AF567059

CHEMICAL THERMODYNAMICS OF URANIUM

CHEMICAL THERMODYNAMICS

Vol. 1. Chemical Thermodynamics of Uranium (H. Wanner and I. Forest, eds.)

CHEMICAL THERMODYNAMICS 1

Chemical Thermodynamics of Uranium

Ingmar GRENTHE (Chairman)
Royal Institute of Technology
Stockholm, Sweden

Jean FUGER
Commission of the European Communities, JRC
European Institute for Transuranium Elements
Karlsruhe, Federal Republic of Germany

Rudy J.M. KONINGS
Netherlands Energy Research Foundation ECN
Petten, The Netherlands

Robert J. LEMIRE
AECL Research, Whiteshell Laboratories
Pinawa, Manitoba, Canada

Anthony B. MULLER
Science Applications International Corporation
McLean, Virginia, USA

Chinh NGUYEN-TRUNG
CREGU
Vandœvre-lès-Nancy, France

Hans WANNER
OECD Nuclear Energy Agency, Data Bank
Issy-les-Moulineaux, France

Edited by
Hans WANNER and Isabelle FOREST
OECD Nuclear Energy Agency, Data Bank
Issy-les-Moulineaux, France

NUCLEAR ENERGY AGENCY
ORGANISATION FOR ECONOMIC CO-OPERATION AND DEVELOPMENT

1992

NORTH-HOLLAND
AMSTERDAM • LONDON • NEW YORK • TOKYO

Published by NORTH-HOLLAND
ELSEVIER SCIENCE PUBLISHERS B.V.
Sara Burgerhartstraat 25
P.O. Box 211, 1000 AE Amsterdam, The Netherlands

Distributors for the United States and Canada:
ELSEVIER SCIENCE PUBLISHING COMPANY INC.
655 Avenue of the Americas
New York, N.Y. 10010, U.S.A.

ISBN: 0 444 89381 4

In order to ensure rapid publication this volume was prepared in camera-ready form by the editors. Time did not allow for the usual extensive editing process of the publisher.

Printed on acid-free paper.

Printed in The Netherlands.

Preface

This is the first volume in a series of critical reviews of the chemical thermodynamics of those elements that are of particular importance in the safety assessment of radioactive waste disposal systems. The NEA Thermochemical Data Base (TDB) project was begun seven years ago by the OECD Nuclear Energy Agency (NEA) as a result of discussions that had taken place during radioactive waste management conferences held in the United States and France. Anthony Muller initiated the project and, at first, he coordinated it at the NEA; in 1986 he was succeeded by Hans Wanner.

In late 1986, two years after the preliminary talks, a meeting was held at which the first drafts of some of the sections in this volume were discussed. This resulted in some major revisions, updates, and also improved documentation of the evaluation and selection procedures. Since then the review team has met twice for extensive discussions.

This volume is a collective effort: each participant contributed sections corresponding to his area of expertise. Robert Lemire has drafted the sections on the oxide and hydroxide compounds and complexes, the uranium nitrides, the solid uranium nitrates and the arsenic-containing uranium compounds, and he has contributed to the sections on uranates in the final stage. He has also written Appendix D and established the procedures used for consistent estimation of entropies. The sections on gaseous and solid uranium halides were originally drafted by Anthony Muller, then commented upon by Jean Fuger, and finally thoroughly revised and extended by Rudy Konings, who joined the review team in 1990, and who has also drafted the section on gaseous uranium oxides and contributed to the sections on uranates in the final stage. Chinh Nguyen-Trung, who joined the review team in 1988, has contributed to the section on solid phosphorus-containing uranium compounds and the one on alkali metal uranates. Hans Wanner has developed the procedures, drafted the sections on uncertainties and on standards and conventions, and has verified and checked the consistency of data submitted by the project participants. My contribution consists of the sections on aqueous complexes (including Section V.2 but excluding the hydroxide complexes) and the parts of the sections on uranium minerals that deal with solubility products. Hans Wanner and I drafted Appendix B on ionic strength corrections.

Hans Wanner has been responsible for editing this volume. He was assisted by his colleagues from the NEA Data Bank: Isabelle Muller (formerly Poirot), succeeded by Isabelle Forest in 1989; Mikazu Yui, succeeded by Kaname Miyahara in 1990; and Pierre Nagel, who has developed the computerized data base system.

I would also like to acknowledge contributions made by Donald Langmuir who participated in our first meeting. Unfortunately, because of insufficient funding, he was unable to take part in subsequent activities.

The project has taken a long time to complete, partly because of lack of proper funding in its initial stages. I am grateful to the NEA Radioactive Waste Management Committee for its assistance in recent years in seeking this financial support. The efforts by Mr. Jean-

Pierre Olivier, Head of the NEA Division for Radiation Protection and Radioactive Waste Management, to establish the TDB project at the NEA, are acknowledged. Dr. Kunihiko Uematsu, Director General of the NEA, has been instrumental in bringing this project to its successful conclusion. Mr. Johnny Rosén, Head of the NEA Data Bank, has always been a source of support and encouragement. On behalf of the review team, I want to thank not only those named above, but also all the colleagues and co-workers who have helped us complete our task.

Stockholm, November 1991 Ingmar Grenthe, Chairman

Acknowledgments

I. Grenthe thanks the Swedish Nuclear Fuel and Waste Management Company (SKB) for financial support.

R.J. Lemire's participation in this project was jointly funded by AECL Research and Ontario Hydro under the auspices of the CANDU Owners Group. He also wishes to express appreciation to P. Taylor and J. Paquette for helpful discussions, and to N.D. Haworth for assistance with the literature searches.

C. Nguyen-Trung's participation in this project was funded by CREGU. He wishes to thank R.E. Mesmer, D.A. Palmer and C.F. Baes, Jr., for helpful discussions and logistic support while he was at the Oak Ridge National Laboratory.

H. Wanner wishes to thank Nathalie Cramer for entering thousands of data into the computerized data base, as well as Séverine Girod and Lionel Morel for their help in formatting and editing the text.

The following eight independent experts have reviewed the contents of the present book according to the peer review procedures (*cf.* [90WAN2]) prepared for this purpose in the framework of the NEA Thermochemical Data Base project. They have viewed and approved the modifications made by the authors according to their comments. The peer review comment records may be obtained on request from the OECD Nuclear Energy Agency.

Prof. S. Ahrland	University of Lund, Lund, Sweden
Dr. C.F. Baes, Jr.	Oak Ridge National Laboratory, Oak Ridge, Tennessee, USA (Retired)
Prof. L. Ciavatta	University of Naples, Naples, Italy
Prof. E.H.P. Cordfunke	Netherlands Energy Research Foundation ECN, Petten, The Netherlands
Prof. F.P. Glasser	University of Aberdeen, Aberdeen, United Kingdom
Prof. F. Grønvold	University of Oslo, Oslo, Norway
Dr. D.L. Hildenbrand	SRI International, Menlo Park, California, USA
Dr. F.J. Pearson, Jr.	Ground-Water Geochemistry, Irving, Texas, USA

We are grateful to I. Puigdomenech (Studsvik AB, Nyköping, Sweden) for the production of the distribution and predominance diagrams, *i.e.*, Figures V.4, V.5, V.6, V.7, V.8, V.17, V.18, V.19, V.20.

In addition, we thank the following for their technical comments and suggestions in the course of the project: P.L. Brown, Australian Nuclear Science and Technology Organization, Lucas Heights, New South Wales, Australia; M.W. Chase, National Institute of Standards and Technology, Gaithersburg, Maryland, USA; K. Czyscinski, Roy F. Weston Inc., Washington, D.C., USA; D. Garvin, National Institute of Standards and Technology,

Gaithersburg, Maryland, USA; I. Khodakovsky, Vernadsky Institute, USSR Academy of Sciences, Moscow, USSR; O. Kubaschewski, Aachen, Germany; L. Maya, Oak Ridge National Laboratory, Oak Ridge, Tennessee, USA; L.R. Morss, Argonne National Laboratory, Argonne, Illinois, USA; T.W. Newton, Los Alamos National Laboratory, Los Alamos, New Mexico, USA; D.K. Nordstrom, US Geological Survey, Menlo Park, California, USA; V.B. Parker, National Institute of Standards and Technology, Gaithersburg, Maryland, USA; S.L. Phillips, Lawrence Berkeley Laboratory, Berkeley, California, USA; M.H. Rand, Harwell Laboratory, Didcot, United Kingdom; J.A. Rard, Lawrence Livermore National Laboratory, Livermore, California, USA; W.J. Ullman, University of Delaware, Lewes, Delaware, USA; P. Vitorge, Commissariat à l'Energie Atomique, Fontenay-aux-Roses, France; E.F. Westrum, Jr., University of Michigan, Ann Arbor, Michigan, USA;

We are indebted to the following for permission to reproduce copyright material: American Institute of Physics and the authors for Figure II.1 (from D.D. Wagman *et al.* [82WAG/EVA]), and John Wiley & Sons, Inc., for Table II.5 (from C.J. Baes, Jr., and R.E. Mesmer [76BAE/MES]).

Paris, January 1992 The Editors

Foreword

The present volume starts a series of detailed expert reviews of the chemical thermodynamics of key elements in nuclear technology and waste management. Volumes on americium, technetium, neptunium and plutonium are currently in progress. The recommended thermodynamic data are the result of a critical assessment of published information. The data network developed at the Data Bank of the OECD Nuclear Energy Agency (NEA), *cf.* Section I.4, ensures consistency not only within the recommended data set on uranium but also among all the data sets to be published in the series.

The NEA Data Bank provides a number of services that may be useful to the reader of this book.

- The recommended data can be obtained on eletronic media (PC diskettes, magnetic tape, or via computer networks) directly from the NEA Data Bank. The special formatting of the data allows easy conversion to any specific formats convenient to the user.

- The NEA Data Bank maintains a library of tested computer programs in various areas. This includes geochemical codes such as PHREEQE, EQ3/6, MINEQL, MINTEQ, PHRQPITZ, *etc.*, in which chemical thermodynamic data like those presented in this book are required as the basic input data. These computer codes can be obtained on request from the NEA Data Bank.

- Short courses on the use of geochemical codes and thermodynamic data have been organized at the NEA Data Bank since 1985, tought by the authors of the codes PHREEQE, MINEQL and EQ3/6. It is planned to give further courses on the codes mentioned above or on new ones, corresponding to the interest expressed by the user's community. Persons interested in further courses are advised to contact the NEA Data Bank.

For requests of data, computer programs, on-line access, and for further information, please write to:

OECD Nuclear Energy Agency
Data Bank
2-12, Rue Jean-Pierre Timbaud
F-92130 Issy-les-Moulineaux
France

Contents

I Introduction 1

I.1 Background 1

I.2 Focus of the review 2

I.3 Review procedure and results 3

I.4 The NEA-TDB data base system 4

I.5 Presentation of the selected data 6

II Standards and Conventions 9

II.1 Symbols, terminology and nomenclature 9

II.1.1 Abbreviations 9

II.1.2 Symbols and terminology 9

II.1.3 Chemical formulae and nomenclature 9

II.1.4 Phase designators 13

II.1.5 Processes 14

II.1.6 Equilibrium constants 14

II.1.7 Order of formulae 20

II.1.8 Reference codes 20

II.2 Units and conversion factors 22

II.3 Standard and reference conditions 24

II.3.1 Standard state 24

II.3.2 Standard state pressure 24

II.3.3 Reference temperature 28

II.4 Fundamental physical constants 28

III Selected uranium data 29

IV Selected auxiliary data 63

V Discussion of data selection 85

V.1 Elemental uranium 85

V.1.1 Uranium metal 85

V.1.2 Uranium gas 86

V.2 Simple uranium aqua ions 86

V.2.1 UO_2^{2+} 86

V.2.2 UO_2^+ 88

V.2.3 U^{4+} 91

V.2.4 U^{3+} 96

V.2.5 U^{2+} 97

V.3 Oxygen and hydrogen compounds and complexes 97

V.3.1 Gaseous uranium oxides 97

V.3.2 Aqueous uranium hydroxide complexes 98

V.3.3 Crystalline and amorphous uranium oxides and hydroxides 131

V.3.4 Uranium hydrides . . . 148
V.4 Group 17 (halogen) compounds and complexes . . . 150
V.4.1 Fluorine compounds and complexes . . . 150
V.4.2 Chlorine compounds and complexes . . . 185
V.4.3 Bromine compounds and complexes . . . 213
V.4.4 Iodine compounds and complexes . . . 226
V.4.5 Mixed halogen compounds . . . 232
V.5 Group 16 (chalcogen) compounds and complexes . . . 236
V.5.1 Sulphur compounds and complexes . . . 236
V.5.2 Selenium compounds and complexes . . . 256
V.5.3 Tellurium compounds . . . 260
V.6 Group 15 compounds and complexes . . . 261
V.6.1 Nitrogen compounds and complexes . . . 261
V.6.2 Phosphorus compounds and complexes . . . 279
V.6.3 Arsenic compounds . . . 301
V.6.4 Antimony compounds . . . 304
V.7 Group 14 compounds and complexes . . . 306
V.7.1 Carbon compounds and complexes . . . 306
V.7.2 Silicon compounds and complexes . . . 334
V.7.3 Lead compounds . . . 336
V.8 Actinide complexes . . . 336
V.8.1 Actinide-actinide interactions . . . 336
V.8.2 Mixed U(VI), Np(VI) and Pu(VI) carbonate complexes . . . 337
V.9 Group 2 (alkaline-earth) compounds . . . 337
V.9.1 Beryllium compounds . . . 337
V.9.2 Magnesium compounds . . . 338
V.9.3 Calcium compounds . . . 340
V.9.4 Strontium compounds . . . 342
V.9.5 Barium compounds . . . 345
V.10 Group 1 (alkali) compounds . . . 347
V.10.1 Lithium compounds . . . 347
V.10.2 Sodium compounds . . . 349
V.10.3 Potassium compounds . . . 358
V.10.4 Rubidium compounds . . . 359
V.10.5 Caesium compounds . . . 360

VI Discussion of auxiliary data selection 365
VI.1 Group 17 (halogen) auxiliary species . . . 366
VI.1.1 Fluorine auxiliary species . . . 366
VI.1.2 Chlorine auxiliary species . . . 368
VI.1.3 Bromine auxiliary species . . . 371
VI.1.4 Iodine auxiliary species . . . 373
VI.2 Group 16 (chalcogen) auxiliary species . . . 374
VI.2.1 Sulphur auxiliary species . . . 374
VI.2.2 Selenium auxiliary species . . . 380
VI.2.3 Tellurium . . . 383
VI.3 Group 15 auxiliary species . . . 384
VI.3.1 Nitrogen auxiliary species . . . 384
VI.3.2 Phosphorus auxiliary species . . . 386
VI.3.3 Arsenic auxiliary species . . . 389
VI.3.4 Antimony . . . 391
VI.4 Group 14 auxiliary species . . . 392
VI.4.1 Carbon auxiliary species . . . 392
VI.4.2 Silicon auxiliary species . . . 394

VI.5 Other auxiliary species . . . 400
VI.5.1 Strontium auxiliary species . . . 400
VI.5.2 Barium auxiliary species . . . 400

VII Reference list **403**

VIII Authors list **485**

IX Formula list **519**

X Alphabetical name list **549**

Appendix A: Discussion of publications **555**

Appendix B: Ionic strength corrections **683**
B.1 The specific ion interaction theory . . . 684
B.1.1 Background . . . 684
B.1.2 Estimation of ion interaction coefficients . . . 688
B.1.3 On the magnitude of ion interaction coefficients . . . 691
B.2 Ion interaction coefficients and equilibrium constants for ion pairs . . . 692
B.3 Tables of ion interaction coefficients . . . 692

Appendix C: Assigned uncertainties **699**
C.1 One source datum . . . 699
C.2 Two or more independent source data . . . 700
C.3 Several data at different ionic strengths . . . 704
C.4 Procedures for data handling . . . 708
C.4.1 Correction to zero ionic strength . . . 708
C.4.2 Propagation of errors . . . 709
C.4.3 Rounding . . . 710
C.4.4 Significant digits . . . 711

Appendix D: The estimation of entropies **713**

List of Tables

II.1 Abbreviations for experimental methods 10
II.2 Symbols and terminology 11
II.3 Abbreviations for chemical processes 15
II.4 Unit conversion factors 22
II.5 Conversion factors of molarity to molality 23
II.6 Reference states for the elements 25
II.7 Fundamental physical constants 28

III.1 Selected uranium data of formation 30
III.2 Selected uranium data of reaction 51
III.3 Selected temperature coefficients of $C_{p,\mathrm{m}}$ 61

IV.1 Selected auxiliary data of formation 64
IV.2 Selected auxiliary data of reaction 80

V.1 Data considered for the evaluation of $\Delta_f H^\circ_m(UO_2^{2+}, aq, 298.15\,K)$ 88
V.2 Experimental redox potentials for the U(VI)/U(V) couple 90
V.3 Experimental redox potentials for the U(VI)/U(IV) couple 93
V.4 Selected data of gaseous uranium oxides 98
V.5 Experimental data for the U(VI) hydroxide system 99
V.6 Experimental data for the U(VI) hydroxide system at different temperatures 104
V.7 Selected formation constants for the U(VI) hydroxide system 107
V.8 Enthalpies and entropies of U(VI) hydroxide species 118
V.9 Experimental data for the U(IV) hydroxide system 121
V.10 Heat capacity functions for uranium oxides 133
V.11 Selected data of uranium oxides and hydroxides 145
V.12 Enthalpy of formation data of $UF_3(g)$ 152
V.13 Enthalpy of sublimation data of $UF_4(g)$ 153
V.14 Enthalpy of formation data of $UF_5(g)$ 155
V.15 Enthalpy of sublimation data of $UF_6(cr)$ 156
V.16 Enthalpy of sublimation data of $UO_2F_2(cr)$ 158
V.17 Experimental data for the U(VI) fluoride system 160
V.18 Experimental data for the U(IV) fluoride system 165
V.19 Enthalpy data of $UF_3(cr)$ and $UF_4(cr)$ 171
V.20 Enthalpies of dehydration of U(VI) fluoride 180
V.21 Enthalpy of sublimation data of $UCl_4(cr)$ 188
V.22 Experimental data for the U(VI) chloride system 193
V.23 Experimental data for the U(IV) chloride system 198
V.24 Enthalpy data of $UCl_3(cr)$ 201
V.25 Enthalpy data of $UCl_4(cr)$ 203
V.26 Enthalpy data of $UO_2Cl_2(cr)$ 209
V.27 Enthalpy of sublimation data of $UBr_4(cr)$ 216
V.28 Experimental data for the U(VI) bromide system 218

V.29 Experimental data for the U(IV) bromide system 219
V.30 Enthalpy data of $UBr_3(cr)$, $UBr_4(cr)$ and $UBr_5(cr)$ 220
V.31 Enthalpy data of $UI_3(cr)$ and $UI_4(cr)$ 230
V.32 Selected thermodynamic parameters on solid uranium sulphides 236
V.33 Experimental data for the U(VI) sulphite system 238
V.34 Experimental data for the U(VI) sulphate system 242
V.35 Experimental data for the U(IV) sulphate system 247
V.36 Experimental data for the U(VI) azide system 266
V.37 Experimental data for the U(VI) nitrate system 268
V.38 Experimental data for the U(IV) nitrate system 271
V.39 Evaluation of the enthalpies of solution of $UO_2(NO_3)_2 \cdot xH_2O(cr)$ 276
V.40 Experimental data for the U(VI)-phosphoric acid system 281
V.41 Selected equilibrium constants in the U(VI)-phosphoric acid system 287
V.42 Experimental data for the U(VI) carbonate system 309
V.43 Selected equilibrium constants in the U(VI) carbonate system 313
V.44 Experimental data for the U(VI) thiocyanate system 330
V.45 Experimental data for the U(IV) thiocyanate system 333
V.46 Enthalpy data of α-Na_2UO_4 351

VI.1 Selected arsenic thermodynamic parameters 391
VI.2 Selected auxiliary data for strontium 400
VI.3 Selected auxiliary data for barium 401

A.1 Analysis of the reported data [69TSY] on the U(VI)-carbonate-hydroxide system . . 614
A.2 Re-evaluation of the U(IV) fluoride data of Ref. [74KAK/ISH] 631
A.3 Reported thermodynamic functions of caesium polyuranates [75JOH] 634
A.4 Models for the U(VI)-carbonate-hydroxide complexes 647
A.5 Test of data in Ref. [82MAY] for the U(VI)-OH-CO_2 system. 657
A.6 Reported equilibrium constants for the U(VI)-carbonate system [84GRE/FER] . . . 664
A.7 Data reported for the U(IV) fluoride system [90AHR/HEF] 678

B.1 Debye-Hückel constants 685
B.2 Preparation of experimental data for the extrapolation to $I = 0$ 689
B.3 Ion interaction coefficients ε for cations 693
B.4 Ion interaction coefficients ε for anions 696
B.5 Ion interaction coefficients ε_1 and ε_1 698

D.1 Contributions to entropies of solids. 715

List of Figures

I.1 Principal schema of the NEA Thermochemical Data Base 5

II.1 Standard order of arrangement 21

V.1 Extrapolation to $I = 0$ for $UO_2^{2+} + e^- \rightleftharpoons UO_2^+$ 91
V.2 Extrapolation to $I = 0$ for $UO_2^{2+} + 2e^- + 4H^+ \rightleftharpoons U^{4+} + 2H_2O(l)$ 94
V.3 Extrapolation to $I = 0$ for $3UO_2^{2+} + 5H_2O(l) \rightleftharpoons (UO_2)_3(OH)_5^+ + 5H^+$ 109
V.4 Predominance diagram of the U(VI) hydroxide system 115
V.5 Solubility and predominance diagram of the U(VI) hydroxide system 116
V.6 Distribution diagram of the U(VI) hydroxide system 116
V.7 Predominance diagram of the U(VI)/U(IV) hydroxide system 128
V.8 Solubility and predominance diagram of the U(VI)/U(IV) hydroxide system 129
V.9 Extrapolation to $I = 0$ for $UO_2^{2+} + HF(aq) \rightleftharpoons UO_2F^+ + H^+$ 162
V.10 Extrapolation to $I = 0$ for $UO_2^{2+} + Cl^- \rightleftharpoons UO_2Cl^+$ 195
V.11 Extrapolation to $I = 0$ for $UO_2^{2+} + 2Cl^- \rightleftharpoons UO_2Cl_2(aq)$ 196
V.12 Extrapolation to $I = 0$ for $U^{4+} + Cl^- \rightleftharpoons UCl^{3+}$ 199
V.13 Extrapolation to $I = 0$ for $U^{4+} + NO_3^- \rightleftharpoons UNO_3^{3+}$ 273
V.14 Extrapolation to $I = 0$ for $U^{4+} + 2NO_3^- \rightleftharpoons U(NO_3)_2^{2+}$ 274
V.15 Extrapolation to $I = 0$ for $UO_2^{2+} + 2CO_3^{2-} \rightleftharpoons UO_2(CO_3)_2^{2-}$ 314
V.16 Extrapolation to $I = 0$ for $UO_2^{2+} + 3CO_3^{2-} \rightleftharpoons UO_2(CO_3)_3^{4-}$ 315
V.17 Distribution diagram of the U(VI) hydroxide-carbonate system at high carbonate concentration 321
V.18 Distribution diagram of the U(VI) hydroxide-carbonate system at low carbonate concentration 322
V.19 Predominance diagram of the U(VI)/U(IV) hydroxide-carbonate system at high carbonate concentration 325
V.20 Solubility and predominance diagram of the U(VI)/U(IV) hydroxide-carbonate system 326
V.21 Extrapolation to $I = 0$ for $Na_4UO_2(CO_3)_3(cr) \rightleftharpoons 4Na^+ + UO_2(CO_3)_3^{4-}$ 328

VI.1 Extrapolation to $I = 0$ for $H^+ + F^- \rightleftharpoons HF(aq)$ 367
VI.2 Extrapolation to $I = 0$ for $HF(aq) + F^- \rightleftharpoons HF_2^-$ 368
VI.3 Extrapolation to $I = 0$ for $Si(OH)_4(aq) \rightleftharpoons SiO(OH)_3^- + H^+$ 395
VI.4 Extrapolation to $I = 0$ for $Si(OH)_4(aq) \rightleftharpoons SiO_2(OH)_2^{2-} + 2H^+$ 396

A.1 Solubility curve of UO_2CO_3 in Na_2CO_3 solutions 570
A.2 Structure models for the complex $(UO_2)_{11}(OH)_{24}(CO_2)_6^{2-}$ 680

B.1 Extrapolation to $I = 0$: An example 690

Chapter I

Introduction

I.1. Background

The modelling of the behaviour of hazardous materials under environmental conditions is among the most important applications of natural and technical sciences for the protection of the environment. In order to assess, for example, the safety of a waste deposit, it is essential to be able to predict the eventual dispersion of its hazardous components in the environment (geosphere, biosphere). For hazardous materials stored in the ground or in geological formations, the only conceivable transport medium is the aqueous phase. It is therefore important to quantitatively predict the reactions that are likely to occur between hazardous waste dissolved or suspended in groundwater, and the surrounding rock material, in order to estimate the quantities of waste that can be transported in the aqueous phase. It is thus essential to know the relative stabilities of the compounds and complexes that may form under the relevant conditions. This information is provided by speciation calculations using chemical thermodynamic data. The local conditions, such as groundwater and rock composition or temperature, may not be constant along the migration paths of hazardous materials, and fundamental thermodynamic data are the indispensable basis for a dynamic modelling of the chemical behaviour of hazardous waste components.

In the field of radioactive waste management, the hazardous material consists to a large extent of actinides and fission products from nuclear reactors. The scientific literature on thermodynamic data, mainly on equilibrium constants and redox potentials in aqueous solution, has been contradictory in a number of cases, especially in the actinide chemistry. A critical and comprehensive review of the available literature is necessary in order to establish a reliable thermochemical data base that fulfils the requirements of a proper modelling of the behaviour of the actinide and fission products in the environment.

The Radioactive Waste Management Committee (RWMC) of the OECD Nuclear Energy Agency recognized the need for an internationally acknowledged, high-quality thermochemical data base for the application in the safety assessment of radioactive waste disposal and undertook the development of the NEA Thermochemical Data Base (TDB) project [85MUL4, 88WAN, 90WAN]. The RWMC assigned a high priority to the critical review of relevant chemical thermodynamic data of compounds and complexes for this area containing the actinides uranium, neptunium, plutonium and americium, as well as the fission product technetium. The present report on uranium thermodynamics is the first volume in the series.

I.2. Focus of the review

The first and most important step in the modelling of chemical reactions is to decide whether they are controlled by chemical thermodynamics or kinetics, or possibly by a combination of the two. This also applies to the modelling of more complex chemical systems and processes, such as waste repositories of various kinds, the processes describing transport of toxic materials in ground and surface water systems, the global geochemical cycles, *etc.*

As outlined in the previous section, the focus of the critical review presented in this report is on the thermodynamic data of uranium relevant to the safety assessment of radioactive waste repositories in the geosphere. This includes the release of waste components from the repository into the geosphere (*i.e.*, its interaction with the waste container and the other near-field materials) and their migration through the geological formations and the various compartments of the biosphere. As groundwaters and porewaters are the transport media for the waste components, the knowledge of the thermodynamics of the corresponding elements in waters of various compositions is of fundamental importance. The present review therefore puts much weight on the assessment of the low-temperature thermodynamics of uranium in aqueous solution and makes independent analyses of the available literature in this area. This also includes the elaboration of a standard method for the analysis of ionic interactions between components dissolved in water (see Appendix B). This method allows the general and consistent use of the selected data for modelling purposes, regardless of the type and composition of the groundwater, within the ionic strength limits given by the experimental data used for the data analyses in the present review.

The interactions between solid compounds, such as the rock materials, and the aqueous solution and its components are as important as the interactions within the aqueous solution, because the solid materials in the geosphere control the chemistry of the groundwater, and they also contribute to the overall solubilities of key elements. The present review therefore also considers the chemical behaviour of solid compounds that are likely to occur, or to be formed, under geological and environmental conditions. It is, however, difficult to assess the relative importance of each of the many known uranium-containing mineral phases for performance assessment purposes, particularly since their interactions with the aqueous phase are in many cases known to be subject to quantitatively unknown kinetic constraints. Moreover, for a large number of these minerals no experimental thermochemical or solubility data exist. This review presents a selection of uranium minerals that may be important in geochemical modelling and for which credible thermodynamic data can be recommended. The most important solid phases for the modelling of uranium released from a nuclear waste repository are the various oxides. The thermodynamic data for the pure phases are of good quality. However, these data can at best be used as guidelines for the modelling of the properties of spent UO_2 fuel, which is not a pure phase. In the same way, the thermodynamic data for pure uranium minerals may be of limited value for geochemical modelling, because most of these are unlikely to form from the uranium released from nuclear waste repositories. Leached uranium is more likely to be found in association with iron(III) oxide hydrates, a common secondary phase in water carrying fractures. If it is considered important, from the performance assessment viewpoint, to include mineral phases other than those presented in this report, an additional review will be envisaged, and its results will be published as an addendum to the present report.

The present review, as described in Chapter V, is a summary and a critical review of the thermodynamic data on compounds and complexes containing uranium, as reported in the available chemical literature up to 1990. A few more recent references are also included.

Although the focus of this review is on uranium, it is necessary to use data on a number of other species during the evaluation process that lead to the recommended data. These so-called auxiliary data are taken from the recent publication of CODATA Key Values [89COX/WAG], and their use is recommended by this review. In many instances the present review requires auxiliary data that are not included in the CODATA list of Key Values. In these cases independent selections are made which are described in Chapter VI. Care has been taken that all the selected thermodynamic data at standard conditions (*cf.* Section II.3) and 298.15 K are internally consistent. For this purpose, a special software has been developed at the NEA Data Bank that is operational in conjunction with the NEA-TDB data base system, *cf.* Section I.4. In order to maintain consistency in the application of these data, it is essential to use these auxiliary data in conjunction with the data on uranium compounds and complexes selected in this review.

The present review does not include any compounds or complexes containing organic ligands. This class of compounds is planned to be the subject of a later review in the NEA-TDB series.

I.3. Review procedure and results

The objective of the present review is to present an assessment of the sources of published thermodynamic data in order to decide on the most reliable values that can be recommended. Previous reviews are not neglected but form a valuable source of critical information on the quality of primary publications. In order to ensure that consistent procedures are used for the evaluation of primary data, a number of guidelines have been developed. They have been updated and improved since 1987, and their most recent versions are available at the NEA [89GRE/WAN, 89WAN2, 89WAN3, 89WAN4]. The procedures are also outlined in the present report, *cf.* Chapter II, Appendix B, Appendix C and Appendix D. A comparatively large number of primary references are discussed separately in Appendix A.

The thermodynamic data selected in the present review (see Chapters III and IV) refer to the reference temperature of 298.15 K and to standard conditions, *cf.* Section II.3. For the modelling of real systems it is, in general, necessary to recalculate the standard thermodynamic data to non-standard state conditions. For aqueous species a procedure for the calculation of the activity factors is thus required. This review uses the approximate specific ion interaction method for the extrapolation of experimental data to the standard state in the data evaluation process. For maximum consistency, this method, as described in Appendix B, should always be used in conjunction with the selected data presented in this review.

The thermodynamic data in the selected set refer to a temperature of 298.15 K (25°C), but they can be recalculated to other temperatures if the corresponding data (enthalpies, entropies, heat capacities) are available. In the case of aqueous species, for which enthalpies of reaction are selected or can be calculated from the selected enthalpies of formation, it is in most cases possible to recalculate equilibrium constants for temperatures up to 100 to 150°C, with an additional uncertainty of about 1 to 2 logarithmic units due to the temperature correction. However, it is important to observe that "new" aqueous species, *i.e.*, species not present in significant amounts at 25°C and therefore not detected, may be significant at higher temperatures, *cf.* Ciavatta, Iuliano and Porto [87CIA/IUL]. Additional high-temperature experiments may therefore be needed in order to ascertain that proper chemical models are used in the modelling of hydrothermal systems. For many species, experimental thermodynamic data are not available to allow a selection of parameters describing the temperature dependence of equilibrium constants and Gibbs energies of formation. A guideline has therefore been developed at the NEA [91PUI/RAR] that gives the user some information on various ways how to estimate the temperature dependence of these thermodynamic parameters.

The thermodynamic data selected in this review are provided with uncertainties representing the 95% confidence level. As discussed in Appendix C, there is no unique way to assign uncertainties, and the assignments made in this review are to a large extent based on the subjective choice by the reviewers, supported by their scientific and technical experience in the corresponding area.

The quality of thermodynamic models cannot be better than the quality of the data they are based on. The quality aspect includes both the numerical values of the thermodynamic data used in the model and the "completeness" of the chemical model used, *e.g.*, the inclusion of all the relevant dissolved chemical species and solid phases. For the user it is important to consider that the selected data set presented in this review (Chapter III) may not be "complete" with respect to all the conceivable systems and conditions; there are gaps in the information, particularly concerning aqueous species that contain more than one kind of ligands, as well as uranium(IV) complexes formed in the range $3 < \mathrm{pH} < 9$. The gaps are pointed out in the various sections of Chapter V, and this information may be used as a basis for the assignment of research priorities.

I.4. The NEA-TDB data base system

A data base system has been developed at the NEA Data Bank that allows the storage of thermodynamic parameters for individual species as well as for reactions. A simplified schema of the NEA-TDB data base system is shown in Figure I.1. The structure of the data base system allows consistent derivation of thermodynamic data

Figure I.1: Principal schema of the NEA Thermochemical Data Base.

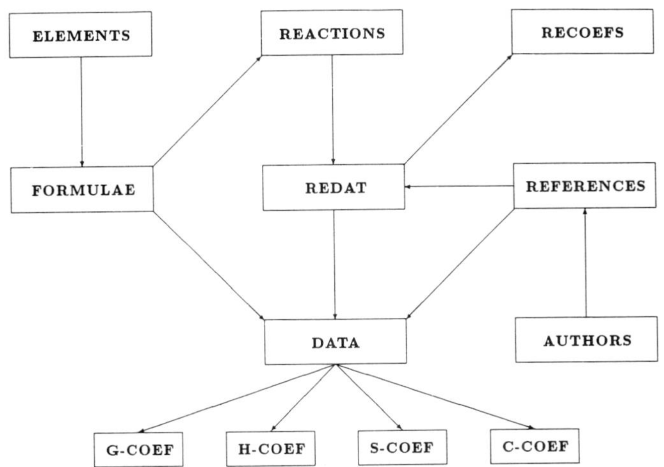

for individual species from reaction data at standard conditions, as well as internal recalculations of data at standard conditions. If a selected value is changed, all the dependent values will be recalculated consistently. The maintenance of consistency of all the selected data, including their uncertainties (*cf.* Appendix C), is ensured by the software developed for this purpose at the NEA Data Bank.

The thermodynamic data and their uncertainties selected for individual species are stored in the DATA record. The following parameters, valid at the reference temperature of 298.15 K, are considered:

$\Delta_f G^\circ_m$	the standard molar Gibbs energy of formation	$(kJ \cdot mol^{-1})$
$\Delta_f H^\circ_m$	the standard molar enthalpy of formation	$(kJ \cdot mol^{-1})$
S°_m	the standard molar entropy	$(J \cdot K^{-1} \cdot mol^{-1})$
$C^\circ_{p,m}$	the standard molar heat capacity	$(J \cdot K^{-1} \cdot mol^{-1})$

As the parameters in this set vary as a function of temperature, provision is made to include the compilation of the coefficients of empirical temperature functions for these thermodynamic data in the records G-COEF, H-COEF, S-COEF and C-COEF, as well as the temperature ranges over which they are valid. In many cases the thermodynamic data measured or calculated at several temperatures were published for a particular species, rather than the deduced temperature functions. In these cases, a non-linear regression method is used to obtain the most significant coefficients of the following empirical function:

$$F(T) = a + b \times T + c \times T^2 + d \times T^{-1} + e \times T^{-2} + f \times \ln T + g \times T \ln T + h \times \sqrt{T} + \frac{i}{\sqrt{T}} + j \times T^3 + k \times T^{-3}. \quad \text{(I.1)}$$

Most temperature variations can be described with three or four parameters, a, b and e being the ones most frequently used. In the present review, only $C^\circ_{p,m}(T)$, *i.e.*, the thermal functions of the heat capacities of individual species, are considered and stored in the record C-COEF. They refer to the relation

$$C^\circ_{p,\mathrm{m}}(T) = a + b \times T + c \times T^2 + d \times T^{-1} + e \times T^{-2} \qquad \text{(I.2)}$$

and are listed in Table III.3.

The pressure dependence of thermodynamic data has not been the subject of critical analysis in the present compilation. The reader interested in higher temperatures and pressures, or the pressure dependency of thermodynamic functions for geochemical applications, is referred to the specialized literature in this area, *e.g.*, [88SHO/HEL, 88TAN/HEL, 89SHO/HEL, 89SHO/HEL2].

Selected standard thermodynamic data referring to chemical reactions (stored in the REACTIONS record in alphanumeric notation) are compiled in the REDAT record. The parameters considered valid at the reference temperature of 298.15 K include:

$\log_{10} K^\circ$	the equilibrium constant of the reaction, logarithmic	
$\Delta_\mathrm{r}G^\circ_\mathrm{m}$	the molar Gibbs energy of reaction	$(\mathrm{kJ \cdot mol^{-1}})$
$\Delta_\mathrm{r}H^\circ_\mathrm{m}$	the molar enthalpy of reaction	$(\mathrm{kJ \cdot mol^{-1}})$
$\Delta_\mathrm{r}S^\circ_\mathrm{m}$	the molar entropy of reaction	$(\mathrm{J \cdot K^{-1} \cdot mol^{-1}})$
$\Delta_\mathrm{r}C^\circ_{p,\mathrm{m}}$	the molar heat capacity of reaction	$(\mathrm{J \cdot K^{-1} \cdot mol^{-1}})$

The temperature functions of these data, if available, are stored in the record RE-COEFS, according to Eq. (I.1).

The literature sources of the data are stored in the REFERENCES record, and each author name has a link to the AUTHORS record for direct retrieval by the author names.

I.5. Presentation of the selected data

The selected data are presented in Chapters III and IV. Unless otherwise indicated, they refer to standard conditions (*cf.* Section II.3) and 298.15 K (25°C) and are provided with an uncertainty that corresponds to the 95% confidence level (see Appendix C). Chapter III contains a table of thermodynamic data for individual compounds and complexes (Table III.1), a table of reaction data (Table III.2) and a table containing thermal functions of the heat capacities of individual species (Table III.3). The selection of these data is discussed in Chapter V. Chapter IV contains, for auxiliary compounds and complexes that do not contain uranium, a table of the thermodynamic data for individual species (Table IV.1) and a table of reaction data (Table IV.2). The selection of the auxiliary data is discussed in Chapter VI. All the selected data presented in Tables III.1, III.2, IV.1 and IV.2 are internally consistent. This consistency is maintained by the internal consistency verification and recalculation software developed at the NEA Data Bank in conjunction with the NEA-TDB data base system, *cf.* Figure I.1. Therefore, for the application of the selected data, the auxiliary data of Chapter IV must be used together with the data in Chapter III to ensure internal consistency of the data set.

It is important to note that Tables III.2 and IV.2 include only those species of which the primary selected data are reaction data. The formation data derived therefrom and listed in Tables III.1 and IV.1 are obtained using auxiliary data, and their uncertainties are propagated accordingly. In order to maintain the uncertainties originally assigned to the selected data in this review, the user is advised to make direct use of the reaction data presented in Tables III.2 and IV.2, rather than taking the derived values in Tables III.1 and IV.1 to calculate the reaction data (which would imply a twofold propagation of the uncertainties and result in reaction data whose uncertainties would be considerably larger than those originally assigned).

Chapter II

Standards and Conventions

This chapter outlines and lists the symbols, terminology and nomenclature, the units and conversion factors, the order of formulae, the standard conditions, and the fundamental physical constants used in this volume. They are derived from international standards and have been specially adjusted for the TDB publications. The assignment and treatment of uncertainties, as well as the propagation of errors and the standard rules for rounding, are described in detail in Appendix C.

II.1. Symbols, terminology and nomenclature

II.1.1. Abbreviations

Abbreviations are mainly used in tables where space is limited. Abbreviations for methods of measurement are kept to a maximum of three characters (except for composed symbols) and are listed in Table II.1.

Other abbreviations may also be used in tables, such as SHE for the standard hydrogen electrode or SCE for the saturated calomel electrode. The abbreviation NHE has been widely used for the "normal hydrogen electrode", which is by definition identical to the SHE. It should nevertheless be noted that NHE customarily refers to a standard state pressure of 1 atm, whereas SHE always refers to a standard state pressure of 0.1 MPa (1 bar) in this review.

II.1.2. Symbols and terminology

The symbols for physical and chemical quantities used in the TDB review follow the recommendations of the International Union of Pure and Applied Chemistry, IUPAC [79WHI2]. They are summarized in Table II.2.

II.1.3. Chemical formulae and nomenclature

This review follows the recommendations made by IUPAC [71JEN, 77FER] on the nomenclature of inorganic compounds and complexes, except for the following items:

Table II.1: Abbreviations for experimental methods.

aix	anion exchange
cal	calorimetry
chr	chromatography
cix	cation exchange
col	colorimetry
con	conductivity
cor	corrected
cou	coulometry
cry	cryoscopy
dis	distribution between two phases
emf	electromotive force, not specified
gl	glass electrode
ise-X	ion selective electrode with ion X stated
ix	ion exchange
kin	rate of reaction
mvd	mole volume determination
nmr	nuclear magnetic resonance
pol	polarography
pot	potentiometry
prx	proton relaxation
qh	quinhydrone electrode
red	emf with redox electrode
rev	review
sp	spectrophotometry
sol	solubility
tc	transient conductivity
tls	thermal lensing spectrophotometry
vlt	voltammetry
?	method unknown to the reviewers

Table II.2: Symbols and terminology.

length	l
height	h
radius	r
diameter	d
wavelength	λ
internal transmittance (transmittance of the medium itself, disregarding boundary or container influence)	τ_{i}
internal transmission density, (decadic absorbance): $\log_{10}(1/T)$	D_{i}
molar (decadic) absorption coefficient: $A/c_{\mathrm{B}}l$	ϵ
volume	V
time	t
frequency	ν
relaxation time	τ
mass	m
density (mass divided by volume)	ρ
pressure	p
relative atomic mass of an element [(a)]	A_{r}
relative molecular mass of a substance [(b)]	M_{r}
amount of substance [(c)]	n
mole fraction of substance B: $n_{\mathrm{B}}/\sum_i n_i$	x_{B}
molality of a solute substance B (amount of B divided by the mass of the solvent) [(d)]	m_{B}
molarity or concentration of a solute substance B (amount of B divided by the volume of the solution) [(e)]	c_{B}, [B]
thermodynamic temperature, absolute temperature	T
Celsius temperature	t
(molar) gas constant	R
Boltzmann constant	k
(molar) entropy	S_{m}
(molar) heat capacity at constant pressure	$C_{p,\mathrm{m}}$
(molar) enthalpy	H_{m}
(molar) Gibbs energy	G_{m}
chemical potential of substance B	μ_{B}
ionic strength: $I_m = \frac{1}{2}\sum_i m_i z_i^2$ or $I_c = \frac{1}{2}\sum_i c_i z_i^2$	I
ion interaction coefficient between substance B_1 and substance B_2	$\varepsilon_{(\mathrm{B_1},\mathrm{B_2})}$

Table II.2 (continued)

partial pressure of substance B: $x_{\mathrm{B}}p$	p_{B}
fugacity of substance B	f_{B}
fugacity coefficient: $f_{\mathrm{B}}/p_{\mathrm{B}}$	$\gamma_{\mathrm{f,B}}$
activity of substance B	a_{B}
activity coefficient, molality basis: $a_{\mathrm{B}}/m_{\mathrm{B}}$	γ_{B}
activity coefficient, concentration basis: $a_{\mathrm{B}}/c_{\mathrm{B}}$	y_{B}
osmotic coefficient	ϕ
stoichiometric coefficient of substance B (negative for reactants, positive for products)	ν_{B}
general equation for a chemical reaction	$0 = \sum_{\mathrm{B}} \nu_{\mathrm{B}}\mathrm{B}$
equilibrium constant [(f)]	K
rate constant	k
Faraday constant	F
charge number of an ion B (positive for cations, negative for anions)	z_{B}
charge number of a cell reaction	n
electromotive force	E
electrolytic conductivity	κ
superscript for standard state [(g)]	$^{\circ}$

(a) The ratio of the average mass per atom of an element to $\frac{1}{12}$ of the mass of an atom of nuclide ^{12}C.

(b) The ratio of the average mass per formula unit of a substance to $\frac{1}{12}$ of the mass of an atom of nuclide ^{12}C.

(c) *cf.* Sections 1.2 and 3.6 of the IUPAC manual [79WHI2].

(d) A solution having a molality equal to $0.1\,\mathrm{mol\cdot kg^{-1}}$ is called a 0.1 molal solution or a 0.1 m solution.

(e) This quantity is called "amount-of-substance concentration" in the IUPAC manual [79WHI2]. A solution with a concentration equal to $0.1\,\mathrm{mol\cdot dm^{-3}}$ is called a 0.1 molar solution or a 0.1 M solution.

(f) Special notations for equilibrium constants are outlined in Section 1.2. In some cases, K_c is used to indicate a concentration constant in molar units, and K_m a constant in molal units.

(g) See Section "Standard conditions".

i) The formulae of coordination compounds and complexes are not enclosed in square brackets [71JEN, Rule 7.21]. No brackets or parentheses are used at all to denote coordination compounds.

ii) The prefixes "oxy-" and "hydroxy-" are retained if used in a general way, *e.g.*, "gaseous uranium oxyfluorides". For specific formula names, however, the IUPAC recommended citation [71JEN, Rule 6.42] is used, *e.g.*, "uranium(IV) difluoride oxide" for $UF_2O(cr)$.

An IUPAC rule that is often not followed by many authors [71JEN, Rules 2.163 and 7.21] is recalled here: The order of arranging ligands in coordination compounds and complexes is the following: Central atom first, followed by ionic ligands and then by the neutral ligands. If there is more than one ionic or neutral ligand, the alphabetical order of the symbols of the ligating atoms determines the sequence of the ligands. For example, $UO_2CO_3(OH)_3^{2-}$ is standard, $UO_2(OH)_3CO_3^{2-}$ is non-standard.

II.1.4. Phase designators

Chemical formulae may refer to different chemical species and are often required to be specified more clearly in order to avoid ambiguities. For example, UF_4 occurs as a gas, a solid, and an aqueous complex. The distinction between the different phases is made by phase designators that immediately follow the chemical formula and appear in parentheses. The only formulae that are not provided with a phase designator are aqueous ions. They are the only charged species in this review since charged gases are not considered. The use of the phase designators is described below.

- The designator (l) is used for pure liquid substances, *e.g.*, $H_2O(l)$.

- The designator (aq) is used for undissociated, uncharged aqueous species, *e.g.*, $U(OH)_4(aq)$, $CO_2(aq)$. Since ionic gases are not considered in this review, all ions may be assumed to be aqueous and are not designed with (aq). If a chemical reaction refers to a medium other than H_2O (*e.g.*, D_2O, 90% ethanol/10% H_2O), then (aq) is replaced by a more explicit designator, *e.g.*, "(in D_2O)" or "(sln)". In the case of (sln), the composition of the solution is described in the text.

- The designator (sln) is used for substances in solution without specifying the actual equilibrium composition of the substance in the solution. Note the difference in the designation of H_2O in Eqs. (II.2) and (II.3). $H_2O(l)$ in Reaction (II.2) indicates that H_2O is present as a pure liquid, *i.e.*, no solutes are present, whereas Reaction (II.3) involves a HCl solution, in which the thermodynamic properties of $H_2O(sln)$ may not be the same as those of the pure liquid $H_2O(l)$. In dilute solutions, however, this difference in the thermodynamic properties of H_2O can be neglected, and $H_2O(sln)$ may be regarded as pure $H_2O(l)$.

  Examples:

$$UOCl_2(cr) + 2HBr(sln) \rightleftharpoons UOBr_2(cr) + 2HCl(sln) \qquad (II.1)$$

$$UO_2Cl_2 \cdot 3H_2O(cr) \rightleftharpoons UO_2Cl_2 \cdot H_2O(cr) + 2H_2O(l) \qquad (II.2)$$
$$UO_3(\gamma) + 2HCl(sln) \rightleftharpoons UO_2Cl_2(cr) + H_2O(sln) \qquad (II.3)$$

- The designators (cr), (am), (vit), and (s) are used for solid substances. (cr) is used when it is known that the compound is crystalline, (am) when it is known that it is amorphous, and (vit) for glassy substances. Otherwise, (s) is used.
- In some cases, more than one crystalline form of the same chemical composition may exist. In such a case, the different forms are distinguished by separate designators that describe the forms more precisely. If the crystal has a mineral name, the designator (cr) is replaced by the first four characters of the mineral name in parentheses, *e.g.*, SiO_2(quar) for quartz and SiO_2(chal) for chalcedony. If there is no mineral name, the designator (cr) is replaced by a Greek letter preceding the formula and indicating the structural phase, *e.g.*, α-UF_5, β-UF_5.

Phase designators are also used in conjunction with thermodynamic symbols to define the state of aggregation of a compound a thermodynamic quantity refers to. The notation is in this case the same as outlined above. In an extended notation (*cf.* [82LAF]) the reference temperature is usually given in addition to the state of aggregation of the composition of a mixture.

Examples:

$\Delta_f G^\circ_m(Na^+, aq, 298.15\,K)$	standard molar Gibbs energy of formation of aqueous Na^+ at 298.15 K
$S^\circ_m(UO_2SO_4 \cdot 2.5H_2O, cr, 298.15\,K)$	standard molar entropy of $UO_2SO_4 \cdot 2.5H_2O$(cr) at 298.15 K
$C^\circ_{p,m}(UO_3, \alpha, 298.15\,K)$	standard molar heat capacity of α-UO_3 at 298.15 K
$\Delta_f H_m(HF, sln, HF \cdot 7.8H_2O)$	enthalpy of formation of HF diluted 1:7.8 with water

II.1.5. Processes

Chemical processes are denoted by the operator Δ, written before the symbol for a property, as recommended by IUPAC [82LAF]. An exception to this rule is the equilibrium constant, *cf.* Section II.1.6. The nature of the process is denoted by annotation of the Δ, *e.g.*, the Gibbs energy of formation, $\Delta_f G_m$, the enthalpy of sublimation, $\Delta_{sub} H_m$, *etc.* The abbreviations of chemical processes are summarized in Table II.3.

The most frequently used symbols for processes are $\Delta_f G$ and $\Delta_f H$, the Gibbs energy and the enthalpy of formation of a compound or complex from the elements in their reference states (*cf.* Table II.6).

II.1.6. Equilibrium constants

The IUPAC has not explicitly defined the symbols and terminology for equilibrium constants of reactions in aqueous solution. The NEA has therefore adopted the con-

Table II.3: Abbreviations used as subscripts of Δ to denote the type of chemical processes.

Subscript of Δ	Chemical process
at	separation of a substance into its constituent gaseous atoms (atomization)
dehyd	elimination of water of hydration (dehydration)
dil	dilution of a solution
f	formation of a compound from its constituent elements
fus	melting (fusion) of a solid
hyd	addition of water of hydration to an unhydrated compound
mix	mixing of fluids
r	chemical reaction (general)
sol	process of dissolution
sub	sublimation (evaporation) of a solid
tr	transfer from one solution or liquid phase to another
trs	transition of one solid phase to another
vap	vaporization (evaporation) of a liquid

ventions that have been used in the work *Stability constants of metal ion complexes* by Sillén and Martell [64SIL/MAR, 71SIL/MAR]. An outline is given in the paragraphs below. Note that, for some simple reactions, there may be different correct ways to index an equilibrium constant. It may sometimes be preferable to indicate the number of the reaction the data refer to, especially in cases where several ligands are discussed that might be confused. For example, for the equilibrium

$$m\mathrm{M} + q\mathrm{L} \rightleftharpoons \mathrm{M}_m\mathrm{L}_q \qquad \text{(II.4)}$$

both $\beta_{q,m}$ and β(II.4) would be appropriate, and $\beta_{q,m}$(II.4) is accepted, too. Note that, in general, K is used for the consecutive or stepwise formation constant, and β is used for the cumulative or overall formation constant. In the following outline, charges are only given for actual chemical species, but are omitted for species containing general symbols (M, L).

II.1.6.1. Protonation of a ligand

$$\mathrm{H}^+ + \mathrm{H}_{r-1}\mathrm{L} \rightleftharpoons \mathrm{H}_r\mathrm{L} \qquad K_{1,r} = \frac{[\mathrm{H}_r\mathrm{L}]}{[\mathrm{H}^+][\mathrm{H}_{r-1}\mathrm{L}]} \qquad \text{(II.5)}$$

$$r\mathrm{H}^+ + \mathrm{L} \rightleftharpoons \mathrm{H}_r\mathrm{L} \qquad \beta_{1,r} = \frac{[\mathrm{H}_r\mathrm{L}]}{[\mathrm{H}^+]^r[\mathrm{L}]} \qquad \text{(II.6)}$$

This notation has been proposed and used by Sillén and Martell [64SIL/MAR], but it has been simplified later by the same authors [71SIL/MAR] from $K_{1,r}$ to K_r. This review retains, for the sake of consistency, *cf.* Eqs. (II.7) and (II.8), the older formulation of $K_{1,r}$.

For the addition of a ligand, the notation shown in Eq.(II.7) is used.

$$\mathrm{HL}_{q-1} + \mathrm{L} \rightleftharpoons \mathrm{HL}_q \qquad K_q = \frac{[\mathrm{HL}_q]}{[\mathrm{HL}_{q-1}][\mathrm{L}]} \qquad \text{(II.7)}$$

Eq. (II.8) refers to the overall formation constant of the species $\mathrm{H}_r\mathrm{L}_q$.

$$r\mathrm{H}^+ + q\mathrm{L} \rightleftharpoons \mathrm{H}_r\mathrm{L}_q \qquad \beta_{q,r} = \frac{[\mathrm{H}_r\mathrm{L}_q]}{[\mathrm{H}^+]^r[\mathrm{L}]^q} \qquad \text{(II.8)}$$

In Eqs. (II.5), (II.6) and (II.8), the second subscript r can be omitted if $r = 1$, as shown in Eq. (II.7).

Example:

$$\mathrm{H}^+ + \mathrm{PO}_4^{3-} \rightleftharpoons \mathrm{HPO}_4^{2-} \qquad \beta_{1,1} = \beta_1 = \frac{[\mathrm{HPO}_4^{2-}]}{[\mathrm{H}^+][\mathrm{PO}_4^{3-}]}$$

$$2\mathrm{H}^+ + \mathrm{PO}_4^{3-} \rightleftharpoons \mathrm{H}_2\mathrm{PO}_4^- + 2\mathrm{H}^+ \qquad \beta_{1,2} = \frac{[\mathrm{H}_2\mathrm{PO}_4^-]}{[\mathrm{H}^+]^2[\mathrm{PO}_4^{3-}]}$$

II.1.6.2. Formation of metal ion complexes

$$\mathrm{ML}_{q-1} + \mathrm{L} \rightleftharpoons \mathrm{ML}_q \qquad K_q = \frac{[\mathrm{ML}_q]}{[\mathrm{ML}_{q-1}][\mathrm{L}]} \qquad \text{(II.9)}$$

$$\mathrm{M} + q\mathrm{L} \rightleftharpoons \mathrm{ML}_q \qquad \beta_q = \frac{[\mathrm{ML}_q]}{[\mathrm{M}][\mathrm{L}]^q} \qquad \text{(II.10)}$$

For the addition of a metal ion, *i.e.*, the formation of polynuclear complexes, the following notation is used, analogous to Eq.(II.5):

$$\mathrm{M} + \mathrm{M}_{q-1}\mathrm{L} \rightleftharpoons \mathrm{M}_q\mathrm{L} \qquad K_{1,q} = \frac{[\mathrm{M}_q\mathrm{L}]}{[\mathrm{M}][\mathrm{M}_{q-1}\mathrm{L}]} \qquad \text{(II.11)}$$

Eq. (II.12) refers to the overall formation constant of a complex $\mathrm{M}_m\mathrm{L}_q$.

$$m\mathrm{M} + q\mathrm{L} \rightleftharpoons \mathrm{M}_m\mathrm{L}_q \qquad \beta_{q,m} = \frac{[\mathrm{M}_m\mathrm{L}_q]}{[\mathrm{M}]^m[\mathrm{L}]^q} \qquad \text{(II.12)}$$

The second index can be omitted if it is equal to 1, *i.e.*, $\beta_{q,m}$ becomes β_q if $m = 1$. The formation constants of mixed ligand complexes are not indexed. In this case, it is necessary to list the chemical reactions considered and to refer the constants to the corresponding reaction numbers.

It has sometimes been customary to use negative values for the indices of the protons to indicate complexation with hydroxide ions, OH^-. This practice is not adopted in this review. If OH^- occurs as a reactant in the notation of the equilibrium, it is treated like a normal ligand L, but in general formulae the index variable n is used instead of q. If H_2O occurs as a reactant to form hydroxide complexes, H_2O is considered as a protonated ligand, HL, so that the reaction is treated as described below in Eqs. (II.13) to (II.15) using n as the index variable. For convenience, no general form is used for the stepwise constants for the formation of the complex $M_mL_qH_r$. In many experiments, the formation constants of metal ion complexes are determined by adding to a metal ion solution a ligand in its protonated form. The complex formation reactions thus involve a deprotonation reaction of the ligand. If this is the case, the equilibrium constant is supplied with an asterisk, as shown in Eqs. (II.13) and (II.14) for mononuclear and in Eq. (II.15) for polynuclear complexes.

$$\mathrm{ML}_{q-1} + \mathrm{HL} \rightleftharpoons \mathrm{ML}_q + \mathrm{H}^+ \qquad {}^*K_q = \frac{[\mathrm{ML}_q][\mathrm{H}^+]}{[\mathrm{ML}_{q-1}][\mathrm{HL}]} \tag{II.13}$$

$$\mathrm{M} + q\mathrm{HL} \rightleftharpoons \mathrm{ML}_q + q\mathrm{H}^+ \qquad {}^*\beta_q = \frac{[\mathrm{ML}_q][\mathrm{H}^+]^q}{[\mathrm{M}][\mathrm{HL}]^q} \tag{II.14}$$

$$m\mathrm{M} + q\mathrm{HL} \rightleftharpoons \mathrm{M}_m\mathrm{L}_q + q\mathrm{H}^+ \qquad {}^*\beta_{q,m} = \frac{[\mathrm{M}_m\mathrm{L}_q][\mathrm{H}^+]^q}{[\mathrm{M}]^m[\mathrm{HL}]^q} \tag{II.15}$$

Examples:

$$\mathrm{UO_2^{2+}} + \mathrm{HF(aq)} \rightleftharpoons \mathrm{UO_2F^+} + \mathrm{H^+} \qquad {}^*K_1 = {}^*\beta_1 = \frac{[\mathrm{UO_2F^+}][\mathrm{H^+}]}{[\mathrm{UO_2^{2+}}][\mathrm{HF(aq)}]}$$

$$3\mathrm{UO_2^{2+}} + 5\mathrm{H_2O(l)} \rightleftharpoons (\mathrm{UO_2})_3(\mathrm{OH})_5^+ + 5\mathrm{H^+} \qquad {}^*\beta_{5,3} = \frac{[(\mathrm{UO_2})_3(\mathrm{OH})_5^+][\mathrm{H^+}]^5}{[\mathrm{UO_2^{2+}}]^3}$$

Note that an asterisk is only assigned to the formation constant if the protonated ligand that is added is deprotonated during the reaction. If a protonated ligand is added and coordinated as such to the metal ion, the asterisk is to be omitted, as shown in Eq. (II.16).

$$\mathrm{M} + q\mathrm{H}_r\mathrm{L} \rightleftharpoons \mathrm{M}(\mathrm{H}_r\mathrm{L})_q \qquad \beta_q = \frac{[\mathrm{M}(\mathrm{H}_r\mathrm{L})_q]}{[\mathrm{M}][\mathrm{H}_r\mathrm{L}]^q} \tag{II.16}$$

Example:

$$\mathrm{UO_2^{2+}} + 3\mathrm{H_2PO_4^-} \rightleftharpoons \mathrm{UO_2(H_2PO_4)_3^-} \qquad \beta_3 = \frac{[\mathrm{UO_2(H_2PO_4)_3^-}]}{[\mathrm{UO_2^{2+}}][\mathrm{H_2PO_4^-}]^3}$$

II.1.6.3. Solubility constants

Conventionally, equilibrium constants involving a solid compound are denoted as "solubility constants" rather than as formation constants of the solid. An index "s" to the equilibrium constant indicates that the constant refers to a solubility process, as shown in Eqs. (II.17) to (II.19).

$$\mathrm{M}_a\mathrm{L}_b(\mathrm{s}) \rightleftharpoons a\mathrm{M} + b\mathrm{L} \qquad K_{\mathrm{s},0} = [\mathrm{M}]^a[\mathrm{L}]^b \tag{II.17}$$

$K_{s,0}$ is the conventional solubility product, and the subscript "0" indicates that the equilibrium reaction involves only uncomplexed aqueous species. If the solubility constant includes the formation of aqueous complexes, a notation analogous to that of Eq. (II.12) is used:

$$\frac{m}{a}\mathrm{M}_a\mathrm{L}_b(\mathrm{s}) \rightleftharpoons \mathrm{M}_m\mathrm{L}_q + \left(\frac{mb}{a} - q\right)\mathrm{L} \qquad K_{\mathrm{s},q,m} = [\mathrm{M}_m\mathrm{L}_q][\mathrm{L}]^{\left(\frac{mb}{a}-q\right)} \tag{II.18}$$

Example:

$$\mathrm{UO_2F_2(cr)} \rightleftharpoons \mathrm{UO_2F^+ + F^-} \qquad K_{\mathrm{s},1,1} = K_{\mathrm{s},1} = [\mathrm{UO_2F^+}][\mathrm{F^-}]$$

Similarly, an asterisk is added to the solubility constant if it simultaneously involves a protonation equilibrium:

$$\frac{m}{a}\mathrm{M}_a\mathrm{L}_b(\mathrm{s}) + \left(\frac{mb}{a} - q\right)\mathrm{H^+} \rightleftharpoons \mathrm{M}_m\mathrm{L}_q + \left(\frac{mb}{a} - q\right)\mathrm{HL}$$

$$^{*}K_{\mathrm{s},q,m} = \frac{[\mathrm{M}_m\mathrm{L}_q][\mathrm{HL}]^{\left(\frac{mb}{a}-q\right)}}{[\mathrm{H^+}]^{\left(\frac{mb}{a}-q\right)}} \tag{II.19}$$

Example:

$$\mathrm{U(HPO_4)_2 \cdot 4H_2O(cr) + H^+} \rightleftharpoons \mathrm{UHPO_4^{2+} + H_2PO_4^- + 4H_2O(l)}$$

$$^{*}K_{\mathrm{s},1,1} = {}^{*}K_{\mathrm{s},1} = \frac{[\mathrm{UHPO_4^{2+}}][\mathrm{H_2PO_4^-}]}{[\mathrm{H^+}]}$$

II.1.6.4. Equilibria involving the addition of a gaseous ligand

A special notation is used for constants describing equilibria that involve the addition of a gaseous ligand, as outlined in Eq. (II.20).

$$\mathrm{ML}_{q-1} + \mathrm{L(g)} \rightleftharpoons \mathrm{ML}_q \qquad K_{\mathrm{p},q} = \frac{[\mathrm{ML}_q]}{[\mathrm{ML}_{q-1}]p_{\mathrm{L}}} \tag{II.20}$$

The subscript "p" can be combined with any other notations given above.

Examples:

$$\mathrm{CO_2(g)} \rightleftharpoons \mathrm{CO_2(aq)} \qquad K_{\mathrm{p}} = \frac{[\mathrm{CO_2(aq)}]}{p_{\mathrm{CO_2}}}$$

$$3UO_2^{2+} + 6CO_2(g) + 6H_2O(l) \rightleftharpoons (UO_2)_3(CO_3)_6^{6-} + 12H^+$$

$$^*\beta_{p,6,3} = \frac{[(UO_2)_3(CO_3)_6^{6-}][H^+]^{12}}{[UO_2^{2+}]^3 p_{CO_2}^6}$$

$$UO_2CO_3(cr) + CO_2(g) + H_2O(l) \rightleftharpoons UO_2(CO_3)_2^{2-} + 2H^+$$

$$^*K_{p,s,2} = \frac{[UO_2(CO_3)_2^{2-}][H^+]^2}{p_{CO_2}}$$

In cases where the subscripts become complicated, it is recommended that K or β be used with or without subscripts, but always followed by the equation number of the equilibrium to which it refers.

II.1.6.5. Redox equilibria

Redox reactions are usually quantified in terms of their electrode (half cell) potential, E. E is identical to the electromotive force (emf) of a redox reaction that involves the standard hydrogen electrode as an electron donor or acceptor. In this review, electrode potentials are given as reduction potentials relative to the standard hydrogen electrode which acts as an electron donor. The standard redox potential, E°, is related to the Gibbs energy change $\Delta_r G_m^\circ$ and the equilibrium constant K as outlined in Eq. (II.21).

$$E^\circ = -\frac{1}{nF}\Delta_r G_m^\circ = \frac{RT}{nF}\ln K \qquad \text{(II.21)}$$

The symbol E° is used for the emf of a galvanic cell relative to the standard hydrogen electrode, *e.g.*, Reaction (II.22).

$$Fe^{3+} + \tfrac{1}{2}H_2(g) \rightleftharpoons Fe^{2+} + H^+ \qquad \text{(II.22)}$$

In the standard hydrogen electrode, $H_2(g)$ is at unit fugacity (an ideal gas at unit pressure, 0.1 MPa), and H^+ is at unit activity. Reaction (II.22) can be formally represented by half cell reactions in which an equal number of electrons are involved, as shown in Eqs. (II.23) and (II.24).

$$Fe^{3+} + e^- \rightleftharpoons Fe^{2+} \qquad \text{(II.23)}$$

$$\tfrac{1}{2}H_2(g) \rightleftharpoons H^+ + e^- \qquad \text{(II.24)}$$

Equilibrium constants may be written for these half cell reactions in the following way:

$$K^\circ(\text{II.23}) = \frac{a_{Fe^{2+}}}{a_{Fe^{3+}} \times a_{e^-}}$$

$$K^\circ(\text{II.24}) = \frac{a_{H^+} \times a_{e^-}}{\sqrt{p_{H_2}}} = 1 \text{ (by definition)} \qquad \text{(II.25)}$$

In addition, $\Delta_r G^\circ_m$(II.24) = $\Delta_r H^\circ_m$(II.24) = $\Delta_r S^\circ_m$(II.24) = 0 by definition, for all temperatures and pressures. The following equations describe the change in the Gibbs energy and redox potential of Reaction (II.22), if p_{H_2} and a_{H^+} are equal to unity:

$$\begin{aligned}\Delta_r G_m(\text{II.22}) &= \Delta_r G^\circ_m(\text{II.22}) + RT\ln\left(\frac{a_{Fe^{2+}}}{a_{Fe^{3+}}}\right)\\ E(\text{II.22}) &= E^\circ(\text{II.22}) - \frac{RT}{nF}\ln\left(\frac{a_{Fe^{2+}}}{a_{Fe^{3+}}}\right)\end{aligned}$$

II.1.7. Order of formulae

To be consistent with CODATA, the data tables are given in "Standard Order of Arrangement" [82WAG/EVA]. This scheme is presented in Figure II.1 below which shows the sequence of the ranks of the elements in this convention. The order follows the ranks of the elements. For uranium, this means that, after elemental uranium and its monoatomic ions (*e.g.*, U^{4+}), the uranium compounds and complexes with oxygen are listed, then those with hydrogen, then those with oxygen and hydrogen, and so on, with decreasing rank of the element and combinations of the elements. Within a class, increasing coefficients of the higher rank elements go before increasing coefficients of the lower rank elements. For example, in the U-O-F class of compounds and complexes, a typical sequence would be UOF_2(cr), UOF_4(cr), UOF_4(g), UO_2F(aq), UO_2F^+, UO_2F_2(aq), UO_2F_2(cr), UO_2F_2(g), $UO_2F_3^-$, $UO_2F_4^{2-}$, $U_2O_3F_6$(cr), *etc.* Formulae with identical stoichiometry are in alphabetical order of their designators.

II.1.8. Reference codes

The references cited in the review are ordered chronologically and alphabetically by the first two authors within each year, as described by CODATA [87GAR/PAR]. A reference code is made up of the final two digits of the year of appearance (if the publication is not from the 20th century, the year will be put in full). The year is followed by the first three letters of the first two authors, separated by a slash. If there are multiple reference codes, a "2" will be added to the second one, a "3" to the third one, and so forth. Reference codes are always enclosed in square brackets.

Examples:

[81HEL/KIR] Helgeson, H.C., Kirkham, D.H., Flowers, G.C., Theoretical prediction of the thermodynamic behaviour of aqueous electrolytes at high temperatures: IV. Calculation of activity coefficients, osmotic coefficients, and apparent molal and standard and relative partial molal properties to 600°C and 5 KB, Am. J. Sci., **281** (1981) 1249-1516.

[88BAR] Barten, H., Comparison of the thermochemistry of uranyl/uranium phosphates and arsenates Thermochim. Acta, **124** (1988) 339-344.

Figure II.1: Standard order of arrangement of the elements and compounds based on the periodic classification of the elements (reproduced by permission from Ref. [82WAG/EVA]).

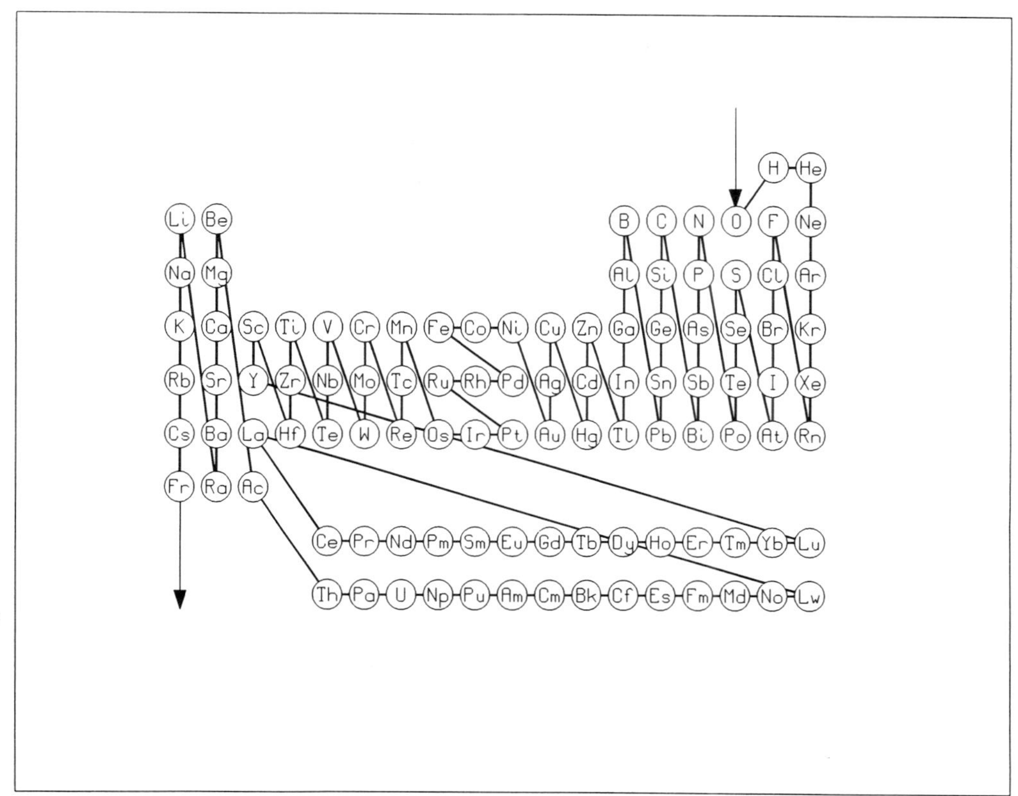

[71AHR/KUL] Ahrland, S., Kullberg, L., Thermodynamics of metal complex formation in aqueous solution: I. A potentiometric study of fluoride complexes of hydrogen, uranium(VI) and vanadium(IV), Acta Chem. Scand., **25** (1971) 3457-3470.

[71AHR/KUL2] Ahrland, S., Kullberg, L., Thermodynamics of metal complex formation in aqueous solution: II. A calorimetric study of fluoride complexes of hydrogen, uranium(VI) and vanadium(IV), Acta Chem. Scand., **25** (1971) 3471-3483.

[1896ALO] Aloy, J., Recherches thermiques sur les composés de l'uranium, C. R. Hebd. Séances Acad. Sci., Ser. C, **122** (1896) 1451-1453.

The assignment of the reference codes is done automatically by the NEA updating programs for the TDB data base. It is therefore possible that multiple reference codes in the TDB data base do not occur in multiple form in the present volume.

The designators "2", "3", *etc.*, are nevertheless retained for reasons of compatibility with the TDB data base.

II.2. Units and conversion factors

Thermodynamic data are given according to the *Système International d'unité* (SI units). The unit of energy is the joule. Some basic conversion factors, also for non-thermodynamic units, are given in Table II.4.

Table II.4: Unit conversion factors.

To convert from (non-SI unit symbol)	to (SI unit symbol)	multiply by
ångström (Å)	metre (m)	1×10^{-10} (exactly)
standard atmosphere (atm)	pascal (Pa)	1.01325×10^{5} (exactly)
bar (bar)	pascal (Pa)	1×10^{5} (exactly)
thermochemical calorie (cal)	joule (J)	4.184 (exactly)
entropy unit (e.u. $\hat{=}$ $\text{cal} \cdot \text{K}^{-1} \cdot \text{mol}^{-1}$)	$\text{J} \cdot \text{K}^{-1} \cdot \text{mol}^{-1}$	4.184 (exactly)

Since a large part of the NEA-TDB project deals with the thermodynamics of aqueous solutions, the units describing the amount of dissolved substance are used very frequently. For convenience, this review uses "M" as an abbreviation of "$\text{mol} \cdot \text{dm}^{-3}$" for molarity, c, and "m" as an abbreviation of "$\text{mol} \cdot \text{kg}^{-1}$" for molality, m. It is often necessary to convert concentration data from molarity to molality and vice versa. This conversion is used for the correction and extrapolation of equilibrium data to zero ionic strength by the specific ion interaction theory which works in molality units (*cf.* Appendix B). This conversion is made n the following way. Molality is defined as m_B moles of substance B dissolved in 1000 grams of pure water. Molarity is defined as c_B moles of substance B dissolved in $(1000\rho - c_B M)$ grams of pure water, where ρ is the density of the solution and M the molar weight of the solute. From this it follows that

$$m_B = \frac{1000 c_B}{1000\rho - c_B M} \qquad \text{(II.26)}$$

Baes and Mesmer [76BAE/MES, *p.*439] give a table with these conversion factors for nine electrolytes and various ionic strengths. Their values are reproduced in Table II.5.

Table II.5: Factors ϱ for the conversion of molarity, c_B, to molality, m_B, of a substance B, in various media (reproduced by permission from Ref. [76BAE/MES]).

	$\varrho = m_B/c_B$ (dm^3 of solution per kg of H_2O)								
c (M)	$NaClO_4$	NaCl	KCl	LiCl	$NaNO_3$	KNO_3	$Mg(ClO_4)_2$	$Ca(ClO_4)_2$	$Ba(ClO_4)_2$
0.1	1.0073	1.0046	1.0057	1.0047	1.0058	1.0068	1.0096	1.0102	1.0108
0.2	1.0118	1.0064	1.0085	1.0065	1.0087	1.0109	1.0167	1.0178	1.0190
0.5	1.0256	1.0120	1.0173	1.0121	1.0178	1.0233	1.0394	1.0422	1.0451
1	1.0499	1.0218	1.0327	1.0218	1.0338	1.0452	1.0821	1.0877	1.0930
1.5	1.0759	1.0323	1.0491	1.0321	1.0509	1.0686	1.1308	1.1396	1.1468
2	1.1037	1.0434	1.0664	1.0428	1.0691	1.0935	1.1863	1.1986	1.2067
3	1.1652	1.0672	1.1041	1.0655	1.1084	1.1481			
4	1.2364	1.0933	1.1457	1.0900	1.152				
5		1.1218		1.1163	1.202				
6				1.1446	1.256				

Examples:

$$\begin{aligned}
1.00\text{ M NaCl} &\hat{=} 1.02\text{ m NaCl}\\
1.00\text{ M NaClO}_4 &\hat{=} 1.05\text{ m NaClO}_4\\
2.00\text{ M KNO}_3 &\hat{=} 2.19\text{ m KNO}_3\\
4.00\text{ M NaClO}_4 &\hat{=} 4.95\text{ m NaClO}_4\\
6.00\text{ M NaNO}_3 &\hat{=} 7.54\text{ m NaNO}_3
\end{aligned}$$

It should be noted that equilibrium constants, unless they are dimensionless, need also to be converted if the concentration scale is changed from molarity to molality or vice versa. For a general equilibrium reaction, $0 = \sum_B \nu_B B$, the equilibrium constants can be expressed either in molarity or molality units, K_c or K_m, respectively:

$$\begin{aligned}
\log_{10} K_c &= \sum_B \nu_B \log_{10} c_B\\
\log_{10} K_m &= \sum_B \nu_B \log_{10} m_B
\end{aligned}$$

With $(m_B/c_B) = \varrho$, or $(\log_{10} m_B - \log_{10} c_B) = \log_{10} \varrho$, the relationship between K_c and K_m becomes very simple, as shown in Eq. (II.27).

$$\log_{10} K_m = \log_{10} K_c + \sum_B \nu_B \log_{10} \varrho \qquad \text{(II.27)}$$

$\sum_B \nu_B$ is the sum of the stoichiometric coefficients of the reaction, and the values of ϱ are the factors for the conversion of molarity to molality as tabulated in Table II.5 for nine different electrolyte media.

II.3. Standard and reference conditions

II.3.1. Standard state

A precise definition of the term "standard state" has been given by IUPAC [82LAF]. The fact that only changes in thermodynamic parameters, but not their absolute values, can be determined experimentally, makes it important to have a well-defined standard state that forms a base line to which the effect of variations can be referred. The IUPAC [82LAF] definition of the standard state has been adopted in the NEA-TDB project. The standard state pressure, $p^\circ = 0.1\,\text{MPa}$ (1 bar), has therefore also been adopted, *cf.* Section II.3.2. The application of the standard state principle to pure substances and mixtures is summarized below. It should be noted that the standard state is always linked to a reference temperature, *cf.* Section II.3.3.

- The standard state for a gaseous substance, whether pure or in a gaseous mixture, is the pure substance at the standard state pressure and in a (hypothetical) state in which it exhibits ideal gas behaviour.
- The standard state for a pure liquid substance is (ordinarily) the pure liquid at the standard state pressure.
- The standard state for a pure solid substance is (ordinarily) the pure solid at the standard state pressure.
- The standard state for a solute B in a solution is a hypothetical solution, at the standard state pressure, in which $m_\text{B} = m^\circ = 1\,\text{mol}\cdot\text{kg}^{-1}$, and in which the activity coefficient γ_B is unity.

It should be emphasized that the use of $^\circ$, *e.g.*, in $\Delta_\text{f}H^\circ_\text{m}$, implies that the compound in question is in the standard state and that the elements are in their reference states. The reference states of the elements at the reference temperature (*cf.* Section II.3.3) are listed in Table II.6.

II.3.2. Standard state pressure

The standard state pressure chosen for all selected data is 0.1 MPa (1 bar) as recommended by the International Union of Pure and Applied Chemistry IUPAC [82LAF]. However, the majority of the thermodynamic data published in the scientific literature and used for the evaluations in this review, refer to the old standard state pressure of 1 "standard atmosphere" (= 0.101325 MPa). The difference between the thermodynamic data for the two standard state pressures is not large and lies in most cases within the uncertainty limits. It is nevertheless essential to make the corrections for the change in the standard state pressure in order to avoid inconsistencies and propagation of errors. In practice the parameters affected by the change between these two standard state pressures are the Gibbs energy and entropy changes of all processes that involve gaseous species. Consequently, changes occur also in the Gibbs energies of formation of species that consist of elements whose reference

Table II.6: Reference states for the elements at the reference temperature of 298.15 K.

O_2	gaseous	Pb	crystalline, cubic
H_2	gaseous	B	β, crystalline, rhombohedral
He	gaseous	Al	crystalline, cubic
Ne	gaseous	Zn	crystalline, hexagonal
Ar	gaseous	Cd	crystalline, hexagonal
Kr	gaseous	Hg	liquid
Xe	gaseous	Cu	crystalline, cubic
F_2	gaseous	Ag	crystalline, cubic
Cl_2	gaseous	Fe	crystalline, cubic
Br_2	liquid	V	crystalline, cubic
I_2	crystalline, orthorhombic	Ti	crystalline, hexagonal
S	crystalline, orthorhombic	U	crystalline, orthorhombic
Se	crystalline, hexagonal	Th	crystalline, cubic
Te	crystalline, hexagonal	Be	crystalline, hexagonal
N_2	gaseous	Mg	crystalline, hexagonal
P	crystalline, cubic ("white")	Ca	crystalline, cubic
As	crystalline, rhombohedral ("gray")	Sr	crystalline, cubic
Sb	crystalline, rhombohedral	Ba	crystalline, cubic
Bi	crystalline, rhombohedral	Li	crystalline, cubic
C	crystalline, hexagonal (graphite)	Na	crystalline, cubic
Si	crystalline, cubic	K	crystalline, cubic
Ge	crystalline, cubic	Rb	crystalline, cubic
Sn	crystalline, tetragonal ("white")	Cs	crystalline, cubic

state is gaseous (H, O, F, Cl, N, and the noble gases). No other parameters are affected significantly. A large part of the following discussion has been taken from the NBS tables of chemical thermodynamic properties [82WAG/EVA], see also Freeman [84FRE].

The following expressions define the effect of pressure on the properties of all substances:

$$\left(\frac{\partial H}{\partial p}\right)_T = V - T\left(\frac{\partial V}{\partial T}\right)_p = V(1-\alpha T) \tag{II.28}$$

$$\left(\frac{\partial C_p}{\partial p}\right)_T = -T\left(\frac{\partial^2 V}{\partial T^2}\right)_p \tag{II.29}$$

$$\left(\frac{\partial S}{\partial p}\right)_T = -V\alpha = -\left(\frac{\partial V}{\partial T}\right)_p \tag{II.30}$$

$$\left(\frac{\partial G}{\partial p}\right)_T = V, \tag{II.31}$$

where
$$\alpha \equiv \frac{1}{V}\left(\frac{\partial V}{\partial T}\right)_p . \tag{II.32}$$

For ideal gases, $V = \frac{RT}{p}$ and $\alpha = \frac{R}{pV} = \frac{1}{T}$. The conversion equations listed below (Eqs. (II.33) to (II.40)) apply to the small pressure change from 1 atm to 1 bar (0.1 MPa). The quantities that refer to the old standard state pressure of 1 atm are assigned the superscript $^{(\text{atm})}$ here, the ones that refer to the new standard state pressure of 1 bar the superscript $^{(\text{bar})}$.

For all substances the change in the enthalpy of formation and the heat capacity is much smaller than the experimental accuracy and can be disregarded. This is exactly true for ideal gases.

$$\Delta_{\text{f}}H^{(\text{bar})}(T) - \Delta_{\text{f}}H^{(\text{atm})}(T) = 0 \tag{II.33}$$
$$C_p^{(\text{bar})}(T) - C_p^{(\text{atm})}(T) = 0 \tag{II.34}$$

For gaseous substances, the entropy difference is

$$\begin{aligned} S^{(\text{bar})}(T) - S^{(\text{atm})}(T) &= R\ln\left(\frac{p^{(\text{atm})}}{p^{(\text{bar})}}\right) \\ &= R\ln 1.01325 \\ &= 0.1094\,\text{J}\cdot\text{K}^{-1}\cdot\text{mol}^{-1}. \end{aligned} \tag{II.35}$$

This is exactly true for ideal gases, as follows from Eq. (II.30) with $\alpha = \frac{R}{pV}$. The entropy change of a reaction or process is thus dependent on the number of moles of gases involved:

$$\begin{aligned} \Delta_{\text{r}}S^{(\text{bar})} - \Delta_{\text{r}}S^{(\text{atm})} &= \delta \times R\ln\left(\frac{p^{(\text{atm})}}{p^{(\text{bar})}}\right) \\ &= \delta \times 0.1094\,\text{J}\cdot\text{K}^{-1}\cdot\text{mol}^{-1}, \end{aligned} \tag{II.36}$$

where δ is the net increase in moles of gas in the process.

Similarly, the change in the Gibbs energy of a process between the two standard state pressures is

$$\begin{aligned} \Delta_{\text{r}}G^{(\text{bar})} - \Delta_{\text{r}}G^{(\text{atm})} &= -\delta \times RT\ln\left(\frac{p^{(\text{atm})}}{p^{(\text{bar})}}\right) \\ &= -\delta \times 0.03263\,\text{kJ}\cdot\text{mol}^{-1} \text{ at } 298.15\,\text{K}. \end{aligned} \tag{II.37}$$

Eq. (II.37) applies also to $\Delta_{\text{f}}G^{(\text{bar})} - \Delta_{\text{f}}G^{(\text{atm})}$, since the Gibbs energy of formation describes the formation process of a compound or complex from the reference states of the elements involved:

$$\Delta_{\text{r}}G^{(\text{bar})} - \Delta_{\text{r}}G^{(\text{atm})} = -\delta \times 0.03263\,\text{kJ}\cdot\text{mol}^{-1} \text{ at } 298.15\,\text{K}. \tag{II.38}$$

The change in the equilibrium constants and cell potentials with the change in the standard state pressure follows from the expression for Gibbs energy changes, Eq. (II.37):

$$\begin{aligned}
\log_{10} K^{(\mathrm{bar})} - \log_{10} K^{(\mathrm{atm})} &= -\frac{\Delta_r G^{(\mathrm{bar})} - \Delta_r G^{(\mathrm{atm})}}{RT \ln 10} \\
&= \delta \times \frac{\ln\left(\frac{p^{(\mathrm{atm})}}{p^{(\mathrm{bar})}}\right)}{\ln 10} = \delta \times \log_{10}\left(\frac{p^{(\mathrm{atm})}}{p^{(\mathrm{bar})}}\right) \\
&= \delta \times 0.005717 \qquad \text{(II.39)}
\end{aligned}$$

$$\begin{aligned}
E^{(\mathrm{bar})} - E^{(\mathrm{atm})} &= -\frac{\Delta_r G^{(\mathrm{bar})} - \Delta_r G^{(\mathrm{atm})}}{nF} \\
&= \delta \times \frac{RT \ln\left(\frac{p^{(\mathrm{atm})}}{p^{(\mathrm{bar})}}\right)}{nF} \\
&= \delta \times \frac{0.0003382}{n} \mathrm{V} \quad \text{at } 298.15\,\mathrm{K}. \qquad \text{(II.40)}
\end{aligned}$$

It should be noted that the standard potential of the hydrogen electrode is equal to 0.00 V exactly, by definition.

$$H^+ + e^- \rightleftharpoons \tfrac{1}{2}H_2(g) \qquad E^\circ \overset{\mathrm{def}}{=} 0.00\,\mathrm{V} \qquad \text{(II.41)}$$

This definition will not be changed, although a gaseous substance, $H_2(g)$, is involved in the process. The change in the potential with pressure for an electrode potential conventionally written as

$$Ag^+ + e^- \rightleftharpoons Ag(cr)$$

should thus be calculated from the balanced reaction that includes the hydrogen electrode,

$$Ag^+ + \tfrac{1}{2}H_2(g) \rightleftharpoons Ag(cr) + H^+.$$

Here $\delta = -0.5$. Hence, the contribution to δ from an electron in a half cell reaction is the same as the contribution of a gas molecule with the stoichiometric coefficient of 0.5. This leads to the same value of δ as the combination with the hydrogen half cell.

Examples:

Reaction	δ	
$Fe(cr) + 2H^+ \rightleftharpoons Fe^{2+} + H_2(g)$	$\delta = 1$	$E^{(\mathrm{bar})} - E^{(\mathrm{atm})} = 0.00017\,\mathrm{V}$
$CO_2(g) \rightleftharpoons CO_2(aq)$	$\delta = -1$	$\log_{10} K^{(\mathrm{bar})} - \log_{10} K^{(\mathrm{atm})} = -0.0057$
$NH_3(g) + \frac{5}{4}O_2(g) \rightleftharpoons NO(g) + \frac{3}{2}H_2O(g)$	$\delta = 0.25$	$\Delta_r G^{(\mathrm{bar})} - \Delta_r G^{(\mathrm{atm})} = -0.008\,\mathrm{kJ \cdot mol^{-1}}$
$\frac{1}{2}Cl_2(g) + 2O_2(g) + e^- \rightleftharpoons ClO_4^-$	$\delta = -3$	$\Delta_f G^{(\mathrm{bar})} - \Delta_f G^{(\mathrm{atm})} = 0.098\,\mathrm{kJ \cdot mol^{-1}}$

II.3.3. Reference temperature

The definitions of standard states given in Section II.3 make no reference to fixed temperature. Hence, it is theoretically possible to have an infinite number of standard states of a substance as the temperature varies. It is, however, convenient to complete the definition of the standard state in a particular context by choosing a reference temperature. As recommended by IUPAC [82LAF], the reference temperature chosen in the NEA-TDB project is $T = 298.15$ K or $t = 25°$C. Where necessary for the discussion, values of experimentally measured temperatures are reported after conversion to the IPTS-68 [69COM]. The relation between the absolute temperature T (K, kelvin) and the Celsius temperature t (°C) is defined by $t = (T - T_0)$ where $T_0 = 273.15$ K.

II.4. Fundamental physical constants

The fundamental physical constants are taken from a recent publication by CODATA [86COD]. Those relevant to this review are listed in Table II.7.

Table II.7: Fundamental physical constants. These values have been taken from CODATA [86COD]. The digits in parentheses are the one-standard-deviation uncertainty in the last digits of the given value.

Quantity	Symbol	Value	Units
speed of light in vacuum	c	299 792 458	$\mathrm{m \cdot s^{-1}}$
permeability of vacuum	μ_o	$4\pi \times 10^{-7}$	$\mathrm{N \cdot A^{-2}}$
		= 12.566 370 614...	$\mathrm{10^{-7}\,N \cdot A^{-2}}$
Planck constant	h	6.626 0755(40)	$\mathrm{10^{-34}\,J \cdot s}$
elementary charge	e	1.602 177 33(49)	$\mathrm{10^{-19}\,C}$
Rydberg constant, $\frac{1}{2}\mu_o ce^2/h$	R_∞	10 973 731.534(13)	$\mathrm{m^{-1}}$
Avogadro constant	N_A	6.022 1367(36)	$\mathrm{10^{23}\,mol^{-1}}$
Faraday constant, $N_\mathrm{A} \times e$	F	96 485.309(29)	$\mathrm{C \cdot mol^{-1}}$
molar gas constant	R	8.314 510(70)	$\mathrm{J \cdot K^{-1} \cdot mol^{-1}}$
Boltzmann constant, R/N_A	k	1.380 658(12)	$\mathrm{10^{-23}\,J \cdot K^{-1}}$
Non-SI units used with SI:			
electron volt, (e/C) J	eV	1.602 177 33(49)	$\mathrm{10^{-19}\,J}$
atomic mass unit, $1\mathrm{u} = m_\mathrm{u} = \frac{1}{12}m(^{12}\mathrm{C})$	u	1.660 5402(10)	$\mathrm{10^{-27}\,kg}$
Bohr Magneton, $\mu_\mathrm{B} = eh/4\pi m_\mathrm{e}$	B.M.	9.274078(36)	$\mathrm{10^{-24}\,J \cdot T^{-1}}$

Chapter III

Selected uranium data

This chapter presents the chemical thermodynamic data set for uranium species which has been selected in this review. Table III.1 contains the recommended thermodynamic data of the uranium compounds and complexes, Table III.2 the recommended thermodynamic data of chemical equilibrium reactions by which the uranium compounds and complexes are formed, and Table III.3 the temperature coefficients of the heat capacity data of Table III.1 where available. Table III.2 contains information only on those reactions for which primary data selections are made in this review. These selected reaction data are used, together with auxiliary data selected in this review, to derive the corresponding formation data in Table III.1. The uncertainties associated with the auxiliary data may in some cases be large, leading to comparatively large uncertainties in the formation data derived in this manner. This is the main reason for including a table for reaction data (Table III.2), where the selected uncertainties are directly based on the experimental accuracies.

The species and reactions in Tables III.1, III.2 and III.3 appear in standard order of arrangement (*cf.* Fig. II.1).

The selected thermal functions of the heat capacities, listed in Table III.3, refer to the relation

$$C^{\circ}_{p,\mathrm{m}}(T) \quad = \quad a + b \times T + c \times T^2 + d \times T^{-1} + e \times T^{-2}. \qquad \text{(III.1)}$$

No references are given in these tables since the selected data are generally not directly attributable to a specific published source. A detailed discussion of the selection procedure is presented in Chapter V.

A warning: The addition of any aqueous species and their data to this internally consistent data base can result in a modified data set which is no longer rigorous and can lead to erroneous results. The situation is similar, to a lesser degree, with the addition of gases and solids.

It should also be noted that the data set presented in this chapter may not be "complete" for all the conceivable systems and conditions. Gaps are pointed out in the various sections of Chapter V.

Table III.1: Selected thermodynamic data for uranium compounds and complexes. All ionic species listed in this table are aqueous species. Unless noted otherwise, all data refer to the reference temperature of 298.15 K and to the standard state, *i.e.*, a pressure of 0.1 MPa and, for aqueous species, infinite dilution ($I = 0$). The uncertainties listed below each value represent total uncertainties and correspond in principal to the statistically defined 95% confidence interval. Values in **bold** typeface are CODATA Key Values and are taken directly from Ref. [89COX/WAG] without further evaluation. Values obtained from internal calculation, *cf.* footnotes (a) and (b), are rounded at the third digit after the decimal point and may therefore not be exactly identical to those given in Chapter V. Systematically, all the values are presented with three digits after the decimal point, regardless of the significance of these digits. The data presented in this table are available on PC diskettes or other computer media from the OECD Nuclear Energy Agency.

Compound	$\Delta_f G_m^\circ$ $(kJ \cdot mol^{-1})$	$\Delta_f H_m^\circ$ $(kJ \cdot mol^{-1})$	S_m° $(J \cdot K^{-1} \cdot mol^{-1})$	$C_{p,m}^\circ$ $(J \cdot K^{-1} \cdot mol^{-1})$
U(cr)	**0.000**	**0.000**	**50.200** **±0.200**	27.660[c] ±0.050
U(g)	**488.400**[a] **±8.000**	**533.000** **±8.000**	**199.790** **±0.100**	23.690[c] ±0.040
U^{3+}	−476.473[b] ±1.810	−489.100[b] ±3.712	−188.170[b] ±13.853	−64.000 ±22.000
U^{4+}	−529.860[b] ±1.765	−591.200 ±3.300	−416.896[a] ±12.553	−48.000 ±15.000
UO(g)	1.871[a] ±17.011	30.500 ±17.000	248.800 ±2.000	42.000[c] ±2.000
UO_2(cr)	**−1031.833**[a] **±1.004**	**−1085.000** **±1.000**	**77.030** **±0.200**	63.600[c] ±0.080
UO_2(g)	−481.064[a] ±20.036	−477.800 ±20.000	266.300 ±4.000	59.500[c] ±2.000
UO_2^+	−961.021 ±1.752	−1025.127[a] ±2.960	−25.000 ±8.000	
UO_2^{2+}	**−952.551**[a] **±1.747**	**−1019.000** **±1.500**	**−98.200** **±3.000**	42.400[c] ±3.000
$UO_{2.25}$(cr)[e]	−1069.125[a] ±1.702	−1128.000 ±1.700	83.530 ±0.170	73.340[c] ±0.150

Table III.1 (continued)

Compound	$\Delta_f G^\circ_m$ (kJ · mol^{-1})	$\Delta_f H^\circ_m$ (kJ · mol^{-1})	S°_m (J · K^{-1} · mol^{-1})	$C^\circ_{p,m}$ (J · K^{-1} · mol^{-1})
β-$UO_{2.25}$[(f)]	−1069.083[(a)] ±1.702	−1127.400 ±1.700	85.400 ±0.200	[(c)]
α-$UO_{2.3333}$			82.170 ±0.500	71.420[(c)] ±0.300
β-$UO_{2.3333}$	−1080.572[(a)] ±1.403	−1142.000 ±1.400	83.510 ±0.200	71.840[(c)] ±0.140
$UO_{2.6667}$(cr)	−1123.157[(a)] ±0.804	−1191.600 ±0.800	94.180 ±0.170	79.310[(c)] ±0.160
$UO_{2.86}\cdot 0.5H_2O$(cr)		−1367.000 ±10.000		
$UO_{2.86}\cdot 1.5H_2O$(cr)		−1666.000 ±10.000		
α-UO_3	−1140.420[(a)] ±3.015	−1217.500 ±3.000	99.400 ±1.000	81.840 ±0.300
β-UO_3	−1142.301[(a)] ±1.307	−1220.300 ±1.300	96.320 ±0.400	81.340[(c)] ±0.160
γ-UO_3	**−1145.739**[(a)] **±1.207**	**−1223.800 ±1.200**	**96.110 ±0.400**	81.670[(c)] ±0.160
δ-UO_3		−1209.000 ±5.000		
ε-UO_3		−1217.200 ±1.300		
UO_3(am)		−1207.900 ±4.000		[(c)]
UO_3(g)	−784.761[(a)] ±15.012	−799.200 ±15.000	309.500 ±2.000	64.500[(c)] ±2.000
β-UH_3	−72.556[(a)] ±0.148	−126.980 ±0.130	63.680 ±0.130	49.290 ±0.080
UOH^{3+}	−763.918[(b)] ±1.798	−830.120[(b)] ±9.540	−199.946[(b)] ±32.521	

Table III.1 (continued)

Compound	$\Delta_f G^\circ_m$ $(kJ \cdot mol^{-1})$	$\Delta_f H^\circ_m$ $(kJ \cdot mol^{-1})$	S°_m $(J \cdot K^{-1} \cdot mol^{-1})$	$C^\circ_{p,m}$ $(J \cdot K^{-1} \cdot mol^{-1})$
$UO_3 \cdot 0.393H_2O(cr)$		−1347.800 ±1.300		
UO_2OH^+	−1160.009[b] ±2.447	−1261.657[a] ±15.107	17.000 ±50.000	
$UO_3 \cdot 0.648H_2O(cr)$		−1424.600 ±1.300		
α-$UO_3 \cdot 0.85H_2O$		−1491.900 ±1.300		
α-$UO_3 \cdot 0.9H_2O$	−1374.560[a] ±2.460	−1506.300 ±1.300	126.000 ±7.000	140.000 ±30.000
$UO_2(OH)_2(aq)$	≥ −1368.038[b]			
β-$UO_2(OH)_2$	−1398.683[a] ±1.765	−1533.800 ±1.300	138.000 ±4.000	141.000[c] ±15.000
γ-$UO_2(OH)_2$		−1531.400 ±1.300		
$U(OH)_4(aq)$	−1452.500 ±8.000	−1655.798[a] ±10.934	40.000 ±25.000	205.000 ±80.000
$UO_2(OH)_3^-$	−1554.377[b] ±2.878			
$UO_3 \cdot 2H_2O(cr)$	−1636.506[a] ±1.705	−1826.100 ±1.700	188.540 ±0.380	172.070[c] ±0.340
$U(OH)_5^-$	> −1621.144[b]			
$UO_2(OH)_4^{2-}$	−1712.746[b] ±11.550			
$UO_4 \cdot 2H_2O(cr)$		−1784.000 ±4.200		
$UO_4 \cdot 4H_2O(cr)$		−2394.800 ±2.100		

Table III.1 (continued)

Compound	$\Delta_f G_m^\circ$ (kJ · mol^{-1})	$\Delta_f H_m^\circ$ (kJ · mol^{-1})	S_m° (J · K^{-1} · mol^{-1})	$C_{p,m}^\circ$ (J · K^{-1} · mol^{-1})
$(UO_2)_2OH^{3+}$	−2126.830[b] ±6.693			
$(UO_2)_2(OH)_2^{2+}$	−2347.302[b] ±3.503	−2572.065[a] ±5.682	−38.000 ±15.000	
$(UO_2)_3(OH)_4^{2+}$	−3738.287[b] ±5.517			
$(UO_2)_3(OH)_5^{+}$	−3954.593[b] ±5.291	−4389.086[a] ±10.394	83.000 ±30.000	
$(UO_2)_3(OH)_7^{-}$	−4340.684[b] ±12.565			
$(UO_2)_4(OH)_7^{+}$	−5345.177[b] ±9.029			
$U_6(OH)_{15}^{9+}$ [g]				
UF(g)	−81.936[a] ±30.148	−52.000 ±30.000	252.000 ±10.000	37.900[c] ±5.000
UF^{3+}	−864.354[b] ±1.964	−932.150[b] ±3.400	−271.814[b] ±12.807	
UF_2(g)	−548.786[a] ±30.148	−530.000 ±30.000	316.000 ±10.000	56.200[c] ±5.000
UF_2^{2+}	−1185.548[b] ±2.401	−1265.400[b] ±3.597	−145.514[b] ±13.132	
UF_3(cr)	−1432.531[a] ±4.702	−1501.400 ±4.700	123.400 ±0.400	95.100[c] ±0.400
UF_3(g)	−1051.998[a] ±15.999	−1054.200[b] ±15.719	347.000 ±10.000	76.200[c] ±5.000
UF_3^{+}	−1497.723[b] ±6.325	−1596.750[b] ±5.540	−43.090[b] ±26.643	
UF_4(aq)	−1802.078[b] ±6.585	−1936.807[b] ±8.713	3.904[b] ±21.374	

Table III.1 (continued)

Compound	$\Delta_f G^\circ_m$ (kJ · mol^{-1})	$\Delta_f H^\circ_m$ (kJ · mol^{-1})	S°_m (J · K^{-1} · mol^{-1})	$C^\circ_{p,m}$ (J · K^{-1} · mol^{-1})
$UF_4(cr)$	−1823.538[(a)] ±4.201	−1914.200 ±4.200	151.700 ±0.200	116.000[(c)] ±0.100
$UF_4(g)$	−1573.537[(a)] ±9.158	−1601.200[(b)] ±9.035	363.000 ±5.000	101.400[(c)] ±3.000
α-UF_5	−1968.688[(a)] ±6.995	−2075.300 ±5.900	199.600 ±12.600	132.200[(c)] ±4.200
β-UF_5	−1970.595[(a)] ±5.635	−2083.200 ±4.200	179.500 ±12.600	132.200[(c)] ±12.000
$UF_5(g)$	−1858.993[(a)] ±15.294	−1910.000 ±15.000	386.100 ±10.000	110.600[(c)] ±5.000
UF_5^-	−2091.650[(b)] ±4.246			
$UF_6(cr)$	−2069.205[(a)] ±1.842	−2197.700 ±1.800	227.600 ±1.300	166.800[(c)] ±0.200
$UF_6(g)$	−2064.500[(a)] ±1.893	−2148.600 ±1.868	376.500 ±1.000	129.500[(c)] ±0.500
UF_6^{2-}	−2384.989[(b)] ±4.629			
$U_2F_9(cr)$	−3812.000 ±17.000	−4015.923[(a)] ±18.016	329.000 ±20.000	251.000[(c)] ±16.700
$U_2F_{10}(g)$	−3860.966[(a)] ±30.148	−4021.000 ±30.000	577.600 ±10.000	234.700 ±5.000
$U_4F_{17}(cr)$	−7464.000 ±30.000	−7849.665[(a)] ±32.284	631.000 ±40.000	485.300[(c)] ±33.000
$UOF_2(cr)$	−1434.127[(a)] ±6.424	−1504.600 ±6.300	119.200 ±4.200	
$UOF_4(cr)$	−1816.264[(a)] ±4.269	−1924.600 ±4.000	195.000 ±5.000	
$UOF_4(g)$	−1703.754[(a)] ±20.221	−1762.000 ±20.000	363.000 ±10.000	108.000[(c)] ±5.000

Table III.1 (continued)

Compound	$\Delta_f G^\circ_m$ (kJ · mol^{-1})	$\Delta_f H^\circ_m$ (kJ · mol^{-1})	S°_m (J · K^{-1} · mol^{-1})	$C^\circ_{p,m}$ (J · K^{-1} · mol^{-1})
UO_2F^+	−1263.128[(b)] ±2.021	−1352.650[(b)] ±1.637	−8.851[(b)] ±3.988	
UO_2F_2(aq)	−1564.800[(b)] ±2.241	−1687.600[(b)] ±1.994	46.272[(b)] ±3.543	
UO_2F_2(cr)	−1557.321[(a)] ±1.307	−1653.500 ±1.300	135.560 ±0.420	103.220[(c)] ±0.420
UO_2F_2(g)	−1318.170[(a)] ±15.349	−1352.500 ±15.056	343.000 ±10.000	86.400[(c)] ±5.000
$UO_2F_3^-$	−1859.338[(b)] ±3.547	−2022.700[(b)] ±2.480	76.961[(b)] ±8.630	
$UO_2F_4^{2-}$	−2145.427[(b)] ±5.166	−2360.110[(b)] ±3.038	71.568[(b)] ±14.189	
$U_2O_3F_6$(cr)	−3372.730[(a)] ±14.801	−3579.200[(b)] ±13.546	324.000 ±20.000	
$U_3O_5F_8$(cr)	−4890.135[(a)] ±9.771	−5192.950[(b)] ±3.929	459.000 ±30.000	304.100[(b)] ±4.143
H_3OUF_6(cr)		−2641.400[(b)] ±3.157		
UOFOH(cr)	−1336.930[(a)] ±12.948	−1426.700 ±12.600	121.000 ±10.000	
$UOFOH{\cdot}0.5H_2O$(cr)	−1458.117[(a)] ±6.970	−1576.100 ±6.300	143.000 ±10.000	
$UOF_2{\cdot}H_2O$(cr)	−1674.474[(a)] ±4.143	−1802.000 ±3.300	161.100 ±8.400	
$UF_4{\cdot}2.5H_2O$(cr)	−2440.282[(a)] ±6.188	−2671.475 ±4.277	263.500 ±15.000	263.700 ±15.000
$UO_2FOH{\cdot}H_2O$(cr)	−1721.700 ±7.500	−1894.500 ±8.400	178.347[(a)] ±37.770	
$UO_2FOH{\cdot}2H_2O$(cr)	−1961.032[(b)] ±8.408	−2190.010[(b)] ±9.392	223.183[(b)] ±38.244	

Table III.1 (continued)

Compound	$\Delta_f G_m^\circ$ $(kJ \cdot mol^{-1})$	$\Delta_f H_m^\circ$ $(kJ \cdot mol^{-1})$	S_m° $(J \cdot K^{-1} \cdot mol^{-1})$	$C_{p,m}^\circ$ $(J \cdot K^{-1} \cdot mol^{-1})$
$UO_2F_2 \cdot 3H_2O(cr)$	−2269.658[(a)] ±6.939	−2534.390[(b)] ±4.398	270.000 ±18.000	
UCl(g)	157.115[(a)] ±22.499	188.200 ±22.300	266.000 ±10.000	43.100[(c)] ±5.000
UCl^{3+}	−670.895[(b)] ±1.918	−777.280[(b)] ±9.586	−391.093[(b)] ±32.788	
$UCl_2(g)$	−182.594[(a)] ±22.499	−163.000 ±22.300	339.000 ±10.000	59.900[(c)] ±5.000
$UCl_3(cr)$	−796.103[(a)] ±2.006	−863.700 ±2.000	158.100 ±0.500	95.100[(c)] ±0.500
$UCl_3(g)$	−535.662[(a)] ±16.079	−537.100 ±15.800	380.000 ±10.000	82.400[(c)] ±5.000
$UCl_4(cr)$	−929.575[(a)] ±2.512	−1018.800 ±2.500	197.100 ±0.800	122.000[(c)] ±0.400
$UCl_4(g)$	−790.174[(a)] ±5.662	−818.100[(b)] ±5.590	402.700 ±3.000	107.700[(c)] ±1.500
$UCl_5(cr)$	−930.115[(a)] ±3.908	−1039.000 ±3.000	242.700 ±8.400	150.600[(c)] ±8.400
$UCl_5(g)$	−831.843[(a)] ±15.294	−882.500 ±15.000	438.000 ±10.000	124.000[(c)] ±5.000
$UCl_6(cr)$	−937.209[(a)] ±3.043	−1066.500 ±3.000	285.800 ±1.700	175.700[(c)] ±4.200
$UCl_6(g)$	−903.587[(a)] ±8.673	−987.500 ±8.544	438.000 ±5.000	147.000[(c)] ±3.000
$U_2Cl_8(g)$	−1639.664[(a)] ±16.750	−1749.600[(b)] ±15.000	624.000 ±25.000	226.000 ±10.000
$U_2Cl_{10}(g)$	−1813.808[(b)] ±10.534	−1967.900[(b)] ±9.948	698.981[(b)] ±11.611	236.000 ±10.000
UOCl(cr)	−785.654[(a)] ±4.890	−833.900 ±4.200	102.500 ±8.400	71.000[(c)] ±5.000

Table III.1 (continued)

Compound	$\Delta_f G^\circ_m$ $(kJ \cdot mol^{-1})$	$\Delta_f H^\circ_m$ $(kJ \cdot mol^{-1})$	S°_m $(J \cdot K^{-1} \cdot mol^{-1})$	$C^\circ_{p,m}$ $(J \cdot K^{-1} \cdot mol^{-1})$
$UOCl_2(cr)$	−998.478[(a)] ±2.701	−1069.300 ±2.700	138.320 ±0.210	95.060[(c)] ±0.420
$UOCl_3(cr)$	−1045.576[(a)] ±8.383	−1140.000 ±8.000	170.700 ±8.400	117.200[(c)] ±4.200
$UO_2Cl(cr)$	−1095.253[(a)] ±8.383	−1171.100 ±8.000	112.500 ±8.400	88.000[(c)] ±5.000
UO_2Cl^+	−1084.738[(b)] ±1.755	−1178.080[(b)] ±2.502	−11.513[(b)] ±7.361	
$UO_2Cl_2(aq)$	−1208.707[(b)] ±2.885	−1338.160[(b)] ±6.188	44.251[(b)] ±21.744	
$UO_2Cl_2(cr)$	−1145.838[(a)] ±1.303	−1243.600 ±1.300	150.540 ±0.210	107.860[(c)] ±0.170
$UO_2Cl_2(g)$	−941.357[(a)] ±15.294	−971.600 ±15.000	377.000 ±10.000	92.300[(c)] ±5.000
$UO_2ClO_3^+$	−963.308[(b)] ±2.239	−1126.900[(b)] ±1.828	60.592[(b)] ±4.562	
$U_2O_2Cl_5(cr)$	−2037.307[(a)] ±4.891	−2197.400 ±4.200	326.300 ±8.400	219.400[(c)] ±5.000
$(UO_2)_2Cl_3(cr)$	−2234.755[(a)] ±2.931	−2404.500 ±1.700	276.000 ±8.000	203.600[(c)] ±5.000
$U_5O_{12}Cl(cr)$	−5517.951[(a)] ±12.412	−5854.400 ±8.600	465.000 ±30.000	
$UO_2Cl_2 \cdot H_2O(cr)$	−1405.003[(a)] ±3.269	−1559.800 ±2.100	192.500 ±8.400	
$UO_2ClOH \cdot 2H_2O(cr)$	−1782.219[(a)] ±4.507	−2010.400 ±1.700	236.000 ±14.000	
$UO_2Cl_2 \cdot 3H_2O(cr)$	−1894.615[(a)] ±3.028	−2164.800 ±1.700	272.000 ±8.400	
$UCl_3F(cr)$	−1146.573[(a)] ±5.155	−1243.000 ±5.000	162.800 ±4.200	118.800 ±4.200

Table III.1 (continued)

Compound	$\Delta_f G_m^\circ$ ($kJ \cdot mol^{-1}$)	$\Delta_f H_m^\circ$ ($kJ \cdot mol^{-1}$)	S_m° ($J \cdot K^{-1} \cdot mol^{-1}$)	$C_{p,m}^\circ$ ($J \cdot K^{-1} \cdot mol^{-1}$)
$UCl_2F_2(cr)$	−1375.967[a] ±5.592	−1466.000 ±5.000	174.100 ±8.400	119.700 ±4.200
$UClF_3(cr)$	−1606.360[a] ±5.155	−1690.000 ±5.000	185.400 ±4.200	120.900 ±4.200
$UBr(g)$	201.772[a] ±17.260	247.000 ±17.000	278.000 ±10.000	44.100[c] ±5.000
UBr^{3+}	−642.044[b] ±2.109			
$UBr_2(g)$	−77.687[a] ±25.177	−31.000 ±25.000	359.000 ±10.000	61.400[c] ±5.000
$UBr_3(cr)$	−673.198[a] ±4.205	−698.700 ±4.200	192.980 ±0.500	105.830[c] ±0.500
$UBr_3(g)$	−401.115[a] ±37.120	−364.000 ±37.000	403.000 ±10.000	85.100[c] ±5.000
$UBr_4(cr)$	−767.479[a] ±3.544	−802.100 ±2.500	238.500 ±8.400	128.000[c] ±4.200
$UBr_4(g)$	−636.182[a] ±8.515	−610.100 ±8.382	442.100 ±5.000	110.300[c] ±3.000
$UBr_5(cr)$	−769.307[a] ±9.205	−810.400 ±8.400	292.900 ±12.600	160.700[c] ±8.000
$UBr_5(g)$	−656.312[a] ±17.441	−637.745 ±17.183	493.000 ±10.000	129.000[c] ±5.000
$UOBr_2(cr)$	−929.648[a] ±8.401	−973.600 ±8.400	157.570 ±0.290	98.000[c] ±0.400
$UOBr_3(cr)$	−901.498[a] ±21.334	−954.000 ±21.000	205.000 ±12.600	120.900[c] ±4.200
UO_2Br^+	−1057.657[b] ±1.759			
$UO_2Br_2(cr)$	−1066.422[a] ±1.808	−1137.400 ±1.300	169.500 ±4.200	116.000[c] ±8.000

Table III.1 (continued)

Compound	$\Delta_f G_m^\circ$ (kJ · mol^{-1})	$\Delta_f H_m^\circ$ (kJ · mol^{-1})	S_m° (J · K^{-1} · mol^{-1})	$C_{p,m}^\circ$ (J · K^{-1} · mol^{-1})
$UO_2BrO_3^+$	−937.077[(b)] ±1.914	−1085.600[(b)] ±1.609	75.697[(b)] ±3.748	
$UO_2Br_2{\cdot}H_2O(cr)$	−1328.644[(a)] ±2.515	−1455.900 ±1.400	214.000 ±7.000	
$UO_2BrOH{\cdot}2H_2O(cr)$	−1744.162[(a)] ±4.372	−1958.200 ±1.300	248.000 ±14.000	
$UO_2Br_2{\cdot}3H_2O(cr)$	−1818.486[(a)] ±5.573	−2058.000 ±1.500	304.000 ±18.000	
$UBr_2Cl(cr)$	−714.389[(a)] ±9.765	−750.600 ±8.400	192.500 ±16.700	
$UBr_3Cl(cr)$	−807.114[(a)] ±9.766	−852.300 ±8.400	238.500 ±16.700	
$UBrCl_2(cr)$	−760.315[(a)] ±9.765	−812.100 ±8.400	175.700 ±16.700	
$UBr_2Cl_2(cr)$	−850.896[(a)] ±9.765	−907.900 ±8.400	234.300 ±16.700	
$UBrCl_3(cr)$	−893.500[(a)] ±9.202	−967.300 ±8.400	213.400 ±12.600	
$UI(g)$	288.010[(a)] ±25.177	341.000 ±25.000	286.000 ±10.000	44.400[(c)] ±5.000
UI^{3+}	−588.719[(b)] ±2.462			
$UI_2(g)$	37.490[(a)] ±25.177	100.000 ±25.000	376.000 ±10.000	61.900[(c)] ±5.000
$UI_3(cr)$	−466.622[(a)] ±4.892	−467.400 ±4.200	221.800 ±8.400	112.100[(c)] ±6.000
$UI_3(g)$	−200.700[(a)] ±25.178	−140.000 ±25.000	428.000 ±10.000	86.000[(c)] ±5.000
$UI_4(cr)$	−513.170[(a)] ±3.836	−518.800 ±2.900	263.600 ±8.400	126.400[(c)] ±4.200

Table III.1 (continued)

Compound	$\Delta_f G_m^\circ$ $(kJ \cdot mol^{-1})$	$\Delta_f H_m^\circ$ $(kJ \cdot mol^{-1})$	S_m° $(J \cdot K^{-1} \cdot mol^{-1})$	$C_{p,m}^\circ$ $(J \cdot K^{-1} \cdot mol^{-1})$
$UI_4(g)$	−370.374[(a)] ±6.507	−308.800 ±5.780	489.000 ±10.000	112.000[(c)] ±5.000
$UO_2IO_3^+$	−1090.305[(b)] ±1.917	−1228.900[(b)] ±1.819	90.959[(b)] ±4.718	
$UO_2(IO_3)_2(aq)$	−1225.718[(b)] ±2.493			
$UO_2(IO_3)_2(cr)$	−1250.206[(b)] ±2.410	−1461.281[(a)] ±3.609	279.000 ±9.000	
$UClI_3(cr)$	−615.789[(a)] ±11.350	−643.800 ±10.000	242.000 ±18.000	
$UCl_2I_2(cr)$	−723.356[(a)] ±11.350	−768.800 ±10.000	237.000 ±18.000	
$UCl_3I(cr)$	−829.877[(a)] ±8.766	−898.300 ±8.400	213.400 ±8.400	
$UBrI_3(cr)$		−589.600 ±10.000		
$UBr_2I_2(cr)$		−660.400 ±10.000		
$UBr_3I(cr)$		−727.600 ±8.400		
$US(cr)$	−320.929[(a)] ±12.600	−322.200 ±12.600	77.990 ±0.210	50.540[(c)] ±0.080
$US_{1.90}(cr)$	−509.470[(a)] ±20.900	−509.900 ±20.900	109.660 ±0.210	73.970[(c)] ±0.130
$US_2(cr)$	−519.241[(a)] ±8.001	−520.400 ±8.000	110.420 ±0.210	74.640[(c)] ±0.130
$US_3(cr)$	−537.253[(a)] ±12.600	−539.600 ±12.600	138.490 ±0.210	95.600[(c)] ±0.250
$U_2S_3(cr)$	−879.786[(a)] ±67.002	−879.000 ±67.000	199.200 ±1.700	133.700[(c)] ±0.800

Table III.1 (continued)

Compound	$\Delta_f G^\circ_m$ $(kJ \cdot mol^{-1})$	$\Delta_f H^\circ_m$ $(kJ \cdot mol^{-1})$	S°_m $(J \cdot K^{-1} \cdot mol^{-1})$	$C^\circ_{p,m}$ $(J \cdot K^{-1} \cdot mol^{-1})$
$U_2S_5(cr)$			243.000 ±25.000	
$U_3S_5(cr)$	−1425.076[(a)] ±100.278	−1431.000 ±100.000	291.000 ±25.000	
USO_4^{2+}	−1311.423[(b)] ±2.113	−1492.540[(b)] ±4.283	−245.591[(b)] ±15.906	
$UO_2SO_3(aq)$	−1477.697[(b)] ±5.563			
$UO_2SO_3(cr)$	−1530.370[(a)] ±12.727	−1661.000 ±12.600	157.000 ±6.000	
$UO_2S_2O_3(aq)$	−1487.827[(b)] ±11.606			
$UO_2SO_4(aq)$	−1714.535[(b)] ±1.800	−1908.840[(b)] ±2.229	46.010[(b)] ±6.173	
$UO_2SO_4(cr)$	−1685.776[(a)] ±2.642	−1845.140 ±0.840	163.200 ±8.400	145.000[(c)] ±3.000
$U(SO_3)_2(cr)$	−1712.826[(a)] ±21.171	−1883.000 ±21.000	159.000 ±9.000	
$U(SO_4)_2(aq)$	−2077.860[(b)] ±2.262	−2377.180[(b)] ±4.401	−69.007[(b)] ±16.158	
$U(SO_4)_2(cr)$	−2084.521[(a)] ±14.070	−2309.600 ±12.600	180.000 ±21.000	
$UO_2(SO_4)_2^{2-}$	−2464.190[(b)] ±1.978	−2802.580[(b)] ±1.972	135.786[(b)] ±4.763	
$U(OH)_2SO_4(cr)$	−1766.223[(b)] ±3.385			
$UO_2SO_4\cdot 2.5H_2O(cr)$	−2298.475 ±1.803	−2607.000 ±0.900	246.056[(a)] ±6.762	
$UO_2SO_4\cdot 3H_2O(cr)$	−2416.561 ±1.811	−2751.500 ±4.600	274.090[(a)] ±16.582	

Table III.1 (continued)

Compound	$\Delta_f G_m^\circ$ (kJ · mol^{-1})	$\Delta_f H_m^\circ$ (kJ · mol^{-1})	S_m° (J · K^{-1} · mol^{-1})	$C_{p,m}^\circ$ (J · K^{-1} · mol^{-1})
$UO_2SO_4 \cdot 3.5H_2O(cr)$	−2535.595 ±1.806	−2901.600 ±0.800	286.524[(a)] ±6.628	
$U(SO_4)_2 \cdot 4H_2O(cr)$	−3033.308[(a)] ±11.433	−3483.200 ±6.300	359.000 ±32.000	
$U(SO_4)_2 \cdot 8H_2O(cr)$	−3987.896[(a)] ±16.735	−4662.600 ±6.300	538.000 ±52.000	
USe(cr)	−276.908[(a)] ±14.600	−275.700 ±14.600	96.520 ±0.210	54.810[(c)] ±0.170
α-USe_2	−427.072[(a)] ±42.000	−427.000 ±42.000	134.980 ±0.250	79.160[(c)] ±0.170
β-USe_2	−427.972[(a)] ±42.179	−427.000 ±42.000	138.000 ±13.000	
$USe_3(cr)$	−451.997[(a)] ±42.305	−452.000 ±42.000	177.000 ±17.000	
$U_2Se_3(cr)$	−721.194[(a)] ±75.002	−711.000 ±75.000	261.400 ±1.700	
$U_3Se_4(cr)$	−988.760[(a)] ±85.752	−983.000 ±85.000	339.000 ±38.000	
$U_3Se_5(cr)$	−1130.611[(a)] ±113.567	−1130.000 ±113.000	364.000 ±38.000	
$UO_2SeO_3(cr)$		−1522.000 ±2.300		
$UO_2SeO_4(cr)$		−1539.300 ±3.300		
UOTe(cr)			111.700 ±1.300	80.600 ±0.800
$UO_2TeO_3(cr)$		−1605.400 ±1.300		
UN(cr)	−265.082[(a)] ±3.001	−290.000 ±3.000	62.430 ±0.220	47.570[(c)] ±0.400

Table III.1 (continued)

Compound	$\Delta_f G_m^\circ$ $(kJ \cdot mol^{-1})$	$\Delta_f H_m^\circ$ $(kJ \cdot mol^{-1})$	S_m° $(J \cdot K^{-1} \cdot mol^{-1})$	$C_{p,m}^\circ$ $(J \cdot K^{-1} \cdot mol^{-1})$
β-$UN_{1.466}$		−362.200 ±2.300		
α-$UN_{1.59}$	−338.202[(a)] ±5.201	−379.200 ±5.200	65.020 ±0.300	
α-$UN_{1.606}$		−381.400 ±5.000		
α-$UN_{1.674}$		−390.800 ±5.000		
α-$UN_{1.73}$	−353.753[(a)] ±7.501	−398.500 ±7.500	65.860 ±0.300	
$UO_2N_3^+$	−619.077[(b)] ±2.705			
$UO_2(N_3)_2(aq)$	−280.867[(b)] ±4.558			
$UO_2(N_3)_3^-$	59.285[(b)] ±6.374			
$UO_2(N_3)_4^{2-}$	412.166[(b)] ±8.302			
UNO_3^{3+}	−649.045[(b)] ±1.960			
$UO_2NO_3^+$	−1065.057[(b)] ±1.990			
$U(NO_3)_2^{2+}$	−764.576[(b)] ±2.793			
$UO_2(NO_3)_2(cr)$	−1106.094[(a)] ±5.675	−1351.000 ±5.000	241.000 ±9.000	
$UO_2(NO_3)_2 \cdot H_2O(cr)$	−1362.965[(a)] ±10.524	−1664.000 ±10.000	286.000 ±11.000	
$UO_2(NO_3)_2 \cdot 2H_2O(cr)$	−1620.500 ±2.000	−1978.700 ±1.700	327.524[(a)] ±8.806	278.000 ±4.000

Table III.1 (continued)

Compound	$\Delta_f G_m^\circ$ $(kJ \cdot mol^{-1})$	$\Delta_f H_m^\circ$ $(kJ \cdot mol^{-1})$	S_m° $(J \cdot K^{-1} \cdot mol^{-1})$	$C_{p,m}^\circ$ $(J \cdot K^{-1} \cdot mol^{-1})$
$UO_2(NO_3)_2 \cdot 3H_2O(cr)$	−1864.693[(a)] ±1.965	−2280.400 ±1.700	367.900 ±3.300	320.100 ±1.700
$UO_2(NO_3)_2 \cdot 6H_2O(cr)$	−2584.212[(a)] ±1.615	−3167.500 ±1.500	505.600 ±2.000	468.000 ±2.500
$UP(cr)$	−265.921[(a)] ±11.101	−269.800 ±11.100	78.280 ±0.420	50.290 ±0.500
$UP_2(cr)$	−294.555[(a)] ±15.031	−304.000 ±15.000	100.700 ±3.200	78.800 ±3.500
$U_3P_4(cr)$	−826.435[(a)] ±26.014	−843.000 ±26.000	259.400 ±2.600	176.900 ±3.600
$UPO_5(cr)$	−1924.713[(a)] ±4.990	−2064.000 ±4.000	137.000 ±10.000	124.000[(c)] ±12.000
$UO_2PO_4^-$	−2053.559[(b)] ±2.504			
$UP_2O_7(cr)$	−2659.272[(a)] ±5.369	−2852.000 ±4.000	204.000 ±12.000	184.000 ±18.000
$(UO_2)_2P_2O_7(cr)$	−3930.002[(a)] ±6.861	−4232.600 ±2.800	296.000 ±21.000	258.000[(c)] ±26.000
$(UO_2)_3(PO_4)_2(cr)$	−5115.975[(a)] ±5.452	−5491.300 ±3.500	410.000 ±14.000	339.000[(c)] ±17.000
$UO_2HPO_4(aq)$	−2089.862[(b)] ±2.777			
$UO_2H_2PO_4^+$	−2108.311[(b)] ±2.378			
$UO_2H_3PO_4^{2+}$	−2106.256[(b)] ±2.504			
$UO_2(H_2PO_4)_2(aq)$	−3254.938[(b)] ±3.659			
$UO_2(H_2PO_4)(H_3PO_4)^+$	−3260.703[(b)] ±3.659			

Table III.1 (continued)

Compound	$\Delta_f G_m^\circ$ (kJ · mol^{-1})	$\Delta_f H_m^\circ$ (kJ · mol^{-1})	S_m° (J · K^{-1} · mol^{-1})	$C_{p,m}^\circ$ (J · K^{-1} · mol^{-1})
$UO_2HPO_4 \cdot 4H_2O(cr)$	−3064.749[b] ±2.414	−3469.968[a] ±7.836	346.000 ±25.000	
$U(HPO_4)_2 \cdot 4H_2O(cr)$	−3844.453[b] ±3.717	−4334.819[a] ±8.598	372.000 ±26.000	460.000 ±50.000
$(UO_2)_3(PO_4)_2 \cdot 4H_2O(cr)$	−6138.967[b] ±6.355	−6739.105[a] ±9.136	589.000 ±22.000	
$(UO_2)_3(PO_4)_2 \cdot 6H_2O(cr)$	−6618.000 ±7.000			
UAs(cr)	−237.908[a] ±8.024	−234.300 ±8.000	97.400 ±2.000	57.900 ±1.200
$UAs_2(cr)$	−252.790[a] ±13.005	−252.000 ±13.000	123.050 ±0.200	79.960 ±0.100
$U_3As_4(cr)$	−725.388[a] ±18.016	−720.000 ±18.000	309.070 ±0.600	187.530 ±0.200
$UAsO_5(cr)$				138.000[c] ±14.000
$UO_2(AsO_3)_2(cr)$	−1944.911[a] ±12.006	−2156.600 ±8.000	231.000 ±30.000	201.000[c] ±20.000
$(UO_2)_2As_2O_7(cr)$	−3130.254[a] ±12.006	−3426.000 ±8.000	307.000 ±30.000	273.000[c] ±27.000
$(UO_2)_3(AsO_4)_2(cr)$	−4310.789[a] ±12.007	−4689.400 ±8.000	387.000 ±30.000	364.000[c] ±36.000
UAsS(cr)			114.500 ±3.400	80.900 ±2.400
UAsSe(cr)			131.900 ±4.000	83.500 ±2.500
UAsTe(cr)			139.900 ±4.200	81.300 ±2.400
USb(cr)	−140.969[a] ±7.711	−138.500 ±7.500	104.000 ±6.000	

Table III.1 (continued)

Compound	$\Delta_f G^\circ_m$ $(kJ \cdot mol^{-1})$	$\Delta_f H^\circ_m$ $(kJ \cdot mol^{-1})$	S°_m $(J \cdot K^{-1} \cdot mol^{-1})$	$C^\circ_{p,m}$ $(J \cdot K^{-1} \cdot mol^{-1})$
$USb_2(cr)$	−173.666[a] ±10.901	−173.600 ±10.900	141.460 ±0.130	80.200 ±0.100
$U_3Sb_4(cr)$	−457.004[a] ±22.602	−451.900 ±22.600	349.800 ±0.400	188.200 ±0.200
$U_4Sb_3(cr)$			379.000 ±20.000	
$UC(cr)$	−98.900 ±3.000	−97.900 ±4.000	59.294[a] ±16.772	50.100[c] ±1.000
α-$UC_{1.94}$	−87.400 ±2.100	−85.324[a] ±2.185	68.300 ±2.000	60.800[c] ±1.500
$U_2C_3(cr)$	−189.317[a] ±10.002	−183.300 ±10.000	137.800 ±0.300	107.400[c] ±2.000
$UO_2CO_3(aq)$	−1535.704[b] ±1.805	−1689.230[b] ±2.512	53.892[b] ±7.455	
$UO_2CO_3(cr)$	−1563.046[b] ±1.805	−1689.646[a] ±1.808	144.200 ±0.300	120.100 ±0.100
$UO_2(CO_3)_2^{2-}$	−2105.044[b] ±2.033	−2350.960[b] ±4.301	188.163[b] ±14.081	
$UO_2(CO_3)_3^{4-}$	−2659.543[b] ±2.123	−3083.890[b] ±4.430	33.852[b] ±14.423	
$UO_2(CO_3)_3^{5-}$	−2586.994[b] ±2.623			
$U(CO_3)_4^{4-}$	−2841.925[b] ±5.958			
$U(CO_3)_5^{6-}$	−3363.431[b] ±5.772	−3987.350[b] ±5.334	−83.051[b] ±25.680	
$(UO_2)_3(CO_3)_6^{6-}$	−6333.284[b] ±8.096	−7171.080[b] ±5.316	228.926[b] ±23.416	
$(UO_2)_2CO_3(OH)_3^-$	−3139.525[b] ±4.517			

Table III.1 (continued)

Compound	$\Delta_f G^\circ_m$ (kJ · mol^{-1})	$\Delta_f H^\circ_m$ (kJ · mol^{-1})	S°_m (J · K^{-1} · mol^{-1})	$C^\circ_{p,m}$ (J · K^{-1} · mol^{-1})
$(UO_2)_3O(OH)_2(HCO_3)^+$	−4100.695 ±5.973			
$(UO_2)_{11}(CO_3)_6(OH)_{12}^{2-}$	−16698.980[(b)] ±22.383			
$USCN^{3+}$	−454.113[(b)] ±4.385	−541.800[(b)] ±9.534	−306.326[(b)] ±35.198	
$U(SCN)_2^{2+}$	−368.776[(b)] ±8.257	−456.400[(b)] ±9.534	−107.175[(b)] ±42.302	
UO_2SCN^+	−867.842[(b)] ±4.558	−939.380[(b)] ±4.272	83.671[(b)] ±19.708	
$UO_2(SCN)_2(aq)$	−774.229[(b)] ±8.770	−857.300[(b)] ±8.161	243.927[(b)] ±39.546	
$UO_2(SCN)_3^-$	−686.438[(b)] ±12.458	−783.800[(b)] ±12.153	394.934[(b)] ±57.938	
$USiO_4(cr)$	−1883.600 ±4.000	−1991.326[(a)] ±5.367	118.000 ±12.000	
$(UO_2)_2(PuO_2)(CO_3)_6^{6-}$[(h)]				
$(UO_2)_2(NpO_2)(CO_3)_6^{6-}$[(h)]				
$Be_{13}U(cr)$	−165.508[(a)] ±17.530	−163.600 ±17.500	180.100 ±3.300	242.300 ±4.200
$MgUO_4(cr)$	−1749.601[(a)] ±1.502	−1857.300 ±1.500	131.950 ±0.170	128.100 ±0.300
$MgU_3O_{10}(cr)$			338.600 ±1.000	305.600 ±5.000
β-$CaUO_4$				124.700 ±2.700
$CaUO_4(cr)$	−1888.706[(a)] ±2.421	−2002.300 ±2.300	121.100 ±2.500	123.800 ±2.500

Table III.1 (continued)

Compound	$\Delta_f G_m^\circ$ (kJ · mol^{-1})	$\Delta_f H_m^\circ$ (kJ · mol^{-1})	S_m° (J · K^{-1} · mol^{-1})	$C_{p,m}^\circ$ (J · K^{-1} · mol^{-1})
Ca_3UO_6(cr)		−3305.400 ±4.100		
α-$SrUO_4$	−1881.355[(a)] ±2.802	−1989.600 ±2.800	153.150 ±0.170	130.620 ±0.200
β-$SrUO_4$		−1990.800 ±2.800		
Sr_2UO_5(cr)		−2635.600 ±3.400		
Sr_3UO_6(cr)		−3263.400 ±3.000		
$Sr_2U_3O_{11}$(cr)		−5242.900 ±4.100		
SrU_4O_{13}(cr)		−5920.000 ±20.000		
$BaUO_3$(cr)		−1690.000 ±10.000		
$BaUO_4$(cr)	−1883.805[(a)] ±3.393	−1993.800 ±3.300	154.000 ±2.500	125.300 ±2.500
Ba_3UO_6(cr)	−3044.951[(a)] ±9.196	−3210.400 ±8.000	298.000 ±15.000	
BaU_2O_7(cr)	−3052.093[(a)] ±6.714	−3237.200 ±5.000	260.000 ±15.000	
$Ba_2U_2O_7$(cr)	−3547.015[(a)] ±7.743	−3740.000 ±6.300	296.000 ±15.000	
Ba_2MgUO_6(cr)		−3245.900 ±6.500		
Ba_2CaUO_6(cr)		−3295.800 ±5.900		
Ba_2SrUO_6(cr)		−3257.300 ±5.700		

Table III.1 (continued)

Compound	$\Delta_f G_m^\circ$ (kJ · mol^{-1})	$\Delta_f H_m^\circ$ (kJ · mol^{-1})	S_m° (J · K^{-1} · mol^{-1})	$C_{p,m}^\circ$ (J · K^{-1} · mol^{-1})
$LiUO_3$(cr)		−1522.300 ±1.800		
Li_2UO_4(cr)	−1853.190[(a)] ±2.215	−1968.200 ±1.300	133.000 ±6.000	
Li_4UO_5(cr)		−2639.400 ±1.700		
$Li_2U_2O_7$(cr)		−3213.600 ±5.300		
$Li_2U_3O_{10}$(cr)		−4437.400 ±4.100		
$NaUO_3$(cr)	−1412.495[(a)] ±1.607	−1494.900 ±1.600	132.840 ±0.400	108.870[(c)] ±0.400
α-Na_2UO_4	−1779.303[(a)] ±3.506	−1897.700 ±3.500	166.000 ±0.500	146.700[(c)] ±0.500
β-Na_2UO_4		−1884.600 ±3.600		
Na_3UO_4(cr)	−1899.909[(a)] ±8.003	−2024.000 ±8.000	198.200 ±0.400	173.000[(c)] ±0.400
Na_4UO_5(cr)		−2456.600 ±1.700		
$Na_2U_2O_7$(cr)	−3011.454[(a)] ±4.015	−3203.800 ±4.000	275.900 ±1.000	227.300[(c)] ±1.000
$Na_6U_7O_{24}$(cr)		−11351.000 ±14.000		
$Na_4UO_2(CO_3)_3$(cr)	−3737.836[(b)] ±2.342			
KUO_3(cr)		−1522.900 ±1.700		
K_2UO_4(cr)	−1798.499[(a)] ±3.248	−1920.700 ±2.200	180.000 ±8.000	

Table III.1 (continued)

Compound	$\Delta_f G_m^\circ$ (kJ · mol^{-1})	$\Delta_f H_m^\circ$ (kJ · mol^{-1})	S_m° (J · K^{-1} · mol^{-1})	$C_{p,m}^\circ$ (J · K^{-1} · mol^{-1})
$K_2U_2O_7(cr)$		−3250.500 ±4.500		
$RbUO_3(cr)$		−1520.900 ±1.800		
$Rb_2UO_4(cr)$	−1800.141[a] ±3.250	−1922.700 ±2.200	203.000 ±8.000	
$Rb_2U_2O_7(cr)$		−3232.000 ±4.300		
$Cs_2UO_4(cr)$	−1805.370[a] ±1.232	−1928.000 ±1.200	219.660 ±0.440	152.760[c] ±0.310
$Cs_2U_2O_7(cr)$	−3022.880[a] ±10.005	−3220.000 ±10.000	327.750 ±0.660	231.200 ±0.500
$Cs_2U_4O_{12}(cr)$	−5251.058[a] ±3.628	−5571.800 ±3.600	526.400 ±1.000	384.000[c] ±1.000

(a) Value calculated internally with the equation $\Delta_f G_m^\circ = \Delta_f H_m^\circ - T\sum_i S_{m,i}^\circ$.
(b) Value calculated internally from reaction data (see Table III.2).
(c) Temperature coefficients of this function are listed in Table III.3.
(d) Mean value for the temperature interval 298 to 473 K.
(e) Stable phase of $UO_{2.25}$ ($\triangleq U_4O_9$) below 348 K.
(f) Stable phase of $UO_{2.25}$ ($\triangleq U_4O_9$) above 348 K. The thermodynamic parameters, however, refer to 298.15 K.
(g) Reaction data are selected for this species at $I = 3$ M ($NaClO_4$), *cf.* Table III.2.
(h) In the absence of recommended thermodynamic data for $PuO_2(CO_3)_3^{4-}$ and $NpO_2(CO_3)_3^{4-}$, only reaction data are selected for $(UO_2)_2(PuO_2)(CO_3)_6^{6-}$ and $(UO_2)_2(NpO_2)(CO_3)_6^{6-}$, *cf.* Table III.2.

Table III.2: Selected thermodynamic data for reactions involving uranium compounds and complexes. All ionic species listed in this table are aqueous species. Reactions are listed only if they were used for primary data selection. The thermodynamic data of formation (see Table III.1) are derived therefrom. Unless noted otherwise, all data refer to the reference temperature of 298.15 K and to the standard state, *i.e.*, a pressure of 0.1 MPa and, for aqueous species, infinite dilution ($I = 0$). The uncertainties listed below each value represent total uncertainties and correspond in principal to the statistically defined 95% confidence interval. Values obtained from internal calculation, *cf.* footnote (a), are rounded at the third digit after the decimal point and may therefore not be exactly identical to those given in Chapter V. Systematically, all the values are presented with three digits after the decimal point, regardless of the significance of these digits. The data presented in this table are available on PC diskettes or other computer media from the OECD Nuclear Energy Agency.

Species	Reaction				
		$\log_{10} K^\circ$	$\Delta_r G_m^\circ$ $(kJ \cdot mol^{-1})$	$\Delta_r H_m^\circ$ $(kJ \cdot mol^{-1})$	$\Delta_r S_m^\circ$ $(J \cdot K^{-1} \cdot mol^{-1})$
U^{3+}	$U^{4+} + e^- \rightleftharpoons U^{3+}$				
		−9.353	53.387	102.100	163.386[(a)]
		±0.070	±0.400	±1.700	±5.857
U^{4+}	$4H^+ + UO_2^{2+} + 2e^- \rightleftharpoons 2H_2O(l) + U^{4+}$				
		9.038	−51.589		
		±0.041	±0.234		
UO_2^+	$UO_2^{2+} + e^- \rightleftharpoons UO_2^+$				
		1.484[(b)]	−8.471		
		±0.022	±0.126		
UOH^{3+}	$H_2O(l) + U^{4+} \rightleftharpoons H^+ + UOH^{3+}$				
		−0.540	3.082	46.910[(a)]	147.000
		±0.060	±0.342	±8.951	±30.000
UO_2OH^+	$H_2O(l) + UO_2^{2+} \rightleftharpoons H^+ + UO_2OH^+$				
		−5.200	29.682		
		±0.300	±1.712		
$UO_2(OH)_2(aq)$	$2H_2O(l) + UO_2^{2+} \rightleftharpoons 2H^+ + UO_2(OH)_2(aq)$				
		≤ -10.300	≥ 58.793		
$UO_2(OH)_3^-$	$3H_2O(l) + UO_2^{2+} \rightleftharpoons 3H^+ + UO_2(OH)_3^-$				
		−19.200	109.594		
		±0.400	±2.283		

Table III.2 (continued)

Species	Reaction	$\log_{10} K^\circ$	$\Delta_r G^\circ_m$ $(kJ \cdot mol^{-1})$	$\Delta_r H^\circ_m$ $(kJ \cdot mol^{-1})$	$\Delta_r S^\circ_m$ $(J \cdot K^{-1} \cdot mol^{-1})$
$U(OH)_5^-$	$H_2O(l) + U(OH)_4(aq) \rightleftharpoons H^+ + U(OH)_5^-$				
		< -12.000	> 68.497		
$UO_2(OH)_4^{2-}$	$4H_2O(l) + UO_2^{2+} \rightleftharpoons 4H^+ + UO_2(OH)_4^{2-}$				
		-33.000	188.365		
		± 2.000	± 11.416		
$(UO_2)_2OH^{3+}$	$H_2O(l) + 2UO_2^{2+} \rightleftharpoons (UO_2)_2OH^{3+} + H^+$				
		-2.700	15.412		
		± 1.000	± 5.708		
$(UO_2)_2(OH)_2^{2+}$	$2H_2O(l) + 2UO_2^{2+} \rightleftharpoons (UO_2)_2(OH)_2^{2+} + 2H^+$				
		-5.620	32.079		
		± 0.040	± 0.228		
$(UO_2)_3(OH)_4^{2+}$	$4H_2O(l) + 3UO_2^{2+} \rightleftharpoons (UO_2)_3(OH)_4^{2+} + 4H^+$				
		-11.900	67.926		
		± 0.300	± 1.712		
$(UO_2)_3(OH)_5^+$	$5H_2O(l) + 3UO_2^{2+} \rightleftharpoons (UO_2)_3(OH)_5^+ + 5H^+$				
		-15.550	88.760		
		± 0.120	± 0.685		
$(UO_2)_3(OH)_7^-$	$7H_2O(l) + 3UO_2^{2+} \rightleftharpoons (UO_2)_3(OH)_7^- + 7H^+$				
		-31.000	176.949		
		± 2.000	± 11.416		
$(UO_2)_4(OH)_7^+$	$7H_2O(l) + 4UO_2^{2+} \rightleftharpoons (UO_2)_4(OH)_7^+ + 7H^+$				
		-21.900	125.006		
		± 1.000	± 5.708		
$U_6(OH)_{15}^{9+}$	$15H_2O(l) + 6U^{4+} \rightleftharpoons 15H^+ + U_6(OH)_{15}^{9+}$				
		-16.900(d)	96.466(d)		
		± 0.600	± 3.425		
UF^{3+}	$F^- + U^{4+} \rightleftharpoons UF^{3+}$				
		9.280	-52.971	-5.600	158.882(a)
		± 0.090	± 0.514	± 0.500	± 2.404
UF_2^{2+}	$2F^- + U^{4+} \rightleftharpoons UF_2^{2+}$				
		16.230	-92.642	-3.500	298.982(a)
		± 0.150	± 0.856	± 0.600	± 3.507

Table III.2 (continued)

Species	Reaction	$\log_{10} K^\circ$	$\Delta_r G_m^\circ$ $(kJ \cdot mol^{-1})$	$\Delta_r H_m^\circ$ $(kJ \cdot mol^{-1})$	$\Delta_r S_m^\circ$ $(J \cdot K^{-1} \cdot mol^{-1})$
$UF_3(g)$	$UF_3(cr) \rightleftharpoons UF_3(g)$				
				447.200	
				±15.000	
UF_3^+	$3F^- + U^{4+} \rightleftharpoons UF_3^+$				
		21.600	−123.294	0.500	415.206[a]
		±1.000	±5.708	±4.000	±23.378
$UF_4(aq)$	$4F^- + U^{4+} \rightleftharpoons UF_4(aq)$				
		25.600	−146.126	−4.206[a]	476.000
		±1.000	±5.708	±7.634	±17.000
$UF_4(g)$	$UF_4(cr) \rightleftharpoons UF_4(g)$				
				313.000	
				±8.000	
UF_5^-	$5F^- + U^{4+} \rightleftharpoons UF_5^-$				
		27.010	−154.174		
		±0.300	±1.712		
$UF_6(g)$	$UF_6(cr) \rightleftharpoons UF_6(g)$				
				49.100	
				±0.500	
UF_6^{2-}	$6F^- + U^{4+} \rightleftharpoons UF_6^{2-}$				
		29.080	−165.990		
		±0.180	±1.027		
UO_2F^+	$F^- + UO_2^{2+} \rightleftharpoons UO_2F^+$				
		5.090	−29.054	1.700	103.149[a]
		±0.130	±0.742	±0.080	±2.503
$UO_2F_2(aq)$	$2F^- + UO_2^{2+} \rightleftharpoons UO_2F_2(aq)$				
		8.620	−49.203	2.100	172.072[a]
		±0.040	±0.228	±0.190	±0.996
$UO_2F_2(g)$	$UO_2F_2(cr) \rightleftharpoons UO_2F_2(g)$				
				301.000	
				±15.000	
$UO_2F_3^-$	$3F^- + UO_2^{2+} \rightleftharpoons UO_2F_3^-$				
		10.900	−62.218	2.350	216.561[a]
		±0.400	±2.283	±0.310	±7.728

Table III.2 (continued)

Species	Reaction			
	$\log_{10} K^\circ$	$\Delta_r G_m^\circ$ $(kJ \cdot mol^{-1})$	$\Delta_r H_m^\circ$ $(kJ \cdot mol^{-1})$	$\Delta_r S_m^\circ$ $(J \cdot K^{-1} \cdot mol^{-1})$
$UO_2F_4^{2-}$	$4F^- + UO_2^{2+} \rightleftharpoons UO_2F_4^{2-}$			
	11.700 ±0.700	−66.784 ±3.996	0.290 ±0.470	224.968[(a)] ±13.494
$U_2O_3F_6(cr)$	$3UOF_4(cr) \rightleftharpoons U_2O_3F_6(cr) + UF_6(g)$			
			46.000 ±6.000	
$U_3O_5F_8(cr)$	$0.5UF_6(g) + 2.5UO_2F_2(cr) \rightleftharpoons U_3O_5F_8(cr)$			
			15.100[(e)] ±2.000	
$H_3OUF_6(cr)$	$6HF(g) + UF_4(cr) + UO_2F_2(cr) \rightleftharpoons 2H_3OUF_6(cr)$			
			−75.300 ±1.700	
$UF_4{\cdot}2.5H_2O(cr)$	$2.5H_2O(l) + UF_4(cr) \rightleftharpoons UF_4{\cdot}2.5H_2O(cr)$			
			−42.700 ±0.800	
$UO_2FOH{\cdot}2H_2O(cr)$	$H_2O(g) + UO_2FOH{\cdot}H_2O(cr) \rightleftharpoons UO_2FOH{\cdot}2H_2O(cr)$			
	1.883[(c)] ±0.666	−10.750 ±3.800	−53.684[(a)] ±4.200	−144.000 ±6.000
$UO_2F_2{\cdot}3H_2O(cr)$	$3H_2O(l) + UO_2F_2(cr) \rightleftharpoons UO_2F_2{\cdot}3H_2O(cr)$			
			−23.400 ±4.200	
UCl^{3+}	$Cl^- + U^{4+} \rightleftharpoons UCl^{3+}$			
	1.720 ±0.130	−9.818 ±0.742	−19.000 ±9.000	−30.797[(a)] ±30.289
$UCl_4(g)$	$UCl_4(cr) \rightleftharpoons UCl_4(g)$			
			200.700 ±5.000	
$UCl_6(g)$	$UCl_6(cr) \rightleftharpoons UCl_6(g)$			
			79.000 ±8.000	
$U_2Cl_8(g)$	$2UCl_4(g) \rightleftharpoons U_2Cl_8(g)$			
			−113.400 ±10.000	

Table III.2 (continued)

Species	Reaction			
	$\log_{10} K^\circ$	$\Delta_r G_m^\circ$ (kJ · mol^{-1})	$\Delta_r H_m^\circ$ (kJ · mol^{-1})	$\Delta_r S_m^\circ$ (J · K^{-1} · mol^{-1})
$U_2Cl_{10}(g)$	$Cl_2(g) + 2UCl_4(cr) \rightleftharpoons U_2Cl_{10}(g)$			
	−7.943[c] ±1.622	45.341[a] ±9.258	69.700 ±8.600	81.700 ±11.500
UO_2Cl^+	$Cl^- + UO_2^{2+} \rightleftharpoons UO_2Cl^+$			
	0.170 ±0.020	−0.970 ±0.114	8.000 ±2.000	30.087[a] ±6.719
$UO_2Cl_2(aq)$	$2Cl^- + UO_2^{2+} \rightleftharpoons UO_2Cl_2(aq)$			
	−1.100 ±0.400	6.279 ±2.283	15.000 ±6.000	29.251[a] ±21.532
$UO_2ClO_3^+$	$ClO_3^- + UO_2^{2+} \rightleftharpoons UO_2ClO_3^+$			
	0.500 ±0.070	−2.854 ±0.400	−3.900 ±0.300	−3.508[a] ±1.676
UBr^{3+}	$Br^- + U^{4+} \rightleftharpoons UBr^{3+}$			
	1.460 ±0.200	−8.334 ±1.142		
$UBr_4(g)$	$UBr_4(cr) \rightleftharpoons UBr_4(g)$			
			192.000 ±8.000	
$UBr_5(g)$	$0.5Br_2(g) + UBr_4(g) \rightleftharpoons UBr_5(g)$			
			−43.100 ±15.000	
UO_2Br^+	$Br^- + UO_2^{2+} \rightleftharpoons UO_2Br^+$			
	0.220 ±0.020	−1.256 ±0.114		
$UO_2BrO_3^+$	$BrO_3^- + UO_2^{2+} \rightleftharpoons UO_2BrO_3^+$			
	0.630 ±0.080	−3.596 ±0.457	0.100 ±0.300	12.397[a] ±1.833
UI^{3+}	$I^- + U^{4+} \rightleftharpoons UI^{3+}$			
	1.250 ±0.300	−7.135 ±1.712		
$UI_4(g)$	$UI_4(cr) \rightleftharpoons UI_4(g)$			
			210.000 ±5.000	

Table III.2 (continued)

Species	Reaction				
		$\log_{10} K^\circ$	$\Delta_r G_m^\circ$ $(kJ \cdot mol^{-1})$	$\Delta_r H_m^\circ$ $(kJ \cdot mol^{-1})$	$\Delta_r S_m^\circ$ $(J \cdot K^{-1} \cdot mol^{-1})$
$UO_2IO_3^+$	$IO_3^- + UO_2^{2+} \rightleftharpoons UO_2IO_3^+$				
		2.000	−11.416	9.800	71.159[a]
		±0.020	±0.114	±0.900	±3.043
$UO_2(IO_3)_2(aq)$	$2IO_3^- + UO_2^{2+} \rightleftharpoons UO_2(IO_3)_2(aq)$				
		3.590	−20.492		
		±0.150	±0.856		
$UO_2(IO_3)_2(cr)$	$2IO_3^- + UO_2^{2+} \rightleftharpoons UO_2(IO_3)_2(cr)$				
		7.880	−44.979		
		±0.100	±0.571		
USO_4^{2+}	$SO_4^{2-} + U^{4+} \rightleftharpoons USO_4^{2+}$				
		6.580	−37.559	8.000	152.805[a]
		±0.190	±1.085	±2.700	±9.759
$UO_2SO_3(aq)$	$SO_3^{2-} + UO_2^{2+} \rightleftharpoons UO_2SO_3(aq)$				
		6.600	−37.673		
		±0.600	±3.425		
$UO_2S_2O_3(aq)$	$S_2O_3^{2-} + UO_2^{2+} \rightleftharpoons UO_2S_2O_3(aq)$				
		2.800	−15.983		
		±0.300	±1.712		
$UO_2SO_4(aq)$	$SO_4^{2-} + UO_2^{2+} \rightleftharpoons UO_2SO_4(aq)$				
		3.150	−17.980	19.500	125.710[a]
		±0.020	±0.114	±1.600	±5.380
$U(SO_4)_2(aq)$	$2SO_4^{2-} + U^{4+} \rightleftharpoons U(SO_4)_2(aq)$				
		10.510	−59.992	32.700	310.889[a]
		±0.200	±1.142	±2.800	±10.142
$UO_2(SO_4)_2^{2-}$	$2SO_4^{2-} + UO_2^{2+} \rightleftharpoons UO_2(SO_4)_2^{2-}$				
		4.140	−23.631	35.100	196.986[a]
		±0.070	±0.400	±1.000	±3.612
$U(OH)_2SO_4(cr)$	$2OH^- + SO_4^{2-} + U^{4+} \rightleftharpoons U(OH)_2SO_4(cr)$				
		31.170	−177.920		
		±0.500	±2.854		
$UO_2SO_4{\cdot}2.5H_2O(cr)$	$2.5H_2O(l) + SO_4^{2-} + UO_2^{2+} \rightleftharpoons UO_2SO_4{\cdot}2.5H_2O(cr)$				
		1.589[c]	−9.070		
		±0.019	±0.110		

Table III.2 (continued)

Species	Reaction			
	$\log_{10} K^\circ$	$\Delta_r G_m^\circ$ (kJ · mol^{-1})	$\Delta_r H_m^\circ$ (kJ · mol^{-1})	$\Delta_r S_m^\circ$ (J · K^{-1} · mol^{-1})
$UO_2SO_4{\cdot}3H_2O(cr)$	$UO_2SO_4{\cdot}3.5H_2O(cr) \rightleftharpoons 0.5H_2O(g) + UO_2SO_4{\cdot}3H_2O(cr)$			
	−0.831	4.743		
	±0.023	±0.131		
$UO_2SO_4{\cdot}3.5H_2O(cr)$	$3.5H_2O(l) + SO_4^{2-} + UO_2^{2+} \rightleftharpoons UO_2SO_4{\cdot}3.5H_2O(cr)$			
	1.585[(c)]	−9.050		
	±0.019	±0.110		
$UO_2N_3^+$	$N_3^- + UO_2^{2+} \rightleftharpoons UO_2N_3^+$			
	2.580	−14.727		
	±0.090	±0.514		
$UO_2(N_3)_2(aq)$	$2N_3^- + UO_2^{2+} \rightleftharpoons UO_2(N_3)_2(aq)$			
	4.330	−24.716		
	±0.230	±1.313		
$UO_2(N_3)_3^-$	$3N_3^- + UO_2^{2+} \rightleftharpoons UO_2(N_3)_3^-$			
	5.740	−32.764		
	±0.220	±1.256		
$UO_2(N_3)_4^{2-}$	$4N_3^- + UO_2^{2+} \rightleftharpoons UO_2(N_3)_4^{2-}$			
	4.920	−28.084		
	±0.240	±1.370		
UNO_3^{3+}	$NO_3^- + U^{4+} \rightleftharpoons UNO_3^{3+}$			
	1.470	−8.391		
	±0.130	±0.742		
$UO_2NO_3^+$	$NO_3^- + UO_2^{2+} \rightleftharpoons UO_2NO_3^+$			
	0.300	−1.712		
	±0.150	±0.856		
$U(NO_3)_2^{2+}$	$2NO_3^- + U^{4+} \rightleftharpoons U(NO_3)_2^{2+}$			
	2.300	−13.128		
	±0.350	±1.998		
$UO_2PO_4^-$	$PO_4^{3-} + UO_2^{2+} \rightleftharpoons UO_2PO_4^-$			
	13.230	−75.517		
	±0.150	±0.856		
$UO_2HPO_4(aq)$	$HPO_4^{2-} + UO_2^{2+} \rightleftharpoons UO_2HPO_4(aq)$			
	7.240	−41.326		
	±0.260	±1.484		

Table III.2 (continued)

Species	Reaction			
	$\log_{10} K^\circ$	$\Delta_r G_m^\circ$ $(kJ \cdot mol^{-1})$	$\Delta_r H_m^\circ$ $(kJ \cdot mol^{-1})$	$\Delta_r S_m^\circ$ $(J \cdot K^{-1} \cdot mol^{-1})$
$UO_2H_2PO_4^+$	$H_3PO_4(aq) + UO_2^{2+} \rightleftharpoons H^+ + UO_2H_2PO_4^+$			
	1.120 ±0.060	−6.393 ±0.342		
$UO_2H_3PO_4^{2+}$	$H_3PO_4(aq) + UO_2^{2+} \rightleftharpoons UO_2H_3PO_4^{2+}$			
	0.760 ±0.150	−4.338 ±0.856		
$UO_2(H_2PO_4)_2(aq)$	$2H_3PO_4(aq) + UO_2^{2+} \rightleftharpoons 2H^+ + UO_2(H_2PO_4)_2(aq)$			
	0.640 ±0.110	−3.653 ±0.628		
$UO_2(H_2PO_4)(H_3PO_4)^+$	$2H_3PO_4(aq) + UO_2^{2+} \rightleftharpoons H^+ + UO_2(H_2PO_4)(H_3PO_4)^+$			
	1.650 ±0.110	−9.418 ±0.628		
$UO_2HPO_4{\cdot}4H_2O(cr)$	$4H_2O(l) + H_3PO_4(aq) + UO_2^{2+} \rightleftharpoons 2H^+ + UO_2HPO_4{\cdot}4H_2O(cr)$			
	2.500 ±0.090	−14.270 ±0.514		
$U(HPO_4)_2{\cdot}4H_2O(cr)$	$4H_2O(l) + 2H_3PO_4(aq) + U^{4+} \rightleftharpoons 4H^+ + U(HPO_4)_2{\cdot}4H_2O(cr)$			
	11.790 ±0.150	−67.298 ±0.856		
$(UO_2)_3(PO_4)_2{\cdot}4H_2O(cr)$	$4H_2O(l) + 2H_3PO_4(aq) + 3UO_2^{2+} \rightleftharpoons (UO_2)_3(PO_4)_2{\cdot}4H_2O(cr) + 6H^+$			
	5.960 ±0.300	−34.020 ±1.712		
$UO_2CO_3(aq)$	$CO_3^{2-} + UO_2^{2+} \rightleftharpoons UO_2CO_3(aq)$			
	9.680 ±0.040	−55.254 ±0.228	5.000 ±2.000	202.092[(a)] ±6.752
$UO_2CO_3(cr)$	$CO_3^{2-} + UO_2^{2+} \rightleftharpoons UO_2CO_3(cr)$			
	14.470 ±0.040	−82.595 ±0.228		
$UO_2(CO_3)_2^{2-}$	$2CO_3^{2-} + UO_2^{2+} \rightleftharpoons UO_2(CO_3)_2^{2-}$			
	16.940 ±0.120	−96.694 ±0.685	18.500 ±4.000	386.363[(a)] ±13.611
$UO_2(CO_3)_3^{4-}$	$3CO_3^{2-} + UO_2^{2+} \rightleftharpoons UO_2(CO_3)_3^{4-}$			
	21.600 ±0.050	−123.294 ±0.285	−39.200 ±4.100	282.052[(a)] ±13.785

Table III.2 (continued)

Species	Reaction			
	$\log_{10} K^\circ$	$\Delta_r G_m^\circ$ $(kJ \cdot mol^{-1})$	$\Delta_r H_m^\circ$ $(kJ \cdot mol^{-1})$	$\Delta_r S_m^\circ$ $(J \cdot K^{-1} \cdot mol^{-1})$
$UO_2(CO_3)_3^{5-}$	$UO_2(CO_3)_3^{4-} + e^- \rightleftharpoons UO_2(CO_3)_3^{5-}$			
	−12.710[b]	72.549		
	±0.270	±1.541		
$U(CO_3)_4^{4-}$	$U(CO_3)_5^{6-} \rightleftharpoons CO_3^{2-} + U(CO_3)_4^{4-}$			
	1.120	−6.393		
	±0.250	±1.427		
$U(CO_3)_5^{6-}$	$5CO_3^{2-} + U^{4+} \rightleftharpoons U(CO_3)_5^{6-}$			
	34.000	−194.073	−20.000	583.845[a]
	±0.900	±5.137	±4.000	±21.838
$(UO_2)_3(CO_3)_6^{6-}$	$6CO_3^{2-} + 3UO_2^{2+} \rightleftharpoons (UO_2)_3(CO_3)_6^{6-}$			
	54.000	−308.234	−62.700	823.526[a]
	±1.000	±5.708	±2.400	±20.768
$(UO_2)_2CO_3(OH)_3^-$	$CO_2(g) + 4H_2O(l) + 2UO_2^{2+} \rightleftharpoons (UO_2)_2CO_3(OH)_3^- + 5H^+$			
	−19.010	108.510		
	±0.500	±2.854		
$(UO_2)_3O(OH)_2(HCO_3)^+$	$CO_2(g) + 4H_2O(l) + 3UO_2^{2+} \rightleftharpoons (UO_2)_3O(OH)_2(HCO_3)^+ + 5H^+$			
	−17.500	99.891		
	±0.500	±2.854		
$(UO_2)_{11}(CO_3)_6(OH)_{12}^{2-}$	$6CO_2(g) + 18H_2O(l) + 11UO_2^{2+} \rightleftharpoons (UO_2)_{11}(CO_3)_6(OH)_{12}^{2-} + 24H^+$			
	−72.500	413.833		
	±2.000	±11.416		
$USCN^{3+}$	$SCN^- + U^{4+} \rightleftharpoons USCN^{3+}$			
	2.970	−16.953	−27.000	−33.698[a]
	±0.060	±0.342	±8.000	±26.857
$U(SCN)_2^{2+}$	$2SCN^- + U^{4+} \rightleftharpoons U(SCN)_2^{2+}$			
	4.260	−24.316	−18.000	21.185[a]
	±0.180	±1.027	±4.000	±13.852
UO_2SCN^+	$SCN^- + UO_2^{2+} \rightleftharpoons UO_2SCN^+$			
	1.400	−7.991	3.220	37.603[a]
	±0.230	±1.313	±0.060	±4.408
$UO_2(SCN)_2(aq)$	$2SCN^- + UO_2^{2+} \rightleftharpoons UO_2(SCN)_2(aq)$			
	1.240	−7.078	8.900	53.590[a]
	±0.550	±3.139	±0.600	±10.720

Table III.2 (continued)

Species	Reaction			
	$\log_{10} K^\circ$	$\Delta_r G_m^\circ$ ($kJ \cdot mol^{-1}$)	$\Delta_r H_m^\circ$ ($kJ \cdot mol^{-1}$)	$\Delta_r S_m^\circ$ ($J \cdot K^{-1} \cdot mol^{-1}$)
$UO_2(SCN)_3^-$	$3SCN^- + UO_2^{2+} \rightleftharpoons UO_2(SCN)_3^-$			
	2.100	−11.987	6.000	60.328[(a)]
	±0.500	±2.854	±1.200	±10.384
$(UO_2)_2(PuO_2)(CO_3)_6^{6-}$	$PuO_2(CO_3)_3^{4-} + 2UO_2(CO_3)_3^{4-} \rightleftharpoons (UO_2)_2(PuO_2)(CO_3)_6^{6-} + 3CO_3^{2-}$			
	−8.800	50.231		
	±0.200	±1.142		
$(UO_2)_2(NpO_2)(CO_3)_6^{6-}$	$NpO_2(CO_3)_3^{4-} + 2UO_2(CO_3)_3^{4-} \rightleftharpoons (UO_2)_2(NpO_2)(CO_3)_6^{6-} + 3CO_3^{2-}$			
	−10.000	57.080		
	±0.100	±0.571		
$Na_4UO_2(CO_3)_3(cr)$	$4Na^+ + UO_2(CO_3)_3^{4-} \rightleftharpoons Na_4UO_2(CO_3)_3(cr)$			
	5.340	−30.481		
	±0.160	±0.913		

(a) Value calculated internally with the equation $\Delta_r G_m^\circ = \Delta_r H_m^\circ - T\Delta_r S_m^\circ$.
(b) Value calculated from a selected standard potential.
(c) Value of $\log_{10} K^\circ$ calculated internally from $\Delta_r G_m^\circ$.
(d) This value is valid at $I = 3$ M ($NaClO_4$).
(e) For the reaction $0.5UF_6(g) + 2.5UO_2F_2(cr) \rightleftharpoons U_3O_5F_8(cr)$, a heat capacity of $\Delta_r C_{p,m}^\circ(298.15\ K) = -(18.7 \pm 4.0)\ J \cdot K^{-1} \cdot mol^{-1}$ is selected.

Table III.3: Selected temperature coefficients for heat capacities marked with [(c)] in Table III.1, according to the form $C^{\circ}_{p,m}(T) = a + bT + cT^2 + dT^{-1} + eT^{-2}$. The functions are valid between the temperatures T_{min} and T_{max} (in K). The values in parentheses represent the power of 10. Units are $J \cdot K^{-1} \cdot mol^{-1}$.

Compound	a	b	c/d [(a)]	e	T_{min}	T_{max}
U(cr)	2.69199(+01)	−2.50203(−03)	2.65558(−05)(c)	−7.69856(+04)	298	941
U(g)	−4.68380(+00)	1.70637(−02)	1.41995(+04)(d)	−2.16503(+06)	298	2500
UO(g)	8.28637(+01)	−6.70169(−02)	3.75787(−05)(c)	−2.15185(+06)	298	1000
UO_2(cr)	6.27740(+01)	3.17400(−02)		−7.69300(+05)	250	600
UO_2(g)	5.78388(+01)	9.01053(−03)	−9.36370(−07)(c)	−7.83110(+04)	298	1000
UO_2^{2+} [(b)]					283	328
$UO_{2.25}$(cr)	1.48760(+03)	−6.97370(+00)	9.73600(−03)(c)	−1.78600(+07)	250	348
β-$UO_{2.25}$	7.90890(+01)	1.36500(−02)		−1.03800(+06)	348	600
α-$UO_{2.3333}$	6.41490(+01)	4.91400(−02)		−6.72000(+05)	237	347
β-$UO_{2.3333}$	6.43380(+01)	4.97900(−02)		−6.55000(+05)	232	346
$UO_{2.6667}$(cr)	8.72760(+01)	2.21600(−02)		−1.24400(+06)	233	600
β-UO_3	8.61700(+01)	2.49840(−02)		−1.09150(+06)	298	678
γ-UO_3	8.81030(+01)	1.66400(−02)		−1.01280(+06)	298	850
UO_3(am)	7.60100(+01)	3.80600(−02)		−2.31000(+05)	400	650
UO_3(g)	6.44292(+01)	3.17328(−02)	−1.51983(−05)(c)	−7.14579(+05)	298	1000
β-$UO_2(OH)_2$	4.18000(+01)	2.00000(−01)		3.53000(+06)	298	473
$UO_3 \cdot 2H_2O$(cr)	8.42380(+01)	2.94592(−01)			298	400
UF(g)	5.33887(+01)	−1.24482(−02)	7.64918(−06)(c)	−1.11051(+06)	298	1000
UF_2(g)	4.27524(+01)	4.37614(−02)	−2.01258(−05)(c)	2.10184(+05)	298	1000
UF_3(cr)	1.06368(+02)	9.17095(−04)		−1.02580(+06)	298	800
UF_3(g)	8.05856(+01)	2.18816(−03)	8.04161(−07)(c)	−4.56379(+05)	298	1000
UF_4(cr)	1.23407(+02)	9.76762(−03)		−9.15376(+05)	298	1200
UF_4(g)	1.05859(+02)	5.48379(−03)	6.30425(−07)(c)	−5.49014(+05)	298	1000
α-UF_5	1.25478(+02)	3.02085(−02)		−1.92464(+05)	298	600
β-UF_5	1.25478(+02)	3.02085(−02)		−1.92464(+05)	298	600
UF_5(g)	1.14197(+02)	3.75552(−02)	−1.64563(−05)(c)	−1.18344(+06)	298	1000
UF_6(cr)	5.23180(+01)	3.83798(−01)			298	337
UF_6(g)	1.41116(+02)	2.94180(−02)	−1.43713(−05)(c)	−1.70571(+06)	298	1000
U_2F_9(cr)	2.35978(+02)	5.99149(−02)		−2.17568(+05)	298	600
U_4F_{17}(cr)	4.53546(+02)	1.18491(−01)		−2.67776(+05)	298	600
UOF_4(g)	1.16063(+02)	2.93016(−02)	−1.42136(−05)(c)	−1.37329(+06)	298	1000
UO_2F_2(cr)	7.72865(+01)	8.69782(−02)			298	400
UO_2F_2(g)	8.98877(+01)	3.10644(−02)	−1.49478(−05)(c)	−1.01918(+06)	298	1000
UCl(g)	5.77709(+01)	−2.74441(−02)	1.87524(−05)(c)	−7.22827(+05)	298	1000
UCl_2(g)	5.92289(+01)	−3.16326(−03)	4.59862(−06)(c)	1.08444(+05)	298	1000
UCl_3(cr)	8.77800(+01)	7.78000(−03)		4.85300(+05)	298	1000
UCl_3(g)	8.55239(+01)	−6.61716(−03)	5.22328(−06)(c)	−1.43347(+05)	298	1000
UCl_4(cr)	1.06859(+02)	4.86448(−02)		−8.99977(+04)	298	800
UCl_4(g)	1.12217(+02)	2.32451(−04)	5.68114(−07)(c)	−4.16473(+05)	298	1000
UCl_5(cr)	1.40164(+02)	3.55640(−02)			298	600
UCl_5(g)	1.25195(+02)	1.79496(−02)	−6.46755(−06)(c)	−5.65216(+05)	298	1000
UCl_6(cr)	1.73427(+02)	3.50619(−02)		−7.40568(+05)	298	452
UCl_6(g)	1.50313(+02)	1.76431(−02)	−6.31723(−06)(c)	−6.93772(+05)	298	1000
UOCl(cr)	7.58140(+01)	1.43510(−02)		−8.28430(+05)	298	900
$UOCl_2$(cr)	9.88121(+01)	2.22099(−02)		−9.22160(+05)	298	700
$UOCl_3$(cr)	1.05269(+02)	3.99154(−02)		−2.09200(+04)	298	900
UO_2Cl(cr)	9.01230(+01)	2.22590(−02)		−7.74040(+05)	298	1000
UO_2Cl_2(cr)	1.15000(+02)	1.82232(−02)		−1.14180(+06)	298	650
UO_2Cl_2(g)	9.45965(+01)	2.28872(−02)	−1.09658(−05)(c)	−7.26861(+05)	298	1000
$U_2O_2Cl_5$(cr)	2.34304(+02)	3.55640(−02)		−2.26773(+06)	298	700
$(UO_2)_2Cl_3$(cr)	2.25936(+02)	3.55640(−02)		−2.92880(+06)	298	900

Table III.3 (continued)

Compound	a	b	c/d[(a)]	e	T_{min}	T_{max}
UBr(g)	5.81507(+01)	−2.83383(−02)	1.90877(−05)(c)	−6.43089(+05)	298	1000
UBr_2(g)	5.96287(+01)	−3.87789(−03)	4.95348(−06)(c)	2.18109(+05)	298	1000
UBr_3(cr)	9.79710(+01)	2.63600(−02)			298	600
UBr_3(g)	8.63269(+01)	−8.05170(−03)	5.93542(−06)(c)	6.20838(+04)	298	1000
UBr_4(cr)	1.19244(+02)	2.97064(−02)			298	700
UBr_4(g)	1.12848(+02)	1.17689(−04)	−1.57354(−07)(c)	−2.32193(+05)	298	1000
UBr_5(cr)	1.50624(+02)	3.34720(−02)			298	400
UBr_5(g)	1.26802(+02)	1.50849(−02)	−5.04742(−06)(c)	−1.67987(+05)	298	1000
$UOBr_2$(cr)	1.10579(+02)	1.36817(−02)		−1.47904(+06)	298	900
$UOBr_3$(cr)	1.30541(+02)	2.05016(−02)		−1.38072(+06)	298	1100
UO_2Br_2(cr)	1.04270(+02)	3.79380(−02)			298	500
UI(g)	5.81857(+01)	−2.84012(−02)	1.91190(−05)(c)	−6.21423(+05)	298	1000
UI_2(g)	5.96972(+01)	−4.00076(−03)	5.01464(−06)(c)	2.63886(+05)	298	1000
UI_3(cr)	1.05018(+02)	2.42672(−02)			298	800
UI_3(g)	8.63983(+01)	−8.17990(−03)	5.99922(−06)(c)	1.33476(+05)	298	1000
UI_4(cr)	1.45603(+02)	9.95792(−03)		−1.97485(+06)	298	720
UI_4(g)	1.13119(+02)	−3.68084(−04)	8.44065(−08)(c)	−7.19505(+04)	298	1000
US(cr)	5.28565(+01)	6.51574(−03)		−3.78288(+05)	298	2000
$US_{1.90}$(cr)	7.80567(+01)	6.63580(−03)		−5.38899(+05)	298	1900
US_2(cr)	7.56635(+01)	8.93702(−03)		−3.27775(+05)	298	1800
US_3(cr)	1.00667(+02)	1.12131(−02)		−7.44752(+05)	298	1100
U_2S_3(cr)	1.29447(+02)	1.46856(−02)			298	2000
UO_2SO_4(cr)	1.12466(+02)	1.08784(−01)			298	820
USe(cr)	5.41473(+01)	7.96237(−03)		−1.52183(+05)	298	800
α-USe_2	7.96932(+01)	8.64868(−03)		−2.64833(+05)	298	800
UN(cr)	5.05400(+01)	1.06600(−02)		−5.23800(+05)	298	1000
UPO_5(cr)	1.10427(+02)	8.52280(−02)		−1.03790(+06)	298	600
$(UO_2)_2P_2O_7$(cr)	2.50669(+02)	1.54299(−01)		−3.40030(+06)	298	600
$(UO_2)_3(PO_4)_2$(cr)	3.26375(+02)	1.96438(−01)		−4.08400(+06)	298	600
$UAsO_5$(cr)	1.30701(+02)	6.42580(−02)		−1.05420(+06)	298	800
$UO_2(AsO_3)_2$(cr)	2.08461(+02)	8.89100(−02)		−3.03460(+06)	298	750
$(UO_2)_2As_2O_7$(cr)	3.07412(+02)	1.05818(−01)		−5.85130(+06)	298	850
$(UO_2)_3(AsO_4)_2$(cr)	3.31151(+02)	2.11360(−01)		−2.68180(+06)	298	850
UC(cr)	5.55714(+01)	9.35943(−03)		−7.32308(+05)	298	600
α-$UC_{1.94}$	6.57748(+01)	2.03745(−02)		−9.86519(+05)	298	600
U_2C_3(cr)	1.14816(+02)	3.10060(−02)		−1.48447(+06)	298	600
$NaUO_3$(cr)	1.15491(+02)	1.91672(−02)		−1.09660(+06)	415	931
α-Na_2UO_4	1.62538(+02)	2.58857(−02)		−2.09660(+06)	618	1165
Na_3UO_4(cr)	1.88901(+02)	2.51788(−02)		−2.08010(+06)	523	1212
$Na_2U_2O_7$(cr)	2.62831(+02)	1.46532(−02)		−3.54904(+06)	390	540
Cs_2UO_4(cr)	1.64881(+02)	1.70232(−02)		−1.52851(+06)	298	1061
$Cs_2U_4O_{12}$(cr)	4.23726(+02)	7.19405(−02)		−5.43750(+06)	298	898

(a) One single column is used for the two coefficients c and d, because they never occur together in the thermal functions presented in this table. The coefficient concerned is indicated in parentheses behind each value.

(b) The thermal function is $C^{\circ}_{p,m}(UO_2^{2+}, aq, T) = \left(350.5 - 0.8722\,T - \frac{5308}{T-190}\right)$ J · K^{-1} · mol^{-1} for 283 K $\leq T \leq$ 328 K.

Chapter IV

Selected auxiliary data

This chapter presents the chemical thermodynamic data for auxiliary compounds and complexes which are selected in this review. Most of these auxiliary species are used in the evaluation of the recommended uranium data in Tables III.1, III.2 and III.3. It is therefore essential to always use these auxiliary data in conjunction with the selected uranium data. The use of other auxiliary data can lead to inconsistencies and erroneous results.

Table IV.1 contains the selected thermodynamic data of the auxiliary species considered in this review and Table IV.2 the selected thermodynamic data of chemical reactions including these auxiliary species. The reason for listing both formation and reaction data is described in Chapter III.

All data in Tables IV.1 and IV.2 refer to a temperature of 298.15 K, the standard state pressure of 0.1 MPa and, for aqueous species and reactions, to the infinite dilution reference state ($I = 0$). The uncertainties listed below each value represent total uncertainties and correspond in principal to the statistically defined 95% confidence interval.

The selection procedure of the auxiliary data is described in Chapter VI.

Note that the values in Tables IV.1 and IV.2 may contain more digits than those listed in Chapter VI, because the data in the present chapter are retrieved directly from the computerized data base and rounded to three digits after the decimal point throughout.

Table IV.1: Selected thermodynamic data for auxiliary compounds and complexes, including the remaining CODATA Key Values [89COX/WAG] of species not containing uranium, as well as other data that were evaluated in Chapter VI and used directly or indirectly for the evaluation of the uranium data presented in Tables III.1 and III.2. All ionic species listed in this table are aqueous species. Unless noted otherwise, all data refer to the reference temperature of 298.15 K and to the standard state, *i.e.*, a pressure of 0.1 MPa and, for aqueous species, infinite dilution ($I = 0$). The uncertainties listed below each value represent total uncertainties and correspond in principal to the statistically defined 95% confidence interval. Values in **bold** typeface are CODATA Key Values and are taken directly from Ref. [89COX/WAG] without further evaluation. Values obtained from internal calculation, *cf.* footnotes (a) and (b), are rounded at the third digit after the decimal point and may therefore not be exactly identical to those given in Chapter VI. Systematically, all the values are presented with three digits after the decimal point, regardless of the significance of these digits. The data presented in this table are available on PC diskettes or other computer media from the OECD Nuclear Energy Agency.

Compound	$\Delta_f G^\circ_m$ (kJ · mol^{-1})	$\Delta_f H^\circ_m$ (kJ · mol^{-1})	S°_m (J · K^{-1} · mol^{-1})	$C^\circ_{p,m}$ (J · K^{-1} · mol^{-1})
O(g)	**231.744**[(a)] ±**0.100**	**249.180** ±**0.100**	**161.059** ±**0.003**	
O_2(g)	0.000	0.000	**205.152** ±**0.005**	
H(g)	**203.276**[(a)] ±**0.006**	**217.998** ±**0.006**	**114.717** ±**0.002**	
H^+	0.000	0.000	0.000	
H_2(g)	0.000	0.000	**130.680** ±**0.003**	
OH^-	**−157.220**[(a)] ±**0.072**	**−230.015** ±**0.040**	**−10.900** ±**0.200**	
H_2O(g)	**−228.582**[(a)] ±**0.040**	**−241.826** ±**0.040**	**188.835** ±**0.010**	
H_2O(l)	**−237.140**[(a)] ±**0.041**	**−285.830** ±**0.040**	**69.950** ±**0.030**	
H_2O_2(aq)		−191.170[(c)] ±0.100		

Table IV.1 (continued)

Compound	$\Delta_f G_m^\circ$ ($kJ \cdot mol^{-1}$)	$\Delta_f H_m^\circ$ ($kJ \cdot mol^{-1}$)	S_m° ($J \cdot K^{-1} \cdot mol^{-1}$)	$C_{p,m}^\circ$ ($J \cdot K^{-1} \cdot mol^{-1}$)
He(g)	**0.000**	**0.000**	**126.153 ±0.002**	
Ne(g)	**0.000**	**0.000**	**146.328 ±0.003**	
Ar(g)	**0.000**	**0.000**	**154.846 ±0.003**	
Kr(g)	**0.000**	**0.000**	**164.085 ±0.003**	
Xe(g)	**0.000**	**0.000**	**169.685 ±0.003**	
F(g)	**62.280**[a] **±0.300**	**79.380 ±0.300**	**158.751 ±0.004**	
F^-	**−281.523**[a] **±0.692**	**−335.350 ±0.650**	**−13.800 ±0.800**	
F_2(g)	**0.000**	**0.000**	**202.791 ±0.005**	
HF(aq)	−299.675[b] ±0.702	−323.150[b] ±0.716	88.000[b] ±1.341	
HF(g)	**−275.400**[a] **±0.700**	**−273.300 ±0.700**	**173.779 ±0.003**	
HF_2^-	−583.709[b] ±1.200	−655.500[b] ±2.221	92.685[b] ±7.260	
Cl(g)	**105.305**[a] **±0.008**	**121.301 ±0.008**	**165.190 ±0.004**	
Cl^-	**−131.217**[a] **±0.117**	**−167.080 ±0.100**	**56.600 ±0.200**	
Cl_2(g)	**0.000**	**0.000**	**223.081 ±0.010**	
ClO^-	−37.670[b] ±0.962			

Table IV.1 (continued)

Compound	$\Delta_f G^\circ_m$ ($kJ \cdot mol^{-1}$)	$\Delta_f H^\circ_m$ ($kJ \cdot mol^{-1}$)	S°_m ($J \cdot K^{-1} \cdot mol^{-1}$)	$C^\circ_{p,m}$ ($J \cdot K^{-1} \cdot mol^{-1}$)
ClO_2^-	10.249[(b)] ±4.044			
ClO_3^-	−7.903[(a)] ±1.342	−104.000 ±1.000	162.300 ±3.000	
ClO_4^-	**−7.890[(a)] ±0.600**	**−128.100 ±0.400**	**184.000 ±1.500**	
HCl(g)	**−95.298[(a)] ±0.100**	**−92.310 ±0.100**	**186.902 ±0.005**	
HClO(aq)	−80.024[(b)] ±0.613			
$HClO_2$(aq)	−0.939[(b)] ±4.043			
Br(g)	**82.379[(a)] ±0.128**	**111.870 ±0.120**	**175.018 ±0.004**	
Br^-	**−103.850[(a)] ±0.167**	**−121.410 ±0.150**	**82.550 ±0.200**	
Br_2(g)	**3.105[(a)] ±0.142**	**30.910 ±0.110**	**245.468 ±0.005**	
Br_2(l)	**0.000**	**0.000**	**152.210 ±0.300**	
BrO^-	−32.140[(b)] ±1.510			
BrO_3^-	19.070[(a)] ±0.634	−66.700 ±0.500	161.500 ±1.300	
HBr(g)	**−53.361[(a)] ±0.166**	**−36.290 ±0.160**	**198.700 ±0.004**	
HBrO(aq)	−81.400 ±1.500			
I(g)	**70.172[(a)] ±0.060**	**106.760 ±0.040**	**180.787 ±0.004**	

Table IV.1 (continued)

Compound	$\Delta_f G_m^\circ$ ($kJ \cdot mol^{-1}$)	$\Delta_f H_m^\circ$ ($kJ \cdot mol^{-1}$)	S_m° ($J \cdot K^{-1} \cdot mol^{-1}$)	$C_{p,m}^\circ$ ($J \cdot K^{-1} \cdot mol^{-1}$)
I^-	**−51.724**[a] **±0.112**	**−56.780** **±0.050**	**106.450** **±0.300**	
$I_2(cr)$	**0.000**	**0.000**	**116.140** **±0.300**	
$I_2(g)$	**19.323**[a] **±0.120**	**62.420** **±0.080**	**260.687** **±0.005**	
IO_3^-	−126.338[a] ±0.779	−219.700 ±0.500	118.000 ±2.000	
HI(g)	**1.700**[a] **±0.110**	**26.500** **±0.100**	**206.590** **±0.004**	
$HIO_3(aq)$	−130.836[b] ±0.797			
S(cr)[d]	**0.000**	**0.000**	**32.054** **±0.050**	22.560 ±0.050
S(g)	**236.689**[a] **±0.151**	**277.170** **±0.150**	**167.829** **±0.006**	
S^{2-}	120.695[b] ±11.610			
$S_2(g)$	**79.686**[a] **±0.301**	**128.600** **±0.300**	**228.167** **±0.010**	
$SO_2(g)$	**−300.095**[a] **±0.201**	**−296.810** **±0.200**	**248.223** **±0.050**	
SO_3^{2-}	−487.473[b] ±4.020			
$S_2O_3^{2-}$	−519.293[b] ±11.345			
SO_4^{2-}	**−744.004**[a] **±0.418**	**−909.340** **±0.400**	**18.500** **±0.400**	
HS^-	**12.243**[a] **±2.115**	**−16.300** **±1.500**	**67.000** **±5.000**	

Table IV.1 (continued)

Compound	$\Delta_f G_m^\circ$ $(kJ \cdot mol^{-1})$	$\Delta_f H_m^\circ$ $(kJ \cdot mol^{-1})$	S_m° $(J \cdot K^{-1} \cdot mol^{-1})$	$C_{p,m}^\circ$ $(J \cdot K^{-1} \cdot mol^{-1})$
$H_2S(aq)$	**−27.648**[a] **±2.115**	**−38.600** **±1.500**	**126.000** **±5.000**	
$H_2S(g)$	**−33.443**[a] **±0.501**	**−20.600** **±0.500**	**205.810** **±0.050**	
HSO_3^-	−528.685[b] ±4.046			
$HS_2O_3^-$	−528.369[b] ±11.377			
$H_2SO_3(aq)$	−539.188[b] ±4.072			
HSO_4^-	**−755.315**[a] **±1.342**	**−886.900** **±1.000**	**131.700** **±3.000**	
$Se(cr)$	0.000	0.000	42.270 ±0.050	25.030 ±0.050
$SeO_2(cr)$		−225.100 ±2.100		
SeO_3^{2-}	−361.570[b] ±1.410			
$H_2Se(aq)$[e]				
$HSeO_3^-$	−409.517[b] ±1.290			
$H_2SeO_3(aq)$	−425.500 ±0.600			
$HSeO_4^-$[f]				
$Te(cr)$	0.000	0.000	49.221 ±0.050	25.550 ±0.100
$N(g)$	**455.537**[a] **±0.400**	**472.680** **±0.400**	**153.301** **±0.003**	

Table IV.1 (continued)

Compound	$\Delta_f G_m^\circ$ (kJ · mol^{-1})	$\Delta_f H_m^\circ$ (kJ · mol^{-1})	S_m° (J · K^{-1} · mol^{-1})	$C_{p,m}^\circ$ (J · K^{-1} · mol^{-1})
N_2(g)	**0.000**	**0.000**	**191.609 ±0.004**	
N_3^-	348.200 ±2.000	275.140 ±1.000	107.710[a] ±7.500	
NO_3^-	**−110.794**[a] **±0.417**	**−206.850 ±0.400**	**146.700 ±0.400**	
HN_3(aq)	321.372[b] ±2.051	260.140[b] ±10.050	147.380[b] ±34.403	
NH_3(aq)	−26.673[b] ±0.305	−81.170[b] ±0.326	109.040[b] ±0.913	
NH_3(g)	**−16.407**[a] **±0.351**	**−45.940 ±0.350**	**192.770 ±0.050**	
NH_4^+	**−79.398**[a] **±0.278**	**−133.260 ±0.250**	**111.170 ±0.400**	
HNO_2(aq)[g]				
P(am)[h]		−7.500 ±2.000		
P(cr)[h]	0.000	0.000	**41.090 ±0.250**	
P(g)	**280.093**[a] **±1.003**	**316.500 ±1.000**	**163.199 ±0.003**	
P_2(g)	**103.468**[a] **±2.006**	**144.000 ±2.000**	**218.123 ±0.004**	
P_4(g)	**24.419**[a] **±0.448**	**58.900 ±0.300**	**280.010 ±0.500**	
PO_4^{3-}	−1025.491[b] ±1.576	−1284.400[b] ±4.085	−220.970[b] ±12.846	
$P_2O_7^{4-}$	−1935.503[b] ±4.563			

Table IV.1 (continued)

Compound	$\Delta_f G^\circ_m$ (kJ · mol^{-1})	$\Delta_f H^\circ_m$ (kJ · mol^{-1})	S°_m (J · K^{-1} · mol^{-1})	$C^\circ_{p,m}$ (J · K^{-1} · mol^{-1})
HPO_4^{2-}	**−1095.985**[(a)] **±1.567**	**−1299.000 ±1.500**	**−33.500 ±1.500**	
$H_2PO_4^-$	**−1137.152**[(a)] **±1.567**	**−1302.600 ±1.500**	**92.500 ±1.500**	
$H_3PO_4(aq)$	−1149.367[(b)] ±1.576	−1294.120[(b)] ±1.616	161.912[(b)] ±2.575	
$HP_2O_7^{3-}$	−1989.158[(b)] ±4.482			
$H_2P_2O_7^{2-}$	−2027.117[(b)] ±4.445			
$H_3P_2O_7^-$	−2039.960[(b)] ±4.362			
$H_4P_2O_7(aq)$	−2045.668[(b)] ±3.299	−2280.210[(b)] ±3.383	274.919[(b)] ±6.954	
As(cr)	0.000	0.000	35.100 ±0.600	24.640 ±0.500
AsO_2^-	−350.022[(a)] ±4.008	−429.030 ±4.000	40.600 ±0.600	
AsO_4^{3-}	−648.360[(a)] ±4.008	−888.140 ±4.000	−162.800 ±0.600	
$As_2O_5(cr)$	−782.449[(a)] ±8.016	−924.870 ±8.000	105.400 ±1.200	116.520 ±0.800
As_4O_6(cubi)[(i)]	−1152.445[(a)] ±16.032	−1313.940 ±16.000	214.200 ±2.400	191.290 ±0.800
As_4O_6(mono)[(j)]	−1154.008[(a)] ±16.041	−1309.600 ±16.000	234.000 ±3.000	
$HAsO_2(aq)$	−402.925[(a)] ±4.008	−456.500 ±4.000	125.900 ±0.600	
$H_2AsO_3^-$	−587.078[(a)] ±4.008	−714.790 ±4.000	110.500 ±0.600	

Table IV.1 (continued)

Compound	$\Delta_f G^\circ_m$ $(kJ \cdot mol^{-1})$	$\Delta_f H^\circ_m$ $(kJ \cdot mol^{-1})$	S°_m $(J \cdot K^{-1} \cdot mol^{-1})$	$C^\circ_{p,m}$ $(J \cdot K^{-1} \cdot mol^{-1})$
$H_3AsO_3(aq)$	−639.681[a] ±4.015	−742.200 ±4.000	195.000 ±1.000	
$HAsO_4^{2-}$	−714.592[a] ±4.008	−906.340 ±4.000	−1.700 ±0.600	
$H_2AsO_4^-$	−753.203[a] ±4.015	−909.560 ±4.000	117.000 ±1.000	
$H_3AsO_4(aq)$	−766.119[a] ±4.015	−902.500 ±4.000	184.000 ±1.000	
$(As_2O_5)_3{\cdot}5H_2O(cr)$		−4248.400 ±24.000		
Sb(cr)	0.000	0.000	45.520 ±0.210	25.260 ±0.200
C(cr)	**0.000**	**0.000**	**5.740 ±0.100**	
C(g)	**671.254[a] ±0.451**	**716.680 ±0.450**	**158.100 ±0.003**	
CO(g)	**−137.168[a] ±0.173**	**−110.530 ±0.170**	**197.660 ±0.004**	
$CO_2(aq)$	**−385.970[a] ±0.270**	**−413.260 ±0.200**	**119.360 ±0.600**	
$CO_2(g)$	**−394.373[a] ±0.133**	**−393.510 ±0.130**	**213.785 ±0.010**	
CO_3^{2-}	**−527.899[a] ±0.390**	**−675.230 ±0.250**	**−50.000 ±1.000**	
HCO_3^-	**−586.845[a] ±0.251**	**−689.930 ±0.200**	**98.400 ±0.500**	
SCN^-	92.700 ±4.000	76.400 ±4.000	144.268[a] ±18.974	
Si(cr)	**0.000**	**0.000**	**18.810 ±0.080**	

Table IV.1 (continued)

Compound	$\Delta_f G_m^\circ$ $(kJ \cdot mol^{-1})$	$\Delta_f H_m^\circ$ $(kJ \cdot mol^{-1})$	S_m° $(J \cdot K^{-1} \cdot mol^{-1})$	$C_{p,m}^\circ$ $(J \cdot K^{-1} \cdot mol^{-1})$
Si(g)	**405.525**[(a)] **±8.000**	**450.000** **±8.000**	**167.981** **±0.004**	
SiO_2(quar)[(k)]	**−856.287**[(a)] **±1.002**	**−910.700** **±1.000**	**41.460** **±0.200**	
$SiO_2(OH)_2^{2-}$	−1175.651[(b)] ±1.265	−1381.960[(b)] ±15.330	−1.487[(b)] ±51.592	
$SiO(OH)_3^-$	−1251.740[(b)] ±1.162	−1431.360[(b)] ±3.743	88.026[(b)] ±13.144	
$Si(OH)_4$(aq)	−1307.735 ±1.156	−1456.960 ±3.163	189.974[(a)] ±11.296	
$Si_2O_3(OH)_4^{2-}$	−2269.878[(b)] ±2.878			
$Si_2O_2(OH)_5^-$	−2332.096[(b)] ±2.878			
$Si_3O_6(OH)_3^{3-}$	−3048.536[(b)] ±3.870			
$Si_3O_5(OH)_5^{3-}$	−3291.955[(b)] ±3.869			
$Si_4O_8(OH)_4^{4-}$	−4075.179[(b)] ±5.437			
$Si_4O_7(OH)_5^{3-}$	−4136.826[(b)] ±4.934			
SiF_4(g)	**−1572.772**[(a)] **±0.814**	**−1615.000** **±0.800**	**282.760** **±0.500**	
Ge(cr)	0.000	0.000	**31.090** **±0.150**	
Ge(g)	331.209[(a)] **±3.000**	**372.000** **±3.000**	**167.904** **±0.005**	
GeO_2(tetr)[(l)]	**−521.404**[(a)] **±1.002**	**−580.000** **±1.000**	**39.710** **±0.150**	

Table IV.1 (continued)

Compound	$\Delta_f G_m^\circ$ $(kJ \cdot mol^{-1})$	$\Delta_f H_m^\circ$ $(kJ \cdot mol^{-1})$	S_m° $(J \cdot K^{-1} \cdot mol^{-1})$	$C_{p,m}^\circ$ $(J \cdot K^{-1} \cdot mol^{-1})$
$GeF_4(g)$	**−1150.017**[(a)]	**−1190.200**	**301.900**	
	±0.584	**±0.500**	**±1.000**	
Sn(cr)	0.000	0.000	**51.180**	
			±0.080	
Sn(g)	**266.223**[(a)]	**301.200**	**168.492**	
	±1.500	**±1.500**	**±0.004**	
Sn^{2+}	**−27.624**[(a)]	**−8.900**	**−16.700**	
	±1.557	**±1.000**	**±4.000**	
SnO(tetr)[(l)]	**−251.914**[(a)]	**−280.710**	**57.170**	
	±0.220	**±0.200**	**±0.300**	
SnO_2(cass)[(m)]	**−515.826**[(a)]	**−577.630**	**49.040**	
	±0.204	**±0.200**	**±0.100**	
Pb(cr)	0.000	0.000	**64.800**	
			±0.300	
Pb(g)	**162.232**[(a)]	**195.200**	**175.375**	
	±0.805	**±0.800**	**±0.005**	
Pb^{2+}	**−24.238**[(a)]	**0.920**	**18.500**	
	±0.399	**±0.250**	**±1.000**	
$PbSO_4(cr)$	**−813.036**[(a)]	**−919.970**	**148.500**	
	±0.447	**±0.400**	**±0.600**	
B(cr)	0.000	0.000	**5.900**	
			±0.080	
B(g)	**521.012**[(a)]	**565.000**	**153.436**	
	±5.000	**±5.000**	**±0.015**	
$B_2O_3(cr)$	**−1194.324**[(a)]	**−1273.500**	**53.970**	
	±1.404	**±1.400**	**±0.300**	
$B(OH)_3(aq)$	**−969.268**[(a)]	**−1072.800**	**162.400**	
	±0.820	**±0.800**	**±0.600**	
$B(OH)_3(cr)$	**−969.667**[(a)]	**−1094.800**	**89.950**	
	±0.820	**±0.800**	**±0.600**	

Table IV.1 (continued)

Compound	$\Delta_f G_m^\circ$ (kJ · mol^{-1})	$\Delta_f H_m^\circ$ (kJ · mol^{-1})	S_m° (J · K^{-1} · mol^{-1})	$C_{p,m}^\circ$ (J · K^{-1} · mol^{-1})
BF_3(g)	−1119.403[a] ±0.803	−1136.000 ±0.800	254.420 ±0.200	
Al(cr)	0.000	0.000	28.300 ±0.100	
Al(g)	289.376[a] ±4.000	330.000 ±4.000	164.554 ±0.004	
Al^{3+}	−491.507[a] ±3.338	−538.400 ±1.500	−325.000 ±10.000	
Al_2O_3(coru)[n]	−1582.257[a] ±1.302	−1675.700 ±1.300	50.920 ±0.100	
AlF_3(cr)	−1431.096[a] ±1.309	−1510.400 ±1.300	66.500 ±0.500	
Zn(cr)	0.000	0.000	41.630 ±0.150	
Zn(g)	94.813[a] ±0.402	130.400 ±0.400	160.990 ±0.004	
Zn^{2+}	−147.203[a] ±0.254	−153.390 ±0.200	−109.800 ±0.500	
ZnO(cr)	−320.479[a] ±0.299	−350.460 ±0.270	43.650 ±0.400	
Cd(cr)	0.000	0.000	51.800 ±0.150	
Cd(g)	77.230[a] ±0.205	111.800 ±0.200	167.749 ±0.004	
Cd^{2+}	−77.733[a] ±0.750	−75.920 ±0.600	−72.800 ±1.500	
CdO(cr)	−228.661[a] ±0.602	−258.350 ±0.400	54.800 ±1.500	
$CdSO_4 \cdot 2.667H_2O$(cr)	−1464.959[a] ±0.810	−1729.300 ±0.800	229.650 ±0.400	

Table IV.1 (continued)

Compound	$\Delta_f G_m^\circ$ ($kJ \cdot mol^{-1}$)	$\Delta_f H_m^\circ$ ($kJ \cdot mol^{-1}$)	S_m° ($J \cdot K^{-1} \cdot mol^{-1}$)	$C_{p,m}^\circ$ ($J \cdot K^{-1} \cdot mol^{-1}$)
Hg(g)	**31.842**[(a)] **±0.054**	**61.380 ±0.040**	**174.971 ±0.005**	
Hg(l)	0.000	0.000	**75.900 ±0.120**	
Hg^{2+}	**164.667**[(a)] **±0.313**	**170.210 ±0.200**	**−36.190 ±0.800**	
Hg_2^{2+}	**153.567**[(a)] **±0.559**	**166.870 ±0.500**	**65.740 ±0.800**	
HgO(mont)[(p)]	**−58.523**[(a)] **±0.154**	**−90.790 ±0.120**	**70.250 ±0.300**	
Hg_2Cl_2(cr)	**−210.725**[(a)] **±0.471**	**−265.370 ±0.400**	**191.600 ±0.800**	
Hg_2SO_4(cr)	**−625.780**[(a)] **±0.411**	**−743.090 ±0.400**	**200.700 ±0.200**	
Cu(cr)	0.000	0.000	**33.150 ±0.080**	
Cu(g)	**297.672**[(a)] **±1.200**	**337.400 ±1.200**	**166.398 ±0.004**	
Cu^{2+}	**65.040**[(a)] **±1.557**	**64.900 ±1.000**	**−98.000 ±4.000**	
$CuSO_4$(cr)	**−662.185**[(a)] **±1.206**	**−771.400 ±1.200**	**109.200 ±0.400**	
Ag(cr)	0.000	0.000	**42.550 ±0.200**	
Ag(g)	**246.007**[(a)] **±0.802**	**284.900 ±0.800**	**172.997 ±0.004**	
Ag^+	**77.096**[(a)] **±0.156**	**105.790 ±0.080**	**73.450 ±0.400**	
AgCl(cr)	**−109.765**[(a)] **±0.098**	**−127.010 ±0.050**	**96.250 ±0.200**	

Table IV.1 (continued)

Compound	$\Delta_f G_m^\circ$ $(kJ \cdot mol^{-1})$	$\Delta_f H_m^\circ$ $(kJ \cdot mol^{-1})$	S_m° $(J \cdot K^{-1} \cdot mol^{-1})$	$C_{p,m}^\circ$ $(J \cdot K^{-1} \cdot mol^{-1})$
Ti(cr)	0.000	0.000	30.720 ±0.100	
Ti(g)	428.403[a] ±3.000	473.000 ±3.000	180.298 ±0.010	
TiO_2(ruti)[q]	−888.767[a] ±0.806	−944.000 ±0.800	50.620 ±0.300	
$TiCl_4$(g)	−726.324[a] ±3.229	−763.200 ±3.000	353.200 ±4.000	
Th(cr)	0.000	0.000	51.800 ±0.500	
Th(g)	560.745[a] ±6.002	602.000 ±6.000	190.170 ±0.050	
ThO_2(cr)	−1169.238[a] ±3.504	−1226.400 ±3.500	65.230 ±0.200	
Be(cr)	0.000	0.000	9.500 ±0.080	
Be(g)	286.202[a] ±5.000	324.000 ±5.000	136.275 ±0.003	
BeO(brom)[r]	−580.090[a] ±2.500	−609.400 ±2.500	13.770 ±0.040	
Mg(cr)	0.000	0.000	32.670 ±0.100	
Mg(g)	112.521[a] ±0.801	147.100 ±0.800	148.648 ±0.003	
Mg^{2+}	−455.375[a] ±1.335	−467.000 ±0.600	−137.000 ±4.000	
MgO(cr)	−569.311[a] ±0.305	−601.600 ±0.300	26.950 ±0.150	
MgF_2(cr)	−1071.051[a] ±1.210	−1124.200 ±1.200	57.200 ±0.500	

Table IV.1 (continued)

Compound	$\Delta_f G^\circ_m$ (kJ · mol^{-1})	$\Delta_f H^\circ_m$ (kJ · mol^{-1})	S°_m (J · K^{-1} · mol^{-1})	$C^\circ_{p,m}$ (J · K^{-1} · mol^{-1})
Ca(cr)	**0.000**	**0.000**	**41.590** **±0.400**	
Ca(g)	**144.020**[(a)] **±0.809**	**177.800** **±0.800**	**154.887** **±0.004**	
Ca^{2+}	**−552.806**[(a)] **±1.050**	**−543.000** **±1.000**	**−56.200** **±1.000**	
CaO(cr)	**−603.296**[(a)] **±0.916**	**−634.920** **±0.900**	**38.100** **±0.400**	
Sr(cr)	0.000	0.000	55.700 ±0.210	
Sr^{2+}	−563.864[(a)] ±0.781	−550.900 ±0.500	−31.500 ±2.000	
SrO(cr)	−559.939[(a)] ±0.914	−590.600 ±0.900	55.440 ±0.500	
$SrCl_2$(cr)	−784.974[(a)] ±0.714	−833.850 ±0.700	114.850 ±0.420	
$Sr(NO_3)_2$(cr)	−783.146[(a)] ±1.018	−982.360 ±0.800	194.600 ±2.100	
Ba(cr)	0.000	0.000	62.420 ±0.840	
Ba^{2+}	−557.656[(a)] ±2.582	−534.800 ±2.500	8.400 ±2.000	
BaO(cr)	−520.394[(a)] ±2.515	−548.100 ±2.500	72.070 ±0.380	
$BaCl_2$(cr)	−806.953[(a)] ±2.514	−855.200 ±2.500	123.680 ±0.250	
Li(cr)	**0.000**	**0.000**	**29.120** **±0.200**	
Li(g)	**126.604**[(a)] **±1.002**	**159.300** **±1.000**	**138.782** **±0.010**	

Table IV.1 (continued)

Compound	$\Delta_f G_m^\circ$ $(kJ \cdot mol^{-1})$	$\Delta_f H_m^\circ$ $(kJ \cdot mol^{-1})$	S_m° $(J \cdot K^{-1} \cdot mol^{-1})$	$C_{p,m}^\circ$ $(J \cdot K^{-1} \cdot mol^{-1})$
Li^+	**−292.918**[a] **±0.109**	**−278.470** **±0.080**	**12.240** **±0.150**	
Na(cr)	0.000	0.000	**51.300** **±0.200**	
Na(g)	**76.964**[a] **±0.703**	**107.500** **±0.700**	**153.718** **±0.003**	
Na^+	**−261.953**[a] **±0.096**	**−240.340** **±0.060**	**58.450** **±0.150**	
K(cr)	0.000	0.000	**64.680** **±0.200**	
K(g)	**60.479**[a] **±0.802**	**89.000** **±0.800**	**160.341** **±0.003**	
K^+	**−282.510**[a] **±0.116**	**−252.140** **±0.080**	**101.200** **±0.200**	
Rb(cr)	0.000	0.000	**76.780** **±0.300**	
Rb(g)	**53.078**[a] **±0.805**	**80.900** **±0.800**	**170.094** **±0.003**	
Rb^+	**−284.009**[a] **±0.153**	**−251.120** **±0.100**	**121.750** **±0.250**	
Cs(cr)	0.000	0.000	**85.230** **±0.400**	
Cs(g)	**49.556**[a] **±1.007**	**76.500** **±1.000**	**175.601** **±0.003**	
Cs^+	**−291.456**[a] **±0.535**	**−258.000** **±0.500**	**132.100** **±0.500**	

Footnotes to Table IV.1:

(a) Value calculated internally with the equation $\Delta_f G^\circ_m = \Delta_f H^\circ_m - T\sum_i S^\circ_{m,i}$.

(b) Value calculated internally from reaction data (see Table IV.2).

(c) From the US NBS [82WAG/EVA]. The uncertainty is estimated in the present review.

(d) Orthorhombic.

(e) In the absence of recommended thermodynamic data for HSe^-, only reaction data are selected for $H_2Se(aq)$, *cf.* Table IV.2.

(f) In the absence of recommended thermodynamic data for SeO_4^{2-}, only reaction data are selected for $HSeO_4^-$, *cf.* Table IV.2.

(g) In the absence of recommended thermodynamic data for NO_2^-, only reaction data are selected for $HNO_2(aq)$, *cf.* Table IV.2.

(h) P(cr) refers to white, crystalline phosphorus and is the reference state for the element phosphorus. P(am) refers to red, amorphous phosphorus, *cf.* Section VI.3.2.

(i) Cubic.

(j) Monoclinic.

(k) Quartz.

(l) Tetragonal.

(m) Cassiterite.

(n) Corundum.

(p) Montroydite, red.

(q) Rutile.

(r) Bromellite.

Table IV.2: Selected thermodynamic data for reactions involving auxiliary compounds and complexes used in the evaluation of the selected uranium data. All ionic species listed in this table are aqueous species. The selection of these data is described in Chapter VI. Reactions are listed only if they were used for primary data selection. The thermodynamic data of formation (see Table IV.1) are derived therefrom. Unless noted otherwise, all data refer to the reference temperature of 298.15 K and to the standard state, *i.e.*, a pressure of 0.1 MPa and, for aqueous species, infinite dilution ($I = 0$). The uncertainties listed below each value represent total uncertainties and correspond in principal to the statistically defined 95% confidence interval. Values obtained from internal calculation, *cf.* footnote (a), are rounded at the third digit after the decimal point and may therefore not be exactly identical to those given in Chapter VI. Systematically, all the values are presented with three digits after the decimal point, regardless of the significance of these digits. The data presented in this table are available on PC diskettes or other computer media from the OECD Nuclear Energy Agency.

Species	Reaction				
		$\log_{10} K^\circ$	$\Delta_r G_m^\circ$ $(kJ \cdot mol^{-1})$	$\Delta_r H_m^\circ$ $(kJ \cdot mol^{-1})$	$\Delta_r S_m^\circ$ $(J \cdot K^{-1} \cdot mol^{-1})$
$HF(aq)$	$F^- + H^+ \rightleftharpoons HF(aq)$				
		3.180	−18.152	12.200	101.800[(a)]
		±0.020	±0.114	±0.300	±1.077
HF_2^-	$F^- + HF(aq) \rightleftharpoons HF_2^-$				
		0.440	−2.512	3.000	18.486[(a)]
		±0.120	±0.685	±2.000	±7.091
ClO^-	$HClO(aq) \rightleftharpoons ClO^- + H^+$				
		−7.420	42.354	19.000	−78.329[(a)]
		±0.130	±0.742	±9.000	±30.289
ClO_2^-	$HClO_2(aq) \rightleftharpoons ClO_2^- + H^+$				
		−1.960	11.188		
		±0.020	±0.114		
$HClO(aq)$	$Cl_2(g) + H_2O(l) \rightleftharpoons Cl^- + H^+ + HClO(aq)$				
		−4.537[(c)]	25.900		
		±0.105	±0.600		
$HClO_2(aq)$	$H_2O(l) + HClO(aq) \rightleftharpoons 2H^+ + HClO_2(aq) + 2e^-$				
		−55.400[(b)]	316.226		
		±0.700	±3.996		
BrO^-	$HBrO(aq) \rightleftharpoons BrO^- + H^+$				
		−8.630	49.260	30.000	−64.600[(a)]
		±0.030	±0.171	±3.000	±10.078

Table IV.2 (continued)

Species	Reaction				
		$\log_{10} K^\circ$	$\Delta_r G_m^\circ$ ($kJ \cdot mol^{-1}$)	$\Delta_r H_m^\circ$ ($kJ \cdot mol^{-1}$)	$\Delta_r S_m^\circ$ ($J \cdot K^{-1} \cdot mol^{-1}$)
$HIO_3(aq)$	$H^+ + IO_3^- \rightleftharpoons HIO_3(aq)$				
		0.788 ±0.029	−4.498 ±0.166		
S^{2-}	$HS^- \rightleftharpoons H^+ + S^{2-}$				
		−19.000 ±2.000	108.453 ±11.416		
SO_3^{2-}	$H_2O(l) + SO_4^{2-} + 2e^- \rightleftharpoons 2OH^- + SO_3^{2-}$				
		−31.400[b] ±0.700	179.233 ±3.996		
$S_2O_3^{2-}$	$3H_2O(l) + 2SO_3^{2-} + 4e^- \rightleftharpoons 6OH^- + S_2O_3^{2-}$				
		−39.200[b] ±1.400	223.755 ±7.991		
HSO_3^-	$H^+ + SO_3^{2-} \rightleftharpoons HSO_3^-$				
		7.220 ±0.080	−41.212 ±0.457	66.000 ±30.000	359.591[a] ±100.632
$HS_2O_3^-$	$H^+ + S_2O_3^{2-} \rightleftharpoons HS_2O_3^-$				
		1.590 ±0.150	−9.076 ±0.856		
$H_2SO_3(aq)$	$H^+ + HSO_3^- \rightleftharpoons H_2SO_3(aq)$				
		1.840 ±0.080	−10.503 ±0.457	16.000 ±5.000	88.891[a] ±16.840
SeO_3^{2-}	$HSeO_3^- \rightleftharpoons H^+ + SeO_3^{2-}$				
		−8.400 ±0.100	47.948 ±0.571	−5.020 ±0.500	−177.654[a] ±2.545
$H_2Se(aq)$	$H^+ + HSe^- \rightleftharpoons H_2Se(aq)$				
		3.800 ±0.300	−21.691 ±1.712		
$HSeO_3^-$	$H_2SeO_3(aq) \rightleftharpoons H^+ + HSeO_3^-$				
		−2.800 ±0.200	15.983 ±1.142	−7.070 ±0.500	−77.319[a] ±4.180
$HSeO_4^-$	$H^+ + SeO_4^{2-} \rightleftharpoons HSeO_4^-$				
		1.800 ±0.140	−10.274 ±0.799	23.800 ±5.000	114.286[a] ±16.983

Table IV.2 (continued)

Species	Reaction			
	$\log_{10} K^\circ$	$\Delta_r G_m^\circ$ $(kJ \cdot mol^{-1})$	$\Delta_r H_m^\circ$ $(kJ \cdot mol^{-1})$	$\Delta_r S_m^\circ$ $(J \cdot K^{-1} \cdot mol^{-1})$
$HN_3(aq)$	$H^+ + N_3^- \rightleftharpoons HN_3(aq)$			
	4.700	−26.828	−15.000	39.671[(a)]
	±0.080	±0.457	±10.000	±33.575
$NH_3(aq)$	$NH_4^+ \rightleftharpoons H^+ + NH_3(aq)$			
	−9.237	52.725	52.090	−2.130[(a)]
	±0.022	±0.126	±0.210	±0.821
$HNO_2(aq)$	$H^+ + NO_2^- \rightleftharpoons HNO_2(aq)$			
	3.210	−18.323	−11.400	23.219[(a)]
	±0.160	±0.913	±3.000	±10.518
PO_4^{3-}	$HPO_4^{2-} \rightleftharpoons H^+ + PO_4^{3-}$			
	−12.350	70.494	14.600	−187.470[(a)]
	±0.030	±0.171	±3.800	±12.758
$P_2O_7^{4-}$	$HP_2O_7^{3-} \rightleftharpoons H^+ + P_2O_7^{4-}$			
	−9.400	53.656		
	±0.150	±0.856		
$H_3PO_4(aq)$	$H^+ + H_2PO_4^- \rightleftharpoons H_3PO_4(aq)$			
	2.140	−12.215	8.480	69.412[(a)]
	±0.030	±0.171	±0.600	±2.093
$HP_2O_7^{3-}$	$H_2P_2O_7^{2-} \rightleftharpoons H^+ + HP_2O_7^{3-}$			
	−6.650	37.958		
	±0.100	±0.571		
$H_2P_2O_7^{2-}$	$H_3P_2O_7^- \rightleftharpoons H^+ + H_2P_2O_7^{2-}$			
	−2.250	12.843		
	±0.150	±0.856		
$H_3P_2O_7^-$	$H_4P_2O_7(aq) \rightleftharpoons H^+ + H_3P_2O_7^-$			
	−1.000	5.708		
	±0.500	±2.854		
$H_4P_2O_7(aq)$	$2H_3PO_4(aq) \rightleftharpoons H_2O(l) + H_4P_2O_7(aq)$			
	−2.790	15.925	22.200	21.045[(a)]
	±0.170	±0.970	±1.000	±4.674
$SiO_2(OH)_2^{2-}$	$Si(OH)_4(aq) \rightleftharpoons 2H^+ + SiO_2(OH)_2^{2-}$			
	−23.140	132.084	75.000	−191.461[(a)]
	±0.090	±0.514	±15.000	±50.340

Table IV.2 (continued)

Species	Reaction			
	$\log_{10} K^\circ$	$\Delta_r G_m^\circ$ $(kJ \cdot mol^{-1})$	$\Delta_r H_m^\circ$ $(kJ \cdot mol^{-1})$	$\Delta_r S_m^\circ$ $(J \cdot K^{-1} \cdot mol^{-1})$
$SiO(OH)_3^-$	$Si(OH)_4(aq) \rightleftharpoons H^+ + SiO(OH)_3^-$			
	−9.810 ±0.020	55.996 ±0.114	25.600 ±2.000	−101.948[(a)] ±6.719
$Si(OH)_4(aq)$	$2H_2O(l) + SiO_2(quar) \rightleftharpoons Si(OH)_4(aq)$			
	−4.000 ±0.100	22.832 ±0.571	25.400 ±3.000	8.613[(a)] ±10.243
$Si_2O_3(OH)_4^{2-}$	$2Si(OH)_4(aq) \rightleftharpoons 2H^+ + H_2O(l) + Si_2O_3(OH)_4^{2-}$			
	−19.000 ±0.300	108.453 ±1.712		
$Si_2O_2(OH)_5^-$	$2Si(OH)_4(aq) \rightleftharpoons H^+ + H_2O(l) + Si_2O_2(OH)_5^-$			
	−8.100 ±0.300	46.235 ±1.712		
$Si_3O_6(OH)_3^{3-}$	$3Si(OH)_4(aq) \rightleftharpoons 3H^+ + 3H_2O(l) + Si_3O_6(OH)_3^{3-}$			
	−28.600 ±0.300	163.250 ±1.712		
$Si_3O_5(OH)_5^{3-}$	$3Si(OH)_4(aq) \rightleftharpoons 3H^+ + 2H_2O(l) + Si_3O_5(OH)_5^{3-}$			
	−27.500 ±0.300	156.971 ±1.712		
$Si_4O_8(OH)_4^{4-}$	$4Si(OH)_4(aq) \rightleftharpoons 4H^+ + 4H_2O(l) + Si_4O_8(OH)_4^{4-}$			
	−36.300 ±0.500	207.202 ±2.854		
$Si_4O_7(OH)_5^{3-}$	$4Si(OH)_4(aq) \rightleftharpoons 3H^+ + 4H_2O(l) + Si_4O_7(OH)_5^{3-}$			
	−25.500 ±0.300	145.555 ±1.712		

(a) Value calculated internally with the equation $\Delta_r G_m^\circ = \Delta_r H_m^\circ - T\Delta_r S_m^\circ$.
(b) Value calculated from a selected standard potential.
(c) Value of $\log_{10} K^\circ$ calculated internally from $\Delta_r G_m^\circ$.

Chapter V

Discussion of data selection

V.1. Elemental uranium

V.1.1. Uranium metal

Three phases of uranium metal exist between room temperature and the melting point of the solid. The temperatures of the $\alpha \rightarrow \beta$ and $\beta \rightarrow \gamma$ transitions were reported by Blumenthal [60BLU] as $(667.7 \pm 1.3)°C$ and $(774.8 \pm 1.6)°C$, respectively. These transition temperatures, corrected from the IPTS-48 to the IPTS-68 [69COM],[1]

$$\begin{array}{lcll} \alpha\text{-U} & \rightleftharpoons & \beta\text{-U} & (941.1 \pm 1.3)\,\text{K} \\ \beta\text{-U} & \rightleftharpoons & \gamma\text{-U} & (1048.5 \pm 1.6)\,\text{K} \end{array}$$

are accepted in the present review, and are essentially the same as those accepted in the IAEA review of the chemical thermodynamics of the actinides [76OET/RAN]. In the present review, the temperature (IPTS-68) of the melting point of γ-U as estimated in the IAEA review [76OET/RAN], (1408 ± 2) K, is accepted. The pure α-phase metal is defined as the uranium reference phase. As such, its Gibbs energy of formation and enthalpy of formation are zero by definition at 298.15 K and 0.1 MPa.

The value of $S^\circ_m(\text{U}, \alpha, 298.15\,\text{K})$ is frequently used as an auxiliary datum in calculations in this report as well as in other chemical thermodynamic calculations involving uranium. α-U has recently been the subject of a number of excellent critical reviews of thermodynamic data. This review recommends the CODATA Key Value [89COX/WAG],

$$S^\circ_m(\text{U}, \alpha, 298.15\,\text{K}) = (50.20 \pm 0.20)\,\text{J} \cdot \text{K}^{-1} \cdot \text{mol}^{-1}.$$

The same value was selected in the IAEA review performed by Oetting, Rand and Ackermann [76OET/RAN]. It was obtained from the heat capacity measurements of Flotow and Lohr [60FLO/LOH] and Flotow and Osborne [66FLO/OSB], which are also selected here.

$$C^\circ_{p,m}(\text{U}, \alpha, 298.15\,\text{K}) = (27.66 \pm 0.05)\,\text{J} \cdot \text{K}^{-1} \cdot \text{mol}^{-1}$$

[1] IPTS = International Practical Temperature Scale

Within the uncertainties stated, the recommended values are the same as the values selected by the US National Bureau of Standards [82WAG/EVA], the US Bureau of Mines [82PAN], the US Geological Survey [78ROB/HEM2], the Gmelin Handbook [83FUG], and the Soviet Academy of Sciences [82GLU/GUR]. The selected temperature function $C_{p,m}(T)$ listed in Table III.3, for the range 298.15 K to 941 K, is that given in Table 1.X of Oetting, Rand and Ackermann [76OET/RAN].

V.1.2. Uranium gas

Although gases are not comprehensively treated in this review, data for ideal monoatomic uranium gas have been selected since they are needed in some thermodynamic calculations. The new CODATA Key Values [89COX/WAG] are adopted,

$$\Delta_f H^\circ_m(\mathrm{U, g, 298.15\,K}) = (533.0 \pm 8.0)\,\mathrm{kJ\cdot mol^{-1}}$$
$$S^\circ_m(\mathrm{U, g, 298.15\,K}) = (199.79 \pm 0.10)\,\mathrm{J\cdot K^{-1}\cdot mol^{-1}}.$$

The recommended value of

$$\Delta_f G^\circ_m(\mathrm{U, g, 298.15\,K}) = (488.4 \pm 8.0)\,\mathrm{kJ\cdot mol^{-1}}$$

is derived from these values.

The heat capacity selected here is from the IAEA review [76OET/RAN], which is based on the observations of Steinhaus [75STE].

$$C^\circ_{p,m}(\mathrm{U, g, 298.15\,K}) = (23.69 \pm 0.04)\,\mathrm{J\cdot K^{-1}\cdot mol^{-1}}$$

The uncertainty estimate on this value is from this review and is consistent with the error estimate by Fuger [83FUG] in the Gmelin Handbook.

Within the uncertainties stated, the selected values are consistent with those selected in the IAEA review [76OET/RAN]. The temperature function $C^\circ_{p,m}(T)$ listed in Table III.3, for the range 298.15 to 2500 K, was obtained by fitting values from Table A.1.6 of Oetting, Rand and Ackermann [76OET/RAN].

V.2. Simple uranium aqua ions

V.2.1. UO_2^{2+}

Extensive and accurate data are available for UO_2^{2+}. Fuger and Oetting [76FUG/OET] made a thorough review of the Gibbs energy and enthalpy of formation of this ion for the IAEA review. In the 15 years since that review appeared, no new data have been published that would materially alter the values they selected.

The $\Delta_f H^\circ_m(\mathrm{UO_2^{2+}, aq, 298.15\,K})$ value selected by Fuger and Oetting [76FUG/OET] is based on two sets of thermochemical cycles, one extrapolated to infinite dilution and one in 0.1 to 1.0 M perchloric acid media. The three cycles leading to values at infinite dilution involve the $\Delta_{sol}H^\circ_m$ of *i)* γ-UO_3 and $UO_2(NO_3)_2\cdot 6H_2O(cr)$ in 6 M HNO_3, *ii)* $UO_2Cl_2(cr)$ in 5.55×10^{-4} M HCl, and *iii)* $UO_2Cl_2\cdot 3H_2O(cr)$ in 4.98 M HCl. The results from these calculations and the sources of the data used, as well as similar information for the $HClO_4$ media cycles are given in Table V.1. Their selected value of $-(1018.8\pm1.7)\,kJ\cdot mol^{-1}$ reflects the results obtained from both sets of calculations. In its critical review, the CODATA Task Group on Key Values for Thermodynamics [78COD] did not consider the values of $\Delta_f H_m(UO_2^{2+}$, dilute $HClO_4)$ in their assessment and gave no weight to the results of cycle 3 in Table V.1. Their value of $-(1019.2\pm2.5)\,kJ\cdot mol^{-1}$ was thus, based only on the solubility data of Cordfunke [66COR] and Bailey and Larson [71BAI/LAR]. The basis for increasing the uncertainty estimate from the value determined by Fuger and Oetting [76FUG/OET] is not known. In their recent revision of Key Values, CODATA [89COX/WAG] included results from some more recent studies [82DEV/YEF, 83FUG/PAR], and adjusted their recommendation to equal the Fuger and Oetting [76FUG/OET] weighted mean from the first set of calculations, although details of their calculations are not totally clear. Because of the thorough IAEA and CODATA reviews, the CODATA [89COX/WAG] value of

$$\Delta_f H_m^\circ(UO_2^{2+}, aq, 298.15\,K) \quad = \quad -(1019.0 \pm 1.5)\,kJ\cdot mol^{-1}$$

is adopted in this review.

Fuger and Oetting [76FUG/OET] based their calculation of $\Delta_f G_m^\circ(UO_2^{2+}$, aq, 298.15 K) on the mean value $\Delta_{sol}H_m^\circ(UO_2(NO_3)_2{\cdot}6H_2O$, cr, 298.15 K) = $(19.58 \pm 0.8)\,kJ\cdot mol^{-1}$ obtained from the analysis of the first cycle in Table V.1. This was combined with $\Delta_{sol}G_m^\circ(UO_2(NO_3)_2{\cdot}6H_2O$, cr, 298.15 K) = $(13.47 \pm 0.21)\,kJ\cdot mol^{-1}$ obtained from the solubility data of Linke [58LIN], $S_m^\circ(UO_2(NO_3)_2{\cdot}6H_2O$, cr, 298.15 K) = $(505.6 \pm 2.1)\,J\cdot K^{-1}\cdot mol^{-1}$ from Kelley and King [61KEL/KIN] and auxiliary data from CODATA [73COD] to yield $S_m^\circ(UO_2^{2+}$, aq, 298.15 K) = $-(97.1\pm3.8)\,J\cdot K^{-1}\cdot mol^{-1}$. CODATA [89COX/WAG] used a similar procedure, incorporating later measurements on the enthalpy of solution of $UO_2(NO_3)_2\cdot 6H_2O(cr)$ [82DEV/YEF] and activity corrections from [79GOL]. They also used the same value of $S_m^\circ(UO_2(NO_3)_2{\cdot}6H_2O, cr, 298.15\,K)$ from Coulter, Pitzer and Latimer [40COU/PIT] that was also adopted by Kelley and King [61KEL/KIN]. The resulting value of $S_m^\circ(UO_2^{2+}, aq, 298.15\,K) = -(98.2 \pm 3.0)\,J\cdot K^{-1}\cdot mol^{-1}$ is practically the same as the value selected by the IAEA reviewers. Simply as an expedient to assure maximum CODATA consistency, the new CODATA [89COX/WAG] value of the entropy of the UO_2^{2+} ion is recommended here:

$$S_m^\circ(UO_2^{2+}, aq, 298.15\,K) \quad = \quad -(98.2 \pm 3.0)\,J\cdot K\cdot mol^{-1}.$$

The Gibbs energy of formation obtained is

$$\Delta_f G_m^\circ(UO_2^{2+}, aq, 298.15\,K) = -(952.55 \pm 1.75)\,kJ\cdot mol^{-1}.$$

The apparent and partial molar heat capacities of aqueous dioxouranium(VI) perchlorate solutions from 10 to 55°C were determined by Hovey, Nguyen-Trung and Tremaine [89HOV/NGU]. These authors also did a review of some earlier estimates of the heat capacity of $UO_2^{2+}(aq)$. This review accepts the conclusions in Ref. [89HOV/NGU] and selects

$$C_{p,m}^\circ(UO_2^{2+}, aq, 298.15\,K) \quad = \quad (42.4 \pm 3.0)\,J\cdot K^{-1}\cdot mol^{-1}$$

Table V.1: Data considered by Fuger and Oetting [76FUG/OET] in selecting $\Delta_f H_m^\circ(UO_2^{2+}, aq, 298.15\,K)$.

Cycle key compounds	Source of data used	Solution	$\Delta_f H_m^\circ(UO_2^{2+}, aq, 298.15\,K)$ $(kJ \cdot mol^{-1})$
γ-UO_3 and $UO_2(NO_3)_2 \cdot 6H_2O(cr)$	[40COU/PIT, 64COR, 66COR, 71BAI/LAR][(a)]	cor $I = 0$	-1019.2 ± 1.5
$UO_2Cl_2(cr)$	[75COR]	cor $I = 0$	-1019.2 ± 1.3
$UO_2Cl_2 \cdot 3H_2O(cr)$	[54LIP/SAM, 73PRI] (after [75COR])	cor $I = 0$	-1018.5 ± 1.7
	Weighted mean:		-1019.0 ± 1.6
γ-UO_3, β-UO_3	[61HOE/SIE, 64COR, 75COR/OUW]	1 M $HClO_4$	-1014.8 ± 2.1
γ-UO_3, $UO_3(am)$	[61HOE/SIE, 64COR]	1 M $HClO_4$	-1016.4 ± 2.1
$UCl_4(cr)$	[47FON, 71KOT]	0.1 m $HClO_4$	-1018.6 ± 2.9
$UCl_4(cr)$	[47FON, 58FON]	0.5 m $HClO_4$	-1024.0 ± 2.9
$UCl_4(cr)$	[47FON, 71FIT/PAV]	0.5 m $HClO_4$	-1021.4 ± 2.9
	Weighted mean (dilute $HClO_4$)		-1018.1 ± 2.9
	Rounded selected value:		-1018.8 ± 1.7

(a) Fuger and Oetting [76FUG/OET] decided not to include the early data of Markétos [12MAR] and de Forcrand [15FOR] or the high ionic strength data of Lipilina and Samoilov [54LIP/SAM] in calculating a mean $\Delta_{sol} H_m^\circ(UO_2(NO_3)_2 \cdot 6H_2O, cr, 298.15\,K)$.

and, for the temperature range 283 to 328 K, the temperature function

$$C_{p,m}^\circ(UO_2^{2+}, aq, T) = \left(350.5 - 0.8722\,T - \frac{5308}{T - 190}\right) J \cdot K^{-1} \cdot mol^{-1}.$$

V.2.2. UO_2^+

The standard Gibbs energy of formation of UO_2^+ is obtained from the standard potential of the reaction

$$UO_2^{2+} + e^- \rightleftharpoons UO_2^+ \qquad (V.1)$$

and the standard Gibbs energy of formation of UO_2^{2+} discussed above. Four precise experimental studies of this standard potential have been performed by using polarographic techniques. Necessary experimental precautions were taken to assure reversibility and to control diffusion potentials in these experiments. In the first case, Kraus and Nelson [49KRA/NEL2] and Kraus, Nelson and Johnson [49KRA/NEL] reported a formal oxidation-reduction potential *vs.* NHE equal to (0.062 ± 0.002) V in 0.1 M KCl. A recalculation to $I = 0$, using the ion interaction coefficients $\varepsilon_{(UO_2^{2+},Cl^-)} = 0.21$ and $\varepsilon_{(UO_2^+,Cl^-)} = 0$, gives the standard potential $E(\text{V.1}, I = 0,$ 298.15 K, 1 atm$) = (0.080 \pm 0.003)$ V. In the second experiment Kritchevsky and Hindman [49KRI/HIN] measured the formal potential of the UO_2^{2+}/UO_2^+ couple in 0.01 M to 3.0 M perchlorate solutions. The formal potentials recalculated in this review to the NHE scale (using $E° = 0.2415$ V for the standard calomel electrode) are given in Table V.2. It is not clear if the electrode and salt bridge system used by Kritchewsky and Hindman [49KRI/HIN] eliminated the diffusion potentials entirely, *cf.* Appendix A. In the third experiment, Kern and Orleman [49KER/ORL] also did a polarographic study of the UO_2^{2+}/UO_2^+ couple. The authors used a cell arrangement which eliminated liquid junction potentials. The formal potential in 0.5 M ClO_4^- is (0.062 ± 0.002) V. Riglet, Vitorge and Grenthe [87RIG/VIT] redetermined the standard potential of the UO_2^{2+}/UO_2^+ couple and determined the ion interaction coefficient $\varepsilon_{(UO_2^+,ClO_4^-)}$. They used the standard polarographic method with a diffusion-free cell and a Ag, AgCl reference electrode. The UO_2^{2+}/UO_2^+ couple was investigated at four different sodium perchlorate solutions (0.5, 1.0, 2.0 and 3.0 M) at (25 ± 1)°C. $E(\text{V.1}, I = 0$, 298.15 K, 1 atm$) = (0.089 \pm 0.002)$ V and $\varepsilon_{(UO_2^+,ClO_4^-)} = (0.28 \pm 0.04)$ were obtained. The published experimental data referring to perchlorate media (see Table V.2) are used to to perform a linear regression to $I = 0$ (see Figure V.1) using the specific ion interaction theory according to Appendix B (The value determined by Capdevila and Vitorge [90CAP/VIT] using cyclic voltammetry is too recent to be included in the evaluation presented in Figure V.1, but it is in very good agreement with the data from the other studies.). The result of this linear regression is $E(\text{V.1}, I = 0$, 298.15 K, 1 atm$) = (0.0879 \pm 0.0013)$ V, and the slope, $\Delta\varepsilon(\text{V.1}) = -(0.20 \pm 0.01)$, leads to the selected ion interaction coefficient $\varepsilon_{(UO_2^+,ClO_4^-)} = (0.26 \pm 0.03)$.

The data of Kraus and Nelson [49KRA/NEL2] and Kraus, Nelson and Johnson [49KRA/NEL] are not very sensitive to errors in the ion interaction coefficients $\varepsilon_{(UO_2^+,Cl^-)}$ and $\varepsilon_{(UO_2^{2+},Cl^-)}$ because of the fairly low ionic strength used. Hence, the correction to $I = 0$ is determined mainly by the Debye-Hückel term. Brand and Cobble [70BRA/COB] determined the standard potential of the NpO_2^{2+}/NpO_2^+ couple. They noted that the difference between the standard potential and the formal potential of this couple, as determined at 1 M $HClO_4$ by Sullivan, Hindman and Zielen [61SUL/HIN], is 0.1 V. They proposed [70BRA/COB, *p.*916] that all reported formal potentials for the actinoid(VI)/actinoid(V) couples should be corrected by this amount. Fuger and Oetting [76FUG/OET] followed this suggestion. This procedure was found to be unsatisfactory in this review since the observed difference may be attributed to experimental error. It should be replaced by an estimation of the activity factors either by the techniques used in this report or by some other technique. The Brand and Cobble [70BRA/COB] correction is much too large, as can be seen

Table V.2: Experimental formal potentials and derived standard potentials for the uranium(VI)/uranium(V) couple at 298.15 K and 1 atm.

Method	Ionic medium	Formal potential (V, *vs.* NHE)	Standard potential (V, *vs.* NHE)	Reference
Redox couple: UO_2^{2+}/UO_2^+, according to $UO_2^{2+} + \frac{1}{2}H_2(g) \rightleftharpoons UO_2^+ + H^+$				
pol	0.1 M Cl^- pH ≈ 3	0.062 ± 0.002	0.080 ± 0.003	[49KRA/NEL]
pol	0.1 M Cl^- pH ≈ 2	0.061 ± 0.001	0.079 ± 0.001	[49KRA/NEL2]
pol	0.1 M ClO_4^-	0.067 ± 0.004		[49KRI/HIN]
	0.15 M ClO_4^-	0.066 ± 0.004		
	0.35 M ClO_4^-	0.058 ± 0.004		
	1.05 M ClO_4^-	0.063 ± 0.004		
	3.05 M ClO_4^-	0.074 ± 0.004		
pol	0.5 M ClO_4^-	0.062 ± 0.002		[49KER/ORL]
pol	0.5 M ClO_4^-	0.062 ± 0.002		[87RIG/VIT]
pol	1.0 M ClO_4^-	0.065 ± 0.002		
pol	2.0 M ClO_4^-	0.071 ± 0.002		
pol	3.0 M ClO_4^-	0.081 ± 0.002		
vlt	1.0 M $HClO_4$	0.060 ± 0.004		[90CAP/VIT]
rev	0		0.163 ± 0.05	[76FUG/OET]
rev	0		0.0879 ± 0.0013	this review

from the UO_2^{2+}/UO_2^+ couple, where accurate data are available at quite low ionic strengths.

The value selected in the present review of the UO_2^{2+}/UO_2^+ couple is the standard potential obtained from the extrapolation shown in Figure V.1, $E^\circ(\text{V.1, 298.15 K}) = (0.0879 \pm 0.0013)\,\text{V}$, which refers, like all the underlying values, to a standard state of 1 atm. Converting this to the standard state pressure of 1 bar preferred in the present review results in the following standard potential and corresponding constant at 298.15 K and $I = 0$,

$$\begin{aligned} E^\circ(\text{V.1}, 298.15\,\text{K}) &= (0.0878 \pm 0.0013)\,\text{V} \\ \log_{10} K^\circ(\text{V.1}, 298.15\,\text{K}) &= 1.484 \pm 0.022. \end{aligned}$$

It is recalled that the equilibrium constant of a redox reaction refers to a cell with the hydrogen electrode as a reference electrode, *cf.* Section II.1.6.5. This value of E° is compared in Table V.2, to the experimental and Fuger and Oetting [76FUG/OET] values. It is used to calculate the selected value of

Figure V.1: Extrapolation to $I = 0$ of experimental formal potential data for the reduction of UO_2^{2+} to UO_2^+ using the specific ion interaction theory. The data refer to $NaClO_4$ media and are taken from [49KRI/HIN] (○), [49KER/ORL] (□) and [87RIG/VIT] (△). All values in this figure refer to a standard state pressure of 1 atm. The shaded area represents the uncertainty range obtained by propagating the resulting uncertainties at $I = 0$ back to $I = 4$ m.

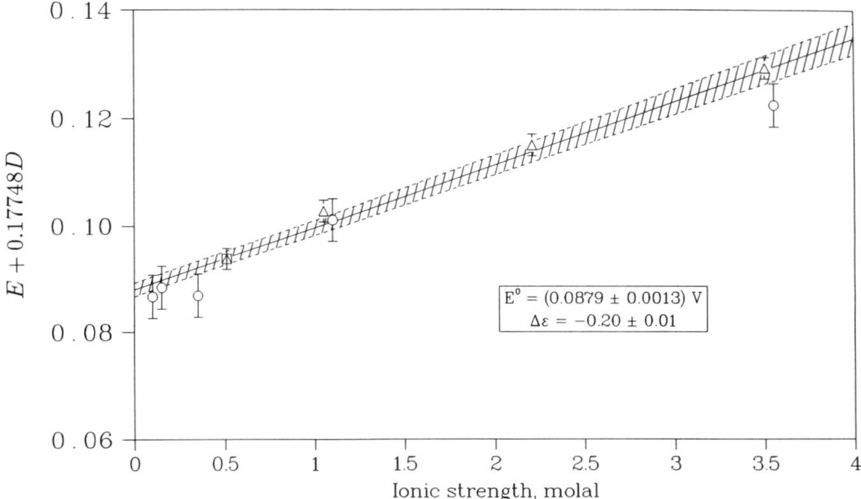

$$\Delta_f G_m^\circ(UO_2^+, aq, 298.15\,K) = -(961.0 \pm 1.8)\,kJ \cdot mol^{-1}$$

from the selected value of $\Delta_f G_m^\circ(UO_2^{2+}, aq, 298.15\,K)$.

In the absence of experimental data, this review selects the entropy value of Fuger and Oetting [76FUG/OET] (who accepted the Brand and Cobble [70BRA/COB] estimate after correcting for a calculational error [76FUG/OET, *p*.35]).

$$S_m^\circ(UO_2^+, aq, 298.15\,K) = -(25 \pm 8)\,J \cdot K^{-1} \cdot mol^{-1}$$

The enthalpy of formation derived from the selected $\Delta_f G_m^\circ$ and S_m° values is

$$\Delta_f H_m^\circ(UO_2^+, aq, 298.15\,K) = -(1025.1 \pm 3.0)\,kJ \cdot mol^{-1}.$$

V.2.3. U^{4+}

The standard Gibbs energy of formation of U^{4+} is obtained from experimental data on the standard potential of the reaction

$$UO_2^{2+} + 4H^+ + 2e^- \rightleftharpoons U^{4+} + 2H_2O(l) \qquad (V.2)$$

and the standard Gibbs energy of formation of UO_2^{2+} discussed in Section V.2.1. The published determinations of this standard potential are summarized in Table V.3. Fuger and Oetting [76FUG/OET] based their selection of the UO_2^{2+}/U^{4+} standard potential on the experimental data of Sobkowski [61SOB] and Sobkowski and Minc [61SOB/MIN], E(V.2) = (0.327 ± 0.001) V, which they readjusted to zero ionic strength by a Debye-Hückel extrapolation, and on the value determined by Nikolaeva [74NIK], E(V.2, I = 0, 298.15 K, 1 atm) = 0.272 V, already thus adjusted. The value E(V.2, I = 0, 298.15 K, 1 atm) = (0.273 ± 0.005) V was selected in the IAEA review [76FUG/OET].

Bruno, Grenthe and Lagerman [86BRU/GRE2] examined the details of the earlier experiments and found these studies to have some shortcomings which are taken into account in this review. These authors measured the formal potential of the UO_2^{2+}/U^{4+} couple over a wide range of ionic strengths. The estimation of activity factors and the extrapolation to I = 0 were made by using both the specific ion interaction theory and the Pitzer method. They proposed a standard potential of E(V.2, I = 0, 298.15 K, 1 atm) = (0.258 ± 0.008) V and a value of $\varepsilon_{(U^{4+},ClO_4^-)} = 1.4$. The latter value is unexpectedly large, especially in view of a recent study of the standard potentials of the MO_2^{2+}/MO_2^+ and M^{4+}/M^{3+} couples performed by Riglet, Robouch and Vitorge [89RIG/ROB]. These authors found that $\varepsilon_{(M^{4+},ClO_4^-)}$ is equal to 0.84 and 1.03 for Np^{4+} and Pu^{4+}, respectively. Bruno, Grenthe and Lagerman [90BRU/GRE] later reinvestigated the UO_2^{2+}/U^{4+} redox system. It seems clear that in the first study of Bruno, Grenthe and Lagerman [86BRU/GRE2] equilibrium was *not* attained at acidities larger than 0.5 M, despite the fact that the redox electrode showed a Nernstian response. The new experimental data were obtained at 45°C where equilibrium was attained much more rapidly than at 25°C. Additional measurements were done at 25°C but allowing a longer time for equilibration. The 25°C data yielded E(V.2, I = 0, 298.15 K, 1 atm) = (0.269 ± 0.008) V, the 45°C data yielded E(V.2, I = 0, 318.15 K, 1 atm) = (0.303 ± 0.008) V. Correction of the latter quantity to 25°C using the known enthalpies of H_2O(l) and H_2(g) and the values of $S^\circ_m(UO_2^{2+}$, aq, 298.15 K) and $S^\circ_m(U^{4+}$, aq, 298.15 K) selected in this review gives E(V.2, I = 0, 298.15 K, 1 atm) = (0.256 ± 0.003) V, which is in fair agreement with the value determined directly at 25°C. The heat capacities for reactants and products are taken from this review.

This review also accepts the discussion of U(VI)/U(IV) redox data in Ref. [90BRU/GRE] and their values of E(V.2) and $\log_{10} K$(V.2) which are based on all the experimental data given in Table V.3. These data are corrected for complex formation using the equilibrium constants and ion interaction coefficients of this review. The linear regression is redone in this review without using (for reasons presented in Appendix B) the value at I = 4.95 m which Bruno, Grenthe and Lagerman [90BRU/GRE] included. A linear regression plot of the data after these corrections is given in Figure V.2. The constant thus obtained, $\log_{10} K^\circ$(V.2, I = 0, 298.15 K, 1 atm) = (9.044 ± 0.041), refers, like all the underlying values, to a standard state pressure of 1 atm. Converting this to the standard state pressure of 1 bar preferred in the present review results in the following constant and corresponding standard

Table V.3: Formal potentials (E') at 25°C and 1 atm, on the normal hydrogen electrode scale, of the UO_2^{2+}/U^{4+} couple in media of various compositions. No correction for activity coefficients and complex formation have been made. The uncertainties reported in the formal potentials refer to the observed variations as the ionic medium is changed in the range given in the second column. The uncertainty reported by Kraus and Nelson [49KRA/NEL2] is discussed by Bruno, Grenthe and Lagerman [90BRU/GRE]. Some of the potentials have been extrapolated to $I = 0$, these are also discussed in the text.

Method	Medium	Formal potential (V, *vs.* NHE)	Reference
Redox couple: U(VI)/U(IV)			
pot	0.05 to 0.5 M H_2SO_4	0.419 ± 0.001	[08LUT/MIC][(a)]
pot	0.05 to 0.5 M H_2SO_4	0.404 ± 0.003	[10TIT][(a)]
pot	0.05 to 0.5 M H_2SO_4	0.420	[43KHL/GUR]
pot	0.2 to 2 m HCl	0.322 to 0.337	[44TAY/SMI]
pot	$\rightarrow$ 1 m HCl	0.310 ± 0.03	[49KRA/NEL2]
pot	$\rightarrow 0$	0.407	[57GUR][(b)]
pot	$\rightarrow 0$	0.329	[61SOB/MIN]
pot	$\rightarrow 0$	0.3288	[61SOB][(c)]
pot	4 M ($NaClO_4$, $HClO_4$)	0.348	[69GRE/VAR]
pot	0.1 to 1.1 M Cl^-	0.304 to 0.344	[70STA][(d)]
pot	$\rightarrow 0$	0.273	[74NIK]
pot	$\rightarrow 0$	0.330	[85GUO/LIU][(e)]
pot	$\rightarrow 0$	0.260 ± 0.003	[86BRU/GRE2]
pot	$\rightarrow 0$	0.2674 ± 0.0015	[90BRU/GRE]
rev	0	0.273 ± 0.005	[76FUG/OET]
rev	0	0.2675 ± 0.0012[(f)]	this review

(a) Measurements performed at approximately 18°C.
(b) Measurements performed in sulphate media.
(c) Data from HCl and H_2SO_4 media also given.
(d) Recalculated using $E' = E_{\text{meas}} - 29.58 \log_{10}\left(\frac{[H^+]^4 \cdot [UO_2^{2+}]}{[U^{4+}]}\right)$.
(e) Measurements performed at 30°C.
(f) Conversion to the standard state pressure of 1 bar yields the value of $E^\circ = (0.2673 \pm 0.0012)$ V recommended by this review.

potential:

$$\begin{aligned} \log_{10} K^\circ(\text{V.2}, 298.15\,\text{K}) &= 9.038 \pm 0.041 \\ E^\circ(\text{V.2}, 298.15\,\text{K}) &= (0.2673 \pm 0.0012)\,\text{V}. \end{aligned}$$

From the resulting value of $\Delta\varepsilon(\text{V.2}) = (0.02 \pm 0.03)$ and the ion interation coefficients of H^+ and UO_2^{2+}, this review obtains, as did Bruno, Grenthe and Lagerman [90BRU/GRE], $\varepsilon_{(U^{4+},ClO_4^-)} = (0.76 \pm 0.06)$, which is in excellent agreement with the values obtained for other +4 cations, *cf.* Appendix B.

Figure V.2: Extrapolation to $I = 0$ of experimental formal potential data for the reduction of UO_2^{2+} to U^{4+} using the specific ion interaction theory. The data are taken from the recent compilation of experimental values by Bruno, Grenthe and Lagerman [90BRU/GRE] for three different aqueous media: perchlorate (○), sulphate (△) and chloride (□). All values in this figure refer to a standard state pressure of 1 atm. The shaded area represents the uncertainty range obtained by propagating the resulting uncertainties at $I = 0$ back to $I = 4$ m.

Equilibrium: $UO_2^{2+} + H_2(g) + 2H^+ \rightleftharpoons U^{4+} + 2H_2O(l)$

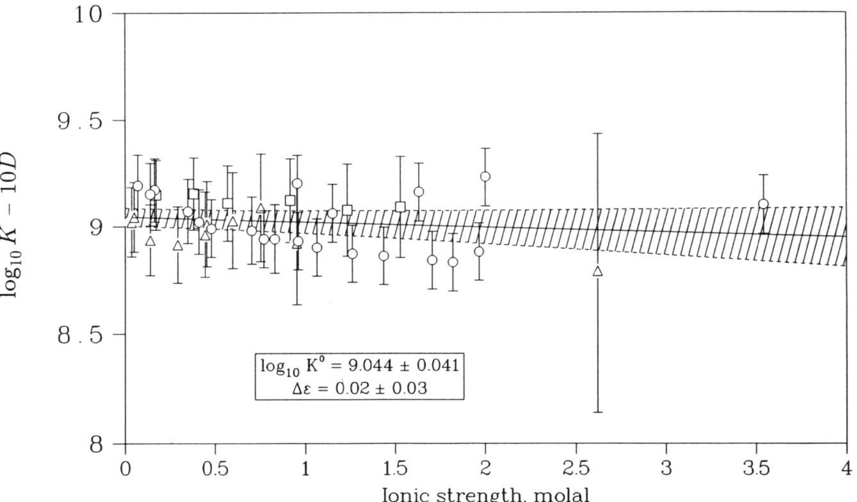

The selected value of $\log_{10} K^\circ$ is used to calculate the selected value of

$$\Delta_f G_m^\circ(U^{4+}, \text{aq}, 298.15\,\text{K}) = -(529.9 \pm 1.8)\,\text{kJ} \cdot \text{mol}^{-1}.$$

Note that the equilibrium constant for the reaction

$$2UO_2^+ + 4H^+ \rightleftharpoons UO_2^{2+} + U^{4+} + 2H_2O(l) \qquad \text{(V.3)}$$

calculated from the selected standard potentials of the UO_2^{2+}/UO_2^+ (Section V.2.2) and UO_2^{2+}/U^{4+} couples, $\log_{10} K^\circ(\text{V.3}) = (6.07 \pm 0.06)$ is in fair agreement with the direct experimental result of Nelson and Kraus [51NEL/KRA], $\log_{10} K^\circ(\text{V.3}) = (6.23 \pm 0.08)$.

The $\Delta_f H_m^\circ(U^{4+}, aq, 298.15\,K)$ value selected by Fuger and Oetting [76FUG/OET] was based on the enthalpy of solution of $UCl_4(cr)$ in various media and on the selected value of $\Delta_f H_m^\circ(UCl_4, cr, 298.15\,K)$. Their selection of $\Delta_{sol} H_m^\circ(UCl_4$, cr, 298.15 K) for the reaction

$$UCl_4(cr) \;\rightarrow\; U^{4+} + 4Cl^- \qquad (V.4)$$

was based on two series of experiments, one in perchloric and one in hydrochloric media. Fuger and Oetting [76FUG/OET] corrected these data to $I = 0$ after including a correction for hydrolysis. From their selected enthalpy of solution according to Reaction (V.4) they derived the selected enthalpy of formation, using data for $\Delta_f H_m^\circ(UCl_4$, cr, 298.15 K), *cf.* Section V.4.2.1.1, and $\Delta_f H_m^\circ(Cl^-$, aq, 298.15 K) which are consistent with the ones selected in this review. The resulting value of the IAEA review [76FUG/OET] is selected here:

$$\Delta_f H_m^\circ(U^{4+}, aq, 298.15\,K) \;=\; -(591.2 \pm 3.3)\,kJ \cdot mol^{-1}.$$

This value is equal to that selected by the US NBS [82WAG/EVA]. By using the selected standard enthalpies of formation, $\Delta_r H_m^\circ(V.5) = -(142.0 \pm 3.9)\,kJ \cdot mol^{-1}$ can be calculated.

$$U^{4+} + \tfrac{1}{2}O_2(g) + H_2O(l) \;\rightleftharpoons\; UO_2^{2+} + 2H^+ \qquad (V.5)$$

This value is in reasonably good agreement with the direct calorimetric determination performed by Grenthe, Spahiu and Olofsson [84GRE/SPA] in 3 M (Na,H)ClO_4, $\Delta_r H_m(V.5, 3\,M\,(Na,H)ClO_4) = -(125 \pm 9)\,kJ \cdot mol^{-1}$.

The entropy calculated from the selected Gibbs energy and enthalpy of formation is

$$S_m^\circ(U^{4+}, aq, 298.15\,K) \;=\; -(416.9 \pm 12.6)\,J \cdot K^{-1} \cdot mol^{-1}.$$

This value is very close to the entropy estimate of Sobkovski [61SOB] $S_m^\circ(U^{4+}$, aq, 298.15 K) $= -416\,J \cdot K^{-1} \cdot mol^{-1}$, based on emf measurements of $\left(\frac{\partial E}{\partial T}\right)_p$.

Due to the absence of $C_{p,m}$ measurements for U^{4+}, this review adopts the Criss-Cobble [64CRI/COB, 64CRI/COB2] estimate of Lemire and Tremaine [80LEM/TRE] for this value. This value, not to be confused with the "absolute" parameter of Criss and Cobble [64CRI/COB], is defined as the mean partial molal heat capacity from 298.15 to 473 K. The assumption that the heat capacity is constant over this temperature range is generally a good one for solids but not for aqueous species. Accordingly, this review assigns a large uncertainty to the selected heat capacity.

$$C_{p,m}|_{298}^{473}\,(U^{4+}, aq) \;=\; -(48 \pm 15)\,J \cdot K^{-1} \cdot mol^{-1}$$

V.2.4. U^{3+}

Thermodynamic data for U^{3+} were discussed in the IAEA review [76FUG/OET]. For the reduction of U^{4+} to U^{3+} according to the reaction

$$U^{4+} + e^- \rightleftharpoons U^{3+} \qquad (V.6)$$

Fuger and Oetting [76FUG/OET] obtained an enthalpy value of

$$\Delta_r H_m^\circ(\text{V.6, 298.15 K}) = (102.1 \pm 1.7)\,\text{kJ} \cdot \text{mol}^{-1},$$

corrected for hydrolysis and referred to $I = 0$. This is based on a measurement by Fontana [58FON] in 0.5 M $HClO_4$. This value, combined with the value of $\Delta_f H_m^\circ(U^{4+}, \text{aq}, 298.15\,\text{K})$ selected in the previous section, yields the selected value of

$$\Delta_f H_m^\circ(U^{3+}, \text{aq, 298.15 K}) = -(489.1 \pm 3.7)\,\text{kJ} \cdot \text{mol}^{-1}.$$

Fuger and Oetting [76FUG/OET] obtained $S_m^\circ(U^{3+}$, aq, 298.15 K) $= -(202.9 \pm 21.8)\,\text{J} \cdot \text{K}^{-1} \cdot \text{mol}^{-1}$ based on the corrected redox data for the U^{4+}/U^{3+} couple from Kritchevsky and Hindman [49KRI/HIN] (E°(V.6, 298.15 K) $= -(0.607 \pm 0.007)\,\text{V}$), the $\Delta_r H_m^\circ$(V.6, 298.15 K) above, and the entropy $S_m^\circ(U^{4+}$, aq, 298.15 K) $= -(417.8 \pm 14.9)\,\text{J} \cdot \text{K}^{-1} \cdot \text{mol}^{-1}$. This review prefers to use the value of E°(V.6, 298.15 K) $= -(0.553 \pm 0.004)\,\text{V}$ from Capdevila and Vitorge [90CAP/VIT], which differs slightly from the value of $-(0.568 \pm 0.010)\,\text{V}$ reported by Riglet, Robouch and Vitorge [89RIG/ROB]. The selected redox potential and equilibrium constant, converted to the 1 bar standard state pressure, are thus

$$\begin{aligned} E^\circ(\text{V.6, 298.15 K}) &= -(0.553 \pm 0.004)\,\text{V} \\ \log_{10} K^\circ(\text{V.6, 298.15 K}) &= -9.35 \pm 0.07. \end{aligned}$$

This results in

$$\Delta_f G_m^\circ(U^{3+}, \text{aq}, 298.15\,\text{K}) = -(476.5 \pm 1.8)\,\text{kJ} \cdot \text{mol}^{-1}$$

which is in agreement with the value $\Delta_f G_m^\circ(U^{3+}, \text{aq}, 298.15\,\text{K}) = -(480.7 \pm 4.6)\,\text{kJ} \cdot \text{mol}^{-1}$ in Ref. [76FUG/OET].

Using the selected enthalpy of reaction and equilibrium constant, the standard entropy is calculated to be

$$S_m^\circ(U^{3+}, \text{aq}, 298.15\,\text{K}) = -(188.2 \pm 13.9)\,\text{J} \cdot \text{K}^{-1} \cdot \text{mol}^{-1}.$$

This value is in better agreement with the value $S_m^\circ(U^{3+}$, aq, 298.15 K) $= -(174.9 \pm 6.4)\,\text{J} \cdot \text{K}^{-1} \cdot \text{mol}^{-1}$, selected by Fuger and Oetting [76FUG/OET] based on analogy with the Pu^{3+} value, than with $-(202.9 \pm 21.8)\,\text{J} \cdot \text{K}^{-1} \cdot \text{mol}^{-1}$, derived from the redox data of Ref. [49KRI/HIN].

As is the case for U^{4+}, the selected value of

$$C_{p,m}|_{298}^{473}\,(U^{3+}, \text{aq}) = -(64 \pm 22)\,\text{J} \cdot \text{K}^{-1} \cdot \text{mol}^{-1}$$

is taken from the Criss-Cobble [64CRI/COB, 64CRI/COB2] estimate of Lemire and Tremaine [80LEM/TRE], and represents the mean value from 298 to 473 K.

V.2.5. U^{2+}

The trivalent cation, U^{3+}, is the lowest oxidation state of uranium which may be generated electrochemically in aqueous media. Although Duflo-Plissonnier and Samhoun [72DUF/SAM] propose a value of $E° = -(1.65 \pm 0.05)$ V for the U^{3+}/U^{2+} couple, there is no evidence for the existence of an aqueous U^{2+} species (see, *e.g.*, Nugent [75NUG]). For this reason and in agreement with Fuger and Oetting [76FUG/OET] this review does not consider this species.

V.3. Oxygen and hydrogen compounds and complexes

V.3.1. Gaseous uranium oxides

The thermodynamics of the gaseous uranium oxides are extremely complex. This is the result of the fact that very high temperatures are required to perform experimental studies, and also because of the lack of information about the electronic level distribution of the molecules, which is of great importance at the high temperatures where formation of the gaseous oxides can be expected.

Since the objective of the present review is to present thermochemical data for radioactive waste management, the gaseous uranium oxides fall outside the scope of the book. However, for scientific reasons, a set of selected values for UO(g), UO_2(g) and UO_3(g) is included in the present review. These data are taken from existing evaluations and are only checked for their consistency with the data presented here.

There are two recent compilations of the thermochemical properties of the gaseous uranium oxides in which a simultaneous analysis of the high-temperature reactions was made. Glushko *et al.* [82GLU/GUR] reviewed information available in 1982 and made a rough estimation of the electronic level distribution. More recently, Fischer [87FIS] presented data for the gaseous uranium oxides which were deduced from a detailed analysis of advanced high-temperature measurements by Breitung and Reil [85BRE/REI], Ohse *et al.* [85OHS/BAB], and Bober and Singer [87BOB/SIN]. These studies were made in the 2150 to 8000 K temperature range and report measurements of the laser evaporation of liquid UO_2 by mass spectrometry. Fischer's [87FIS] approach involved the adjustment of the electronic level density distribution to fit the high-temperature equilibria, using fixed values for the enthalpies of formation.

The recommendations from the two compilations differ significantly. The evaluation by Glushko *et al.* [82GLU/GUR] is adopted here for the following reasons:

- Fischer's [87FIS] evaluation is solely based on a restricted number of high-temperature mass spectrometric measurements, whereas Glushko *et al.* [82GLU/GUR] analysed a larger number of well-defined experimental investigations at lower temperatures.

- The high-temperature measurements by Breitung and Reil [85BRE/REI], Ohse *et al.* [85OHS/BAB] and Bober and Singer [87BOB/SIN] were made at conditions where temperature measurement techniques and mass spectrometric sampling are subjected to considerable uncertainties.

The selected values are summarized in Table V.4.

Table V.4: Selected thermodynamic parameters on the gaseous uranium oxide compounds at 298.15 K and 1 bar. The Gibbs energies of formation are calculated from the selected enthalpies of formation and entropies.

	$\Delta_f G_m^\circ$ ($kJ \cdot mol^{-1}$)	$\Delta_f H_m^\circ$ ($kJ \cdot mol^{-1}$)	S_m° ($J \cdot K^{-1} \cdot mol^{-1}$)	$C_{p,m}^\circ$ ($J \cdot K^{-1} \cdot mol^{-1}$)
$UO(g)$	1.9 ± 17.0	30.5 ± 17.0	$248.8 \pm 2.0^{(a)}$	42.0 ± 2.0
$UO_2(g)$	-481.1 ± 20.0	-477.8 ± 20.0	266.3 ± 4.0	59.5 ± 2.0
$UO_3(g)$	-784.8 ± 15.0	-799.2 ± 15.0	309.5 ± 2.0	64.5 ± 2.0

(a) It should be mentioned that a new value for the entropy of UO(g) has been published recently, S_m°(UO, g, 298.15 K) = $(252.0 \pm 0.1)\ J \cdot K^{-1} \cdot mol^{-1}$ [90GUR], which is based on new spectroscopic data from the Soviet group.

V.3.2. Aqueous uranium hydroxide complexes

V.3.2.1. U(VI) hydroxide complexes

The hydrolysis of dioxouranium(VI) has been the subject of extensive study. The data were reviewed previously by Sylva and Davidson [79SYL/DAV2], Tripathi [84TRI], and Baes and Mesmer [76BAE/MES], among others. Hydrolysis constant values given in the literature for temperatures near 298.15 K are summarized in Table V.5 and those for higher temperatures in Table V.6. These hydrolysis constants $^*\beta_{n,m}$ ($^*\beta^\circ_{n,m}$ defined as $^*\beta_{n,m}$ at $I = 0$) are the equilibrium constants for reactions of the form

$$m\mathrm{UO}_2^{2+} + n\mathrm{H_2O(l)} \rightleftharpoons (\mathrm{UO_2})_m(\mathrm{OH})_n^{2m-n} + n\mathrm{H}^+. \qquad \text{(V.7)}$$

The values in Tables V.5 and V.6 correspond to those given in the original papers cited.[2] However, it is obvious that in some cases analysis of the experimental data was based on erroneous assumptions concerning the major species present in the solutions studied. In other cases, values of constants were given for species representing no more than a few percent of the total uranium in solution in any of the experiments. These species and constants may be correct, or there may be different species that could be proposed and used to fit the data equally well. The results may

[2] See also Footnote 7 in Section V.4.2.1.2.a on the formation of ternary complexes of in the U(VI)-H_2O-Cl^- system, and Footnote 8 in Section V.5.1.3.1.a on the formation of ternary complexes of the general form $(UO_2)_m(OH)_n(SO_4)_q$.

Table V.5: Experimental equilibrium data for the uranium(VI) hydroxide system, according to the equilibria

$$m\mathrm{UO_2^{2+}} + n\mathrm{H_2O(l)} \rightleftharpoons (\mathrm{UO_2})_m(\mathrm{OH})_n^{2m-n} + n\mathrm{H^+}.$$

$n:m$	t (°C)	Method	Ionic strength (medium)	$\log_{10}{}^*\beta_{n,m}$	Reference
1:1	25?	col	self	$-4.50^{(a)}$	[47GUI]
2:2				$-4.95^{(a)}$	
1:1	25	pol	var	-4.09	[47HAR/KOL]
2:2	20-25	pot	self	-5.87	[47MAC/LON]
2:2	25?	pot	var	-5.91	[47SUT]
3:4				-12.51	
2:2	25?	pot	0.06 M ($Ba(NO_3)_2$)	-5.36	[48FAU, 54FAU]
2:2	25?	pot	0.60 M ($Ba(NO_3)_2$)	-5.97	
1:1	20	pot	1.0 M ($NaClO_4$)	$-4.77^{(b)}$	[49AHR]
2:2				-6.10	
5:3				-16.74	
2:2	25	pot	self (0.03 M)	-5.99	[49SUT]
3:4				$-13.3^{(c)}$	
5:3				-16.8	
6:3				-23.32	
7:3				-30.7	
8:3				-41.7	
9:3				-52.7	
2:2	25	pot	self (0.03 M)	-5.94	
3:4				-12.9	
5:3				-16.4	
1:1	25	sol	0.0	-4.14	[55GAY/LEI]
1:1	25?	sp	0.0	-4.19	[55KOM/TRE]
1:1	25	dis	0.1 M ($NaClO_4$)	-4.20	[55RYD]
2:1				-9.40	
2:2	?	pot	self (SO_4^{2-})	-5.06	[56ORB/BAR]
2:4				-1.26	
1:1	25	pot	0.035 M ($BaCl_2$)	-5.82 ± 0.80	[57HEA/WHI]
2:2				-6.16	
1:1	25	pot	0.35 M ($BaCl_2$)	-5.40 ± 0.80	
2:2				-5.82	
1:1	20	sol	$I = 0$ cor	$< -4.$	[58BRU]
2:2	25	pot	0.16 M (KNO_3)	-6.12 ± 0.40	[58LI/DOO]
$\frac{(2+2n)}{(2+n)}$				$(-6.1 - 6.35n)$	
1:2	25	pot	1.0 M (self/$NaClO_4$)	-3.66	[59HIE/SIL]
2:2				-6.02	
1:2	25	pot	3.0 M (self/$NaClO_4$)	-3.68	
2:2				-6.31	
4:3				-12.6	
1:1	25	pot	0.1 M (KNO_3)	-6.10	[60GUS/RIC]
2:2				-5.84 ± 0.30	
6:4				-17.6	

Table V.5 (continued)

$n:m$	t (°C)	Method	Ionic strength (medium)	$\log_{10}{}^*\beta_{n,m}$	Reference
1:1	20	dis	0.1 M ($NaClO_4$)	−5.0	[60STA]
2:1				−10.5	
3:1				−17.1	
2:2	25	pot	4.5 M (Na_2SO_4)	−8.20	[61PET, 63DUN/HIE]
4:3				−16.21	
5:3				−22.13	
6:4				−24.56	
8:5				−32.30	
1:1	25	pot	0.5 M (KNO_3)	-5.70 ± 0.60	[62BAE/MEY]
2:2				-5.92 ± 0.08	
5:3				-16.22 ± 0.30	
2:2	25?	sp	0.5 M (?NO_3)	−6.2	[62NIK/PAR]
5:3	25?	ix	0.5 M (?NO_3)	−16.	
1:1	25	pot	1.0 M (NaCl)	−5.60	[62RUS/JOH]
2:2				-6.17 ± 0.20	
3:4				-12.33 ± 0.40	
5:3				-17.00 ± 0.40	
2:2	25	?	0.1 M ($NaNO_3$)	−6.2	[63BAR/SOM, 76BAE/MES]
2:2	25	?	0.1 M ($NaClO_4$)	−6.2	
1:2	25	pot	1.0 M (KNO_3)	−4.16	[63DUN/HIE]
2:2				-5.96 ± 0.10	
4:3				-12.79 ± 1.00	
5:3				-16.21 ± 0.30	
2:2	25	pot	3.0 M (NaCl)	-6.64 ± 0.10	[63DUN/HIE, 63DUN/SIL]
4:3				-12.54 ± 0.40	
5:3				-18.07 ± 0.40	
6:4				−19.96	
7:4				−24.91	
1:2	25	pot	4.5 M ($Mg(ClO_4)_2$)	−3.81	
2:2				−6.25	
4:3				−13.33	
5:3				−17.18	
6:4				−20.18	
1:2	25	pot	4.5 M ($Ca(ClO_4)_2$)	−3.96	[63DUN/HIE, 63HIE/ROW]
2:2				−6.20	
4:3				−13.44	
5:3				−16.91	
1:2	25	pot	3.0 M ($NaClO_4$)	−3.7 [(d)]	
2:2				-6.02 ± 0.10	
4:3				-13.83 ± 1.00	
5:3				-16.54 ± 0.30	
1:1	27	pot	self (NO_3^-)	−4.59	[63POZ/STE]
1:1	16	pot	self (NO_3^-)	−5.05	

Table V.5 (continued)

$n:m$	t (°C)	Method	Ionic strength (medium)	$\log_{10}{}^*\beta_{n,m}$	Reference
2:2	25	sp	1.0 M ($NaClO_4$)	−5.94	[63RUS/JOH]
5:3				−16.41	
2:2	25	pot	1.0 M ($NaClO_4$)	-5.91 ± 0.30	
5:3				-16.43 ± 0.60	
1:2	22-25	sp	0.1 M ($NaClO_4$)	−1.9	[64BAR/SOM]
2:2				−6.28	
2:2	25	pot	0.1 M ($NaClO_4$)	-6.09 ± 0.40	
2:1	25?	pot	var ($NaNO_3$)	−8.49	[65ISR]
2:2				−4.8	
3:1				−16.15	
5:2				−34.06	
6:4				−14.76	
1:1	25	kin	self	−6.0	[67COL/EYR]
2:2	25	cal	3.0 M ($NaClO_4$)	-6.02 ± 0.30	[68ARN/SCH]
5:3				-16.54 ± 0.60	
2:2	25	pot	0.2 M ($NaClO_4$)	-5.92 ± 0.30	[68OST/CAM]
5:3				-16.16 ± 0.60	
1:1	25	pot	4.5 M ($Mg(NO_3)_2$)	−5.38	[68SCH/FRY]
2:2				−6.34	
5:3				−17.37	
1:1	25	pot	7.5 M ($Mg(NO_3)_2$)	−5.53	
2:2				−6.52	
5:3				−17.76	
1:1	25	pot	0.1 M ($NaClO_4$)	−4.39	[69TSY]
1:2				−2.22	
2:2				−6.09	
5:3				−15.64	
7:3				−24.03	
1:1	25	pot	0.5 M (KNO_3)	−5.70	[69VAN/OST]
2:2				-5.95 ± 0.10	
5:3				-16.36 ± 0.30	
2:2	25	pot	3.0 M ($NaClO_4$)	-6.17 ± 0.20	[72MAE/AMA]
3:4				-12.92 ± 1.00	
5:3				-17.04 ± 0.40	
1:1	25	sol	self	−5.08[(e)]	[72NIK/SER]
1:2				−12.28[(e)]	
2:2				−5.73[(e)]	
1:1	20	ix, sp	0.5 M ($NaNO_3$)	−3.43	[73DAV/EFR, 83DAV/EFR]
2:2	22	pot	var ($NaClO_4$)	−6.0	[73MAV, 74MAV]
3:4				−13.0	
1:1	30	sp	0.5 M ($NaClO_4$)	−4.00	[76GHO/MUK]
1:1	25?	pot/dis	$I=0$ cor (ClO_4^-)	−4.03	[77VOL/BEL]
1:1	25?	pot/dis	$I=0$ cor (TcO_4^-)	−4.20	
1:1	25?	tc	self (ClO_4^-)	−5.2	[78SCH/SUL]

Table V.5 (continued)

$n:m$	t (°C)	Method	Ionic strength (medium)	$\log_{10}{}^*\beta_{n,m}$	Reference
2:2	25	pot	3.0 M ($NaClO_4$)	-6.0	[79CIA/FER]
5:3				-16.6	
2:2				$-6.02\pm0.10^{(f)}$	
5:3				$-16.80\pm0.60^{(f)}$	
1:2	25	pot	0.5 M ($NaClO_4$)	-3.81	[79LAJ/PAR]
2:2				-6.034 ± 0.100	
4:3				-13.17 ± 0.50	
5:3				-16.778 ± 0.300	
6:4				-18.91	
2:2	25	pot	3.0 M (KCl)	-6.30	[79MIL/ELK]
2:3				-11.20	
6:4				-17.85	
2:2	25	pot	0.1 M (Et_4NClO_4)	-5.63 ± 0.30	[79SPI/ARN]
5:3				-15.87 ± 0.60	
1:1	25	pot	0.1 M (KNO_3)	-5.50 ± 0.40	[79SYL/DAV2]
2:2				-5.89 ± 0.06	
3:4				-12.31 ± 0.40	
5:3				-16.46 ± 0.20	
7:4				-22.76	
1:1	25	pot	0.0	-5.1 ± 0.4	[80DON/LAN]
1:1	25	pot	0.2 M ($NaNO_3$)	-4.8	[80PON/DOU]
2:2				-5.83	
5:3				-18.96	
2:2	25	pot	0.024 M ($NaClO_4$)	-5.64 ± 0.40	[81VAI/MAK]
5:3				-15.94 ± 0.80	
2:2	25	pot	0.105 M ($NaClO_4$)	-5.85 ± 0.30	
5:3				-16.32 ± 0.60	
2:2	25	pot	0.254 M ($NaClO_4$)	-5.89 ± 0.30	
5:3				-16.46 ± 0.60	
2:2	25	pot	0.506 M ($NaClO_4$)	-5.97 ± 0.30	
5:3				-16.51 ± 0.60	
2:2	25	pot	1.005 M ($NaClO_4$)	-6.06 ± 0.30	
5:3				-16.67 ± 0.60	
2:2	25	pot	2.003 M ($NaClO_4$)	-6.16 ± 0.30	
5:3				-16.79 ± 0.60	
2:2	25	pot	0.1 M ($NaClO_4$)	-5.89	[82MAY]
5:3				-16.19	
2:2	25	pot	3.0 M ($NaNO_3$)	-6.09 ± 0.30	[82MIL/SUR]
5:3				-16.72 ± 0.60	
2:2	25	pot	0.5 M ($NaNO_3$)	-6.01	
3:4				-12.24	
2:2	25	pot	1.0 M ($NaNO_3$)	-6.07	
3:4				-12.31	
2:2	25	pot	1.5 M ($NaNO_3$)	-6.10	
3:4				-12.40	

Table V.5 (continued)

$n:m$	t (°C)	Method	Ionic strength (medium)	$\log_{10}{}^*\beta_{n,m}$	Reference
2:2	25	pot	2.0 M ($NaNO_3$)	−6.13	[82MIL/SUR]
3:4				−12.41	
2:2	25	pot	2.5 M ($NaNO_3$)	−6.13	
3:4				−12.48	
2:2	25	pot	3.0 M ($NaNO_3$)	−6.13	
5:3				−16.65	
1:3	25	pot	0.1 M (KNO_3)	−4.5	[82OVE/LUN]
2:2				-5.95 ± 0.08	
3:4				-12.5 ± 0.4	
5:3				-16.54 ± 0.30	
1:1	25	dis	0.0	−5.89	[83CAC/CHO2]
1:1			0.05 M ($NaClO_4$)	−6.04	
1:1			0.1 M ($NaClO_4$)	−6.09	
1:1			0.4 M ($NaClO_4$)	−6.20	
1:1			0.7 M ($NaClO_4$)	−6.09	
1:1			1.0 M ($NaClO_4$)	−6.03	
2:2	25	pot	0.1 M (KNO_3)	−6.45	[84KOT/EVS]
5:3				−17.29	
7:4				−23.12	
2:2	20	pot/sp	< 0.005 M (self)	$\sim -6.$	[87VIL]
3:2				~ -10.3	
19:9				~ -69.7	
5:3	25	sol	0.5 M ($NaClO_4$)	−16.54	[89BRU/SAN]
7:3				−31.9	
3:1				−19.69	
2:2	25	pot	0.1 M (KNO_3)	−6.45	[84KOT/EVS]
3:1	25	sol	0.5 M ($NaClO_4$)	-19.09 ± 0.40	[90BRU/SAN]
2:2	25	pot	0.5 M ($NaClO_4$)	−6.03	[91GRE/LAG]
5:3				−16.37	

(a) Plus hydroxonitrato complexes.
(b) As recalculated in Ref. [62RUS/JOH].
(c) From Ref. [49SUT, Fig. 6]. Probably incorrect in the text of Ref. [49SUT].
(d) Fixed value used during the refinement of the other hydrolysis constants.
(e) Smoothed values from $F(T)$.
(f) As recalculated as part of this review, see Appendix A.

Table V.6: Experimental equilibrium data for the uranium(VI) hydroxide system at different temperatures, according to the equilibria: $m\mathrm{UO}_2^{2+} + n\mathrm{H_2O(l)} \rightleftharpoons (\mathrm{UO_2})_m(\mathrm{OH})_n^{2m-n} + n\mathrm{H}^+$.

↓ $n:m$	$\log_{10}{}^*\beta_{n,m}$										Method	Ionic strength (medium)	Reference
	37°C	40°C	50°C	70°C	90°C	94.4°C	100°C	125°C	150°C	200°C			
1:1		−5.09									pot	0.035 M ($BaCl_2$)	[57HEA/WHI]
						−4.19					pot	0.5 m (KNO_3)	[62BAE/MEY]
			−4.47	−4.40	−3.57		−3.38	−2.94			pot	?? (self)	[71NIK][(a)]
			−4.36	−3.85	−3.40		−3.20	−2.73	−2.31	−1.61	pot	$I = 0$ corr	[72NIK/SER][(b)]
2:2		−5.92									pot	0.035 M ($BaCl_2$)	[57HEA/WHI]
							−4.27		−3.84		sol	$I = 0$ corr	[60LIE/STO][(c)]
						−4.51					pot	0.5 m (KNO_3)	[62BAE/MEY]
							−4.9		−4.1	−3.8	con	$I = 0$	[67RYZ/NAU]
			−5.21	−4.79	−4.42		−4.25	−3.85			pot	?? (self)	[71NIK][(a)]
			−4.97	−4.45	−3.98		−3.76	−3.27	−2.84	−2.11	pot	$I = 0$ corr	[72NIK/SER][(b)]
	−5.693										pot	0.15 M (NaCl)	[86DES/KRA]
2:1							−7.83	−7.05	−5.98		pot	~ 0.01 (self)	[71NIK]
			−11.24				−9.56		−8.28	−7.23	sol	$I = 0$ corr	[72NIK/SER]
4:3	−11.499										pot	0.15 M (NaCl)	[86DES/KRA]
5:3						−12.74					pot	0.5 m (KNO_3)	[62BAE/MEY]
	−16.001										pot	0.15 M (NaCl)	[86DES/KRA]
7:4	−21.027										pot	0.15 M (NaCl)	[86DES/KRA]

(a) These are fitted constants based on experimental work reported in Ref. [68NIK/ANT].
(b) These are fitted (and, above 125°C, extrapolated) constants based on a reanalysis of experimental work reported in Ref. [68NIK/ANT].
(c) From the fitting function [60LIE/STO, Table 1 and Eq. (9)].

also be mere artifacts of systematic experimental errors (*e.g.*, incorrect standardization of pH electrodes for solutions containing a high concentration of supporting electrolyte). The ionic strength at which the measurements were performed is particularly important for the minor species. The ionic medium must necessarily change during the experiment. Hence, some changes in the activity factors will occur and these will affect the minor species to a large extent. There is always the risk of obtaining "artificial" complexes when doing a computer refinement of the minor species. No computer program can provide safeguards against this. Hence, an element of chemical reasoning is always required when discussing the stoichiometric composition of these species. Discussions of some of these problems have been given by Baes and Meyer [62BAE/MEY], Sylva and Davidson [79SYL/DAV2], Baes and Mesmer [76BAE/MES], and Wanner [84WAN]. Few papers give sufficient experimental details to enable a reviewer to recalculate the raw data in terms of a different set of species.

Error limits given in the literature for hydrolysis constants generally reflect the precision of agreement between experimental data and values generated using a set of fitted parameters (equilibrium constants), but do not include contributions from systematic experimental errors, or errors from an incorrect choice of species (model error). For this reason, it is necessary to evaluate the reliability of each study and assign uncertainties to the reported parameters.

The lowest uncertainties are given to values based on good quality emf data obtained using properly calibrated electrodes, and derived considering a wide range of possible species. Since different species are important for different experimental conditions, the uncertainties assigned to the constants from a single paper are not necessarily the same for all species. Only rarely was it possible to directly use the uncertainties estimated by the original authors. A brief assessment of each reference considered in Tables V.5 and V.6 is given in Appendix A, except for a few papers for which all values are rejected.

In many early studies (*e.g.*, [55KOM/TRE, 55RYD, 63POZ/STE, 76GHO/MUK]) the predominance of the monomeric hydrolysis species UO_2OH^+ was assumed for conditions under which the major species were later found to be polymeric. In such studies the reported values for $^*\beta_1$ are inevitably too large. Similarly, the species $(UO_2)_3(OH)_4^{2+}$ and other "core-and-links" [54SIL, 54SIL2] species were often assumed without considering the possibility of species such as $(UO_2)_3(OH)_5^+$. In this latter case, values of $^*\beta_{2,2}$ obtained simultaneously from the data were probably influenced only slightly, but the $^*\beta_{4,3}$ values cannot be considered to be correct.

In the present review of hydrolysis data, measurements in chloride and nitrate media are reinterpreted by considering values (Sections V.4.2.1.2.a and V.6.1.3.1.a) of the first and second association constants of UO_2^{2+} with chloride and the first association constant of UO_2^{2+} with nitrate. In principle the values of the apparent complexation constants can be calculated using the values of $\log_{10}\beta_n^\circ$ and the appropriate interaction coefficients. However, the interaction coefficients of UO_2Cl^+ with Cl^- and $UO_2NO_3^+$ with NO_3^- are not known, although values are calculated for the interaction coefficients of these complex cations with ClO_4^-. No attempt is made to use literature activity coefficient data for aqueous solutions of UO_2Cl_2 or $UO_2(NO_3)_2$ [59ROB/STO] to extract sets of association constants with interaction coefficients.

The resulting parameters would almost certainly be highly correlated. In the absence of a better procedure, the value of each $\Delta\varepsilon$ for simple association reactions in a chloride or nitrate medium is set equal to the corresponding value for a perchlorate media of the same ionic strength.

From the calculated formation constants for the chloride and nitrate complexes, the ratio $[UO_2^{2+}] : [UO_2L^+]$, where $L^- = Cl^-$ or NO_3^-, is calculated. Thus, the apparent hydrolysis constants reported in the original paper can be used to calculate apparent constants in the absence of chloride or nitrate complex.

Using the "method of least squares", the data from ClO_4^-, Cl^- and NO_3^- media can then be treated simultaneously to obtain values for a hydrolysis constant at $I = 0$ and the ion interaction coefficient differences, $\Delta\varepsilon(n : m, ClO_4^-)$, $\Delta\varepsilon(n : m, Cl^-)$ and $\Delta\varepsilon(n : m, NO_3^-)$. This properly weights the value for $I = 0$, and should give a more precise and consistent set of ion interaction coefficients than separately treating data from solutions with supporting electrolytes having different anions.

Analysis using the alternative model assuming association and a single common hydrolysis interaction coefficient $\Delta\varepsilon$ for all anions interacting with a specific hydrolysis species gives a much poorer fit to the literature data. This situation is quite different from that found in examining other systems, *e.g.*, the UO_2^{2+}/U^{4+} couple discussed in Section V.2.3. The different values of $\Delta\varepsilon$ probably are, in part, the result of the assumption that $\Delta\varepsilon$'s for the complexation reactions are identical to those in perchlorate media.

Another treatment of the data (not explicitly including association constants for complexation of Cl^- or NO_3^- with UO_2^{2+}, but instead, treating association as part of the "interaction" described by the $\Delta\varepsilon$'s) does not significantly improve the goodness of fit.

No data from experiments at $I > 3.5$ m are used in the analysis. The selected values of the equilibrium constants and ion interaction coefficients are given in Table V.7.

V.3.2.1.1. The major polymeric U(VI) hydrolysis species at 298.15 K

The vast majority of experimental work on dioxouranium(VI) hydrolysis was done in aqueous, slightly acidic ($2 < pH < 5$) media with total uranium concentrations above 10^{-4} M. For this range of conditions there is a general consensus that the dimer, $(UO_2)_2(OH)_2^{2+}$, is a major species. In a curve-fitting analysis of potentiometric (and other) data the value of $^*\beta_{2,2}$ is not strongly correlated with other hydrolysis constants. Also, it is not greatly affected by changes on the concentration and nature of 1:1 supporting electrolytes. The change in Debye-Hückel terms between 0.1 M $NaClO_4$ and 3.0 M $NaClO_4$ results in a difference of only $\sim$0.25 in $\log_{10}{^*\beta_{2,2}}$. Also, the ion interaction coefficients tend to be such as to minimize medium effect on the value of $^*\beta_{2,2}$. Thus, literature values have a tendency to appear consistent with one another. Although there is a scarcity of reliable data in chloride media, the use of data from studies in perchlorate and nitrate solutions to determine $^*\beta_{2,2}$ allows a reasonable estimate to be made for $\Delta\varepsilon(V.7, n = 2, m = 2, Cl^-)$. Bruno [87BRU] calculated ion interaction coefficients for several polymeric hydrolysis cations of beryllium, and

Table V.7: Selected formation constants for the uranium(VI) hydroxide system at 25°C and $I = 0$, according to the equilibria

$$m\mathrm{UO}_2^{2+} + n\mathrm{H_2O(l)} \rightleftharpoons (\mathrm{UO_2})_m(\mathrm{OH})_n^{2m-n} + n\mathrm{H}^+.$$

Species	$\log_{10}{}^*\beta^\circ_{n,m}$	$\Delta\varepsilon_{(n:m,\mathrm{ClO_4^-})}$	$\Delta\varepsilon_{(n:m,\mathrm{Cl^-})}$	$\Delta\varepsilon_{(n:m,\mathrm{NO_3^-})}$
$(UO_2)_2(OH)_2^{2+}$	-5.62 ± 0.04	-0.07 ± 0.02	0.01 ± 0.03	-0.29 ± 0.07
$(UO_2)_3(OH)_4^{2+}$	-11.9 ± 0.3	0.1 ± 0.2	-0.40 ± 0.15	-0.4 ± 1.0
$(UO_2)_3(OH)_5^+$	-15.55 ± 0.12	-0.23 ± 0.07	0.03 ± 0.13	-0.6 ± 0.2
UO_2OH^+	-5.2 ± 0.3	-0.4 ± 3.7		0.1 ± 1.4
$UO_2(OH)_2(aq)$	$\leq -10.3^{(a)}$			
$UO_2(OH)_3^-$	$-19.2 \pm 0.4^{(a,c)}$	$-0.13 \pm 0.08^{(b)}$		
$UO_2(OH)_4^{2-}$	$-33 \pm 2^{(a)}$			
$(UO_2)_2OH^{3+}$	$-2.7 \pm 1.0^{(a)}$			
$(UO_2)_3(OH)_7^-$	$-31 \pm 2^{(a,c)}$			
$(UO_2)_4(OH)_7^+$	$-21.9 \pm 1.0^{(a)}$			

(a) Estimated.

(b) Estimated value referring to $\Delta\varepsilon_{(n:m,\mathrm{Na^+})}$.

(c) Sandino [91SAN] recently determined the equilibrium constants for the formation of $UO_2(OH)_3^-$ and $(UO_2)_3(OH)_7^-$ from solubility measurements of schoepite and $(UO_2)_3(PO_4)_2 \cdot 4H_2O(cr)$. She obtained $\log_{10}{}^*\beta_3^\circ(m = 1, n = 3, 298.15\,\mathrm{K}) = -(19.99 \pm 0.32)$ and $-(19.18 \pm 0.27)$, as well as $\log_{10}{}^*\beta^\circ_{7,3}(m = 3, n = 7, 298.15\,\mathrm{K}) = -(31.54 \pm 0.44)$, where the uncertainties are twice the standard deviations resulting from the least-squares refinement. These values are in good agreement with the selected values presented in this table.

found the interaction coefficients to be similar to those reported [80CIA] for "hard" cations of charge +2. The constants found for $\Delta\varepsilon(\mathrm{V.7}, n = 2, m = 2, \mathrm{X}^-)$ are more consistent with the ion interaction coefficients for 3+ ions, as might have been expected considering the effective charge in dioxouranium(VI) ions.

Two tri-uranyl species, $(UO_2)_3(OH)_5^+$ and $(UO_2)_3(OH)_4^{2+}$, are also reasonably well established [76BAE/MES, 79SYL/DAV2]. Nevertheless, when experimental conditions are chosen so that neither of these is a major species, the two hydrolysis constants from curve-fitting analyses can be highly correlated with each other. Correlation can also occur with the formation constant for the monomer UO_2OH^+.

An extreme example is found in a paper by Milic and Suranji [82MIL/SUR]. The species $(UO_2)_3(OH)_4^{2+}$ and $(UO_2)_3(OH)_5^+$ would account for less than 15% of the uranium in any of the $NaNO_3$ solutions considered. For the 3.0 M $NaNO_3$ medium, for which Milic and Suranji [82MIL/SUR] assumed the existence of $(UO_2)_3(OH)_5^+$, but not $(UO_2)_3(OH)_4^{2+}$, their own extrapolated value for $^*\beta_{4,3}$ (from experiments in media with different nitrate concentrations) would suggest slightly greater concentrations of $(UO_2)_3(OH)_4^{2+}$ than $(UO_2)_3(OH)_5^+$. The species $(UO_2)_3(OH)_4^{2+}$ is most firmly established for chloride media at 25°C. Baes and Meyer [62BAE/MEY] found no evidence for it in 0.5 m KNO_3 solution, but Sylva and Davidson [79SYL/DAV2] concluded it exists as a minor species in 0.1 M KNO_3. The calculated uncertainty in the ion interaction coefficient $\Delta\varepsilon(\text{V.7}, n = 4, m = 3, NO_3^-)$ is very large; nevertheless, based on the fitted constants in Table V.7, the (4,3) species would only have been a minor species in the solutions used by Baes and Meyer [62BAE/MEY]. Thus, mathematically, there is no real discrepancy; nevertheless, problems remain as to whether structures in agreement with experimental evidence and within reasonable geometrical restrictions can be proposed for both $(UO_2)_3(OH)_4^{2+}$ and $(UO_2)_3(OH)_5^+$. Aberg [70ABE] used X-ray diffraction to elucidate the structure of uranium hydrolytic species in a solution of dioxouranium(VI) chloride (~ 3 M). Her data indicate that the trimeric species consists of a triangular array of uranyl groups, linked by hydroxide groups bridging each pair of uranium atoms with a single oxygen atom in the middle of the triangle, *i.e.*, a structure very similar to the one found in $[(UO_2)_3O(OH)_3(H_2O)_6]NO_3{\cdot}4H_2O$. The proposed $(UO_2)_3O(OH)_3^+$ structure, which is stoichiometrically equivalent to $(UO_2)_3(OH)_5^+$ if coordinated water molecules are not considered, was proposed for the trimeric species in solutions in which $(UO_2)_3(OH)_4^{2+}$ had been found to be a major solution species [63DUN/SIL]. However, if the species were $(UO_2)_3(OH)_4^{2+}$, addition of an extra proton to the $(UO_2)_3O(OH)_3^+$ structure would be required—one possibility being protonation of the central oxygen as proposed by Evans [63EVA]. It is not clear why protonation of this structure would be especially favoured in chloride (as opposed to perchlorate) media, and it also remains uncertain that the hydrolysis species in such concentrated solutions of uranium have the same structures as those in more dilute solutions of uranium containing high chloride concentrations. Some authors [76BAE/MES] suggested that $(UO_2)_3(OH)_4^{2+}$ may actually be $(UO_2)_3(OH)_4X^+$, where X is an anion. The structure $(UO_2)_3O(OH)_2X^+$ would be possible if the anion replaced one of the bridging hydroxide groups in a $(UO_2)_3O(OH)_3^+$ structure (see also Ref. [71ABE]). If this is so, the procedure used in this review to obtain $^*\beta_{4,3}$ is incorrect, as a slightly different formation constant at $I = 0$ would be expected for each anion. There are insufficient data available to reach a firm conclusion about the nature of this species.

A problem is found in the analysis of the reported values for $^*\beta_{5,3}$, especially those from studies in nitrate media. Attempts to combine the data from studies in perchlorate, chloride and nitrate media lead to a poor fit to the data from essentially all of the studies in nitrate media (Figure V.3). If the fitting is done omitting the somewhat suspect value of Milic and Suranji [82MIL/SUR] (3.33 m $NaNO_3$, see Appendix A), the calculated value of $\log_{10}{}^*\beta_{5,3}$ becomes significantly more negative (~ −15.7) and $\Delta\varepsilon(\text{V.7}, n = 5, m = 3, NO_3^-) = -1.2$ is much more negative than $\Delta\varepsilon(\text{V.7}, n = 5, m = 3, Cl^-) = -0.02$. This difference in the ion interaction coefficients is unexpectedly large. It would have been expected, on the basis of values for other ion interaction coefficients (see also Refs. [80CIA, 87BRU]), that the ion

Figure V.3: Extrapolation to $I = 0$ of experimental data for the formation of $(UO_2)_3(OH)_5^+$ using the specific ion interaction theory. The data refer to three different aqueous media, perchlorate (○), nitrate (△) and chloride (□), and they have been corrected for the formation of UO_2^{2+} chloride and nitrate complexes in these media.

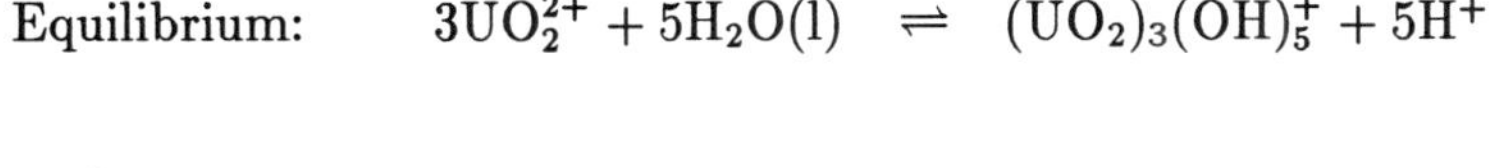

Equilibrium: $3UO_2^{2+} + 5H_2O(l) \rightleftharpoons (UO_2)_3(OH)_5^+ + 5H^+$

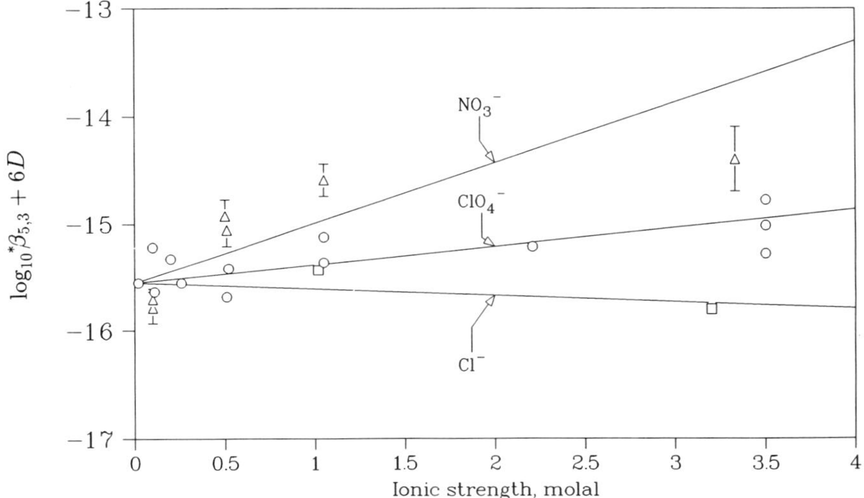

interaction coefficient differences ($\Delta\varepsilon$) for the hydrolysis reaction of UO_2^{2+} to form $(UO_2)_3(OH)_5^+$ in perchlorate, chloride and nitrate media would not differ by more than ~ 0.3. Thus, there are difficulties with the results in nitrate media in addition to those specific to the analysis of Milic and Suranji [82MIL/SUR]. If $\Delta\varepsilon(\text{V.7}, n = 5, m = 3, NO_3^-) = \Delta\varepsilon_1 + \Delta\varepsilon_2 \log_{10} I$ is used, the values $\Delta\varepsilon_1 = -(1.10 \pm 0.29)$ and $\Delta\varepsilon_2 = (1.16 \pm 0.57)$ are obtained, and the value of $\log_{10}{}^*\beta^\circ_{5,3} = -(15.74 \pm 0.15)$ and the values and uncertainties of the other ion interaction coefficients are similar to those obtained by omission of the value from Ref. [82MIL/SUR]. Possible reasons for the scatter include undetected precipitation at high pH values in some of the studies, or the omission of more hydrolysed species (*e.g.*, $(UO_2)_4(OH)_7^+$) in models used for analysis of data from solutions of high pH and total uranium concentration. In this review it is finally decided to simply include the results of Ref. [82MIL/SUR] and assume the ion interaction coefficient difference, $\Delta\varepsilon(\text{V.7}, n = 5, m = 3, NO_3^-)$, to be independent of the ionic strength. It is also possible, but less likely, that the source of inconsistencies in the ion interaction coefficient differences could be an undetected major experimental error in the work of Dunsmore, Hietanen and Sillén [63DUN/HIE] ($[Cl^-] = 3.2$ m), or an error in their data analysis.

It should be mentioned that Viljoen [87VIL] recently presented an alternate hydrolysis scheme (omitting the species $(UO_2)_3(OH)_5^+$) that is discussed in more detail in Appendix A.

V.3.2.1.2. The first monomeric U(VI) hydrolysis species: UO_2OH^+

The existence of UO_2OH^+ and the equilibrium constant for its formation have been the subject of debate for almost forty years. With the values for the formation constants of $(UO_2)_2(OH)_2^{2+}$, $(UO_2)_3(OH)_4^{2+}$, and $(UO_2)_3(OH)_5^+$, as established above, it is clear that studies using total uranium concentrations above 10^{-4} M that did not include polymeric species in their data analysis [47HAR/KOL, 55GAY/LEI, 55KOM/TRE, 73DAV/EFR, 76GHO/MUK, 78SCH/SUL, 83DAV/EFR] must be eliminated from further consideration. Careful potentiometric studies [62BAE/MEY, 79SYL/DAV2] gave values of the formation constant of $10^{-5.5}$ and $10^{-5.7}$ in 0.1 and 0.5 m nitrate media, respectively. Dongarra and Langmuir [80DON/LAN] gave a somewhat higher value for $^*\beta_1^\circ$ at $I = 0$. Hearne and White [57HEA/WHI] neglected the formation of $(UO_2)_3(OH)_5^+$ as did Gustafson, Richard and Martell [60GUS/RIC], although the latter group did include $(UO_2)_4(OH)_6^{2+}$. The reason for the large value obtained by Tsymbal [69TSY] is not clear. In all these studies UO_2OH^+ is a minor species.

Rydberg [55RYD], who used an extraction technique, ignored polymer formation at total uranium concentrations below 10^{-3} M. Even in the recent extraction work of Caceci and Choppin [83CAC/CHO2], from which $^*\beta_1 \approx 10^{-6.1}$ was obtained using ^{233}U at tracer levels, it can be shown that $(UO_2)_3(OH)_5^+$ should be significant for solutions with pH > 6. Stary [60STA], working at slightly higher uranium concentrations (and at 20°C), considered other monomeric uranium hydrolysis species. This procedure resulted in $^*\beta_1 = 10^{-5.0}$, which is in fair agreement with the potentiometric work. To some extent, the inclusion of $UO_2(OH)_2(aq)$ and $UO_2(OH)_3^-$ in his analysis compensated for the omission of polymeric species in the hydrolysis scheme.

Cole *et al.* [67COL/EYR] determined rate constants for the hydrolysis equilibrium

$$UO_2^{2+} + H_2O(l) \rightleftharpoons UO_2OH^+ + H^+ \qquad (V.8)$$

using the dissociation field effect relaxation method. Their experiments indicate $^*\beta_1$ is in the range of 10^{-5} to 10^{-6}.

Nikitin *et al.* [72NIK/SER] reported $^*\beta_1^\circ = 10^{-5.1}$ at 25°C based on extrapolation of the equilibrium constants determined between 50 and 150°C in the potentiometric study of Nikolaeva, Antipina and Pastukova [68NIK/ANT]. Unfortunately, $(UO_2)_3(OH)_5^+$ was not included in the set of species assumed by these researchers; furthermore, the uncertainty in the extrapolation is reasonably large. Nikitin *et al.* [72NIK/SER] also reported solubility measurements for $UO_3{\cdot}2H_2O(cr)$ at 25°C (pH 6.2 to 8.2). The use of the hydrolysis constants selected in this review with the set of Gibbs energies of formation for UO_2^{2+} and $UO_3 \cdot 2H_2O$ given by Nikitin *et al.* [72NIK/SER] leads to calculated solubilities that are much higher than the experimental values, and to the conclusion that $(UO_2)_3(OH)_5^+$ is the major solution species. Thus, the value calculated by Nikitin *et al.* [72NIK/SER] for $^*\beta_1^\circ$ is probably incorrect.

The value of $^*\beta_1^\circ$ is not really well defined by the experimental data, and estimation of activity coefficients for species such as UO_2OH^+ using sparse data is not a clear-cut procedure. This is particularly true for nitrate media, where data for possible analogue species such as UO_2F^+ are also lacking. In their review, Baes and Mesmer [76BAE/MES] selected a value of $\log_{10}{^*\beta_1^\circ} = -(5.8 \pm 0.2)$. This is smaller than the value determined experimentally by all groups except Gustafson, Richard and Martell [60GUS/RIC] and Caceci and Choppin [83CAC/CHO2], and since Ref. [76BAE/MES] was prepared, new data have been published [79SYL/DAV, 80DON/LAN] that suggest slightly more positive values of $\log_{10}{^*\beta_1^\circ}$. This review thus selects a value of

$$\log_{10}{^*\beta_1^\circ}(\text{V.8}, 298.15\,\text{K}) = -(5.2 \pm 0.3),$$

and considers medium effects responsible for several smaller values found, for example, by Baes and Meyer [62BAE/MEY].[3] The values calculated in this review for $\Delta\varepsilon(\text{V.8}, ClO_4^-)$ and $\Delta\varepsilon(\text{V.8}, NO_3^-)$ have extremely large uncertainties (resulting from the sparse data, most of which were obtained at relatively low ionic strengths). Assuming only Debye-Hückel effects are important, *i.e.*, $\Delta\varepsilon(\text{V.8}, ClO_4^-) = \Delta\varepsilon(\text{V.8}, NO_3^-) = 0.0$, has essentially no effect on the weighted average value of $\log_{10}{^*\beta_1^\circ}$, and a maximum effect on $\log_{10}{^*\beta_1}$ of 0.09 for constants calculated for the media used in the experimental work.

V.3.2.1.3. Neutral and anionic U(VI) hydrolysis species

Several authors have hypothesized neutral and/or anionic hydrolysis species of dioxouranium(VI) in an attempt to fit experimental data. The study of species in neutral and alkaline solutions of dioxouranium(VI) is complicated by the formation of very insoluble uranate solids of varying compositions, and by formation of very strong carbonate complexes $UO_2(CO_3)_3^{4-}$ and $UO_2(CO_3)_2^{2-}$ (*cf.* Section V.7.1.2.1.a). Musikas [72MUS] attempted to circumvent the first problem by working with solutions containing the tetrabutylammonium ion as the sole cation. He concluded that in very basic solutions a species with an OH:U ratio approaching 4 is obtained above pH 13. In alkaline solutions of lower pH, polymeric uranium species (OH:U ratios between 2 and 4) predominate even in solutions with total uranium concentrations less than 6×10^{-3} M. Maya and Begun [81MAY/BEG] obtained Raman spectroscopic evidence that confirmed the existence of at least two dioxouranium(VI) hydrolysis species in 0.1 M tetramethylammonium hydroxide solution. These species were shown to have spectra different from those of $(UO_2)_2(OH)_2^{2+}$ and $(UO_2)_3(OH)_5^+$.

Sutton [49SUT] and Tsymbal [69TSY] proposed the species $(UO_2)_3(OH)_7^-$ on the basis of potentiometric studies in slightly acidic solutions. The values of $^*\beta_{7,3}$ given by these authors differ by six orders of magnitude. Reanalysis of Tsymbal's data gives $^*\beta_{7,3} = 10^{-28}$, a more reasonable value than the 10^{-24} originally reported. If $^*\beta_{7,3}$

[3] A paper by Choppin and Mathur [91CHO/MAT], received after the draft of this section was completed, reports $\log_{10}{^*\beta_1} = -5.9$ in 0.1 M $NaClO_4$, suggesting $\log_{10}{^*\beta_1^\circ}$ may be somewhat more negative than recommended in the present review.

were as large as 10^{-24}, then $(UO_2)_3(OH)_7^-$ would be a very important species under the experimental conditions for many other studies. Sutton [49SUT] also proposed the species $(UO_2)_3(OH)_6(aq)$ and $(UO_2)_3(OH)_8^{2-}$ but there is little experimental evidence for their existence. Bruno and Sandino [89BRU/SAN] invoked $UO_2(OH)_3^-$ and $(UO_2)_3(OH)_7^-$ to explain their results for the solubility of schoepite between pH values of 7 and 9, and Sandino [91SAN] calculated a formation constant for $UO_2(OH)_3^-$ in their study of phosphate complexation of dioxouranium(VI). Recently, based on a scoping study, Viljoen [87VIL] proposed $(UO_2)_9(OH)_{19}^-$ ($\log_{10}\beta_{19,9} \approx -69.7$) as the major polymeric anionic hydrolysis species of dioxouranium(VI).

In a search for a useable value for ${}^*\beta^\circ_{7,3}$, Tripathi [84TRI] reanalysed the data of Pongi, Double and Hurwic [80PON/DOU], Tsymbal [69TSY], and Sylva and Davidson [79SYL/DAV2]. He obtained $\log_{10}{}^*\beta_{7,3} = -28.84$ (0.2 M $NaNO_3$), $\log_{10}{}^*\beta_{7,3} = -27.9$ (0.1 M $NaClO_4$) and -28.9 (0.1 M $NaClO_4$), respectively. As noted by Tripathi [84TRI] and in Appendix A of this review, there are serious problems in using the data from any of these studies to obtain ${}^*\beta_{7,3}$. Tripathi's selected value is $\log_{10}{}^*\beta^\circ_{7,3} = -(28.34 \pm 0.03)$ based on an extended Debye-Hückel extrapolation to $I = 0$. If this value is correct, it is expected that $(UO_2)_3(OH)_7^-$ would be a major uranium species in solutions with a total uranium concentration $\sim 10^{-2}$ M and a $[CO_3]_T$:$[U]_T$ ratio near 3.0, *i.e.*, in solutions similar to those used in several studies designed to determine formation constants of carbonate complexes of UO_2^{2+} (*e.g.*, [56MCC/BUL, 69TSY, 82MAY]). Except for Tsymbal's work [69TSY], no evidence for $(UO_2)_3(OH)_7^-$ has been claimed in such studies. Reanalysis of the data from Ref. [82MAY] provide no further support for this species. Thus, the formation constant for $(UO_2)_3(OH)_7^-$ is probably smaller than estimated by Tripathi [84TRI]. Bruno and Sandino [89BRU/SAN] report a value of $\log_{10}{}^*\beta^\circ_{7,3} = -(31.4 \pm 0.1)$, but do not provide the raw data, nor do they discuss details of their model selection (see Appendix A); more recent data were reported by Sandino [91SAN]. In view of the strong qualitative evidence for a polymeric anionic hydrolysis species, this review accepts the existence of $(UO_2)_3(OH)_7^-$, and a value of

$$\log_{10}{}^*\beta^\circ_{7,3}(\text{V.7}, m = 3, n = 7, 298.15\,\text{K}) = -31 \pm 2.$$

The consistent Gibbs energy of formation is calculated to be

$$\Delta_f G^\circ_m((UO_2)_3(OH)_7^-, \text{aq}, 298.15\,\text{K}) = -(4340.7 \pm 12.6)\,\text{kJ} \cdot \text{mol}^{-1}.$$

At very low total solution concentrations of uranium, it would be expected that monomeric species $UO_2(OH)_n^{2-n}$ would predominate over polymeric species. However, no direct evidence for such species has been found in neutral and weakly basic solutions. Data from solubility [72NIK/SER] and potentiometric studies of solutions at temperatures above 25°C [76NIK] were interpreted using schemes including the species $UO_2(OH)_2(aq)$. However, these authors neglected one or more polymeric species (*e.g.*, $(UO_2)_3(OH)_5^+$) that may be important for their experimental conditions. Furthermore, these studies provide no real evidence that the species more highly hydrolysed than UO_2OH^+ or $(UO_2)_2(OH)_2^{2+}$ is really $UO_2(OH)_2(aq)$.

Stary's work [60STA], done using an extraction method, indicates the probable formation of a neutral hydrolysis species (20°C, $[U]_T \leq 2 \times 10^{-4}$ M, pH ≈ 6) and an anionic species in more basic aqueous solution. There are insufficient data to determine if the neutral and anionic species are monomeric (as proposed by Stary [60STA]) or polymeric.

If it is assumed that $(UO_2)_3(OH)_7^-$ and $UO_2(OH)_4^{2-}$ are the only major species at $\overline{n} = 3.0$ in Musikas' study [72MUS], a rough value of $\log_{10}{}^*\beta_4^\circ = -(33 \pm 2)$ is calculated using the value of $\log_{10}{}^*\beta_{7,3}^\circ$ accepted in this review. There is no unambiguous evidence to confirm the existence of $UO_2(OH)_2(aq)$, nevertheless, an upper limit can be assigned to the formation constant for this species. The maximum value for the equilibrium constant of Reaction (V.9),

$$UO_3 \cdot 2H_2O(cr) \quad \rightarrow \quad UO_2(OH)_2(aq) + H_2O(l), \qquad (V.9)$$

that is compatible with the 25°C solubility data of Nikitin *et al.* [72NIK/SER] is $\log_{10} K(V.9) = -5.5$. Thus, a limiting value of $\log_{10}{}^*\beta_2^\circ \leq -10.3$ is selected in this review, leading to $\Delta_f G_m^\circ(UO_2(OH)_2, aq, 298.15\,K) \geq -1368\,kJ \cdot mol^{-1}$.[4]

The value of $\log_{10}{}^*\beta_3^\circ$ must be of the order of -20, or smaller, to permit agreement with the qualitative study of Musikas [72MUS]. It appears that $\log_{10}{}^*\beta_3 = -(19.09 \pm 0.27)$ from the phosphate complexation study of Sandino [91SAN] is a well defined value. The value was corrected by Sandino [91SAN] to $I = 0$ using the ion interaction coefficients $\varepsilon_{(UO_2^{2+},ClO_4^-)} = 0.46$, $\varepsilon_{(UO_2(OH)_3^-,\ Na^+)} = -0.09$ and $\varepsilon_{(H^+,ClO_4^-)} = 0.14$. This extrapolation resulted in $\log_{10}{}^*\beta_3^\circ = -(19.18 \pm 0.29)$. This value of the equilibrium constant is accepted in this review, but the uncertainty is increased to ± 0.4 to reflect uncertainties in the model selection.

$$\log_{10}{}^*\beta_3^\circ(V.7, m = 1, n = 3, 298.15\,K) \quad = \quad -19.2 \pm 0.4$$

From this, the value

$$\Delta_f G_m^\circ(UO_2(OH)_3^-, aq, 298.15\,K) \quad = \quad -(1554.4 \pm 2.9)\,kJ \cdot mol^{-1}$$

is calculated. None of these values is inconsistent with potentiometric data obtained in acidic solutions.

V.3.2.1.4. Other polymeric cationic U(VI) hydrolysis species

Although several additional polymeric cations have been proposed as hydrolysis species in the dioxouranium(VI) system, only $(UO_2)_2OH^{3+}$, $(UO_2)_4(OH)_6^{2+}$ and $(UO_2)_4(OH)_7^+$ have been reported repeatedly (see Table V.5). The formation of $(UO_2)_2OH^{3+}$ was initially reported from potentiometric studies in which high uranium concentrations were used [63DUN/HIE, 63HIE/ROW]. More recently, however, Frei and Wendt [70FRE/WEN] presented good kinetic evidence for the existence of $(UO_2)_2OH^{3+}$ in a 0.5 M nitrate medium. At 3°C they determined $\log_{10}{}^*\beta_{1,2} =$

[4] A paper by Choppin and Mathur [91CHO/MAT], received after the draft of this section of the review was completed, reports $\log_{10}{}^*\beta_2 = -12.4$ in 0.1 M $NaClO_4$, suggesting a value of $\log_{10}{}^*\beta_2^\circ = -12$ and, hence, $\Delta_f G_m^\circ(UO_2(OH)_2$, aq, 298.15 K$) = -1360\,kJ \cdot mol^{-1}$. This is consistent with $UO_2(OH)_2(aq)$ being less stable than the (upper) stability limit proposed in the present review.

$\log_{10}{}^*\beta_{2,2} + 2.89$. If $\log_{10}{}^*\beta^\circ_{2,2} = -5.62$ at 25°C and $I = 0$, and a similar equation is used, the value $\log_{10}{}^*\beta_{1,2} \approx -2.7$ is obtained. The values of ${}^*\beta_{1,2}(\approx 10^{-4})$ at high ionic strength [63DUN/HIE, 63HIE/ROW] are reasonably compatible with this value. However, Tsymbal [69TSY] reported a value almost an order of magnitude larger (from potentiometric measurements in 0.1 M $NaClO_4$, see Appendix A). If ${}^*\beta^\circ_{1,2}$ (and ${}^*\beta_{1,2}$) were as large as 10^{-2}, it would be expected the species would be important in several other studies. This review selects the value

$$\log_{10}{}^*\beta^\circ_{1,2}(\text{V.7}, m = 2, n = 1, 298.15\,\text{K}) = -2.7 \pm 1.0.$$

This value is used to calculate the Gibbs energy of formation,

$$\Delta_f G^\circ_m((UO_2)_2OH^{3+}, \text{aq}, 298.15\,\text{K}) = -(2126.8 \pm 6.7)\,\text{kJ}\cdot\text{mol}^{-1}.$$

Both $(UO_2)_4(OH)_6^{2+}$ and $(UO_2)_4(OH)_7^+$ have been proposed on the basis of potentiometric studies [60GUS/RIC, 63DUN/HIE, 79LAJ/PAR, 79MIL/ELK, 79SYL/DAV2, 84KOT/EVS]. It is apparent that at least one polymeric species aside from $(UO_2)_2(OH)_2^{2+}$, $(UO_2)_3(OH)_4^{2+}$ and $(UO_2)_3(OH)_5^+$ must be invoked to explain pH titration curves above pH 4 to 5. However, it should be noted that it is difficult to envisage reasonable structures (*e.g.*, see Ref. [65ISR]) for species such as $(UO_2)_4(OH)_n^{8-n}$ ($n = 6$ and 7), and constants for these species should be regarded with skepticism. It also seems that none of the groups proposing $(UO_2)_4(OH)_6^{2+}$ or $(UO_2)_4(OH)_7^+$ considered $(UO_2)_3(OH)_7^-$ as a possible species. The careful work of Sylva and Davidson [79SYL/DAV2] provides support for $(UO_2)_4(OH)_7^+$ in the hydrolysis scheme in preference to $(UO_2)_4(OH)_6^{2+}$ or to including both species. Deschenes *et al.* [86DES/KRA] reached similar conclusions from their study at 37°C (see below), although their value for ${}^*\beta_{7,4}$ is almost two orders of magnitude larger than that found by Sylva and Davidson for conditions that were not extremely different. Tripathi's reanalysis of the data [84TRI] shows that $\log_{10}{}^*\beta_{7,4}$ is not changed by more than 0.1 if $(UO_2)_3(OH)_7^-$ is added to the model (even with a larger value of ${}^*\beta_{7,3}$ than chosen in this review). In this review, Sylva and Davidson's [79SYL/DAV2] $\log_{10}{}^*\beta_{7,4} \approx -22.76$ in 0.1 M KNO_3 is crudely corrected to $I = 0$ using the extended Debye-Hückel portion of the medium correction used in this review. This results in the selected value

$$\log_{10}{}^*\beta^\circ_{7,4}(\text{V.7}, m = 4, n = 7, 298.15\,\text{K}) = -21.9 \pm 1.0,$$

where the uncertainty is an estimate.

From this value, the Gibbs energy of formation is calculated,

$$\Delta_f G^\circ_m((UO_2)_4(OH)_7^+, \text{aq}, 298.15\,\text{K}) = -(5345.2 \pm 9.0)\,\text{kJ}\cdot\text{mol}^{-1}.$$

As an illustration, solubility, predominance and distribution diagrams of the dioxouranium(VI) hydroxide system in the range $4.5 < \text{pH} < 10$ are presented in Figures V.4, V.5 and V.6. For a correct interpretation of these figures, the following comments may be useful:

- Figure V.5 represents an experimental modelling situation. It shows that $UO_2(OH)_2(aq)$ is the predominant species in the range $5.1 < pH < 9.7$. In the range $6 < pH < 9$ schoepite precipitates if the total uranium concentration exceeds $10^{-5.5}$ M. The only experimental study performed in this region is the solubility study of Nikitin *et al.* [72NIK/SER]. From the discussion in Appendix A it is clear that this study is not very precise, and therefore only an upper limit for the formation constant of $UO_2(OH)_2(aq)$ can be given.

- Figures V.4 and V.6 do not represent a real experimental modelling situation because the calculations were made by assuming that no solid phase is formed. This was done in order to illustrate the areas of predominance and relative concentrations of the various dioxouranium(VI) species in a homogeneous aqueous phase.

Figure V.4: Predominance diagram of the dioxouranium(VI) hydroxide system at 25°C in the range $4.5 < pH < 10$. The precipitation of solid phases is suppressed.

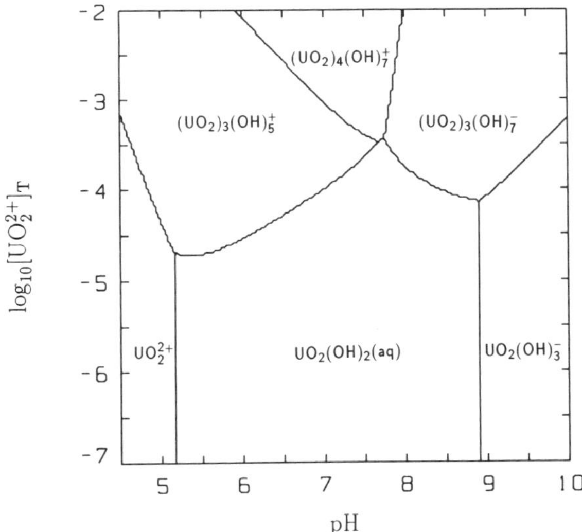

V.3.2.1.5. Temperature dependence of U(VI) hydrolysis constants

Only a few studies have been done of dioxouranium(VI) hydrolysis at temperatures outside the range of 20 to 30°C (Table V.6). There is agreement that as temperature increases the formation of monomeric hydrolysis species, particularly UO_2OH^+, is enhanced at the expense of the polymeric species. However, there is considerable disagreement as to the major species and the values of their formation constants. In general, the results at high temperatures are based on very few experimental measurements. The main exceptions are the potentiometric study by Baes and Meyer

Figure V.5: Solubility and predominance diagram of the dioxouranium(VI) hydroxide system at 25°C in the range 4.5 < pH < 10. The solubility limiting phase under the conditions covered in the graph is schoepite, $UO_3 \cdot 2H_2O(cr)$.

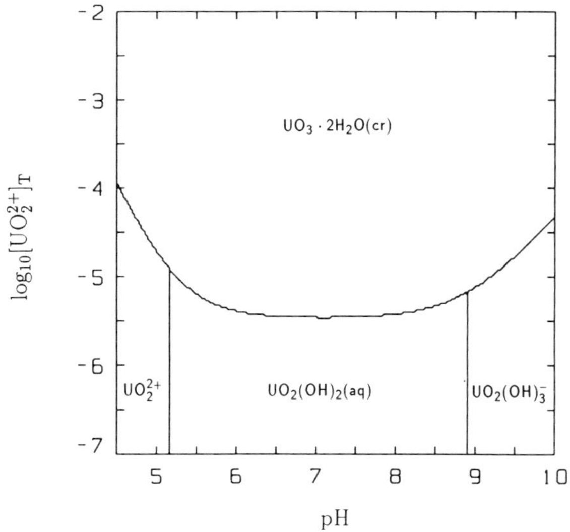

Figure V.6: Distribution diagram of the dioxouranium(VI) hydroxide system at 25°C in the range 4.5 < pH < 10. The precipitation of solid phases is suppressed.

$$[UO_2^{2+}]_T \quad = \quad 10^{-5}\ M$$

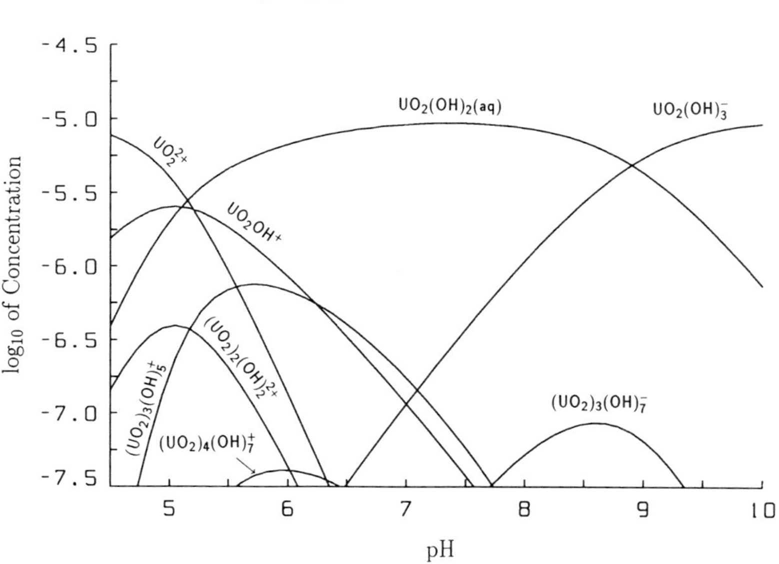

[62BAE/MEY] and the solubility work of Lietzke and Stoughton [60LIE/STO]. The only reliable calorimetric study of uranium hydrolysis [68ARN/SCH] is for a medium of 3 M $NaClO_4$(298.15 K).

Baes and Meyer [62BAE/MEY] analysed their potentiometric data in terms of three species, UO_2OH^+, $(UO_2)_2(OH)_2^{2+}$, and $(UO_2)_3(OH)_5^+$, but Nikolaeva [71NIK, 76NIK] omitted the trimer in analysis of data from experiments for the temperature range 50 to 150°C, as did Hearne and White [57HEA/WHI] for their results at 40°C. Deschenes *et al.* [86DES/KRA] concluded that the major species at 37°C (0.15 M NaCl) are $(UO_2)_2(OH)_2^{2+}$, $(UO_2)_3(OH)_4^{2+}$, $(UO_2)_3(OH)_5^+$ and $(UO_2)_4(OH)_7^+$. The $(UO_2)_3(OH)_5^+$ species may not be significant above 100°C, but omission of this species in the analyses of data for 50, 70 and 90°C by Nikitin *et al.* [72NIK/SER] and Nikolaeva [71NIK, 76NIK] is unfortunate. However, it can be shown that $UO_2(OH)_2$(aq), if it has a formation constant in the range suggested in these papers, would not be a major species in the solutions of Baes and Meyer at 94°C [62BAE/MEY] (because of the relatively high total uranium concentrations used by this group).

The values for $^*\beta_1$ and $^*\beta_{2,2}$ in Ref. [72NIK/SER] are fitted values based on a reanalysis of the data in an unavailable report by Nikolaeva, Antipina and Pastukova [68NIK/ANT]. The values for these constants given in Refs. [71NIK] and [76NIK] are apparently fitted values (not those actually reported as being calculated from the raw data for each separate temperature) from that same report. Values for $^*\beta_{2,1}$ were obtained from the data for 100, 125 and 150°C in Ref. [71NIK]. It is not clear if these are the values criticised in Ref. [72NIK/SER], or whether the data in Table 1 of Ref. [71NIK] are from a different set of experiments.

Ryzhenko, Naumov and Goglev [67RYZ/NAU] measured the conductance of reasonably concentrated aqueous solutions (> 0.01 M) of dioxouranium(VI) nitrate at temperatures from 100 to 200°C. At these high total uranium concentrations the concentrations of the monomeric hydrolysis species were assumed small, and the data were analysed by including only the dimer in the hydrolysis model. At higher temperatures, at which the approximations in the conductance study are more likely to be valid, the authors found values for $^*\beta_{2,2}$ that are in reasonable agreement with those of Lietzke and Stoughton [60LIE/STO] (based on differential solubility measurements). From comparison with the other reported constants, the value for $^*\beta_{2,2}$ at 100°C from Ryzhenko, Naumov and Goglev [67RYZ/NAU] seems too small.

Enthalpies and entropies of reaction, and calculated entropies of the hydrolysis species are shown in Table V.8. The data in Table V.6 are fitted to the equation

$$\log_{10}{}^*\beta_{n,m}(T) = \left(\frac{298.15}{T}\right)\log_{10}{}^*\beta_{n,m}(298.15\,\mathrm{K}) + \Delta_r S_m\left(\frac{T-298.15}{RT\ln(10)}\right). \quad (V.10)$$

Except where noted, the equilibrium constant at 25°C is fixed using previously selected values for the Gibbs energies at 25°C. The use of this equation implies that $\Delta_r C_{p,m}$ is zero for each reaction. This is a very crude assumption; however, in no case does the precision of the available data warrant the use of an extra fitting parameter. Because of the sparse experimental detail in many of the papers, the fact that in several cases only fitted rather than raw data are available, and the use of incomplete

Table V.8: Enthalpies and entropies of the reaction $m\mathrm{UO}_2^{2+} + n\mathrm{H_2O(l)} \rightleftharpoons (\mathrm{UO_2})_m(\mathrm{OH})_n^{2m-n} + n\mathrm{H}^+$ and standard entropies of $(\mathrm{UO_2})_m(\mathrm{OH})_n^{2m-n}$.

$n:m$	t (°C)	$\Delta_r H_m^\circ$ (a) (kJ · mol^{-1})	$\Delta_r S_m^\circ$ (a) (J · K^{-1} · mol^{-1})	S_m° (J · K^{-1} · mol^{-1})	Reference
1:1	25-94	46 ± 17 (b)	44 ± 42 (b)	16 (c)	[62BAE/MEY]
	25-125	53 (d)	76 (d)	48 (d)	[71NIK]
		(50)	(70)	(42)	
	25-125	58 (d)	94 (d)	66 (d)	[72NIK/SER] *ex* [71NIK]
		(53)	(82)	(54)	
2:2	100-150	37 (d)	15 (d,e)	−42 (d)	[60LIE/STO]
	25-94	43 ± 4 (b)	30 ± 17 (b)	−27 (c)	[62BAE/MEY]
	150-200	28 (d)	−14 (d)	−70 (d)	[67RYZ/NAU]
	25	39.7 ± 0.3 (f)	18.0 ± 5.8 (f)	−39 (c)	[68ARN/SCH]
	25-125	40 (d)	26 (d)	−30 (d)	[71NIK]
		(41)	(38)	(−18)	
	25-125	54 (d)	73 (d)	16 (d)	[72NIK/SER] *ex* [71NIK]
		(56)	(78)	(21)	
2:1	100-150	122 (d)	174 (d)	216 (d)	[71NIK]
		(112)	(149)	(191)	
	50-200	20 (a)	−54 (a)	69	[72NIK/SER]
5:3	25-94	105 ± 13 (b)	42 ± 42 (b)	97 (c)	[62BAE/MEY]
	25	102 ± 4 (f)	25 ± 18 (f)	80 (c)	[68ARN/SCH]

(a) All enthalpies and entropies of reaction are for the appropriate hydrolysis reaction, Eq. (V.7), except for the solubility work of Nitkitin *et al.* [72NIK/SER] where the parameters are for the dissolution of β-$\mathrm{UO_3} \cdot \mathrm{H_2O}$ to form $\mathrm{UO_2(OH)_2(aq)}$ (values from Ref. [72NIK/SER] are not compatible with the limiting value at 25°C, see Appendix A).

(b) The value of $\Delta_r S_m$ or $\Delta_r H_m$ is calculated from the equilibrium constants at the two temperatures for an aqueous medium of 0.5 m $\mathrm{KNO_3}$. The value used for the equilibrium constant at 25°C is not linked to the value selected in this review. Uncertainties are twice those reported by the authors.

(c) Using the approximation that $\Delta_r S_m(I \neq 0) = \Delta_r S_m^\circ$.

(d) From a fit of the data including the selected value of the hydrolysis constant at 25°C as a fixed point. Values in parentheses are from fitting the data without this constraint.

(e) The value in the original paper (in the text above Table 3 in Ref. [60LIE/STO]) appears to be in error.

(f) For a medium of 3.0 M $\mathrm{NaClO_4}$.

(or incorrect) sets of hydrolysis species in the data analyses, it is difficult to properly assign uncertainties to the values in Tables V.6 and V.8.

The values resulting from the reanalysis by Nikitin *et al.* [72NIK/SER] of the data of Nikolaeva [68NIK/ANT, 71NIK] differ markedly from the values reported by Nikolaeva [71NIK, 76NIK]. Thus, the uncertainties in the values from these experiments are at least equal to these differences: $18\,J\cdot K^{-1}\cdot mol^{-1}$ in $S^\circ_m(UO_2OH^+$, aq, 298.15 K); $46\,J\cdot K^{-1}\cdot mol^{-1}$ in $S^\circ_m((UO_2)_2(OH)_2^{2+}$, aq, 298.15 K). It should also be noted that the reanalysis tends, in both cases, to move the value of the entropy of the hydrolysis species further from other reported values.

The calorimetric study [68ARN/SCH] and the potentiometric study of Baes and Meyer [62BAE/MEY] are in fairly close agreement considering the different media. For $(UO_2)_2(OH)_2^{2+}$ and $(UO_2)_3(OH)_5^+$ the appropriately weighted average values of the entropies from these studies are used in this review. The uncertainties reflect the lack of correction for medium effects. For UO_2OH^+ only the value from Ref. [62BAE/MEY] is considered useable because of the sparse data and poor model used in Ref. [71NIK]. The uncertainty in the entropy was raised to reflect the lack of confirmation of this value. No entropy is selected in this review for $UO_2(OH)_2(aq)$. The selected values for the entropies are

$$\begin{aligned}
S^\circ_m(UO_2OH^+, aq, 298.15\,K) &= (17 \pm 50)\,J\cdot K^{-1}\cdot mol^{-1}\\
S^\circ_m((UO_2)_2(OH)_2^{2+}, aq, 298.15\,K) &= -(38 \pm 15)\,J\cdot K^{-1}\cdot mol^{-1}\\
S^\circ_m((UO_2)_3(OH)_5^{+}, aq, 298.15\,K) &= (83 \pm 30)\,J\cdot K^{-1}\cdot mol^{-1}.
\end{aligned}$$

They should not be used for extrapolations above 150°C, and if used to calculate equilibrium constants or Gibbs energies of reaction at temperatures other than 25°C, they should not be used in combination with heat capacities for these or other species.

V.3.2.1.6. *Oxouranium(VI), UO^{4+}*

The species UO^{4+} [68NEM/PAL] is not credited in this review (see Appendix A) as it has not been otherwise confirmed.

V.3.2.1.7. *Solvolysis in D_2O (l)*

Maeda *et al.* [70MAE/KAK, 72MAE/AMA] studied the solvolysis of UO_2^{2+} in D_2O (3 M $NaClO_4$). From these papers, and the values selected in this review for the hydrolysis constants of UO_2^{2+} (in H_2O), values of $\log_{10}{}^*\beta^\circ_{2,2} = -(6.3 \pm 0.2)$, $\log_{10}{}^*\beta^\circ_{5,3} = -(17.2 \pm 0.4)$, and $\log_{10}{}^*\beta^\circ_{4,3} = -(13.2 \pm 1.0)$ are estimated for solvolysis of UO_2^{2+} in D_2O.

V.3.2.2. *U(V) hydroxide complexes*

No aqueous models which need to call upon UO_2^+ hydroxide species have been proposed for interpreting experimental data. The regions in which UO_2^+ has been proposed as a significant species are at pH < 5. By analogy with NpO_2^+, no hydrolysis of UO_2^+ would be expected under these conditions. In higher pH regions, UO_2^+ hydroxide species are not expected to be found at significant concentrations because of disproportionation of dioxouranium(V). Therefore, this review finds no credible UO_2^+ hydroxide species.

V.3.2.3. U(IV) hydroxide complexes

Hydrolysis of the U^{4+} ion is extensive except in strongly acidic solutions, and precipitation of extremely insoluble uranium dioxide or hydroxide occurs readily from uranium(IV) solutions as pH is increased. Even in strongly basic solutions (pOH < 2), the equilibrium solution concentration of uranium over such solids remains very low. These factors have limited the number of reliable studies of the hydrolysis species and their equilibrium constants ${}^*\beta_{n,m}$ for the reactions

$$m\mathrm{U}^{4+} + n\mathrm{H_2O(l)} \rightleftharpoons \mathrm{U}_m(\mathrm{OH})_n^{4m-n} + n\mathrm{H}^+. \qquad (V.11)$$

V.3.2.3.1. UOH^{3+}

Information about the (1,1) monomeric hydrolysis species UOH^{3+} has primarily been derived from studies of acidic solutions of uranium(IV). There is little to be added to the summary given by Baes and Mesmer [76BAE/MES]. The spectrophotometric study of Kraus and Nelson [50KRA/NEL] remains the best available source of data on which to base an estimate of ${}^*\beta_1^\circ$(V.11, $m = 1, n = 1$) at 25°C. The value found spectrophotometrically by Sullivan and Hindman [59SUL/HIN] (ClO_4^-, 2 M) agrees well with the earlier work. The values of Nikolaeva [78NIK] (0.25, 0.36 and 0.48 m $HClO_4$) are smaller than those found by Kraus and Nelson [55KRA/NEL], while the value of Betts [55BET] ($\log_{10}{}^*\beta_1 = -(1.1 \pm 0.1)$ at 24.7°C and $I = 0.19$ M) is markedly larger. Hietanen's potentiometric study [56HIE, 56HIE2] was done at higher ionic strength, and the reanalysis of her data by Baes and Mesmer [76BAE/MES] gave a value of $\log_{10}{}^*\beta_1 = -(2.1 \pm 0.3)$ (see Appendix A). A recent thermal lensing spectrophotometric study by Grenthe, Bidoglio and Omenetto [89GRE/BID] ($I = 3$ M, $(Na^+, H^+)ClO_4$) with low total uranium concentrations such that monomeric hydrolysis species were predominant, gave results that were consistent with those of Kraus and Nelson [50KRA/NEL]. The formation constant for UOH^{3+} reported by Grenthe, Bidoglio and Omenetto [89GRE/BID] for the high ionic strength medium, $\log_{10}{}^*\beta_1 = -(1.65 \pm 0.05)$, is in only fair agreement with the value from Hietanen's study. The value of $\log_{10}{}^*\beta_1^\circ = -(0.51 \pm 0.03)$ and of the ion interaction coefficient difference reported by Grenthe, Bidoglio and Omenetto [89GRE/BID], rely primarily on the results of Kraus and Nelson [50KRA/NEL] in perchlorate medium.

The values of $\log_{10}{}^*\beta_{n,m}$ in Table V.9 correspond to those given in the original papers, although it is clear that there are problems with some of the data analyses. A brief assessment of each reference is given in Appendix A. For the formation of UOH^{3+},

$$\mathrm{U}^{4+} + \mathrm{H_2O(l)} \rightleftharpoons \mathrm{UOH}^{3+} + \mathrm{H}^+, \qquad (V.12)$$

Table V.9: Experimental equilibrium data for the uranium(IV) hydroxide system, according to the equilibria

$$m\mathrm{U}^{4+} + n\mathrm{H_2O(l)} \rightleftharpoons \mathrm{U}_m(\mathrm{OH})_n^{4m-n} + n\mathrm{H}^+.$$

$n:m$	Method	Medium	t (°C)	$\log_{10}{}^*\beta_{n,m}$ [(a)]	Reference
1:1	sp	2.00 M (Na/$HClO_4$)	25	-1.63 ± 0.09	[50KRA/NEL]
1:1		1.01 M (Na/$HClO_4$)		-1.56 ± 0.09	
1:1		0.55 M (Na/$HClO_4$)		-1.45 ± 0.09	
1:1		0.52 M (Na/$HClO_4$)		-1.51 ± 0.09	
1:1		0.52 M (Na/$HClO_4$)		-1.50 ± 0.09	
1:1		0.50 M (Na/$HClO_4$)		-1.54 ± 0.09	
1:1		0.27 M (Na/$HClO_4$)		-1.36 ± 0.09	
1:1		0.12 M (Na/$HClO_4$)		-1.29 ± 0.09	
1:1		0.12 M (Na/$HClO_4$)		-1.23 ± 0.09	
1:1		0.11 M (Na/$HClO_4$)		-1.28 ± 0.09	
1:1		0.11 M (Na/$HClO_4$)		-1.22 ± 0.09	
1:1		0.063 M (Na/$HClO_4$)		-1.14 ± 0.17	
1:1		0.045 M (Na/$HClO_4$)		-1.10 ± 0.17	
1:1		0.035 M (Na/$HClO_4$)		-1.00 ± 0.17	
1:1		0.033 M (Na/$HClO_4$)		-1.07 ± 0.17	
1:1		0.017 M (Na/$HClO_4$)		-0.92 ± 0.17	
1:1		2.00 M (Na/HCl)		-1.82	
1:1		1.01 M (Na/HCl)		-1.73	
1:1		0.67 M (Na/HCl)		-1.59	
1:1		0.53 M (Na/HCl)		-1.63	
1:1		0.52 M (Na/HCl)		-1.58	
1:1		0.51 M (Na/HCl)		-1.59	
1:1		0.27 M (Na/HCl)		-1.49	
1:1		0.12 M (Na/HCl)		-1.36	
1:1		0.12 M (Na/HCl)		-1.35	
1:1		0.11 M (Na/HCl)		-1.34	
1:1	sp	0.065 M (Na/HCl)	25	-1.21	
1:1		0.045 M (Na/HCl)		-1.17	
1:1		0.040 M (Na/HCl)		-1.10	
1:1		0.035 M (Na/HCl)		-1.12	
1:1		0.033 M (Na/HCl)		-1.14	
1:1		0.025 M (Na/HCl)		-1.02	
1:1		0.024 M (Na/HCl)		-1.01	
1:1	dis	0.0	25	-0.68[(b)]	
1:1	sp	0.19 M (Na/$HClO_4$)	15.2	-1.4 ± 0.2	[55BET]
1:1	sp	0.19 M (Na/$HClO_4$)	20.4	-1.2 ± 0.2	
1:1	sp	0.19 M (Na/$HClO_4$)	24.7	-1.1 ± 0.2	
1:1	sp	0.5 M (Na/$HClO_4$)	10.0	-1.90 ± 0.09	[55KRA/NEL]
1:1	sp	0.5 M (Na/$HClO_4$)	43.0	-1.00 ± 0.09	
1:1	pot	3 M ($NaClO_4$)	25	-2.0	[56HIE]
$3n:(1+n)$				$-1.2 - 3.4n$	

Table V.9 (continued)

$n:m$	Method	Medium	t (°C)	$\log_{10}{}^*\beta_{n,m}$ [(a)]	Reference
1:1	pot	3 M ($NaClO_4$)	25	–1.95[(c)]	[56HIE2]
$3n$:(1+n)				$-1.22-3.4n$	
1:1	sp	2.0 M ($Na/HClO_4$)	25	-1.68 ± 0.2	[59SUL/HIN]
1:1	pot, sol	self (UF_4 soln)	25	–2.64	[61NIK/LUK]
2:1				–6.40	
3:1				–12.3	
4:1				–16.8	
1:1	prx	self?	25?	–1.52	[63VDO/ROM]
1:1	sp	1.00 M ($Na/HClO_4$)	20	–1.57	[64MCK/WOO]
1:1	sp	0.0	25	–1.11[(d)]	[67SOL/TSV]
2:1	prx	self?	25?	–1.80[(e)]	[69VDO/STE]
1:1	sp	0.5 M (NH_4/HCl)	25?	–1.56	[75DAV/EFR]
2:1				–4.11	
1:1		0.5 M ($Na/HClO_4$)		–1.74	
2:1				–4.16	
1:1	sp	var/self	25?	–1.28	
1:1	pot	3 M ($NaClO_4$)	25	-2.0 ± 0.3[(c)]	[76BAE/MES]
15:6				-16.9 ± 0.6	
1:1	sp	0.25 m ($HClO_4$)	25	-1.44 ± 0.2	[78NIK]
1:1		0.36 m ($HClO_4$)		-1.59 ± 0.2	
1:1		0.48 m ($HClO_4$)		-1.70 ± 0.2	
1:1		0.36 m ($HClO_4$)		-1.00 ± 0.2	
1:1		0.48 m ($HClO_4$)		-1.07 ± 0.2	
1:1		0.51 m ($HClO_4$)		-1.21 ± 0.2	
1:1		0.65 m ($HClO_4$)		-1.31 ± 0.2	
1:1		0.75 m ($HClO_4$)		-1.44 ± 0.2	
1:1		0.86 m ($HClO_4$)		-1.54 ± 0.2	
1:1		0.99 m ($HClO_4$)		-1.68 ± 0.2	
1:1		1.09 m ($HClO_4$)		-1.72 ± 0.2	
1:1		0.51 m ($HClO_4$)	75	-0.56 ± 0.2	
1:1		0.65 m ($HClO_4$)		-0.70 ± 0.2	
1:1		0.75 m ($HClO_4$)		-0.74 ± 0.2	
1:1		0.86 m ($HClO_4$)		-0.82 ± 0.2	
1:1		0.99 m ($HClO_4$)		-0.93 ± 0.2	
1:1		1.09 m ($HClO_4$)		-1.00 ± 0.2	
1:1		0.51 m ($HClO_4$)	100	-0.25 ± 0.2	
1:1		0.65 m ($HClO_4$)		-0.36 ± 0.2	
1:1		0.75 m ($HClO_4$)		-0.39 ± 0.2	
1:1		0.86 m ($HClO_4$)		-0.42 ± 0.2	
1:1		0.99 m ($HClO_4$)		-0.50 ± 0.2	
1:1		1.09 m ($HClO_4$)		-0.53 ± 0.2	
1:1		1.20 m ($HClO_4$)		-0.55 ± 0.2	
1:1		0.65 m ($HClO_4$)	125	-0.21 ± 0.2	
1:1		0.75 m ($HClO_4$)		-0.23 ± 0.2	
1:1		0.86 m ($HClO_4$)		-0.23 ± 0.2	
1:1		0.99 m ($HClO_4$)		-0.31 ± 0.2	

Table V.9 (continued)

$n:m$	Method	Medium	t (°C)	$\log_{10}{}^*\beta_{n,m}$ [(a)]	Reference
1:1		1.09 m ($HClO_4$)		-0.33 ± 0.2	[78NIK]
1:1		1.20 m ($HClO_4$)		-0.40 ± 0.2	
1:1		1.68 m ($HClO_4$)		-0.56 ± 0.2	
1:1	sp	1.09 m ($HClO_4$)	150	-0.09 ± 0.2	
1:1		1.20 m ($HClO_4$)		-0.28 ± 0.2	
1:1		1.68 m ($HClO_4$)		-0.43 ± 0.2	
3:1	sol	0.5 M ($NaClO_4$)	25	-1.1	[87BRU/CAS]
1:1	tls	3 M ($NaClO_4$)	25	-1.65 ± 0.1	[89GRE/BID]

(a) Uncertainty estimated in this review (see Appendix A).
(b) Using only the perchlorate data.
(c) Data from Hietanen [56HIE].
(d) A recalculation of the data in Kraus and Nelson [50KRA/NEL].
(e) A recalculation of the data in Vdovenko, Romanov and Shcherbakov [63VDO/ROM].

the selected value is obtained from a linear regression to $I = 0$, excluding the data from the studies in chloride media of Kraus and Nelson [50KRA/NEL] (see Appendix A).

$$\log_{10}{}^*\beta_1^\circ(\text{V.12}, 298.15\,\text{K}) = -0.54 \pm 0.06$$

The uncertainty is therefore a statistical uncertainty. Here, as elsewhere in this review, the inherent uncertainty in assuming the Debye-Hückel limiting law and specific ion interaction theory to be applicable to equilibria involving such highly charged species, was disregarded. From the slope of the regression line, $\Delta\varepsilon(\text{V.12}) = -(0.14 \pm 0.05)$ is obtained, which leads to the new ion interaction coefficient $\varepsilon_{(\text{UOH}^{3+},\text{ClO}_4^-)} = (0.48 \pm 0.08)$.

The selected constant corresponds to a Gibbs energy of reaction of $(3.1 \pm 0.3)\,\text{kJ}\cdot\text{mol}^{-1}$ for Reaction (V.12). Using the selected value of $\Delta_f G_m^\circ(U^{4+}, \text{aq}, 298.15\text{ K})$ and the CODATA [89COX/WAG] value of $\Delta_f G_m^\circ(H_2O, \text{l}, 298.15\,\text{K})$ a value of

$$\Delta_f G_m^\circ(\text{UOH}^{3+}, \text{aq}, 298.15\,\text{K}) = -(763.9 \pm 1.8)\,\text{kJ}\cdot\text{mol}^{-1}$$

is obtained.

Kraus and Nelson [50KRA/NEL, 55KRA/NEL] measured the first hydrolysis constant of U^{4+} at (10.0 ± 0.2)°C, (25.0 ± 0.2)°C and (43.0 ± 0.3)°C in perchlorate media ($I = 0.5$ M). Betts [55BET] also measured the equilibrium constant of Reaction (V.12) spectrophotometrically at 15.2, 20.4 and 24.7°C ($I = 0.19$ M (Na,H)ClO_4). Values of $\Delta_r H_m(\text{V.12})$ and $\Delta_r S_m(\text{V.12})$ obtained in the two studies are in good agreement (see Appendix A). The values of $\Delta_r H_m$(V.12) are also similar to the value of 46 kJ · mol^{-1} reported by Fontana [47FON], but as noted by Betts [55BET] this agreement is probably fortuitous (see Appendix A). Nikolaeva [78NIK] measured the hydrolysis constant in perchloric acid solutions over a much wider range of temperatures, 25 to 150°C. The enthalpies and entropies of reaction from the data in this paper ($I = 0.5$ to 1.0 m) agree well with those obtained from the earlier studies over more limited temperature ranges [55BET, 55KRA/NEL] ($I = 0.19$ and 0.5 M). The values

$$\Delta_r H_m^\circ(\text{V.12, 298.15 K}) = (46.9 \pm 9.0)\,\text{kJ}\cdot\text{mol}^{-1}$$
$$\Delta_r S_m^\circ(\text{V.12, 298.15 K}) = (147 \pm 30)\,\text{J}\cdot\text{K}^{-1}\cdot\text{mol}^{-1}$$

are selected, based on a weighted average of results, extrapolated to $I = 0$ (see Appendix A), from the studies of Nikolaeva [78NIK] and Kraus and Nelson [55KRA/NEL]. In the calculations the equilibrium constant at 25°C is constrained to the value selected in this review, and a value of 0.0 J · K^{-1} · mol^{-1} is assigned to $\Delta_r C_{p,m}^\circ$. The uncertainty in $\Delta_r H_m^\circ$(V.12) is based on the estimated uncertainties in $\Delta_r S_m^\circ$(V.12) and $\log_{10}{}^*\beta_1^\circ$.

The following enthalpy of formation and entropy values are derived:

$$\Delta_f H_m^\circ(\text{UOH}^{3+}, \text{aq}, 298.15\,\text{K}) = -(830.1 \pm 9.6)\,\text{kJ}\cdot\text{mol}^{-1}$$
$$S_m^\circ(\text{UOH}^{3+}, \text{aq}, 298.15\,\text{K}) = -(200 \pm 33)\,\text{J}\cdot\text{K}^{-1}\cdot\text{mol}^{-1}.$$

V.3.2.3.2. $U(OH)_4(aq)$ and $U(OH)_5^-$

Considerable effort has been expended on the determination of the solubility of uranium dioxide, "amorphous" uranium dioxide, "hydrous uranium(IV) oxide" and/or uranium(IV) hydroxide [57GAY/LEI, 60STE/GAL, 81TRE/CHE, 83RYA/RAI, 85PAR/POH, 87BRU/CAS, 88PAR/POH, 90RAI/FEL] in basic solutions (*cf.* Section V.3.3.2). In theory, such measurements should establish whether these solids are amphoteric, and the minimum solubility (as a function of pH) should give a limit for the stability of the neutral aqueous species, the monomeric form of which would be $U(OH)_4$(aq). In practice, difficulties in characterization of the surface of the solid, problems both with sorption and with removal of fine particles from solution and the use of analytical methods of insufficient sensitivity, have given solubility limits rather than actual solubilities.

There is no clear evidence for the amphoterism of UO_2(cr). The work of Gayer and Leider [57GAY/LEI] was properly criticized by Tremaine *et al.* [81TRE/CHE] and Ryan and Rai [83RYA/RAI]. Tremaine *et al.* [81TRE/CHE] proposed a value for $^*\beta_5$(V.11, $m = 1, n = 5$), but they were working near the limit of their analytical method. The data of Ryan and Rai [83RYA/RAI] show some increase in uranium(IV) solubility at high pH, but the authors specifically excluded an anionic species in their data analysis. Parks and Pohl [88PAR/POH] found the solubility of UO_2(cr) to be independent of pH above pH $\approx$ 4 for temperatures from 100° to 300°C. This review includes only a limiting value for $U(OH)_5^-$ in the selected data set. The species, if it exists, is apparently of minor importance.

If UO_2 is not amphoteric, then it can be assumed that the low uranium concentrations in basic solutions reported by Parks and Pohl [88PAR/POH], Tremaine *et al.* [81TRE/CHE], and Ryan and Rai [83RYA/RAI] represent upper limits on the equilibrium concentration of $U(OH)_4(aq)$ (or some neutral polymer thereof). Tremaine *et al.* [81TRE/CHE] used a well-characterized UO_2 solid, and measured the solubility in a flow system at a series of temperatures in the presence of hydrogen gas (to ensure the uranium remained in the +4 oxidation state). Ryan and Rai [83RYA/RAI] used other chemical reductants to produce an uncharacterized "hydrous uranium(IV) oxide" at room temperature. It is not clear if the holding reductants could have affected their results by increasing the rate of ripening of the precipitated solid, or by sorption of uranium species. Even lower solubilities were found by Parks and Pohl [85PAR/POH] using a static system (500 bars $H_2(g)$) and $UO_2(cr)$ from the same batch studied by Tremaine *et al.* [81TRE/CHE]. It must be concluded that the values obtained in Ref. [81TRE/CHE] merely reflect the lower limits of their method of uranium analysis. Tremaine *et al.* [81TRE/CHE] and Parks and Pohl [88PAR/POH] assumed that hydrogen at high temperatures ($\sim$ 300°C) ensured that the surface of the uranium oxide in contact with water was $UO_{2.00}$, and Parks and Pohl [88PAR/POH] did experiments to prove that the addition of dioxouranium(VI) to acidic solutions did not significantly increase the measured uranium solubility. However, hydrogen reduction of neutral dioxouranium(VI) solutions was reported [86BRU/GRE] to give a solid of composition near U_3O_8, and the kinetics of further reduction of the uranium solid may be slow.

The experiments of Parks and Pohl [88PAR/POH] showed changes in uranium solubility of less than an order of magnitude as the temperature was raised from 100 to 300°C. Indeed, recent papers by Red'kin *et al.* [89RED/SAV] and Dubessy *et al.* [87DUB/RAM] suggest that the solubility of UO_2 does not change substantially with temperature even in the temperature range 300 to 600°C. The solubility was also independent of the hydrogen ion concentration from pH $\approx$ 4 to pOH $\approx$ 2. The authors averaged the results (pH $>$ 4) for all temperatures to obtain an average solubility of $10^{-(9.47\pm0.56)}$ M. If the only major uranium solution species under these conditions is $U(OH)_4(aq)$, the assumption of a temperature independent solubility is equivalent to assuming $\Delta_r C_{p,m}(V.13) = 0.0\,J\cdot K^{-1}\cdot mol^{-1}$ at all temperatures, and $\Delta_r S^\circ_m(V.13) = -181.3\,J\cdot K^{-1}\cdot mol^{-1}$.

$$UO_2(cr) + 2H_2O(l) \rightleftharpoons U(OH)_4(aq) \qquad (V.13)$$

Use of these values with the values accepted for $UO_2(cr)$ (Section V.3.3.2.1) and water [89COX/WAG] results in a calculated value of $S^\circ_m(U(OH)_4, aq, 298.15\ K) = 36\,J\cdot K^{-1}\cdot mol^{-1}$ and an average value of $C^\circ_{p,m}(U(OH)_4, aq, 298.15\ K) \approx 220\,J\cdot K^{-1}\cdot mol^{-1}$. These are reasonable values for a neutral aqueous species.

If the solubility of the same solid were markedly higher (by more than one to two orders of magnitude) at 25°C, and the solution species assumed to be the same, the entropy of $U(OH)_4(aq)$ would, unreasonably, be markedly negative. Therefore, the solubility of the Parks and Pohl solid at 25°C is probably not greater than the average 100 to 300°C value by more than an order of magnitude. Choosing the uncertainties to reflect this, $\Delta_r G_m^\circ$(V.13, 298.15 K) is calculated to be $(54.1 \pm 6.5)\,\text{kJ}\cdot\text{mol}^{-1}$, and hence, $\Delta_f G_m^\circ(\text{U(OH)}_4, \text{aq}, 298.15\,\text{K}) = -(1452.1 \pm 6.6)\,\text{kJ}\cdot\text{mol}^{-1}$.

Bruno *et al.* [87BRU/CAS] measured the solubility of a so-called amorphous (actually partly crystalline) form of UO_2 at 25°C in 0.5 M $NaClO_4$. The solubility of this material was $10^{-(4.4\pm0.4)}$ M, independent of pH between pH values of 5.5 and 10.0. If the predominant uranium solution species is again $U(OH)_4(aq)$, the solubility in the neutral pH range should also be relatively independent of the ionic medium. For the reaction

$$UO_2(cr) + 4H^+ \rightleftharpoons U^{4+} + 2H_2O(l), \qquad (V.14)$$

the solubility product, $K_{s,0} = [U^{4+}]/[H^+]^4$, (determined potentiometrically in 3 M $NaClO_4$ [86BRU/FER]) for a similarly prepared solid was reported as $10^{(0.5\pm0.6)}$. The nature of the precipitated uranium oxide solids is discussed in Section V.3.3.2.2. The same authors calculated that the solubility product for a more crystalline form of UO_2 would be shifted from $10^{-1.2}$ to $10^{-1.6}$ on transfer from 3 M $NaClO_4$ to pure water [86BRU/FER]. If the same correction is applied to the value for the "amorphous" UO_2, the solubility product in pure water is calculated to be $^*K_{s,0}^\circ$(V.14, 298.15 K) = $10^{(0.1\pm0.7)}$. Thus, $\Delta_f G_m^\circ(UO_2$, "am", 298.15 K$) = -(1003.6 \pm 4.4)\,\text{kJ}\cdot\text{mol}^{-1}$. If the solid phase controlling the results in the potentiometric and solubility studies is the same, $\Delta_r G_m^\circ$ for the dissolution of UO_2("am") to form $U(OH)_4(aq)$ is $(25.1 \pm 2.3)\,\text{kJ}\cdot\text{mol}^{-1}$ and $\Delta_f G_m^\circ(\text{U(OH)}_4, \text{aq}, 298.15\,\text{K}) = -(1452.8 \pm 5.0)\,\text{kJ}\cdot\text{mol}^{-1}$.

The agreement in the values of $\Delta_f G_m^\circ(\text{U(OH)}_4, \text{aq}, 298.15\,\text{K})$ is unexpectedly good, especially considering the difficulties in characterising an amorphous solid as the one used by Bruno *et al.* [87BRU/CAS]. The weighted average value,

$$\Delta_f G_m^\circ(\text{U(OH)}_4, \text{aq}, 298.15\,\text{K}) = -(1452.5 \pm 8.0)\,\text{kJ}\cdot\text{mol}^{-1},$$

is selected in this review, the uncertainty being increased to allow for uncertainties in the nature of the solids and for compatibility with the values for other hydrolysis species (see below). In this review, to reflect the temperature independence of the experimentally measured solubilities of UO_2 [88PAR/POH] (also see Appendix A and Ref. [88LEM]), $\Delta_r C_{p,m}$(V.13) is assumed to be $0\,\text{J}\cdot\text{K}^{-1}\cdot\text{mol}^{-1}$ and

$$S_m^\circ(\text{U(OH)}_4,\ \text{aq},\ 298.15\ \text{K}) = (40 \pm 25)\,\text{J}\cdot\text{K}^{-1}\cdot\text{mol}^{-1}$$

is estimated. It follows that

$$C_{p,m}^\circ(\text{U(OH)}_4,\ \text{aq},\ 298.15\ \text{K}) = (205 \pm 80)\,\text{J}\cdot\text{K}^{-1}\cdot\text{mol}^{-1}.$$

The uncertainty in $C_{p,m}^\circ$(U(OH)$_4$, aq, 298.15 K) is selected to reflect the estimated uncertainty in S_m°(U(OH)$_4$, aq, 298.15 K). However, the values of S_m°(U(OH)$_4$, aq, 298.15 K), $C_{p,m}^\circ$(U(OH)$_4$, aq, 298.15 K) and $\Delta_f G_m^\circ$(U(OH)$_4$, aq, 298.15 K) are correlated. It should be noted that there is no direct evidence in either study that the neutral aqueous species is actually monomeric.

The limit

$$\Delta_f G_m^\circ(\text{U(OH)}_5^-, \text{aq}, 298.15\,\text{K}) > -1621\,\text{kJ}\cdot\text{mol}^{-1}$$

is based on the assumption that $a_{U(OH)_5^-} < a_{U(OH)_4(aq)}$ at pH $\leq$ 12 (*i.e.*, $\Delta_r G_m^\circ >$ 68.5 kJ $\cdot$ mol^{-1} for $U(OH)_4(aq) + H_2O(l) \rightleftharpoons U(OH)_5^- + H^+$).

As this review was in its final stages of preparation, Rai, Felmy and Ryan [90RAI/FEL] published a paper on the solubility of "$UO_2 \cdot xH_2O$(am)". This appears to be an excellent study, however, the results imply a value of $\Delta_f G_m^\circ(U(OH)_4$, aq, 298.15 K) that is approximatively 40 kJ $\cdot$ mol^{-1} more positive than the value selected above, and very low equilibrium uranium concentrations of 10^{-14} M in non-complexing aqueous solutions with pH values between 4 and 12 over crystalline UO_2. This is completely inconsistent with the earlier work of Parks and Pohl [88PAR/POH]. Further experimental work, perhaps using single crystals of UO_2, will be required to resolve this discrepancy. Solubility measurements for UO_2(cr) at low pH could prove useful.

V.3.2.3.3. Other U(IV) hydrolysis species

Hydrolysis species intermediate between UOH^{3+} and $U(OH)_4$(aq) have been proposed by several authors [61NIK/LUK, 63VDO/ROM, 75DAV/EFR, 76BAE/MES, 89GRE/BID]. Baes and Mesmer's [76BAE/MES] reanalysis of Hietanen's [56HIE] data indicated $U_6(OH)_{15}^{9+}$ as well as another species with the ratio $(n : m) > 2.5$. This is consistent with a recent study on the hydrolysis of Th^{4+} [83BRO/ELL] which suggested a model involving the (1,1), (15,6) and (12,4) species. In this review the value

$$\log_{10}{}^*\beta_{15,6}\ (\text{V.11}, n = 15, m = 6,\ 3\,\text{M NaClO}_4) \quad = \quad -16.9 \pm 0.6$$

from the reanalysis of Baes and Mesmer [76BAE/MES] is accepted as reasonably representing U(IV) hydrolysis polymerisation (although other models may be at least as satisfactory [83BRO/ELL]). However, considering the high charge and large size of this species, no attempt is made to correct the formation constant to standard ($I = 0$) conditions. On the basis of the work of Baes *et al.* [65BAE/MEY, 76BAE/MES] for the corresponding thorium system, the entropy of such a species would be expected to be large and negative (~ -600 J $\cdot$ K^{-1} $\cdot$ mol^{-1}).

It might also be expected that other monomeric hydrolysis species would be involved at low uranium concentrations. However, no unambiguous evidence for the formation of such species in acidic solutions has been reported. Values proposed by Baes and Mesmer [76BAE/MES] for ${}^*\beta_2(\text{V.11}, m = 1, n = 2)$ and ${}^*\beta_3(\text{V.11}, m = 1, n = 3)$ were based on an early value for ${}^*\beta_5^\circ(\text{V.11}, m = 1, n = 5)$ that now appears to be incorrect. Recently, Grenthe, Bidoglio and Omenetto [89GRE/BID] published data that are used to derive a limiting value of $\log_{10}{}^*\beta_2 \leq -4.5$ in a 3 M $NaClO_4$ medium. The data of Parks and Pohl [88PAR/POH] also suggest the existence of some hydrolysis species with a hydroxide to uranium ratio of less than four. The authors offered the alternative explanations of formation of UOH^{3+} or $U(OH)_2^{2+}$ in acidic solutions at temperatures from 100 to 300°C.

The measurements at 25°C of Bruno *et al.* [87BRU/CAS] indicated the solubility of UO_2("am") (I = 0.5 M, $NaClO_4$) increases slowly with decreasing pH between pH values of 5 and 2. This increase was interpreted as indicating that the predominant species is $U(OH)_3^+$ (or a polymer thereof). On the basis of this species being monomeric and the only significant uranium(IV) hydrolysis species aside from $U(OH)_4(aq)$, the equilibrium constant for the formation of $U(OH)_3^+$ would be $\log_{10}{}^*\beta_3 = -(1.1 \pm 0.1)$ ($I = 0.5$ M, $NaClO_4$) or $\log_{10}{}^*\beta_3 = -(0.6 \pm 0.7)$ at $I = 0$ (see Appendix A). As shown in Appendix A such large values for ${}^*\beta_3$ are incompatible with the experimental values of ${}^*\beta_1$.

As an illustration, solubility and predominance diagrams of the dioxouranium(VI)/ uranium(IV) hydroxide system in the ranges $4.5 < \text{pH} < 10$ and $-5 < \text{pe} < 10$ are presented in Figures V.7 and V.8. It should be mentioned that Figure V.7 does not represent a real experimental situation because the calculations were made by assuming that no solid phase is formed. This is done in order to illustrate the areas of predominance and relative concentrations of the various dioxouranium(VI) species in a homogeneous aqueous phase. On the contrary, Figure V.8 does represent an experimental modelling situation. It shows that a total uranium concentration of 10^{-5} M can only be maintained at pH < 5.3 and pe > 5, or at pH > 9.2 and pe >~2.

Figure V.7: Predominance diagram of the dioxouranium(VI)/uranium(IV) hydroxide system at 25°C in the range 4.5 < pH < 10, as a function of the redox potential, pe. The precipitation of solid phases is suppressed.

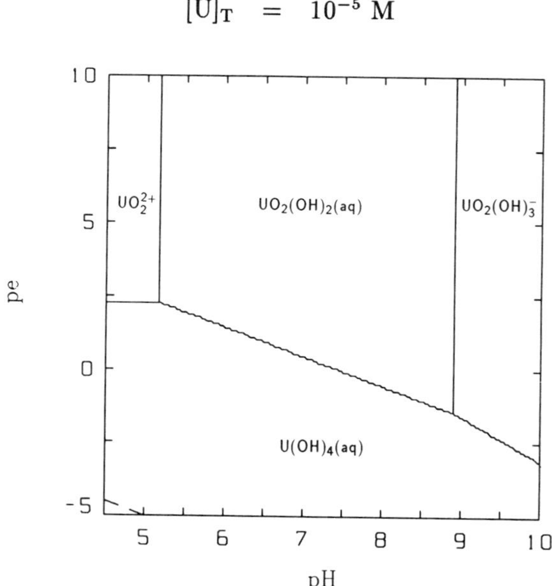

Figure V.8: Solubility and predominance diagram of the dioxouranium(VI)/uranium(IV) hydroxide system at 25°C in the range 4.5 < pH < 10, as a function of the redox potential. The solubility limiting phases are indicated on the graph. Schoepite is $UO_3 \cdot 2H_2O(cr)$, uraninite is $UO_2(cr)$.

$$[U]_T = 10^{-5} \text{ M}$$

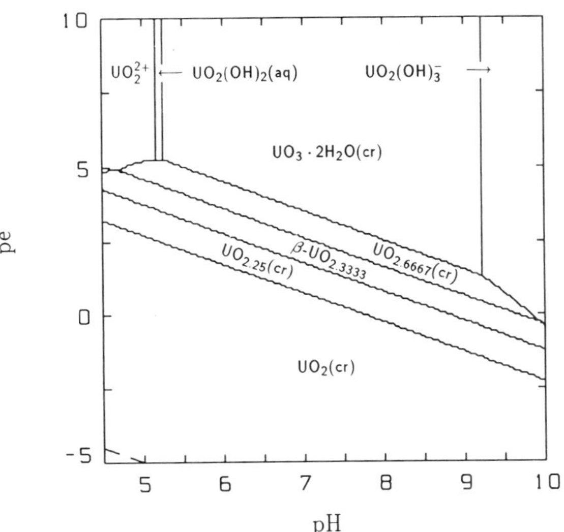

A potential inconsistency:

Values for $\Delta_f G_m^\circ(UO_2, cr, 298.15\,K)$ and $\Delta_f G_m^\circ(H_2O, l, 298.15\,K)$ as recommended by CODATA [89COX/WAG], and for $\Delta_f G_m^\circ(U^{4+}, aq, 298.15\,K)$ as recommended in this review lead to a value of the solubility product of $UO_2(cr)$, $^*K_{s,0}^\circ = (a_{U^{4+}}/a_{H^+}^4) = 10^{-4.8}$. This differs slightly from the $10^{-4.6}$ that is consistent with the compilation of Fuger and Oetting [76FUG/OET], primarily because of a small difference in the selected value of $\Delta_f G_m^\circ(U^{4+}, aq, 298.15\,K)$.

If all hydrolysis species for U^{4+} are neglected except for UOH^{3+} and $U(OH)_4(aq)$, the following equations are relevant for hydrolysis:

$$^*K_{s,0}^\circ = \frac{a_{U^{4+}}}{a_{H^+}^4} \qquad \text{(V.15)}$$

$$^*\beta_4^\circ = \frac{a_{U(OH)_4(aq)} \times a_{H^+}^4}{a_{U^{4+}}} \qquad \text{(V.16)}$$

$$^*\beta_1^\circ = \frac{a_{UOH^{3+}} \times a_{H^+}}{a_{U^{4+}}} \qquad \text{(V.17)}$$

Therefore,

$$\frac{{}^*\beta_4^\circ}{{}^*\beta_1^\circ} = \frac{a_{\mathrm{U(OH)_4(aq)}} \times a^3_{\mathrm{H^+}}}{a_{\mathrm{UOH^{3+}}}} \quad \text{(V.18)}$$

and since

$${}^*\beta_1^\circ\, {}^*K^\circ_{\mathrm{s,0}} = \frac{a_{\mathrm{UOH^{3+}}}}{a^3_{\mathrm{H^+}}} \quad \text{(V.19)}$$

therefore,

$$\log_{10} a_{\mathrm{U(OH)_4(aq)}} = \log_{10}\left(\frac{{}^*\beta_4^\circ}{{}^*\beta_1^\circ}\right) + \log_{10}{}^*\beta_1^\circ + \log_{10}{}^*K^\circ_{\mathrm{s,0}} \quad \text{(V.20)}$$

Using the values selected above, $\log_{10} a_{\mathrm{U(OH)_4(aq)}} = -(9.5 \pm 1.0)$, $\log_{10}{}^*K^\circ_{\mathrm{s,0}} = -(4.8 \pm 0.2)$, and $\log_{10}{}^*\beta_1^\circ = -(0.54 \pm 0.06)$, then

$$\log_{10}\left(\frac{{}^*\beta_4^\circ}{{}^*\beta_1^\circ}\right) = -4.2,$$

and equal concentrations of UOH^{3+} and $U(OH)_4(aq)$ would exist at pH ≈ 1.4 in the hypothetical solution (hypothetical because the low pH value implies $I \gg 0$).

It is difficult to extend this calculation accurately to real solutions. The highly charged species, U^{4+} and UOH^{3+}, will be stabilized at moderate ionic strengths (≤ 3 M) [88LEM]. At $I \approx 0.5$ M, Kraus and Nelson found $\log_{10}{}^*\beta_1 = -1.5$ in perchlorate media. If, from this, $f_\pm = 0.69$ is estimated, equal concentrations of UOH^{3+} and $U(OH)_4(aq)$ should coexist near pH $= 1.8$, and significant concentrations of $U(OH)_4(aq)$ are calculated to have been present in parts of the pH range used for the hydrolysis experiments. There has been no direct experimental evidence that $U(OH)_4(aq)$ occurs in such solutions.

Furthermore, the calculations suggest neither $U(OH)_2^{2+}$ nor $U(OH)_3^+$ are likely to be predominant species at any value of pH, or they would also have been present in significant quantities during experiments to measure ${}^*\beta_1$ in acidic media. Yet the results of Bruno *et al.* [87BRU/CAS] suggest some hydrolysis species (albeit perhaps polymeric) besides UOH^{3+} and $U(OH)_4(aq)$ is significant between pH 2 and 5.

The simplest explanation is that the experimental solubility of UO_2("am") [86BRU/FER, 87BRU/CAS] reflects dissolution of material ("fines" or material at grain boundaries) less stable than the solid that effectively sets the redox potential, and that the material used by Parks and Pohl [88PAR/POH] is less stable than $UO_2(cr)$. Therefore, the stability of the neutral species $U(OH)_4(aq)$ has been overestimated by orders of magnitude, and the agreement in $\Delta_f G^\circ_m(U(OH)_4, aq, 298.15\ K)$ calculated from the two studies is coincidental.

Indeed, the recent results of Rai, Felmy and Ryan [90RAI/FEL] on the solubility of "$UO_2 \cdot xH_2O(am)$" suggest a much lower stability of $U(OH)_4$. Unresolved inconsistencies remain with the work of Parks and Pohl [88PAR/POH], and further experimental work, perhaps on single crystals of UO_2, will be required to resolve these discrepancies.

Other possible reasons include underestimation of the stabilization of U^{4+} and UOH^{3+} in high ionic strength solutions, misinterpretation of data concerning the value for the first hydrolysis constant or, least likely, large errors in the selected values for $\Delta_f G^\circ_m(U^{4+}, aq, 298.15\,K)$ and/or $\Delta_f G^\circ_m(UO_2, cr, 298.15\,K)$.

V.3.2.3.4. UOH^{3+} in D_2O

Sullivan and Hindman [59SUL/HIN] determined the first solvolysis constant for U^{4+} in both H_2O and D_2O. By comparison of their values for $^*\beta_1$ ($I = 2$ M, (Na, H)ClO_4) in the two solvents and the value for $\log_{10}{^*\beta^\circ_1}$ selected here, the value $\log_{10}{^*\beta^\circ_1} = -(0.6 \pm 0.3)$ is estimated for formation of UOH^{3+} in D_2O.

V.3.2.4. U(III) hydroxide complexes

No credible data for U(III) hydroxo species are found by this review.

V.3.2.5. Polynuclear complexes containing different metal ions

Hydroxide and oxide ions are efficient bridge-forming ligands as indicated by the formation of a large number of polynuclear complexes. Most polynuclear systems studied to date contain only one type of metal ion. However, polynuclear systems containing different metal ions are also known. Toth *et al.* [84TOT/FRI, 86SHE/TOT] have studied the formation of such complexes between U(VI) and Th(IV), Hf(IV) and Zr(IV). However, no equilibrium constants are available. Such mixed complexes may also be formed containing atoms of a given actinide in different oxidation states, *e.g.*, U(IV) and U(VI).

V.3.3. Crystalline and amorphous uranium oxides and hydroxides

There have been many studies on the thermodynamic properties of uranium oxides. A substantial proportion of the experimental work concerns properties at much higher temperatures than need to be considered for nuclear fuel waste management. This review restricts itself primarily to data for temperatures ≤ 600 K.

V.3.3.1. U(VI) oxides and hydroxides

V.3.3.1.1. γ-UO_3(cr)

The accepted values

$$\begin{aligned}
\Delta_f H^\circ_m(UO_3, \gamma, 298.15\,K) &= -(1223.8 \pm 1.2)\,kJ\cdot mol^{-1} \\
S^\circ_m(UO_3, \gamma, 298.15\,K) &= (96.11 \pm 0.40)\,J\cdot K^{-1}\cdot mol^{-1}
\end{aligned}$$

are those selected by CODATA [89COX/WAG]. The enthalpy value selected by CODATA is based on a recalculation of the results of Fitzgibbon, Pavone and Holley [67FIT/PAV] for $UO_{2.993}$ using a more recent value for $\Delta_f H^\circ_m(U_3O_8)$ [69HUB/HOL]. The correction to stoichiometric UO_3 was apparently done as in Ref. [67FIT/PAV].

The CODATA value is in good agreement with that of Cordfunke and Aling [65COR/ALI, 75COR/OUW]. As pointed out by Cordfunke, Ouweltjes and Prins [75COR/OUW], the somewhat different value for $\Delta_f H_m^\circ(\gamma\text{-}UO_3)$ based on the heat of solution experiments of Vidavskii, Byakhova and Ippolitova [65VID/BYA] relies on the value for $\Delta_f H_m^\circ$(HF, aq, 298.15 K), which is not firmly established. The CODATA entropy is directly from the low-temperature heat capacity work of Westrum [66WES] (recently described in more detail by Cordfunke and Westrum [88COR/WES]). Jones, Gordon and Long [52JON/GOR] reported $S_m^\circ(UO_3, cr, 298.15\,K) = (98.6\pm0.3)\,J\cdot K^{-1}\cdot mol^{-1}$ based on heat capacity measurements for a sample of UO_3 that was probably predominantly γ-UO_3. This earlier value was apparently assigned a weight of zero in the CODATA evaluation.

The Gibbs energy of formation is calculated from the selected enthalpy of formation and entropy.

$$\Delta_f G_m^\circ(UO_3, \gamma, 298.15\,K) = -(1145.7 \pm 1.2)\,kJ\cdot mol^{-1}$$

The experimental heat capacities determined by Jones, Gordon and Long [52JON/GOR] are greater than those of Cordfunke and Westrum at all temperatures (by 2 to 4%), and show considerable scatter above 200 K. There is reason to believe that the sample of γ-UO_3 used by Jones, Gordon and Long [52JON/GOR] contained a small amount of water [47MOO/KEL, 88COR/WES]. Moore and Kelley [47MOO/KEL] reported enthalpy data (to 885 K) for a UO_3 sample that had been dried at 550 to 600°C in a stream of oxygen. The heat capacity parameters derived by Pankratz [82PAN] were based on the fit of the drop calorimetry data of Moore and Kelley [47MOO/KEL] to a temperature dependent function constrained by the Westrum's heat capacity value [66WES] for γ-UO_3 at 298.15 K. For temperatures to 700 K, heat capacities calculated using this function differ by less than 2% from those calculated using the function reported by Cordfunke and Westrum [88COR/WES] based on a drop calorimetry study (347 to 693 K). The Cordfunke and Westrum [88COR/WES] parameters differ somewhat from those briefly noted earlier by Cordfunke, Ouweltjes and Prins [75COR/OUW].

The γ-UO_3 sample used by Cordfunke and Westrum [88COR/WES] was better characterized than the sample used by Moore and Kelley [47MOO/KEL]. The two sets of enthalpy data can be used to calculate heat capacities between 300 and 350 K that match the results of the adiabatic calorimetry in this temperature range within 1%, however, agreement is markedly better with the values based on Moore and Kelley [47MOO/KEL, 82PAN]. From 400 to 700 K the Cordfunke and Westrum function gives calculated heat capacities that are approximately 1% lower than those calculated using the function of Pankratz [82PAN]. In this review the enthalpy data of Moore and Kelley [47MOO/KEL] and Cordfunke and Westrum [88COR/WES] have been equally weighted to obtain

$$\begin{aligned} H_m^\circ(T) - H_m^\circ(298.15\,K) &= 88.103\,T + (8.320\times10^{-3})\,T^2 \\ &\quad +(10.128\times10^5)\,T^{-2} - 30404 \end{aligned} \qquad (V.21)$$

(using the same constraints as in Refs. [82PAN] and [88COR/WES]) and hence the heat capacity function in Table V.10. It should be stressed that these equations will not give reasonable values for the enthalpy difference nor $C^\circ_{p,m}(\gamma\text{-}UO_3)$ for temperatures below 298.15 K, however, the uncertainty in $C^\circ_{p,m}(\gamma\text{-}UO_3)$ in the range 298.15 to 700 K is only 1%. The value

$$C^\circ_{p,m}(\gamma\text{-}UO_3, 298.15\,K) = (81.67 \pm 0.16)\,J{\cdot}K^{-1}{\cdot}mol^{-1}$$

selected in this review is the value from Ref. [88COR/WES]. The uncertainty is estimated as twice the 0.1% uncertainty indicated by the authors.

Table V.10: Temperature dependent heat capacity functions of the form $C^\circ_{p,m}(T) = a + 10^{-3}bT + 10^{-3}cT^2 + 10^5 e\,T^{-2}$.(a,b)

Compound	T-range (K) T_{min}	T_{max}	a	b	c	e
UO_2(cr)	250	600	62.774	31.74		−7.693
$UO_{2.25}$(cr)	250	348	1487.6	−6973.7	9.736	−178.6
β-$UO_{2.25}$	348	600	79.089	13.65		−10.38
α-$UO_{2.3333}$	237	347	64.149	49.14		−6.72
β-$UO_{2.3333}$	232	346	64.338	49.79		−6.55
$UO_{2.6667}$(cr)	233	600	87.276	22.16		−12.44
β-UO_3	298	678	86.170	24.984		−10.915
γ-UO_3	298	850	88.103	16.640		−10.128
UO_3(am)	400	650	76.01	38.06		−2.31
β-$UO_2(OH)_2$ ($\equiv \beta$-$UO_3{\cdot}H_2O$)	298	473	41.8	200		35.3
$UO_3 \cdot 2H_2O$(cr)	298	400	84.238	294.592		

(a) The units of $C^\circ_{p,m}$ are $J \cdot K^{-1} \cdot mol^{-1}$.
(b) Values of the heat capacity calculated from the function parameters in this table generally differ less than 2% from the experimental values. Exceptions may occur near transition temperatures ($U_4O_9 \equiv 4UO_{2.25}$, $U_3O_8 \equiv 3UO_{2.6667}$) or near the limits of each temperature range. Parameters for β-$UO_2(OH)_2$ are estimated (see text). In all cases values of $C^\circ_{p,m}$(298.15 K) from Table V.11 are to be preferred to values calculated for 298.15 K using the parameters in Table V.10.

V.3.3.1.2. Other forms of anhydrous UO_3(cr)

The enthalpies of formation of these solids are obtained from the differences between the enthalpies of solution and the values found for the enthalpy of solution of γ-UO_3 in the same medium. Using the selected value for $\Delta_f H^\circ_m(UO_3, \gamma, 298.15\,K)$ and the appropriately weighted (according to the authors' stated uncertainties) enthalpy data from the work of Cordfunke [64COR] and Hoekstra and Siegel [61HOE/SIE], the following selected values are obtained:

$$\Delta_f H_m^\circ(UO_3, \alpha, 298.15\,K) = -(1217.5 \pm 3.0)\,kJ \cdot mol^{-1}$$
$$\Delta_f H_m^\circ(UO_3, \varepsilon, 298.15\,K) = -(1217.2 \pm 1.3)\,kJ \cdot mol^{-1}.$$

As noted by Loopstra and Cordfunke [66LOO/COR], α-UO_3 has a disordered structure and samples may not have the exact stoichiometry $UO_{3.00}$. In the present review, this problem is considered when estimating the uncertainty in $\Delta_f H_m^\circ(UO_3$, α, 298.15 K). The heat of solution of α-UO_3 in a $Ce(SO_4)_2/H_2SO_4$ mixture, reported by Cordfunke, Prins and van Vlaanderen [68COR/PRI], is not used in estimating $\Delta_f H_m^\circ(UO_3, \alpha, 298.15\,K)$ and is not consistent with the selected value (see Appendix A). The difference $[\Delta_f H_m^\circ(UO_3, \gamma, 298.15\,K) - \Delta_f H_m^\circ(UO_3, \beta, 298.15\,K)] = -(3.5 \pm 0.2)\,kJ \cdot mol^{-1}$ is calculated from data from the same two sources as well as data from Cordfunke, Ouweltjes and Prins [75COR/OUW]. Hence,

$$\Delta_f H_m^\circ(UO_3, \beta, 298.15\,K) = -(1220.3 \pm 1.3)\,kJ \cdot mol^{-1}.$$

There is a difference of $10\,kJ \cdot mol^{-1}$ between the values of $\Delta_f H_m^\circ(UO_3, \delta, 298.15\,K)$ recalculated from the experiments of Hoekstra and Siegel [61HOE/SIE] and Cordfunke [64COR]. This review accepts the value from the latter source, an estimate based on the heat of solution of a mixture of α-UO_3 and δ-UO_3, and obtains

$$\Delta_f H_m^\circ(UO_3, \delta, 298.15\,K) = -(1209 \pm 5)\,kJ \cdot mol^{-1}.$$

The selected entropy value,

$$S_m^\circ(UO_3, \alpha, 298.15\,K) = (99.4 \pm 1.0)\,J \cdot K^{-1} \cdot mol^{-1}$$

is based on low temperature heat capacity measurements reported by Westrum [66WES]. However, insufficient data were provided to permit calculation of the temperature dependence of the heat capacities of this solid above 250 K. The value of $S_m^\circ(UO_3$, $\alpha)$ is greater than the values for $S_m^\circ(UO_3$, $\gamma)$ or $S_m^\circ(UO_3$, $\beta)$, reflecting the greater structural disorder in α-UO_3 [66LOO/COR]. The uncertainty is an estimate (see above).

For β-UO_3, the following value is selected in this review:

$$S_m^\circ(UO_3, \beta, 298.15\,K) = (96.32 \pm 0.40)\,J \cdot K^{-1} \cdot mol^{-1}.$$

This is the value reported by Cordfunke and Westrum [88COR/WES] based on low temperature heat capacity measurements. The selected parameters for the temperature dependence of the heat capacity, (*cf.* Table V.10), and enthalpy difference for β-UO_3 are those reported in the same paper [88COR/WES], based on drop calorimetry experiments (404.9 to 678.4 K). As is found for γ-UO_3 (Section V.3.3.1.1), the differences between the values of $C_{p,m}^\circ(\beta$-$UO_3)$ determined by adiabatic calorimetry and from extrapolation using the drop calorimetry results [88COR/WES] are less than 1% in the temperature range 300 to 346 K.

The Gibbs energies of formation for α-UO_3 and β-UO_3 are calculated from the selected enthalpies of formation and entropies.

$$\Delta_f G_m^\circ(UO_3, \alpha, 298.15\,K) = -(1140.4 \pm 3.0)\,kJ \cdot mol^{-1}$$
$$\Delta_f G_m^\circ(UO_3, \beta, 298.15\,K) = -(1142.3 \pm 1.3)\,kJ \cdot mol^{-1}$$

The values

$$C_{p,m}^\circ(UO_3, \alpha, 298.15\,K) = (81.84 \pm 0.30)\,J \cdot K^{-1} \cdot mol^{-1}$$
$$C_{p,m}^\circ(UO_3, \beta, 298.15\,K) = (81.34 \pm 0.16)\,J \cdot K^{-1} \cdot mol^{-1}$$

selected in this review are the values from Refs. [66WES] and [88COR/WES], respectively. The uncertainty in $C_{p,m}^\circ(UO_3, \beta, 298.15\text{ K})$ is estimated as twice the 0.1% uncertainty indicated in Ref. [88COR/WES], and a slightly larger uncertainty is assigned to the value for α-UO_3 based on possible uncertainties in the stoichiometry [66LOO/COR].

V.3.3.1.3. UO_3 (am)

The difference between the enthalpies of solution of UO_3(am) and γ-UO_3 was reported as $(15.5\pm2.5)\,kJ\cdot mol^{-1}$ by Hoekstra and Siegel [61HOE/SIE] and $(15.7\pm1.0)\,kJ\cdot mol^{-1}$ by Cordfunke [64COR] (see Appendix A). The difference between the enthalpies of solution of β-UO_3 and UO_3(am) is $(13.8 \pm 2.4)\,kJ \cdot mol^{-1}$ according to Hoekstra and Siegel [61HOE/SIE]. Vidavskii and Ippolitova [71VID/IPP] measured the enthalpies of solution (into aqueous HF) of γ-UO_3 and a mixture of UO_3(am) and α-UO_3. From their data, $[\Delta_f H_m^\circ(UO_3, \text{am}, 298.15\text{ K}) - \Delta_f H_m^\circ(UO_3, \gamma, 298.15\text{ K})] = (13.9\pm0.8)\,kJ \cdot mol^{-1}$ is obtained (*cf.* Appendix A). This enthalpy difference is somewhat smaller than those found in Refs. [64COR] and [61HOE/SIE]. Because of the impure nature of the amorphous oxide used by Vidavskii and Ippolitova [71VID/IPP], their data are not used for the calculation of $\Delta_f H_m^\circ(UO_3, \text{am}, 298.15\,K)$ in this review. Using the values of $\Delta_f H_m^\circ(UO_3, \gamma, 298.15\text{ K})$ and of the enthalpy difference $[\Delta_f H_m^\circ(UO_3, \gamma, 298.15\text{ K}) - \Delta_f H_m^\circ(UO_3, \beta, 298.15\text{ K})] = -(3.5 \pm 0.2)\,kJ \cdot mol^{-1}$ selected above, the selected standard enthalpy of formation is obtained,

$$\Delta_f H_m^\circ(UO_3, \text{am}, 298.15\,K) = -(1207.9 \pm 4.0)\,kJ \cdot mol^{-1},$$

based on the results of Cordfunke [64COR] and Hoekstra and Siegel [61HOE/SIE]. The uncertainty in this value has been increased to allow for variability in the particle size of the amorphous solid (see also the discussion on β-UO_3 in Ref. [64COR]).

The values of Popov, Gal'chenko and Senin [58POP/GAL] for the heat capacity of UO_3(am) from 392 to 673 K are less than 3% greater than those obtained using the parameters for γ-UO_3 from Table V.10 for 400 to 600 K. At 700 K the difference is approximately 5%. As there are no low-temperature heat capacity measurements available for the amorphous solid, the parameters in Table V.10 based on the results of Popov *et al.* [58POP/GAL] should only be used for the temperature range 400 to 650 K.

V.3.3.1.4. U(VI) peroxides

Cordfunke and Aling [63COR/ALI] measured the enthalpy of precipitation of $UO_4{\cdot}4H_2O(cr)$ to be $-(31.21 \pm 0.29)\,kJ \cdot mol^{-1}$. Using $\Delta_f H^\circ_m(H_2O_2, aq, 298.15\ K) = -(191.17 \pm 0.10)\,kJ \cdot mol^{-1}$ from [82WAG/EVA] (the uncertainty is estimated in the present review), the value

$$\Delta_f H^\circ_m(UO_4{\cdot}4H_2O, cr, 298.15\ K) = -(2394.8 \pm 2.1)\,kJ \cdot mol^{-1}$$

is calculated. The enthalpy of dehydration of the tetrahydrate to form the dihydrate,

$$UO_4 \cdot 4H_2O(cr) \rightleftharpoons UO_4 \cdot 2H_2O(cr) + 2H_2O(g), \qquad (V.22)$$

was calculated to be $(117.02 \pm 1.00)\,kJ \cdot mol^{-1}$ from vapour pressure measurements [66COR]. Using this value,

$$\Delta_f H^\circ_m(UO_4{\cdot}2H_2O, cr, 298.15\ K) = -(1784.0 \pm 4.2)\,kJ \cdot mol^{-1}$$

is calculated. The selected uncertainties for these enthalpies of formation are those proposed by Cordfunke and O'Hare [78COR/OHA].

No entropy or Gibbs energy data are available for these compounds.

V.3.3.1.5. U(VI) hydrated oxides and hydroxides

As discussed by Hoekstra and Siegel [73HOE/SIE], there are several forms of hydrated UO_3 (and/or hydrated dioxouranium(VI) hydroxides) that can form at temperatures below 300°C. These include a dihydrate, at least two stoichiometric monohydrates, a hypostoichiometric "monohydrate" (α-$UO_3 \cdot (0.8 \pm 0.1)H_2O$) and a lower hydrate variously described as $UO_3{\cdot}0.5H_2O$ [56DAW/WAI] or $U_3O_8(OH)_2$ [73HOE/SIE] ($\triangleq UO_3{\cdot}0.33H_2O$). The interconversion of the various phases was described [73HOE/SIE], and it is clear that several forms must have similar thermodynamic stability when in contact with liquid water at temperatures ≤ 100°C.

The papers of Santalova *et al.* [71SAN/VID] and Vidavskii, Byakhova and Ippolitova [65VID/BYA] lead to a value of $-(30.04 \pm 1.01)\,kJ \cdot mol^{-1}$ for the enthalpy of hydration of γ-UO_3 to the dihydrate (schoepite), based on the differences in the enthalpies of solution of the two solids in dilute aqueous hydrofluoric acid (see Appendix A). The recent work of Tasker *et al.* [88TAS/OHA] reports a similarly determined experimental value of $-(29.93 \pm 0.52)\,kJ \cdot mol^{-1}$. These compare with $-(31.2_4 \pm 0.6)\,kJ \cdot mol^{-1}$ based on the enthalpies of solution in 6 M HNO_3 [64COR, 65PAR] and $\Delta_{tr}H_m[H_2O, H_2O(l) \rightarrow 6\ M\ HNO_3(aq)] = -0.5_4\,kJ \cdot mol^{-1}$ [65PAR, 76FUG/OET]. The weighted average value for the enthalpy differences from measurements in aqueous HF, and the difference from measurements in aqueous HNO_3, are not consistent within the estimated uncertainties. In this review the average enthalpy difference, $-(30.6 \pm 1.2)\,kJ \cdot mol^{-1}$ is used. When combined with the CODATA [89COX/WAG] values for $\Delta_f H^\circ_m(UO_3, \gamma, 298.15\,K)$ and $\Delta_f H^\circ_m(H_2O, l, 298.15\,K)$, this results in

$$\Delta_f H_m^\circ(UO_3 \cdot 2H_2O, cr, 298.15\,K) = -(1826.1 \pm 1.7)\,kJ \cdot mol^{-1}.$$

The $\Delta_f H_m^\circ$ value for $UO_3 \cdot 2H_2O(cr)$ is similar to that calculated from the much earlier enthalpy of solution work of de Forcrand [15FOR]. It should be noted that for the reaction

$$UO_3(am) + 2H_2O(l) \rightarrow UO_3 \cdot 2H_2O(cr), \qquad (V.23)$$

the value of $\Delta_{hyd} H_m^\circ$ calculated from the values selected in this review is $-(46.5 \pm 0.4)\,kJ \cdot mol^{-1}$, which is in reasonable agreement with the value of $-(45.9 \pm 2.5)\,kJ \cdot mol^{-1}$ obtained directly by Drobnic and Kolar [66DRO/KOL] (see Appendix A). Some of the difference can be attributed to differences in the method of synthesis of the amorphous solids.

The selected values of

$$S_m^\circ(UO_3 \cdot 2H_2O, cr, 298.15\,K) = (188.54 \pm 0.38)\,J \cdot K^{-1} \cdot mol^{-1}$$

and

$$C_{p,m}^\circ(UO_3 \cdot 2H_2O, cr, 298.15\,K < T < 400) = (84.238 + 0.294592\,T)\,J \cdot K^{-1} \cdot mol^{-1}$$

were recently reported by Tasker *et al.* [88TAS/OHA]. These values are the only available direct experimental values for these parameters, and they are selected in this review.

The Gibbs energy of formation,

$$\Delta_f G_m^\circ(UO_3 \cdot 2H_2O, cr, 298.15\,K) = -(1636.5 \pm 1.7)\,kJ \cdot mol^{-1},$$

is calculated from the selected enthalpy of formation and entropy. This value of $\Delta_f G_m^\circ(UO_3 \cdot 2H_2O, cr, 298.15\,K)$ is preferred to the substantially different value calculated by Lemire and Tremaine [80LEM/TRE], based on the results of the solubility study of Gayer and Leider [55GAY/LEI, 76BAE/MES] (also see below).[5]

The selected values of

$$\Delta_f H_m^\circ(UO_2(OH)_2 \equiv UO_3 \cdot H_2O, \beta, 298.15\,K) = -(1533.8 \pm 1.3)\,kJ \cdot mol^{-1}$$
$$\Delta_f H_m^\circ(UO_2(OH)_2 \equiv UO_3 \cdot H_2O, \gamma, 298.15\,K) = -(1531.4 \pm 1.3)\,kJ \cdot mol^{-1}$$

are based on the differences between the enthalpies of solution of the hydrated oxides and $\Delta_{sol} H_m(UO_3, \gamma, 298.15\,K)$ reported in Ref. [64COR]. The compound ε-$UO_3 \cdot H_2O$ of Cordfunke [64COR] is apparently the same as γ-$UO_3 \cdot H_2O$ of Hoekstra and Siegel [73HOE/SIE]. The enthalpies of formation are calculated using the CODATA Key Value for $\Delta_f H_m^\circ(UO_3, \gamma, 298.15\,K)$ and $\Delta_{tr} H[H_2O, H_2O(l) \rightarrow 6\ M\ HNO_3(aq)] = -0.54\,kJ \cdot mol^{-1}$ [65PAR, 76FUG/OET]. In a recent review [78COR/OHA] the

[5] Sandino [91SAN] recently determined the solubility product for schoepite and reported $\log_{10}{}^*K_s^\circ = (5.96 \pm 0.18)$ for the reaction $UO_3 \cdot 2H_2O(cr) + 2H^+ \rightleftharpoons UO_2^{2+} + 3H_2O(l)$, where the uncertainty is twice the standard deviation resulting from the least-squares refinement of the solubility data. Sandino also reported the solubility product of an amorphous phase, $\log_{10}{}^*K_s^\circ = (6.33 \pm 0.17)$.

values for the enthalpies of formation of the β- and γ-hydroxides (and for $\Delta_f H_m^\circ(UO_3{\cdot}0.85H_2O, \alpha, 298.15\ K)$, see below) appear to have been calculated without considering this enthalpy of transfer. The $\Delta_f H_m^\circ$ value for β-$UO_2(OH)_2$ is similar to that calculated from the much earlier enthalpy of solution work of de Forcrand [15FOR]. The enthalpies of solution obtained by Pissarjewsky [00PIS] and especially those of Aloy [1896ALO] give more negative values.

The temperature dependent heat capacity of β-$UO_2(OH)_2$ was estimated by Nikitin *et al.* [72NIK/SER] using the heat capacities of $Ba(OH)_2$ [60KEL], $BaSO_4$ [60KEL] and UO_2SO_4 [64OWE/MAY]. An error in this calculation was corrected by Lemire and Tremaine [80LEM/TRE], and their value,

$$C_{p,m}^\circ(UO_2(OH)_2, \beta, 298.15\,K) \quad = \quad (141 \pm 15)\,J\cdot K^{-1}\cdot mol^{-1},$$

and function for the temperature dependence of the heat capacity (Table V.10) are selected in this review. The uncertainty in $C_{p,m}^\circ(UO_2(OH)_2, \beta, 298.15\,K)$ is estimated in this review.

Robins [66ROB] suggested that, based on precipitation studies, $UO_3\cdot 2H_2O(cr)$ becomes unstable with respect to β-$UO_2(OH)_2$ at a temperature between 40 and 100°C. Hoekstra and Siegel [73HOE/SIE] described the occasional conversion of $UO_3\cdot$ $2H_2O$ to β-$UO_2(OH)_2$ on digestion in aqueous dioxouranium(VI) nitrate solutions at 55°C.

$$\beta\text{-}UO_2(OH)_2 + H_2O(l) \quad \rightarrow \quad UO_3\cdot 2H_2O(cr), \qquad (V.24)$$

Thus, it is reasonable to assume that the Gibbs energy of dehydration of $UO_3\cdot$ $2H_2O(cr)$ to β-$UO_2(OH)_2$ (in contact with liquid water) is zero at some transformation temperature, T_{trs}, near this value. Assuming T_{trs} is (330 ± 30) K, a calculation similar to that of Lemire and Tremaine [80LEM/TRE] using values selected in this review (including the values of $C_{p,m}^\circ(UO_3{\cdot}2H_2O, cr, 298.15\ K)$, $C_{p,m}^\circ(UO_3{\cdot}H_2O, cr, 298.15\ K)$) and $C_{p,m}^\circ(H_2O, l, 298.15\ K)$ [82WAG/EVA], results in the value

$$S_m^\circ(UO_2(OH)_2, \beta, 298.15\,K) \quad = \quad (138 \pm 4)\,J\cdot K^{-1}\cdot mol^{-1}.$$

The result is not greatly affected by the choice of the transition temperature for Reaction (V.24), but is strongly influenced by the uncertainties in the difference $\Delta_f H_m^\circ(UO_2(OH)_2, \beta, 298.15\ K) - \Delta_f H_m^\circ(UO_3{\cdot}2H_2O, cr, 298.15\ K)$. Within the stated uncertainties, the selected entropy is the same as the value estimated by Robins [66ROB].

From the selected enthalpy and entropy,

$$\Delta_f G_m^\circ(UO_2(OH)_2, \beta, 298.15\,K) \quad = \quad -(1398.7 \pm 1.8)\,kJ\cdot mol^{-1}.$$

is calculated. Hoekstra and Siegel [73HOE/SIE] suggested that γ-$UO_2(OH)_2$ is a metastable phase, and in this review no value is selected for $S_m^\circ(UO_2(OH)_2, \gamma, 298.15\ K)$ and, hence, no value for $\Delta_f G_m^\circ(UO_2(OH)_2, \gamma, 298.15\ K)$.

Cordfunke [64COR] also reported heat of solution data for α-$UO_3\cdot 0.85H_2O$ and, using the same auxiliary data as for the β- and γ-hydroxides above,

$$\Delta_f H_m^\circ(UO_3 \cdot 0.85H_2O, \alpha, 298.15\,K) \quad = \quad -(1491.9 \pm 1.3)\,kJ \cdot mol^{-1}$$

is calculated. O'Hare, Lewis and Nguyen [88OHA/LEW] recently reported that the enthalpy of solution (in aqueous HF) of samples of the α-trioxide hydrate with the composition $UO_3 \cdot 0.9H_2O$ is $(25.27 \pm 0.51)\,kJ \cdot mol^{-1}$ more endothermic than the heat of solution of γ-UO_3 in the same medium [88TAS/OHA]. Using the CODATA Key Value for $\Delta_f H_m^\circ(UO_3, \gamma, 298.15\,K)$, a value of

$$\Delta_f H_m^\circ(UO_3 \cdot 0.9H_2O, \alpha, 298.15\,K) \quad = \quad -(1506.3 \pm 1.3)\,kJ \cdot mol^{-1}$$

is calculated. The difference between the values for the 0.9 hydrate and the 0.85 hydrate is within $0.5\,kJ \cdot mol^{-1}$ of the enthalpy of formation of 0.05 moles of liquid water (*i.e.*, $\Delta_r H_m$(V.25, 298.15 K) is essentially zero within the uncertainty limits for the measurements).

$$\alpha\text{-}UO_3 \cdot 0.90H_2O \quad \rightarrow \quad \alpha\text{-}UO_3 \cdot 0.85H_2O + 0.05H_2O(l) \qquad \text{(V.25)}$$

The difference in the enthalpies of formation of α-$UO_3 \cdot 0.85H_2O$ and β-$UO_2(OH)_2$, as calculated from the enthalpies of solution, is not consistent with the large $(-5.69\,kJ \cdot mol^{-1})$ heat measured [72TAY/KEL] for the transformation of β-$UO_2(OH)_2$ to α-$UO_3 \cdot H_2O$ at 5°C. This suggests either that the measured enthalpy of formation of β-$UO_2(OH)_2$ is incorrect, or more probably, that the α-phase observed by Taylor, Kelley and Downer [72TAY/KEL] is not simply α-$UO_3 \cdot 0.9H_2O$ with an extra 0.1 moles of water per mole of uranium [73HOE/SIE].

For the conversion of α-$UO_3 \cdot 0.9H_2O$ to $UO_3 \cdot 2H_2O$(cr)

$$\alpha\text{-}UO_3 \cdot 0.9H_2O + 1.1H_2O(g) \quad \rightarrow \quad UO_3 \cdot 2H_2O(cr), \qquad \text{(V.26)}$$

O'Hare, Lewis and Nguyen [88OHA/LEW] assumed the equilibrium decomposition pressure (p) approaches 101.325 kPa (the reference pressure p°), at a temperature (T) near 373 K, and derived the approximate value $S_m^\circ(UO_3 \cdot 0.9H_2O$, α, 298.15 K) = $125\,J \cdot K^{-1} \cdot mol^{-1}$ from the calculated value of $\Delta_r H_m$(V.26) and an estimate of the temperature independent term in the expression

$$\log_{10}\left(\frac{p}{p^\circ}\right) \quad = \quad \frac{a}{T} + b. \qquad \text{(V.27)}$$

Although Tasker *et al.* [88TAS/OHA] indicated that in a sealed capsule, transformation of $UO_3 \cdot 2H_2O$(cr) to α-$UO_3 \cdot H_2O$ occurs at temperatures between 400 K and 425 K, there are reports [73HOE/SIE] of at least partial conversion of $UO_3 \cdot 2H_2O$(cr) in contact with liquid water to α-$UO_3 \cdot H_2O$ at lower temperatures. Thus, it is reasonable to assume that the Gibbs energy of dehydration of $UO_3 \cdot 2H_2O$(cr) (in contact with liquid water) is zero at some lower transformation temperature, T_{trs}. Assuming T_{trs} is (360 ± 30) K and $\Delta_r C_{p,m} = (50 \pm 30)\,J \cdot K^{-1} \cdot mol^{-1}$, a calculation similar to that reported above for β-$UO_2(OH)_2$ leads to a value of

$$S_m^\circ(UO_3 \cdot 0.9H_2O, \alpha, 298.15\,K) \quad = \quad (126 \pm 7)\,J \cdot K^{-1} \cdot mol^{-1}.$$

From the selected enthalpy and entropy, the value

$$\Delta_f G_m^\circ(UO_3 \cdot 0.9H_2O, \alpha, 298.15\,K) = -(1374.6 \pm 2.5)\,kJ \cdot mol^{-1}$$

is calculated.

Santalova *et al.* [72SAN/VID] reported enthalpies of solution of $UO_3 \cdot 0.393H_2O(cr)$ and $UO_3 \cdot 0.648H_2O(cr)$ in aqueous HF that are, respectively, $(11.71 \pm 0.52)\,kJ \cdot mol^{-1}$ and $(15.61 \pm 0.57)\,kJ \cdot mol^{-1}$ more endothermic than $\Delta_{sol}H_m^\circ(UO_3, \gamma, 298.15\,K)$ as determined by the same research group for the same medium [65VID/BYA]. From these enthalpy differences, the following selected enthalpies of formation are obtained:

$$\Delta_f H_m^\circ(UO_3 \cdot 0.393H_2O, cr, 298.15\,K) = -(1347.8 \pm 1.3)\,kJ \cdot mol^{-1}$$
$$\Delta_f H_m^\circ(UO_3 \cdot 0.648H_2O, cr, 298.15\,K) = -(1424.6 \pm 1.3)\,kJ \cdot mol^{-1}.$$

In this case the dehydration reaction

$$UO_3 \cdot 0.648H_2O(cr) \rightarrow UO_3 \cdot 0.393H_2O(cr) + 0.255H_2O(l) \qquad (V.28)$$

is clearly endothermic at 298.15 K. No entropies or Gibbs energies are estimated in this review for these lower hydrates.

O'Hare, Lewis and Nguyen [88OHA/LEW] have recently compared experimental solubility data for uranium trioxide hydrates, *e.g.*, [55GAY/LEI, 58BRU, 67GRY/KOR], with calculated values based on values for the Gibbs energies of the hydrated oxides (similar to those selected in this review). In general the agreement is not particularly good, partially, as suggested by O'Hare, Lewis and Nguyen [88OHA/LEW], because of "the difficulty of achieving equilibrium between sparingly soluble compounds and aqueous solution", *i.e.*, the presence of active sites on the solid surface in the solubility experiments and poor characterization of the solids. Also, in most of the solubility papers, the hydrolysis scheme was greatly oversimplified, leading to an overestimate of the uranium present as UO_2^{2+} in the solutions (see also Section V.3.2.1.3).

V.3.3.2. U(IV) oxides

V.3.3.2.1. Crystalline UO_2 (uraninite)

Values for the enthalpy of formation and entropy of UO_2 at 298.15 K and 1 bar were assessed by CODATA [89COX/WAG]. From these,

$$\Delta_f G_m^\circ(UO_2, cr, 298.15\,K) = -(1031.8 \pm 1.0)\,kJ \cdot mol^{-1}$$

is calculated.

The selected enthalpy of formation,

$$\Delta_f H_m^\circ(UO_2, cr, 298.15\,K) = -(1085.0 \pm 1.0)\,kJ \cdot mol^{-1},$$

is based on the enthalpy of oxidation of UO_2(cr) to U_3O_8(cr) as measured by Huber and Holley [69HUB/HOL]. The entropy value is the value calculated from low temperature heat capacity measurements by Huntzicker and Westrum [71HUN/WES]. These values are accepted in this review, and are combined with the data of Grønvold *et al.* [70GRO/KVE] (304 to 600 K) and Huntzicker and Westrum [71HUN/WES] (192 to 346 K) to obtain parameters to describe the variation of the heat capacity of solid UO_2 over the temperature range 250 to 600 K. The data were adjusted to account for the slight hyperstoichiometry of the samples of UO_{2+y} as discussed by Grønvold *et al.* [70GRO/KVE]. This was particularly necessary in the region of the $UO_{2.25}$ λ-transition near 350 K. Engel [69ENG] did heat capacity measurements for UO_2 (300 to 1100 K), however, only curve-fitted results were reported. These are given zero weight in this review. The value

$$C^{\circ}_{p,\mathrm{m}}(\mathrm{UO_2, cr, 298.15\,K}) = (63.60 \pm 0.08)\,\mathrm{J\cdot K^{-1}\cdot mol^{-1}}$$

selected in this review is the value from Ref. [71HUN/WES]. The uncertainty is estimated as twice the 0.06% uncertainty indicated by the authors. As noted above (Section V.3.2.3.3) the values recommended in this review lead to a value of $10^{-4.8}$ for the solubility product for UO_2(cr). Experimental solubility product values are discussed in Sections V.3.2.3.2 and V.3.3.2.2.

V.3.3.2.2. "Hydrous" or "amorphous" UO_2

Uranium dioxide produced by the reduction of U(VI) solutions, or by the hydrolysis of U(IV) salts, is generally microcrystalline [60RAF/KAN]. Crystallite sizes can be as small as 1 nm. Formation of larger crystals is slow even at higher temperatures. There is some spectroscopic evidence for the formation of a stable hydrate or hydroxides [70BAR, 84FUL/SMY], and freshly precipitated and filtered UO_2 may also have varying amounts of adsorbed water. Nikolaeva and Pirozhkov [78NIK/PIR] reported partial dehydration of the oxide between 125 and 150°C. Thus, the terms "hydrous" or "amorphous" UO_2 as used in the literature do not refer to unique materials, but rather to a range of solids with differing thermodynamic stabilities.

A number of solubility studies have been reported for "precipitated" or "hydrous" UO_2 [57GAY/LEI, 60GAL/STE, 60STE/GAL, 83RYA/RAI, 87BRU/CAS]. In principle, the relative solubilities of this material and of well-characterized crystalline UO_2 should yield a value for $\Delta_f G^{\circ}_{m}(\mathrm{UO_2, am, 298.15\,K})$. In practice, there are considerable difficulties in ensuring the precipitated solids contain only uranium(IV), the "crystalline" material does not have a partially oxidised surface layer or that the aqueous "saturated" solutions are sufficiently free of oxygen and carbon dioxide. Furthermore, the solubility would be expected to be dependent on the size of the particles of a precipitated microcrystalline solid.

Stepanov and Galkin [60STE/GAL] obtained $K_{s,0} = 10^{-(52.0\pm0.4)}$ (20°C) from the pH of initial precipitaion of solid from a uranium(IV) solution. At best, this represents an extreme upper limit ($\log_{10}{}^*K_{s,0} < 4.3$, 20°C) for an active form of hydrated oxide.

As this review was in its final stages of preparation, Rai, Felmy and Ryan [90RAI/FEL] published a paper on the solubility of "$UO_2 \cdot x H_2O$(am)". The solid used in that work was apparently X-ray amorphous both before and after the solubility measurements, and of comparable stability to the solid used by Stepanov and Galkin [60STE/GAL].

Nikolaeva and Pirozhkov [78NIK/PIR] determined the potential of solutions of dioxouranium(VI) over a precipitated uranium(IV) oxide, and reported $\log_{10}{}^{*}K_{s,0} = -(3.2 \pm 0.2)$ at 25°C. However, neglect of dioxouranium(VI) hydrolysis, and probable junction potential problems [86BRU/GRE2] leave doubts as to the uncertainty that should be assigned to this value. Recently, Bruno *et al.* [86BRU/FER] used a similar method, with more careful control of the pH and under conditions that were much less likely to cause oxidation, for three different samples of UO_2 in 3 M aqueous $NaClO_4$. An "amorphous" (actually very weakly crystalline) form was reported to have a value of $\log_{10}{}^{*}K_{s,0} = (0.5 \pm 0.3)$, while a ripened precipitate gave $\log_{10}{}^{*}K_{s,0} = -(1.2 \pm 0.1)$. The latter result was identical to the value obtained for a ceramic pellet (UO_2(cr)) electrode for which the potential may have been set by "surface active" material. Thus a difference from the "thermodynamic" value of approximately three orders of magnitude is found, even if ionic strength corrections are made [86BRU/FER]. Measurements with well-characterized single crystals would be of interest.

Based on solubility measurements using freshly precipitated material in acidic media [60STE/GAL, 90RAI/FEL], the value of $\Delta_f G^\circ_m(UO_2$, "am" or "hyd", 298.15 K) can be as great as $-977\,\text{kJ}\cdot\text{mol}^{-1}$. As mentioned in Section V.3.2.3.2, $\Delta_f G^\circ_m(UO_2$, "am", 298.15 K) $= -1003.6\,\text{kJ}\cdot\text{mol}^{-1}$ for the aged "amorphous" solid described by Bruno *et al.* [86BRU/FER, 87BRU/CAS], based on electrochemical measurements (although, as noted by Rai, Felmy and Ryan [90RAI/FEL], the measured solubility of this solid for a pH value near 2 is similar to that reported [90RAI/FEL] for freshly precipitated material). More crystalline precipitates of UO_2 from aqueous solutions (*e.g.*, as in Ref. [86BRU/FER]), even those with adsorbed water, undoubtedly have more negative values of $\Delta_f G^\circ_m$(298.15 K), although none have been shown to have values as negative as that recommended for $\Delta_f G^\circ_m(UO_2$, cr, 298.15 K) in Section V.3.3.2.1.

When required, the Gibbs energy of formation of an "amorphous" or "hydrated" form of UO_2 must be based on measurements for the particular solid being considered. No specific value for $\Delta_f G^\circ_m(UO_2$, "am", 298.15 K) or $\Delta_f G^\circ_m(UO_2$, "hyd", 298.15 K) is recommended in the present review.

V.3.3.3. *Mixed valence oxides*

V.3.3.3.1. $UO_{2.6667}$ ($\equiv \frac{1}{3}U_3O_8$)

The CODATA [89COX/WAG] Key Value of

$$\Delta_f H^\circ_m(UO_{2.6667} \equiv \tfrac{1}{3}U_3O_8, \text{cr}, 298.15\,\text{K}) = -(1191.6 \pm 0.8)\,\text{kJ}\cdot\text{mol}^{-1},$$

is accepted in this review. The value is based primarily on the experiments of Huber and Holley [69HUB/HOL] to determine the heat of combustion of uranium metal to U_3O_8. These measurements are in satisfactory agreement with, but more accurate than, earlier measurements by Huber, Holley and Meierkord [52HUB/HOL] and Popov and Ivanov [57POP/IVA].

The CODATA [89COX/WAG] Key Value for

$$S^\circ_m(UO_{2.6667} \equiv \frac{1}{3}U_3O_8, cr, 298.15\,K) = (94.18 \pm 0.17)\,J \cdot K^{-1} \cdot mol^{-1},$$

based on the work of Westrum and Grønvold [59WES/GRO], is also accepted in this review.

The Gibbs energy of formation is calculated from the selected enthalpy of formation and entropy.

$$\Delta_f G^\circ_m(UO_{2.6667} \equiv \frac{1}{3}U_3O_8, cr, 298.15\,K) = -(1123.2 \pm 0.8)\,kJ \cdot mol^{-1}$$

There have been a number of studies of the heat capacity of $UO_{2.6667}$ [58POP/GAL, 59WES/GRO, 68GIR/WES, 77INA/SHI]. The results of Popov, Gal'chenko and Senin [58POP/GAL] appear to be less accurate than those from the other studies. For temperatures between 300 and 350 K the heat capacity values of Girdhar and Westrum [68GIR/WES] are systematically smaller (by as much as 1%) than those reported by Westrum and Grønvold [59WES/GRO]. The values obtained by Inaba, Shimizu and Naito [77INA/SHI] also are systematically larger than those of Girdhar and Westrum over a large temperature range above 300 K.

Inaba *et al.* [77INA/SHI, 82NAI/INA] showed that several small λ-type heat capacity transitions occur above 450 K. The transition temperatures are apparently dependent on the thermal history of the sample, on the conditions of the measurement and, of course, the exact U:O ratio. (For a discussion of nonstoichiometry in U_3O_8 see the paper by Fujino, Tagawa and Adachi [81FUJ/TAG].) The transition enthalpy and entropy reported by Girdhar and Westrum [68GIR/WES] for the transition near 483 K (57.2 J per mole of uranium and $0.12\,J \cdot K^{-1}$ per mole of uranium, respectively) are only about 40% of the values determined by Inaba *et al.* [77INA/SHI, 82NAI/INA]. The transition heat and entropy for the transition at 568 K as reported by Inaba are similar in magnitude to those for the 483 K transition. Naito, Tsuji and Ohya [83NAI/TSU] examined the relation of phase transitions, as indicated by electrical conductivity and heat capacity measurements, to the conversion of the crystal structure of U_3O_{8-z} from orthorhombic to hexagonal. Within the temperature range of interest (< 600 K) there is little error ($< 0.3\,kJ \cdot mol^{-1}$) introduced in calculated values of $\Delta_f G^\circ_m(UO_{2.6667})$ by ignoring the transitions at 483 and 568 K, and using a single smooth heat capacity function over the temperature range 250 to 600 K. Therefore, all data from Refs. [59WES/GRO, 68GIR/WES, 77INA/SHI] for the temperature range 232 to 600 K are weighted using uncertainties proportional to those indicated by the the authors (Appendix A) to obtain the parameters given in Table V.10. It should be emphasised that this fitting function will not give accurate values for $C^\circ_{p,m}(UO_{2.6667})$ near the transition temperatures. The value

$$C^\circ_{p,m}(UO_{2.6667} \equiv \frac{1}{3}U_3O_8, cr, 298.15\,K) = (79.31 \pm 0.16)\,J \cdot K^{-1} \cdot mol^{-1}$$

selected in this review is the value from Ref. [59WES/GRO]. The uncertainty is estimated as twice the 0.1% uncertainty indicated by the authors.

V.3.3.3.2. $UO_{2.3333}$ ($\equiv \frac{1}{3} U_3O_7$)

In the present review, thermodynamic data are selected for this stoichiometry based on the analyses for the α- and β-forms in the original papers. It is recognized that the sample of the α-form [62WES/GRO2] was probably primarily (tetragonal) $U_{16}O_{37}$ (*i.e.*, $UO_{2.31}$) [70HOE/SIE]. Also, samples of the β-form [62WES/GRO2, 67FIT/PAV] may have been material similar in structure to the monoclinic U_8O_{19} (*i.e.*, $UO_{2.375}$) reported by Hoekstra, Siegel and Gallagher [70HOE/SIE].

The selected enthalpy of formation for β-$UO_{2.3333}$,

$$\Delta_f H^\circ_m(UO_{2.3333} \equiv \tfrac{1}{3}U_3O_7, \beta, 298.15\,K) = -(1142.0 \pm 1.4)\,kJ \cdot mol^{-1},$$

is from a re-evaluation of the enthalpy of solution data of Fitzgibbon, Pavone and Holley [67FIT/PAV] (see also Fuger [72FUG]).

The entropy of α-$UO_{2.3333}$, $(82.17 \pm 0.50)\,J \cdot K^{-1} \cdot mol^{-1}$, is the value of Westrum and Grønvold [62WES/GRO2] (from low temperature heat capacity measurements) less the contribution of 0.38 $J{\cdot}K^{-1}{\cdot}mol^{-1}$ originally attributed to a small transition at 30.5 K, but later ascribed to UO_2 impurity in the α-$UO_{2.3333}$ sample [66WES] (UO_2 has a sharp antiferromagnetic-paramagnetic transition at 30.3 to 30.5 K [66WES, 71HUN/WES, 83KHA]). The rather large uncertainty in the entropy, estimated in this review, reflects the impurity of the sample of α-$UO_{2.33}$ used for the low-temperature heat capacity study [62WES/GRO2]. The selected entropy for β-$UO_{2.3333}$,

$$S^\circ_m(UO_{2.3333} \equiv \tfrac{1}{3}U_3O_7, \beta, 298.15\,K) = (83.51 \pm 0.20)\,J \cdot K^{-1} \cdot mol^{-1},$$

is that reported by Westrum and Grønvold [62WES/GRO2], and the uncertainty in the entropy ($\pm 0.20\,J \cdot K^{-1} \cdot mol^{-1}$) is estimated in this review.

The Gibbs energy of formation is calculated from the selected enthalpy of formation and entropy.

$$\Delta_f G^\circ_m(UO_{2.3333} \equiv \tfrac{1}{3}U_3O_7, \beta, 298.15\,K) = -(1080.6 \pm 1.4)\,kJ \cdot mol^{-1}$$

Heat capacities of both α-$UO_{2.3333}$ and β-$UO_{2.3333}$ from the same source are fitted to give parameters for the temperature dependence over the temperature range noted in Table V.10. The experimental heat capacity data of Mukaibo *et al.* [62MUK/NAI] for an unspecified form of U_3O_7 (373 to 673 K) are inadequate to help extend the range of the heat capacity functions. If necessary, it would probably introduce less error to use the parameters in Table V.10 even above 350 K, especially for β-$UO_{2.3333}$. In the same paper Mukaibo *et al.* [62MUK/NAI] reported values for $\Delta_f H^\circ_m(U_3O_7)$ from differential thermal analysis data for successive oxidation of UO_2 to U_3O_7 and U_3O_8. Recalculation using values of $\Delta_f H^\circ_m(UO_2)$ and $\Delta_f H^\circ_m(UO_{2.6667})$ from Table V.11 leads, respectively, to $-1137\,kJ \cdot mol^{-1}$ and $-1146\,kJ \cdot mol^{-1}$ for $\Delta_f H^\circ_m(UO_{2.3333})$, in rough agreement with the calorimetric study for β-$UO_{2.3333}$. The values of $\Delta_f H^\circ_m(UO_{2.3333})$ would be consistent if the enthalpies of the oxidation reactions were slightly more exothermic than the experimental enthalpies of reaction reported in the DTA study.

Table V.11: Selected thermodynamic data of uranium oxides and hydroxides at 298.15 K.

Compound	$\Delta_f G^\circ_m$ (kJ · mol^{-1})	$\Delta_f H^\circ_m$ (kJ · mol^{-1})	S°_m (J · K^{-1} · mol^{-1})	$C^\circ_{p,m}$ (J · K^{-1} · mol^{-1})
$UO_2(cr)$	-1031.8 ± 1.0	-1085.0 ± 1.0	77.03 ± 0.20	63.60 ± 0.08
$UO_{2.25}(cr)$[(a)] ($\equiv \frac{1}{4}$-$U_4O_9(cr)$)	-1069.1 ± 1.7	-1128.0 ± 1.7	83.53 ± 0.17	73.34 ± 0.15
β-$UO_{2.25}$[(b)] ($\equiv \frac{1}{4}\beta$-U_4O_9)	-1069.1 ± 1.7	-1127.4 ± 1.7	85.4 ± 0.2	
α-$UO_{2.3333}$ ($\equiv \frac{1}{3}\alpha$-U_3O_7)			82.17 ± 0.50	71.42 ± 0.30
β-$UO_{2.3333}$ ($\equiv \frac{1}{3}\beta$-U_3O_7)	-1080.6 ± 1.4	-1142.0 ± 1.4	83.51 ± 0.20	71.84 ± 0.14
$UO_{2.6667}(cr)$ ($\equiv \frac{1}{3}U_3O_8(cr)$)	-1123.2 ± 0.8	-1191.6 ± 0.8	94.18 ± 0.17	79.31 ± 0.16
$UO_{2.86} \cdot 0.5H_2O$		-1367 ± 10		
$UO_{2.86} \cdot 1.5H_2O$		-1666 ± 10		
α-UO_3	-1140.4 ± 3.0	-1217.5 ± 3.0	99.4 ± 1.0	81.84 ± 0.30
β-UO_3	-1142.3 ± 1.3	-1220.3 ± 1.3	96.32 ± 0.40	81.34 ± 0.16
γ-UO_3	-1145.7 ± 1.2	-1223.8 ± 1.2	96.11 ± 0.40	81.67 ± 0.16
δ-UO_3		-1209 ± 5		
ε-UO_3		-1217.2 ± 1.3		
$UO_3(am)$		-1207.9 ± 4.0		
$UO_3 \cdot 0.393H_2O(cr)$		-1347.8 ± 1.3		
$UO_3 \cdot 0.648H_2O(cr)$		-1424.6 ± 1.3		
α-$UO_3 \cdot 0.85H_2O$		-1491.9 ± 1.3		
α-$UO_3 \cdot 0.9H_2O$	-1374.6 ± 2.5	-1506.3 ± 1.3	126 ± 7	140 ± 30
β-$UO_2(OH)_2$ ($\equiv \beta$-$UO_3 \cdot H_2O$)	-1398.7 ± 1.8	-1533.8 ± 1.3	138 ± 4	141 ± 15
γ-$UO_2(OH)_2$ ($\equiv \gamma$-$UO_3 \cdot H_2O \equiv \varepsilon$-$UO_3 \cdot H_2O$)		-1531.4 ± 1.3		
$UO_3 \cdot 2H_2O(cr)$	-1636.5 ± 1.7	-1826.1 ± 1.7	188.54 ± 0.38	172.07 ± 0.34

(a) Stable phase below 348 K.
(b) Stable phase above 348 K. The thermodynamic parameters, however, refer to 298.15 K.

The values

$$C^{\circ}_{p,\mathrm{m}}(\mathrm{UO_{2.3333}} \equiv \tfrac{1}{3}\mathrm{U_3O_7}, \alpha, 298.15\,\mathrm{K}) = (71.42 \pm 0.30)\,\mathrm{J \cdot K^{-1} \cdot mol^{-1}}$$
$$C^{\circ}_{p,\mathrm{m}}(\mathrm{UO_{2.3333}} \equiv \tfrac{1}{3}\mathrm{U_3O_7}, \beta, 298.15\,\mathrm{K}) = (71.84 \pm 0.14)\,\mathrm{J \cdot K^{-1} \cdot mol^{-1}}$$

selected in this review are the values from Ref. [62WES/GRO2]. The uncertainty is estimated for β-$UO_{2.3333}$ as twice the 0.1% uncertainty indicated by the authors. The larger uncertainty in $C^{\circ}_{p,\mathrm{m}}(\alpha\text{-}\mathrm{UO_{2.3333}})$ is chosen to reflect the presence of UO_2 as an impurity in the sample used by Westrum and Grønvold [62WES/GRO2].

V.3.3.3.3. $UO_{2.25}$ ($\equiv \frac{1}{4}U_4O_9$)

A recalculation of the enthalpy of solution data of Fitzgibbon, Pavone and Holley [67FIT/PAV] (see also Fuger [72FUG]) results in a value of $-(1128.3 \pm 1.5)\,\mathrm{kJ \cdot mol^{-1}}$ for the enthalpy of formation of $UO_{2.25}$ at 298.15 K. This is essentially identical to the value proposed by Glushko *et al.* [82GLU/GUR] who reported $-(4512 \pm 7)\,\mathrm{kJ \cdot mol^{-1}}$ for U_4O_9 (*i.e.*, $-(1128.0 \pm 1.7)\,\mathrm{kJ \cdot mol^{-1}}$ for $UO_{2.25}$), based on the same data. In the present review, rather than tabulate a "new" value that does not differ significantly from the value in the earlier assessment, the value

$$\Delta_{\mathrm{f}}H^{\circ}_{\mathrm{m}}(\mathrm{UO_{2.25}} \equiv \tfrac{1}{4}\mathrm{U_4O_9},\ \mathrm{cr},\ 298.15\ \mathrm{K}) = -(1128.0 \pm 1.7)\,\mathrm{kJ \cdot mol^{-1}}$$

is selected. It should be mentioned that in this review, values are selected for the composition $UO_{2.25}$, although recently Willis [87WIL] concluded from neutron-diffraction data that the structure of β-U_4O_9 is actually consistent with the composition $UO_{2.25-y}$, $y = 0.015625$.

The selected entropy,

$$S^{\circ}_{\mathrm{m}}(\mathrm{UO_{2.25}} \equiv \tfrac{1}{4}\mathrm{U_4O_9},\ \mathrm{cr},\ 298.15\ \mathrm{K}) = (83.53 \pm 0.17)\,\mathrm{J \cdot K^{-1} \cdot mol^{-1}}$$

is from the work of Osborne, Westrum and Lohr [57OSB/WES] with a correction of $-0.450\,\mathrm{J \cdot K^{-1}}$ per mole of $UO_{2.25}$ ($-0.430\,\mathrm{cal \cdot K^{-1}}$ per mole of U_4O_9) applied on the basis of the low-temperature heat capacity data of Flotow, Osborne and Westrum [68FLO/OSB]. There are several sets of heat capacity data for uranium oxides with a 1:2.250 [65GOT/NAI, 65WES/TAK, 70GRO/KVE, 73INA/NAI, 82SET/MAT]. These indicate that a λ-type transition to β-U_4O_9 occurs near 348 K. The exact temperature of the transition apparently depends to some extent on the thermal history of the oxide sample [73INA/NAI].

The Gibbs energy of formation, valid at 298.15 K, is calculated from the selected enthalpy of formation and entropy.

$$\Delta_{\mathrm{f}}G^{\circ}_{\mathrm{m}}(\mathrm{UO_{2.25}} \equiv \tfrac{1}{4}\mathrm{U_4O_9},\ \mathrm{cr},\ 298.15\ \mathrm{K}) = -(1069.1 \pm 1.7)\,\mathrm{kJ \cdot mol^{-1}}$$

Heat capacity data for the temperature range 235 to 348 K are fitted to an equation of the form

$$C^{\circ}_{p,\mathrm{m}}(\mathrm{UO_{2.25}} \equiv \tfrac{1}{4}\mathrm{U_4O_9}, \mathrm{cr}, T) = a + bT + cT^2 + eT^{-2}. \qquad (\mathrm{V.29})$$

Data are adjusted to $UO_{2.25}$ (by assuming the heat capacities of the oxides are proportional to their molecular weights) and weighted using uncertainties proportional to those ascribed to the data by the authors of the original papers. Eq. (V.29) gives a reasonable fit between 250 and 348 K, and allows calculation of $S^\circ_m(T)$ and $\Delta_f G^\circ_m(T)$ in that temperature range. The fit is poorer near 348 K, but no simple function can be found that is more satisfactory. The parameters (Table V.10) are highly correlated (and if fewer digits are retained for these parameters, the calculated values of $C^\circ_{p,m}(T)$ are significantly affected).

The sharp decrease in the heat capacity of $UO_{2.25}$ immediately above 348 K cannot be easily described by a simple function of temperature. To facilitate calculation of chemical thermodynamic parameters for $UO_{2.25}$ at higher temperatures, values of S°_m and $\Delta_f H^\circ_m$ are calculated for the (hypothetical) high temperature form (β-U_4O_9) at 298.15 K. All available experimental heat capacity data for 190 K to 600 K, excluding those in the region of the λ-transition (275 to 400 K), are appropriately weighted and fitted to a function of the form

$$C^\circ_{p,m}(UO_{2.25} \equiv \tfrac{1}{4}U_4O_9, \beta, T) \quad = \quad a + bT + eT^{-2}. \qquad \text{(V.30)}$$

$a = (74.27 \pm 0.41)$, $b = (0.02153 \pm 0.00078)$, $e = -(8.227 \pm 0.115)$. (It should be noted that the data of Inaba and Naito [73INA/NAI] for 405 to 470 K are systematically 1 to 4% greater than predicted by Eq. (V.30)). Using

$$S^\circ_m(T_2) \quad = \quad S^\circ_m(T_1) + a\ln\left(\frac{T_2}{T_1}\right) + b(T_2 - T_1) - e\left(\frac{T_2^2 - T_1^2}{2}\right) \qquad \text{(V.31)}$$

a value of $S^\circ_m(UO_{2.25}, \text{cr}, 298.15\,\text{K}) = (83.4 \pm 0.2)\,\text{J}\cdot\text{K}^{-1}\cdot\text{mol}^{-1}$ is calculated for $UO_{2.25}$($\frac{1}{4}$ α-U_4O_9) if there were no λ-type transition. The value for $\Delta_f H^\circ_m$(298.15 K) for α-U_4O_9 (within the uncertainty selected by this review) would be unchanged from the selected value for $UO_{2.25}$. The values of the enthalpy and entropy of the λ-type transition at 348 K are calculated to be $0.63_6\,\text{kJ}\cdot\text{mol}^{-1}$ and $1.9_7\,\text{J}\cdot\text{K}^{-1}\cdot\text{mol}^{-1}$, respectively, from the weighted results of Gotoo and Naito [65GOT/NAI] Inaba and Naito [73INA/NAI], Westrum, Takahashi and Grønvold [65WES/TAK] and Grønvold *et al.* [70GRO/KVE]. Thus, for the higher temperature form of $UO_{2.25}$, the selected values are

$$\Delta_f H^\circ_m(UO_{2.25} \equiv \tfrac{1}{4}U_4O_9, \beta,\ 298.15\ \text{K}) \quad = \quad -(1127.4 \pm 1.7)\,\text{kJ}\cdot\text{mol}^{-1}$$
$$S^\circ_m(UO_{2.25} \equiv \tfrac{1}{4}U_4O_9, \beta,\ 298.15\ \text{K}) \quad = \quad (85.4 \pm 0.2)\,\text{J}\cdot\text{K}^{-1}\cdot\text{mol}^{-1}.$$

The Gibbs energy of formation, valid at 298.15 K, is calculated from the selected enthalpy of formation and entropy.

$$\Delta_f G^\circ_m(UO_{2.25} \equiv \tfrac{1}{4}U_4O_9, \beta,\ 298.15\ \text{K}) \quad = \quad -(1069.1 \pm 1.7)\,\text{kJ}\cdot\text{mol}^{-1}$$

To calculate the Gibbs energy of $UO_{2.25}$ from 348 to 675 K these values should be used in conjunction with the heat capacity function parameters calculated from experimental heat capacity data for 390 to 675 K (*cf.* Table V.10). Uncertainties in $\Delta_f G^\circ_m(UO_{2.25})$ introduced by this procedure are within the experimental uncertainties, even below 370 K.

The value and uncertainty limit of

$$C^\circ_{p,m}(UO_{2.25} \equiv \tfrac{1}{4}U_4O_9, \text{cr}, 298.15 \text{ K}) = (73.34 \pm 0.15) \text{ J} \cdot \text{K}^{-1} \cdot \text{mol}^{-1}$$

are from the work of Osborne *et al.* [57OSB/WES, 68FLO/OSB].

V.3.3.3.4. Other mixed valence oxides

An oxide with a composition $UO_{2.92}$ was obtained on decomposition of amorphous UO_3 in air at 450 to 550°C and also by oxidation of U_3O_8 at 500°C (40 atm O_2) [61HOE/SIE]. Although the enthalpy of solution of this solid in 6 M HNO_3 was reported [64COR], no information was provided concerning the species reduced on oxidation of the solid to U(VI) in solution. As a result, no value for $\Delta_f H^\circ_m(UO_{2.92}$, cr, 298.15 K) is selected in this review.

There are several sets of heat capacity data (T < 600 K) reported for non-stoichiometric oxides (U_3O_{8-y} [70MAR/CIO, 82NAI/INA], U_4O_{9-y} [70MAR/CIO, 73INA/NAI, 82SET/MAT], $UO_{2.5}$ [71MAR/CIO]). These data are not critically evaluated in this review.

V.3.3.3.5. Hydrates of mixed valence oxides

Cordfunke, Prins and van Vlaanderen [68COR/PRI] reported enthalpies of solution for $UO_{2.86} \cdot 0.5H_2O$(cr) and $UO_{2.86} \cdot 1.5H_2O$(cr) in a mixture of $Ce(SO_4)_2$ and H_2SO_4, and in a later publication [78COR/OHA] these enthalpies of solution were used to calculate enthalpies of formation for the two solid hydrates. These values for the enthalpies of formation,

$$\Delta_f H^\circ_m(UO_{2.86} \cdot 0.5H_2O, \text{cr}, 298.15 \text{ K}) = -(1367 \pm 10) \text{ kJ} \cdot \text{mol}^{-1}$$
$$\Delta_f H^\circ_m(UO_{2.86} \cdot 1.5H_2O, \text{cr}, 298.15 \text{ K}) = -(1666 \pm 10) \text{ kJ} \cdot \text{mol}^{-1}$$

are accepted in this review. The estimation of the uncertainties is discussed in Appendix A. The X-ray pattern of $UO_{2.86} \cdot 1.5H_2O$(cr), prepared photochemically from solution, was found [68COR/PRI] to be identical to that for the mineral ianthinite, $UO_{2.833} \cdot 1.76H_2O$(cr) [59GUI/PRO].

V.3.4. Uranium hydrides

Flotow, Haschke and Yamauchi [84FLO/HAS] recently reviewed the chemical thermodynamics of the actinide hydrides for the IAEA. This review accepts their findings which are briefly summarized below.

Three phases are known in the uranium-hydrogen system: α-UH_r, α-UH_3 and β-UH_3. α-UH_r represents hydrogen dissolved in α-U at the H/U atomic ratio r. Mulford, Ellinger and Zachariasen [54MUL/ELL] and Caillat, Coriou and Perio [53CAI/COR] independently found some preparations of β-UH_3 to contain a second crystalline phase of the same stoichiometry, but no way is known to synthesize the pure substance. Above 473 K α-UH_3 transforms to β-UH_3, but the reverse has never been observed, even after cycling between room temperature and 4.2 K. Abraham and Flotow [55ABR/FLO] found the enthalpy of transition to be negligible, so all results here apply only to β-UH_3.

Abraham and Flotow [55ABR/FLO] determined the enthalpy of formation of β-UH_3 with all three isotopes of hydrogen using a reaction calorimetric technique. They found the values

$$\begin{aligned}\Delta_f H_m^\circ(UH_3, \beta, 298.15\,K) &= -(126.98 \pm 0.13)\,kJ \cdot mol^{-1}\\ \Delta_f H_m^\circ(UD_3, \beta, 298.15\,K) &= -(129.79 \pm 0.13)\,kJ \cdot mol^{-1}\\ \Delta_f H_m^\circ(UT_3, \beta, 298.15\,K) &= -(130.29 \pm 0.21)\,kJ \cdot mol^{-1}\end{aligned}$$

which are accepted here.

The heat capacity of β-UH_3 was measured by Flotow and coworkers from 1.4 to 23 K [67FLO/OSB] and from 5 to 350 K [59FLO/LOH]. The combined results and thermal function are accepted here, as they were by the IAEA [84FLO/HAS].

$$C_{p,m}^\circ(UH_3, \beta, 298.15\,K) = (49.29 \pm 0.08)\,J \cdot K^{-1} \cdot mol^{-1}$$

The heat capacity of β-UD_3 was measured by Abraham *et al.* [60ABR/OSB] from 5 to 350 K. These results are also accepted.

$$C_{p,m}^\circ(UD_3, \beta, 298.15\,K) = (64.98 \pm 0.08)\,J \cdot K^{-1} \cdot mol^{-1}$$

No low-temperature heat capacity measurement exists for β-UT_3. Flotow, Haschke and Yamauchi [84FLO/HAS] estimated

$$C_{p,m}^\circ(UT_3, \beta, 298.15\,K) = (74.43 \pm 0.75)\,J \cdot K^{-1} \cdot mol^{-1}$$

and the corresponding thermal function from the data for β-UH_3 and β-UD_3 using a method which sums heat capacity contributions [84FLO/HAS, *p.*7].

The thermal functions of heat capacity lead to the accepted entropy values

$$\begin{aligned}S_m^\circ(UH_3, \beta, 298.15\,K) &= (63.68 \pm 0.13)\,J \cdot K^{-1} \cdot mol^{-1}\\ S_m^\circ(UD_3, \beta, 298.15\,K) &= (71.76 \pm 0.13)\,J \cdot K^{-1} \cdot mol^{-1}\\ S_m^\circ(UT_3, \beta, 298.15\,K) &= (79.08 \pm 0.79)\,J \cdot K^{-1} \cdot mol^{-1}.\end{aligned}$$

Corresponding consistent Gibbs energy values are calculated. The required entropy of D_2(g), *i.e.*, $^2H_2$(g), is taken from the US NBS tables [82WAG/EVA], $S_m^\circ(D_2$, g, 298.15 K) = $(144.96 \pm 0.02)\,J \cdot K^{-1} \cdot mol^{-1}$. The entropy of T_2(g), *i.e.*, $^3H_2$(g), is taken from the Russian tables [78GLU/GUR] but is converted from the 1 atm to the 1 bar standard state pressure, $S_m^\circ(T_2$, g, 298.15 K) = $(153.326 \pm 0.020)\,J \cdot K^{-1} \cdot mol^{-1}$. The uncertainties of these entropies are estimated by this review.

$$\begin{aligned}\Delta_f G_m^\circ(UH_3, \beta, 298.15\,K) &= -(72.56 \pm 0.15)\,kJ \cdot mol^{-1}\\ \Delta_f G_m^\circ(UD_3, \beta, 298.15\,K) &= -(71.39 \pm 0.15)\,kJ \cdot mol^{-1}\\ \Delta_f G_m^\circ(UT_3, \beta, 298.15\,K) &= -(70.33 \pm 0.32)\,kJ \cdot mol^{-1}\end{aligned}$$

It should be mentioned that the data on β-UD_3 and β-UT_3 are not included in Table III.1 of Chapter III, since no isotopic distinctions are made in the computerized data base of the NEA Thermochemical Data Base project.

V.4. Group 17 (halogen) compounds and complexes

V.4.1. Fluorine compounds and complexes

V.4.1.1. Gaseous uranium fluorides

V.4.1.1.1. UF(g) and UF_2(g)

Lau and Hildenbrand [82LAU/HIL] as well as Gorokhov, Smirnov and Khodeev [84GOR/SMI] studied molecular exchange reactions for these species by mass spectrometry. The former authors studied the reactions of the lower uranium fluorides with BaF(g):

$$\mathrm{U(g) + BaF(g) \rightleftharpoons UF(g) + Ba(g)} \quad (V.32)$$
$$\mathrm{UF(g) + BaF(g) \rightleftharpoons UF_2(g) + Ba(g)} \quad (V.33)$$
$$\mathrm{UF_2(g) + BaF(g) \rightleftharpoons UF_3(g) + Ba(g)} \quad (V.34)$$

Third-law analysis of Reaction (V.32) gives $\Delta_r H^\circ_m$(V.32, 298.15 K) $= -(87.8\pm0.2)$ kJ · mol^{-1} and, hence, $\Delta_f H^\circ_m$(UF, g, 298.15 K) $= -(58.8 \pm 22.8)$ kJ · mol^{-1}. Similarily, $\Delta_r H^\circ_m$(V.33, 298.15 K) $= -(8.5 \pm 0.6)$ kJ · mol^{-1} is obtained, corresponding to [$\Delta_f H^\circ_m$(UF, g, 298.15 K) $-$ $\Delta_f H^\circ_m$(UF_2, g, 298.15 K)] $= (512.5\pm29.2)$ kJ · mol^{-1}, and $\Delta_r H^\circ_m$(V.34, 298.15 K) $= -(36.1 \pm 0.3)$ kJ · mol^{-1}, giving $\Delta_f H^\circ_m$(UF_2, g, 298.15 K) $= -(514.1 \pm 26.2)$ kJ · mol^{-1}, when combined with the selected values for UF(g) and UF_3(g).

Gorokhov, Smirnov and Khodeev [84GOR/SMI] studied the reactions

$$\mathrm{2UF_3(g) \rightleftharpoons UF_4(g) + UF_2(g)} \quad (V.35)$$
$$\mathrm{2UF_2(g) \rightleftharpoons UF(g) + UF_3(g)} \quad (V.36)$$

By third-law analysis $\Delta_r H^\circ_m$(V.35, 298.15 K) $= -(21.70\pm0.34)$ kJ · mol^{-1} is obtained, giving $\Delta_f H^\circ_m$(UF_2, g, 298.15 K) $= -(528.9 \pm 26.2)$ kJ · mol^{-1} when combined with the selected values for UF_4(g) and UF_3(g). Similarily, $\Delta_r H^\circ_m$(V.36, 298.15 K) $= -(30.0 \pm 3.2)$ kJ · mol^{-1} is obtained.

Recently, Hildenbrand and Lau [91HIL/LAU2] performed mass spectrometric measurements for the reaction

$$\mathrm{U(g) + UF_2(g) \rightleftharpoons 2UF(g)} \quad (V.37)$$

from which $\Delta_r H^\circ_m$(V.37, 298.15 K) $= -(80.0 \pm 0.4)$ kJ · mol^{-1} is obtained by third-law analysis. Combining the results for Eqs. (V.36) and (V.37) gives $\Delta_f H^\circ_m$(UF_2, g, 298.15 K) $= -(532\pm30)$ kJ·mol^{-1} and $\Delta_f H^\circ_m$(UF, g, 298.15 K) $= -(39\pm30)$ kJ·mol^{-1}, with uncertainties estimated by this review. These values are in good agreement with the results cited above.

For the selection of $\Delta_f H^\circ_m$(UF, g, 298.15 K), this review gives more weight to the value derived from Reaction (V.32) using well defined auxiliary data, and the value

$$\Delta_f H^\circ_m(\mathrm{UF, g, 298.15\ K}) = -(52 \pm 30)\ \mathrm{kJ \cdot mol^{-1}}$$

is recommended.

For UF_2(g), the concordant results from the studies of Gorokhov, Smirnov and Khodeev [84GOR/SMI] and Hildenbrand and Lau [91HIL/LAU2] are preferred, and the value of

$$\Delta_f H_m^\circ(UF_2, g, 298.15\ K) = -(530 \pm 30)\ kJ \cdot mol^{-1}$$

is selected.

The thermal functions of UF_2(g) and UF(g) are calculated from molecular constants estimated by Glushko *et al.* [82GLU/GUR].

$$S_m^\circ(UF, g, 298.15\ K) = (252 \pm 10)\ J \cdot K^{-1} \cdot mol^{-1}$$
$$S_m^\circ(UF_2, g, 298.15\ K) = (316 \pm 10)\ J \cdot K^{-1} \cdot mol^{-1}$$

$$C_{p,m}^\circ(UF, g, 298.15\ K) = (37.9 \pm 5.0)\ J \cdot K^{-1} \cdot mol^{-1}$$
$$C_{p,m}^\circ(UF_2, g, 298.15\ K) = (56.2 \pm 5.0)\ J \cdot K^{-1} \cdot mol^{-1}$$

The Gibbs energies of formation are calculated from the selected enthalpies of formation and entropies.

$$\Delta_f G_m^\circ(UF, g, 298.15\ K) = -(82 \pm 30)\ kJ \cdot mol^{-1}$$
$$\Delta_f G_m^\circ(UF_2, g, 298.15\ K) = -(549 \pm 30)\ kJ \cdot mol^{-1}$$

V.4.1.1.2. UF_3 (g)

The enthalpy of formation of UF_3(g) is calculated from the enthalpy of sublimation obtained by third-law analysis of the vapour pressure data measurements (1252 to 1469 K) by Gorokhov, Smirnov and Khodeev [84GOR/SMI]. For the equilibrium

$$UF_3(cr) \rightleftharpoons UF_3(g) \qquad (V.38)$$

this review evaluates the selected value with an estimated uncertainty:

$$\Delta_{sub} H_m^\circ(V.38, 298.15\ K) = (447.2 \pm 15.0)\ kJ \cdot mol^{-1}.$$

The enthalpy of formation of UF_3(cr) selected in this review is used to derive

$$\Delta_f H_m^\circ(UF_3, g, 298.15\ K) = -(1054.2 \pm 15.7)\ kJ \cdot mol^{-1}.$$

This value is in reasonable agreement with the enthalpies of formation derived from the mass spectrometric measurements by Zmbov [69ZMB] and Lau and Hildenbrand [82LAU/HIL] for the reaction

$$UF_4(g) + Ca(g) \rightleftharpoons UF_3(g) + CaF(g) \qquad (V.39)$$

which are given in Table V.12. The results of the analysis of Reaction (V.39) are, however, considered less reliable since the uncertainties in the auxilairy data are considerably larger. Roy *et al.* [82ROY/PRA] studied the vapour pressure of UF_3(cr) in a hydrogen stream. A third-law analysis of these vapour pressure data (1229 to

Table V.12: Enthalpy of formation values for $UF_3(g)$ at 298.15 K, as evaluated from third-law analysis of experimental data.

Reference	Reaction investigated	$\Delta_f H_m^\circ(UF_3, g, 298.15\ K)$ $(kJ \cdot mol^{-1})$
[69ZMB]	$UF_4(g) + Ca(g) \rightleftharpoons UF_3(g) + CaF(g)$	-1066.4 ± 17.8
[82LAU/HIL]	$UF_4(g) + Ca(g) \rightleftharpoons UF_3(g) + CaF(g)$	-1088.3 ± 17.8
[82ROY/PRA]	$UF_3(cr) \rightleftharpoons UF_3(g)$	-1136.3 ± 15.7
[84GOR/SMI]	$UF_3(cr) \rightleftharpoons UF_3(g)$	-1054.2 ± 15.7

1367 K) yields $\Delta_{sub}H_m^\circ(V.38, 298.15\ K) = (365.1 \pm 1.2)\ kJ \cdot mol^{-1}$; however, the results show a clear temperature dependence. The value significantly differs from the results of Gorokhov, Smirnov and Khodeev [84GOR/SMI] which may be due to the poor characterization of the equilibria occuring in the experiments.

Values for the heat capacity and entropy of $UF_3(g)$ are calculated from estimated molecular parameters given by Glushko *et al.* [82GLU/GUR].

$$S_m^\circ(UF_3, g, 298.15\ K) = (347 \pm 10)\ J \cdot K^{-1} \cdot mol^{-1}$$
$$C_{p,m}^\circ(UF_3, g, 298.15\ K) = (76.2 \pm 5.0)\ J \cdot K^{-1} \cdot mol^{-1}$$

The Gibbs energy of formation is calculated from the selected enthalpy of formation and entropy.

$$\Delta_f G_m^\circ(UF_3, g, 298.15\ K) = -(1052.0 \pm 16.0)\ kJ \cdot mol^{-1}$$

V.4.1.1.3. UF_4 (g)

The enthalpy of formation of $UF_4(g)$ is calculated from the enthalpy of sublimation and the enthalpy of formation of the solid. The former value is obtained from a third-law analysis of the vapour pressure data for the reaction

$$UF_4(cr) \rightleftharpoons UF_4(g) \quad (V.40)$$

as listed in Table V.13. The studies are in reasonable agreement except for the results of Altman [43ALT] and Akishin and Khodeev [61AKI/KHO]. The value

$$\Delta_{sub}H_m^\circ(V.40, 298.15\ K) = (313.0 \pm 8.0)\ kJ \cdot mol^{-1}$$

is selected here, where the uncertainty reflects the uncertainties in the thermal function. This leads to the enthalpy of formation of

$$\Delta_f H_m^\circ(UF_4, g, 298.15\ K) = -(1601.2 \pm 9.0)\ kJ \cdot mol^{-1}.$$

Table V.13: Enthalpy of sublimation values for $UF_4(cr) \rightleftharpoons UF_4(g)$ at 298.15 K, as evaluated from third-law analysis of experimental data.

Reference	T-range (K)	$\Delta_{sub}H^{o}_{m}$(V.40, 298.15 K)[(a)] (kJ · mol^{-1})
[43ALT]	1091-1188	363.5 ± 4.3
[47JOH]	1120-1275	306.8 ± 0.7
[59POP/KOS]	1148-1273	318.4 ± 1.0
[60LAN/BLA]	1291-1575	313.9 ± 0.2
[61AKI/KHO]	950-1040	344.0 ± 3.0
[70CHU/CHO]	828-1280	309.7 ± 0.5
[77HIL]	980-1130	313.1 ± 0.4
[80NAG/BHU]	1169-1427	314.5 ± 0.4
(b)	1060-1137	308.0 ± 1.6

(a) The indicated errors represent the statistical uncertainties.
(b) Khodeev, Yu.S., *et al.*, cited by Glushko *et al.* [82GLU/GUR].

The heat capacity and entropy are calculated from molecular parameters (including estimated electronic energy levels up to 22000 cm^{-1}) given by Glushko *et al.* [82GLU/GUR]. These are in good agreement with experimental data from recent infrared and electron diffraction studies. The infrared study [83KOV/CHI] located the U-F streching frequency at 550 cm^{-1}; the electron diffraction data by Girichev *et al.* [83GIR/PET] indicated a distorted tetrahedral structure of C_{3v} or a D_{2d} symmetry. Glushko *et al.* [82GLU/GUR] selected a C_{2v} configuration from earlier measurements by Ezhov, Akishin and Rambidi [69EZH/AKI].

$$S^{o}_{m}(UF_4, g, 298.15\ K) = (363.0 \pm 5.0)\ J \cdot K^{-1} \cdot mol^{-1}$$
$$C^{o}_{p,m}(UF_4, g, 298.15\ K) = (101.4 \pm 3.0)\ J \cdot K^{-1} \cdot mol^{-1}$$

The Gibbs energy of formation is calculated from the selected enthalpy of formation and entropy.

$$\Delta_f G^{o}_{m}(UF_4, g, 298.15\ K) = -(1573.5 \pm 9.2)\ kJ \cdot mol^{-1}$$

V.4.1.1.4. UF_5 (g)

There are a number of recent studies on the enthalpy of formation of $UF_5(g)$. Bondarenko *et al.* [87BON/KOR] studied the reaction

$$UF_4(cr) + UF_6(g) \rightleftharpoons 2UF_5(g) \qquad (V.41)$$

from 663 to 809 K by mass spectrometry (also reported by Borshchevskii *et al.* [88BOR/BOL]). However, the $UF_5(g)$ pressure was not measured directly, but calculated from the pressure of the dimer, using the equilibrium constant from [79KLE/HIL] which is also selected by the present review (see Section V.4.1.1.6). From their data $\Delta_r H_m^\circ(\text{V.41, 0 K}) = (245.6 \pm 1.5)\,\text{kJ}\cdot\text{mol}^{-1}$ is obtained by third-law analysis, and, hence, $\Delta_f H_m^\circ(\text{UF}_5\text{, g, 298.15 K}) = -(1908.7 \pm 10.0)\,\text{kJ}\cdot\text{mol}^{-1}$ by combining this value with auxiliary data from this assessment, the uncertainty being raised to account for uncertainties in the thermal functions of $UF_5(g)$ and in the pressure calibration.

Smirnov and Gorokhov [84SMI/GOR] studied the reaction

$$\text{UO}_2\text{F}_2(\text{g}) + 2\text{UF}_5(\text{g}) \rightleftharpoons 3\text{UF}_4(\text{g}) + \text{O}_2(\text{g}) \qquad (\text{V.42})$$

by mass spectrometry. A third-law analysis by the authors (no experimental data were reported) gave $\Delta_r H_m^\circ(\text{V.42, 0 K}) = (352 \pm 40)\,\text{kJ}\cdot\text{mol}^{-1}$, which leads to $\Delta_f H_m^\circ(\text{UF}_5\text{, g, 298.15 K}) = -(1906 \pm 40)\,\text{kJ}\cdot\text{mol}^{-1}$.

Lau, Brittain and Hildenbrand [85LAU/BRI] found $\Delta_f H_m^\circ(\text{UF}_5\text{, g, 298.15 K}) = -(1894.5 \pm 15)\,\text{kJ}\cdot\text{mol}^{-1}$ from an analysis of the reaction

$$\tfrac{5}{2}\text{UO}_2\text{F}_2(\text{cr}) \rightleftharpoons \text{UF}_5(\text{g}) + \tfrac{1}{2}\text{O}_2(\text{g}) + \tfrac{1}{2}\text{U}_3\text{O}_8(\text{cr}) \qquad (\text{V.43})$$

This value could not be recalculated since no experimental data were reported. Gorokhov, Smirnov and Khodeev [84GOR/SMI] studied three different equilibria involving $UF_5(g)$. From their data for the reaction

$$\text{UF}_4(\text{g}) + \text{CuF}_2(\text{g}) \rightleftharpoons \text{UF}_5(\text{g}) + \text{CuF}(\text{g}), \qquad (\text{V.44})$$

$\Delta_r H_m^\circ(\text{V.44, 298.15 K}) = -(44.7 \pm 0.6)\,\text{kJ}\cdot\text{mol}^{-1}$ is obtained by third-law analysis and, hence, $\Delta_f H_m(\text{UF}_5\text{, g, 298.15 K}) = -(1900.2 \pm 23.8)\,\text{kJ}\cdot\text{mol}^{-1}$. For the reactions

$$\text{UF}_4(\text{cr}) + \text{UF}_6(\text{g}) \rightleftharpoons 2\text{UF}_5(\text{g}) \qquad (\text{V.45})$$

$$\tfrac{4}{3}\text{UF}_{4.25}(\text{cr}) + \text{UF}_6(\text{g}) \rightleftharpoons \tfrac{7}{3}\text{UF}_5(\text{g}), \qquad (\text{V.46})$$

$\Delta_r H_m^\circ(\text{V.45, 298.15 K}) = (239.1 \pm 1.5)\,\text{kJ}\cdot\text{mol}^{-1}$ and $\Delta_r H_m^\circ(\text{V.46, 298.15 K}) = (266.5 \pm 2.1)\,\text{kJ}\cdot\text{mol}^{-1}$ are obtained, respectively, which lead to $\Delta_f H_m^\circ(\text{UF}_5\text{, g, 298.15 K}) = -(1911.9 \pm 10.0)\,\text{kJ}\cdot\text{mol}^{-1}$ and $-(1928.1 \pm 12.3)\,\text{kJ}\cdot\text{mol}^{-1}$, respectively, when combined with auxiliary values from the present assessment. Similar to Bondarenko *et al.* [87BON/KOR], the UF_5 pressure was obtained from measurement of the dimer pressure.

The results of these studies, as summarized in Table V.14, are considerably different from the study by Hildenbrand [77HIL] of the reaction

$$\text{Ag}(\text{g}) + \text{UF}_5(\text{g}) \rightleftharpoons \text{AgF}(\text{g}) + \text{UF}_4(\text{g}) \qquad (\text{V.47})$$

for which $\Delta_r H_m^\circ(\text{V.47, 298.15 K}) = (66.8 \pm 0.2)\,\text{kJ}\cdot\text{mol}^{-1}$ is obtained by third-law method. Recently, Hildenbrand and Lau [91HIL/LAU2] restudied this equilibrium.

Table V.14: Enthalpy of formation values for $UF_5(g)$ at 298.15 K, as evaluated from third-law analysis of experimental data.

Reference	Reaction investigated	$\Delta_f H_m^\circ(UF_5, g, 298.15 K)$ ($kJ \cdot mol^{-1}$)
[77HIL]	$UF_5(g) + Ag(g) \rightleftharpoons AgF(g) + UF_4(g)$	-1945.4 ± 12.8
[83LEI]	$UF_5(cr) \rightleftharpoons UF_5(g)$	-1918.5
[84SMI/GOR]	$UO_2F_2(g) + 2UF_5(g) \rightleftharpoons 3UF_5(g) + O_2(g)$	-1906 ± 40
[84GOR/SMI]	$UF_4(cr) + UF_6(g) \rightleftharpoons 2UF_5(g)$	-1911.9 ± 10.0
	$\frac{3}{4}UF_{4.25}(cr) + UF_6(g) \rightleftharpoons \frac{7}{3}UF_5(g)$	-1928.1 ± 12.3
	$UF_4(g) + CuF_2(g) \rightleftharpoons UF_5(g) + CuF(g)$	-1900.2 ± 23.8
[85LAU/BRI]	$\frac{5}{2}UO_2F_2 \rightleftharpoons UF_5(g) + \frac{1}{2}O_2(g) + \frac{1}{2}U_3O_8(cr)$	-1894.5 ± 15.0
[87BON/KOR]	$UF_4(cr) + UF_6(g) \rightleftharpoons 2UF_5(g)$	-1908.7 ± 10.0
[91HIL/LAU2]	$UF_5(g) + Ag(g) \rightleftharpoons AgF(g) + UF_4(g)$	-1946.5 ± 12.8

From these data $\Delta_r H_m^\circ(V.47, 298.15 K) = (67.9 \pm 0.5)\,kJ \cdot mol^{-1}$ is obtained by third-law method, in good agreement with the earlier value. Combining these values with the recent value for $\Delta_f H_m^\circ(AgF, g, 298.15 K) = (7.5 \pm 6)\,kJ \cdot mol^{-1}$ from Hildenbrand and Lau [91HIL/LAU2], the values $\Delta_f H_m^\circ(UF_5, g, 298.15 K) = -(1945.4 \pm 12.8)\,kJ \cdot mol^{-1}$ and $-(1946.5 \pm 12.8)\,kJ \cdot mol^{-1}$ are obtained, respectively.

Also shown in Table V.14 is the value proposed by Leitnaker [83LEI], which was calculated from the data for the sublimation reaction

$$UF_5(cr) \rightleftharpoons UF_5(g). \qquad (V.48)$$

This value was derived from a correction of the vapour pressure measurements of Wolf *et al.* [65WOL/POS] applying corrections for dimerization and dissociation processes.

From the table it is clear that considerable variation exists in the resulting data. This is mainly due to the many difficulties involved in the experiments and uncertainties in the auxiliary data, however, obvious sources of errors cannot be found in the studies. As a best value

$$\Delta_f H_m^\circ(UF_5, g, 298.15\,K) = -(1910 \pm 15)\,kJ \cdot mol^{-1}$$

is suggested, primarily based on the equilibrium studies in the U-F system to preserve the internal consistency between the U-F species as well as the vapour pressure data. More work is still required to resolve the discrepancies.

The thermal functions of $UF_5(g)$ are calculated from molecular parameters given by Glushko *et al.* [82GLU/GUR]:

$$S_m^\circ(UF_5, g, 298.15 K) = (386.1 \pm 10.0)\,J \cdot K^{-1} \cdot mol^{-1}$$
$$C_{p,m}^\circ(UF_5, g, 298.15 K) = (110.6 \pm 5.0)\,J \cdot K^{-1} \cdot mol^{-1}$$

The Gibbs energy of formation is calculated from the selected enthalpy of formation and entropy.

$$\Delta_f G^\circ_m(UF_5, g, 298.15 K) = -(1859 \pm 15) kJ \cdot mol^{-1}$$

V.4.1.1.5. UF_6 (g)

The enthalpy of formation of UF_6(g) is calculated from the enthalpy of sublimation,

$$UF_6(cr) \rightleftharpoons UF_6(g) \quad (V.49)$$

and the enthalpy of formation of the solid. The former value is obtained by a third-law analysis of the concordant vapour pressure measurements listed in Table V.15. Not included in the analysis are the vapour pressure measurements by Ruff and Heinzelman [11RUF/HEI], which seriously deviate. In addition, Masi [49MAS] determined the enthalpy of sublimation by direct calorimetric measurements to be 49.67 $kJ \cdot mol^{-1}$ at 298.15 K. This value is somewhat higher than that from the vapour pressure measurements.

Table V.15: Enthalpy of sublimation values for $UF_6(cr) \rightleftharpoons UF_6(g)$ at 298.15 K, as evaluated from third-law analysis of experimental data.

Reference	T-range (K)	$\Delta_{sub}H^\circ_m(V.49, 298.15 K)$[(a)] ($kJ \cdot mol^{-1}$)
[41AMP/THO]	294-329	48.96 ± 0.04
[41WIL/LYN]	259-285	48.88 ± 0.07
[48AMP/MUL]	286-303	49.14 ± 0.02
[48WEI/CRI]	273-336	49.09 ± 0.05
[50KIG]	273-308	49.23 ± 0.20
[53LLE]	273-337	49.20 ± 0.02
[53OLI/MIL]	250-320	49.12 ± 0.07
[78MEI/HEI]	250-320	49.12 ± 0.07

(a) The indicated errors represent the statistical uncertainties.

The selected value is

$$\Delta_{sub}H^\circ_m(V.49, 298.15 K) = (49.1 \pm 0.5) kJ \cdot mol^{-1},$$

where the uncertainty reflects the uncertainties in the thermal function. This leads to the enthalpy of formation of

$$\Delta_f H_m^\circ(UF_6, g, 298.15\,K) = -(2148.6 \pm 1.9)\,kJ \cdot mol^{-1}.$$

The thermal functions of $UF_6(g)$ are calculated from the vibrational parameters summarised by Aldridge *et al.* [85ALD/BRO], from a number of recent high-resolution spectroscopic studies and $r(\text{U-F}) = (0.19962 \pm 0.00007)$ nm as derived by the same authors from an analysis of the rotational spectrum of the ν_3 stret- ching frequency of $UF_6(g)$.

$$S_m^\circ(UF_6,\ g,\ 298.15\ K) = (376.5 \pm 1.0)\,J \cdot K^{-1} \cdot mol^{-1}$$
$$C_{p,m}^\circ(UF_6,\ g,\ 298.15\ K) = (129.5 \pm 0.5)\,J \cdot K^{-1} \cdot mol^{-1}$$

The Gibbs energy of formation is calculated from the selected enthalpy of formation and entropy.

$$\Delta_f G_m^\circ(UF_6,\ g,\ 298.15\ K) = -(2064.5 \pm 1.9)\,kJ \cdot mol^{-1}$$

V.4.1.1.6. $U_2F_{10}(g)$

Hildenbrand, Gurvich and Yungman [85HIL/GUR] calculated $\Delta_f H_m^\circ(U_2F_{10},\ g,\ 298.15\ K) = -(4042 \pm 27)\,kJ \cdot mol^{-1}$ from the enthalpy of dimerisation of $UF_5(g)$ reported by Kleinschmidt and Hildenbrand [79KLE/HIL]. Based on the same data, Leitnaker [83LEI] also obtained $\Delta_f H_m^\circ(U_2F_{10}, g, 298.15\,K) = -4000\,kJ{\cdot}mol^{-1}$ by analogy. This review selects the average of the estimated enthalpies of formation reported by Leitnaker [83LEI] and Hildenbrand, Gurvich and Yungman [85HIL/GUR],

$$\Delta_f H_m^\circ(U_2F_{10}, g, 298.15\,K) = -(4021 \pm 30)\,kJ \cdot mol^{-1}.$$

The uncertainties are assigned by this review.

Leitnaker [83LEI] used $UF_5(g)$ as an analogue to $U_2F_{10}(g)$ and calculated

$$C_{p,m}^\circ(U_2F_{10}, g, 298.15\,K) = (234.7 \pm 5.0)\,J \cdot K^{-1} \cdot mol^{-1}$$

based on the ratio of the contribution of the translation, rotational and vibrational energy per degree of freedom in the classical approximation. This value is selected by this review with an estimated uncertainty.

From the $UF_5(g)$ transport data of Wolf, Posey and Rapp [65WOL/POS], Leitnaker [83LEI] obtained a value of

$$S_m^\circ(U_2F_{10}, g, 298.15\,K) = (577.6 \pm 10.0)\,J \cdot K^{-1} \cdot mol^{-1},$$

which is selected by this review. The uncertainty is estimated to $\pm 10\,J \cdot K^{-1} \cdot mol^{-1}$ due to the comparatively large uncertainty of the electronic contribution.

The selected Gibbs energy of formation is calculated from the enthalpy of formation and the entropy.

$$\Delta_f G_m^\circ(U_2F_{10},\ g,\ 298.15\ K) = -(3861 \pm 30)\,kJ \cdot mol^{-1}$$

V.4.1.1.7. $UO_2F_2(g)$

The enthalpy of formation of $UO_2F_2(g)$ is derived by third-law method from the three vapour pressure studies listed in Table V.16. This review selects

$$\Delta_{sub}H_m^\circ(V.50,\ 298.15\ K) = (301 \pm 15)\ kJ \cdot mol^{-1}$$

for the sublimation reaction

$$UO_2F_2(cr) \rightleftharpoons UO_2F_2(g). \qquad (V.50)$$

This value is supported by the second-law results of Lau, Brittain and Hildenbrand [85LAU/BRI], $\Delta_{sub}H_m^\circ(V.50,\ 298.15\ K) = (302.5 \pm 6.3)\ kJ \cdot mol^{-1}$. The selected uncertainty is estimated to reflect the uncertainties in the auxiliary thermal functions of the gas. Combining this value with the selected value for the enthalpy of formation of the solid $UO_2F_2(cr)$ from Section V.4.1.3.2.d, results in

$$\Delta_f H_m^\circ(UO_2F_2,\ g,\ 298.15\ K) = -(1352.5 \pm 15.1)\ kJ \cdot mol^{-1}.$$

Table V.16: Enthalpy of sublimation values for $UO_2F_2(cr) \rightleftharpoons UO_2F_2(g)$ at 298.15 K, as evaluated from third-law analysis of experimental data.

Reference	T-range (K)	$\Delta_{sub}H_m^\circ(V.50, 298.15\ K)$[(a)] (kJ · mol^{-1})
[69KNA/LOS]	1033-1132	296.8 ± 3.5
[84SMI/GOR]	929-1094	302.3 ± 2.2
[85LAU/BRI]	957-1000	301.3 ± 0.1

(a) The indicated errors represent the statistical uncertainties.

The values for the heat capacity and entropy of $UO_2F_2(g)$ are calculated from estimated molecular parameters given by Glushko *et al.* [82GLU/GUR]:

$$S_m^\circ(UO_2F_2,\ g,\ 298.15\ K) = (343 \pm 10)\ J \cdot K^{-1} \cdot mol^{-1}$$
$$C_{p,m}^\circ(UO_2F_2,\ g,\ 298.15\ K) = (86.4 \pm 5.0)\ J \cdot K^{-1} \cdot mol^{-1}.$$

The selected Gibbs energy of formation is calculated from the enthalpy of formation and the entropy.

$$\Delta_f G_m^\circ(UO_2F_2,\ g,\ 298.15\ K) = -(1318.2 \pm 15.3)\ kJ \cdot mol^{-1}$$

V.4.1.1.8. UOF_4 (g)

Smirnov and Gorokhov [84SMI/GOR] and Lau *et al.* [85LAU/BRI] identified $UOF_4(g)$ as a minor decomposition product from the sublimation of $UO_2F_2(cr)$, in addition to $UO_2F_2(g)$, $UF_5(g)$, $O_2(g)$ and $U_3O_8(cr)$. From the results by Smirnov and Gorokhov [84SMI/GOR] for the reaction

$$UO_2F_2(g) + 2UF_5(g) + \tfrac{1}{2}O_2(g) \rightleftharpoons 3UOF_4(g) \qquad (V.51)$$

a value of $\Delta_r H_m^\circ(V.51, 298.15\text{ K}) = -(121.3 \pm 1.9)\,\text{kJ}\cdot\text{mol}^{-1}$ is derived by third-law analysis, giving $\Delta_f H_m^\circ(UOF_4, g, 298.15\text{ K}) = -(1764 \pm 14)\,\text{kJ}\cdot\text{mol}^{-1}$. Lau, Brittain and Hildenbrand [85LAU/BRI] reported $\Delta_r H_m^\circ(V.52, 298.15\text{ K}) = (356 \pm 20)\,\text{kJ}\cdot\text{mol}^{-1}$ by third-law and $(381.2 \pm 10)\,\text{kJ}\cdot\text{mol}^{-1}$ by second-law analysis for the reaction

$$2UO_2F_2(cr) \rightleftharpoons UOF_4(g) + \tfrac{1}{6}O_2(g) + \tfrac{1}{3}U_3O_8(cr). \qquad (V.52)$$

These values correspond to $\Delta_f H_m^\circ(UOF_4, g, 298.15\text{ K}) = -(1759 \pm 20)\,\text{kJ}\cdot\text{mol}^{-1}$ and $-(1734 \pm 10)\,\text{kJ}\cdot\text{mol}^{-1}$, respectively. The weighted average of the two values obtained by third-law analysis,

$$\Delta_f H_m^\circ(UOF_4, g, 298.15\text{ K}) = -(1762 \pm 20)\,\text{kJ}\cdot\text{mol}^{-1},$$

is selected by the present review.

The values for the heat capacity and entropy of $UOF_4(g)$ are calculated from estimated molecular parameters given by Glushko *et al.* [82GLU/GUR].

$$S_m^\circ(UOF_4, g, 298.15\text{ K}) = (363 \pm 10)\,\text{J}\cdot\text{K}^{-1}\cdot\text{mol}^{-1}$$
$$C_{p,m}^\circ(UOF_4, g, 298.15\text{ K}) = (108 \pm 5)\,\text{J}\cdot\text{K}^{-1}\cdot\text{mol}^{-1}$$

The selected Gibbs energy of formation is calculated from the enthalpy of formation and the entropy.

$$\Delta_f G_m^\circ(UOF_4, g, 298.15\text{ K}) = -(1704 \pm 20)\,\text{kJ}\cdot\text{mol}^{-1}$$

V.4.1.2. Aqueous uranium fluorides

V.4.1.2.1. Aqueous U(VI) fluorides

Equilibrium studies have been reported on the two major kinds of reactions:

$$UO_2^{2+} + q\text{HF(aq)} \rightleftharpoons UO_2F_q^{2-q} + qH^+ \qquad (V.53)$$
$$UO_2^{2+} + qF^- \rightleftharpoons UO_2F_q^{2-q}. \qquad (V.54)$$

A summary of the equilibrium data available for these reactions is presented in Table V.17. The data for the formation of UO_2F^+ according to Reaction (V.53) are used to determine $\log_{10}{}^*\beta_1^\circ$, the equilibrium constant at zero ionic strength, and $\Delta\varepsilon(V.53, n = 1)$ according to the specific ion interaction theory (*cf.* Appendix B).

Table V.17: Experimental equilibrium data for the uranium(VI) fluoride system.

Method	Ionic Medium	t (°C)	$\log_{10} K^{(a)}$	$\log_{10} K^{\circ(a)}$	$\log_{10} \beta_q^{\circ(b)}$	Reference
$UO_2^{2+} + HF(aq) \rightleftharpoons UO_2F^+ + H^+$						
dis	0.05 M $HClO_4$	25	$1.71 \pm 0.10^{(c)}$			[54DAY/POW]
dis	0.20 M $NaClO_4$ +0.05 M $HClO_4$	25	$1.57 \pm 0.10^{(c)}$			
dis	0.45 M $NaClO_4$ +0.05 M $HClO_4$	25	$1.38 \pm 0.10^{(c)}$			
dis	0.95 M $NaClO_4$ +0.05 M $HClO_4$	25	$1.43 \pm 0.10^{(c)}$			
dis	1.95 M $NaClO_4$ +0.05 M $HClO_4$	25	$1.41 \pm 0.10^{(c)}$			
sp	0.65 M	25	1.18			[61CON/PAU]
cix	2.10 M $HClO_4$	25?	$1.31 \pm 0.11^{(c)}$			[68KRY/KOM3]
cix	1.04 M $HClO_4$	25?	$1.52 \pm 0.11^{(c)}$			
cix	0.51 M $HClO_4$	25?	$1.54 \pm 0.11^{(c)}$			
cix	0.20 M $HClO_4$	25?	1.57			
red	3.40 M $NaClO_4$ +0.60 M $HClO_4$	25	1.48 ± 0.09			[69GRE/VAR]
dis	2.00 M $HClO_4$	25	$1.56 \pm 0.10^{(c)}$			[76PAT/RAM]
aix	HCl + HF var.	25	1.18			[82GRE/HAS]
$UO_2^{2+} + 2HF(aq) \rightleftharpoons UO_2F_2(aq) + 2H^+$						
cix	0.20 M $HClO_4$	25?	1.8 ± 0.2	1.90 ± 0.2	8.3 ± 0.2	[68KRY/KOM3]
$UO_2^{2+} + 3HF(aq) \rightleftharpoons UO_2F_3^- + 3H^+$						
cix	2.0 M $HClO_4$	25	1.68	1.68	11.2	[68KRY/KOM3]
$UO_2^{2+} + F^- \rightleftharpoons UO_2F^+$						
aix	var.	25	$4.32^{(e)}$			[50MOO/KRA]
qh	1.00 M $NaClO_4$	20				[54AHR/LAR]
qh	1.00 M $NaClO_4$	20	$4.55 \pm 0.05^{(c)}$		5.15 ± 0.08	[56AHR/LAR]
sp	~ 0	25?	4.77			[61KUT]
sol	var	25?	4.0			[61TAN/DEI]
nmr	0.5 M $NaClO_4$	25	$4.66 \pm 0.10^{(c)}$		5.25 ± 0.10	[69VDO/STE2]
qh, ise-F^-	1.00 M $NaClO_4$	25	$4.54 \pm 0.05^{(c)}$		5.14 ± 0.08	[71AHR/KUL]
ise-F^-	1.0 M NaCl	25	$4.52 \pm 0.06^{(d)}$		5.13 ± 0.09	[77ISH/KAO]
rev		25			5.1	[78LAN]
rev		25			5.1 ± 0.3	[80LEM/TRE]
ise-F^-	1.00 M $NaClO_4$	21	$4.56 \pm 0.05^{(c)}$		5.17 ± 0.08	[85SAW/CHA]
$UO_2^{2+} + 2F^- \rightleftharpoons UO_2F_2(aq)$						
qh	1.00 M $NaClO_4$	20				[54AHR/LAR]
qh	1.00 M $NaClO_4$	20	$7.89 \pm 0.05^{(c)}$		8.56 ± 0.07	[56AHR/LAR]
sol	var	25?	8.7			[61TAN/DEI]
qh, ise-F^-	1.00 M $NaClO_4$	25	$7.98 \pm 0.05^{(c)}$		8.64 ± 0.07	[71AHR/KUL]
ise-F^-	1.0 M NaCl	25	$7.90 \pm 0.06^{(d)}$		8.60 ± 0.08	[77ISH/KAO]
rev		25			9.0	[78LAN]
rev		25			9.0 ± 0.4	[80LEM/TRE]
ise-F^-	1.00 M $NaClO_4$	21	$7.99 \pm 0.05^{(c)}$		8.66 ± 0.07	[85SAW/CHA]

Table V.17 (continued)

Method	Ionic Medium	t (°C)	$\log_{10} K^{(a)}$	$\log_{10} K^{\circ(a)}$	$\log_{10} \beta_q^{\circ(b)}$	Reference
$UO_2^{2+} + 3F^- \rightleftharpoons UO_2F_3^-$						
qh	1.00 M $NaClO_4$	20	10.49 ± 0.09			[54AHR/LAR]
qh	1.00 M $NaClO_4$	20	$10.46 \pm 0.05^{(c)}$		11.09 ± 0.10	[56AHR/LAR]
qh, ise-F^-	1.00 M $NaClO_4$	25	$10.41 \pm 0.05^{(c)}$		11.04 ± 0.10	[71AHR/KUL]
ise-F^-	1.0 M NaCl	25	$9.96 \pm 0.07^{(d)}$		10.64 ± 0.11	[77ISH/KAO]
rev		25			11.3	[78LAN]
rev		25			11.3 ± 0.4	[80LEM/TRE]
ise-F^-	1.00 M $NaClO_4$	21	$10.34 \pm 0.06^{(c)}$		10.96 ± 0.10	[85SAW/CHA]
....						
$UO_2^{2+} + 4F^- \rightleftharpoons UO_2F_4^{2-}$						
qh	1.00 M $NaClO_4$	20	11.85 ± 0.06			[54AHR/LAR]
qh	1.00 M $NaClO_4$	20	$11.81 \pm 0.05^{(c)}$		11.90 ± 0.12	[56AHR/LAR]
qh, ise-F^-	1.00 M $NaClO_4$	25	$11.89 \pm 0.05^{(c)}$		11.98 ± 0.12	[71AHR/KUL]
ise-F^-	1.0 M NaCl	25	$10.94 \pm 0.12^{(d)}$		11.11 ± 0.15	[77ISH/KAO]
rev		25			12.6	[78LAN]
rev		25			12.6 ± 0.4	[80LEM/TRE]
....						
$H^+ + UO_2F_3^- \rightleftharpoons HUO_2F_3(aq)$						
qh	var	25	4.1			[72CHA/BAN]

(a) Refers to the reactions indicated, $\log_{10} K$ in the ionic medium and at the temperature given in the table, $\log_{10} K^\circ$ (in molal units) at $I = 0$ and 298.15 K.
(b) $\log_{10} \beta_q^\circ$ refers to the formation reactions $UO_2^{2+} + qF^- \rightleftharpoons UO_2F_q^{2-q}$ at $I = 0$ and 298.15 K.
(c) Uncertainties estimated by this review as described in Appendix A.
(d) This constant is corrected for Cl^- complexing as described in Appendix A.
(e) See comments in Appendix A.

The data from Refs. [76PAT/RAM, 68KRY/KOM3, 54DAY/POW] are used for the weighted linear regression. As described in Appendix C, the uncertainties $\sigma(\log_{10}{}^*\beta_1^\circ)$ assigned to each accepted value of $\log_{10}{}^*\beta_1^\circ$ (see Table V.17) are used in the weighting procedure. Ref. [69GRE/VAR] refers to such a high ionic strength (4 M) that the validity of the specific ion interaction theory is doubtful. Although this value falls close to the regression line ($\log_{10}{}^*\beta_1$(V.53, 4 M $NaClO_4$) $= (1.48 \pm 0.09)$ *vs.* (1.47 ± 0.21) calculated by extrapolating the regression line to $I_m = 4.95$ m) it is not used for the linear extrapolation to $I = 0$ in Figure V.9. The results obtained are $\log_{10}{}^*\beta_1^\circ = (1.84 \pm 0.06)$ and $\Delta\varepsilon(\text{V.53}, q = 1) = -(0.03 \pm 0.04)$. The uncertainties represent the 95% confidence level, see also Figure V.9. From the value of $\Delta\varepsilon$ and the ion interaction coefficients $\varepsilon_{(UO_2^{2+},ClO_4^-)}$ and $\varepsilon_{(H^+,ClO_4^-)}$ given in Table B.1, this review obtains $\varepsilon_{(UO_2F^+,ClO_4^-)} = (0.29 \pm 0.05)$.

By combining the equilibrium constant of Reaction (V.53) with the equilibrium

Figure V.9: Extrapolation to $I = 0$ of experimental data for the formation of UO_2F^+ using the specific ion interaction theory. The data refer to $NaClO_4$ media and are taken from [54DAY/POW] (○), [68KRY/KOM3] (□) and [76PAT/RAM] (△). The shaded area represents the uncertainty range obtained by propagating the resulting uncertainties at $I = 0$ back to $I = 3$ m.

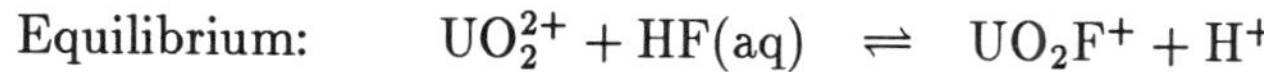

$$\text{Equilibrium:} \quad UO_2^{2+} + HF(aq) \rightleftharpoons UO_2F^+ + H^+$$

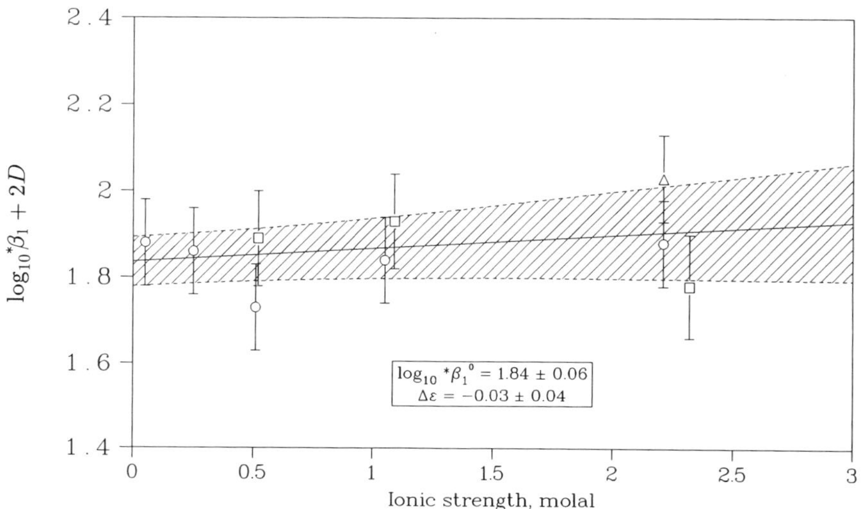

constant of the reaction $H^+ + F^- \rightleftharpoons HF(aq)$ listed in Chapter VI, $\log_{10} K^\circ = (3.18 \pm 0.02)$, this review obtains $\log_{10} \beta_1^\circ = (5.02 \pm 0.06)$.

Refs. [56AHR/LAR, 69VDO/STE2, 71AHR/KUL, 77ISH/KAO, 85SAW/CHA] provide a second set of data for the determination of $\log_{10} \beta_1^\circ$ according to Reaction (V.54). The values at $I = 0$ are calculated by using the specific ion interaction theory with $\varepsilon_{(UO_2^{2+},ClO_4^-)} = (0.46 \pm 0.03)$, $\varepsilon_{(F^-,Na^+)} = (0.02 \pm 0.02)$ and $\varepsilon_{(UO_2F^+,ClO_4^-)} = (0.29 \pm 0.05)$, resulting in $\Delta\varepsilon(V.54, n = 1) = -(0.19 \pm 0.06)$. The concentration constants are converted to molality units. The results are given in Table V.17. From this second set a weighted average of $\log_{10} \beta_1^\circ = (5.16 \pm 0.04)$ is obtained. The data in the first set cluster around $\log_{10} \beta_1^\circ = 5.02$ and those in the second set around $\log_{10} \beta_1^\circ = 5.16$. The difference is not large, nevertheless a systematic error between the two sets is indicated. The first data set refers to $HClO_4$ solutions and the second to $NaClO_4$ solutions. Hence, a possible cause of the discrepancy may be a medium effect which is not taken into account by the approximate specific ion interaction theory. The value selected for $\log_{10} \beta_1^\circ$ in this review is therefore the unweighted average of the two sets, and its uncertainty is assigned such that it spans the range of uncertainty of the two single sets, as described in Appendix C:

$$\log_{10} \beta_1^\circ(V.54,\ q = 1,\ 298.15\ K) = 5.09 \pm 0.13.$$

There are not enough precise equilibrium data at different ionic strengths to allow an independent determination of both the equilibrium constants at $I = 0$ and the ion interaction coefficients for $UO_2F_3^-$ and $UO_2F_4^{2-}$. The data given in Table V.17 are thus corrected to $I = 0$ after conversion to molality constants, using $\varepsilon_{(UO_2F_3^-,Na^+)} \approx \varepsilon_{(UO_2F_3^-,H^+)} = (0.00 \pm 0.05)$ and $\varepsilon_{(UO_2F_4^{2-},Na^+)} = -(0.08 \pm 0.06)$. This leads to $\Delta\varepsilon(V.54, q = 2) = -(0.50 \pm 0.05)$, $\Delta\varepsilon(V.54, q = 3) = -(0.52 \pm 0.08)$ and $\Delta\varepsilon(V.54, q = 4) = -(0.62 \pm 0.10)$ in sodium perchlorate media. The estimates are based on analogies with the values listed in Table B.4 for ions of charge −1 and −2, respectively. The $\Delta\varepsilon$ values in sodium chloride medium used in Ref. [77ISH/KAO] are assumed to be the same as in perchlorate medium, *cf.* Appendix A. The data of Krylov, Komarov and Pushlenkov [68KRY/KOM3] are not included in the calculation of the selected formation constant of $UO_2F_3^-$ because of the large experimental uncertainties in the concentration range where this complex is formed. The value for $\log_{10}\beta_2^\circ$ [68KRY/KOM3] also deviates from the other data in Table V.17. It is considered as an outlier and is given zero weight, *cf.* Appendix A. The data of Ahrland and Larsson [54AHR/LAR] were re-evaluated later by Ahrland, Larsson and Rosengren [56AHR/LAR], and they are therefore not included in the calculation of the recommended set of equilibrium constants. The values of $\log_{10}\beta_q(q = 1,2)$ from Ishiguro, Kao and Kakihana [77ISH/KAO], corrected for Cl^- complex formation, *cf.* Appendix A, are in good agreement with other determinations.

A weighted average of the values of $\log_{10}\beta_2^\circ$ given in Table V.17 leads to the following selected value:

$$\log_{10}\beta_2^\circ(V.54,\ q = 2) \quad = \quad 8.62 \pm 0.04.$$

The values of $\log_{10}\beta_3^\circ$ and $\log_{10}\beta_4^\circ$ calculated from the data of Ishiguro, Kao and Kakihana [77ISH/KAO] are significantly different from the other data. This review finds no reason to discard these values, and they are therefore used to calculate the selected $\log_{10}\beta_3^\circ$ and $\log_{10}\beta_4^\circ$ and their uncertainties according to the guideline outlined in Appendix C for discrepant data, that is, the four values of $\log_{10}\beta_3^\circ$ and the three values of $\log_{10}\beta_4^\circ$ are used to calculate unweighted averages and uncertainties are assigned that span the range of uncertainty of all the values involved.

$$\log_{10}\beta_3^\circ(V.54,\ q = 3) \quad = \quad 10.9 \pm 0.4$$
$$\log_{10}\beta_4^\circ(V.54,\ q = 4) \quad = \quad 11.7 \pm 0.7$$

The $\log_{10}\beta_q^\circ$ values selected in this review are in good agreement with those proposed by Langmuir [78LAN] and Lemire and Tremaine [80LEM/TRE]. The latter authors give $\log_{10}\beta_1^\circ = (5.41 \pm 0.13)$, $\log_{10}\beta_2^\circ = (8.56 \pm 0.10)$, $\log_{10}\beta_3^\circ = (11.0 \pm 0.1)$, and $\log_{10}\beta_4^\circ = (12.25 \pm 0.10)$.

The recommended Gibbs energies of formation for the species $UO_2F_q^{2-q}$, $q = 1$ to 4, are calculated by using the values of $\Delta_f G_m^\circ$ for UO_2^{2+} and F^- selected in this review.

$$\Delta_f G_m^\circ(UO_2F^+, aq, 298.15\,K) \quad = \quad -(1263.1 \pm 2.0)\,kJ \cdot mol^{-1}$$
$$\Delta_f G_m^\circ(UO_2F_2, aq, 298.15\,K) \quad = \quad -(1564.8 \pm 2.2)\,kJ \cdot mol^{-1}$$
$$\Delta_f G_m^\circ(UO_2F_3^-, aq, 298.15\,K) \quad = \quad -(1859.3 \pm 3.5)\,kJ \cdot mol^{-1}$$
$$\Delta_f G_m^\circ(UO_2F_4^{2-}, aq, 298.15\,K) \quad = \quad -(2145.4 \pm 5.2)\,kJ \cdot mol^{-1}$$

The stepwise enthalpy of formation values determined by Ahrland and Kullberg [71AHR/KUL2] are accepted and recalculated for the reactions (V.54), *cf.* Appendix A. The selected enthalpies of formation are derived therefrom.

$$\begin{aligned}
\Delta_r H^\circ_m(\text{V.54}, q=1) &= (1.70 \pm 0.08)\,\text{kJ}\cdot\text{mol}^{-1}\\
\Delta_r H^\circ_m(\text{V.54}, q=2) &= (2.10 \pm 0.19)\,\text{kJ}\cdot\text{mol}^{-1}\\
\Delta_r H^\circ_m(\text{V.54}, q=3) &= (2.35 \pm 0.31)\,\text{kJ}\cdot\text{mol}^{-1}\\
\Delta_r H^\circ_m(\text{V.54}, q=4) &= (0.29 \pm 0.47)\,\text{kJ}\cdot\text{mol}^{-1}
\end{aligned}$$

$$\begin{aligned}
\Delta_f H^\circ_m(\text{UO}_2\text{F}^+, \text{aq}, 298.15\,\text{K}) &= -(1352.7 \pm 1.6)\,\text{kJ}\cdot\text{mol}^{-1}\\
\Delta_f H^\circ_m(\text{UO}_2\text{F}_2, \text{aq}, 298.15\,\text{K}) &= -(1687.6 \pm 2.0)\,\text{kJ}\cdot\text{mol}^{-1}\\
\Delta_f H^\circ_m(\text{UO}_2\text{F}_3^-, \text{aq}, 298.15\,\text{K}) &= -(2022.7 \pm 2.5)\,\text{kJ}\cdot\text{mol}^{-1}\\
\Delta_f H^\circ_m(\text{UO}_2\text{F}_4^{2-}, \text{aq}, 298.15\,\text{K}) &= -(2360.1 \pm 3.0)\,\text{kJ}\cdot\text{mol}^{-1}
\end{aligned}$$

The data of Ahrland and Kullberg [71AHR/KUL2] refer to $I = 1.00$ M ($NaClO_4$). This review assumes that these enthalpies of reaction are valid also at $I = 0$. Fuger *et al.* [83FUG/PAR] independently evaluated the standard enthalpy of formation of UO_2F_2(aq) from $\Delta_{sol}H_m(UO_3, \gamma)$ determinations [65VID/BYA, 71VID/IPP, 77COR/OUW, 78JOH/OHA] and obtained $\Delta_f H^\circ_m(UO_2F_2$, aq, 298.15 K) = −(1689.5 ± 2.0) $\text{kJ}\cdot\text{mol}^{-1}$, which is consistent with the value selected here. This agreement is not surprising in view of the small enthalpies of reaction of the various (V.54) reactions.

Using the selected reaction data, the entropies are calculated via $\Delta_r S^\circ_m$.

$$\begin{aligned}
S^\circ_m(\text{UO}_2\text{F}^+, \text{aq}, 298.15\,\text{K}) &= -(8.9 \pm 4.0)\,\text{J}\cdot\text{K}^{-1}\cdot\text{mol}^{-1}\\
S^\circ_m(\text{UO}_2\text{F}_2, \text{aq}, 298.15\,\text{K}) &= (46.3 \pm 3.5)\,\text{J}\cdot\text{K}^{-1}\cdot\text{mol}^{-1}\\
S^\circ_m(\text{UO}_2\text{F}_3^-, \text{aq}, 298.15\,\text{K}) &= (77.0 \pm 8.6)\,\text{J}\cdot\text{K}^{-1}\cdot\text{mol}^{-1}\\
S^\circ_m(\text{UO}_2\text{F}_4^{2-}, \text{aq}, 298.15\,\text{K}) &= (72 \pm 14)\,\text{J}\cdot\text{K}^{-1}\cdot\text{mol}^{-1}
\end{aligned}$$

The mean value $C^\circ_{p,m}|^{473}_{298}(UO_2F^+, aq)$ calculated by Lemire and Tremaine [80LEM/TRE] is not likely to be a good approximation to $C^\circ_{p,m}(UO_2F^+$, aq, 298.15 K).

V.4.1.2.2. *Aqueous U(V) fluorides*

No information exists on aqueous species of the form $UO_2F_q^{1-q}$, presumably due to the limited stability range of dioxouranium(V) in aqueous media. The analogous system with NpO_2^+ has been studied though, *e.g.*, Refs. [84CHO/RAO, 85SAW/RIZ]. This review has chosen not to select a formation constant for UO_2F(aq) by analogy to the corresponding neptunium(V) system.

V.4.1.2.3. *Aqueous U(IV) fluorides*

The available experimental data are compiled in Table V.18 and refer to the equilibria

$$\text{U}^{4+} + q\text{HF(aq)} \rightleftharpoons \text{UF}_q^{4-q} + q\text{H}^+. \qquad \text{(V.55)}$$

Table V.18: Experimental equilibrium data for the uranium(IV) fluoride system.

Method	Ionic Medium	t (°C)	$\log_{10} K^{(a)}$	$\log_{10} K^{\circ(a)}$	$\log_{10} \beta_q^{\circ(b)}$	Reference
$U^{4+} + HF(aq) \rightleftharpoons UF^{3+} + H^+$						
dis	$I = 2$ M	25	6			[55DAY/WIL]
nmr	2 M $HClO_4$	20	5.35			[63VDO/ROM2]
red	3.4 M $NaClO_4$ 0.6 M $HClO_4$	25	$5.37 \pm 0.03^{(c)}$	6.24 ± 0.25	9.42 ± 0.25	[69GRE/VAR]
ise-F^-	4 M $HClO_4$	20	$5.49 \pm 0.02^{(d)}$	6.36 ± 0.25	9.54 ± 0.25	[69NOR]
ise-F^-	1 M (Na,H)Cl	25	$5.02 \pm 0.11^{(e)}$	6.10 ± 0.15	9.28 ± 0.15	[74KAK/ISH]
ise-F^-	1 M $HClO_4$	25	4.82 ± 0.16	5.91 ± 0.17	9.09 ± 0.17	[76CHO/UNR]
....						
$UF^{3+} + Al^{3+} \rightleftharpoons U^{4+} + AlF^{2+}$						
sp	var.	20	2.6		10.2	[66VDO/ROM]
....						
$U^{4+} + 2HF(aq) \rightleftharpoons UF_2^{2+} + 2H^+$						
dis	$I = 2$ M	25	8			[55DAY/WIL]
nmr	2 M $HClO_4$	20	8.80			[63VDO/ROM2]
red	3.4 M $NaClO_4$ 0.6 M $HClO_4$	25	$8.29 \pm 0.04^{(c)}$	10.01 ± 0.50	16.37 ± 0.50	[69GRE/VAR]
ise-F^-	4 M $HClO_4$	20	$8.64 \pm 0.03^{(d)}$	10.36 ± 0.50	16.72 ± 0.50	[69NOR]
ise-F^-	1 M (Na,H)Cl	25	$7.98 \pm 0.11^{(e)}$	9.80 ± 0.15	16.16 ± 0.16	[74KAK/ISH]
....						
$U^{4+} + 3HF(aq) \rightleftharpoons UF_3^{+} + 3H^+$						
nmr	2 M $HClO_4$	20	12.3			[63VDO/ROM2]
red	3.4 M $NaClO_4$ 0.6 M $HClO_4$	25	9.45			[69GRE/VAR]
ise-F^-	4 M $HClO_4$	20	$10.57 \pm 0.05^{(d)}$	12.52 ± 0.50	22.06 ± 0.50	[69NOR]
ise-F^-	1 M (Na,H)Cl	25	$9.48 \pm 0.12^{(e)}$	11.69 ± 0.16	21.23 ± 0.17	[74KAK/ISH]
....						
$U^{4+} + 4HF(aq) \rightleftharpoons UF_4(aq) + 4H^+$						
nmr	2 M $HClO_4$	20	15.9			[63VDO/ROM2]
ise-F^-	1 M (Na,H)Cl	25	$10.64 \pm 0.20^{(e)}$	12.89 ± 0.22	25.61 ± 0.23	[74KAK/ISH]
....						
$UF_4 \cdot 2.5H_2O(cr) \rightleftharpoons UF_2^{2+} + 2F^- + 2.5H_2O(l)$						
sol	0.12 M ($HClO_4$)	25	-12.45 ± 0.04	-13.14 ± 0.04		[60SAV/BRO]
....						
$UF_4 \cdot 2.5H_2O(cr) \rightleftharpoons UF_3^{+} + F^- + 2.5H_2O(l)$						
sol	0.12 M ($HClO_4$)	25	-8.24 ± 0.09	-8.47 ± 0.09		[60SAV/BRO]
$UF_4 \cdot 2.5H_2O(cr) \rightleftharpoons UF_4(aq) + 2.5H_2O(l)$						
sol	0.12 M ($HClO_4$)	25	-3.96 ± 0.03	-3.96 ± 0.03		[60SAV/BRO]
....						

Table V.18 (continued)

Method	Ionic Medium	t (°C)	$\log_{10} K^{(a)}$	$\log_{10} K^{\circ(a)}$	$\log_{10} \beta_q^{\circ(b)}$	Reference
$UF_4 \cdot 2.5H_2O(cr) + F^- \rightleftharpoons UF_5^- + 2.5H_2O(l)$						
sol	0.12 M ($HClO_4$)	25	-2.39 ± 0.25	-2.39 ± 0.25	27.01 ± 0.31	[60SAV/BRO]
....						
$UF_4 \cdot 2.5H_2O(cr) + 2F^- \rightleftharpoons UF_6^{2-} + 2.5H_2O(l)$						
sol	0.12 M ($HClO_4$)	25	-0.08 ± 0.08	-0.32 ± 0.08	29.08 ± 0.20	[60SAV/BRO]

(a) Refers to the reactions indicated, $\log_{10} K$ in the ionic medium and at the temperature given in the table, $\log_{10} K^\circ$ (in molal units) at $I = 0$ and 298.15 K.
(b) $\log_{10} \beta_q^\circ$ (in molal units) refers to the formation reactions $U^{4+} + qF^- \rightleftharpoons UF_q^{4-q}$ at $I = 0$ and 298.15 K.
(c) Uncertainties estimated by this review as described in Appendix A.
(d) These values are corrected to 298.15 K as described in Appendix A.
(e) The constants reported in Ref. [74KAK/ISH] are corrected for chloride complex formation (see Appendix A). For the correction to $I = 0$ the formation constants $\log_{10} \beta_q$ reported in Ref. [74KAK/ISH] are used, rather than the values of $\log_{10}{}^*\beta_q$ listed here (see Appendix A).

There are fairly few measurements of Reaction (V.55) at different ionic strengths, and one cannot simultaneously determine $\log_{10}{}^*\beta_q^\circ$ and the corresponding $\Delta\varepsilon$ values from the specific ion interaction theory. The value of $\Delta\varepsilon = -(0.14 \pm 0.05)$ found for the reaction $U^{4+} + H_2O(l) \rightleftharpoons UOH^{3+} + H^+$ (*cf.* discussion in Section V.3.2.3.1) is used as an estimate of $\Delta\varepsilon(V.55, q = 1)$. Published equilibrium data for the formation of ThF^{3+}, UF^{3+} and NpF^{3+} are also compatible with this value of $\Delta\varepsilon$.

The selected formation constants $\log_{10} \beta_q^\circ$ according to the reactions

$$U^{4+} + qF^- \rightleftharpoons UF_q^{4-q} \qquad (V.56)$$

are derived from the values of the corrected data listed in Table V.18. For the conversion of $\log_{10}{}^*\beta_q^\circ$ to $\log_{10} \beta_q^\circ$, $\log_{10} K^\circ = (3.18 \pm 0.02)$ is used for the formation of HF(aq), *cf.* Chapter VI.

The weighted average value of $\log_{10} \beta_1^\circ$ is calculated from the corrected values of Refs. [69GRE/VAR, 69NOR, 74KAK/ISH, 76CHO/UNR]:

$$\log_{10} \beta_1^\circ(V.56,\ q = 1,\ 298.15\ K) = 9.28 \pm 0.09.$$

The equilibrium constants for the formation of UF_q^{4-q}, $q = 2$ to 4, are corrected to $I = 0$ by using the experimental or estimated ion interaction coefficients $\varepsilon_{(U^{4+},ClO_4^-)} = (0.76 \pm 0.06)$, $\varepsilon_{(UF_2^{2+},ClO_4^-)} \approx \varepsilon_{(M^{2+},ClO_4^-)} = (0.3 \pm 0.1)$ and $\varepsilon_{(UF_3^+,ClO_4^-)} \approx \varepsilon_{(M^+,ClO_4^-)} = (0.1 \pm 0.1)$, with the additional assumption that the uncertainty in $\Delta\varepsilon$ should be at most ± 0.10 (see Appendix B). This upper limit is reasonable as judged by the span of the experimental $\Delta\varepsilon$ values. The value of $\log_{10}{}^*\beta_3$ for the formation of UF_3^+ given by Grenthe and Varfeldt [69GRE/VAR] is not considered reliable because of the small amounts of this complex formed even at the highest fluoride concentrations used in [69GRE/VAR]. The value of $\log_{10}\beta_1^\circ = 10.2$ calculated from the data of Vdovenko, Romanov and Shcherbakov [66VDO/ROM] is not credited because of some experimental shortcomings, *cf.* Appendix A.

$\log_{10}\beta_2^\circ$ is the weighted average of the values from Refs. [69GRE/VAR, 69NOR, 74KAK/ISH].

$$\log_{10}\beta_2^\circ(\text{V.56}, q=2, 298.15\ \text{K}) = 16.23 \pm 0.15$$

The uncertainty ranges of the two reliable values for $\log_{10}\beta_3^\circ$ [69NOR, 74KAK/ISH] do not overlap, and these values are therefore not consistent. This review selects the unweighted average of these values and assigns an uncertainty that spans the uncertainty ranges of both values, *cf.* Appendix C.

$$\log_{10}\beta_3^\circ(\text{V.56}, q=3, 298.15\ \text{K}) = 21.6 \pm 1.0$$

The only reliable value of $\log_{10}\beta_4^\circ$ [74KAK/ISH] is selected here. However, in view of the disagreement of $\log_{10}\beta_3^\circ$ from this paper [74KAK/ISH] with that from Ref. [69NOR], this review suggests that $\log_{10}\beta_4^\circ$ may be affected with a similar uncertainty as $\log_{10}\beta_3^\circ$. The uncertainty is therefore raised to ± 1.0, as obtained for $\log_{10}\beta_3^\circ$.

$$\log_{10}\beta_4^\circ(\text{V.56}, q=4, 298.15\ \text{K}) = 25.6 \pm 1.0$$

The only experimental study of anionic uranium(IV) fluoride complexes is a solubility study by Savage and Browne [60SAV/BRO]. From their values of $\log_{10}K_{s,q}$, $q = 2$ to 4, recalculated to $I = 0$ (see Appendix A), and the values of $\log_{10}\beta_2^\circ$, $\log_{10}\beta_3^\circ$ and $\log_{10}\beta_4^\circ$ selected above, the solubility product of $UF_4{\cdot}2.5H_2O(cr)$ can be calculated. The following values are obtained: $\log_{10}K_{s,0}^\circ = -(29.38 \pm 0.19)$, $-(30.07 \pm 1.00)$, and $-(29.56 \pm 1.00)$. The weighted average of these three values, $\log_{10}K_{s,0}^\circ = -(29.40 \pm 0.16)$, is used, together with the values of $\log_{10}K_{s,5}^\circ$ and $\log_{10}K_{s,6}^\circ$ from Ref. [60SAV/BRO], to evaluate

$$\log_{10}\beta_5^\circ(\text{V.56}, q=5, 298.15\ \text{K}) = 27.01 \pm 0.30$$
$$\log_{10}\beta_6^\circ(\text{V.56}, q=6, 298.15\ \text{K}) = 29.08 \pm 0.18,$$

cf. Appendix A. It should be noted that the solubility product $\log_{10}K_{s,0}^\circ = -(29.40 \pm 0.16)$, as used here for the evaluation of $\log_{10}\beta_5^\circ$ and $\log_{10}\beta_6^\circ$ (see Ref. [60SAV/BRO] in Appendix A), is different from the value of $-(33.5 \pm 1.2)$ which can be calculated from $\Delta_f G_m^\circ(UF_4{\cdot}2.5H_2O, \text{cr}, 298.15\ \text{K})$ recommended by this review (*cf.* discussion in Section V.4.1.3.1.b). The proposed values of $\log_{10}\beta_q^\circ$ differ partly from the ones proposed in reviews by Langmuir [78LAN] and Lemire and Tremaine [80LEM/TRE]. Both these reviews gave the same numerical values of $\log_{10}\beta_q^\circ$, Lemire and Tremaine [80LEM/TRE] also gave estimates of the uncertainty. They proposed for Reactions (V.56): $\log_{10}\beta_1^\circ = (9 \pm 1)$, $\log_{10}\beta_2^\circ = (14 \pm 1)$, $\log_{10}\beta_3^\circ = (19 \pm 2)$, $\log_{10}\beta_4^\circ = (24 \pm 2)$, $\log_{10}\beta_5^\circ = (25 \pm 3)$, and $\log_{10}\beta_6^\circ = (28 \pm 3)$.

It should be mentioned that relatively strong U(IV) fluoride complexes appear to be formed in aqueous solution above 300°C [89RED/SAV].

The standard Gibbs energies of formation of the first six UF_q^{4-q} species are calculated from the $\log_{10}\beta_q^\circ$ (V.56) values using $\Delta_f G_m^\circ(U^{4+}, aq, 298.15\,K)$ selected in Section V.2.3 and the CODATA [89COX/WAG] value of $\Delta_f G_m^\circ(F^-, aq, 298.15\,K)$, converted to the 0.1 MPa standard state pressure:

$$\begin{aligned}
\Delta_f G_m^\circ(UF^{3+}, aq, 298.15\,K) &= -(864.4 \pm 2.0)\,kJ\cdot mol^{-1}\\
\Delta_f G_m^\circ(UF_2^{2+}, aq, 298.15\,K) &= -(1185.5 \pm 2.4)\,kJ\cdot mol^{-1}\\
\Delta_f G_m^\circ(UF_3^{+}, aq, 298.15\,K) &= -(1497.7 \pm 6.3)\,kJ\cdot mol^{-1}\\
\Delta_f G_m^\circ(UF_4, aq, 298.15\,K) &= -(1802.1 \pm 6.6)\,kJ\cdot mol^{-1}\\
\Delta_f G_m^\circ(UF_5^{-}, aq, 298.15\,K) &= -(2091.7 \pm 4.2)\,kJ\cdot mol^{-1}\\
\Delta_f G_m^\circ(UF_6^{2-}, aq, 298.15\,K) &= -(2385.0 \pm 4.6)\,kJ\cdot mol^{-1}
\end{aligned}$$

There are two experimental studies of enthalpies of reaction of the type (V.55). Choppin and Unrein [76CHO/UNR] reported $\Delta_r H_m(V.56, q=1) = (0.75 \pm 4.00)\,kJ\cdot mol^{-1}$. Their study is discussed in Appendix A and re-evaluated by this review. The resulting value is $\Delta_r H_m(V.56, q=1) = (0.8 \pm 4.0)\,kJ\cdot mol^{-1}$. Ahrland, Hefter and Norén [90AHR/HEF] made a calorimetric study of the reactions (V.55) in 4 M $HClO_4$ and reported enthalpies of Reaction (V.55, $q = 1$ to 3). They used the enthalpy of protonation of F^- from Kozlov *et al.* [73KOZ/BLO] to obtain $\Delta_r H_m(V.56, q = 1$ to $3)$. This review selects the $\Delta_r H_m$ values from Ahrland, Hefter and Norén [90AHR/HEF] and assumes that they do not depend strongly on the ionic strength, *i.e.*, $\Delta_r H_m^\circ(V.56, I=0) = \Delta_r H_m(V.56, 4\text{ M } HClO_4)$.

$$\begin{aligned}
\Delta_r H_m^\circ(V.56, q=1, 298.15\,K) &= -(5.6 \pm 0.5)\,kJ\cdot mol^{-1}\\
\Delta_r H_m^\circ(V.56, q=2, 298.15\,K) &= -(3.5 \pm 0.6)\,kJ\cdot mol^{-1}\\
\Delta_r H_m^\circ(V.56, q=3, 298.15\,K) &= (0.5 \pm 4.0)\,kJ\cdot mol^{-1}
\end{aligned}$$

The corresponding enthalpies of formation of the various UF_q^{4-q} complexes are

$$\begin{aligned}
\Delta_f H_m^\circ(UF^{3+}, aq, 298.15\,K) &= -(932.2 \pm 3.4)\,kJ\cdot mol^{-1}\\
\Delta_f H_m^\circ(UF_2^{2+}, aq, 298.15\,K) &= -(1265.4 \pm 3.6)\,kJ\cdot mol^{-1}\\
\Delta_f H_m^\circ(UF_3^{+}, aq, 298.15\,K) &= -(1596.8 \pm 5.5)\,kJ\cdot mol^{-1}.
\end{aligned}$$

Using the selected reaction data, the entropies are calculated via $\Delta_r S_m^\circ$.

$$\begin{aligned}
S_m^\circ(UF^{3+}, aq, 298.15\,K) &= -(272 \pm 13)\,J\cdot K^{-1}\cdot mol^{-1}\\
S_m^\circ(UF_2^{2+}, aq, 298.15\,K) &= -(146 \pm 13)\,J\cdot K^{-1}\cdot mol^{-1}\\
S_m^\circ(UF_3^{+}, aq, 298.15\,K) &= -(43 \pm 27)\,J\cdot K^{-1}\cdot mol^{-1}
\end{aligned}$$

Ahrland, Hefter and Norén [90AHR/HEF] did not give an enthalpy of reaction for $UF_4(aq)$. This review therefore estimates $\Delta_r S_m(V.57) = (67 \pm 10)\,J\cdot K^{-1}\cdot mol^{-1}$ for the stepwise reaction

$$UF_3^+ + F^- \rightleftharpoons UF_4(aq). \qquad (V.57)$$

This estimate is based on the experimental data in Ref. [90AHR/HEF] for the Zr^{4+}, Hf^{4+}, Th^{4+} and U^{4+} fluoride systems, which indicate that the value of $\Delta_r S_m$ for the stepwise reactions, $MF_{q-1}^{4-q+1} + F^- \rightleftharpoons MF_q^{4-q}$, do not vary much with the M^{4+} ion for a given value of q. The estimated value of $\Delta_r S_m$(V.57) given above is the average of the reported Zr^{4+} and Hf^{4+} data for the reaction $MF_3^+ + F^- \rightleftharpoons MF_4(aq)$. From this value, this review estimates $\Delta_r S_m(\text{V.56}, q = 4, 4\text{ M HClO}_4) = (476\pm17)\,\text{J}\cdot\text{K}^{-1}\cdot\text{mol}^{-1}$ and assumes that this value is equal to $\Delta_r S_m^\circ$, which is certainly an approximation. The selected value is therefore

$$\Delta_r S_m^\circ(\text{V.56}, q = 4, 298.15\text{ K}) = (476 \pm 17)\,\text{J}\cdot\text{K}^{-1}\cdot\text{mol}^{-1}.$$

The entropy of $UF_4(aq)$ is then calculated to be

$$S_m^\circ(\text{UF}_4, \text{aq}, 298.15\,\text{K}) = (4 \pm 21)\,\text{J}\cdot\text{K}^{-1}\cdot\text{mol}^{-1}.$$

Using the selected reaction data, $\Delta_f H_m^\circ$ is calculated via $\Delta_r H_m^\circ$.

$$\Delta_f H_m^\circ(\text{UF}_4, \text{aq}, 298.15\,\text{K}) = -(1936.8 \pm 8.7)\,\text{kJ}\cdot\text{mol}^{-1}$$

This review concludes that the enthalpies of reaction in the U^{4+}-F^- system are small, hence the corresponding equilibria are not strongly influenced by temperature.

Although no heat capacity data are available for any of these species, Lemire and Tremaine [80LEM/TRE] estimated the mean heat capacity between 298 and 473 K of the charged UF_q^{4-q} species using a modified Criss Cobble treatment. These mean values should be regarded as approximate values, suitable only for crude calculations of changes in equilibrium constants over this temperature range. They are not likely to be good approximations to the true values at 298.15 K. Because of the uncertainties discussed above and the fact that no heat capacity value is available for the F^- ion, this review does not select any heat capacity values for the uranium(IV) fluoride complexes. Experimental studies by Luk'yanychev and Nikolaev [62LUK/NIK, 63LUK/NIK, 63LUK/NIK2], and Vdovenko and Romanov [67VDO/ROM], are given zero weight for reasons discussed in Appendix A.

V.4.1.3. Solid uranium fluorides

V.4.1.3.1. Binary uranium fluorides and their hydrates

a) UF_3 (cr) and UF_4 (cr)

The selection of enthalpy of formation data for UF_3(cr) and UF_4(cr) presents the problem most feared by critical reviewers of chemical thermodynamic data: different methods lead to apparently equally reliable values which are appreciably different. For the enthalpy of formation of UF_3(cr), critical compilations have reported values as different as $-(1480\pm20)\,\text{kJ}\cdot\text{mol}^{-1}$ [72FUG] and $-(1509\pm4)\,\text{kJ}\cdot\text{mol}^{-1}$ [82FUG]; and for the enthalpy of formation of UF_4(cr), $-(1920.5\pm3.0)\,\text{kJ}\cdot\text{mol}^{-1}$ [82GLU/GUR] and $-(1914.2\pm4.2)\,\text{kJ}\cdot\text{mol}^{-1}$ [83FUG/PAR]. The difference between the latter two compilations is related to the use of different unpublished results of fluorine combustion studies at Argonne National Laboratory. Glushko *et al.* [78GLU/GUR] cited a report by Wijbenga, Cordfunke and Johnson, in which they gave $\Delta_f H_m^\circ$(UF_4, cr, 298.15 K)

$= -(1921.4 \pm 4.6)\,\mathrm{kJ\cdot mol^{-1}}$. Wijbenga [81WIJ] described the earlier experiments and obtained for two different samples $\Delta_f H_m^\circ(\mathrm{UF_4}$, cr, 298.15 K) $= -(1916.3 \pm 3.8)$ and $-(1916.2 \pm 2.6)\,\mathrm{kJ\cdot mol^{-1}}$. Fuger *et al.* [83FUG] also cited this report but argued that a correction for HF(g) formed from H impurity had to be applied. In addition, they had access to preliminary data from a more recent fluorine combustion study published later by Johnson [85JOH]. Johnson obtained $\Delta_f H_m^\circ(\mathrm{UF_4}$, cr, 298.15 K) $= -(1910.6 \pm 2.0)\,\mathrm{kJ\cdot mol^{-1}}$; he also discussed Wijbenga's [81WIJ] experiments and concluded that her value was subject to various interpretations due to uncertainties in the analyses of the $\mathrm{UF_4}$ samples. Fuger *et al.* [83FUG/PAR] presented a review and detailed discussion of the available experimental data. A summary of published data on both $\mathrm{UF_4}$(cr) and $\mathrm{UF_3}$(cr) is presented in Table V.19.

The value selected in the IAEA review [83FUG/PAR] is a weighted average of the values published by Johnson [85JOH] (which were based on the fluorination of $\mathrm{UF_4}$(cr) to $\mathrm{UF_6}$(cr) and the selected enthalpy of formation of $\mathrm{UF_6}$(cr) and of Cordfunke and Ouweltjes [81COR/OUW2] (which were based on calorimetric measurements and the selected enthalpies of formation of HF(aq), $\mathrm{U_3O_8}$(cr) and γ-$\mathrm{UO_3}$). It is thus a combination of the best of the fluoride system and the oxide system data. This review accepts the value selected by the IAEA [83FUG/PAR],

$$\Delta_f H_m^\circ(\mathrm{UF_4}, \mathrm{cr}, 298.15\ \mathrm{K}) = -(1914.2 \pm 4.2)\,\mathrm{kJ\cdot mol^{-1}},$$

even though it is recognized that further measurements are needed to resolve the discrepancies in the experimental values.

The difference $[\Delta_f H_m^\circ(\mathrm{UF_3}$, cr, 298.15 K) $-$ $\Delta_f H_m^\circ(\mathrm{UF_4}$, cr, 298.15 K)] is less subject to experimental and calculational uncertainties. A value of $(412.8 \pm 2.0)\,\mathrm{kJ\cdot mol^{-1}}$ is selected for this difference from the data in Table V.19. This is used to obtain the selected value of

$$\Delta_f H_m^\circ(\mathrm{UF_3}, \mathrm{cr}, 298.15\,\mathrm{K}) = -(1501.4 \pm 4.7)\,\mathrm{kJ\cdot mol^{-1}}.$$

The uncertainty intervals are estimated in order to better reflect the uncertainties associated with estimating the absolute values of the parameters, but when used in calculations, they tend to over-estimate the uncertainty associated with processes which can be directly defined with greater confidence. This must be taken into account when using the uncertainty intervals presented here. It should be mentioned that the relationships $\mathrm{UCl_4}$-$\mathrm{UF_4}$, $\mathrm{UCl_4}$-$\mathrm{UO_2Cl_2}$ and $\mathrm{UCl_4}$-$\mathrm{UO_3}$ require further experimental work.

The selected temperature functions $C_{p,m}^\circ(\mathrm{UF_3}$, cr, T) and the values of $C_{p,m}^\circ S_m^\circ$ at 298.15 K are taken from Cordfunke, Konings and Westrum [89COR/KON]. The data of Burns, Osborne and Westrum [60BUR/OSB] (1.3 to 20 K) are combined with the consistent values of Osborne, Westrum and Lohr [55OSB/WES] to obtain the values at 298.15 K selected for $\mathrm{UF_4}$(cr). The $C_{p,m}^\circ$ function of $\mathrm{UF_4}$(cr) is taken from the assessment by Glushko *et al.* [82GLU/GUR].

Table V.19: Data considered for the selection of $\Delta_f H^\circ_m(UF_3$, cr, 298.15 K) and $\Delta_f H^\circ_m(UF_4$, cr, 298.15 K).

Reference	$\Delta_f H^\circ_m(UF_3)$ $(kJ \cdot mol^{-1})$	$\Delta_f H^\circ_m(UF_4)^{(a)}$ $(kJ \cdot mol^{-1})$	$\Delta_f H^\circ_m(UF_3 - UF_4)^{(b)}$ $(kJ \cdot mol^{-1})$
[58AGR]		$-1920.9 \pm 8.4^{(q)}$	
[58BRI]		$-1912.0^{(r)}$	
[60MAL/GAG, 80HU/NI]		$-1892.8^{(c)}$	
[66HEU/EGA]	$-1496.2 \pm 6.3^{(i)}$	$-1906.6 \pm 6.3^{(i)}$	410.4 ± 4.2
	$-1492.8 \pm 6.3^{(j)}$	$-1902.5 \pm 6.3^{(j)}$	409.7 ± 4.2
[67HAY](s)	$-1494.0 \pm 3.1^{(g)}$	$-1898.1 \pm 0.7^{(g)}$	404.1 ± 2.1
	$-1505.7 \pm 2.5^{(h)}$	$-1909.6 \pm 0.5^{(h)}$	403.9 ± 2.1
[67MAR/BON]	$-1498.7 \pm 6.3^{(i)}$	$-1909.6 \pm 6.3^{(i)}$	410.9 ± 4.2
	$-1494.5 \pm 6.3^{(j)}$	$-1904.1 \pm 6.3^{(j)}$	409.6 ± 4.2
[68KHA]		$-1892.4 \pm 0.9^{(d)}$	407.9 ± 5.0
[70KHA/KRI]	$-1484.9^{(e)}$		
	$-1483.9^{(f)}$		
[80HU/NI]		$-1895.8 \pm 4.6^{(o)}$	
[80SUG/CHI]		$-1884.9 \pm 2.9^{(n)}$	
[80VOL/SUG]		$-1891.6 \pm 4.6^{(m)}$	
[81COR/OUW2]	$-1509.6 \pm 5.4^{(k)}$	$-1921.3 \pm 4.2^{(k)}$	411.7 ± 3.3
[81WIJ]	$-1502.4 \pm 6.4^{(l)}$	$-1916.7 \pm 2.1^{(l)}$	413.8 ± 6.0
[85JOH]		$-1910.6 \pm 2.0^{(p)}$	

(a) Values recalculated by Fuger *et al.* [83FUG/PAR].
(b) $\Delta_f H^\circ_m(UF_3, cr, 298.15\ K) - \Delta_f H^\circ_m(UF_4, cr, 298.15\ K)$ in $kJ \cdot mol^{-1}$.
(c) Solution calorimetry at 293 K, dependent on UCl_4(cr) and HF(sln).
(d) Solution calorimetry at 323 K, dependent on UCl_4(cr) and HF(sln).
(e) Solution calorimetry at 323 K, dependent on UO_2Cl_2(cr), (α-UO_3,UCl_4)HF(sln) and $FeCl_2$(cr)/$FeCl_3$(cr).
(f) Solution calorimetry at 323 K, dependent on UCl_4(cr), HF(sln) and $FeCl_2$(cr)/$FeCl_3$(cr).
(g) Fluoridation of α-U and UF_3(cr) to UF_6(g).
(h) Fluoridation of UF_3(cr) to UF_6(g) with selected $\Delta_f H^\circ_m(UF_6, g, 298.15\ K) = -(2147.4 \pm 1.7)\ kJ \cdot mol^{-1}$.
(i) emf measurement at 873 K, dependent on AlF_3(cr).
(j) emf measurement at 873 K, dependent on MgF_2(cr).
(k) Solution calorimetry at 298.15 K K, dependent on γ-UO_3, HF(sln) and U_3O_8(cr).
(l) Fluoride bomb calorimetry at 298.15 K, dependent on selected $\Delta_f H^\circ_m(UF_6, cr, 298.15\ K) = -(2197.7 \pm 1.8)\ kJ \cdot mol^{-1}$.
(m) Obtained from Ref. [67MAR/BON], $4UF_3(cr) \rightleftharpoons 3UF_4(cr) + U(cr)$ with $\Delta_r H^\circ_m = (265 \pm 17)\ kJ \cdot mol^{-1}$.
(n) Dependent on UCl_4(cr) and KF(cr).
(o) Dependent on UCl_4(cr) and HF(sln).
(p) With $\Delta_f H^\circ_m(UF_6, cr, 298.15\ K) = -(2197.7 \pm 1.8)\ kJ \cdot mol^{-1}$.
(q) Transpiration, dependent on UF_6(g) and α-UF_5, recalculated by Fuger *et al.* [83FUG/PAR].
(r) Equilibrium at 620 to 955 K, third law, dependent on UO_2(cr) and HF(g).
(s) Unpublished paper.

$$S^\circ_m(UF_3, cr, 298.15\,K) = (123.4 \pm 0.4)\,J \cdot K^{-1} \cdot mol^{-1}$$
$$S^\circ_m(UF_4, cr, 298.15\,K) = (151.7 \pm 0.2)\,J \cdot K^{-1} \cdot mol^{-1}$$

$$C^\circ_{p,m}(UF_3, cr, 298.15\,K) = (95.1 \pm 0.4)\,J \cdot K^{-1} \cdot mol^{-1}$$
$$C^\circ_{p,m}(UF_4, cr, 298.15\,K) = (116.0 \pm 0.1)\,J \cdot K^{-1} \cdot mol^{-1}$$

The Gibbs energy of formation values for these two compounds are calculated from the corresponding $\Delta_f H^\circ_m$ and S°_m values.

$$\Delta_f G^\circ_m(UF_3, cr, 298.15\,K) = -(1432.5 \pm 4.7)\,J \cdot K^{-1} \cdot mol^{-1}$$
$$\Delta_f G^\circ_m(UF_4, cr, 298.15\,K) = -(1823.5 \pm 4.2)\,J \cdot K^{-1} \cdot mol^{-1}$$

b) UF_4 (cr) hydrates

With the exception of $UF_4 \cdot 2.5H_2O$(cr), UF_4(cr) hydrates are poorly defined and generally metastable compounds. The following hydrates have been identified (by Gagarinskii and Khanaev [67GAG/KHA] and Gagarinskii and Mashirev [59GAG/MAS], except for the monohydrate, which was reported by Katz and Rabinowitch [51KAT/RAB]):

$UF_4 \cdot 0.5H_2O$ (orthorhombic)
$UF_4 \cdot 0.5H_2O$ (cubic)
$UF_4 \cdot 0.5H_2O$ (monoclinic)
$UF_4 \cdot H_2O$(cr)
$UF_4 \cdot 1.33H_2O$ (monoclinic)
$UF_4 \cdot 1.5H_2O$ (cubic)
$UF_4 \cdot 2H_2O$ (cubic)
$UF_4 \cdot 2.5H_2O$ (orthorhombic)

Only the 1-, 1.33-, 1.5- and 2.5-hydrates have thermodynamic data reported (only enthalpy data in all cases). Katz and Rabinowitch [51KAT/RAB] reported two reactions which lead to a mean value of $-2224\,kJ \cdot mol^{-1}$ for the enthalpy of formation of the monohydrate. The discrepancy in the two values ($75\,kJ \cdot mol^{-1}$), the lack of experimental detail available and the lack of structural characterization of the solid in such a potentially complex system cause this review not to accept this value and not to select data for this compounds. The experiments of Gagarinskii and Khanaev [67GAG/KHA] for the reaction

$$UF_4(cr) + xH_2O(l) \rightarrow UF_4 \cdot xH_2O(cr) \qquad (V.58)$$

lead to $\Delta_r H_m$(V.58, $x = 1.33$) $= -(31.1 \pm 0.3)\,kJ \cdot mol^{-1}$ and $\Delta_r H_m$(V.58, $x = 1.5$) $= -(21.3 \pm 0.2)\,kJ \cdot mol^{-1}$ at 323 K. This review accepts the conclusion of Fuger *et al.* [83FUG/PAR] that these compounds were not sufficiently well defined in the experiment to allow consistent enthalpy of formation selections to be made.

Stable, orthorhombic $UF_4 \cdot 2.5H_2O$(cr) has been studied by several researchers [50POP/GAG, 57POP/KOS, 60MAL/GAG, 67GAG/KHA]. This review accepts the conclusion of Fuger *et al.* [83FUG/PAR] that incomplete sample characterization and the possibility of the formation of lower hydrates cast doubt on the results of all studies except that of Gagarinskii and Khanaev [67GAG/KHA]. Fuger *et al.* [83FUG/PAR] adjusted $\Delta_r H_m$(V.58, $x = 2.5$) $= -(43.9 \pm 0.2)$ kJ · mol^{-1} at 323 K [67GAG/KHA] to

$$\Delta_r H^\circ_m(\text{V.58}, x = 2.5, 298.15\text{ K}) = -(42.7 \pm 0.8)\text{ kJ}\cdot\text{mol}^{-1}.$$

In combination with the value of $\Delta_f H^\circ_m(\text{UF}_4$, cr, 298.15 K) selected above and $\Delta_f H^\circ_m(\text{H}_2\text{O}$, l, 298.15 K) from CODATA [89COX/WAG], this leads to the selected enthalpy of formation of

$$\Delta_f H^\circ_m(\text{UF}_4\cdot 2.5\text{H}_2\text{O}, \text{cr}, 298.15\text{ K}) = -(2671.5 \pm 4.3)\text{ kJ}\cdot\text{mol}^{-1}.$$

The solubility data of Savage and Browne [60SAV/BRO] are used to calculate the solubility constant of $\text{UF}_4\cdot 2.5\text{H}_2\text{O}$(cr). These data, reduced to zero ionic strength by using the specific ion interaction theory (*cf.* Appendix B), appear at the bottom of Table V.18. The solubility data of Luk'yanychev and Nikolaev [63LUK/NIK] are re-evaluated by this review, *cf.* Appendix A. The solubility product thus obtained, $\log_{10} K^\circ_{s,0} = -(25.7 \pm 0.3)$, is much lower than the value proposed by the authors, $\log_{10} K^\circ_{s,0} \approx -21.2$, and in better agreement with the value of Savage and Browne [60SAV/BRO]. This review does not credit the value obtained from the data in Ref. [63LUK/NIK] in view of the experimental shortcomings. A value of $\log_{10} K^\circ_{s,0} = -(29.40 \pm 0.16)$ is obtained from data in Ref. [60SAV/BRO] for use in the evaluation of higher complex formation equilibria as described in Section V.4.1.2.3. Selected auxiliary data for $\Delta_f G^\circ_m$ from this review are used to calculate $\Delta_f G^\circ_m(\text{UF}_4\cdot 2.5\text{H}_2\text{O}, \text{cr}, 298.15\text{ K}) = -(2416.6 \pm 3.4)$ kJ · mol^{-1}.

This value differs from the one proposed by Fuger *et al.* [83FUG/PAR], $\Delta_f G^\circ_m(\text{UF}_4\cdot 2.5\text{H}_2\text{O}, \text{cr}, 298.15\text{ K}) = -(2437 \pm 6)$ kJ · mol^{-1}, which is based on $\Delta_f H^\circ_m(\text{UF}_4\cdot 2.5\text{H}_2\text{O}$, cr, 298.15 K) and $S^\circ_m(\text{UF}_4\cdot 2.5\text{H}_2\text{O}, \text{cr}, 298.15\text{ K}) = (251 \pm 8)$ J · K^{-1} · mol^{-1}, estimated from the measured $S^\circ_m(\text{UF}_4, \text{cr}, 298.15\text{ K})$ plus an estimate from 2.5 H_2O (40 J · K^{-1} · mol^{-1} per molecule of water). This review prefers to use the method described in Appendix D to estimate

$$S^\circ_m(\text{UF}_4\cdot 2.5\text{H}_2\text{O}, \text{cr}, 298.15\text{ K}) = (263.5 \pm 15.0)\text{ J}\cdot\text{K}^{-1}\cdot\text{mol}^{-1}.$$

This value is in fair agreement with the estimate of Fuger *et al.* [83FUG/PAR].

From the enthalpy of formation and the entropy, this review obtains

$$\Delta_f G^\circ_m(\text{UF}_4\cdot 2.5\text{H}_2\text{O}, \text{cr}, 298.15\text{ K}) = -(2440.3 \pm 6.2)\text{ kJ}\cdot\text{mol}^{-1}.$$

From this selected value, a solubility product of $\log_{10} K^\circ_{s,0} = -(33.5 \pm 1.2)$ can be calculated which is incompatible with the ones obtained experimentally [60SAV/BRO, 63LUK/NIK]. Since the solid phase used for the solubility measurements was not characterized, this review prefers, for the crystalline compound, to rely on the Gibbs energy based on the experimental $\Delta_f H^\circ_m$ and the entropy estimate which can be made with a good degree of confidence.

In the absence of direct measurements, this review accepts the heat capacity equation used by Lemire and Tremaine [80LEM/TRE] and selects

$$C^\circ_{p,m}(UF_4 \cdot 2.5H_2O, cr, 298.15\,K) = (263.7 \pm 15.0)\,J \cdot K^{-1} \cdot mol^{-1}.$$

c) UF_5 (cr)

Crystalline uranium(V) pentafluoride exists in two allotropic forms. Both have tetragonal crystal structures, but the low-temperature β-phase has eight molecules per unit cell while the α-phase, which is the stable form above 398 K (in the presence of UF_6(g) to avoid disproportionation), has only two molecules per unit cell [51KAT/RAB].

α-UF_5: Fuger *et al.* [83FUG/PAR] selected a value of $\Delta_f H^\circ_m(UF_5, \alpha, 298.15\,K)$ which is essentially identical to the $-(2075.5 \pm 6.7)\,kJ \cdot mol^{-1}$ obtained by O'Hare, Malm and Eller [82OHA/MAL] from their solution-calorimetric experiments. This is consistent with the earlier results of Agron [58AGR] when recalculated using the values of $\Delta_f H^\circ_m(UF_6, g, 298.15\,K)$ and $\Delta_f H^\circ_m(UF_4, cr, 298.15\,K)$ selected in this review. The value selected here is

$$\Delta_f H^\circ_m(UF_5, \alpha, 298.15\,K) = -(2075.3 \pm 5.9)\,kJ \cdot mol^{-1},$$

where the uncertainty is estimated by this review.

Katz and Rabinowitch [51KAT/RAB] report that Brickwedde, Hodge and Scott [51BRI/HOD] measured the heat capacity of α-UF_5 on a sample which contained only 83% UF_5 by weight, the remainder being UF_4 and UO_2F_2. In the absence of additional data, Fuger *et al.* [83FUG/PAR] adjusted the resulting entropy value by $+11.3\,J \cdot K^{-1} \cdot mol^{-1}$ to obtain a value which is consistent with Agron's [58AGR] dissociation measurements for α-UF_5, β-UF_5, $UF_{4.5}$(cr) and $UF_{4.25}$(cr). This value,

$$S^\circ_m(UF_5, \alpha, 298.15\,K) = (199.6 \pm 12.6)\,J \cdot K^{-1} \cdot mol^{-1},$$

is accepted by this review.

The Gibbs energy of formation is calculated from the selected enthalpy of formation and entropy.

$$\Delta_f G^\circ_m(UF_5, \alpha, 298.15\,K) = -(1968.7 \pm 7.0)\,kJ \cdot mol^{-1}$$

The selected heat capacity value of

$$C^\circ_{p,m}(UF_5, \alpha, 298.15\,K) = (132.2 \pm 4.2)\,J \cdot K^{-1} \cdot mol^{-1}$$

is interpolated from the data in Katz and Rabinowitch [51KAT/RAB] between 275 and 300 K. The selected function $C_{p,m}(UF_5, \alpha, T)$ is from Knacke, Lossman and Müller [69KNA/LOS2], who did not distinguish between α- and β-phases. This lack of distinction should influence less their results for α-UF_5, which should be within the uncertainty stated at 298.15 K.

β-UF_5: The value selected by Fuger *et al.* [83FUG/PAR] and accepted here,

$$\Delta_f H^\circ_m(UF_5, \beta, 298.15\,K) = -(2083.2 \pm 4.2)\,kJ \cdot mol^{-1},$$

is also based on the $-(2083 \pm 6)\,\text{kJ}\cdot\text{mol}^{-1}$ value calculated from solution calorimetry data by O'Hare, Malm and Eller [82OHA/MAL]. The observed difference of $\Delta_f H_m^\circ(\text{UF}_5, \beta, 298.15\,\text{K}) - \Delta_f H_m^\circ(\text{UF}_5, \alpha, 298.15\,\text{K}) = 7.9\,\text{kJ}\cdot\text{mol}^{-1}$ is consistent with Parker's [80PAR] value for the β-$\text{UF}_5 \rightarrow \alpha$-$\text{UF}_5$ transition enthalpy of $(7.9 \pm 2)\,\text{kJ}\cdot\text{mol}^{-1}$ based on the earlier measurements of Agron [58AGR].

Fuger *et al.* [83FUG/PAR] calculated a β-$\text{UF}_5 \rightarrow \alpha$-$\text{UF}_5$ transition entropy of $20.1\,\text{J}\cdot\text{K}^{-1}\cdot\text{mol}^{-1}$ at 298.15 K from the data of Agron [58AGR], which leads to the selected

$$S_m^\circ(\text{UF}_5, \beta, 298.15\,\text{K}) = (179.5 \pm 12.6)\,\text{J}\cdot\text{K}^{-1}\cdot\text{mol}^{-1},$$

when combined with $S_m^\circ(\text{UF}_5, \alpha, 298.15\,\text{K})$ selected above.

The Gibbs energy of formation is calculated from the selected enthalpy of formation and entropy.

$$\Delta_f G_m^\circ(\text{UF}_5, \beta, 298.15\,\text{K}) = -(1970.6 \pm 5.6)\,\text{kJ}\cdot\text{mol}^{-1}$$

In the absence of direct experimental data, the assumption $C_{p,m}(\text{UF}_5, \alpha, 298.15\text{ K}) = C_{p,m}(\text{UF}_5, \beta, 298.15\text{ K})$ is made, but the uncertainty associated with the selected value for the β-form is increased to $12\,\text{J}\cdot\text{K}^{-1}\cdot\text{mol}^{-1}$.

$$C_{p,m}^\circ(\text{UF}_5, \beta, 298.15\,\text{K}) = (132.2 \pm 12.0)\,\text{J}\cdot\text{K}^{-1}\cdot\text{mol}^{-1}$$

The same function from Knacke, Lossman and Müller [69KNA/LOS2] is used for $C_{p,m}^\circ(\text{UF}_5, \beta, T)$ as is used for α-UF_5.

d) UF_6 (cr)

The selected enthalpy of formation of UF_6(cr) is from a recent fluorine combustion study by Johnson [79JOH].

$$\Delta_f H_m^\circ(\text{UF}_6, \text{cr}, 298.15\text{ K}) = -(2197.7 \pm 1.8)\,\text{kJ}\cdot\text{mol}^{-1}$$

This value is in good agreement with $\Delta_f H_m^\circ(\text{UF}_6, \text{cr}, 298.15\text{ K}) = -(2197.0 \pm 1.7)\,\text{kJ}\cdot\text{mol}^{-1}$, as given by Parker [80PAR], which involved a correction of $-10.46\,\text{kJ}\cdot\text{mol}^{-1}$, obtained by simultaneous analysis of other uranium halides, oxides, chalcogenides and pnictides. This review prefers the direct calorimetric value.

This review accepts the entropy and heat capacity values for UF_6(cr) reported by Brickwedde, Hodge and Scott [48BRI/HOD],

$$\begin{aligned} S_m^\circ(\text{UF}_6, \text{cr}, 298.15\,\text{K}) &= (227.6 \pm 1.3)\,\text{J}\cdot\text{K}^{-1}\cdot\text{mol}^{-1} \\ C_{p,m}^\circ(\text{UF}_6, \text{cr}, 298.15\,\text{K}) &= (166.8 \pm 0.2)\,\text{J}\cdot\text{K}^{-1}\cdot\text{mol}^{-1}. \end{aligned}$$

The Gibbs energy of formation is calculated from the selected enthalpy of formation and entropy.

$$\Delta_f G_m^\circ(\text{UF}_6, \text{cr}, 298.15\,\text{K}) = -(2069.2 \pm 1.8)\,\text{kJ}\cdot\text{mol}^{-1}$$

The $C^\circ_{p,\mathrm{m}}(T)$ equations given by Glushko *et al.* [82GLU/GUR] are accepted in this review.

e) $U_2F_9(cr)$ and $U_4F_{17}(cr)$

These compounds, often written $UF_{4.5}$(cr) and $UF_{4.25}$(cr), are considered intermediates between UF_4(cr) and UF_6(cr), as is UF_5(cr). All three of these compounds tend to disproportionate to UF_6(g) and the next lower fluoride solid [63RAN/KUB]. Rand and Kubaschewski [63RAN/KUB] estimated the entropies of $UF_{4.5}$(cr) and $UF_{4.25}$(cr) in two ways: by Latimer's method [52LAT] and by linearly interpolating between the values for entropies of UF_4(cr) and UF_5(cr) obtained from Agron [58AGR]. A value of $S^\circ_\mathrm{m}(\mathrm{UF_5, cr, 298.15\,K}) = 188.3\,\mathrm{J\cdot K^{-1}\cdot mol^{-1}}$ was used for the undifferentiated pentafluoride. Rand and Kubaschewski [63RAN/KUB] selected values intermediate between these two estimates, which they found most consistent with Agron's [58AGR] data. The values accepted by Rand and Kubaschewski [63RAN/KUB], and adopted by Fuger *et al.* [83FUG/PAR], are almost indistinguishable from those obtained by linear interpolation between $S^\circ_\mathrm{m}(\mathrm{UF_4, cr, 298.15\ K}) = (151.7 \pm 0.2)\,\mathrm{J\cdot K^{-1}\cdot mol^{-1}}$ and $S^\circ_\mathrm{m}(\mathrm{UF_5}, \beta, \mathrm{298.15\ K}) = (180 \pm 13)\,\mathrm{J\cdot K^{-1}\cdot mol^{-1}}$ selected here. The values of Fuger *et al.* [83FUG/PAR] are selected in this review.

$$S^\circ_\mathrm{m}(\mathrm{U_2F_9, cr, 298.15\,K}) = (329 \pm 20)\,\mathrm{J\cdot K^{-1}\cdot mol^{-1}}$$
$$S^\circ_\mathrm{m}(\mathrm{U_4F_{17}, cr, 298.15\,K}) = (631 \pm 40)\,\mathrm{J\cdot K^{-1}\cdot mol^{-1}}$$

Fuger *et al.* [83FUG/PAR] calculated $\Delta_\mathrm{r}G^\circ_\mathrm{m}(T)$ functions from the disproportionation pressures given by Agron [58AGR]. These, in turn, were used to calculate standard enthalpies of formation. This review accepts these values.

$$\Delta_\mathrm{f}G^\circ_\mathrm{m}(\mathrm{U_2F_9, cr, 298.15\,K}) = -(3812 \pm 17)\,\mathrm{kJ\cdot mol^{-1}}$$
$$\Delta_\mathrm{f}G^\circ_\mathrm{m}(\mathrm{U_4F_{17}, cr, 298.15\,K}) = -(7464 \pm 30)\,\mathrm{kJ\cdot mol^{-1}}$$

The enthalpy of formation values are calculated from the selected Gibbs energy of formation and entropy values.

$$\Delta_\mathrm{f}H^\circ_\mathrm{m}(\mathrm{U_2F_9, cr, 298.15\,K}) = -(4016 \pm 18)\,\mathrm{kJ\cdot mol^{-1}}$$
$$\Delta_\mathrm{f}H^\circ_\mathrm{m}(\mathrm{U_4F_{17}, cr, 298.15\,K}) = -(7850 \pm 32)\,\mathrm{kJ\cdot mol^{-1}}$$

This review accepts $C^\circ_{p,\mathrm{m}}$ and for both compounds from the IAEA review [83FUG/PAR] based on Knacke, Lossman and Müller [69KNA/LOS2].

$$C^\circ_{p,\mathrm{m}}(\mathrm{U_2F_9, cr, 298.15\,K}) = (251.0 \pm 16.7)\,\mathrm{J\cdot K^{-1}\cdot mol^{-1}}$$
$$C^\circ_{p,\mathrm{m}}(\mathrm{U_4F_{17}, cr, 298.15\,K}) = (485.3 \pm 33.0)\,\mathrm{J\cdot K^{-1}\cdot mol^{-1}}.$$

The $C^\circ_{p,\mathrm{m}}(T)$ data tabulated by the IAEA [83FUG/PAR] are fitted between 298.15 and 600 K to obtain the selected $C^\circ_{p,\mathrm{m}}(T)$ function listed in Table III.3.

V.4.1.3.2. Other U(VI) fluorides

a) UOF_4 (cr)

All published thermodynamic data on UOF_4(cr) can be traced to the solution-calorimetry work of O'Hare and Malm [82OHA/MAL2]. Using CODATA compatible auxiliary data also selected in this review, these authors calculated $\Delta_f H^\circ_m(UOF_4, cr, 298.15 K)$ and estimated $S^\circ_m(UOF_4, cr, 298.15 K)$, based on the auxiliary entropy data reported by Parker [80PAR]. This review accepts these values, but increases the uncertainty of the enthalpy of formation to $4\,kJ \cdot mol^{-1}$ to reflect that the selection is based on a single determination.

$$\begin{array}{lcl} \Delta_f H^\circ_m(UOF_4, cr, 298.15\,K) & = & -(1924.6 \pm 4.0)\,kJ \cdot mol^{-1} \\ S^\circ_m(UOF_4, cr, 298.15\,K) & = & (195 \pm 5)\,J \cdot K^{-1} \cdot mol^{-1} \end{array}$$

Note that the enthalpy of Reaction (V.59),

$$2UOF_4(cr) \rightleftharpoons UO_2F_2(cr) + UF_6(g), \tag{V.59}$$

$\Delta_r H^\circ_m = (46.3 \pm 5.5)\,kJ \cdot mol^{-1}$, is very close to the enthalpy of sublimation of UF_6(cr), $\Delta_{sub} H^\circ_m = (50.3 \pm 4.0)\,kJ \cdot mol^{-1}$, indicating that from a thermodynamic point of view, UOF_4(cr) behaves as a very loosely bound 1:1 complex of UO_2F_2(cr) and UF_6(cr) [82OHA/MAL2].

The Gibbs energy of formation is calculated from the selected enthalpy of formation and entropy.

$$\Delta_f G^\circ_m(UOF_4, cr, 298.15\,K) = -(1816.3 \pm 4.3)\,kJ \cdot mol^{-1},$$

No heat capacity or thermal function data are available for this species.

b) $U_2O_3F_6$ (cr)

The data for this species also originate in the work of O'Hare and Malm [82OHA/MAL2]. Based on the measured enthalpy of Reaction (V.59) above, these authors estimated

$$\Delta_r H^\circ_m(V.60) = (46 \pm 6)\,kJ \cdot mol^{-1}$$

for the reaction

$$3UOF_4(cr) \rightleftharpoons U_2O_3F_6(cr) + UF_6(g). \tag{V.60}$$

This, combined with the selected enthalpies for UOF_4(cr) and UF_6(g), leads to the selected value of

$$\Delta_f H^\circ_m(U_2O_3F_6, cr, 298.15\,K) = -(3579.2 \pm 13.5)\,kJ \cdot mol^{-1}.$$

O'Hare and Malm [82OHA/MAL2] also estimated $S^\circ_m(U_2O_3F_6, cr, 298.15\,K) = (355 \pm 10)\,J \cdot K^{-1} \cdot mol^{-1}$ using a Latimer-like method. No details were given, and in this review an approximate value of the entropy is estimated as the sum of the entropies of γ-UO_3 and UF_6(cr),

$$S^\circ_m(U_2O_3F_6, cr, 298.15\,K) = (324 \pm 20)\,J \cdot K^{-1} \cdot mol^{-1}.$$

These selected results are combined with selected auxiliary data to obtain the standard Gibbs energy of formation.

$$\Delta_f G^\circ_m(U_2O_3F_6, cr, 298.15\,K) = -(3372.7 \pm 14.8)\,kJ \cdot mol^{-1}$$

No heat capacity or thermal function data are available for this species.

c) $U_3O_5F_8(cr)$

Otey and Le Doux [67OTE/DOU] deduced from a differential thermal analysis study that $\Delta_r H_m(V.61) = -7.5\,kJ \cdot mol^{-1}$ at 681 K.

$$U_3O_5F_8(cr) \rightleftharpoons 2\tfrac{1}{2}UO_2F_2(cr) + \tfrac{1}{2}UF_6(g) \quad (V.61)$$

O'Hare and Malm [82OHA/MAL2] assumed that this value is not appreciably different at 298.15 K and give $\Delta_r H^\circ_m(V.61) \approx -8\,kJ \cdot mol^{-1}$. This review prefers to make an estimate of $\Delta_r C^\circ_{p,m}(V.61)$ and to calculate $\Delta_r H^\circ_m$(V.61, 298.15 K) from the corresponding value at 681 K. It is assumed that $C^\circ_{p,m}(U_3O_8F_8$, cr, 298.15 K) $= 2.5C^\circ_{p,m}(UO_2F_2$, cr, 298.15 K) + $0.5C^\circ_{p,m}(UF_6$, cr, 298.15 K). Hence $\Delta_r C^\circ_{p,m}(V.61)$ $= 0.5[C^\circ_{p,m}(UF_6$, g, 298.15 K) − $C^\circ_{p,m}(UF_6$, cr, 298.15 K)], leading to the selected

$$\Delta_r C^\circ_{p,m}(V.61, 298.15\,K) = (18.7 \pm 4.0)\,J \cdot K^{-1} \cdot mol^{-1}.$$

The value of $\Delta_r H^\circ_m(V.61)$ at 298.15 K is then

$$\Delta_r H^\circ_m(V.61, 298.15\,K) = -(15.1 \pm 2.0)\,kJ \cdot mol^{-1},$$

where the uncertainty is assigned by this review.

From this value, and the selected standard enthalpies of formation, this review obtains

$$\Delta_f H^\circ_m(U_3O_5F_8, cr, 298.15\,K) = -(5193.0 \pm 3.9)\,kJ \cdot mol^{-1}.$$

The entropy of $U_3O_5F_8(cr)$, estimated in this review as approximately equal to the sum of the entropies of γ-UO_3, $UF_6(cr)$ and $UO_2F_2(cr)$, *cf.* Paragraph d) below, is

$$S^\circ_m(U_3O_5F_8, cr, 298.15\,K) = (459 \pm 30)\,J \cdot K^{-1} \cdot mol^{-1}.$$

The Gibbs energy of formation is calculated from the selected enthalpy of formation and entropy.

$$\Delta_f G^\circ_m(U_3O_5F_8, cr, 298.15\,K) = -(4890.1 \pm 9.8)\,kJ \cdot mol^{-1}$$

The heat capacity is calculated from the value of $\Delta_r C^\circ_{p,m}$(V.61, 298.15 K) selected above:

$$C^\circ_{p,m}(U_3O_5F_8, cr, 298.15\,K) = (304.1 \pm 4.1)\,J \cdot K^{-1} \cdot mol^{-1}.$$

d) UO_2F_2 *(cr) and its hydrates*

UO_2F_2(cr): Fuger *et al.* [83FUG/PAR] accepted the enthalpy of solution measurements of Popov, Kostylev and Karpova [57POP/KOS] (combined with $\Delta_{sol}H_m(UO_3, \gamma)$ from Vidavskii, Byakhova and Ippolitova [65VID/BYA]) and of Cordfunke and Ouweltjes [77COR/OUW] (combined with $\Delta_{sol}H_m(UO_3, \gamma)$ from Cordfunke, Ouweltjes and Prins [76COR/OUW]). Their selected value of

$$\Delta_f H_m^\circ(UO_2F_2, cr, 298.15\,K) = -(1653.5 \pm 1.3)\,kJ \cdot mol^{-1}$$

is a weighted average of these results and is accepted in this review. It has been further supported experimentally by the more recent calorimetric results of O'Hare and Malm [82OHA/MAL2] who reported $-(1654.8 \pm 2.1)\,kJ \cdot mol^{-1}$, based on the hydrolysis of UF_6. The high-temperature decomposition results of Knacke, Lossman and Müller [69KNA/LOS] are inconsistent with the room-temperature solubility data and are therefore not considered. The selected value is consistent with a standard enthalpy of solution of $\Delta_{sol}H_m^\circ(UO_2F_2, cr, 298.15\,K) = -36\,kJ \cdot mol^{-1}$ obtained by Fuger *et al.* [83FUG/PAR].

Wacker and Cheney [47WAC/CHE] performed heat capacity measurements on UO_2F_2(cr) from 13 to 418 K. Their entropy and heat capacity data are accepted here, as they were in the IAEA review [83FUG/PAR].

$$S_m^\circ(UO_2F_2, cr, 298.15\,K) = (135.56 \pm 0.42)\,J \cdot K^{-1} \cdot mol^{-1}$$
$$C_{p,m}^\circ(UO_2F_2, cr, 298.15\,K) = (103.22 \pm 0.42)\,J \cdot K^{-1} \cdot mol^{-1}$$

The temperature function $C_{p,m}(UO_2F_2, cr, T)$ is based on a least-squares fit of the enthalpy measurements by Cordfunke, Muis and Prins [79COR/MUI], which is constrained to fit the selected $C_{p,m}^\circ$ value at 298.15 K. The agreement between the low- and high-temperature values is found satisfactory, contrary to the opinion of Cordfunke, Muis and Prins [79COR/MUI].

The Gibbs energy of formation is calculated from the selected enthalpy of formation and entropy.

$$\Delta_f G_m^\circ(UO_2F_2, cr, 298.15\,K) = -(1557.3 \pm 1.3)\,kJ \cdot mol^{-1}$$

UO_2F_2(cr) hydrates: The following five hydrates of dioxouranium(VI) fluoride have enthalpy data reported in the literature, even though it should be mentioned that uncertainties exist as to the existence of these hydrates:

$UO_2F_2 \cdot H_2O$ (cr)
$UO_2F_2 \cdot 1.6H_2O$ (cr)
$UO_2F_2 \cdot 2H_2O$ (cr)
$UO_2F_2 \cdot 3H_2O$ (cr)
$UO_2F_2 \cdot 4H_2O$ (cr)

Table V.20: Enthalpies of dehydration of dioxouranium(VI) fluoride hydrates, $UO_2F_2 \cdot xH_2O(cr)$ to $UO_2F_2(cr)$.

	$\Delta_{dehyd}H^\circ_m(UO_2F_2 \cdot xH_2O \rightarrow UO_2F_2, 298.15\,K)$ (kJ · mol^{-1})	
x	Suponitskii *et al.* [71SUP/TSV]	Tsvetkov *et al.* [71TSV/SEL]
1		8.47
1.6	21.21	
2	22.43	16.61
3	23.68[(a)]	22.84[(a)]
4	24.14	31.55

(a) Used to calculate $\Delta_f H^\circ_m(UO_2F_2 \cdot 3H_2O$, cr, 298.15 K).

The available enthalpy data for the dehydration reaction are summarized in Table V.20. The values from Suponitskii *et al.* [71SUP/TSV] are derived from enthalpy of solution measurements. Those from Tsvetkov *et al.* [71TSV/SEL] are from tensimetric measurements of enthalpies of reaction.

This review concurs with Fuger *et al.* [83FUG/PAR] that only the data for the trihydrate are sufficiently consistent for a selection to be based upon them (values estimated using the effective charge model of Kaganyuk, Kyskin and Kazin [83KAG/KYS] are not used).[6] This review accepts the value of Fuger *et al.* [83FUG/PAR] of

$$\Delta_{dehyd}H^\circ_m(V.62, 298.15\text{ K}) = (23.4 \pm 4.2)\,\text{kJ} \cdot \text{mol}^{-1}$$

for the dehydration of the trihydrate,

$$UO_2F_2 \cdot 3H_2O(cr) \rightleftharpoons UO_2F_2(cr) + 3H_2O(l), \qquad (V.62)$$

and calculates

$$\Delta_f H^\circ_m(UO_2F_2 \cdot 3H_2O, \text{cr}, 298.15\,\text{K}) = -(2534.4 \pm 4.4)\,\text{kJ} \cdot \text{mol}^{-1}.$$

The selected entropy value is calculated here based on the selected $S^\circ_m(UO_2F_2$, cr, 298.15 K) with an increased uncertainty, plus an estimated $44.7\,\text{J} \cdot \text{K}^{-1} \cdot \text{mol}^{-1}$, additional contribution per mole of water of hydration, *cf.* Appendix D.

[6] It should be mentioned that a recent paper by Lychev, Mikhalev and Suglobov [90LYC/MIK] provides some experimental information on dioxouranium(VI) fluoride hydrates.

$$S^{\circ}_{m}(UO_2F_2 \cdot 3H_2O, cr, 298.15\,K) = (270 \pm 18)\,J \cdot K^{-1} \cdot mol^{-1}$$

The Gibbs energy of formation is calculated from the selected enthalpy of formation and entropy.

$$\Delta_f G^{\circ}_{m}(UO_2F_2 \cdot 3H_2O, cr, 298.15\,K) = -(2269.7 \pm 6.9)\,kJ \cdot mol^{-1}$$

No heat capacity or thermal function data are available for this species.

e) UO_2FOH (cr) and its hydrates

Tsvetkov *et al.* [71TSV/SEL] observed the existence of two hydrated salts of variable composition of the form $UO_2F_{2-y}(OH)_y{\cdot}2H_2O(cr)$ and $UO_2F_{2-y}(OH)_y{\cdot}H_2O(cr)$ in the UO_3-HF-H_2O system (where y varies between 0.3 and 1.2). For the case where $y = 1$ they reported [73TSV/SEL] the dehydration pressure and decomposition pressure as a function of temperature of Reaction (V.63) for 303 to 379 K and of Reaction (V.64) for 303 to 408 K.

$$UO_2FOH \cdot 2H_2O(cr) \rightleftharpoons UO_2FOH \cdot H_2O(cr) + H_2O(g) \quad (V.63)$$
$$UO_2FOH \cdot H_2O(cr) \rightleftharpoons \tfrac{1}{2}UO_2F_2(cr) + \tfrac{1}{2}UO_3 \cdot H_2O(cr) + H_2O(g) \quad (V.64)$$

Using the Clausius-Claperon equation, Fuger *et al.* [83FUG/PAR] calculated the $\Delta_r H^{\circ}_{m}$ and $\Delta_r G^{\circ}_{m}$ values of these reactions from these pressure data. By combining $\Delta_r H^{\circ}_{m}$(V.64) with the selected formation properties of UO_2F_2(cr) and $UO_3 \cdot H_2O$(cr), they obtained their selected values of

$$\Delta_f G^{\circ}_{m}(UO_2FOH \cdot H_2O, cr, 298.15\,K) = -(1721.7 \pm 7.5)\,kJ \cdot mol^{-1}$$
$$\Delta_f H^{\circ}_{m}(UO_2FOH \cdot H_2O, cr, 298.15\,K) = -(1894.5 \pm 8.4)\,kJ \cdot mol^{-1},$$

which are also accepted in this review, even though it should be mentioned that some uncertainty exists as to the exact composition of this compound.

The standard entropy is calculated from the selected $\Delta_f G^{\circ}_{m}$ and $\Delta_f H^{\circ}_{m}$ values.

$$S^{\circ}_{m}(UO_2FOH \cdot H_2O, cr, 298.15\,K) = (178 \pm 38)\,J \cdot K^{-1} \cdot mol^{-1}$$

This value is in excellent agreement with the value estimated according to Appendix D, $S^{\circ}_{m}(UO_2FOH \cdot H_2O, cr, 298.15\,K) = (176.2 \pm 10.0)\,J \cdot K^{-1} \cdot mol^{-1}$. No heat capacity data for this species are available.

Performing a similar calculation with $\Delta_r H_m$(V.63) and the selected data for $UO_2FOH \cdot H_2O$(cr), Fuger *et al.* [83FUG/PAR] found $\Delta_r S_m$(V.63) = 119.2 kJ per mole of H_2O(g). They found this to be unacceptably low, since $\Delta_{dehyd} S^{\circ}_{m}$ is normally about 146.5 to 150.5 kJ per mole of H_2O(g). Therefore, rather than accepting these results, Fuger *et al.* [83FUG/PAR] recalculated $\Delta_r H_m$(V.63) at the mean temperature (340 K) using the measured $\Delta_r G_m(V.63) = 4.64\,kJ \cdot mol^{-1}$ and an estimated $\Delta_r S_m(V.63) = 146.4\,J \cdot K^{-1} \cdot mol^{-1}$ to obtain a more reasonable value of $\Delta_r H_m(V.63) = 54.4\,kJ \cdot mol^{-1}$ and the function (V.65), valid between 298.15 and 380 K.

$$\Delta_r G_m(V.63, T) = (54.4 - 146.4 \times 10^{-3}\,T)\,(kJ \cdot mol^{-1}) \quad (V.65)$$

The selected Gibbs energy of reaction at 298.15 K is then, with an estimated uncertainty,

$$\Delta_r G_m^\circ(\text{V.63, 298.15 K}) = (10.75 \pm 3.80)\,\text{kJ}\cdot\text{mol}^{-1}.$$

Using the method in Appendix D, this review estimates

$$\Delta_r S_m^\circ(\text{V.63, 298.15 K}) = (144.0 \pm 6.0)\,\text{J}\cdot\text{K}^{-1}\cdot\text{mol}^{-1},$$

in good agreement with [83FUG/PAR]. This value is preferred for consistency and combined with the selected data for the monohydrate. This leads to the selected standard entropy. $\Delta_r S_m^\circ$ is used, together with the selected Gibbs energy of reaction, to derive, via $\Delta_r H_m^\circ$, the standard enthalpy of formation.

$$\begin{aligned}\Delta_f G_m^\circ(\text{UO}_2\text{FOH}\cdot 2\text{H}_2\text{O, cr, 298.15 K}) &= -(1961.0 \pm 8.4)\,\text{kJ}\cdot\text{mol}^{-1}\\ \Delta_f H_m^\circ(\text{UO}_2\text{FOH}\cdot 2\text{H}_2\text{O, cr, 298.15 K}) &= -(2190.0 \pm 9.4)\,\text{kJ}\cdot\text{mol}^{-1}.\\ S_m^\circ(\text{UO}_2\text{FOH}\cdot 2\text{H}_2\text{O, cr, 298.15 K}) &= (223 \pm 38)\,\text{J}\cdot\text{K}^{-1}\cdot\text{mol}^{-1}\end{aligned}$$

No heat capacity data for this compound are available.

There is no evidence for the existence of the anhydrous UO_2OHF(cr).

f) $UO_2(HF_2)_2 \cdot 4H_2O$(cr)

Kaganyuk, Kyskin and Kazin [83KAG/KYS] predicted the enthalpy of formation of $UO_2(HF_2)_2 \cdot 4H_2O$(cr) based on an effective charge model. Their value compares well with the one they cited as an experimental value from Galkin [71GAL]. This latter handbook was not available to the reviewers either in the original or in translation. Therefore, no data for $UO_2(HF_2)_2 \cdot 4H_2O$(cr) are selected in this review.

V.4.1.3.3. Other U(V) fluorides

a) H_3OUF_6(cr)

H_3OUF_6(cr) is the only thermodynamically characterized uranium(V) solid fluoride other than UF_5(cr). Morss [86MOR], who denoted this solid as "$(H_3O^+)(UF_6^-)$", took its enthalpy of formation from a thermal decomposition study by Masson *et al.* [76MAS/DES] according to the reaction

$$2\text{H}_3\text{OUF}_6(\text{cr}) \rightarrow \text{UF}_4(\text{cr}) + \text{UO}_2\text{F}_2(\text{cr}) + 6\text{HF(g)}. \qquad \text{(V.66)}$$

From calibration of thermograms, Masson *et al.* [76MAS/DES] obtained

$$\Delta_r H_m^\circ(\text{V.66, 298.15 K}) = (75.3 \pm 1.7)\,\text{kJ}\cdot\text{mol}^{-1}.$$

From this value, the following standard enthalpy of formation is calculated using auxiliary data selected in this review:

$$\Delta_f H_m^\circ(\text{H}_3\text{OUF}_6\text{, cr, 298.15 K}) = -(2641.4 \pm 3.2)\,\text{kJ}\cdot\text{mol}^{-1}.$$

b) UOF_3 (cr)

Although Krestov [72KRE] reporte thermodynamic data for UOF_3(cr), no clear source for the data were presented. No other reference to this compound was found. Therefore, no data for UOF_3(cr) are selected in this review.

V.4.1.3.4. Other U(IV) fluorides

a) UOF_2 (cr) and its hydrates

UOF_2 (cr): Vdovenko, Romanov and Solntseva [69VDO/ROM] measured the vapour pressure for the dehydration reaction (V.67) between 283 and 363 K.

$$UOF_2 \cdot H_2O(cr) \quad \rightarrow \quad UOF_2(cr) + H_2O(g) \qquad (V.67)$$

Their data lead to $\Delta_r H^\circ_m(V.67) = 46.4\,kJ \cdot mol^{-1}$ and $\Delta_r S^\circ_m(V.67) = 116.3 J \cdot K^{-1} \cdot mol^{-1}$. Fuger *et al.* [83FUG/PAR] pointed out that these values are unreasonably low, $\Delta_{dehyd} S^\circ_m$ normally being expected to be in the range of 146.5 to 150.5 kJ per mole of H_2O(g). Therefore, rather than using the pressure function, Fuger *et al.* [83FUG/PAR] used the observed value of $\Delta_r G^\circ_m$(V.67, 298.15 K) $= 11.80\,kJ \cdot mol^{-1}$ [69VDO/ROM] and the estimated value of $\Delta_r S^\circ_m$(V.67, 298.15 K) $= 146.4\,kJ \cdot K^{-1} \cdot mol^{-1}$ to obtain a more reasonable value of $\Delta_r H^\circ_m(V.67) = 55.6\,kJ \cdot mol^{-1}$. This, combined with the selected value of $\Delta_f H^\circ_m(UOF_2 \cdot H_2O, cr, 298.15\,K)$ selected in this review, leads to the selected value of

$$\Delta_f H^\circ_m(UOF_2, cr, 298.15\,K) \quad = \quad -(1504.6 \pm 6.3)\,kJ \cdot mol^{-1}$$

which is consistent with the selected auxiliary data in this review.

This review also accepts

$$S^\circ_m(UOF_2, cr, 298.15\,K) \quad = \quad (119.2 \pm 4.2)\,J \cdot K^{-1} \cdot mol^{-1}$$

estimated on the basis of comparison with the selected values for UCl_4(cr), UF_4(cr), $UOCl_2$(cr), UO_2Cl_2(cr), UO_2F_2(cr), UO_2(cr) and γ-UO_3 [83FUG/PAR].

The Gibbs energy of formation is calculated from the selected enthalpy of formation and entropy.

$$\Delta_f G^\circ_m(UOF_2, cr, 298.15\,K) \quad = \quad -(1434.1 \pm 6.4)\,kJ \cdot mol^{-1}$$

No heat capacity data for this species are available.

$UOF_2 \cdot H_2O$ (cr): Only the monohydrate of UOF_2(cr) has been thermodynamically characterized. Vdovenko, Romanov and Solntseva [69VDO/ROM] measured the enthalpy of the reaction

$$UCl_4(cr) + 2H_2O(sln) + 2HF(sln) \quad \rightleftharpoons \quad UOF_2 \cdot H_2O(cr) + 4HCl(sln) \qquad (V.68)$$

in 40% HF and 5 M HCl solution. Fuger *et al.* [83FUG/PAR] used the data of Mal'tsev, Gagarinskii and Popov [60MAL/GAG] at corresponding concentration of $\Delta_{sol}H^{\circ}_{m}(UCl_4$, cr, in HCl(sln)) to calculate $\Delta_r H_m(V.68) = -(195.3 \pm 0.8)\,kJ \cdot mol^{-1}$. They reduced this to the selected value of

$$\Delta_f H^{\circ}_{m}(UOF_2 \cdot H_2O, cr, 298.15\,K) = -(1802.0 \pm 3.3)\,kJ \cdot mol^{-1}$$

using enthalpy of formation data for HF, HCl and H_2O at the appropriate concentrations.

The accepted entropy value is based on $S^{\circ}_{m}(UOF_2, cr, 298.15\,K)$ and estimated entropy of dehydration of $UOF_2 \cdot H_2O(cr)$.

$$S^{\circ}_{m}(UOF_2 \cdot H_2O, cr, 298.15\,K) = (161.1 \pm 8.4)\,J \cdot K^{-1} \cdot mol^{-1}$$

The Gibbs energy of formation is calculated from the selected enthalpy of formation and entropy.

$$\Delta_f G^{\circ}_{m}(UOF_2 \cdot H_2O, cr, 298.15\,K) = -(1674.5 \pm 4.1)\,kJ \cdot mol^{-1}$$

No heat capacity data for this species are available.

b) UOFOH(cr) and its hydrates

UOFOH(cr): Uranium(IV) oxide fluoride hydroxide is often written in the structurally more meaningful form UOF(OH)(cr), but the notation UOFOH(cr) is preferred by the convention in this review (*cf.* Section II.1.3 and IUPAC [71JEN] Rule 2.163). The data for UOFOH(cr), whose composition is subject to some uncertainties, are derived from those of the better characterized hemihydrate discussed below. Vdovenko, Romanov and Solntseva [70VDO/ROM2] measured the vapour pressure of the dehydration reaction (V.69) between 280 and 363 K.

$$UOFOH \cdot \tfrac{1}{2}H_2O(cr) \rightleftharpoons UOFOH(cr) + \tfrac{1}{2}H_2O(g) \qquad (V.69)$$

As was the case for UOF_2(cr), Fuger *et al.* [83FUG/PAR] rejected their resulting $\Delta_r H_m(V.69)$ and $\Delta_r S_m(V.69)$ as too low. They preferred to assume $\Delta_r S^{\circ}_{m}(V.69, 298.15\,K) = 73\,J \cdot K^{-1} \cdot mol^{-1}$ and to accept the measured $\Delta_r G^{\circ}_{m}(V.69, 298.15\,K) = 5.8\,kJ \cdot mol^{-1}$ to obtain $\Delta_r H^{\circ}_{m}(V.69, 298.15\,K) = 27.6\,kJ \cdot mol^{-1}$. This, combined with the selected $\Delta_f H^{\circ}_{m}(UOFOH \cdot 0.5H_2O, cr, 298.15\,K)$ from below, results in the selected value of

$$\Delta_f H^{\circ}_{m}(UOFOH, cr, 298.15\,K) = -(1426.7 \pm 12.6)\,kJ \cdot mol^{-1}$$

which is also accepted here.

The entropy estimate is made from $S^{\circ}_{m}(UOF_2, cr, 298.15\,K)$ by assuming the appropriate Latimer contributions when substituting OH^- for F^-. The uncertainty is higher because of the ambiguity of the effective charge of the central ion. The selected value is

$$S^{\circ}_{m}(UOFOH, cr, 298.15\,K) = (121 \pm 10)\,J \cdot K^{-1} \cdot mol^{-1}.$$

The Gibbs energy of formation is calculated from the selected enthalpy of formation and entropy.

$$\Delta_f G^\circ_m(\text{UOFOH, cr, 298.15 K}) = -(1336.9 \pm 12.9)\,\text{kJ}\cdot\text{mol}^{-1}$$

No heat capacity data for this species are available.

UOFOH · 0.5H₂O(cr): Only the hemihydrate of UOFOH(cr) has been prepared [70VDO/ROM]. Vdovenko, Romanov and Solntseva [70VDO/ROM2] measured the enthalpy of solution of this compound and of UCl_4(cr) in 5 M HCl at 293 K. Fuger *et al.* [83FUG/PAR] used this value and auxiliary data from Mal'tsev, Gagarinskii and Popov [60MAL/GAG] at corresponding concentrations to calculate $\Delta_r H_m(\text{V.70}) = -(143.2 \pm 0.8)\,\text{kJ}\cdot\text{mol}^{-1}$.

$$\text{UCl}_4(\text{cr}) + \text{HF(sln)} + 2.5\text{H}_2\text{O(sln)} \rightleftharpoons \text{UOFOH}\cdot0.5\text{H}_2\text{O(cr)} + 4\text{HCl(sln)} \quad (\text{V.70})$$

They assumed $\Delta_r C^\circ_{p,m}(\text{V.70}) = -502\,\text{J}\cdot\text{K}^{-1}\cdot\text{mol}^{-1}$ to correct this value to 298.15 K and combined the result with enthalpy of formation data for HF(sln), HCl(sln) and H_2O(sln) at the appropriate concentration to obtain the selected value of

$$\Delta_f H^\circ_m(\text{UOFOH}\cdot0.5\text{H}_2\text{O, cr, 298.15 K}) = -(1576.1 \pm 6.3)\,\text{kJ}\cdot\text{mol}^{-1}.$$

The selected entropy value,

$$S^\circ_m(\text{UOFOH}\cdot0.5\text{H}_2\text{O, cr, 298.15 K}) = (143 \pm 10)\,\text{J}\cdot\text{K}^{-1}\cdot\text{mol}^{-1},$$

is estimated in this review using the experimental entropy of $S^\circ_m(\text{UOF}_2\text{, cr, 298.15 K})$ and substituting F^- for OH^- and adding the entropy contribution of $0.5H_2O$.

The Gibbs energy of formation is calculated from the selected enthalpy of formation and entropy.

$$\Delta_f G^\circ_m(\text{UOFOH}\cdot0.5\text{H}_2\text{O, cr, 298.15 K}) = -(1458.1 \pm 7.0)\,\text{kJ}\cdot\text{mol}^{-1}$$

No heat capacity data for this species are available.

V.4.2. Chlorine compounds and complexes

V.4.2.1. Uranium chlorides

V.4.2.1.1. Gaseous uranium chlorides

a) UCl(g) and UCl₂(g)

The enthalpies of formation of these species are derived from the mass spectrometric molecular study of the following molecular exchange reactions by Lau and Hildenbrand [84LAU/HIL].

$$\text{UCl}_2(\text{g}) + \text{CaCl(g)} \rightleftharpoons \text{UCl}_3(\text{g}) + \text{Ca(g)} \quad (\text{V.71})$$

$$\text{UCl(g)} + \text{CaCl(g)} \rightleftharpoons \text{UCl}_2(\text{g}) + \text{Ca(g)} \quad (\text{V.72})$$

$$\text{U(g)} + \text{UCl}_2(\text{g}) \rightleftharpoons 2\text{UCl(g)} \quad (\text{V.73})$$

By third-law analysis the values $\Delta_r H_m^\circ$(V.71, 298.15 K) = $-(90.2 \pm 0.4)$ kJ · mol^{-1}, $\Delta_r H_m^\circ$(V.72, 298.15 K) = $-(67.9 \pm 0.3)$ kJ · mol^{-1}, and $\Delta_r H_m^\circ$(V.73, 298.15 K) = (4.6 ± 0.5) kJ · mol^{-1} are obtained, respectively. These values differ significantly from the second-law values given Lau and Hildenbrand [84LAU/HIL], but the third-law values are preferred by the present review in order to maintain consistency with the heat capacity and entropy data. Combining these values with auxiliary data for Ca(g), CaCl(g) and the selected value for UCl_3(g), Reaction (V.71) leads to $\Delta_f H_m^\circ$(UCl_2, g, 298.15 K) = $-(165.4 \pm 21.9)$ kJ·mol^{-1} and solving the thermochemical cycles of Reactions (V.72) and (V.73) results in $\Delta_f H_m^\circ$(UCl, g, 298.15 K) = $-(188.2 \pm 22.3)$ kJ · mol^{-1} and $\Delta_f H_m^\circ$(UCl_2, g, 298.15 K) = $-(161.2 \pm 22.3)$ kJ · mol^{-1}.

The present review selects

$$\begin{aligned} \Delta_f H_m^\circ(\mathrm{UCl, g, 298.15\ K}) &= (188.2 \pm 22.3)\ \mathrm{kJ \cdot mol^{-1}} \\ \Delta_f H_m^\circ(\mathrm{UCl_2, g, 298.15\ K}) &= -(163.0 \pm 22.3)\ \mathrm{kJ \cdot mol^{-1}} \end{aligned}$$

The thermal functions of UCl and UCl_2(g) are calculated from molecular constants estimated by Gurvich and Dorofeeva [84GUR/DOR].

$$\begin{aligned} S_m^\circ(\mathrm{UCl, g, 298.15\ K}) &= (266 \pm 10)\ \mathrm{J \cdot K^{-1} \cdot mol^{-1}} \\ S_m^\circ(\mathrm{UCl_2, g, 298.15\ K}) &= (339 \pm 10)\ \mathrm{J \cdot K^{-1} \cdot mol^{-1}} \end{aligned}$$

$$\begin{aligned} C_{p,m}^\circ(\mathrm{UCl, g, 298.15\ K}) &= (43.1 \pm 5.0)\ \mathrm{J \cdot K^{-1} \cdot mol^{-1}} \\ C_{p,m}^\circ(\mathrm{UCl_2, g, 298.15\ K}) &= (59.9 \pm 5.0)\ \mathrm{J \cdot K^{-1} \cdot mol^{-1}} \end{aligned}$$

The Gibbs energies of formation are calculated from the selected enthalpies of formation and entropies.

$$\begin{aligned} \Delta_f G_m^\circ(\mathrm{UCl, g, 298.15\ K}) &= -(157.1 \pm 22.5)\ \mathrm{kJ \cdot mol^{-1}} \\ \Delta_f G_m^\circ(\mathrm{UCl_2, g, 298.15\ K}) &= -(182.6 \pm 22.5)\ \mathrm{kJ \cdot mol^{-1}} \end{aligned}$$

b) UCl_3 (g)

Values for the enthalpy of formation of UCl_3(g) are derived from mass-spectrometric studies of molecular exchange reactions as well as vapour pressure measurements. Lau and Hildenbrand [84LAU/HIL] studied the reaction

$$\mathrm{UCl_3(g) + CuCl(g) \rightleftharpoons UCl_4(g) + Cu(g)} \qquad \text{(V.74)}$$

from which $\Delta_r H_m^\circ$(V.74, 298.15 K) = $-(34.7 \pm 0.5)$ kJ · mol^{-1} is derived by the third-law method and, hence, $\Delta_f H_m^\circ$(UCl_3, g, 298.15 K) = $-(537.1 \pm 15.8)$ kJ · mol^{-1}, the absolute uncertainty being estimated by the present review.

Tentative vapour pressure for the reaction

$$\mathrm{UCl_3(cr) \rightleftharpoons UCl_3(g)} \qquad \text{(V.75)}$$

was measured by Altman [43ALT]. By third-law analysis, $\Delta_{sub} H_m^\circ$(V.75, 298.15 K) = (366 ± 14) kJ · mol^{-1} and $\Delta_f H_m^\circ$(UCl_3, g, 298.15 K) = $-(496 \pm 14)$ kJ · mol^{-1} are obtained.

Since the vapour pressure data are subject to considerable uncertainties, the value selected by the present review is solely based on the results of Lau and Hildenbrand [84LAU/HIL]:

$$\Delta_f H_m^\circ(\mathrm{UCl_3, g, 298.15\ K}) = -(537.1 \pm 15.8)\,\mathrm{kJ \cdot mol^{-1}}.$$

The heat capacity and entropy of UCl_3(g) are calculated from molecular constants estimated by Hildenbrand, Gurvich and Yungman [85HIL/GUR].

$$S_m^\circ(\mathrm{UCl_3, g, 298.15\ K}) = (380 \pm 10)\,\mathrm{J \cdot K^{-1} \cdot mol^{-1}}$$
$$C_{p,m}^\circ(\mathrm{UCl_3, g, 298.15\ K}) = (82.4 \pm 5.0)\,\mathrm{J \cdot K^{-1} \cdot mol^{-1}}$$

The Gibbs energy of formation is calculated from the selected enthalpy of formation and entropy.

$$\Delta_f G_m^\circ(\mathrm{UCl_3, g, 298.15\ K}) = -(535.7 \pm 16.1)\,\mathrm{kJ \cdot mol^{-1}}$$

c) UCl_4 (g)

The enthalpy of formation of UCl_4(g) is a key parameter in the evaluation of the UCl_q(g) series. A large number of experimental determinations of the vapour pressure of UCl_4 have been reported; the third-law enthalpies of sublimation are summarized in Table V.21 (after [91COR/KON]). For the sublimation reaction

$$\mathrm{UCl_4(cr)} \rightleftharpoons \mathrm{UCl_4(g)} \qquad \text{(V.76)}$$

the value of

$$\Delta_{sub} H_m^\circ(\mathrm{V.38, 298.15\ K}) = (200.7 \pm 5.0)\,\mathrm{kJ \cdot mol^{-1}}$$

is selected here, as did Cordfunke and Konings [91COR/KON]. The uncertainty is estimated taking into account the uncertainties in the auxiliary thermal functions of the gas. Combining this value with the selected enthalpy of formation for UCl_4(cr), the following value is obtained:

$$\Delta_f H_m^\circ(\mathrm{UCl_4, g, 298.15\ K}) = -(818.1 \pm 5.6)\,\mathrm{kJ \cdot mol^{-1}}.$$

The highly deviating results of Sawlewicz and Siekierski [75SAW/SIE], who did not give any numerical data, are not considered in the analysis.

The thermal functions of UCl_4(g) are calculated from molecular constants as obtained by Gurvich and Dovofeeva [84GUR/DOR] from an analysis of the then available data. Gurvich and Dovofeeva assumed the molecular structure of UCl_4(g) to be a distorted tetrahedron of C_{2v} symmetry in analogy with UF_4(g). This assumption was recently confirmed by electron diffraction measurements by Ezhov and co-workers [88EZH/KOM, 89BAZ/KOM]. A small correction to the moment of inertia, to account for the new geometric parameters is applied. From these data the selected values are calculated.

Table V.21: Enthalpy of sublimation values for $UCl_4(cr) \rightleftharpoons UCl_4(g)$ at 298.15 K, as evaluated from third-law analysis of experimental data.

Reference	T-range (K)	$\Delta_{sub}H^\circ_m$(V.76, 298.15 K) ($kJ \cdot mol^{-1}$)[(a)]
[42JEN/AND]	970-1103	196.3 ± 1.6
[45DAV]	743-775	200.4 ± 0.4[(b)]
[45DAV/STR]	640-731	200.6 ± 0.7[(b)]
[46GRE2]	874-1044	196.4 ± 1.0
[48MUE]	629-703	201.6 ± 0.4[(b)]
[48THO/SCH]	663-783	199.8 ± 0.6[(b)]
[56SHC/VAS]	631-708	200.6 ± 0.5[(b)]
[58JOH/BUT]	723-948	192.6 ± 1.8
[58YOU/GRA]	864-1064	196.3 ± 0.3
[68CHO/CHU]	648-748	198.1 ± 0.5
[75HIL/CUB]	575-654	201.7 ± 1.7[(b)]
[78SIN/PRA]	763-931	197.7 ± 1.8
[90COR/KON]	699-842	201.3 ± 0.3[(b)]
[91HIL/LAU]	588-675	200.4 ± 0.2[(b)]

(a) The indicated errors represent the statistical uncertainties.
(b) Values considered in the determination of the selected weighted mean.

$$S^\circ_m(\mathrm{UCl_4, g, 298.15\ K}) = (402.7 \pm 3.0)\,\mathrm{J \cdot K^{-1} \cdot mol^{-1}}.$$
$$C^\circ_{p,m}(\mathrm{UCl_4, g, 298.15\ K}) = (107.7 \pm 1.5)\,\mathrm{J \cdot K^{-1} \cdot mol^{-1}}.$$

The Gibbs energy of formation is calculated from the selected enthalpy of formation and entropy.

$$\Delta_f G^\circ_m(\mathrm{UCl_4, g, 298.15\ K}) = -(790.2 \pm 5.7)\,\mathrm{kJ \cdot mol^{-1}}$$

d) UCl_5 (g)

Lau and Hildenbrand [84LAU/HIL] studied the equilibrium

$$\mathrm{UCl_4(g)} + \tfrac{1}{2}\mathrm{Cl_2(g)} \rightleftharpoons \mathrm{UCl_5(g)} \qquad (V.77)$$

by mass spectrometry. By second-law method they derived $\Delta_r H^\circ_m$(V.77, 298.15 K) $= -(70.3 \pm 6.3)\,kJ \cdot mol^{-1}$. This review obtains for the third-law enthalpy of reaction $\Delta_r H^\circ_m$(V.77, 298.15 K) $= -(64.4 \pm 1.3)\,kJ \cdot mol^{-1}$, however, exhibiting a slight temperature trend. The latter value is accepted by the present review, but the uncertainty is raised to $\pm15\,\mathrm{kJ\cdot mol^{-1}}$ to reflect the uncertainties in the thermal functions, resulting in

$$\Delta_f H_m^\circ(\mathrm{UCl_5, g, 298.15\ K}) = -(882.5 \pm 15.0)\,\mathrm{kJ\cdot mol^{-1}}$$

The heat capacity and entropy of $UCl_5(g)$ are calculated from molecular constants estimated by Hildenbrand, Gurvich and Yungman [85HIL/GUR]. The uncertainties are assigned by the present review.

$$S_m^\circ(\mathrm{UCl_5, g, 298.15\ K}) = (438 \pm 10)\,\mathrm{J\cdot K^{-1}\cdot mol^{-1}}$$
$$C_{p,m}^\circ(\mathrm{UCl_5, g, 298.15\ K}) = (124 \pm 5)\,\mathrm{J\cdot K^{-1}\cdot mol^{-1}}$$

The Gibbs energy of formation is calculated from the selected enthalpy of formation and entropy.

$$\Delta_f G_m^\circ(\mathrm{UCl_5, g, 298.15\ K}) = -(831.8 \pm 15.3)\,\mathrm{kJ\cdot mol^{-1}}$$

e) UCl_6 (g)

There are only few experimental data on $UCl_6(g)$: Johnson *et al.* [58JOH/BUT] and Altman, Lipkin and Weissman [43ALT/LIP] reported vapour pressure measurements for this compound, but the latter authors observed dissociation of UCl_6 above 413 K. Third-law analysis of the two experimental studies gives $\Delta_{sub}H_m^\circ(\mathrm{V.78, 298.15\ K}) = (81.4 \pm 1.0)\,\mathrm{kJ\cdot mol^{-1}}$ and $(76.6 \pm 0.8)\,\mathrm{kJ\cdot mol^{-1}}$, respectively, for the reaction

$$\mathrm{UCl_6(cr)} \rightleftharpoons \mathrm{UCl_6(g)}, \qquad (V.78)$$

slight temperature trends being observed in both studies. The value

$$\Delta_{sub}H_m^\circ(\mathrm{V.38, 298.15\ K}) = (79.0 \pm 8.0)\,\mathrm{kJ\cdot mol^{-1}}$$

is selected, where the uncertainty reflects the uncertainties in the thermal functions. Combining this with the enthalpy of formation of $UCl_6(cr)$ results in

$$\Delta_f H_m^\circ(\mathrm{UCl_6, g, 298.15\ K}) = -(987.5 \pm 8.5)\,\mathrm{kJ\cdot mol^{-1}}.$$

The heat capacity and entropy of $UCl_6(g)$ are calculated from estimated molecular parameters for an octahedral structure.

$$S_m^\circ(\mathrm{UCl_6, g, 298.15\ K}) = (438 \pm 5)\,\mathrm{J\cdot K^{-1}\cdot mol^{-1}}$$
$$C_{p,m}^\circ(\mathrm{UCl_6, g, 298.15\ K}) = (147 \pm 3)\,\mathrm{J\cdot K^{-1}\cdot mol^{-1}}$$

The Gibbs energy of formation is calculated from the selected enthalpy of formation and entropy.

$$\Delta_f G_m^\circ(\mathrm{UCl_6, g, 298.15\ K}) = -(903.6 \pm 8.7)\,\mathrm{kJ\cdot mol^{-1}}$$

f) $U_2Cl_8(g)$

All available data in the literature on this compound are derived from two experiments on the dimerization reaction (V.79) by Binnewies and Schäfer [74BIN/SCH], who found $\Delta_r H_m(\text{V.79}) = -102.1\,\text{kJ}\cdot\text{mol}^{-1}$ at approximately 745 K.

$$2\text{UCl}_4(\text{g}) \rightleftharpoons \text{U}_2\text{Cl}_8(\text{g}) \qquad (\text{V.79})$$

Cordfunke and Kubaschewski [84COR/KUB] estimated the entropy and heat capacity of $U_2Cl_8(g)$ from an empirical relationship between these values for mono- and dimeric molecules. This leads to $\Delta_r S_m(\text{V.79, 745 K}) = -136\,\text{J}\cdot\text{K}^{-1}\cdot\text{mol}^{-1}$ and the selected reaction enthalpy,

$$\Delta_r H_m^\circ(\text{V.79, 298.15 K}) = -(113.4 \pm 10.0)\,\text{kJ}\cdot\text{mol}^{-1}.$$

Combining this value with the selected value of $\Delta_f H_m^\circ(\text{UCl}_4, \text{g } 298.15\text{ K})$, the resulting enthalpy of formation is obtained.

$$\Delta_f H_m^\circ(\text{U}_2\text{Cl}_8, \text{g}, 298.15\,\text{K}) = -(1749.6 \pm 15.0)\,\text{kJ}\cdot\text{mol}^{-1}$$

The entropy and its uncertainty estimated by Cordfunke and Kubaschewski [84COR/KUB] are selected here:

$$S_m^\circ(\text{U}_2\text{Cl}_8, \text{g}, 298.15\,\text{K}) = (624 \pm 25)\,\text{J}\cdot\text{K}^{-1}\cdot\text{mol}^{-1}$$

The Gibbs energy of formation is calculated from the selected enthalpy of formation and entropy.

$$\Delta_f G_m^\circ(\text{U}_2\text{Cl}_8, \text{g}, 298.15\,\text{K}) = -(1639.7 \pm 16.8)\,\text{kJ}\cdot\text{mol}^{-1}$$

The $C_{p,m}^\circ$ value and its temperature function estimated by Cordfunke and Kubaschewski [84COR/KUB] are also adopted in this review with an estimated uncertainty.

$$C_{p,m}^\circ(\text{U}_2\text{Cl}_8, \text{g}, 298.15\,\text{K}) = (226 \pm 10)\,\text{J}\cdot\text{K}^{-1}\cdot\text{mol}^{-1}$$

g) $U_2Cl_{10}(g)$

Gruen and McBeth [69GRU/MCB] found $\Delta_r G_m(\text{V.80}) = (63.3 - 0.0644\,T)\,\text{kJ}\cdot\text{mol}^{-1}$ for the temperature range from 450 to 650 K,

$$2\text{UCl}_4(\text{cr}) + \text{Cl}_2(\text{g}) \rightleftharpoons \text{U}_2\text{Cl}_{10}(\text{g}), \qquad (\text{V.80})$$

i.e., $\Delta_r H_m^\circ(\text{V.80, 450 to 650 K}) = (63.3 \pm 8.0)\,\text{kJ}\cdot\text{mol}^{-1}$ and $\Delta_r S_m^\circ(\text{V.80, 450 to 650 K}) = (64.4 \pm 8.0)\,\text{J}\cdot\text{K}^{-1}\cdot\text{mol}^{-1}$, assuming these parameters to be constant over the temperature range indicated. The uncertainties are estimated by this review. Fuger *et al.* [83FUG/PAR] assumed $\Delta_r C_{p,m}(\text{V.80}) = 0$ and therefore selected the above values also for 298.15 K. Since the IAEA review [83FUG/PAR] was published, Cordfunke and Kubaschewski [84COR/KUB] estimated the heat capacity of $U_2Cl_{10}(g)$ from 298.15 to 2000 K (the value in Table 2 of Ref. [84COR/KUB] given for $UCl_5(g)$ is assumed to represent $\frac{1}{2}U_2Cl_{10}(g)$ as $UCl_5(g)$ is not discussed in this paper). The 298.15 K value is also accepted by this review with an estimated uncertainty.

$$C^\circ_{p,m}(U_2Cl_{10}, g, 298.15\,K) = (236 \pm 10)\,J \cdot K^{-1} \cdot mol^{-1}$$

From the selected value and $C^\circ_{p,m}(Cl_2, g, 298.15\,K) = (33.949 \pm 0.002)\,J \cdot K^{-1} \cdot mol^{-1}$ from CODATA [87GAR/PAR] and $C^\circ_{p,m}(UCl_4, cr, 298.15\,K) = (122.0 \pm 0.4)\,J \cdot K^{-1} \cdot mol^{-1}$ selected in Section V.4.2.1.3.b one obtains $\Delta_r C^\circ_{p,m}(V.80) = -(42 \pm 10)\,J \cdot K^{-1} \cdot mol^{-1}$. Assuming this value to be constant from 298.15 to 450 K, the values $\Delta_r H_m(V.80)$ and $\Delta_r S_m(V.80)$ can be corrected from 450 K to 298.15 K and $\Delta_r G^\circ_m(V.80)$ calculated from these. This results in the selected values of

$$\begin{aligned} \Delta_r G^\circ_m(V.80,\ 298.15\ K) &= (45.3 \pm 9.3)\,kJ \cdot mol^{-1} \\ \Delta_r H^\circ_m(V.80,\ 298.15\ K) &= (69.7 \pm 8.6)\,kJ \cdot mol^{-1} \\ \Delta_r S^\circ_m(V.80,\ 298.15\ K) &= (81.7 \pm 11.5)\,J \cdot K^{-1} \cdot mol^{-1}. \end{aligned}$$

Use of the auxiliary data from Chapter VI and Section V.4.2.1.3.b allows the calculation of the formation data,

$$\begin{aligned} \Delta_f G^\circ_m(U_2Cl_{10}, g, 298.15\,K) &= -(1813.8 \pm 10.5)\,kJ \cdot mol^{-1} \\ \Delta_f H^\circ_m(U_2Cl_{10}, g, 298.15\,K) &= -(1967.9 \pm 9.9)\,kJ \cdot mol^{-1} \\ S^\circ_m(U_2Cl_{10}, g, 298.15\,K) &= (699.0 \pm 11.6)\,J \cdot K^{-1} \cdot mol^{-1}. \end{aligned}$$

h) $UO_2Cl_2(g)$

The equilibrium

$$U_3O_8(cr) + 3Cl_2(g) \rightleftharpoons 3UO_2Cl_2(g) + O_2(g) \qquad (V.81)$$

was investigated by Cordfunke and Prins [74COR/PRI] from 1133 to 1328 K. A third-law analysis of the data, using estimated quatities for $UO_2Cl_2(g)$, yields $\Delta_r H^\circ_m(V.81, 298.15\ K) = (660 \pm 0.7)\,kJ \cdot mol^{-1}$. Estimating the absolute uncertainty to be $\pm 45\,kJ \cdot mol^{-1}$, the value

$$\Delta_f H^\circ_m(UO_2Cl_2,\ g,\ 298.15\ K) = -(971.6 \pm 15.0)\,kJ \cdot mol^{-1}$$

is obtained. Reaction (V.81) was also investigated by Kangro [63KAN], but their results are significantly different. Since the authors did not report any primary data, their results are not considered by the present review.

The heat capacity and entropy of $UO_2Cl_2(g)$ are calculated from estimated molecular parameters. $\Delta_f G^\circ_m$ is calculated from the selected values of $\Delta_f H^\circ_m$ and S°_m.

$$\begin{aligned} S^\circ_m(UO_2Cl_2,\ g,\ 298.15\ K) &= (377 \pm 10)\,J \cdot K^{-1} \cdot mol^{-1} \\ C^\circ_{p,m}(UO_2Cl_2,\ g,\ 298.15\ K) &= (92.3 \pm 5.0)\,J \cdot K^{-1} \cdot mol^{-1} \\ \Delta_f G^\circ_m(UO_2Cl_2,\ g,\ 298.15\ K) &= -(941.4 \pm 15.3)\,kJ \cdot mol^{-1} \end{aligned}$$

V.4.2.1.2. Aqueous uranium chlorides

a) Aqueous U(VI) chlorides

The dioxouranium(VI) chloride complexes are very weak. This accounts for a large part of the scattering of the experimental data summarized in Table V.22.

Davies and Monk [57DAV/MON] and Awasthi and Sundaresan [81AWA/SUN] made spectrophotometric studies of the reaction

$$UO_2^{2+} + Cl^- \rightleftharpoons UO_2Cl^+ \qquad (V.82)$$

which cover a wide range of ionic strengths. After conversion to molality units, these data and those of Bednarczyk and Fidelis [78BED/FID] and Ahrland [51AHR] are used to determine the value of $\Delta\varepsilon$ for Reaction (V.82). The data of Ahrland [51AHR] are corrected to 25°C using the enthalpy value selected below. The experimental data are given as plots of $(\log_{10}\beta_1+4D)$ *vs.* m_{Cl^-} in Figure V.10. The emf determination of Ref. [51AHR] seems to be an outlier and has been omitted in the evaluation process. The weighted linear regression, for which values from Refs. [51AHR, 57DAV/MON, 78BED/FID, 81AWA/SUN] are used as shown in Figure V.10, yields $\log_{10}\beta_1^\circ(V.82) = (0.17 \pm 0.02)$ and $\Delta\varepsilon(V.82) = -(0.25 \pm 0.02)$. Seven values of $\log_{10}\beta_1$ at different ionic strengths are calculated from the equation given by Awasthasi and Sundaresan [81AWA/SUN] and are included in the linear regression in order not to give these data too low a weight in comparison with the seven experimental determinations of Davies and Monk [57DAV/MON]. From $\Delta\varepsilon = -(0.25 \pm 0.02)$, and the values given in Appendix B for $\varepsilon_{(Cl^-,H^+)} = (0.12 \pm 0.01)$ and $\varepsilon_{(UO_2^{2+},ClO_4^-)} = (0.46 \pm 0.03)$, one obtains $\varepsilon_{(UO_2Cl^+,ClO_4^-)} = (0.33 \pm 0.04)$, which is very close to the value obtained for $\varepsilon_{(UO_2F^+,ClO_4^-)}$, *cf.* Appendix B.

There are only two reliable experimental studies of the formation reaction

$$UO_2^{2+} + 2Cl^- \rightleftharpoons UO_2Cl_2(aq). \qquad (V.83)$$

Awasthi and Sundaresan [81AWA/SUN] used a spectrophotometric method and Bednarczyk and Fidelis [78BED/FID] a chromatographic method. The ionic strength dependence given in Ref. [81AWA/SUN] seems to conform to the specific ion interaction theory, which is rather unexpected in view of the large medium changes necessary to study these weak complexes.

It is in practice impossible to distinguish between complex formation and ionic strength effects in these cases. In order to verify the ion interaction coefficients, however, this review does an extrapolation to $I = 0$ using the value of Ref. [78BED/FID] and five values at different ionic strengths (see Table V.22) calculated with the function given in Ref. [81AWA/SUN]. The regression is shown in Figure V.11 and yielded $\log_{10}\beta_2^\circ = -(1.07\pm0.35)$ and $\Delta\varepsilon(V.83) = -(0.62\pm0.17)$. It is interesting to note that using the ion interaction coefficients listed in Appendix B for a pure $HClO_4$ medium leads to $\Delta\varepsilon(V.83) = -(0.70 \pm 0.04)$, which is compatible with the value obtained from the linear regression. This is an indication that the ionic strength dependence

Table V.22: Experimental equilibrium data for the uranium(VI) chloride system.

Method	Ionic Medium	t (°C)	$\log_{10}\beta_q^{(a)}$	$\log_{10}\beta_q^{\circ(a)}$	Reference
$UO_2^{2+} + Cl^- \rightleftharpoons UO_2Cl^+$					
emf	1 M (Na, H)(Cl, ClO_4)	20	-0.10 ± 0.11		[51AHR]
sp	1 M (Na, H)(Cl, ClO_4)	20	-0.30 ± 0.26		
sp, emf	var.	25		0.38	[51NEL/KRA]
dis	2 M (Na, H)(Cl, ClO_4)	10	-0.24		[54DAY/POW]
dis	2 M (Na, H)(Cl, ClO_4)	25	-0.06		
dis	2 M (Na, H)(Cl, ClO_4)	40	0.06		
sp	0.852 M (Cl^-, ClO_4^-)	25	-0.42 ± 0.04[b,c]		[57DAV/MON]
	0.651 M (Cl^-, ClO_4^-)	25	-0.34 ± 0.04[b,c]		
	0.552 M (Cl^-, ClO_4^-)	25	-0.42 ± 0.04[b,c]		
	0.403 M (Cl^-, ClO_4^-)	25	-0.41 ± 0.04[b,c]		
	0.303 M (Cl^-, ClO_4^-)	25	-0.39 ± 0.04[b,c]		
	0.254 M (Cl^-, ClO_4^-)	25	-0.35 ± 0.04[b,c]		
sp	0.05-0.12 M Na(Cl, ClO_4)	25		0.22 ± 0.03	[57BAL/DAV]
sp	1.238 M $NaClO_4$	25	1.64		[60HEF/AMI]
cry	sat. KNO_3($\sim$ 2 M)	-3	0.26		[62FAU/CRE]
emf	0.54 M Na(NO_3, ClO_4)	25	0.38		[67OHA/MOR]
	0.82 M Na(NO_3, ClO_4)	25	0.15		
	1.06 M Na(NO_3, ClO_4)	25	-0.05		
emf	0.05-0.5 M	15		1.29[e]	[67OHA/MOR2]
		25		1.19[e]	
		35		1.11[e]	
dis	8 M H^+	20	-0.09		[70LAH/KNO]
cix	0.60 M H(ClO_4, Cl)	RT?	0.0		[74BUN]
	2.00 M H(ClO_4, Cl)	RT?	0.0		
mvd	1 M Ca(Cl, ClO_4)$_2$	RT?	-0.60		[74JED]
	2 M Ca(Cl, ClO_4)$_2$	RT?	-0.60		
	4 M Ca(Cl, ClO_4)$_2$	RT?	-0.52		
	6 M Ca(Cl, ClO_4)$_2$	RT?	-0.46		
	8 M Ca(Cl, ClO_4)$_2$	RT?	-0.22		
cix	var.	RT?	1.59		[75ALY/ABD]
cix	var.	25	0.34		[76SOU/SHA]
pot	0.03 M KCl	50-150	$\sim$ 1.7 at 25°C		[77NIK2]
dis, chr	2 M H(ClO_4, Cl)	25	-0.12 ± 0.10[c]		[78BED/FID]
rev		25		0.2	[78LAN]
rev		25		2.0 ± 1.0	[80LEM/TRE]

Table V.22 (continued)

Method	Ionic Medium	t (°C)	$\log_{10}\beta_q^{(a)}$	$\log_{10}\beta_q^{o(a)}$	Reference
sp	0.1 M $H(Cl, ClO_4)$	25	-0.17 ± 0.10[c,f]		[81AWA/SUN]
	0.2 M $H(Cl, ClO_4)$	25	-0.25 ± 0.10[c,f]		
	0.5 M $H(Cl, ClO_4)$	25	-0.32 ± 0.10[c,f]		
	1 M $H(Cl, ClO_4)$	25	-0.31 ± 0.10[c,f]		
	1.5 M $H(Cl, ClO_4)$	25	-0.24 ± 0.10[c,f]		
	2 M $H(Cl, ClO_4)$	25	-0.15 ± 0.10[c,f]		
	2.5 M $H(Cl, ClO_4)$	25	-0.06 ± 0.10[c,f]		
	3 M $H(Cl, ClO_4)$	25	0.04 ± 0.10[c,f]		
	various constant values of I	25		0.23 ± 0.02[d]	
	various constant values of I	35		0.27 ± 0.02[d]	
	various constant values of I	45		0.31 ± 0.02[d]	

....

$UO_2^{2+} + 2Cl^- \rightleftharpoons UO_2Cl_2(aq)$

Method	Ionic Medium	t (°C)	$\log_{10}\beta_q^{(a)}$	$\log_{10}\beta_q^{o(a)}$	Reference
dis, chr	2 M $H(ClO_4, Cl)$	25	-0.64 ± 0.30[c]		[78BED/FID]
sp	0.10 M HCl	25	-1.78 ± 0.30[c,f]		[81AWA/SUN]
	0.5 M $H(Cl, ClO_4)$	25	-1.91 ± 0.30[c,f]		
	1 M $H(Cl, ClO_4)$	25	-1.76 ± 0.30[c,f]		
	2 M $H(Cl, ClO_4)$	25	-1.29 ± 0.30[c,f]		
	3 M $H(Cl, ClO_4)$	25	-0.75 ± 0.30[c,f]		
	various constant values of I	25		-1.20 ± 0.03[d]	
	various constant values of I	35		-1.10 ± 0.03[d]	
	various constant values of I	45		-1.00 ± 0.03[d]	

(a) Refers to the reactions indicated, $\log_{10}\beta_q$ in the ionic medium and at the temperature given in the table, $\log_{10}\beta_q^o$ (in molal units) at $I = 0$ and 298.15 K.
(b) Recalculated from primary data given in Ref. [57DAV/MON, Table 2].
(c) Uncertainties estimated by this review as described in Appendix A.
(d) Calculated at $I = 0$ with the formula given by the authors. This value refers to the experimental temperature.
(e) Values valid at $I = 0$, but at the temperature of the experiments.
(f) Calculated with the ionic strength function given by the authors.

in a mixed chloride/perchlorate medium is close to that in a pure perchlorate medium if complex formation is properly taken into account.[7]

Figure V.10: Extrapolation to $I = 0$ of experimental data for the formation of UO_2Cl^+ using the specific ion interaction theory. The data refer to mixed Cl^-/ClO_4^- media and are taken from [51AHR] (◇), [57DAV/MON] (○), [78BED/FID] (□) and [81AWA/SUN] (△). The shaded area represents the uncertainty range obtained by propagating the resulting uncertainties at $I = 0$ back to $I = 4$ m.

$$\text{Equilibrium:} \quad UO_2^{2+} + Cl^- \rightleftharpoons UO_2Cl^+$$

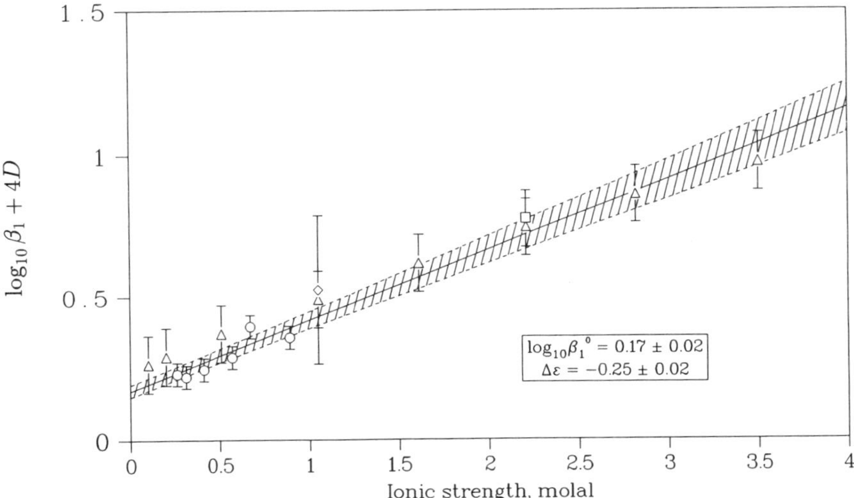

Summarizing, the values selected by this review of the formation constants of UO_2Cl^+ and $UO_2Cl_2(aq)$, respectively, are

$$\log_{10}\beta_1^\circ(\text{V.82}, 298.15\ \text{K}) = 0.17 \pm 0.02$$
$$\log_{10}\beta_2^\circ(\text{V.83}, 298.15\ \text{K}) = -1.1 \pm 0.4.$$

The value proposed for $\log_{10}\beta_1^\circ$ is very close to the one given in a review by Langmuir [78LAN], but differs considerably from the value $\log_{10}\beta_1^\circ = (2\pm1)$ proposed by Lemire and Tremaine [80LEM/TRE]. The results of the qualitative study by Stankov, Čeleda and Jedunáková [73STA/CEL] do not contradict the equilibrium constants selected in this review. The selected $\log_{10}\beta_q^\circ$ values are used to calculate the standard Gibbs energies of reaction, which in turn are combined with the $\Delta_f G_m^\circ$ values of UO_2^{2+} and Cl^- selected in this review to obtain the recommended $\Delta_f G_m^\circ$ values for UO_2Cl^+ and $UO_2Cl_2(aq)$.

7 It should be mentioned that Dunsmore, Hietanen and Sillén [63DUN/HIE] discussed the possible formation of ternary complexes in the dioxouranium(VI)-water-chloride system. However, such complexes, if formed at all, are much weaker than the corresponding sulphate complexes as discussed in Footnote 8 in Section V.5.1.3.1.a.

Figure V.11: Extrapolation to $I = 0$ of experimental data for the formation of $UO_2Cl_2(aq)$ using the specific ion interaction theory. The data refer to mixed Cl^-/ClO_4^- media and are taken from [78BED/FID] (□) and [81AWA/SUN] (△). The shaded area represents the uncertainty range obtained by propagating the resulting uncertainties at $I = 0$ back to $I = 4$ m.

Equilibrium: $UO_2^{2+} + 2Cl^- \rightleftharpoons UO_2Cl_2(aq)$

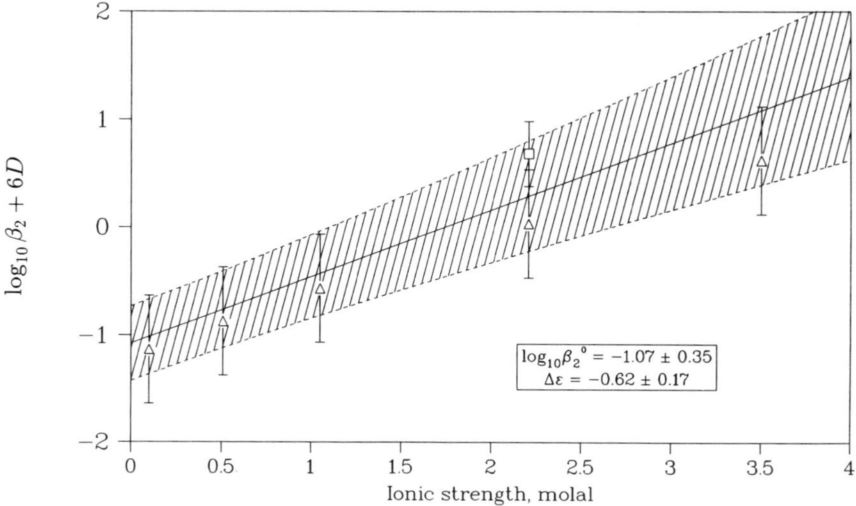

$$\Delta_f G_m^\circ(UO_2Cl^+, aq, 298.15\,K) = -(1084.7 \pm 1.8)\,kJ \cdot mol^{-1}$$
$$\Delta_f G_m^\circ(UO_2Cl_2, aq, 298.15\,K) = -(1208.7 \pm 2.9)\,kJ \cdot mol^{-1}$$

The enthalpy data for the UO_2^{2+}-Cl^- system from Ohashi and Morozumi [67OHA/MOR2] are inconsistent with those obtained by Awasthi and Sundarsan [81AWA/SUN] and by Nikolaeva [77NIK2], and are therefore not considered in this review. The data from Nikolaeva [77NIK2] would have to be extrapolated out of their experimental temperature range to obtain $\Delta_f H_m^\circ(UO_2Cl^+, aq, 298.15\,K)$, which can lead to appreciable error. The values selected here are therefore based exclusively on the data of Awasthi and Sundarsan [81AWA/SUN], who found

$$\Delta_r H_m^\circ(V.82,\ 298.15\ K) = (8 \pm 2)\,kJ \cdot mol^{-1}$$
$$\Delta_r H_m^\circ(V.83,\ 298.15\ K) = (15 \pm 6)\,kJ \cdot mol^{-1}.$$

Combining this with selected auxiliary enthalpy data for UO_2^{2+} and Cl^- yields the enthalpies of formation selected in this review.

$$\Delta_f H_m^\circ(UO_2Cl^+, aq, 298.15\,K) = -(1178.1 \pm 2.5)\,kJ \cdot mol^{-1}$$
$$\Delta_f H_m^\circ(UO_2Cl_2, aq, 298.15\,K) = -(1338.2 \pm 6.2)\,kJ \cdot mol^{-1}$$

Using the selected reaction data, the entropies are calculated via $\Delta_r S_m^\circ$:

$$S^\circ_m(UO_2Cl^+, aq, 298.15\,K) = -(11.5 \pm 7.4)\,J \cdot K^{-1} \cdot mol^{-1}$$
$$S^\circ_m(UO_2Cl_2, aq, 298.15\,K) = (44 \pm 23)\,J \cdot K^{-1} \cdot mol^{-1}$$

No heat capacity data are available for $UO_2Cl_2(aq)$. The value estimated by Lemire and Tremaine [80LEM/TRE] for the mean value of $C^\circ_{p,m}|^{473}_{298}$ (UO_2Cl^+, aq) is not likely to be a good approximation to the value $C^\circ_{p,m}(UO_2Cl^+, aq, 298.15\,K)$.

b) Aqueous U(V) chlorides

No aqueous species of the form $UO_2Cl_q^{1-q}$ have been identified.

c) Aqueous U(IV) chlorides

There are fairly few studies of chloride complexes of uranium(IV). The existing data are compiled in Table V.23 and refer to the reactions

$$U^{4+} + qCl^- \rightleftharpoons UCl_q^{4-q}. \quad (V.84)$$

The most precise studies are those of Ahrland and Larsson [54AHR/LAR2] and of Day, Wilhite and Hamilton [55DAY/WIL]. Sobkowski [61SOB] did an extensive number of emf-studies covering a broad range of ionic strengths. This review corrects these results for the formation of UO_2Cl^+, *cf.* Appendix A. This review uses the data from Refs. [54AHR/LAR, 55DAY/WIL, 61SOB, 63VDO/ROM2] to calculate $\log_{10}\beta_1^\circ$ and $\Delta\varepsilon$ using the specific ion interaction theory. A plot of $(\log_{10}\beta_1 + 8D)$ *vs.* $m_{ClO_4^-}$ is shown in Figure V.12. The results of the linear regression are

$$\log_{10}\beta_1^\circ(V.84, q=1, 298.15\ K) = 1.72 \pm 0.13$$

and $\Delta\varepsilon = -(0.29 \pm 0.08)$. This value agrees reasonably well with $\Delta\varepsilon$ used for the analogous fluoride complexation reaction (V.56). From $\Delta\varepsilon$ a new ion interaction coefficient is obtained, $\varepsilon_{(UCl^{3+},ClO_4^-)} = (0.59 \pm 0.10)$. Large changes in the ionic composition of the test solution have to be made in order to study the complex formation reactions. Hence, the accuracy of both $\log_{10}\beta_1^\circ$ and $\Delta\varepsilon$ may be lower than indicated by the values obtained from Figure V.12. The data of Bunus [74BUN], Kraus and Nelson [50KRA/NEL], and Nikolaeva [77NIK] fall outside. This may be due to experimental shortcomings in all cases, *cf.* Appendix A.

The recommended value of

$$\Delta_f G^\circ_m(UCl^{3+}, aq, 298.15\,K) = -(670.9 \pm 1.9)\,kJ \cdot mol^{-1}$$

is obtained from the selected $\log_{10}\beta_1^\circ$ and selected auxiliary data for U^{4+}, and Cl^- from CODATA [89COX/WAG].

The value of $\log_{10}\beta_1^\circ$ selected in this review is in fair agreement with the value proposed by Langmuir, $\log_{10}\beta_1^\circ = 1.3$ [78LAN], but not with the value $\log_{10}\beta_1^\circ = (3 \pm 1)$ proposed by Lemire and Tremaine [80LEM/TRE].

No reliable value of $\log_{10}\beta_2$ can be obtained from Day, Wilhite and Hamilton [55DAY/WIL], the only experimental work to address this species. This reflects the

Table V.23: Experimental equilibrium data for the uranium(IV) chloride system.

Method	Ionic Medium	t (°C)	$\log_{10}\beta_q^{(a)}$	$\log_{10}\beta_q^{\circ(a)}$	Reference
$U^{4+} + Cl^- \rightleftharpoons UCl^{3+}$					
emf	var.	25		0.85 ± 0.13	[50KRA/NEL]
emf	$1\ M\ (H,Na)(Cl,ClO_4)$	20	0.30 ± 0.11		[54AHR/LAR2]
dis	$(H,Na)(Cl,ClO_4)$	10	$0.18 \pm 0.15^{(b)}$		[55DAY/WIL]
	$1\ M\ H^+, I = 2\ M$	25	$0.26 \pm 0.15^{(b)}$		
		40	$0.52 \pm 0.15^{(b)}$		
aix	HCl var, > 5.5 M	25	(c)		[56KRA/MOO]
emf	$0.25\ M\ H(Cl,ClO_4)$	25	$0.60 \pm 0.20^{(b)}$		[61SOB]
	$0.64\ M\ H(Cl,ClO_4)$	25	$0.47 \pm 0.20^{(b)}$		
	$1\ M\ H(Cl,ClO_4)$	25	$0.40 \pm 0.20^{(b)}$		
	$1.44\ M\ H(Cl,ClO_4)$	25	$0.33 \pm 0.20^{(b)}$		
	$2.25\ M\ H(Cl,ClO_4)$	25	$0.26 \pm 0.20^{(b)}$		
nmr	$\sim 2.5\ M\ H(Cl,ClO_4)$	20	$0.78 \pm 0.22^{(b)}$		[63VDO/ROM3]
cix	$2.06\ M\ HClO_4$	25	−0.51		[74BUN]
	$3\ M\ HClO_4$	25	−0.12		
	$3.93\ M\ HClO_4$	25	0.01		
emf	0.5 to 4.0 M	50-150		$2.6^{(d)}$	[77NIK]
rev		25		1.3	[78LAN]
rev		25		3.0 ± 1.0	[80LEM/TRE]
....					
$U^{4+} + 2Cl^- \rightleftharpoons UCl_2^{2+}$					
dis	$(H,Na)(Cl,ClO_4)$ $1\ M\ H^+, I = 2\ M$	25	0.06		[55DAY/WIL]

(a) Refers to the reactions indicated, $\log_{10}\beta$ in the ionic medium and at the temperature given in the table, $\log_{10}\beta^\circ$ (in molal units) at $I = 0$ and 298.15 K.
(b) Uncertainties estimated by this review as described in Appendix A.
(c) This study contains evidence for anionic complexes.
(d) This value has been calculated for 25°C by the formula given by the author.

Figure V.12: Extrapolation to $I = 0$ of experimental data for the formation of UCl^{3+} using the specific ion interaction theory. The data refer to mixed Cl^-/ClO_4^- media and are taken from [54AHR/LAR2] (□), [55DAY/WIL] (○), [61SOB] (△) and [63VDO/ROM3] (◇). The shaded area represents the uncertainty range obtained by propagating the resulting uncertainties at $I = 0$ back to $I = 3$ m.

$$\text{Equilibrium:} \qquad U^{4+} + Cl^- \rightleftharpoons UCl^{3+}$$

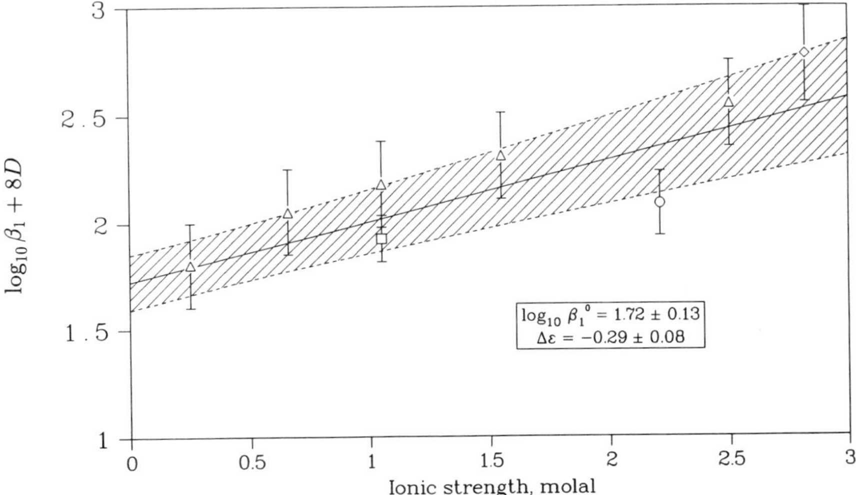

general difficulty in determining accurate values for the stability constants of weak complexes. Because of large variations in the composition of test solutions, it is also difficult to assure constant activity factors in the equilibrium experiments.

It should be mentioned that relatively strong U(IV) chloride complexes appear to be formed in aqueous solution above 300°C [89RED/SAV].

From the equilibrium data reported by Day, Wilhite and Hamilton [55DAY/WIL] for Reaction (V.84, $q = 1$) at $I = 2$ M, this review calculates the selected value of

$$\Delta_r H_m^\circ(\text{V.84}, q = 1, 298.15\ K) = -(19 \pm 9)\,\text{kJ}\cdot\text{mol}^{-1},$$

and uses it to calculate the selected enthalpy of formation,

$$\Delta_f H_m^\circ(\text{UCl}^{3+}, \text{aq}, 298.15\,\text{K}) = -(777.3 \pm 9.6)\,\text{kJ}\cdot\text{mol}^{-1}.$$

Using the selected reaction data, the entropy is calculated via $\Delta_r S_m^\circ$.

$$S_m^\circ(\text{UCl}^{3+}, \text{aq}, 298.15\,\text{K}) = -(391 \pm 33)\,\text{J}\cdot\text{K}^{-1}\cdot\text{mol}^{-1}$$

The mean value $C_{p,m}^\circ|_{298}^{473}(\text{UCl}^{3+}, \text{aq})$ calculated by Lemire and Tremaine [80LEM/TRE] is not likely to be a good approximation to $C_{p,m}^\circ(\text{UCl}^{3+}, \text{aq}, 298.15\,\text{K})$.

d) Aqueous U(III) chlorides

Shiloh and Marcus [65SHI/MAR] reported a value of $\beta_1 = (1.3 \pm 0.3) \times 10^{-3}$ at 25°C for the formation of UCl^{2+} in aqueous solutions of LiCl. Uranium(III) is not stable in aqueous solutions and is rapidly oxidized. The value of β_1 was determined in solutions of such high ionic strength ($I > 6$ M) that the specific ion interaction theory cannot be used to estimate a value of β_1°. Hence, no standard Gibbs energy of formation is reported in this review.

V.4.2.1.3. Solid uranium chlorides

The chemical thermodynamic properties of the binary uranium chloride solids are reliably reviewed in a recent IAEA report [83FUG/PAR]. Since the objective of this review is to extend and complete the IAEA work in areas most relevant to nuclear technologies, that work has only been reviewed and no extensive re-evaluation of these compounds is performed here. Note that although Krestov [72KRE] compiled thermodynamic data for UCl_2(cr), insufficient basis exists for this review to credit this compound.

a) UCl_3 (cr)

Fuger *et al.* [83FUG/PAR] performed a detailed review of the thermodynamic properties of uranium(III) chloride for the IAEA. Table V.24 summarizes the data upon which their selection was made. The recommended value,

$$\Delta_f H_m^\circ(UCl_3, cr, 298.15\,K) = -(863.7 \pm 2.0)\,kJ \cdot mol^{-1},$$

is the weighed mean of the results Fuger *et al.* [83FUG/PAR] derived from Fontana [58FON] and Barkelew [45BAR], plus the recent results by Cordfunke, Ouweltjes and Prins [82COR/OUW] who reported $\Delta_f H_m^\circ(UCl_3$, cr, 298.15 K) $= -(862.1 \pm 3.2)\,kJ \cdot mol^{-1}$. This remains unchanged when recalculated with auxiliary data from the present review. The selected value is in good agreement with $\Delta_f H_m^\circ(UCl_3$, cr, 298.15 K) $= -(863.6 \pm 7)\,kJ \cdot mol^{-1}$ which they [83FUG/PAR] estimated from the trend of $[\Delta_f H_m^\circ(MCl_3$, cr, 298.15 K) $- \Delta_f H_m^\circ(M^{3+}$, aq, 298.15 K)] for plutonium and americium. A more recent solution calorimetry determination [87SUG/CHI] resulted in a value of $\Delta_f H_m^\circ(UCl_3$, cr, 298.15 K) $= -(861.4 \pm 2.9)\,kJ \cdot mol^{-1}$ (after recalculation using $\Delta_f H_m^\circ(UCl_4$, cr, 298.15 K) selected in this review), which is in agreement with the selected value.

The values for the entropy and heat capacity at 298.15 K are taken from the low-temperature heat capacity measurements reported by Cordfunke, Konings and Westrum [89COR/KON].

$$S_m^\circ(UCl_3, cr, 298.15\,K) = (158.1 \pm 0.5)\,J \cdot K^{-1} \cdot mol^{-1}.$$
$$C_{p,m}(UCl_3, cr, 298.15\,K) = (95.10 \pm 0.50)\,J \cdot K^{-1} \cdot mol^{-1}.$$

Table V.24: Data considered for the selection of $\Delta_f H_m^\circ(UCl_3, cr, 298.15 K)$.

Reference	Reaction	$\Delta_f H_m^\circ(UCl_3, cr, 298.15\,K)$ $(kJ \cdot mol^{-1})$
[44ALT] and [45GRE]	$UCl_4(cr) + \frac{1}{2}H_2(g) \rightleftharpoons UCl_3(cr) + HCl(g)$	-866.1[(a)] -857.7[(b)]
[45BAR]	$UCl_3(cr) + \frac{1}{2}Cl_2(g) \rightleftharpoons UCl_4(cr)$	-859.4 ± 8.4[(a)]
[58FON]	$UCl_3(cr) + \frac{1}{2}UO_2^{2+} + 2H^+ \rightleftharpoons \frac{3}{2}U^{4+} + 3Cl^- + H_2O(l)$	-867.3 ± 5.4[(c)]
[58FON]	$UCl_3(cr) \rightleftharpoons U^{3+} + 3Cl^-$	-864.0 ± 5.4[(d)]
[58FON]	$UCl_3(cr) + HCl(sln) \rightleftharpoons UCl_4(cr) + \frac{1}{2}H_2(g)$	-866.1 ± 3.8[(e)]
[82COR/OUW]	$UCl_3(cr) + HCl(sln) \rightleftharpoons UCl_4(cr) + \frac{1}{2}H_2(g)$	-862.1 ± 3.2[(f)]
[87SUG/CHI]	$UCl_3(cr) + HCl(sln) \rightleftharpoons UCl_4(cr) + \frac{1}{2}H_2(g)$	-861.4 ± 2.9[(g)]

(a) Data reported by MacWood [58MAC]. The auxiliary value used for $\Delta_f H_m^\circ(UCl_4$, cr, 298.15 K) is the one selected here.

(b) Using $\Delta_r S_m^\circ = 169.5\,J \cdot K \cdot mol^{-1}$, after MacWood [58MAC].

(c) In 0.5 M $HClO_4$. Auxiliary data on $\Delta_f H_m^\circ(U^{4+}$, 0.5 M $HClO_4$) and $\Delta_f H_m^\circ(UO_2^{2+}$, 0.5 M $HClO_4$) from Ref. [76FUG/OET]. $\Delta_f H_m^\circ(Cl^-$, 0.5 M $HClO_4$) = $\Delta_f H_m^\circ(Cl^-$, 0.5 M HCl) assumed.

(d) In 0.5 M $HClO_4$. Auxiliary data on $\Delta_f H_m^\circ(U^{3+}$, aq, 298.15 K) is selected value [76FUG/OET].

(e) In 0.5 M $HClO_4$. Reaction above rearranged to depend on UCl_4(cr) [76FUG/OET].

(f) In 1.511 M H_2SO_4.

(g) In 4 M HCl + 0.005 M Na_2SiF_6 + $1.5 \cdot 10^{-6}$ M H_2PtCl_6. Recalculated using $\Delta_f H_m^\circ(UCl_4$, cr, 298.15 K) selected in this review.

These are preferred to the earlier measurements of Ferguson and Prather [44FER/PRA] tabulated by Katz and Rabinowich [51KAT/RAB].

These selected values at 298.15 K were combined with the thermal function fit by Cordfunke, Konings and Westrum [89COR/KON] to heat content measurements relative to 273 K (in the range 273 to 998 K) [47GIN/COR] to obtain the $C_{p,m}(UCl_3, cr, T)$ functions selected in this review.

The Gibbs energy of formation is calculated from the selected enthalpy of formation and entropy.

$$\Delta_f G^\circ_m(UCl_3, cr, 298.15\,K) = -(796.1 \pm 2.0)\,kJ \cdot mol^{-1}$$

b) UCl_4 (cr)

Fuger *et al.* [83FUG/PAR] performed a detailed review of the thermodynamic properties of uranium(IV) chloride. They pointed out that the value of $\Delta_f H^\circ_m(UCl_4$, cr, 298.15 K) seemed well established until 1971 [63RAN/KUB, 52LAT, 58BRE/BRO, 57KAT/SEA]. Early measurements supported a value of $\Delta_f H^\circ_m(UCl_4$, cr, 298.15 K) $= -1050\,kJ \cdot mol^{-1}$. The experimental basis for this value is given in the first four lines of Table V.25, which all represent the results of enthalpy of solution measurements of U(cr) and UCl_4(cr) in HCl solutions. Fitzgibbon, Pavone and Holley [71FIT/PAV] and Cordfunke, Ouweltjes and Prins [76COR/OUW] obtained significantly different results by using different methods. In attempts to verify the results of the earlier experiments, Suglobova and Chirkst [78SUG/CHI] and Thakur, Ahmad and Prasad [78THA/AHM] obtained values disturbingly similar to the original measurements (*cf.* Table V.25). After extensive analysis, Fuger *et al.* [83FUG/PAR] discarded these determinations, as well as that of Suglobova and Chirkst [80SUG/CHI], because of their implications in altering the internal consistency of the selected uranium thermochemical network (particularly γ-UO_3, UO_2Cl_2(cr), and UF_6(cr)). This value is further supported by the more recent calorimetric result of O'Hare [85OHA2] who obtained $\Delta_f H^\circ_m(UCl_4$, cr, 298.15 K) $= -(1022.8 \pm 3.8)\,kJ \cdot mol^{-1}$ from the oxidation of UCl_4 to UO_2Cl_2 in solution by XeO_3. Recalculating this value with auxiliary data from the present review gives $\Delta_f H^\circ_m(UCl_4$, cr, 298.15 K) $= -(1018.5 \pm 1.8)\,kJ \cdot mol^{-1}$. Because of the key position of UCl_4(cr) in the thermodynamical network for the U-O-Cl system, the value selected by Fuger *et al.* [83FUG/PAR] is also selected in this review.

$$\Delta_f H^\circ_m(UCl_4,\ cr,\ 298.15\ K) = -(1018.8 \pm 2.5)\,kJ \cdot mol^{-1}.$$

Incidentally, one should note that the above results by Suglobova and Chirkst [87SUG/CHI] points, as already noted [83FUG/PAR, 85OHA2], to the difficult situation encountered where one tries to reconciliate selected values for the enthalpies of formation of UF_4(cr) and UCl_4(cr).

As did Fuger *et al.* [83FUG/PAR], this review accepts the following values from the smoothed temperature functions for $S^\circ_m(UCl_4, cr, T)$ and $C^\circ_{p,m}(UCl_4, cr, T)$ tabulated by Katz and Rabinowitch [51KAT/RAB] of the low-temperature specific heat measurements of Ferguson and Prather [44FER/PRA].

Table V.25: Data considered for the selection of $\Delta_f H_m^\circ(UCl_4$, cr, 298.15 K).

Reference	Reaction	$\Delta_f H_m^\circ(UCl_4, cr, 298.15\,K)$ $(kJ \cdot mol^{-1})$
[28BIL/FEN]	$U(cr) + 2Cl_2(g) \rightleftharpoons UCl_4(cr)$	-1038.5[(a)]
[45BAR]	$U(cr) + 2Cl_2(g) \rightleftharpoons UCl_4(cr)$	-1045.2[(b)]
[61ARG/MER]	$U(cr) + 2Cl_2(g) \rightleftharpoons UCl_4(cr)$	-1048.1[(c)]
[69SMI/THA]	$U(cr) + 2Cl_2(g) \rightleftharpoons UCl_4(cr)$	-1050.2[(c)]
[71FIT/PAV]	$U(cr) + 2Cl_2(g) \rightleftharpoons UCl_4(cr)$	-1017.6 ± 2.9[(d,i)]
		-1019.1 ± 4.6[(e,i)]
[71FIT/PAV]	$UO_2(cr) + 4HCl(sln) \rightleftharpoons UCl_4(cr) + 2H_2O(sln)$	-1017.6 ± 2.9[(f,i)]
[76COR/OUW]	$UO_2(cr) + 4HCl(sln) \rightleftharpoons UCl_4(cr) + 2H_2O(sln)$	-1019.1 ± 4.6[(f,i)]
[78SUG/CHI]	$U(cr) + 2Cl_2(g) \rightleftharpoons UCl_4(cr)$	-1052.3 ± 3.8[(g)]
	$U(cr) + 2Cl_2(g) \rightleftharpoons UCl_4(cr)$	-1024.7 ± 4.2[(d)]
[80SUG/CHI]	$UCl_4(cr) + 6NaF(cr) + 2FeCl_3(cr) \rightleftharpoons UF_6(cr) + 6NaCl(cr) + 2FeCl_2(cr)$	-1042.2 ± 2.9[(h)]
[85OHA2]	$UCl_4(cr) + \left(\frac{1}{3}XeO_3 + H_2O\right)(sln) \rightleftharpoons UO_2Cl_2(cr) + 2HCl(sln) + \frac{2}{3}Xe(g)$	-1022.8 ± 3.8

(a) Net reaction, at 298.15 K in 8.13 M HCl, from dissolution of U(cr) and UCl_4(cr).
(b) Net reaction, at 273.15 K in 12 M HCl/$FeCl_3$ solution from dissolution of U(cr) and UCl_4(cr). Fuger *et al.* [83FUG/PAR] revised $\Delta_f H_m^\circ$(HCl, sln) in this medium and temperature from -138.1 to $-136.8\,kJ \cdot mol^{-1}$ using selected auxiliary data.
(c) Fuger *et al.* [83FUG/PAR] revised the earlier value of $\Delta_f H_m^\circ(UCl_4, cr, 298.15\,K)$ of $-1050\,kJ \cdot mol^{-1}$ to account for erroneous $\Delta_f H_m^\circ$(HCl, sln) and erroneous enthalpy of mixing.
(d) From $\Delta_{sol} H_m(U, cr)$ and $\Delta_{sol} H_m(UCl_4, cr)$ in 4.35 M HCl.
(e) From $\Delta_{sol} H_m(U, cr)$ and $\Delta_{sol} H_m(UCl_4, cr)$ in 6.83 M HCl.
(f) The thermodynamic cycles for this determination were presented in detail in Table 8.XIV of Ref. [83FUG/PAR].
(g) From $\Delta_{sol} H_m(U, cr)$ and $\Delta_{sol} H_m(UCl_4, cr)$ in 4 M HCl with Na_2SiF_6 to prevent hydrolysis of U(III) in the dissolution of U(cr), and with H_2PtCl_6 to accelerate the oxidation of U(III).
(h) From $\Delta_{sol} H_m(UCl_4, cr)$ and $\Delta_{sol} H_m(UCl_6, cr)$ in 2% HCl, 0.5% $FeCl_3$ and 1.5% H_3BO_3.
(i) Values considered in the determination of the selected weighted mean.

$$S^\circ_m(\mathrm{UCl_4, cr, 298.15\,K}) = (197.1 \pm 0.8)\,\mathrm{J \cdot K^{-1} \cdot mol^{-1}}$$
$$C^\circ_{p,m}(\mathrm{UCl_4, cr, 298.15\,K}) = (122.0 \pm 0.4)\,\mathrm{J \cdot K^{-1} \cdot mol^{-1}}$$

The temperature function $C^\circ_{p,m}(T)$ accepted by the IAEA [83FUG/PAR] and based on Katz and Rabinowitch [51KAT/RAB] is also accepted by this review.

The Gibbs energy of formation is calculated from the selected enthalpy of formation and entropy.

$$\Delta_f G^\circ_m(\mathrm{UCl_4, cr, 298.15\,K}) = -(929.6 \pm 2.5)\,\mathrm{kJ \cdot mol^{-1}}$$

c) UCl_5 *(cr) and* UCl_6 *(cr)*

Fuger *et al.* [83FUG/PAR] pointed out the discrepancy between $\Delta_f H^\circ_m(\mathrm{UCl_5}$, cr, 298.15 K) = $-1062.3\,\mathrm{kJ \cdot mol^{-1}}$ and $\Delta_f H^\circ_m(\mathrm{UCl_6}$, cr, 298.15 K) = $-1101.6\,\mathrm{kJ \cdot mol^{-1}}$ recalculated from MacWood [58MAC] (after the experimental results of Barkelew [45BAR]) compared to $\Delta_f H^\circ_m(\mathrm{UCl_5}$, cr, 298.15 K) = $-1036.8\,\mathrm{kJ \cdot mol^{-1}}$ and $\Delta_f H^\circ_m(\mathrm{UCl_6}$, cr, 298.15 K) = $-1054.4\,\mathrm{kJ \cdot mol^{-1}}$ calculated after Gross, Hayman and Wilson [71GRO/HAY]. By adjusting the auxiliary data for the enthalpy of oxidation of $FeCl_2$(aq) to $FeCl_3$(aq) in the experimental solution from $-84.5\,\mathrm{kJ \cdot mol^{-1}}$ to what they consider to be the more reasonable value of $-79.5\,\mathrm{kJ \cdot mol^{-1}}$, they obtained $\Delta_f H^\circ_m(\mathrm{UCl_5}$, cr, 298.15 K) = $-1057.3\,\mathrm{kJ \cdot mol^{-1}}$ and $\Delta_f H^\circ_m(\mathrm{UCl_6}$, cr, 298.15 K) = $-1091.6\,\mathrm{kJ \cdot mol^{-1}}$. Based on this analysis, Fuger *et al.* [83FUG/PAR] provisionally selected the weighted averages which are also accepted here. These selections are not in agreement with the more recent results of Cordfunke, Ouweltjes and Prins [82COR/OUW] based on enthalpy of solution measurements. After a small correction for the energy equivalent of the calorimeter [83COR/OUW] these authors gave $\Delta_f H^\circ_m(\mathrm{UCl_5}$, cr, 298.15 K) = $-(1041.5 \pm 1.9)\,\mathrm{kJ \cdot mol^{-1}}$ and $\Delta_f H^\circ_m(\mathrm{UCl_6}$, cr, 298.15 K) = $-(1068.3 \pm 1.9)\,\mathrm{kJ \cdot mol^{-1}}$. Since the results of Cordfunke, Ouweltjes and Prins [82COR/OUW] for UCl_5(cr) are in good agreement with the direct chlorination experiments by Gross, Hayman and Wilson [71GRO/HAY], this review rejects the results of Macwood [58MAC] and selects

$$\Delta_f H^\circ_m(\mathrm{UCl_5,\ cr,\ 298.15\ K}) = -(1039.0 \pm 3.0)\,\mathrm{kJ \cdot mol^{-1}}$$
$$\Delta_f H^\circ_m(\mathrm{UCl_6,\ cr,\ 298.15\ K}) = -(1066.5 \pm 3.0)\,\mathrm{kJ \cdot mol^{-1}}.$$

The first value is the average of the concordant values by Cordfunke, Ouweltjes and Prins [83COR/OUW] and Gross, Hayman and Wilson [71GRO/HAY], the second value is solely based on the results of Cordfunke, Ouweltjes and Prins [83COR/OUW]. The uncertainties are assigned in the present review.

This review accepts the values for $C_{p,m}(\mathrm{UCl_6}$, cr, $T)$ and $S_m(\mathrm{UCl_6}$, cr, $T)$ derived by Fuger *et al.* [83FUG/PAR] as tabulated from 0 to 350 K by Katz and Rabinowitch [51KAT/RAB] from the measurements of Ferguson and Rand [44FER/RAN]. The selected values at 298.15 K were obtained from these functions.

$$S^\circ_m(\mathrm{UCl_6, cr, 298.15\,K}) = (285.8 \pm 1.7)\,\mathrm{J \cdot K^{-1} \cdot mol^{-1}}$$
$$C^\circ_{p,m}(\mathrm{UCl_6, cr, 298.15\,K}) = (175.7 \pm 4.2)\,\mathrm{J \cdot K^{-1} \cdot mol^{-1}}$$

The selected temperature function $C^\circ_{p,m}(T)$ is also from the IAEA review [83FUG/PAR].

Since similar data are unavailable for UCl_5(cr), the values for UCl_5(cr) at 298.15 K were estimated by Fuger *et al.* [83FUG/PAR] from the selected values for UCl_4(cr) and UCl_6(cr).

$$S^\circ_m(\mathrm{UCl_5, cr, 298.15\,K}) = (242.7 \pm 8.4)\,\mathrm{J \cdot K^{-1} \cdot mol^{-1}}$$
$$C^\circ_{p,m}(\mathrm{UCl_5, cr, 298.15\,K}) = (150.6 \pm 8.4)\,\mathrm{J \cdot K^{-1} \cdot mol^{-1}}$$

The Gibbs energies of formation are calculated from the selected enthalpies of formation and entropies.

$$\Delta_f G^\circ_m(\mathrm{UCl_5, cr, 298.15\,K}) = -(930.1 \pm 3.9)\,\mathrm{kJ \cdot mol^{-1}}$$
$$\Delta_f G^\circ_m(\mathrm{UCl_6, cr, 298.15\,K}) = -(937.2 \pm 3.0)\,\mathrm{kJ \cdot mol^{-1}}$$

Further experimental work on UCl_5(cr) and UCl_6(cr) and on their relationship to UCl_4(cr) is clearly needed.

d) UOCl(cr)

UOCl(cr) is the only thermodynamically well-characterized uranium(III) chloride other than UCl_3(cr). The only thermochemical data available to Fuger *et al.* [83FUG/PAR] for UOCl(cr) was the equilibrium constant (as a function of temperature) determination by Gregory [45GRE] tabulated by Katz and Rabinowitch [51KAT/RAB]. These data lead to $\Delta_f H^\circ_m(\mathrm{UOCl, cr, 298.15\,K}) = -892\,\mathrm{kJ \cdot mol^{-1}}$ which they [83FUG/PAR] found inconsistent (too negative) with what would be expected from $\Delta_f H^\circ_m$(PuOCl, cr, 298.15 K) $-$ $\Delta_f H^\circ_m$($PuCl_3$, cr, 298.15 K), $\Delta_f H^\circ_m$(PuOBr, cr, 298.15 K) $-$ $\Delta_f H^\circ_m$($PuBr_3$, cr, 298.15 K), and $\Delta_f H^\circ_m$(AmOCl, cr, 298.15 K) $-$ $\Delta_f H^\circ_m$($AmCl_3$, cr, 298.15 K). No data were therefore selected for UOCl(cr) in the IAEA review. Since Ref. [83FUG/PAR] appeared, Cordfunke, Ouweltjes and van Vlaanderen [83COR/OUW2] calculated the enthalpy of formation of UOCl(cr) from enthalpy of solution measurements and auxiliary data consistent with [89COX/WAG]. This value, corrected for a small error in the energy equivalent of the calorimeter [83COR/OUW], is selected in this review.

$$\Delta_f H^\circ_m(\mathrm{UOCl, cr, 298.15\,K}) = -(833.9 \pm 4.2)\,\mathrm{kJ \cdot mol^{-1}}$$

In the absence of experimental data, S°_m(UOCl, cr, 298.15 K) and $C^\circ_{p,m}$(UOCl, cr, T) estimated by Cordfunke and Kubaschewski [84COR/KUB] were selected.

$$S^\circ_m(\mathrm{UOCl, cr, 298.15\,K}) = (102.5 \pm 8.4)\,\mathrm{J \cdot K^{-1} \cdot mol^{-1}}$$
$$C^\circ_{p,m}(\mathrm{UOCl, cr, 298.15\,K}) = (71 \pm 5)\,\mathrm{J \cdot K^{-1} \cdot mol^{-1}}$$

The Gibbs energy of formation is calculated from the selected enthalpy of formation and entropy.

$$\Delta_f G^\circ_m(\mathrm{UOCl, cr, 298.15\,K}) = -(785.7 \pm 4.9)\,\mathrm{kJ \cdot mol^{-1}}$$

e) $UOCl_2$ (cr)

$UOCl_2$(cr) is the only thermodynamically well-characterized uranium(IV) chloride other than UCl_4(cr). Fuger *et al.* [83FUG/PAR] accepted the evaluation by Rand and Kubaschewski [63RAN/KUB] for this compound. The enthalpy of formation is based on the measurements of Davidson, MacWood and Streetor [45DAV/MAC] and of Shchukarev *et al.* [58SHC/VAS2] in the range 700 to 800 K of the equilibrium

$$2UOCl_2(cr) \rightleftharpoons UO_2(cr) + UCl_4(g). \qquad (V.85)$$

They obtained $\Delta_f H^\circ_m(UOCl_2, cr, 298.15\,K) = -1066.5\,kJ \cdot mol^{-1}$ using $\Delta_r C_{p,m}(V.85) = -25\,J \cdot K^{-1} \cdot mol^{-1}$ to adjust from 750 K. Knacke, Müller and van Rensen
[72KNA/MUE] studied the same equilibrium in the range 700 to 1023 K. Using the same $\Delta_r C_{p,m}(V.85)$ to adjust from 807 K, Fuger *et al.* [83FUG/PAR] found the essentially identical value of $\Delta_f H^\circ_m(UOCl_2, cr, 298.15\,K) = -(1066.9 \pm 2.5)\,kJ \cdot mol^{-1}$. Since then, the enthalpy of formation of $UOCl_2$(cr) at room temperature has been determined by solution calorimetry by Cordfunke *et al.* [83COR/OUW]. From a solution cycle based upon UCl_4(cr) they obtained

$$\Delta_f H^\circ_m(UOCl_2, cr, 298.15\,K) = -(1069.3 \pm 2.7)\,kJ \cdot mol^{-1}$$

which is selected here.

The selected entropy and heat capacity values were obtained from the heat capacity measurements of Greenberg and Westrum [56GRE/WES].

$$S^\circ_m(UOCl_2, cr, 298.15\,K) = (138.32 \pm 0.21)\,J \cdot K^{-1} \cdot mol^{-1}$$
$$C^\circ_{p,m}(UOCl_2, cr, 298.15\,K) = (95.06 \pm 0.42)\,J \cdot K^{-1} \cdot mol^{-1}$$

The selected temperature function $C^\circ_{p,m}(T)$ is based on the these values and the high-temperature (298.15 to 800 K) heat capacity functions of Cordfunke and Kubaschewski [84COR/KUB] who cited unpublished measurements by Cordfunke.

The Gibbs energy of formation is calculated from the selected enthalpy of formation and entropy.

$$\Delta_f G^\circ_m(UOCl_2, cr, 298.15\,K) = -(998.5 \pm 2.7)\,kJ \cdot mol^{-1}$$

f) $UOCl_3$ (cr)

Shchukarev *et al.* [58SHC/VAS2] measured the enthalpies of reaction of $UOCl_3$(cr) and UCl_4(cr) in a 2% HCl, 0.5% $FeCl_3$ solution. Fuger *et al.* [83FUG/PAR] combined this with selected auxiliary data and with the enthalpy of chlorination of $FeCl_2$(cr) to $FeCl_3$(cr) to obtain $\Delta_r H^\circ_m(V.86) = (129.3 \pm 0.9)\,kJ \cdot mol^{-1}$.

$$UOCl_3(cr) + FeCl_2(cr) + 2HCl(sln) \rightleftharpoons UCl_4(cr) + FeCl_3(cr) + H_2O(sln) \qquad (V.86)$$

Combining this with the selected $\Delta_f H_m^\circ$(UCl_4, cr, 298.15 K) leads to $\Delta_f H_m^\circ$($UOCl_3$, cr, 298.15 K) = $-(1161.1 \pm 3.3)\,kJ \cdot mol^{-1}$.

However, the purity of the sample used by Shchukarev *et al.* [58SHC/VAS2] is questionable, and the derived value for the enthalpy of formation is not in accordance with the phase relations in the U-O-Cl system, as discussed by Cordfunke and Kubaschewski [84COR/KUB]. These authors argued that the value should be about $14.6\,kJ \cdot mol^{-1}$ less negative to fit the known phase relations. This review accepts the value suggested by Cordfunke and Kubaschewski [84COR/KUB], which is rounded to

$$\Delta_f H_m^\circ(UOCl_3, cr, 298.15\ K) = -(1140 \pm 8)\,kJ \cdot mol^{-1}.$$

Fuger *et al.* [83FUG/PAR] estimated S_m°($UOCl_3$, cr, 298.15 K) = $(171.5 \pm 8.4)\,J \cdot K^{-1} \cdot mol^{-1}$ from the entropies of UCl_3(cr), UCl_4(cr), $UOCl_2$(cr) and UO_2Cl_2(cr). Cordfunke and Kubaschewski [84COR/KUB] estimated S_m°($UOCl_3$, cr, 298.15 K) = $(169.9 \pm 8.4)\,J \cdot K^{-1} \cdot mol^{-1}$. This review selects the average of these two values,

$$S_m^\circ(UOCl_3, cr, 298.15\ K) = (170.7 \pm 8.4)\,J \cdot K^{-1} \cdot mol^{-1}.$$

The Gibbs energy of formation is calculated from the selected enthalpy of formation and entropy.

$$\Delta_f G_m^\circ(UOCl_3, cr, 298.15\,K) = -(1045.6 \pm 8.4)\,kJ \cdot mol^{-1}$$

This review accepts the heat capacity from Fuger *et al.* [83FUG/PAR] estimated by Barin and Knacke [73BAR/KNA]. The $C_{p,m}^\circ(T)$ data tabulated by the IAEA [83FUG/PAR] are fitted between 298.15 and 600 K to obtain the selected $C_{p,m}^\circ(T)$ function listed in Table III.3.

$$C_{p,m}^\circ(UOCl_3, cr, 298.15\,K) = (117.2 \pm 4.2)\,J \cdot K^{-1} \cdot mol^{-1}$$

is obtained.

g) UO_2Cl(cr)

Although UO_2Cl(cr) is known, its thermodynamic properties have not yet been determined experimentally. Cordfunke and Kubaschewski [84COR/KUB] used the mass spectrometric data of Münstermann [76MUE] to calculate $\Delta_f H_m^\circ$(UO_2Cl, cr, 298.15 K) = $-1171.1\,kJ \cdot mol^{-1}$. This is somewhat more negative than the value they obtained additively, so a comparatively large uncertainty is assigned to the estimated value selected by this review,

$$\Delta_f H_m^\circ(UO_2Cl, cr, 298.15\,K) = -(1171.1 \pm 8.0)\,kJ \cdot mol^{-1}.$$

Cordfunke and Kubaschewski [84COR/KUB] used a modified form of Latimer's method [52LAT] to estimate the entropy which is selected by this review.

$$S_m^\circ(UO_2Cl, cr, 298.15\,K) = (112.5 \pm 8.4)\,J \cdot K^{-1} \cdot mol^{-1}$$

The Gibbs energy of formation is calculated from the selected enthalpy of formation and entropy.

$$\Delta_f G^\circ_m(UO_2Cl, cr, 298.15\,K) = -(1095.3 \pm 8.4)\,kJ \cdot mol^{-1}.$$

The thermal function and the 298.15 K value of $C^\circ_{p,m}$ estimated by Cordfunke and Kubaschewski [84COR/KUB] are also accepted here.

$$C^\circ_{p,m}(UO_2Cl, cr, 298.15\,K) = (88 \pm 5)\,J \cdot K^{-1} \cdot mol^{-1}$$

h) UO_2Cl_2 (cr) and its hydrates

Four reliable measurements leading to $\Delta_f H^\circ_m(UO_2Cl_2$, cr, 298.15 K) through three different thermochemical cycles were evaluated by Fuger *et al.* [83FUG/PAR] and are summarized in Table V.26. The agreement among the values dependent on the selected enthalpies of formation of γ-UO_3, UCl_4(cr) and UF_4(cr) is reassuring. Since then, O'Hare [85OHA2] measured the enthalpy of formation of UO_2Cl_2(cr) relative to UO_2F_2(cr). Recalculating his result with the selected value for the latter compound gives $\Delta_f H^\circ_m(UO_2Cl_2$, cr, 298.15 K) $= -(1243.6 \pm 1.3)\,kJ \cdot mol^{-1}$, in good agreement with the earlier results. The selected value obtained in the IAEA report [83FUG/PAR],

$$\Delta_f H^\circ_m(UO_2Cl_2, cr, 298.15\,K) = -(1243.6 \pm 1.3)\,kJ \cdot mol^{-1},$$

is adopted in this review. No new value is proposed because of the key position of UO_2Cl_2(cr) in the thermodynamic network of the U-O-Cl system.

This review also accepts the entropy and heat capacity selections of Fuger *et al.* [83FUG/PAR], obtained from the low-temperature calorimetric measurements of Greenberg and Westrum [56GRE/WES3].

$$S^\circ_m(UO_2Cl_2, cr, 298.15\,K) = (150.54 \pm 0.21)\,J \cdot K^{-1} \cdot mol^{-1}$$
$$C^\circ_{p,m}(UO_2Cl_2, cr, 298.15\,K) = (107.86 \pm 0.17)\,J \cdot K^{-1} \cdot mol^{-1}$$

The selected temperature function $C^\circ_{p,m}(T)$ was generated by Fuger *et al.* [83FUG/PAR] from high-temperature (392 to 690 K) heat capacity measurements of Cordfunke, Muis and Prins [79COR/MUI], which they preferred over those of Prins [73PRI].

The Gibbs energy of formation is calculated from the selected enthalpy of formation and entropy.

$$\Delta_f G^\circ_m(UO_2Cl_2, cr, 298.15\,K) = -(1145.8 \pm 1.3)\,kJ \cdot mol^{-1}$$

The monohydrate and trihydrate of UO_2Cl_2(cr) have been thermodynamically characterized by Prins [73PRI], and by Shchukarev, Vasilkova and Drozdova [58SHC/VAS]. The measurements of Lipilina and Samoilov [54LIP/SAM] who measured the enthalpy of solution of the trihydrate, gave a range consistent with that of Prins [73PRI]. Cordfunke [66COR] combined his equilibrium vapour pressure data

Table V.26: Data considered for the selection of $\Delta_f H^\circ_m(UO_2Cl_2, cr, 298.15\,K)$.

Reference	Reaction	$\Delta_f H^\circ_m(UO_2Cl_2)$ $(kJ \cdot mol^{-1})$
[76COR/OUW]	γ-$UO_3 + 2\,HCl(sln) \rightleftharpoons UO_2Cl_2(cr) + 2\,H_2O(sln)$	-1243.2 ± 1.3[a]
[76COR/OUW]	$UCl_4(cr) + 2FeCl_3(cr) + 2H_2O(sln) \rightleftharpoons UO_2Cl_2(cr) + 2FeCl_2(cr) + 4HCl(sln)$	-1246.4 ± 3.8[b]
[58SHC/VAS2]	$UCl_4(cr) + 2FeCl_3(cr) + 2H_2O(sln) \rightleftharpoons UO_2Cl_2(cr) + 2FeCl_2(cr) + 4HCl(sln)$	-1221.7 ± 8.4[b,c] -1245.5 ± 4.2[b,d]
[70KHA/KRI]	$UO_2Cl_2(cr) + 3FeCl_2(cr) + 3HF(sln) + HCl(sln) \rightleftharpoons UF_3(cr) + 3FeCl_3(cr) + 2H_2O(sln)$	-1242.6 ± 4.2[e]

(a) In 5.545 M HCl at 298.15 K.
(b) In 1.98% (mass) HCl, 0.5% (mass) $FeCl_3$ at 298.15 K.
(c) Auxiliary data consistent with NBS [82WAG/EVA].
(d) Using data for enthalpy of dissolution of $FeCl_2(cr)$ and $FeCl_3(cr)$ rather than own value for $FeCl_3$ only.
(e) In boric + hydrofluoric + hydrochloric acid solutions at 323.15 K corrected to 298.15 K by Fuger *et al.* [83FUG/PAR].

with those of Kapustinskii and Baranova [52KAP/BAR] for the dehydration of the trihydrate to the monohydrate and found an enthalpy inconsistent with those of Refs. [73PRI] and [58SHC/VAS] and an unreasonably high entropy of dehydration. For these reasons, Fuger *et al.* [83FUG/PAR] accepted the dehydration values of Prins [73PRI] and calculated enthalpy of formation values consistent with the selected $\Delta_f H^\circ_m(UO_2Cl_2, cr, 298.15\,K)$.

$$\Delta_f H^\circ_m(UO_2Cl_2 \cdot H_2O, cr, 298.15\,K) = -(1559.8 \pm 2.1)\,kJ \cdot mol^{-1}$$
$$\Delta_f H^\circ_m(UO_2Cl_2 \cdot 3H_2O, cr, 298.15\,K) = -(2164.8 \pm 1.7)\,kJ \cdot mol^{-1}$$

The entropy values were estimated by Fuger *et al.* [83FUG/PAR].

$$S^\circ_m(UO_2Cl_2 \cdot H_2O, cr, 298.15\,K) = (192.5 \pm 8.4)\,J \cdot K^{-1} \cdot mol^{-1}$$
$$S^\circ_m(UO_2Cl_2 \cdot 3H_2O, cr, 298.15\,K) = (272.0 \pm 8.4)\,J \cdot K^{-1} \cdot mol^{-1}$$

The Gibbs energies of formation are calculated from the selected enthalpies of formation and entropies.

$$\Delta_f G^\circ_m(UO_2Cl_2 \cdot H_2O, cr, 298.15\,K) = -(1405.0 \pm 3.3)\,kJ \cdot mol^{-1}$$
$$\Delta_f G^\circ_m(UO_2Cl_2 \cdot 3H_2O, cr, 298.15\,K) = -(1894.6 \pm 3.0)\,kJ \cdot mol^{-1}$$

No heat capacity data are available for these two hydrated compounds.

i) $(UO_2)_2Cl_3(cr)$

Cordfunke, Prins and van Vlaanderen [77COR/PRI] measured the enthalpy of solution of $(UO_2)_2Cl_3(cr)$ in sulphuric acid. Combining this with the data of Cordfunke and Ouweltjes [77COR/OUW] for the enthalpy of solution of γ-UO_3 in the same solution yields $\Delta_r H_m(V.87) = -(82.2 \pm 1.3)\,kJ \cdot mol^{-1}$ in 1.505 M H_2SO_4.

$$3\gamma\text{-}UO_3 + UCl_4(cr) + 2HCl(g) \rightleftharpoons 2(UO_2)_2Cl_3(cr) + H_2O(sln) \qquad (V.87)$$

Combining this with the selected values for the enthalpies of formation of γ-UO_3 and $UCl_4(cr)$ and with selected auxiliary data (after correction for dilution effects) yields the value of

$$\Delta_f H_m^\circ((UO_2)_2Cl_3, cr, 298.15\,K) = -(2404.5 \pm 1.7)\,kJ \cdot mol^{-1}$$

obtained by Fuger *et al.* [83FUG/PAR] and selected here. This is equal to the value chosen by Cordfunke and Kubaschewski [84COR/KUB]. In the absence of experimental data, the entropy, heat capacity and its temperature function estimated by Cordfunke and Kubaschewski [84COR/KUB] are selected here.

$$S_m^\circ((UO_2)_2Cl_3, cr, 298.15\,K) = (276 \pm 8)\,J \cdot K^{-1} \cdot mol^{-1}$$
$$C_{p,m}^\circ((UO_2)_2Cl_3, cr, 298.15\,K) = (203.6 \pm 5.0)\,J \cdot K^{-1} \cdot mol^{-1}$$

The Gibbs energy of formation is calculated from the selected enthalpy of formation and entropy.

$$\Delta_f G_m^\circ((UO_2)_2Cl_3, cr, 298.15\,K) = -(2234.8 \pm 2.9)\,kJ \cdot mol^{-1}$$

j) $U_2O_2Cl_5(cr)$

Cordfunke, Ouweltjes and van Vlaanderen [83COR/OUW2] determined the enthalpy of solution of $U_2O_2Cl_5(cr)$ calorimetrically. This is reduced to

$$\Delta_f H_m^\circ(U_2O_2Cl_5, cr, 298.15\,K) = -(2197.4 \pm 4.2)\,kJ \cdot mol^{-1}$$

using selected auxiliary data.

In the absence of experimental data, the entropy and heat capacity values, as well as the temperature function of the heat capacity estimated by Cordfunke and Kubaschewski [84COR/KUB] are selected.

$$S_m^\circ(U_2O_2Cl_5, cr, 298.15\,K) = (326.3 \pm 8.4)\,J \cdot K^{-1} \cdot mol^{-1}$$
$$C_{p,m}^\circ(U_2O_2Cl_5, cr, 298.15\,K) = (219.4 \pm 5.0)\,J \cdot K^{-1} \cdot mol^{-1}$$

The Gibbs energy of formation is calculated from the selected enthalpy of formation and entropy.

$$\Delta_f G_m^\circ(U_2O_2Cl_5, cr, 298.15\,K) = -(2037.3 \pm 4.9)\,kJ \cdot mol^{-1}$$

k) $U_5O_{12}Cl(cr)$

Cordfunke *et al.* [85COR/VLA] calorimetrically determined the enthalpy of solution of orthorombic $U_5O_{12}Cl(cr)$, which is a uranium(V) compound. This is reduced to the selected value of

$$\Delta_f H^\circ_m(U_5O_{12}Cl, cr, 298.15\,K) = -(5854.4 \pm 8.6)\,kJ \cdot mol^{-1}.$$

using selected auxiliary data.

In the absence of experimental data, the entropy was estimated by these authors by comparison with other uranium oxide chlorides as

$$S^\circ_m(U_5O_{12}Cl, cr, 298.15\,K) = (465 \pm 30)\,J \cdot K^{-1} \cdot mol^{-1}.$$

The Gibbs energy of formation is calculated from the selected enthalpy of formation and entropy.

$$\Delta_f G^\circ_m(U_5O_{12}Cl, cr, 298.15\,K) = -(5518.0 \pm 12.4)\,kJ \cdot mol^{-1}$$

l) $UO_2ClOH \cdot 2H_2O(cr)$

Only the dihydrate of $UO_2ClOH(cr)$ has been thermodynamically characterized. Prins [73PRI] measured $\Delta_{sol}H_m$ of both $UO_2Cl_2 \cdot 3H_2O(cr)$ and $UO_2ClOH \cdot 2H_2O(cr)$ in 5.55 M HCl. Combining these values and the selected value of $\Delta_f H^\circ_m(UO_2Cl_2 \cdot 3H_2O, cr, 298.15\,K)$, Fuger *et al.* [83FUG/PAR] calculated the selected value of

$$\Delta_f H^\circ_m(UO_2ClOH \cdot 2H_2O, cr, 298.15\,K) = -(2010.4 \pm 1.7)\,kJ \cdot mol^{-1}.$$

This review estimates, according to Appendix D, the following value for the entropy:

$$S^\circ_m(UO_2ClOH \cdot 2H_2O, cr, 298.15\,K) = (236 \pm 14)\,J \cdot K^{-1} \cdot mol^{-1}.$$

The Gibbs energy of formation estimate is calculated from the estimated entropy and selected enthalpy of formation data.

$$\Delta_f G^\circ_m(UO_2ClOH \cdot 2H_2O, cr, 298.15\,K) = -(1782.2 \pm 4.5)\,kJ \cdot mol^{-1}$$

No heat capacity data are available for this compound.

V.4.2.2. Uranium chlorites

V.4.2.2.1. Aqueous uranium chlorites

a) Aqueous U(VI) and U(V) chlorites

Gordon and Kern [64GOR/KER] studied the formation of chlorite complexes of dioxouranium(VI). Their data indicate the formation of a weak complex $UO_2ClO_2^+$. This review estimates, from this study, a formation constant of $\log_{10}\beta_1 \approx -0.5$ at $I = 1.0$ M and 25°C. This value is very uncertain, and no selection is made by this review.

No aqueous chlorite complexes of UO_2^+ are identified.

b) Aqueous U(IV) chlorites

Buchacek and Gordon [72BUC/GOR] found that the rate law for the oxidation of uranium(IV) by $HClO_2$ could be interpreted assuming the formation of a weak complex according to the reaction $U^{4+} + HClO_2(aq) \rightleftharpoons UOClOH^{4+}$ with $K = (4.13 \pm 0.25)$, *cf.* Appendix A. However, the complex is an intermediate which is rapidly oxidized to UO_2^{2+}. This review therefore does not include any equilibrium data for uranium(IV)-chlorite species.

V.4.2.2.2. Solid uranium chlorites

No evaluation of thermodynamic data on solid uranium chlorites of any oxidation has been made in the literature.

V.4.2.3. Uranium chlorates

V.4.2.3.1. Aqueous uranium chlorates

Khalili, Choppin and Rizkalla [88KHA/CHO] determined the equilibrium constant for Reaction (V.88)

$$UO_2^{2+} + ClO_3^- \rightleftharpoons UO_2ClO_3^+ \qquad (V.88)$$

and gave $\beta_1(V.88) = (1.2 \pm 0.2)$ at 25°C. The correction to $I = 0$ is done by this review (see Appendix A) and results in

$$\log_{10} \beta_1^\circ(V.88, 298.15\ K) = 0.50 \pm 0.07,$$

which is selected here.

The Gibbs energy of formation is calculated using the selected auxiliary data given in Section V.2.1 and Chapter VI.

$$\Delta_f G_m^\circ(UO_2ClO_3^+, aq, 298.15\,K) = -(963.3 \pm 2.2)\,kJ \cdot mol^{-1}$$

The enthalpy value from Ref. [88KHA/CHO] is tentatively accepted here, assuming the value reported for a 0.1 M $NaClO_4$ medium can also be used at $I = 0$.

$$\Delta_r H_m^\circ(V.88,\ 298.15\ K) = -(3.9 \pm 0.3)\,kJ \cdot mol^{-1}$$

The resulting standard enthalpy of formation is

$$\Delta_f H_m^\circ(UO_2ClO_3^+, aq, 298.15\,K) = -(1126.9 \pm 1.8)\,kJ \cdot mol^{-1}.$$

Using the selected reaction data, the entropy is calculated via $\Delta_r S_m^\circ$:

$$S_m^\circ(UO_2ClO_3^+, aq, 298.15\,K) = (60.6 \pm 4.6)\,J \cdot K^{-1} \cdot mol^{-1}.$$

No data are available on the aqueous chlorate complexes of either UO_2^+ or U^{4+}.

V.4.2.3.2. Solid uranium chlorates

No evaluation of thermodynamic data on solid uranium chlorates of any oxidation state has been made in the literature.

V.4.2.4. Uranium perchlorates

V.4.2.4.1. Aqueous uranium perchlorates

No data are available on the aqueous perchlorate complexes of either UO_2^{2+} or UO_2^+.

Vdovenko, Romanov and Shcherbakov [63VDO/ROM3] studied the perchlorate complexation of U^{4+}. They found that these complexes are very weak and reported $\log_{10} \beta_1(\text{V.89}) = -0.9$ in 2 M $HClO_4$.

$$U^{4+} + ClO_4^- \rightleftharpoons UClO_4^{3+} \qquad (\text{V.89})$$

It is difficult to distinguish between the effects of complex formation and variations in the activity coefficients under conditions where such weak complexes may be formed. Because of the very low stability of this species, and the absence of experimental confirmation of this study, no data for $UClO_4^{3+}$ are selected in this review.

A warning: If a value for this constant is included in a data base, it will be incompatible with the values calculated from data selected in this review, except (strictly) at $I = 0$.

V.4.2.4.2. Solid uranium perchlorates

No solid perchlorates of uranium at any oxidation state are credited in this review.

V.4.3. Bromine compounds and complexes

V.4.3.1. Uranium bromides

V.4.3.1.1. Gaseous uranium bromides

a) UBr(g) and UBr₂(g)

The enthalpies of formation of these species are derived from an analysis of the mass spectrometric study by Lau and Hildenbrand [87LAU/HIL] of the following molecular exchange reactions:

$$\begin{aligned} U(g) + Br(g) &\rightleftharpoons UBr(g) && (\text{V.90}) \\ U(g) + UBr_2(g) &\rightleftharpoons 2UBr(g) && (\text{V.91}) \\ UBr(g) + UBr_3(g) &\rightleftharpoons 2UBr_2(g) && (\text{V.92}) \\ UBr_2(g) + UBr_4(g) &\rightleftharpoons 2UBr_3(g). && (\text{V.93}) \end{aligned}$$

A third-law analysis of the equilibria yields the following values:

$$\Delta_r H^\circ_m(\text{V.90, 298.15 K}) = -(398.0 \pm 0.5)\,\text{kJ}\cdot\text{mol}^{-1}$$
$$\Delta_r H^\circ_m(\text{V.91, 298.15 K}) = -(12.3 \pm 1.0)\,\text{kJ}\cdot\text{mol}^{-1}$$
$$\Delta_r H^\circ_m(\text{V.92, 298.15 K}) = (47.6 \pm 0.9)\,\text{kJ}\cdot\text{mol}^{-1}$$
$$\Delta_r H^\circ_m(\text{V.93, 298.15 K}) = -(83.2 \pm 0.5)\,\text{kJ}\cdot\text{mol}^{-1}$$

Our selected value for the enthalpy of formation of UBr(g) is solely based on Eq. (V.90), which is subjected to the least uncertainties in the auxiliary data.

$$\Delta_f H^\circ_m(\text{UBr, g, 298.15 K}) = (247 \pm 17)\,\text{kJ}\cdot\text{mol}^{-1}$$

The second-law values given by Lau and Hildenbrand [87LAU/HIL] for the above differ considerably, however, they do not include the electronic contributions to the thermal functions. The enthalpy of formation of UBr_2(g) can now be obtained from Eq. (V.91), for which $\Delta_f H^\circ_m(\text{UBr}_2\text{, g, 298.15 K}) = -(27 \pm 35)\,\text{kJ}\cdot\text{mol}^{-1}$ is obtained. Solving the thermochemical cycles for Reactions (V.92) and (V.93) yields $\Delta_f H^\circ_m(\text{UBr}_2\text{, g, 298.15 K}) = -(35 \pm 37)\,\text{kJ}\cdot\text{mol}^{-1}$ and $\Delta_f H^\circ_m(\text{UBr}_3\text{, g, 298.15 K}) = -(364\pm37)\,\text{kJ}\cdot\text{mol}^{-1}$, respectively. The two values for UBr_2(g) are in good agreement and the avarage value is recommended:

$$\Delta_f H^\circ_m(\text{UBr}_2\text{, g, 298.15 K}) = -(31 \pm 25)\,\text{kJ}\cdot\text{mol}^{-1}$$

The thermal functions of UBr and UBr_2(g) are calculated from molecular constants estimated by Hildenbrand, Gurvich and Yungman [85HIL/GUR]; the electronic contribution is assumed to be identical to the corresponding uranium chlorides.

$$S^\circ_m(\text{UBr, g, 298.15 K}) = (278 \pm 10)\,\text{J}\cdot\text{K}^{-1}\cdot\text{mol}^{-1}$$
$$S^\circ_m(\text{UBr}_2\text{, g, 298.15 K}) = (359 \pm 10)\,\text{J}\cdot\text{K}^{-1}\cdot\text{mol}^{-1}$$

$$C^\circ_{p,m}(\text{UBr, g, 298.15 K}) = (44.1 \pm 5.0)\,\text{J}\cdot\text{K}^{-1}\cdot\text{mol}^{-1}$$
$$C^\circ_{p,m}(\text{UBr}_2\text{, g, 298.15 K}) = (61.4 \pm 5.0)\,\text{J}\cdot\text{K}^{-1}\cdot\text{mol}^{-1}$$

The Gibbs energies of formation are calculated from the selected enthalpies of formation and entropies.

$$\Delta_f G^\circ_m(\text{UBr, g, 298.15 K}) = (202 \pm 17)\,\text{kJ}\cdot\text{mol}^{-1}$$
$$\Delta_f G^\circ_m(\text{UBr}_2\text{, g, 298.15 K}) = -(78 \pm 25)\,\text{kJ}\cdot\text{mol}^{-1}$$

b) UBr_3 (g)

Values for the enthalpy of formation of UBr_3(g) are derived from mass-spectrometric studies of molecular exchange reactions as well as vapour pressure measurements. Lau and Hildenbrand [84LAU/HIL] studied the reactions between the lower valent UBr_q species ($q = 1$ to 4). Analysis of the equilibria as discussed in Paragraph a) yields $\Delta_f H^\circ_m(\text{UBr}_3\text{, g, 298.15 K}) = -(364 \pm 37)\,\text{kJ}\cdot\text{mol}^{-1}$. Vapour pressure measurements for the reaction

$$\text{UBr}_3(\text{cr, l}) \rightleftharpoons \text{UBr}_3(\text{g}) \qquad \text{(V.94)}$$

were measured by Altman [43ALT2] and by Webster [43WEB], which are in poor agreement. However, Webster used a mixture of U(cr) and UBr_3(cr) to suppress dissociation, and his results are therefore rejected here. By third-law analysis of Altman's results $\Delta_{sub}H^{\circ}_{m}(V.94, 298.15\ K) = (289.3 \pm 1.2)\ kJ \cdot mol^{-1}$ is derived, leading to $\Delta_f H^{\circ}_{m}(UBr_3, g, 298.15\ K) = -(400.4 \pm 6.0)\ kJ \cdot mol^{-1}$, the absolute uncertainty being estimated here. This value is in poor agreement with the results of Lau and Hildenbrand. The selected value of

$$\Delta_f H^{\circ}_{m}(UBr_3, g, 298.15\ K) = -(364 \pm 37)\ kJ \cdot mol^{-1}$$

is based on the results of Lau and Hildenbrand [84LAU/HIL], as Altman's measurements for UF_4 and UCl_3 [43ALT] are also deviating significantly.

The heat capacity and entropy of UBr_3(g) are calculated from molecular constants for the vibrational and rotational parameters, as estimated by Hildenbrand, Gurvich and Yungman [85HIL/GUR]. The electronic contribution was assumed to be identical to UCl_3(g).

$$S^{\circ}_{m}(UBr_3, g, 298.15\ K) = (403 \pm 10)\ J \cdot K^{-1} \cdot mol^{-1}$$
$$C^{\circ}_{p,m}(UBr_3, g, 298.15\ K) = (85.1 \pm 5.0)\ J \cdot K^{-1} \cdot mol^{-1}$$

The Gibbs energy of formation is calculated from the selected enthalpy of formation and entropy.

$$\Delta_f G^{\circ}_{m}(UBr_3, g, 298.15\ K) = -(401 \pm 37)\ kJ \cdot mol^{-1}$$

c) UBr_4 (g)

The enthalpy of formation of UBr_4(g) is derived from experimental vapour pressure data (*cf.* Table V.27) for the sublimation equilibrium

$$UBr_4(cr) \rightleftharpoons UBr_4(g). \qquad (V.95)$$

A third-law analysis shows the reasonable agreement between the studies, and the value of

$$\Delta_{sub}H^{\circ}_{m}(V.95, 298.15\ K) = (192.0 \pm 8.0)\ kJ \cdot mol^{-1}$$

is selected here. Most weight is given to the effusion study by Hildenbrand and Lau [91HIL/LAU] and the boiling point study by Prasad *et al.* [79PRA/NAG], which are considered to be the most reliable. The uncertainty is raised to reflect the uncertainties in the thermal functions of the gaseous and condensed phases. Combining this value with the enthalpy of formation of the solid, gives

$$\Delta_f H^{\circ}_{m}(UBr_4, g, 298.15\ K) = -(610.1 \pm 8.4)\ kJ \cdot mol^{-1}.$$

Table V.27: Enthalpy of sublimation values for $UBr_4(cr) \rightleftharpoons UBr_4(g)$ at 298.15 K, as evaluated from third-law analysis of experimental data.

Reference	T-range (K)	$\Delta_{sub}H^{\circ}_{m}(V.95, 298.15\,K)$[(a)] ($kJ\cdot mol^{-1}$)
[42THO/SCH]	573-723	190.1 ± 0.9
[44NOT/POW]	723-898	192.1 ± 1.7
[46GRE3]	815-1033	189.1 ± 0.5
		188.5 ± 0.4
[79PRA/NAG]	759-1004	191.4
[91HIL/LAU]	579-693	191.91 ± 0.04

(a) The indicated errors represent the statistical uncertainties.

The heat capacity and entropy of $UBr_4(g)$ are calculated for a distorted tetrahedral structure as recently established by electron diffraction measurements by Ezhov *et al.* [89EZH/BAZ]. The authors also estimated the vibrational frequencies of the molecule using the only experimental value for the U-Br stretching frequency [83KOV/CHI], $\nu(\text{U-Br}) = 230\ cm^{-1}$, as a guidance. In the present calculation the data of Ezhov *et al.* are combined with estimated electronic energy levels obtained by comparison with UCl_4 and from the matrix isolation electronic study by Clifton, Grün and Ron [69CLI/GRU].

$$S^{\circ}_{m}(UBr_4,\ g,\ 298.15\ K) = (442.1 \pm 5.0)\ J\cdot K^{-1}\cdot mol^{-1}$$
$$C^{\circ}_{p,m}(UBr_4,\ g,\ 298.15\ K) = (110.3 \pm 3.0)\ J\cdot K^{-1}\cdot mol^{-1}$$

The Gibbs energy of formation is calculated from the selected enthalpy of formation and entropy.

$$\Delta_f G^{\circ}_{m}(UBr_4,\ g,\ 298.15\ K) = -(636.2 \pm 8.5)\ kJ\cdot mol^{-1}$$

d) $UBr_5(g)$

Lau and Hildenbrand [87LAU/HIL] studied the equilibrium

$$UBr_4(g) + \tfrac{1}{2}Br_2(g) \rightleftharpoons UBr_5(g) \qquad (V.96)$$

by mass spectrometry. They argued that the second-law enthalpy of reaction should be used since the electronic level distribution of $UBr_5(g)$ is not known, and they derived $\Delta_r H^{\circ}_{m}(V.96,\ 298.15\ K) = -(70.3 \pm 6.3)\ kJ\cdot mol^{-1}$. Assuming the electronic level distribution in $UBr_5(g)$ to be the same as in $UCl_5(g)$ [84GUR/DOR], this review obtains, for the third-law enthalpy of reaction, $\Delta_r H_m^\circ$(V.96, 298.15 K) = $-(43.1 \pm 2.2)\,kJ\cdot mol^{-1}$. The latter value is selected here to obtain consistency with the selected values for the heat capacity and entropy, but its uncertainty is raised to $\pm 15\,kJ\cdot mol^{-1}$:

$$\Delta_r H_m^\circ(\text{V.96, 298.15 K}) = -(43.1 \pm 15.0)\,\text{kJ}\cdot\text{mol}^{-1}.$$

Using the selected enthalpy of formation of UBr_4(g), this results in resulting in

$$\Delta_f H_m^\circ(\text{UBr}_5\text{, g, 298.15 K}) = -(637.7 \pm 17.2)\,\text{kJ}\cdot\text{mol}^{-1}.$$

The entropy and heat capacity of UBr_5(g) are calculated from molecular constants estimated by Lau and Hildenbrand [87LAU/HIL] for a square pyramidal structure. The electronic energy level distribution is assumed to be the same as for UCl_5 [84GUR/DOR]. However, this should not affect the values at 298.15 K. This review thus obtains the values

$$S_m^\circ(\text{UBr}_5\text{, g, 298.15 K}) = (493 \pm 10)\,\text{J}\cdot\text{K}^{-1}\cdot\text{mol}^{-1}$$
$$C_{p,m}^\circ(\text{UBr}_5\text{, g, 298.15 K}) = (129 \pm 5)\,\text{J}\cdot\text{K}^{-1}\cdot\text{mol}^{-1}.$$

The uncertainties are assigned by the present review.

The Gibbs energy of formation is calculated from the selected enthalpy of formation and entropy.

$$\Delta_f G_m^\circ(\text{UBr}_5\text{, g, 298.15 K}) = -(656.3 \pm 17.4)\,\text{kJ}\cdot\text{mol}^{-1}$$

V.4.3.1.2. Aqueous uranium bromides

a) Aqueous U(VI) bromides

The results of the two reliable experimental studies of the equilibrium

$$\text{UO}_2^{2+} + \text{Br}^- \rightleftharpoons \text{UO}_2\text{Br}^+ \qquad \text{(V.97)}$$

are presented in Table V.28. The data of Ahrland [51AHR] are assumed to be valid also at 25°C, and the primary data measured by Davies and Monk [57DAV/MON] are re-evaluated, *cf.* Appendix A. These data are extrapolated to $I = 0$ (after conversion to molality units) using the specific ion interaction theory assuming that $\Delta\varepsilon$(V.97) = $-(0.25 \pm 0.02)$ as evaluated for the corresponding chloride reaction. This leads to a weighted mean value of

$$\log_{10}\beta_1^\circ(\text{V.97, 298.15 K}) = 0.22 \pm 0.02.$$

The selected Gibbs energy of formation,

$$\Delta_f G_m^\circ(\text{UO}_2\text{Br}^+\text{, aq, 298.15 K}) = -(1057.7 \pm 1.8)\,\text{kJ}\cdot\text{mol}^{-1},$$

Table V.28: Experimental equilibrium data for the uranium(VI) bromide system.

Method	Ionic Medium	t (°C)	$\log_{10}\beta^{(a)}$	$\log_{10}\beta^{\circ(b)}$	Reference
$UO_2^{2+} + Br^- \rightleftharpoons UO_2Br^+$					
emf	1 M Na(Br, ClO_4)	20	-0.30 ± 0.17	0.24 ± 0.17	[51AHR]
sp	0.532 M Na(Br, ClO_4)	25	$-0.37\pm0.04^{(c)}$	0.20 ± 0.05	[57DAV/MON]
	0.367 M Na(Br, ClO_4)	25	$-0.36\pm0.04^{(c)}$	0.19 ± 0.04	
	0.317 M Na(Br, ClO_4)	25	$-0.30\pm0.04^{(c)}$	0.24 ± 0.04	
	0.217 M Na(Br, ClO_4)	25	$-0.24\pm0.04^{(c)}$	0.26 ± 0.04	
	0.168 M Na(Br, ClO_4)	25	$-0.28\pm0.04^{(c)}$	0.19 ± 0.04	
mvd	1 M Na(Br, ClO_4)	25?	-0.70		[74JED]
	2 M Na(Br, ClO_4)	25?	-0.70		
	3 M Na(Br, ClO_4)	25?	-0.52		
	5 M Na(Br, ClO_4)	25?	-0.40		

(a) $\log_{10}\beta$ refers to the ionic strength and temperature given in the table.
(b) $\log_{10}\beta^\circ$ (in molal units) refers to the equilibrium constant corrected to $I = 0$ at 298.15 K.
(c) Recalculated from primary data given in Ref. [57DAV/MON, Table 3].

is obtained from the auxiliary data selected in this review.

No enthalpy data are available for this species.

Although data have been estimated for higher order dioxouranium(VI) bromide species [87BRO/WAN], they are not credited in this review since they are not experimentally based. No other enthalpy, entropy or heat capacity data are available.

b) Aqueous U(V) bromides

No aqueous dioxouranium(V) bromide species were identified.

c) Aqueous U(IV) bromides

Reliable experimental data are available only for the equilibrium

$$U^{4+} + Br^- \rightleftharpoons UBr^{3+}. \qquad (V.98)$$

Assuming $\Delta\varepsilon$(V.98) to be the same as in the corresponding chloride reaction, $\Delta\varepsilon(V.98) = -(0.29\pm0.08)$, the equilibrium data of Vdovenko, Romanov and Shcherbakov [63VDO/ROM3] and Ahrland and Larsson [54AHR/LAR2] listed in Table V.29 are corrected to $I = 0$ using the specific ion interaction theory, after conversion to molality constants. These values refer to 20°C, but they are assumed to be valid also at 25°C. The resulting weighted mean of these two values is selected here:

$$\log_{10} \beta_1^\circ(\text{V.98}, 298.15 \text{ K}) = 1.46 \pm 0.20.$$

This constant leads to the selected Gibbs energy of formation,

$$\Delta_f G_m^\circ(\text{UBr}^{3+}, \text{aq}, 298.15\,\text{K}) = -(642.0 \pm 2.1)\,\text{kJ} \cdot \text{mol}^{-1}.$$

Although values have been proposed for higher aqueous uranium(IV) bromides, these species are not credited in this review.

Table V.29: Experimental equilibrium data for the uranium(IV) bromide system.

Method	Ionic Medium	t (°C)	$\log_{10} \beta_1^{(a)}$	$\log_{10} \beta_1^{\circ(b)}$	Reference
$U^{4+} + Br^- \rightleftharpoons UBr^{3+}$					
emf	1 M $(H, Na)(Br, ClO_4)$	20	0.18 ± 0.14	1.50 ± 0.16	[54AHR/LAR2]
nmr	2.5 M $H(ClO_4, Br)$	20	$0.30 \pm 0.13^{(c)}$	1.37 ± 0.26	[63VDO/ROM3]

(a) $\log_{10} \beta_1$ refers to the ionic strength and temperature given in the table.
(b) $\log_{10} \beta_1^\circ$ (in molal units) refers to the equilibrium constant corrected to $I = 0$ and assumed to be valid at 298.15 K.
(c) Uncertainty estimated in this review, *cf.* Appendix A.

d) Aqueous U(III) bromides

Shiloh and Marcus [65SHI/MAR] report the formation of UBr^{2+}, with $\beta_1 = (1.1 \pm 0.2) \times 10^{-4}$ at 25°C in very concentrated LiBr solutions ($I > 6$ M). This value cannot be extrapolated to $I = 0$, and no value for the standard Gibbs energy of formation of UBr^{2+} is therefore reported in this review.

V.4.3.1.3. Solid uranium bromides

In their recent review for the IAEA, Fuger *et al.* [83FUG/PAR] thoroughly reviewed the available data on the three binary uranium bromide solid compounds. The basis for their selections is presented in Table V.30. Their selected values are accepted in this review and their analysis is summarized here. Note that although Krestov [72KRE] has compiled thermodynamic data on UBr_2(cr) and UBr_6(cr), insufficient basis exists for this review to credit these compounds.

Table V.30: Enthalpy data considered for $UBr_3(cr)$, $UBr_4(cr)$ and $UBr_5(cr)$.

Reference	Reaction	$\Delta_f H_m^\circ(UBr_q, cr, 298.15 K)$ $(kJ \cdot mol^{-1})$
		$q = 3$:
[44ALT, 45GRE]	$UBr_4(cr) + \frac{1}{2}H_2(g) \rightleftharpoons UBr_3(cr) + HBr(g)$	-698.7[c]
[45BAR]	$UBr_3(cr) + 3HCl(sln) \rightleftharpoons UCl_3(cr) + 3HBr(sln)$	-700.4 ± 0.8[a]
[45BAR]	$UBr_3(cr) + FeCl_3(aq) + HBr(sln) \rightleftharpoons UBr_4(cr) + FeCl_2(aq) + HCl(sln)$	-693.7[b]
[78SUG/CHI2]	$UBr_3(cr) + 4HCl(sln) \rightleftharpoons UCl_4(cr) + 3HBr(sln) + \frac{1}{2}H_2(g)$	-695.0 ± 3.3[d,e]
		$q = 4$:
[45BAR]	$UBr_4(cr) + 4HCl(sln) \rightleftharpoons UCl_4(cr) + 4HBr(sln)$	-802.5 ± 8.4[f,i]
[59SHC/VAS]	$UBr_4(cr) + 4HCl(sln) \rightleftharpoons UCl_4(cr) + 4HBr(sln)$	-805.6 ± 3.3[f,h]
[73FUG/BRO]	$UBr_4(cr) + 4HCl(sln) \rightleftharpoons UCl_4(cr) + 4HBr(sln)$	-802.1 ± 2.9[f,g]
[73VDO/SUG]	$UBr_4(cr) + 4HCl(sln) \rightleftharpoons UCl_4(cr) + 4HBr(sln)$	-800.8 ± 2.5[f,g]
[86SUG/CHI]	$UBr_4(cr) + 4HCl(sln) \rightleftharpoons UCl_4(cr) + 4HBr(sln)$	-799.9 ± 3.1[h,m]
		$q = 5$:
[73BLA/IHL]	$UBr_5(cr) \rightleftharpoons UBr_4(cr) + \frac{1}{2}Br_2(g)$	-811.7[j,k]
		-809.2[j,l]

(a) Data reported by MacWood [58MAC]. In 12 M HCl/10% $FeCl_3$ solution. Fuger *et al.* [83FUG/PAR] used auxiliary data for HCl and HBr appropriate for the solution.
(b) Same data as above, related to $UBr_4(cr)$ through its enthalpy of solution.
(c) Based on equilibrium measurements over the range 648 to 798 K.
(d) From enthalpy of solution in 4 M HCl containing 0.005 M Na_2SiF_6 and 1.5×10^{-6} M H_2PtCl_6, combined with the enthalpy of solution of $UCl_4(cr)$ in the same medium.
(e) The value reported by Suglobova and Chirkst [78SUG/CHI2] was not considered in the IAEA selection. This, however, does not lead to a revised value for this review.
(f) From the enthalpy of solution of $UBr_4(cr)$ and $UCl_4(cr)$ and the selected value of $\Delta_f H_m^\circ(UCl_4, cr)$, and auxiliary data appropriate for the solution.
(g) In 1 M HCl solution.
(h) In 2% HCl/0.5% $FeCl_3$ solution.
(i) In 12 M HCl/10% $FeCl_3$ solution.
(j) From thermal decomposition data obtained in the range 298 to 400 K. Above 383 K, $UBr_5(cr)$ decomposes irreversibly to $UBr_4(cr)$ and $Br_2(l)$.
(k) Third law treatment of the vapour pressure data below 383 K.
(l) Second law treatment of the same data.
(m) Recalculated using the value of $\Delta_f H_m^\circ(UCl_4, cr, 298.15 K)$ selected in this review. This result does not, however, lead to a revision of the selected value for $\Delta_f H_m^\circ(UBr_4, cr, 298.15 K)$.

a) UBr_3 (cr)

Fuger *et al.* [83FUG/PAR] based their selection of $\Delta_r H_m^\circ(UBr_3, cr, 298.15\,K)$ on the enthalpy of solution measurements of Barkelew [45BAR] and the temperature dependence data of the hydrogen reduction reaction of UBr_4(cr) to UBr_3(cr) of Altman [44ALT] and Gregory [45GRE], all reported by MacWood [58MAC]. The two results are consistent and lead to a selected value of

$$\Delta_f H_m^\circ(UBr_3, cr, 298.15\ K) = -(698.7 \pm 4.2)\,kJ \cdot mol^{-1}.$$

The more recent solution calorimetry result by Suglobova and Chirkst [86SUG/CHI] yields $\Delta_f H_m^\circ(UBr_3, cr, 298.15\ K) = -(686.1 \pm 3.3)\,kJ \cdot mol^{-1}$ when recalculated using $\Delta_f H_m^\circ(UCl_4, cr, 298.15\ K)$ selected in this review. This rather discordant value is not used in the present evaluation.

Low-temperature heat capacity measurements of UBr_3(cr) were reported by Cordfunke, Konings and Westrum [89COR/KON]. Their measurements result in the following recommended values:

$$\begin{aligned} S_m^\circ(UBr_3, cr, 298.15\ K) &= (192.98 \pm 0.50)\,J \cdot K^{-1} \cdot mol^{-1} \\ C_{p,m}(UBr_3, cr, 298.15\ K) &= (105.83 \pm 0.50)\,J \cdot K^{-1} \cdot mol^{-1}. \end{aligned}$$

The Gibbs energy of formation is calculated from the selected enthalpy of formation and entropy.

$$\Delta_f G_m^\circ(UBr_3, cr, 298.15\,K) = -(673.2 \pm 4.2)\,kJ \cdot mol^{-1}$$

The $C_{p,m}^\circ(T)$ function is from the IAEA evaluation [83FUG/PAR], slightly adjusted to fit the selected low-temperature data.

b) UBr_4 (cr)

All four of the experiments upon which the IAEA [83FUG/PAR] selected value of $\Delta_f H_m^\circ(UBr_4, cr, 298.15\ K)$ is based are determinations of the enthalpies of solution of UCl_4(cr) and UBr_4(cr) in the same medium. As can be seen from the summary in Table V.30, the various authors used different aqueous HCl solutions. The value proposed by Fuger *et al.* [83FUG/PAR] is selected here:

$$\Delta_f H_m^\circ(UBr_4, cr, 298.15\,K) = -(802.1 \pm 2.5)\,kJ \cdot mol^{-1}.$$

This selection is further supported by the more recent solution calorimetry result of Suglobova and Chirkst [86SUG/CHI], which yields $\Delta_f H_m^\circ(UBr_4, cr, 298.15\ K) = -(799.9 \pm 3.1)\,kJ \cdot mol^{-1}$ after recalculation with $\Delta_f H_m^\circ(UCl_4, cr, 298.15\ K)$ selected in this review.

The selected heat capacity function from Fuger *et al.* [83FUG/PAR] was based on Krestov [72KRE]. The selected values of

$$\begin{aligned} S_m^\circ(UBr_4, cr, 298.15\,K) &= (238.5 \pm 8.4)\,J \cdot K^{-1} \cdot mol^{-1} \\ C_{p,m}^\circ(UBr_4, cr, 298.15\,K) &= (128.0 \pm 4.2)\,J \cdot K^{-1} \cdot mol^{-1} \end{aligned}$$

are derived from this function.

The Gibbs energy of formation at 298.15 K is derived from the selected enthalpy of formation and entropy.

$$\Delta_f G^\circ_m(\text{UBr}_4, \text{cr}, 298.15\,\text{K}) = -(767.5 \pm 3.5)\,\text{kJ}\cdot\text{mol}^{-1}$$

c) UBr_5 (cr)

As did Fuger *et al.* [83FUG/PAR], this review selects

$$\Delta_f H^\circ_m(\text{UBr}_5, \text{cr}, 298.15\,\text{K}) = -(810.4 \pm 8.4)\,\text{kJ}\cdot\text{mol}^{-1}$$

which is the average of the second-law and third-law treatment of the thermal decomposition data (between 298.15 and 400 K) of Blair and Ihle [73BLA/IHL]. The estimated entropy and heat capacity selected are taken from Krestov [72KRE].

$$S^\circ_m(\text{UBr}_5, \text{cr}, 298.15\,\text{K}) = (292.9 \pm 12.6)\,\text{J}\cdot\text{K}^{-1}\cdot\text{mol}^{-1}$$
$$C^\circ_{p,m}(\text{UBr}_5, \text{cr}, 298.15\,\text{K}) = (160.7 \pm 8.0)\,\text{J}\cdot\text{K}^{-1}\cdot\text{mol}^{-1}$$

The heat capacity function tabulated by Fuger *et al.* [83FUG/PAR] and based on Krestov [72KRE] is also selected.

The Gibbs energy of formation is calculated from the selected enthalpy of formation and entropy.

$$\Delta_f G^\circ_m(\text{UBr}_5, \text{cr}, 298.15\,\text{K}) = -(769.3 \pm 9.2)\,\text{kJ}\cdot\text{mol}^{-1}$$

d) UOBr(cr)

The only thermodynamic data available for UOBr(cr) were tabulated by Katz and Rabinovitch [51KAT/RAB] from results of Gregory [45GRE] on the reduction of $UOBr_2$(cr) by hydrogen. Fuger *et al.* [83FUG/PAR] analyzed these data together with those reported by the same author for UOCl(cr). They found the value $\Delta_f H^\circ_m$(UOBr, cr, 298.15 K) $= -849\,\text{kJ}\cdot\text{mol}^{-1}$ inconsistent (too negative) with what would be expected from the known differences, [$\Delta_f H^\circ_m$(PuOCl, cr, 298.15 K) $-$ $\Delta_f H^\circ_m$($PuCl_3$, cr, 298.15 K)], [$\Delta_f H^\circ_m$(PuOBr, cr, 298.15 K) $-$ $\Delta_f H^\circ_m$($PuBr_3$, cr, 298.15 K)], and [$\Delta_f H^\circ_m$(AmOCl, cr, 298.15 K) $-$ $\Delta_f H^\circ_m$($AmCl_3$, cr, 298.15 K)]. They therefore rejected the results of Gregory [45GRE] for UOBr(cr) as well as for UOCl(cr). This review adopts the same position.

e) $UOBr_2$ (cr)

The only solid uranium(IV) bromide, other than UBr_4(cr), for which reliable chemical thermodynamic data are available, is $UOBr_2$(cr). Müller [48MUE] reported the unpublished equilibrium vapour pressure measurements obtained by Gregory [45GRE] for $UBr_4(g)$ over $UOBr_2(cr)$ in the range 710 to 960 K. From this function it can be calculated that $\Delta_r H_m(V.99) = 207.9\,kJ\cdot mol^{-1}$ at 835 K.

$$2UOBr_2(cr) \rightleftharpoons UBr_4(g) + UO_2(cr) \qquad (V.99)$$

Fuger *et al.* [83FUG/PAR] showed that $\Delta_r S^\circ_m(V.99,\ 298.15\ K) = 184\,J\cdot K^{-1}\cdot mol^{-1}$ derived from these experiments is inconsistent with $\Delta_r S^\circ_m(V.99,\ 298.15\ K) = 221.7\,J\cdot K^{-1}\cdot mol^{-1}$ obtained from the selected standard entropies of the products and reactants in Reaction (V.99) (for $S^\circ_m(UOBr_2,\ cr,\ 298.15\ K)$, see below). They therefore re-analyzed the experimental data with the latter $\Delta_r S^\circ_m(V.99,\ 298.15\ K)$ value as a constraint and obtained $\Delta_f H^\circ_m(UOBr_2,\ cr,\ 298.15\ K) = -970.7\,kJ\cdot mol^{-1}$. They also re-evaluated the experimental results of Barkelew [45BAR] on the oxidation of $UOBr_2(cr)$ and $UOCl_2(cr)$ in a 12 M HCl, 10% $FeCl_3$ solution. From this they could obtain $\Delta_f H^\circ_m(UOBr_2,\ cr,\ 298.15\ K) = -979.5\,kJ\cdot mol^{-1}$, which is in reasonable agreement with the result from the re-analysis of Gregory's [45GRE] data. Fuger *et al.* [83FUG/PAR] selected

$$\Delta_f H^\circ_m(UOBr_2, cr, 298.15\,K) = -(973.6 \pm 8.4)\,kJ\cdot mol^{-1},$$

which is also adopted in this review.

The selected entropy and heat capacity values at 298.15 K are adopted, as in the IAEA review [83FUG/PAR], from the heat capacity measurements of Greenberg and Westrum [56GRE/WES] from 5 to 339 K.

$$S^\circ_m(UOBr_2, cr, 298.15\,K) = (157.57 \pm 0.29)\,J\cdot K^{-1}\cdot mol^{-1}$$
$$C^\circ_{p,m}(UOBr_2, cr, 298.15\,K) = (98.0 \pm 0.4)\,J\cdot K^{-1}\cdot mol^{-1}$$

The $C^\circ_{p,m}(T)$ data tabulated by the IAEA [83FUG/PAR] are fitted between 298.15 and 600 K to obtain the selected $C^\circ_{p,m}(T)$ function listed in Table III.3.

The Gibbs energy of formation is calculated from the selected enthalpy of formation and entropy.

$$\Delta_f G^\circ_m(UOBr_2, cr, 298.15\,K) = -(929.6 \pm 8.4)\,kJ\cdot mol^{-1}$$

f) $UOBr_3$ (cr)

The only solid uranium(V) bromide, other than $UBr_5(cr)$, for which reliable chemical thermodynamic data are available is $UOBr_3(cr)$. The assessed values for the enthalpy of formation of this compound are based on an experiment from Shchukarev, Vasilkova and Drozdova [58SHC/VAS]. Those appearing before 1982 are equal to about $-987\,kJ\cdot mol^{-1}$, based on the evaluation of Rand and Kubaschewski [63RAN/KUB] while those after that date equal $-954\,kJ\cdot mol^{-1}$, apparently based on preliminary values from the IAEA [83FUG/PAR] review. The IAEA review [83FUG/PAR] re-evaluated the available experimental data [58SHC/VAS, 58SHC/VAS2] on the relative solubility of $UOBr_3(cr)$ and $UOCl_3(cr)$ in a (2% HCl, 0.5% $FeCl_3$) solution. Combination of this result with the rather uncertain selected value of $\Delta_f H^\circ_m(UOCl_3, cr, 298.15\,K)$ leads to

$$\Delta_f H_m^\circ(\mathrm{UOBr_3, cr, 298.15\,K}) = -(954 \pm 21)\,\mathrm{kJ \cdot mol^{-1}},$$

which is accepted in this review.

The selected entropy was estimated by Fuger *et al.* [83FUG/PAR] based on the entropies of similar and related chloride and bromide compounds.

$$S_m^\circ(\mathrm{UOBr_3, cr, 298.15\,K}) = (205.0 \pm 12.6)\,\mathrm{J \cdot K^{-1} \cdot mol^{-1}}$$

The heat capacity derived from the $C_{p,m}(\mathrm{UOBr_3, cr}, T)$ function estimated by Barin, Knacke and Kubaschewski [77BAR/KNA] was accepted by Fuger *et al.* [83FUG/PAR] and is also selected here.

$$C_{p,m}^\circ(\mathrm{UOBr_3, cr, 298.15\,K}) = (120.9 \pm 4.2)\,\mathrm{J \cdot K^{-1} \cdot mol^{-1}}$$

The $C_{p,m}^\circ(T)$ data tabulated by the IAEA [83FUG/PAR] are fitted between 298.15 and 600 K to obtain the selected $C_{p,m}^\circ(T)$ function listed in Table III.3.

The Gibbs energy of formation is calculated from the selected enthalpy of formation and entropy.

$$\Delta_f G_m^\circ(\mathrm{UOBr_3, cr, 298.15\,K}) = -(901.5 \pm 21.3)\,\mathrm{kJ \cdot mol^{-1}}$$

g) UO_2Br_2 (cr) and its hydrates

Prins, Cordfunke and Ouweltjes [78PRI/COR] and Shchukarev, Vasilkova and Drozdova [58SHC/VAS] measured the enthalpy of solution of UO_2Br_2(cr). As did Fuger *et al.* [83FUG/PAR], this review accepts the arguments of Parker [80PAR], and rejects the results from the latter experiment, which lead to $\Delta_f H_m^\circ(\mathrm{UO_2Br_2}$, cr, 298.15 K) $= -1128.8\,\mathrm{kJ \cdot mol^{-1}}$. The former results lead to the selected value of

$$\Delta_f H_m^\circ(\mathrm{UO_2Br_2, cr, 298.15\,K}) = -(1137.4 \pm 1.3)\,\mathrm{kJ \cdot mol^{-1}}$$

which is consistent with the enthalpies of formation selected for the monohydrate and trihydrate.

This review accepts the entropy estimate,

$$S_m^\circ(\mathrm{UO_2Br_2, cr, 298.15\,K}) = (169.5 \pm 4.2)\,\mathrm{J \cdot K^{-1} \cdot mol^{-1}},$$

made by Fuger *et al.* [83FUG/PAR] based on the selected entropies of $UOCl_2$(cr), $UOBr_2$(cr) and UO_2Cl_2(cr). Cordfunke, Muis and Prins [79COR/MUI] measured $[H_m^\circ(T) - H_m^\circ(298.15\,\mathrm{K})]$ in the range 345 to 454 K, from which Fuger *et al.* [83FUG/PAR] estimated the selected

$$C_{p,m}^\circ(\mathrm{UO_2Br_2, cr, 298.15\,K}) = (116 \pm 8)\,\mathrm{J \cdot K^{-1} \cdot mol^{-1}}.$$

The uncertainty has been assigned by this review. The $C_{p,m}^\circ(T)$ data tabulated by the IAEA [83FUG/PAR] are fitted between 298.15 and 600 K to obtain the selected $C_{p,m}^\circ(T)$ function listed in Table III.3.

The Gibbs energy of formation is calculated from the selected enthalpy of formation and entropy.

$$\Delta_f G_m^\circ(UO_2Br_2, cr, 298.15\,K) = -(1066.4 \pm 1.8)\,kJ \cdot mol^{-1}$$

As for UO_2Br_2(cr), the selection of the enthalpy of formation for the monohydrate and trihydrate is based on the enthalpy of solution results of Prins, Cordfunke and Ouweltjes [78PRI/COR] which yield

$$\Delta_f H_m^\circ(UO_2Br_2 \cdot H_2O, cr, 298.15\,K) = -(1455.9 \pm 1.4)\,kJ \cdot mol^{-1}$$
$$\Delta_f H_m^\circ(UO_2Br_2 \cdot 3H_2O, cr, 298.15\,K) = -(2058.0 \pm 1.5)\,kJ \cdot mol^{-1}.$$

The selected entropies were estimated by Fuger *et al.* [83FUG/PAR] from the $S_m^\circ(UO_2Br_2, cr, 298.15\,K)$ estimate by assuming an entropy of hydration of $39.75\,J \cdot K^{-1}$ per mole of H_2O. This review uses a value of $(44.7 \pm 4.0)\,J \cdot K^{-1} \cdot mol^{-1}$, *cf.* Appendix D.

$$S_m^\circ(UO_2Br_2 \cdot H_2O, cr, 298.15\,K) = (214 \pm 7)\,J \cdot K^{-1} \cdot mol^{-1}$$
$$S_m^\circ(UO_2Br_2 \cdot 3H_2O, cr, 298.15\,K) = (304 \pm 18)\,J \cdot K^{-1} \cdot mol^{-1}$$

The Gibbs energy of formation values are calculated from the selected enthalpy of formation and entropy values.

$$\Delta_f G_m^\circ(UO_2Br_2 \cdot H_2O, cr, 298.15\,K) = -(1328.6 \pm 2.5)\,kJ \cdot mol^{-1}$$
$$\Delta_f G_m^\circ(UO_2Br_2 \cdot 3H_2O, cr, 298.15\,K) = -(1818.5 \pm 5.6)\,kJ \cdot mol^{-1}$$

No heat capacity or thermal function data are available.

h) $UO_2BrOH \cdot 2H_2O$(cr)

Prins, Cordfunke and Ouweltjes [78PRI/COR] measured the enthalpy of solution of γ-UO_3 and $UO_2BrOH{\cdot}2H_2O$(cr) in the same HBr solution. Using auxiliary data for HBr(sln) and H_2O(sln) appropriate to that solution, Fuger *et al.* [83FUG/PAR] calculated the enthalpy of the reaction

$$\gamma\text{-}UO_3 + HBr(aq) + 2H_2O(l) \rightleftharpoons UO_2BrOH \cdot 2H_2O(cr). \qquad (V.100)$$

Using the selected $\Delta_f H_m^\circ(UO_3, \gamma, 298.15\,K)$ they obtained

$$\Delta_f H_m^\circ(UO_2BrOH \cdot 2H_2O, cr, 298.15\,K) = -(1958.2 \pm 1.3)\,kJ \cdot mol^{-1}.$$

This review estimates, according to Appendix D, the following value for the entropy:

$$S_m^\circ(UO_2BrOH \cdot 2H_2O, cr, 298.15\,K) = (248 \pm 14)\,J \cdot K^{-1} \cdot mol^{-1}.$$

The Gibbs energy of formation estimate is calculated from the estimated entropy and selected enthalpy of formation data.

$$\Delta_f G_m^\circ(UO_2BrOH \cdot 2H_2O, cr, 298.15\,K) = -(1744.2 \pm 4.4)\,kJ \cdot mol^{-1}$$

No heat capacity data are available for this compound.

V.4.3.2. Uranium bromates

V.4.3.2.1. Aqueous uranium bromates

Khalili, Choppin and Rizkalla [88KHA/CHO] have determined the equilibrium constant for Reaction (V.101)

$$UO_2^{2+} + BrO_3^- \rightleftharpoons UO_2BrO_3^+ \qquad (V.101)$$

and give β_1(V.101) = (1.6 ± 0.3) in 0.1 M $NaClO_4$ and at 25°C. The correction to $I = 0$ is done by this review (see Appendix A) and results in

$$\log_{10} \beta_1^\circ(\text{V.101}, 298.15\,\text{K}) = 0.63 \pm 0.08,$$

which is selected by this review.

The Gibbs energy of formation is calculated using the selected auxiliary data given in Section V.2.1 and Chapter VI.

$$\Delta_f G_m^\circ(UO_2BrO_3^+, \text{aq}, 298.15\,\text{K}) = -(937.1 \pm 1.9)\,\text{kJ} \cdot \text{mol}^{-1}$$

The enthalpy value from Ref. [88KHA/CHO] is tentatively accepted here, assuming the value reported for a 0.1 M $NaClO_4$ medium can also be used at $I = 0$.

$$\Delta_r H_m^\circ(\text{V.101, 298.15 K}) = (0.1 \pm 0.3)\,\text{kJ} \cdot \text{mol}^{-1}$$

The resulting standard enthalpy of formation is

$$\Delta_f H_m^\circ(UO_2BrO_3^+, \text{aq}, 298.15\,\text{K}) = -(1085.6 \pm 1.6)\,\text{kJ} \cdot \text{mol}^{-1}.$$

Using the selected reaction data, the standard entropy is calculated via the entropy of reaction, $\Delta_r S_m^\circ$:

$$S_m^\circ(UO_2BrO_3^+, \text{aq}, 298.15\,\text{K}) = (75.7 \pm 3.7)\,\text{J} \cdot \text{K}^{-1} \cdot \text{mol}^{-1}.$$

No data are available on the aqueous bromate complexes of either UO_2^+ or U^{4+}.

V.4.3.2.2. Solid uranium bromates

No data on solid uranium bromates of any oxidation state are reported in the literature.

V.4.4. Iodine compounds and complexes

Iodine may exist as either iodide or iodate in natural systems. Mainly uranium iodides have been studied experimentally, but thermodynamic data are also selected for two aqueous iodates and one solid iodate, *cf.* Sections V.4.4.2.1 and V.4.4.2.2.

V.4.4.1. Uranium iodides

V.4.4.1.1. Gaseous uranium iodides

a) UI(g), UI_2(g) and UI_3(g)

No experimental data exist for the these species. Estimated values were reported by Hildenbrand, Gurvich and Yungman [85HIL/GUR], who estimated an average bond dissociation energy from the selected value for the enthalpy of formation of UI_4(g) as recommended by Fuger *et al.* [83FUG/PAR]. Since selected value for the enthalpy of formation of UI_4(g) is in good agreement with the value proposed by Fuger *et al.*, the estimates by Hildenbrand, Gurvich and Yungman for the enthalpies of formation of UI_3(g), UI_2(g) and UI(g) are selected here:

$$\begin{aligned}
\Delta_f H_m^\circ(\text{UI, g, 298.15 K}) &= (341 \pm 25)\ \text{kJ}\cdot\text{mol}^{-1}\\
\Delta_f H_m^\circ(\text{UI}_2\text{, g, 298.15 K}) &= (100 \pm 25)\ \text{kJ}\cdot\text{mol}^{-1}\\
\Delta_f H_m^\circ(\text{UI}_3\text{, g, 298.15 K}) &= -(140 \pm 25)\ \text{kJ}\cdot\text{mol}^{-1}
\end{aligned}$$

The thermal functions of UI(g), UI_2(g) and UI_3(g) are calculated from molecular constants estimated by Hildenbrand, Gurvich and Yungman [85HIL/GUR]; the electronic contribution is assumed to be identical to the corresponding uranium chlorides.

$$\begin{aligned}
S_m^\circ(\text{UI, g, 298.15 K}) &= (286 \pm 10)\ \text{J}\cdot\text{K}^{-1}\cdot\text{mol}^{-1}\\
S_m^\circ(\text{UI}_2\text{, g, 298.15 K}) &= (376 \pm 10)\ \text{J}\cdot\text{K}^{-1}\cdot\text{mol}^{-1}\\
S_m^\circ(\text{UI}_3\text{, g, 298.15 K}) &= (428 \pm 10)\ \text{J}\cdot\text{K}^{-1}\cdot\text{mol}^{-1}
\end{aligned}$$

$$\begin{aligned}
C_{p,m}^\circ(\text{UI, g, 298.15 K}) &= (44.4 \pm 5.0)\ \text{J}\cdot\text{K}^{-1}\cdot\text{mol}^{-1}\\
C_{p,m}^\circ(\text{UI}_2\text{, g, 298.15 K}) &= (61.9 \pm 5.0)\ \text{J}\cdot\text{K}^{-1}\cdot\text{mol}^{-1}\\
C_{p,m}^\circ(\text{UI}_3\text{, g, 298.15 K}) &= (86.0 \pm 5.0)\ \text{J}\cdot\text{K}^{-1}\cdot\text{mol}^{-1}
\end{aligned}$$

The Gibbs energies of formation are calculated from the selected enthalpies of formation and entropies.

$$\begin{aligned}
\Delta_f G_m^\circ(\text{UI, g, 298.15 K}) &= (288 \pm 25)\ \text{kJ}\cdot\text{mol}^{-1}\\
\Delta_f G_m^\circ(\text{UI}_2\text{, g, 298.15 K}) &= (37 \pm 25)\ \text{kJ}\cdot\text{mol}^{-1}\\
\Delta_f G_m^\circ(\text{UI}_3\text{, g, 298.15 K}) &= -(201 \pm 25)\ \text{kJ}\cdot\text{mol}^{-1}
\end{aligned}$$

b) UI_4(g)

There is only little information on the enthalpy of formation of UI_4(g). Vapour pressure studies were performed by Schelberg and Thompson [42SCH/THO] for UI_4(cr) by effusion method from 573 to 683 K, and by Gregory [46GRE] for UI_4(l) by boiling point technique from 823 to 837 K at a known pressure of I_2(g) to prevent dissociation. The two sets of measurements are not in agreement; third-law analyses yield $\Delta_{sub}H_m^\circ$(V.102, 298.15 K) $= (207.7 \pm 1.2)\ \text{kJ}\cdot\text{mol}^{-1}$ and $(203.7 \pm 0.6)\ \text{kJ}\cdot\text{mol}^{-1}$, respectively, for the reaction

$$\text{UI}_4(\text{cr}) \rightleftharpoons \text{UI}_4(\text{g}). \qquad \text{(V.102)}$$

The analysis of these equilibria is complicated by the dissociation of $UI_4(g)$ to $UI_3(g)$ and iodine gas, which is also evident from the temperature dependence of the third-law enthalpies of sublimation. The dissociation pressure of iodine in equilibrium with $UI_4(g)$ and $UI_3(g)$ is about one order of magnitude lower at 573 K but increases with temperature. Since the measurements of Schelberg and Thompson [42SCH/THO] were made at the lowest temperatures, where the effect of dissociation is the smallest, preference is given to their results. The selected enthalpy of sublimation, is estimated from their results,

$$\Delta_{\text{sub}}H_{\text{m}}^{\circ}(\text{V.102, 298.15 K}) = (210 \pm 5)\,\text{kJ}\cdot\text{mol}^{-1}.$$

Combining the enthalpy of sublimation derived from their study with the enthalpy of formation of the solid, the present review obtains:

$$\Delta_{\text{f}}H_{\text{m}}^{\circ}(\text{UI}_4\text{, g, 298.15 K}) = -(308.8 \pm 5.8)\,\text{kJ}\cdot\text{mol}^{-1}$$

The heat capacity and entropy of $UI_4(g)$ are calculated from molecular constants estimated by Hildenbrand, Gurvich and Yungman [85HIL/GUR] by analogy to heavy metal iodides. The electronic contribution, not considered by Hildenbrand, Gurvich and Yungman, is assumed to be the same as for $UBr_4(g)$:

$$S_{\text{m}}^{\circ}(\text{UI}_4\text{, g, 298.15 K}) = (489 \pm 10)\,\text{J}\cdot\text{K}^{-1}\cdot\text{mol}^{-1}$$
$$C_{p,\text{m}}^{\circ}(\text{UI}_4\text{, g, 298.15 K}) = (112 \pm 5)\,\text{J}\cdot\text{K}^{-1}\cdot\text{mol}^{-1}.$$

The Gibbs energy of formation is calculated from the selected enthalpy of formation and entropy.

$$\Delta_{\text{f}}G_{\text{m}}^{\circ}(\text{UI}_4\text{, g, 298.15 K}) = -(370.4 \pm 6.5)\,\text{kJ}\cdot\text{mol}^{-1}$$

V.4.4.1.2. Aqueous uranium iodides

a) Aqueous U(VI) and U(V) iodides

Davies and Monk [57DAV/MON] did not find any indication for ion pair formation between UO_2^{2+} and I^-. Jedináková [74JED] investigated this system and concluded that the data she obtained in 7 M NaI solution might be interpreted in terms of UO_2I^+ formation. Insufficient information is available to support this conclusion.

There are no published experimental data on aqueous iodide complexes of UO_2^+.

b) Aqueous U(IV) iodides

The only available experimental study on aqueous uranium(IV) iodides is that of Vdovenko, Romanov and Shcherbakov [63VDO/ROM3]. Their equilibrium constant for the reaction

$$U^{4+} + I^- \rightleftharpoons UI^{3+} \qquad \text{(V.103)}$$

is consistent with the trend of the other known first halide complexes of uranium(IV). This review therefore adopts their value of $\log_{10}\beta_1(\text{V.103}) = (0.18 \pm 0.14)$ at $I = 2.5$ M

(Na,H)ClO_4) and 20°C and corrects it to $I = 0$ by using $\Delta\varepsilon(\text{V.103}) \approx \Delta\varepsilon(\text{V.84}, q = 1) = -(0.29 \pm 0.08)$. The resulting value, assumed to be valid also at 25°C, is selected here.

$$\log_{10} \beta_1^\circ(\text{V.103}, 298.15\text{ K}) = 1.25 \pm 0.30$$

An increased uncertainty is chosen by this review to reflect the fact that only one experimental study is available.

The Gibbs energy of formation is calculated from the selected enthalpy of formation and entropy.

$$\Delta_f G_m^\circ(\text{UI}^{3+}, \text{aq}, 298.15\,\text{K}) = -(588.7 \pm 2.5)\,\text{kJ} \cdot \text{mol}^{-1}$$

No enthalpy, entropy or heat capacity data are available for this species. Although data have been estimated for higher aqueous uranium(IV) iodides, these species are not credited in this review.

V.4.4.1.3. Solid uranium iodides

Only two solids in the U-O-H-I system are sufficiently well characterized for chemical thermodynamic data selection to be possible.

The selected data, for both UI_3(cr) and UI_4(cr) are adopted from Fuger *et al.* [83FUG/PAR]. The basis for their enthalpy of formation selections is summarized in Table V.31.

$$\Delta_f H_m^\circ(\text{UI}_3, \text{cr}, 298.15\,\text{K}) = -(467.4 \pm 4.2)\,\text{kJ} \cdot \text{mol}^{-1}$$
$$\Delta_f H_m^\circ(\text{UI}_4, \text{cr}, 298.15\,\text{K}) = -(518.8 \pm 2.9)\,\text{kJ} \cdot \text{mol}^{-1}$$

The selected standard entropy of UI_4(cr) was estimated by Fuger *et al.* [83FUG/PAR], and the selected heat capacity of UI_4(cr) was obtained by them from an extrapolation of the measurements of Popov, Galchenko and Senin [59POP/GAL].

$$S_m^\circ(\text{UI}_4, \text{cr}, 298.15\,\text{K}) = (263.6 \pm 8.4)\,\text{J} \cdot \text{K}^{-1} \cdot \text{mol}^{-1}$$
$$C_{p,m}^\circ(\text{UI}_4, \text{cr}, 298.15\,\text{K}) = (126.4 \pm 4.2)\,\text{J} \cdot \text{K}^{-1} \cdot \text{mol}^{-1}$$

The entropy of UI_3(cr) is estimated by the present review from the experimental data for UF_3(cr), UCl_3(cr) and UBr_3(cr):

$$S_m^\circ(\text{UI}_3, \text{cr}, 298.15\,\text{K}) = (221.8 \pm 8.4)\,\text{J} \cdot \text{K}^{-1} \cdot \text{mol}^{-1}$$

This value is in reasonable agreement with that calaulated by Fuger *et al.* [83FUG/PAR] from the decomposition pressure measurements (*cf.* Table V.31) of MacWood [58MAC], $S_m^\circ(\text{UI}_3, \text{cr}, 298.15\text{ K}) = (227.7 \pm 5.0)\,\text{kJ} \cdot \text{mol}^{-1}$.

The heat capacity estimated by Krestov [72KRE] was accepted in the IAEA review [83FUG/PAR]. The uncertainty is assigned by this review.

$$C_{p,m}^\circ(\text{UI}_3, \text{cr}, 298.15\,\text{K}) = (112.1 \pm 6.0)\,\text{J} \cdot \text{K}^{-1} \cdot \text{mol}^{-1}$$

The Gibbs energies of formation are calculated from the selected enthalpies of formation and entropies.

Table V.31: Enthalpy data considered for $UI_3(cr)$ and $UI_4(cr)$.

Reference	Reaction	$\Delta_f H^\circ_m(UI_q$, cr, 298.15 K) (kJ · mol^{-1})
		$q = 3$:
[45BAR]	$UI_4(cr) \rightleftharpoons UI_3(cr) + \frac{1}{2}I_2(cr)$	-467.4 ± 4.2[(a)]
[58MAC]	$UI_4(cr) \rightleftharpoons UI_3(cr) + \frac{1}{2}I_2(cr)$	-472.8 ± 5.0[(b)]
[68TVE/CHA]	$U(cr) + \frac{3}{2}I_2(g) \rightleftharpoons UI_3(cr)$	-461.1 ± 4.2[(c)]
		$q = 4$:
[73FUG/BRO]	$UI_4(cr) + 4HCl(sln) \rightleftharpoons UCl_4(cr) + 4HI(sln)$	-521.3 ± 2.9[(d)]
	$UI_4(cr) + 4HBr(sln) \rightleftharpoons UBr_4(cr) + 4HI(sln)$	-516.7 ± 2.9[(e)]

(a) Reduction in 12 M HCl, reported by MacWood [58MAC].
(b) Second law value from decomposition pressure measurements in the range 523 to 660 K.
(c) emf cell measurements at 643 K.
(d) Dissolution in 1 M HCl using enthalpy of dissolution of $UCl_4(cr)$ determined in the same solution and using suitable auxiliary data.
(e) Dissolution in 6 M HCl using enthalpy of dissolution of $UBr_4(cr)$ determined in the same solution and using suitable auxiliary data.

$$\Delta_f G^\circ_m(UI_3, cr, 298.15\,K) = -(466.6 \pm 4.9)\,kJ \cdot mol^{-1}$$
$$\Delta_f G^\circ_m(UI_4, cr, 298.15\,K) = -(513.2 \pm 3.8)\,kJ \cdot mol^{-1}$$

The $C^\circ_{p,m}(T)$ data tabulated by the IAEA [83FUG/PAR] are fitted between 298.15 and 600 K to obtain the selected $C^\circ_{p,m}(T)$ function listed in Table III.3.

V.4.4.2. Uranium iodates

V.4.4.2.1. Aqueous uranium iodates

Khalili, Choppin and Rizkalla [88KHA/CHO] determined the formation constant of the complex $UO_2IO_3^+$ and give $\beta_1(V.104, q = 1) = (37.6 \pm 1.0)$ at 25°C. The correction to $I = 0$ is done by this review (see Appendix A) and results in $\log_{10}\beta_1^\circ = (2.00 \pm 0.02)$. Klygin, Smirnova and Nikol'skaya [59KLY/SMI2] did not include the complex $UO_2IO_3^+$ in the interpretation of their solubility study. This review therefore re-evaluates this study from which the solubility product of $UO_2(IO_3)_2(cr)$ and the equilibrium constants for the reactions

$$UO_2^{2+} + qIO_3^- \rightleftharpoons UO_2(IO_3)_q^{2-q} \qquad (V.104)$$

are obtained, *cf.* Appendix A. The solubility product is discussed in Section V.4.4.2.2. The value of $\log_{10}\beta_1^\circ(V.104, q = 1) = (2.05 \pm 0.14)$ obtained is in good agreement with the one found in Ref. [88KHA/CHO]. This review selects the weighted average of these two values, and accepts the value for $\log_{10}\beta_2^\circ(\text{V.104}, q = 2)$ evaluated from the data in Ref. [59KLY/SMI2].

$$\begin{aligned}\log_{10}\beta_1^\circ(\text{V.104}, q = 1, 298.15\,\text{K}) &= 2.00 \pm 0.02\\ \log_{10}\beta_2^\circ(\text{V.104}, q = 2, 298.15\,\text{K}) &= 3.59 \pm 0.15\end{aligned}$$

No value for $\log_{10}\beta_3$ is credited in view of the large changes in the ionic medium that occurred during the experiment in the concentration range where this complex is formed.

From the equilibrium constants selected above and the standard Gibbs energies of formation of the reactants, this review obtains

$$\begin{aligned}\Delta_f G_m^\circ(\text{UO}_2\text{IO}_3^+, \text{aq}, 298.15\,\text{K}) &= -(1090.3 \pm 1.9)\,\text{kJ}\cdot\text{mol}^{-1}\\ \Delta_f G_m^\circ(\text{UO}_2(\text{IO}_3)_2, \text{aq}, 298.15\,\text{K}) &= -(1225.7 \pm 2.5)\,\text{kJ}\cdot\text{mol}^{-1}.\end{aligned}$$

The enthalpy of Reaction (V.104, $q = 1$) from Ref [88KHA/CHO] is tentatively accepted here, assuming the value reported at $I = 0.1$ M is not significantly different at $I = 0$,

$$\Delta_r H_m^\circ(\text{V.104},\ q = 1,\ 298.15\ \text{K}) = (9.8 \pm 0.9)\,\text{kJ}\cdot\text{mol}^{-1},$$

which leads to the selected

$$\Delta_f H_m^\circ(\text{UO}_2\text{IO}_3^+, \text{aq}, 298.15\,\text{K}) = -(1228.9 \pm 1.8)\,\text{kJ}\cdot\text{mol}^{-1}.$$

Using these selected values, the standard entropy is calculated to be

$$S_m^\circ(\text{UO}_2\text{IO}_3^+, \text{aq}, 298.15\,\text{K}) = (91.0 \pm 4.7)\,\text{J}\cdot\text{K}^{-1}\cdot\text{mol}^{-1}.$$

No heat capacity data are available on aqueous dioxouranium(VI) iodates. No aqueous uranium iodates of either dioxouranium(V) or uranium(IV) are identified in this review.

It should be noted that the dioxouranium(VI) iodate complexes are considerably more stable than the corresponding chlorate and bromate complexe. This indicates that iodate forms inner-sphere complexes with UO_2^{2+}, while chlorate and bromate form outer-sphere complexes.

V.4.4.2.2. Solid uranium iodates

Klygin, Smirnova and Nikol'skaya [59KLY/SMI2] measured the solubility product of $\text{UO}_2(\text{IO}_3)_2(\text{cr})$ at 25°C and 60°C in a 0.2 M NH_4Cl aqueous solution.

$$\text{UO}_2(\text{IO}_3)_2(\text{cr}) \rightleftharpoons \text{UO}_2^{2+} + 2\text{IO}_3^- \qquad \text{(V.105)}$$

This review reinterprets the measurements using a different chemical model (see Appendix A). After correction to zero ionic strength using the specific ion interaction approach, the standard solubility product obtained is

$$\log_{10} K^\circ_{s,0}(\text{V.105}, 298.15\text{ K}) = -7.88 \pm 0.10.$$

The selected Gibbs energy of formation is calculated using the selected auxiliary data in Chapter VI, resulting in

$$\Delta_f G^\circ_m(UO_2(IO_3)_2, \text{cr}, 298.15\,\text{K}) = -(1250.2 \pm 2.4)\,\text{kJ} \cdot \text{mol}^{-1}.$$

No reliable enthalpy of formation for $UO_2(IO_3)_2$(cr) can be extracted from the available experimental information in Ref. [59KLY/SMI2].

An entropy estimate according to the method in Appendix D results in

$$S^\circ_m(UO_2(IO_3)_2, \text{cr}, 298.15\,\text{K}) = (279 \pm 9)\,\text{J} \cdot \text{K}^{-1} \cdot \text{mol}^{-1}.$$

The enthalpy of formation is calculated from the selected $\Delta_f G^\circ_m$ and S°_m values.

$$\Delta_f H^\circ_m(UO_2(IO_3)_2, \text{cr}, 298.15\,\text{K}) = -(1461.3 \pm 3.6)\,\text{kJ} \cdot \text{mol}^{-1}$$

V.4.4.3. Uranium periodates

V.4.4.3.1. Aqueous uranium periodates

Hutin [71HUT, 75HUT] studied equilibria between dioxouranium(VI) and various periodate species. This review finds that these studies give no conclusive evidence about the stoichiometries of the complexes proposed. No other equilibrium data seem to be available about these systems.

V.4.4.3.2. Solid uranium periodates

Hutin [75HUT] observed the formation of $H(UO_2)_2IO_6$(cr) by precipitation. However, the author gave no chemical analysis of the solid, and there is not enough information available for the extraction of any thermodynamic data.

V.4.5. Mixed halogen compounds

V.4.5.1. Chlorofluoride compounds

V.4.5.1.1. Gaseous and aqueous uranium chlorofluorides

No experimental or estimated thermodynamic data have been published on gaseous or aqueous chlorofluoride species of any oxidation state of uranium.

V.4.5.1.2. Solid uranium chlorofluorides

Uranium(IV) chlorofluorides have been known since the Manhattan Project era, and some evidence of uranium(V) compounds also exists [51KAT/RAB], but no data are available on uranium(VI) compounds. Sufficient data upon which to base selections are available only for the U(IV)-Cl-F system.

Maslov [64MAS] calculated S°_m and $C_{p,m}$ for the range 0 to 350 K for all three compounds intermediate to UCl_4(cr) and UF_4(cr). As was done in the Gmelin [83FUG] and IAEA [83FUG/PAR] compilations, this review accepts these values at 298.15 K.

$$S^\circ_m(UCl_3F, cr, 298.15\,K) = (162.8 \pm 4.2)\,J \cdot K^{-1} \cdot mol^{-1}$$
$$S^\circ_m(UCl_2F_2, cr, 298.15\,K) = (174.1 \pm 8.4)\,J \cdot K^{-1} \cdot mol^{-1}$$
$$S^\circ_m(UClF_3, cr, 298.15\,K) = (185.4 \pm 4.2)\,J \cdot K^{-1} \cdot mol^{-1}$$

$$C^\circ_{p,m}(UCl_3F, cr, 298.15\,K) = (118.8 \pm 4.2)\,J \cdot K^{-1} \cdot mol^{-1}$$
$$C^\circ_{p,m}(UCl_2F_2, cr, 298.15\,K) = (119.7 \pm 4.2)\,J \cdot K^{-1} \cdot mol^{-1}$$
$$C^\circ_{p,m}(UClF_3, cr, 298.15\,K) = (120.9 \pm 4.2)\,J \cdot K^{-1} \cdot mol^{-1}$$

Both S°_m and $C^\circ_{p,m}$ vary linearly between the UCl_4(cr) and UF_4(cr) end compositions. Because of this relationship, and since $\Delta_f H^\circ_m$ varies linearly in the uranium(IV) bromochloride system (*cf.* Section V.4.5.2.2), this review estimates the enthalpies of formation of the three intermediate chlorofluoride compounds, UCl_3F(cr), UCl_2F_2(cr) and $UClF_3$(cr), with a linear model, based on the selected values of $\Delta_f H^\circ_m(UCl_4, cr, 298.15\,K)$ and $\Delta_f H^\circ_m(UF_4, cr, 298.15\,K)$.

$$\Delta_f H^\circ_m(UCl_3F, cr, 298.15\,K) = -(1243 \pm 5)\,kJ \cdot mol^{-1}$$
$$\Delta_f H^\circ_m(UCl_2F_2, cr, 298.15\,K) = -(1466 \pm 5)\,kJ \cdot mol^{-1}$$
$$\Delta_f H^\circ_m(UClF_3, cr, 298.15\,K) = -(1690 \pm 5)\,kJ \cdot mol^{-1}$$

The Gibbs energies of formation are calculated from the selected enthalpies of formation and entropies.

$$\Delta_f G^\circ_m(UCl_3F, cr, 298.15\,K) = -(1147 \pm 5)\,kJ \cdot mol^{-1}$$
$$\Delta_f G^\circ_m(UCl_2F_2, cr, 298.15\,K) = -(1376 \pm 6)\,kJ \cdot mol^{-1}$$
$$\Delta_f G^\circ_m(UClF_3, cr, 298.15\,K) = -(1606 \pm 5)\,kJ \cdot mol^{-1}$$

V.4.5.2. *Bromochloride compounds*

V.4.5.2.1. *Gaseous and aqueous uranium bromochlorides*

The only experimental study of gaseous mixed uranium bromochlorides in the study of Alikhanyan *et al.* [88ALI/MAL], involving $UBrCl_3$(g), UBr_2Cl_2(g) and UBr_3Cl(g). Some of the other thermodynamic data reported in this study are at variance with the data in Ref. [85LAU/HIL], indicating the presence of systematic errors. The data for the uranium bromochlorides are therefore not considered by this review. For this reason no thermodynamic data are reported on the UBr_qCl_{4-q} compounds. No aqueous mixed uranium bromochloride compounds are known.

V.4.5.2.2. *Solid uranium bromochlorides*

Binary uranium(III) and uranium(IV) bromochloride compounds are known. Uranium(V) bromochlorides may exist [51KAT/RAB], but no experimental data are available for them.

a) U(IV) bromochlorides

MacWood [58MAC] reported the enthalpy of solution of the three uranium(IV) bromochloride complexes intermediate to UBr_4(cr) and UCl_4(cr) in 12 M HCl and 10% $FeCl_3$(aq). From these values, Fuger *et al.* [83FUG/PAR] calculated their selected enthalpies of formation for UBr_3Cl(cr), UBr_2Cl_2(cr) and $UBrCl_3$(cr), which are also accepted here.

$$\begin{aligned}
\Delta_f H^\circ_m(\mathrm{UBr_3Cl, cr, 298.15\,K}) &= -(852.3 \pm 8.4)\,\mathrm{kJ\cdot mol^{-1}}\\
\Delta_f H^\circ_m(\mathrm{UBr_2Cl_2, cr, 298.15\,K}) &= -(907.9 \pm 8.4)\,\mathrm{kJ\cdot mol^{-1}}\\
\Delta_f H^\circ_m(\mathrm{UBrCl_3, cr, 298.15\,K}) &= -(967.3 \pm 8.4)\,\mathrm{kJ\cdot mol^{-1}}
\end{aligned}$$

MacWood [58MAC] also reported the equilibrium constants for the hydrogen reduction of various uranium(III) and uranium(IV) bromochloride complexes [44ALT] and also for the exchange equilibrium reactions with HBr(g) in the range 600 to 773 K. Fuger *et al.* [83FUG/PAR] used these data to estimate the entropies of these compounds at standard state.

$$\begin{aligned}
S^\circ_m(\mathrm{UBr_3Cl, cr, 298.15\,K}) &= (238.5 \pm 16.7)\,\mathrm{J\cdot K^{-1}\cdot mol^{-1}}\\
S^\circ_m(\mathrm{UBr_2Cl_2, cr, 298.15\,K}) &= (234.3 \pm 16.7)\,\mathrm{J\cdot K^{-1}\cdot mol^{-1}}\\
S^\circ_m(\mathrm{UBrCl_3, cr, 298.15\,K}) &= (213.4 \pm 12.6)\,\mathrm{J\cdot K^{-1}\cdot mol^{-1}}
\end{aligned}$$

These estimates are selected in this review.

The Gibbs energy of formation values are calculated from the selected enthalpy of formation and entropy values.

$$\begin{aligned}
\Delta_f G^\circ_m(\mathrm{UBr_3Cl, cr, 298.15\,K}) &= -(807.1 \pm 9.8)\,\mathrm{kJ\cdot mol^{-1}}\\
\Delta_f G^\circ_m(\mathrm{UBr_2Cl_2, cr, 298.15\,K}) &= -(850.9 \pm 9.8)\,\mathrm{kJ\cdot mol^{-1}}\\
\Delta_f G^\circ_m(\mathrm{UBrCl_3, cr, 298.15\,K}) &= -(893.5 \pm 9.2)\,\mathrm{kJ\cdot mol^{-1}}
\end{aligned}$$

No heat capacity data are available for these species.

b) U(III) bromochlorides

The same sources of data are used and procedures followed for $UBrCl_2$(cr) and UBr_2Cl(cr) as for the uranium(IV) bromochlorides in Section V.4.5.2.2.a above.

$$\begin{aligned}
\Delta_f G^\circ_m(\mathrm{UBr_2Cl, cr, 298.15\,K}) &= -(714.4 \pm 9.8)\,\mathrm{kJ\cdot mol^{-1}}\\
\Delta_f G^\circ_m(\mathrm{UBrCl_2, cr, 298.15\,K}) &= -(760.3 \pm 9.8)\,\mathrm{kJ\cdot mol^{-1}}
\end{aligned}$$

$$\begin{aligned}
\Delta_f H^\circ_m(\mathrm{UBr_2Cl, cr, 298.15\,K}) &= -(750.6 \pm 8.4)\,\mathrm{kJ\cdot mol^{-1}}\\
\Delta_f H^\circ_m(\mathrm{UBrCl_2, cr, 298.15\,K}) &= -(812.1 \pm 8.4)\,\mathrm{kJ\cdot mol^{-1}}
\end{aligned}$$

$$\begin{aligned}
S^\circ_m(\mathrm{UBr_2Cl, cr, 298.15\,K}) &= (192.5 \pm 16.7)\,\mathrm{J\cdot K^{-1}\cdot mol^{-1}}\\
S^\circ_m(\mathrm{UBrCl_2, cr, 298.15\,K}) &= (175.7 \pm 16.7)\,\mathrm{J\cdot K^{-1}\cdot mol^{-1}}
\end{aligned}$$

V.4.5.3. Chloroiodide compounds

Experimental enthalpy data on uranium chloroiodide species are limited to the enthalpy of solution determination of $UCl_3I(cr)$ of MacWood [58MAC]. This was obtained in 12 M HCl in equilibrium with $I_2(cr)$ at 273 K. Fuger *et al.* [83FUG/PAR] assumed $\Delta_r C_{p,m}$(V.106) is negligible for the reaction

$$UCl_3(cr) + \tfrac{1}{2}I_2(cr) \rightleftharpoons UCl_3I(cr) \qquad (V.106)$$

and converted $\Delta_r H^\circ_m$(V.106, 298.15 K) = 36.4 kJ · mol^{-1} to the selected value of

$$\Delta_f H^\circ_m(UCl_3I, cr, 298.15\,K) = -(898.3 \pm 8.4)\,kJ \cdot mol^{-1}.$$

Fuger *et al.* [83FUG/PAR] used data presented by MacWood [58MAC] on the dissociation pressure of the equilibrium

$$UCl_3I(cr) \rightleftharpoons UCl_3(cr) + \tfrac{1}{2}I_2(g) \qquad 472\text{ to }475\,K \qquad (V.107)$$

to obtain

$$S^\circ_m(UCl_3I, cr, 298.15\,K) = (213.4 \pm 8.4)\,J \cdot K^{-1} \cdot mol^{-1},$$

assuming $\Delta_r C^\circ_{p,m}$(V.107) = −8.4 J·K^{-1}·mol^{-1}. $\Delta_f H^\circ_m$ is consistent with a linear model for the change of these parameters between $UCl_4(cr)$ and $UI_4(cr)$. This review extends this relationship to interpolate $\Delta_f H^\circ_m$ for the remaining two intermediate chloroiodide solids, $UCl_2I_2(cr)$ and $UClI_3(cr)$, both of which are known [51KAT/RAB].

$$\Delta_f H^\circ_m(UCl_2I_2, cr, 298.15\,K) = -(768.8 \pm 10.0)\,kJ \cdot mol^{-1}$$
$$\Delta_f H^\circ_m(UClI_3, cr, 298.15\,K) = -(643.8 \pm 10.0)\,kJ \cdot mol^{-1}$$

$$S^\circ_m(UCl_2I_2, cr, 298.15\,K) = (237 \pm 18)\,J \cdot K^{-1} \cdot mol^{-1}$$
$$S^\circ_m(UClI_3, cr, 298.15\,K) = (242 \pm 18)\,J \cdot K^{-1} \cdot mol^{-1}$$

The Gibbs energies of formation of all three compounds are calculated from the selected enthalpies of formation and entropies.

$$\Delta_f G^\circ_m(UCl_3I, cr, 298.15\,K) = -(829.9 \pm 8.8)\,kJ \cdot mol^{-1}$$
$$\Delta_f G^\circ_m(UCl_2I_2, cr, 298.15\,K) = -(723.4 \pm 11.3)\,kJ \cdot mol^{-1}$$
$$\Delta_f G^\circ_m(UClI_3, cr, 298.15\,K) = -(615.8 \pm 11.4)\,kJ \cdot mol^{-1}$$

No heat capacity data are available for any of these compounds.

V.4.5.4. Bromoiodide compounds

Experimental data on the uranium bromochloride species are limited to the enthalpy of solution determinations of $UBr_3I(cr)$ of MacWood [58MAC]. These were obtained in 12 M HCl in equilibrium with $I_2(cr)$ at 273 K. Fuger *et al.* [83FUG/PAR] assumed $\Delta_r C_{p,m}$(V.108) is negligible for the temperature correction from 273 to 298.15 K for the reaction

$$UCl_3I(cr) + 3HBr(sln) \rightleftharpoons UBr_3I(cr) + 3HCl(sln). \qquad (V.108)$$

Using auxiliary data appropriate to the solution, Fuger *et al.* [83FUG/PAR] obtained the selected value of

$$\Delta_f H_m^\circ(\text{UBr}_3\text{I, cr, 298.15 K}) = -(727.6 \pm 8.4)\,\text{kJ}\cdot\text{mol}^{-1}$$

from these data. This value is consistent with a linear model for the change in $\Delta_f H_m^\circ$ between UBr_4(cr) and UI_4(cr). This review uses such a model to estimate the enthalpies of formation of UBr_2I_2(cr) and $UBrI_3$(cr), which are known to exist [51KAT/RAB].

$$\Delta_f H_m^\circ(\text{UBr}_2\text{I}_2\text{, cr, 298.15 K}) = -(660.4 \pm 10.0)\,\text{kJ}\cdot\text{mol}^{-1}$$
$$\Delta_f H_m^\circ(\text{UBrI}_3\text{, cr, 298.15 K}) = -(589.6 \pm 10.0)\,\text{kJ}\cdot\text{mol}^{-1}$$

No other experimental data for uranium(IV) bromoiodides are available.

V.5. Group 16 (chalcogen) compounds and complexes

V.5.1. Sulphur compounds and complexes

V.5.1.1. Uranium sulphides

The chemical thermodynamic parameters for the solid uranium sulphides were recently reviewed by Grønvold *et al.* [84GRO/DRO]. This review makes no attempt to re-evaluate data assessed in that reference, and for the most part accepts the values selected therein. The selected values for the solid uranium sulphide compounds are listed in Table V.32.

Table V.32: Selected thermodynamic parameters on the solid uranium sulphide compounds at 298.15 K. The Gibbs energies of formation are calculated from the selected enthalpies of formation and entropies.

Compound	$\Delta_f G_m^\circ$ ($kJ\cdot mol^{-1}$)	$\Delta_f H_m^\circ$ ($kJ\cdot mol^{-1}$)	S_m° ($J\cdot K^{-1}\cdot mol^{-1}$)	$C_{p,m}^\circ$ ($J\cdot K^{-1}\cdot mol^{-1}$)
US(cr)	-320.9 ± 12.6	-322.2 ± 12.6	77.99 ± 0.21	50.54 ± 0.08
U_2S_3(cr)	-880 ± 67	-879 ± 67	199.2 ± 1.7	133.7 ± 0.8[a]
U_3S_5(cr)	-1425 ± 100	-1431 ± 100	291 ± 25	
$US_{1.90}$(cr)	-509.5 ± 20.9	-509.9 ± 20.9	109.66 ± 0.21	73.97 ± 0.13
US_2(cr)	-519.2 ± 8.0	-520.4 ± 8.0	110.42 ± 0.21	74.64 ± 0.13
U_2S_5(cr)			243 ± 25	
US_3(cr)	-537.3 ± 12.6	-539.6 ± 12.6	138.49 ± 0.21	95.60 ± 0.25

(a) The value given in Ref. [84GRO/DRO, *p.*36] for $US_{1.50}$(cr) is misprinted as 16.9 instead of 16.0 $cal\cdot K^{-1}\cdot mol^{-1}$.

Since the review of Grønvold *et al.* was prepared, Settle and O'Hare [84SET/OHA] have published their study of the enthalpy of formation of $US_{1.992}$, and reported $\Delta_f H_m^\circ(US_{1.992}, cr, 298.15\,K) = -519.7\,kJ\cdot mol^{-1}$. This value differs somewhat from values previously presented in preliminary reports of the same work [75OHA/ADE,

84GRO/DRO]. Using the value selected in this review for $\Delta_f H_m^\circ(UF_6$, cr, 298.15 K), $-2197.0\,kJ \cdot mol^{-1}$ (instead of $-2197.7\,kJ \cdot mol^{-1}$ used by Settle and O'Hare [84SET/OHA]), $\Delta_f H_m^\circ(US_{1.992}, cr, 298.15\,K) = -519.0\,kJ \cdot mol^{-1}$ is calculated. From this, using the procedure used by Settle and O'Hare [84SET/OHA], the value

$$\Delta_f H_m^\circ(US_2, cr, 298.15\,K) = -(520.4 \pm 8.0)\,kJ \cdot mol^{-1}$$

is calculated. A linear extrapolation based on the value for $\Delta_f H_m^\circ(US_{1.992}$, cr, 298.15 K) and the value of $\Delta_f H_m^\circ$(US, cr, 298.15 K) [84GRO/DRO] gives a similar result. This value and the corresponding Gibbs energy of formation, calculated using the entropy of Ref. [84GRO/DRO],

$$\Delta_f G_m^\circ(US_2, cr, 298.15\,K) = -(519.2 \pm 8.0)\,kJ \cdot mol^{-1}$$

are selected in this review in preference to the values from the IAEA review [84GRO/DRO]. Using this revised value for $\Delta_f H_m^\circ(US_2$, cr, 298.15 K), analysis identical to that of Grønvold *et al.* [84GRO/DRO] leads to

$$\begin{aligned}\Delta_f H_m^\circ(US_{1.90}, \text{ cr, } 298.15\text{ K}) &= -(509.9 \pm 20.9)\,kJ \cdot mol^{-1}\\ \Delta_f H_m^\circ(US_3, \text{ cr, } 298.15\text{ K}) &= -(539.6 \pm 12.6)\,kJ \cdot mol^{-1}.\end{aligned}$$

From these values and the selected entropies this review obtains

$$\begin{aligned}\Delta_f G_m^\circ(US_{1.90}, \text{ cr, } 298.15\text{ K}) &= -(509.5 \pm 20.9)\,kJ \cdot mol^{-1}\\ \Delta_f G_m^\circ(US_3, \text{ cr, } 298.15\text{ K}) &= -(537.3 \pm 12.6)\,kJ \cdot mol^{-1}.\end{aligned}$$

It should also be mentioned that, as discussed in Ref. [84GRO/DRO], the heat capacity of $US_{1.5}$(cr) ($\frac{1}{2}U_2S_3$) appears to rise sharply in a temperature range near room temperature. Therefore, in Ref. [84GRO/DRO], the value selected for $C_{p,m}(US_{1.5}$, cr, 298.15 K) is not the same as that calculated from the selected function for the calculation of high-temperature heat capacities. This same discrepancy exists between the values for U_2S_3(cr) in Tables III.1 and III.3 of this review.

V.5.1.2. Uranium sulphites

V.5.1.2.1. Aqueous uranium sulphites

a) Aqueous U(VI) sulphites

The only quantitative studies available on the equilibria in the dioxouranium(VI)-sulphite system are those of Klygin and Kolyada [59KLY/KOL] and of Zakharova and Orlova [67ZAK/ORL]. These data are presented in Table V.33. The data of Zakharova and Orlova [67ZAK/ORL] were obtained at 23°C, and the numerical values are assumed to be the same at 25°C. These data demonstrate the formation of two sulphite complexes of dioxouranium(VI), UO_2SO_3(aq) and $UO_2(SO_3)_2^{2-}$, according to Eq. (V.109).

$$UO_2^{2+} + qSO_3^{2-} \rightleftharpoons UO_2(SO_3)_q^{2-2q} \qquad (V.109)$$

Table V.33: Experimental equilibrium data for the uranium(VI) sulphite system.

Method	Ionic Medium	t (°C)	$\log_{10} K^{(a)}$	$\log_{10} K^{\circ\,(b)}$	Reference
$UO_2^{2+} + SO_3^{2-} \rightleftharpoons UO_2SO_3(aq)$					
sol	var. ~ 0.02 M	25	$6.0 \pm 0.2^{(c)}$	6.5 ± 0.3	[59KLY/KOL]
emf	0.02 to 0.04 M	23	$5.8 \pm 0.3^{(c)}$	6.4 ± 0.3	[67ZAK/ORL]
sp	0.1 M NH_4ClO_4	23	$6.2 \pm 0.2^{(c)}$	7.0 ± 0.2	
$UO_2SO_3 \cdot 4.5H_2O(cr) + SO_3^{2-} \rightleftharpoons UO_2(SO_3)_2^{2-} + 4.5H_2O(l)$					
sol	var. ~ 0.02 M	25	$-1.5 \pm 0.1^{(d)}$	-1.5 ± 0.1	[59KLY/KOL]
sol	1 M NaCl	23	1.0		[67ZAK/ORL]
$UO_2SO_3 \cdot 4.5H_2O(cr) \rightleftharpoons UO_2^{2+} + SO_3^{2-} + 4.5H_2O(l)$					
sol	var. ~ 0.02 M	25	$-8.6 \pm 0.2^{(d)}$	-9.1 ± 0.3	[59KLY/KOL]

(a) $\log_{10} K$ refers to the reactions indicated in the experimental ionic medium and at the temperature given in the table.
(b) $\log_{10} K^\circ$ (in molal units) refers to the reactions indicated at $I = 0$ and 298.15 K.
(c) These uncertainties have been estimated by this review, as described in Appendix A.
(d) See comments in Appendix A.

The most precise study is the spectrophotometric study of the formation of $UO_2SO_3(aq)$ given in Ref. [67ZAK/ORL]. The solubility studies by Klygin and Kolyada [59KLY/KOL] and Zakharova and Orlova [67ZAK/ORL] are less satisfactory, but they demonstrate the formation of $UO_2(SO_3)_2^{2-}$ very clearly. The various equilibrium constants are recalculated to $I = 0$ by this review, after conversion to molality constants, by using the ion interaction coefficients listed in the tables in Appendix B. This results in $\Delta\varepsilon(\text{V.109}, q = 1) = -(0.38 \pm 0.06)$ for perchlorate medium, and $\Delta\varepsilon = (0.13 \pm 0.05)$ for the second reaction listed in Table V.33 for NaCl medium.

The results after correction to $I = 0$ are shown in Table V.33. The three values are not strictly consistent, and this review selects the unweighted average with an uncertainty that spans the uncertainty range of all three values:

$$\log_{10} \beta_1^\circ(\text{V.109}, q = 1, 298.15\,\text{K}) = 6.6 \pm 0.6.$$

From the solubility data of $UO_2SO_3 \cdot 4.5H_2O(cr)$ in Ref. [59KLY/KOL], this review calculates $\log_{10}(\beta_2 \times K_{s,0}) = -(1.5 \pm 0.1)$, *cf.* Appendix A. The value of the solubility product of $UO_2SO_3 \cdot 4.5H_2O(cr)$ is taken from Ref. [59KLY/KOL]:

$\log_{10} K^\circ_{s,0} = \log_{10}(a_{UO_2^{2+}} \times a_{SO_3^{2-}}) = -(9.1 \pm 0.3)$. This is in good agreement with $\log_{10} K^\circ_{s,0} = -(9.23 \pm 0.16)$, obtained from the product $\log_{10}(\beta^\circ_1 \times K^\circ_{s,0}) = -(2.49 \pm 0.04)$ [67ZAK/ORL] and the value of $\log_{10} \beta^\circ_1$ selected in this review. The value of $\log_{10} \beta^\circ_2 = (7.6 \pm 0.3)$ obtained from [59KLY/KOL] results in a ratio between the two first stepwise equilibrium constants equal to 8×10^6 as compared to 1.4×10^3 for sulphate. This review finds no compelling chemical evidence for such a large difference between the two ligands and does not consider the quantitative value of $\log_{10} \beta^\circ_2$ sufficiently good to include in the data base. However, the review accepts the ***qualitative*** evidence that anionic sulphite complexes may be formed.

Combining $\log_{10} \beta^\circ_1$ with the selected auxiliary data leads to the selected Gibbs energy of formation:

$$\Delta_f G^\circ_m(UO_2SO_3, aq, 298.15\,K) = -(1477.7 \pm 5.6)\,kJ \cdot mol^{-1}.$$

No enthalpy, entropy or heat capacity data are available.

b) Aqueous U(V) and U(IV) sulphites

No experimental information is available for any aqueous dioxouranium(V) sulphites. Formation of aqueous uranium(IV) sulphite complexes was reported in a qualitative study by Rosenheim and Kelmy [32ROS/KEL]. However, no experimental chemical thermodynamic data on these species are available.

V.5.1.2.2. Solid uranium sulphites

The only known uranium sulphite solids are $UO_2SO_3(cr)$, $UO_2SO_3 \cdot 4.5H_2O(cr)$ and $U(SO_3)_2(cr)$.

a) UO_2SO_3 (cr)

This review accepts Fuger's [83FUG] selected value of

$$\Delta_f H^\circ_m(UO_2SO_3, cr, 298.15\,K) = -(1661.0 \pm 12.6)\,kJ \cdot mol^{-1}$$

based on unpublished experimental data determined by Brandenburg [78BRA] in 1.5 M H_2SO_4/0.035 M $Ca(SO_4)_2$.

The selected entropy,

$$S^\circ_m(UO_2SO_3, cr, 298.15\,K) = (157 \pm 6)\,J \cdot K^{-1} \cdot mol^{-1},$$

is estimated by this review as described in Appendix D.

The Gibbs energy of formation is calculated from the selected enthalpy of formation and entropy.

$$\Delta_f G^\circ_m(UO_2SO_3, cr, 298.15\,K) = -(1530.4 \pm 12.7)\,kJ \cdot mol^{-1}$$

b) $UO_2SO_3 \cdot 4.5H_2O(cr)$

Fuger [83FUG] selected a value of $\Delta_f H_m^\circ(UO_2SO_3 \cdot 4.5H_2O, cr, 298.15 K) = -(3009.6 \pm 4.2)$ kJ · mol^{-1} based on unpublished data from the thesis of Brandenburg [78BRA].

The Gibbs energy of formation can be calculated from the data of Klygin and Kolyada [59KLY/KOL] and Zakharova and Orlova [67ZAK/ORL]. Using $\log_{10} K^\circ_{s,0} = -(9.1 \pm 0.3)$, *cf.* Section V.5.1.2.1.a, leads to $\Delta_f G_m^\circ(UO_2SO_3 \cdot 4.5H_2O, cr, 298.15 K) = -(2559.1 \pm 4.7)$ kJ · mol^{-1}. On the other hand, by estimating the entropy as $S_m^\circ(UO_2SO_3 \cdot 4.5H_2O, cr, 298.15 K) = (358 \pm 19)$ kJ · mol^{-1} and using the value of Fuger [83FUG] for the standard enthalpy of reaction of the solid, a Gibbs energy of formation of $\Delta_f G_m^\circ(UO_2SO_3 \cdot 4.5H_2O, cr, 298.15 K) = -(2625.9 \pm 7.1)$ kJ · mol^{-1} is obtained. The two values differ so much that this review does not find it meaningful to select a standard free energy of formation for $UO_2SO_3 \cdot 4.5H_2O(cr)$. Additional experiments are needed to resolve this problem.

c) $U(SO_3)_2(cr)$

Erdös [62ERD] used a correlation method to predict a value of $\Delta_f H_m^\circ(U(SO_3)_2, cr, 298.15 K) = -(1907 \pm 13)$ kJ · mol^{-1}. With another method [62ERD2] he obtained $\Delta_r H_m^\circ(V.110) = 183.6$ kJ · mol^{-1} for the reaction

$$U(SO_3)_2(cr) \rightleftharpoons UO_2(cr) + 2SO_2(g). \qquad (V.110)$$

From this value Cordfunke and O'Hare [78COR/OHA] obtained $\Delta_f H_m^\circ(U(SO_3)_2, cr, 298.15 K) = -1862$ kJ · mol^{-1}. Their selected value of

$$\Delta_f H_m^\circ(U(SO_3)_2, cr, 298.15 K) = -(1883 \pm 21) \text{ kJ} \cdot \text{mol}^{-1}$$

is a rounded average of these two estimates. Using selected auxiliary data from CODATA [89COX/WAG] and from this review to repeat these calculations yields a value insignificantly different from that of the IAEA review [78COR/OHA]. Their value is therefore accepted here.

This review has repeated the estimation by Cordfunke and O'Hare [78COR/OHA] to obtain of $S_m^\circ(U(SO_3)_2, cr, 298.15 K)$, *cf.* Appendix D and found

$$S_m^\circ(U(SO_3)_2, cr, 298.15 K) = (159 \pm 9) \text{ J} \cdot \text{K}^{-1} \cdot \text{mol}^{-1}.$$

The Gibbs energy of formation is calculated from the selected enthalpy of formation and entropy.

$$\Delta_f G_m^\circ(U(SO_3)_2, cr, 298.15 K) = -(1713 \pm 21) \text{ J} \cdot \text{K}^{-1} \cdot \text{mol}^{-1}$$

There are insufficient data available to select a value for the heat capacity.

V.5.1.3. Uranium sulphates

V.5.1.3.1. Aqueous uranium sulphates

a) Aqueous U(VI) sulphates

The dioxouranium(VI)-sulphate system has been extensively investigated with many different experimental methods. The results of these studies are compiled in Table V.34. Conclusive quantitative evidence exists for the formation of $UO_2SO_4(aq)$ and $UO_2(SO_4)_2^{2-}$. The recent paper of Yang and Pitzer [89YAN/PIT] on the investigation of activity coefficients in dioxouranium(VI) sulphate solutions has to be discarded because it does not consider the complex formation between UO_2^{2+} and SO_4^{2-}, *cf.* Appendix A.[8]

Some of the data listed in Table V.34 for Reaction (V.111)

$$UO_2^{2+} + SO_4^{2-} \rightleftharpoons UO_2SO_4(aq) \qquad (V.111)$$

were determined at temperatures other than 298.15 K. These are recalculated to 25°C using $\Delta_r H_m^\circ(V.111) = (19.5\pm1.6)\,kJ\cdot mol^{-1}$ selected below. The various experimental data with their estimated uncertainties are given in Table V.34. Key papers are discussed in some detail in Appendix A.

[8] A comment on the formation of ternary U(VI)-OH^--SO_4^{2-} complexes: One of the authors (I. Grenthe) is currently investigating this system experimentally. A reinterpretation of the hydrolysis data of Peterson [61PET] in 1.5 M Na_2SO_4 solution indicates clearly that comparatively strong ternary complexes are formed. The equilibrium constants reported by Peterson are conditional constants that refer to the reactions

$$m\text{“}UO_2^{2+}\text{”} + nH_2O(l) + qSO_4^{2-} \rightleftharpoons (\text{“}UO_2\text{”})_m(OH)_n(SO_4)_q^{2m-n-2q} + nH^+,$$

where $[\text{“}UO_2^{2+}\text{”}] = [UO_2^{2+}] + [UO_2SO_4(aq)] + [UO_2(SO_4)_2^{2-}]$. By using $\log_{10}\beta_1 \approx 2$ and $\log_{10}\beta_2 \approx 3$ for the formation of $UO_2SO_4(aq)$ and $UO_2(SO_4)_2^{2-}$ in this medium, the hydrolysis constants $\log_{10}{}^*\beta_{2,2}$, $\log_{10}{}^*\beta_{4,3}$ and $\log_{10}{}^*\beta_{8,5}$ reported by Peterson are recalculated. The original constants reported by Peterson [61PET], the values recalculated by I. Grenthe as explained above, as well as the constants in 3 M $NaClO_4$ are listed below.

	Values reported by Peterson [61PET] (1.5 M Na_2SO_4)	Values of [61PET] recalc. by I. Grenthe (1.5 M Na_2SO_4)	Values recommended by this review (3 M $NaClO_4$)
$\log_{10}{}^*\beta_{2,2}$	−8.17	−1.51	−6.1
$\log_{10}{}^*\beta_{4,3}$	−16.20	−5.96	−13.2
$\log_{10}{}^*\beta_{8,5}$	−32.14	−15.36	not formed

The large difference between the constants in the last two columns indicates that ternary complexes are predominant in the solutions used by Peterson. However, it is not possible to determine the number of bonded sulphates in these complexes because the sulphate concentration was not varied in Peterson's experiments. It is also observed that the stoichiometric coefficients m and n are different in the two media, the complex $(UO_2)_5(OH)_8^{2+}$ is not formed in perchlorate media.

Table V.34: Experimental equilibrium data for the uranium(VI) sulphate system.

Method	Ionic Medium	t (°C)	$\log_{10} K$[(a)]	$\log_{10} \beta_q^{\circ}$[(b)]	Reference
$UO_2^{2+} + SO_4^{2-} \rightleftharpoons UO_2SO_4(aq)$					
qh	1 M $NaClO_4$	20	1.70 ± 0.09	3.02 ± 0.11	[51AHR2]
sp	1 M $NaClO_4$	20	1.75 ± 0.05	3.07 ± 0.09	
con	$I < 0.005$ M	25	3.23 ± 0.07		[54BRO/BUN]
sp	$I = 0.3$ M	25	2.95		[57DAV/MON]
sol	var	25-200	2.72		[60LIE/STO]
sp	1 M $NaClO_4$	25	1.85 ± 0.10[(c)]	3.12 ± 0.12	[60MAT]
cix	1 M $NaClO_4$	32	1.63 ± 0.20[(c)]	2.82 ± 0.21	[61BAN/TRI]
emf	var 0.001-0.03 M	24.7	3.8		[63POZ/STE]
dis	0.150 M NH_4ClO_4	25	2.26 ± 0.05[(c)]	3.20 ± 0.05	[67WAL]
	0.100 M NH_4ClO_4	25	2.34 ± 0.05[(c)]	3.18 ± 0.05	
	0.075 M NH_4ClO_4	25	2.44 ± 0.05[(c)]	3.20 ± 0.05	
	0.050 M NH_4ClO_4	25	2.51 ± 0.05[(c)]	3.17 ± 0.05	
	0.025 M NH_4ClO_4	25	2.65 ± 0.05[(c)]	3.13 ± 0.05	
	0.010 M NH_4ClO_4	25	2.80 ± 0.05[(c)]	3.15 ± 0.05	
	0.150 M NH_4ClO_4	35	2.34		
	0.100 M NH_4ClO_4	35	2.47		
	0.050 M NH_4ClO_4	35	2.62		
	0.150 M NH_4ClO_4	50	2.50		
con	$I < 0.02$ M	25	3.35		[70NIK]
	$I < 0.02$ M	50	[(d)]		
	$I < 0.02$ M	70	[(d)]		
	$I < 0.02$ M	90	[(d)]		
gl	$I \sim 0.001$ M	25	2.93		[71NIK]
rev	0	25		2.72	[78LAN]
rev	0	25		2.9 ± 0.5	[80LEM/TRE]
....					
$UO_2^{2+} + 2SO_4^{2-} \rightleftharpoons UO_2(SO_4)_2^{2-}$					
qh	1 M $NaClO_4$	20	2.54 ± 0.19	3.90 ± 0.24	[51AHR2]
sp	1 M $NaClO_4$	20	2.65 ± 0.05	4.00 ± 0.15	
sol	var	25-200	4.20		[60LIE/STO]
sp	1 M $NaClO_4$	25	2.40 ± 0.20[(c)]	3.65 ± 0.25	[60MAT]
cix	1 M $NaClO_4$	32	3.8		[61BAN/TRI]
dis	0.15 M NH_4ClO_4	25	3.32 ± 0.10[(c)]	4.26 ± 0.10	[67WAL]
	0.15 M NH_4ClO_4	35	3.44		
	0.15 M NH_4ClO_4	50	3.67		
aix	var	25?	4.4		[73MAJ]
rev	0	25		4.16	[78LAN]
....					

Table V.34 (continued)

Method	Ionic Medium	t (°C)	$\log_{10} K^{(a)}$	$\log_{10} \beta_q^{\circ(b)}$	Reference
$UO_2^{2+} + 3SO_4^{2-} \rightleftharpoons UO_2(SO_4)_3^{4-}$					
qh	1 M $NaClO_4$	20	3.4		[51AHR2]
aix	var	25?	3.4		[73MAJ]
....					
$UO_2SO_4(aq) + SO_4^{2-} \rightleftharpoons UO_2(SO_4)_2^{2-}$					
dis	$I = 1$ M $NaClO_4$	25	0.78 ± 0.17	0.76 ± 0.17	[58ALL]
....					
$UO_2^{2+} + HSO_4^- \rightleftharpoons UO_2SO_4(aq) + H^+$					
sp	2.00 M $HClO_4$ $I = 2.65$ M	25	0.70 ± 0.06[c]	2.66 ± 0.14	[49BET/MIC]
dis	2.00 M $HClO_4$	10	0.79 ± 0.15[c]		[54DAY/POW]
dis	2.00 M $HClO_4$	25	0.81 ± 0.15[c]	2.99 ± 0.18	
dis	2.00 M $HClO_4$	40	0.81 ± 0.15[c]		
dis	2.00 M $HClO_4$	25	0.88 ± 0.06[c]	3.07 ± 0.12	[76PAT/RAM]
....					
$UO_2^{2+} + 2HSO_4^- \rightleftharpoons UO_2(SO_4)_2^{2-} + 2H^+$					
dis	2.00 M $HClO_4$	25	1.23 ± 0.13[c]	4.57 ± 0.26	[76PAT/RAM]

(a) $\log_{10} K$ refers to the reactions indicated in the ionic medium and at the temperature given in the table.
(b) $\log_{10} \beta_q^\circ$ (molal units) refers to the formation reactions $UO_2^{2+} + qSO_4^{2-} \rightleftharpoons UO_2(SO_4)_q^{2-2q}$ at $I = 0$ and 298.15 K.
(c) Uncertainties estimated by this review as described in Appendix A.
(d) See comments in Appendix A.

After conversion to molality constants, the correction to $I = 0$ is done with the specific ion interaction theory with $\Delta\varepsilon(V.111) = -(0.34 \pm 0.07)$, calculated with the ion interaction coefficients in Appendix B. The results shown in Table V.34 agree very well. Equilibria of the type

$$UO_2^{2+} + HSO_4^- \rightleftharpoons UO_2SO_4(aq) + H^+ \qquad (V.112)$$

were studied in acid solutions [49BET/MIC, 54DAY/POW, 76PAT/RAM]. The data from these experiments are corrected to zero ionic strength using the specific ion interaction theory with $\Delta\varepsilon(V.112) = -(0.33 \pm 0.04)$ (see discussion in Appendix A for more details). The resulting equilibrium constants at $I = 0$ and 298.15 K are combined with that of the reaction $SO_4^{2-} + H^+ \rightleftharpoons HSO_4^-$, $\log_{10} K_1^\circ = (1.98 \pm 0.05)$ (see Chapter VI) to obtain the equilibrium constant for Reaction (V.111). These values of $\log_{10}\beta_1^\circ$ are listed in Table V.34. The values are consistent with the set of values obtained from the experiments in less acid solutions. This review selects the weighted average of the 13 values obtained from Refs. [49BET/MIC, 51AHR2, 54DAY/POW, 60MAT, 61BAN/TRI, 67WAL, 76PAT/RAM],

$$\log_{10}\beta_1^\circ(V.111, 298.15\,K) = 3.15 \pm 0.02.$$

The Gibbs energy of formation is calculated by using the auxiliary data selected by this review.

$$\Delta_f G_m^\circ(UO_2SO_4, aq, 298.15\,K) = -(1714.5 \pm 1.8)\,kJ \cdot mol^{-1}$$

The equilibrium constant for Reaction (V.113)

$$UO_2^{2+} + 2SO_4^{2-} \rightleftharpoons UO_2(SO_4)_2^{2-} \qquad (V.113)$$

are evaluated from the data in Refs. [51AHR2, 60MAT, 67WAL]. They are corrected to $I = 0$ by using the estimates $\varepsilon_{(UO_2(SO_4)_2^{2-},NH_4^+)} \approx \varepsilon_{(SO_4^{2-},NH_4^+)} \approx \varepsilon_{(SO_4^{2-},Na^+)} = -(0.12 \pm 0.06)$, *i.e.*, $\Delta\varepsilon(V.113) = -(0.34 \pm 0.14)$. The results are shown in Table V.34. The values from Ref. [51AHR2] are corrected to 25°C with $\Delta_r H_m^\circ(V.113) = (35.1 \pm 1.0)\,kJ \cdot mol^{-1}$, *cf.* Appendix A. Patil and Ramakrishna [76PAT/RAM] determined $\log_{10}{}^*\beta_2(V.114)$ in acid solution.

$$UO_2^{2+} + 2HSO_4^- \rightleftharpoons UO_2(SO_4)_2^{2-} + 2H^+ \qquad (V.114)$$

From their value this review calculates $\log_{10}\beta_2^\circ(V.113) = (4.57 \pm 0.26)$, which is not inconsistent with the value obtained from less acid solutions, and this review selects the weighted average of the five values obtained from Refs. [51AHR2, 60MAT, 67WAL, 76PAT/RAM],

$$\log_{10}\beta_2^\circ(V.113, 298.15\,K) = 4.14 \pm 0.07.$$

From the data of Allen [58ALL] this review calculates the stepwise equilibrium constant for the reaction $UO_2SO_4(aq) + SO_4^{2-} \rightleftharpoons UO_2(SO_4)_2^{2-}$, $\log_{10} K_2^\circ = (0.76 \pm 0.17)$. This constant is independent of the ionic strength and is in fair agreement with the value $\log_{10} K_2^\circ = (1.00 \pm 0.07)$ calculated from the values of $\log_{10}\beta_1^\circ$ and $\log_{10}\beta_2^\circ$ selected above.

The Gibbs energy of formation is calculated by using the auxiliary data selected by this review.

$$\Delta_f G^\circ_m(UO_2(SO_4)_2^{2-}, aq, 298.15\,K) = -(2464.2 \pm 2.0)\,kJ \cdot mol^{-1}$$

There is some evidence to suggest the existence of $UO_2(SO_4)_3^{4-}$ (*cf.* Ahrland and Kullberg [71AHR/KUL3] and Majchrzak [73MAJ]). However, this complex is very weak (if formed at all), and large changes in the ionic medium occurred in the range where it is formed. Allen [58ALL] concluded that the present evidence suggested only very low or negligible proportions of the trisulphate complex at sulphate molarities below one. Formation data for this species were often tabulated (*e.g.*, Refs. [78ALL/BEA, 81TUR/WHI, 82JEN, 86WAN, 87BRO/WAN]). Based on these data, the species is weak, at most. This review does not assign any thermodynamic data to this species as its formation has not been unambiguously demonstrated.

The enthalpy changes for the reactions (V.111) and (V.113) were obtained calorimetrically by Bailey and Larson [71BAI/LAR] at $I = 0$, by Ahrland and Kullberg [71AHR/KUL3] at $I = 1$ M, and by Ullman and Schreiner [86ULL/SCH] at ~0.005 M $< I <$ ~0.015 M. The agreement between the $\Delta_r H_m$ values for Reaction (V.111) is fair, $(20.8 \pm 0.3)\,kJ \cdot mol^{-1}$, $(18.23 \pm 0.17)\,kJ \cdot mol^{-1}$, and $(19.6 \pm 0.7)\,kJ \cdot mol^{-1}$, respectively, indicating that $\Delta_r H_m$(V.111) is not strongly dependent on the ionic strength. This is confirmed by the value of $\Delta_r H_m$(V.111) obtained at $I = 0.15$ M by Wallace [67WAL] using $\log_{10} K(T)$ data. Wallace [67WAL] found $\Delta_r H_m$(V.111) = $(21 \pm 2)\,kJ \cdot mol^{-1}$. The enthalpy value from Day and Powers [54DAY/POW], $\Delta_r H_m$(V.111) = $10\,kJ \cdot mol^{-1}$, differs considerably from the other authors and is not credited in this review.

This review selects the unweighted average of the data reported by Bailey and Larson [71BAI/LAR], Ahrland and Kullberg [71AHR/KUL3] and Ullman and Schreiner [86ULL/SCH], where the uncertainty is chosen to span the uncertainty range of all the values involved.

$$\Delta_r H^\circ_m(\text{V.111}, 298.15\text{ K}) = (19.5 \pm 1.6)\,kJ \cdot mol^{-1}$$

The standard enthalpy of formation is calculated by using the auxiliary data selected by this review.

$$\Delta_f H^\circ_m(UO_2SO_4, aq, 298.15\,K) = -(1908.8 \pm 2.2)\,kJ \cdot mol^{-1}$$

The enthalpy of Reaction (V.113) was determined by Ahrland and Kullberg [71AHR/KUL3], Wallace [67WAL] and Ullman and Schreiner [86ULL/SCH]. The values of $\Delta_r H_m$(V.113) = $(35.11 \pm 0.40)\,kJ \cdot mol^{-1}$, $(29 \pm 4)\,kJ \cdot mol^{-1}$ and $(37.8 \pm 2.0)\,kJ \cdot mol^{-1}$ are in fair agreement. This review selects the value of Ahrland and Kullberg [71AHR/KUL3] as the most precise one, but assigns a larger uncertainty.

$$\Delta_r H^\circ_m(\text{V.113}, 298.15\text{ K}) = (35.1 \pm 1.0)\,kJ \cdot mol^{-1}.$$

The enthalpy of formation is calculated by using the auxiliary data selected by this review.

$$\Delta_f H^\circ_m(UO_2(SO_4)_2^{2-}, aq, 298.15\,K) = -(2802.6 \pm 2.0)\,kJ \cdot mol^{-1}$$

Using the selected reaction data, the entropies are calculated via $\Delta_r S^\circ_m$:

$$S^\circ_m(UO_2SO_4, aq, 298.15\,K) = (46.0 \pm 6.2)\,J \cdot K^{-1} \cdot mol^{-1}$$
$$S^\circ_m(UO_2(SO_4)_2^{2-}, aq, 298.15\,K) = (135.8 \pm 4.8)\,J \cdot K^{-1} \cdot mol^{-1}.$$

The data selected for $UO_2SO_4(aq)$ and $UO_2(SO_4)_2^{2-}$ are in fair agreement with the values proposed in other reviews, *cf.* Refs. [80LEM/TRE] and [78LAN].

For the heat capacity of $UO_2SO_4(aq)$, Lemire and Tremaine [80LEM/TRE] estimated a mean value of $C^\circ_{p,m}|^{473}_{298}$ $(UO_2SO_4, aq) = 354\,J \cdot K^{-1} \cdot mol^{-1}$ based on the temperature dependence of the complexation reaction (V.113).[9]

b) Aqueous U(V) sulphates

No experimental information is available on any aqueous dioxouranium(V) sulphates.

c) Aqueous U(IV) sulphates

The uranium(IV)-sulphate system has been studied in strongly acidic solutions to avoid hydrolysis. All available experimental data are compiled in Table V.35. They refer to reactions of the type

$$U^{4+} + qHSO_4^- \rightleftharpoons U(SO_4)_q^{4-2q} + qH^+. \qquad (V.115)$$

The data are limited. Only those of Betts and Leigh [50BET/LEI] (recalculated by Sullivan and Hindman [52SUL/HIN]) and Day, Wilhite and Hamilton [55DAY/WIL] are accepted by this review. The data do not allow an independent determination of $\log_{10}{}^*\beta^\circ_q$ and $\Delta\varepsilon$. This review uses experimental and estimated ion interaction coefficients, $\varepsilon_{(U^{4+},ClO_4^-)} = (0.76 \pm 0.06)$ and $\varepsilon_{(USO_4^{2+},ClO_4^-)} = (0.3 \pm 0.1)$, *i.e.*, $\Delta\varepsilon(V.115, q = 1) = -(0.31 \pm 0.12)$ and $\Delta\varepsilon(V.115, q = 2) = -(0.46 \pm 0.08)$, to calculate $\log_{10}{}^*\beta^\circ_q$. The study of Rao and Pai [69RAO/PAI] is precise, but the correction to $I = 0$ of these data from a mixed $(H^+, Na^+)(ClO_4^-, NO_3^-)$ ionic medium of high ionic strength is not sufficiently accurate to permit their use. The study of Vdovenko, Romanov and Shcherbakov [63VDO/ROM3] was also made in a mixed ionic medium which makes it difficult to perform a correction to $I = 0$. However, the equilibrium constants reported in Refs. [69RAO/PAI] and [63VDO/ROM3] are of the same magnitude as the constants selected in this review. From Refs. [52SUL/HIN] and [55DAY/WIL] this review obtains, after conversion to molality constants and correction to $I = 0$, values of $\log_{10}{}^*\beta^\circ_1(V.115, q = 1)$ and $\log_{10}{}^*\beta^\circ_2(V.115, q = 2)$. Combining these equilibrium constants with the protonation constant for SO_4^{2-} (*cf.* Chapter VI) leads to the equilibrium constants for Reactions (V.116), as shown in Table V.35.

$$U^{4+} + qSO_4^{2-} \rightleftharpoons U(SO_4)_q^{4-2q} \qquad (V.116)$$

This review selects the weighted average values from Refs. [52SUL/HIN] and [55DAY/WIL] for both $\log_{10}\beta^\circ_1$ and $\log_{10}\beta^\circ_2$:

[9] In a recent publication, Nguyen-Trung and Hovey [90NGU/HOV] reported the standard state heat capacity function of $UO_2SO_4(aq)$ from 283 to 328 K, $C^\circ_{p,m}(UO_2SO_4, T) = (-2773.1 + 13.636 \times T - 0.015322 \times T^2)\,J \cdot K^{-1} \cdot mol^{-1}$.

Table V.35: Experimental equilibrium data for the uranium(IV) sulphate system.

Method	Ionic Medium	t (°C)	$\log_{10} K^{(a)}$	$\log_{10} {}^*\beta_q^{o(b)}$	$\log_{10} \beta_q^{o(c)}$	Reference
$U^{+4} + HSO_4^- \rightleftharpoons USO_4^{2+} + H^+$						
dis	2 M $HClO_4$	25	2.53			[50BET/LEI]
dis	2 M $HClO_4$	25	$2.41 \pm 0.05^{(d)}$	4.54 ± 0.27	6.52 ± 0.27	[52SUL/HIN]
dis	$(H^+, Na^+)ClO_4^-$ 1 M H^+, $I = 2$ M	10	$2.63 \pm 0.05^{(d)}$			[55DAY/WIL]
dis	$(Na^+, H^+)ClO_4^-$ 1 M H^+, $I = 2$ M	25	$2.52 \pm 0.05^{(d)}$	4.65 ± 0.27	6.63 ± 0.27	
dis	$(Na^+, H^+)ClO_4^-$ 1 M H^+, $I = 2$ M	40	$2.38 \pm 0.05^{(d)}$			
dis	(Na^+, H^+) (ClO_4^-, NO_3^-) $[H^+] = 1.00$ M $[NO_3^-] = 1.00$ M $I = 3.8$ M	25?	$2.36 \pm 0.06^{(d)}$			[69RAO/PAI]
....						
$U^{+4} + 2HSO_4^- \rightleftharpoons U(SO_4)_2(aq) + 2H^+$						
dis	2 M $HClO_4$	25	4.93			[50BET/LEI]
dis	2 M $HClO_4$	25	3.72 ± 0.25	6.46 ± 0.31	10.42 ± 0.33	[52SUL/HIN]
dis	$(H^+, Na^+)ClO_4^-$ 1 M H^+, $I = 2$ M	10	$3.97 \pm 0.15^{(d)}$			[55DAY/WIL]
dis	$(Na^+, H^+)ClO_4^-$ 1 M H^+, $I = 2$ M	25	$3.87 \pm 0.15^{(d)}$	6.60 ± 0.23	10.56 ± 0.25	
dis	$(Na^+, H^+)ClO_4^-$ 1 M H^+, $I = 2$ M	40	$3.76 \pm 0.15^{(d)}$			
dis	(Na^+, H^+) (ClO_4^-, NO_3^-) $[H^+] = 1.00$ M $[NO_3^-] = 1.00$ M $I = 3.8$ M	25?	$3.73^{(e)} \pm 0.06^{(d)}$			[69RAO/PAI]
....						
$U^{+4} + SO_4^{2-} \rightleftharpoons USO_4^{2+}$						
nmr	2 M $HClO_4$	20	1.70			[63VDO/ROM3]
pot	2 M $NaClO_4$	25	3.8 ± 0.1		6.8 ± 0.3	[77NIK/TSV]
pot	1.8 M $NaClO_4$	25	4.0 ± 0.3		7.0 ± 0.4	
pot	1.6 M $NaClO_4$	25	4.1 ± 0.3		7.1 ± 0.4	
pot	1.4 M $NaClO_4$	25	4.3 ± 0.3		7.3 ± 0.3	
pot	1.1 M $NaClO_4$	25	4.4 ± 0.3		7.4 ± 0.3	
pot	0.4 M $NaClO_4$	25	5.1 ± 0.2		7.6 ± 0.2	
rev	0	25			5.46	[78LAN]
rev	0	25			5 ± 1	[80LEM/TRE]

Table V.35 (continued)

Method	Ionic Medium	t (°C)	$\log_{10} K^{(a)}$	$\log_{10} {}^*\beta_q^{o(b)}$	$\log_{10} \beta_q^{o(c)}$	Reference
$U^{+4} + 2SO_4^{2-} \rightleftharpoons U(SO_4)_2(aq)$						
pot	2 M $NaClO_4$	25	4.9 ± 0.1		9.3 ± 0.3	[77NIK/TSV]
pot	1.8 M $NaClO_4$	25	5.2 ± 0.2		9.6 ± 0.3	
pot	1.6 M $NaClO_4$	25	5.3 ± 0.1		9.7 ± 0.2	
pot	1.4 M $NaClO_4$	25	5.4 ± 0.3		9.8 ± 0.4	
pot	1.1 M $NaClO_4$	25	5.7 ± 0.3		10.1 ± 0.3	
pot	0.4 M $NaClO_4$	25	6.8 ± 0.3		10.6 ± 0.3	
rev	0	25			9.75	[78LAN]
rev	0	25			10 ± 3	[80LEM/TRE]

(a) $\log_{10} K$ refers to the reactions indicated in the ionic medium and at the temperature given in the table.
(b) $\log_{10} {}^*\beta_q^o$ (in molal units) refers to the formation reactions involving the deprotonation of sulphate, $U^{4+} + q HSO_4^- \rightleftharpoons U(SO_4)_q^{4-2q} + qH^+$, at $I = 0$ and 298.15 K.
(c) $\log_{10} \beta_q^o$ (in molal units) refers to the formation reactions $U^4 + + qSO_4^{2-} \rightleftharpoons U(SO_4)_q^{4-2q}$ at $I = 0$ and 298.15 K.
(d) Uncertainties estimated by this review as described in Appendix A.
(e) See comments in Appendix A.

$$\log_{10} \beta_1^o(\text{V.116}, q = 1, 298.15\ K) = 6.58 \pm 0.19$$
$$\log_{10} \beta_2^o(\text{V.116}, q = 2, 298.15\ K) = 10.51 \pm 0.20$$

This review does not find it justified to take the data of Nikolaeva and Tsveldub [77NIK/TSV] into account, *cf.* comments in Appendix A.

The standard Gibbs energies of formation are obtained by using auxiliary data for U^{4+} and SO_4^{2-} from Section V.2.3 and Chapter VI, respectively.

$$\Delta_f G_m^o(USO_4^{2+}, aq, 298.15\,K) = -(1311.4 \pm 2.1)\,kJ \cdot mol^{-1}$$
$$\Delta_f G_m^o(U(SO_4)_2, aq, 298.15\,K) = -(2077.9 \pm 2.3)\,kJ \cdot mol^{-1}$$

These values are in good agreement with those obtained in earlier reviews [80LEM/TRE, 78LAN].

The value $\Delta_r H_m(\text{V.116}, q = 1, I = 2\ \text{M}) = 42.3\,kJ \cdot mol^{-1}$ determined by Nikolaeva and Tsveldub [77NIK/TSV], based on the measurements of the temperature dependence of $\log_{10} \beta_1$ in the range 298 to 398 K, appears very high in comparison with corresponding enthalpy data for Th^{4+} [59ZIE], Np^{4+} [54SUL/HIN] and Pu^{4+} [83NAS/CLE2]. This review therefore does not consider the enthalpy data from Ref. [77NIK/TSV], but prefers the values obtained by recalculating the data reported by Day, Wilhite and Hamilton [55DAY/WIL] at $I = 2$ M and 10, 25 and 40°, *cf.* Appendix A. The values are combined with the selected enthalpy of the sulphate protonation reaction, resulting in

$$\begin{aligned}\Delta_r H_m^\circ(\text{V.116}, q = 1, 298.15\text{ K}) &= (8.0 \pm 2.7)\,\text{kJ}\cdot\text{mol}^{-1}\\ \Delta_r H_m^\circ(\text{V.116}, q = 2, 298.15\text{ K}) &= (32.7 \pm 2.8)\,\text{kJ}\cdot\text{mol}^{-1}.\end{aligned}$$

The selected $\Delta_f H_m^\circ$ values of USO_4^{2+} and $UO_2(SO_4)_2(aq)$ are obtained by using auxiliary data selected in this review.

$$\begin{aligned}\Delta_f H_m^\circ(\text{USO}_4^{2+}, \text{aq}, 298.15\,\text{K}) &= -(1492.5 \pm 4.3)\,\text{kJ}\cdot\text{mol}^{-1}\\ \Delta_f H_m^\circ(\text{U(SO}_4)_2, \text{aq}, 298.15\,\text{K}) &= -(2377.2 \pm 4.4)\,\text{kJ}\cdot\text{mol}^{-1}\end{aligned}$$

Using the selected reaction data, the entropies are calculated via $\Delta_r S_m^\circ$.

$$\begin{aligned}S_m^\circ(\text{USO}_4^{2+}, \text{aq}, 298.15\,\text{K}) &= -(246 \pm 16)\,\text{J}\cdot\text{K}^{-1}\cdot\text{mol}^{-1}\\ S_m^\circ(\text{U(SO}_4)_2, \text{aq}, 298.15\,\text{K}) &= -(69 \pm 16)\,\text{J}\cdot\text{K}^{-1}\cdot\text{mol}^{-1}\end{aligned}$$

No experimental heat capacity data are available. A mean value of $C_{p,m}^\circ|_{298}^{473}$ (USO_4^{2+}, aq) = $(121 \pm 20)\,\text{J}\cdot\text{K}^{-1}\cdot\text{mol}^{-1}$ was calculated by Lemire and Tremaine [80LEM/TRE] but no value for 298.15 K is accepted in this review.

No species above the uranium(IV) disulphate ion are credited in this review.

V.5.1.3.2. Solid uranium sulphates

a) UO_2SO_4 (cr) and its hydrates

The dioxouranium(VI) sulphate hydrates $UO_2SO_4 \cdot qH_2O(cr)$ with $q = 0.5$, 1, 2, 2.5, 3 and 3.5 have been identified [69COR, 72COR]. Only the 2.5-hydrate is thermally stable at room temperature [72COR]. No chemical thermodynamic data are available for the metastable phases $UO_2SO_4 \cdot 0.5H_2O(cr)$ and $UO_2SO_4 \cdot 2H_2O(cr)$, or for the monohydrates α-, β- and γ-$UO_2SO_4 \cdot H_2O$.

UO_2SO_4 (cr): Two forms of $UO_2SO_4(cr)$ exist. The orthorhombic α-phase is always metastable with respect to the monoclinic β-phase, with which it can coexist over a wide temperature range [72COR]. No data are available for the α-phase.

The enthalpy of solution of $UO_2SO_4(cr)$ was first measured by Favre and Silbermann [1853FAV/SIL]. This review accepts the finding of Cordfunke and O'Hare [78COR/OHA] that the works of Beck [28BEC] and Owens and Mayer [64OWE/MAY] must be disregarded due to the poorly defined reactions studied. Cordfunke [72COR] measured the enthalpy of solution of β-UO_3 and $UO_2SO_4(cr)$ in 1 M H_2SO_4. He found $\Delta_{sol}H_m^\circ(UO_3, \beta, 298.15\text{ K})$ and $\Delta_{sol}H_m^\circ(UO_2SO_4, \text{cr}, 298.15\text{ K})$ to be $-(88.7\pm0.3)$ and $-(63.1\pm0.2)\,\text{kJ}\cdot\text{mol}^{-1}$, respectively, in that medium. Combining these values with $\Delta_f H_m^\circ$ of $H_2O(l)$ and $H_2SO_4(aq)$ in this medium, Cordfunke and O'Hare [78COR/OHA] obtained $\Delta_f H_m^\circ(UO_2SO_4, \text{cr}, 298.15\text{ K}) = -(1845.6\pm0.8)\,\text{kJ}\cdot\text{mol}^{-1}$. This differs somewhat from Cordfunke's [72COR] own calculated value of $\Delta_f H_m^\circ(UO_2SO_4, \text{cr}, 298.15\text{ K}) = -1851\,\text{kJ}\cdot\text{mol}^{-1}$, which was not obtained with auxiliary data for 1 M H_2SO_4. In a more recent work, Cordfunke and Ouweltjes

[77COR/OUW2] repeated their experiment in 1.5 M H_2SO_4, obtaining $-(84.2 \pm 0.2)$ and $-(62.4 \pm 0.2)\,kJ \cdot mol^{-1}$ for $\Delta_{sol}H^\circ_m(UO_3, \gamma, 298.15\,K)$ and $\Delta_{sol}H^\circ_m(UO_2SO_4$, cr, 298.15 K), respectively, which leads to $\Delta_{sol}H^\circ_m(UO_2SO_4$, cr, 298.15 K) $= -(1844.7 \pm 1.0)\,kJ \cdot mol^{-1}$. Cordfunke and O'Hare [78COR/OHA] averaged these two values of $\Delta_f H^\circ_m(UO_2SO_4$, cr, 298.15 K) to obtain their selected value of

$$\Delta_f H^\circ_m(UO_2SO_4, cr, 298.15\,K) \quad = \quad -(1845.14 \pm 0.84)\,kJ \cdot mol^{-1}$$

which is accepted in this review.

Experimental data are not available for determining either $\Delta_f G^\circ_m$ or S°_m. For this reason, Latimer's method [52LAT] was used by Cordfunke and O'Hare [78COR/OHA] to estimate S°_m. Their selected value, which is essentially the same as the value calculated as per Appendix D of this review, is accepted here.

$$S^\circ_m(UO_2SO_4, cr, 298.15\,K) \quad = \quad (163.2 \pm 8.4)\,J \cdot K^{-1} \cdot mol^{-1}$$

The Gibbs energy of formation is calculated from the selected enthalpy of formation and entropy.

$$\Delta_f G^\circ_m(UO_2SO_4, cr, 298.15\,K) \quad = \quad -(1685.8 \pm 2.6)\,kJ \cdot mol^{-1}$$

Calorimetric determination of $C_{p,m}$ from near 0 K to room temperature, to obtain S°_m, or measurement of the solubility product of this solid, seems desirable.

Neither the characterization of UO_2SO_4(cr) nor the calorimetry performed by Owens and Mayer [64OWE/MAY] are in question. These measurements, performed to 869 K, were extrapolated to 298.15 K to obtain the selected

$$C^\circ_{p,m}(UO_2SO_4, cr, 298.15\,K) \quad = \quad (145 \pm 3)\,J \cdot K^{-1} \cdot mol^{-1}.$$

The uncertainty is estimated in this review. The $C^\circ_{p,m}(T)$ data tabulated by the IAEA [78COR/OHA] are fitted between 298.15 and 600 K to obtain the selected $C^\circ_{p,m}(T)$ function listed in Table III.3.

$UO_2SO_4 \cdot 2.5H_2O$(cr): This is the only stable hydrated UO_2SO_4 compound. The thermodynamics were reviewed by the IAEA [78COR/OHA].

The heat of solution of 1 M $UO_2SO_4 \cdot 2.5H_2O$(cr) in H_2SO_4 was determined by Cordfunke [72COR], $\Delta sol H_m(V.117, 1\ MH_2SO_4, 298.15\,K) = -(15.90 \pm 0.29)\,kJ \cdot mol^{-1}$.

$$UO_2SO_4 \cdot 2.5H_2O(cr) \rightleftharpoons UO_2^{2+}(sln) + SO_4^{2-}(sln) + 2.5H_2O(sln) \qquad (V.117)$$

By combining this value with $\Delta_{sol}H_m(UO_2SO_4$, cr, 1 M H_2SO_4, 298.15 K) $= -(63.1 \pm 0.2)\,kJ \cdot mol^{-1}$ (see above), and $\Delta_f H_m(H_2O$, sln, 1 M H_2SO_4, 298.15 K) $= -285.85\,kJ \cdot mol^{-1}$ [78COR/OHA, *p*.34], an enthalpy of formation of $\Delta_f H^\circ_m(UO_2SO_4 \cdot 2.5H_2O$, cr, 298.15 K) $= -(2607.0 \pm 0.9)\,kJ \cdot mol^{-1}$ is obtained. This value is consistent with the one selected in the IAEA review [78COR/OHA], $\Delta_f H^\circ_m(UO_2SO_4 \cdot 2.5H_2O$, cr, 298.15 K) $= -(2607.5 \pm 0.8)\,kJ \cdot mol^{-1}$.

Cordfunke and O'Hare [78COR/OHA] used the unpublished solubility data of Cordfunke [75COR2] to calculate $\Delta_r G^\circ_m(V.117) = (9.07 \pm 0.11)\,kJ \cdot mol^{-1}$. The measured solubility is high (4.32 m at 298.15 K), and the corrections for non-ideality were made by using mean activity coefficient data from Robinson and Stokes [59ROB/STO], $\gamma_\pm = 0.0453$ and $a_{H_2O} = 0.854$. These data were recently used by Yang and Pitzer [89YAN/PIT] and seem to be reliable. The uncertainty reported by Cordfunke [75COR2] seems small, but this review has no basis for a re-evaluation.

$$\Delta_r G^\circ_m(\text{V.117, 298.15 K}) = (9.07 \pm 0.11)\,\text{kJ}\cdot\text{mol}^{-1}$$

From the $\Delta_f G^\circ_m(UO_2SO_4{\cdot}2.5H_2O$, cr, 298.15 K) derived from the Gibbs energy of reaction and the $\Delta_f H^\circ_m(UO_2SO_4{\cdot}2.5H_2O$, cr, 298.15 K), the value $S^\circ_m(UO_2SO_4{\cdot}2.5H_2O$, cr, 298.15 K) $= (246.1 \pm 6.8)\,\text{J}\cdot\text{K}^{-1}\cdot\text{mol}^{-1}$ is obtained. An estimation of the entropy using Latimer's method (Appendix D) yields $(274 \pm 16)\,\text{J}\cdot\text{K}^{-1}\cdot\text{mol}^{-1}$. The difference between the two values is significant, and an independent calorimetric determination of the entropy is desirable. This review selects the values listed below, where the uncertainties in the enthalpy of formation and the entropy might be underestimated.

$$\begin{aligned}
\Delta_f G^\circ_m(UO_2SO_4\cdot 2.5H_2O, \text{cr}, 298.15\,\text{K}) &= -(2298.5 \pm 1.8)\,\text{kJ}\cdot\text{mol}^{-1}\\
\Delta_f H^\circ_m(UO_2SO_4\cdot 2.5H_2O, \text{cr}, 298.15\,\text{K}) &= -(2607.0 \pm 0.9)\,\text{kJ}\cdot\text{mol}^{-1}\\
S^\circ_m(UO_2SO_4\cdot 2.5H_2O, \text{cr}, 298.15\,\text{K}) &= (246.1 \pm 6.8)\,\text{J}\cdot\text{K}^{-1}\cdot\text{mol}^{-1}
\end{aligned}$$

$UO_2SO_4\cdot 3H_2O$(cr): This is not a stable compound as it decomposes rapidly to $UO_2SO_4\cdot 2.5H_2O$(cr) at room temperature [72COR]. Cordfunke [72COR] measured the vapor pressure as a function of temperature (287 to 313 K) for the partial dehydration reaction

$$UO_2SO_4\cdot 3.5H_2O(\text{cr}) \rightleftharpoons UO_2SO_4\cdot 3H_2O(\text{cr}) + 0.5H_2O(\text{g}) \qquad (\text{V.118})$$

to be

$$\log_{10} p_{H_2O}(\text{atm}) = -(2651 \pm 14)\,T^{-1} + (7.224 \pm 0.001).$$

Cordfunke and O'Hare [78COR/OHA] used this to find the value of $\Delta_r H^\circ_m$(V.118, 298.15 K) $= (25.40 \pm 0.13)\,\text{kJ}\cdot\text{mol}^{-1}$, from which this review obtains $\Delta_f H^\circ_m(UO_2SO_4{\cdot}3H_2O$, cr, 298.15 K) $= -(2755.3 \pm 0.8)\,\text{kJ}\cdot\text{mol}^{-1}$. Note that this value is only $1.0\,\text{kJ}\cdot\text{mol}^{-1}$ from the average of the $\Delta_f H^\circ_m$ selected values of $UO_2SO_4{\cdot}3.5H_2O$(cr) and $UO_2SO_4{\cdot}2.5H_2O$(cr). Cordfunke also made an experimentally based estimate of the enthalpy of solution of $UO_2SO_4{\cdot}3H_2O$(cr), from which he calculated $\Delta_f H^\circ_m(UO_2SO_4{\cdot}3H_2O$, cr, 298.15 K) $= -2747.6\,\text{kJ}\cdot\text{mol}^{-1}$. This review selects the unweighted average of these two values,

$$\Delta_f H^\circ_m(UO_2SO_4\cdot 3H_2O, \text{cr}, 298.15\,\text{K}) = -(2751.5 \pm 4.6)\,\text{kJ}\cdot\text{mol}^{-1},$$

which is very close to the value selected in the IAEA review [78COR/OHA], $\Delta_f H^\circ_m(UO_2SO_4{\cdot}3H_2O$, cr, 298.15 K) $= -(2748.9 \pm 4.2)\,\text{kJ}\cdot\text{mol}^{-1}$.

Cordfunke and O'Hare [78COR/OHA] did not make a selection of a Gibbs energy of formation. In this review, $0.5\log_{10} p_{H_2O}(\text{V.118}) = \log_{10} K(\text{V.118})$ is used, after correction to the 0.1 MPa standard state pressure, as the selected equilibrium constant:

$$\log_{10} K^\circ(\text{V.118}, 298.15\text{ K}) = -0.831 \pm 0.023.$$

Using $\Delta_f G^\circ_m(UO_2SO_4 \cdot 3.5H_2O, cr, 298.15\,K)$ selected below, the Gibbs energy of formation is obtained:

$$\Delta_f G^\circ_m(UO_2SO_4 \cdot 3H_2O, cr, 298.15\,K) = -(2416.6 \pm 1.8)\,kJ \cdot mol^{-1}.$$

From the derived reaction entropy, $\Delta_r S^\circ_m(\text{V.118}) = (69.28 \pm 0.62)\,J \cdot K^{-1} \cdot mol^{-1}$, the selected entropy value is obtained,

$$S^\circ_m(UO_2SO_4 \cdot 3H_2O, cr, 298.15\,K) = (274 \pm 17)\,J \cdot K^{-1} \cdot mol^{-1}.$$

This value is consistent with the estimated entropy $S^\circ_m(UO_2SO_4{\cdot}3H_2O$, cr, 298.15 K) $= (297 \pm 19)\,J{\cdot}K^{-1}{\cdot}mol^{-1}$. Although Owens and Mayer [64OWE/MAY] reported $C_{p,m}$ data for $UO_2SO_4 \cdot 3H_2O(cr)$ from measurements from 373 to 470 K, Cordfunke and O'Hare [78COR/OHA] convincingly argued that the phase they studied was actually $UO_2SO_4 \cdot 2.5H_2O(cr)$ contaminated with water. Thus, no $C_{p,m}$ data are selected.

Some older publications reporting investigations on $UO_2SO_4 \cdot 3H_2O(cr)$ were discussed by Cordfunke and O'Hare [78COR/OHA] and found unreliable.

$UO_2SO_4 \cdot 3.5H_2O(cr)$: Cordfunke [72COR] reported an enthalpy of solution of $\Delta_{sol}H_m$(V.119, 1 M H_2SO_4, 298.15 K) $= -(7.4 \pm 0.3)\,kJ \cdot mol^{-1}$ for this unstable hydrate.

$$UO_2SO_4 \cdot 3.5H_2O(cr) \rightleftharpoons UO_2^{2+} + SO_4^{2-} + 3.5H_2O(l) \qquad (\text{V.119})$$

By using the same auxiliary data as for β-UO_2SO_4 discussed above, Cordfunke and O'Hare [78COR/OHA] found

$$\Delta_f H^\circ_m(UO_2SO_4 \cdot 3.5H_2O, cr, 298.15\,K) = -(2901.6 \pm 0.8)\,kJ \cdot mol^{-1}.$$

Cordfunke and O'Hare [78COR/OHA] used the unpublished solubility data of Cordfunke [75COR2] to calculate

$$\Delta_r G^\circ_m(\text{V.119, 298.15 K}) = (9.05 \pm 0.11)\,kJ \cdot mol^{-1},$$

using mean activity coefficient data from Robinson and Stokes [59ROB/STO]. This review finds the reported uncertainty small but accepts it because of lack of experimental details.

The Gibbs energy of formation derived therefrom is

$$\Delta_f G^\circ_m(UO_2SO_4 \cdot 3.5H_2O, cr, 298.15\,K) = -(2535.6 \pm 1.8)\,kJ \cdot mol^{-1}.$$

The standard entropy is calculated from the selected $\Delta_f G^\circ_m$ and $\Delta_f H^\circ_m$ values.

$$S^\circ_m(UO_2SO_4 \cdot 3.5H_2O, cr, 298.15\,K) = (286.5 \pm 6.6)\,J \cdot K^{-1} \cdot mol^{-1}$$

This value is only in fair agreement with the entropy value that can be estimated using the method described in Appendix D, $S^\circ_m(UO_2SO_4 \cdot 3.5H_2O$, cr, 298.15 K) = $(319 \pm 22)\,J \cdot K^{-1} \cdot mol^{-1}$. The entropy increase is approximately $35\,J \cdot K^{-1} \cdot mol^{-1}$ for each water in the UO_2SO_4 hydrates, whereas the one used in estimations according to Latimer's method (Appendix D) is $45\,J \cdot K^{-1} \cdot mol^{-1}$.

No $C_{p,m}$ data are available for $UO_2SO_4 \cdot 3.5H_2O$(cr).

b) $U(SO_4)_2$(cr) and its hydrates

$U(SO_4)_2$(cr): The IAEA value [78COR/OHA] of the standard enthalpy of formation is accepted by this review,

$$\Delta_f H^\circ_m(U(SO_4)_2, cr, 298.15\,K) = -(2309.6 \pm 12.6)\,kJ \cdot mol^{-1}.$$

This value was also selected by Morss [86MOR].

All reported values of $S^\circ_m(U(SO_4)_2$, cr, 298.15 K) are estimates, *cf.* [78COR/OHA]. This review selects the Cordfunke and O'Hare [78COR/OHA] estimate,

$$S^\circ_m(U(SO_4)_2, cr, 298.15\,K) = (180 \pm 21)\,J \cdot K^{-1} \cdot mol^{-1},$$

which is in fair agreement with the estimate of $(167 \pm 9)\,J \cdot K^{-1} \cdot mol^{-1}$ based on the method described in Appendix D.

The Gibbs energy of formation is calculated from the selected enthalpy of formation and entropy.

$$\Delta_f G^\circ_m(U(SO_4)_2, cr, 298.15\,K) = -(2085 \pm 14)\,kJ \cdot mol^{-1}$$

No heat capacity data are available for this compound.

$U(SO_4)_2 \cdot 4H_2O$(cr) and $U(SO_4)_2 \cdot 8H_2O$(cr): The standard enthalpies of formation from Cordfunke and O'Hare [78COR/OHA] are accepted by this review,

$$\Delta_f H^\circ_m(U(SO_4)_2 \cdot 4H_2O, cr, 298.15\,K) = -(3483.2 \pm 6.3)\,kJ \cdot mol^{-1}$$
$$\Delta_f H^\circ_m(U(SO_4)_2 \cdot 8H_2O, cr, 298.15\,K) = -(4662.6 \pm 6.3)\,kJ \cdot mol^{-1}.$$

Differences in the auxiliary data were considered small compared to the other uncertainties. The standard entropies are estimated by this review, *cf.* Appendix D.

$$S^\circ_m(U(SO_4)_2 \cdot 4H_2O, cr, 298.15\,K) = (359 \pm 32)\,J \cdot K^{-1} \cdot mol^{-1}$$
$$S^\circ_m(U(SO_4)_2 \cdot 8H_2O, cr, 298.15\,K) = (538 \pm 52)\,J \cdot K^{-1} \cdot mol^{-1}$$

The estimates are based on the [78COR/OHA] entropy of $U(SO_4)_2$(cr).

The Gibbs energies of formation are calculated from the enthalpies of formation and the entropies.

$$\Delta_f G^\circ_m(U(SO_4)_2 \cdot 4H_2O, cr, 298.15\,K) = -(3033.3 \pm 11.4)\,kJ \cdot mol^{-1}$$
$$\Delta_f G^\circ_m(U(SO_4)_2 \cdot 8H_2O, cr, 298.15\,K) = -(3987.9 \pm 16.7)\,kJ \cdot mol^{-1}$$

c) Other uranium sulphates

$U_2(SO_4)_3(cr)$: Wilcox and Bromley [63WIL/BRO] calculated a value for $\Delta_f H^\circ_m(U_2(SO_4)_3$, cr, 298.15 K). This solid has not been identified and is therefore not considered in this review.

$U(OH)_2SO_4(cr)$: When uranium(IV) sulphate solutions are hydrolyzed, a sparingly soluble oxo or hydroxo sulphate is formed. Several compositions have been proposed for this solid: $UOSO_4 \cdot 3H_2O$ [1842RAM], $UOSO_4 \cdot 2H_2O$ [1848PEL], $4UOSO_4 \cdot UO_2 \cdot 8H_2O$ [01CON], and $3UOSO_4 \cdot U(SO_4)_2 \cdot xH_2O$ [06GLI/LIB]. Lundgren [53LUN] determined the structure and composition of $U(OH)_2SO_4$(cr) (orthorhombic, $a = 11.575 \times 10^{-10}$, $b = 5.926 \times 10^{-10}$, $c = 6.969 \times 10^{-10}$ m). This review considers Lundgren's chemical and structural characterization of "basic uranium(IV) sulphate" as the most precise one.

The solubility product of "basic uranium(IV) sulphate" was determined by Stepanov and Galkin [62STE/GAL]. This review accepts this solubility product but increases its uncertainty to ±0.5 logarithmic units, assuming that Stepanov and Galkin [62STE/GAL] used the same solid phase as Lundgren [53LUN], *i.e.*, that the solubility product corresponds to the hypothetical reaction

$$U(OH)_2SO_4(cr) \rightleftharpoons U^{4+} + 2OH^- + SO_4^{2-}. \qquad (V.120)$$

The selected solubility product is

$$\log_{10} K^\circ_{s,0}(V.120, 298.15\ K) = -31.17 \pm 0.50.$$

The slope of the straight line "solubility *vs.* pH" given in Ref. [62STE/GAL] is in agreement with the stoichiometry of Reaction (V.120). From the solubility product and the standard Gibbs energy of the corresponding aqueous species this review obtains

$$\Delta_f G^\circ_m(U(OH)_2SO_4, cr, 298.15\,K) = -(1766.2 \pm 3.4)\,kJ \cdot mol^{-1}.$$

Measurements of the magnetic susceptibility and heat capacity were done by Blaise *et al.* [79BLA/LAG]. The experimental data reveal a heat capacity anomaly at 21 K, but no experimental data at higher temperatures were reported.

No other thermodynamical data are available on this compound.

Zippeites: The uranium sulphates were among the first naturally occurring uranium minerals to be recognized and were known in the early part of the 19th century. Zippeites are basic dioxouranium(VI) sulphates first found in Joachimsthal. Systematic work on the zippeite group was made by Frondel *et al.* [76FRO/ITO]. Haacke and Williams [79HAA/WIL] determined solubility products and standard Gibbs energies of formation of $M_2(UO_2)_6(SO_4)_3(OH)_{10} \cdot 8H_2O$, M = Mg, Co, Ni and Zn at

25°C and $I \approx 0$. This work is discussed in Appendix A, and the resulting values are not selected by this review. The solubility products for the reactions

$$\begin{aligned} M_2(UO_2)_6(SO_4)_3(OH)_{10} \cdot 8H_2O(cr) \rightleftharpoons\ & 2M^{2+} + 6UO_2^{2+} + 3SO_4^{2-} + 10OH^- \\ & +8H_2O(l) \end{aligned} \qquad (V.121)$$

reported by Haacke and Williams [79HAA/WIL] are not consistent with their $\Delta_f G_m^\circ$ values of the zippeites. These derived Gibbs energies were accepted by Hemingway [82HEM], but no discussion was given about the discrepancy with the solubility products. Further experimental studies with pH variations are needed to obtain reliable information on the behaviour of the zippeites in aqueous solutions.

V.5.1.4. *Uranium thiosulphates*

V.5.1.4.1. *Aqueous uranium thiosulphates*

a) Aqueous U(VI) thiosulphates

The only quantitative information about aqueous uranium thiosulphate complexes is the study by Melton and Amis [63MEL/AMI], who examined the reaction

$$UO_2^{2+} + S_2O_3^{2-} \rightleftharpoons UO_2S_2O_3(aq). \qquad (V.122)$$

They reported $\log_{10}\beta_1(V.122, 298.15\ K) = (2.04 \pm 0.07)$ at $I = 0.0638$ M. This review tentatively accepts this value, (although confirmation of the results from another study would be useful) and corrects it to $I = 0$ using the Debye-Hückel term only. The selected value thus obtained is

$$\log_{10}\beta_1^\circ(V.122, 298.15\ K) = 2.8 \pm 0.3,$$

where the uncertainty is an estimate.

From the $\Delta_f G_m^\circ$ values of UO_2^{2+} and $S_2O_3^{2-}$ selected in this review, the selected value of $\Delta_f G_m^\circ(UO_2S_2O_3, aq, 298.15\,K)$ is obtained.

$$\Delta_f G_m^\circ(UO_2S_2O_3, aq, 298.15\,K) = -(1487.8 \pm 11.6)\,kJ \cdot mol^{-1}$$

The corresponding enthalpy change reported [63MEL/AMI] was $\Delta_r H_m(V.122) = -77.4\,kJ \cdot mol^{-1}$. This value is not credited here because of the large uncertainty associated with the small temperature range of the experiment (15 to 25°C).

b) Aqueous U(V) and U(IV) thiosulphates

No experimental information is available on thiosulphate aqueous complexes of either dioxouranium(V) or uranium(IV).

V.5.1.4.2. Solid uranium thiosulphates

Gabelica [77GAB] examined the method of preparing $UO_2S_2O_3 \cdot H_2O(cr)$ described by Klygin and Kolyada [60KLY/KOL] but found no evidence for the formation of a stoichiometric compound. The solid formed seems to be a mixture indicating decomposition of thiosulphate into sulphite and elemental sulphur. This review finds no reliable evidence for the formation of solid uranium thiosulphate compounds.

V.5.2. Selenium compounds and complexes

V.5.2.1. Uranium selenides

V.5.2.1.1. USe(cr)

With the exception of some early estimates (*e.g.*, [62WES/GRO, 68KAR/KAR]), all tabulated values of $\Delta_f H^\circ_m$(USe, cr, 298.15 K) appear to originate with the calorimetric measurements of Baskin and Smith [70BAS/SMI]. Their value was based upon their estimate of $\Delta_f H^\circ_m(USe_2$, cr, 298.15 K) described and accepted in the following section. Their value is accepted in this review, as it was by the IAEA review [84GRO/DRO], by Barin and Knacke [77BAR/KNA] and by Mills [74MIL].

$$\Delta_f H^\circ_m(\text{USe, cr, 298.15 K}) = -(275.7 \pm 14.6)\,\text{kJ} \cdot \text{mol}^{-1}$$

The entropy and heat capacity values accepted by Grønvold, Drowart and Westrum [84GRO/DRO] were from the measurements of Takahashi and Westrum [65TAK/WES]. These values are also accepted by this review.

$$S^\circ_m(\text{USe, cr, 298.15 K}) = (96.52 \pm 0.21)\,\text{J} \cdot \text{K}^{-1} \cdot \text{mol}^{-1}$$
$$C^\circ_{p,m}(\text{USe, cr, 298.15 K}) = (54.81 \pm 0.17)\,\text{J} \cdot \text{K}^{-1} \cdot \text{mol}^{-1}$$

The $C^\circ_{p,m}(T)$ data tabulated by the IAEA [84GRO/DRO] are fitted between 298.15 and 600 K to obtain the selected $C^\circ_{p,m}(T)$ function listed in Table III.3.

The Gibbs energy of formation is calculated from the selected enthalpy of formation and entropy.

$$\Delta_f G^\circ_m(\text{USe, cr, 298.15 K}) = -(276.9 \pm 14.6)\,\text{kJ} \cdot \text{mol}^{-1}$$

V.5.2.1.2. USe_2(cr)

Khodadad [57KHO] reported an α-phase of USe_2(cr) with tetragonal structure, a β-phase with orthorhombic structure and a γ-phase which is hexagonal. Thermodynamic data are available for only the first two of these phases.

a) α-USe_2

The selected standard enthalpy of formation was estimated by Baskin and Smith [70BAS/SMI] on the assumption that $\Delta_f H^\circ_m$ values of uranium selenides are a linear function of the Se:U ratio of the compound. The uncertainty in the value accepted here is from Ref. [84GRO/DRO] who also accepted this estimate in their IAEA review.

$$\Delta_f H^\circ_m(USe_2, \alpha, 298.15\,K) = -(427 \pm 42)\,kJ \cdot mol^{-1}$$

Westrum and Grønvold [70WES/GRO] reported heat capacity measurements from 5 to 350 K resulting in $C^\circ_{p,m}(USe_2, \alpha, 298.15\ K) = (79.16 \pm 0.17)\,J \cdot K^{-1} \cdot mol^{-1}$ and $S^\circ_m(USe_2, \alpha, 298.15\ K) = (133.84 \pm 0.25)\,J \cdot K^{-1} \cdot mol^{-1}$. The former value was accepted by the IAEA review [84GRO/DRO] while the latter was adjusted by $1.17\,J \cdot K^{-1} \cdot mol^{-1}$ to account for a structural disorder in the α-phase. The two values are selected here.

$$S^\circ_m(USe_2, \alpha, 298.15\,K) = (134.98 \pm 0.25)\,J \cdot K^{-1} \cdot mol^{-1}$$
$$C^\circ_{p,m}(USe_2, \alpha, 298.15\,K) = (79.16 \pm 0.17)\,J \cdot K^{-1} \cdot mol^{-1}$$

The $C^\circ_{p,m}(T)$ data tabulated by the IAEA [84GRO/DRO] are fitted between 298.15 and 600 K to obtain the selected $C^\circ_{p,m}(T)$ function listed in Table III.3.

The Gibbs energy of formation is calculated from the selected enthalpy of formation and entropy.

$$\Delta_f G^\circ_m(USe_2, \alpha, 298.15\,K) = -(427.1 \pm 42.0)\,kJ \cdot mol^{-1}$$

b) β-USe$_2$

The enthalpy of formation of this phase is within the uncertainty of the estimate for $\Delta_f H^\circ_m(USe_2, \alpha, 298.15\,K)$ above. The value tentatively accepted by the IAEA review [84GRO/DRO] is adopted here.

$$\Delta_f H^\circ_m(USe_2, \beta, 298.15\,K) = -(427 \pm 42)\,kJ \cdot mol^{-1}$$

Grønvold, Drowart and Westrum [84GRO/DRO] applied the estimation scheme of Westrum and Grønvold [62WES/GRO] to β-USe_2 to obtain the value selected in the IAEA review and accepted in this review.

$$S^\circ_m(USe_2, \beta, 298.15\,K) = (138 \pm 13)\,J \cdot K^{-1} \cdot mol^{-1}$$

The Gibbs energy of formation is calculated from the selected enthalpy of formation and entropy.

$$\Delta_f G^\circ_m(USe_2, \beta, 298.15\,K) = -(428.0 \pm 42.2)\,kJ \cdot mol^{-1}$$

V.5.2.1.3. USe$_3$(cr)

The IAEA review [84GRO/DRO] used the selenium pressure data obtained by Sev-ast'yanov, Slovyanskikh and Ellert [71SEV/SLO] over decomposing USe_2(cr) (at 1023 to 1178 K) to derive the Gibbs energy of decomposition of USe_3(cr) at 1100 K according to the reaction

$$USe_3(cr) \rightarrow USe_2(cr) + \tfrac{1}{2}Se_2(g). \qquad (V.123)$$

This result is combined with the thermodynamic data for USe_2(cr), USe_3(cr), Se(cr) and Se_2(g) selected in Ref. [84GRO/DRO] to obtain

$$\Delta_f H_m^\circ(USe_3, cr, 298.15\,K) = -(452 \pm 42)\,kJ \cdot mol^{-1}.$$

Westrum and Grønvold's [62WES/GRO] Latimer rule estimate,

$$S_m^\circ(USe_3, cr, 298.15\,K) = (177 \pm 17)\,J \cdot K^{-1} \cdot mol^{-1},$$

accepted by the IAEA review [84GRO/DRO], is also selected here.

The Gibbs energy of formation is calculated from the selected enthalpy of formation and entropy.

$$\Delta_f G_m^\circ(USe_3, cr, 298.15\,K) = -(452 \pm 42)\,kJ \cdot mol^{-1}$$

V.5.2.1.4. U_2Se_3 (cr)

The enthalpy of formation estimated for U_2Se_3(cr) by Mills [74MIL] was adopted by the IAEA review [84GRO/DRO] and is accepted here.

$$\Delta_f H_m^\circ(U_2Se_3, cr, 298.15\,K) = -(711 \pm 75)\,kJ \cdot mol^{-1}$$

Heat capacity measurements by Lagnier, Suski and Wojakowski [75LAG/SUS] were integrated by Lagnier and collegues in 1978 and reported to the authors of the IAEA review [84GRO/DRO] who obtained the value of

$$S_m^\circ(U_2Se_3, cr, 298.15\,K) = (261.4 \pm 1.7)\,J \cdot K^{-1} \cdot mol^{-1}$$

selected here.

Results reported in the IAEA review [84GRO/DRO] suggest a sharp maximum in the heat capacity of U_2Se_3(cr) near 300 K. In the absence of other more detailed experimental data, this review does not select any value for $C_{p,m}^\circ(U_2Se_3, cr, 298.15\,K)$.

The Gibbs energy of formation is calculated from the selected enthalpy of formation and entropy.

$$\Delta_f G_m^\circ(U_2Se_3, cr, 298.15\,K) = -(721 \pm 75)\,kJ \cdot mol^{-1}$$

V.5.2.1.5. U_3Se_4 (cr)

The values selected here are accepted from the IAEA review [84GRO/DRO]. The $\Delta_f H_m^\circ$ value was obtained by Baskin and Smith [70BAS/SMI] and is dependent on the $\Delta_f H_m^\circ(USe_2, \alpha, 298.15\,K)$ estimate accepted above.

$$\Delta_f H_m^\circ(U_3Se_4, cr, 298.15\,K) = -(983 \pm 85)\,kJ \cdot mol^{-1}$$

The selected standard entropy value was estimated by Westrum and Grønvold [62WES/GRO] using a modified Latimer's rule method which added a magnetic contribution.

$$S_m^\circ(U_3Se_4, cr, 298.15\,K) = (339 \pm 38)\,J \cdot K^{-1} \cdot mol^{-1}$$

The Gibbs energy of formation is calculated from the selected enthalpy of formation and entropy.

$$\Delta_f G_m^\circ(U_3Se_4, cr, 298.15\,K) = -(989 \pm 86)\,kJ \cdot mol^{-1}$$

V.5.2.1.6. $U_3Se_5(cr)$

The standard enthalpy of formation of U_3Se_5(cr) estimated by Mills [74MIL] was adopted by the IAEA review [84GRO/DRO] and is accepted here.

$$\Delta_f H_m^\circ(U_3Se_5, cr, 298.15\,K) = -(1130 \pm 113)\,kJ \cdot mol^{-1}$$

The selected standard entropy value was estimated by Westrum and Grønvold [62WES/GRO] using a modified Latimer's rule method which added a magnetic contribution.

$$S_m^\circ(U_3Se_5, cr, 298.15\,K) = (364 \pm 38)\,J \cdot K^{-1} \cdot mol^{-1}$$

The Gibbs energy of formation is calculated from the selected enthalpy of formation and entropy.

$$\Delta_f G_m^\circ(U_3Se_5, cr, 298.15\,K) = -(1131 \pm 114)\,kJ \cdot mol^{-1}$$

V.5.2.2. Uranium selenites

V.5.2.2.1. Aqueous uranium selenites

Verma and Khandelwal [73VER/KHA] made a conductometric study of the interactions between U(VI) and selenous acid. They proposed the formation of UO_2SeO_3(aq) and $(UO_2)_2(OH)_2SeO_3$(aq), but there is no information to substantiate this claim and no attempts are made to interpret the experiments with a quantitative chemical model.

V.5.2.2.2. Solid uranium selenites

The only uranium selenite compound for which reliable chemical thermodynamic data are available is dioxouranium(VI) selenite. Cordfunke and Ouweltjes [77COR/OUW2] calorimetrically determined the enthalpy of solution of α-UO_2SeO_3 and calculated $\Delta_f H_m^\circ(UO_2SeO_3, cr, 298.15\,K) = -(1522.1 \pm 0.8)\,kJ \cdot mol^{-1}$. The uncertainty in $\Delta_f H_m^\circ(SeO_2$, cr, 298.15 K) was underestimated in Ref. [77COR/OUW2]. Using auxiliary data from the present review, the selected value

$$\Delta_f H_m^\circ(UO_2SeO_3, cr, 298.15\,K) = -(1522.0 \pm 2.3)\,kJ \cdot mol^{-1}.$$

is calculated. Note that the reference in an earlier review, [83FUG], should also be to Cordfunke and Ouweltjes [77COR/OUW2] as the data source.

No other thermodynamic parameters for this compound have data available, and a calculation of the Gibbs energy of formation is therefore not possible.

V.5.2.3. *Uranium selenates*

V.5.2.3.1. *Aqueous uranium selenates*

No aqueous dioxouranium(VI) selenate complexes are credited in this review. Note that Fuger [83FUG] reported the enthalpy of formation of a solution of UO_2SeO_4 in $500(H_2SO_4 + 34.7H_2O)$.

V.5.2.3.2. *Solid uranium selenates*

The only uranium selenate compound for which reliable thermodynamic data are available is dioxouranium(VI) selenate. The selected value of the enthalpy of formation of α-UO_2SeO_4,

$$\Delta_f H^\circ_m(UO_2SeO_4, cr, 298.15\,K) = -(1539.3 \pm 3.3)\,kJ \cdot mol^{-1},$$

is recalculated from the data in Ref. [77COR/OUW2] using the value for $\Delta_f H^\circ_m(UO_2SO_4, \beta, 298.15\text{ K})$ accepted in this review.

No entropy or Gibbs energy data appear to be available for α-UO_2SeO_4.

V.5.3. *Tellurium compounds*

V.5.3.1. *Uranium tellurides*

V.5.3.1.1. *Binary uranium tellurides*

The thermochemistry of these compounds was reviewed recently by the IAEA [84GRO/DRO]. There are few reliable measurements on these systems and most of the values selected in that review were based on estimates. No analysis of these systems is done in this review and no data are selected. The cautious use of the values in the IAEA review [84GRO/DRO] is recommended.

V.5.3.1.2. *UOTe(cr)*

The values for $S^\circ_m(\text{UOTe, cr, 298.15 K})$ and $C^\circ_{p,m}(\text{UOTe, cr, 298.15 K})$ from the specific heat measurements by Staliński, Niemic and Biegański [63STA/NIE] are selected. It should be noted that both Morss [86MOR] and Fuger [83FUG] reported the values from this source, but Morss [86MOR] reported $S^\circ_m(\text{UOTe, cr, 298.15 K})$ as 116.7 rather than 111.67 $J \cdot K^{-1} \cdot mol^{-1}$, which is a typographical error. The uncertainties adopted here are from Fuger [83FUG].

$$S^\circ_m(UOTe, cr, 298.15\,K) = (111.7 \pm 1.3)\,J \cdot K^{-1} \cdot mol^{-1}$$
$$C^\circ_{p,m}(UOTe, cr, 298.15\,K) = (80.6 \pm 0.8)\,J \cdot K^{-1} \cdot mol^{-1}$$

No other thermodynamic data are available for this compound.

It should be mentioned that Morss' [86MOR] citation of S°_m and $C^\circ_{p,m}$ values for $UOTe_3$(cr) from Blaise [79BLA] is erroneous. Blaise did not mention $UOTe_3$(cr) but cited values for UTe_3(cr).

V.5.3.2. Uranium tellurites

V.5.3.2.1. Aqueous uranium tellurites

No aqueous uranium tellurites are known.

V.5.3.2.2. Solid uranium tellurites

The only uranium tellurite compound for which reliable chemical thermodynamic data are available is dioxouranium(VI) tellurite (schmitterite). Brandenburg [78BRA] reported

$$\Delta_f H_m^\circ(UO_2TeO_3, cr, 298.15\,K) = -(1605.4 \pm 1.3)\,kJ \cdot mol^{-1}.$$

No data are available to define other thermodynamic parameters for this compound.

V.6. Group 15 compounds and complexes

V.6.1. Nitrogen compounds and complexes

V.6.1.1. Uranium nitrides

An extensive discussion of the complex U-N system [86WEI] is beyond the scope of the present review. The review by Tagawa [74TAG] provides a thorough discussion of phase relationships in the uranium-nitrogen system.

V.6.1.1.1. UN(cr)

Values for the enthalpy of formation of uranium mononitride, UN(cr), have been reported in a large number of papers and reports. Many of these values result from second and third law analysis of high-temperature measurements of the nitrogen pressure over UN(cr) [63CAV/BON, 63OLS/MUL, 64BUG/BAU, 65VOZ/CRE, 68INO/LEI, 69GIN, 76IKE/TAM, 80PRI/COR] for the equilibrium reaction

$$2UN(cr) \rightleftharpoons N_2(g) + 2U(l), \qquad (V.124)$$

or of nitrogen and/or uranium pressures [69ALE/OGD, 71HOE, 80KHR/LYU] for the equilibrium

$$2UN(cr) \rightleftharpoons N_2(g) + 2U(g). \qquad (V.125)$$

As has been noted elsewhere, interpretation of the data from experiments using reaction (V.124) is difficult because the dissolution of N_2(g) in U(l) significantly reduces the activity of the uranium [69GIN, 72OET/LEI, 80PRI/COR]. For experiments involving reaction (V.125), a value for the enthalpy of sublimation of uranium is required, *i.e.*, $\Delta_f H_m^\circ$(U, g, 298.15 K). However, the CODATA Key Value

[89COX/WAG], $\Delta_f H_m^\circ(\text{U, g, 298.15 K}) = (533 \pm 8)\,\text{kJ}\cdot\text{mol}^{-1}$, has a very large uncertainty. There is even some question [87MAT/OHS] as to whether the vaporization of UN(cr) is congruent.

Tagawa [74TAG] and Prins, Cordfunke and Depaus [80PRI/COR] recalculated much of the earlier data for equilibrium (V.124). As stated in the latter reference, "in most cases there is a large discrepancy between the second-law and third-law calculations. The third-law calculations of $\Delta_f H_m^\circ(298\text{ K})$ agree better among themselves, especially the more recent investigations". The weighted average of the third-law results reviewed in Ref. [80PRI/COR] for $\Delta_f H_m^\circ(\text{UN, cr, 298.15 K})$ is $-(283.6\pm0.7)\,\text{kJ}\cdot\text{mol}^{-1}$, and is markedly less negative than the results from calorimetric studies discussed below.

Also, a number values based on calorimetry have been reported. Neumann, Kroeger and Haebler [32NEU/KRO] reported $\Delta_f H_m^\circ(\text{UN, cr, 298.15 K}) = -(286.6 \pm 2.1)\,\text{kJ}\cdot\text{mol}^{-1}$ based on reaction of an impure sample of uranium with a large excess of nitrogen (700°C, 2.53 MPa) in a bomb calorimeter. Gross, Hayman and Clayton [62GRO/HAY] used a hot-zone calorimeter to measure the heat of reaction of nitrogen with an excess of uranium powder (750 to 800°C), and obtained an average value of $\Delta_f H_m^\circ(\text{UN, cr, 298.15 K}) = -(291.2 \pm 3.3)\,\text{kJ}\cdot\text{mol}^{-1}$. Hubbard [62HUB] reported $-5361\,\text{kJ}\cdot\text{mol}^{-1}$ for the oxidation of UN(cr) in a bomb calorimeter according to

$$6\text{UN(cr)} + 8\text{O}_2\text{(g)} \rightleftharpoons 2\text{U}_3\text{O}_8\text{(cr)} + 3\text{N}_2\text{(g)} \qquad \text{(V.126)}$$

and, hence, $\Delta_f H_m^\circ(\text{UN, cr, 298.15 K})$ would be $-(298.1 \pm 2.9)\,\text{kJ}\cdot\text{mol}^{-1}$ based on the value for $UO_{2.6667}$ in the present review. O'Hare *et al.* [68OHA/SET] reported an unpublished value, $\Delta_f H_m^\circ(\text{UN, cr, 298.15 K}) = -(301.2 \pm 5.0)\,\text{kJ}\cdot\text{mol}^{-1}$, determined by Fredrickson using the same method. Of these four studies, only that of Gross, Hayman and Clayton [62GRO/HAY] is credited in the present review. The others are rejected either because of the impure solid used [32NEU/KRO], or because details of the samples and experiments are lacking.

O'Hare *et al.* [68OHA/SET] reported the heat of combustion of $UN_{0.965}$ with fluorine. Oetting and Leitnaker [72OET/LEI] claimed that results of the uranium analyses for $UN_{0.965}$ in Ref. [68OHA/SET] were probably high, and hence the nitrogen content had been underestimated. They suggested the composition was probably closer to $U_{0.9957}$. If the data from O'Hare *et al.* [68OHA/SET] are recalculated based on this assumption, and auxiliary data consistent with the present review are used, $\Delta_f H_m^\circ(\text{UN, cr, 298.15 K}) = -(304.3 \pm 4.6)\,\text{kJ}\cdot\text{mol}^{-1}$ is obtained. Johnson and Cordfunke [81JOH/COR] did similar measurements using samples of uranium nitride having the composition $UN_{0.997}$. Recalculation of the results in the latter paper, using auxiliary data consistent with the present review, results in $\Delta_f H_m^\circ(\text{UN}_{0.997}\text{, cr, 298.15 K}) = -(289.5 \pm 2.2)\,\text{kJ}\cdot\text{mol}^{-1}$.

In the present review, the selected value

$$\Delta_f H_m^\circ(\text{UN, cr, 298.15 K}) = -(290.0 \pm 3.0)\,\text{kJ}\cdot\text{mol}^{-1}$$

is based on the weighted average of the calorimetric studies of Johnson and Cordfunke [81JOH/COR] and Gross, Hayman and Clayton [62GRO/HAY]. The estimated uncertainty is increased slightly to reflect the severe experimental problems encountered in obtaining the values and the probable substoichiometry of the experimental samples.

Using adiabatic calorimetry, Westrum and Barber [66WES/BAR] measured the heat capacity of UN for temperatures between 5 and 346 K. Similar measurements (11 to 320 K) were reported by Counsell, Dell and Martin [66COU/DEL] (for $UN_{1.01}$). For temperatures above 80 K, the values from the latter study are slightly smaller than those reported by Westrum and Barber [66WES/BAR], with the difference increasing to 1% at 300 K. Based on these data, the values $S^\circ_m(\mathrm{UN, cr, 298.15\ K}) = 62.63\ \mathrm{J \cdot K^{-1} \cdot mol^{-1}}$ [66WES/BAR] and $S^\circ_m(\mathrm{UN_{1.01}, cr, 298.15\ K}) = 62.22\ \mathrm{J \cdot K^{-1} \cdot mol^{-1}}$ [66COU/DEL] were reported. Low-temperature heat capacity experiments by Rudigier, Ott and Vogt [85RUD/OTT, 85RUD/OTT2] did not reveal any additional contributions to the entropy.

The average entropy value

$$S^\circ_m(\mathrm{UN, cr, 298.15\ K}) = (62.43 \pm 0.22)\ \mathrm{J \cdot K^{-1} \cdot mol^{-1}}$$

is accepted in the present review. The uncertainty is calculated assuming an uncertainty of 0.5% in each of the reported entropy values.

The accepted heat capacity value

$$C^\circ_{p,m}(\mathrm{UN, cr, 298.15\ K}) = (47.57 \pm 0.40)\ \mathrm{J \cdot K^{-1} \cdot mol^{-1}}$$

is based on the average of the values of Westrum and Barber [66WES/BAR] and Counsell, Dell and Martin [66COU/DEL], and the uncertainty is assigned in the present review. The thermal function for uranium mononitride for temperatures above 298.15 K was recently assessed by Matsui and Ohse [87MAT/OHS] and Hayes, Thomas and Peddicord [90HAY/THO]. As the primary temperature range of interest in the present review is limited to 600 K, the heat capacity function proposed by Matsui and Ohse [87MAT/OHS] for the temperature range $298.15 < T/\mathrm{K} < 1000$ is selected,

$$C^\circ_{p,m}(\mathrm{UN, cr}, T) = (50.54 + 1.066 \times 10^{-2}T - 5.238 \times 10^{5}T^{-2})\ \mathrm{J \cdot K^{-1} \cdot mol^{-1}}.$$

Although the value generated by this function at 298.15 K is not identical to the value selected above, the function is consistent with $C^\circ_{p,m}(\mathrm{UN, cr, 298.15\ K})$ within the stated uncertainties.

The value

$$\Delta_f G^\circ_m(\mathrm{UN, cr, 298.15\ K}) = -(265.1 \pm 3.0)\ \mathrm{kJ \cdot mol^{-1}}$$

is calculated from the selected $\Delta_f H^\circ_m$ and S°_m values.

V.6.1.1.2. *"α-U_2N_3" (α-$UN_{1.5+x}$)*

O'Hare *et al.* [68OHA/SET] reported heats for combustion of $UN_{1.51}$ and $UN_{1.69}$ with fluorine. Later, Johnson and Cordfunke [81JOH/COR] did similar measurements using samples of uranium nitrides having compositions $UN_{1.606}$ and $UN_{1.674}$.

Recalculation of the results in the latter paper, using auxiliary data consistent with the present review, results in $\Delta_f H_m^\circ(UN_{1.606}, \alpha, 298.15 \text{ K}) = -(381.4 \pm 2.3)\,\text{kJ}\cdot\text{mol}^{-1}$ and $\Delta_f H_m^\circ(UN_{1.674}, \text{cr}, 298.15 \text{ K}) = -(390.8 \pm 2.3)\,\text{kJ}\cdot\text{mol}^{-1}$. These values are approximately $10\,\text{kJ}\cdot\text{mol}^{-1}$ more positive than the enthalpies of formation that are recalculated using the data of O'Hare *et al.* [68OHA/SET]. The differences were attributed [81JOH/COR] to less complete combustion of samples in the earlier study. Also, Oetting and Leitnaker [72OET/LEI] claimed that results of the uranium analyses in Ref. [68OHA/SET] may be high (at least for UN), and hence the nitrogen content may be underestimated. Tagawa [74TAG] also pointed out that the $UN_{1.51}$ sample was probably a mixture of $U_{1.54}$ plus UN. In light of these difficulties, the results for $UN_{1.51}$ and $UN_{1.69}$ in Ref. [68OHA/SET] are rejected in the present review.

Gross, Hayman and Clayton [62GRO/HAY] measured the enthalpy of reaction of $N_2(g)$ with UN(cr). As pointed out by O'Hare *et al.* [68OHA/SET], the value for $\Delta_r H_m^\circ$ from this work probably refers to formation of a higher nitride than stoichiometric $UN_{1.5}$. However, if the reaction was

$$\frac{2}{1+2x}\text{UN(cr)} + 0.5\text{N}_2\text{(g)} \rightleftharpoons \frac{2}{1+2x}\text{UN}_{1.5+x}, \qquad \text{(V.127)}$$

no reasonable value of x can be chosen that allows the heat of reaction to be rationalized with both the value of $\Delta_f H_m^\circ(\text{UN, cr}, 298.15 \text{ K})$ selected in the present review and the enthalpies of combustion for α-$UN_{1.5+x}$ measured by Johnson and Cordfunke [81JOH/COR] (or by O'Hare *et al.* [68OHA/SET]). Instead, the value for $\Delta_f H_m^\circ(UN_{1.5+x}, \alpha, 298.15 \text{ K})$ calculated from Gross, Hayman and Clayton [62GRO/HAY] is much more positive than the value from the heat of combustion measurements. Lapat and Holden [64LAP/HOL] showed, from vapour pressure measurements, that the heat of solution of $N_2(g)$ in $UN_{1.5+x}$ appears to vary with the value of x, and reported the heat of nitration of UN to be $-238\,\text{kJ}\cdot\text{mol}^{-1}$. This value is quite similar to the value reported by Gross, Hayman and Clayton [62GRO/HAY]. In the present review, only the values reported by Johnson and Cordfunke [81JOH/COR] (based on bomb-calorimetry) are accepted, however, the uncertainties are increased to reflect the lack of confirmatory values.

$$\Delta_f H_m^\circ(UN_{1.606}, \alpha, 298.15 \text{ K}) = -(381.4 \pm 5.0)\,\text{kJ}\cdot\text{mol}^{-1}$$
$$\Delta_f H_m^\circ(UN_{1.674}, \alpha, 298.15 \text{ K}) = -(390.8 \pm 5.0)\,\text{kJ}\cdot\text{mol}^{-1}$$

Counsell, Dell and Martin [66COU/DEL] measured the heat capacity of two samples of α-$UN_{1.5+x}$, with compositions $UN_{1.59}$ and $UN_{1.73}$, at temperatures from 11 to 320 K. Based on these measurements,

$$S_m^\circ(UN_{1.59}, \alpha, 298.15 \text{ K}) = (65.02 \pm 0.30)\,\text{J}\cdot\text{K}^{-1}\cdot\text{mol}^{-1}$$
$$S_m^\circ(UN_{1.73}, \alpha, 298.15 \text{ K}) = (65.86 \pm 0.30)\,\text{J}\cdot\text{K}^{-1}\cdot\text{mol}^{-1},$$

were reported. These are accepted in the present review, and uncertainties of 0.5% are assigned to the values.

Linear extrapolation of the selected enthalpy values to the compositions of α-U_2N_3 used by Counsell, Dell and Martin would lead to

$$\Delta_f H^\circ_m(UN_{1.59}, \alpha, 298.15 \text{ K}) = -(379.2 \pm 5.2) \text{ kJ} \cdot \text{mol}^{-1}$$
$$\Delta_f H^\circ_m(UN_{1.73}, \alpha, 298.15 \text{ K}) = -(398.5 \pm 7.5) \text{ kJ} \cdot \text{mol}^{-1},$$

which are also accepted in this review with the increased uncertainties reflecting extra uncertainty from the extrapolation procedure. From these and the selected entropies, the values

$$\Delta_f G^\circ_m(UN_{1.59}, \alpha, 298.15 \text{ K}) = -(338.2 \pm 5.2) \text{ kJ} \cdot \text{mol}^{-1}$$
$$\Delta_f G^\circ_m(UN_{1.73}, \alpha, 298.15 \text{ K}) = -(353.8 \pm 7.5) \text{ kJ} \cdot \text{mol}^{-1},$$

are calculated.

V.6.1.1.3. *β-U_2N_3 (β-$UN_{1.5}$)*

This compound is difficult to prepare in a pure form at 298.15 K. Johnson and Cordfunke [81JOH/COR] reported $\Delta_f H^\circ_m(UN_{1.466}, \beta, 298.15 \text{ K}) = -(362.9 \pm 2.3) \text{ kJ} \cdot \text{mol}^{-1}$ based on the energy of combustion of samples in a fluorine bomb calorimeter. Recalculation using auxiliary data consistent with the present review results in

$$\Delta_f H^\circ_m(UN_{1.466}, \beta, 298.15 \text{ K}) = -(362.2 \pm 2.3) \text{ kJ} \cdot \text{mol}^{-1},$$

which is accepted. No experimental data are available to provide values for the entropy or heat capacity data of this solid.

V.6.1.2. *Uranium azides*

The earliest study of azide complexes of dioxouranium(VI) was by Sherif and Awad [61SHE/AWA]. These authors reported the equilibrium constant for the formation of $UO_2N_3^+$. In two other studies [62SHE/AWA, 62SHE/AWA2] they presented evidence for the formation of higher azide complexes but no quantitative information was given. Nair, Prabhu and Vartak [61NAI/PRA] used the same experimental method, but found an equilibrium constant that is more than ten times larger. This might have been due to decomposition of HN_3 present in not negligible amounts in the test solutions of Nair, Prabhu and Vartak [61NAI/PRA]. Chierice and Neves [83CHI/NEV] did a potentiometric study of the aqueous dioxouranium(VI) azide system and reported equilibrium data for the formation of $UO_2(N_3)_q^{2-q}$, where $q = 1$ to 6, according to Eq. (V.128).

$$UO_2^{2+} + qN_3^- \rightleftharpoons UO_2(N_3)_q^{2-q} \qquad (V.128)$$

The available experimental data are compiled in Table V.36. A graphical re-evaluation of the experimental data in Ref. [83CHI/NEV] is made in this review in order to estimate the uncertainties of the various $\log_{10}\beta_q$ values. This review finds $\log_{10}\beta_1 = (2.11 \pm 0.01)$, $\log_{10}\beta_2 = (4.03 \pm 0.07)$, $\log_{10}\beta_3 = (5.48 \pm 0.04)$ and $\log_{10}\beta_4 = (5.39 \pm 0.09)$. These values differ from the ones proposed by the authors ($\log_{10}\beta_1 = 2.14$, $\log_{10}\beta_2 = 3.92$, $\log_{10}\beta_3 = 5.71$, $\log_{10}\beta_4 = 5.85$, $\log_{10}\beta_5 = 6.36$, and $\log_{10}\beta_6 = 7.08$). This review does not consider the existence of $UO_2(N_3)_5^{3-}$ and

Table V.36: Experimental equilibrium data for the dioxouranium(VI) azide system.

Method	Ionic Medium	t (°C)	$\log_{10}\beta_q^{(a)}$	$\log_{10}\beta_q^{\circ(b)}$	Reference
$UO_2^{2+} + N_3^- \rightleftharpoons UO_2N_3^+$					
sp	0.3 M $NaClO_4$	~ 25	3.49		[61NAI/PRA]
sp	0.096 M $NaClO_4$	25	2.31 ± 0.04	2.58 ± 0.10(c)	[61SHE/AWA]
sp	0.04 M	25	2.64		[62SHE/AWA]
sp	(qualitative information)				[62SHE/AWA2]
gl	2 M $NaClO_4$	25	2.11 ± 0.01(d)	2.65 ± 0.22	[83CHI/NEV]
$UO_2^{2+} + 2N_3^- \rightleftharpoons UO_2(N_3)_2(aq)$					
gl	2 M $NaClO_4$	25	4.03 ± 0.07(d)	4.33 ± 0.23	[83CHI/NEV]
$UO_2^{2+} + 3N_3^- \rightleftharpoons UO_2(N_3)_3^-$					
gl	2 M $NaClO_4$	25	5.48 ± 0.04(d)	5.74 ± 0.22	[83CHI/NEV]
$UO_2^{2+} + 4N_3^- \rightleftharpoons UO_2(N_3)_4^{2-}$					
gl	2 M $NaClO_4$	25	5.39 ± 0.09(d)	4.92 ± 0.24	[83CHI/NEV]
$UO_2^{2+} + 5N_3^- \rightleftharpoons UO_2(N_3)_5^{3-}$					
gl	2 M $NaClO_4$	25	6.36		[83CHI/NEV]
$UO_2^{2+} + 6N_3^- \rightleftharpoons UO_2(N_3)_6^{4-}$					
gl	2 M $NaClO_4$	25	7.08		[83CHI/NEV]

(a) $\log_{10}\beta_q$ refers to the formation reaction indicated, in the ionic medium and at the temperature given in the table.
(b) $\log_{10}\beta_q^\circ$ (in molal units) refers to the formation reaction at $I = 0$ and 298.15 K.
(c) Uncertainty estimated by this review, *cf.* Appendix A.
(d) Result of a re-evaluation of the experimental data by this review, *cf.* Appendix A.

$UO_2(N_3)_6^{4-}$ to be adequately proven. In fact, it is doubtful if there is room for six azide ligands in the coordination sphere of UO_2^{2+}. This review prefers the re-evaluated equilibrium constants to the values given by the authors.

After conversion to molality constants, the equilibrium constants are corrected to $I = 0$ by using the estimated ion interaction coefficients $\varepsilon_{(UO_2N_3^+,ClO_4^-)} = (0.3 \pm 0.1)$, $\varepsilon_{(UO_2(N_3)_3^-,Na^+)} \approx \varepsilon_{(N_3^-,Na^+)} = (0.0 \pm 0.1)$, and $\varepsilon_{(UO_2(N_3)_4^{2-},Na^+)} = -(0.1 \pm 0.1)$, with the additional assumption that the uncertainty in $\Delta\varepsilon$ should be at most ± 0.10 (see Appendix B). This upper limit is reasonable as judged by the span of the experimental $\Delta\varepsilon$ values. These estimates are based on the ion interaction coefficients of similar complexes of the same charge type. They result in $\Delta\varepsilon(\text{V.128}, q = 1) = -(0.16 \pm 0.10)$, $\Delta\varepsilon(\text{V.128}, q = 2) = -(0.46 \pm 0.10)$, $\Delta\varepsilon(\text{V.128}, q = 3) = -(0.46 \pm 0.10)$, and $\Delta\varepsilon(\text{V.128}, q = 4) = -(0.56 \pm 0.10)$, for sodium perchlorate medium. The selected value for $\log_{10} \beta_1^\circ$ is the weighted average between the values from Refs. [61SHE/AWA] and [83CHI/NEV], corrected to $I = 0$ as shown in Table V.36. For $q > 1$ the values evaluated from Ref. [83CHI/NEV] are accepted.[10]

$$\begin{aligned}
\log_{10} \beta_1^\circ(\text{V.128}, q = 1, 298.15\,\text{K}) &= 2.58 \pm 0.09 \\
\log_{10} \beta_2^\circ(\text{V.128}, q = 2, 298.15\,\text{K}) &= 4.33 \pm 0.23 \\
\log_{10} \beta_3^\circ(\text{V.128}, q = 3, 298.15\,\text{K}) &= 5.74 \pm 0.22 \\
\log_{10} \beta_4^\circ(\text{V.128}, q = 4, 298.15\,\text{K}) &= 4.92 \pm 0.24
\end{aligned}$$

The selected Gibbs energies of formation are calculated from these selected equilibrium constants using selected auxiliary data.

$$\begin{aligned}
\Delta_f G_m^\circ(UO_2N_3^+, \text{aq}, 298.15\,\text{K}) &= -(619.1 \pm 2.7)\,\text{kJ} \cdot \text{mol}^{-1} \\
\Delta_f G_m^\circ(UO_2(N_3)_2, \text{aq}, 298.15\,\text{K}) &= -(280.9 \pm 4.6)\,\text{kJ} \cdot \text{mol}^{-1} \\
\Delta_f G_m^\circ(UO_2(N_3)_3^-, \text{aq}, 298.15\,\text{K}) &= (59.3 \pm 6.4)\,\text{kJ} \cdot \text{mol}^{-1} \\
\Delta_f G_m^\circ(UO_2(N_3)_4^{2-}, \text{aq}, 298.15\,\text{K}) &= (412.2 \pm 8.3)\,\text{kJ} \cdot \text{mol}^{-1}
\end{aligned}$$

There are no enthalpy, entropy or heat capacity data available for any of the dioxouranium(VI) azide complexes.

V.6.1.3. Uranium nitrates

V.6.1.3.1. Aqueous uranium nitrates

a) Aqueous U(VI) nitrates

The dioxouranium(VI) nitrate complexes are weak, and it is therefore difficult to distinguish between complex formation and changes in the activity factors of the solutes caused by the (often large) changes in solute concentration. Among the data listed in Table V.37, this review only finds the data of Ahrland [51AHR] and Day and Powers [54DAY/POW] useful for determining $\log_{10} \beta_1^\circ$(V.129, 298.15 K).

---

[10] The reviewer (L.C.) has noted that by using a modified specific ion interaction theory [90CIA] the following values are obtained: $\log_{10} \beta_1^\circ(\text{V.128}, q = 1, 298.15\,\text{K}) = 2.45$ and $\log_{10} \beta_2^\circ(\text{V.128}, q = 2, 298.15\,\text{K}) = 4.54$.

Table V.37: Experimental equilibrium data for the uranium(VI) nitrate system.

Method	Ionic Medium	t (°C)	$\log_{10} K_q^{(a)}$	$\log_{10} K_q^{\circ(a)}$	Reference
$UO_2^{2+} + NO_3^- \rightleftharpoons UO_2NO_3^+$					
sp	2 M $HClO_4$, $I = 5.4$ M	25	−0.68		[49BET/MIC]
	2 M $HClO_4$, $I = 7$ M	25	−0.57		
qh	1 M $(Na,H)(Cl^-,ClO_4)$	20	-0.30 ± 0.17	0.38 ± 0.18	[51AHR]
dis	2 M $Na(Cl,ClO_4)$	10	−0.52		[54DAY/POW]
	2 M $Na(Cl,ClO_4)$	25	$-0.62 \pm 0.09^{(b)}$	0.01 ± 0.26	
	2 M $Na(Cl,ClO_4)$	40	−0.77		
cix	1 M $NaClO_4$	32	−1.4		[61BAN/TRI]
gl	$I = 0.54$ M	25	−0.43		[67OHA/MOR]
	$I = 0.82$ M	25	−0.70		
	$I = 1.06$ M	25	−0.72		
dis	8 M $HClO_4$	20	−0.47		[70LAH/KNO]
emf	3 M H^+	30	−0.48		[85GUO/LIU]
....					
$UO_2^{2+} + 2NO_3^- \rightleftharpoons UO_2(NO_3)_2(aq)$					
cix	1 M $NaClO_4$	32	−1.4		[61BAN/TRI]
sp	var. 0.59 to 11.1 M	20	-1.66 ± 0.16		[70KLY/KOL]
dis	8 M $HClO_4$	20	−1.5		[70LAH/KNO]
....					
$UO_2^{2+} + 3NO_3^- \rightleftharpoons UO_2(NO_3)_3^-$					
cix	1 M $NaClO_4$	32	0.5		[61BAN/TRI]
dis	6 M HNO_3	20	−1.65		[70LAH/KNO]
....					
$UO_2^{2+} + 3NO_3^- + H^+ \rightleftharpoons HUO_2(NO_3)_3(aq)$					
sp	var. 0.59 to 11.1 M	20	-1.74 ± 0.15		[70KLY/KOL]

(a) Refers to the reactions indicated, $\log_{10} K_q$ in the ionic medium and at the temperature given in the table, $\log_{10} K_q^\circ$ (in molal units) at $I = 0$ and 298.15 K.
(b) Uncertainty estimated by this review, *cf.* Appendix A.

$$UO_2^{2+} + NO_3^- \rightleftharpoons UO_2NO_3^+ \qquad (V.129)$$

The data of Betts and Michaels [49BET/MIC] were obtained at such a high ionic strength ($I = 5.4$ and 7 M) that correction to $I = 0$ is not feasible. The data of Ohashi and Morozumi [67OHA/MOR] and Guorong, Liufang and Chengfa [85GUO/LIU] are discussed in Appendix A and are not considered reliable. The equilibrium constants in Refs. [54DAY/POW, 51AHR] are converted to molality constants and corrected to $I = 0$ by using the ion interaction coefficients listed in Appendix B, assuming that $\varepsilon_{(UO_2NO_3^+,ClO_4^-)} \approx \varepsilon_{(UO_2Cl^+,ClO_4^-)} = (0.33 \pm 0.04)$, yielding $\Delta\varepsilon(V.129) = -(0.09 \pm 0.06)$. It should be mentioned that the experimental data used for the selection of $\log_{10} \beta_1^\circ$ refer to slightly different temperatures. No reliable enthalpy data are available in the literature. This review assumes that the enthalpy change in Reaction (V.129) is similar to that of the corresponding chloride reaction, and that the value reported at 20°C [51AHR] does not differ significantly from that at 25°C. From the corrected data listed in Table V.37 of Refs. [51AHR, 54DAY/POW] this review selects the weighted average,

$$\log_{10} \beta_1^\circ(V.129, 298.15\,K) = (0.30 \pm 0.15).$$

The values of $\Delta_f G_m^\circ$ selected in this review for UO_2^{2+} and NO_3^- are used to obtain the selected value of

$$\Delta_f G_m^\circ(UO_2NO_3^+, aq, 298.15\,K) = -(1065.1 \pm 2.0)\,kJ \cdot mol^{-1}.$$

There is no reliable quantitative information on higher dioxouranium(VI) nitrato complexes. Lahr and Knoch [70LAH/KNO] found evidence for the formation of $UO_2(NO_3)_2(aq)$ with $\log_{10} \beta_2 = 0.03$. However, very large changes in the composition of the solution were made in this study (0 to 8 M HNO_3), and the data might be explained by activity factor variations as well. The same is true for the spectrophotometric study of Klygin, Kolyada and Smirnova [70KLY/KOL], who found no evidence for $UO_2NO_3^+$, but found evidence instead for $UO_2(NO_3)_2(aq)$ and $HUO_2(NO_3)_3(aq)$. The cation exchange study of Banerjea and Tripathi [61BAN/TRI] is not precise enough to extract information on nitrate complexes, *cf.* Appendix A. Lahr and Knoch [70LAH/KNO] also reported a value of $\log_{10} \beta_3$ for the formation of $UO_2(NO_3)_3^-$. This value is also considered unreliable by this review.

There is no reliable enthalpy information available on the U(VI) nitrate system. The data of Day and Powers [54DAY/POW] are not sufficiently well documented to justify their selection. No entropy or heat capacity data are available.

b) Aqueous U(V) nitrates

No aqueous species of dioxouranium(V) have been identified.

c) Aqueous U(IV) nitrates

The uranium(IV) nitrate system was studied by several investigators using spectrophotometry, distribution measurements and potentiometry. The data for the reactions (V.130) are compiled in Table V.38.

$$U^{4+} + qNO_3^- \rightleftharpoons U(NO_3)_q^{4-q} \quad (V.130)$$

The most precise studies are those of McKay and Woodhead [64MCK/WOO] and Ermolaev and Krot [62ERM/KRO]. Guorong, Liufang and Chengfa [85GUO/LIU] obtained results by potentiometry that are in good agreement with the first two studies. However, this review points out some experimental shortcomings, *cf.* Appendix A, and therefore discards their [85GUO/LIU] data. Although the data of Rao and Pai [69RAO/PAI] are in fair agreement with those of the previous investigators, they are not taken into account by this review, *cf.* Appendix A.

The data of Lahr and Knoch [70LAH/KNO], which were obtained at I = 8 M (H^+), also support the formation of the higher nitrate complexes, $U(NO_3)_3^+$ and $U(NO_3)_4$(aq). Due to the ionic strength and the large variations in the composition of the test solution it is unlikely that the activity coefficients are constant. Hence, it is difficult to distinguish between complex formation and a medium dependent variation of the activity coefficients of the ionic species. This is a general difficulty in all systems where weak complexes are formed.

In the spectrophotometric studies [62ERM/KRO, 64MCK/WOO], and in the cation exchange study [62ERM/KRO], one auxiliary parameter for each complex (molar absorptivity or distribution coefficient) has to be determined along with the equilibrium constants.

There is, in general, a large covariation between the equilibrium constant and the auxiliary parameter for a given complex. This makes it difficult to determine higher complexes with any degree of confidence. The data of Ermolaev and Krot [62ERM/KRO] were criticized by McKay and Woodhead [64MCK/WOO]. This review considers the equilibrium constants for the formation of UNO_3^{3+} and $U(NO_3)_2^{2+}$ as reliable (*cf.* details in Appendix A). There is no doubt that there is well supported evidence for the formation of higher complexes at high nitrate concentrations [61WIL/KED, 66KOC/SCH]. However, because of the large changes in the ionic medium, this review finds it impossible to propose any numerical values for the equilibrium constants of these complexes.

The equilibrium constants in Refs. [62ERM/KRO] and [64MCK/WOO] were measured at several different ionic strengths. Those at $I \leq 3.5$ m are used, after correction to molality constants, to evaluate $\log_{10}\beta_q^\circ$ and $\Delta\varepsilon$ for the formation of UNO_3^{3+} and $U(NO_3)_2^{2+}$ using the specific ion interaction theory as described in Appendix B. The corresponding plots are shown in Figures V.13 and V.14. The following formation constants result from this extrapolation:

$$\log_{10}\beta_1^\circ(\text{V.130}, q = 1, 298.15\,\text{K}) = 1.47 \pm 0.13$$
$$\log_{10}\beta_2^\circ(\text{V.130}, q = 2, 298.15\,\text{K}) = 2.30 \pm 0.35.$$

Table V.38: Experimental equilibrium data for the uranium(IV) nitrate system.

Method	Ionic Medium	t (°C)	$\log_{10}\beta_q^{(a)}$	Reference
$U^{4+} + NO_3^- \rightleftharpoons UNO_3^{3+}$				
cix	3.5 M $H(ClO_4, NO_3)$	26.5	$0.30 \pm 0.20^{(b)}$	[62ERM/KRO]
sp	3.5 M $H(ClO_4, NO_3)$	26.5	$0.36 \pm 0.20^{(b)}$	
	3.0 M $H(ClO_4, NO_3)$	26.5	$0.28 \pm 0.10^{(b)}$	
	2.5 M $H(ClO_4, NO_3)$	26.5	$0.20 \pm 0.10^{(b)}$	
	2.0 M $H(ClO_4, NO_3)$	26.5	$0.20 \pm 0.10^{(b)}$	
sp	1.0 M $H(ClO_4, NO_3)$	20	$0.18 \pm 0.15^{(b)}$	[64MCK/WOO]
	3.0 M $Li(ClO_4, NO_3)$			
	1.0 M $H(ClO_4, NO_3)$	20	$0.20 \pm 0.15^{(b)}$	
	2.0 M $Li(ClO_4, NO_3)$			
	1.0 M $H(ClO_4, NO_3)$	20	$0.06 \pm 0.10^{(b)}$	
	1.0 M $Li(ClO_4, NO_3)$			
	1.0 M $H(ClO_4, NO_3)$	20	$0.04 \pm 0.10^{(b)}$	
dis	3.8 M H^+	25?	0.08	[69RAO/PAI]
dis	8 M H^+	20	−0.08	[70LAH/KNO]
pot	3 M H^+	30	0.22	[85GUO/LIU]
....				
$U^{4+} + 2NO_3^- \rightleftharpoons U(NO_3)_2^{2+}$				
cix	3.5 M $H(ClO_4, NO_3)$	26.5	$0.42 \pm 0.20^{(b)}$	[62ERM/KRO]
sp	3.5 M $H(ClO_4, NO_3)$	26.5	$0.47 \pm 0.20^{(b)}$	
	3.0 M $H(ClO_4, NO_3)$	26.5	$0.31 \pm 0.20^{(b)}$	
	2.5 M $H(ClO_4, NO_3)$	26.5	$0.20 \pm 0.20^{(b)}$	
	2.0 M $H(ClO_4, NO_3)$	26.5	$0.18 \pm 0.20^{(b)}$	
sp	1.0 M $H(ClO_4, NO_3)$	20	$0.78 \pm 0.30^{(b)}$	[64MCK/WOO]
	3.0 M $Li(ClO_4, NO_3)$			
	1.0 M $H(ClO_4, NO_3)$	20	$0.30 \pm 0.30^{(b)}$	
	2.0 M $Li(ClO_4, NO_3)$			
	1.0 M $H(ClO_4, NO_3)$	20	$0.00 \pm 0.30^{(b)}$	
	1.0 M $Li(ClO_4, NO_3)$			
	1.0 M $H(ClO_4, NO_3)$	20	$-0.30 \pm 0.30^{(b)}$	
dis	8 M H^+	20	0.58	[70LAH/KNO]
pot	3 M H^+	30	0.18	[85GUO/LIU]
....				

Table V.38 (continued)

Method	Ionic Medium	t (°C)	$\log_{10} \beta_q^{(a)}$	Reference
$U^{4+} + 4NO_3^- \rightleftharpoons U(NO_3)_4(aq)$				
cix	3.5 M $H(ClO_4, NO_3)$	26.5	0.38	[62ERM/KRO]
sp	3.5 M $H(ClO_4, NO_3)$	26.5	0.42	
	3.0 M $H(ClO_4, NO_3)$	26.5	0.17	
	2.5 M $H(ClO_4, NO_3)$	26.5	0.41	
	2.0 M $H(ClO_4, NO_3)$	26.5	−0.018	
dis	8 M H^+	20	−0.43	[70LAH/KNO]
pot	3 M H^+	30	−0.17	[85GUO/LIU]
....				
$U^{4+} + 3NO_3^- \rightleftharpoons U(NO_3)_3^+$				
cix	3.5 M $H(ClO_4, NO_3)$	27	0.18	[62ERM/KRO]
sp	3.5 M $H(ClO_4, NO_3)$	27	0.18	
	3.0 M $H(ClO_4, NO_3)$	27	−0.15	
	2.5 M $H(ClO_4, NO_3)$	27	−0.01	
	2.0 M $H(ClO_4, NO_3)$	27	−0.46	
dis	8 M H^+	20	−0.30	[70LAH/KNO]
pot	3 M H^+	30	−0.40	[85GUO/LIU]

(a) Refers to the formation reaction indicated at the ionic strength and temperature given in the table.

(b) Uncertainties estimated by this review as described in Appendix A.

Figure V.13: Extrapolation to $I = 0$ of experimental data for the formation of UNO_3^{3+} using the specific ion interaction theory. The data refer to mixed perchlorate/nitrate media and are taken from [62ERM/KRO] (△) and [64MCK/WOO] (○). The shaded area represents the uncertainty range obtained by propagating the resulting uncertainties at $I = 0$ back to $I = 4$ m.

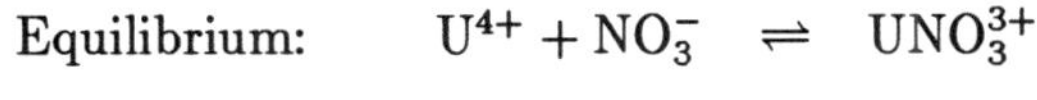

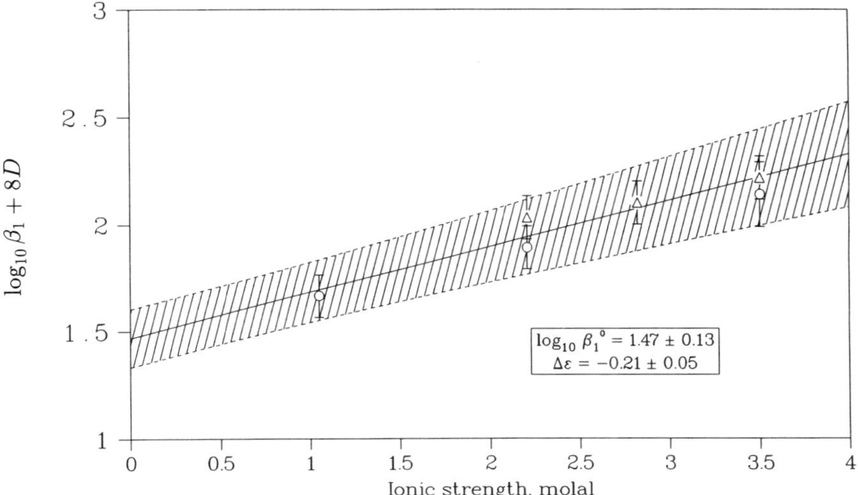

By combining the resulting $\Delta\varepsilon$ values (see Figures V.13 and V.14) with the ion interaction coefficients $\varepsilon_{(U^{4+},ClO_4^-)}$ and $\varepsilon_{(NO_3^-,H^+)} \approx \varepsilon_{(NO_3^-,Li^+)}$, two new coefficients are obtained: $\varepsilon_{(UNO_3^{3+},ClO_4^-)} = (0.62 \pm 0.08)$ and $\varepsilon_{(U(NO_3)_2^{2+},ClO_4^-)} = (0.49 \pm 0.14)$. It should be mentioned that the experimental data used for the analysis in Figures V.13 and V.14 refer to slightly different temperatures. However, no enthalpy data are available in the literature that would allow a correction to $t = 25°C$. This review assumes that the enthalpy values of the reactions of U^{4+} with nitrate do not differ much from those of the corresponding reactions with chloride. The different temperatures will then not significantly affect the extrapolation procedure.

The selected $\Delta_f G_m^\circ$ values are calculated using auxiliary data selected in this review.

$$\begin{aligned}\Delta_f G_m^\circ(UNO_3^{3+}, aq, 298.15\,K) &= -(649.0 \pm 2.0)\,kJ \cdot mol^{-1}\\ \Delta_f G_m^\circ(U(NO_3)_2^{2+}, aq, 298.15\,K) &= -(764.6 \pm 2.8)\,kJ \cdot mol^{-1}\end{aligned}$$

No enthalpy, entropy or heat capacity data are available for uranium(IV) nitrate species.

Figure V.14: Extrapolation to $I = 0$ of experimental data for the formation of $U(NO_3)_2^{2+}$ using the specific ion interaction theory. The data refer to mixed perchlorate/nitrate media and are taken from [62ERM/KRO] (△) and [64MCK/WOO] (○). The shaded area represents the uncertainty range obtained by propagating the resulting uncertainties at $I = 0$ back to $I = 4$ m.

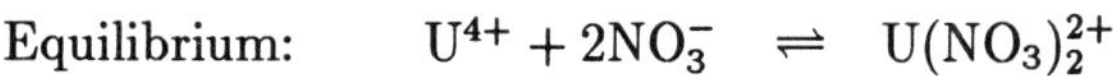

$$\text{Equilibrium:} \quad U^{4+} + 2NO_3^- \rightleftharpoons U(NO_3)_2^{2+}$$

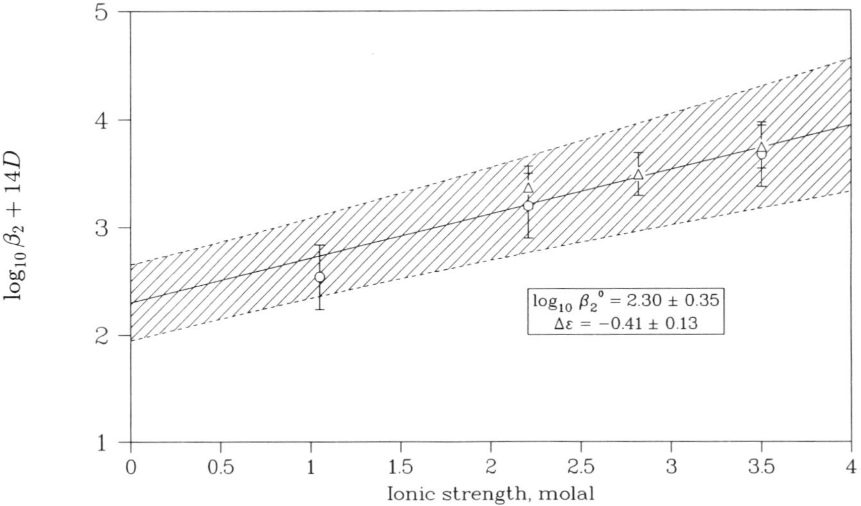

V.6.1.3.2. Solid uranium nitrates

The thermodynamic data for solid dioxouranium(VI) nitrates were reviewed by Rand and Kubaschewski [63RAN/KUB] (in the early 1960s) and by Cordfunke and O'Hare [78COR/OHA] (in the late 1970s). In the CODATA review [89COX/WAG], the values for the hexahydrate were reanalysed (based, in part, on the work of Fuger and Oetting [76FUG/OET]) and were incorporated in the selection of values for $\Delta_f G_m^\circ(UO_2^{2+})$. Thus, from the point of view of maintaining consistency with the CODATA compilation, the values of $\Delta_f H_m^\circ(UO_2(NO_3)_2 \cdot 6H_2O, \text{cr}, 298.15\text{ K})$ and $S_m^\circ(UO_2(NO_3)_2 \cdot 6H_2O, \text{cr}, 298.15\text{ K})$ are fixed. The values for the other hydrates are then related to those for the hexahydrate.

a) $UO_2(NO_3)_2 \cdot 6H_2O(cr)$

In the CODATA review [89COX/WAG],

$$\Delta_f H_m^\circ(UO_2(NO_3)_2 \cdot 6H_2O, \text{cr}, 298.15\text{ K}) = -(3167.5 \pm 1.5)\text{ kJ} \cdot \text{mol}^{-1}$$

is based on Cordfunke's measurements of the enthalpies of solution of $UO_2(NO_3)_2{\cdot}6H_2O(cr)$ and γ-UO_3 in aqueous HNO_3 [64COR]. Although

$\Delta_f H^\circ_m(UO_2(NO_3)_2{\cdot}6H_2O$, cr, 298.15 K) is not listed in the CODATA table of Key Values, the key CODATA value for $\Delta_f H^\circ_m(UO_2^{2+}$, aq, 298.15 K) is linked to $\Delta_f H^\circ_m(UO_2(NO_3)_2{\cdot}6H_2O$, cr, 298.15 K). The CODATA value of $\Delta_f H^\circ_m(UO_2^{2+}$, aq, 298.15 K) is the average of the values calculated in two different ways. The first is from $\Delta_f H^\circ_m(UO_2(NO_3)_2{\cdot}6H_2O$, cr, 298.15 K) and the enthalpy of solution of $UO_2(NO_3)_2{\cdot}6H_2O$(cr) in H_2O(l), $(19.35 \pm 0.20)\,kJ \cdot mol^{-1}$ (based on the value from Fuger and Oetting [76FUG/OET] as calculated from earlier studies, and the later work of Devina *et al.* [82DEV/YEF]). The second is from $\Delta_f H^\circ_m(UO_2Cl_2$, cr, 298.15 K) and the enthalpy of solution of UO_2Cl_2(cr) in H_2O(l). There do not appear to be any more recently published experimental values for $\Delta_f H^\circ_m(UO_2(NO_3)_2{\cdot}6H_2O$, cr, 298.15 K), and for the sake of consistency the CODATA value for $\Delta_f H^\circ_m(UO_2(NO_3)_2{\cdot}6H_2O$, cr, 298.15 K) is selected in this review.

In calculating $S^\circ_m(UO_2^{2+}$, aq, 298.15 K) the value

$$S^\circ_m(UO_2(NO_3)_2{\cdot}6H_2O, \text{cr}, 298.15\text{ K}) = (505.6 \pm 2.0)\,\text{J} \cdot \text{K}^{-1} \cdot \text{mol}^{-1}$$

was used by CODATA [89COX/WAG], based on the low-temperature heat capacity measurements of Coulter *et al.* [40COU/PIT]. Again, for the sake of consistency, the same value is selected in this review. It is noted that the estimated uncertainty in the entropy [89COX/WAG] is essentially the same as that selected in the IAEA review [78COR/OHA], but is approximately $0.5\,J \cdot K^{-1} \cdot mol^{-1}$ less than the uncertainty estimated in the original paper [40COU/PIT].

The Gibbs energy of formation is calculated from the selected enthalpy of formation and entropy.

$$\Delta_f G^\circ_m(UO_2(NO_3)_2{\cdot}6H_2O, \text{cr}, 298.15\text{ K}) = -(2584.2 \pm 1.6)\,\text{kJ} \cdot \text{mol}^{-1}$$

The selection of the value, based on Ref. [40COU/PIT],

$$C^\circ_{p,m}(UO_2(NO_3)_2{\cdot}6H_2O, \text{cr}, 298.15\text{ K}) = (468.0 \pm 2.5)\,\text{J} \cdot \text{K}^{-1} \cdot \text{mol}^{-1},$$

is discussed in Appendix A.

b) $UO_2(NO_3)_2 \cdot 3H_2O$(cr)

The value $\Delta_f H^\circ_m(UO_2(NO_3)_2{\cdot}3H_2O$, cr, 298.15 K) $= -(1280.0 \pm 1.6)\ J{\cdot}K^{-1}{\cdot}mol^{-1}$ was recalculated using CODATA compatible auxiliary data from the enthalpies of solution of γ-UO_3 and $UO_2(NO_3)_2 \cdot 3H_2O$(cr) in 6 M HNO_3 reported by Cordfunke [64COR]. Also, using the selected value of $\Delta_f H^\circ_m(UO_2(NO_3)_2 \cdot 6H_2O$, cr, 298.15 K), the value for $\Delta_f H^\circ_m(UO_2(NO_3)_2 \cdot 3H_2O$, cr, 298.15 K) can be calculated from the differences in the enthalpies of solution of $UO_2(NO_3)_2 \cdot 6H_2O$(cr) and $UO_2(NO_3)_2 \cdot 3H_2O$(cr) in aqueous media. A summary of the available data is given in Table V.39. Because only differences in the enthalpies of solution are required, corrections for the enthalpies of dilution of $UO_2(NO_3)_2$(aq) are not needed if the final solutions from any one set of results have roughly equal uranium concentrations (assuming $\Delta_{tr} H^\circ_m(H_2O, \text{sln} \rightarrow H_2O, \text{l}) = 0\,kJ \cdot mol^{-1}$).

Table V.39: Summary of the differences in the enthalpies of solution of $UO_2(NO_3)_2 \cdot 6H_2O(cr)$ and $UO_2(NO_3)_2 \cdot xH_2O(cr)$, where $x = 2, 3$.

Medium ($U:H_2O$)	$\Delta_{sol}H^\circ_m(UO_2(NO_3)_2 \cdot 6H_2O$, cr, 298.15 K) $-\Delta_{sol}H^\circ_m(UO_2(NO_3)_2 \cdot xH_2O$, cr, 298.15 K)			
	T(°C)	$x = 3$	$x = 2$	Reference
H_2O (1:?)	?	28	42.6	[12MAR]
H_2O (1:1000)	12	28.45	43.93	[15FOR]
	25(cor)	29.46	45.38	
H_2O (1:180)	25	29.87	45.40	[52KAT/SIM]
H_2O (1:120)	18	30.12	43.72$^{(a)}$	[58VDO/MAL]
	25(cor)	30.67		
H_2O (1:200)	25	30.07	45.23	[65FEN/LI]

(a) Based on a single measurement for the dihydrate at a $U:H_2O$ ratio of (1:220), not (1:120).

The value

$$C^\circ_{p,m}(UO_2(NO_3)_2 \cdot 3H_2O, \text{cr}, 298.15\,\text{K}) = (320.1 \pm 1.7)\,\text{J} \cdot \text{K}^{-1} \cdot \text{mol}^{-1}$$

is selected in this review from the work of Feng, Zhang and Zhang [82FEN/ZHA]. Enthalpy of solution values from de Forcrand [15FOR] and Vdovenko and Mal'tseva [58VDO/MAL] are corrected to 25°C using the value $C^\circ_{p,m}(UO_2(NO_3)_2 \cdot 6H_2O$, cr, 298.15 K) selected above and $C^\circ_{p,m}(H_2O$, l, 298.15 K) [82WAG/EVA]. The data from the work of Markétos [12MAR] are not judged to be sufficiently precise to be used. The average of the other results (at, or corrected to, 25°C), $(30.0 \pm 1.0)\,\text{kJ} \cdot \text{mol}^{-1}$, is used to calculate $\Delta_f H^\circ_m(UO_2(NO_3)_2 \cdot 3H_2O, \text{cr}, 298.15\,\text{K}) = -(2280.0 \pm 1.7)\,\text{kJ} \cdot \text{mol}^{-1}$.

A third method of estimating $\Delta_f H^\circ_m(UO_2(NO_3)_2 \cdot 3H_2O$, cr, 298.15 K) is based on the temperature dependence of the equilibrium vapour pressure of water over mixtures of $UO_2(NO_3)_2 \cdot 6H_2O(cr)$ and $UO_2(NO_3)_2 \cdot 3H_2O(cr)$. Cordfunke [66COR] and Cordfunke and O'Hare [78COR/OHA] analysed data measured by several authors [29GER/FRE, 52KAP/BAR, 54WEN/KIR, 57KIN/PFE, 59VDO/SOK, 66COR] for the equilibrium

$$UO_2(NO_3)_2 \cdot 6H_2O(cr) \rightleftharpoons UO_2(NO_3)_2 \cdot 3H_2O(cr) + 3H_2O(g). \quad \text{(V.131)}$$

Such data also provide a value of $\Delta_f S^\circ_m(UO_2(NO_3)_2 \cdot 3H_2O$, cr, 298.15 K). In this review the vapour pressure data from Refs. [52KAP/BAR, 54WEN/KIR,

59VDO/SOK, 66COR] (see Appendix A) are reanalysed by fitting to the equation

$$p_{H_2O} = \exp\{-\frac{1}{3RT}(\Delta_r H_m^\circ(298.15\,K) - T\Delta_r S_m^\circ(298.15\,K) + (T-298.15)\Delta_r C_{p,m} - T\Delta_r C_{p,m}\ln(T/298.15))\}. \quad (V.132)$$

A value of $\Delta_r C_{p,m}$(V.131) = $-47.1\,J\cdot K^{-1}\cdot mol^{-1}$ (calculated using $C^\circ_{p,m}(H_2O$, g, 298.15 K) = $33.6\,J\cdot K^{-1}\cdot mol^{-1}$ [82WAG/EVA] and the selected heat capacity values for the hexahydrate and the trihydrate) is assumed to be independent of temperature within the range of interest, and is held constant. The calculated values are $\Delta_r H_m^\circ$(V.131) = $(159.7\pm1.9)\,kJ\cdot mol^{-1}$ and $\Delta_r S_m^\circ$(V.131) = $(422.7\pm6.0)\,J\cdot K^{-1}\cdot mol^{-1}$ leading to $\Delta_f H_m^\circ(UO_2(NO_3)_2\cdot3H_2O$, cr, 298.15 K) = $-(2282.3\pm2.4)\,kJ\cdot mol^{-1}$ and $S_m^\circ(UO_2(NO_3)_2\cdot3H_2O$, cr, 298.15 K) = $(362.1\pm6.3)\,J\cdot K^{-1}\cdot mol^{-1}$. The calculated value for $\Delta_f H_m^\circ$ differs from both values calculated from the enthalpies of solution data, although the difference is within the sum of the uncertainties. Similarly, agreement is marginal with the value of $S_m^\circ(UO_2(NO_3)_2\cdot3H_2O$, cr, 298.15 K) = $(369.4\pm3.8)\,J\cdot K^{-1}\cdot mol^{-1}$ reported by Feng, Zhang and Zhang [82FEN/ZHA] based on heat capacity measurements between 62 and 303 K.

The values from the different types of experiments are combined and appropriately weighted while maintaining the correlation in the $\Delta_r S_m^\circ$ and $\Delta_r H_m^\circ$ values imposed by the vapour pressure measurements.

The values $\Delta_r S_m^\circ = (428.5\pm2.5)\,J\cdot K^{-1}\cdot mol^{-1}$ and $\Delta_r H_m^\circ = (161.6\pm0.8)\,kJ\cdot mol^{-1}$ are obtained, leading to the selected values

$$\Delta_f H_m^\circ(UO_2(NO_3)_2\cdot3H_2O, cr, 298.15\,K) = -(2280.4\pm1.7)\,kJ\cdot mol^{-1}$$
$$S_m^\circ(UO_2(NO_3)_2\cdot3H_2O, cr, 298.15\,K) = (367.9\pm3.3)\,J\cdot K^{-1}\cdot mol^{-1}.$$

The Gibbs energy of formation is calculated from the selected enthalpy of formation and entropy.

$$\Delta_f G_m^\circ(UO_2(NO_3)_2\cdot3H_2O, cr, 298.15\,K) = -(1864.7\pm2.0)\,kJ\cdot mol^{-1}$$

c) $UO_2(NO_3)_2\cdot 2H_2O(cr)$

For the dihydrate, the value

$$C^\circ_{p,m}(UO_2(NO_3)_2\cdot2H_2O, cr, 298.15\,K) = (278\pm4)\,J\cdot K^{-1}\cdot mol^{-1}$$

is selected in this review from the unpublished work of Morss and Cobble as reported by Cordfunke and O'Hare [78COR/OHA]. Using this and the data in Table V.39, the average difference between the enthalpies of solution of $UO_2(NO_3)_2\cdot6H_2O$(cr) and $UO_2(NO_3)_2\cdot2H_2O$(cr) is $(45.34\pm1.00)\,kJ\cdot mol^{-1}$. The uncertainty is an estimate. Neither the value of Markétos [12MAR] nor that of Vdovenko and Mal'tseva [58VDO/MAL] is used in calculating the average. The Vdovenko and Mal'tseva [58VDO/MAL] value for the $\Delta_{sol}H_m^\circ(UO_2(NO_3)_2\cdot2H_2O$, cr, 298.15 K) cannot be used directly because the final solution from the experiment with this salt did not have the same uranium nitrate concentration as those from their other measurements.

From the average difference, the value $\Delta_f H_m^\circ(UO_2(NO_3)_2{\cdot}2H_2O$, cr, 298.15 K) = $-(1978.8 \pm 1.8)\,kJ \cdot mol^{-1}$ is calculated.

As described above for the trihydrate, water vapour pressure data [54WEN/KIR, 59VDO/SOK, 66COR], in this case for trihydrate/dihydrate conversion, can be used to obtain $\Delta_r H_m^\circ(V.133) = (60.5 \pm 4.0)\,kJ \cdot mol^{-1}$ and $\Delta_r S_m^\circ(V.133) = (150.2 \pm 12.6)\,J \cdot K^{-1} \cdot mol^{-1}$ for the equilibrium

$$UO_2(NO_3)_2 \cdot 3H_2O(cr) \rightleftharpoons UO_2(NO_3)_2 \cdot 2H_2O(cr) + H_2O(g), \qquad (V.133)$$

and $\Delta_f H_m^\circ(UO_2(NO_3)_2{\cdot}2H_2O$, cr, 298.15 K) = $-(1978.0 \pm 4.3)\,kJ \cdot mol^{-1}$. The weighted average of the values for the enthalpy of formation is

$$\Delta_f H_m^\circ(UO_2(NO_3)_2{\cdot}2H_2O, \text{cr}, 298.15\text{ K}) = -(1978.7 \pm 1.7)\,kJ \cdot mol^{-1},$$

and using this and the pressure data

$$\Delta_f G_m^\circ(UO_2(NO_3)_2{\cdot}2H_2O, \text{cr}, 298.15\text{ K}) = -(1620.5 \pm 2.0)\,kJ \cdot mol^{-1},$$

is calculated.

From $\Delta_r H_m^\circ(V.133)$ and $\Delta_r G_m^\circ(V.133)$, the value $\Delta_r S_m^\circ(V.133) = (148 \pm 8)\,J \cdot K^{-1} \cdot mol^{-1}$ is calculated, and hence,

$$S_m^\circ(UO_2(NO_3)_2{\cdot}2H_2O, \text{cr}, 298.15\text{ K}) = (327.5 \pm 8.8)\,J \cdot K^{-1} \cdot mol^{-1}.$$

d) $UO_2(NO_3)_2 \cdot H_2O(cr)$

Although there is reasonable evidence that this compound exists [15FOR, 68SMI, 70CHO], at least as an intermediate compound in the thermal decomposition of higher hydrates of dioxouranium(VI) nitrate, its isolation as a pure compound has proven difficult. Therefore, only limited thermodynamic data have been reported for this compound.

De Forcrand [15FOR] reported a value for the enthalpy of solution in water that can be compared to values for the enthalpies of solution of the other hydrated dioxouranium(VI) nitrates. From this (see Appendix A), $\Delta_f H_m^\circ(UO_2(NO_3)_2{\cdot}H_2O$, cr, 298.15 K) = $-1663.8\,kJ \cdot mol^{-1}$ can be calculated. This value is quite similar to the average value of $-1663.6\,kJ \cdot mol^{-1}$ that can be calculated from the DTA data of Smith [68SMI]. The value

$$\Delta_f H_m^\circ(UO_2(NO_3)_2{\cdot}H_2O, \text{cr}, 298.15\text{ K}) = -(1664 \pm 10)\,kJ \cdot mol^{-1}$$

is selected in this review. The large uncertainty is chosen to reflect the absence of data based on a more adequately characterized solid.

Using the method in Appendix D, the entropy is estimated to be

$$S_m^\circ(UO_2(NO_3)_2{\cdot}H_2O, \text{cr}, 298.15\text{ K}) = (286 \pm 11)\,J \cdot K^{-1} \cdot mol^{-1}.$$

The Gibbs energy of formation is calculated from the selected enthalpy of formation and entropy.

$$\Delta_f G_m^\circ(UO_2(NO_3)_2{\cdot}H_2O, cr, 298.15\ K) = -(1363.0 \pm 10.5)\ kJ \cdot mol^{-1}$$

e) $UO_2(NO_3)_2(cr)$

This compound is difficult to synthesize in a pure form. Enthalpies of solution reported by Markétos [12MAR] and by Vdovenko, Koval'skaya and Kovaleva [57VDO/KOV] appear to be based on experiments using impure samples. Analysis of data from de Forcrand [15FOR] and Smith [68SMI] in the same manner as done above for the monohydrate results in the values of $\Delta_f H_m^\circ(UO_2(NO_3)_2$, cr, 298.15 K) of $-1347.3\ kJ \cdot mol^{-1}$ and $-1354.7\ kJ \cdot mol^{-1}$, respectively. Vdovenko *et al.* [63VDO/SUG] and Kanevskii *et al.* [74KAN/ZAR] reported similar values for the enthalpy of solution of well-characterized samples of the anhydrous salt in water. These values were corrected for hydrolysis by Cordfunke and O'Hare [78COR/OHA] and an average $\Delta_{sol} H_m^\circ = 82.0\ kJ{\cdot}mol^{-1}$ was calculated. Using this value, $\Delta_f H_m^\circ(UO_2(NO_3)_2$, cr, 298.15 K) $= -1350.6\ kJ \cdot mol^{-1}$ is calculated, a number consistent with the other studies. In this review

$$\Delta_f H_m^\circ(UO_2(NO_3)_2, cr, 298.15\ K) = -(1351 \pm 5)\ kJ \cdot mol^{-1}$$

is selected, based primarily on the results in Refs. [74KAN/ZAR, 63VDO/SUG], with the estimated uncertainty reflecting the difficulties in preparing the pure compound.

Using the method in Appendix D, the entropy is estimated to be

$$S_m^\circ(UO_2(NO_3)_2, cr, 298.15\ K) = (241 \pm 9)\ J \cdot K^{-1} \cdot mol^{-1}.$$

The Gibbs energy of formation is calculated from the selected enthalpy of formation and entropy.

$$\Delta_f G_m^\circ(UO_2(NO_3)_2, cr, 298.15\ K) = -(1106 \pm 6)\ kJ \cdot mol^{-1}$$

V.6.2. Phosphorus compounds and complexes

V.6.2.1. Aqueous uranium phosphorus species

V.6.2.1.1. The uranium-phosphoric acid system

The experimental studies of equilibria in the uranium-phosphoric acid system are complicated by the formation of a number of sparingly soluble solid phases and the formation of ternary complexes of the type $MH_r(PO_4)_q$, where $M = UO_2^{2+}$ or U^{4+}.

There are few precise studies available in the literature, and most of them refer to solutions of low pH and fairly high concentrations of phosphoric acid. Dongarra and Langmuir [80DON/LAN] covered the pH region up to 4.7. The only experimental study which extends into the pH range encountered in ground and surface waters is the recent thesis by Sandino [91SAN], where the solubility of $(UO_2)_3(PO_4)_2{\cdot}4H_2O(cr)$ was measured in the range $6 < -\log_{10}[H^+] < 9$. The experimental studies have not always been carried out over a sufficiently large hydrogen ion concentration range to allow a determination of the proton content of the complexes.

a) Complex formation in the U(VI)-H_3PO_4 system

Baes, Schreyer and Lesser [53BAE/SCH2], Baes [56BAE] and Thamer [57THA] studied the equilibria of dioxouranium(VI) in phosphoric acid solutions of varying acidity. Baes [56BAE] reported equilibrium constants for the formation of $UO_2H_3PO_4^{2+}$, $UO_2H_2PO_4^+$, $UO_2(H_2PO_4)(H_3PO_4)^+$, and $UO_2(H_2PO_4)_2(aq)$, *cf.* Table V.40. Thamer [57THA] reported the same conditional equilibrium constants as Baes [56BAE] in 1 M $HClO_4$, but found no evidence for the formation of complexes containing H_3PO_4 as a ligand. Marcus [58MAR] identified the formation of several species of this type, and his results are in fair agreement with those of Baes [56BAE]. This review accepts the evidence presented by Baes [56BAE] and Marcus [58MAR] for the formation of complexes containing H_3PO_4 as a ligand. However, Marcus' data [58MAR] are not used in the evaluation of the selected constants because they are so-called "mixed" equilibrium constants, *cf.* Appendix A. A recent study by Mathur [91MAT] yielded somewhat different results, because an incomplete model was used by the author for the interpretation of his measurements, *cf.* Appendix A.

Several solubility studies are available on the dioxouranium(VI)-phosphoric acid system. The experimental conditions were in general optimized for the determination of solubility products, and it is difficult to make an unambiguous assignment of species from these data alone. Schreyer and Baes [55SCH/BAE] measured the solubilities of $(UO_2)_3(PO_4)_2 \cdot 6H_2O(cr)$, $UO_2HPO_4 \cdot 4H_2O(cr)$ and $UO_2(H_2PO_4)_2 \cdot 3H_2O(cr)$ in phosphoric acid solutions of various concentrations. The *increase* in solubility with increasing concentrations of H_3PO_4 for the first two phases indicates that complexes with two or more phosphates per dioxouranium(VI) are formed. The *decrease* in solubility of $UO_2(H_2PO_4)_2{\cdot}3H_2O(cr)$ with increasing concentration of phosphoric acid indicates that no significant amounts of complexes with three or more phosphates per dioxouranium(VI) are formed at phosphoric acid concentrations < 10 M.

Moskvin, Shelyakina and Perminov [67MOS/SHE] performed a solubility study where they claimed evidence for the formation of $UO_2HPO_4(aq)$ and $UO_2(HPO_4)_2^{2-}$. These data were discussed by Tripathi [84TRI] and are also reinterpreted by this review, *cf.* Appendix A. They give support to the formation of $UO_2(H_2PO_4)_2(aq)$, but there is no conclusive evidence for the formation of complexes containing HPO_4^{2-} as a ligand. Dongarra and Langmuir [80DON/LAN] interpreted a potentiometric study of the UO_2^{2+}-phosphoric acid system in terms of the formation of $UO_2(HPO_4)_2^{2-}$. This interpretation was criticized by Tripathi [84TRI]. This review agrees with the arguments presented by Tripathi [84TRI] and does therefore not consider the results from Ref. [80DON/LAN].

Hence, this review considers only the following equilibria in acidic solution, with H_3PO_4 and $H_2PO_4^-$ as ligands:

$$UO_2^{2+} + H_3PO_4(aq) \rightleftharpoons UO_2H_2PO_4^+ + H^+ \quad (V.134)$$

$$UO_2^{2+} + H_3PO_4(aq) \rightleftharpoons UO_2H_3PO_4^{2+} \quad (V.135)$$

$$UO_2^{2+} + 2H_3PO_4(aq) \rightleftharpoons UO_2(H_2PO_4)_2(aq) + 2H^+ \quad (V.136)$$

$$UO_2^{2+} + 2H_3PO_4(aq) \rightleftharpoons UO_2(H_2PO_4)(H_3PO_4)^+ + H^+. \quad (V.137)$$

Table V.40: Experimental equilibrium data for the uranium(VI)-phosphoric acid system.

Method	Ionic medium	t (°C)	$\log_{10} K^{(a)}$	$\log_{10} K^{\circ(a)}$	Reference
$UO_2^{2+} + PO_4^{3-} \rightleftharpoons UO_2PO_4^-$					
sol	$I = 0.5$ M ($NaClO_4$)	25	11.28 ± 0.09	$13.23 \pm 0.15^{(b)}$	[91SAN]
$UO_2^{2+} + HPO_4^{2-} \rightleftharpoons UO_2HPO_4(aq)$					
sol	$I = 0.5$ M ($NaClO_4$)	25	6.00 ± 0.14	$7.24 \pm 0.26^{(b)}$	[91SAN]
$UO_2^{2+} + H_3PO_4(aq) \rightleftharpoons UO_2H_2PO_4^+ + H^+$					
sp	$I = 1.0$ M, $[H^+] = 0.1$ M and 1.0 M	25	$0.72 \pm 0.04^{(c)}$	$1.12 \pm 0.06^{(c)}$	[56BAE]
sp	$I = 1.0$ M, $[H^+] = 0.5$ M and 1.0 M	25	1.19 ± 0.04		[57THA]
aix	var.	25	$0.9 \pm 0.1^{(d)}$		[58MAR]
dis	var.	25		1.32 ± 0.02	[91MAT]
$UO_2^{2+} + H_3PO_4(aq) \rightleftharpoons UO_2H_3PO_4^{2+}$					
sp	$I = 1.0$ M, $[H^+] = 0.1$ M and 1.0 M	25	$0.76 \pm 0.15^{(c)}$	$0.76 \pm 0.15^{(c)}$	[56BAE]
$UO_2^{2+} + 2H_3PO_4(aq) \rightleftharpoons UO_2(H_rPO_4)_2^{2r-4} + (6-2r)H^+$					
dis	$I = 1.0$ M, $[H^+] = 0.5$ M and 1.0 M	25	1.34 ± 0.08		[57THA]
sol	$I = 0.5$ M (HNO_3)	25	$1.54 \pm 0.09^{(c)}$		[67MOS/SHE]
$UO_2^{2+} + 2H_3PO_4(aq) \rightleftharpoons UO_2(HPO_4)_2^{2-} + 4H^+$					
pot	0	25		$18.3 \pm 0.2^{(d)}$	[80DON/LAN]
$UO_2^{2+} + 2H_3PO_4(aq) \rightleftharpoons UO_2(H_2PO_4)_2(aq) + 2H^+$					
sp	$I = 1$ M, $[H^+] = 0.1$ M and 1.0 M	25	$0.41 \pm 0.10^{(c)}$	$0.64 \pm 0.11^{(c)}$	[56BAE]
aix	var.	25	$1.3 \pm 0.1^{(d)}$		[58MAR]
dis	var.	25		1.47 ± 0.02	[91MAT]

Table V.40 (continued)

Method	Ionic medium	t (°C)	$\log_{10} K^{(a)}$	$\log_{10} K^{\circ(a)}$	Reference
$UO_2^{2+} + 2H_3PO_4(aq) \rightleftharpoons UO_2(H_2PO_4)(H_3PO_4)^+ + H^+$					
sp	$I = 1$ M, $[H^+] = 0.1$ M and 1.0 M	25	$1.33 \pm 0.10^{(c)}$	$1.65 \pm 0.11^{(c)}$	[56BAE]
$UO_2^{2+} + 2H_3PO_4(aq) \rightleftharpoons UO_2(H_3PO_4)_2^{2+}$					
aix	var.	25	$3.9 \pm 0.1^{(d)}$		[58MAR]
$UO_2^{2+} + 3H_3PO_4(aq) \rightleftharpoons UO_2(H_2PO_4)_3^- + 3H^+$					
aix	var.	25	$1.1 \pm 0.1^{(d)}$		[58MAR]
$UO_2^{2+} + 3H_3PO_4(aq) \rightleftharpoons UO_2(H_2PO_4)(H_3PO_4)_2^+ + H^+$					
aix	var.	25	$4.8 \pm 0.1^{(d)}$		[58MAR]
$UO_2^{2+} + 3H_3PO_4(aq) \rightleftharpoons UO_2(H_3PO_4)_3^{2+}$					
aix	var.	25	$5.3 \pm 0.1^{(d)}$		[58MAR]
$UO_2HPO_4(cr) + 2H^+ \rightleftharpoons UO_2^{2+} + H_3PO_4(aq)$					
sol	$I = 0.32$ M (HNO_3)	20	$-2.14 \pm 0.05^{(c)}$	$-2.42 \pm 0.05^{(c)}$	[65VES/PEK]
sol	var.	25		$-2.59 \pm 0.05^{(c)}$	[67MOS/SHE]
$(UO_2)_3(PO_4)_2 \cdot 6H_2O(cr) + 6H^+ \rightleftharpoons 3UO_2^{2+} + 2H_3PO_4(aq) + 6H_2O(l)$					
sol	$I = 0.01$ M (HNO_3)	25	$-5.27 \pm 0.30^{(c)}$	$-5.42 \pm 0.30^{(c)}$	[61KAR]
$(UO_2)_3(PO_4)_2 \cdot 4H_2O(cr) + 6H^+ \rightleftharpoons 3UO_2^{2+} + 2H_3PO_4(aq) + 4H_2O(l)$					
sol	$I = 0.32$ M (HNO_3)	20	$-5.13 \pm 0.30^{(c)}$	$-5.96 \pm 0.30^{(c)}$	[65VES/PEK]
$(UO_2)_3(PO_4)_2 \cdot 4H_2O(cr) \rightleftharpoons 3UO_2^{2+} + 2PO_4^{3-} + 4H_2O(l)$					
sol	$I = 0.5$ M ($NaClO_4$)	25	-48.48 ± 0.12	-53.28 ± 0.13	[91SAN]

Table V.40 (continued)

Method	Ionic medium	t (°C)	$\log_{10} K^{(a)}$	$\log_{10} K^{\circ(a)}$	Reference
$MUO_2PO_4 \cdot xH_2O(cr) + 3H^+ \rightleftharpoons M^+ + UO_2^{2+} + H_3PO_4(aq) + xH_2O(l)$					
sol (M = Na)	$I = 0.2$ M (NO_3^-)	20	$-1.66 \pm 0.09^{(c)}$	$-1.94 \pm 0.09^{(c)}$	[65VES/PEK]
(M = K)		20	$-3.00 \pm 0.15^{(c)}$	$-3.28 \pm 0.15^{(c)}$	
(M = Rb)		20	$-3.23 \pm 0.20^{(c)}$	$-3.51 \pm 0.20^{(c)}$	
(M = Cs)		20	$-2.90 \pm 0.20^{(c)}$	$-3.18 \pm 0.20^{(c)}$	
(M = NH_4)		20	$-3.82 \pm 0.10^{(c)}$	$-4.10 \pm 0.10^{(c)}$	

(a) Refers to the reactions indicated, $\log_{10} K$ in the ionic medium and at the temperature given in the table, $\log_{10} K^\circ$ (in molal units) at $I = 0$ and 298.15 K.
(b) Uncertainty estimated in this review.
(c) Value obtained from a reinterpretation of the experimental data, *cf.* Appendix A.
(d) Uncertainty reported in the paper, but the value is discarded in this review, *cf.* Appendix A.

The evaluation of the data given in Table V.40 for Reactions (V.134) to (V.137) yields the following selected equilibrium constants:

$$\begin{aligned}
\log_{10} K^\circ(\text{V.134}, 298.15\text{ K}) &= 1.12 \pm 0.06\\
\log_{10} K^\circ(\text{V.135}, 298.15\text{ K}) &= 0.76 \pm 0.15\\
\log_{10} K^\circ(\text{V.136}, 298.15\text{ K}) &= 0.64 \pm 0.11\\
\log_{10} K^\circ(\text{V.137}, 298.15\text{ K}) &= 1.65 \pm 0.11.
\end{aligned}$$

From these constants the Gibbs energies of formation are obtained by using auxiliary data selected in the present review.

$$\begin{aligned}
\Delta_f G^\circ_m(UO_2H_2PO_4^+, \text{aq}, 298.15\,\text{K}) &= -(2108.3 \pm 2.4)\,\text{kJ}\cdot\text{mol}^{-1}\\
\Delta_f G^\circ_m(UO_2H_3PO_4^{2+}, \text{aq}, 298.15\,\text{K}) &= -(2106.3 \pm 2.5)\,\text{kJ}\cdot\text{mol}^{-1}\\
\Delta_f G^\circ_m(UO_2(H_2PO_4)_2, \text{aq}, 298.15\,\text{K}) &= -(3254.9 \pm 3.7)\,\text{kJ}\cdot\text{mol}^{-1}\\
\Delta_f G^\circ_m(UO_2(H_2PO_4)(H_3PO_4)^+, \text{aq}, 298.15\,\text{K}) &= -(3260.7 \pm 3.7)\,\text{kJ}\cdot\text{mol}^{-1}
\end{aligned}$$

In neutral to basic solutions, a solubility study of $(UO_2)_3(PO_4)_2{\cdot}4H_2O(cr)$ in the pH range between 6 and 9 was published recently by Sandino [91SAN]. Equilibrium data were reported for the formation of $UO_2HPO_4(aq)$, $UO_2PO_4^-$ and $UO_2(OH)_3^-$, in addition to the solubility product of the solid phase. The constant reported for $UO_2(OH)_3^-$ is consistent with the value selected in Section V.3.2.1.3. This review accepts Sandino's formation constants for $UO_2PO_4^-$ and $UO_2HPO_4(aq)$ according to the reactions

$$UO_2^{2+} + PO_4^{3-} \rightleftharpoons UO_2PO_4^-, \quad \text{(V.138)}$$

$$UO_2^{2+} + HPO_4^{2-} \rightleftharpoons UO_2HPO_4(aq) \quad \text{(V.139)}$$

but increases the uncertainties, *cf.* Appendix A.

$$\log_{10} K^\circ(\text{V.138}, 298.15\text{K}) = 13.23 \pm 0.15$$
$$\log_{10} K^\circ(\text{V.139}, 298.15\text{K}) = 7.24 \pm 0.26$$

The following selected Gibbs energies of formation are derived therefrom:

$$\Delta_f G^\circ_m(UO_2PO_4^-, \text{aq}, 298.15\,\text{K}) = -(2053.6 \pm 2.5)\,\text{kJ} \cdot \text{mol}^{-1}$$
$$\Delta_f G^\circ_m(UO_2HPO_4, \text{aq}, 298.15\,\text{K}) = -(2089.9 \pm 2.8)\,\text{kJ} \cdot \text{mol}^{-1}.$$

In view of the importance of the phosphate system for the modelling of dioxouranium(VI) in the environment, it is highly desirable to have additional experimental verification of the dioxouranium(VI) phosphate system in the neutral and alkaline pH ranges.

Marković and Pavković [83MAR/PAV] performed a solubility study of $UO_2HPO_4 \cdot 4H_2O$(cr) and $(UO_2)_3(PO_4)_2 \cdot 8H_2O$(cr) in acid solutions of phosphoric acid. This study is discussed at length in Appendix A. The authors proposed a slightly different chemical model from the one selected in this review, and also somewhat different equilibrium constants. A recalculation shows that the equilibrium constants selected in this review give an excellent fit to the experimental data; the fit seems to be even better than the one claimed by Marković and Pavković, although no statistical analysis of this is made by this review. The conclusion of this review is that the results of Ref. [83MAR/PAV] confirm the selected set of equilibrium data.

b) Solubility equilibria in the U(VI)-H_3PO_4 system

The determinations of the solubility products of the dioxouranium(VI) phosphate phases were carried out in acid solutions. The measured solubilities must in all cases be corrected for the formation of aqueous phosphate complexes. This was not always done in the literature, *e.g.*, [56CHU/STE, 61CHU/ALY, 61KAR, 67MOS/SHE]. The solubility products depend on the choice of the chemical model and the numerical values of the complex formation constants; an example is given by Veselý, Pekarek and Abbrent [65VES/PEK]. Tripathi [84TRI] reviewed earlier data. The data of Veselý, Pekarek and Abbrent [65VES/PEK] and of Moskvin, Shelyakina and Perminov [67MOS/SHE] are used in this review to determine the selected equilibrium constant for the reaction

$$UO_2HPO_4 \cdot 4H_2O(\text{cr}) + 2H^+ \rightleftharpoons UO_2^{2+} + H_3PO_4(\text{aq}) + 4H_2O(\text{l}). \quad (\text{V.140})$$

The primary data are recalculated by using the equilibrium constants for the dioxouranium(VI)-phosphoric acid system selected in this review, recalculated to the appropriate ionic strengths, *cf.* Appendix A for more details. The resulting selected value is

$$\log_{10}{}^*K^\circ_s(\text{V.140}, 298.15\text{ K}) = -2.50 \pm 0.09.$$

The solubility constant for the reaction

$$(UO_2)_3(PO_4)_2 \cdot 4H_2O(\text{cr}) + 6H^+ \rightleftharpoons 3UO_2^{2+} + 2H_3PO_4(\text{aq}) + 4H_2O(\text{l}) \quad (\text{V.141})$$

is selected from Veselý, Pekarek and Abbrent [65VES/PEK]. Chukhlantsev and Alyamovskaya [61CHU/ALY] measured the solubility of "$(UO_2)_3(PO_4)_2$(s)" but did not take complex formation into account in their evaluation of the solubility product. Their reported value is therefore too high and has to be discarded, *cf.* Appendix A. Karpov [61KAR] studied the same reaction using a solid phase with six waters of hydration instead of four. This review reinterprets the original data in Refs. [65VES/PEK] and [61KAR] using the protonation constants of phosphate and the stability constants of the dioxouranium(VI) phosphate complexes selected in this review. The two values are in fair agreement, $\log_{10}{}^*K_s^\circ(\text{V.141}) = -(5.96 \pm 0.30)$ [65VES/PEK] and $\log_{10}{}^*K_s^\circ(\text{V.141}) = -(5.42 \pm 0.30)$ [61KAR], respectively. The experimental procedures are described in more detail in Ref. [65VES/PEK] than in Ref. [61KAR]. Further, Schreyer and Baes [54SCH/BAE] discussed the state of hydration of $(UO_2)_3(PO_4)_2$(cr), indicating a close similarily between the tetra- and hexahydrate. For these reasons, this review prefers the value obtained from the data in Ref. [65VES/PEK]. The selected equilibrium constant is

$$\log_{10}{}^*K_s^\circ(\text{V.141}, 298.15\text{ K}) \quad = \quad -5.96 \pm 0.30.$$

The selected equilibrium constants $\log_{10}{}^*K_s^\circ(\text{V.140})$ and $\log_{10}{}^*K_s^\circ(\text{V.141})$ are used to calculate the selected Gibbs energies of formation of the two solids, *cf.* Sections V.6.2.2.10.c and V.6.2.2.5.a, respectively.

Sandino [91SAN] recently determined the solubility product of a well defined $(UO_2)_3(PO_4)_2 \cdot 4H_2O$(cr) phase. She reported a value of $\log_{10} K_{s,0}$(V.142, 0.5 M $NaClO_4$) = $-(48.48 \pm 0.12)$ for the reaction

$$(UO_2)_3(PO_4)_2 \cdot 4H_2O(\text{cr}) \quad \rightleftharpoons \quad 3UO_2^{2+} + 2PO_4^{3-} + 4H_2O(\text{l}) \qquad (\text{V.142})$$

Recalculation to $I = 0$ for the reaction involving phosphoric acid instead of phosphate results in $\log_{10}{}^*K_{s,0}^\circ$(V.141, 298.15) = $-(9.88 \pm 0.16)$, which is much lower than the value reported from investigations at lower ionic strength and the value selected in this review, $\log_{10}{}^*K_{s,0}^\circ$(V.141, 298.15) = $-(5.96 \pm 0.30)$. The most obvious reason for this discrepancy is a difference in the crystallinity between the two phases. Additional determinations of the solubility product over a range of ionic strengths are desirable. This review prefers the solubility product determinations made at lower ionic strength because they were performed under conditions which are closer to the ones used in modelling uranium migration in non-saline ground and surface water systems.

A check of the consistency of the solubility products can be provided by measurements of the phosphoric acid concentration in a system where $UO_2HPO_4 \cdot 4H_2O$(cr) and $(UO_2)_3(PO_4)_2 \cdot 4H_2O$(cr) are in equilibrium. Such measurements were done by Schreyer and Baes [54SCH/BAE], Hoffmann [72HOF] and Weigel, *cf.* [85WEI, Figure 7]. At equilibrium, $\log_{10}[H_3PO_4] = 3\log_{10}{}^*K_s^\circ(\text{V.140}) - \log_{10}{}^*K_s^\circ(\text{V.141})$. From the solubility constants selected in this review, $\log_{10}[H_3PO_4] = -(1.54 \pm 0.40)$ is obtained, *i.e.*, $[H_3PO_4] = (0.029 \pm 0.027)$ M. This value is consistent with the experimental values of 0.014 M [54SCH/BAE] and 0.006 to 0.012 M [85WEI].

The solubility of $UO_2(H_2PO_4)_2 \cdot 3H_2O$(cr) was measured in concentrated phosphoric acid solutions by Schreyer and Baes [54SCH/BAE] but no equilibrium constant was given for the reaction

$$UO_2(H_2PO_4)_2 \cdot 3H_2O(cr) + 2H^+ \rightleftharpoons UO_2^{2+} + 2H_3PO_4(aq) + 3H_2O(l). \quad (V.143)$$

This constant is estimated by this review from the phosphoric acid concentration at equilibrium between $UO_2HPO_4 \cdot 4H_2O(cr)$ and $UO_2(H_2PO_4)_2 \cdot 3H_2O(cr)$ given in Ref. [54SCH/BAE]. At this point the following relation is valid: $\log_{10}[H_3PO_4] = \log_{10}{}^*K_s^\circ(V.143) - \log_{10}{}^*K_s^\circ(V.140) \approx 0.8$. The ionic strength at the intersection is around 0.5 m. Using the selected value for $\log_{10}{}^*K_s^\circ(V.140)$ above, this review tentatively proposes

$$\log_{10}{}^*K_s^\circ(V.143, 298.15\,K) = -1.7$$

as an approximate value for the solubility constant of $UO_2(H_2PO_4)_2 \cdot 3H_2O(cr)$. No uncertainty estimate is made. Note that this phase is not thermodynamically stable at low phosphoric acid concentrations. It should be mentioned that $UO_2HPO_4{\cdot}4H_2O(cr)$ is an excellent ion exchanger, *cf.* Refs. [65PEK/VES, 80KOB/KOL]. Veselý, Pekárek and Abbrent [65VES/PEK] determined the solubility constants for the reactions

$$MUO_2PO_4 \cdot xH_2O(cr) + 3H^+ \rightleftharpoons M^+ + UO_2^{2+} + H_3PO_4(aq) + xH_2O(l) \quad (V.144)$$

where $M^+ = Na^+, K^+, Rb^+, Cs^+$ and NH_4^+. The original data are corrected for the formation of dioxouranium(VI) phosphate complexes using the stability constants selected by this review, *cf.* Appendix A. The results are given in Table V.40.

Muto [65MUT] synthesized $H_2(UO_2)_2(PO_4)_2 \cdot 10H_2O(cr)$ by a cation exchange reaction with decahydrated Ca-autunite in acidic solution, as well as by precipitation from a mixture of phosphoric acid and dioxouranium(VI) chloride solution, as represented by Eq. (V.145).

$$2UO_2^{2+} + 2H_3PO_4(aq) + 10H_2O(aq) \rightleftharpoons H_2(UO_2)_2(PO_4)_2 \cdot 10H_2O(cr) + 4H^+ \quad (V.145)$$

Muto, Hirono and Kurata [65MUT, 65MUT/HIR] experimentally determined the constant of Reaction (V.145) starting from both directions. Details of the experimental procedure of the precipitation reaction [65MUT/HIR] were not reported, and this review therefore does not accept these data. The selected equilibrium constants in the dioxouranium(VI)-phosphoric acid system are summarized in Table V.41.

c) The aqueous U(IV)-H_3PO_4 system

There are few experimental studies of equilibria in the uranium(IV)-phosphate system. However, Zebroski, Alter and Heumann [51ZEB/ALT] performed a precise study of the corresponding thorium system. They clearly demonstrated that both H_3PO_4 and $H_2PO_4^-$ act as ligands in the hydrogen ion concentration range 0.25 to 2.00 M. The equilibria in the corresponding plutonium(IV) system were investigated by Denotkina, Moskvin and Shevchenko [60DEN/MOS]. These authors interpreted

Table V.41: Selected equilibrium constants in the dioxouranium(VI)-phosphoric acid system at 298.15 K and $I = 0$.

Reaction			$\log_{10} K^\circ$
$UO_2^{2+} + PO_4^{3-}$	$\rightleftharpoons$	$UO_2PO_4^-$	13.23 ± 0.15
$UO_2^{2+} + HPO_4^{2-}$	$\rightleftharpoons$	$UO_2HPO_4(aq)$	7.24 ± 0.26
$UO_2^{2+} + H_3PO_4(aq)$	$\rightleftharpoons$	$UO_2H_2PO_4^+ + H^+$	1.12 ± 0.06
$UO_2^{2+} + H_3PO_4(aq)$	$\rightleftharpoons$	$UO_2H_3PO_4^{2+}$	0.76 ± 0.15
$UO_2^{2+} + 2H_3PO_4$	$\rightleftharpoons$	$UO_2(H_2PO_4)_2(aq) + 2H^+$	0.64 ± 0.11
$UO_2^{2+} + 2H_3PO_4$	$\rightleftharpoons$	$UO_2(H_2PO_4)(H_3PO_4)^+ + H^+$	1.65 ± 0.11
Solubility constants:			
$UO_2HPO_4 \cdot 4H_2O(cr) + 2H^+$	$\rightleftharpoons$	$UO_2^{2+} + H_3PO_4(aq) + 4H_2O(l)$	-2.50 ± 0.09
$(UO_2)_3(PO_4)_2 \cdot 4H_2O(cr) + 6H^+$	$\rightleftharpoons$	$3UO_2^{2+} + 2H_3PO_4(aq) + 4H_2O(l)$	-5.96 ± 0.30
$UO_2(H_2PO_4)_2 \cdot 3H_2O(cr) + 2H^+$	$\rightleftharpoons$	$UO_2^{2+} + 2H_3PO_4(aq) + 3H_2O(l)$	-1.7[(a)]

(a) This value is not selected in this review but is given as a guideline.

their results assuming that the complex formation occurs with HPO_4^{2-} as the ligand. There is no experimental evidence for this assumption and this review therefore prefers the chemical model proposed in Ref. [51ZEB/ALT] and assumes this to be valid for all the actinide(IV) ions in the acidity range 0.25 to 2.00 M.

The uranium(IV)-phosphate system was discussed by Moskvin, Essen and Bukhtiyarova [67MOS/ESS]. Their experimental data were taken from a study of Markov, Vernyi and Vinogradov [64MAR/VER]. As the experiments refer to a constant acidity it is not possible to determine the degree of protonation of the ligand; however, the data clearly indicate that a maximum of four phosphate groups are bonded per U(IV). This review recalculates the data in Ref. [67MOS/ESS] (see Appendix A) and finds $\log_{10}{}^*K_s(V.146) = -(9.96 \pm 0.15)$ for the reaction

$$U(HPO_4)_2 \cdot 4H_2O(cr) + 4H^+ \rightleftharpoons U^{4+} + 2H_3PO_4(aq) + 4H_2O(l). \qquad (V.146)$$

Using the specific ion interaction theory to correct to $I = 0$ results in the selected value of

$$\log_{10}{}^*K_s^\circ(V.146, 298.15\ K) = -11.79 \pm 0.15.$$

This corresponds to a solubility constant of $\log_{10}{}^*K_s^\circ = -(30.49 \pm 0.16)$ for the reaction $U(HPO_4)_2(s) \rightleftharpoons U^{4+} + 2HPO_4^{2-}$, compared to $\log_{10}{}^*K_s^\circ = -26.8$ given in Ref. [67MOS/ESS].

This review finds it impossible to obtain any reliable information on the composition of the aqueous uranium(IV) phosphate complexes and the numerical values of their formation constants. However, there is no doubt that very stable uranium(IV) phosphate complexes are formed, *e.g.*, see [55SCH, 56BAE2], and that additional investigations are needed. Some guidance as to the complexes that might be formed can be obtained from a recent study of the thorium(IV) phosphoric acid system by Elyahyaoui *et al.* [90ELY/BRI].

V.6.2.1.2. *Aqueous uranium pyrophosphates*

a) *Aqueous U(VI) pyrophosphates*

The complexation of dioxouranium(VI) ions with pyrophosphate ligands was investigated by Drăgulescu, Julean and Vîlceanu [65DRA/JUL, 70DRA/VIL]. These authors used five analytical methods (potentiometry, conductometry, spectrophotometry, radiometry, radioelectrophoresis) to detect uranium complexes in sodium pyrophosphate/uranium nitrate solutions in the pH range 1 to 11. For reasons given in Appendix A, no value is selected for the complexes $(UO_2)_mH_r(P_2O_7)_q^{2m+r-4q}$ reported by Drăgulescu, Julean and Vîlceanu [65DRA/JUL]. Qualitative information on the formation of pyrophosphate and polyphosphate complexes has been provided by van Wazer and Campanella [50WAZ/CAM], *cf.* Appendix A.

b) *Aqueous U(IV) pyrophosphates*

Only one neutral complex, $UP_2O_7(aq)$, was suggested by Merkusheva *et al.* [67MER/SKO]. These authors measured the solubility of $UP_2O_7 \cdot 20H_2O(cr)$ in (Na, H)ClO_4 (I = 0.1 M) solutions in the range $1 \leq pH \leq 2.68$ and obtained a solubility curve nearly independent of pH. Based on this result, Merkusheva *et al.* [67MER/SKO] interpreted the solubility of $UP_2O_7 \cdot 20H_2O(cr)$ according to the reaction

$$UP_2O_7 \cdot 20H_2O(cr) \rightleftharpoons UP_2O_7(aq) + 20H_2O \qquad (V.147)$$

and calculated the constants for the two equilibria

$$UP_2O_7(cr) \rightleftharpoons U^{4+} + P_2O_7^{4-} \qquad (V.148)$$
$$U^{4+} + P_2O_7^{4-} \rightleftharpoons UP_2O_7(aq). \qquad (V.149)$$

They obtained $K(V.148, I = 0.1) = (1.35 \pm 0.32) \times 10^{-24}$ and $K(V.149, I = 0.1) = (1.17 \pm 0.34) \times 10^{19}$. The suggestion of the presence of $UP_2O_7(aq)$ as a unique species seems to be logical but incomplete. It has been shown that pyrophosphate ligands may form many complexes with dioxouranium(VI) in acid solutions [65DRA/JUL, 70DRA/VIL, 73DRA/VIL]. It is thus probable that other neutral species such as $U(H_2P_2O_7)_2(aq)$, $U(OH)HP_2O_7(aq)$, $U(OH)_2H_2P_2O_7(aq)$ and $U(OH)_3H_3P_2O_7(aq)$ could predominate individually or exist together without changing the slope of the solubility line of $UP_2O_7(cr)$. Furthermore, the solid phase was apparently not characterized after its equilibration with (Na,H)ClO_4 solutions, and it is not clear that the solid is effectively hydrated with $20H_2O$. In this review, it is decided that the data of Merkusheva *et al.* [67MER/SKO] are insufficient to permit the calculation of unambiguous thermodynamic parameters.

V.6.2.2. Solid uranium phosphorus compounds

V.6.2.2.1. Uranium phosphides

Uranium forms the following solid phosphides: UP(cr), U_3P_4(cr), UP_2(cr) and possibly U_2P(s).

a) UP(cr)

Counsell *et al.* [67COU/DEL] used an adiabatic calorimeter to measure the heat capacity of UP(cr) at low temperature (11 to 320 K). Using an improved laser flash method, Yokokawa, Takahashi and Mukaibo [75YOK/TAK] determined the heat capacity of this compound in the temperature range 80 to 1080 K. Moser and Kruger [67MOS/KRU] used a laser as a heat pulse source for heat capacity measurements and obtained a nearly constant heat capacity value of $C_{p,\mathrm{m}}$(UP, cr, 298 to 973 K) $= 50.63\,\mathrm{J\cdot K^{-1}\cdot mol^{-1}}$. Results reported in Ref. [67MOS/KRU] are not accepted in this review because it is unlikely that the heat capacity value of UP(cr) remains constant over a large temperature range (298 to 973 K). The $C^\circ_{p,\mathrm{m}}(\mathrm{UP, cr, 298.15\,K})$ values reported in Refs. [67COU/DEL] and [75YOK/TAK] differ by 1.4%. Yokokawa, Takahashi and Mukaibo [75YOK/TAK] used an accurate improved laser flash method and did measurements on pure samples containing less UO_2 ($UP_{1.032} + 0.0042UO_2$) than those prepared by Counsell *et al.* [67COU/DEL] ($UP_{1.007} + 0.0175UO_2$). For these reasons, this review accepts the heat capacity value reported by Yokokawa, Takahashi and Mukaibo [75YOK/TAK] with an uncertainty of about ±1%.

$$C^\circ_{p,\mathrm{m}}(\mathrm{UP, cr, 298.15\,K}) = (50.29 \pm 0.50)\,\mathrm{J\cdot K^{-1}\cdot mol^{-1}}$$

The entropy of UP(cr) was determined by Counsell *et al.* [67COU/DEL] and recalculated with new heat capacity data for temperatures > 80 K by Yokokawa, Takahashi and Mukaibo [75YOK/TAK]. Both groups reported the same value,

$$S^\circ_{\mathrm{m}}(\mathrm{UP, cr, 298.15\,K}) = (78.28 \pm 0.42)\,\mathrm{J\cdot K^{-1}\cdot mol^{-1}},$$

which is accepted in this review. The value actually refers to a solid with a composition closer to $UP_{1.007}$.

The enthalpy of formation of UP(cr) has been estimated and experimentally determined by many authors. Niessen and de Boer [81NIE/BOE] used the semi-empirical model of Miedema, de Châtel and de Boer [80MIE/CHA] to estimate $\Delta_f H^\circ_{\mathrm{m}}$(UP, cr, 298.15 K) $= -312\,\mathrm{kJ\cdot mol^{-1}}$. O'Hare *et al.* [68OHA/SET] obtained $\Delta_f H^\circ_{\mathrm{m}}$(UP, cr, 298.15 K) $= -(316.0 \pm 6.0)\,\mathrm{kJ\cdot mol^{-1}}$ by fluorine and oxygen bomb calorimetry. Reishus and Gundersen [67REI/GUN] studied the vaporization of UP(cr) in the temperature range 2073 to 2423 K and derived $\Delta_f H^\circ_{\mathrm{m}}$(UP, cr, 298.15 K) $= -(318 \pm 17)\,\mathrm{kJ\cdot mol^{-1}}$. Using direct-reaction calorimetry, Baskin and

Smith [70BAS/SMI] determined $\Delta_f H_m^\circ$(UP, cr, 298.15 K) = $-(262.3\pm10.9)\,\mathrm{kJ\cdot mol^{-1}}$ which refers to red phosphorus rather than to the crystalline, white phosphorus selected as a reference phase in this review. The US NBS [82WAG/EVA] selected the value $-268\,\mathrm{kJ\cdot mol^{-1}}$. This review does not accept the value reported in Ref. [68OHA/SET] because errors arising from non-stoichiometry, contamination and incomplete reaction in the fluorine and oxygen bomb and in the vaporisation of UP(cr) could be very significant. The value estimated by Niessen and de Boer [81NIE/BOE] is similar to those in Refs. [67REI/GUN] and [68OHA/SET]. Due to the apparent simplicity of the direct-reaction calorimetry method, and for reasons of internal consistency with U_3P_4(cr) and UP_2(cr), this review accepts the value from direct-reaction calorimetry [70BAS/SMI], corrected to the standard state of phosphorus in this review:

$$\Delta_f H_m^\circ(\mathrm{UP, cr, 298.15\,K}) = -(269.8 \pm 11.1)\,\mathrm{kJ\cdot mol^{-1}}.$$

It is, however, recognized that further measurements are needed for the enthalpy of formation of UP(cr) to resolve the discrepancies in the experimental values.

Using the entropy and enthalpy values selected above, the Gibbs energy of formation is calculated:

$$\Delta_f G_m^\circ(\mathrm{UP, cr, 298.15\,K}) = -(265.9 \pm 11.1)\,\mathrm{kJ\cdot mol^{-1}}.$$

This value is in good agreement with the value of $-264\,\mathrm{kJ\cdot mol^{-1}}$ selected by the US NBS [82WAG/EVA] and of $-(266.1 \pm 10.9)\,\mathrm{kJ\cdot mol^{-1}}$ reported by Fuger [83FUG].

b) U_3P_4 (cr)

The heat capacity of U_3P_4(cr) was measured in an adiabatic calorimeter by Counsell *et al.* [67COU/DEL] at low temperature (11 to 320 K) and by Staliński, Bigański and Troć [66STA/BIE] in the temperature range 22.5 to 349.0 K. The $C_{p,m}^\circ$(U_3P_4, cr, 298.15 K) values reported in Refs. [67COU/DEL] and [66STA/BIE] differ by 2.1%. This review accepts the mean value and estimates the uncertainty,

$$C_{p,m}^\circ(\mathrm{U_3P_4, cr, 298.15\,K}) = (176.9 \pm 3.6)\,\mathrm{J\cdot K^{-1}\cdot mol^{-1}}.$$

The entropy of U_3P_4(cr) was determined by Counsell *et al.* [67COU/DEL] and Staliński, Biegański and Troć [66STA/BIE]. S_m°(U_3P_4, cr, 298.15 K) values obtained by both groups differ only by 0.7%. This review selects the mean value and assigns the uncertainty,

$$S_m^\circ(\mathrm{U_3P_4, cr, 298.15\,K}) = (259.4 \pm 2.6)\,\mathrm{J\cdot K^{-1}\cdot mol^{-1}}.$$

The heat capacity and entropy values obtained by Counsell *et al.* [67COU/DEL] refer to a solid of composition $UP_{1.35}$. Analytical data were not provided in the paper of Staliński, Biegański and Troć [66STA/BIE].

The enthalpy of formation of U_3P_4(cr) was experimentally determined by Dogu, Val and Accary [68DOG/VAL] and Baskin and Smith [70BAS/SMI]. Using the differential thermal analysis at temperatures up to 800 K, Dogu, Val and Accary

[68DOG/VAL] obtained $\Delta_f H^\circ_m(U_3P_4$, cr, 298.15 K) = $-(1113 \pm 105)$ kJ · mol^{-1}. Baskin and Smith [70BAS/SMI] reported the value $\Delta_f H^\circ_m(U_3P_4$, cr, 298.15 K) = $-(813 \pm 25)$ kJ · mol^{-1} from accurate direct-reaction calorimetry experiments. The result obtained in Ref. [68DOG/VAL] is not considered reliable because of possible errors arising from non-stoichiometry and sample contamination (phosphorus containing up to 2400 ppm of oxygen). This review prefers the value of Baskin and Smith [70BAS/SMI] and corrects it to the standard state of phosphorus used in this review.

$$\Delta_f H^\circ_m(U_3P_4, \text{cr}, 298.15\,\text{K}) = -(843 \pm 26)\,\text{kJ} \cdot \text{mol}^{-1}$$

This value is consistent with the value of -837 kJ · mol^{-1} selected by the US NBS [82WAG/EVA].

The Gibbs energy of formation is calculated from the selected enthalpy of formation and entropy.

$$\Delta_f G^\circ_m(U_3P_4, \text{cr}, 298.15\,\text{K}) = -(826 \pm 26)\,\text{kJ} \cdot \text{mol}^{-1}.$$

This value is is not inconsistent with the value of -820 kJ · mol^{-1} selected by US NBS [82WAG/EVA] and of $-(827 \pm 25)$ kJ · mol^{-1} reported by Fuger [83FUG].

c) UP_2 (cr)

Staliński, Biegański and Troć [67STA/BIE] and Blaise *et al.* [78BLA/FOU, 79BLA] have reported heat capacity and entropy determinations for UP_2(cr). $C^\circ_{p,m}(UP_2$, cr, 298.15 K) and $S^\circ_m(UP_2$, cr, 298.15 K) values reported in Refs. [67STA/BIE] and [79BLA] differ by 3.2% and 2.2%, respectively. This review selects the average values of $C^\circ_{p,m}$ and S°_m reported by Staliński, Biegański and Troć [67STA/BIE] and Blaise *et al.* [78BLA/FOU, 79BLA].

$$C^\circ_{p,m}(UP_2, \text{cr}, 298.15\,\text{K}) = (78.8 \pm 3.5)\,\text{J} \cdot \text{K}^{-1} \cdot \text{mol}^{-1}$$
$$S^\circ_m(UP_2, \text{cr}, 298.15\,\text{K}) = (100.7 \pm 3.2)\,\text{J} \cdot \text{K}^{-1} \cdot \text{mol}^{-1}$$

A large discrepancy between estimated and experimental data is observed for the enthalpy of formation of UP_2(cr). Dogu, Val and Accary [68DOG/VAL] used differential thermal analysis at temperatures up to 800 K to determine $\Delta_f H^\circ_m(UP_2$, cr, 298.15 K) = $-(418 \pm 50)$ kJ·mol^{-1}. However, this value is not considered in this review for the same reason as the value for U_3P_4(cr). Using the direct-reaction calorimetry, Baskin and Smith [70BAS/SMI] determined $\Delta_f H^\circ_m$ for UP(cr) and U_3P_4(cr) and estimated $\Delta_f H^\circ_m(UP_2$, cr, 298.15 K) = -289 kJ · mol^{-1}. This estimation was based on the assumption that the $\Delta_f H^\circ_m$ values in the U-P system vary linearly with the phosphorus/uranium ratio. Niessen and de Boer [81NIE/BOE] estimated $\Delta_f H^\circ_m(UP_2$, cr, 298.15 K) = -372 kJ · mol^{-1} using the semi-empirical model of Miedema, de Châtel and de Boer [80MIE/CHA]. The US NBS [82WAG/EVA] selected the value $\Delta_f H^\circ_m(UP_2$, cr, 298.15 K) = -305.0 kJ · mol^{-1} but did not indicate the source of data and the evaluation procedure. As for UP(cr) and U_3P_4(cr), this review prefers the value from Ref. [70BAS/SMI] with an estimated uncertainty, and corrected to the standard state of phosphorus selected in this review.

$$\Delta_f H_m^\circ(UP_2, cr, 298.15\,K) = -(304 \pm 15)\,kJ \cdot mol^{-1}.$$

The Gibbs energy of formation is calculated from the selected enthalpy of formation and entropy.

$$\Delta_f G_m^\circ(UP_2, cr, 298.15\,K) = -(295 \pm 15)\,kJ \cdot mol^{-1}$$

This value is less negative than $-294.6\,kJ \cdot mol^{-1}$ proposed by Fuger [83FUG] and $-297\,kJ \cdot mol^{-1}$ selected by the US NBS [82WAG/EVA].

d) $U_2P(cr)$

Niessen and de Boer [81NIE/BOE] estimated $\Delta_f H_m^\circ(U_2P$, s, 298.15 K$) = -375\,kJ \cdot mol^{-1}$. Since there is no experimental proof for the existence of $U_2P(s)$, this review does not accept this estimate.

V.6.2.2.2. Uranium hypophosphites

Three solid dioxouranium(VI) hypophosphites have been identified: $UO_2(H_2PO_2)_2(cr)$, $UO_2(H_2PO_2)_2{\cdot}H_2O(cr)$ and $UO_2(H_2PO_2)_2{\cdot}3H_2O(cr)$. The synthesis of $UO_2(H_2PO_2)_2(cr)$ was described by Pascal [61PAS]. The monohydrate was obtained by Rammelsberg [1872RAM]. The trihydrate was synthesized by Rosenheim and Trewendt [22ROS/TRE]. X-ray diffraction data were reported only for the monohydrate [86POW: file 14-584], and no study on its structure has been done. No thermodynamic data appear to have been published for these three solid compounds.

Two solid uranium(IV) hypophosphites, $U(H_2PO_2)_4(s)$ and $U(H_2PO_2)_4{\cdot}H_3PO_2(s)$, have been reported [61PAS]. No thermodynamic data have been published for these two compounds.

V.6.2.2.3. Uranium phosphites

Two solid dioxouranium(VI) phosphites, $UO_2HPO_3(cr)$ [37MON] and $UO_2HPO_3 \cdot 3H_2O(cr)$ [37CHR/KRA, 37CHR/KRA2], have been synthesized. No thermodynamic data appear to have been reported for these two solid compounds.

Pascal [61PAS] reported two solid phases of uranium(IV), $U(HPO_3)_2{\cdot}4H_2O(cr)$ and $U(H_2PO_3)_2HPO_3{\cdot}5H_2O(cr)$. No thermodynamic data are available for these two compounds.

V.6.2.2.4. Uranium metaphosphates

Barten and Cordfunke [80BAR/COR2] reported the synthesis of three solid phases denoted α-, β- and γ-$UO_2(PO_3)_2$. Cordfunke and Ouweltjes [85COR/OUW] estimated the enthalpy of formation of "$UO_2(PO_3)_2(cr)$" to $\Delta_f H_m^\circ(UO_2(PO_3)_2$, cr, 298.15 K$) = -(2973 \pm 3)\,kJ \cdot mol^{-1}$. The value was obtained by analogy with the enthalpy of solution of the corresponding arsenate [82COR/OUW2]. The composition of the solid phases α-, β- and γ-$UO_2(PO_3)_2$(cr) synthesized by Barten and Cordfunke [80BAR/COR2] cannot be verified. Powder X-ray data offered for the characterization of these phases are assigned a low reliability because of uncertain composition, lack of indexing and inadequate range of intensities [86POW: files 35-207, 33-1430, 33-1431]. The analogy with the enthalpy of solution of the corresponding arsenate [82COR/OUW2, 85COR/OUW] is not evident because the X-ray diffraction patterns of α-$UO_2(AsO_3)_2$ [86POW: file 33-1415] and β-$UO_2(AsO_3)_2$ [86POW: file 33-1417] also indicate lack of indexing and uncertain composition.

For these reasons, this review does not accept the enthalpy data reported by Cordfunke and Ouweltjes [85COR/OUW].

V.6.2.2.5. Uranium orthophosphates

a) U(VI) orthophosphates

The main compound of dioxouranium(VI) orthophosphates has the formula $(UO_2)_3(PO_4)_2$(cr). Four hydrated species have been synthesized to date: $(UO_2)_3(PO_4)_2{\cdot}5H_2O$(cr) and $(UO_2)_3(PO_4)_2{\cdot}H_2O$(cr) by Barten and Cordfunke [80BAR/COR2], and $(UO_2)_3(PO_4)_2{\cdot}4H_2O$(cr) and $(UO_2)_3(PO_4)_2{\cdot}6H_2O$(cr) by Schreyer and Baes [54SCH/BAE]. Three anhydrous solid phases exist: α-, β- and γ-$(UO_2)_3(PO_4)_2$. Among these solid phases, only the poorly crystalline $(UO_2)_3(PO_4)_2$(cr) and the hexahydrated compound were thermodynamically studied.

$(UO_2)_3(PO_4)_2$(cr): Cordfunke and Ouweltjes [85COR/OUW] determined

$$\Delta_f H^\circ_m((UO_2)_3(PO_4)_2, \text{cr}, 298.15\,\text{K}) = -(5491.3 \pm 3.5)\,\text{kJ}\cdot\text{mol}^{-1}.$$

This experimental value is accepted by this review.

Using Latimer's method (*cf.* Appendix D) this review estimates

$$S^\circ_m((UO_2)_3(PO_4)_2, \text{cr}, 298.15\,\text{K}) = (410 \pm 14)\,\text{J}\cdot\text{K}^{-1}\cdot\text{mol}^{-1}.$$

This is similar to values recently estimated by Langmuir [78LAN] and Barten [88BAR3], $S^\circ_m((UO_2)_3(PO_4)_2, \text{cr}, 298.15\,\text{K}) = 406\,\text{J}\cdot\text{K}^{-1}\cdot\text{mol}^{-1}$.

The Gibbs energy of formation is calculated from the selected enthalpy of formation and entropy.

$$\Delta_f G^\circ_m((UO_2)_3(PO_4)_2, \text{cr}, 298.15\,\text{K}) = -(5116.0 \pm 5.5)\,\text{J}\cdot\text{K}^{-1}\cdot\text{mol}^{-1}$$

Langmuir [78LAN] calculated the value $\Delta_f H^\circ_m((UO_2)_3(PO_4)_2$, cr, 298.15 K) = $-6008.2\,\text{kJ}\cdot\text{mol}^{-1}$. This value was accepted by Phillips *et al.* [85PHI/PHI, 88PHI/HAL]. In fact, the values provided by Langmuir [78LAN] and by Naumov, Ryzhenko and Khodakovsky [71NAU/RYZ] were based on solubility data for hydrated $(UO_2)_3(PO_4)_2{\cdot}xH_2O$, and the use of the simple formula "$(UO_2)_3(PO_4)_2$" was merely a formalism. In addition, as pointed out by Alwan and Williams [80ALW/WIL], the enthalpy of formation of "$(UO_2)_3(PO_4)_2$" in Ref. [78LAN] was miscalculated.

Cordfunke and Prins [86COR/PRI] measured the enthalpy content of $(UO_2)_3(PO_4)_2$(cr) in the temperature range 298 to 771 K and obtained the following equation:

$$H^\circ_T - H^\circ_{298} = (326.375\,T + 0.098219\,T^2 + 4.0840 \times 10^6\,T^{-1} - 119737.5)\,\mathrm{J \cdot mol^{-1}}. \qquad \text{(V.150)}$$

By differentiation of Eq. (V.150), Cordfunke and Prins [86COR/PRI] calculated the heat capacity of $(UO_2)_3(PO_4)_2$(cr). Its value was reported by Barten [88BAR3] and is accepted in this review.

$$C^\circ_{p,\mathrm{m}}((\mathrm{UO_2})_3(\mathrm{PO_4})_2, \mathrm{cr}, 298.15\,\mathrm{K}) = (339 \pm 17)\,\mathrm{J \cdot K^{-1} \cdot mol^{-1}}$$

$(UO_2)_3(PO_4)_2 \cdot xH_2O(cr)$: The solubility equilibria of $(UO_2)_3(PO_4)_2{\cdot}xH_2O$(cr) are discussed in Section V.6.2.1.1.b. The equilibrium constant selected there for the solubility of $(UO_2)_3(PO_4)_2{\cdot}4H_2O$(cr) in a phosphoric acid solution is used to calculate the Gibbs energy of formation of

$$\Delta_\mathrm{f} G^\circ_\mathrm{m}((\mathrm{UO_2})_3(\mathrm{PO_4})_2 \cdot 4\mathrm{H_2O}, \mathrm{cr}, 298.15\,\mathrm{K}) = -(6139.0 \pm 6.4)\,\mathrm{kJ \cdot mol^{-1}}.$$

Since there are no experimental data available on the entropy or enthalpy of formation, this review estimates the entropy (*cf.* Appendix D) and obtains

$$S^\circ_\mathrm{m}((\mathrm{UO_2})_3(\mathrm{PO_4})_2 \cdot 4\mathrm{H_2O}, \mathrm{cr}, 298.15\,\mathrm{K}) = (589 \pm 22)\,\mathrm{J \cdot K^{-1} \cdot mol^{-1}},$$

where the uncertainty is an estimate.

The standard enthalpy of formation is calculated from the selected Gibbs energy of formation and entropy.

$$\Delta_\mathrm{f} H^\circ_\mathrm{m}((\mathrm{UO_2})_3(\mathrm{PO_4})_2 \cdot 4\mathrm{H_2O}, \mathrm{cr}, 298.15\,\mathrm{K}) = -(6739.1 \pm 9.1)\,\mathrm{kJ \cdot mol^{-1}}$$

From the observation of Schreyer and Baes [54SCH/BAE] that the hexahydrate transforms to the corresponding tetrahydrate at a temperature less than 100°C, and from Kobets, Kolevich and Umreiko [78KOB/KOL] that the tetra- and hexahydrates are both stable at room temperature, this review estimates that the Gibbs energy of formation cannot differ from the sum of the value of the tetrahydrate plus that of two moles of liquid water by more than a few $\mathrm{kJ \cdot mol^{-1}}$. Accordingly, this review selects

$$\Delta_\mathrm{f} G^\circ_\mathrm{m}((\mathrm{UO_2})_3(\mathrm{PO_4})_2 \cdot 6\mathrm{H_2O}, \mathrm{cr}, 298.15\,\mathrm{K}) = -(6618 \pm 7)\,\mathrm{kJ \cdot mol^{-1}}.$$

b) U(IV) orthophosphate

The synthesis of the compound $U_3(PO_4)_4$(cr) was described by Pascal [61PAS]. No experimental determinations of the thermodynamic data of $U_3(PO_4)_4$(cr) have been made. Allard and Beall [78ALL/BEA] predicted the solubility product according to the reaction

$$\mathrm{U_3(PO_4)_4(cr)} \rightleftharpoons 3\mathrm{U^{4+}} + 4\mathrm{PO_4^{3-}} \qquad \text{(V.151)}$$

to be $\log_{10} K^\circ_{s,0}$(V.151) $= -90$. This value appears to be an estimate based on analogy with the corresponding compound of cerium(IV). This review does not consider this estimate reliable, *cf.* discussion of Ref. [61CHU/ALY] in Appendix A.

V.6.2.2.6. *Uranium pyrophosphates*

a) *U(VI) pyrophosphates*

$(UO_2)_2P_2O_7$(cr): The enthalpy of formation of $(UO_2)_2P_2O_7$(cr) was determined by means of solution calorimetry by Cordfunke and Ouweltjes [85COR/OUW]. This review accepts the rounded value and reported uncertainty.

$$\Delta_f H^\circ_m((UO_2)_2P_2O_7, \text{cr}, 298.15\,\text{K}) = -(4232.6 \pm 2.8)\,\text{kJ} \cdot \text{mol}^{-1}$$

Barten [88BAR3] measured the oxygen pressure p_{O_2} for the reactions

$$(UO_2)_3(PO_4)_2(\text{cr}) \rightleftharpoons \tfrac{1}{3}U_3O_8(\text{cr}) + 2UPO_5(\text{cr}) + \tfrac{2}{3}O_2(\text{g}) \qquad (V.152)$$

$$(UO_2)_2P_2O_7(\text{cr}) \rightleftharpoons 2UPO_5(\text{cr}) + \tfrac{1}{2}O_2(\text{g}) \qquad (V.153)$$

in the temperature ranges 1089 to 1280 K and 979 to 1130 K, respectively. He then calculated the temperature dependence of the Gibbs energy of Reaction (V.152) using the relation

$$\Delta_r G^\circ_m(V.152, T) = -2.3RT \log_{10}(p_{O_2})^n, \qquad (V.154)$$

where n is the stoichiometric coefficient of oxygen gas. The derivation of this equation leads to the mean difference $[S^\circ_m((UO_2)_3(PO_4)_2$, cr, 298.15 K) $-$ $2S^\circ_m(UPO_5$, cr, 298.15 K)] = $(135.4 \pm 2.7)\,\text{J} \cdot \text{K}^{-1} \cdot \text{mol}^{-1}$. Using this result and the entropy value of $(UO_2)_3(PO_4)_2$(cr), Barten [88BAR3] calculated $S^\circ_m(UPO_5$, cr, 298.15 K) = $(135.2 \pm 10)\,\text{J} \cdot \text{K}^{-1} \cdot \text{mol}^{-1}$, which corresponds to $S^\circ_m(UPO_5$, cr, 298.15 K) = $(137 \pm 10)\,\text{J} \cdot \text{K}^{-1} \cdot \text{mol}^{-1}$ using the entropy of $(UO_2)_3(PO_4)_2$(cr) selected in this review. The same type of calculation was done for Reaction (V.153) and led to the difference $[2S^\circ_m(UPO_5$, cr, 298.15 K) $-$ $S^\circ_m((UO_2)_2P_2O_7$, cr, 298.15 K)] = $-(21.55 \pm 9.02)\,\text{J} \cdot \text{K}^{-1} \cdot \text{mol}^{-1}$. Using the value of the entropy obtained above for UPO_5(cr), Barten [88BAR3] obtained $S^\circ_m((UO_2)_2P_2O_7$, cr, 298.15 K) = $(292 \pm 21)\,\text{J} \cdot \text{K}^{-1} \cdot \text{mol}^{-1}$, which is corrected (to be consistent with the data selected in this review) to the selected value of

$$S^\circ_m((UO_2)_2P_2O_7, \text{cr}, 298.15\,\text{K}) = (296 \pm 21)\,\text{J} \cdot \text{K}^{-1} \cdot \text{mol}^{-1}.$$

It should be mentioned that Barten estimated the entropy by a second method to be $S^\circ_m((UO_2)_2P_2O_7$, cr, 298.15 K) = $(277 \pm 17)\,\text{J} \cdot \text{K}^{-1} \cdot \text{mol}^{-1}$, and he finally selected $S^\circ_m((UO_2)_2P_2O_7$, cr, 298.15 K) = $(284 \pm 13)\,\text{J} \cdot \text{K}^{-1} \cdot \text{mol}^{-1}$.

The Gibbs energy of formation is calculated from the selected enthalpy of formation and entropy.

$$\Delta_f G^\circ_m((UO_2)_2P_2O_7, \text{cr}, 298.15\,\text{K}) = -(3930 \pm 6.9)\,\text{kJ} \cdot \text{mol}^{-1}$$

The heat capacity of $(UO_2)_2P_2O_7$(cr) was reported by Barten [88BAR3]. This review accepts the rounded value and conservatively assigns an uncertainty of ±10% to it.

$$C^\circ_{p,m}((UO_2)_2P_2O_7, \text{cr}, 298.15\,\text{K}) = (258 \pm 26)\,\text{J} \cdot \text{K}^{-1} \cdot \text{mol}^{-1}$$

$(UO_2)_2P_2O_7 \cdot 4H_2O$(cr): Pascal [61PAS] reported the synthesis of the compound $(UO_2)_2P_2O_7{\cdot}4H_2O$(cr) from the reaction of $Na_4P_2O_7$(cr) and an aqueous solution of $UO_2(NO_3)_2$. No information on experimental thermodynamic data on this solid phase is available.

$UO_2H_2P_2O_7$(cr): $UO_2H_2P_2O_7$(cr) was synthesized by Barten and Cordfunke [80BAR/COR2]. No thermodynamic data on this compound have been published.

b) Other uranium pyrophosphate compounds

UPO_5(cr) ($\frac{1}{2}U_2O_3P_2O_7$(cr)): UPO_5(cr) was synthesized by Cordfunke and Ouweltjes [85COR/OUW] by heating $(UO_2)_2P_2O_7$(cr) at 1275 K for 20 hours. They determined the enthalpy of formation of this compound by means of solution calorimetry. This review selects the rounded value obtained by Cordfunke and Ouweltjes [85COR/OUW] and doubles the uncertainty in order to be more consistent with other cycles studied by the authors.

$$\Delta_f H^\circ_m(\mathrm{UPO_5, cr, 298.15\,K}) = -(2064 \pm 4)\,\mathrm{kJ \cdot mol^{-1}}$$

The entropy of UPO_5(cr) was derived by combining the oxygen pressure measurements for Reaction (V.152) and the corresponding relation (V.154, $n = 0.66667$) [88BAR3]. This review selects the value reported in Ref. [88BAR3] after correction to be consistent with the data selected in this review.

$$S^\circ_m(\mathrm{UPO_5, cr, 298.15\,K}) = (137 \pm 10)\,\mathrm{J \cdot K^{-1} \cdot mol^{-1}}$$

The Gibbs energy of formation is calculated from the selected enthalpy of formation and entropy.

$$\Delta_f G^\circ_m(\mathrm{UPO_5, cr, 298.15\,K}) = -(1924.7 \pm 5.0)\,\mathrm{kJ \cdot mol^{-1}}$$

The heat capacity of UPO_5(cr) was calculated by Barten [88BAR3]. This review adopts his rounded value with an uncertainty of ±10%.

$$C^\circ_{p,m}(\mathrm{UPO_5, cr, 298.15\,K}) = (124 \pm 12)\,\mathrm{J \cdot K^{-1} \cdot mol^{-1}}$$

UP_2O_7(cr): The enthalpy of formation of UP_2O_7(cr) was first determined by Cordfunke and Ouweltjes [85COR/OUW] using solution calorimetry. The recalculation by Cordfunke and Prins [86COR/PRI] led to the value selected by this review,

$$\Delta_f H^\circ_m(\mathrm{UP_2O_7, cr, 298.15\,K}) = -(2852 \pm 4)\,\mathrm{kJ \cdot mol^{-1}}.$$

The entropy of UP_2O_7(cr) was determined by Barten [86BAR] by combining the oxygen pressure measurements for Reaction (V.155),

$$\mathrm{UO_2(PO_3)_2(cr)} \rightleftharpoons \mathrm{UP_2O_7(cr)} + \tfrac{1}{2}\mathrm{O_2(g)}, \qquad \text{(V.155)}$$

by deriving Eq. (V.154) with $n = 0.5$. The derivation of Eq. (V.154, $n = 0.5$) leads to the difference [S°_m($UO_2(PO_3)_2$, cr, 298.15 K) − S°_m(UP_2O_7, cr, 298.15 K)] = $-(1.5 \pm 7.8)\,\mathrm{J \cdot K^{-1} \cdot mol^{-1}}$. This leads to the selected value of

$$S^\circ_m(UP_2O_7, cr, 298.15\,K) = (204 \pm 12)\,J \cdot K^{-1} \cdot mol^{-1}.$$

The selected auxiliary data allow the calculation of

$$\Delta_f G^\circ_m(UP_2O_7, cr, 298.15\,K) = -(2659.3 \pm 5.4)\,J \cdot K^{-1} \cdot mol^{-1},$$

which corresponds to a solubility product of $\log_{10} K^\circ_{s,0} = -(33.6 \pm 1.5)$. This value would predict a solubility that is considerably lower than measured for "$UP_2O_7 \cdot 20H_2O$(cr)" in the available dissolution experiments [67MER/SKO]. It is possible that the discrepancy is due to differences in the crystallinity, on which no information has been published in Ref. [67MER/SKO]. This review accepts the calorimetric information and the $\Delta_f G^\circ_m$ value above calculated therefrom. The value for

$$C^\circ_{p,m}(UP_2O_7, cr, 298.15\,K) = (184 \pm 18)\,J \cdot K^{-1} \cdot mol^{-1}$$

given by Barten [86BAR] is accepted by this review. The uncertainty is estimated by this review.

$UP_2O_7 \cdot xH_2O$(cr): Two hydrated uranium(IV) pyrophosphates have been reported in the literature. Pascal [67PAS] described $UP_2O_7 \cdot 6H_2O$(cr), for which no thermodynamic data are available. Merkusheva *et al.* [67MER/SKO] prepared a compound whose analysis corresponded to the composition $UP_2O_7 \cdot 20H_2O$(cr). There is insufficient information for making a firm conclusion about the degree of hydration. The solubility data are discussed in Appendix A, where this review concludes that they cannot be used for the unambiguous assignments of thermodynamic parameters.

V.6.2.2.7. Uranium hydrogen phosphates

a) U(VI) hydrogen phosphates

The compounds with the composition $H_2(UO_2)_2(PO_4)_2 \cdot xH_2O$(cr), sometimes referred to as $UO_2HPO_4 \cdot \frac{x}{2}H_2O$(cr), are discussed in Section V.6.2.2.10.c.

b) U(IV) hydrogen phosphates

In the experimental study of the UO_2-P_2O_5-H_2O system, Schreyer [55SCH] synthesized $U(HPO_4)_2 \cdot 4H_2O$(cr) and $U(HPO_4)_2 \cdot 6H_2O$(cr), and he suggested the formation of $UOHPO_4$(cr). Two other compounds, $U(HPO_4)_2$(cr) and $U(HPO_4)_2 \cdot 2H_2O$(cr), were identified by Schreyer and Phillips [56SCH/PHI].

$U(HPO_4)_2$(cr) was synthesized as a metastable solid phase [56SCH/PHI]. $U(HPO_4)_2 \cdot 2H_2O$(cr) was synthesized by Schreyer and Phillips [56SCH/PHI]. $U(HPO_4)_2 \cdot 6H_2O$(cr) was precipitated from a 10.9 M phosphate solution by Schreyer [55SCH]. Its solubility in phosphoric acid solution [55SCH] provides insufficient information to extract a solubility product.

The only phase for which reliable solubility data exist is the tetrahydrate, $U(HPO_4)_2 \cdot 4H_2O$(cr). As discussed in Section V.6.2.1.1.c, this review selects the equilibrium constant $\log_{10} {}^*K^\circ_s(V.146) = -(11.79 \pm 0.15)$, based on recalculated data of

Moskvin, Essen and Bukhtiyarova [67MOS/ESS]. The Gibbs energy obtained from this constant is

$$\Delta_f G^\circ_m(U(HPO_4)_2 \cdot 4H_2O, cr, 298.15\,K) = -(3844.5 \pm 3.7)\,kJ \cdot mol^{-1}.$$

Using Latimer's method this review estimates the entropy and its uncertainty as

$$S^\circ_m(U(HPO_4)_2 \cdot 4H_2O, cr, 298.15\,K) = (372 \pm 26)\,J \cdot K^{-1} \cdot mol^{-1}.$$

The Gibbs energy of formation is calculated from the selected enthalpy of formation and entropy.

$$\Delta_f H^\circ_m(U(HPO_4)_2 \cdot 4H_2O, cr, 298.15\,K) = -(4334.8 \pm 8.6)\,kJ \cdot mol^{-1}$$

Lemire and Tremaine [80LEM/TRE] estimated the heat capacity of $U(HPO_4)_2{\cdot}4H_2O(cr)$. This review accepts their value and assigns an uncertainty of $\pm 50\,J \cdot K^{-1} \cdot mol^{-1}$ to it.

$$C^\circ_{p,m}(U(HPO_4)_2 \cdot 4H_2O, cr, 298.15\,K) = (460 \pm 50)\,kJ \cdot mol^{-1}$$

V.6.2.2.8. *Uranium dihydrogen phosphates*

Two hydrated dioxouranium(VI) compounds are known. $UO_2(H_2PO_4)_2{\cdot}3H_2O(cr)$ was synthesized by Schreyer and Baes [54SCH/BAE], and $UO_2(H_2PO_4)_2{\cdot}H_2O(cr)$ was prepared by Barten and Cordfunke [80BAR/COR2]. The solubility of the trihydrate in phosphoric acid solutions was measured [54SCH/BAE], but the results obtained are not sufficient to select a solubility product or any other thermodynamic data. No other thermodynamic data are available for these compounds.

V.6.2.2.9. *Uranium mixed phosphates*

In the experimental study of the UO_2-P_2O_5-H_2O system, Schreyer [55SCH] synthesized $U(HPO_4)_2H_3PO_4{\cdot}H_2O(cr)$ from a solution containing more than 9.8 M total phosphate. The solubility measurements of this compound in phosphoric acid solutions [55SCH] do not provide sufficient information (pH, forms of dissolved uranium and phosphate species) to extract a solubility product. According to Pascal [61PAS], $U(HPO_4)_2H_3PO_4{\cdot}H_2O(cr)$ can be written as $UHPO_4(H_2PO_4)_2{\cdot}H_2O(cr)$.

V.6.2.2.10. *Other uranium phosphorus compounds*

a) $H_2(UO_2)_3(HPO_3)_4 \cdot 12H_2O(cr)$

Rammelsberg [1872RAM], cited by Pascal [67PAS], reported the synthesis of a solid phosphite compound of the composition $H_2(UO_2)_3(HPO_3)_4 \cdot 12H_2O(cr)$ by addition of $(NH_4)_2UO_4(cr)$ into a solution of H_3PO_3. No information on the crystallographic and thermodynamic data of this compound is available.

b) $H_{11}(UO_2)_2(PO_4)_5(cr)$

Staritzky and Cromer [56STA/CRO] reported the synthesis of a solid compound with the suggested formula $H_{11}(UO_2)_2(PO_4)_5(cr)$ [86POW: file 9-160]. The stoichiometry of this compounds is doubtful, and the compound in question may well be $UO_2(H_2PO_4)_2 \cdot 3H_2O(cr)$ because of the similar synthesis conditions and crystallographic data [54SCH/BAE, 86POW: file 36-91].

c) $H_2(UO_2)_2(PO_4)_2 \cdot xH_2O(cr)$

The H-autunite system was studied by numerous authors [55ROS, 56DUN, 57WEI/HAR, 58FRO, 58NEC, 65MUT, 67MOS/SHE, 73BEL, 76WEI/HOF]. However, the hydration number is still poorly known. Muto [65MUT] reported the synthesis of the decahydrated H-autunite from the decahydrated Ca-autunite by a cation exchange reaction in acidic solution, but he did not provide the crystallographic data of the product obtained. Three other phases of H-autunite were also identified by means of X-ray diffractometry [65MUT]. The three basal spacings, $d = 8.6 \times 10^{-10}$, 9.1×10^{-10} and 10.5×10^{-10} m for the (001) plane, were attributed to a compound formulated as "$UO_2HPO_4 \cdot 2H_2O(cr)$" by Dunn [56DUN, 86POW: file 13-61], the compound $H_2(UO_2)_2(PO_4)_2 \cdot 8H_2O(cr)$ [55ROS, 58FRO, 73BEL: file 9-296], and an unidentified dioxouranium(VI) phosphate compound, respectively.

Belova [73BEL] reported crystallographic data of a solid compound formulated as $(H_3O)_2(UO_2)_2(PO_4)_2 \cdot 6H_2O(cr)$ with $d = 8.53 \times 10^{-10}$ m for the (001) plane [86POW, file 26-887]. A systematic investigation of the H-autunite system using X-ray diffractometry and/or gravimetry and Karl Fischer titration for water determination by Weigel and Hoffmann [76WEI/HOF] indicated the existence of three hydrated species, $H_2(UO_2)_2(PO_4)_2 \cdot xH_2O(cr)$ ($x = 1$, 4 and 8), and its anhydrous compound. Only the octahydrated H-autunite was measured by X-ray diffractometry [76WEI/HOF, 86POW: file 29-670]. From an re-examination of the crystallographic data of the solid phases described above, this review concludes:

- The phases $H_2(UO_2)_2(PO_4)_2 \cdot xH_2O(cr)$ ($x = 6, 8$) crystallize in the tetragonal system [55ROS, 57WEI/HAR, 73BEL].

- The X-ray diffractometry pattern of the compound "$UO_2HPO_4 \cdot 2H_2O(cr)$" [56DUN] does not conform to that of the hexa- and octahydrated H-autunite cited above, and there is no information available on the crystallographic system. Therefore, one cannot affirm if this solid phase has the basal spacing $d = 8.6 \times 10^{-10}$ m as reported by Muto [65MUT].

- The two solid phases $(H_3O)_2(UO_2)_2(PO_4)_2{\cdot}6H_2O(cr)$ [73BEL, 86POW: file 26-887] and $H_2(UO_2)_2(PO_4)_2{\cdot}8H_2O(cr)$ [76WEI/HOF, 86POW: file 29-670] have similar crystallographic data with $d = (8.55 \pm 0.02) \times 10^{-10}$ m for the (001) plane, whereas the species $H_2(UO_2)_2(PO_4)_2 \cdot 8H_2O(cr)$ reported in Ref. [6037] has a larger value of $d = 9.032 \times 10^{-10}$ m for the (001) plane.

Hence, the compound "$UO_2HPO_4 \cdot 2H_2O$(cr)" studied by Dunn [56DUN] is not accepted in this review as a mineral in the H-autunite family. This review concludes that Belova [73BEL] and Weigel and Hoffmann [76WEI/HOF] studied the same solid species, $H_2(UO_2)_2(PO_4)_2 \cdot 8H_2O$(cr). The sample with the large value of $d = 9.032 \times 10^{-10}$ m for the (001) plane measured by Ross [55ROS] seems to be the H-autunite containing more that eight waters of hydration.

This review accepts five H-autunite phases: One anhydrous species, $H_2(UO_2)_2(PO_4)_2$(cr), and four hydrated compounds, $H_2(UO_2)_2(PO_4)_2 \cdot xH_2O$(cr) ($x$ = 2, 4, 8 and 10). See discussion in Section V.6.2.1.1.b) and under Ref. [65VES/PEK] in Appendix A.

$H_2(UO_2)_2(PO_4)_2 \cdot 10H_2O$(cr): This compound is discussed in Section V.6.2.1.1.b) and for reasons stated there this review does not select any thermodynamic data for this compound.

$H_2(UO_2)_2(PO_4)_2 \cdot 8H_2O$(cr): This review considers the thermodynamic properties of this compound to be identical to those of $UO_2HPO_4 \cdot 4H_2O$(cr).

Thermodynamic data are obtained from the solubility products as discussed in Section V.6.2.1.1.b. The selected solubility of $UO_2HPO_4 \cdot 4H_2O$(cr) in a phosphoric acid solution gives a Gibbs energy of formation of

$$\Delta_f G^\circ_m(UO_2HPO_4 \cdot 4H_2O, cr, 298.15\,K) = -(3064.7 \pm 2.4)\,kJ \cdot mol^{-1}.$$

The entropy of $UO_2HPO_4 \cdot 4H_2O$ is estimated in this review to be

$$S^\circ_m(UO_2HPO_4 \cdot 4H_2O, cr, 298.15\,K) = (346 \pm 25)\,J \cdot K^{-1} \cdot mol^{-1}.$$

The enthalpy of formation is calculated from the selected Gibbs energy of formation and entropy.

$$\Delta_f H^\circ_m(UO_2HPO_4 \cdot 4H_2O, cr, 298.15\,K) = -(3470.0 \pm 7.8)\,kJ \cdot mol^{-1}$$

$H_2(UO_2)_2(PO_4)_2 \cdot 4H_2O$(cr): This solid compound was found by Weigel and Hoffmann [76WEI/HOF]. No information on the thermodynamic data of this species are available.

$H_2(UO_2)_2(PO_4)_2 \cdot 2H_2O$(cr): This solid compound was also identified by Weigel and Hoffmann [76WEI/HOF] but no thermodynamic data have been reported.

$H_2(UO_2)_2(PO_4)_2$(cr): The anhydrous solid phase of H-autunite was identified by Weigel and Hoffmann [76WEI/HOF]. Naumov, Ryzhenko and Khodakovsky [71NAU/RYZ] reported $\Delta_f G^\circ_m$("HUO_2PO_4", cr, 298.15 K) = $-(2111.7 \pm 8.4)\,kJ \cdot mol^{-1}$. Later, Langmuir [78LAN] has proposed $\Delta_f G^\circ_m(H_2(UO_2)_2(PO_4)_2$, cr, 298.15 K) = $-4217.5\,kJ \cdot mol^{-1}$ estimated from experimental solubility data of $H_2(UO_2)_2(PO_4)_2 \cdot 8H_2O$(cr) [67MOS/SHE]. These values do not refer to the anhydrous phase. No values for the anhydrous phase are accepted in this review.

V.6.3. Arsenic compounds

V.6.3.1. Uranium arsenides

V.6.3.1.1. UAs(cr)

The enthalpy of formation for UAs(cr) was determined by Baskin and Smith [70BAS/SMI], and the value obtained by them is accepted in this review.

$$\Delta_f H_m^\circ(\text{UAs, cr, 298.15 K}) = -(234.3 \pm 8.0)\,\text{kJ}\cdot\text{mol}^{-1}$$

Heat capacities were measured for the temperature range 5 to 300 K by Blaise *et al.* [80BLA/TRO], and for the ranges 0.12 K $< T <$ 1 K and 1.5 K $< T <$ 11 K by Rudigier, Ott and Vogt [85RUD/OTT, 85RUD/OTT2]. Blaise *et al.* [80BLA/TRO] derived values of $S_m^\circ(\text{UAs, cr, 298.15 K}) = 97.44\,\text{J}\cdot\text{K}^{-1}\cdot\text{mol}^{-1}$, and $C_{p,m}^\circ(\text{UAs, cr, 298.15 K}) = 57.91\,\text{J}\cdot\text{K}^{-1}\cdot\text{mol}^{-1}$. They also estimated errors in the heat capacities above 15 K to be less than 2%. The low-temperature experiments did not reveal any significant additional contributions to the entropy. Therefore, this review selects the 298.15 K values of S_m°(UAs, cr, 298.15 K) and $C_{p,m}^\circ$(UAs, cr, 298.15 K) from Blaise *et al.* [80BLA/TRO].

$$S_m^\circ(\text{UAs, cr, 298.15 K}) = (97.4 \pm 2.0)\,\text{J}\cdot\text{K}^{-1}\cdot\text{mol}^{-1}$$
$$C_{p,m}^\circ(\text{UAs, cr, 298.15 K}) = (57.9 \pm 1.2)\,\text{J}\cdot\text{K}^{-1}\cdot\text{mol}^{-1}$$

The Gibbs energy of formation is calculated from the selected enthalpy of formation and entropy.

$$\Delta_f G_m^\circ(\text{UAs, cr, 298.15 K}) = -(237.9 \pm 8.0)\,\text{kJ}\cdot\text{mol}^{-1}$$

V.6.3.1.2. U_3As_4(cr)

The enthalpy of formation for U_3As_4(cr) was determined by Baskin and Smith [70BAS/SMI], and the value obtained by them is accepted in this review.

$$\Delta_f H_m^\circ(\text{U}_3\text{As}_4\text{, cr, 298.15 K}) = -(720 \pm 18)\,\text{kJ}\cdot\text{mol}^{-1}$$

Heat capacities were measured for the temperature range 5 to 950 K by Alles *et al.* [77ALL/FAL] and for 5 to 300 K by Blaise *et al.* [80BLA/LAG]. The values obtained in the latter study are generally slightly larger than those from Ref. [77ALL/FAL]. The authors of Ref. [80BLA/LAG] indicate that the earlier results of Alles *et al.* [77ALL/FAL] (which have less scatter) are to be preferred for $T >$ 200 K. $S_m^\circ(\text{U}_3\text{As}_4\text{, cr, 298.15 K})$ values from the two sets of experiments differ by only 0.6%. This review has selected the entropy and heat capacity values of Alles *et al.* [77ALL/FAL].

$$S_m^\circ(\text{U}_3\text{As}_4\text{, cr, 298.15 K}) = (309.07 \pm 0.60)\,\text{J}\cdot\text{K}^{-1}\cdot\text{mol}^{-1}$$
$$C_{p,m}^\circ(\text{U}_3\text{As}_4\text{, cr, 298.15 K}) = (187.53 \pm 0.20)\,\text{J}\cdot\text{K}^{-1}\cdot\text{mol}^{-1}$$

The Gibbs energy of formation is calculated from the selected enthalpy of formation and entropy.

$$\Delta_f G_m^\circ(\text{U}_3\text{As}_4\text{, cr, 298.15 K}) = -(725 \pm 18)\,\text{kJ}\cdot\text{mol}^{-1}$$

V.6.3.1.3. UAs_2 (cr)

Heat capacity measurements for UAs_2(cr) were reported by Grønvold *et al.* [75WES/SOM, 78GRO/ZAK] and by Blaise *et al.* [78BLA/FOU]. Heat capacity values for UAs_2(cr) from the latter reference are generally greater than those reported by Grønvold *et al.* Also, the shapes of the heat capacity curves for 5 to 300 K in the two studies differ slightly, although the experimental values of the Néel temperature differ by only 2 K. There is no value of $S^{\circ}_{m}(UAs_2, cr, 298.15\,K)$ reported in Ref. [78BLA/FOU], nor are sufficient data available in that reference to allow a reviewer to easily calculate the entropy. Furthermore, Grønvold *et al.* [78GRO/ZAK] reported results obtained using two different types of calorimeters, and these gave essentially the same results for the heat capacity near 300 K. For these reasons this review accepts the entropy and heat capacity values from Ref. [78GRO/ZAK].

$$S^{\circ}_{m}(UAs_2, cr, 298.15\,K) = (123.05 \pm 0.20)\,J \cdot K^{-1} \cdot mol^{-1}$$
$$C^{\circ}_{p,m}(UAs_2, cr, 298.15\,K) = (79.96 \pm 0.10)\,J \cdot K^{-1} \cdot mol^{-1}$$

Baskin and Smith [70BAS/SMI] suggested that the enthalpy of formation of UAs_2(cr) can be estimated by assuming a linear relationship between $\Delta_f H^{\circ}_{m}$ and y for compounds UAs_y(cr) ($y = 1$ to 2). They showed this functional dependence holds for the corresponding sulphur and antimony systems. This review accepts the use of this method with the uncertainty of ±5% claimed by Baskin and Smith [70BAS/SMI] and, based on the values for $\Delta_f H^{\circ}_{m}(UAs, cr, 298.15\,K)$ and $\Delta_f H^{\circ}_{m}(UAs_{1.33}, cr, 298.15\,K)$ ($= \frac{1}{3}\Delta_f H^{\circ}_{m}(U_3As_4, cr, 298.15\,K)$) selected below, obtains

$$\Delta_f H^{\circ}_{m}(UAs_2, cr, 298.15\,K) = -(252 \pm 13)\,kJ \cdot mol^{-1}.$$

The Gibbs energy of formation is calculated from the selected enthalpy of formation and entropy.

$$\Delta_f G^{\circ}_{m}(UAs_2, cr, 298.15\,K) = -(253 \pm 13)\,kJ \cdot mol^{-1}$$

V.6.3.1.4. Ternary uranium arsenides

Blaise *et al.* [80BLA/LAG2] reported heat capacity measurements (1 to 300 K) for UAsS(cr), UAsSe(cr) and UAsTe(cr). From these, the authors calculated heat capacities and entropies for the compounds at 298 K. This review accepts these values and assigns an uncertainty of ±3% as suggested by the authors.

$$S^{\circ}_{m}(UAsS, cr, 298.15\,K) = (114.5 \pm 3.4)\,J \cdot K^{-1} \cdot mol^{-1}$$
$$S^{\circ}_{m}(UAsSe, cr, 298.15\,K) = (131.9 \pm 4.0)\,J \cdot K^{-1} \cdot mol^{-1}$$
$$S^{\circ}_{m}(UAsTe, cr, 298.15\,K) = (139.9 \pm 4.2)\,J \cdot K^{-1} \cdot mol^{-1}$$

$$C^{\circ}_{p,m}(UAsS, cr, 298.15\,K) = (80.9 \pm 2.4)\,J \cdot K^{-1} \cdot mol^{-1}$$
$$C^{\circ}_{p,m}(UAsSe, cr, 298.15\,K) = (83.5 \pm 2.5)\,J \cdot K^{-1} \cdot mol^{-1}$$
$$C^{\circ}_{p,m}(UAsTe, cr, 298.15\,K) = (81.3 \pm 2.4)\,J \cdot K^{-1} \cdot mol^{-1}$$

No enthalpy or Gibbs energy data for these ternary compounds appear to have been reported.

V.6.3.2. Uranium arsenates

V.6.3.2.1. U(VI) arsenates

Chemical thermodynamic data for several anhydrous dioxouranium(VI) arsenates were recently reported by Barten, Cordfunke and Ouweltjes [82COR/OUW2, 85BAR/COR, 86BAR, 88BAR2]. The values $\Delta_f H^\circ_m((UO_2)_3(AsO_4)_2$, cr, 298.15 K) = $-(4689.4 \pm 2.8)$ kJ · mol^{-1}, $\Delta_f H^\circ_m((UO_2)_2As_2O_7$, cr, 298.15 K) = $-(3426.0 \pm 2.1)$ kJ · mol^{-1} and $\Delta_f H^\circ_m(UO_2(AsO_3)_2$, cr, 298.15 K) = $-(2156.6 \pm 1.3)$ kJ · mol^{-1} [82COR/OUW2], based on the value of $\Delta_f H^\circ_m(H_3AsO_4$, aq, 298.15 K) [68WAG/EVA], were decreased by 20.7 kJ·mol^{-1} in Barten's thesis [86BAR]. It appears that the "correction" was itself in error, or that a different thermodynamic cycle was used to obtain the new result (see Appendix A). Indeed, the values of $\Delta_f H^\circ_m(H_3AsO_4$, aq, 298.15 K) and $\Delta_f H^\circ_m((As_2O_5)_3 \cdot 5H_2O$, cr, 298.15 K) are probably too negative [65BEE/MOR, 71MIN/GIL, 88BAR/BUR], and not too positive as would be indicated by the direction of Barten's "correction". It should also be mentioned that the composition of these solids may not be proven, as the supporting X-ray diffraction patterns of α-$UO_2(AsO_3)_2$ [86POW, file 33-1415] and β-$UO_2(AsO_3)_2$ [86POW, file 33-1417] are of uncertain quality. The values from Ref. [82COR/OUW2] are accepted in this review, but the uncertainties are estimated as ± 8 kJ · mol^{-1}.

$$\Delta_f H^\circ_m((UO_2)_3(AsO_4)_2, \text{cr}, 298.15\,\text{K}) = -(4689.4 \pm 8.0)\,\text{kJ} \cdot \text{mol}^{-1}$$
$$\Delta_f H^\circ_m((UO_2)_2As_2O_7, \text{cr}, 298.15\,\text{K}) = -(3426.0 \pm 8.0)\,\text{kJ} \cdot \text{mol}^{-1}$$
$$\Delta_f H^\circ_m(UO_2(AsO_3)_2, \text{cr}, 298.15\,\text{K}) = -(2156.6 \pm 8.0)\,\text{kJ} \cdot \text{mol}^{-1}$$

Barten [85BAR/COR, 86BAR] reported the vapour pressures of As_4O_{10}(g) over the three solids as functions of temperature and oxygen partial pressure. The data for p_{O_2} = 1 atm and the "enthalpy content" functions were used to calculate values for the entropies and heat capacities of the solids. (It should be noted that Ref. [86BAR] is the only source for some of the needed auxiliary data, such as $S^\circ_m(As_4O_{10}$, g, 298.15 K).) Some calculations and estimates rest on data that may be inconsistent with the arsenic data base in this review.

The values

$$S^\circ_m((UO_2)_3(AsO_4)_2, \text{cr}, 298.15\,\text{K}) = (387 \pm 30)\,\text{J} \cdot \text{K}^{-1} \cdot \text{mol}^{-1}$$
$$S^\circ_m((UO_2)_2As_2O_7, \text{cr}, 298.15\,\text{K}) = (307 \pm 30)\,\text{J} \cdot \text{K}^{-1} \cdot \text{mol}^{-1}$$
$$S^\circ_m(UO_2(AsO_3)_2, \text{cr}, 298.15\,\text{K}) = (231 \pm 30)\,\text{J} \cdot \text{K}^{-1} \cdot \text{mol}^{-1}$$

are selected, based on auxiliary data from this review (also see Appendix A). The large uncertainties are chosen to reflect the uncertainties in the values for $S^\circ_m(As_4O_{10}$, g, 298.15 K), $\Delta_f H^\circ_m(As_4O_{10}$, g, 298.15 K) and the selected enthalpies of formation.

From the accepted values of $\Delta_f H^\circ_m$ and S°_m, the Gibbs energies are calculated.

$$\Delta_f G^\circ_m((UO_2)_3(AsO_4)_2, \text{cr}, 298.15\,\text{K}) = -(4310.8 \pm 12.0)\,\text{kJ} \cdot \text{mol}^{-1}$$
$$\Delta_f G^\circ_m((UO_2)_2As_2O_7, \text{cr}, 298.15\,\text{K}) = -(3130.3 \pm 12.0)\,\text{kJ} \cdot \text{mol}^{-1}$$
$$\Delta_f G^\circ_m(UO_2(AsO_3)_2, \text{cr}, 298.15\,\text{K}) = -(1944.9 \pm 12.0)\,\text{kJ} \cdot \text{mol}^{-1}$$

The heat capacities at 298.15 K reported in Ref. [88BAR2] are accepted with uncertainties as assigned in this review.

$$C^\circ_{p,m}((UO_2)_3(AsO_4)_2, cr, 298.15\,K) = (364 \pm 36)\ J \cdot K^{-1} \cdot mol^{-1}$$
$$C^\circ_{p,m}((UO_2)_2As_2O_7, cr, 298.15\,K) = (273 \pm 27)\ J \cdot K^{-1} \cdot mol^{-1}$$
$$C^\circ_{p,m}(UO_2(AsO_3)_2, cr, 298.15\,K) = (201 \pm 20)\ J \cdot K^{-1} \cdot mol^{-1}$$

The heat capacity functions are given in Table III.3.

V.6.3.2.2. $UAsO_5$ (cr)

Barten [86BAR, 87BAR] reported values for $\Delta_f H^\circ_m$, S°_m, $\Delta_f G^\circ_m$ and $C^\circ_{p,m}$ for $UAsO_5$(cr) at 298.15 K and as a function of temperature. For reasons outlined in Appendix A, only the heat capacity as a function of temperature is accepted in this review, $C^\circ_{p,m}(UAsO_5, cr, 298.15\ K) = 130.701 + 64.258 \times 10^{-3}T - 10.542 \times 10^5 T^{-2}$. Hence, the value at 298.15 K is

$$C^\circ_{p,m}(UAsO_5, cr, 298.15\,K) = (138 \pm 14)\,J \cdot K^{-1} \cdot mol^{-1}.$$

V.6.4. Antimony compounds

Chiotti *et al.* [81CHI/AKH] reported four phases in the U-Sb system. Some data are available for USb(cr), USb_2(cr) and U_3Sb_4(cr). This review estimates the thermodynamic properties of U_4Sb_3(cr).

V.6.4.1. USb(cr)

Baskin and Smith [70BAS/SMI] obtained

$$\Delta_f H^\circ_m(USb,\ cr,\ 298.15\ K) = -(138.5 \pm 7.5)\,kJ \cdot mol^{-1}$$

by direct reaction calorimetry. This value is accepted in this review.

The entropy is estimated according to the third method described in Appendix D based on the entropies of USb_2(cr) and U_3Sb_4(cr).

$$S^\circ_m(USb,\ cr,\ 298.15\ K) = (104 \pm 6)\,J \cdot K^{-1} \cdot mol^{-1}$$

The Gibbs energy of formation is calculated from the selected enthalpy of formation and entropy.

$$\Delta_f G^\circ_m(USb,\ cr,\ 298.15\ K) = -(141.0 \pm 7.7)\,kJ \cdot mol^{-1}$$

This value is slightly more negative than $\Delta_f H^\circ_m$, as would be expected for such an alloy.

V.6.4.2. USb_2 (cr)

Baskin and Smith [70BAS/SMI] obtained

$$\Delta_f H^\circ_m(USb_2,\ cr,\ 298.15\ K) = -(173.6 \pm 10.9)\,kJ \cdot mol^{-1}$$

by direct reaction calorimetry. This value is accepted in this review.

Grønvold *et al.* [78GRO/ZAK] measured the heat capacity of tetragonal USb_2(cr) by adiabatic shield calorimetry from 5 to 750 K. They obtained

$$C^{\circ}_{p,m}(USb_2, cr, 298.15\ K) = (80.2 \pm 0.1)\ J \cdot K^{-1} \cdot mol^{-1}$$

and calculated

$$S^{\circ}_{m}(USb_2, cr, 298.15\ K) = (141.46 \pm 0.13)\ J \cdot K^{-1} \cdot mol^{-1}.$$

These values and their thermal functions are accepted here.

The Gibbs energy of formation is calculated from the selected enthalpy of formation and entropy.

$$\Delta_f G^{\circ}_{m}(USb_2, cr, 298.15\ K) = -(173.7 \pm 10.9)\ kJ \cdot mol^{-1}$$

V.6.4.3. U_3Sb_4(cr)

Baskin and Smith [70BAS/SMI] obtained

$$\Delta_f H^{\circ}_{m}(U_3Sb_4, cr, 298.15\ K) = -(451.9 \pm 22.6)\ kJ \cdot mol^{-1}$$

by direct reaction calorimetry. This value is accepted here.

Alles *et al.* [77ALL/FAL] measured the heat capacity of U_3Sb_4(cr) by adiabatic shield calorimetry from 5 to 950 K. They obtained

$$C^{\circ}_{p,m}(U_3Sb_4, cr, 298.15\ K) = (188.2 \pm 0.2)\ J \cdot K^{-1} \cdot mol^{-1}$$

and calculated

$$S^{\circ}_{m}(U_3Sb_4, cr, 298.15\ K) = (349.8 \pm 0.4)\ J \cdot K^{-1} \cdot mol^{-1}.$$

These values and their thermal functions are accepted here.

The Gibbs energy of formation is calculated from the selected enthalpy of formation and entropy.

$$\Delta_f G^{\circ}_{m}(U_3Sb_4, cr, 298.15\ K) = -(457.0 \pm 22.6)\ kJ \cdot mol^{-1}.$$

V.6.4.4. U_4Sb_3(cr)

The entropy is estimated according to the third method described in Appendix D, based on the entropies of USb_2(cr) and U_3Sb_4(cr).

$$S^{\circ}_{m}(U_4Sb_3, cr, 298.15\ K) = (379 \pm 20)\ J \cdot K^{-1} \cdot mol^{-1}$$

Based on the stability and high melting point of U_4Sb_3, it is probable that the alloy has an enthalpy of formation per mole of atoms that is similar to the values for $\frac{1}{2}$USb (-69.3 kJ · mol^{-1}) and $\frac{1}{7}U_3Sb_4$ (-64.6 kJ · mol^{-1}). However, considering the large uncertainties involved, no values for $\Delta_f H^{\circ}_{m}(U_4Sb_3, cr)$ and $\Delta_f G^{\circ}_{m}(U_4Sb_3, cr)$ are recommended in the present review.

V.7. Group 14 compounds and complexes

V.7.1. *Carbon compounds and complexes*

V.7.1.1. *Uranium carbides*

Holley, Rand and Storms [84HOL/RAN] reported some of the results of the IAEA Carbide Panel, whose unpublished 1978 report addressed the chemical thermodynamics of actinide carbides. According to the phase diagram developed by the panel [84HOL/RAN, Figure 6.3], UC(cr) is the only uranium carbide stable up to about 1100 K. β-UC_2 is stable only in carbon-rich systems at very high temperature, transforming to α-UC_2 below about 2050 K. α-UC_2 is metastable below 1790 K, and U_2C_3(cr) probably becomes unstable below 1200 K. Nevertheless, because of their slow decomposition rates, α-UC_2 and U_2C_3(cr) exist as metastable compounds at room temperature.

V.7.1.1.1. *UC(cr)*

A sizeable and varied set of information exists for face-centered cubic UC(cr). Samples may vary in composition due to residual dissolved uranium or carbon. This review accepts the results of the selection process performed by Holley, Rand and Storms [84HOL/RAN]. It should be mentioned that there is some evidence [71TET/HUN] that the results are rather composition-sensitive. The uncertainties assigned in this review are based on the discussions given in the IAEA review [84HOL/RAN]. The standard entropy is calculated from the selected $\Delta_f G_m^\circ$ and $\Delta_f H_m^\circ$ values. The selected values are thus:

$$\begin{aligned}
\Delta_f G_m^\circ(\mathrm{UC, cr, 298.15\,K}) &= -(98.9 \pm 3.0)\,\mathrm{kJ \cdot mol^{-1}} \\
\Delta_f H_m^\circ(\mathrm{UC, cr, 298.15\,K}) &= -(97.9 \pm 4.0)\,\mathrm{kJ \cdot mol^{-1}} \\
S_m^\circ(\mathrm{UC, cr, 298.15\,K}) &= (59.3 \pm 16.8)\,\mathrm{J \cdot K^{-1} \cdot mol^{-1}}.
\end{aligned}$$

The $C_{p,m}^\circ(T)$ values tabulated in the IAEA review [84HOL/RAN] are fitted between 298.15 and 600 K. The resulting parameters are tabulated in Table III.3. The value at 298.15 K, with an uncertainty estimated by this review, is

$$C_{p,m}^\circ(\mathrm{UC, cr, 298.15\,K}) = (50.1 \pm 1.0)\,\mathrm{J \cdot K^{-1} \cdot mol^{-1}}.$$

V.7.1.1.2. *U_2C_3(cr)*

Limited experimental data exist for the body-centered cubic sesquicarbide. This review accepts the enthalpy and entropy values selected by the IAEA review [84HOL/RAN]. The uncertainties are estimated in this review.

$$\begin{aligned}
\Delta_f H_m^\circ(\mathrm{U_2C_3, cr, 298.15\,K}) &= -(183.3 \pm 10.0)\,\mathrm{kJ \cdot mol^{-1}} \\
S_m^\circ(\mathrm{U_2C_3, cr, 298.15\,K}) &= (137.8 \pm 0.3)\,\mathrm{J \cdot K^{-1} \cdot mol^{-1}}
\end{aligned}$$

The Gibbs energy of formation is calculated from the selected enthalpy of formation and entropy.

$$\Delta_f G^\circ_m(U_2C_3, cr, 298.15\,K) = -(189.3 \pm 10.0)\,kJ \cdot mol^{-1}$$

This value is inconsistent with $\Delta_f G^\circ_m(U_2C_3, cr, 298.15\,K) = -204.3\,kJ \cdot mol^{-1}$ given in [84HOL/RAN, Table A.6.7], although it is consistent with the enthalpy of formation and entropy values selected in the table at the beginning of Section 6.4 of Ref. [84HOL/RAN].

The $C^\circ_{p,m}(T)$ values tabulated in the IAEA review [84HOL/RAN] are refitted between 298.15 and 600 K. The resulting parameters are tabulated in Table III.3. The value at 298.15 K, with an uncertainty estimated by this review, is

$$C^\circ_{p,m}(U_2C_3, cr, 298.15\,K) = (107.4 \pm 2.0)\,J \cdot K^{-1} \cdot mol^{-1}.$$

V.7.1.1.3. α-$UC_{1.94}$ (α-UC_2)

There are insufficient data on face-centered cubic β-UC_2 to select values for its thermodynamic parameters at 298.15 K. Tetragonal α-UC_2 normally shows a composition in the range $UC_{1.84}$ to $UC_{1.96}$. As was done by Holley, Rand and Storms [84HOL/RAN], this review refers to a nominal stoichiometry of $UC_{1.94}$(cr) as do the values in the table at the beginning of Section 6.4 of Ref. [84HOL/RAN]. This review accepts the results of the selection process performed by Holley, Rand and Storms [84HOL/RAN]. The uncertainty of the selected Gibbs energy of formation is from the IAEA review [84HOL/RAN]. The uncertainty assigned to the entropy by this review incorporates a contribution reflecting the uncertainty about the presence of structural zero-point entropy in $UC_{1.94}$ and $UC_{1.90}$.

$$\Delta_f G^\circ_m(UC_{1.94}, \alpha, 298.15\,K) = -(87.4 \pm 2.1)\,kJ \cdot mol^{-1}$$
$$S^\circ_m(UC_{1.94}, \alpha, 298.15\,K) = (68.3 \pm 2.0)\,J \cdot K^{-1} \cdot mol^{-1}$$

The enthalpy of formation is calculated from the selected Gibbs energy of formation and entropy.

$$\Delta_f H^\circ_m(UC_{1.94}, \alpha, 298.15\,K) = -(85.3 \pm 2.2)\,kJ \cdot mol^{-1}$$

In the US NBS Tables [82WAG/EVA] a randomization entropy of $2.72\,J \cdot K^{-1} \cdot mol^{-1}$ (ambiguously stated in [82WAG/EVA] to be 0.65, which is the value in $cal \cdot K^{-1} \cdot mol^{-1}$, not $J \cdot K^{-1} \cdot mol^{-1}$) is included in $S^\circ_m(UC_{1.94}$, cr, 298.15 K). This randomization entropy does not appear to be included in the values listed in the IAEA review [84HOL/RAN, Table A.6.8], nor is it included in the values selected in the present review.

The $C^\circ_{p,m}(T)$ values tabulated in the IAEA review [84HOL/RAN] are refitted between 298.15 and 600 K. The resulting parameters are tabulated in Table III.3. The value at 298.15 K, with an uncertainty estimated by this review, is

$$C^\circ_{p,m}(UC_{1.94}, \alpha, 298.15\,K) = (60.8 \pm 1.5)\,J \cdot K^{-1} \cdot mol^{-1}.$$

V.7.1.2. Uranium carbonates

V.7.1.2.1. The aqueous uranium carbonate system

There is an extensive literature on equilibria in the uranium-carbonate system. Reviews of these data were made by Langmuir [78LAN], Lemire and Tremaine [80LEM/TRE], Robel [83ROB] and Tripathi [84TRI]. The available information is summarized in Table V.42, where some experimental information is also given. The equilibrium constants are corrected to $I = 0$ by using the specific ion interaction theory, *cf.* Appendix B. Two different methods are used:

- When experimental data over a range of ionic strengths are available, a linear regression as described in Appendix B and illustrated in Figures V.15 and V.16 is used.
- When data from only one or a few ionic strengths are available, the correction to $I = 0$ is made by using the specific ion interaction theory and the ion interaction coefficients given in Appendix B. The constants obtained in this way are also listed in Table V.42.

Details of their determination (or estimation) are given in the appropriate parts of Appendix A, where the individual experimental studies are discussed in more detail. Studies of the carbonate complexes of uranium are seldom straightforward. Reasons for this include:

- The carbonate ion is a fairly strong base. Hence, the experimental measurements must often be made in a pH-range where extensive hydrolysis may occur and where the formation of ternary uranium-hydroxide-carbonate complexes may be expected.
- The high proton affinity of the carbonate ion may result in the formation of bicarbonate complexes in addition to hydroxide and carbonate complexes.
- Sparingly soluble oxide-hydrates and carbonates may form, which may complicate the experimental studies.
- Uranium(IV) and dioxouranium(V) carbonate complexes are strong reducing agents, and are thus easily oxidized by traces of oxygen.

The aqueous uranium carbonate system is formally a three-component system, where the complexes may be denoted (omitting the charges) by $M_m(OH)_n(CO_2)_q$. Note that $(OH)CO_2^-$ is equivalent to HCO_3^-, and $(OH)_2CO_2^{2-}$ is equivalent to $CO_3^{2-} + H_2O$. It is possible to obtain a unique chemical model in terms of the stoichiometric coefficients m, n and q. However, in order to obtain the chemical composition in terms of coordinated OH^-, HCO_3^- and CO_3^{2-}, additional chemical information is necessary. Such information may be obtained from additional spectroscopic [87BRU/GRE, 88FER/GLA] or structural [83ABE/FER] information. Chemical knowledge of preferred coordination numbers and coordination geometries and the magnitude of the equilibrium constants may also give the required additional information.

Table V.42: Experimental equilibrium data for the dioxouranium(VI) carbonate system.

Method	Ionic Medium	t (°C)	$\log_{10} K^{(a)}$	$\log_{10} K^{\circ(a)}$	Reference
$UO_2^{2+} + CO_3^{2-} \rightleftharpoons UO_2CO_3(aq)$					
emf, gl	0.1 M ($NaClO_4$)	25	$9.30 \pm 0.20^{(b,d)}$	10.13 ± 0.20	[69TSY]
sol, sp	0.03 M	25	$9.44 \pm 0.30^{(d)}$	10.00 ± 0.30	[72SER/NIK]
pol	var.	25	2.8		[73ALM/GAR]
sol	0	25		10.0	[76PIR/NIK]
emf, gl	3.0 M ($NaClO_4$)	25	$9.01 \pm 0.04^{(b)}$	9.51 ± 0.15	[79CIA/FER]
sol	0.5 M ($NaClO_4$)	25	$8.39 \pm 0.10^{(b,c)}$	9.57 ± 0.10	[84GRE/FER]
	3.0 M ($NaClO_4$)	25	$8.89 \pm 0.10^{(b,c)}$	9.38 ± 0.17	
emf, gl	0.5 m ($NaClO_4$)	25	8.54 ± 0.05	9.72 ± 0.05	[91GRE/LAG]
rev	0	25		10.1	[78LAN]
rev	0	25		10.1 ± 0.4	[80LEM/TRE]
$UO_2^{2+} + 2CO_3^{2-} \rightleftharpoons UO_2(CO_3)_2^{2-}$					
sol	0	25		14.6	[56MCC/BUL]
sol	0.2 M (NH_4NO_3)	25	15.6		[60BAB/KOD]
emf, gl	0.1 M ($NaClO_4$)	25	$16.2 \pm 0.3^{(b,d)}$		[69TSY]
sol	0.03 M	25	$16.7 \pm 0.3^{(d)}$		[72SER/NIK]
pol	var.	25	4.0		[73ALM/GAR]
dis	0.1 M ($NaNO_3$)	25[e]	$16.22 \pm 0.30^{(d,e)}$		[77SCA]
emf, gl	0.1 M ($NaClO_4$)	25	$16.15 \pm 0.30^{(d)}$		[82MAY]
sol	0.5 M ($NaClO_4$)	25	$15.56 \pm 0.15^{(b,c)}$		[84GRE/FER]
	3.0 M ($NaClO_4$)	25	$16.20 \pm 0.15^{(b,c)}$		
emf, gl	0.5 m ($NaClO_4$)	25	14.93 ± 0.30		[91GRE/LAG]
rev	0	25		17.0	[78LAN]
rev	0	25		17.1 ± 0.4	[80LEM/TRE]
$UO_2^{2+} + 3CO_3^{2-} \rightleftharpoons UO_2(CO_3)_3^{4-}$					
sol, sp	0	25		18.3	[56MCC/BUL]
sol	1.0	25	22.8		[59KLY/SMI]
sol	0.2 M (NH_4NO_3)	25	20.7		[60BAB/KOD]
emf, gl	0.1 M ($NaClO_4$)	25	$21.5 \pm 0.3^{(b,d)}$		[69TSY]
pol	var.	25	7.7		[73ALM/GAR]
dis	0.1 M ($NaNO_3$)	25[e]	$21.58 \pm 0.06^{(d,e)}$		[75CIN/SCA]
dis	0.1 M ($NaNO_3$)	25[e]	$21.74 \pm 0.30^{(d,e)}$		[77SCA]
emf, gl	0.1 M ($NaClO_4$)	25	$21.80 \pm 0.10^{(c,d)}$		[82MAY]
sol	0.5 M ($NaClO_4$)	25	$21.76 \pm 0.15^{(b,c)}$		[84GRE/FER]
	3.0 M ($NaClO_4$)	25	$22.61 \pm 0.15^{(b,c)}$		
emf, gl	0.5 m ($NaClO_4$)	25	22.30 ± 0.11		[91GRE/LAG]
rev	0	25		21.4	[78LAN]
rev	0	25		21.4 ± 0.4	[80LEM/TRE]

Table V.42 (continued)

Method	Ionic Medium	t (°C)	$\log_{10} K^{(a)}$	$\log_{10} K^{\circ(a)}$	Reference
$UO_2(CO_3)_2^{2-} + CO_3^{2-} \rightleftharpoons UO_2(CO_3)_3^{4-}$					
ix	0.5 M ($NaNO_3$)	20(?)	7.00 ± 0.04		[62PAR/NIK]
pot	~0.1 M (H,Na)NO_3	20(?)	5.52 ± 0.22		[62PAR/NIK2]
tls	0.5 M ($NaClO_4$)	25	6.35 ± 0.05		[91BID/CAV]
$3UO_2^{2+} + 6CO_3^{2-} \rightleftharpoons (UO_2)_3(CO_3)_6^{6-}$					
emf, gl	0.1 M ($NaClO_4$)	25	$54.8 \pm 0.3^{(b,d)}$	54.7 ± 0.3	[69TSY]
emf, gl	3.0 M ($NaClO_4$)	25	$56.49 \pm 0.10^{(g)}$	54.10 ± 0.23	[81CIA/FER2]
sol	0.5 M ($NaClO_4$)	25	$54.33 \pm 0.30^{(b,c)}$	53.97 ± 0.30	[84GRE/FER]
	3.0 M ($NaClO_4$)	25	$56.23 \pm 0.30^{(b,c)}$	53.84 ± 0.37	
emf	0.5 m ($NaClO_4$)	25	53.82 ± 0.17	53.46 ± 0.17	[91GRE/LAG]
$2UO_2^{2+} + CO_2(g) + 4H_2O(l) \rightleftharpoons (UO_2)_2CO_3(OH)_3^- + 5H^+$					
dis	0.1 M ($NaClO_4$)	25	-18.63 ± 0.08	-18.43 ± 0.08	[82MAY]
emf	0.5 m ($NaClO_4$)	25	-19.40 ± 0.11	-19.16 ± 0.15	[91GRE/LAG]
$3UO_2^{2+} + CO_2(g) + 4H_2O(l) \rightleftharpoons (UO_2)_3O(OH)_2HCO_3^+ + 5H^+$					
emf, gl	3.0 M ($NaClO_4$)	25	$-16.6 \pm 0.2^{(b)}$	-17.5 ± 0.5	[79CIA/FER]
$11UO_2^{2+} + 6CO_2(g) + 18H_2O(l) \rightleftharpoons (UO_2)_{11}(CO_3)_6(OH)_{12}^{2-} + 24H^+$					
emf, gl	3.0 M ($NaClO_4$)	25	$-72.14 \pm 0.13^{(b)}$		[79CIA/FER]
emf, gl	0.1 M ($NaClO_4$)	25	$-72.6 \pm 0.3^{(b,d)}$		[69TSY]
emf, gl	0.5 m ($NaClO_4$)	25	-72.48 ± 0.31		[91GRE/LAG]
$UO_2(CO_3)_3^{4-} + e^- \rightleftharpoons UO_2(CO_3)_3^{5-}$					
pot	1 M (Na_2CO_3)	25	-8.32	$-0.492\ V^{(f)}$	[69CAJ/PRA]
emf	3.0 M ($NaClO_4$)	25	-8.851 ± 0.005	$-(0.5236 \pm 0.0003)\ V^{(f)}$	[83FER/GRE]
vlt	0.5 M ($NaClO_4$)	25	-10.77 ± 0.17	$-(0.637 \pm 0.010)\ V^{(f)}$	[90RIG]
	1.0 M ($NaClO_4$)	25	-10.02 ± 0.17	$-(0.593 \pm 0.010)\ V^{(f)}$	
	2.0 M ($NaClO_4$)	25	-8.91 ± 0.17	$-(0.527 \pm 0.010)\ V^{(f)}$	
	3.0 M ($NaClO_4$)	25	-8.59 ± 0.17	$-(0.508 \pm 0.010)\ V^{(f)}$	

Table V.42 (continued)

Method	Ionic Medium	t (°C)	$\log_{10} K$(a)	$\log_{10} K^\circ$(a)	Reference
$UO_2^+ + 3CO_3^{2-} \rightleftharpoons UO_2(CO_3)_3^{5-}$					
emf	3.0 M ($NaClO_4$)	25		6.54 ± 0.49(i)	[83FER/GRE]
vlt	1.0 M ($HClO_4$)	25		6.6 ± 0.3	[90CAP/VIT]
vlt	0.4 to 3.0 M ($NaClO_4$)	25(?)		7.1 ± 0.2	[90RIG]
....					
$UO_2(CO_3)_3^{4-} + 2CO_2(g) + 2e^- \rightleftharpoons U(CO_3)_5^{6-}$					
emf	3.0 M ($NaClO_4$)	25	-9.44 ± 0.03	$-(0.279 \pm 0.001)$ V(f)	[83CIA/FER]
sp, emf	3.0 M ($NaClO_4$)	25	-9.44 ± 0.03	$-(0.279 \pm 0.001)$ V(f)	
....					
$U(CO_3)_4^{4-} + CO_3^{2-} \rightleftharpoons U(CO_3)_5^{6-}$					
sp	0.5 M ($NaClO_4$)	25	1.95 ± 0.15		[90BRU/GRE]
	1.0 M ($NaClO_4$)	25	2.15 ± 0.15		
	2.0 M ($NaClO_4$)	25	2.90 ± 0.15		
	3.0 M ($NaClO_4$)	25	3.50 ± 0.15		
....					
$UO_2CO_3(cr) \rightleftharpoons UO_2^{2+} + CO_3^{2-}$					
sol	0.0002 to 0.02 M	25	-14.26(j)	-14.26 ± 0.30(d)	[72SER/NIK]
sol	0.01 M(h)	25	-14.15 ± 0.08	-14.50 ± 0.17	[76NIK2]
	0.01 M(h)	50	-14.84 ± 0.08		
	0.01 M(h)	75	-15.04 ± 0.08		
	0.01 M(h)	100	-15.22 ± 0.08		
	0.01 M(h)	125	-16.14 ± 0.08		
	0.01 M(h)	150	-17.4(?)		
sol	0.5 M ($NaClO_4$)	25	-13.31 ± 0.03(b,c)	-14.48 ± 0.04	[84GRE/FER]
	3.0 M ($NaClO_4$)	25	-13.94 ± 0.03(b,c)	-14.38 ± 0.14	

(a) Refers to the reactions indicated, $\log_{10} K$ in the ionic medium and at the temperature given in the table, $\log_{10} K^\circ$ (in molal units) at $I = 0$ and 298.15 K.
(b) Re-evaluated in this review, *cf.* Appendix A.
(c) The reported constant is corrected for the different protonation constant of carbonate used in this review (Chapter VI), *cf.* Appendix A.
(d) Uncertainties estimated by this review.
(e) Original value corrected from 20 to 25°C using the enthalpy value selected in this review.
(f) Measured potential E *vs.* NHE.
(g) Value calculated by this review from the constant for the equilibrium $3UO_2(CO_3)_3^{4-} + 3CO_2(g) + 3H_2O(l) \rightleftharpoons (UO_2)_3(CO_3)_6^{6-} + 6HCO_3^-$.
(h) Ionic strength assumed to be similar to that used in Ref. [72SER/NIK].
(i) Value recalculated to $I = 0$ by this review, *cf.* Appendix A.
(j) Value refers to $I = 0$.

The most practical way to study systems of this type is to reduce them to two components by working at a constant partial pressure of $CO_2(g)$. At constant p_{CO_2}, the equilibrium constant for the reaction

$$m\mathrm{M}^{z+} + n\mathrm{H_2O(l)} + q\mathrm{CO_2(g)} \rightleftharpoons \mathrm{M}_m(\mathrm{OH})_n(\mathrm{CO_2})_q^{mz-n} + n\mathrm{H}^+ \qquad \text{(V.156)}$$

can be written

$$\frac{[\mathrm{M}_m(\mathrm{OH})_n(\mathrm{CO_2})_q^{mz-n}][\mathrm{H}^+]^n}{[\mathrm{M}^{z+}]^m} = {}^*\beta_{q,n,m} \times (p_{\mathrm{CO_2}})^q = {}^*K_{n,m}. \qquad \text{(V.157)}$$

The stoichiometric coefficients m and n and the conditional equilibrium constants ${}^*K_{n,m}$ can be determined by standard methods, *cf.* [61ROS/ROS]. The coefficient q and the values of ${}^*\beta_{q,n,m}$ are determined from the p_{CO_2} dependence of the conditional constants ${}^*K_{n,m}$, *cf.* [81CIA/FER2]. This review prefers to describe the chemical composition in the U-H_2O-$CO_2(g)$ system in terms of coordinated HCO_3^-/CO_3^{2-}, rather than using the "CO_2-notation". The chemical composition of the complexes formed in the U-H_2O-CO_2 system were either determined by ^{13}C-NMR or derived from structural considerations (*cf.* [91GRE/LAG] in Appendix A).

The published experimental studies cover a wide range of pH, p_{CO_2} and uranium concentration values. In individual studies the concentration ranges investigated may be much smaller. In this case the correct assessment of the minor species is difficult or impossible. Details are given in Appendix A. The reported studies cover a wide range of experimental techniques, solubility, ion exchange, spectrophotometry and emf-methods. The most precise of these are in general the emf-methods.

Most of the measurements were carried out in media of constant and often high ionic strength. This is the most precise method to establish the equilibrium model (*i.e.*, the chemical composition of all the complexes formed, *cf.* [61ROS/ROS]). Some investigators took protolysis constants for the $CO_2(g)$-$H_2O(l)$ system from the literature, usually for $I = 0$, and used these constants at the ionic strength of their study. This introduced a systematic error in their equilibrium constants. Nevertheless, data of this type can in general be used to deduce a correct chemical equilibrium model for the major species.

The numerical values of the carbonate equilibrium constants are to some extent influenced by the numerical values of the formation constants of major hydrolysis species such as $(UO_2)_2(OH)_2^{2+}$ and $(UO_2)_3(OH)_5^+$. In most cases, this correlation is small because the binary hydroxide complexes are often minor species.

The set of selected constants is calculated from the data in Table V.42, recalculated to zero ionic strength. A summary of the selected values is given in Table V.43.

Values selected in the reviews of Langmuir [78LAN] and of Lemire and Tremaine [80LEM/TRE] are also given in Table V.42 for comparison.

Standard enthalpies of formation of the various uranium carbonate complexes are obtained from selected standard enthalpies of formation of UO_2^{2+} and CO_3^{2-} and measured enthalpies of reaction. These were obtained either by direct calorimetric measurements or from the temperature dependence of the corresponding equilibrium constant. In general, the second method leads to a much lower accuracy than the

Table V.43: Selected equilibrium constants in the dioxouranium(VI) carbonate system at 298.15 K and $I = 0$, including the mixed hydroxide-carbonate complexes.

Reaction			$\log_{10} K^\circ$
$UO_2^{2+} + CO_3^{2-}$	$\rightleftharpoons$	$UO_2CO_3(aq)$	9.68 ± 0.04
$UO_2^{2+} + 2CO_3^{2-}$	$\rightleftharpoons$	$UO_2(CO_3)_2^{2-}$	16.94 ± 0.12
$UO_2^{2+} + 3CO_3^{2-}$	$\rightleftharpoons$	$UO_2(CO_3)_3^{4-}$	21.60 ± 0.05
$3UO_2^{2+} + 6CO_3^{2-}$	$\rightleftharpoons$	$(UO_2)_3(CO_3)_6^{6-}$	54.0 ± 1.0
$2UO_2^{2+} + CO_2(g) + 4H_2O(l)$	$\rightleftharpoons$	$(UO_2)_2CO_3(OH)_3^- + 5H^+$	-19.01 ± 0.50
$3UO_2^{2+} + CO_2(g) + 4H_2O(l)$	$\rightleftharpoons$	$(UO_2)_3O(OH)_2HCO_3^+ + 5H^+$	-17.5 ± 0.5
$11UO_2^{2+} + 6CO_2(g) + 18H_2O(l)$	$\rightleftharpoons$	$(UO_2)_{11}(CO_3)_6(OH)_{12}^{2-} + 24H^+$	-72.5 ± 2.0
Solubility product:			
$UO_2CO_3(cr)$	$\rightleftharpoons$	$UO_2^{2+} + CO_3^{2-}$	-14.47 ± 0.04

calorimetric method. It is often practical to perform the calorimetric experiments in a bomb calorimeter in order to have a well-defined state of the gas phase in the system. Because of the low pressure, there is no problem in converting the measured $\Delta_r U$ values to the corresponding $\Delta_r H$ quantities, *cf.* [84GRE/SPA].

a) Major U(VI) carbonate complexes

The stoichiometric compositions of the three mononuclear U(VI) carbonate complexes, $UO_2CO_3(aq)$, $UO_2(CO_3)_2^{2-}$ and $UO_2(CO_3)_3^{4-}$, are well established. The same is true for the trimer $(UO_2)_3(CO_3)_6^{6-}$. The equilibrium constants reported for Reactions (V.158) through (V.160), or the reactions from which they are derived, are given in Table V.42.

$$UO_2^{2+} + CO_3^{2-} \rightleftharpoons UO_2CO_3(aq) \quad (V.158)$$
$$UO_2^{2+} + 2CO_3^{2-} \rightleftharpoons UO_2(CO_3)_2^{2-} \quad (V.159)$$
$$UO_2^{2+} + 3CO_3^{2-} \rightleftharpoons UO_2(CO_3)_3^{4-} \quad (V.160)$$

There are only few experimental data for Reaction (V.158), and this review therefore corrects each value to $I = 0$ using the specific ion interaction theory and the interaction coefficients listed in Appendix B, *i.e.*, $\Delta\varepsilon(V.158) = -(0.41 \pm 0.04)$. The six values of $\log_{10} K^\circ(V.158)$ from Refs. [69TSY, 72SER/NIK, 79CIA/FER, 84GRE/FER, 91GRE/LAG] listed in Table V.42 are used to calculate the weighted average which is selected here.[11]

$$\log_{10} \beta_1^\circ(V.158, 298.15\,K) = 9.68 \pm 0.04$$

[11] The reviewer (L.C.) has noted that by using a modified specific ion interaction theory [90CIA] the extrapolation to $I = 0$ of the six values accepted by this review results in the following

An equally large number of experimental data are available for the equilibria (V.159) and (V.160). The data from Refs. [69TSY, 72SER/NIK, 77SCA, 82MAY, 84GRE/FER], as listed in Table V.42, for these two equilibria are corrected to molality constants and to 25°C where necessary (using the $\Delta_r H_m^\circ$ values selected below) and used in a linear regression to evaluate the corresponding values of $\log_{10}\beta^\circ$ and $\Delta\varepsilon$. The plots are shown in Figures V.15 and V.16. The following constants resulting from these regressions are thus selected:[12]

$$\log_{10}\beta_2^\circ(\text{V.159}, 298.15\,\text{K}) = 16.94 \pm 0.12$$
$$\log_{10}\beta_3^\circ(\text{V.160}, 298.15\,\text{K}) = 21.60 \pm 0.05.$$

Figure V.15: Extrapolation to $I = 0$ of experimental data for the formation of $UO_2(CO_3)_2^{2-}$ using the specific ion interaction theory. The data are taken from [69TSY] (○), [72SER/NIK] (◇), [82MAY] (▽) and [84GRE/FER] (□) ($NaClO_4$ media), as well as [77SCA] (△) ($NaNO_3$ medium). The shaded area represents the uncertainty range obtained by propagating the resulting uncertainties at $I = 0$ back to $I = 4$ m.

Equilibrium: $UO_2^{2+} + 2CO_3^{2-} \rightleftharpoons UO_2(CO_3)_2^{2-}$

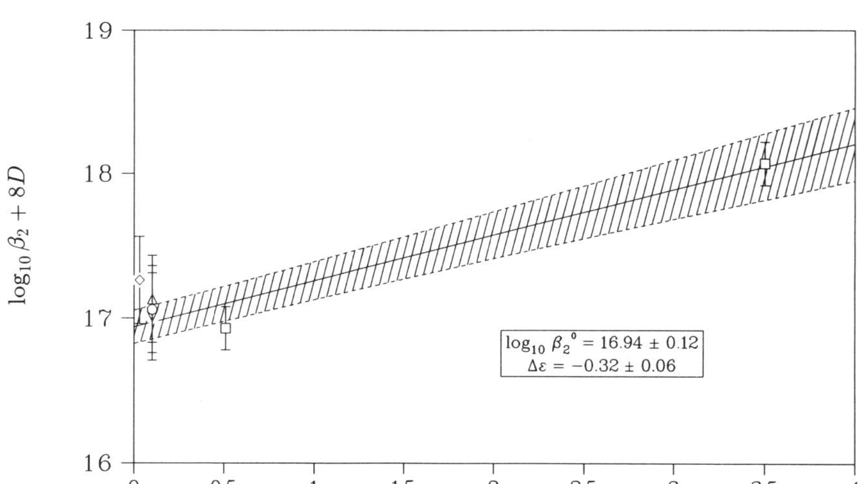

values for $\log_{10}\beta_1^\circ$: 9.99 (extrapolated from 0.03 m), 10.16 (from 0.1 m), 9.69 (from 0.513 m), 9.81 (from 0.513 m), 10.23 (from 3.5 m), and 10.11 (from 3.5 m), with an average value of $\log_{10}\beta_1^\circ = (9.95 \pm 0.25)$.

[12] The reviewer (L.C.) has noted that by using a modified specific ion interaction theory [90CIA] the extrapolation to $I = 0$ of the six values accepted by this review results in the following values for $\log_{10}\beta_2^\circ$: 17.25 (extrapolated from 0.03 m), 17.05 (from 0.1 m), 17.22 (from 0.513 m), 16.96 (from 0.513 m), and 17.29 (from 3.5 m), with an average value of $\log_{10}\beta_2^\circ = (17.15 \pm 0.16)$.

Figure V.16: Extrapolation to $I = 0$ of experimental data for the formation of $UO_2(CO_3)_3^{4-}$ using the specific ion interaction theory. The data are taken from [69TSY] (○), [82MAY] (▽) and [84GRE/FER] (□) ($NaClO_4$ media), as well as [75CIN/SCA] (◇) and [77SCA] (△) ($NaNO_3$ media). The shaded area represents the uncertainty range obtained by propagating the resulting uncertainties at $I = 0$ back to $I = 4$ m.

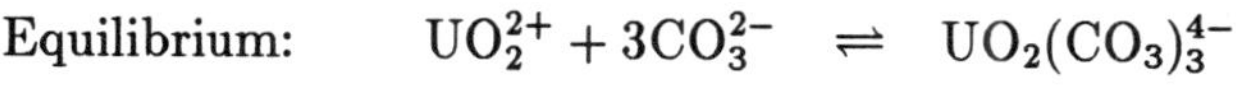

$$\text{Equilibrium:} \quad UO_2^{2+} + 3CO_3^{2-} \rightleftharpoons UO_2(CO_3)_3^{4-}$$

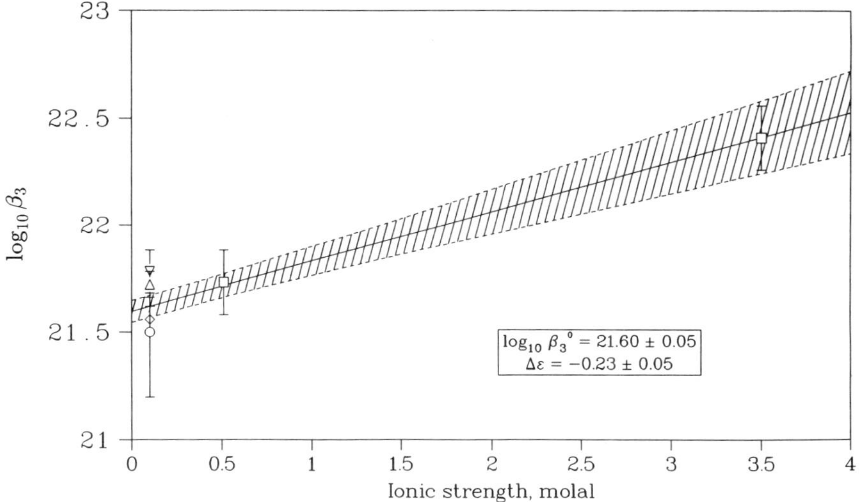

The resulting slopes are $\Delta\varepsilon(\text{V.159}) = -(0.32 \pm 0.06)$ and $\Delta\varepsilon(\text{V.160}) = -(0.23 \pm 0.05)$, from which two new ion interaction coefficients can be deduced: $\varepsilon_{(UO_2(CO_3)_2^{2-},Na^+)} = (0.04 \pm 0.09)$ and $\varepsilon_{(UO_2(CO_3)_3^{4-},Na^+)} = (0.08 \pm 0.11)$. $\log_{10}\beta_3^\circ(\text{V.160})$ is a key value because it is used to deduce the equilibrium constants for the formation of $UO_2(CO_3)_3^{5-}$ and $U(CO_3)_5^{6-}$.

A recent thermal lensing study by Bidoglio *et al.* [91BID/CAV] of the equilibrium

$$UO_2(CO_3)_2^{2-} + CO_3^{2-} \rightleftharpoons UO_2(CO_3)_3^{4-} \qquad \text{(V.161)}$$

resulted in $\log_{10} K_3(\text{V.161, 0.5 M NaClO}_4) = (6.35 \pm 0.05)$. By using the selected $\log_{10}\beta_2^\circ(\text{V.159})$ and $\log_{10}\beta_3^\circ(\text{V.160})$ which, converted to $I = 0.5$ M ($NaClO_4$), are (15.70 ± 0.12) and (21.72 ± 0.06), respectively, the value $\log_{10} K_3(\text{V.161, 0.5 M NaClO}_4) = (6.02 \pm 0.14)$ is obtained, which is in fair agreement with the value given by Bidoglio *et al.* [91BID/CAV]. However, this comparison also shows that the uncertainties of the values obtained in this review by linear regression, *cf.* Figures V.15 and V.16, may be somewhat underestimated.

The trinuclear complex $(UO_2)_3(CO_3)_6^{6-}$ was first identified by Ciavatta *et al.* [81CIA/FER2], and the equilibrium constants for its formation were determined by

Ciavatta *et al.* [81CIA/FER2], Ferri, Grenthe and Salvatore [81FER/GRE], Grenthe *et al.* [84GRE/FER], and recently by Grenthe and Lagerman [91GRE/LAG]. In the latter study $(UO_2)_3(CO_3)_6^{6-}$ is a minor species, reaching at most 20% of the total U(VI) concentration in one of the titrations. This might explain the slight difference between the value from Ref. [91GRE/LAG] and the others reported in Table V.42. The value of Ref. [91GRE/LAG] is not used to calculate the selected value, but is well within the reported uncertainty limits.

Previous experimental data [69TSY] are reinterpreted by this review and are shown to confirm the existence of $(UO_2)_3(CO_3)_6^{6-}$. The available experimental equilibrium constants for the reaction

$$3UO_2^{2+} + 6CO_3^{2-} \rightleftharpoons (UO_2)_3(CO_3)_6^{6-} \qquad \text{(V.162)}$$

are used to deduce the equilibrium constant at $I = 0$ and the ion interaction coefficient $\varepsilon_{((UO_2)_3(CO_3)_6^{6-},Na^+)}$. The Debye-Hückel terms cancel between reactants and products, and the ionic strength dependence of $\log_{10}\beta_{6,3}$(V.162) is determined only by $\Delta\varepsilon$(V.162). The five available determinations, corrected to $I = 0$, are not consistent (*cf.* Table V.42), and this review therefore selects the unweighted average of these values and assigns an uncertainty that spans the range of uncertainty of all the values involved.

$$\log_{10}\beta^{\circ}_{6,3}(\text{V.162}, 298.15\,\text{K}) = 54.0 \pm 1.0$$

The new ion interaction coefficient derived is $\varepsilon_{((UO_2)_3(CO_3)_6^{6-},Na^+)} = (0.55 \pm 0.11)$. This value deviates considerably from the values of other ions with a large negative charge *cf.* Appendix B, Table B.4. The equilibrium constant for the polymerization reaction $3UO_2(CO_3)_2^{2-} \rightleftharpoons (UO_2)_3(CO_3)_6^{6-}$ increases strongly with increasing ionic strength; hence, the trinuclear complex is a minor species at low ionic strength and at low total concentration of U(VI). It is interesting to observe that the solubility data of Blake *et al.* [56BLA/COL] clearly indicate the formation of $(UO_2)_3(CO_3)_6^{6-}$, even though this was not observed by them [56BLA/COL], *cf.* Appendix A.

There are very few precise enthalpy data available for the species formed in the uranium carbonate system. Sergeyeva *et al.* [72SER/NIK] and Piroshkov and Nikolaeva [76PIR/NIK] estimated $\Delta_r H^{\circ}_m$(V.158, 298.15 K). The results of these two estimates, $\Delta_r H^{\circ}_m(\text{V.158}) = 4.6\,\text{kJ}\cdot\text{mol}^{-1}$ and $\Delta_r H^{\circ}_m(\text{V.158}) = -10.9\,\text{kJ}\cdot\text{mol}^{-1}$, respectively, are not consistent. More experimental details were given by Sergeyeva *et al.* [72SER/NIK] than in Ref. [76PIR/NIK], and the procedures used there seem to be satisfactory. This review thus accepts a value of

$$\Delta_r H^{\circ}_m(\text{V.158, 298.15 K}) = (5 \pm 2)\,\text{kJ}\cdot\text{mol}^{-1}.$$

The enthalpy of reaction (V.159) is based on determinations of $\log_{10} K$(V.159) at 25 and 50°C by Sergeyeva *et al.* [72SER/NIK]. Assuming a $\Delta_r C_{p,m}(\text{V.159}) = 0$, the selected value of

$$\Delta_r H^{\circ}_m(\text{V.159, 298.15 K}) = (18.5 \pm 4.0)\ \text{kJ}\cdot\text{mol}^{-1}$$

is obtained. The uncertainty is an estimate (*cf.* Appendix A). In Langmuir's previous analysis, based on the temperature dependence of $\log_{10} K$(V.159), derivation of the equation in [78LAN, Table 2] leads to $\Delta_r H_m^\circ$(V.159, 298.15 K) = 16.3 kJ · mol^{-1}, although $\Delta_f H_m^\circ(UO_2(CO_3)_2^{2-}$, aq, 298.15 K) in [78LAN, Table 1] would suggest $\Delta_r H_m^\circ$(V.159, 298.15 K) = 14.6 kJ · mol^{-1}. The reason for the discrepancy is not clear.

Devina *et al.* [83DEV/YEF] measured the enthalpy of Reaction (V.160) calorimetrically as a function of ionic strength and temperature. They reported $\Delta_r H_m^\circ$(V.160, 298.15 K) = $-(40.62 \pm 0.13)$ kJ · mol^{-1} at $I = 0$. The ionic strength dependence of this quantity is small around 25°C. Grenthe, Spahiu and Olofsson [84GRE/SPA] reported $\Delta_r H_m$(V.160) = $-(35.9 \pm 0.8)$ kJ · mol^{-1} in 3 M $NaClO_4$, while Bauman and de Arman [79BAU/ARM] found $\Delta_r H_m$(V.160) = -37.6 kJ · mol^{-1} in 0.1 M $NaClO_4$. Only an abstract of this paper is available, and therefore this study is not taken into account in the selection procedure. The calorimetric study by Schreiner *et al.* [85SCH/FRI] is recalculated in this review taking the sulphate complexation into account, *cf.* Appendix A, to give $\Delta_r H_m^\circ$(V.160) = $-(38.0 \pm 2.2)$ kJ · mol^{-1}. Ullman and Schreiner reported $\Delta_r H_m^\circ$(V.160, 298.15 K) = $-(42.1 \pm 1.3)$ kJ · mol^{-1}. This review selects the unweighted average of the values reported by Devina *et al.* [83DEV/YEF], Grenthe, Spahiu and Olofsson [84GRE/SPA], the recalculated value from Schreiner *et al.* [85SCH/FRI], and the value from Ullman and Schreiner [88ULL/SCH],

$$\Delta_r H_m^\circ(\text{V.160}, 298.15\,\text{K}) = -(39.2 \pm 4.1)\,\text{kJ} \cdot \text{mol}^{-1}.$$

The uncertainty is assigned to span the range of uncertainty of the three values. This value is in good agreement with the one proposed by Langmuir [78LAN].

The selected enthalpy of formation of $(UO_2)_3(CO_3)_6^{6-}$ is obtained from calorimetric enthalpy data for the reaction

$$UO_2^{2+} + 2CO_3^{2-} \rightleftharpoons \tfrac{1}{3}(UO_2)_3(CO_3)_6^{6-}, \qquad \text{(V.163)}$$

based on the experimental work of Grenthe, Spahiu and Olofsson [84GRE/SPA] and Ullman and Schreiner [88ULL/SCH] as reinterpreted by this review. The weighted average is $\Delta_r H_m^\circ$(V.163, 298.15 K) = $-(20.9 \pm 0.8)$ kJ · mol^{-1}. The selected value for Reaction (V.162) is

$$\Delta_r H_m^\circ(\text{V.162}, 298.15\text{K}) = -(62.7 \pm 2.4)\,\text{kJ} \cdot \text{mol}^{-1}.$$

The standard Gibbs energies and enthalpies of the carbonate complexes are calculated using the necessary auxiliary data selected in this review. The entropies are obtained from the derived reaction entropies.

$$\begin{aligned}
\Delta_f G_m^\circ(UO_2CO_3, \text{aq}, 298.15\,\text{K}) &= -(1535.7 \pm 1.8)\,\text{kJ} \cdot \text{mol}^{-1}\\
\Delta_f G_m^\circ(UO_2(CO_3)_2^{2-}, \text{aq}, 298.15\,\text{K}) &= -(2105.0 \pm 2.0)\,\text{kJ} \cdot \text{mol}^{-1}\\
\Delta_f G_m^\circ(UO_2(CO_3)_3^{4-}, \text{aq}, 298.15\,\text{K}) &= -(2659.5 \pm 2.1)\,\text{kJ} \cdot \text{mol}^{-1}\\
\Delta_f G_m^\circ((UO_2)_3(CO_3)_6^{6-}, \text{aq}, 298.15\,\text{K}) &= -(6333.3 \pm 8.1)\,\text{kJ} \cdot \text{mol}^{-1}
\end{aligned}$$

$$\begin{aligned}
\Delta_f H^\circ_m(UO_2CO_3, aq, 298.15\,K) &= -(1689.2 \pm 2.5)\,kJ\cdot mol^{-1}\\
\Delta_f H^\circ_m(UO_2(CO_3)_2^{2-}, aq, 298.15\,K) &= -(2351.0 \pm 4.3)\,kJ\cdot mol^{-1}\\
\Delta_f H^\circ_m(UO_2(CO_3)_3^{4-}, aq, 298.15\,K) &= -(3083.9 \pm 4.4)\,kJ\cdot mol^{-1}\\
\Delta_f H^\circ_m((UO_2)_3(CO_3)_6^{6-}, aq, 298.15\,K) &= -(7171.1 \pm 5.3)\,kJ\cdot mol^{-1}
\end{aligned}$$

$$\begin{aligned}
S^\circ_m(UO_2CO_3, aq, 298.15\,K) &= (53.9 \pm 7.5)\,J\cdot K^{-1}\cdot mol^{-1}\\
S^\circ_m(UO_2(CO_3)_2^{2-}, aq, 298.15\,K) &= (188.2 \pm 14.1)\,J\cdot K^{-1}\cdot mol^{-1}\\
S^\circ_m(UO_2(CO_3)_3^{4-}, aq, 298.15\,K) &= (33.9 \pm 14.4)\,J\cdot K^{-1}\cdot mol^{-1}\\
S^\circ_m((UO_2)_3(CO_3)_6^{6-}, aq, 298.15\,K) &= (228.9 \pm 23.4)\,J\cdot K^{-1}\cdot mol^{-1}
\end{aligned}$$

b) Mixed U(VI) hydroxide-carbonate complexes

These complexes are often minor species, and there are several different proposals for their composition. Ciavatta *et al.* [79CIA/FER] suggested that the complexes $(UO_2)_3(OH)_5CO_2^+$, which is the same as $(UO_2)_3O(OH)_2(HCO_3)^+$, and a large polynuclear species, $(UO_2)_{11}(CO_3)_6(OH)_{12}^{2-}$, exist in solution. Maya [82MAY] proposed $(UO_2)_2CO_3(OH)_3^-$, and Tsymbal [69TSY] suggested the formation of $UO_2CO_3OH^-$.

This review, and Grenthe and Lagerman [91GRE/LAG], have tested several different chemical models in order to come to a conclusion about the stoichiometry of the minor species. The study of Ciavatta *et al.* [79CIA/FER] is optimized for the determination of UO_2CO_3(aq) and $(UO_2)_2(OH)_2^{2+}$, while all other species are minor. The large polynuclear complex represents, at most, 15% of the total uranium in the solutions. Nevertheless, it is well defined because of its strong effect on the shape of the $Z(-\log_{10}[H^+])$ curve [79CIA/FER, Figure 1]. The formation of this complex was also well established in the [91GRE/LAG] study, and its inclusion significantly improves the Tsymbal model [69TSY]. Details about the reinterpretation are given in Appendix A, where it is also shown that the proposed stoichiometry is not incompatible with known structural features of U(VI) carbonate solids and complexes.

The conclusion of this review is thus that there is good evidence for the formation of a highly polynuclear mixed hydroxide-carbonate complex. The equilibrium constants do not vary much between 0.1 M $NaClO_4$ and 3.0 M $NaClO_4$, and the average value is $\log_{10}{}^*K^\circ(V.164) = -72.5$.

$$11UO_2^{2+} + 6CO_2(g) + 18H_2O(l) \rightleftharpoons (UO_2)_{11}(CO_3)_6(OH)_{12}^{2-} + 24H^+ \qquad (V.164)$$

This review does not find it meaningful to extrapolate these data to $I = 0$ because the result is very sensitive even to small model errors due to the very large Debye-Hückel term. This review selects

$$\log_{10}{}^*K^\circ(V.164, 298.15\ K) = -72.5 \pm 2.0,$$

where the large uncertainty reflects the difficulty of making extrapolations to $I = 0$.

The complex $(UO_2)_3(OH)_5CO_2^+ \triangleq (UO_2)_3O(OH)_2(HCO_3)^+$, first identified by Ciavatta *et al.* [79CIA/FER], was a minor species, making up at most 5% of the total uranium concentration, in all of their titrations. The same is also the case in the recent study of Grenthe and Lagerman [91GRE/LAG]. It is interesting to observe that this species has about the same value for its formation constant in both studies. The stoichiometry is also chemically reasonable, because the complex might be formed by the addition of CO_2 to one of the bridging OH groups in $(UO_2)_3(OH)_5^+ \triangleq (UO_2)_3O(OH)_3^+$ according to

$$(UO_2)_3O(OH)_3^+ + CO_2(g) \rightleftharpoons (UO_2)_3O(OH)_2(HCO_3)^+, \qquad (V.165)$$

cf. Bruno and Grenthe [87BRU/GRE] for the discussion of an analogous reaction between $Be_3(OH)_3^{3+}$ and CO_2. The complex $(UO_2)_3O(OH)_2(HCO_3)^+$ is never a predominant complex, and it is difficult to establish its existence experimentally. However, as a guideline for the reaction

$$3UO_2^{2+} + CO_2(g) + 4H_2O(l) \rightleftharpoons (UO_2)_3O(OH)_2(HCO_3)^+ + 5H^+, \qquad (V.166)$$

this review proposes the weighted average of the two values from Refs. [79CIA/FER, 91GRE/LAG], corrected to $I = 0$, *cf.* Table V.42,

$$\log_{10}{}^*K^\circ(V.166, 298.15\text{ K}) = -17.5 \pm 0.5.$$

The Gibbs energy of formation derived therefrom is

$$\Delta_f G_m^\circ((UO_2)_3O(OH)_2(HCO_3)^+, aq, 298.15\text{ K}) = -(4100.7 \pm 6.0)\text{ kJ} \cdot \text{mol}^{-1}.$$

Maya [82MAY] made extensive studies of the dioxouranium(VI)-carbonate system and presents evidence for $(UO_2)_2CO_3(OH)_3^-$ as a major complex in addition to $UO_2(CO_3)_3^{4-}$. Maya also used additional experimental methods, such as pulse polarography [82MAY2], where he assigned to $(UO_2)_2CO_3(OH)_3^-$ a peak at -0.59V *vs.* SCE. A Raman study [81MAY/BEG] was also interpreted in terms of this complex. It is not clear from the published data how the stoichiometry was obtained from the experiments in Refs. [81MAY/BEG, 82MAY2], and this review concludes that, although these studies indicate that a mixed complex is formed, they are not conclusive as to its composition.

In order to test the possible formation of $(UO_2)_2CO_3(OH)_3^-$, Grenthe and Lagerman [91GRE/LAG] designed an experiment where larger amounts of this complex might be formed. These data strongly support the formation of a mixed complex of the stoichiometry suggested by Maya [82MAY], and Grenthe and Lagerman [91GRE/LAG] obtained $\log_{10}{}^*K(V.167, 0.5\text{ M NaClO}_4) = -(19.42 \pm 0.11)$ for the reaction

$$2UO_2^{2+} + CO_2(g) + 4H_2O(l) \rightleftharpoons (UO_2)_2CO_3(OH)_3^- + 5H^+. \qquad (V.167)$$

Grenthe and Lagerman [91GRE/LAG] also tried to reinterpret the data of Ciavatta *et al.* [79CIA/FER] using the complex $(UO_2)_2CO_3(OH)_3^-$. The results were not conclusive due to the small amounts of this complex formed under the experimental conditions used.

The data for Reaction (V.167) of Maya [82MAY] and Grenthe and Lagerman [91GRE/LAG] from Table V.42 are recalculated to $I = 0$ using $\Delta\varepsilon(V.167) = -(0.22 \pm 0.07)$ obtained by assuming $\varepsilon_{((UO_2)_2CO_3(OH)_3^-, Na^+)} \approx (0.00 \pm 0.05)$. The resulting values are $\log_{10}{}^*K^\circ(V.167) = -(18.43\pm0.08)$ [82MAY] and $\log_{10}{}^*K^\circ(V.167) = -(19.01\pm0.13)$ [91GRE/LAG]. There is a fairly large difference between the two values, where Maya's value corresponds to such a large stability of $(UO_2)_2CO_3(OH)_3^-$ that it is not compatible with the data in Refs. [91GRE/LAG] and [79CIA/FER, 81CIA/FER2]. This is clearly a case of conflicting evidence where additional experimental information is necessary to resolve the issue. However, this review has more experimental data available from the study of Grenthe and Lagerman [91GRE/LAG] than from that of Maya [82MAY] and therefore prefers the former set and selects the value

$$\log_{10}{}^*K^\circ(V.167, 298.15K) = -19.01 \pm 0.50,$$

where the uncertainty is increased to reflect the fact that there may be an unresolved issue on the numerical value of the equilibrium constant.

The Gibbs energy of formation derived therefrom is

$$\Delta_f G_m^\circ((UO_2)_2CO_3(OH)_3^-, aq, 298.15\,K) = -(3139.5 \pm 4.5)\,kJ \cdot mol^{-1}.$$

The data of Tsymbal [69TSY] were criticized by Tripathi [84TRI]. Tsymbal's hydrolysis data are discussed in Appendix A and Section V.3.2.1.3. There is clearly a mistake in Tsymbal's value of the equilibrium constant for the complex $(UO_2)_3(OH)_7^-$. The data also indicate the presence of a systematic error in the total concentration of protons in the dioxouranium(VI) stock solution. However, the carbonate data appear to be of good quality and are therefore reinterpreted here. The data from Tsymbal are tested by using four different chemical models for the most important mixed complexes, *cf.* Appendix A. It is difficult to make a choice between these three proposals because of the small amounts of these minor complexes present in the test solutions.

Double, Pongi and Hurwic [80DOU/PON] used the inflection points in titration curves to propose the composition of a large number of U(VI) carbonate complexes. This method is not reliable since a test of a chemical model can only be made by a comparison between an experimental variable and the corresponding quantity calculated from the model.

Based on these considerations, the selected data listed in Table V.43 were derived. Only the $\log_{10} K$ value of the $(UO_2)_3(OH)_5CO_2^+$ reaction can be extrapolated to $I = 0$. The experimental data on the large polynuclear species are not sufficient for an extrapolation to $I = 0$. Because of its large Debye-Hückel term, such an extrapolation is extremely susceptible to error. No reliable enthalpy, entropy or heat capacity data are currently available for either of these minor species.

As an illustration, distribution diagrams of the dioxouranium(VI) hydroxide-carbonate system in the range $4.5 < pH < 10$ are presented in Figures V.17 and V.18 for two different total carbonate concentrations.

c) Mixed U(VI), Np(VI) and Pu(VI) carbonate complexes

Carbonate is an excellent bridging ligand and the formation of polynuclear carbonate complexes containing one type of metal ion is well known. Such complexes may also

Figure V.17: Distribution diagram of the dioxouranium(VI) hydroxide-carbonate system at 25°C in the range 4.5 < pH < 10.0. The precipitation of solid phases is suppressed.

$$[UO_2^{2+}]_T = 1 \times 10^{-5}\ M$$
$$[CO_3^{2-}]_T = 2 \times 10^{-3}\ M$$

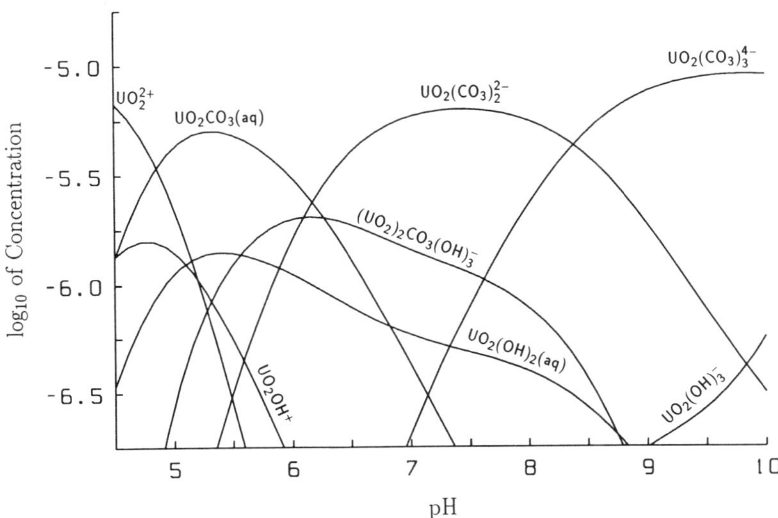

contain two or more different types of metal ions. Known examples are carbonate complexes containing UO_2^{2+}, NpO_2^{2+} and PuO_2^{2+} as metal ions. These complexes are discussed in Section V.8.2.

d) U(V) carbonate complexes

Only one dioxouranium(V) carbonate complex, $UO_2(CO_3)_3^{5-}$, was identified in aqueous solution. Information about this species was obtained by using various electrochemical techniques. Caja and Pradvic [69CAJ/PRA] measured the redox potential of the reaction

$$UO_2(CO_3)_3^{4-} + e^- \rightleftharpoons UO_2(CO_3)_3^{5-} \qquad (V.168)$$

using chronopotentiometry in 1 M Na_2CO_3 medium. They found a redox potential of $E = -0.760$ V *vs.* SCE, *i.e.*, -0.492 V *vs.* NHE. This value is not used by this review to evaluate the standard potential at $I = 0$ because of the uncertainties in the estimation of the liquid junction potential. However, the value found by Caja and Pradvic [69CAJ/PRA] agrees well with that found by other authors [83FER/GRE, 90RIG]. Ferri, Grenthe and Salvatore [83FER/GRE] found $E°(V.168) = -(0.5236 \pm 0.0003)$ V from emf-measurements in 3 M $NaClO_4$. Riglet [90RIG] studied this redox

Figure V.18: Distribution diagram of the dioxouranium(VI) hydroxide-carbonate system at 25°C in the range 4.5 < pH < 10.0. The precipitation of solid phases is suppressed.

$$[UO_2^{2+}]_T = 1.0 \times 10^{-5}\ M$$
$$[CO_3^{2-}]_T = 1.5 \times 10^{-4}\ M$$

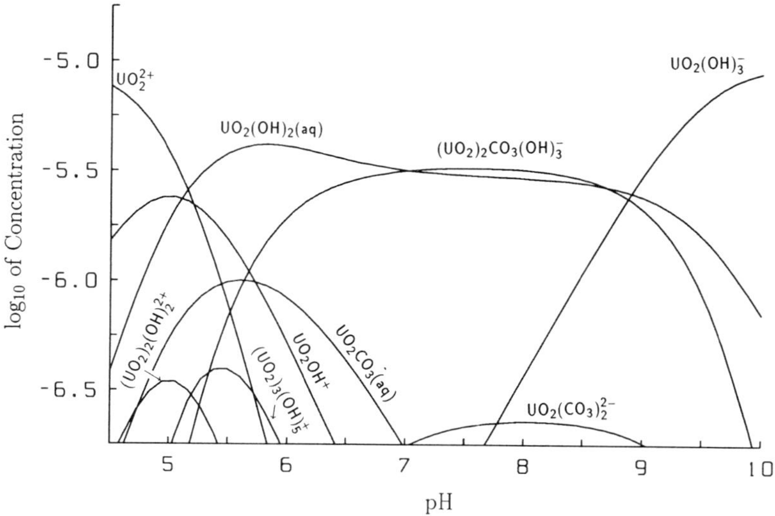

couple using polarography and cyclic voltammetry at four different ionic strengths. There is a difference of 0.0156 V between the data of Riglet [90RIG] and Ferri *et al.* [83FER/GRE]. The latter data are much more precise, while those of Riglet may include a systematic error due to non-reversible electrode reactions *cf.* [90RIG, *pp*.99-106]. This error is assumed to be independent of the ionic strength, and this review therefore selects the data of Riglet [90RIG] because they provide the only possibility to determine both $\log_{10} K^\circ$(V.168) and $\Delta\varepsilon$(V.168). This review increases the uncertainty to ±0.016 V in order to make the uncertainty range consistent with the value of Ferri *et al.* [83FER/GRE]. The recommended values for the standard potential and the corresponding $\log_{10} K^\circ$ value are

$$E^\circ(\text{V.168, 298.15 K}) = -(0.752 \pm 0.016)\ \text{V}$$
$$\log_{10} K^\circ(\text{V.168, 298.15 K}) = -12.71 \pm 0.27.$$

From the value of $\Delta\varepsilon(\text{V.168}) = -(0.61 \pm 0.10)$ obtained by Riglet [90RIG], this review calculates the new ion interaction coefficient $\varepsilon_{(UO_2(CO_3)_3^{5-},Na^+)} = -(0.66 \pm 0.14)$. The large negative value of $\varepsilon_{(UO_2(CO_3)_3^{5-},Na^+)}$ indicates a fairly strong ion pairing with Na^+. Riglet [90RIG] estimated $\beta_1 = (1.4 \pm 0.5)$ for the ionic pairing constant. The strong ionic pairing makes the extrapolation of the data of Ferri, Grenthe and Salvatore [83FER/GRE] uncertain; therefore, their value is not included in the selection of $\log_{10} K^\circ$(V.169).

The standard Gibbs energy of formation of $UO_2(CO_3)_3^{5-}$ is derived from the selected formation constant:

$$\Delta_f G_m^\circ(UO_2(CO_3)_3^{5-}, aq, 298.15\,K) = -(2587.0 \pm 2.6)\,kJ \cdot mol^{-1}.$$

By combining the selected values of $\log_{10} K^\circ$(V.168), $\log_{10} K^\circ$(V.160) and $\log_{10} K^\circ$(V.1), this review calculates the formation constant according to the reaction

$$UO_2^+ + 3CO_3^{2-} \rightleftharpoons UO_2(CO_3)_3^{5-} \quad (V.169)$$

to be $\log_{10} \beta_3^\circ$(V.169) $= (7.41 \pm 0.27)$. This value differs somewhat from the one given by Riglet [90RIG] because this review uses a different constant for the formation of $UO_2(CO_3)_3^{4-}$. The value reported by Capdevila and Vitorge [90CAP/VIT], $\log_{10} \beta_3^\circ$(V.169) $= (6.6 \pm 0.3)$, is significantly different from the value selected here. The reason for this discrepancy is not clear, *cf.* Appendix A.

Riglet [90RIG] also studied the carbonate complexes of Np(V) and Pu(V) and reported equilibrium constants for the formation of $NpO_2CO_3^-$ and $NpO_2(CO_3)_2^{3-}$, $\log_{10} \beta_1^\circ(NpO_2CO_3^-) = (4.7 \pm 0.2)$ and $\log_{10} \beta_2^\circ(NpO_2(CO_3)_2^{3-}) = (6.2 \pm 0.4)$. These equilibrium constants can be used as guidelines for the estimation of the corresponding constants for U(V).

e) U(IV) carbonate complexes

There is considerably less information about the carbonate complexes of U(IV) and U(V) than about U(VI). The uranium(IV) carbonate complexes were only investigated in solutions of rather high bicarbonate concentration. The chemical composition and the equilibrium constant of the limiting complex $U(CO_3)_5^{6-}$ are well established. However, the magnitude of the equilibrium constant depends on the value of the standard potential of UO_2^{2+}/U^{4+}. This value is discussed in Section V.2.3.

No information is available on the composition and equilibrium constants of U(IV) carbonate complexes in acid solutions. However, from the studies on the corresponding Th(IV) system by Bruno *et al.* [87BRU/CAS2] and Grenthe and Lagerman [91GRE/LAG2], this review concludes that mixed hydroxide carbonate/bicarbonate complexes of U(IV) are likely to be formed at pH < 7, complexes that will affect both the speciation and the solubility of uranium(IV). Experimental information on the complexes formed and their stability constants is badly needed. Such information should probably first be obtained for Th(IV), where the experimental difficulties seem less formidable.

The earliest studies on the U(IV) carbonate systems were those by McClaine, Bullwinkel and Huggins [56MCC/BUL], Stabrowski [60STA2] and by Cukman, Caja and Pravdic [68CUK/CAJ]. The first of these is a spectrophotometric study, where evidence for the formation of carbonate complexes was presented, but without proof of the suggested stoichiometry. The second is a polarographic study, where the author proposed extensive hydrolysis of U(IV), U(V) and U(VI) in the 1 M $Na_2CO_3/NaHCO_3$ solutions studied. This assumption is erroneous. The oxidation of U(VI) was assumed to take place via $U(CO_3)_4(OH)_2^{6-}$ and $U(CO_3)_5^{6-}$, but no evidence was presented for these compositions. Cukman, Caja and Pravdic [68CUK/CAJ] proposed the formation of a large number of mixed hydroxide/carbonate species: $UCO_3(OH)_2(aq)$, $U(CO_3)_2(OH)_2^{2-}$, $U(CO_3)_2(OH)_3^{4-}$ and $U(CO_3)_4(OH)_2^{6-}$, to account for the kinetics of the oxidation of U(IV) and U(VI) in sodium bicarbonate solutions. No experimental support was given for this choice. The only quantitative study available in the literature is that of Ciavatta *et al.* [83CIA/FER], who determined the redox potential of the reaction

$$UO_2(CO_3)_3^{4-} + 2e^- + 2CO_2(g) \rightleftharpoons U(CO_3)_5^{6-}. \qquad (V.170)$$

Ciavatta *et al.* [83CIA/FER] also used auxiliary data for the standard potentials of the UO_2^{2+}/U^{4+} couple and β_3 for the formation of $UO_2(CO_3)_3^{4-}$ to calculate the equilibrium constant for the reaction

$$U^{4+} + 5CO_3^{2-} \rightleftharpoons U(CO_3)_5^{6-}. \qquad (V.171)$$

The auxiliary data are revised by this review, see Appendix A for details. By using these data, corrected to $I = 3$ M ($NaClO_4$), this review calculates $\log_{10}\beta_5^\circ(V.171, I = 3\text{ M}) = (36.86 \pm 0.55)$. Since the Debye-Hückel terms cancel, the correction of this value to $I = 0$ depends only on $\Delta\varepsilon(V.171)$ which is taken equal to $-(0.82 \pm 0.19)$ calculated from the values in Appendix B, and $\varepsilon_{(U(CO_3)_5^{6-},Na^+)} = -(0.27 \pm 0.15)$ as evaluated below. The selected value thus calculated is

$$\log_{10}\beta_5^\circ(V.171, 298.15\text{ K}) = 34.0 \pm 0.9.$$

The dissociation of the limiting complex $U(CO_3)_5^{6-}$ to $U(CO_3)_4^{4-}$ was studied by Bruno, Grenthe and Robouch [89BRU/GRE] at 25°C and in varying ionic media.

$$U(CO_3)_4^{4-} + CO_3^{2-} \rightleftharpoons U(CO_3)_5^{6-} \qquad (V.172)$$

These data were used by the authors to estimate

$$\log_{10} K_5^\circ(V.172, 298.15\text{ K}) = -1.12 \pm 0.25$$

and $\Delta\varepsilon(V.172) = -(0.13 \pm 0.11)$, which are accepted by this review. By using $\varepsilon_{(U(CO_3)_4^{4-},Na^+)} = (0.09 \pm 0.10)$, a value of $\varepsilon_{(U(CO_3)_5^{6-},Na^+)} = -(0.27 \pm 0.15)$ is obtained. The negative value of $\varepsilon_{(U(CO_3)_5^{6-},Na^+)}$ indicates ion pairing with Na^+, *cf.* [89BRU/GRE, *p.*225].

The selected equilibrium constants are used to calculate the consistent standard Gibbs energies of formation.

$$\Delta_f G_m^\circ(U(CO_3)_4^{4-}, \text{aq}, 298.15\,\text{K}) = -(2841.9 \pm 6.0)\,\text{kJ}\cdot\text{mol}^{-1}$$
$$\Delta_f G_m^\circ(U(CO_3)_5^{6-}, \text{aq}, 298.15\,\text{K}) = -(3363.4 \pm 5.8)\,\text{kJ}\cdot\text{mol}^{-1}$$

The only experimental enthalpy data for carbonate complexes of uranium(IV) refer to the Reaction (V.171) studied by Grenthe, Spahiu and Olofsson [84GRE/SPA]. The value

$$\Delta_r H_m^\circ(V.171, 298.15\,\text{K}) = -(20 \pm 4)\,\text{kJ}\cdot\text{mol}^{-1}$$

is selected here and is assigned a large uncertainty because of the low concentration of uranium(IV) that had to be used in order to avoid precipitation.

The enthalpy of formation of $U(CO_3)_5^{6-}$ is derived therefrom.

$$\Delta_f H_m^\circ(U(CO_3)_5^{6-}, aq, 298.15\,K) = -(3987.4 \pm 5.3)\,kJ \cdot mol^{-1}$$

Using the selected reaction data, the entropy is calculated via $\Delta_r S_m^\circ$.

$$S_m^\circ(U(CO_3)_5^{6-}, aq, 298.15\,K) = -(83 \pm 26)\,J \cdot K^{-1} \cdot mol^{-1}$$

As an illustration, predominance and solubility diagrams of the dioxouranium(VI)/uranium(IV) hydroxide-carbonate system in the ranges $4.5 < pH < 10$ and $-5 < pe < 10$ are presented in Figures V.19 and V.20. It should be mentioned that Figure V.19 does not represent a real experimental situation because the calculations were made by assuming that no solid phase is formed. This figure is included to illustrate the areas of predominance and relative concentrations of the various dioxouranium(VI) species in a homogeneous aqueous phase. On the contrary, Figure V.20 does represent an experimental modelling situation. A comparison of Figure V.20 with Figure V.8 shows the stabilising effect of carbonate on the solution species of uranium due to the formation of stable aqueous complexes.

Figure V.19: Predominance diagram of the dioxouranium(VI)/uranium(IV) hydroxide-carbonate system at 25°C in the range $4.5 < pH < 10.0$, as a function of the redox potential, pe. The precipitation of solid phases is suppressed.

$$[U]_T = 1.0 \times 10^{-5}\,M$$
$$[CO_3^{2-}]_T = 2.0 \times 10^{-3}\,M$$

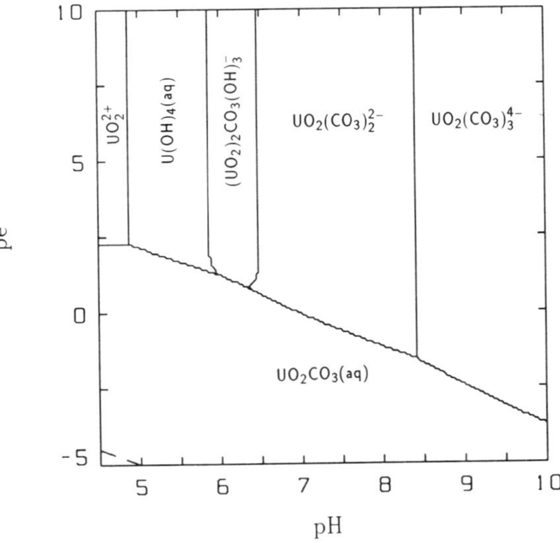

Figure V.20: Solubility and predominance diagram of the dioxouranium(VI)/ uranium(IV) hydroxide-carbonate system at 25°C in the range 4.5 < pH < 10.0, as a function of the redox potential. The solubility limiting phases are indicated on the graph. Schoepite is $UO_3 \cdot 2H_2O(cr)$, uraninite is $UO_2(cr)$.

$$\begin{aligned} [U]_T &= 1.0 \times 10^{-5} \text{ M} \\ [CO_3^{2-}]_T &= 2.0 \times 10^{-3} \text{ M} \end{aligned}$$

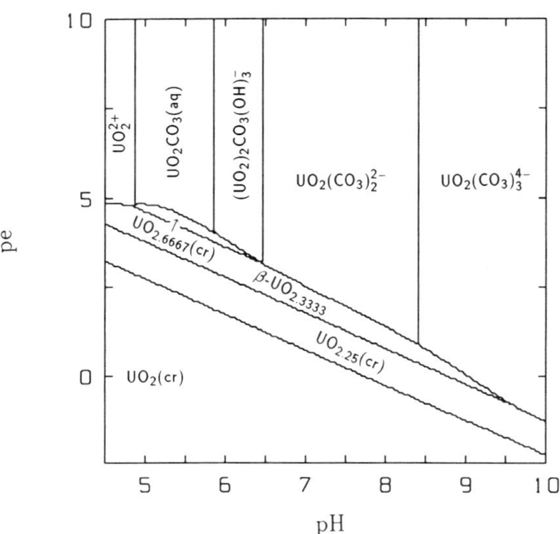

V.7.1.2.2. Solid uranium carbonates

a) $UO_2CO_3(cr)$

The only known stable solid in the U-CO_2-H_2O system is the simple dioxouranium(VI) carbonate $UO_2CO_3(cr)$. When naturally ocurring, this yellow orthorhombic mineral is called rutherfordine [83FLE].

The only three reliable determinations of the solubility product for the reaction

$$UO_2CO_3(cr) \rightleftharpoons UO_2^{2+} + CO_3^{2-} \tag{V.173}$$

found in the literature were by Sergeyeva *et al.* [72SER/NIK], Nikolaeva [76NIK2] and Grenthe *et al.* [84GRE/FER]. The $\log_{10} K_{s,0}$(V.173) values are corrected to zero ionic strength using the specific ion interaction coefficients in Appendix B. These values, along with the original values in ionic media, are also given in Table V.42. This review calculates the weighted average of the four values, leading to the selected value of

$$\log_{10} K^\circ_{s,0}(\text{V.173, 298.15 K}) = -(14.47 \pm 0.04).$$

This value, combined with the selected values of $\Delta_f G^\circ_m$ for UO_2^{2+} and for CO_3^{2-}, leads to the selected

$$\Delta_f G^\circ_m(UO_2CO_3, cr, 298.15\,K) = -(1563.0 \pm 1.8)\,kJ \cdot mol^{-1}.$$

This is a refinement of the value selected by the IAEA [78COR/OHA], $\Delta_f G^\circ_m(UO_2CO_3$, cr, 298.15 K) = $-(1561.9 \pm 3.3)\,kJ \cdot mol^{-1}$, which was based only on Sergeyeva *et al.* [72SER/NIK].

The entropy and heat capacity of UO_2CO_3(cr) were determined experimentally by Gurevich *et al.* [87GUR/SER] whose values for these two quantities are selected by this review.

$$S^\circ_m(UO_2CO_3, cr, 298.15\,K) = (144.2 \pm 0.3)\,J \cdot K^{-1} \cdot mol^{-1}$$
$$C^\circ_{p,m}(UO_2CO_3, cr, 298.15\,K) = (120.1 \pm 0.1)\,J \cdot K^{-1} \cdot mol^{-1}$$

The enthalpy of formation is calculated from the selected Gibbs energy of formation and entropy.

$$\Delta_f H^\circ_m(UO_2CO_3, cr, 298.15\,K) = -(1689.6 \pm 1.8)\,kJ \cdot mol^{-1}$$

Earlier estimates [78COR/OHA] of the entropy value and the standard enthalpy of formation are in good agreement with the more precise values selected here.

Some other dioxouranium(VI) carbonates were described by Bagnall, Jacob and Potter [83BAG/JAC]. These phases are $UO_2CO_3 \cdot H_2O$(cr), $UO_2CO_3 \cdot xH_2O$(cr) ($x \leq 2$, joliotite, structure characterized), $UO_2CO_3 \cdot 2.2H_2O$(cr) and $UO_2CO_3 \cdot 2.5H_2O$(cr). No thermodynamic data are available for any of these compounds.

b) Other uranium carbonates

$M^I_{2y}M^{II}_{2-y}UO_2(CO_3)_3 \cdot xH_2O(cr)$: A large number of solids of the general composition $M^I_{2y}M^{II}_{2-y}UO_2(CO_3)_3 \cdot xH_2O$, where M^I and M^{II} are cations of charge +1 and +2, respectively, were synthesized and characterized by X-ray diffractions. All of them contain the $UO_2(CO_3)_3^{4-}$ ion. Thermodynamic data exist for $Na_4UO_2(CO_3)_3$(cr), $Ca_2UO_2(CO_3)_3 \cdot 10H_2O$(cr) (liebigite), $CaMgUO_2(CO_3)_3 \cdot 12H_2O$(cr) (swartzite), $Mg_2UO_2(CO_3)_3 \cdot 18H_2O$(cr) (bayleyite), and $CaNa_2UO_2(CO_3)_3 \cdot 6H_2O$(cr) (andersonite). The crystallographic characteristics of these and other phases containing $UO_2(CO_3)_3^{4-}$ were reviewed by Weigel [85WEI].

The solubility product of $Na_4UO_2(CO_3)_3$(cr) was measured by Blake *et al.* [56BLA/COL] at different ionic strengths and in different media. This review uses the six values reported for $NaClO_4$ media up to $I = 3$ M for the reaction

$$Na_4UO_2(CO_3)_3(cr) \rightleftharpoons 4Na^+ + UO_2(CO_3)_3^{4-} \qquad (V.174)$$

to make an extrapolation to $I = 0$ (*cf.* Figure V.21). The resulting selected solubility constant is

$$\log_{10} K^\circ_{s,3}(V.174, 298.15\,K) = -5.34 \pm 0.16.$$

Figure V.21: Extrapolation to $I = 0$ of experimental data for the solubility constant of $Na_4UO_2(CO_3)_3(cr)$ using the specific ion interaction theory. The data are taken from Blake *et al.* [56BLA/COL] and refer to an aqueous $NaClO_4$ medium. The shaded area represents the uncertainty range obtained by propagating the resulting uncertainties at $I = 0$ back to $I = 4$ m.

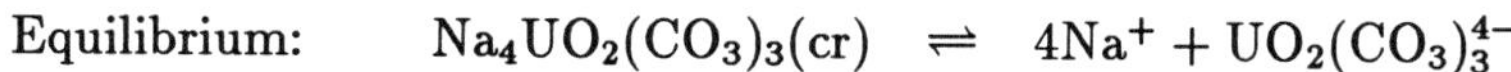

$$\text{Equilibrium:} \quad Na_4UO_2(CO_3)_3(cr) \rightleftharpoons 4Na^+ + UO_2(CO_3)_3^{4-}$$

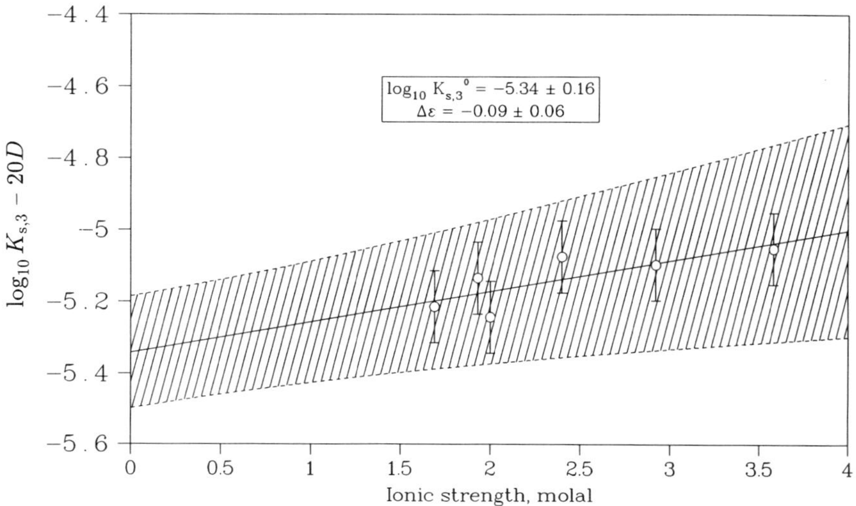

It should be mentioned that the resulting slope, $\Delta\varepsilon(\text{V.174}) = -(0.09 \pm 0.06)$, leads to $\varepsilon_{(UO_2(CO_3)_3^{4-},Na^+)} = -(0.13 \pm 0.07)$ which differs significantly from the value (0.08 ± 0.11) obtained from the regression in Figure V.16 and selected by this review.

The Gibbs energies and enthalpies of formation of liebigite, swartzite, bayleyite and andersonite were reported by Alwan and Williams [80ALW/WIL] based on solubility determinations at five different temperatures in the range 0 to 25°C. There are no experimental details in Ref. [80ALW/WIL] and no information about the source of auxiliary data used to obtain the published standard enthalpies and Gibbs energies of formation. There is an obvious error in the published data for andersonite, which results in a solubility product for this compound that differs widely from the others. This is not in agreement with the data in Figure 1 of Ref. [80ALW/WIL]. This review therefore does not select any data for liebigite, swartzite, bayleyite and andersonite.

It should be mentioned that Weigel [85WEI, Table 5] listed the stoichiometry and crystallographic characteristics of many more compounds of the general composition $M^{I}_{2y}M^{II}_{2-y}UO_2(CO_3)_3{\cdot}xH_2O(cr)$ $(y = 0 \text{ to } 2)$.

$M_2UO_2(CO_3)_2 \cdot xH_2O(cr)$ and $MUO_2(CO_3)_2 \cdot xH_2O(cr)$: Solids of the above composition were listed in the compilation of Weigel [85WEI, Table 1]. Only the two cal-

cium compounds $CaUO_2(CO_3)_2{\cdot}5H_2O(cr)$ and $CaUO_2(CO_3)_2{\cdot}3H_2O(cr)$ were crystallographically characterized. No thermodynamic data exist, only a statement that the compounds $CaUO_2(CO_3)_2{\cdot}10H_2O(cr)$, $BaUO_2(CO_3)_2{\cdot}2H_2O(cr)$ and $M_2UO_2(CO_3)_2{\cdot}xH_2O(cr)$ are very insoluble in water. The original information comes from Ref. [66CHE], which is not available to the reviewers.

Basic U(VI) carbonate phases: A number of basic dioxouranium(VI) carbonate phases have been described in the literature. A survey was given in the Gmelin Handbook [83BAG/JAC]. Structure characterizations were made of the compounds $MUO_2(CO_3)OH{\cdot}3H_2O(cr)$, M = NH_4, $\frac{1}{2}$Ba, Ag, Tl; $M(UO_2)_2(CO_3)_3OH{\cdot}5H_2O(cr)$, M = NH_4, $\frac{1}{2}$Ba, Ag, Tl; and of the mineral rabbitite (see Section V.9.3.3.2). No thermodynamic data have been reported for any of these compounds.

Mixed U(VI) carbonate phases: Bagnall, Jacob and Potter [83BAG/JAC] reported structure data for the compounds $Ca_3NaUO_2(CO_3)_3F(SO_4){\cdot}4H_2O(cr)$, $Ca_3NaUO_2(CO_3)_3F(SO_4){\cdot}10H_2O(cr)$ (schroeckingerite), as well as for peroxo-carbonatodioxouranate(VI) compounds, and fluoride containing phases, *i.e.*, $K_3UO_2(CO_3)F_3(cr)$, $CsUO_2(CO_3)F{\cdot}H_2O(cr)$ and $BaUO_2(CO_3)F_2{\cdot}2H_2O(cr)$. No thermodynamic data are available for any of these compounds.

V.7.1.3. Uranium bicarbonates

No simple U(IV), U(V) or U(IV) bicarbonate aqueous species or solids are known. The oxy- and hydroxy-complexes $UO_2CO_3OH^-$ and $(UO_2)_3(OH)_5CO_2^+ \triangleq (UO_2)_3O(OH)_2(HCO_3)^+$ are treated as carbonate species and are discussed in Section V.7.1.2.1.b. Bagnall, Jacob and Potter [83BAG/JAC] reported the existence of four uranium(IV) fluorobicarbonate phases, $M_2[UF_4(HCO_3)_2]$, M = Na, NH_4. However, no structural data have been reported for any of these compounds.

V.7.1.4. Uranium thiocyanates

V.7.1.4.1. Aqueous uranium thiocyanates

a) Aqueous U(VI) thiocyanate complexes

Several accurate studies of equilibria and enthalpy changes in the dioxouranium(VI) thiocyanate system have been reported. The equilibrium data of the reactions are compiled in Table V.44. The most precise ones are those of Ahrland [49AHR2] and Ahrland and Kullberg [71AHR/KUL3]. The key references are discussed in Appendix A. The experimental values of $\log_{10}\beta_q$(V.175) for the formation reactions

$$UO_2^{2+} + qSCN^- \;=\; UO_2(SCN)_q^{2-q}, \qquad (V.175)$$

where q = 1, 2 and 3, are corrected to $I = 0$ using the specific ion interaction theory. $\Delta\varepsilon(V.175, q = 1) = -(0.25 \pm 0.02)$ is adopted from the corresponding reaction with Cl^-, $\Delta\varepsilon(V.175, q = 2) = -(0.56 \pm 0.04)$ is calculated from the values listed

Table V.44: Experimental equilibrium data for the dioxouranium(VI) thiocyanate system.

Method	Ionic Medium	t (°C)	$\log_{10}\beta_q^{(a)}$	$\log_{10}\beta_q^{o(b)}$	Reference
$UO_2^{2+} + SCN^- \rightleftharpoons UO_2SCN^+$					
sp	1 M $NaClO_4$	20	0.76 ± 0.03	1.32 ± 0.05	[49AHR2]
sp	1 M $NaClO_4$	25	0.93 ± 0.03	1.49 ± 0.05	[57DAV/MON]
cix	1 M $NaClO_4$	32	-1.3		[61BAN/TRI]
sp	4 M $NaNO_3$	22	0.71 ± 0.03		[64VAS/MUK]
	2.5 M $NaNO_3$	22	0.72 ± 0.03	1.17 ± 0.07	
cal	1.0 M (?H)ClO_4	25	0.75 ± 0.03	1.31 ± 0.10	[71AHR/KUL3]
pot	0.1 M KNO_3	25	1.19 ± 0.03	1.60 ± 0.03	[71POL/BAR]
emf	0.33 M KNO_3	21	0.90 ± 0.10	1.49 ± 0.10	[75BAR/SEN]
....					
$UO_2^{2+} + 2SCN^- \rightleftharpoons UO_2(SCN)_2(aq)$					
sp	1 M $NaClO_4$	20	0.74 ± 0.08	1.43 ± 0.14	[49AHR2]
cix	1 M $NaClO_4$	32	1.05		[61BAN/TRI]
sp	4 M $NaNO_3$	22	0.72 ± 0.11		[64VAS/MUK]
	2.5 M $NaNO_3$	22	0.70 ± 0.11	0.85 ± 0.18	
cal	1.0 M $HClO_4$	25	0.72 ± 0.10	1.41 ± 0.11	[71AHR/KUL3]
pot	0.1 M KNO_3	25	2.76 ± 0.12		[71POL/BAR]
....					
$UO_2^{2+} + 3SCN^- \rightleftharpoons UO_2(SCN)_3^-$					
sp	1 M $NaClO_4$	20	1.71 ± 0.14	2.36 ± 0.16	[49AHR2]
cix	1 M $NaClO_4$	32	1.08		[61BAN/TRI]
cal	1.0 M $HClO_4$	25	1.18 ± 0.18	1.83 ± 0.21	[71AHR/KUL3]

(a) The values of $\log_{10}\beta_q$ are the equilibrium constants for the formation of $UO_2(SCN)_q^{2-q}$ at the temperature and the ionic strength given in the table
(b) $\log_{10}\beta_q^o$ is the corresponding value corrected to $I = 0$ at 298.15 K.
(c) Corrections for the formation of $UO_2NO_3^+$ have been made by using $\log_{10}\beta_1^o(UO_2NO_3^+) = 0.0$.

in Appendix B, and $\Delta\varepsilon(\text{V.175}, q = 3) = -(0.61 \pm 0.07)$ is calculated assuming that $\varepsilon_{(\text{UO}_2(\text{SCN})_3^-,\text{Na}^+)} \approx \varepsilon_{(\text{UO}_2\text{F}_3^-,\text{Na}^+)} = (0.00 \pm 0.05)$. The experiments performed in nitrate media are corrected for NO_3^- complexation of UO_2^{2+} using the selected formation constant in this review in the relevant ionic medium. The uncertainties in the constants turn out to be rather small, and the constants do not seem to belong to the same parent distributions for all the cases of $q = 1$, 2 and 3. This review therefore calculates the unweighted averages for the three constants and assigns an uncertainty to each that spans the ranges of the uncertainties for the constant.[13]

$$\begin{aligned}
\log_{10}\beta_1^\circ(\text{V.175}, q = 1, 298.15\,\text{K}) &= 1.40 \pm 0.23\\
\log_{10}\beta_2^\circ(\text{V.175}, q = 2, 298.15\,\text{K}) &= 1.24 \pm 0.55\\
\log_{10}\beta_3^\circ(\text{V.175}, q = 3, 298.15\,\text{K}) &= 2.1 \pm 0.5
\end{aligned}$$

The only source of enthalpy information on the dioxouranium(VI) thiocyanate system is the calorimetric study of Ahrland and Kullberg [71AHR/KUL3]. This is a precise study performed at $I = 1$ M ($NaClO_4$). This review estimates that the ionic strength dependence of the reported values is small, similarly to the cases of the UO_2^{2+} sulphate and carbonate complexes. The enthalpies of reaction selected at $I = 0$ are therefore assumed to be the same as those reported by Ahrland and Kullberg [71AHR/KUL3] for $I = 1$ M ($NaClO_4$). The uncertainties reported in Ref. [71AHR/KUL3] are doubled by this review to account for the uncertainty in the assumptions made.

$$\begin{aligned}
\Delta_r H_m^\circ(\text{V.175}, q = 1, 298.15\text{ K}) &= (3.22 \pm 0.06)\,\text{kJ}\cdot\text{mol}^{-1}\\
\Delta_r H_m^\circ(\text{V.175}, q = 2, 298.15\text{ K}) &= (8.9 \pm 0.6)\,\text{kJ}\cdot\text{mol}^{-1}\\
\Delta_r H_m^\circ(\text{V.175}, q = 3, 298.15\text{ K}) &= (6.0 \pm 1.2)\,\text{kJ}\cdot\text{mol}^{-1}
\end{aligned}$$

This review uses the US NBS data [82WAG/EVA] for SCN^- (*cf.* Chapter VI) to calculate the thermodynamic data of formation of the dioxouranium(VI) thiocyanate complexes. The entropies are calculated via $\Delta_r S_m^\circ$ from the selected equilibrium constants and enthalpies of reaction. The selected data are listed below.

$$\begin{aligned}
\Delta_f G_m^\circ(\text{UO}_2\text{SCN}^+, \text{aq}, 298.15\,\text{K}) &= -(867.8 \pm 4.6)\,\text{kJ}\cdot\text{mol}^{-1}\\
\Delta_f G_m^\circ(\text{UO}_2(\text{SCN})_2, \text{aq}, 298.15\,\text{K}) &= -(774.2 \pm 8.8)\,\text{kJ}\cdot\text{mol}^{-1}\\
\Delta_f G_m^\circ(\text{UO}_2(\text{SCN})_3^-, \text{aq}, 298.15\,\text{K}) &= -(686.4 \pm 12.5)\,\text{kJ}\cdot\text{mol}^{-1}
\end{aligned}$$

$$\begin{aligned}
\Delta_f H_m^\circ(\text{UO}_2\text{SCN}^+, \text{aq}, 298.15\,\text{K}) &= -(939.4 \pm 4.3)\,\text{kJ}\cdot\text{mol}^{-1}\\
\Delta_f H_m^\circ(\text{UO}_2(\text{SCN})_2, \text{aq}, 298.15\,\text{K}) &= -(857.3 \pm 8.2)\,\text{kJ}\cdot\text{mol}^{-1}\\
\Delta_f H_m^\circ(\text{UO}_2(\text{SCN})_3^-, \text{aq}, 298.15\,\text{K}) &= -(783.8 \pm 12.2)\,\text{kJ}\cdot\text{mol}^{-1}
\end{aligned}$$

$$\begin{aligned}
S_m^\circ(\text{UO}_2\text{SCN}^+, \text{aq}, 298.15\,\text{K}) &= (84 \pm 20)\,\text{J}\cdot\text{K}^{-1}\cdot\text{mol}^{-1}\\
S_m^\circ(\text{UO}_2(\text{SCN})_2, \text{aq}, 298.15\,\text{K}) &= (244 \pm 40)\,\text{J}\cdot\text{K}^{-1}\cdot\text{mol}^{-1}\\
S_m^\circ(\text{UO}_2(\text{SCN})_3^-, \text{aq}, 298.15\,\text{K}) &= (395 \pm 58)\,\text{J}\cdot\text{K}^{-1}\cdot\text{mol}^{-1}
\end{aligned}$$

[13] The reviewer (L.C.) has noted that by using a modified specific ion interaction theory [90CIA] the following values are obtained: $\log_{10}\beta_1^\circ(\text{V.175}, q = 1, 298.15\,\text{K}) = 1.38$ and $\log_{10}\beta_2^\circ(\text{V.175}, q = 2, 298.15\,\text{K}) = 1.54$.

b) Aqueous U(V) thiocyanate complexes

No experimental information is available on aqueous dioxouranium(V) thiocyanate complexes.

c) Aqueous U(IV) thiocyanate complexes

There are only two experimental determinations of equilibrium constants in the uranium(IV) thiocyanate system, a potentiometric study by Ahrland and Larsson [54AHR/LAR2] and an extraction study by Day, Wilhite and Hamilton [55DAY/WIL]. The agreement between these two investigations is satisfactory (see Table V.45).

$$U^{4+} + q\mathrm{SCN}^- \rightleftharpoons \mathrm{U(SCN)}_q^{4-q} \qquad \text{(V.176)}$$

The experimental data for $q = 1$ and 2 are corrected to $I = 0$ using the specific ion interaction theory. $\Delta\varepsilon(\text{V.176}, q = 1) = -(0.13 \pm 0.05)$ is adopted from the corresponding reaction with Cl^-, and $\Delta\varepsilon(\text{V.176}, q = 2) = -(0.56 \pm 0.14)$ is calculated from the values listed in Appendix B, assuming that $\varepsilon_{(\mathrm{U(SCN)}_2^{2+},\mathrm{ClO}_4^-)} \approx \varepsilon_{(\mathrm{UF}_2^{2+},\mathrm{ClO}_4^-)} = (0.3 \pm 0.1)$. The corrected values agree well, and this review selects the weighted averages of the data from Refs. [54AHR/LAR2, 55DAY/WIL].[14]

$$\begin{aligned}\log_{10}\beta_1^\circ(\text{V.176}, q = 1, 298.15\,\mathrm{K}) &= 2.97 \pm 0.06\\ \log_{10}\beta_2^\circ(\text{V.176}, q = 2, 298.15\,\mathrm{K}) &= 4.26 \pm 0.18\end{aligned}$$

The Gibbs energies of formation are derived using the auxiliary data selected in this review.

$$\begin{aligned}\Delta_f G_m^\circ(\mathrm{USCN}^{3+}, \mathrm{aq}, 298.15\,\mathrm{K}) &= -(454.1 \pm 4.4)\,\mathrm{kJ \cdot mol^{-1}}\\ \Delta_f G_m^\circ(\mathrm{U(SCN)}_2^{2+}, \mathrm{aq}, 298.15\,\mathrm{K}) &= -(368.8 \pm 8.3)\,\mathrm{kJ \cdot mol^{-1}}\end{aligned}$$

Ahrland and Larsson [54AHR/LAR2] suggested the formation of $\mathrm{U(SCN)}_3^+$ and proposed $\log_{10}\beta_3^\circ(\text{V.176}, q = 3, \mathrm{I} = 1.0\ \mathrm{M}) = 2.2$. On the other hand, Day, Wilhite and Hamilton [55DAY/WIL], who used even higher SCN^- concentrations, found no evidence for this complex. This review concurs with Ahrland and Larsson [54AHR/LAR2] that a $\mathrm{U(SCN)}_3^+$ complex may form. However, the quantitative information given by them is insufficient to select reliable data for this species.

Day, Wilhite and Hamilton [55DAY/WIL] determined the enthalpy changes for the Reactions (V.176) from the temperature dependence of the corresponding equilibrium constants. This review accepts their results with the uncertainties estimated to reflect the assumption that the values for $I = 2$ M are valid at $I = 0$.

[14] The reviewer (L.C.) has noted that by using a modified specific ion interaction theory [90CIA] the following values are obtained: $\log_{10}\beta_1^\circ(\text{V.176}, q = 1, 298.15\,\mathrm{K}) = 2.69$ and $\log_{10}\beta_2^\circ(\text{V.176}, q = 2, 298.15\,\mathrm{K}) = 4.27$.

Table V.45: Experimental equilibrium data for the uranium(IV) thiocyanate system.

Method	Ionic Medium	t (°C)	$\log_{10}\beta_q^{(a)}$	$\log_{10}\beta_q^{\circ(b)}$	Reference
$U^{4+} + SCN^- \rightleftharpoons USCN^{3+}$					
emf	0.6 M $HClO_4$ +0.4 M $NaClO_4$	20	1.49 ± 0.03	2.97 ± 0.06	[54AHR/LAR2]
dis	1.00 M $HClO_4$	10	1.78		[55DAY/WIL]
	1.00 M $NaClO_4$	25	$1.49 \pm 0.20^{(c)}$	2.97 ± 0.21	
	1.00 M $NaClO_4$	40	1.30		
....					
$U^{4+} + 2SCN^- \rightleftharpoons U(SCN)_2^{2+}$					
emf	0.6 M $HClO_4$ +0.4 M $NaClO_4$	20	1.95 ± 0.17	4.20 ± 0.22	[54AHR/LAR2]
dis	1.00 M $HClO_4$	10	2.30		[55DAY/WIL]
	1.00 M $NaClO_4$	25	$2.11 \pm 0.30^{(c)}$	4.36 ± 0.30	
	1.00 M $NaClO_4$	40	1.98		
....					
$U^{4+} + 3SCN^- \rightleftharpoons U(SCN)_3^{+}$					
emf	0.6 M $HClO_4$ +0.4 M $NaClO_4$	20	2.2		[54AHR/LAR2]

(a) $\log_{10}\beta_q$ refers to the equilibrium and the ionic strength given in the table.
(b) $\log_{10}\beta_q^\circ$ is the corresponding value corrected to $I = 0$ at 298.15 K.
(c) Uncertainty estimated by this review.

$$\Delta_r H_m^\circ(\text{V.176}, q=1, 298.15\text{ K}) = -(27 \pm 8)\text{ kJ}\cdot\text{mol}^{-1}$$
$$\Delta_r H_m^\circ(\text{V.176}, q=2, 298.15\text{ K}) = -(18 \pm 4)\text{ kJ}\cdot\text{mol}^{-1}$$

These values are used to derive the enthalpies of formation of the two complexes $USCN^{3+}$ and $U(SCN)_2^{2+}$:

$$\Delta_f H_m^\circ(\text{USCN}^{3+}, \text{aq}, 298.15\text{ K}) = -(541.8 \pm 9.5)\text{ kJ}\cdot\text{mol}^{-1}$$
$$\Delta_f H_m^\circ(\text{U(SCN)}_2^{2+}, \text{aq}, 298.15\text{ K}) = -(456.4 \pm 9.5)\text{ kJ}\cdot\text{mol}^{-1}.$$

Using the selected reaction data, the entropies are calculated via $\Delta_r S_m^\circ$.

$$S_m^\circ(\text{USCN}^{3+}, \text{aq}, 298.15\text{ K}) = -(306 \pm 35)\text{ J}\cdot\text{K}^{-1}\cdot\text{mol}^{-1}$$
$$S_m^\circ(\text{U(SCN)}_2^{2+}, \text{aq}, 298.15\text{ K}) = -(107 \pm 42)\text{ J}\cdot\text{K}^{-1}\cdot\text{mol}^{-1}$$

V.7.1.4.2. Solid uranium thiocyanates

No thermodynamic data are available on any solid uranium thiocyanate compounds.

V.7.2. Silicon compounds and complexes

V.7.2.1. Aqueous uranium silicates

The only experimental information on uranium silicate complexes refers to the reaction

$$UO_2^{2+} + Si(OH)_4(aq) \rightleftharpoons UO_2SiO(OH)_3^+ + H^+ \qquad (V.177)$$

with $\log_{10}{}^*K(V.177) = -(1.98 \pm 0.13)$ at 25°C and in 0.2 M $NaClO_4$ solution [71POR/WEB]. The experimental data clearly indicate that a complex formation takes place. However, it was not demonstrated by Porter and Weber [71POR/WEB] that the ligand really is $SiO(OH)_3^-$. The method of preparing the ligand solution indicates that considerable quantities of polysilicates might be present. From the known value of the first dissociation constant of $Si(OH)_4(aq)$, *cf.* Chapter VI, it can be calculated that $\log_{10} K(V.178) = -1.98 + 9.55 = 7.57$.

$$UO_2^{2+} + SiO(OH)_3^- \rightleftharpoons UO_2SiO(OH)_3^+ \qquad (V.178)$$

This value is unexpectedly large and indicates that the ligand must have a larger negative charge and that a chelate complex must be formed. Both these facts suggest that the complex contains a polynuclear silicate ligand. The experimental information on aqueous uranium silicate species is not sufficiently precise to define a stoichiometry and to evaluate an equilibrium constant. However, aqueous uranium silicate complexes may well exist in groundwater systems, and additional experimental investigations are necessary to decide the type of complexes formed and their stabilities.

V.7.2.2. Solid uranium silicates

V.7.2.2.1. $(UO_2)_2SiO_4 \cdot 2H_2O(cr)$

In the uranium(VI)-silicate mineral group, $(UO_2)_2SiO_4 \cdot 2H_2O(cr)$, soddyite, is the only compound reported [81STO/SMI]. Thermodynamic data for this solid phase have not been reported.

V.7.2.2.2. $USiO_4(cr)$

Coffinite, $USiO_4(cr)$, is a relatively abundant mineral in reduced sedimentary uranium ore deposits. This mineral generally forms small crystals ($< 5\mu m$), and is almost always associated with amorphous $USiO_4$, uraninite and auxiliary minerals. Coffinite minerals have been synthesized only with difficulty because many particular conditions are necessary: reducing media, basic pH ($7 < pH < 10$), solutions rich in dissolved silicate. Coffinite minerals are always obtained in association with $UO_2(cr)$ and $SiO_2(cr)$ [59FUC/HOE]. Therefore, it is very difficult to determine thermodynamic data for pure coffinite experimentally. Brookins [75BRO] was the first author to estimate a value for the Gibbs energy of formation of coffinite, $\Delta_f G_m^\circ(USiO_4, cr, 298.15\,K) = -1907.9\,kJ \cdot mol^{-1}$. From this value the aqueous equilibrium between coffinite and uraninite according to Eq. (V.179)

$$USiO_4(cr) + 2H_2O(l) \rightleftharpoons UO_2(cr) + Si(OH)_4(aq) \qquad (V.179)$$

should be established at $10^{-6.9}$ M dissolved silica (8 ppb as SiO_2). This value is much lower than the average value (17 ppm) found in most groundwaters. Consequently, according to Brookin's estimate, coffinite should be more stable than uraninite in natural waters. This result seems to be unrealistic because uraninite is more abundant than coffinite. Langmuir [78LAN] assumed an average silica concentration of 10^{-3} M (60 ppm as SiO_2) for the coffinite-uraninite equilibrium according to Eq. (V.179). This assumption (with auxiliary data from Ref. [78LAN]) leads to $\Delta_f G_m^\circ(USiO_4, cr, 298.15\,K) = -1882.4\,kJ \cdot mol^{-1}$ [80LAN/CHA] (the value of $-1891.2\,kJ \cdot mol^{-1}$ reported by Langmuir [78LAN] was in error). This estimated value [80LAN/CHA] seems to be more reasonable than that suggested by Brookins [75BRO]. Therefore, this review accepts Langmuir's assumption, and calculates

$$\Delta_f G_m^\circ(USiO_4, cr, 298.15\,K) = -(1883.6 \pm 4.0)\,kJ \cdot mol^{-1}$$

based on auxiliary data presented in Tables III.1 and IV.1. The uncertainty is estimated by this review.

Langmuir [78LAN] estimated the entropy of this compound as the sum of the values for quartz and $UO_2(cr)$. Using data for these two compounds as selected in this review, this entropy value is reestimated and an uncertainty of $\pm 10\%$ is assigned to it.

$$S_m^\circ(USiO_4, cr, 298.15\,K) = (118 \pm 12)\,J \cdot K^{-1} \cdot mol^{-1}$$

The enthalpy of formation is calculated from the selected Gibbs energy of formation and entropy.

$$\Delta_f H_m^\circ(USiO_4, cr, 298.15\,K) = -(1991.3 \pm 5.4)\,kJ \cdot mol^{-1}$$

V.7.2.2.3. $USiO_4$ (am)

Amorphous $USiO_4$ is often observed together with crystalline coffinite and uraninite in several reduced sedimentary uranium ore deposits. No experimental thermodynamic data of amorphous $USiO_4$ appeared to be reported. Galloway and Kaiser [80GAL/KAI] estimated the Gibbs energy difference between crystalline and amorphous $USiO_4$ to be similar to that between crystalline and amorphous UO_2. In this way, using $\Delta_f G_m^\circ(UO_2, cr, 298.15\,K) - \Delta_f G_m^\circ(UO_2, am, 298.15\,K) = -31.8\,kJ \cdot mol^{-1}$, they calculated [80GAL/KAI] $\Delta_f G_m^\circ(USiO_4, am, 298.15\,K) = -1850.6\,kJ \cdot mol^{-1}$.

As discussed in Section V.3.3.2.2, the value of the Gibbs energy of formation of an "amorphous" form of UO_2 is specific to a particular sample, and may be 20 to 50 kJ·mol^{-1} less stable than UO_2(cr). However, the evidence for such a large difference in Gibbs energies between crystalline and amorphous $USiO_4$ is not conclusive, and in this review no chemical thermodynamic parameters for $USiO_4$(am) are selected.

V.7.3. Lead compounds

In the lead-uranium(VI)-silicate group, the only mineral identified is kasolite, $PbUO_2SiO_4 \cdot H_2O$(cr). The exact formula of this compound was determined by Rosenzweig and Ryan [77ROS/RYA]. Thermodynamic data of this mineral have not been reported.

V.8. Actinide complexes

V.8.1. Actinide-actinide interactions

The tendency of actinide(V) cations, MO_2^+, to interact with certain cations (mostly multicharged) was first identified by Sullivan, Hindman and Zielen [61SUL/HIN], who found that, in acid solution, NpO_2^+ ions formed complexes with UO_2^{2+} ions. In the years following this discovery, many publications appeared and cation-cation interactions of the actinides were studied by an array of techniques, mainly absorption spectrophotometry, proton spin relaxation, potentiometric techniques and later Raman spectroscopy. All relevant studies before 1977 were extensively reviewed by [79FRO/RYK].

Interactions of UO_2^+-UO_2^{2+} (1:1 complexes) were reported [65NEW/BAK] using spectrophotometry over the temperature range 273 to 298.15 K at $I = 2$ M ($NaClO_4$/$HClO_4$), and these results yielded values for the enthalpy and entropy changes due to the association.

The interaction of NpO_2^+ with cations was studied very extensively, and quantitative data appeared, among others, on the NpO_2^+-UO_2^{2+} system [61SUL/HIN, 79MAD/GUI, 82GUI/BEG]. All these studies, carried out over a range of ionic strengths and media, concluded that weak complexes are formed, with no reported value above $\log_{10}\beta_1(\text{V.180}) = 0.4$.

$$MO_2^+ + M'O_2^{2+} \rightleftharpoons MM'O_4^{3+} \qquad (\text{V.180})$$

However, even restricted to perchlorate media, the data are too scattered to warrant generalization. Even comparing UO_2^+-UO_2^{2+} with NpO_2^+-UO_2^{2+} or NpO_2^+-NpO_2^{2+} is difficult. For $I = 2$ M ($NaClO_4$/$HClO_4$), $\log_{10}\beta_1(\text{V.180}, UO_2^+\text{-}UO_2^{2+}) = (1.22 \pm 0.02)$ was reported [65NEW/BAK], while other authors found $\log_{10}\beta_1$(V.180, NpO_2^+-UO_2^{2+}, 3 M (Na, Mg, H)ClO_4) $= -(0.16 \pm 0.01)$ [61SUL/HIN] and $\log_{10}\beta_1$(V.180, NpO_2^+-UO_2^{2+}, 7 M (Mg, H)ClO_4) $= (0.48 \pm 0.02)$ [79MAD/GUI]. It is interesting to note the identification of a NpO_2^+-NpO_2^+ dimeric species with $\log_{10}\beta_1 = -(0.09 \pm 0.03)$ [82GUI/BEG] observed by Raman spectroscopy in perchlorate media with $[NpO_2^+] >$ 0.2 M.

Similar complexes of AmO_2^+ with UO_2^{2+} and NpO_2^{2+} [81GUI/HOB] were reported for $I = 10$ M (perchlorate) and in media of variable ionic strength for the species AmO_2^+-UO_2^{2+}.

As it is clear from this discussion, no data can be recommended for any of these associations of cations.

V.8.2. Mixed U(VI), Np(VI) and Pu(VI) carbonate complexes

Carbonate is an excellent bridging ligand and the formation of polynuclear carbonate complexes containing one type of metal ion is well known. Such complexes may also contain two or more different types of metal ions. Known examples are carbonate complexes containing UO_2^{2+}, NpO_2^{2+} and PuO_2^{2+} as metal ions. One example is $(UO_2)_2(MO_2)(CO_3)_6^{6-}$, M = Np and Pu, studied by Grenthe, Riglet and Vitorge [86GRE/RIG]. They determined the equilibrium constants for the reactions

$$2UO_2(CO_3)_3^{4-} + MO_2(CO_3)_3^{4-} \rightleftharpoons (UO_2)_2(MO_2)(CO_3)_6^{6-} + 3CO_3^{2-} \quad M = Np, Pu \qquad (V.181)$$

at (22 ± 1)°C in 3 M $NaClO_4$ medium and reported $\log_{10} K(V.181) = -11.63$, -10.0 and -8.8 for M = U, Np and Pu, respectively. The extrapolation of these values to $I = 0$ is facilitated by the fact that the Debye-Hückel terms cancel. By assuming the same $\Delta\varepsilon$ for M = U, Np and Pu this review obtains

$$\log_{10} K^\circ(V.181, M = Np, 298.15\,K) = -10.0 \pm 0.1$$
$$\log_{10} K^\circ(V.181, M = Pu, 298.15\,K) = -8.8 \pm 0.2$$

where the uncertainties estimated by the authors [86GRE/RIG] are accepted by this review.

The standard Gibbs energies of formation of these complexes will be reported in the upcoming volumes on chemical thermodynamics of neptunium and plutonium, to be published in the NEA-TDB series.

V.9. Group 2 (alkaline-earth) compounds

V.9.1. Beryllium compounds

The only beryllium compound for which thermodynamic data have been reported to date is the crystalline binary alloy $Be_{13}U(cr)$.[15] All tabulated enthalpy data [63RAN/KUB, 81CHI/AKH, 82WAG/EVA, 83FUG, 86MOR] can be traced to the acid solution calorimetry determination of Ivanov and Tumbakov [59IVA/TUM]. Chiotti *et al.* [81CHI/AKH] recalculated their results with CODATA compatible auxiliary data to obtain

$$\Delta_f H_m^\circ(Be_{13}U, cr, 298.15\,K) = -(163.6 \pm 17.5)\,kJ \cdot mol^{-1}.$$

[15] Note that $UBe_3(cr)$ reported by Morss [86MOR] is a typographical error for $Be_{13}U(cr)$.

They reported the heat capacity measurements of both Farkas and Eldridge [68FAR/ELD] and Samarukov *et al.* [74SAM/KOS], which are in good agreement ($\Delta < 7\,\mathrm{J\cdot K^{-1}\cdot mol^{-1}}$) at 298.15 K, but show no preference. This review accepts

$$C^{\circ}_{p,\mathrm{m}}(\mathrm{Be_{13}U, cr, 298.15\,K}) = (242.3 \pm 4.2)\,\mathrm{J\cdot K^{-1}\cdot mol^{-1}}$$

from Samarukov *et al.* [74SAM/KOS], as well as the corresponding value

$$S^{\circ}_{\mathrm{m}}(\mathrm{Be_{13}U, cr, 298.15\,K}) = (180.1 \pm 3.3)\,\mathrm{J\cdot K^{-1}\cdot mol^{-1}},$$

since the data from Farkas and Eldridge [68FAR/ELD] at high temperatures are somewhat higher than expected from a smooth extrapolation of the low-temperature data.

The Gibbs energy of formation is calculated from the selected enthalpy of formation and entropy.

$$\Delta_{\mathrm{f}}G^{\circ}_{\mathrm{m}}(\mathrm{Be_{13}U, cr, 298.15\,K}) = -(165.5 \pm 17.5)\,\mathrm{kJ\cdot mol^{-1}}$$

Systematics on the alkaline-earth uranates [85FUG] indicate that $BeUO_4(cr)$ is not stable towards its decomposition to the oxides.

V.9.2. *Magnesium compounds*

V.9.2.1. *Magnesium uranates*

V.9.2.1.1. *$MgUO_4$ (cr)*

O'Hare *et al.* [77OHA/BOE] reported the heats of reaction of $MgUO_4(cr)$ with 12 M HNO_3 and with an HNO3/HF mixture. From these and CODATA consistent auxiliary data (see Appendix A) the recommended value

$$\Delta_{\mathrm{f}}H^{\circ}_{\mathrm{m}}(\mathrm{MgUO_4, cr, 298.15\,K}) = -(1857.3 \pm 1.5)\,\mathrm{kJ\cdot mol^{-1}},$$

is calculated based on the weighted average of the results from the two different experimental cycles.

The heat capacity and entropy values selected by Cordfunke and O'Hare [78COR/OHA] are $S^{\circ}_{\mathrm{m}}(\mathrm{MgUO_4, cr, 298.15\,K}) = 131.9\,\mathrm{J\cdot K^{-1}\cdot mol^{-1}}$ and $C^{\circ}_{p,\mathrm{m}}(\mathrm{MgUO_4, cr, 298.15\,K}) = 128.1\,\mathrm{J\cdot K^{-1}\cdot mol^{-1}}$. These extensively quoted data were based on the unpublished value of Westrum and Magadanz [75WES/MAG], which was subsequently superseded by the low-temperature measurements of Westrum, Zainel and Jakes [80WES/ZAI]. The entropy value

$$S^{\circ}_{\mathrm{m}}(\mathrm{MgUO_4, cr, 298.15\,K}) = (131.95 \pm 0.17)\,\mathrm{J\cdot K^{-1}\cdot mol^{-1}}$$

in the latter reference is virtually identical, and is accepted in this review. However, two different values for the heat capacity are given. One, from Ref. [80WES/ZAI, Table III], is $C^{\circ}_{p,\mathrm{m}}(\mathrm{MgUO_4, cr, 300.0\,K}) = 128.4\,\mathrm{J\cdot K^{-1}\cdot mol^{-1}}$. The other, from [80WES/ZAI, Table VI], is $C^{\circ}_{p,\mathrm{m}}(\mathrm{MgUO_4, cr, 298.15\,K}) = 133.4\,\mathrm{J\cdot K^{-1}\cdot mol^{-1}}$. These values are inconsistent. The smaller value is in closer agreement with data from high-temperature enthalpy increments [67JAK/SCH, 77OHA/BOE]. Also, the authors showed in their figures that the heat capacity of the magnesium compound falls between those of the calcium and strontium compounds. The values from their Table III (but not their Table VI) are consistent with this, and in this review it is concluded that the larger value (133.4 $kJ \cdot mol^{-1}$) is a typographical error. The correction from 300 to 298.15 K is estimated as $-(0.3 \pm 0.2)\,J \cdot K^{-1} \cdot mol^{-1}$, and hence,

$$C^{\circ}_{p,m}(MgUO_4, cr, 298.15\,K) = (128.1 \pm 0.3)\,J \cdot K^{-1} \cdot mol^{-1}$$

where the uncertainty is an estimate. This is in agreement with the value as selected by other reviewers, *e.g.*, [82WAG/EVA, 83FUG, 88PHI/HAL].

The Gibbs energy of formation is calculated from the selected enthalpy of formation and entropy.

$$\Delta_f G^{\circ}_m(MgUO_4, cr, 298.15\,K) = -(1749.6 \pm 1.5)\,kJ \cdot mol^{-1}$$

V.9.2.1.2. MgU_2O_6 *(cr) and* MgU_2O_7 *(cr)*

Golt'sev *et al.* [81GOL/TRE] estimated $\Delta_f H^{\circ}_m(MgU_2O_6, cr, 298.15\,K) = 3064\,kJ \cdot mol^{-1}$ and $\Delta_f H^{\circ}_m(MgU_2O_7, cr, 298.15\,K) = 3190\,kJ \cdot mol^{-1}$. This review assumes these values were intended to be negative. No confirmation of the existence of these compounds is available to this review, so they are not credited here.

V.9.2.1.3. MgU_3O_{10} *(cr)*

Cordfunke and O'Hare [78COR/OHA] also accepted the unpublished entropy and heat capacity values of Westrum and Magadanz [75WES/MAG]. Unlike the results for $MgUO_4$(cr), these do not appear to have ever been published. They are accepted here as

$$S^{\circ}_m(MgU_3O_{10}, cr, 298.15\,K) = (338.6 \pm 1.0)\,J \cdot K^{-1} \cdot mol^{-1}$$
$$C^{\circ}_{p,m}(MgU_3O_{10}, cr, 298.15\,K) = (305.6 \pm 5.0)\,J \cdot K^{-1} \cdot mol^{-1}.$$

Uncertainties are adjusted to be consistent with the discussion on $MgUO_4$(cr) above. No Gibbs energy or enthalpy data are available.

V.9.2.1.4. $Mg_3U_3O_{10}$ *(cr)*

The value reported by Fuger [83FUG] contains a typographical error, reporting MgU_3O_{10}(cr) as $Mg_3U_3O_{10}$(cr). The latter compound is not known.

V.9.2.2. *Magnesium-uranium sulphates*

Magnesium zippeite, $(Mg_2(UO_2)_6(SO_4)_3(OH)_{10} \cdot 8H_2O$(cr), is discussed with the other zippeites in Section V.5.1.3.2.c.

V.9.2.3. Magnesium-uranium carbonate: Bayleyite

Bayleyite is a hydrated magnesium-dioxouranium(VI) carbonate of the composition $Mg_2UO_2(CO_3)_3{\cdot}18H_2O(cr)$. The solubility study performed by Alwan and Williams [80ALW/WIL] is discussed under the solid uranium carbonates (Section V.7.1.2.2.b). The values reported there are not accepted in this review. No other thermodynamic data are available on bayleyite.

V.9.3. Calcium compounds

V.9.3.1. Calcium uranates

V.9.3.1.1. $CaUO_4$ (cr)

The chemical thermodynamic properties of rhombohedral [86WEI] calcium uranate have been extensively reported in the scientific literature. The values in the recent western literature are essentially the same, based on two precise studies. Early results from the Soviet Union, *e.g.*, [61IPP/FAU, 72KRE], are somewhat more positive and are not considered here.

O'Hare, Boerio and Hoekstra [76OHA/BOE] performed solution calorimetric measurements on well-characterized samples of $CaUO_4$(cr). Using auxiliary data consistent with the present review (see Appendix A), the value

$$\Delta_f H^\circ_m(\mathrm{CaUO_4, cr, 298.15\,K}) = -(2002.3 \pm 2.3)\,\mathrm{kJ \cdot mol^{-1}}$$

is calculated.

Westrum, Zainel and Jakes [80WES/ZAI] measured the low-temperature heat capacity of $CaUO_4$(cr) by adiabatic calorimetry from 5 K to 350 K, obtaining

$$C^\circ_{p,m}(\mathrm{CaUO_4, cr, 298.15\,K}) = (123.8 \pm 2.5)\,\mathrm{J \cdot K^{-1} \cdot mol^{-1}}$$

and calculating

$$S^\circ_m(\mathrm{CaUO_4, cr, 298.15\,K}) = (121.1 \pm 2.5)\,\mathrm{J \cdot K^{-1} \cdot mol^{-1}}.$$

These data are accepted here, with uncertainties increased to reflect absolute uncertainty as well as experimental error (*cf.* Appendix C).

Note that due to a typographical error in the abstract of Ref. [80WES/ZAI], Morss [86MOR] incorrectly cited these results. This error has been perpetuated by Phillips *et al.* [88PHI/HAL2], who reported values that are inconsistent with the defining equation, $\Delta_f G^\circ_m = \Delta_f H^\circ_m - T \sum_i S^\circ_{m,i}$.

The solution calorimetry data of O'Hare, Boerio and Hoekstra [76OHA/BOE] are combined with the entropy from [80WES/ZAI] to obtain the selected value of

$$\Delta_f G^\circ_m(\mathrm{CaUO_4, cr, 298.15\,K}) = -(1888.7 \pm 2.4)\,\mathrm{kJ \cdot mol^{-1}}.$$

This value is in fair agreement with the two Gibbs energy values of Tagawa, Fujino and Yamashita [79TAG/FUJ].

Leonidov, Rezukhina and Bereznikova [60LEO/REZ] report a discontinuity in the heat capacity of $CaUO_4$(cr) between 1022 and 1027 K. Naumov, Ryzhenko and Khodakovsky [71NAU/RYZ] report both as α-(low-temperature) phases, though no crystallographic description is given of either. Although this report accepted a heat capacity value at 298.15 K which is $5.8\,\text{J}\cdot\text{K}^{-1}\cdot\text{mol}^{-1}$ lower than the one reported in Ref. [60LEO/REZ], the difference of $0.92\,\text{J}\cdot\text{K}^{-1}\cdot\text{mol}^{-1}$ between the phases is accepted here. The selected values above refer to an α-phase, and this review therefore selects for the high-temperature phase (designated β-phase here) the value

$$C^\circ_{p,\text{m}}(\text{CaUO}_4, \beta, 298.15\,\text{K}) = (124.7 \pm 2.7)\,\text{J}\cdot\text{K}^{-1}\cdot\text{mol}^{-1}.$$

No other thermodynamic data are available for the β-phase.

V.9.3.1.2. CaU_2O_6(cr) and CaU_2O_7(cr)

All reported values for these compounds are derived from the enthalpy of formation estimated by Gol'tsev *et al.* [81GOL/TRE]. Neither the dioxouranium(V) nor dioxouranium(VI) oxide are well known [86WEI]. For this reason, this review does not credit the values for the compounds CaU_2O_6(cr) and CaU_2O_7(cr), similarly to the corresponding magnesium analogues, *cf.* Section V.9.2.1.2.

Note that Ref. [85PHI/PHI] contains a typographical error, reporting CaU_2O_7(cr) data as $Ca_2U_2O_7$(cr).

V.9.3.1.3. Ca_3UO_6(cr)

The enthalpy of formation of Ca_3UO_6(cr) was determined by Morss *et al.* [83MOR/WIL2] from solution calorimetry measurements on two well-characterized samples. The enthalpy of reaction data from this work is used with auxiliary data consistent with the present review (see Appendix A), to calculate

$$\Delta_f H^\circ_\text{m}(\text{Ca}_3\text{UO}_6, \text{cr}, 298.15\,\text{K}) = -(3305.4 \pm 4.1)\,\text{kJ}\cdot\text{mol}^{-1}$$

which is accepted here.

V.9.3.2. Calcium-uranium carbonates

V.9.3.2.1. Zellerite

Zellerite is a hydrated calcium-dioxouranium(VI) carbonate of the composition $CaUO_2(CO_3)_2\cdot 5H_2O$(cr). No thermodynamic data are available on zellerite.

V.9.3.2.2. Liebigite

Liebigite is a hydrated calcium-dioxouranium(VI) carbonate of the composition $Ca_2UO_2(CO_3)_3\cdot 10H_2O$(cr). The solubility study performed by Alwan and Williams [80ALW/WIL] is discussed under the solid uranium carbonates (Section V.7.1.2.2.b). The values reported there are not accepted in this review. No other thermodynamic data are available on liebigite.

V.9.3.3. Calcium-magnesium-uranium compounds

V.9.3.3.1. Swartzite

Swartzite is a hydrated calcium-magnesium-dioxouranium(VI) carbonate of the composition $CaMgUO_2(CO_3)_3{\cdot}12H_2O(cr)$. The solubility study performed by Alwan and Williams [80ALW/WIL] is discussed under the solid uranium carbonates (Section V.7.1.2.2.b). The values reported there are not accepted in this review. No other thermodynamic data are available on swartzite.

V.9.3.3.2. Rabbitite

Rabbitite is a hydrated calcium-magnesium-dioxouranium(VI) carbonate of the composition $Ca_3Mg_3(UO_2)_2(CO_3)_6(OH)_4{\cdot}18H_2O(cr)$. Hemingway [82HEM] estimated its Gibbs energy of formation using the method of Chen [75CHE], and obtained $\Delta_f G^\circ_m(Ca_3Mg_3(UO_2)_2(CO_3)_6(OH)_4{\cdot}18H_2O, cr, 298.15\ K) = -(13525 \pm 30)\ kJ \cdot mol^{-1}$. No experimental data are available to confirm this estimate, and this review does not select this value.

V.9.4. Strontium compounds

V.9.4.1. $SrUO_4(cr)$

Strontium uranate is known as both a rhombohedral (α-$SrUO_4$) and orthorombic (β-$SrUO_4$) crystal.

V.9.4.1.1. α-$SrUO_4$

Cordfunke and O'Hare [78COR/OHA] used the enthalpy of solution data of Cordfunke and Loopstra [67COR/LOO] for the reaction

$$\begin{aligned}\alpha\text{-}SrUO_4 + 4HNO_3(sln) &\rightleftharpoons Sr(NO_3)_2(sln) + UO_2(NO_3)_2(sln)\\ &\quad +2H_2O(sln) \qquad (V.182)\end{aligned}$$

in 6 M HNO_3 to calculate $\Delta_f H^\circ_m(SrUO_4, \alpha, 298.15\,K) = -(1985.3 \pm 2.1)\,kJ \cdot mol^{-1}$. This is based on $\Delta_f H^\circ_m(Sr(NO_3)_2, cr, 298.15\,K) = -(978.2 \pm 0.8)\,kJ \cdot mol^{-1}$ [71PAR/WAG] and auxiliary data for 6 M HNO_3 of unspecified origin. Morss *et al.* [83MOR/WIL2] recalculated this value with auxiliary data that was compatible with CODATA [89COX/WAG] (except for some of the uncertainties) and revised strontium data (that was almost the same as, but not identical to, that proposed in [84BUS/PLU]) to obtain $\Delta_f H^\circ_m(SrUO_4, \alpha, 298.15\,K) = -(1989.2 \pm 3.2)\,kJ \cdot mol^{-1}$. In this review, using data for the strontium species from Busenberg, Plummer and Parker [84BUS/PLU] ($\Delta_f H^\circ_m(Sr(NO_3)_2, cr, 298.15\,K) = -(982.36 \pm 0.80)\,kJ \cdot mol^{-1}$) and other auxiliary data that are compatible with CODATA,

$$\Delta_f H^\circ_m(SrUO_4, \alpha, 298.15\,K) = -(1989.6 \pm 2.8)\,kJ \cdot mol^{-1}$$

is calculated, which is recommended by this review.

Westrum, Zainel and Jakes [80WES/ZAI] measured the low-temperature heat capacity of α-$SrUO_4$, obtaining

$$S^\circ_m(SrUO_4, \alpha, 298.15\,K) = (153.15 \pm 0.17)\,J \cdot K^{-1} \cdot mol^{-1}$$
$$C^\circ_{p,m}(SrUO_4, \alpha, 298.15\,K) = (130.62 \pm 0.20)\,J \cdot K^{-1} \cdot mol^{-1}$$

which are recommended by this review. Note that the uncertainty in $C^\circ_{p,m}$ is estimated in this review.

The Gibbs energy of formation is calculated from the selected enthalpy of formation and entropy.

$$\Delta_f G^\circ_m(SrUO_4, \alpha, 298.15\,K) = -(1881.4 \pm 2.8)\,kJ \cdot mol^{-1}$$

V.9.4.1.2. β*-SrUO*$_4$

Cordfunke and Loopstra [67COR/LOO] also determined enthalpy of solution data for β-$SrUO_4$. In this review the enthalpy of formation is recalculated using a cycle similar to that described by Morss *et al.* [83MOR/WIL2] for Sr_3UO_6, but using auxiliary data from this review (see previous section on α-$SrUO_4$), obtaining

$$\Delta_f H^\circ_m(SrUO_4, \beta, 298.15\,K) = -(1990.8 \pm 2.8)\,kJ \cdot mol^{-1}.$$

Tagawa, Fujino and Yamashita [79TAG/FUJ] estimated S°_m($SrUO_4$, β, 298.15 K) = 155 $J \cdot K^{-1} \cdot mol^{-1}$ using data for strontium molybdates(VI). This review assigns this estimate an uncertainty of a magnitude which would make the result indistinguishable from the S°_m of α-$SrUO_4$, and therefore does not accept the estimate. Note that Morss [86MOR] reported values of S°_m and $C^\circ_{p,m}$ for β-$SrUO_4$, but these data actually pertain to α-$SrUO_4$. Thus, no additional reliable thermodynamic data are available for β-$SrUO_4$.

V.9.4.2. *SrU*$_4$*O*$_{13}$*(cr)*

Cordfunke and O'Hare [78COR/OHA] extrapolated the strontium uranate solubility data of Cordfunke and Loopstra [67COR/LOO] to SrU_4O_{13}(cr). Auxiliary data from this review are used (as for α-$SrUO_4$, above) to recalculate the value based on Ref. [67COR/LOO], obtaining

$$\Delta_f H^\circ_m(SrU_4O_{13}, cr, 298.15\,K) = -(5920 \pm 20)\,kJ \cdot mol^{-1}.$$

Note that Fuger [83FUG] performed a similar recalculation with unspecified auxiliary data, obtaining the same value.

No additional thermodynamic data are available.

V.9.4.3. Sr_2UO_5(cr)

Cordfunke and Loopstra [67COR/LOO] also determined the enthalpy of solution of Sr_2UO_5(cr). Auxiliary data from this review are used (as for α-$SrUO_4$, above) to recalculate the value, obtaining

$$\Delta_f H_m^\circ(Sr_2UO_5, cr, 298.15\,K) = -(2635.6 \pm 3.4)\,kJ \cdot mol^{-1}.$$

Note that Fuger [83FUG] performed a similar recalculation with unspecified auxiliary data, obtaining $\Delta_f H_m^\circ(Sr_2UO_5, cr, 298.15\,K) = -(2632.6 \pm 2.5)\,kJ \cdot mol^{-1}$. No additional thermodynamic data are available for this compound.

V.9.4.4. $Sr_2U_3O_{11}$(cr)

Cordfunke and Loopstra [67COR/LOO] also determined the enthalpy of solution of $Sr_2U_3O_{11}$(cr). Auxiliary data from this review are used (as for α-$SrUO_4$, above) to recalculate the value, obtaining

$$\Delta_f H_m^\circ(Sr_2U_3O_{11}, cr, 298.15\,K) = -(5242.9 \pm 4.1)\,kJ \cdot mol^{-1}.$$

Note that Fuger [83FUG] performed a similar recalculation with unspecified auxiliary data, obtaining $\Delta_f H_m^\circ(Sr_2U_3O_{11}, cr, 298.15\,K) = -(5240.9 \pm 3.3)\,kJ \cdot mol^{-1}$.

No additional primary thermodynamic data are available for this compound.

V.9.4.5. Sr_3UO_6(cr)

Cordfunke and Loopstra [67COR/LOO] also determined the enthalpy of solution of Sr_3UO_6(cr) in 6 M HNO_3. This value was recalculated by Morss *et al.* [83MOR/WIL2]. Using the same method, but slightly different auxiliary data (see the section on α-$SrUO_4$), this review does a further recalculation, obtaining $\Delta_f H_m^\circ(Sr_3UO_6, cr, 298.15\,K) = -(3262.7 \pm 4.3)\,kJ \cdot mol^{-1}$. Morss *et al.* [83MOR/WIL2] measured the enthalpy of solution of Sr_3UO_6(cr) in 1.0 M HCl, obtaining $\Delta_f H_m^\circ(Sr_3UO_6, cr, 298.15\,K) = -(3263.4 \pm 4.4)\,kJ \cdot mol^{-1}$. Using CODATA compatible auxiliary data the value $\Delta_f H_m^\circ(Sr_3UO_6, cr, 298.15\,K) = -(3264.0 \pm 4.2)\,kJ \cdot mol^{-1}$ is obtained from a recalculation in this review. The weighted average of the enthalpies from the two different sets of experiments is accepted here:

$$\Delta_f H_m^\circ(Sr_3UO_6, cr, 298.15\,K) = -(3263.4 \pm 3.0)\,kJ \cdot mol^{-1}.$$

The good agreement between the recalculated values of Morss *et al.* [83MOR/WIL2] and of Cordfunke and Loopstra [67COR/LOO] lends confidence in the recalculated values for the other strontium uranates in this section.

V.9.5. *Barium compounds*

V.9.5.1. *Barium uranates*

V.9.5.1.1. $BaUO_3$ *(cr)*

The only reported measurement to date of the thermodynamic properties of cubic $BaUO_3$(cr) is the solution calorimetry work done by Williams, Morss and Choi [84WIL/MOR]. They performed six enthalpy of solution measurements on two samples, $BaUO_{3.20}$(cr) and $BaUO_{3.06}$(cr), to obtain

$$\Delta_f H_m^\circ(BaUO_3, cr, 298.15\,K) = -(1690 \pm 10)\,kJ \cdot mol^{-1}$$

with CODATA compatible auxiliary data. This value is accepted here.

V.9.5.1.2. $BaUO_4$ *(cr)*

For the IAEA review, Cordfunke and O'Hare [78COR/OHA] accepted the experimentally based enthalpy of formation and estimated entropy of $BaUO_4$(cr) reported by O'Hare, Boerio and Hoekstra [76OHA/BOE]. Recalculation with more recent auxiliary data (see Appendix A) gives

$$\Delta_f H_m^\circ(BaUO_4, cr, 298.15\,K) = -(1993.8 \pm 3.3)\,kJ \cdot mol^{-1}.$$

Cordfunke and O'Hare [78COR/OHA] accepted $C_{p,m}^\circ$($BaUO_4$, cr, 298.15 K) = 133.5 $J \cdot K^{-1} \cdot mol^{-1}$ from the high-temperature heat capacity measurements of Leonidov, Rezukhina and Bereznikova [60LEO/REZ]. Westrum, Zainel and Jakes [80WES/ZAI] obtained a significantly different value of $C_{p,m}^\circ$($BaUO_4$, cr, 298.15 K) = 143.84 $J \cdot K^{-1} \cdot mol^{-1}$ from low-temperature adiabatic calorimetry measurements on a sample with no explicit chemical analysis reported. O'Hare, Flotow and Hoekstra [80OHA/FLO] performed similar experiments on a well-characterized sample, obtaining $C_{p,m}^\circ$($BaUO_4$, cr, 298.15 K) = $(125.27 \pm 0.25)\,J \cdot K^{-1} \cdot mol^{-1}$. They explain the discrepancy with Westrum, Zainel and Jakes [80WES/ZAI] by questioning the purity of their sample. The last value is consistent with the selected $C_{p,m}^\circ$ values selected for the other alkaline-earth monourantes, and is accepted here along with its low-temperature function. O'Hare *et al.* [80OHA/FLO] point out that this merges smoothly with the high-temperature curve from Leonidov, Rezukhina and Bereznikova [60LEO/REZ]. The value of S_m°($BaUO_4$, cr, 298.15 K) = $(153.97 \pm 0.31)\,J \cdot K^{-1} \cdot mol^{-1}$ was obtained from the $C_{p,m}(T)$ curve. The uncertainty of both values was raised to $\pm 2.5\,J \cdot K^{-1} \cdot mol^{-1}$ by this review and the accepted values were rounded accordingly to

$$C_{p,m}^\circ(BaUO_4, cr, 298.15\,K) = (125.3 \pm 2.5)\,J \cdot K^{-1} \cdot mol^{-1}$$
$$S_m^\circ(BaUO_4, cr, 298.15\,K) = (154.0 \pm 2.5)\,J \cdot K^{-1} \cdot mol^{-1}.$$

The Gibbs energy of formation is calculated from the selected enthalpy of formation and entropy.

$$\Delta_f G_m^\circ(BaUO_4, cr, 298.15\,K) = -(1883.8 \pm 3.4)\,kJ \cdot mol^{-1}$$

V.9.5.1.3. $BaU_2O_7(cr)$ and $Ba_2U_2O_7(cr)$

Cordfunke and Ouweltjes [88COR/OUW] measured the enthalpy of solution of BaU_2O_7(cr) and $Ba_2U_2O_7$(cr) in ($HCl+0.0419FeCl_3+70.66H_2O$). Using the auxiliary data described in Appendix A,

$$\begin{aligned}\Delta_f H^\circ_m(BaU_2O_7, \text{cr}, 298.15\text{ K}) &= -(3237.2 \pm 5.0)\,\text{kJ}\cdot\text{mol}^{-1}\\ \Delta_f H^\circ_m(Ba_2U_2O_7, \text{cr}, 298.15\text{ K}) &= -(3740.0 \pm 6.3)\,\text{kJ}\cdot\text{mol}^{-1}\end{aligned}$$

are obtained.

No experimental data for the entropy are available. Cordfunke and IJdo [88COR/IJD] estimated the entropies from phase diagram calculations, combined with experimental evidence. They obtained

$$\begin{aligned}S^\circ_m(BaU_2O_7, \text{cr}, 298.15\text{ K}) &= (260 \pm 15)\,\text{J}\cdot\text{K}^{-1}\cdot\text{mol}^{-1}\\ S^\circ_m(Ba_2U_2O_7, \text{cr}, 298.15\text{ K}) &= (296 \pm 15)\,\text{J}\cdot\text{K}^{-1}\cdot\text{mol}^{-1}.\end{aligned}$$

These values are selected and the uncertainties estimated by this review.

These values allow the calculation of the Gibbs energies of formation:

$$\begin{aligned}\Delta_f G^\circ_m(BaU_2O_7, \text{cr}, 298.15\text{ K}) &= -(3052.1 \pm 6.7)\,\text{kJ}\cdot\text{mol}^{-1}\\ \Delta_f G^\circ_m(Ba_2U_2O_7, \text{cr}, 298.15\text{ K}) &= -(3547.0 \pm 7.7)\,\text{kJ}\cdot\text{mol}^{-1}.\end{aligned}$$

V.9.5.1.4. $Ba_3UO_6(cr)$

Morss [82MOR] tabulates the thermodynamic properties of several alkaline-earth actinide(VI) oxides, citing Morss *et al.* [81MOR/FAH] as the source. However, Ref. [81MOR/FAH] contains no numerical results. Apparently Morss [82MOR] had access to preliminary results which were later published in detailed papers. Such is the case with Ba_3UO_6(cr). The preliminary result have been cited by some compilations (*e.g.*, [83FUG, 85PHI/PHI, 88PHI/HAL2]). The experimental work was published by Morss *et al.* [83MOR/WIL2] who determined the enthalpy of solution of Ba_3UO_6(cr) in 1 M HCl. These measurements are accepted and recalculated using the auxiliary data selected in this review. The selected value is

$$\Delta_f H^\circ_m(Ba_3UO_6, \text{cr}, 298.15\text{ K}) = -(3210.4 \pm 8.0)\,\text{kJ}\cdot\text{mol}^{-1}.$$

As for BaU_2O_7(cr) and $Ba_2U_2O_7$(cr), Cordfunke and IJdo [88COR/IJD] estimated the entropy from phase diagram calculations.

$$S^\circ_m(Ba_3UO_6, \text{cr}, 298.15\text{ K}) = (298 \pm 15)\,\text{J}\cdot\text{K}^{-1}\cdot\text{mol}^{-1}$$

The Gibbs energy of formation is calculated from the selected enthalpy of formation and entropy.

$$\Delta_f G^\circ_m(Ba_3UO_6, \text{cr}, 298.15\text{ K}) = -(3045.0 \pm 9.2)\,\text{kJ}\cdot\text{mol}^{-1}$$

V.9.5.2. Barium alkaline-earth compounds

Enthalpy of formation data have been reported for three barium alkaline-earth compounds, the perovskite-type oxides of the form Ba_2MUO_6(cr) with M = Mg, Ca and Sr.

No entropy or Gibbs energy values are available for these compounds.

V.9.5.2.1. Ba_2MgUO_6(cr)

Based on the enthalpy of reaction of Ba_2MgUO_6(cr) with 1 M HCl(aq) reported by Morss, Fuger and Jenkins [82MOR/FUG] and auxiliary data discussed in Appendix A,

$$\Delta_f H^\circ_m(\mathrm{Ba_2MgUO_6, cr, 298.15\,K}) = -(3245.9 \pm 6.5)\,\mathrm{kJ \cdot mol^{-1}}$$

is calculated and accepted in this review.

V.9.5.2.2. Ba_2CaUO_6(cr)

The value of the heat of solution of Ba_2CaUO_6(cr) in 1 M HCl(aq) was reported by Gens *et al.* [85GEN/FUG] from fourteen replicate measurements on two well-characterized samples. From this and auxiliary data in Appendix A, the recommended value

$$\Delta_f H^\circ_m(\mathrm{Ba_2CaUO_6, cr, 298.15\,K}) = -(3295.8 \pm 5.9)\,\mathrm{kJ \cdot mol^{-1}}$$

is calculated.

V.9.5.2.3. Ba_2SrUO_6(cr)

The value of the heat of solution of Ba_2CaUO_6(cr) in 1 M HCl(aq) was reported by Gens *et al.* [85GEN/FUG] from 22 replicate measurements on four well-characterized samples. From this and auxiliary data in Appendix A, the recommended value

$$\Delta_f H^\circ_m(\mathrm{Ba_2SrUO_6, cr, 298.15\,K}) = -(3257.3 \pm 5.7)\,\mathrm{kJ \cdot mol^{-1}}$$

is calculated.

V.10. Group 1 (alkali) compounds

V.10.1. Lithium compounds

V.10.1.1. $LiUO_3$(cr)

From calorimetric measurements of the enthalpy of solution of $LiUO_3$(cr) in a sulphuric acid/ceric sulphate solution, Cordfunke and Ouweltjes [81COR/OUW] determined the enthalpy of formation of $LiUO_3$(cr). After minor corrections to the auxiliary data (*cf.* Appendix A), the value of the enthalpy of formation accepted in this review is

$$\Delta_f H_m^\circ(\text{LiUO}_3, \text{cr}, 298.15\text{ K}) = -(1522.3 \pm 1.8)\text{ kJ}\cdot\text{mol}^{-1}.$$

Data on S_m° and $\Delta_f G_m^\circ$ of $LiUO_3$(cr) have not been published.

V.10.1.2. Li_2UO_4 (cr)

The enthalpy of formation of Li_2UO_4(cr) was first determined by Ippolitova, Faustova and Spitsyn [61IPP/FAU]. These authors obtained the value $\Delta_f H_m^\circ(\text{Li}_2\text{UO}_4, \text{cr}, 298.15\text{ K}) = -1953.9\text{ kJ}\cdot\text{mol}^{-1}$ but did not report details on the experimental conditions. Later, O'Hare and Hoekstra [74OHA/HOE4] measured the enthalpy of dissolution of Li_2UO_4(cr) in HCl (1 M) according to the reaction

$$\text{Li}_2\text{UO}_4(\text{cr}) + 4\text{HCl}(\text{sln}) \rightleftharpoons 2\text{LiCl}(\text{sln}) + \text{UO}_2\text{Cl}_2(\text{sln}) + 2\text{H}_2\text{O}(\text{sln}) \quad (\text{V.183})$$

and obtained $\Delta_r H_m(\text{V.183}) = -(174.77 \pm 0.08)\text{ kJ}\cdot\text{mol}^{-1}$. These authors combined this measurement with appropriate auxiliary thermodynamic data including the enthalpies of formation of LiCl(cr), H_2O(l), Li^+(aq), Cl^-(aq), H_2(g), Cl_2(g) and UO_2Cl_2(cr) and obtained $\Delta_f H_m^\circ(\text{Li}_2\text{UO}_4, \text{cr}, 298.15\text{ K}) = -(1938.49\pm3.51)\text{ kJ}\cdot\text{mol}^{-1}$. The value of $\Delta_f H_m^\circ(\text{UO}_2\text{Cl}_2, \text{cr}, 298.15\text{ K})$ was in error, as was the heat of solution of this compound in the aqueous HCl. This probably resulted from the reaction of atmospheric moisture with the UO_2Cl_2(cr).

Using a more recent value of $\Delta_f H_m^\circ(\text{UO}_2\text{Cl}_2, \text{cr}, 298.15\text{ K}) = -(1243.5 \pm 0.8)\text{ kJ}\cdot\text{mol}^{-1}$ determined by O'Hare, Boerio and Hoekstra [76OHA/BOE], Cordfunke and O'Hare [78COR/OHA] recalculated $\Delta_f H_m^\circ(\text{Li}_2\text{UO}_4, \text{cr}, 298.15\text{ K}) = -(1961.5 \pm 1.3)\text{ kJ}\cdot\text{mol}^{-1}$. Wagman *et al.* [82WAG/EVA] suggested the value $\Delta_f H_m^\circ(\text{Li}_2\text{UO}_4, \alpha, 298.15\text{ K}) = -1971.9\text{ kJ}\cdot\text{mol}^{-1}$ without any explanation, and this last value was also selected by Morss [86MOR]. Recently, Cordfunke, Ouweltjes and Prins [85COR/OUW2] reported $\Delta_f H_m^\circ(\text{Li}_2\text{UO}_4, \text{cr}, 298.15\text{ K}) = -(1967.54\pm2.00)\text{ kJ}\cdot\text{mol}^{-1}$ based on dissolution of Li_2UO_4 in 1.505 M H_2SO_4(aq). Because of the lack of information on the experimental procedure and calculation in the publications of Ippolitova, Faustova and Spitsyn [61IPP/FAU] the results of this study are not accepted. Recalculation of $\Delta_f H_m^\circ(\text{Li}_2\text{UO}_4, \text{cr}, 298.15\text{ K})$ from the other two calorimetric studies has been done using CODATA consistent auxiliary data (see Appendix A). The value $\Delta_f H_m^\circ(\text{Li}_2\text{UO}_4, \text{cr}, 298.15\text{ K}) = -(1967.4 \pm 1.5)\text{ kJ}\cdot\text{mol}^{-1}$ is obtained from the sulphate cycle, and $\Delta_f H_m^\circ(\text{Li}_2\text{UO}_4, \text{cr}, 298.15\text{ K}) = -(1971.2 \pm 3.0)\text{ kJ}\cdot\text{mol}^{-1}$ from the chloride cycle. The results are in marginal agreement, and the weighted average

$$\Delta_f H_m^\circ(\text{Li}_2\text{UO}_4, \text{cr}, 298.15\text{ K}) = -(1968.2 \pm 1.3)\text{ kJ}\cdot\text{mol}^{-1}$$

is accepted in the present review.

O'Hare and Hoekstra [74OHA/HOE4] estimated $S_m^\circ(\text{Li}_2\text{UO}_4, \text{cr}, 298.15\text{ K}) = (132.6 \pm 6.3)\text{ J}\cdot\text{K}^{-1}\cdot\text{mol}^{-1}$ based on Latimer's method. This review accepts the rounded value.

$$S_m^\circ(\text{Li}_2\text{UO}_4, \text{cr}, 298.15\text{ K}) = (133 \pm 6)\text{ J}\cdot\text{K}^{-1}\cdot\text{mol}^{-1}$$

The Gibbs energy of formation is calculated from the selected enthalpy of formation and entropy.

$$\Delta_f G_m^\circ(\text{Li}_2\text{UO}_4, \text{cr}, 298.15\text{ K}) = -(1853.2 \pm 2.2)\text{ kJ}\cdot\text{mol}^{-1}$$

V.10.1.3. Li_4UO_5(cr)

Cordfunke, Ouweltjes and Prins [85COR/OUW2] recently reported $\Delta_f H_m^\circ(Li_4UO_5$, cr, 298.15 K) = $-(2639.7 \pm 3.7)\,kJ \cdot mol^{-1}$. Based on heats of solution from this reference and slightly different auxiliary data (see Appendix A), the value

$$\Delta_f H_m^\circ(Li_4UO_5, \text{cr}, 298.15\ \text{K}) = -(2639.4 \pm 1.7)\,\text{kJ} \cdot \text{mol}^{-1},$$

is accepted in the present review. It is presumed that a private communication of the same experimental data [85COR/OUW2] was the source for the value $\Delta_f H_m^\circ(Li_4UO_5$, cr, 298.15 K) = $-2640.5\,kJ{\cdot}mol^{-1}$ in the compilation of Wagman *et al.* [82WAG/EVA].

No data on the entropy and the Gibbs energy of formation of Li_4UO_5(cr) appear to be available.

V.10.1.4. $Li_2U_2O_7$(cr)

The enthalpy of formation of $Li_2U_2O_7$(cr) was determined from five measurements of the enthalpy of solution of $Li_2U_2O_7$(cr) in HCl (1 M) by Judge, Brown and Fuger as cited by Fuger [85FUG]. Based on the reported average value of the enthalpy of solution and CODATA consistent auxiliary data (Appendix A), the value

$$\Delta_f H_m^\circ(Li_2U_2O_7, \text{cr}, 298.15\ \text{K}) = -(3213.6 \pm 5.3)\,\text{kJ} \cdot \text{mol}^{-1}$$

is calculated.

Data on the entropy and the Gibbs energy of formation of $Li_2U_2O_7$(cr) seem to be unavailable.

V.10.1.5. $Li_2U_3O_{10}$(cr)

Cordfunke, Ouweltjes and Prins [85COR/OUW2] recently reported the enthalpy of solution of $Li_2U_3O_{10}$(cr) in 1.505 M H_2SO_4(aq). From this and CODATA consistent auxiliary data (Appendix A), the value

$$\Delta_f H_m^\circ(Li_2U_3O_{10}, \text{cr}, 298.15\ \text{K}) = -(4437.4 \pm 4.1)\,\text{kJ} \cdot \text{mol}^{-1}$$

is calculated.

No data on the entropy and the Gibbs energy of formation of $Li_2U_3O_{10}$(cr) appear to be available.

V.10.2. Sodium compounds

Sodium monouranates, polyuranates and peroxyuranates of dioxouranium(VI) are known, as are monouranates of dioxouranium(V).

V.10.2.1. Sodium monouranates

V.10.2.1.1. Sodium monouranates(VI)

Four solid species have been identified: α-Na_2UO_4, β-Na_2UO_4, α-Na_4UO_5 and β-Na_4UO_5. Thermodynamic data are available only for the two polymorphs of Na_2UO_4 and β-Na_4UO_5.

a) α-Na_2UO_4

The heat of formation of α-Na_2UO_4 was calculated by Rand and Kubaschewski [63RAN/KUB] to be $\Delta_f H_m^\circ(Na_2UO_4, \alpha, 298.15\,K) = -(2050 \pm 21)\,kJ \cdot mol^{-1}$, based on the combustion calorimetry data of Mixter [12MIX]. (The value would be $-2045\,kJ \cdot mol^{-1}$ based on current auxiliary data.) Later, Ippolitova, Faustova and Spitsyn [61IPP/FAU] reported the value of $-2000\,kJ \cdot mol^{-1}$. Neither of these studies distinguished between the α- and β-forms, and the samples of Mixter [12MIX] were not well characterized. The enthalpy of solution of Na_2UO_4 in approximately 1 M HCl(aq) reported by Ippolitova *et al.* [61IPP/FAU] is more than 70 $kJ \cdot mol^{-1}$ less exothermic than later reported for the reaction in the same medium by other authors [76OHA/HOE, 85TSO/BRO]. The reason for the large difference is not clear, but it appears that the value in Ref. [61IPP/FAU] is erroneous.

Cordfunke and Loopstra [71COR/LOO] determined the enthalpy of dissolution of α-Na_2UO_4 in 6 M HNO_3:

$$\alpha\text{-}Na_2UO_4 + 4HNO_3(sln) \rightleftharpoons UO_2(NO_3)_2(sln) + 2NaNO_3(sln) + 2H_2O(sln). \qquad (V.184)$$

From the value obtained, $\Delta_r H_m(V.184) = -(192.8 \pm 1.4)\,kJ \cdot mol^{-1}$, these authors reported $\Delta_f H_m^\circ(Na_2UO_4, \alpha, 298.15\,K) = -1874.4\,kJ \cdot mol^{-1}$. Recalculation (see Appendix A) yields $\Delta_f H_m^\circ(Na_2UO_4, \alpha, 298.15\,K) = -(1892.4 \pm 2.3)\,kJ \cdot mol^{-1}$. Independently, O'Hare and Hoekstra [73OHA/HOE2] measured the enthalpy of dissolution of α-Na_2UO_4 in 0.1 M HCl according to the reaction

$$\alpha\text{-}Na_2UO_4 + 4HCl(sln) \rightleftharpoons UO_2Cl_2(sln) + 2NaCl(sln) + 2H_2O(sln). \qquad (V.185)$$

They obtained $\Delta_r H_m(V.185) = -(175.5 \pm 0.8)\,kJ \cdot mol^{-1}$ and derived $\Delta_f H_m^\circ(Na_2UO_4, \alpha, 298.15\,K) = -(1864.2 \pm 3.6)\,kJ \cdot mol^{-1}$. Recalculation by Cordfunke and O'Hare [78COR/OHA] using different auxiliary data (see Appendix A) gave the value $\Delta_f H_m^\circ(Na_2UO_4, \alpha, 298.15\,K) = -(1887.8 \pm 1.7)\,kJ \cdot mol^{-1}$. Later, Cordfunke *et al.* [82COR/MUI] remeasured the enthalpy of the dissolution of α-Na_2UO_4 in sulphuric acid,

$$\alpha\text{-}Na_2UO_4 + 4H_2SO_4(sln) \rightarrow UO_2SO_4(sln) + 2Na_2SO_4(sln) + 2H_2O(sln), \qquad (V.186)$$

and obtained $\Delta_r H_m(V.186) = -(182.95 \pm 0.44)\,kJ \cdot mol^{-1}$. From this value, Cordfunke *et al.* [82COR/MUI] calculated $\Delta_f H_m^\circ(Na_2UO_4, \alpha, 298.15\,K) = -(1897.27 \pm 1.06)\,kJ \cdot$

mol^{-1}. According to Cordfunke *et al.* [82COR/MUI], the difference between results obtained from the work in sulphate solution and the earlier work in chloride solution [73OHA/HOE2] was due, for the most part, to the evaluation of the enthalpy of solution of UO_2Cl_2 in 0.1 M HCl(aq). The solid phase, UO_2Cl_2(cr), used for the experiments of O'Hare and Hoekstra [73OHA/HOE2] was probably partially hydrated. Cordfunke *et al.* [82COR/MUI] reported a new value ($-(107.70 \pm 0.8)\,kJ \cdot mol^{-1}$) for this enthalpy of solution, and found the recalculated value based on the data in Ref. [73OHA/HOE2] to be consistent with the value determined from dissolution in aqueous sulphuric acid.

Tso *et al.* [85TSO/BRO] prepared high purity α-Na_2UO_4 by thermal decomposition of $Na_2UO_2(C_2O_4)\cdot 4H_2O$(cr) at 720 to 760°C for 19.5 to 22 hours, and remeasured the enthalpy of dissolution of α-Na_2UO_4 in 1 M HCl. From the value obtained, $\Delta_{sol}H_m = -(171.9 \pm 0.4)\,kJ \cdot mol^{-1}$, these authors determined $\Delta_f H^\circ_m(Na_2UO_4, \alpha, 298.15\ K) = -(1901.3 \pm 2.3)\,kJ \cdot mol^{-1}$. The enthalpy of solution in this work is in reasonable agreement with the value, $-(172.6 \pm 0.1)\,kJ \cdot mol^{-1}$, reported by O'Hare, Hoekstra and Fredrickson [76OHA/HOE] for the enthalpy of solution, also in 1 M HCl(aq). The values, recalculated in this review, are summarised in Table V.46. The values do not agree within the stated uncertainties. As suggested by Tso *et al.* [85TSO/BRO] there still may be unidentified inconsistencies in the auxiliary data used in the various thermodynamic data. Certainly similar problems are found in comparing the data for other uranates when such data have been derived from dissolution experiments using different media. There also may be differences in the samples although, for some uranates, it has been shown that similar enthalpy of solution values are obtained with the samples used for different studies [85FUG]. In the absence of any better method the unweighted average of the five values in Table (V.46) is selected.

Table V.46: Recalculated values of the enthalpy of formation of α-Na_2UO_4.

medium	$\Delta_r H^\circ_m$ (a) ($kJ \cdot mol^{-1}$)	$\Delta_f H^\circ_m$ ($kJ \cdot mol^{-1}$)	Reference
6 M HNO_3	-192.8 ± 1.4	-1892.4 ± 2.3	[71COR/LOO]
0.1 M HCl	-175.5 ± 0.7	-1896.8 ± 1.2	[73OHA/HOE2]
1.0 M HCl	-172.6 ± 0.1	-1900.5 ± 2.3	[76OHA/HOE]
1.0 M HCl	-171.9 ± 0.4	-1901.3 ± 2.3	[85TSO/BRO]
1.5 M H_2SO_4	-183.0 ± 0.4	-1897.4 ± 1.1	[82COR/MUI]

(a) The enthalpies of reaction refer, depending on the composition of the medium, to Eqs. (V.184), (V.185), and (V.186), respectively.

$$\Delta_f H^\circ_m(Na_2UO_4, \alpha, 298.15\,K) = -(1897.7 \pm 3.5)\,kJ \cdot mol^{-1}.$$

The uncertainty is assigned by this review. For comparison, Wagman *et al.* [82WAG/EVA] suggested the value $\Delta_f H_m^\circ(Na_2UO_4, \alpha, 298.15\,K) = -1893.3\,kJ\cdot mol^{-1}$ (probably based on an earlier value for the enthalpy of solution of UO_2Cl_2), and Morss [86MOR] selected the value $\Delta_f H_m^\circ(Na_2UO_4, \alpha, 298.15\,K) = -(1897 \pm 4)\,kJ \cdot mol^{-1}$ from results reported in Refs. [82COR/MUI, 78COR/OHA, 85FUG].

Osborne, Flotow and Hoekstra [74OSB/FLO] determined the entropy and the heat capacity of α-Na_2UO_4 using low-temperature calorimetry. They reported an entropy of $(166.02 \pm 0.33)\,J \cdot K^{-1} \cdot mol^{-1}$ and a heat capacity of $(146.65 \pm 0.29)\,J \cdot K^{-1} \cdot mol^{-1}$. These two values were adopted in both papers [78COR/OHA, 82COR/MUI]. This review accepts the rounded values obtained by Osborne, Flotow and Hoekstra [74OSB/FLO] and assigns uncertainties as follows:

$$S_m^\circ(Na_2UO_4, \alpha, 298.15\,K) = (166.0 \pm 0.5)\,J \cdot K^{-1} \cdot mol^{-1}$$
$$C_{p,m}^\circ(Na_2UO_4, \alpha, 298.15\,K) = (146.7 \pm 0.5)\,J \cdot K^{-1} \cdot mol^{-1}.$$

The Gibbs energy of formation of α-Na_2UO_4, recalculated in this review using the selected data, gives the value

$$\Delta_f G_m^\circ(Na_2UO_4, \alpha, 298.15\,K) = -(1779.3 \pm 3.5)\,kJ \cdot mol^{-1}.$$

This value is in good agreement with values from other recent assessments. The value $-1778.3\,kJ \cdot mol^{-1}$ was suggested by Cordfunke *et al.* [82COR/MUI] and $-(1779 \pm 2)\,kJ \cdot mol^{-1}$ by Morss [86MOR]. The apparent good agreement with the value $-1777.72\,kJ \cdot mol^{-1}$ in Wagman *et al.* [82WAG/EVA] seems to be the result of a calculation error in that reference.

Fredrickson and O'Hare [76FRE/OHA] reported measurements of enthalpy increments of α-Na_2UO_4 from 618 to 1165 K. This review accepts the (fitted) heat capacity function obtained in that paper:

$$C_{p,m}^\circ(T) = (162.538 + 25.8857 \times 10^{-3}\,T - 20.966 \times 10^5\,T^{-2})\,J \cdot K^{-1} \cdot mol^{-1}.$$

The earlier work of Faustova, Ippolitova and Spitsyn [61FAU/IPP] was criticized elsewhere [74OSB/FLO], and the values from that source are not accepted in this review.

b) β-Na_2UO_4

Cordfunke and Loopstra [71COR/LOO] determined the enthalpy of dissolution of β-Na_2UO_4 in 6 M HNO_3 and obtained $\Delta_r H_m = -(199.0 \pm 1.0)\,kJ \cdot mol^{-1}$. However, they indicated their sample was impure, containing some $Na_2U_2O_7$, and the difference between their enthalpies of solution of α-Na_2UO_4 and β-Na_2UO_4 in the same medium was only $(6.23 \pm 1.71)\,kJ \cdot mol^{-1}$.

O'Hare, Hoekstra and Fredrickson [76OHA/HOE] measured the enthalpy of solution of β-Na_2UO_4 in 1 M HCl, obtaining a value of $-(186.48 \pm 0.13)\,kJ \cdot mol^{-1}$. Thus, using their measured value of the enthalpy of solution of α-Na_2UO_4 in the same medium (above), the difference between the enthalpies of solution of α-Na_2UO_4 and β-Na_2UO_4 was found to be $(13.86 \pm 0.15)\,kJ \cdot mol^{-1}$. This is in only fair agreement with $(12.3 \pm 0.6)\,\text{kJ}\cdot\text{mol}^{-1}$ obtained by Tso *et al.* [85TSO/BRO] for the difference in the heats of solution of high-purity samples of the two solids in the same medium.

Based on the considerations above, it does not appear that the work of Cordfunke and Loopstra [71COR/LOO] can be used to obtain a precise value for $\Delta_f H_m^\circ(\beta\text{-}Na_2UO_4)$. Using the auxiliary data in this review, the values $\Delta_f H_m^\circ(\beta\text{-}Na_2UO_4) = -(1886.7 \pm 2.3)\,\text{kJ}\cdot\text{mol}^{-1}$ and $\Delta_f H_m^\circ(\beta\text{-}Na_2UO_4) = -(1889.0 \pm 2.3)\,\text{kJ}\cdot\text{mol}^{-1}$ are calculated from the data of Refs. [76OHA/HOE] and [85TSO/BRO], respectively.

However, both these results were obtained from enthalpies of solution in the same medium (1.0 M HCl(aq)). There is a large variation in the value of $\Delta_f H_m^\circ(\alpha\text{-}Na_2UO_4)$ that appears to depend on the medium used for the enthalpy of solution measurements (probably a result of problems with the requisite auxiliary data). Therefore, in this review, the selected value of

$$\Delta_f H_m^\circ(Na_2UO_4,\ \beta,\ 298.15\ \text{K}) = -(1884.6 \pm 3.6)\,\text{kJ}\cdot\text{mol}^{-1}$$

is calculated by adding the average difference of the enthalpies of solution for the α- and β-phases $(13.1 \pm 0.8)\,\text{kJ}\cdot\text{mol}^{-1}$ to the value selected above for $\Delta_f H_m^\circ(\alpha\text{-}Na_2UO_4)$. This value is slightly more positive than the value calculated directly from any of the experimental heats of solution. Further experimental work is required, probably involving dissolution of identical samples in different media, with particular attention to auxiliary data.

No data on the entropy and the Gibbs energy of formation $\beta\text{-}Na_2UO_4$ appear to be available. Values for the enthalpy increments of $\beta\text{-}Na_2UO_4$ for 1198 to 1273 K were reported by Fredrickson and O'Hare [76FRE/OHA]. However, in the absence of values for thermodynamic parameters for lower temperatures (except for $\Delta_f H_m^\circ(Na_2UO_4,\ \beta,\ 298.15\ \text{K})$), these enthalpy increments have not been included in the selected data set in this review.

c) Na_4UO_5(cr)

Cordfunke and Loopstra [71COR/LOO] measured the heat of the reaction

$$\begin{aligned} Na_4UO_5(cr) + 6HNO_3(sln) \rightleftharpoons\ & 4NaNO_3(sln) + UO_2(NO_3)_2(sln) \\ & +3H_2O(sln) \end{aligned} \qquad (V.187)$$

in 6 M HNO_3(aq), and reported $\Delta_r H_m(V.187) = -(418.9 \pm 2.1)\,\text{kJ}\cdot\text{mol}^{-1}$. They calculated the heat of formation to be $\Delta_f H_m^\circ(Na_4UO_5,\ \text{cr},\ 298.15\ \text{K}) = -2414.2\,\text{kJ}\cdot\text{mol}^{-1}$. Recalculation using more recent auxiliary data (Appendix A) gives $\Delta_f H_m^\circ(Na_4UO_5,\ \text{cr},\ 298.15\ \text{K}) = -(2456.2 \pm 2.1)\,\text{kJ}\cdot\text{mol}^{-1}$. Tso *et al.* [85TSO/BRO] reported $\Delta_r H_m(V.187) = -(386.6 \pm 0.6)\,\text{kJ}\cdot\text{mol}^{-1}$ for the enthalpy of solution of Na_4UO_5 in 1 M HCl(aq), and using auxiliary data consistent with this review, calculated $\Delta_f H_m^\circ(Na_4UO_5,\ \text{cr},\ 298.15\ \text{K}) = -(2457.3 \pm 2.8)\,\text{kJ}\cdot\text{mol}^{-1}$. In this review the weighted average,

$$\Delta_f H_m^\circ(Na_4UO_5, \text{cr}, 298.15\,\text{K}) = -(2456.6 \pm 1.7)\,\text{kJ}\cdot\text{mol}^{-1},$$

is selected.

No entropy or Gibbs energy data seem to be available for this compound.

V.10.2.1.2. Sodium monouranates(V)

a) $NaUO_3(cr)$

Values for $\Delta_f H^\circ_m(NaUO_3)$ have been measured by King *et al.* [71KIN/SIE], Cordfunke and Loopstra [71COR/LOO], O'Hare and Hoekstra [74OHA/HOE3] and Cordfunke and Ouweltjes [81COR/OUW]. After recalculation using auxiliary data from this review (Appendix A), the values from Cordfunke and Loopstra [71COR/LOO], $\Delta_f H^\circ_m(NaUO_3, cr, 298.15\ K) = -(1497.6\pm4.9)\ kJ\cdot mol^{-1}$, and Cordfunke and Ouweltjes [81COR/OUW], $\Delta_f H^\circ_m(NaUO_3, cr, 298.15\ K) = -(1494.6\pm1.7)\ kJ\cdot mol^{-1}$ are in good agreement. These values are also apparently in good agreement with the results of King *et al.* [71KIN/SIE]. These results were unavailable to the reviewers. According to Cordfunke and O'Hare [78COR/OHA] the final value from that reference has a large uncertainty because the samples used contained 20 to 25% UO_2. The recalculated result (Appendix A) from the work of O'Hare and Hoekstra [74OHA/HOE3], $\Delta_f H^\circ_m(NaUO_3, cr, 298.15\ K) = -1520.5\ kJ\cdot mol^{-1}$, does not agree with the other values. In this review, the selected value

$$\Delta_f H^\circ_m(NaUO_3, cr, 298.15\ K) = -(1494.9\pm1.6)\ kJ\cdot mol^{-1}$$

is calculated from the weighted average of the values recalculated from the results of Cordfunke and Loopstra [71COR/LOO] and Cordfunke and Ouweltjes [81COR/OUW].

Lyon *et al.* [77LYO/OSB] measured the heat capacity of $NaUO_3(cr)$ from 5 to 350 K by aneroid adiabatic calorimetry. In this review the values from this work are accepted and slightly larger uncertainties assigned:

$$\begin{aligned} S^\circ_m(NaUO_3, cr, 298.15\ K) &= (132.84\pm0.40)\ J\cdot K^{-1}\cdot mol^{-1} \\ C^\circ_{p,m}(NaUO_3, cr, 298.15\ K) &= (108.87\pm0.40)\ J\cdot K^{-1}\cdot mol^{-1}. \end{aligned}$$

Cordfunke *et al.* [82COR/MUI] reported measurements of enthalpy increments of $NaUO_3(cr)$ from 415 to 931 K. This review accepts the heat capacity function obtained by taking the temperature derivative of the fitting equation in that paper.

$$C^\circ_{p,m}(T) = (115.491 + 19.1672\times10^{-3}\,T - 10.966\times10^{5}\,T^{-2})\ J\cdot K^{-1}\cdot mol^{-1}$$

The Gibbs energy of formation is calculated from the selected enthalpy of formation and entropy.

$$\Delta_f G^\circ_m(NaUO_3, cr, 298.15\ K) = -(1412.5\pm1.6)\ kJ\cdot mol^{-1}$$

b) $Na_3UO_4(cr)$

The enthalpy of solution and oxidation of $Na_3UO_4(cr)$ in aqueous HCl containing XeO_3 was measured by O'Hare *et al.* [72OHA/SHI], and from that value $\Delta_f H^\circ_m(Na_3UO_4, cr, 298.15\ K) = -1998.7\ kJ\cdot mol^{-1}$ was calculated. The value was recalculated in the same manner as for the value of $NaUO_3(cr)$ [74OHA/HOE3] (see

Appendix A), using auxiliary data consistent with those used in this review. The resulting value is $\Delta_f H^\circ_m(Na_3UO_4$, cr, 298.15 K) $= -(2024.1 \pm 2.5)\,kJ\cdot mol^{-1}$. An inconsistent value of $\Delta_f H^\circ_m(NaUO_3$, cr, 298.15 K) was determined by XeO_3 oxidation in aqueous HCl by the same group [74OHA/HOE3] (see above). This review agrees with Cordfunke and O'Hare [78COR/OHA] that the solution calorimetry result for the enthalpy of formation is to be preferred to a result from mass spectrometry [72BAT/SHI]. However, although the result based on calorimetry is selected, the uncertainty has been increased.

$$\Delta_f H^\circ_m(Na_3UO_4, cr, 298.15\,K) = -(2024 \pm 8)\,kJ\cdot mol^{-1}$$

Osborne and Flotow [72OSB/FLO] measured the heat capacity of Na_3UO_4(cr) from 6.3 to 348.1 K with an adiabatic calorimeter. This review accepts the values from this work:

$$S^\circ_m(Na_3UO_4, cr, 298.15\,K) = (198.2 \pm 0.4)\,J\cdot K^{-1}\cdot mol^{-1}$$
$$C^\circ_{p,m}(Na_3UO_4, cr, 298.15\,K) = (173.0 \pm 0.4)\,J\cdot K^{-1}\cdot mol^{-1}.$$

Fredrickson and Chasanov [72FRE/CHA] reported measurements of enthalpy increments of Na_3UO_4(cr) from 523 to 1212 K. This review accepts the (fitted) heat capacity function obtained in that paper.

$$C^\circ_{p,m}(NaUO_3, cr, T) = (188.901 + 25.1788 \times 10^{-3}\,T \quad 20.801 \times 10^5\,T^{-2})\,J\cdot K^{-1}\cdot mol^{-1}$$

$$\Delta_f G^\circ_m(Na_3UO_4, cr, 298.15\,K) = -(1900 \pm 8)\,kJ\cdot mol^{-1}$$

V.10.2.2. *Sodium polyuranates*

V.10.2.2.1. $Na_2U_2O_7$(cr)

The first determination of enthalpy of formation of $Na_2U_2O_7$(cr) was done by Cordfunke and Loopstra [71COR/LOO]. These authors measured the enthalpy of reaction of $Na_2U_2O_7$(cr) in 6 M HNO_3.

$$Na_2U_2O_7(cr) + 6HNO_3(sln) \rightleftharpoons 2UO_2(NO_3)_2(sln) + 2NaNO_3(sln) + 3H_2O(sln) \qquad (V.188)$$

They obtained $\Delta_r H_m(V.188) = -(184.35 \pm 1.13)\,kJ\cdot mol^{-1}$ and derived $\Delta_f H^\circ_m(Na_2U_2O_7$, cr, 298.15 K) $= -3179.8\,kJ\cdot mol^{-1}$. Recalculation by Cordfunke and O'Hare [78COR/OHA] using the enthalpy of reaction and recent thermodynamic data gave the value $-(3194.5 \pm 2.1)\,kJ\cdot mol^{-1}$. Recalculation using auxiliary data consistent with this review gives the value $\Delta_f H^\circ_m(Na_2U_2O_7$, cr, 298.15 K) $= -(3196.1 \pm 3.9)\,kJ\cdot mol^{-1}$ (see Appendix A). Later, Cordfunke *et al.* [82COR/MUI] measured the enthalpy of reaction of $Na_2U_2O_7$(cr) in 1.5 M H_2SO_4.

$$Na_2U_2O_7(cr) + 3H_2SO_4(sln) \rightleftharpoons 2UO_2(SO_4)(sln) + 2Na_2SO_4(sln) + 3H_2O(sln) \qquad (V.189)$$

They obtained $\Delta_r H_m(V.189) = -(193.53 \pm 0.37)\,kJ \cdot mol^{-1}$ and calculated $\Delta_f H_m^\circ(Na_2U_2O_7, cr, 298.15\ K) = -(3194.8 \pm 1.8)\,kJ \cdot mol^{-1}$. This value is in good agreement with the previous determination based on Ref. [71COR/LOO], and Cordfunke *et al.* [82COR/MUI] suggested the mean value $-(3194.7 \pm 1.4)\,kJ \cdot mol^{-1}$. Tso *et al.* [85TSO/BRO] prepared high-purity $Na_2U_2O_7$(cr) by thermal decomposition of $NaUO_2(CH_3COO)_3$ at 650 to 800°C in oxygen atmosphere for up to 22 hours and measured the enthalpy of solution of $Na_2U_2O_7$(cr) in 1 M HCl. From the result, $-(171.8 \pm 1.0)\,kJ \cdot mol^{-1}$, these authors derived $\Delta_f H_m^\circ(Na_2U_2O_7, cr, 298.15\ K) = -(3203.8 \pm 2.8)\,kJ \cdot mol^{-1}$. O'Hare, cited as "personal communication 1984" by Tso *et al.* [85TSO/BRO], obtained the value $-(172.4 \pm 2.0)\,kJ \cdot mol^{-1}$ for the enthalpy of solution in 1 M HCl, of part of $Na_2U_2O_7$(cr) used by Cordfunke *et al.* [82COR/MUI]. This result is in good agreement with data reported by Tso *et al.* [85TSO/BRO] in the same medium, $-(171.8 \pm 1.0)\,kJ \cdot mol^{-1}$. Wagman *et al.* [82WAG/EVA] selected the value $\Delta_f H_m^\circ(Na_2U_2O_7, cr, 298.15\ K) = -(3196.2 \pm 3.0)\,kJ \cdot mol^{-1}$, which is very close to the value suggested by Cordfunke *et al.* [82COR/MUI]. Fuger [85FUG] and Morss [86MOR] preferred the data reported by Tso *et al.* [85TSO/BRO].

Because of the good agreement between data reported by Tso *et al.* [85TSO/BRO] and O'Hare as cited in Ref. [85TSO/BRO], this review accepts the value obtained by Tso *et al.* [85TSO/BRO] with the assigned uncertainty:

$$\Delta_f H_m^\circ(Na_2U_2O_7, cr, 298.15\,K) = -(3203.8 \pm 4.0)\,kJ \cdot mol^{-1}.$$

The only source of heat capacity and entropy data is from Cordfunke *et al.* [82COR/MUI]. From calorimetric measurements these authors obtained the value $C_{p,m}^\circ(Na_2U_2O_7, cr, 298.15\,K) = (227.26 \pm 0.68)\,J \cdot K^{-1} \cdot mol^{-1}$ and determined $S_m^\circ(Na_2U_2O_7, cr, 298.15\,K) = (275.86 \pm 0.83)\,J \cdot K^{-1} \cdot mol^{-1}$. This review accepts rounded values based on those reported by Cordfunke *et al.* [82COR/MUI].

$$C_{p,m}^\circ(Na_2U_2O_7, cr, 298.15\,K) = (227.3 \pm 1.0)\,J \cdot K^{-1} \cdot mol^{-1}$$
$$S_m^\circ(Na_2U_2O_7, cr, 298.15\,K) = (275.9 \pm 1.0)\,J \cdot K^{-1} \cdot mol^{-1}$$

The Gibbs energy of formation is calculated from the selected enthalpy of formation and entropy.

$$\Delta_f G_m^\circ(Na_2U_2O_7, cr, 298.15\,K) = -(3011.5 \pm 4.0)\,kJ \cdot mol^{-1}$$

Cordfunke *et al.* [82COR/MUI] also reported measurements of enthalpy increments of α-$Na_2U_2O_7$ from 390 to 540 K, and β-$Na_2U_2O_7$ from 681 to 926 K. The temperature of the phase transition is not well established, and this review accepts only the heat capacity function for the α-form, obtained by taking the temperature derivative of the fitting equation in Ref. [82COR/MUI].

$$C_{p,m}^\circ(Na_2U_2O_7, cr, T) = (262.831 + 14.6532 \times 10^{-3}\,T - 35.4904 \times 10^5\,T^{-2})\,J \cdot K^{-1} \cdot mol^{-1}$$

V.10.2.2.2. $Na_2U_2O_7 \cdot 1.5H_2O(cr)$

Wagman *et al.* [82WAG/EVA] reported the enthalpy of formation of this compound, $\Delta_f H_m^\circ(Na_2U_2O_7 \cdot 1.5H_2O, cr, 298.15\ K) = -3653.51\,kJ \cdot mol^{-1}$. The source of the datum appears to be a paper by Pissarjewsky [00PIS]. In that paper there are no detailed analytical data on the composition of the salt. The value appears reasonable compared to the value for the anhydrous salt, but the lack of experimental details makes it impossible to assign an uncertainty. The value is not selected in this review.

V.10.2.2.3. $Na_6U_7O_{24}(cr)$

Cordfunke and Loopstra [71COR/LOO] determined the heat of the reaction

$$Na_6U_7O_{24}(cr) + 20HNO_3(sln) \rightleftharpoons 6NaNO_3(sln) + 7UO_2(NO_3)_2(sln) + 10H_2O(sln) \quad (V.190)$$

by calorimetry, and obtained $\Delta_r H_m(V.190) = -(85.1 \pm 0.3)\,kJ \cdot mol^{-1}$. These authors calculated the enthalpy of formation of $Na_6U_7O_{24}(cr)$ and obtained $\Delta_f H_m^\circ(Na_6U_7O_{24}, cr, 298.15\ K) = -10799\,kJ \cdot mol^{-1}$. Recalculation in this review (Appendix A), using more recent auxiliary data, gives

$$\Delta_f H_m^\circ(Na_6U_7O_{24}, cr, 298.15\,K) = -(11351 \pm 14)\,kJ \cdot mol^{-1}.$$

This value is accepted in this review (this is essentially the same value selected earlier by Cordfunke and O'Hare [78COR/OHA] and by Morss (misprinted as $-(1135.5 \pm 8)\,kJ \cdot mol^{-1}$ in Ref. [82MOR])). The rather large uncertainty reflects uncertainties in the auxiliary data.

No entropy or Gibbs energy data seem to be available for this compound.

V.10.2.3. Sodium carbonates

The compounds $Na_4UO_2(CO_3)_3(cr)$, on which Blake *et al.* [56BLA/COL] reported a number of solubility products in different media, is described in Section V.7.1.2.2.b. The Gibbs energy of formation that results from the selected solubility constant is

$$\Delta_f G_m^\circ(Na_4UO_2(CO_3)_3, cr, 298.15\,K) = -(3737.8 \pm 2.3)\,kJ \cdot mol^{-1}.$$

V.10.2.4. Calcium-sodium compounds

V.10.2.4.1. Andersonite

Andersonite is a hydrated calcium-sodium-dioxouranium(VI) carbonate of the composition $CaNa_2UO_2(CO_3)_3 \cdot 6H_2O(cr)$. The solubility study performed by Alwan and Williams [80ALW/WIL] is discussed under the solid uranium carbonates (Section V.7.1.2.2.b). There is an obvious error in the $\Delta_f G_m^\circ$ value for andersonite as can be seen from the corresponding solubility product. The value reported in this study is therefore not accepted in this review. No other thermodynamic data are available on andersonite.

V.10.3. Potassium compounds

V.10.3.1. KUO_3 (cr)

Cordfunke and Ouweltjes [81COR/OUW] reported calorimetric measurements of enthalpy of solution of KUO_3(cr) in a sulphuric acid/ceric sulphate solution. From these and auxiliary data in Appendix A, the standard enthalpy of formation,

$$\Delta_f H^\circ_m(KUO_3, cr, 298.15 K) = -(1522.9 \pm 1.7) kJ \cdot mol^{-1},$$

is calculated.

No data on S°_m and $\Delta_f G^\circ_m$ of KUO_3(cr) appear to be available.

V.10.3.2. K_2UO_4 (cr)

Ippolitova, Faustova and Spitsyn [61IPP/FAU] measured the enthalpy of solution of K_2UO_4(cr) in HCl (approximately 1 M) according to the reaction

$$K_2UO_4(cr) + 4HCl(sln) \rightleftharpoons 2KCl(sln) + UO_2Cl_2(sln) + 2H_2O(sln) \quad (V.191)$$

and obtained $\Delta_r H^\circ_m$(V.191, 298.15 K) = $-(141.08 \pm 0.42)$ kJ · mol^{-1}. Later, O'Hare and Hoekstra [74OHA/HOE4] remeasured calorimetrically the enthalpy of reaction of carefully prepared and characterized K_2UO_4(cr) in aqueous HCl (1 M) and obtained the value $\Delta_r H^\circ_m$(V.191, 298.15 K) = $-(176.02 \pm 0.21)$ kJ · mol^{-1}. From this measurement and auxiliary data of the enthalpies of solution in acid of potassium chloride and of dioxouranium(VI) chloride, these authors derived the standard enthalpy of formation of K_2UO_4(cr): $\Delta_f H^\circ_m$(K_2UO_4, cr, 298.15 K) = $-(1888.62 \pm 3.47)$ kJ · mol^{-1}. The value suggested by Cordfunke and O'Hare [78COR/OHA] and apparently that proposed by Wagman *et al.* [82WAG/EVA] are based on this second enthalpy of solution measurement [74OHA/HOE4], but incorporate markedly different auxiliary data (*e.g.*, values for $\Delta_f H^\circ_m$ of UO_2Cl_2(sln)). The heats of reaction reported by Ippolitova, Faustova and Spitsyn [61IPP/FAU] for several uranates are found to be inconsistent with values from later investigations [78COR/OHA]. Also, insufficient details were provided by these authors to allow proper incorporation of newer data for the enthalpies of dilution. Only the enthalpies of solution of K_2UO_4 and KCl(cr) reported by O'Hare and Hoekstra [74OHA/HOE] are accepted in the present review. From these and auxiliary data discussed in Appendix A, the value

$$\Delta_f H^\circ_m(K_2UO_4, cr, 298.15 K) = -(1920.7 \pm 2.2) kJ \cdot mol^{-1}$$

is calculated.

O'Hare and Hoekstra [74OHA/HOE4] used Latimer's method to estimate the entropy of K_2UO_4(cr). This review accepts this rounded value with the large uncertainty.

$$S^\circ_m(K_2UO_4, cr, 298.15 K) = (180 \pm 8) J \cdot K^{-1} \cdot mol^{-1}$$

The Gibbs energy of formation is calculated from the selected enthalpy of formation and entropy.

$$\Delta_f G^\circ_m(K_2UO_4, cr, 298.15 K) = -(1798.5 \pm 3.2) kJ \cdot mol^{-1}$$

V.10.3.3. $K_2U_2O_7(cr)$

The standard enthalpy of formation of $K_2U_2O_7$(cr) was determined from eight calorimetric measurements of the enthalpy of solution of $K_2U_2O_7$(cr) in aqueous HCl (1 M) by Judge, Brown and Fuger as cited by Fuger [85FUG]. From the reported average enthalpy of solution and auxiliary data discussed in Appendix A the following value for the enthalpy of formation of $K_2U_2O_7$(cr) is calculated:

$$\Delta_f H_m^\circ(K_2U_2O_7, \text{cr}, 298.15 \text{ K}) = -(3250.5 \pm 4.5) \text{ kJ} \cdot \text{mol}^{-1}.$$

Data on the entropy and the Gibbs energy of formation of $K_2U_2O_7$(cr) seem to be unavailable.

V.10.4. Rubidium compounds

V.10.4.1. $RbUO_3(cr)$

Cordfunke and Ouweltjes [81COR/OUW] calorimetrically measured the enthalpy of solution of $RbUO_3$(cr) in a sulphuric acid/ceric sulphate solution. From this and auxiliary data discussed in Appendix A, the value

$$\Delta_f H_m^\circ(RbUO_3, \text{cr}, 298.15 \text{ K}) = -(1520.9 \pm 1.8) \text{ kJ} \cdot \text{mol}^{-1}$$

is calculated.

No data on S_m° and $\Delta_f G_m^\circ$ of $RbUO_3$(cr) appear to be available.

V.10.4.2. $Rb_2UO_4(cr)$

Ippolitova, Faustova and Spitsyn [61IPP/FAU] reported the enthalpy of solution of Rb_2UO_4(cr) in approximately 1 M HCl according to the reaction

$$Rb_2UO_4(\text{cr}) + 4HCl(\text{sln}) \rightleftharpoons 2RbCl(\text{sln}) + UO_2Cl_2(\text{sln}) + 2H_2O(\text{l}) \quad (\text{V.192})$$

and obtained $\Delta_r H_m^\circ(\text{V.192}, 298.15 \text{ K}) = -(139.83 \pm 0.21) \text{ kJ} \cdot \text{mol}^{-1}$. The enthalpy of the same reaction was remeasured calorimetrically by O'Hare and Hoekstra [74OHA/HOE4] using carefully prepared and characterized Rb_2UO_4(cr). This gave $\Delta_r H_m^\circ(\text{V.192}, 298.15 \text{ K}) = -(172.80 \pm 0.21) \text{ kJ} \cdot \text{mol}^{-1}$. From this measurement and auxiliary enthalpy data, O'Hare and Hoekstra [74OHA/HOE4] derived the standard enthalpy formation of Rb_2UO_4(cr), $\Delta_f H_m^\circ(Rb_2UO_4, \text{cr}, 298.15 \text{ K}) = -(1891.17 \pm 3.56) \text{ kJ} \cdot \text{mol}^{-1}$. The value suggested by Cordfunke and O'Hare [78COR/OHA] and apparently that proposed by Wagman *et al.* [82WAG/EVA] are based on this second enthalpy of solution measurement [74OHA/HOE4], but incorporate markedly different auxiliary data, *e.g.*, values for $\Delta_f H_m^\circ$ of UO_2Cl_2(sln). The enthalpy of solution reported by Ippolitova, Faustova and Spitsyn [61IPP/FAU] is rejected in the present review for the same reasons discussed above with respect to their measurements for K_2UO_4(cr). From the enthalpies of solution of Rb_2UO_4(cr) and RbCl(cr) reported by O'Hare and Hoekstra [74OHA/HOE] and auxiliary data discussed in Appendix A, the value

$$\Delta_f H_m^\circ(Rb_2UO_4, cr, 298.15 K) = -(1922.7 \pm 2.2) kJ \cdot mol^{-1}$$

is calculated.

Based on Latimer's method, O'Hare and Hoekstra [74OHA/HOE4] estimated the entropy of $Rb_2UO_4(cr)$. This review accepts their rounded value with an estimated uncertainty,

$$S_m^\circ(Rb_2UO_4, cr, 298.15 K) = (203 \pm 8) J \cdot K^{-1} \cdot mol^{-1}.$$

The Gibbs energy of formation is calculated from the selected enthalpy of formation and entropy.

$$\Delta_f G_m^\circ(Rb_2UO_4, cr, 298.15 K) = -(1800.1 \pm 3.3) kJ \cdot mol^{-1}$$

V.10.4.3. $Rb_2U_2O_7(cr)$

Judge, Brown and Fuger, as cited by Fuger [85FUG], measured the enthalpy of solution of $Rb_2U_2O_7(cr)$ in aqueous HCl (1 M). From this and auxiliary data discussed in Appendix A, the standard enthalpy of formation of $Rb_2U_2O_7(cr)$ is calculated to be

$$\Delta_f H_m^\circ(Rb_2U_2O_7, cr, 298.15 K) = -(3232.0 \pm 4.3) kJ \cdot mol^{-1}.$$

This value is accepted in this review.

No data on S_m° and $\Delta_f G_m^\circ$ of $Rb_2U_2O_7(cr)$ appear to be available.

V.10.5. Caesium compounds

V.10.5.1. Caesium monouranate

Two closely concordant values of $\Delta_f H_m^\circ(Cs_2UO_4, cr, 298.15 K)$, based on different thermodynamic cycles, have been reported by Cordfunke, Ouweltjes and Prins [86COR/OUW]. Recalculation using auxiliary data consistent with the present review (as described in Appendix A) gives $\Delta_f H_m^\circ(Cs_2UO_4, cr, 298.15 K) = -(1928.5 \pm 2.4) kJ \cdot mol^{-1}$ for the value based on dissolution of Cs_2UO_4 in 0.5 M HF(aq), and $\Delta_f H_m^\circ(Cs_2UO_4, cr, 298.15 K) = -(1927.1 \pm 1.7) kJ \cdot mol^{-1}$ for the value based on dissolution of Cs_2UO_4 in 1.505 M H_2SO_4(aq). Recalculation of the earlier experimental data of O'Hare and Hoekstra [74OHA/HOE2] (see Appendix A) leads to $\Delta_f H_m^\circ(Cs_2UO_4, cr, 298.15 K) = -(1929.4 \pm 2.5) kJ \cdot mol^{-1}$. The value recommended in the present review,

$$\Delta_f H_m^\circ(Cs_2UO_4, cr, 298.15 K) = -(1928.0 \pm 1.2) kJ \cdot mol^{-1},$$

is based on the weighted average of the three independent values. All these data are considered to be more reliable than the earlier values of Ippolitova, Faustova and Spitsyn [61IPP/FAU] and Johnson and Steidl [72JOH/STE].

The entropy and heat capacity at room temperature are derived from the low-temperature adiabatic heat capacity measurements (5 to 350 K) by Osborne *et al.* [76OSB/BRL].

$$S^\circ_m(Cs_2UO_4, cr, 298.15 K) = (219.66 \pm 0.44)\,J \cdot K^{-1} \cdot mol^{-1}$$
$$C_{p,m}(Cs_2UO_4, cr, 298.15 K) = (152.76 \pm 0.31)\,J \cdot K^{-1} \cdot mol^{-1}$$

The selected $C_{p,m}(T)$ function is also taken from the assessment by Cordfunke *et al.* [90COR/KON] and is based on drop calorimetric measurements by Fredrickson and O'Hare [76FRE/OHA].

The Gibbs energy of formation is calculated from the selected enthalpy of formation and entropy.

$$\Delta_f G^\circ_m(Cs_2UO_4, cr, 298.15 K) = -(1805.4 \pm 1.2)\,kJ \cdot mol^{-1}$$

V.10.5.2. *Caesium polyuranates*

V.10.5.2.1. *$Cs_2U_2O_7$(cr)*

The enthalpy of reaction of $Cs_2U_2O_7$(cr) in aqueous HCl (1 M) according to the reaction

$$Cs_2U_2O_7(cr) + 6HCl(sln) \rightleftharpoons 2CsCl(sln) + 2UO_2Cl_2(sln) + 3H_2O(sln) \quad (V.193)$$

was measured in a solution calorimeter by O'Hare and Hoekstra [75OHA/HOE2], $\Delta_r H^\circ_m$(V.193, 298.15 K) = $-(192.83 \pm 0.43)\,kJ \cdot mol^{-1}$. From this measurement and auxiliary data of the enthalpies of solution of CsCl in a HCl solution containing dioxouranium(VI) ions, O'Hare and Hoekstra [75OHA/HOE2] deduced the standard enthalpy of formation, $\Delta_f H^\circ_m(Cs_2U_2O_7$, cr, 298.15 K) = $-(3156.0 \pm 6.7)\,kJ \cdot mol^{-1}$.

Recalculating this value with auxiliary data from the present review gives $\Delta_f H^\circ_m(Cs_2U_2O_7$, cr, 298.15 K) = $-(3220.1 \pm 4.5)\,kJ \cdot mol^{-1}$. Calorimetric measurements of the enthalpy of the Reaction (V.193), using carefully characterized $Cs_2U_2O_7$(cr), by Judge, Brown and Fuger, as cited by Fuger [85FUG], yielded a different value, $\Delta_r H^\circ_m$(V.193, 298.15 K) = $-(187.4 \pm 1.6)\,kJ \cdot mol^{-1}$, leading to (on recalculation using auxiliary data from the present review) the substantially more negative value $\Delta_f H^\circ_m(Cs_2U_2O_7$, cr, 298.15 K) = $-(3225.5 \pm 4.7)\,kJ \cdot mol^{-1}$. The the two enthalpies of solution do not agree within the cited uncertainties. The average value of $\Delta_r H^\circ_m$(V.193, 298.15 K) is $-(190.1 \pm 4.3)\,kJ \cdot mol^{-1}$, and hence, based on the two chloride results, $\Delta_f H^\circ_m(Cs_2U_2O_7$, cr, 298.15 K) = $-(3222.8 \pm 6.2)\,kJ \cdot mol^{-1}$ is calculated.

The value from the earlier work [75OHA/HOE2] is in marginal agreement with $\Delta_f H^\circ_m(Cs_2U_2O_7$, cr, 298.15 K) = $-(3215.3 \pm 1.9)\,kJ \cdot mol^{-1}$ as determined by Cordfunke, from the enthalpy of solution in aqueous H_2SO_4 (unpublished results cited in Ref. [90COR/KON]). However, if CODATA consistent auxiliary data are used (Appendix A) it is probable the value based on this experiment would be even less negative ($\Delta_f H^\circ_m(Cs_2U_2O_7$, cr, 298.15 K) = $-(3214.3 \pm 2.1)\,kJ \cdot mol^{-1}$) and agreement is poorer.

It would appear further experiments may be needed to resolve the discrepancies. Although the sulphate cycle is less likely to be subject to errors in the auxiliary data, the problem appears to lie in part with the actual enthalpies of solution (*i.e.*, the purity and handling of the solids). The discrepancies are somewhat similar to, but more severe than, those noted in Section V.10.2.2.1 for $Na_2U_2O_7(cr)$. The results from different cycles for the monourantates (*i.e.*, Li_2UO_4, Cs_2UO_4 and even Na_2UO_4) appear more concordant. Based on the available values, this review selects

$$\Delta_f H^\circ_m(Cs_2U_2O_7, \text{cr}, 298.15 \text{ K}) = -(3220 \pm 10) \text{ kJ} \cdot \text{mol}^{-1}.$$

The entropy is derived from low-temperature heat capacity measurements by O'Hare, Flotow and Hoekstra [81OHA/FLO] as

$$S^\circ_m(Cs_2U_2O_7, \text{cr}, 298.15 \text{ K}) = (327.75 \pm 0.66) \text{ J} \cdot \text{K}^{-1} \cdot \text{mol}^{-1}.$$

The Gibbs energy of formation is calculated from the selected enthalpy of formation and entropy.

$$\Delta_f G^\circ_m(Cs_2U_2O_7, \text{cr}, 298.15 \text{ K}) = -(3023 \pm 10) \text{ kJ} \cdot \text{mol}^{-1}$$

The low-temperature (5 to 350 K) heat capacity was measured by adiabatic calorimetry [81OHA/FLO] and the following result was reported:

$$C^\circ_{p,m}(Cs_2U_2O_7, \text{cr}, 298.15 \text{ K}) = (231.2 \pm 0.5) \text{ J} \cdot \text{K}^{-1} \cdot \text{mol}^{-1}.$$

This review accepts this value.

V.10.5.2.2. $Cs_2U_4O_{12}(cr)$

The compound $Cs_2U_4O_{12}(cr)$ which may be written as $Cs_2O \cdot UO_2 \cdot 3UO_3(cr)$ can exist as three polymorphs: α, β and γ. The α-polymorph is stable up to 898 K, the β-polymorph from 898 to 968 K, and the γ-polymorph at higher temperatures.

Cordfunke and Westrum [80COR/WES] measured the enthalpy of solution of α-$Cs_2U_4O_{12}$ in 1.505 M H_2SO_4 + 0.0350 M $Ce(SO_4)_2$ solution by calorimetry according to Reaction (V.194).

$$\alpha\text{-}Cs_2U_4O_{12} + 4H_2SO_4(\text{sln}) + 2Ce(SO_4)_2(\text{sln}) \rightleftharpoons Cs_2SO_4(\text{sln}) + 4UO_2SO_4(\text{sln}) + 4H_2O(\text{sln}) + Ce_2(SO_4)_3(\text{sln}) \quad (V.194)$$

From the value obtained, $\Delta_r H^\circ_m(V.194, 298.15 \text{ K}) = -(479.9 \pm 2.3) \text{ kJ} \cdot \text{mol}^{-1}$, Cordfunke and Westrum [80COR/WES] calculated the standard enthalpy of formation of α-$Cs_2U_4O_{12}$: $\Delta_f H^\circ_m(Cs_2U_4O_{12}, \alpha, 298.15 \text{ K}) = -(5573.92 \pm 3.56) \text{ kJ} \cdot \text{mol}^{-1}$. In the present review, a different cycle less prone to experimental error is used. The cycle incorporates data for $U_3O_8(cr)$ rather than $UO_2(cr)$ (as was suggested previously by Cordfunke and Ouweltjes [81COR/OUW] for MUO_3 uranates). From the enthalpy of reaction of α-$Cs_2U_4O_{12}$ from Cordfunke and Westrum [80COR/WES] and auxiliary data discussed in Appendix A, the value

$$\Delta_f H^\circ_m(Cs_2U_4O_{12}, \alpha, 298.15 \text{ K}) = -(5571.8 \pm 3.6) \text{ kJ} \cdot \text{mol}^{-1}$$

is calculated.

The entropy of α-$Cs_2U_4O_{12}$ was estimated by Johnson [75JOH]: $S^\circ_m(Cs_2U_4O_{12}, \alpha, 298.15\ K) = 522.6\ J \cdot K^{-1} \cdot mol^{-1}$. This value is in agreement with the experimental value determined from calorimetric measurements by Cordfunke and Westrum [80COR/WES], $S^\circ_m(Cs_2U_4O_{12}, \alpha, 298.15\ K) = 526.43\ J \cdot K^{-1} \cdot mol^{-1}$. This review accepts the rounded value reported by Cordfunke and Westrum [80COR/WES] with an estimated uncertainty,

$$S^\circ_m(Cs_2U_4O_{12}, \alpha, 298.15\ K) = (526.4 \pm 1.0)\ J \cdot K^{-1} \cdot mol^{-1}.$$

Only one calorimetric study to measure the heat capacity of α-$Cs_2U_4O_{12}$ has been reported [80COR/WES]. The temperature function for 298.15 to 898 K from that study is accepted in this review,

$$C^\circ_{p,m}(T) = (423.726 + 0.0719405\,T - 54.375 \times 10^5\,T^{-2})\ kJ \cdot mol^{-1},$$

as is the 298.15 K value, to which an uncertainty of $\pm 1.0\ J \cdot K^{-1} \cdot mol^{-1}$ is assigned:

$$C^\circ_{p,m}(Cs_2U_4O_{12}, \alpha, 298.15\ K) = (384.0 \pm 1.0)\ J \cdot K^{-1} \cdot mol^{-1}.$$

The Gibbs energy of formation is calculated from the selected enthalpy of formation and entropy.

$$\Delta_f G^\circ_m(Cs_2U_4O_{12}, \alpha, 298.15\ K) = -(5251.1 \pm 3.6)\ kJ \cdot mol^{-1}$$

V.10.5.2.3. Other caesium polyuranates

Thermodynamic functions of six caesium polyuranates, $Cs_4U_5O_{17}(cr)$, $Cs_2U_4O_{13}(cr)$, $Cs_2U_5O_{16}(cr)$, $Cs_2U_7O_{22}(cr)$, $Cs_2U_{16}O_{49}(cr)$ and $Cs_2U_4O_{12}(cr)$, were calculated by Johnson [75JOH] using a graphical method. For reasons discussed in Appendix A, the values estimated by Johnson [75JOH] are not accepted in this review.

Chapter VI

Discussion of auxiliary data selection

A number of auxiliary thermodynamic data are used in the present review of uranium thermodynamics. Among these auxiliary data are the entropies of the reference states of each element (see Table II.6) and thermodynamic data of chemical species that contain no uranium but occur in reactions from which the data for uranium compounds and complexes selected in this review are derived. Further, a large number of ligands that form metal ion complexes are protolytes, usually anions of weak acids or neutral molecules such as ammonia. The protonation constants of these ligands are needed as auxiliary data, not only for the evaluation of the experimental data, but also for speciation calculations by the users of the selected data. This chapter contains a discussion of the selection of such auxiliary data. However, the set of auxiliary data selected here and listed in Tables IV.1 and IV.2 is not claimed to be complete, and further selections will be made as needed in subsequent volumes of this series.

Many of the equilibrium constants (particularly the protonation constants) reported in this chapter have been determined with a very high precision by emf methods. A precision of ± 0.002 units in $\log_{10} K$ is not unusual. The accuracy of these constants can be judged by comparing the values reported by different investigators. The accuracy, too, is excellent in most cases, but always several times lower than the precision. As in Chapter V, the uncertainties assigned in the present chapter reflect the accuracy and not the precision of the selected values. Unless otherwise indicated, the uncertainties represent the largest difference between the selected value and the literature values accepted in this review. This uncertainty may be many times larger than the precision of an individual study, but still small in comparison with the uncertainties in the formation constants of metal ion complexes.

When available, the CODATA [89COX/WAG] Key Values, converted to refer to a standard state pressure of 0.1 MPa, are used as auxiliary data. However, the list of these high-quality data is limited, and selections of thermodynamic data for additional species are made by this review. The selected equilibrium data for auxiliary reactions are listed in Table IV.2.

VI.1. Group 17 (halogen) auxiliary species

VI.1.1. Fluorine auxiliary species

The fluoride ion is the only moderately strong base among the halides. No CODATA Key Values are available for HF(aq). The equilibrium constants for Reactions (VI.1) and (VI.2)

$$H^+ + F^- \rightleftharpoons HF(aq) \quad (VI.1)$$

$$HF(aq) + F^- \rightleftharpoons HF_2^- \quad (VI.2)$$

were discussed by Bond and Hefter [80BON/HEF]. By using their compiled equilibrium constants in $NaClO_4$ media at 25°C, judged as "recommended" by the authors, this review performs an extrapolation to $I = 0$ using the specific ion interaction theory (see Figure VI.1) after transformation of the concentration constants to molality constants. The resulting selected value is

$$\log_{10} K_1^\circ(\text{VI.1, 298.15 K}) = 3.18 \pm 0.02.$$

This value agrees with the average selected by Bond and Hefter [80BON/HEF] for $I = 0$, $\log_{10} K_1^\circ(\text{VI.1}) = (3.19 \pm 0.05)$. The value of $\Delta\varepsilon$ obtained from the extrapolation to $I = 0$, as shown in Figure VI.1, $\Delta\varepsilon = -\varepsilon_{(H^+,ClO_4^-)} - \varepsilon_{(F^-,Na^+)} = -(0.16 \pm 0.01)$, does not quite agree with the $\Delta\varepsilon = -(0.10 \pm 0.01)$ calculated from the data of Ciavatta [80CIA]. Since the extrapolation shown in Figure VI.1 is based on 19 experimental points ranging from 0.51 to 3.50 m solutions, the slope in the straight line is determined with fairly high accuracy, as the uncertainty obtained for $\Delta\varepsilon$ indicates. One of the two ion interaction coefficients given by Ciavatta [80CIA], $\varepsilon_{(H^+,ClO_4^-)}$ or $\varepsilon_{(F^-,Na^+)}$, must therefore be erroneous. It is likely that this is the case for $\varepsilon_{(F^-,Na^+)}$. The new value calculated from $\Delta\varepsilon = -(0.16 \pm 0.01)$ is then $\varepsilon_{(F^-,Na^+)} = (0.02 \pm 0.02)$, compared to -0.04 given by Ciavatta [80CIA]. This assumption is supported by the magnitude of the sodium interaction coefficients of OH^-, Cl^- and Br^- which are all in the range of 0.03 to 0.05. The $\Delta_f G_m^\circ$ value for HF(aq) is calculated by using the CODATA [89COX/WAG] value for F^-, giving the selected Gibbs energy value of

$$\Delta_f G_m^\circ(\text{HF, aq, 298.15 K}) = -(299.68 \pm 0.70)\,\text{kJ} \cdot \text{mol}^{-1}.$$

The data for $\log_{10} K_2$(VI.2) compiled by Bond and Hefter [80BON/HEF] were judged only "tentative" by the authors, which means that they assigned an uncertainty of ± 0.2 to each value. This review extrapolates the data with $I \leq 3.5$ m to $I = 0$ (see Figure VI.2) and obtains $\Delta\varepsilon = -(0.13 \pm 0.06)$ and

$$\log_{10} K_2^\circ(\text{VI.2, 298.15 K}) = 0.44 \pm 0.12.$$

From this value of $\Delta\varepsilon$ a new ion interaction coefficient can be derived, $\varepsilon_{(HF_2^-,Na^+)} = -(0.11 \pm 0.06)$, which is obtained by using $\varepsilon_{F^-,Na^+)} = (0.02 \pm 0.02)$ evaluated above. The $\Delta_f G_m^\circ$ value for HF_2^- is calculated by using $\Delta_f G_m^\circ$(HF, aq, 298.15 K) obtained above, yielding

Figure VI.1: Extrapolation to $I = 0$ of experimental data for the formation of HF(aq) in $NaClO_4$ media using the specific ion interaction theory. The data are those from the compilation of Bond and Hefter [80BON/HEF, *p.*19] referring to perchlorate media. The shaded area represents the uncertainty range obtained by propagating the resulting uncertainties at $I = 0$ back to $I = 4$ m.

Equilibrium: $H^+ + F^- \rightleftharpoons HF(aq)$

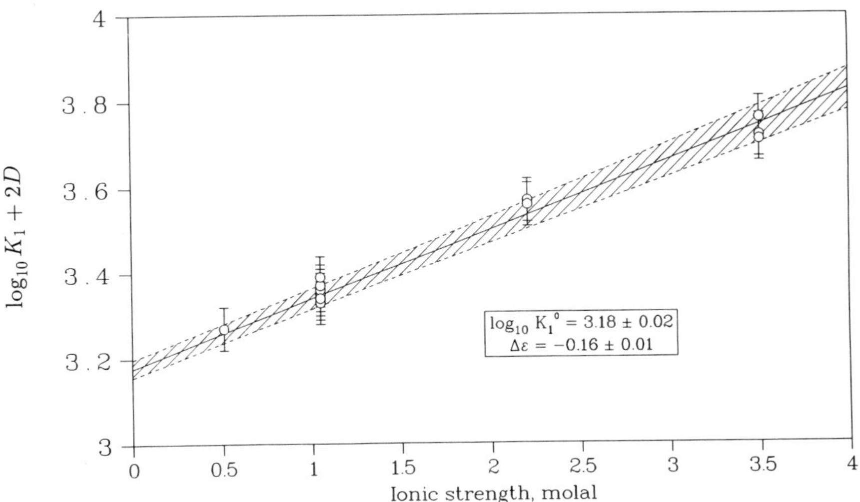

$$\Delta_f G^\circ_m(HF_2^-, aq, 298.15\,K) = -(583.7 \pm 1.2)\,kJ \cdot mol^{-1}.$$

Ahrland and Kullberg [71AHR/KUL2] made, at 25°C and $I = 1$ M ($NaClO_4$), a calorimetric determination of $\Delta_r H_m(VI.1) = (12.17 \pm 0.03)\,kJ \cdot mol^{-1}$. In addition, enthalpies of Reaction (VI.1) were compiled by Sillén and Martell [64SIL/MAR, 71SIL/MAR]. The five values they reported lie between $\Delta_r H^\circ_m(VI.1) = 10.9$ (in 0.5 M $NaClO_4$) and $13.3\,kJ \cdot mol^{-1}$, are thus not incompatible with the experimental value of Ahrland and Kullberg [71AHR/KUL2]. Perrin [82PER, *p.*49] gave $\log_{10} K^\circ_1$ values at different temperatures from three sources. Integration of the data of Broene and de Vries [47BRO/VRI] from 15 to 35°C, of Wooster [38WOO] from 0 to 25°C, and of Ellis [63ELL] from 25 to 50°C yields $\Delta_r H^\circ_m(VI.1) = 12.7$, 12.5, and $16.2\,kJ \cdot mol^{-1}$, respectively. This review selects the calorimetric value reported by Ahrland and Kullberg [71AHR/KUL2] and increases its uncertainty to reflect the assumption that the influence of the ionic strength is negligible.

$$\Delta_r H^\circ_m(VI.1, 298.15\ K) = (12.2 \pm 0.3)\,kJ \cdot mol^{-1}$$

From this, the following enthalpy of formation of HF(aq) is obtained:

$$\Delta_f H^\circ_m(HF, aq, 298.15\ K) = -(323.15 \pm 0.72)\,kJ \cdot mol^{-1}.$$

Figure VI.2: Extrapolation to $I = 0$ of experimental data for the formation of HF_2^- in $NaClO_4$ media using the specific ion interaction theory. The data are those from the compilation of Bond and Hefter [80BON/HEF, *p.*20] referring to perchlorate media. The shaded area represents the uncertainty range obtained by propagating the resulting uncertainties at $I = 0$ back to $I = 4$ m.

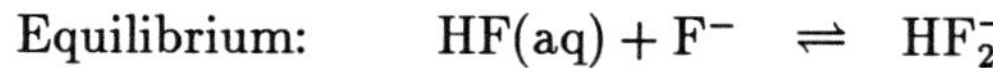

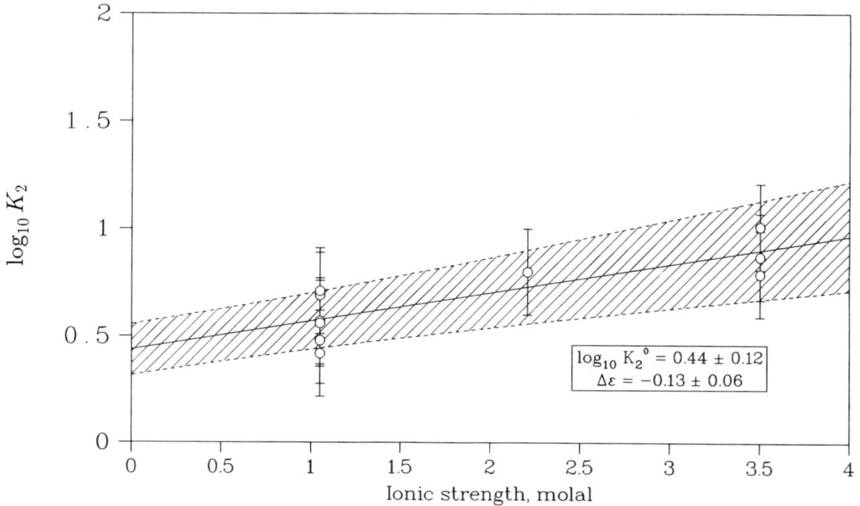

This value is in fair agreement with the one proposed by the US NBS [82WAG/EVA], $\Delta_f H_m^\circ(\mathrm{HF, aq, 298.15\,K}) = -320.08\,\mathrm{kJ \cdot mol^{-1}}$.

Ahrland and Kullberg [71AHR/KUL2] in addition obtained $\Delta_r H_m(\mathrm{VI.2}) = (2.93 \pm 0.28)\,\mathrm{kJ \cdot mol^{-1}}$ at 25°C and $I = 1$ M ($NaClO_4$). This review assumes that $\Delta_r H_m$ is not significantly dependent on the ionic strength. This assumption is reflected by the uncertainty assigned to the selected value,

$$\Delta_r H_m^\circ(\mathrm{VI.2,\ 298.15\ K}) \quad = \quad (3 \pm 2)\,\mathrm{kJ \cdot mol^{-1}}.$$

The enthalpy of formation of HF_2^- is found to be

$$\Delta_f H_m^\circ(\mathrm{HF_2^-, aq, 298.15\,K}) \quad = \quad -(655.5 \pm 2.2)\,\mathrm{kJ \cdot mol^{-1}}.$$

The value proposed by the US NBS [82WAG/EVA], $\Delta_f H_m^\circ = -649.94\,\mathrm{kJ \cdot mol^{-1}}$, is slightly different from the one selected here.

VI.1.2. Chlorine auxiliary species

CODATA [89COX/WAG] Key Values exist for the following chlorine species: Cl(g), Cl_2(g), Cl^- and ClO_4^-. For various calculations, standard thermodynamic data on other chlorine species are required. Their selection is described below.

VI.1.2.1. Hypochlorous acid

Protonation constants of hypochlorite to form HClO(aq) according to Eq. (VI.3) have been tabulated by Perrin [82PER].

$$H^+ + ClO^- \rightleftharpoons HClO(aq) \qquad (VI.3)$$

The published data fall within two groups, one with $\log_{10} K_1^\circ = (7.52 \pm 0.03)$, the other with $\log_{10} K_1^\circ = (7.32 \pm 0.02)$. The first value was obtained by six different groups [66MOR, 72PAL/LVO, 57CAR, 40HAG, 38SHI/GLA, 51MUU/REF], the second one by three groups [58FLI, 63OSI/BYN, 69STO/JAC]. The documentation of the data from the first group is better than that from the second. Nevertheless, this review does not find it justified to prefer one value over the other and therefore selects the unweighted average of the two values,

$$\log_{10} K_1^\circ(\text{VI.3, 298.15 K}) = 7.42 \pm 0.13.$$

The selected value of $\Delta_r G_m^\circ$(VI.6) (*cf.* Section VI.1.2.4) allows the calculation of $\Delta_f G_m^\circ(\text{HClO, aq, 298.15 K})$, which can be used with $\log_{10} K^\circ$(VI.3) to calculate $\Delta_f G_m^\circ(\text{ClO}^-\text{, aq, 298.15 K})$:

$$\begin{aligned} \Delta_f G_m^\circ(\text{HClO, aq, 298.15 K}) &= -(80.0 \pm 0.6)\,\text{kJ} \cdot \text{mol}^{-1} \\ \Delta_f G_m^\circ(\text{ClO}^-\text{, aq, 298.15 K}) &= -(37.67 \pm 0.96)\,\text{kJ} \cdot \text{mol}^{-1}. \end{aligned}$$

The values given by Latimer [52LAT] are in good agreement with the ones calculated here.

The enthalpies of reaction evaluated from these references are $\Delta_r H_m^\circ(\text{VI.3}) = (15 \pm 2)\,\text{kJ} \cdot \text{mol}^{-1}$ from Refs. [66MOR, 57CAR] and $\Delta_r H_m^\circ(\text{VI.3}) = (22.2 \pm 6)\,\text{kJ} \cdot \text{mol}^{-1}$ from Refs. [58FLI, 63OSI/BYN, 69STO/JAC]. This review selects the unweighted average,

$$\Delta_r H_m^\circ(\text{VI.3, 298.15 K}) = (19 \pm 9)\,\text{kJ} \cdot \text{mol}^{-1}.$$

VI.1.2.2. Chlorous acid

Perrin [82PER] reported three values for the protonation constant of chlorite to form $HClO_2$(aq) according to Eq. (VI.4).

$$H^+ + ClO_2^- \rightleftharpoons HClO_2(aq) \qquad (VI.4)$$

They agree well, and this review selects the average value,

$$\log_{10} K_1^\circ(\text{VI.4, 298.15 K}) = 1.96 \pm 0.02.$$

The selected value of E° for Reaction (VI.7) (*cf.* Section VI.1.2.4) allows the calculation of $\Delta_f G_m^\circ(\text{HClO}_2\text{, aq, 298.15 K})$, which can be combined with $\log_{10} K_1^\circ$(VI.4) to give $\Delta_f G_m^\circ(\text{ClO}_2^-\text{, aq, 298.15 K})$:

$$\begin{aligned} \Delta_f G_m^\circ(\text{HClO}_2\text{, aq, 298.15 K}) &= -(0.94 \pm 4.04)\,\text{kJ} \cdot \text{mol}^{-1} \\ \Delta_f G_m^\circ(\text{ClO}_2^-\text{, aq, 298.15 K}) &= (10.25 \pm 4.04)\,\text{kJ} \cdot \text{mol}^{-1}. \end{aligned}$$

VI.1.2.3. Chlorate

This review accepts the assessment made by Johnson, Smith, Appelman and Hubbard [70JOH/SMI] but increases the uncertainty of S°_m slightly to be consistent with the entropy selected by Kelley [40KEL] and Kelley and King [61KEL/KIN]. The selected values are

$$S^\circ_m(ClO_3^-, aq, 298.15\,K) = (162.3 \pm 3.0)\,J\cdot K^{-1}\cdot mol^{-1}$$

and

$$\Delta_f H^\circ_m(ClO_3^-, aq, 298.15\,K) = -(104 \pm 1)\,kJ\cdot mol^{-1}.$$

The Gibbs energy of formation is calculated from the selected enthalpy of formation and entropy.

$$\Delta_f G^\circ_m(ClO_3^-, aq, 298.15\,K) = -(7.9 \pm 1.3)\,kJ\cdot mol^{-1}$$

VI.1.2.4. Redox data involving chlorine species

The reduction of gaseous chlorine according to Reaction (VI.5),

$$\tfrac{1}{2}Cl_2(g) + e^- \rightleftharpoons Cl^-, \qquad \text{(VI.5)}$$

is fairly rapid. From the CODATA [89COX/WAG] values, converted to refer to a standard state pressure of 0.1 MPa, the following recommended values are calculated:

$$\begin{aligned}\log_{10} K^\circ(\text{VI.5}, 298.15\,K) &= 22.99 \pm 0.02\\ E^\circ(\text{VI.5}, 298.15\,K) &= (1.360 \pm 0.001)\,V\\ \Delta_r H^\circ_m(\text{VI.5}, 298.15\,K) &= -(167.08 \pm 0.10)\,kJ\cdot mol^{-1}.\end{aligned}$$

This review tentatively accepts the Gibbs energy given by Latimer [52LAT] for Reaction (VI.6) and estimates an uncertainty of $\pm 0.6\,kJ\cdot mol^{-1}$ in $\Delta_r G^\circ_m$.

$$Cl^- + HClO(aq) + H^+ \rightleftharpoons Cl_2(g) + H_2O(l) \qquad \text{(VI.6)}$$

$$\begin{aligned}\Delta_r G^\circ_m(\text{VI.6}, 298.15\,K) &= -25.9 \pm 0.6\\ \log_{10} K^\circ(\text{VI.6}, 298.15\,K) &= 4.54 \pm 0.11\end{aligned}$$

The redox equilibria between the oxyacids of chlorine are in general slow and cannot be studied by equilibrium methods. They have mostly been calculated from the enthalpies of formation and the ionic entropies of the corresponding chlorine containing species. Both hypochlorous acid/hypochlorite and chlorous acid/chlorite are good oxidizing agents. For the couple chlorite/hypochlorite this review tentatively accepts the standard potential proposed by Latimer [52LAT]. Since the value obtained here (*cf.* Section VI.1.2.1) for $\Delta_f G^\circ_m$(HClO, aq, 298.15 K) is the same as the one selected by Latimer [52LAT], but the one for $\Delta_f G^\circ_m(ClO^-$, aq, 298.15 K) is not, this review uses the potential reported in acid solution. The uncertainty is estimated by this review.

$$HClO_2(aq) + 2H^+ + 2e^- \rightleftharpoons HClO(aq) + H_2O(l) \qquad \text{(VI.7)}$$

$$E^\circ(\text{VI.7}, 298.15\text{ K}) = (1.64 \pm 0.02)\text{ V}$$
$$\log_{10} K^\circ(\text{VI.7}, 298.15\text{ K}) = 55.4 \pm 0.7$$

The standard potentials of the couples chlorate/chlorite and perchlorate/ chlorate can now be calculated based on the selections made above for all the participating species.

$$ClO_3^- + 3H^+ + 2e^- \rightleftharpoons HClO_2(aq) + H_2O(l) \quad \text{(VI.8)}$$
$$ClO_4^- + 2H^+ + 2e^- \rightleftharpoons ClO_3^- + H_2O(l) \quad \text{(VI.9)}$$

$$E^\circ(\text{VI.8}, 298.15\text{ K}) = (1.19 \pm 0.02)\text{ V}$$
$$E^\circ(\text{VI.9}, 298.15\text{ K}) = (1.23 \pm 0.01)\text{ V}$$
$$\log_{10} K^\circ(\text{VI.8}, 298.15\text{ K}) = 40.3 \pm 0.7$$
$$\log_{10} K^\circ(\text{VI.9}, 298.15\text{ K}) = 41.55 \pm 0.26$$
$$\Delta_r H_m^\circ(\text{VI.9}, 298.15\text{ K}) = -(261.7 \pm 1.1)\text{ kJ} \cdot \text{mol}^{-1}$$

VI.1.3. Bromine auxiliary species

CODATA [89COX/WAG] Key Values exist for the following bromine species: Br(g), Br_2(g), Br_2(l), Br^- and HBr(g). For various calculations, standard thermodynamic data on other bromine species are required. Their selection is briefly described below.

VI.1.3.1. Hypobromous acid

Perrin [82PER] reported protonation constants of hypobromite to HBrO(aq) according to Reaction (VI.10) from four different sources [64FLI/MIS, 56KEL/TAR, 34SHI, 38SHI/GLA].

$$H^+ + BrO^- \rightleftharpoons HBrO(aq) \quad \text{(VI.10)}$$

The data agree well, and this review selects the average after correction of the data reported for 22°C [34SHI] and 20°C [38SHI/GLA] to 25°C by using the temperature function given in Ref. [64FLI/MIS]. The selected value is

$$\log_{10} K_1^\circ(\text{VI.10}, 298.15\text{ K}) = 8.63 \pm 0.03.$$

The Gibbs energy of formation of HBrO(aq) is calculated in Section VI.1.3.3 from data for the disproportionation reaction (VI.12), $\Delta_f G_m^\circ$(HBrO, aq, 298.15 K) = $-(81.4 \pm 1.5)$ kJ·mol^{-1}. The protonation constant selected here allows the calculation of the Gibbs energy of formation of BrO^-:

$$\Delta_f G_m^\circ(BrO^-, aq, 298.15\text{ K}) = -(32.14 \pm 1.51)\text{ kJ} \cdot \text{mol}^{-1}.$$

The enthalpy of Reaction (VI.10) is evaluated from the data in Ref. [64FLI/MIS]. This review obtains the selected

$$\Delta_r H_m^\circ(\text{VI.10}, 298.15\text{ K}) = -(30 \pm 3)\text{ kJ} \cdot \text{mol}^{-1},$$

which is an average referring to the temperature interval 10 to 50°C. The uncertainty is based on an estimated uncertainty of ±0.03 in $\log_{10} K_1^\circ$(VI.10).

VI.1.3.2. Bromate

This review accepts the assessment made by Johnson, Smith, Appelman and Hubbard [70JOH/SMI]. The selected values are

$$S^{\circ}_{m}(BrO_3^-, aq, 298.15\,K) = (161.5 \pm 1.3)\,J \cdot K^{-1} \cdot mol^{-1}$$
$$\Delta_f H^{\circ}_{m}(BrO_3^-, aq, 298.15\,K) = -(66.7 \pm 0.5)\,kJ \cdot mol^{-1}.$$

The entropy value is consistent with the one selected earlier by Kelley and King [61KEL/KIN].

The Gibbs energy of formation is calculated from the selected enthalpy of formation and entropy.

$$\Delta_f G^{\circ}_{m}(BrO_3^-, aq, 298.15\,K) = (19.07 \pm 0.63)\,kJ \cdot mol^{-1}$$

VI.1.3.3. Redox data involving bromine species

The reduction of liquid bromine according to Reaction (VI.11)

$$\tfrac{1}{2}Br_2(l) + e^- \rightleftharpoons Br^- \qquad (VI.11)$$

is fairly rapid. From the CODATA [89COX/WAG] values that exist for bromide ion, the reaction data recommended are

$$\log_{10} K^{\circ}(VI.11,\ 298.15\ K) = 18.19 \pm 0.03$$
$$E^{\circ}(VI.11,\ 298.15\ K) = (1.076 \pm 0.002)\,V$$
$$\Delta_r H^{\circ}_{m}(VI.11,\ 298.15\ K) = -(121.41 \pm 0.15)\,kJ \cdot mol^{-1}.$$

The values given by Latimer [52LAT] for the hypobromite/bromide and hypobromite/bromine couples are based on a review by Liebhafsky [34LIE] on the equilibrium constant of the disproportionation reaction of aqueous bromine:

$$Br_2(aq) + H_2O(l) \rightleftharpoons HBrO(aq) + Br^- + H^+. \qquad (VI.12)$$

Liebhafsky [34LIE] selected $\log_{10} K^{\circ}(VI.12) = -8.24$. This review uses this value, with an estimated uncertainty of ± 0.20, to calculate $\Delta_f G^{\circ}_{m}$ of HBrO(aq). The Gibbs energy of formation of Br_2(aq) needed for this calculation is obtained by averaging the value given by Latimer [52LAT], $4.09\,kJ \cdot mol^{-1}$, and the value obtained by Wu, Birky and Hepler [63WU/BIR] from solubility measurements of Br_2(l), $5.77\,kJ \cdot mol^{-1}$, leading to $\Delta_f G^{\circ}_{m}(Br_2, aq, 298.15\,K) = (4.9 \pm 1.0)\,kJ \cdot mol^{-1}$. The uncertainty is chosen so high that it also includes the US NBS [82WAG/EVA] value, $3.93\,kJ \cdot mol^{-1}$. The resulting Gibbs energy of formation of hypobromous acid is

$$\Delta_f G^{\circ}_{m}(HBrO, aq, 298.15\,K) = -(81.4 \pm 1.5)\,kJ \cdot mol^{-1}.$$

The standard potential of the bromate/bromide couple can be calculated from the data selected for BrO_3^- (see Section VI.1.3.2) and the CODATA [89COX/WAG] values after correction to 0.1 MPa.

$$BrO_3^- + 3H_2O(l) + 6e^- \rightleftharpoons Br^- + 6OH^- \qquad (VI.13)$$

$$E^{\circ}(VI.13,\ 298.15\ K) = (0.6130 \pm 0.0014)\ V$$
$$\log_{10} K^{\circ}(VI.13,\ 298.15\ K) = 62.17 \pm 0.14$$
$$\Delta_r H^{\circ}_{m}(VI.13,\ 298.15\ K) = -(577.31 \pm 0.59)\,kJ \cdot mol^{-1}$$

VI.1.4. Iodine auxiliary species

CODATA [89COX/WAG] Key Values exist for the following iodine species: I(g), I_2(g), I_2(cr), I^- and HI(g). For various calculations, standard thermodynamic data on other iodine species are required. Their selection is briefly described below.

VI.1.4.1. Iodic acid

Perrin [82PER] reported six values [67PET/PRU, 41LI/LO, 39NAI/RIC, 68HAL/SAB, 33FUO/KRA, 55LEI] of the protonation constant of iodate to form iodic acid, HIO_3(aq), at 25°C, according to Reaction (VI.14).

$$H^+ + IO_3^- \rightleftharpoons HIO_3(aq) \qquad (VI.14)$$

Two of these values [33FUO/KRA, 55LEI] are based on data from Kraus and Parker [22KRA/PAR]. The agreement between the values reported is excellent, and this review selects the average,

$$\log_{10} K_1^\circ(\text{VI.14, 298.15 K}) = 0.788 \pm 0.029.$$

The value of $\log_{10} K_1^\circ$(VI.14) = (0.77 ± 0.03) reported by Smith and Martell [76SMI/MAR] is in good agreement with the value proposed by this review.

Li and Lo [41LI/LO] report equilibrium constants also at 30 and 35°C. However, these data are not sufficiently precise to allow a determination of the corresponding enthalpy of reaction. Smith and Martell [76SMI/MAR] reported the enthalpy of reaction, $\Delta_r H_m^\circ(\text{VI.14}) = 10.0\,\text{kJ}\cdot\text{mol}^{-1}$, and in the first supplement [82MAR/SMI] they reported $\Delta_r H_m^\circ(\text{VI.14}) = 2.7\,\text{kJ}\cdot\text{mol}^{-1}$. This review does not consider any of these data sufficiently well documented to warrant their inclusion in this data base.

VI.1.4.2. Iodate

As for the chlorate and bromate species, this review accepts the assessment made by Johnson, Smith, Appelman and Hubbard [70JOH/SMI]. The selected values are

$$S_m^\circ(IO_3^-, \text{aq}, 298.15\,\text{K}) = (118 \pm 2)\,\text{J}\cdot\text{K}^{-1}\cdot\text{mol}^{-1}$$
$$\Delta_f H_m^\circ(IO_3^-, \text{aq}, 298.15\,\text{K}) = -(219.7 \pm 0.5)\,\text{kJ}\cdot\text{mol}^{-1}.$$

The Gibbs energy of formation is calculated from the selected enthalpy of formation and entropy.

$$\Delta_f G_m^\circ(IO_3^-, \text{aq}, 298.15\,\text{K}) = -(126.34 \pm 0.78)\,\text{kJ}\cdot\text{mol}^{-1}$$

VI.1.4.3. Redox data involving iodine species

Iodine species participate in a large number of redox processes. The thermodynamic data of the couple I_2(cr)/I^- according to Reaction (VI.15) can be calculated from the CODATA [89COX/WAG] values corrected to the standard state pressure of 0.1 MPa.

$$I_2(cr) + 2e^- \rightleftharpoons 2I^- \qquad (VI.15)$$

$$E^\circ(\text{VI.15, 298.15 K}) = (0.5361 \pm 0.0012)\ \text{V}$$
$$\log_{10} K^\circ(\text{VI.15, 298.15 K}) = 18.12 \pm 0.04$$
$$\Delta_r H^\circ_m(\text{VI.15, 298.15 K}) = -(113.56 \pm 0.10)\ \text{kJ} \cdot \text{mol}^{-1}$$

VI.2. Group 16 (chalcogen) auxiliary species

VI.2.1. Sulphur auxiliary species

VI.2.1.1. Elemental sulphur

The entropy of S(cr) is a CODATA [89COX/WAG] Key Value. As S(cr) is the elemental reference state of sulphur, the Gibbs energy of formation and enthalpy of formation are by definition equal to zero. This review accepts the heat capacity evaluated in the IAEA review [84GRO/DRO].

$$C^\circ_{p,m}(\text{S, cr, 298.15 K}) = (22.56 \pm 0.05)\ \text{J} \cdot \text{K}^{-1} \cdot \text{mol}^{-1}$$

VI.2.1.2. Sulphide auxiliary species

Hydrogen sulphide is a weak dibasic acid, and only the first dissociation constant can be measured in aqueous solution. The sulphide ion, S^{2-}, is a much stronger base than the hydroxide ion, OH^-. Hence, no measurable quantities of S^{2-} can be be found in aqueous solution. Solubility products of sparingly soluble metal ion sulphides are always determined by using reactions involving HS^- or $H_2S(aq)$, *e.g.*,

$$\text{MS(s)} + \text{H}^+ \rightleftharpoons \text{M}^{2+} + \text{HS}^-, \qquad \text{(VI.16)}$$

for which the solubility product is $^*K_s = ([\text{M}^{2+}][\text{HS}^-])/[\text{H}^+]$. $^*K_s$ can usually be determined experimentally with high precision, but this is not the case for reactions such as

$$\text{MS(s)} \rightleftharpoons \text{M}^{2+} + \text{S}^{2-}. \qquad \text{(VI.17)}$$

It is unfortunate that many publications and compilations report solubility products according to Eq. (VI.17), rather than solubility constants according to Eq. (VI.16). In order to obtain the solubility product, the experimentally determined value of $^*K_s$(VI.16) is combined with the poorly known dissociation constant of HS^-, *cf.* Eq. (VI.19), to calculate a value for K_s(VI.17) which cannot be more accurate than the dissociation constant of HS^-.

In most thermochemical calculations, *e.g.*, of solubilities and speciation, it is sufficient to use Eq. (VI.16), because S^{2-} is not present in measurable amounts in aqueous solutions. If these calculations are made by using Eqs. (VI.17) and (VI.19) instead, it is essential that the equilibrium constants are consistent, *i.e.*, that the same numerical value is used for K(VI.19) for relating $^*K_s$(VI.16) to K_s(VI.17). This review recommends special caution when using compiled solubility products of metal ion sulphides. For the few cases where a reliable value for the second dissociation constant is desirable, this review therefore tries to assess the various estimates for this constant that have been made in the past.

The first dissociation constant of $H_2S(aq)$ refers to the reaction

$$H_2S(aq) \rightleftharpoons HS^- + H^+ \qquad (VI.18)$$

and can be calculated from the CODATA [89COX/WAG]. The value obtained is $\log_{10} K^\circ(VI.18) = -(6.99 \pm 0.52)$. The uncertainty obtained in this way is certainly too large. The average of the nine equilibrium constants reported by Perrin [82PER] at 25°C is $\log_{10} K^\circ(VI.18) = -6.96$, and the largest deviation of a single value from the average is 0.17. This review considers this deviation as an appropriate uncertainty of the selected value calculated from the CODATA values.

$$\log_{10} K^\circ(VI.18, 298.15\ K) = -(6.99 \pm 0.17)$$

The enthalpy of dissociation is obtained from the CODATA [89COX/WAG] values.

$$\Delta_r H_m^\circ(VI.18, 298.15\ K) = (22.3 \pm 2.1)\ kJ \cdot mol^{-1}.$$

The second dissociation constant refers to Reaction (VI.19).

$$HS^- \rightleftharpoons S^{2-} + H^+ \qquad (VI.19)$$

The value of the dissociation constant, $\log_{10} K(VI.19)$, has been the subject of controversy for more than fifty years. Recent investigations by Myers [86MYE] and Licht and Manassen [87LIC/MAN] confirmed the results of Giggenbach [71GIG], Ellis and Giggenbach [71ELL/GIG] and Meyer *et al.* [83MEY/WAR]. Schoonen and Barnes [88SCH/BAR] used a flocculation method to estimate the second dissociation constant of hydrogen sulphide. They reported $\log_{10} K(VI.19) = -(18.51 \pm 0.56)$ at 20°C and $I = 0$. Dubois, Lelieur and Lepoutre [88DUB/LEL] considered the value of $\log_{10} K(VI.19) = -(17.6 \pm 0.3)$ as reliable. However, this value was obtained from measurements in very concentrated solutions of NaOH and does not refer to $I = 0$. This review accepts the value recommended by Myers [86MYE] as a best estimate.

$$\log_{10} K^\circ(VI.19, 298.15\ K) = -19 \pm 2$$

The standard Gibbs energy of formation that now can be calculated,

$$\Delta_f G_m^\circ(S^{2-}, aq, 298.15\ K) = (120.7 \pm 11.6)\ kJ \cdot mol^{-1},$$

is significantly different from those in many other data compilations [80BEN/TEA, 80COD, 82WAG/EVA, 69HEL, 52LAT, 78ROB/HEM2], where $\Delta_f G_m^\circ(S^{2-}$, aq, 298.15 K) lies mostly between 85 and 93 kJ · mol^{-1}, indicating that these authors chose values between −13 and −14 for $\log_{10} K^\circ(VI.19)$.

VI.2.1.3. Polysulphide auxiliary species

A number of oligomeric sulphur species may form in highly alkaline aqueous solution. Dubois, Lelieur and Lepoutre [88DUB/LEL] recently evaluated some of these equilibria in liquid ammonia and discussed equilibrium constants for the dissociation reactions of S_4^{2-}, S_3^{2-} and S_2^{2-} in aqueous solution, calculated from data of Licht,

Hodes and Manassen [86LIC/HOD]. However, these authors [86LIC/HOD] reported that these constants cannot be corrected to $I = 0$ since they were obtained in solutions of high ionic strength (2 M). These data are therefore not selected here. If they are used as a guideline in connection with data selected in this review, one has to be careful not to use constants reported for equilibria involving S^{2-}, *e.g.*, the reaction $2S_2^{2-} \rightleftharpoons S_3^{2-} + S^{2-}$ [88DUB/LEL], but rather those involving HS^-, in order to avoid inconsistencies with different values used for $\log_{10} K$(VI.19). Giggenbach [74GIG] provided evidence that the polysulphide species are not protonated in alkaline solution.

VI.2.1.4. Thiosulphate auxiliary species

Only the first protonation constant of thiosulphate can be studied experimentally in aqueous solution:

$$S_2O_3^{2-} + H^+ \rightleftharpoons HS_2O_3^-. \qquad \text{(VI.20)}$$

Perrin [82PER] tabulated, for 25°C and $I = 0$, protonation constants for Reaction (VI.20) originating from three experimental determinations: $\log_{10} K_1^\circ(\text{VI.20}) = 1.74$, 1.46, and 1.56. This review selects their unweighted average and assigns an uncertainty that spans the range of all three values:

$$\log_{10} K_1^\circ(\text{VI.20, 298.15 K}) = 1.59 \pm 0.15.$$

This value is consistent with the value $\log_{10} K_1^\circ = (1.6 \pm 0.1)$ reported by Smith and Martell [76SMI/MAR].

The Gibbs energy of formation of $S_2O_3^{2-}$ is calculated from the standard potential selected for the sulphite/thiosulphate couple, *cf.* Section VI.2.1.7, and the $\Delta_f G_m^\circ$ value of SO_3^{2-}, selected in Section VI.2.1.5. $\Delta_f G_m^\circ(HS_2O_3^-, \text{aq}, 298.15\,\text{K})$ follows then from $\log_{10} K^\circ$(VI.20):

$$\begin{aligned} \Delta_f G_m^\circ(S_2O_3^{2-}, \text{aq}, 298.15\,\text{K}) &= -(519.3 \pm 11.3)\,\text{kJ}\cdot\text{mol}^{-1} \\ \Delta_f G_m^\circ(HS_2O_3^-, \text{aq}, 298.15\,\text{K}) &= -(528.4 \pm 11.4)\,\text{kJ}\cdot\text{mol}^{-1}. \end{aligned}$$

VI.2.1.5. Sulphite auxiliary species

Sulphurous acid is a medium strength diprotic acid. The stepwise protonation reactions to form sulphurous acid from sulphite are the following:

$$SO_3^{2-} + H^+ \rightleftharpoons HSO_3^- \qquad \text{(VI.21)}$$
$$HSO_3^- + H^+ \rightleftharpoons H_2SO_3(\text{aq}) \qquad \text{(VI.22)}$$

A survey of the protonation constants at $I = 0$ was published by Perrin [82PER] who reported $\log_{10} K_1^\circ(\text{VI.21}) = 7.205$, 7.205, 7.18, and 7.30. The data published by Krunchak *et al.* [73KRU/ROD] at eight different ionic strengths are used by this review to obtain an estimate of the uncertainty involved. The correction of each value to $I = 0$ is done by using the specific ion interaction method and neglecting the $\Delta\varepsilon\, m$ term because of the low ionic strengths used by the authors. This review finds an uncertainty of ±0.05 in $\log_{10} K_1^\circ$(VI.21) of Krunchak *et al.* [73KRU/ROD]. Tartar and Garretson [41TAR/GAR, Tables IV and V] reported uncertainties in their $\log_{10} K_1^\circ$(VI.21) of ±0.03 and ±0.07, calculated from their data in bromide and chloride solutions, respectively. This review recommends the unweighted average of the $\log_{10} K_1^\circ$(VI.21) values compiled by Perrin and mentioned above, with an estimated uncertainty of ±0.08 such that the range of all the four values is spanned:

$$\log_{10} K_1^\circ(\text{VI.21, 298.15 K}) = 7.22 \pm 0.08.$$

Smith and Martell [76SMI/MAR] selected $\log_{10} K_1^\circ(\text{VI.21}) = (7.18 \pm 0.03)$, which is in agreement with the value selected in this review.

The enthalpy change for Reaction (VI.21) was derived by Krunchak *et al.* [73KRU/ROD] from experimental determinations of $\log_{10} K_1$(VI.21) at different temperatures and after extrapolation to $I = 0$. The values found at 25 and 30°C differ by $32\,\text{kJ}\cdot\text{mol}^{-1}$ which indicates that $\Delta_r H_m^\circ$ is strongly temperature-dependent, or that the uncertainty in the experiment is large. This review prefers the latter interpretation and therefore assigns a large uncertainty to the selected value:

$$\Delta_r H_m^\circ(\text{VI.21, 298.15 K}) = (66 \pm 30)\,\text{kJ}\cdot\text{mol}^{-1}.$$

The value given by Smith and Martell [76SMI/MAR], $\Delta_r H_m^\circ(\text{VI.21}) \approx 13\,\text{kJ}\cdot\text{mol}^{-1}$, is only an approximation and they consider it very uncertain.

Equilibrium constants for the second protonation reaction at $I = 0$ and 25°C were also compiled by Perrin [82PER] who reported $\log_{10} K_{1,2}^\circ(\text{VI.22}) = 1.89, 1.82, 1.76, 1.86, 1.764, 1.90, 1.86, 1.86$, and 1.89. This review recommends the unweighted average of these values, $\log_{10} K_{1,2}^\circ(\text{VI.22}) = 1.84$. The uncertainty estimates, based on the experimental data in Refs. [41TAR/GAR] and [34JOH/LEP], are ±0.06 and ±0.02, respectively. This review selects an uncertainty that spans the range of all the values given above:

$$\log_{10} K_{1,2}^\circ(\text{VI.22, 298.15 K}) = 1.84 \pm 0.08.$$

The value selected by Smith and Martell [76SMI/MAR] is $\log_{10} K_{1,2}^\circ = (1.91 \pm 0.02)$, which is not inconsistent with the one selected by this review.

The enthalpy change $\Delta_r H_m^\circ$ for Reaction (VI.22) is taken from the data of Johnstone and Leppla [34JOH/LEP] who reported $\Delta_r H_m^\circ(\text{VI.22}) = 16\,\text{kJ}\cdot\text{mol}^{-1}$ from the temperature variation of $\log_{10} K_{1,2}^\circ$(VI.22). This review estimates the uncertainty to $\pm5\,\text{kJ}\cdot\text{mol}^{-1}$. The selected value is thus

$$\Delta_r H_m^\circ(\text{VI.22, 298.15 K}) = (16 \pm 5)\,\text{kJ}\cdot\text{mol}^{-1}.$$

As a comparison, Smith and Martell [76SMI/MAR] gave $\Delta_r H_m^\circ(\text{VI.22}) = 16.7\,\text{kJ}\cdot\text{mol}^{-1}$, which agrees very well with the one selected here.

The Gibbs energy of formation of SO_3^{2-} is calculated from the selected standard potential of the couple SO_4^{2-}/SO_3^{2-}, *cf.* Section VI.2.1.7. The values of $\log_{10} K_1^\circ$(VI.21) and $\log_{10} K_{1,2}^\circ$(VI.22) selected above allow the calculation of the Gibbs energy of formation for both HSO_3^- and $H_2SO_3(aq)$:

$$\begin{aligned}\Delta_f G_m^\circ(SO_3^{2-}, aq, 298.15\,K) &= -(487.47 \pm 4.02)\,kJ \cdot mol^{-1}\\ \Delta_f G_m^\circ(HSO_3^-, aq, 298.15\,K) &= -(528.69 \pm 4.05)\,kJ \cdot mol^{-1}\\ \Delta_f G_m^\circ(H_2SO_3, aq, 298.15\,K) &= -(539.19 \pm 4.07)\,kJ \cdot mol^{-1}.\end{aligned}$$

These values are in good agreement with those recommended by the US NBS [82WAG/EVA].

VI.2.1.6. Sulphate auxiliary species

The present review selects the CODATA [89COX/WAG] values for SO_4^{2-} and HSO_4^-, which are used to calculate the equilibrium constant for the reaction

$$SO_4^{2-} + H^+ \rightleftharpoons HSO_4^-. \qquad \text{(VI.23)}$$

The value obtained is $\log_{10} K_1^\circ(\text{VI.23}) = (1.98 \pm 0.25)$ at 298.15 K, where the uncertainty is based on the uncertainties calculated for $\Delta_f G_m^\circ(SO_4^{2-}$, aq, 298.15 K) and $\Delta_f G_m^\circ(HSO_4^-$, aq, 298.15 K) from $\Delta_f H_m^\circ$ and S_m° of these species given by CODATA [89COX/WAG]. The selection of the CODATA values [89COX/WAG, *p.*34] is based on $\Delta_r G_m^\circ(\text{VI.23}) = -(11.340 \pm 0.050)\,kJ \cdot mol^{-1}$, which corresponds to $\log_{10} K_1^\circ(\text{VI.23}) = (1.987 \pm 0.009)$. Since the values selected in the present review are consistent with the values presented in the selected data tables of the CODATA review [89COX/WAG], the value of $\log_{10} K_1^\circ(\text{VI.23}) = 1.98$ is maintained. However, its derived uncertainty of ± 0.25 is the result of a series of error propagations, and this review therefore examines the range of uncertainty that may realistically be expected from direct measurements of $\log_{10} K_1(\text{VI.23})$.

Several compilations of such data have been published in the past, one of the more recent ones by Perrin [82PER] who gave a dozen values that refer to 25°C and $I = 0$. The unweighted average of these values is $\log_{10} K_1^\circ(\text{VI.23}) = 1.98$, which is identical with the value calculated using the CODATA [89COX/WAG] values. The uncertainty in $\log_{10} K_1^\circ(\text{VI.23})$ can be estimated based on discussions given in Refs. [58NAI/NAN, 64DUN/NAN, 66MAR/JON, 78MON/AMI]. Nair and Nancollas [58NAI/NAN] and Dunsmore and Nancollas [64DUN/NAN] reported an uncertainty of ± 0.004 in $\log_{10} K_1^\circ(\text{VI.23})$, where the main source of the uncertainty seems to be the extrapolation procedure to $I = 0$. Marshall and Jones [66MAR/JON] obtained $\log_{10} K_1^\circ(\text{VI.23})$ from solubility data and reported an uncertainty of ± 0.004, too. Monk and Amira [78MON/AMI] also estimated an uncertainty of ± 0.004 in $\log_{10} K_1^\circ(\text{VI.23})$ obtained from emf data. Pitzer, Roy and Silvester [77PIT/ROY] discussed the thermodynamics of sulphuric acid and reinterpreted previous experimental data by using Pitzer's virial equations [73PIT]. The authors found that two sets of virial coefficients and $\log_{10} K(\text{VI.23}, 298.15\text{ K})$ gave essentially the same fit. Pitzer, Roy and Silvester preferred the set with $\log_{10} K(\text{VI.23}, 298.15\text{ K}) = 1.979$, which is the same as the CODATA value. Several authors (*e.g.*, Dunsmore and Nancollas [64DUN/NAN]) pointed out that the value of $\log_{10} K_1^\circ(\text{VI.23})$ is strongly dependent on the method used to extrapolate the experimental data to $I = 0$.[1] For

[1] A paper by Dickson *et al.* [90DIC/WES], received after the draft of this section was completed,

this reason, the present review assigns a comparatively large uncertainty of ± 0.05 to the recommended value.

$$\log_{10} K_1^\circ(\text{VI.23, 298.15 K}) = 1.98 \pm 0.05$$

The selected enthalpy change in the protonation reaction of sulphate is

$$\Delta_r H_m^\circ(\text{VI.23, 298.15 K}) = (22.4 \pm 1.1)\,\text{kJ} \cdot \text{mol}^{-1}$$

calculated from the CODATA [89COX/WAG] values. This value is in good agreement with the values given by Nair and Nancollas [58NAI/NAN], $\Delta_r H_m^\circ(\text{VI.23}) = (23.4 \pm 0.8)\,\text{kJ} \cdot \text{mol}^{-1}$, Ahrland and Kullberg [71AHR/KUL3], $\Delta_r H_m^\circ(\text{VI.23}) = (23.47 \pm 0.25)\,\text{kJ} \cdot \text{mol}^{-1}$ for a 1 M $NaClO_4$ solution, and Smith and Martell [76SMI/MAR], $\Delta_r H_m^\circ(\text{VI.23}) = (22.6 \pm 1.2)\,\text{kJ} \cdot \text{mol}^{-1}$, but differs considerably from the one measured by Marshall and Jones [66MAR/JON], $\Delta_r H_m^\circ(\text{VI.23}) = -16.1\,\text{kJ} \cdot \text{mol}^{-1}$. The analysis in Ref. [77PIT/ROY] gives $\Delta_r H_m^\circ(\text{VI.23, 298.15 K}) = 23.5\,\text{kJ} \cdot \text{mol}^{-1}$ for the set preferred by the authors. This value is in agreement with the CODATA value and with most of the calorimetric data.

Additional protons can only be added to HSO_4^- at very high acidities and/or very low water activities. These conditions are not encountered in ordinary aquatic systems, and no equilibrium data are therefore selected for these systems.

VI.2.1.7. Redox data involving sulphur species

Redox equilibria between sulphur containing species are in general slow, and equilibrium is only attained in hydrothermal systems ($t > 200$°C). The standard potential of the couple SO_4^{2-}/S(cr), for example, can be calculated from the CODATA [89COX/WAG] values, but it has no practical significance. Redox equilibria between polysulphide species are in general more rapid, and such equilibria may determine the redox potential of waters containing sulphide on the intrusion of oxygen gas [86LIC/HOD].

This review selects a small number of standard potentials that allow the calculation of the standard Gibbs energies of formation of the species described in Sections VI.2.1.4 and VI.2.1.5.

The standard potential given by Latimer [52LAT] for the sulphate/sulphite couple, according to the reaction

$$SO_4^{2-} + H_2O(l) + 2e^- \rightleftharpoons SO_3^{2-} + 2OH^-, \qquad \text{(VI.24)}$$

in basic solution is accepted with an estimated uncertainty of ± 0.02 V.

$$E^\circ(\text{VI.24, 298.15 K}) = -(0.93 \pm 0.02)\ \text{V}$$
$$\log_{10} K^\circ(\text{VI.24, 298.15 K}) = -31.4 \pm 0.7$$

reports measurements of the dissociation constant of HSO_4^- in aqueous sodium chloride solutions up to 250°C. The ionic strength was varied from 0.1 to 5 m. The results, coupled with isopiestic, calorimetric and solubility studies, may be used to develop an internally consistent model for speciation calculations in the system $Na^+/H^+/HSO_4^-/SO_4^{2-}/Cl^-$ in aqueous solutions over a wide range of temperatures and salinities.

This value is used to calculate $\Delta_f G_m^\circ$ of SO_3^{2-}, *cf.* Section VI.2.1.5.

Similarly, this review accepts Latimer's [52LAT] standard potential for the sulphite/thiosulphate couple, according to the reaction

$$2SO_3^{2-} + 3H_2O(l) + 4e^- \rightleftharpoons S_2O_3^{2-} + 6OH^-, \qquad (VI.25)$$

in basic solution, also with an estimated uncertainty of ±0.02 V.

$$E^\circ(\text{VI.25, 298.15 K}) = -(0.58 \pm 0.02)\ \text{V}$$
$$\log_{10} K^\circ(\text{VI.25, 298.15 K}) = -39.2 \pm 1.4$$

The $\Delta_f G_m^\circ$ value of $S_2O_3^{2-}$ is calculated by this review, see Section VI.2.1.4.

The standard potential for the reduction of rhombic sulphur to sulphide can be calculated from the CODATA [89COX/WAG] values, after correction to the standard state pressure of 0.1 MPa, for either HS^- or $H_2S(aq)$. The standard potential and equilibrium constant of the $S(cr)/HS^-$ couple according to the reaction

$$S(cr) + H^+ + 2e^- \rightleftharpoons HS^- \qquad (VI.26)$$

thus calculated are

$$E^\circ(\text{VI.26, 298.15 K}) = (0.0634 \pm 0.0110)\ \text{V}$$
$$\log_{10} K^\circ(\text{VI.25, 298.15 K}) = 2.145 \pm 0.371.$$

VI.2.2. Selenium auxiliary species

This review does not make an independent survey of the literature on auxiliary thermodynamic data for species containing selenium. For aqueous species, the compilations reported by Perrin [82PER], Baes and Mesmer [76BAE/MES] and Latimer [52LAT] are used. Cowan [88COW] made a review of selenium thermodynamic data, including a recalculation of solution chemical data to $I = 0$ using the Davies equation. This review prefers the specific ion interaction method (*cf.* Appendix B). The selected values in the Battelle report [88COW] are on the whole in good agreement with the ones selected by this review.

VI.2.2.1. Elemental selenium

This review accepts the standard entropy and heat capacity values evaluated in the IAEA review [84GRO/DRO]. As Se(cr) is the elemental reference state of selenium, the Gibbs energy of formation and enthalpy of formation are by definition equal to zero.

$$S_m^\circ(\text{Se, cr, 298.15 K}) = (42.27 \pm 0.05)\ \text{J} \cdot \text{K}^{-1} \cdot \text{mol}^{-1}$$
$$C_{p,m}^\circ(\text{Se, cr, 298.15 K}) = (25.03 \pm 0.05)\ \text{J} \cdot \text{K}^{-1} \cdot \text{mol}^{-1}$$

VI.2.2.2. Selenide auxiliary species

Only very few determinations of the dissociation constants of hydrogen selenide, $H_2Se(aq)$, have been made [82PER]. $H_2Se(aq)$ is a stronger acid than $H_2S(aq)$, and the selenide ion, Se^{2-}, is a very strong base whose concentration in aqueous solution is very small. Three values for the dissociation constant according to the reaction

$$H_2Se(aq) \rightleftharpoons HSe^- + H^+ \quad \text{(VI.27)}$$

were tabulated by Perrin [82PER]. They are $\log_{10} K(\text{VI.27}) = -3.73$, -3.77 and -3.89, but they do not refer to $I = 0$ and only limited information is available on the ionic media. Latimer [52LAT] accepts $\log_{10} K(\text{VI.27}) = -3.73$, and Baes and Mesmer [76BAE/MES] listed $\log_{10} K(\text{VI.27}) = -3.73$, and -3.89. It is difficult to make a selection of $\log_{10} K^\circ(\text{VI.27})$ on this basis. Nevertheless, this review tentatively accepts the average of the three values reported by Perrin [82PER] and assigns an uncertainty of ± 0.3 to account for the medium effects.

$$\log_{10} K^\circ(\text{VI.27, 298.15 K}) = -3.8 \pm 0.3.$$

This value is in agreement with the one selected from the Gibbs energy values given by the US NBS [82WAG/EVA], $\log_{10} K^\circ(\text{VI.27}) = -3.82$, and the value $\log_{10} K^\circ(\text{VI.27}) = -3.69$ proposed by Cowen [88COW]. This review does not accept the value $\Delta_r H_m^\circ(\text{VI.27}) = -3.3\,\text{kJ} \cdot \text{mol}^{-1}$ given in Ref. [88COW], as this value is not based on direct experimental data. Furthermore, the present review does not accept the thermodynamic data for the selenide ion. HSe^- is such a weak acid, *cf.* HS^-, that no reliable experimental measurements of the equilibrium $HSe^- \rightleftharpoons H^+ + Se^{2-}$ can be made.

There is not enough information available for the selection of a value for the second dissociation constant of hydrogen selenide.

VI.2.2.3. Selenite auxiliary species

Like sulphurous acid, selenious acid is a medium strength diprotic acid. The stepwise protonation reactions to form selenious acid from selenite are the following:

$$SeO_3^{2-} + H^+ \rightleftharpoons HSeO_3^- \quad \text{(VI.28)}$$
$$HSeO_3^- + H^+ \rightleftharpoons H_2SeO_3(aq) \quad \text{(VI.29)}$$

Baes and Mesmer [76BAE/MES] gave $\log_{10} K_1^\circ(\text{VI.28}) = 8.50$ and $\log_{10} K_{1,2}^\circ(\text{VI.29}) = 2.75$, whereas the averages of the data tabulated by Perrin [82PER] at 25°C and $I = 0$ are $\log_{10} K_1^\circ(\text{VI.28}) = 8.34$ and $\log_{10} K_{1,2}^\circ(\text{VI.29}) = 2.77$. The agreement between the various values available is good. This review selects the following values:

$$\log_{10} K_1^\circ(\text{VI.28, 298.15 K}) = 8.4 \pm 0.1$$
$$\log_{10} K_{1,2}^\circ(\text{VI.29, 298.15 K}) = 2.8 \pm 0.2.$$

These values are neither in agreement with those calculated from the Gibbs energies of formation tabulated by the US NBS [82WAG/EVA], $\log_{10} K_1^\circ(\text{VI.28}) = 7.30$ and

$\log_{10} K^{\circ}_{1,2}(\text{VI.29}) = 2.57$, whose sources are unknown, nor with the values of Ref. [88COW], $\log_{10} K^{\circ}(\text{VI.28}) = 8.04$, $\log_{10} K^{\circ}(\text{VI.29}) = 2.46$. The present review does not accept any of the polynuclear selenite species proposed by Barcza and Sillén [71BAR/SIL], as they were made without taking proper account of the liquid junction potentials arising when a large part of the ionic medium is exchanged for the reactants.

The value of $\Delta_f G^{\circ}_m(H_2SeO_3, \text{aq}, 298.15\text{ K}) = -(425.5 \pm 0.6)\,\text{kJ}\cdot\text{mol}^{-1}$ selected in Section VI.2.2.5 allows the calculation of the standard Gibbs energies of hydrogen selenite and selenite:

$$\begin{aligned} \Delta_f G^{\circ}_m(HSeO_3^-, \text{aq}, 298.15\,\text{K}) &= -(409.5 \pm 1.3)\,\text{kJ}\cdot\text{mol}^{-1} \\ \Delta_f G^{\circ}_m(SeO_3^{2-}, \text{aq}, 298.15\,\text{K}) &= -(361.6 \pm 1.4)\,\text{kJ}\cdot\text{mol}^{-1}. \end{aligned}$$

This review accepts, in agreement with the review by Cowen [88COW], the enthalpies of reaction given by Arnek and Barcza [72ARN/BAR].

$$\begin{aligned} \Delta_r H^{\circ}_m(\text{VI.28, 298.15 K}) &= (5.02 \pm 0.50)\,\text{kJ}\cdot\text{mol}^{-1} \\ \Delta_r H^{\circ}_m(\text{VI.29, 298.15 K}) &= (7.07 \pm 0.50)\,\text{kJ}\cdot\text{mol}^{-1} \end{aligned}$$

The uncertainties are estimated by this review.

The following selected entropies of reaction are then calculated:

$$\begin{aligned} \Delta_r S^{\circ}_m(\text{VI.28, 298.15 K}) &= (177.7 \pm 2.5)\,\text{J}\cdot\text{K}^{-1}\cdot\text{mol}^{-1} \\ \Delta_r S^{\circ}_m(\text{VI.29, 298.15 K}) &= (77.3 \pm 4.2)\,\text{J}\cdot\text{K}^{-1}\cdot\text{mol}^{-1}. \end{aligned}$$

VI.2.2.4. Selenate auxiliary species

Selenic acid strongly resembles sulphuric acid in its properties, their dissociation constants being very similar. The only protonation reaction that needs to be considered in aqueous solution is thus the protonation of selenate according to Reaction (VI.30).

$$SeO_4^{2-} + H^+ \rightleftharpoons HSeO_4^- \qquad \text{(VI.30)}$$

There are two experimental studies of equilibrium (VI.30) by Nair [64NAI] and Pamfilov and Agafonova [50PAM/AGA]. Nair [64NAI] used the Davies equation (see Appendix B) to correct to $I = 0$ and obtained $\log_{10} K^{\circ}_1(\text{VI.30}) = 1.66$. Covington and Dobson [65COV/DOB] reanalysed Nair's [64NAI] experiment using a Debye-Hückel expression with two different ion size parameters and obtained $\log_{10} K^{\circ}_1(\text{VI.30}) = (1.70 \pm 0.01)$ and (1.78 ± 0.01), respectively. Covington and Dobson [65COV/DOB] also reanalysed the measurements made by Pamfilov and Agafonova [50PAM/AGA] and obtained $\log_{10} K^{\circ}_1(\text{VI.30}) = (1.92 \pm 0.02)$. This review concurs with Covington and Dobson [65COV/DOB] that there is a discrepancy between these values that cannot be resolved due to insufficient experimental information on the measurements of Pamfilov and Agafonova [50PAM/AGA]. Ghosh and Nair [70GHO/NAI] studied the temperature dependence of Reaction (VI.30) and obtained an equilibrium constant at 25°C of $\log_{10} K^{\circ}_1(\text{VI.30}) = (1.745 \pm 0.019)$ which is in good agreement with Nair's [64NAI] value. For the reasons outlined above, the value selected here is provided with a comparatively large uncertainty:

$$\log_{10} K^\circ(\text{VI.30, 298.15 K}) = 1.80 \pm 0.14.$$

The value calculated from the Gibbs energies of formation published by the US NBS [82WAG/EVA] is $\log_{10} K_1^\circ(\text{VI.30}) = 1.91$, which is consistent with the one selected here. The selected value is also in excellent agreement with the value proposed by Cowen [88COW], $\log_{10} K^\circ(\text{VI.30}) = 1.75$.

The enthalpy of reaction from Ghosh and Nair [70GHO/NAI] is accepted in this review with an estimated uncertainty:

$$\Delta_r H_m^\circ(\text{VI.30, 298.15 K}) = (23.8 \pm 5.0)\,\text{kJ}\cdot\text{mol}^{-1}.$$

The selected entropy of reaction is then calculated to be

$$\Delta_r S_m^\circ(\text{VI.30, 298.15 K}) = (114 \pm 17)\,\text{J}\cdot\text{K}^{-1}\cdot\text{mol}^{-1}.$$

VI.2.2.5. Redox data involving selenium species

Redox reactions involving species containing selenium are in general slow. The information available on the $Se(cr)/Se^{2-}$ couple is insufficient for the selection of a reliable standard potential. This review only considers the standard potential of the couple $H_2SeO_3(aq)/Se(cr)$. The equilibrium of the selenate/selenite couple is very slow and is not included in the selected set of auxiliary data.

Schott, Swift and Yost [28SCH/SWI] determined the potential of the couple $H_2SeO_3(aq)/Se(cr)$ with iodine and iodide, according to Reaction (VI.31).

$$H_2SeO_3(aq) + 4I^- + 4H^+ = Se(cr) + 2I_2(cr) + 3H_2O(l) \qquad \text{(VI.31)}$$

They obtained $\log_{10} K^\circ(\text{VI.31}) = 13.84$ to which this review assigns an uncertainty of ± 0.10. By using the CODATA [89COX/WAG] values corrected to the 0.1 MPa standard state pressure, this review obtains

$$\Delta_f G_m^\circ(H_2SeO_3, \text{aq}, 298.15\,\text{K}) = -(425.5 \pm 0.6)\,\text{kJ}\cdot\text{mol}^{-1}.$$

VI.2.2.6. Solid selenium auxiliary compounds

This review uses the value

$$\Delta_f H_m^\circ(SeO_2, \text{cr}, 298.15\,\text{K}) = -(225.1 \pm 2.1)\,\text{kJ}\cdot\text{mol}^{-1}$$

based on the evaluation of Mills [74MIL].

VI.2.3. Tellurium

This review accepts the standard entropy and heat capacity values evaluated in the IAEA review [84GRO/DRO]. As Te(cr) is the elemental reference state of tellurium, the Gibbs energy of formation and enthalpy of formation are by definition equal to zero.

$$S_m^\circ(\text{Te, cr}, 298.15\,\text{K}) = (49.221 \pm 0.050)\,\text{J}\cdot\text{K}^{-1}\cdot\text{mol}^{-1}$$
$$C_{p,m}^\circ(\text{Te, cr}, 298.15\,\text{K}) = (25.55 \pm 0.10)\,\text{J}\cdot\text{K}^{-1}\cdot\text{mol}^{-1}$$

VI.3. Group 15 auxiliary species

VI.3.1. Nitrogen auxiliary species

Protonation constants of four nitrogen species are briefly discussed here. These include ammonia, $NH_3(aq)$, azide, N_3^-, nitrite, NO_2^-, and nitrate, NO_3^-.

VI.3.1.1. Ammonium

The protonation equilibrium of ammonia according to the reaction

$$NH_3(aq) + H^+ \rightleftharpoons NH_4^+ \qquad \text{(VI.32)}$$

has been extensively studied. From the tabulation of Perrin [82PER] this review obtained the selected value

$$\log_{10} K_1^\circ(\text{VI.32, 298.15 K}) = 9.237 \pm 0.022,$$

which agrees well with the results of other evaluations, *e.g.*, $\log_{10} K_1^\circ(\text{VI.32}) = 9.244$ proposed by Martell and Smith [82MAR/SMI].

The standard Gibbs energy of formation calculated for aqueous ammonia is

$$\Delta_f G_m^\circ(NH_3, aq, 298.15\,K) = -(26.67 \pm 0.30)\,kJ \cdot mol^{-1}.$$

For the enthalpy of reaction this review accepts the value given by Martell and Smith [82MAR/SMI],

$$\Delta_r H_m^\circ(\text{VI.32, 298.15 K}) = -(52.09 \pm 0.21)\,kJ \cdot mol^{-1}.$$

The standard enthalpy of formation calculated for aqueous ammonia is

$$\Delta_f H_m^\circ(NH_3, aq, 298.15\,K) = -(81.17 \pm 0.33)\,kJ \cdot mol^{-1}.$$

VI.3.1.2. Hydroazoic acid

The protonation constant of azide to form hydroazoic acid according to the reaction

$$N_3^- + H^+ = HN_3(aq) \qquad \text{(VI.33)}$$

is evaluated from the values referring to $I = 0$ given by Perrin [82PER]. The data given at 20 and 22°C are corrected to 25°C using the enthalpy of reaction given by Martell and Smith [82MAR/SMI], $\Delta_r H_m^\circ(\text{VI.33}) = 15\,kJ \cdot mol^{-1}$. The selected value is

$$\log_{10} K_1^\circ(\text{VI.33, 298.15 K}) = 4.70 \pm 0.08$$

which is in good agreement with the one proposed by Martell and Smith [82MAR/SMI], $\log_{10} K_1^\circ(\text{VI.33}) = (4.65 \pm 0.03)$.

The enthalpy of reaction given by Martell and Smith [82MAR/SMI] and adopted here,

$$\Delta_r H_m^\circ(\text{VI.33, 298.15 K}) = -(15 \pm 10)\,\text{kJ}\cdot\text{mol}^{-1},$$

is very uncertain and should only be used for estimates.

This review adopts the values of $\Delta_f G_m^\circ$ and $\Delta_f H_m^\circ$ of N_3^- recommended by the US NBS [82WAG/EVA]. The uncertainties are estimated by this review. These values are used to calculate the $\Delta_f G_m^\circ$ and $\Delta_f H_m^\circ$ values for $HN_3(aq)$.

$$\begin{aligned}\Delta_f G_m^\circ(N_3^-, \text{aq}, 298.15\,\text{K}) &= (348.20 \pm 2.00)\,\text{kJ}\cdot\text{mol}^{-1}\\ \Delta_f G_m^\circ(HN_3, \text{aq}, 298.15\,\text{K}) &= (321.37 \pm 2.05)\,\text{kJ}\cdot\text{mol}^{-1}\end{aligned}$$

$$\begin{aligned}\Delta_f H_m^\circ(N_3^-, \text{aq}, 298.15\,\text{K}) &= (275.14 \pm 1.00)\,\text{kJ}\cdot\text{mol}^{-1}\\ \Delta_f H_m^\circ(HN_3, \text{aq}, 298.15\,\text{K}) &= (260.14 \pm 10.05)\,\text{kJ}\cdot\text{mol}^{-1}\end{aligned}$$

VI.3.1.3. Nitrous acid

The experimental determination of the protonation constant of nitrite according to the reaction

$$NO_2^- + H^+ = HNO_2(aq) \qquad \text{(VI.34)}$$

is complicated by the kinetic instability of $HNO_2(aq)$. Nevertheless, there are some accurate experimental determinations which were tabulated by Perrin [82PER]. This review selects the average of the values that refer to 25°C and $I = 0$ and obtains the selected value of

$$\log_{10} K_1^\circ(\text{VI.34, 298.15 K}) = 3.21 \pm 0.16.$$

The value is in good agreement with the one proposed by Martell and Smith [82MAR/SMI], $\log_{10} K_1^\circ(\text{VI.34}) = 3.15$.

An estimate of the enthalpy of reaction is obtained from $\log_{10} K(T)$ data given by Tummavuori and Lumme [68TUM/LUM] and Schmid, Marchgraber and Dunkl [37SCH/MAR]. These authors reported $\Delta_r H_m^\circ(\text{VI.34}) = -12.2\,\text{kJ}\cdot\text{mol}^{-1}$ and $-10.5\,\text{kJ}\cdot\text{mol}^{-1}$, respectively. This review selects the average,

$$\Delta_r H_m^\circ(\text{VI.34, 298.15 K}) = -(11.4 \pm 3.0)\,\text{kJ}\cdot\text{mol}^{-1}.$$

The uncertainty is chosen such that the value is not inconsistent with the one proposed by Martell and Smith [82MAR/SMI], $\Delta_r H_m^\circ(\text{VI.34}) = -8.4\,\text{kJ}\cdot\text{mol}^{-1}$.

VI.3.1.4. Nitric acid

Nitric acid is a strong electrolyte. Undissociated molecules of $HNO_3(aq)$ are found in solutions of very high acidity (> 2 M), and the protonation constant of NO_3^- to form $HNO_3(aq)$ is around $\log_{10} K_1 \approx -1.4$, *cf.* Perrin [82PER]. This review does not consider data obtained under such extreme conditions where it is difficult or impossible to disentangle effects of incomplete dissociation and activity factor variations. Therefore no value for the protonation constant of nitrate is selected in this review. For solutions with pH > 0.5 nitric acid can be considered as completely dissociated.

VI.3.2. Phosphorus auxiliary species

Values for the phosphorus species in this review are relative to the crystalline, white form. Where needed to interpret data, the enthalpy of formation of red, amorphous phosphorus is selected from the US NBS tables [82WAG/EVA].

$$\Delta_f H_m^\circ(\mathrm{P, am, red, 298.15\,K}) = -(7.5 \pm 2.0)\,\mathrm{kJ \cdot mol^{-1}}.$$

Phosphorus forms a number of oxyacids but only those of P(V) are discussed here.

VI.3.2.1. Phosphoric acid

The dissociation constants of phosphoric acid, $H_3PO_4(aq)$, have been studied experimentally over a period of more than eighty years. The data considered in this review for the first dissociation constant of $H_3PO_4(aq)$ according to Reaction (VI.35)

$$\mathrm{H_3PO_4(aq)} \rightleftharpoons \mathrm{H_2PO_4^- + H^+} \qquad \text{(VI.35)}$$

are those reported by Perrin [82PER] for 25°C and $I = 0$, with the addition of the precise data reported by Mesmer and Baes [74MES/BAE] for Reaction (VI.36).

$$\mathrm{H_3PO_4(aq) + OH^-} \rightleftharpoons \mathrm{H_2PO_4^- + H_2O(l)} \qquad \text{(VI.36)}$$

Note that the data reported by Mesmer and Baes [74MES/BAE] for Reactions (VI.36) and (VI.38) were erroneously represented by Perrin [82PER] as referring to the third and second dissociation constant of phosphoric acid (pK_3 and pK_2), respectively. This review selects

$$\log_{10} K^\circ(\mathrm{VI.35,\ 298.15\ K}) = -2.14 \pm 0.03,$$

where the uncertainty represents the largest deviation of the reported values from the average selected here. This value is in excellent agreement with the value obtained by Pitzer and Silvester [76PIT/SIL], $\log_{10} K^\circ(\mathrm{VI.35}) = 2.146$, as well as with that selected by Martell and Smith [82MAR/SMI], $\log_{10} K^\circ(\mathrm{VI.35}) = (2.148 \pm 0.002)$,

The standard Gibbs energy of formation of phosphoric acid is calculated from $\log_{10} K^\circ(\mathrm{VI.35})$ and the CODATA [89COX/WAG] value for $H_2PO_4^-$:

$$\Delta_f G_m^\circ(\mathrm{H_3PO_4, aq, 298.15\,K}) = -(1149.4 \pm 1.6)\,\mathrm{kJ \cdot mol^{-1}}.$$

The enthalpy of Reaction (VI.35) is calculated from the data reported by Mesmer and Baes [74MES/BAE] for Reaction (VI.36). By using the CODATA [89COX/WAG] value for the enthalpy of the reaction $H^+ + OH^- \rightleftharpoons H_2O(l)$, this review obtains $\Delta_r H_m^\circ(\mathrm{VI.35}) = -(8.48 \pm 0.34)\,\mathrm{kJ \cdot mol^{-1}}$. Martell and Smith [82MAR/SMI] reported $\Delta_r H_m^\circ(\mathrm{VI.35}) = -(7.95 \pm 0.42)\,\mathrm{kJ \cdot mol^{-1}}$. This review prefers the value of Mesmer and Baes [74MES/BAE] but assigns a slightly larger uncertainty:

$$\Delta_r H_m^\circ(\mathrm{VI.35,\ 298.15\ K}) = -(8.48 \pm 0.60)\,\mathrm{kJ \cdot mol^{-1}}.$$

This value is used to calculate the standard enthalpy of formation of $H_3PO_4(aq)$,

$$\Delta_f H_m^\circ(H_3PO_4, aq, 298.15\,K) = -(1294.1 \pm 1.6)\,kJ \cdot mol^{-1}.$$

The second dissociation constant of phosphoric acid according to the reaction

$$H_2PO_4^- \rightleftharpoons HPO_4^{2-} + H^+ \qquad (VI.37)$$

can be calculated from the CODATA [89COX/WAG] values and results in $\log_{10} K^\circ(VI.37) = (7.212 \pm 0.388)$, where the uncertainty is based on the uncertainties calculated for $\Delta_f G_m^\circ(H_2PO_4^-$, aq, 298.15 K) and $\Delta_f G_m^\circ(HPO_4^{2-}$, aq, 298.15 K) from $\Delta_f H_m^\circ$ and S_m° of these species given by CODATA [89COX/WAG]. The derived uncertainty of ± 0.388 is thus the result of a series of error propagations. This review reduces the uncertainty to ± 0.013 based on the span of the values from accurate experimental data reported by Perrin [82PER]. The selected value is

$$\log_{10} K^\circ(VI.37, 298.15\,K) = -7.212 \pm 0.013.$$

The value proposed by Martell and Smith [82MAR/SMI], $\log_{10} K^\circ(VI.37) = -(7.199 \pm 0.002)$, is consistent with the one selected here.

The enthalpy of Reaction (VI.37) calculated from the CODATA [89COX/WAG] values is $\Delta_r H_m^\circ(VI.35) = (3.6 \pm 2.1)\,kJ \cdot mol^{-1}$. Mesmer and Baes [74MES/BAE] reported data for the reaction

$$H_2PO_4^- + OH^- \rightleftharpoons HPO_4^{2-} + H_2O(l). \qquad (VI.38)$$

The equilibrium constant, erroneously represented by Perrin [82PER], is used above for the selection of the uncertainty in $\log_{10} K^\circ(VI.37)$. By using the CODATA [89COX/WAG] value for the enthalpy of the reaction $H^+ + OH^- \rightleftharpoons H_2O(l)$, the enthalpy data obtained by Mesmer and Baes [74MES/BAE] lead to $\Delta_r H_m^\circ(VI.37) = (3.20 \pm 0.22)\,kJ \cdot mol^{-1}$, which is in good agreement with the CODATA value. Martell and Smith [82MAR/SMI] reported $\Delta_r H_m^\circ(VI.37) = (3.77 \pm 0.42)\,kJ \cdot mol^{-1}$. This review reduces the uncertainty in $\Delta_r H_m^\circ(VI.37)$ to $\pm 1.0\,kJ \cdot mol^{-1}$:

$$\Delta_r H_m^\circ(VI.37, 298.15\,K) = (3.6 \pm 1.0)\,kJ \cdot mol^{-1}.$$

The third dissociation constant of phosphoric acid refers to Reaction (VI.39).

$$HPO_4^{2-} \rightleftharpoons PO_4^{3-} + H^+ \qquad (VI.39)$$

This review considers the determinations of Bjerrum and Unmack [29BJE/UNM] and of Vanderzee and Quist [61VAN/QUI] as the most precise ones and selects the average of the values reported in these two studies,

$$\log_{10} K^\circ(VI.39, 298.15\,K) = -12.35 \pm 0.03.$$

The value of Ghosh, Ghosh and Prasad [80GHO/GHO], $\log_{10} K^\circ(VI.39) = -11.958$, differs considerably from the other two values. One reason for this may be that the modified Davies equation used to extrapolate the experimental data to zero ionic strength is not satisfactory for the highly charged PO_4^{3-} ion. Martell and Smith [82MAR/SMI] reported $\log_{10} K^\circ(VI.39) = -(12.35 \pm 0.02)$ which is in excellent agreement with the value selected here.

The derived standard Gibbs energy of formation of the phosphate ion is

$$\Delta_f G_m^\circ(PO_4^{3-}, aq, 298.15\,K) = -(1025.5 \pm 1.6)\,kJ \cdot mol^{-1}.$$

Martell and Smith [82MAR/SMI] reported an enthalpy of $\Delta_r H_m^\circ(VI.39) = (14.6 \pm 3.8)\,kJ \cdot mol^{-1}$. The value obtained by Ghosh, Ghosh and Prasad [80GHO/GHO] is $\Delta_r H_m^\circ(VI.39) = 5.2\,kJ \cdot mol^{-1}$. In an earlier work, Latimer, Pitzer and Smith [38LAT/PIT] reported $\Delta_r S_m^\circ(VI.39) = -180\,J \cdot K^{-1} \cdot mol^{-1}$, leading to $\Delta_r H_m^\circ(VI.39) = 16.8\,kJ \cdot mol^{-1}$ by using the $\log_{10} K(VI.39)$ value selected here. Since the value of $\log_{10} K^\circ(VI.39)$ differs considerably from the one selected in this review, the enthalpy of reaction from Martell and Smith [82MAR/SMI] is preferred over the one reported in Ref. [80GHO/GHO].

$$\Delta_r H_m^\circ(VI.39,\ 298.15\ K) = (14.6 \pm 3.8)\,kJ \cdot mol^{-1}$$

The standard enthalpy of formation is calculated to be

$$\Delta_f H_m^\circ(PO_4^{3-}, aq, 298.15\,K) = -(1284.4 \pm 4.1)\,kJ \cdot mol^{-1}.$$

The entropy is obtained from $\Delta_r S_m^\circ$ via the selected equilibrium constant and enthalpy of reaction.

$$S_m^\circ(PO_4^{3-}, aq, 298.15\,K) = -(221 \pm 13)\,J \cdot K^{-1} \cdot mol^{-1}$$

VI.3.2.2. Polyphosphoric acids

At higher concentrations and temperatures, phosphoric acid tends to form oligomers. However, their concentrations are negligible at phosphate concentrations less than 0.045 m and temperatures below 200°C [74MES/BAE]. This review only considers the pyrophosphates among the polyphosphates. The dimerization constant according to Reaction (VI.40),

$$2H_3PO_4(aq) \rightleftharpoons H_4P_2O_7(aq) + H_2O(l), \qquad (VI.40)$$

is calculated from the Gibbs energy values published by the US NBS [82WAG/EVA],

$$\log_{10} K^\circ(VI.40,\ 298.15\ K) = -2.79 \pm 0.17.$$

The uncertainty is derived by this review from an estimated uncertainty of $\pm 1.0\,kJ \cdot mol^{-1}$ in $\Delta_r G_m^\circ(VI.40)$.

The enthalpy of reaction is also calculated from Ref. [82WAG/EVA]. The selected value and the estimated uncertainty are

$$\Delta_r H_m^\circ(VI.40,\ 298.15\ K) = (22.2 \pm 1.0)\,kJ \cdot mol^{-1}.$$

For the dissociation constants of pyrophosphoric acid, $H_4P_2O_7(aq)$, the average values of the data tabulated by Perrin [82PER] from different sources at 25°C and $I = 0$ are selected by this review. The uncertainties represent the largest deviations of the single determinations from the selected average values.

$$\begin{array}{lcll}
H_4P_2O_7(aq) & \rightleftharpoons & H_3P_2O_7^- + H^+ & \log_{10} K^\circ = -1.0 \pm 0.5 \\
H_3P_2O_7^- & \rightleftharpoons & H_2P_2O_7^{2-} + H^+ & \log_{10} K^\circ = -2.25 \pm 0.15 \\
H_2P_2O_7^{2-} & \rightleftharpoons & HP_2O_7^{3-} + H^+ & \log_{10} K^\circ = -6.65 \pm 0.10 \\
HP_2O_7^{3-} & \rightleftharpoons & P_2O_7^{4-} + H^+ & \log_{10} K^\circ = -9.40 \pm 0.15
\end{array}$$

The Gibbs energies of formation are calculated from the selected $\log_{10} K^\circ$(VI.40) and the $\log_{10} K^\circ$ values above, using $\Delta_f G_m^\circ(H_3PO_4, aq, 298.15\ K)$ selected in Section VI.3.2.1.

$$\begin{array}{lcl}
\Delta_f G_m^\circ(H_4P_2O_7, aq, 298.15\,K) & = & -(2045.7 \pm 3.3)\,kJ \cdot mol^{-1} \\
\Delta_f G_m^\circ(H_3P_2O_7^-, aq, 298.15\,K) & = & -(2040.0 \pm 4.4)\,kJ \cdot mol^{-1} \\
\Delta_f G_m^\circ(H_2P_2O_7^{2-}, aq, 298.15\,K) & = & -(2027.1 \pm 4.4)\,kJ \cdot mol^{-1} \\
\Delta_f G_m^\circ(HP_2O_7^{3-}, aq, 298.15\,K) & = & -(1989.2 \pm 4.5)\,kJ \cdot mol^{-1} \\
\Delta_f G_m^\circ(P_2O_7^{4-}, aq, 298.15\,K) & = & -(1935.5 \pm 4.6)\,kJ \cdot mol^{-1}
\end{array}$$

The standard enthalpy of formation of $H_4P_2O_7(aq)$ is calculated from the selected value for $\Delta_r H_m^\circ$(VI.40) above and $\Delta_f H_m^\circ(H_3PO_4, aq, 298.15\,K)$ selected in Section VI.3.2.1. The entropy is derived from $\log_{10} K$(VI.40) and $\Delta_r H_m^\circ$(VI.40) via $\Delta_r S_m^\circ$(VI.40).

$$\begin{array}{lcl}
\Delta_f H_m^\circ(H_4P_2O_7, aq, 298.15\,K) & = & -(2280.2 \pm 3.4)\,kJ \cdot mol^{-1} \\
S_m^\circ(H_4P_2O_7, aq, 298.15\,K) & = & (274.9 \pm 7.0)\,J \cdot K^{-1} \cdot mol^{-1}
\end{array}$$

VI.3.3. Arsenic auxiliary species

There are no arsenic compounds at all in the CODATA compilation of Key Values [89COX/WAG], nor in the extensive work by Glushko *et al.* [78GLU/GUR, 82GLU/GUR, *etc.*]. Furthermore, there are several major problems with the chemical thermodynamic databases that are found in the literature for this element, its compounds and aqueous solution species. Although the US NBS values [82WAG/EVA] are internally consistent, the published tables were last revised in 1964. Since then several relevant papers [65BEE/MOR, 71MIN/GIL, 80JOH/PAP, 82COR/OUW2, 88BAR/BUR] on the enthalpy of formation of arsenic compounds have appeared in the literature. Also, the entropy value used for α-As in the US NBS [82WAG/EVA] Tables, $35.1\,J \cdot K^{-1} \cdot mol^{-1}$, differs from the value $35.7\,J \cdot K^{-1} \cdot mol^{-1}$ recommended in later compilations [73HUL/DES, 82PAN]. According to Barten [86BAR], a partial re-evaluation of the arsenic data has been done at US NBS, but this has not been published.

The tables of Naumov, Ryzhenko and Khodakovsky [71NAU/RYZ] use a slightly different entropy ($35.6\,J \cdot K^{-1} \cdot mol^{-1}$) for "As(cr)", assumed to be α-As, and the enthalpy of formation values for the arsenic oxides do not incorporate corrections [71MIN/GIL, 88BAR/BUR] to the key paper of Beezer, Mortimer and Tyler [65BEE/MOR]. Also, these tables are not as extensive as the US NBS tables. The analysis of Sadiq and Lindsay [81SAD/LIN] is similar in many respects to that of Naumov, Ryzhenko and Khodakovsky [71NAU/RYZ], but it used the older entropy value for α-As. Barten [86BAR] reported a database purportedly compatible with the US NBS tables [82WAG/EVA], but with revised values for $\Delta_f H_m^\circ(As_2O_5, cr, 298.15\,K)$

and $S^\circ_m(\text{As}, \alpha, 298.15\,\text{K})$. This database is not complete, and may not be internally consistent.

Although needed, a complete reanalysis of the chemical thermodynamic data for arsenic species is not within the scope of the current review. In general, for key (non-arsenic) species, the US NBS [82WAG/EVA] values do not differ greatly from the CODATA [89COX/WAG] values. In the interest of attempting to maintain a reasonable approximation to internal consistency, only values from the US NBS compilation [82WAG/EVA] are used as auxiliary data for arsenic species. The notation of the chemical formulae is, for reasons of consistency, also adopted from the US NBS tables. Where possible, an indication is given (usually in Appendix A) as to how problems with the arsenic data base affect the selected values.

Uncertainties are not provided for the values in the US NBS tables [82WAG/EVA]. It would appear that $\Delta_f H^\circ_m(\text{As}_2\text{O}_5, \text{cr}, 298.15\,\text{K})$ may be as much as $8\,\text{kJ}\cdot\text{mol}^{-1}$ more positive than the tabulated value. As noted above, $S^\circ_m(\text{As}, \alpha, 298.15\text{ K})$ may be $0.6\text{ J}\cdot\text{K}^{-1}\cdot\text{mol}^{-1}$ in error. This considered, the values given below are assigned uncertainties of $\pm 4\,\text{kJ}\cdot\text{mol}^{-1}$ per arsenic atom in the formula for $\Delta_f H^\circ_m$ and $\Delta_f G^\circ_m$ values, and $\pm 0.6\text{ J}\cdot\text{K}^{-1}\cdot\text{mol}^{-1}$ for the values of S°_m. These large uncertainties do not necessarily apply, of course, to reactions between species listed in the Table below. Recognizing that uncertainties for values in the Tables should be "within 8 to 80 units of the last (right-most) digit" [82WAG/EVA], consistent, smaller uncertainties can be calculated for some reactions (*e.g.*, the protonation reactions).

These data allow the calculation of the equilibrium constants for the following protonation reactions:

$$\text{AsO}_2^- + \text{H}^+ \rightleftharpoons \text{HAsO}_2(\text{aq}) \quad \text{(VI.41)}$$
$$\text{AsO}_4^{3-} + \text{H}^+ \rightleftharpoons \text{HAsO}_4^{2-} \quad \text{(VI.42)}$$
$$\text{HAsO}_4^{2-} + \text{H}^+ \rightleftharpoons \text{H}_2\text{AsO}_4^- \quad \text{(VI.43)}$$
$$\text{H}_2\text{AsO}_4^- + \text{H}^+ \rightleftharpoons \text{H}_3\text{AsO}_4(\text{aq}). \quad \text{(VI.44)}$$

It should be mentioned that H_2AsO_3^- is identical with $\text{AsO}_2^- + \text{H}_2\text{O(l)}$, and $\text{H}_3\text{AsO}_3(\text{aq})$ is identical with $\text{HAsO}_2(\text{aq}) + \text{H}_2\text{O(l)}$. The derived constants for the above protonation equilibria are

$$\log_{10} K^\circ(\text{VI.41, 298.15 K}) = 9.23$$
$$\log_{10} K_1^\circ(\text{VI.42, 298.15 K}) = 11.60$$
$$\log_{10} K_2^\circ(\text{VI.43, 298.15 K}) = 6.76$$
$$\log_{10} K_3^\circ(\text{VI.44, 298.15 K}) = 2.24.$$

As mentioned above, the uncertainties in these equilibrium constants may be smaller than calculated from the uncertainties of the $\Delta_f G^\circ_m$ values listed in Table VI.1.

Table VI.1: Selected thermodynamic data for arsenic, arsenic compounds and solution species.

Species	$\Delta_f G^\circ_m$ (kJ·mol^{-1})	$\Delta_f H^\circ_m$ (kJ·mol^{-1})	S°_m (J·K^{-1}·mol^{-1})	$C^\circ_{p,m}$ (J·K^{-1}·mol^{-1})
α-As	0.0	0.0	35.1 ± 0.6	24.64 ± 0.50
AsO_2^-	−350.02 ± 4.01	−429.03 ± 4.00	40.6 ± 0.6	
AsO_4^{3-}	−648.36 ± 4.01	−888.14 ± 4.00	−162.8 ± 0.6	
As_2O_5(cr)	−782.45 ± 8.02	−924.87 ± 8.00	105.4 ± 1.2	116.52 ± 0.80
As_4O_6(cubi)	−1152.45 ± 16.03	−1313.94 ± 16.00	214.2 ± 2.4	191.29 ± 0.80
As_4O_6(mono)	−1154.0 ± 16.0	−1309.6 ± 16.0	234. ± 3.	
$HAsO_2$(aq)	−402.9 ± 4.0	−456.5 ± 4.0	125.9 ± 0.6	
$HAsO_4^{2-}$	−714.59 ± 4.01	−906.34 ± 4.00	−1.7 ± 0.6	
$H_2AsO_3^-$	−587.08 ± 4.01	−714.79 ± 4.00	110.5 ± 0.6	
$H_2AsO_4^-$	−753.20 ± 4.02	−909.56 ± 4.00	117. ± 1.	
H_3AsO_3(aq)	−639.7 ± 4.0	−742.2 ± 4.0	195. ± 1.	
H_3AsO_4(aq)[(a)]	−766.1 ± 4.0	−902.5 ± 4.0	184. ± 1.	
H_3AsO_4(sln)[(a)]		−904.6 ± 4.00		
$(As_2O_5)_3\cdot 5H_2O$(cr)		−4248.4 ± 24.0		

(a) H_3AsO_4(aq) in the present review is defined as "H_3AsO_4(ao)" in Ref. [82WAG/EVA] where H_3AsO_4(aq) is defined as $\frac{1}{2}(As_2O_5(aq) + 3H_2O(l))$, here noted as "$H_3AsO_4$(sln)". It is not obvious exactly what conditions the value for "H_3AsO_4(sln)" refers to, and this formula is therefore not included in Table IV.1. For other species in the table "aq" is defined as "ao" in Ref. [82WAG/EVA].

VI.3.4. Antimony

The entropy and heat capacity values of crystalline antimony are accepted from Robie, Hemingway and Fisher [78ROB/HEM2].

$$S^\circ_m(\text{Sb, cr, 298.15 K}) = (45.52 \pm 0.21)\ \text{J}\cdot\text{K}^{-1}\cdot\text{mol}^{-1}$$
$$C^\circ_{p,m}(\text{Sb, cr, 298.15 K}) = (25.26 \pm 0.20)\ \text{J}\cdot\text{K}^{-1}\cdot\text{mol}^{-1}$$

These values are consistent with the values reported by other authors [61KEL/KIN, 63HUL/ORR, 71NAU/RYZ, 77BAR/KNA, 79KUB/ALC, 82PAN, 82WAG/EVA, 85PAS].

VI.4. Group 14 auxiliary species

VI.4.1. Carbon auxiliary species

VI.4.1.1. Carbonate auxiliary species

Carbonic acid occurs in aqueous solution as a mixture of $H_2CO_3(aq)$ and $CO_2(aq)$. The equilibrium

$$CO_2(aq) + H_2O(l) \rightleftharpoons H_2CO_3(aq) \qquad (VI.45)$$

lies rather far to the left. In most experimental studies "carbonic acid" is defined as the sum of the concentrations of $H_2CO_3(aq)$ and $CO_2(aq)$, due to the difficulty of analytically distinguishing between the two forms. This convention is adopted in this review, and aqueous carbonic acid is consistently referred to as $CO_2(aq)$.

CODATA [89COX/WAG] values are available for $CO_2(g)$, $CO_2(aq)$, CO_3^{2-}, HCO_3^-, and $H_2O(l)$. From the $\Delta_f H_m^\circ$ and S_m° values given, the values of $\Delta_f G_m^\circ$ and, consequently, all the relevant equilibrium constants and enthalpy changes can be calculated. The twofold propagation of errors during this procedure, however, leads to uncertainties in the resulting equilibrium constants that are significantly higher than those obtained from experimental determinations of the constants. For the equilibria

$$CO_3^{2-} + H^+ \rightleftharpoons HCO_3^- \qquad (VI.46)$$
$$HCO_3^- + H^+ \rightleftharpoons CO_2(aq) + H_2O(l) \qquad (VI.47)$$
$$CO_2(g) \rightleftharpoons CO_2(aq) \qquad (VI.48)$$

the following values are calculated based on the CODATA [89COX/WAG] Key Values, $\log_{10} K_1^\circ(VI.46) = 10.329$, $\log_{10} K_{1,2}^\circ(VI.47) = 6.354$, and $\log_{10} K_p^\circ(VI.48) = -1.472$. The uncertainties obtained from the propagation of the uncertainties of the underlying CODATA values, are ± 0.081, ± 0.065 and ± 0.053, respectively. They are reestimated in this review by using experimental values for the corresponding equilibrium constants. Perrin [82PER] gave the following values for Reaction (VI.46), taken from different sources, at $I = 0$ and 25°C: $\log_{10} K_1^\circ(VI.46) = 10.329$, 10.32, and 10.32. This review selects an uncertainty of ± 0.02. The rounded selected value is thus

$$\log_{10} K_1^\circ(VI.46, 298.15\ K) = 10.33 \pm 0.02.$$

The equilibrium constant of Reaction (VI.47) has been determined by a large number of investigators, and compilations of data were given by Sillén and Martell [64SIL/MAR] and Perrin [82PER]. The agreement between different investigators is excellent, and the precision in the experimental results is generally very good. The values reported by Perrin [82PER] for $I = 0$ and 25°C are used to estimate the uncertainty in $\log_{10} K_{1,2}^\circ$. They are $\log_{10} K_{1,2}^\circ = 6.351$, 6.352, 6.349, 6.366, 6.35, 6.35, 6.38, and 6.364. These results are consistent with the value calculated from the CODATA [89COX/WAG] values, and this review assigns an uncertainty of ± 0.02. This leads to the following rounded selected value:

$$\log_{10} K^\circ_{1,2}(\text{VI.47, 298.15 K}) = 6.35 \pm 0.02.$$

Reaction (VI.48) describes the equilibrium of CO_2 between the gaseous and the aqueous phase. The equilibrium constant of this reaction, K_p, is identical with the Henry's law constant, K_H. Buch, as cited by Stumm and Morgan [81STU/MOR, *p.*204], evaluated experimental solubility data of $CO_2(g)$ and tabulated values of $\log_{10} K^\circ_p$ *vs.* temperature. He reported $\log_{10} K^{\circ(\text{atm})}_p = -1.47$ ($M \cdot atm^{-1}$) at 25°C which, corrected to the 0.1 MPa standard state pressure, becomes $\log_{10} K^\circ_p = -1.48$ ($M{\cdot}bar^{-1}$). Buch also reported values for seawater of different salinities. Nilsson, Rengemo and Sillén [58NIL/REN] and Frydman *et al.* [58FRY/NIL] reported for 25°C referring to 1 atm standard state pressure, $\log_{10} K_p = -(1.51 \pm 0.01)$ in a 1 m $NaClO_4$ and $-(1.55 \pm 0.01)$ in a 3.5 m $NaClO_4$ solution, *i.e.*, $\log_{10} K_p = -(1.52 \pm 0.01)$ and $-(1.56 \pm 0.01)$, respectively, after correction to the 0.1 MPa standard state pressure referred to in this review. Based on the experimental data in Refs. [58NIL/REN] and [58FRY/NIL], and taking into account the uncertainty associated with the correction to $I = 0$, this review assigns an uncertainty of ± 0.02 to the rounded value derived from the CODATA [89COX/WAG] Key Values:

$$\log_{10} K^\circ_p(\text{VI.48, 298.15 K}) = -1.47 \pm 0.02.$$

In chemical systems with a constant and known partial pressure of $CO_2(g)$ it is sometimes convenient to define the protonation constants according to Reactions (VI.49) and (VI.50).

$$CO_3^{2-} + 2H^+ \rightleftharpoons CO_2(g) + H_2O(l) \qquad \text{(VI.49)}$$
$$HCO_3^- + H^+ \rightleftharpoons CO_2(g) + H_2O(l) \qquad \text{(VI.50)}$$

The corresponding equilibrium constants can be calculated from the equilibrium constants of Reactions (VI.46) to (VI.48):

$$\log_{10} K^\circ(\text{VI.49, 298.15 K}) = 18.15 \pm 0.04$$
$$\log_{10} K^\circ(\text{VI.50, 298.15 K}) = 7.82 \pm 0.04.$$

The standard enthalpies of formation of $CO_2(g)$, $CO_2(aq)$, HCO_3^- and CO_3^{2-} were evaluated by CODATA [89COX/WAG]. They allow the calculation of the selected $\Delta_r H^\circ_m$ for Reactions (VI.46) to (VI.48). The uncertainties are calculated from the uncertainties of the CODATA values.

$$\Delta_r H^\circ_m(\text{VI.46, 298.15 K}) = -(14.70 \pm 0.32)\,\text{kJ} \cdot \text{mol}^{-1}$$
$$\Delta_r H^\circ_m(\text{VI.47, 298.15 K}) = -(9.16 \pm 0.29)\,\text{kJ} \cdot \text{mol}^{-1}$$
$$\Delta_r H^\circ_m(\text{VI.48, 298.15 K}) = -(19.75 \pm 0.24)\,\text{kJ} \cdot \text{mol}^{-1}$$

As a comparison, Smith and Martell [76SMI/MAR] reported $\log_{10} K^\circ_1(\text{VI.46}) = (10.329 \pm 0.01)$, $\log_{10} K^\circ_{1,2}(\text{VI.47}) = (6.352 \pm 0.01)$, and $\log_{10} K^{\circ(\text{atm})}_p(\text{VI.48}) = -(1.464 \pm 0.01)$ at 10^5 Pa, corresponding to $\log_{10} K^\circ_p(\text{VI.48}) = -(1.470 \pm 0.01)$. These values agree very well with the ones recommended in this review. This is also true for the various enthalpies of reaction.

VI.4.1.2. Thiocyanate auxiliary species

This review does not make an independent assessment of the chemical thermodynamic data of the thiocyanate ion or of thiocyanic acid. However, in order to provide auxiliary data for the derivation of the thermodynamic data of formation from reaction data, this review selects the US NBS [82WAG/EVA] Gibbs energy and enthalpy of formation. The entropy is derived using the selected values of $\Delta_f H_m^\circ$ and $\Delta_f G_m^\circ$.

$$\begin{aligned}
\Delta_f G_m^\circ(\mathrm{SCN^-, aq, 298.15\,K}) &= (92.7 \pm 4.0)\,\mathrm{kJ\cdot mol^{-1}}\\
\Delta_f H_m^\circ(\mathrm{SCN^-, aq, 298.15\,K}) &= (76.4 \pm 4.0)\,\mathrm{kJ\cdot mol^{-1}}\\
S_m^\circ(\mathrm{SCN^-, aq, 298.15\,K}) &= (144 \pm 19)\,\mathrm{J\cdot K^{-1}\cdot mol^{-1}}
\end{aligned}$$

VI.4.2. Silicon auxiliary species

Only the oxidation state +IV of silicon needs to be taken into account in aqueous systems. The solution chemistry of Si(IV) is complicated by the formation of polynuclear species and the fact that reactions involving polysilicates are very slow. These problems occur at pH > 10. The following discussion is based on the review by Baes and Mesmer [76BAE/MES] and some more recent studies by Busey and Mesmer [77BUS/MES] and Sjöberg *et al.* [83SJO/HAE, 81SJO/NOR, 77BUS/MES].

VI.4.2.1. Mononuclear silicate auxiliary species

A compilation of dissociation constants of $Si(OH)_4(aq)$ for the equilibria (VI.51) and (VI.52) was published by Baes and Mesmer [76BAE/MES] who estimated $\log_{10} K^\circ(\mathrm{VI.51}) = -(9.86 \pm 0.07)$ and $\log_{10} K^\circ(\mathrm{VI.52}) = -(22.92 \pm 0.06)$.

$$\mathrm{Si(OH)_4(aq) \rightleftharpoons SiO(OH)_3^- + H^+} \qquad \text{(VI.51)}$$

$$\mathrm{Si(OH)_4(aq) \rightleftharpoons SiO_2(OH)_2^{2-} + 2H^+} \qquad \text{(VI.52)}$$

Since that publication Sjöberg *et al.* [83SJO/HAE, 81SJO/NOR] and Busey and Mesmer [77BUS/MES] published new measurements of Si(IV) hydrolysis at different ionic strengths and temperatures. The data for $I \neq 0$ and $I < 3.5$ m reported by Baes and Mesmer [76BAE/MES] and the more recent ones in Refs. [77BUS/MES], 81SJO/NOR, 83SJO/HAE] are used for an extrapolation to $I = 0$ with the specific ion interaction theory after reassigning the uncertainties given by the authors (see Figures VI.3 and VI.4). The following results are obtained:

$$\begin{aligned}
\log_{10} K^\circ(\mathrm{VI.51,\ 298.15\ K}) &= -9.81 \pm 0.02\\
\log_{10} K^\circ(\mathrm{VI.52,\ 298.15\ K}) &= -23.14 \pm 0.09.
\end{aligned}$$

These values agree well with the results of Sjöberg *et al.* [83SJO/HAE] who obtained $\log_{10} K^\circ(\mathrm{VI.51}) = -9.842$ and $\log_{10} K^\circ(\mathrm{VI.52}) = -23.27$, those of Busey and Mesmer [77BUS/MES] who obtained $\log_{10} K^\circ(\mathrm{VI.51}) = -9.82$, and the value of Baes and Mesmer [81BAE/MES], $\log_{10} K^\circ(\mathrm{VI.51}) = -(9.825 \pm 0.033)$.

From the slopes $\Delta\varepsilon$ of the straight lines in Figures VI.3 and VI.4 (the uncertainties of which are increased by a factor of three in order to account for the scattering of the data points), $\Delta\varepsilon = \varepsilon_{(\mathrm{SiO(OH)_3^-,Na^+})} + \varepsilon_{(\mathrm{H^+,Cl^-})} = (0.04 \pm 0.03)$ and

Figure VI.3: Extrapolation to $I = 0$ of experimental data for the formation of $SiO(OH)_3^-$ using the specific ion interaction theory. Data are taken from [76BAE/MES] (△), [77BUS/MES] (□), [81SJO/NOR] (◇) and [83SJO/HAE] (○), and they refer to $NaClO_4$ (△) and NaCl media (□, ◇, ○), respectively. The shaded area represents the uncertainty range obtained by propagating the resulting uncertainties at $I = 0$ back to $I = 4$ m.

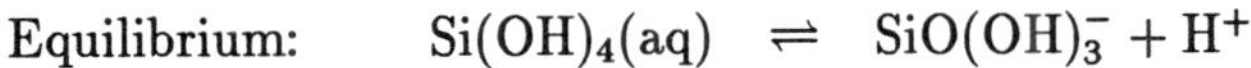

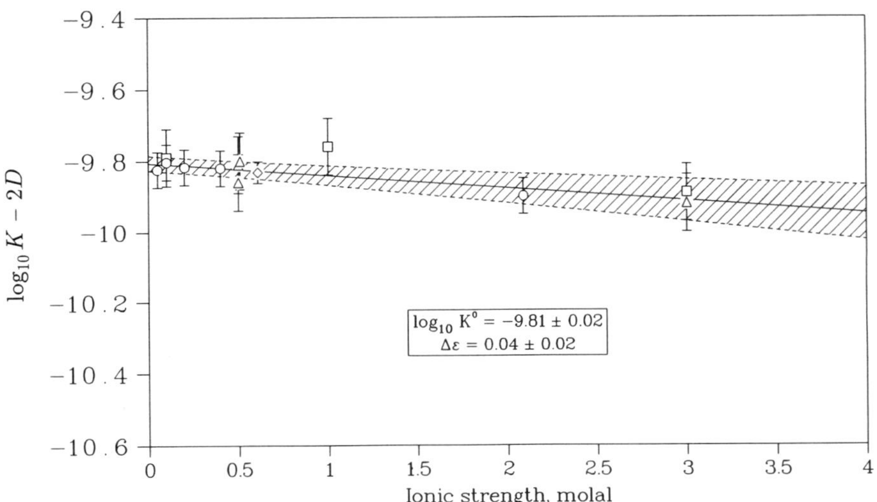

$\Delta\varepsilon = \varepsilon_{(SiO_2(OH)_2^{2-},Na^+)} + 2\varepsilon_{(H^+,Cl^-)} = (0.14 \pm 0.07)$, and $\varepsilon_{(H^+,Cl^-)} = (0.12 \pm 0.01)$ given by Ciavatta [80CIA], this review obtains two new ion interaction coefficients: $\varepsilon_{(SiO(OH)_3^-,Na^+)} = -(0.08 \pm 0.03)$ and $\varepsilon_{(SiO_2(OH)_2^{2-},Na^+)} = -(0.10 \pm 0.07)$. The first value is more negative than would be expected from comparison with other ion interaction coefficients for species of the same charge and similar size (see Table B.4). This may indicate complex formation between the sodium ion and $SiO(OH)_3^-$ and $SiO_2(OH)_2^{2-}$, respectively, as discussed by Busey and Mesmer [77BUS/MES] and Sjöberg *et al.* [83SJO/HAE].

The standard Gibbs energy of formation of $Si(OH)_4(aq)$ is evaluated from the solubility of quartz, according to the equilibrium

$$SiO_2(\text{quartz}) + 2H_2O(l) \rightleftharpoons Si(OH)_4(aq). \qquad (VI.53)$$

Van Lier, de Bruyn and Overbeek [60LIE/BRU] determined the solubility of quartz at 25 and 90°C and found values of 1.8×10^{-4} m (298.15 K) and 9.0×10^{-4} m (363.15 K), respectively. From these two values, $\log_{10} K_s^\circ(\text{VI.53}, 298.15\,\text{K}) = -3.74$ and $\Delta_r H_m^\circ(\text{VI.53}, 298.15\text{ K}) = 22.3\,\text{kJ}\cdot\text{mol}^{-1}$ are obtained, assuming $\Delta_r C_{p,m}^\circ(\text{VI.53}) = 0$. Morey, Fournier and Rowe [62MOR/FOU] measured the solubility of quartz

Figure VI.4: Extrapolation to $I = 0$ of experimental data for the formation of $SiO_2(OH)_2^{2-}$ using the specific ion interaction theory. The data are taken from [76BAE/MES] (△), [81SJO/NOR] (◇) and [83SJO/HAE] (○), and they refer to $NaClO_4$ (△) and NaCl media (△ (lower point at I = 0.5 m), □, ◇, ○), respectively. The shaded area represents the uncertainty range obtained by propagating the resulting uncertainties at $I = 0$ back to $I = 4$ m.

$$\text{Equilibrium:} \quad Si(OH)_4(aq) \rightleftharpoons SiO_2(OH)_2^{2-} + 2H^+$$

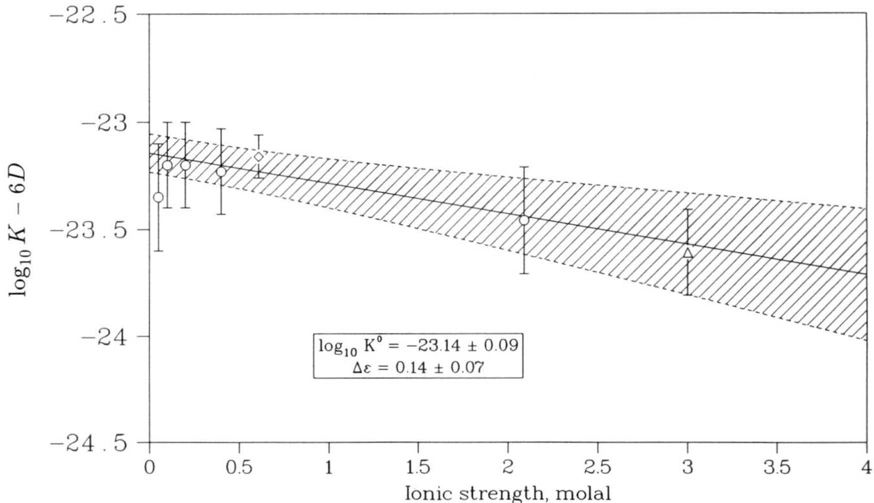

at seven different temperatures between 25 and 240°C. In an experiment that was carried out over a period of 530 days, the solubility equilibrated at 6 ppm SiO_2, *i.e.*, at a concentration of $10^{-4.0}$ m. The authors explained that the difference between this value and the one measured by van Lier, de Bruyn and Overbeek [60LIE/BRU] may be a result of the greater surface energy of the fine-grained quartz used in Ref. [60LIE/BRU]. This review accepts the results of Morey, Fournier and Rowe [62MOR/FOU] and estimates the uncertainty to be ±0.1 in $\log_{10} K_s^\circ$.

$$\log_{10} K_s^\circ(\text{VI.53}, 298.15\,\text{K}) = -4.0 \pm 0.1$$

From the temperature variation of the solubility of quartz between 25 and 240°C [62MOR/FOU], the enthalpy of reaction is derived,

$$\Delta_r H_m^\circ(\text{VI.53}, 298.15\,\text{K}) = (25.4 \pm 3.0)\,\text{kJ}\cdot\text{mol}^{-1}.$$

The uncertainty of $\pm 0.6\,\text{kJ}\cdot\text{mol}^{-1}$ given in Ref. [62MOR/FOU] reflects the precision of the experiment. This review estimates the uncertainty in the selected enthalpy of formation to be $\pm 3.0\,\text{kJ}\cdot\text{mol}^{-1}$.

This allows the derivation of the thermodynamic parameters of $Si(OH)_4(aq)$.

$$\begin{aligned}
\Delta_f G^\circ_m(\mathrm{Si(OH)_4, aq, 298.15\,K}) &= -(1307.7 \pm 1.2)\,\mathrm{kJ \cdot mol^{-1}} \\
\Delta_f H^\circ_m(\mathrm{Si(OH)_4, aq, 298.15\,K}) &= -(1457.0 \pm 3.2)\,\mathrm{kJ \cdot mol^{-1}} \\
S^\circ_m(\mathrm{Si(OH)_4, aq, 298.15\,K}) &= (190 \pm 11)\,\mathrm{J \cdot K^{-1} \cdot mol^{-1}}
\end{aligned}$$

From $\log_{10} K^\circ$(VI.51) and from $\log_{10} K^\circ$(VI.52), the Gibbs energies of formation of the deprotonated species are calculated.

$$\begin{aligned}
\Delta_f G^\circ_m(\mathrm{SiO(OH)_3^-, aq, 298.15\,K}) &= -(1251.7 \pm 1.2)\,\mathrm{kJ \cdot mol^{-1}} \\
\Delta_f G^\circ_m(\mathrm{SiO_2(OH)_2^{2-}, aq, 298.15\,K}) &= -(1175.7 \pm 1.3)\,\mathrm{kJ \cdot mol^{-1}}
\end{aligned}$$

The enthalpies of the formation of $SiO(OH)_3^-$ and $SiO_2(OH)_2^{2-}$ were determined by Busey and Mesmer [77BUS/MES] for temperatures from 0 to 300°C. They reported $\Delta_r H^\circ_m = -(30.2 \pm 1.1)\,\mathrm{kJ \cdot mol^{-1}}$ for the association reaction $\mathrm{Si(OH)_4(aq) + OH^- \rightleftharpoons SiO(OH)_3^- + H_2O(l)}$, and $\Delta_r H_m = -(36.4 \pm 6.3)\,\mathrm{kJ \cdot mol^{-1}}$ for the overall association reaction $\mathrm{Si(OH)_4(aq) + 2OH^- \rightleftharpoons SiO_2(OH)_2^{2-} + 2H_2O(l)}$, the latter referring to a temperature range of 60 to 200°C and a solution of 1 m NaCl. By using the CODATA [89COX/WAG] values to calculate the ionization product of water, the following values are obtained:

$$\begin{aligned}
\Delta_r H^\circ_m(\mathrm{VI.51,\ 298.15\ K}) &= (25.6 \pm 2.0)\,\mathrm{kJ \cdot mol^{-1}} \\
\Delta_r H^\circ_m(\mathrm{VI.52,\ 298.15\ K}) &= (75 \pm 15)\,\mathrm{kJ \cdot mol^{-1}}.
\end{aligned}$$

The uncertainties are assigned by this review. The uncertainty in the second enthalpy value is chosen comparatively high to account for the non-standard conditions the original value refers to.

The enthalpies of formation are then calculated to be

$$\begin{aligned}
\Delta_f H^\circ_m(\mathrm{SiO(OH)_3^-,\ aq,\ 298.15\ K}) &= -(1431.4 \pm 3.7)\,\mathrm{kJ \cdot mol^{-1}} \\
\Delta_f H^\circ_m(\mathrm{SiO_2(OH)_2^{2-},\ aq,\ 298.15\ K}) &= -(1382 \pm 15)\,\mathrm{kJ \cdot mol^{-1}}.
\end{aligned}$$

The entropies are derived via $\Delta_r S^\circ_m$ from the selected equilibrium constants and enthalpies of reaction.

$$\begin{aligned}
S^\circ_m(\mathrm{SiO(OH)_3^-,\ aq,\ 298.15\ K}) &= (88 \pm 13)\,\mathrm{J \cdot K^{-1} \cdot mol^{-1}} \\
S^\circ_m(\mathrm{SiO_2(OH)_2^{2-},\ aq,\ 298.15\ K}) &= -(1 \pm 52)\,\mathrm{J \cdot K^{-1} \cdot mol^{-1}}
\end{aligned}$$

Following Baes and Mesmer [76BAE/MES], this review does not credit the conductance measurements of Ryzhenko and Khitarov [68RYZ/KHI] and their determinations of the dissociation constants for the reactions $\mathrm{SiO_2(OH)_2^{2-} \rightleftharpoons SiO_3OH^{3-} + H^+}$ and $\mathrm{SiO_3OH^{3-} \rightleftharpoons SiO_4^{4-} + H^+}$.

VI.4.2.2. Polynuclear silicate auxiliary species

Polynuclear Si(IV) species are only significant at pH > 10 and at total concentrations of dissolved silica larger than 10^{-3} M. The solubility of silica increases strongly with increasing temperature, but the significance of polynuclear species tends to decrease with increasing temperature [77BUS/MES]. This review does not make an independent evaluation of equilibrium constants and stoichiometries of the polynuclear silicate species. Baes and Mesmer [76BAE/MES] proposed $\log_{10} K^\circ(\text{VI.54}) = -(13.44 \pm 0.2)$ and $\log_{10} K^\circ(\text{VI.55}) = -35.80$.

$$4\text{Si(OH)}_4(\text{aq}) \rightleftharpoons \text{Si}_4\text{O}_6(\text{OH})_6^{2-} + 2\text{H}^+ + 4\text{H}_2\text{O(l)} \quad (\text{VI.54})$$
$$4\text{Si(OH)}_4(\text{aq}) \rightleftharpoons \text{Si}_4\text{O}_8(\text{OH})_4^{4-} + 4\text{H}^+ + 4\text{H}_2\text{O(l)} \quad (\text{VI.55})$$

Lagerström [59LAG] found $\log_{10} K(\text{VI.55}, 3\ \text{M NaClO}_4, 298.15\ \text{K}) = -32.48$. Sjöberg, Öhmann and Ingri [85SJO/OEH] made a combined potentiometric and Si-NMR study of polysilicates at $11.0 \leq \text{pH} \leq 12.2$ in 0.6 m NaCl and found that a large number of polynuclear silicate species are formed. They proposed the following set of equilibria, in addition to Eq. (VI.55):

$$2\text{Si(OH)}_4(\text{aq}) \rightleftharpoons \text{Si}_2\text{O}_3(\text{OH})_4^{2-} + 2\text{H}^+ + \text{H}_2\text{O(l)} \quad (\text{VI.56})$$
$$2\text{Si(OH)}_4(\text{aq}) \rightleftharpoons \text{Si}_2\text{O}_2(\text{OH})_5^{-} + \text{H}^+ + \text{H}_2\text{O(l)} \quad (\text{VI.57})$$
$$3\text{Si(OH)}_4(\text{aq}) \rightleftharpoons \text{Si}_3\text{O}_6(\text{OH})_3^{3-} + 3\text{H}^+ + 3\text{H}_2\text{O(l)} \quad (\text{VI.58})$$
$$3\text{Si(OH)}_4(\text{aq}) \rightleftharpoons \text{Si}_3\text{O}_5(\text{OH})_5^{3-} + 3\text{H}^+ + 2\text{H}_2\text{O(l)} \quad (\text{VI.59})$$
$$4\text{Si(OH)}_4(\text{aq}) \rightleftharpoons \text{Si}_4\text{O}_7(\text{OH})_5^{3-} + 3\text{H}^+ + 4\text{H}_2\text{O(l)} \quad (\text{VI.60})$$

The species $\text{Si}_4\text{O}_7(\text{OH})_5^{3-}$ has a proposed cyclic structure, while the geometry of the other tetrameric species is unknown. The equilibrium constants refer to 25°C and a 0.6 m NaCl medium. There is not enough experimental information that could be used for the correction to $I = 0$ with the specific ion interaction theory. However, it is important to obtain some guidance about the silicate speciation in strongly alkaline solutions, and therefore an attempt is made to provide a silicate data set that is consistently usable in the range from $I = 0$ to the ionic strength reported, $I = 0.6$ m. This review compares these reactions with the dissociation reactions of phosporic acid, $\text{H}_3\text{PO}_4(\text{aq})$. Reactions (VI.58) to (VI.60) belong to the same charge class as the dissociation of $\text{H}_3\text{PO}_4(\text{aq})$ to PO_4^{3-}, Reaction (VI.56) can be compared with the dissociation of $\text{H}_3\text{PO}_4(\text{aq})$ to HPO_4^{2-}, and Reaction (VI.57) with the dissociation of $\text{H}_3\text{PO}_4(\text{aq})$ to $\text{H}_2\text{PO}_4^{-}$. Although the silicate species appear larger in size than the corresponding phosphate species, probably the best approach that is available. From the ion interaction coefficients for H_2PO_4^-, HPO_4^{2-}, and PO_4^{3-}, given by Ciavatta (see Table B.4), the following $\Delta\varepsilon$ values are obtained for $I = 0.6$ m NaCl: $\Delta\varepsilon = (0.09 \pm 0.06)$ for Reaction (VI.56), $\Delta\varepsilon = (0.04 \pm 0.04)$ for Reaction (VI.57), and $\Delta\varepsilon = (0.11 \pm 0.03)$ for Reactions (VI.58) to (VI.60). It can be seen that the correction factors from $I = 0.6$ m(NaCl) to $I = 0$ that are due to the specific ion interaction term are very small, leading to corrections of less than 0.07 logarithmic units in the equilibrium constants, which is within the uncertainty limits. For consistency reasons, this review nevertheless includes the specific ion interaction term in the correction procedure and suggests using the same ion interaction coefficients for the polynuclear silicates as are tabulated in Table B.4 for the phosphate species of the corresponding charge. As is usual for these kinds of reactions, the Debye-Hückel term gives a significant contribution to the correction factor to $I = 0$, *i.e.*, -1.09 logarithmic units for Reaction (VI.56), -0.36 for Reaction (VI.57), and -2.19 for Reactions (VI.58) to

(VI.60). For Reaction (VI.55), the value of $\Delta\varepsilon(\text{VI.55}) = (0.22 \pm 0.10)$ is estimated using $\varepsilon_{(\text{H}^+,\text{Cl}^-)} = 0.12$ and $\varepsilon_{(\text{Si}_4\text{O}_8(\text{OH})_4^{4-},\text{Na}^+)} \approx \varepsilon_{(\text{P}_2\text{O}_7^{4-},\text{Na}^+)} = -0.26$. The correction to $I = 0$ of the value reported by Sjöberg, Öhmann and Ingri [85SJO/OEH] results in $\log_{10} K^\circ(\text{VI.55}) = -36.34$, compared with -36.7 based on the value reported by Lagerström [59LAG], and with -35.8 reported by Baes and Mesmer [76BAE/MES]. This review selects the value of Ref. [85SJO/OEH] that fits with the values selected for the other polynuclear species, however, the uncertainty is assigned such that the values of Refs. [59LAG] and [76BAE/MES] are not inconsistent. This review selects the following constants for the formation of polynuclear silicates at $I = 0$:

$$\begin{aligned}
\log_{10} K^\circ(\text{VI.55, 298.15 K}) &= -36.3 \pm 0.5 \\
\log_{10} K^\circ(\text{VI.56, 298.15 K}) &= -19.0 \pm 0.3 \\
\log_{10} K^\circ(\text{VI.57, 298.15 K}) &= -8.1 \pm 0.3 \\
\log_{10} K^\circ(\text{VI.58, 298.15 K}) &= -28.6 \pm 0.3 \\
\log_{10} K^\circ(\text{VI.59, 298.15 K}) &= -27.5 \pm 0.3 \\
\log_{10} K^\circ(\text{VI.60, 298.15 K}) &= -25.5 \pm 0.3
\end{aligned}$$

Based on the selection of $\Delta_f G_m^\circ(\text{Si(OH)}_4, \text{aq}, 298.15\,\text{K}) = -(1309 \pm 3)\,\text{kJ}\cdot\text{mol}^{-1}$ described in Section VI.4.2.1, the $\Delta_f G_m^\circ$ values of the hydrolysis species formed in Reactions (VI.51) to (VI.60) can be calculated. The following values are obtained:

$$\begin{aligned}
\Delta_f G_m^\circ(\text{Si}_2\text{O}_3(\text{OH})_4^{2-}, \text{aq}, 298.15\,\text{K}) &= -(2269.9 \pm 2.9)\,\text{kJ}\cdot\text{mol}^{-1} \\
\Delta_f G_m^\circ(\text{Si}_2\text{O}_2(\text{OH})_5^{-}, \text{aq}, 298.15\,\text{K}) &= -(2332.1 \pm 2.9)\,\text{kJ}\cdot\text{mol}^{-1} \\
\Delta_f G_m^\circ(\text{Si}_3\text{O}_6(\text{OH})_3^{3-}, \text{aq}, 298.15\,\text{K}) &= -(3048.5 \pm 3.9)\,\text{kJ}\cdot\text{mol}^{-1} \\
\Delta_f G_m^\circ(\text{Si}_3\text{O}_5(\text{OH})_5^{3-}, \text{aq}, 298.15\,\text{K}) &= -(3292.0 \pm 3.9)\,\text{kJ}\cdot\text{mol}^{-1} \\
\Delta_f G_m^\circ(\text{Si}_4\text{O}_8(\text{OH})_4^{4-}, \text{aq}, 298.15\,\text{K}) &= -(4075.2 \pm 5.4)\,\text{kJ}\cdot\text{mol}^{-1} \\
\Delta_f G_m^\circ(\text{Si}_4\text{O}_7(\text{OH})_5^{3-}, \text{aq}, 298.15\,\text{K}) &= -(4136.8 \pm 4.9)\,\text{kJ}\cdot\text{mol}^{-1}
\end{aligned}$$

There is no information about the enthalpy changes of these reactions.

In a recent paper, Zarubin and Nemkina [90ZAR/NEM] described a solubility study of amorphous silica in alkaline sodium chloride solutions (1 M and 3 M) at 25°C. The charged soluble species are dominated by polynuclear species of the charge -1. It is not clear to this review how the authors arrived at the chemical species proposed. However, their proposed equilibrium constants for Reaction (VI.57), $\log_{10} K(\text{VI.57}) = -(7.76 \pm 0.08)$, and $-(7.67 \pm 0.09)$, in 1 M and 3 M NaCl, respectively, are in good agreement with the value selected in this review.

Eikenberg [90EIK] recently made a comparison of experimental solubilities of amorphous silica in the range $7 < \text{pH} < 11.5$. The constants used by Eikenberg were from a draft version of this report which contained an erroneous value of $\log_{10} K^\circ(\text{VI.55})$. Eikenberg found that the NEA selected data fitted the experimental solubilities very well up to pH ≈ 10.5. At higher pH the calculated solubilities were much higher than the experimental values. Eikenberg's analysis indicated that this was due to a value of $\log_{10} K(\text{VI.55})$ which was too high. By using the proper selected value the fit between experimental and calculated solubilities is satisfactory up to pH ≈ 11. The good agreement between the experimental solubility data and the solubilities calculated from main potentiometric data gives additional support for the constants selected in the present review.

VI.5. Other auxiliary species

VI.5.1. Strontium auxiliary species

There are no CODATA [89COX/WAG] values for strontium, its compounds or solution species. Busenberg, Plummer and Parker [84BUS/PLU] recommended a set of values for the enthalpies of formation and entropies at 298.15 K for a number of strontium species, and listed calculated values for the Gibbs energies of formation for the 1 atm standard state. These values are fully compatible with the auxiliary values later selected by CODATA [89COX/WAG]. In this review, values based on Ref. [84BUS/PLU] are used for calculations involving strontium species.

The values for the Gibbs energies of formation for the 1 bar standard state in Table IV.1 of the present review are calculated from the enthalpies of formation and the entropies.

The selected values are summarized in Table VI.2.

Table VI.2: Selected auxiliary thermodynamic data for strontium.

Species	$\Delta_f G_m^\circ$ $(\text{kJ}\cdot\text{mol}^{-1})$	$\Delta_f H_m^\circ$ $(\text{kJ}\cdot\text{mol}^{-1})$	S_m° $(\text{J}\cdot\text{K}^{-1}\cdot\text{mol}^{-1})$
Sr(cr)	0.0	0.0	55.70 ± 0.21
SrO(cr)	-559.94 ± 0.91	-590.60 ± 0.90	55.44 ± 0.50
Sr^{2+}(aq)	-563.86 ± 0.78	-550.90 ± 0.50	-31.5 ± 2.0
$SrCl_2$(cr)	-784.97 ± 0.71	-833.85 ± 0.70	114.85 ± 0.42
$Sr(NO_3)_2$(cr)	-783.15 ± 1.02	-982.36 ± 0.80	194.6 ± 2.1

VI.5.2. Barium auxiliary species

There are no CODATA [89COX/WAG] values for barium, its compounds or solution species. Busenberg and Plummer [86BUS/PLU] and Cordfunke, Konings and Ouweltjes [90COR/KON2] recently discussed the inconsistencies in chemical thermodynamic data bases for barium. In both cases new experimental data were used that were not considered by the other group. In the present review the values of the enthalpies of formation of BaO(cr), Ba(cr) and Ba^{2+}(aq) proposed by Cordfunke, Konings and Ouweltjes [90COR/KON2] are accepted. The value for BaO(cr) is essentially the same as that proposed by Busenberg and Plummer [86BUS/PLU], while those for $BaCl_2$(cr) and Ba^{2+} are slightly closer to those from the US NBS Tables [82WAG/EVA].

The values for the entropies of BaO(cr), Ba(cr) and Ba^{2+}(aq) proposed by Busenberg and Plummer [86BUS/PLU] are used, and the entropy of $BaCl_2$(cr) from Ref. [66GOO/WES] is accepted with an estimated uncertainty of 0.2%. Uncertainties in the enthalpies of formation and in $S^\circ_m(Ba^{2+}$, aq, 298.15 K) are estimated in the present review.

The values for the Gibbs energies of formation for the 1 bar standard state in Table IV.1 of the present review are calculated from the enthalpies of formation and the entropies.

The selected values are summarized in Table VI.3.

Table VI.3: Selected auxiliary thermodynamic data for barium.

Species	$\Delta_f G^\circ_m$ $(kJ \cdot mol^{-1})$	$\Delta_f H^\circ_m$ $(kJ \cdot mol^{-1})$	S°_m $(J \cdot K^{-1} \cdot mol^{-1})$
Ba(cr)	0.0	0.0	62.42 ± 0.84
BaO(cr)	-520.4 ± 2.5	-548.1 ± 2.5	72.07 ± 0.38
Ba^{2+}(aq)	-557.7 ± 2.6	-534.8 ± 2.5	8.4 ± 2.0
$BaCl_2$(cr)	-807.0 ± 2.5	-855.2 ± 2.5	123.68 ± 0.25

Chapter VII

Reference list

[1842RAM] Rammelsberg, S., Pogg. Ann., **56** (1842) 132, cited by Stepanov, M.A., Galkin, N.P., Solubility product of basic uranium(IV) sulphate, Russ. J. Inorg. Chem., **7** (1962) 506-508.

[1848PEL] Peligot, E., Annalen der Chemie, **43** (1848) 313, cited by Stepanov, M.A., Galkin, N.P., Solubility product of basic uranium(IV) sulphate, Russ. J. Inorg. Chem., **7** (1962) 506-508.

[1853FAV/SIL] Favre, P.A., Silbermann, J.T., Recherches sur les quantités de chaleur degagées dans les actions chimiques et moléculaires, Ann. Chim. Phys., **37** (1853) 406-509, in French.

[1872RAM] Rammelsberg, C.F., Monatsber. Preuss. Akad. Wiss., **407** (1872) 53, cited by Pascal, P., Nouveau traité de chimie minérale: Tome XV. Uranium et transuraniens, Quatrième fascicule: Uranium (compléments), Paris: Masson et Cie., 1967, 1213p, in French.

[1896ALO] Aloy, J., Recherches thermiques sur les composés de l'uranium, C. R. Hebd. Séances Acad. Sci., Ser. C, **122** (1896) 1541-1543, in French.

[00PIS] Pissarjewsky, L., Die Überuran-, Übermolybdän- und Überwolframsäuren und entsprechende Säuren, Z. Anorg. Allg. Chem., **24** (1900) 108-122, in German.

[01CON] De Conink, O., Bull. Cl. Sci., Acad. R. Belg., (1901) 485, cited by Stepanov, M.A., Galkin, N.P., Solubility product of basic uranium(IV) sulphate, Russ. J. Inorg. Chem., **7(5)** (1962) 506-508.

[06GLI/LIB] Gliolitti, F., Liberi, G., Gazz. Chim. Ital., **36** (1906) 467, cited by Stepanov, M.A., Galkin, N.P., Solubility product of basic uranium(IV) sulphate, Russ. J. Inorg. Chem., **7** (1962) 506-508.

[08LUT/MIC] Luther, R., Michie, A.C., Das elektromotorische Verhalten von Uranyl-Uranogemengen, Z. Elektrochem., **14(51)** (1908) 826-829, in German.

[10TIT] Titlestad, N., Photo-Voltaketten mit Urano- und Uranylsulfat, Z. Phys. Chem., **72** (1910) 257-307, in German.

[11RUF/HEI] Ruff, O., Heinzelmann, A., Über das Uranhexafluorid, Z. Anorg. Chem., **72** (1911) 63-84, in German.

[12MAR] Markétos, M., Sur les nitrates anhydres d'uranyle et de zinc, C. R. Séances Hebd. Acad. Sci., **155** (1912) 210-213, in French.

[12MIX] Mixter, W.G., Heat of formation of the oxides of vanadium and uranium, and eighth paper on heat of combination of acidic oxides with sodium oxide, Am. J. Sci. and Arts 2, **34** (1912) 141-156.

[15FOR] De Forcrand, R., Etude des hydrates du nitrate d'uranyle et de l'anhydride uranique, Ann. de Chim. (Paris), **9(3)** (1915) 5-48, in French.

[22BRO] Brønsted, J.N., Studies on solubility. IV. The principle of the specific interaction of ions, J. Am. Chem. Soc., **44** (1922) 877-898.

[22BRO2] Brønsted, J.N., Calculation of the osmotic and activity functions in solutions of uni-univalent salts, J. Am. Chem. Soc., **44** (1922) 938-948.

[22KRA/PAR] Kraus, C.A., Parker, H.C., The conductance of aqueous solutions of iodic acid and the limiting value of the equivalent conductance of the hydrogen ion, J. Am. Chem. Soc., **44** (1922) 2429-2449.

[22ROS/TRE] Rosenheim, A., Trewendt, G., Komplexe Uranyl-Hypophosphite, Ber. Dtsch. Chem. Ges., **7B** (1922) 1957-1960, in German.

[28BEC] Beck, G., Thermodynamische Beziehungen zur Konstitution von Verbindungen drei- und mehrwertiger Elemente, Z. Anorg. Allg. Chem., **174** (1928) 31-41, in German.

[28BIL/FEN] Biltz, W., Fendius, C., Beiträge zur systematischen Verwandtschaftslehre: 48. Die Bildungswärmen von UCl_4, UCl_3 und UO_3, Z. Anorg. Allg. Chem., **176** (1928) 49-63, in German.

[28JOB] Job, P., Recherches sur la formation de complexes minéraux en solution, et sur leur stabilité, Ann. de Chim. (Paris), **9** (1928) 113-203, in French.

[28SCH/SWI] Schott, H.F., Swift, E.H., Yost, D.M., The reduction potential of selenium and the free energy of aqueous selenous acid, J. Am. Chem. Soc., **50** (1928) 721-727.

[29BJE/UNM] Bjerrum, N., Unmack, A., Kgl. Danske Videnskab Selskab, Mat-fys Medd., **9** (1929), cited by Perrin, D.D., Ionisation constants of inorganic acids and bases in aqueous solution, 2*nd* ed., IUPAC Chemical Data Series **29**, Oxford: Pergamon Press, 1982, 180p.

[29GER/FRE] Germann, F.E.E., Frey, P.R., The hydrates of uranyl nitrate, their respective vapor pressures, and the relative stability of the lower hydrate, J. Colorado-Wyoming Acad. Sci., **34** (1929) 54-55.

[32MAI/KEL] Maier, C.G., Kelley, K.K., An equation for the representation of high-temperature heat content data, J. Am. Chem. Soc., **54** (1932) 3243-3246.

[32NEU/KRO] Neumann, B., Kröger, C., Haebler, H., Über die Bildungswärmen der Nitride: IV. Uran-, Thor- und Lanthannitrid, Z. Anorg. Allg. Chem., **207** (1932) 145-149, in German.

[32ROS/KEL] Rosenheim, A., Kelmy, M., Über die Verbindungen des vierwertigen Urans, Z. Anorg. Allg. Chem., **206** (1932) 31-34, in German.

[33FUO/KRA] Fuoss, R.M., Kraus, C.A., Properties of electrolytic solutions: II. The evaluations of Λ_0 and of K for incompletely dissociated electrolytes, J. Am. Chem. Soc., **55** (1933) 476-488.

[34JOH/LEP] Johnstone, H.F., Leppla, P.W., The solubility of sulfur dioxide at low partial pressures: The ionization constant and heat of ionization of sulfurous acid, J. Am. Chem. Soc., **56** (1934) 2233-2238.

[34LIE] Liebhafsky, H.A., The equilibrium constant of the bromine hydrolysis and its variation with temperature, J. Am. Chem. Soc., **56** (1934) 1500-1505.

[34SHI] Shilov, E.A., Zh. Fiz. Khim., **5** (1934) 654, cited by Perrin, D.D., Ionisation constants of inorganic acids and bases in aqueous solution, 2*nd* ed., IUPAC Chemical Data Series **29**, Oxford: Pergamon Press, 1982, 180p.

[35GUG] Guggenheim, E.A., The specific thermodynamic properties of aqueous solutions of strong electrolytes, Philos. Mag., **19** (seventh series) (1935) 588-643.

[36SCA] Scatchard, G., Concentrated solutions of strong electrolytes, Chem. Rev., **19** (1936) 309-327.

[37CHR/KRA] Chrétien, A., Kraft, J., Phosphites d'uranyle et phosphite neutre d'uranium tétravalent, Bull. Soc. Chim. Fr., (1937) 1182-1183, in French.

[37CHR/KRA2] Chrétien, A., Kraft, J., Sur les phosphites d'uranyle, C. R. Hebd. Séances Acad. Sci., (1937) 1736-1738, in French.

[37KIE] Kielland, J., Individual activity coefficients of ions in aqueous solutions J. Am. Chem. Soc., **59** (1937) 1675.

[37MON] Montignie, E., Etude de quelques sels d'uranyle, Bull. Soc. Chim. Fr., (1937) 1142-1144, in French.

[37SCH/MAR] Schmid, H., Marchgraber, R., Dunkl, F., Z. Elektrochem., **43** (1937) 337, cited by Perrin, D.D., Ionisation constants of inorganic acids and bases in aqueous solution, 2*nd* ed., IUPAC Chemical Data Series **29**, Oxford: Pergamon Press, 1982, 180p.

[38LAT/PIT] Latimer, W.M., Pitzer, K.S., Smith, W.V., The entropies of aqueous ions, J. Am. Chem. Soc., **60** (1938) 1829-1831.

[38SHI/GLA] Shilov, E.A., Gladtchikova, J.N., On the calculation of the dissociation constants of hypohalogenous acids from kinetic data, J. Am. Chem. Soc., **60** (1938) 490-492.

[38WOO] Wooster, C.B., J. Am. Chem. Soc., **60** (1938) 1609, cited by Perrin, D.D., Ionization constants of inorganic acids and bases in aqueous solution, 2*nd* ed., IUPAC Chemical Data Series **29**, Oxford: Pergamon Press, 1982, 180p.

[39NAI/RIC] Naidich, S., Ricci, J.E., Solubility of barium iodate monohydrate in solutions of uni-univalent electrolytes at 25°C, and the calculation of the dissociation constant of iodic acid from solubility data, J. Am. Chem. Soc., **612** (1939) 3268-3273.

[40COU/PIT] Coulter, L.V., Pitzer, K.S., Latimer, W.M., Entropies of large ions: The heat capacity, entropy and heat of solution of K_2PtCl_6, $(CH_3)_4NI$ and $UO_2(NO_3)_2{\cdot}6H_2O$, J. Am. Chem. Soc., **62** (1940) 2845-2851.

[40HAG] Hagisawa, H., Bull. Inst. Phys. Chem. Res. (Tokyo), **19** (1940) 1220, cited by Perrin, D.D., Ionisation constants of inorganic acids and bases in aqueous solution, 2*nd* ed., IUPAC Chemical Data Series **29**, Oxford: Pergamon Press, 1982, 180p.

[40KEL] Kelley, K.K., Contributions to the data on theoretical metallurgy: IX. The entropies of inorganic substancies. Revision (1940) of data and methods of calculation, US Bureau of Mines Bulletin, **434** (1948) 115p.

[41AMP/THO] Amphlett, C.B., Thomas, L.F., Reports B-18 and B-27, 1941, cited by Katz, J.J., Rabinowitch, E., The chemistry of uranium, New York: Dover Publications, 1951, 608p.

[41LI/LO] Li, N.C., Lo, Y.T., Solubility studies: III. The ionization constant of iodic acid at 25, 30 and 35°C, J. Am. Chem. Soc., **63** (1941) 397-399.

[41TAR/GAR] Tartar, H.V., Garretson, H.H., The thermodynamic ionization constants of sulfurous acid at 25°C, J. Am. Chem. Soc., **63** (1941) 808-816.

[41VOS/COO] Vosburgh, W.C., Cooper, G.R., Complex ions: I. The identification of complex ions in solution by spectrophotometric measurements, J. Am. Chem. Soc., **63** (1941) 437-442.

[41WIL/LYN] Williamson, A.T., Lynch,., Report B-18, 1941, cited by Katz, J.J., Rabinowitch, E., The chemistry of uranium, New York: Dover Publications, 1951, 608p.

[42JEN/AND] Jenkins, F.A., Anderson, O.E., Reports RL 4.6.3 and 4.6.4, September 1942, cited by Müller, M.E., The vapor pressure of uranium halides, Report AECD-2029, US Atomic Energy Commission, Oak Ridge, Tennesse, USA, 1948, 118p.

[42SCH/THO] Schelberg, A., Thompson, R.W., Unpublished report A-809, 1942, cited by Katz, J.J., Rabinowitch, E., The chemistry of uranium, New York: Dover Publications, 1951, 608p.

[42THO/SCH] Thompson, R.W., Schelberg, A., Unpublished report A-179, 1942, cited by Katz, J.J., Rabinowitch, E., The density of uranium, New York: Dover Publications, 1951, 608p.

[43ALT/LIP] Altman, D., Lipkin, D., Weissman, S., Unpublished report RL-4.6.22, 1943, cited by Katz, J.J., Rabinowitch, E., The chemistry of uranium, New York: Dover Publications, 1951, 608p.

[43ALT] Altman, D., Unpublished report RL-4.6.156, 1943, cited by Katz, J.J., Rabinowitch, E., The chemistry of uranium, New York: Dover Publications, 1951, 608p.

[43ALT2] Altman, D., Unpublished report RL-4.6.202, 1943, cited by Katz, J.J., Rabinowitch, E., The chemistry of uranium, New York: Dover Publications, 1951, 608p.

[43KHL/GUR] Khlopin, V.G., Gurevich, A.M., Bull. Acad. Sci. URSS, (1943) 271, cited by Bruno, J., Grenthe, I., Lagerman, B., On the UO_2^{2+}/U^{4+} redox potential, Acta Chem. Scand., **44** (1990) 896-901.

[43WEB] Webster, R.A., Unpublished report CK-873, 1943, cited by Katz, J.J., Rabinowitch, E., The chemistry of uranium, New York: Dover Publications, 1951, 608p.

[44ALT] Altman, D., Unpublished report RL-4.6.276, 1944, cited by MacWood, G.E., Thermodynamic properties of uranium compounds, in: Chemistry of uranium: Collected papers, US Atomic Energy Commission, Oak Ridge, Tennessee, USA, 1958, *pp.*543-609.

[44FER/PRA] Ferguson, W.J., Prather, J., Unpublished reports A-3143, 1944, cited by MacWood, G.E., Thermodynamic properties of uranium compounds, in: Chemistry of uranium, US Atomic Energy Commission, Oak Ridge, Tennessee, USA, 1958, *pp.*543-609.

[44FER/RAN] Ferguson, W.J., Rand, R.D., Unpublished data, 1944, cited by Katz, J.J., Rabinowich, E., The chemistry of uranium, New York: Dover Publications, 1951, 608p.

[44NOT/POW] Nottorf, R., Powell, J., The vapor pressure of UBr_4, in: Chemical Research 10 May to 10 June, 1944 (Spedding, F.H., Wilhelm, H.A., *eds.*), Ames Project Report CC-1504 (A-2087), 1944, *pp.*9-17.

[44TAY/SMI] Taylor, J.K., Smith, E.R., Report A-1972 (1944), cited by Bruno, J., Grenthe, I., Lagerman, B., On the redox potential of U(VI)/U(IV), Acta Chem. Scand., **44** (1990) 896-901.

[45BAR] Barkelew, C., Unpublished reports RL-4.6.906 and RL-4.6.929, 1945, cited by MacWood, G.E., Thermodynamic properties of uranium compounds, in: Chemistry of uranium, US Atomic Energy Commission, Oak Ridge, Tennessee, USA, 1958, *pp.*543-609.

[45BRE/BRO] Brewer, L., Bromley, L.A., Gilles, P.W., Lofgren, N.L., The thermodynamic properties and equilibria at high temperatures of uranium halides, oxides, nitrides, and carbides, Report MDDC-1543, US Atomic Energy Commission, Oak Ridge, Tennessee, USA, 1945, 81p.

[45DAV/MAC] Davidson, P., MacWood, G.E., Streeter, I., Unpublished report RL-4.6.932, 1945, cited by MacWood, G.E., Thermodynamic properties of uranium compounds, in: Chemistry of uranium: Collected papers, US Atomic Energy Commission, Oak Ridge, Tennessee, USA, 1958, *pp.*543-609.

[45DAV/STR] Davidson, P.H., Streeter, I., The equilibrium vapor pressure of UCl_4 above $UOCl_2$, Report RL 4.6.920, June 1945, cited by Müller, M.E., The vapor pressure of uranium halides, Report AECD-2029, US Atomic Energy Commission, Oak Ridge, Tennessee, USA, 1948, 118p.

[45DAV] Davidson, P.H., The flow of UCl_4 vapor through tubes of cylindrical cross-section, Report RL 4.6.931, August 1945, cited by Müller, M.E., The vapor pressure of uranium halides, Report AECD-2029, U.S. Atomic Energy Commission, Oak Ridge, Tennessee, USA, 1948, 118p.

[45GRE] Gregory, N.W., Unpublished reports RL-4.6.928 and RL-4.6.940, 1945, cited by MacWood, G.E., Thermodynamic properties of uranium compounds, in: Chemistry of uranium, US Atomic Energy Commission, Oak Ridge, Tennessee, USA, 1958, *pp.*543-609.

[46GRE] Gregory, N.W., The vapor pressure of uranium tetraiodide, Report AECD-3342, US Atomic Energy Commission, Oak Ridge, Tennessee, USA, 1946.

[46GRE2] Gregory, N.W., Applications of a diaphragm gauge for the investigation of vapor pressures and chemical equilibria at high temperatures: I. The vapor pressure and gas density of UCl_4, Report BC-16, June 1946, cited by Müller, M.E., The vapor pressure of uranium halides, Report AECD-2029, US Atomic Energy Commission, Oak Ridge, Tennessee, USA, 1948, 118p.

[46GRE3] Gregory, N.W., The vapour pressure of UCl_3Br, UCl_2Br_2, and UBr_4, Report BC-3, May 1946, cited by Müller, M.E., The vapor pressure of uranium halides, Report AECD-2029, U.S. Atomic Energy Commission, Oak Ridge, Tennessee, USA, 118p.

[47BRO/VRI] Broene, H.H., de Vries, T., The thermodynamics of aqueous hydrofluoric acid solutions, J. Am. Chem. Soc., **69** (1947) 1644-1646.

[47FON] Fontana, B.J., The heat of solution of uranium tetrachloride in aqueous perchloric acid solutions, Report MDDC-1452, US Atomic Energy Commission, Oak Ridge, Tennessee, USA, 1947, 2p.

[47GIN/COR] Ginnings, D.C., Corruccini, R.J., Heat capacities at high temperatures of uranium, uranium trichloride, and uranium tetrachloride, J. Res. Nat. Bur. Stand., **39** (1947) 309-316.

[47GUI] Guiter, H., Hydrolyse du nitrate d'uranyle, Bull. Soc. Chim. Fr., (1947) 64-67, in French.

[47HAR/KOL] Harris, W.E., Kolthoff, I.M., The polarography of uranium: III. Polarography in very weakly acid, neutral or basic solution, J. Am. Chem. Soc., **69** (1947) 446-451.

[47JOH] Johnsson, K.O., The vapor pressure of uranium tetrafluoride, Report Y-42, Clinton Engineer Works, Carbon and Carbon Chemicals Corporation, Y-12 Plant, Oak Ridge, Tennessee, USA, 1947, 19p.

[47MAC/LON] MacInnes, D.A., Longsworth, L.G., The measurement and interpretation of pH and conductance values of aqueous solutions of uranyl salts, Report MDDC-9911, US Atomic Energy Commission, Oak Ridge, Tennessee, USA, March 1947, 10p.

[47MOO/KEL] Moore, G.E., Kelley, K.K., High-temperature heat contents of uranium, uranium dioxide and uranium trioxide, J. Am. Chem. Soc., **69** (1947) 2105-2107.

[47SUT] Sutton, J., Ionic species in uranyl solutions, CRC-325, National Research Council of Canada, Chalk River, Ontario, Mar 1947, 11p.

[47WAC/CHE] Wacker, P.F., Cheney, R.K., Specific heat, enthalpy and entropy of uranyl fluoride, J. Res. Nat. Bur. Stand., **39** (1947) 317-320.

[48AMP/MUL] Amphlett, C.B., Mullinger, L.W., Thomas, L.F., Some physical properties of uranium hexafluoride, Trans. Faraday Soc., **44** (1948) 927-938.

[48BRI/HOD] Brickwedde, F.G., Hodge, H.J., Scott, R.B., The low temperature heat capacities, enthalpies and entropies of UF_4 and UF_6, J. Chem. Phys., **16** (1948) 429-436.

[48FAU] Faucherre, J., Sur la condensation des ions basiques dans l'hydrolyse des nitrates d'aluminium et d'uranyle, C. R. Hebd. Séances Acad. Sci., **227** (1948) 1367-1369, in French.

[48MUE] Müller, M.E., The vapor pressure of uranium halides, Report AECD-2029, U.S. Atomic Energy Commission, Oak Ridge, Tennessee, USA, 1948, 118p.

[48THO/SCH] Thompson, R.W., Schelberg, A., OSRD Proj. SSRC-5, Princeton Report 24. A-809, cited by Müller, M.E., The vapor pressure of uranium halides, Report AECD-2029, US Atomic Energy Commission, Oak Ridge, Tennessee, USA, 1948, 118p.

[48WEI/CRI] Weinstock, B., Crist, R.H., The vapor pressure of uranium hexafluoride, J. Chem. Phys., **16** (1948) 436-441.

[49AHR] Ahrland, S., On the complex chemistry of the uranyl ion: I. The hydrolysis of the six-valent uranium in aqueous solutions, Acta Chem. Scand., **3** (1949) 374-400.

[49AHR2] Ahrland, S., On the complex chemistry of the uranyl ion: III. The complexity of uranyl thiocyanate: An extinctiometric investigation, Acta Chem. Scand., **3** (1949) 1067-1076.

[49BET/MIC] Betts, R.H., Michels, R.K., Ionic association in aqueous solutions of uranyl sulphate and uranyl nitrate, J. Chem. Soc., **S58** (1949) 286-294.

[49KER/ORL] Kern, D.M.H., Orlemann, E.F., The potential of the uranium(V), uranium(VI) couple and the kinetics of uranium(V) disproportionation in perchlorate media, J. Am. Chem. Soc., **71** (1949) 2102-2106.

[49KRA/NEL] Kraus, K.A., Nelson, F., Johnson, G.L., Chemistry of aqueous uranium(V) solutions: I. Preparation and properties. Analogy between uranium(V), neptunium(V) and plutonium(V), J. Am. Chem. Soc., **71** (1949) 2510-2517.

[49KRA/NEL2] Kraus, K.A., Nelson, F., Chemistry of aqueous uranium(V) solutions: II. Reaction of uranium pentachloride and water. Thermodynamic stability of UO_2^+. Potential of U(IV)/(V), U(IV)/(VI) and U(V)/U(VI) couples, J. Am. Chem. Soc., **71** (1949) 2517-2522.

[49KRI/HIN] Kritchevsky, E.S., Hindman, J.C., The potentials of the uranium three-four and five-six couples in perchloric and hydrochloric acids, J. Am. Chem. Soc., (1949) 2096-2106.

[49MAS] Masi, J.F., J. Chem. Phys., **17** (1949) 755, cited by Fuger, F., Parker, V.B., Hubbard, W.N., Oetting, F.L., The chemical thermodynamics of actinide elements and compounds: Part 8. The actinide halides, Vienna: International Atomic Energy Agency, 1983, 267p.

[49SUT] Sutton, J., The hydrolysis of the uranyl ion: Part I., J. Chem. Soc., **S57** (1949) 275-286.

[50BET/LEI] Betts, R.H., Leigh, R.M., Ionic species of tetravalent uranium in perchloric and sulphuric acids, Can. J. Res., **28** (1950) 514-525.

[50KIG] Kigoshi, K., Bull. Chem. Soc. Japan, **23** (1950) 67.

[50KRA/NEL] Kraus, K.A., Nelson, F., Hydrolytic behavior of metal ions: I. The acid constants of uranium(IV) and plutonium(IV), J. Am. Chem. Soc., **72** (1950) 3901-3906.

[50MOO/KRA] Moore, G.E., Kraus, K.A., Anionic exchange studies of uranium(VI) in HCl-HF mixtures: Separation of Pa(VI) and U(VI), Report ORNL-795, Oak Ridge National Laboratory, Oak Ridge, Tennessee, USA, 1950, *pp.*16-17.

[50PAM/AGA] Pamfilov, A.V., Agafonova, A.L., Zh. Fiz. Khim., **24** (1950) 1147, cited by Perrin D.D., Ionisation constants of inorganic acids and bases in aqueous solution, 2*nd* ed., IUPAC Chemical Data Series **29**, Oxford: Pergamon Press, 1982, 180p.

[50POP/GAG] Popov, M.M., Gagarinskii, Yu.V., Composition and dissociation pressure of stable crystallohydrate of uranium tetrafluoride, Russ. J. Inorg. Chem., **2** (1950) 1-8.

[50WAZ/CAM] Van Wazer, J.R., Campanella, D.A., Structure and properties of the condensed phosphates: IV. Complex ion formation in polyphosphate solutions, J. Am. Chem. Soc., **72** (1950) 655-663.

[51AHR] Ahrland, S., On the complex chemistry of the uranyl ion: VI. The complexity of uranyl chloride, bromide and nitrate, Acta Chem. Scand., **5** (1951) 1271-1282.

[51AHR2] Ahrland, S., On the complex chemistry of the uranyl ion: V. The complexity of uranyl sulfate, Acta Chem. Scand., **5** (1951) 1151-1167.

[51BRI/HOD] Brickwedde, F.G., Hodge, H.J., Scott, R.B., Unpublished NBS Report A-607, cited by Katz, J.J., Rabinowitch, E. The chemistry of uranium, New York: Dover Publications, 1951, 608p.

[51FRO] Fronæus, S., The use of cation exchangers for the quantitative investigation of complex systems, Acta Chem. Scand., **5** (1951) 859-871.

[51KAT/RAB] Katz, J.J., Rabinowitch, E., The chemistry of uranium, New York: Dover Publications, 1951, 608p.

[51MUU/REF] Muus, J., Refn, I., Asmussen, R.W., Trans. Dan. Acad. Tech. Sci., **23** (1951), cited by Perrin, D.D., Ionisation constants of inorganic acids and bases in aqueous solution, 2*nd* ed., IUPAC Chemical Data Series **29**, Oxford: Pergamon Press, 1982, 180p.

[51NEL/KRA] Nelson, F., Kraus, K.A., Chemistry of aqueous uranium(V) solutions: III. The uranium(IV)-(V)-(VI) equilibrium in perchlorate and chloride solutions, J. Am. Chem. Soc., **73** (1951) 2157-2161.

[51ZEB/ALT] Zebroski, E.L., Alter, H.W., Heumann, F.K., Thorium complexes with chloride, fluoride, nitrate, phosphate and sulfate, J. Am. Chem. Soc., **73** (1951) 5646-5650.

[52HUB/HOL] Huber, E.J., Jr., Holley, C.E., Jr., Meierkord, E.H., The heats of combustion of thorium and uranium, J. Am. Chem. Soc., **74** (1952) 3406-3408.

[52JON/GOR] Jones, W.M., Gordon, J., Long, E.A., The heat capacities of uranium, uranium trioxide and uranium dioxide from 15 K to 300 K, J. Chem. Phys., **20** (1952) 695-699.

[52KAP/BAR] Kapustinsky, A.F., Baranova, L.I., Free energy and heat of dehydration of crystal-hydrates of uranyl nitrate and chloride, Izv. Akad. Nauk SSSR, Otdel. Khim. Nauk, **6** (1952) 1122-1124, in Russian; Engl. transl.: Bull. Acad. Sci. USSR, **6** (1952) 981-983.

[52KAT/SIM] Katzin, L.I., Simon, D.M., Ferraro, J.R., Heat of solution of uranyl nitrates hydrates in water and in organic solvents, J. Am. Chem. Soc., **74** (1952) 1191-1194.

[52LAT] Latimer, W.M., The oxidation states of the elements and their potentials in aqueous solutions, 2*nd* ed., New York: Prentice-Hall Inc., 1952, 392p.

[52SUL/HIN] Sullivan, J.C., Hindman, J.C., An analysis of the general mathematical formulations for the calculation of association constants on complex ion systems, J. Am. Chem. Soc., **74** (1952) 6091-6096.

[53BAE/SCH2] Baes, C.F., Jr., Schreyer, J.M., Lesser, J.M., The chemistry of uranium(VI) orthophosphate solutions: Part I. A spectrophotometric investigation of uranyl phosphate complex formation in perchloric acid solutions, Report AECD-3596, US Atomic Energy Commission, Oak Ridge, Tennessee, USA, 1953.

[53CAI/COR] Caillat, R., Coriou, H., Perio, P., Sur une nouvelle variété d'hydrure d'uranium, C. R. Hebd. Séances Acad. Sci., (1953) 812-815, in French.

[53LLE] Llewellyn, D.R., Some physical properties of uranium hexafluoride, J. Chem. Soc., (1953) 28-36.

[53LUN] Lundgren, G., The crystal structure of $U_6O_4(OH)_4(SO_4)_6$, Ark. för Kemi, **34** (1953) 349-363.

[53OLI/MIL] Oliver, G.D., Milton, H.T., Grisard, J.W., The vapor pressure and critical constants of uranium hexafluoride, J. Am. Chem. Soc., **75** (1953) 2827-2829.

[54AHR/HIE] Ahrland, S., Hietanen, S., Sillén, L.G., Studies on the hydrolysis of metal ions, Acta Chem. Scand., **8** (1954) 1907-1916.

[54AHR/LAR] Ahrland, S., Larsson, R., The complexity of uranyl fluoride, Acta Chem. Scand., **8** (1954) 354-366.

[54AHR/LAR2] Ahrland, S., Larsson, R., The complexity of uranium(IV) chloride, bromide and thiocyanate, Acta Chem. Scand., **8** (1954) 137-150.

[54BRO/BUN] Brown, R.D., Bunger, W.B., Marshall, W.L., Secoy, C.H., The electrical conductivity of uranyl sulfate in aqueous solution, J. Am. Chem. Soc., **76** (1954) 1532-1535.

[54COU] Coughlin, J.P., Contributions to the data on theoretical metallurgy: XII. Heats and free energies of formation of inorganic oxides, US Bureau of Mines Bulletin, **542** (1954) 1-80.

[54DAY/POW] Day, R.A., Jr., Powers, R.M., Extraction of uranyl ion from some aqueous salt solutions with 2-thenoyltrifluoroacetone, J. Am. Chem. Soc., **76** (1954) 3895-3897.

[54FAU] Faucherre, J., Sur la constitution des ions basiques métalliques, Bull. Soc. Chim. Fr., Sér. 5, **21(2)** (1954) 253-267, in French.

[54HIN] Hindman, J.C., Ionic and molecular species of plutonium in solution, in: The actinide elements (Seaborg, G.T., Katz, J.J., *eds.*), New York: McGraw-Hill, 1954, 301-370.

[54HOE/KAT] Hoekstra, H.R., Katz, J.J., The chemistry of uranium, in: The actinide elements (Seaborg, G.T., Katz, J.J., *eds.*), 1*st* ed., New York: McGraw-Hill, 1954, *p*.174.

[54LIP/SAM] Lipilina, I.I., Samoilov, O.Ya., Thermochemical investigation of the nature of dilute aqueous solutions of uranyl chloride and nitrate, Dokl. Akad. Nauk SSSR, **98** (1954) 95-102, in Russian.

[54MUL/ELL] Mulford, R.N.R., Ellinger, F.H., Zachariasen, W.H., A new form of uranium hydride, J. Am. Chem. Soc, **76** (1954) 297-298.

[54SCH/BAE] Schreyer, J.M., Baes, C.F., Jr., The solubility of uranium(VI) orthophosphates in phosphoric acid solutions, J. Am. Chem. Soc., **76** (1954) 354-357.

[54SIL] Sillén, L.G., On equilibria in systems with polynuclear complex formation: I. Methods for deducing the composition of the complexes from experimental data, "core+links" complexes, Acta Chem. Scand., **8** (1954) 299-317.

[54SIL2] Sillén, L.G., On equilibria in systems with polynuclear complex formation: II. Testing simple mechanisms which give "core+links" complexes of composition $B(A_tB)_n$, Acta Chem. Scand., **8** (1954) 318-335.

[54SUL/HIN] Sullivan, J.C., Hindman, J.C., Thermodynamics of the neptunium(IV) sulfate complex ions, J. Am. Chem. Soc., **77** (1954) 5931-5934.

[54WEN/KIR] Wendolkowski, W.S., Kirslis, S.S., Thermal decomposition of uranyl nitrate hexahydrate, Report K-1086, K-25 Plant, Carbide and Carbon Chemicals Company, Union Carbide and Carbon Corporation, Oak Ridge, Tennessee, USA, 1954, 23p.

[55ABR/FLO] Abraham, B.M., Flotow, H.E., The heats of formation of uranium hydride, uranium deuteride and uranium tritride at 25°C, J. Am. Chem. Soc., **77** (1955) 1446-1448.

[55BET] Betts, R.H., Heat of hydrolysis of uranium(IV) in perchloric acid solutions, Can. J. Chem., **33** (1955) 1775-1779.

[55DAY/WIL] Day, R.A., Jr., Wilhite, R.N., Hamilton, F.D., Stability of complexes of uranium(IV) with chloride, sulfate and thiocyanate, J. Am. Chem. Soc., **77** (1955) 3180-3182.

[55GAY/LEI] Gayer, K.H., Leider, H., The solubility of uranium trioxide, $UO_3{\cdot}H_2O$, in solutions of sodium hydroxide and perchloric acid at 25°C, J. Am. Chem. Soc., **77** (1955) 1448-1550.

[55KOM/TRE] Komar, N.P., Tretyak, Z.A., Investigation of complex compounds of the uranyl ion which are of importance in analytical chemistry, J. Anal. Chem. (Engl. Transl.), **10** (1955) 223-229.

[55KRA/NEL] Kraus, K.A., Nelson, F., Hydrolytic behavior of metal ions: IV. The acid constant of uranium(IV) as a function of temperature, J. Am. Chem. Soc., **77** (1955) 3721-3722.

[55LEI] Leist, M., Z. Phys. Chem. Leipzig, **205** (1955) 16, cited by Perrin, D.D., Ionisation constants of inorganic acids and bases in aqueous solution, 2*nd* ed., IUPAC Chemical Data Series **29**, Oxford: Pergamon Press, 1982, 180p.

[55MAL/SHE] Mallett, M.W., Sheipline, V.M., Carbides, in: The reactor handbook, Report AECD-3647, Vol. 3: Materials, Sec. 1: General properties, US Atomic Energy Commission, Washington, D.C.: US Government Printing Office, 1955, *pp.*113-122.

[55MAL/SHE2] Mallett, M.W., Sheipline, V.M., Hydrides, in: The reactor handbook, Report AECD-3647, Vol. 3: Materials, Sec. 1: General properties, US Atomic Energy Commission, Washington, D.C.: US Government Printing Office, 1955, *pp.*155-168.

[55OSB/WES] Osborne, D.W., Westrum, E.F., Jr., Lohr, H.R., The heat capacity of uranium tetrafluoride from 5 to 300 K, J. Am. Chem. Soc., **77** (1955) 2737-2739.

[55ROS] Ross, V., Studies of uranium minerals: XXI. Synthetic hydrogen-autunite, Am. Mineral., **40** (1955) 917-918.

[55RYD] Rydberg, J., Studies on the extraction of metal complexes: XII-B. The formation of composit, mononuclear complexes. Part B. Studies on the thorium and uranium(VI)-acetylacetone-H_2O-organic solvent systems; Ark. för Kemi, **8(14)** (1955) 113-140.

[55SCH/BAE] Schreyer, J.M., Baes, C.F., Jr., The solubility of $UO_2HPO_4{\cdot}4H_2O$ in perchloric acid solutions, J. Phys. Chem., **59** (1955) 1179-1181.

[55SCH] Schreyer, J.M., The solubility of uranium(IV) orthophosphates in phosphoric acid solutions, J. Am. Chem. Soc., **77** (1955) 2972-2974.

[56AHR/LAR] Ahrland, S., Larsson, R., Rosengren, K., On the complex chemistry of the uranyl ion: VIII. The complexity of uranyl fluoride, Acta Chem. Scand., **10** (1956) 705-718.

[56BAE] Baes, C.F., Jr., A spectrophotometric investigation of uranyl phosphate complex formation in perchloric acid solution, J. Phys. Chem., **60** (1956) 878-883.

[56BAE2] Baes, C.F., Jr., The reduction of uranium(VI) by iron(II) in phosphoric acid solution, J. Phys. Chem., **60** (1956) 805-806.

[56BLA/COL] Blake, C.A., Coleman, C.F., Brown, K.B., Hill, D.G., Lowrie, R.S., Schmitt, J.M., Studies in the carbonate-uranium system, J. Am. Chem. Soc., **78** (1956) 5978-5983.

[56CHU/STE] Chukhlantsev, V.G., Stepanov, S.I., Solubility of uranyl and thorium phosphates, Zh. Neorg. Khim., **1** (1956) 487-494, in Russian, CA 51:47c; Engl. transl.: J. Inorg. Chem. USSR, **1** (1956) 135-141.

[56DAW/WAI] Dawson, J.K., Wait, E., Alcock, K., Chilton, D.R., Some aspects of the system uranium trioxide-water, J. Chem. Soc., (1956) 3531-3540.

[56DUN] Dunn, H.W., X-ray diffraction data for some uranium compounds, Report ORNL-2092, Oak Ridge National Laboratory, Oak Ridge, Tennessee, USA, 1956.

[56GRE/WES] Greenberg, E., Westrum, E.F., Jr., Heat capacity and thermodynamic functions of uranium(IV) oxychloride and uranium(IV) oxybromide from 10 to 350 K, J. Am. Chem. Soc., **78** (1956) 5144-5447.

[56GRE/WES3] Greenberg, E., Westrum, E.F., Jr., Heat capacity and thermodynamic functions of uranyl chloride from 6 to 350 K, J. Am. Chem. Soc., **78** (1956) 4526-4528.

[56HIE] Hietanen, S., Studies on the hydrolysis of metal ions: 17. The hydrolysis of the uranium(IV) ion, U^{4+}, Acta Chem. Scand., **10** (1956) 1531-1546.

[56HIE2] Hietanen, S., The mechanism of thorium(IV) and uranium(IV) hydrolysis, Rec. Trav. Chim. Pays-Bas, **75** (1956) 711-715.

[56KEL/TAR] Kelley, C.M., Tartar, H.V., On the system: Bromine-water, J. Am. Chem. Soc., **78** (1956) 5752-5756.

[56KRA/MOO] Kraus, K.A., Moore, G.E., Nelson, F., Anion-exchange studies: XXI. Th(IV) and U(IV) in hydrochloric acid, separation of thorium, protactinium and uranium, J. Am. Chem. Soc., **78** (1956) 2692-2694.

[56MCC/BUL] McClaine, L.A., Bullwinkel, E.P., Huggins, J.C., The carbonate chemistry of uranium: Theory and applications, in: Proc. International Conf. on the Peaceful Uses of Atomic Energy, held 8-20 August, 1955, in Geneva, Switzerland, Vol. **VIII**, New York: United Nations, 1956, *pp.*26-37.

[56ORB/BAR] Orban, E., Barnett, M.K., Boyle, J.S., Heiks, J.R., Jones, L.V., Physical properties of aqueous uranyl sulphate solutions from 20 to 90°C, J. Am. Chem. Soc., **60** (1956) 413-415.

[56SCH/PHI] Schreyer, J.M., Phillips, L.R., The solubility of uranium(IV) orthophosphates in perchloric acid solutions, J. Phys. Chem., **60** (1956) 588-591.

[56SHC/VAS] Shchukarev, S.A., Vasil'kova, I.V., Efimov, A.I., The disproportionation of uranium trichloride, Zh. Neorg. Khim., **1(12)** (1956) 2652-2656, in Russian; Engl. Transl.: Russ. J. Inorg. Chem., **1(12)** (1956) 4-9.

[56STA/CRO] Staritzky, E., Cromer, D.T., Hendecahydrogen diuranyl pentaphosphate, $H_{11}(UO_2)_2(PO_4)_5$, Anal. Chem., **28** (1956) 1354-1355.

[57BAL/DAV] Bale, W.D., Davies, E.W., Morgans, D.B., Monk, C.B., The study of some ion-pairs by spectrophotometry, Discuss. Faraday Soc., **24** (1957) 94-102.

[57CAR] Caramazza, R., Gazz. Chim. Ital., **87** (1957) 1507, cited by Perrin, D.D., Ionisation constants of inorganic acids and bases in aqueous solution, 2*nd* ed., IUPAC Chemical Data Series **29**, Oxford: Pergamon Press, 1982, 180p.

[57COL/APP] Coleman, R.G., Appleman, D.E., Umohoite from the Lucky MC Mine, Wyoming, Am. Mineral., **42** (1957) 657-660.

[57DAV/MON] Davies, E.W., Monk, C.B., Spectrophotometric studies of electrolytic dissociation: Part 4. Some uranyl salts in water, Trans. Faraday Soc., **53** (1957) 442-449.

[57GAY/LEI] Gayer, K.H., Leider, H., The solubility of uranium(IV) hydroxide in solutions of sodium hydroxide and perchloric acid at 25°C, Can. J. Chem., **35(5)** (1957) 5-7.

[57GUR] Gurevich, A.M., Trudy Radievogo Inst. im. V.G. Khopina, **6** (1957) 88. cited by Bruno, J., Grenthe, I., Lagerman, B., On the redox potential of U(VI)/U(IV), Acta Chem. Scand., **44** (1990) 896-901.

[57HEA/WHI] Hearne, J.A., White, A.G., Hydrolysis of the uranyl ion, J. Chem. Soc., (1957) 2168-2174.

[57KAT/SEA] Katz, J.J., Seaborg, G.T., The chemistry of the actinide elements, London: Metheun & Co. Ltd., 1957, *pp.*406-475.

[57KHO] Khodadad, P., Sur les séléniures de l'uranium tétravalent, C. R. Séances Hebd. Acad. Sci., **245** (1957) 934-936, in French.

[57KIN/PFE] King, A.J., Pfeiffer, R., Zeek, W., Thermal stability of the hydrates of uranyl nitrate, Final Report NYO-6313, US Atomic Energy Commission, 1957.

[57OSB/WES] Osborne, D.W., Westrum, E.F., Jr., Lohr, H.R., The heat capacity and thermodynamic functions of tetrauranium nonoxide from 5 to 310 K, J. Am. Chem. Soc., **79** (1957) 529-530.

[57POP/IVA] Popov, M.M., Ivanov, M.I., Heats of formation of PuO_2 and U_3O_8, Sov. J. At. Energy, **2** (1957) 439-443.

[57POP/KOS] Popov, M.M., Kostylev, F.A., Karpova, T.F., The heat of formation of uranyl fluoride and the heats of reactions of hexa- and tetra-fluorides of uranium with water, J. Inorg. Chem. USSR, **2** (1957) 9-15.

[57THA] Thamer, B.J., Spectrophotometric and solvent-extraction studies of uranyl phosphate complexes, J. Am. Chem. Soc., **79** (1957) 4298-4305.

[57VDO/KOV] Vdovenko, V.M., Koval'skaya, M.P., Kovaleva, T.V., Investigations of the complexes of uranyl nitrate with diethyl ether, J. Inorg. Chem. USSR, **II(7)** (1957) 359-367.

[57WEI/HAR] Weiss, A., Hartl, K., Hofmann, U., Zur Kenntnis von Mono-hydrogen-uranylphosphat $HUO_2PO_4{\cdot}4H_2O$ und Mono-hydrogen-uranylarsenat $HUO_2AsO_4{\cdot}4H_2O$, Z. Naturforsch., **12B** (1957) 669-671.

[58AGR] Agron, P.A., Thermodynamics of intermediate uranium fluorides from measurements of the disproportionation pressures, in: Chemistry of uranium, collected papers (Katz, J.J., Rabinowitch, E., *eds.*), US Atomic Energy Commission, Oak Ridge, Tennessee, USA, 1958, *pp.*610-626.

[58ALL] Allen, K.A., The uranyl sulfate complexes from Tri-n-octylamine extraction equilibria, J. Am. Chem. Soc., **80** (1958) 4133-4137.

[58BEL] Belova, L.N., Arsenuranylite, the arsenic analogue of phosphuranylite, Zapiski Vses. Mineralog. Obshch., 87 (1958) 589-602, in Russian, Abstract in Am. Mineral., 44 (1959) 208.

[58BRE/BRO] Brewer, L., Bromley, L.A., Gilles, P.W., Lofgren, N.L., Thermodynamic properties and equilibriums of uranium halides, oxides, nitrides and carbides at high temperatures, in: Chemistry of uranium, collected papers (Katz, J.J., Rabinowitch, E., *eds.*), US Atomic Energy Commission, Oak Ridge, Tennessee, USA, 1958, *pp.*219-268.

[58BRI] Briggs, G.G., Chemical equilibria and reaction rates for hydrofluorination of UO_2 from "ammonium diuranate" and from UO_3, NLCO-720, National Lead Co. of Ohio, Cincinnati, Ohio, 1958, 72p.

[58BRU] Brusilovskii, S.A., Investigation of the precipitation of hexavalent uranium hydroxide Proc. Acad. Sci. USSR, **120** (1958) 343-347.

[58CON/WAH] Connick, R.E., Wahl, A.C., The acetate complexes of uranyl ion and solubility of sodium uranyl acetate, in: Chemistry of uranium, collected papers Katz, J.J., Rabinowitch, E., *eds.*), US Atomic Energy Commission, Oak Ridge, Tennessee, USA, 1958, *pp.*154-160.

[58FLI] Flis, I.E., Russ. J. Appl. Chem., **31** (1958) 1183, cited by Perrin, D.D., Ionisation constants of inorganic acids and bases in aqueous solution, 2*nd* ed., IUPAC Chemical Data Series **29**, Oxford Pergamon Press, 1982, 180p.

[58FON] Fontana, B.J., Heats of reaction of the aqueous uranium oxidation-reduction couples, in: Chemistry of uranium, collected papers (Katz, J.J., Rabinowitch, E., *eds.*), US Atomic Energy Commission, Oak Ridge, Tennessee, USA, 1958, *pp.*289-299.

[58FRO] Frondel, C., Systematic mineralogy of uranium and thorium, US Geological Survey Bull. **1064**, (1958) 400p.

[58FRY/NIL] Frydman, M., Nilsson, G., Rengemo, T., Sillén, L.G., Some solution equilibria involving calcium sulfite and carbonate: III. The acidity constants of H_2CO_3 and H_2SO_3, and $CaCO_3+CaSO_3$ equilibria in $NaClO_4$ medium at 25°C, Acta Chem. Scand., **12** (1958) 868-872.

[58IVA/TUM] Ivanov, M.I., Tumbakov, V.A., Podolshaia, N.S., The heats of formation of UAl_2, UAl_3 and UAl_4, Sov. At. Energy, **4** (1958) 1007-1011.

[58JOH/BUT] Johnson, O., Butler, T., Newton, A.S., Preparation, purification, and properties of anhydrous uranium chlorides, in: Chemistry of uranium (Katz, J.J., Rabinowitch, E., *eds.*), Vol. **1**, Oak Ridge, Tennessee, USA, 1958, *pp.*1-28.

[58LI/DOO] Li, N.C., Doody, E., White, J.M., Copper(II), nickel and uranyl complexes of some amino acids, Am. J. Sci., **80** (1958) 5901-5903.

[58LIN] Linke, W.F., Solubilities: Inorganic and metal-organic compounds: Vol. I. A-Ir, Princeton, New Jersey: Van Nostrand Co., 1958, 1487p.

[58MAC] MacWood, G.E., Thermodynamic properties of uranium compounds, in: Chemistry of uranium: Collected papers (Katz, J.J., Rabinowitch, E., *eds.*), US Atomic Energy Commission, Oak Ridge, Tennessee, USA, 1958, *pp.*543-609.

[58MAR] Marcus, Y., Anion exchange of metal complexes: The uranyl phosphate system, in: Proc. 2*nd* United Nations International Conference on the Peaceful Uses of Atomic Energy, held 1-13 September, 1958, in Geneva, Switzerland, Vol. **3**, Geneva: United Nations, 1958, *pp.*465-471.

[58NAI/NAN] Nair, V.S.K., Nancollas, G.H., Thermodynamic properties of ion association: Part V. Dissociation of the bisulphate ion, J. Chem. Soc., (1958) 4144-4147.

[58NEC] Neckrasova, Z.A., titel unknown, in: Proc 2nd int'l conf on the peaceful uses of atomic energy held 1-13 Sep 1958 in Geneva, Geneva: United Nations, **2** (1958) 286.

[58NIL/REN] Nilsson, G., Rengemo, T., Sillén, L.G., Some solution equilibria involving calcium sulfite and carbonate: I. Simple solubility equilibria of CO_2, SO_2, $CaCO_3$, and $CaSO_4$, Acta Chem. Scand., **12** (1958) 868-872.

[58POP/GAL] Popov, M.M., Gal'chenko, G.L., Senin, M.D., The heat capacities of UO_2, U_3O_8 and UO_3 at high temperatures, J. Inorg. Chem. USSR, **8** (1958) 18-22.

[58SHC/VAS] Shchukarev, S.A., Vasil'kova, I.V., Drozdova, V.M., The heats of formation of uranyl bromide, uranium bromide and monoxytribromide, Zh. Neorg. Khim., **3(12)** (1958) 2651-2653, in Russian; Engl. transl.: J. Inorg. Chem. USSR, **3(12)** (1958) 75-78.

[58SHC/VAS2] Shchukarev, S.A., Vasil'kova, I.V., Martynova, N.S., Mal'tsev, Y.G., Heats of formation of uranyl chloride and uranium monoxy-trichloride,

Zh. Neorg. Khim., **3(12)** (1958) 2647-2650, in Russian; Engl. transl.: J. Inorg. Chem. USSR, **3(12)** (1958) 70-74.

[58VDO/MAL] Vodvenko, V.M., Mal'tseva, L.A., Determination of the heats of solution of crystalline hydrates of uranyl nitrate hydrates in water and in diethyl ether, Trudy Radievogo Inst. Im. V.G. Khlopina, **8** (1958) 25-29.

[58YOU/GRA] Young, H.S., Grady, H.F., Physical constants of uranium tetrachloride, in: Chemistry of uranium, Vol. **2** (Katz, J.J., Rabinowitch, E., *eds.*), Oak Ridge, Tennessee, USA, 1958, *pp.*749-756.

[59FLO/LOH] Flotow, H.E., Lohr, H.R., Abraham, B.M., Osborne, D.W., The heat capacity and thermodynamic functions of β-uranium hydride from 5 to 350 K, J. Am. Chem. Soc., **81** (1959) 3529-3533.

[59FUC/HOE] Fuchs, L.H., Hoekstra, H.R., The preparation and properties of uranium(IV) silicate, Am. Mineral., **44** (1959) 1057-1063.

[59GAG/MAS] Gagarinskii, Yu.V., Mashirev, V.P., Crystalline hydrates of uranium tetrafluoride, Russ. J. Inorg. Chem., **4** (1959) 565-568.

[59GUI/PRO] Guillemin, C., Protas, J., Ianthinite et wyarite, Bull. Soc. Fr. Minér. Crist., **82** (1959) 80-86, in French.

[59HIE/SIL] Hietanen, S., Sillén, L.G., Studies on the hydrolysis of metal ions: 24. Hydrolysis of the uranyl ion, UO_2^{2+}, in perchlorate self-medium, Acta Chem. Scand., **13** (1959) 1828-1838.

[59IVA/TUM] Ivanov, M.I., Tumbakov, V.A., Heat of formation of UBe_{13}, Sov. J. At. Energy, **7** (1959) 559-561.

[59KLY/KOL] Klygin, A.E., Kolyada, N.S., A study of the UO_2SO_3-$(NH_4)_2SO_3$-H_2O system by the solubility method, Russ. J. Inorg. Chem., **4** (1959) 101-103.

[59KLY/SMI] Klygin, A.E., Smirnova, I.D., The dissociation constant of the $UO_2(CO_3)_3^{4-}$ ion, Russ. J. Inorg. Chem., **4** (1959) 16-18.

[59KLY/SMI2] Klygin, A.E., Smirnova, I.D., Nikol'skaya, N.A., Equilibria in the $UO_2(IO_3)_2$-KIO_3-H_2O system, Russ. J. Inorg. Chem., **4** (1959) 754-756.

[59LAG] Lagerström, G., Equilibrium studies of polyanions: III. Silicate ions in $NaClO_4$ medium, Acta Chem. Scand., **13** (1959) 722-736.

[59POP/GAL] Popov, M.M., Gal'chenko, G.L., Senin, M.D., Specific heats and heats of fusion of UCl_4 and UI_4 and heat of transformation of UI_4, Russ. J. Inorg. Chem., **4** (1959) 560-562.

[59POP/KOS] Popov, M.M., Kostylev, F.A., Zubova, N.V., Vapour pressure of uranium tetrafluoride, Russ. J. Inorg. Chem., **4** (1959) 770-771.

[59ROB/STO] Robinson, R.A., Stokes, R.H., Electrolyte solutions, London: Butterworths, 2*nd ed.*, 1959, 559p.

[59SHC/VAS] Shchukarev, S.A., Vasil'kova, I.V., Drozdova, V.M., Martynova, N.S., III. The thermochemistry of solid uranium oxyhalides and the substitution principle, Russ. J. Inorg. Chem., **4** (1959) 13-15.

[59SUL/HIN] Sullivan, J.C., Hindman, J.C., The hydrolysis of neptunium(IV), J. Phys. Chem., **63** (1959) 1332-1333.

[59VDO/SOK] Vdovenko, V.M., Sokolov, A.P., Crystallisation of uranyl nitrate, Radiokhimiya, **1** (1959) 117-120, in Russian.

[59WES/GRO] Westrum, E.F., Jr., Grønvold, F., Low temperature heat capacity and thermodynamic functions of triuranium octoxide, J. Am. Chem. Soc., **81** (1959) 1777-1780.

[59ZIE] Zielen, A.J., Thermodynamics of the sulfate complexes of thorium, J. Am. Chem. Soc., **81** (1959) 5022-5028.

[60ABR/OSB] Abraham, B.M., Osborne, D.W., Flotow, H.E., Marcus, R.B., The heat capacity and thermodynamic functions of β-uranium deuteride from 5 to 350 K, J. Am. Chem. Soc., **82** (1960) 1064-1068.

[60BAB/KOD] Babko, A.K., Kodenskaya, V.S., Equilibria in solutions of uranyl carbonate complexes, Russ. J. Inorg. Chem., **5** (1960) 1241-1244.

[60BLU] Blumenthal, B., The transformation temperatures of high-purity uranium, J. Nucl. Mater., **2** (1960) 23-30.

[60BUR/OSB] Burns, J.H., Osborne, D.W., Westrum, E.F., Jr., Heat capacity of uranium tetrafluoride from 1.3 to 20 K and the thermodynamic functions to 300 K calorimeter for the range 0.8 to 20 K, J. Chem. Phys., **33** (1960) 387-394.

[60DEN/MOS] Denotkina, R.G., Moskvin, A.I., Shevchenko, V.B., The solubility product of plutonium(IV) hydrogen phosphate and its solubility in various acids, Russ. J. Inorg. Chem., **5** (1960) 387-389.

[60FLO/LOH] Flotow, H.E., Lohr, H.R., Heat capacity and thermodynamic functions of U from 5 to 350 K, J. Phys. Chem., **64** (1960) 904-906.

[60GAL/STE] Galkin, N.P., Stepanov, M.A., Solubility of uranium(IV) hydroxide in sodium hydroxide, Sov. At. Energy, **8** (1960) 231-233.

[60GUS/RIC] Gustafson, R.L., Richard, C., Martell, A.E., Polymerization of uranyl-tiron chelates, J. Am. Chem. Soc., **82** (1960) 1526-1534.

[60HEF/AMI] Hefley, J.D., Amis, E.S., A spectrophotometric study of the complexes formed between uranyl and chloride ions in water and water-ethanol solvents, J. Phys. Chem., **64** (1960) 870-872.

[60KEL] Kelley, K.K., Contributions to the data on theoretical metallurgy: XIII. High-temperature heat-content, heat-capacity and entropy data for the elements and inorganic compounds, US Bureau of Mines Bulletin, **584** (1960) 232p.

[60KLY/KOL] Klygin, A.E., Kolyada, N.S., Uranyl thiosulphate, Russ. J. Inorg. Chem., **5** (1960) 562-563.

[60LAN/BLA] Langer, S., Blankenship, F.F., The vapour pressure of uranium tetrafluoride, J. Inorg. Nucl. Chem., **14** (1960) 26-31.

[60LEO/REZ] Leonidov, V.Ya., Rezukhina, T.N., Bereznikova, I.A., Heat capacity of calcium and barium uranates(VI) at high temperatures, Zh. Fiz. Khim., **34(8)** (1960) 1862-1865, Engl. transl: Russ. J. Phys. Chem., **34(8)** (1960) 885-886.

[60LIE/BRU] Van Lier, J.A., de Bruyn, P.L., Overbeek, J.T.G., The solubility of quartz, J. Phys. Chem., **64** (1960) 1675-1682.

[60LIE/STO] Lietzke, M.H., Stoughton, R.W., The solubility of silver sulfate in electrolyte solutions: Part 7. Solubility in uranyl sulfate solutions, J. Phys. Chem., **64** (1960) 816-820.

[60MAL/GAG] Mal'tsev, V.A., Gagarinskii, Yu.V., Popov, M.M., The heat of formation of uranium tetrafluoride, Zh. Neorg. Khim., **5(1)** (1960) 228-229, in Russian; Engl. transl.: Russ. J. Inorg. Chem., **5(1)** (1960) 109-110.

[60MAT] Matsuo, S., Sulphate complexes of uranium(VI), J. Chem. Soc. Japan, **81** (1960) 833-836, in Japanese.

[60RAF/KAN] Rafalskii, R.P., Kandykin, Yu.M., Experimental data on uranium dioxide crystallization in hydrothermal conditions, Geol. Rudn. Mestorozhdenii, **1** (1960) 98-106, in Russian, Chem. Abstr. 55:1302h.

[60RYA] Ryan, J.L., Species involved in the anion-exchange absorption of quadrivalent actinide nitrates, J. Phys. Chem., **64** (1960) 1375-1385.

[60SAV/BRO] Savage, A.W., Jr., Browne, J.C., The solubility of uranium(IV) fluoride in aqueous fluoride solutions, J. Am. Chem. Soc., **82** (1960) 4817-4821.

[60STA] Starý, J., Untersuchungen über die Extraktion des U(VI)-Komplexes mit Benzoylaceton, Coll. Czech. Chem. Commun., **25** (1960) 890-896, in German.

[60STA2] Stabrovskii, A.I., Polarography of uranium compounds in carbonate and bicarbonate solutions, Russ. J. Inorg. Chem., **5** (1960) 389-394.

[60STE/GAL] Stepanov, M.A., Galkin, N.P., The solubility product of the hydroxide of tetravalent uranium, Sov. At. Energy, **9** (1960) 817-821.

[61AKI/KHO] Akishin, P.A., Khodeev, Yu.S., Determination of the heat of sublimation of uranium tetrafluoride by the mass-spectroscopic method, Russ. J. Phys. Chem., **35** (1961) 574-575.

[61ALC/GRI] Alcock, C.B., Grieveson, P., A thermodynamic study of the compounds of uranium with silicon, germanium, tin and lead, J. Inst. Met., **90** (1961/62) 304-310.

[61ARG/MER] Argue, G.R., Mercer, E.E., Cobble, J.W., An ultrasensitive thermistor microcalorimeter and heats of solution of neptunium, uranium and uranium tetrachloride, J. Phys. Chem., **65** (1961) 2041-2048.

[61BAN/TRI] Banerjea, D., Tripathi, K.K., Association of uranium(VI) with anions in aqueous perchloric acid medium, J. Inorg. Nucl. Chem., **18** (1961) 199-206.

[61CHU/ALY] Chukhlantsev, V.G., Alyamovskaya, K.V., Solubility product of uranyl, beryllium and cerium phosphates, Izv. Vuz. Khim. Tekhnol., **4** (1961) 359-363, in Russian.

[61CON/PAU] Connick, R.E., Paul, A.D., The fluoride complexes of silver and stannous ions in aqueous solution, J. Phys. Chem., **65** (1961) 1216-1220.

[61FAU/IPP] Faustova, D.G., Ippolitova, V.I., Spitsyn, V.I., Determination of true thermal capacities of uranates of alkali elements and calcium, in: Investigations in the field of uranium chemistry (Spitsyn, V.I., *ed.*), Moscow University, Moscow, USSR, 1961; Engl. transl.: ANL-Trans-33, Argonne National Laboratory, Argonne, Illinois, USA, *pp.*173-175.

[61HOE/SIE] Hoekstra, H.R., Siegel, S., The uranium-oxygen system: U_3O_8-UO_3, J. Inorg. Nucl. Chem., **18** (1961) 154-165.

[61IPP/FAU] Ippolitova, E.A., Faustova, D.G., Spitsyn, V.I., Determination of enthalpies of formation of monouranates of alkali elements and calcium, in: Investigations in the field of uranium chemistry (Spitsyn, V.I., *ed.*), Moscow University, Moscow, USSR, 1961; Engl. transl.: ANL-Trans-33, Argonne National Laboratory, Argonne, Illinois, USA, *pp.*170-172.

[61KAR] Karpov, V.I., The solubility of triuranyl phosphate, Russ. J. Inorg. Chem., **6** (1961) 271-272.

[61KEL/KIN] Kelley, K.K., King, E.G., Contributions to the data on theoretical metallurgy: XIV. Entropies of the data on theoretical metallurgy, US Bureau of Mines Bulletin, **592** (1961) 149p.

[61KHO/SPI] Khomakov, K.G., Spitsyn, V.I., Zhvanko, S.A., True thermal capacity of uranoso-uranic oxide, in: Investigations in the field of uranium chemistry (Spitsyn, V.I., *ed.*), Moscow University, Moscow, USSR, 1961, in Russian; Engl. transl.: Report ANL-Trans-33, Argonne National Laboratory, Argonne, Illinois, USA, *pp.*165-169.

[61KUT] Kuteinikov, A.F., Spectroscopic investigation of the stability of the complex compound of uranium(VI) fluorine, Radiokhimiya, **3** (1961) 706-711, in Russian.

[61NAI/PRA] Nair, B.K.S., Prabhu, L.H., Vartak, D.G., Spectrophotometric investigation of uranyl azide complex in aqueous solution, J. Sci. Indust. Res., Sect. B, **20** (1961) 489-492.

[61NIK/LUK] Nikolaev, N.S., Luk'yanychev, Yu.A., The hydrolysis of uranium tetrafluoride, Sov. At. Energy, **11** (1961) 704-706.

[61PAS] Pascal, P., Nouveau traité de chimie minérale: Tome XV. Uranium et transuraniens, Deuxième fascicule: Combinaisons de l'uranium, Paris: Masson et Cie., 1961, 639p., in French.

[61PET] Peterson, A., Studies on the hydrolysis of metal ions: 32. The uranyl ion, UO_2^{2+}, in Na_2SO_4 medium, Acta Chem. Scand., **15** (1961) 101-120.

[61PIT/BRE] Pitzer, K.S., Brewer, L., Thermodynamics, *2nd* ed., New York: McGraw-Hill, 1961, 723p.

[61ROS/ROS] Rossotti, F.J.C., Rossotti, H., The determination of stability constants and other equilibrium constants in solution, New York: McGraw-Hill, 1961.

[61SHE/AWA] Sherif, F.G., Awad, A.M., The structure of uranyl azide: Its instability constant in aqueous solutions, J. Inorg. Nucl. Chem., **19** (1961) 94-100.

[61SOB/MIN] Sobkowski, J., Minc, S., Oxidation-reduction potential of UO_2^{2+}-U^{4+} system in perchloric acid, J. Inorg. Nucl. Chem., **19** (1961) 101-106.

[61SOB] Sobkowski, J., The oxidation-reduction potential of UO_2^{2+}-U^{4+} system: II. The influence of $HClO_4$, HCl, H_2SO_4 and of temperature on the oxidation potential of UO_2^{2+}-U^{4+}, J. Inorg. Nucl. Chem., **23** (1961) 81-90.

[61SUL/HIN] Sullivan, J.C., Hindman, J.C., Zielen, A.J., Specific interaction between Np(V) and U(VI) in aqueous perchloric acid media, J. Am. Chem. Soc., **83** (1961) 3373-3378.

[61TAN/DEI] Tanayev, I.V., Deichman, E.N., The properties of uranium hydroxyfluoride in solutions, Radiokhimiya, **3** (1961) 712-718, in Russian.

[61VAN/QUI] Vanderzee, C.E., Quist, A.S., The third dissociation constant of orthophosphoric acid, J. Phys. Chem., **65** (1961) 118-123.

[61WIL/KED] Wilson, A.S., Keder, W.E., The extraction of uranium(IV) from nitric acid by tri-*n*-octylamine, J. Inorg. Nucl. Chem., **18** (1961) 259-262.

[62ALC/GRI] Alcock, C.B., Grieveson, P., Study of uranium borides and carbides by means of the Knudsen effusion technique, in: Thermodynamics of nuclear materials, Proc. Symp. held 21-25 May, 1962, in Vienna, Vienna: International Atomic Energy Agency, 1962, *pp.*563-579.

[62BAE/MEY] Baes, C.F., Jr., Meyer, N.J., Acidity measurements at elevated temperatures: I. Uranium(VI) hydrolysis at 25 and 94°C, Inorg. Chem., **1** (1962) 780-789.

[62DAV] Davies, C.W., Ion association, Washington, D.C.: Butterworths, 1962.

[62ERD] Erdös, E., Thermodynamic properties of sulphites: I. Standard heats of formation, Coll. Czech. Chem. Commun., **27** (1962) 1428-1437.

[62ERD2] Erdös, E., Thermodynamic properties of sulphites: II. Absolute entropies, heat capacities and dissociation pressures, Coll. Czech. Chem. Commun., **27** (1962) 2273-2283.

[62ERM/KRO] Ermolaev, N.P., Krot, N.N., Some data on the behavior of uranium(IV) in nitric acid solutions, Sov. Radiochem., **4** (1962) 600-606.

[62FAU/CRE] Faucherre, J., Crego, A., Détermination cryoscopique de constantes de dissociation de complexes peu stables, Bull. Soc. Chim. Fr., (1962) 1820-1824, in French.

[62GRO/HAY] Gross, P., Hayman, C., Clayton, H., Heats of formation of uranium silicides and nitrides, in: Thermodynamics of nuclear materials, Proc. Symp. held 21-25 May, 1962, in Vienna, Vienna: International Atomic Energy Agency, 1962, *pp.*653-665.

[62HOS/GAR] Hostetler, P.B., Garrels, R.M., Transportation and precipitation of U and V at low temperature with special reference to sandstone-type U deposits, Econ. Geol., **57(2)** (1962) 137-167.

[62HUB/HOL] Huber, E.J., Jr., Holley, C.E., Jr., The thermodynamic properties of the actinide carbides including new measurements of the heats of formation of some thorium, uranium and plutonium carbides, in: Thermodynamics of nuclear materials, Proc. Symp. held 21-25 May, 1962, in Vienna, Vienna: International Atomic Energy Agency, 1962, *pp.*581-599.

[62HUB] Hubbard, W.N., Thermodynamics studies at Argonne laboratory, Report TID-15554, US Atomic Energy Commission, 1962, 5p.

[62LUK/NIK] Luk'yanychev, Yu.A., Nikolaev, N.S., Solubility of tetravalent uranium hydroxide in hydrofluoric acid solutions, Sov. At. Energy, **13** (1962) 779-781.

[62MOR/FOU] Morey, G.W., Fournier, R.O., Rowe, J.J., The solubility of quartz in water in the temperature interval from 25 to 300 C, Geochim. Cosmochim. Acta, 26 (1962) 1029-1043.

[62MUK/NAI] Mukaibo, T., Naito, K., Sato, K., Uchijima, T., The heat of formation of U_3O_7, in: Thermodynamics of nuclear materials, Proc. Symp. held 21-25 May, 1962, in Vienna, Vienna: International Atomic Energy Agency, 1962, *pp.*723-734.

[62MUK/NAI2] Mukaibo, T., Naito, K., Sato, K., Uchijima, T., The measurement of the specific heat of uranium carbides above room temperatures, in: Thermodynamics of nuclear materials, Proc. Symp. held 21-25 May, 1962, in Vienna, Vienna: International Atomic Energy Agency, 1962, *pp.*645-650.

[62NIK/PAR] Nikolaeva, N.M., Paramonova, V.I., Kolychev, V.B., Hydrolysis of uranyl in nitrate solutions, Izv. Sib. Otd. Akad. Nauk. SSSR, **3** (1962), in Russian; Engl. transl.: Report ORNL-TR-417, Oak Ridge National Laboratory, Oak Ridge, Tennessee, USA, 1965, 15p.

[62PAR/NIK] Paramonova, V.I., Nikolaeva, N.M., Use of ion exchange to study the state of a substance in solution: VIII. Study of uranyl carbonate solutions by ion exchange, Sov. Radiochem., **4** (1962) 72-76.

[62PAR/NIK2] Paramonova, V.I., Nikol'skii, B.P., Nikolaeva, N.M., Interaction of uranyl nitrate solutions and alkali-metal carbonates, Russ. J. Inorg. Chem., **7** (1962) 528-532.

[62RUS/JOH] Rush, R.M., Johnson, J.S., Kraus, K.A., Hydrolysis of uranium(VI): Ultracentrifugation and acidity measurements in chloride solutions, Inorg. Chem., **1** (1962) 378-386.

[62SHE/AWA] Sherif, F.G., Awad, A.M., Spectrophotometric determination of uranium(VI) by the azide reaction, Anal. Chim. Acta, **26** (1962) 235-241.

[62SHE/AWA2] Sherif, F.G., Awad, A.M., Re-evaluation of the continuous variation method as applied to the uranyl azide complexes, J. Inorg. Nucl. Chem., **24** (1962) 179-182.

[62STE/GAL] Stepanov, M.A., Galkin, N.P., Solubility product of basic uranium(IV) sulphate, Russ. J. Inorg. Chem., **7** (1962) 506-508.

[62STO] Storms, E.K., A mass spectometric study of the vapour pressure of U(g) and UC_2(g) over various compositinos in the uranium-carbon system, in: Thermodynamics of nuclear materials, Proc. Symp. held 21-25 May, 1962, in Vienna, Vienna: International Atomic Energy Agency, 1962, *pp*.309-343.

[62VAS] Vasil'ev, V.P., Influence of ionic strength on the instability constants of complexes, Russ. J. Inorg. Chem., **7** (1962) 924-927.

[62WES/GRO] Westrum, E.F., Jr., Grønvold, F., Chemical thermodynamics of the actinide element chalcogenides, in: Thermodynamics of nuclear materials, Proc. Symp. held 21-25 May, 1962, in Vienna, Vienna: International Atomic Energy Agency, 1962, *pp*.3-37.

[62WES/GRO2] Westrum, E.F., Jr., Grønvold, F., Triuranium heptaoxides: Heat capacities and thermodynamic properties of α- and β-U_3O_7 from 5 to 350 K, J. Phys. Chem. Solids, **23** (1962) 39-53.

[63ACH/ALC] Achachinsky, V.V., Alcock, C.B., Holley, C.E., Jr., Kubaschewski, O., Livey, D.T., Mukaibo, T., Neckel, A., Nowotny, H., Rand, M.H., Reschetnikov, F., Thorn, R.J., Westrum, E.F., Jr., Hara, R., The uranium-carbon and plutonium-carbon system, Technical Report Series No. 14, Vienna: International Atomic Energy Agency, 1963, 44p.

[63BAR/SOM] Bartušek, M., Sommer, L., Ph.D. Thesis, Institut für Analytische Chemie der J.E. Purkyne Universität, Brno, Czechoslovakia, 1963, cited by Baes, C.F., Jr., Mesmer, R.F., The hydrolysis of cations, New York: Wiley & Sons, 1976, 177p.

[63CAV/BON] Cavett, A.D., Bonham, R.W., High temperature decomposition of uranium mononitride, in: Summary technical report No. NLCO-885, (Burgett, R., *ed.*), National Lead Company of Ohio, Cincinnati, Ohio, USA, April-June, 1963, *pp.*17-18.

[63COR/ALI] Cordfunke, E.H.P., Aling, P., Rec. Trav. Chim. Pays-Bas, **82** (1963) 257.

[63DUN/HIE] Dunsmore, H.S., Hietanen, S., Sillén, L.G., Studies on the hydrolysis of metal ions: 46. Uranyl ion, UO_2^{2+}, in chloride, perchlorate, nitrate and sulfate media, Acta Chem. Scand., **17** (1963) 2644-2656.

[63DUN/SIL] Dunsmore, H.S., Sillén, L.G., Studies on the hydrolysis of metal ions: 47. Uranyl ion in 3 M (Na)Cl medium, Acta Chem. Scand., **17** (1963) 2657-2663.

[63ELL] Ellis, A.J., The effect of temperature on the ionization of hydrofluoric acid. J. Chem. Soc., (1963) 4300-4304.

[63EVA] Evans, H.T., Jr., Uranyl ion coordination, Science, **141** (1963) 154-158.

[63HIE/ROW] Hietanen, S., Row, B.R.L., Sillén, L.G., Studies on the hydrolysis of metal ions: 48. The uranyl ion in sodium, magnesium and calcium perchlorate medium, Acta Chem. Scand., **17** (1963) 2735-2749.

[63HUL/ORR] Hultgren, R., Orr, R.L., Anderson, P.D., Kelley, K.K., Selected values of thermodynamic properties of metals and alloys, New York: J. Wiley & Sons, 1963, 963p.

[63KAN] Kangro, W., Z. Erzbergbau Metallhüttenwes., **16** (1963) 107, cited by Fuger, J., Parker, V.B., Hubbard, W.N., Oetting, F.L., The chemical thermodynamics of actinide elements and compounds, Part 8: The actinide halides, Vienna: International Atomic Energy Agency, 1983, 267p.

[63LUK/NIK] Luk'yanychev, Yu.A., Nikolaev, N.S., The solubility of uranium tetrafluoride in aqueous solutions of acids, Sov. At. Energy, **15** (1963) 1184-1187.

[63LUK/NIK2] Luk'yanychev, Yu.A., Nikolaev, N.S., Solubility product of uranium tetrafluoride, Russ. J. Inorg. Chem., **8** (1963) 927-928.

[63MEL/AMI] Melton, S.L., Amis, E.S., A spectrophotometric study of the complexes formed between uranyl and thiosulfate ions in water and water-ethanol solvents, Anal. Chem., **35** (1963) 1626-1630.

[63OLS/MUL] Olson, W.M., Mulford, R.N.R., The decomposition pressure and melting point of uranium mononitride, J. Phys. Chem., **67** (1963) 952-954.

[63OSI/BYN] Osinska-Tanevska, S.M., Bynyaeva, M.K., Mishchenko, K.P., Flis, I.E., Russ. J. Appl. Chem., **36** (1963) 1162, cited by Perrin, D.D., Ionisation constants of inorganic acids and bases in aqueous solution, 2*nd* ed., IUPAC Chemical Data Series **29**, Oxford: Pergamon Press, 1982, 180p.

[63POZ/STE] Pozharskii, B.G., Sterlingova, T.N., Petrova, A.E., Hydrolysis and complex formation of uranyl in mineral acid solutions, Russ. J. Inorg. Chem., **8** (1963) 831-839.

[63RAN/KUB] Rand, M.H., Kubaschewski, O., The thermodynamical properties of uranium compounds, New York: J. Wiley & Sons, 1963, 95p.

[63RUS/JOH] Rush, R.M., Johnson, J.S., Hydrolysis of uranium(VI): Absorption spectra of chloride and perchlorate solutions, J. Phys. Chem., **67** (1963) 821-825.

[63SET/FED] Settle, J.L., Feder, H.M., Hubbard, W.N., Fluorine bomb calorimetry: VI. The enthalpy of formation of uranium hexafluoride, J. Phys. Chem., **67** (1963) 1892-1895.

[63STA/NIE] Staliński, B., Niemic, J., Biegański, Z., Heat capacity and thermodynamical functions of antiferromagnetic uranium oxytelluride, UOTe, within the range 21 to 362 K, Bull. Acad. Polon. Sci., Ser. Sci. Chim., **9** (1963) 267-270.

[63VDO/ROM] Vdovenko, V.M., Romanov, G.A., Shcherbakov, V.A., Hydrolysis of the ion U(IV), Sov. Radiochem., **5** (1963) 119-120.

[63VDO/ROM2] Vdovenko, V.M., Romanov, G.A., Shcherbakov, V.A., Study of the complex-formation or uranium(IV) with the fluoride ion by the proton resonance method, Sov. Radiochem., **5** (1963) 538-541.

[63VDO/ROM3] Vdovenko, V.M., Romanov, G.A., Shcherbakov, V.A., Proton resonance study of complex formation of U(IV) with halide, sulfate, and perchlorate ions, Sov. Radiochem., **5** (1963) 624-627.

[63VDO/SUG] Vdovenko, V.M., Suglobova, I.G., Suglobov, D.N., Datyuk, Yu.V., Heat of solution of uranyl nitrate and some of its complex compounds, Sov. Radiochem., **5** (1963) 699-701.

[63WIC/BLO] Wicks, C.E., Block, F.E., Thermodynamic properties of 65 elements, their oxides, halides, carbides and nitrides, US Bureau of Mines Bulletin, **605** (1963) 146p.

[63WIL/BRO] Wilcox, D.E., Bromley, L.A., Computer estimation of heat and free energy of formation for simple inorganic compounds, Ind. Eng. Chem., **55(7)** (1963) 32-39.

[63WU/BIR] Wu, C.H., Birky, M.M., Hepler, L.G., Thermochemistry of some bromine and iodine species in aqueous solution, J. Phys. Chem., **67** (1963) 1202-1205.

[64AND/COU] Andon, R.J.L., Counsell, J.F., Martin, J.F., Hedger, H.J., Thermodynamic properties of uranium compounds: Part 1. Low-temperature heat capacity and entropy of three uranium carbides, Trans. Faraday Soc., **60** (1964) 1030-1037.

[64BAR/SOM] Bartušek, M., Sommer, L., Über die Hydrolyse von UO_2^{2+} in verdünnten Lösungen, Z. Phys. Chem. Leipzig, **226** (1964) 309-332, in German.

[64BUG/BAU] Bugl, J., Bauer, A.A., Phase and thermodynamic studies. In development of uranium mononitride, Report BMI-X-10083 (Keller, D.L., *ed.*), Batelle Memorial Institute, Columbus, Ohio, USA, 1964, *pp.*3-5.

[64COR] Cordfunke, E.H.P., Heats of formation of some hexavalent uranium compounds, J. Phys. Chem., **68** (1964) 3353-3356.

[64CRI/COB] Criss, C.M., Cobble, J.W., The thermodynamic properties of high temperature aqueous solutions: IV. Entropies of the ions up to 200°C and the correspondence principle, J. Am. Chem. Soc., **86** (1964) 5385-5390.

[64CRI/COB2] Criss, C.M., Cobble, J.W., The thermodynamic properties of high temperature aqueous solutions: V. The calculation of ionic heat capacities up to 200°C. Entropies and heat capacities above 200°C, J. Am. Chem. Soc., **86** (1964) 5390-5393.

[64DUN/NAN] Dunsmore, H.S., Nancollas, G.H., Dissociation of the bisulphate ion, J. Phys. Chem., **68** (1964) 1579-1581.

[64FLI/MIS] Flis, I.E., Mishchenko, K.P., Pusenok, G.I., Izv. Vysshikh Uchebn. Zavedenii Khim. i Khim. Tekhnol., **7** (1964) 764, cited by Perrin, D.D., Ionisation constants of inorganic acids and bases in aqueous solution, 2*nd* ed., IUPAC Chemical Data Series **29**, Oxford: Pergamon Press, 1982, 180p.

[64GOR/KER] Gordon, G., Kern, D.M.H., Observations on the complex between uranyl and chlorite ions, Inorg. Chem., **3** (1964) 1055-1056.

[64LAP/HOL] Lapat, P.E., Holden, R.B., Thermodynamics of the uranium-nitrogen system, in: Proc. International Symp. on compounds of interest nuclear reactor technology (Waber, T.J., Chiotti, P., Miner, W.N., *eds.*), Nuclear Metallurgy, Vol. **X**, IMD Special Report No. 13, Ann Arbor, Michigan, USA, 1964, *pp.*225-236.

[64MAR/VER] Markov, V.K., Vernyi, E.A., Vinogradov, A.V., Methods for the determination of uranium, Moscow: Atomizdat, 1964, *p.*92, cited by Moskvin, A.I., Essen, L.N., Bukhtiyarova, T.N., The formation of thorium(IV) and uranium(IV) complexes in phosphate solutions, Russ. J. Inorg. Chem., **12** (1967) 1794-1795.

[64MAS] Maslov, P.G., Thermodynamic characteristics of the uranium halides $UClF_3$, UCl_2F_2, and $UClF_3$, Zh. Neorg. Khim., **9(9)** 2076-2078, in Russian; Engl. transl.: Russ. J. Inorg. Chem., **9(9)** (1964) 1122-1124.

[64MCK/WOO] McKay, H.A.C., Woodhead, J.L., A spectrophotometric study of the nitrate complexes of uranium(IV), J. Chem. Soc., (1964) 717-723.

[64NAI] Nair, V.S.K., Dissociation of the biselenate ion, J. Inorg. Nucl. Chem., **26** (1964) 1911-1917.

[64NOR/WIN] Norman, J.H., Winchell, P., Mass spectrometry - Knudsen cell studies of the vaporization of uranium dicarbide, J. Phys. Chem., **68** (1964) 3802-3805.

[64OWE/MAY] Owens, B.B., Mayer, S.W., The thermodynamic properties of uranyl sulphate, J. Inorg. Nucl. Chem., **26** (1964) 501-507.

[64SIL/MAR] Sillén, L.G., Martell, A.E., Stability constants of metal-ion complexes, Special Publ. No. 17, Chemical Society, London, 1964, 754p.

[64VAS/MUK] Vasil'ev, V.P., Mukhina, P.S., Equilibria of uranyl complexes in aqueous thiocyanate solutions, Izv. Vyssh. Ucheb. Zaved. Khim., **7** (1964) 711-714, in Russian, Chem. Abstr. **62** (1965) 7172c.

[65BAE/MEY] Baes, C.F., Jr., Meyer, N.J., Roberts, C.E., The hydrolysis of thorium(IV) at 0 and 95°C, Inorg. Chem., **4** (1965) 518-527.

[65BEE/MOR] Beezer, A.E., Mortimer, C.T., Tyler, E.G., Heats of formation and bond energies: Part III. Arsenic tribromide, arsenious and arsenic oxides, and aqueous solutions of sodium arsenite and sodium arsenate, J. Chem. Soc., (1965) 4471-4478.

[65COR/ALI] Cordfunke, E.H.P., Aling, P., System $UO_3 + U_3O_8$: Dissociation pressure of γ-UO_3, Trans. Faraday Soc., **61** (1965) 50-53.

[65COV/DOB] Covington, A.K., Dobson, J.V., The dissociation constant of the biselenate ion at 25°C, J. Inorg. Nucl. Chem., **27** (1965) 1435-1436.

[65DRA/JUL] Drăgulescu, C., Julean, I., Vîlceanu, N., Sur les pyrophosphates complexes d'uranyle, Rev. Roum. Chim., **10** (1965) 809-819, in French.

[65FEN/LI] Feng, S.Y., Li, H.Y., Measurement of the heat of solution and calculation of the heat of formation of hydrated uranyl nitrates in water and in alcohol, Yuan Tzu Neng, **3** (1965) 233-238, from Communist Chinese Sci. Abstr., Chem., **6(4)** (1965).

[65GOT/NAI] Gotoo, K., Naito, K., Study of U_4O_9: Part I. An anomaly of the heat capacity near the room temperature, J. Phys. Chem. Solids, **26** (1965) 1673-1677.

[65ISR] Israeli, Y.J., Le mécanisme d'hydrolyse de l'ion uranyle, Bull. Soc. Chim. Fr., (1965) 193-196, in French.

[65MUT/HIR] Muto, T., Hirono, S., Kurata, H., Some aspects of fixation of uranium from natural waters, Mining Geol., **15(74)** (1965) 287-298, in Japanese; Engl. transl.: Report NSJ-Tr 91, Japanese Atomic Energy Research Institute, Tokai-Mura, Japan, 27p.

[65MUT] Muto, T., Thermochemical stability of ningyoite, Mineral. J., **4** (1965) 245-274.

[65NEW/BAK] Newton, T.W., Baker, F.B., A uranium(V)-uranium(VI) complex and its effect on the uranium(V) disproportionation rate, Inorg. Chem., **4** (1965) 1166-1170.

[65PAR] Parker, V.B., Thermal properties of aqueous uni-univalent electrolytes, National Standard Reference Data Series, National Bureau of Standards **2**, Washington, D.C.: US Government printing office, 1965.

[65PEK/VES] Pekárek, V., Veselý, V., A study on uranyl phosphates: II. Sorption properties of some 1- to 4-valent cations on uranyl hydrogen phosphate heated to various temperatures, J. Inorg. Nucl. Chem., **27** (1965) 1151-1158.

[65SHI/MAR] Shiloh, M., Marcus, Y., A spectrophotometric study of trivalent actinide complexes in solution: I. Uranium, Israel J. Chem., **3** (1965) 123-131.

[65TAK/WES] Takahashi, Y., Westrum, E.F., Jr., Uranium monoselenide. Heat capacity and thermodynamic properties from 5 to 350 K, J. Phys. Chem., **69** (1965) 3618-3621.

[65VES/PEK] Veselý, V., Pekárek, V., Abbrent, M., A study of uranyl phosphates: III. Solubility products of uranyl hydrogen phosphate, uranyl orthophosphate and some alkali uranyl phosphates, J. Inorg. Nucl. Chem., **27** (1965) 1159-1166.

[65VID/BYA] Vidavskii, L.M., Byakhova, N.I., Ippolitova, E.A., Enthalpy of the reaction of γ-UO_3 with hydrofluoric acid and enthalpy of formation of γ-UO_3, Zh. Neorg. Khim., **10(7)** (1965) 1746-1747, in Russian; Engl. transl.: Russ. J. Inorg. Chem., **10** (1965) 953-954.

[65VOZ/CRE] Vozzella, P.A., de Crescente, M.A., Thermodynamic properties of uranium mononitride, Report PWAC-479, Pratt & Whitney Aircraft, Hiddletown, Connecticut, USA, 1965, 16p.

[65WES/SUI] Westrum, E.F., Jr., Suits, E., Lonsdale, H.K., Uranium monocarbide and hypostoichiometric dicarbide: Heat capacities and thermodynamic properties from 5 to 350 K, in: Thermophysical properties at extreme temperatures and pressures (Gratch, S., *ed.*), New York: Am. Soc. Mech. Eng., 1965, *pp.*156-161.

[65WES/TAK] Westrum, E.F., Jr., Takahashi, Y., Grønvold, F., λ-type thermal anomaly in tetrauranium enneaoxide at 348 K, J. Phys. Chem., **69** (1965) 3192-3193.

[65WOL/POS] Wolf, A.S., Posey, J.C., Rapp, K.E., α-uranium pentafluoride: I. Characterization, Inorg. Chem., **4** (1965) 751-757.

[66BEH/EGA] Behl, W.K., Egan, J.J., Standard molar free energy of formation of uranium dicarbide and uranium sesquicarbide by electromotive force measurements, J. Electrochem. Soc., **113** (1966) 376-378.

[66CHE] Chernyayev, I.I. (editor), Complex compounds of uranium, Israel Program for Scientific Translations, Jerusalem, 1966, cited by Weigel, F., The carbonates, phosphates and arsenates of the hexavalent and pentavalent actinides, in: Handbook on the Physics and Chemistry of the Actinides, Vol **3** (Freeman, A.J., Keller, C., *eds.*), Amsterdam: Elsevier Science Publishers BV, 1985, *pp.*243-288.

[66COR] Cordfunke, E.H.P., Thermodynamic properties of hexavalent uranium compounds, in: Thermodynamics, Vol. **2**, Proc. Symp. held 22-27 July, 1965, in Vienna, Austria, Vienna: International Atomic Energy Agency, 1966, *pp.*483-495.

[66COU/DEL] Counsell, J.F., Dell, R.M., Martin, J.F., Thermodynamic properties of uranium compounds: Part 2. Low-temperature heat capacity and entropy of three uranium nitrides, Trans. Faraday Soc., **62** (1966) 1736-1747.

[66DRO/KOL] Drobnic, M., Kolar, D., Calorimetric determination of the enthalpy of hydration of UO_3, J. Inorg. Nucl. Chem., **28** (1966) 2833-2835.

[66FLO/OSB] Flotow, H.E., Osborne, D.W., Heat capacity of α-uranium from 1.7 to 25 K, Phys. Rev., **151** (1966) 564-570.

[66GOO/WES] Goodman, R.M., Westrum, E.F., Jr., Barium chloride heat capacities and thermodynamic properties from 5 to 350 K , J. Chem. Eng. Data, **11** (1966) 294-295.

[66GUG] Guggenheim, E.A., Applications of statistical mechanics, Oxford: Clarendon Press, 1966, 211p.

[66HEU/EGA] Heus, R.J., Egan, J.J., Free energies of formation of some inorganic fluorides by solid state electromotive force measurements, Z. Phys. Chem. (Leipzig), **49** (1966) 38-43.

[66KOC/SCH] Koch, G., Schwind, E., Extraktion von Uran(IV)-Nitrat durch Tricaprylmethylammoniumnitrat, J. Inorg. Nucl. Chem., **28** (1966) 571-575, in German.

[66LOO/COR] Loopstra, B.O., Cordfunke, E.H.P., On the structure of α-UO_3, Rec. Trav. Chim. Pays-Bas, **85** (1966) 135-142.

[66MAR/JON] Marshall, W.L., Jones, E.V., Second dissociation constant of sulfuric acid from 25 to 350°C evaluated from solubilities of calcium sulfate in sulfuric acid solutions, J. Phys. Chem., **70** (1966) 4028-4040.

[66MIT/MAL] Mitra, R.R., Malhotra, H.C., Jain, D.V.S., Dissociation equilibria in pyrophosphates and kinetics of degradation, Trans. Faraday Soc., **62** (1966) 167-172.

[66MOR] Morris, J.C., The acid ionization constant of HOCl from 5 to 35°C, J. Phys. Chem., **70** (1966) 3798-3805.

[66ROB] Robins, R.G., Hydrolysis of uranyl nitrate solutions at elevated temperatures, J. Inorg. Nucl. Chem., **28**, (1966) 119-123.

[66SHI/NAZ] Shilin, I.V., Nazarov, V.K., Complex formation of neptunium(IV) with nitrate and chloride ions, Sov. Radiochem., **8** (1966) 474-478.

[66STA/BIE] Staliński, B., Biegański, Z., Troć, R., Low-temperature heat capacity and thermodynamical functions of ferromagnetic uranium phosphide U_3P_4, Phys. Status Solidi, **17** (1966) 837-841.

[66VDO/ROM] Vdovenko, V.M., Romanov, G.A., Shcherbakov, V.A., A study of fluoro-complexes of uranium(IV) in aluminium salt solutions, Russ. J. Inorg. Chem., **11** (1966) 139-141.

[66WES/BAR] Westrum, E.F., Jr., Barber, C.M., Uranium mononitride: Heat capacity and thermodynamic properties from 5 to 350 K, J. Chem. Phys., **45** (1966) 635-639.

[66WES] Westrum, E.F., Jr., Recent developments in the chemical thermodynamics of the uranium chalcogenides, in: Thermodynamics, Vol. **2**, Proc. Symp. held 22-27 July, 1965, in Vienna, Vienna: International Atomic Energy Agency, 1966, *pp.*497-510.

[67AHR] Ahrland, S., Enthalpy and entropy changes by formation of different types of complexes, Helv. Chim. Acta, **50** (1967) 306-318.

[67BAR] Bartušek, M., Über die Methoden zum Studium der Uranylhydrolyse, Institut für Analytische Chemie der J.E. Purkyne Universität, Brno, Czechoslovakia, **480** (1967) 51-62, in German.

[67COL/EYR] Cole, D.L., Eyring, E.M., Rampton, D.T., Silzars, A., Jensen, R.P., Rapid reaction rates in a uranyl ion hydrolysis equilibrium, J. Phys. Chem., **71** (1967) 2771-2775.

[67COR/LOO] Cordfunke, E.H.P., Loopstra, B.O., Preparation and properties of the uranates of calcium and strontium, J. Inorg. Nucl. Chem., **29** (1967) 51-57.

[67COU/DEL] Counsell, J.F., Dell, R.M., Junkison, A.R., Martin, J.F., Thermodynamic properties of uranium compounds: Part 3. Low-temperature heat capacity and entropy of UP and U_2P_4, Trans. Faraday Soc., **63** (1967) 72-79.

[67FIT/PAV] Fitzgibbon, G.C., Pavone, D., Holley, C.E., Jr., Enthalpies of solution and formation of some uranium oxides, J. Chem. Eng. Data, **12** (1967) 122-125.

[67FLO/OSB] Flotow, H.E., Osborne, D.W., β-Uranium hydride: Heat capacities from 1.4 to 23 K and some derived properties, Phys. Rev., **164** (1967) 755-758.

[67GAG/KHA] Gagarinskii, Yu.V., Khanaev, E.I., Thermochemical study of uranium tetrafluoride crystal hydrates and their dehydration products, Russ. J. Inorg. Chem., **12** (1967) 54-56.

[67GRY/KOR] Gryzin, Yu.I., Koryttsev, K.Z., A study of the behaviour of UO_3 and its hydrates in solutions with an electrode of the third kind, Russ. J. Inorg. Chem., **12** (1967) 50-53.

[67HAY] Hayman, H.J.G., Paper No. 7, Thermodynamik Symposium Heidelberg (Schafer, K.L., *ed.*), 1967, cited by Fuger, J., Parker, V.b., Hubbard, W.N., Oetting, F.L., The chemical thermodynamics of the actinide elements: Part 8. The actinide halides, Vienna: International Atomic Energy Agency, 1983, 267p.

[67ING/KAK] Ingri, N., Kakolowicz, W., Sillén, L.G., Warnquist, B., High-speed computers as a supplement to graphical methods-V. HALTAFALL, a general program for calculating the composition of equilibrium mixtures, Talanta, **14** (1967) 1261.

[67JAK/SCH] Jakes, D., Schauer, V., Heat capacity of some uranates, Proc. Brit. Ceram. Soc., **8** (1967) 123-125.

[67MAR/BON] Markin, T.L., Bones, R.L., Wheeler, V.J., Galvanic cells and gas equilibration techniques, Proc. Br. Ceram. Soc., **8** (1967) 51-66.

[67MER/SKO] Merkusheva, S.A., Skorik, N.A., Kumok, V.N., Serebrennikov, V.V., Thorium and uranium(IV) pyrophosphates, Sov. Radiochem, **9** (1967) 683-685.

[67MOS/ESS] Moskvin, A.I., Essen, L.N., Bukhtiyarova, T.N., The formation of thorium(IV) and uranium(IV) complexes in phosphate solutions, Russ. J. Inorg. Chem., **12** (1967) 1794-1795.

[67MOS/KRU] Moser, J.B., Kruger, O.L., Thermal conductivity and heat capacity of the monocarbide, monophosphide and monosulphide of uranium, J. Appl. Phys., **38** (1967) 3215-3222.

[67MOS/SHE] Moskvin, A.I., Shelyakina, A.M., Perminov, P.S., Solubility product of uranyl phosphate and the composition and dissociation constants of uranyl phosphato-complexes, Zh. Neorg. Khim., **12** (1967) 3319-3326, in Russian; Engl. transl.: Russ. J. Inorg. Chem., **12** (1967) 1756-1760.

[67OHA/MOR] Ohashi, H., Morozumi, T., Electrometric determination of stability constants of uranyl chloride and uranyl nitrate complexes with pCl-Stat, J. At. Energy Soc. Japan, **9** (1967) 65-71, in Japanese, Chem. Abstr. **67** (1967) 111876.

[67OHA/MOR2] Ohashi, H., Morozumi, T., Temperature dependence of the stability constants of uranylchloro complexes, J. At. Energy Soc. Japan, **9(4)** (1967) 200-201, in Japanese, Chem. Abstr. **69** (1968) 80999v.

[67OTE/DOU] Otey, M.G., Le Doux, R.A., $U_3O_5F_8$—A new compound in the U-O-F system, J. Inorg. Nucl. Chem., **29** (1967) 2249-2256.

[67PAS] Pascal, P., Nouveau traité de chimie minérale: Tome XV. Uranium et transuraniens, Quatrième fascicule: Uranium (compléments), Paris: Masson et Cie., 1967, 1213p., in French.

[67PET/PRU] Pethybridge, A.D., Prue, J.E., Equilibria in aqueous solutions of iodic acid, Trans. Faraday Soc., **63** (1967) 2019-2033.

[67REI/GUN] Reishus, J.W., Gundersen, G.E., Vaporization of uranium monophosphide, Chemical Engineering Division, Semiannual Report ANL-7375, Argonne National Laboratory, Argonne, Illinois, USA, 1967, *pp.*103-106.

[67RYZ/NAU] Ryzhenko, B.N., Naumov, G.B., Goglev, V.S., Hydrolysis of uranyl ions at elevated temperatures, Geochem. Internat., **4** (1967) 363-367.

[67SOL/TSV] Solovkin, A.S., Tsvetkova, Z.N., Ivantsov, A.I., Thermodynamic constants for complex formation by U^{4+} with OH^- ions and by Zr^{4+} with OH^- and NO_3^- ions, Russ. J. Inorg. Chem., **12** (1967) 326-330.

[67STA/BIE] Staliński, B., Biegański, Z., Troć, R., Low temperature heat capacity and thermodynamical functions of antiferromagnetic uranium diphosphide UP_2, Bull. Acad. Polon. Sci., Ser. Sci. Chim., **16(6)** (1967) 257-260.

[67STO/HUB] Storms, E.K., Huber, E.J., Jr., The heat of formation of uranium carbide, J. Nucl. Mater., **23** (1967) 19-24.

[67STO] Storms, E.K., The refractory carbides, New York: Academic Press, 1967.

[67VDO/ROM] Vdovenko, V.M., Romanov, G.A., Solubility of uranium tetrafluoride in solutions of aluminium salts, Sov. Radiochem., **9** (1967) 375-377.

[67WAL] Wallace, R.M., Determination of stability constants by Donnan membrane equilibrium: The uranyl sulphate complexes, J. Phys. Chem., **71** (1967) 1271-1276.

[67ZAK/ORL] Zakharova, F.A., Orlova, M.M., The formation of uranyl sulphito-complexes, Russ. J. Inorg. Chem., **12** (1967) 1596-1599.

[68AHR] Ahrland, S., Thermodynamics of complex formation between hard and soft acceptors and donors, Struct. Bonding (Berlin), **5** (1968) 118-149.

[68ARN/SCH] Arnek, R., Schlyter, K., Thermochemical studies of hydrolytic reactions: 8. A recalculation of calorimetric data on uranyl hydrolysis, Acta Chem. Scand., **22** (1968) 1331-1333.

[68BRO] Brown, D., Halides of the lanthanides and actinides, London: Wiley-Interscience, 1968.

[68CHO/CHU] Choporov, D.Ya., Chudinov, É.T., Melting point and saturated vapor pressure of neptunium tetrachloride, Radiokhimiya, **10** (1968) 221-227. in Russian; Engl. transl.: Sov. Radiochem., **10** (1968) 208-213.

[68COR/PRI] Cordfunke, E.H.P., Prins, G., van Vlaanderen, P., Preparation and properties of the violet "U_3O_8 hydrate", J. Inorg. Nucl. Chem., **30** (1968) 1745-1750.

[68COU/MAR] Counsell, J.F., Martin, J.F., Dell, R.M., Junkison, A.R., Thermodynamic properties of uranium compounds: Part 4. Low-temperature heat capacities and entropies of Us-UP solid solutions, in: Thermodynamics of nuclear materials 1967, Proc. Symp. held 1967 in Vienna, Vienna: International Atomic Energy Agency, 1968, *pp.*385-394.

[68CUK/CAJ] Cukman, D., Caja, J., Pravdic, V., The electrochemical oxidation of uranium(IV) in sodium bicarbonate solutions, J. Electroanal. Chem., **19** (1968) 267-274.

[68DOG/VAL] Dogu, A., Val, C., Accary, A., Application d'une méthode quantitative d'analyse thermique différentielle à la détermination de U_3P_4 et UP_2, J. Nucl. Mater., **28** (1968) 271-279, in French.

[68FAR/ELD] Farkas, M., Eldridge, E., Heat contents and specific heats of some uranium-bearing fuels, J. Nucl. Mater., **27** (1968) 94.

[68FLO/OSB] Flotow, H.E., Osborne, D.W., Westrum, E.F., Jr., Heat capacity of tetrauranium enneaoxide U_4O_9 from 1.6 to 24 K: The magnetic contribution to the entropy, J. Chem. Phys., **49** (1968) 2438-2442.

[68GIR/WES] Girdhar, H.L., Westrum, E.F., Jr., λ-type thermal anomaly in triuranium octaoxide at 482.7 K, J. Chem. Eng. Data, **13** (1968) 531-533.

[68HAL/SAB] Haladjan, J., Sabbah, R., Bianco, P., J. Chim. Phys. Physico-Chim. Biol., 65 (1968) 1751, CA-70-51218c, cited by Perrin, D.D., Ionisation constants of inorganic acids and bases in aqueous solution, 2nd ed., IUPAC Chemical Data Series 29, Oxford: Pergamon Press, 1982, 180p.

[68HOL/STO] Holley, C.E., Jr., Storms, E.K., Actinide carbides: A review of the thermodynamic properties, in: Thermodynamics of nuclear materials 1967, Proc. Symp. held 1967 in Vienna, Vienna: International Atomic Energy Agency, 1968, *pp.*397-426.

[68INO/LEI] Inouye, H., Leitnaker, J.M., Equilibrium nitrogen pressures and thermodynamic properties of UN, J. Am. Ceram. Soc., **51** (1968) 6-9.

[68KAR/KAR] Karapet'yants, M.K., Karapet'yants, M.L., The fundamental thermodynamic constants of inorganic and organic compounds, Khimiya, Moscow, USSR, 1968, cited by Krestov, G.A., Thermochemistry of compounds of rare-earth and actinide elements, Atomizdat, Moscow, USSR, 1972; Engl. transl.: Report AEC-tr-7505, US Atomic Energy Commission, 253p.

[68KHA] Khanaev, E.I., Enthalpy of formation of uranium tetrafluoride, Bull. Acad. Polon. Sci., Ser. Sci. Chim., **4** (1968) 449-451.

[68KRY/KOM] Krylov, V.N., Komarov, E.V., Pushlenkov, M.F., Complex formation of Pu(VI) with the fluoride ion in solutions of $HClO_4$, Sov. Radiochem., **10** (1968) 705-707.

[68KRY/KOM3] Krylov, V.N., Komarov, E.V., Pushlenkov, M.F., Complex formation of U(VI) with the fluoride ion in solution of $HClO_4$, Sov. Radiochem., **10(6)** (1968) 708-710.

[68LEB/NIC2] Lebedev, V.A., Nichkov, I.F., Raspopin, S.P., Kapitonov, V.I., An electromotive force study of the thermodynamics of the intermetallic compounds of bismuth and uranium, Russ. J. Phys. Chem., **42(3)** (1968) 365-367.

[68MAR/KUD] Martynova, N.S., Kudryashova, Z.P., Vasilkova, I.V., Enthalpy of formation of K_2UCl_6, $KUCl_5$, Na_2UCl_6, and $KNaUCl_6$, Sov. At. Energy, **25** (1968) 1000-1001.

[68NEM/PAL] Nemodruk, A.A., Palei, P.N., Glukova, L.P., Karyakin, A.V., Muradova, G.A., The decomposition of the uranyl group in strong acid solutions, Dokl. Chem. (Engl. Transl.), **180** (1968) 510-513.

[68NIK/ANT] Nikolaeva, N.M., Antipina, V.A., Pastukova, E.D., Investigation of hydrolysis of uranyl nitrate at elevated temperatures, Deposited document No. 395, VINITI, Moscow, 1968, in Russian.

[68OHA/SET] O'Hare, P.A.G., Settle, J.L., Feder, H.M., Hubbard, W.N., The thermochemistry of some uranium compounds, in: Thermodynamics of nuclear materials 1967, Proc symp held 1967 in Vienna, Vienna: International Atomic Energy Atomic, 1968, *pp.*265-278.

[68OST/CAM] Ostacoli, G., Campi, E., Gennaro, M.C., Complessi di alcuni acidi diidrossibenzoici con lo ione uranile in soluzione acquosa, Gazz. Chim. Ital., **98** (1968) 301-315, in Italian.

[68ROB/WAL] Robie, R.A., Waldbaum, D.R., Thermodynamic properties of minerals and related substances at 298.15 K (25°C) and one atmosphere (1.013 bars) pressure and at higher temperatures, US Geological Survey Bull., Nr. 1259, 1968, 17p.

[68RYZ/KHI] Ryzhenko, B.N., Khitarov, N.I., Silica in aqueous solutions, Geochem. Int., **5** (1968) 791-795.

[68SCH/FRY] Schedin, U., Frydman, M., Studies on the hydrolysis of metal ions: 59. The uranyl ion in magnesium nitrate medium, Acta Chem. Scand., **22** (1968) 115-127.

[68SMI] Smith, W.H., Thermal dehydration of uranyl nitrate hydrates, J. Inorg. Nucl. Chem., **30** (1968) 1761-1768.

[68TUM/LUM] Tummavuori, J., Lumme, P., Protolysis of nitrous acid in aqueous sodium nitrate and sodium nitrite soultions at different temperatures, Acta Chem. Scand., **22** (1968) 2003-2011.

[68TVE/CHA] Tveekrem, J.O., Chandrasekharaiah, M.S., The standard free energy of formation of UI_3 from EMF measurements on a solid electrolyte galvanic cell, J. Electrochem. Soc., **115** (1968) 1021-1023.

[68WAG/EVA] Wagman, D.D., Evans, W.H., Parker, V.B., Halow, I., Bailey, S.M., Schumm, R.H., Selected values of chemical thermodynamic properties: Tables for the first thirty-four elements in the standard order of arrangement, U.S. National Bureau of Standards Technical Note 270-3, Washington, D.C., USA, 1968.

[69ALE/OGD] Alexander, C.A., Ogden, J.S., Pardue, W.M., Volatilization characteristics of uranium mononitride, J. Nucl. Mater., **31** (1969) 13-24.

[69BEV] Bevington, P.R., Data reduction and error analysis for the physical sciences, New York: McGraw-Hill, 1969, 336p.

[69CAJ/PRA] Caja, J., Pravdic, V., Contribution to the electrochemistry of uranium(V) in carbonate solutions, Croat. Chem. Acta, **41** (1969) 213-222.

[69CHI/KAT] Chiotti, P., Kateley, J.A., Thermodynamic properties of uranium-aluminum alloys, J. Nucl. Mater., **32** (1969) 135-145.

[69CLI/GRU] Clifton, J.R., Gruen, D.M., Ron, A., Electronic absorption spectra of matrix-isolated uranium tetrachloride and tetrabromide molecules, J. Chem. Phys., **51** (1969) 224.

[69COM] Comité International des Poids et des Mesures, The International Practical Temperature Scale of 1968, Metrologia, **5(2)** (1969) 35-47.

[69COR] Cordfunke, E.H.P., The system uranyl sulphate - water: I. Preparation and characterization of the phases in the system, J. Inorg. Nucl. Chem., **31** (1969) 1327-1335.

[69ENG] Engel, T.K., The heat capacities of Al_2O_3, UO_2 and PuO_2 from 300 to 1100 K, J. Nucl. Mater., **31** (1969) 211-214.

[69EZH/AKI] Ezhov, Yu.S., Akishin, P.A., Rambidi, N.G., Zh. Struct. Khim., **10** (1969) 571, cited by Fuger, J., Parker, V.B., Hubbard, W.N., Oetting, F.L., The chemical thermodynamics of actinide elements and compounds: Part 8. The actinide halides, Vienna: International Atomic Energy Agency, 1983, 267p.

[69FLO/OSB] Flotow, H.E., Osborne, D.W., O'Hare, P.A.G., Settle, J.L., Mrazek, F.C., Hubbard, W.N., Uranium diboride: Preparation, enthalpy of formation at 298.15 K, heat capacity from 1 to 350 K, and some derived thermodynamic properties, J. Chem. Phys., **51** (1969) 583-592.

[69GIN] Gingerich, K.A., Vaporization of uranium mononitride and heat of sublimation of uranium, J. Chem. Phys., **51** (1969) 4433-4439.

[69GRE/VAR] Grenthe, I., Varfeldt, J., A Potentiometric study of fluoride complexes of uranium(IV) and uranium(VI) using the U(VI)/U(IV) redox couple, Acta Chem. Scand., **23** (1969) 988-998.

[69GRU/MCB] Gruen, D.M., McBeth, R.L., Vapor complexes of uranium pentachloride and uranium tetrachloride with aluminium chloride: The nature of gaseous uranium pentachloride, Inorg. Chem., **8** (1969) 2625-2633.

[69HEL] Helgeson, H.C., Thermodynamics of hydrothermal systems at elevated temperatures and pressures, Am. J. Sci., **267** (1969) 729-804.

[69HUB/HOL] Huber, E.J., Jr., Holley, C.E., Jr., Enthalpies of formation of triuranium octaoxide and uranium dioxide, J. Chem. Thermodyn., **1** (1969) 267-272.

[69HUB/HOL3] Huber, E.J., Jr., Holley, C.E., Jr., Witteman, W.G., Enthalpy of formation of uranium sesquicarbide, J. Chem. Thermodyn., **1** (1969) 579-587.

[69KNA/LOS] Knacke, O., Lossmann, G., Müller, F., Zur thermischen Dissoziation und Sublimation von UO_2F_2, Z. Anorg. Allg. Chem., **371** (1969) 32-39, in German.

[69KNA/LOS2] Knacke, O., Lossmann, G., Müller, F., Zustandsdiagramme zum System Uran-Sauerstoff-Fluor, Z. Anorg. Allg. Chem., **370** (1969) 91-103, in German.

[69MAC] MacLeod, A.C., High-temperature thermodynamic properties of uranium dicarbide, J. Inorg. Nucl. Chem., **31** (1969) 715-725.

[69MOS] Moskvin, A.I., Complex formation of the actinides with anions of acids in aqueous solutions, Sov. Radiochem., **11(4)** (1969) 447-449.

[69NOR] Norén, B., A solvent extraction and potentiometric study of fluoride complexes of thorium(IV) and uranium(IV), Acta Chem. Scand., **23** (1969) 931-942.

[69RAO/PAI] Rao, C.L., Pai, S.A., A study of nitrate and sulphate complexes of uranium(IV), Radiochim. Acta, **12** (1969) 135-140.

[69SIL/WAR] Sillén, L.G., Warnquist, B., High-speed computers as a supplement to graphical methods: 6. A strategy for two-level LETAGROP adjustment of common and "group" parameters. Some features that avoid divergence, Ark. för Kemi, **31** (1969) 315-339.

[69SMI/THA] Smith, B.C., Thakur, L., Wassef, M.A., Thermochemical studies of thorium(IV) and uranium(IV) chlorides, Indian J. Chem., **7** (1969) 1154-1157.

[69STO/JAC] Stojkovic, D., Jacksic, M.M., Nikolic, B.Z., Glas. Hem. Drus. Beograd, **34** (1969) 171, cited by Perrin, D.D., Ionisation constants of inorganic acids and bases in aqueous solution, 2*nd* ed., IUPAC Chemical Data Series **29**, Oxford: Pergamon Press, 1982, 180p.

[69TSY] Tsymbal, C., Contribution à la chimie de l'uranium(VI) en solution, Ph.D. thesis, Report CEA-R-3479, Université de Grenoble, Grenoble, France, 1969, 97p, in French.

[69VAN/OST] Vanni, A., Ostacoli, G., Roletto, E., Complessi formati dallo ione uranile in soluzione acquosa con acidi bicarbossilici della serie satura, Ann. Chim. Rome, **59** (1969) 847-859, in Italian.

[69VDO/ROM] Vdovenko, V.M., Romanov, G.A., Solntseva, L.V., Heat of formation of UOF_2, Radiokhimiya, **11(4)** (1969) 466-468, in Russian; Engl. transl.: Sov. Radiochem., **11(4)** (1969) 455-457.

[69VDO/STE] Vdovenko, V.M., Stebunov, O.B., Relaxation processes during complex formation: IV. Determination of the stability constants of the complexes from the data of the proton relaxation method, Sov. Radiochem., **11** (1969) 625-629.

[69VDO/STE2] Vdovenko, V.M., Stebunov, O.B., Relaxation processes during complex formation: V. Determination of the stability constants of strong complexes by the NMR method in the case of unknown equilibrium concentrations of the ligand, Sov. Radiochem., **11** (1969) 630-632.

[69ZMB] Zmbov, K.F., in: Proc. 1*st* International Conf. on calorimetry and thermodynamics, Polish Scientific Publisher, Warsaw, Poland, 1969, *p*.423, cited by Fuger, J., Parker, V.B., Hubbard, W.N., Oetting, F.L., The chemical thermodynamics of actinide elements and compounds: Part 8. The actinide halides, Vienna: International Atomic Energy Agency, 1983, 267p.

[70ABE] Aberg, M., On the structures of the predominant hydrolysis products of uranyl(VI) in solution, Acta Chem. Scand., **24** (1970) 2901-2915.

[70BAR] Baran, V., U(IV)-U(VI) hydrolytic precipitate, Inorg. Nucl. Chem. Letters, **6** (1970) 651-655.

[70BAS/SMI] Baskin, Y., Smith, S.D., Enthalpy-of-formation data on compounds of uranium with groups VA and VIA elements, J. Nucl. Mater., **37** (1970) 209-222.

[70BRA/COB] Brand, J.R., Cobble, J.W., The thermodynamic functions of neptunium(V) and neptunium(VI), Inorg. Chem., **9** (1970) 912-917.

[70CHO] Chottard, G., Décomposition thermique du nitrate d'uranyle hexahydraté. Etude des intermédiaires de cette décomposition, CEA-R-3717, (1970) 54p.

[70CHU/CHO] Chudinov, É.G., Choporov, D.Ya., Sublimation of americium tetrafluoride, Sov. At. Energy, **28** (1970) 71-73.

[70DRA/VIL] Drăgulescu, C., Vîlceanu, N., Julean, I., Untersuchung der Bildung anionischer Komplexe, Z. Anorg. Allg. Chem., **376** (1970) 196-204, in German.

[70FRE/BAR] Fredrickson, D.R., Barnes, R.D., Chasanov, M.G., The enthalpy of uranium diboride from 1300 to 2300 K by drop calorimetry, High Temp. Sci., **2** (1970) 299-301.

[70FRE/WEN] Frei, V., Wendt, H., Fast ionic reactions in solution: VIII. The fission and formation kinetics of the dimeric isopolybase $(UO_2OH)_2^{2+}$, Ber. Bunsen-Ges. **74** (1970) 593-599.

[70GHO/NAI] Ghosh, R., Nair, V.S.K., Studies on metal complexes in aqueous solution: III. The biselenate ion and transition metal selenates, J. Inorg. Nucl. Chem., **32** (1970) 3041-3051.

[70GIN] Gingerich, K.A., Gaseous metal borides: II. Mass-spectrometric evidence for the molecules UB_2, UB and CeB and predicted stability of gaseous diborides of electropositive transition metals, J. Chem. Phys., **53** (1970) 746-748.

[70GRO/KVE] Grønvold, F., Kveseth, N.J., Sveen, A., Tichy, J., Thermodynamics of the UO_{2+x} phase: I. Heat capacities of $UO_{2.017}$ and $UO_{2.254}$ from 300 to 1000 K and electronic contributions, J. Chem. Thermodyn., **2** (1970) 665-679.

[70HOE/SIE] Hoekstra, H.R., Siegel, S., Gallagher, F.X., The uranium-oxygen system at high pressure, J. Inorg. Nucl. Chem., **32** (1970) 3237-3248.

[70JOH/SMI] Johnson, G.K., Smith, P.N., Appleman, E.H., Hubbard, W.N., The thermodynamic properties of perbromate and bromate ions, Inorg. Chem., **9** (1970) 119-125.

[70KHA/KRI] Khanaev, E.I., Kripin, L.A., Standard enthalpy of the formation of uranium trifluoride, Sov. Radiochem., **12** (1970) 158-160.

[70KLY/KOL] Klygin, A.E., Kolyada, N.S., Smirnova, I.D., Spectrophotometric investigation of complex formation in the uranyl nitrate - nitric acid - water system, Russ. J. Inorg. Chem., **15** (1970) 1719-1721.

[70LAH/KNO] Lahr, H., Knoch, W., Bestimmung von Stabilitätskonstanten einiger Aktinidenkomplexe: II. Nitrat- und Chloridkomplexe von Uran, Neptunium, Plutonium und Americium, Radiochim. Acta, **13** (1970) 1-5, in German.

[70MAE/KAK] Maeda, M., Kakihana, H., The hydrolysis of the uranyl ion in heavy water, Bull. Chem. Soc. Japan, **43** (1970) 1097-1100.

[70MAR/CIO] Marchidan, D.I., Ciopec, M., High temperature enthalpy and related thermodynamic functions of some materials from the uranium-oxygen system, Rev. Roum. Chim., **15** (1970) 1287-1301.

[70NIK] Nikolaeva, N.M., Complexing in the solutions of uranyl sulphate at elevated temperatures, Izv. Sib. Otd. Akad. Nauk SSSR Ser. Khim. Nauk, (1970) 62-66, in Russian.

[70STA] Stabrovskii, A.I., Elektrokhimiya, **10** (1970) 1471, cited by Bruno, J., Grenthe, I., Lagerman, B., The redox potential of U(VI)/U(IV), Acta Chem. Scand., **44** (1990) 896-901.

[70VDO/ROM] Vdovenko, V.M., Romanov, G.A., Malinin, G.V., Solntseva, L.V., Production and study of some physicochemical properties of $UOF(OH)\cdot 0.5H_2O$, Sov. Radiochem., **12** (1970) 727-729.

[70VDO/ROM2] Vdovenko, V.M., Romanov, G.A., Solntseva, L.V., Heat of formation of $UOF(OH)\cdot 0.5H_2O$, Radiokhimiya, **12(5)** (1970) 764-766, in Russian; Engl. transl.: Sov. Radiochem., **12** (1970) 730-731.

[70WAL] Walker, J.B., A note on the thermodynamic parameters of hydrofluoric acid in aqueous media, J. Inorg. Nucl. Chem., **32** (1970) 2793-2796.

[70WES/GRO] Westrum, E.F., Jr., Grønvold, F., Uranium chalcogenides: III. Heat capacities and thermodynamic properties of $US_{1.9}$ and USe_2 from 5 to 350 K, J. Inorg. Nucl. Chem., **32** (1970) 2169-2177.

[71ABE] Aberg, M., On the crystal structure of a tetranuclear hydroxo complex of uranyl(VI), Acta Chem. Scand., **25** (1971) 368-369.

[71AHR/KUL] Ahrland, S., Kullberg, L., Thermodynamics of metal complex formation in aqueous solution: I. A potentiometric study of fluoride complexes of hydrogen, uranium(VI) and vanadium(IV), Acta Chem. Scand., **25** (1971) 3457-3470.

[71AHR/KUL2] Ahrland, S., Kullberg, L., Thermodynamics of metal complex formation in aqueous solution: II. A calorimetric study of fluoride complexes of hydrogen, uranium(VI) and vanadium(IV), Acta Chem. Scand., **25** (1971) 3471-3483.

[71AHR/KUL3] Ahrland, S., Kullberg, L., Thermodynamics of metal complex formation in aqueous solution: III. A calorimetric study of hydrogen sulphate and uranium(VI) sulphate, acetate, and thiocyanate complexes, Acta Chem. Scand., **25** (1971) 3677-3691.

[71ALE/OGD] Alexander, C.A., Ogden, J.S., Pardue, W.M., Thermophysical properties of (U,Pu)N, in: Plutonium 1970 and other actinides, Proc. 4*th* International Conf. held 5-9 October, 1970, in Santa Fe, New Mexico, USA, Nuclear Metallurgy, Vol. **17**, Part **I**, 1971, *pp*.95-103.

[71BAI/LAR] Bailey, A.R., Larson, J.W., Heats of dilution and the thermodynamics of dissociation of uranyl and vanadyl sulfates, J. Phys. Chem., **75** (1971) 2368-2372.

[71BAR/SIL] Barcza, L., Sillén, L.G., Equilibrium studies of polyanions: 19. Polyselenite equilibria in various ionic media, Acta Chem. Scand., **25** (1971) 1250-1260.

[71COR/LOO] Cordfunke, E.H.P., Loopstra, B.O., Sodium uranates: Preparation and thermochemical properties, J. Inorg. Nucl. Chem., **33** (1971) 2427-2436.

[71ELL/GIG] Ellis, A.J., Giggenbach, W.F., Hydrogen sulphide ionization and sulphur hydrolysis in high temperature solutions, Geochim. Cosmochim. Acta, **35** (1971) 247-260.

[71FIT/PAV] Fitzgibbon, G.C., Pavone, D., Holley, C.E., Jr., Enthalpy of formation of uranium tetrachloride, J. Chem. Thermodyn., **3** (1971) 151-162.

[71FUG/BRO] Fuger, J., Brown, D., Thermodynamics of the actinide elements: Part III. Heats of formation of the dicaesium actinide hexachloro-complexes, J. Chem. Soc. A, (1971) 841-846.

[71GAL] Galkin, N.P. (editor), Basic properties of inorganic fluorides, Moscow: Atomizdat, 1971, cited by Kaganyuk, D.S., Kyskin, V.I., Kazin, I.V., Calculation of enthalpies of formation for radioelement compounds, Sov. Radiochem., **25** (1983) 65-69.

[71GIG] Giggenbach, W.F., Optical spectra of highly alkaline sulfide solutions and the second dissociation constant of hydrogen sulfide, Inorg. Chem., **10** (1971) 1333-1338.

[71GRO/HAY] Gross, P., Hayman, C., Wilson, G.L., Bildungsenthalpie von Uranpentachlorid, Monatsh. Chem., **102** (1971) 924-927, in German.

[71HOE] Hoenig, C.L., Phase equilibria, vapor pressure, and kinetic studies in the uranium-nitrogen system, J. Amer. Ceramic Soc., **54** (1971) 391-398.

[71HUN/WES] Huntzicker, J.J., Westrum, E.F., Jr., The magnetic transition, heat capacity and thermodynamic properties of uranium dioxide from 5 to 350 K, J. Chem. Thermodyn., **3** (1971) 61-76.

[71HUT] Hutin, M.F., Chimie analytique. Sur l'existence de complexes uranyl-periodiques. Application analytique, C. R. Hebd. Séances Acad. Sci., Ser. C, **273** (1971) 739-740, in French.

[71JEN] Jensen, K.A. (chairman), Nomenclature of inorganic chemistry, 2*nd* ed., IUPAC Commission on Nomenclature of Inorganic Chemistry, Oxford: Pergamon Press, 1971, 110p.

[71KIN/SIE] King, E.G., Siemans, R.E., Richardson, D.W., Report USBM-RC-1516, US Department of the Interior, Bureau of Mines, Albany Metallurgical Research Centre, Albany, Oregon, USA, 1971.

[71KOT] Kothari, P.S., Design, construction and operation of an isoperibol solution calorimeter: Uranyl-uranium(IV) system, Ph.D. thesis in physical chemistry, Wayne State University, Detroit, Michigan, USA, 1971, 115p.

[71MAR/CIO] Marchidan, D.I., Ciopec, M., High temperature enthalpy and related thermodynamic function of UO_{2+x}, Rev. Roum. Chim., **16** (1971) 1145-1154.

[71MIN/GIL] Minor, J.I., Gillikland, A.A., Wagman, D.D., Report 10860, US National Bureau of Standards, Washington, D.C., 1971, cited by Barnes, D.S., Burkinshaw, P.N., Mortimer, C.T., Enthalpy of combustion of triphenylarsine oxide, Thermochim. Acta, **131** (1988) 107-113.

[71NAU/RYZ] Naumov, G.B., Ryzhenko, B.N., Khodakovsky, I.L., Handbook of thermodynamic data, Moscow: Atomizdat, 1971, in Russian; Engl. transl.: Report WRD-74-01 (Soleimani, G.J., Barnes I., Speltz, V., *eds.*), US Geological Survey, Menlo Park, California, USA, 1974, 328p.

[71NIK] Nikolaeva, N.M., The study of hydrolysis and complexing of uranyl ions in sulphate solutions at elevated temperatures, Izv. Sib. Otd. Akad. Nauk SSSR, Ser. Khim. Nauk, **7(3)** (1971) 61-67, translated from the Russian: ORNL/TR-88/1, Oak Ridge National Laboratory, Oak Ridge, Tennessee, USA, 1988, 14p.

[71PAR/WAG] Parker, V.B., Wagman, D.D., Evans, W.H., Technical note 270-6, National Bureau of Standards, Washington, D.C., USA, 1971, cited by Cordfunke, E.H.P., O'Hare, P.A.G., The chemical thermodynamics of actinide elements and compounds: Part 3. Miscellaneous actinide compounds, Vienna: International Atomic Energy Agency, 1978, 83p.

[71POL/BAR] Polasek, M., Bartušek, M., Thiocyanate complexes of bivalent metals in dilute aqueous solutions, Scr. Fac. Sci. Natur. Univ. Purkynianae Brun., **1(9)** (1971) 109-118, Chem. Abstr. **77** (1972) 131395v.

[71POR/WEB] Porter, R.A., Weber, Jr., The interaction of silicic acid with iron(III) and uranyl ions in dilute aqueous solutions, J. Inorg. Nucl. Chem., **33** (1971) 2443-2449.

[71SAN/VID] Santalova, N.A., Vidavskii, L.M., Dunaeva, K.M., Ippolitova, E.A., Enthalpy of formation of uranium trioxide dihydrate, Sov. Radiochem., **13** (1971) 608-612.

[71SEV/SLO] Sevast'yanov, V.G., Slovyanskikh, V.K., Ellert, G.V., Equilibria in the US_x-Br_2 and USe_x-Br_2 system, Russ. J. Inorg. Chem., **16** (1971) 1776-1779.

[71SIL/MAR] Sillén, L.G., Martell, A.E., Stability constants of metal-ion complexes, Suppl. No. 1, Special Publ. No. 25, London: The Chemical Society, 1971, 865p.

[71SUP/TSV] Suponitskii, Yu.L., Tsvetkov, A.A., Seleznev, V.P., Karapet'yants, M.K., Sudarikov, B.N., Gromov, B.V. Enthalpy of formation of crystaline hydrates of uranyl fluoride, Russ. J. Phys. Chem., **45** (1971) 558-559.

[71TET/HUN] Tetenbaum, M., Hunt, P.D., High-temperature thermodynamic properties of hypo- and hyper-stoichiometric uranium carbides, J. Nucl. Mater., **40** (1971) 104-112.

[71TSV/SEL] Tsvetkov, A.A., Seleznev, V.P., Sudarikov, B.N., Gromov, B.V., Tensimetric study of the hydrates of ammonium tetraoxopentafluorodiuranate, Russ. J. Inorg. Chem., **16** (1971) 1229-1231.

[71VID/IPP] Vidavskii, L.M., Ippolitova, E.A., Standard enthalpy of formation of the amorphous modification of uranium trioxide, Sov. Radiochem., **13** (1971) 306-308.

[72ARN/BAR] Arnek, R., Barcza, L., Thermochemical stidues of hydrolytic reactions: 11. Polyselenite equilibria in various ionic media, Acta Chem. Scand., **26** (1972) 213-217.

[72BAT/SHI] Battles, J.E., Shinn, W.A., Blackburn, P.E., Thermodynamic investigation of trisodium uranium(V) oxide (Na_3UO_4): IV. Mass spectrometric study of the Na+U+O system, J. Chem. Thermodyn., **4** (1972) 425-439.

[72BUC/GOR] Buchacek, R., Gordon, G., The kinetics and mechanism of the oxidation-reduction reaction between uranium(IV) and chlorine(III) in the presence of phenol in aqueous acid solution, Inorg. Chem., **11** (1972) 2154-2160.

[72CHA/BAN] Chakravorti, M.C., Bandyopadhyay, N., Fluoro complexes of hexavalent uranium-IV: Complexes of the series $UO_2F_3^-$ and $(UO_2)_2F_5^-$, J. Inorg. Nucl. Chem., **34** (1972) 2867-2874.

[72COR] Cordfunke, E.H.P., The system uranyl sulphate-water: II. Phase relationships and thermochemical properties of the phases in the system UO_3-SO_3-H_2O, J. Inorg. Nucl. Chem., **34** (1972) 1551-1561.

[72DUF/SAM] Duflo-Plissonnier, M., Samhoun, K., Etude de la réduction U(III)-U(0) en solution aqueuse par polarographie radiochimique, Radiochem. Radioanal. Lett., **12(2/3)** (1972) 131-138, in French.

[72FRE/CHA] Fredrickson, D.R., Chasanov, M.G., Thermodynamic investigation of trisodium uranium(V) oxide (Na_3UO_4): III. Enthalpy to 1200 K by drop calorimetry, J. Chem. Thermodyn., **4** (1972) 419-423.

[72FUG] Fuger, J., Thermodynamic properties of simple actinide compounds, in: Lanthanides and actinides, MIT Int. Rev. Sci., Inorg. Chem. Ser. I, **7** (Bagnall, K.W., *ed.*), London: Butterworths, 1972, *pp.*157-210.

[72GLE/KLY] Glevob, V.A., Klygin, A.E., Smirnova, I.D., Kolyada, N.S., Hydrolysis of quadrivalent uranium, Russ. J. Inorg. Chem., **17** (1972) 1740-1742.

[72HOF] Hoffmann, G., Über Phosphate und Arsenate von Uran(VI) und Neptunium(VI), Ph.D. thesis, University of Munich, Munich, Federal Republic of Germany, 1972, in German.

[72JOH/STE] Johnson, I., Steidl, D.V., U-Pu-O fuel-fission product interaction, in: Chemical Engineering Division, Fuels and Materials Chemistry, Semiannual Report July-December 1971 (Blackburn, P.E., Johnson, C.E., Johnson, I., Martin, A.E., Tetenbaum, M., Crouthamel, C.E., Tevebaugh, A.D., Vogel, R.C., *eds.*), Argonne National Laboratory, Argonne, Illinois, USA, April 1972, *pp.*16-18.

[72KNA/MUE] Knacke, O., Müller, F., van Rensen, E., Thermische Dissoziation von $ThOCl_2$ sowie $UOCl_2$ und UO_2Cl_2, Z. Phys. Chem. Leipzig, **80** (1972) 91-100, in German.

[72KRE] Krestov, G.A., Thermochemistry of compounds of rare-earth and actinide elements, Atomizdat, Moscow, USSR, 1972; Engl. transl.: Report AEC-tr-7505, US Atomic Energy Commission, 253p.

[72MAE/AMA] Maeda, M., Amaya, T., Ohtaki, H., Kakihana, H., Mixed solvent deuterium isotope effects on the hydrolysis of UO_2^{2+} and Be^{2+} ions, Bull. Chem. Soc. Jpn., **45** (1972) 2464-2468.

[72MUS] Musikas, C., Formation d'uranates solubles par hydrolyse des ions uranyle(VI), Radiochem. Radioanal. Letters, **11(5)** (1972) 307-316, in French.

[72NIK/SER] Nikitin, A.A., Sergeyeva, E.I., Khodakovsky, I.L., Naumov, G.B., Hydrolysis of uranyl in the hydrothermal region, Geokhimiya, **3** (1972) 297-307, in Russian; Engl. transl.: AECL translation Nr. 3554, Atomic Energy of Canada Ltd., Pinawa, Manitoba, Canada, 21p.

[72OET/LEI] Oetting, F.L., Leitnaker, J.M., The chemical thermodynamic properties of nuclear materials: I. Uranium mononitride, J. Chem. Thermodyn., **4** (1972) 199-211.

[72OET/NAV] Oetting, F.L., Navratil, J.D., The chemical thermodynamics of nuclear materials: II. High temperature properties of uranium carbides, J. Nucl. Mater., **45** (1972/73) 271-283.

[72OHA/SHI] O'Hare, P.A.G., Shinn, W.A., Mrazek, F.C., Martin, A.E., Thermodynamic investigation of trisodium uranium(V) oxide (Na_3UO_4): I. Preparation and enthalpy of formation, J. Chem. Thermodyn., **4** (1972) 401-409.

[72OSB/FLO] Osborne, D.W., Flotow, H.E., Thermodynamic investigation of trisodium uranium(V) oxide (Na_3UO_4): II. Heat capacity, entropy, and enthalpy increment from 5 to 350 K. Gibbs energy of formation at 298.15 K, J. Chem. Thermodyn., **4** (1972) 411-418.

[72PAL/LVO] Pal'chevskii, V.V., L'vova, T.I., Krunchak, V.G., Issled. Otd. Proized. Polufabrikat., Ochetki Prom. Stokon, (1972) 166, cited by Perrin, D.D., Ionisation constants of inorganic acids and bases in aqueous solution, 2*nd* ed., IUPAC Chemical Data Series **29**, Oxford: Pergamon Press, 1982, 180p.

[72SAN/VID] Santalova, N.A., Vidavskii, L.M., Dunaeva, K.M., Ippolitova, E.A., Enthalpy of formation of uranium trioxide hemihydrate, Sov. Radiochem., **14** (1972) 741-745.

[72SER/NIK] Sergeyeva, E.I., Nikitin, A.A., Khodakovsky, I.L., Naumov, G.B., Vernadskiy, V.I., Experimental investigation of equilibria in the system UO_3-CO_2-H_2O in 25-200°C temperature interval, Geochem. Int., **11** (1972) 900-910.

[72TAY/KEL] Taylor, J.C., Kelly, J.W., Downer, B., A study of the $\beta \rightarrow \alpha$ phase transition in $UO_2(OH)_2$ by dilatometric, microcalorimetric and X-ray diffraction techniques, J. Solid State Chem., **5** (1972) 291-299.

[72TSV/SEL] Tsvetkov, A.A., Seleznev, V.P., Sudarikov, B.N., Gromov, B.V., Equilibrium diagram of the uranyl fluoride-water system, Zh. Neorg. Khim., **17** (1972) 2020-2023, in Russian; Engl. transl.: Russ. J. Inorg. Chem., **17(7)** (1972) 1048-1050.

[73ALM/GAR] Almagro, V., Garcia, F.S., Sancho, J., Estudio polarografico del ion UO_2^{2+} en medio carbonato sodico, An. Quim., **69** (1973) 709-716, in Spanish.

[73BAR/KNA] Barin, I., Knacke, O., Thermochemical properties of inorganic substances, Berlin: Springer-Verlag, 1973, 921p.

[73BEL] Belova, L.N., Voprosy prikladnoi radiogeologii, Moscow, (1973) 174-177, cited in Powder Diffraction File, alphabetic index: Inorganic phases, International Centre for Diffraction Data, Swarthmore, Pennsylvania, USA, 1986.

[73BLA/IHL] Blair, A., Ihle, H., The thermal decomposition and thermodynamic properties of uranium pentabromide, J. Inorg. Nucl. Chem., **35** (1973) 3795-3803.

[73COD] CODATA recommended Key Values for thermodynamics 1973, Committee on Data for Science and Technology (CODATA), Bulletin Nr. 10, Paris: International Council of Scientific Unions, 1973, 10p.

[73DAV/EFR] Davydov, Yu.P., Efremenkov, V.M., Investigation of the hydrolysis properties of hexavalent uranium, Vest. Akad. Nauk Belaruskai SSR Ser. Fiz. Energ., **(4)** (1973) 21-25, in Russian.

[73DRA/VIL] Drăgulescu, C., Vîlceanu, N., Uranyl-pyrophosphate complexes: III. The monopyrophosphato-pentauranyl cation complex, Rev. Roum. Chim., **18** (1973) 49-52.

[73FUG/BRO] Fuger, J., Brown, D., Thermodynamics of the actinide elements: Part IV. Heats and free energies of formation of the tetrachlorides, tetrabromides, and tetraiodides of thorium, uranium and neptunium, J. Chem. Soc. Dalton Trans., (1973) 428-434.

[73HOE/SIE] Hoekstra, H.R., Siegel, S., The uranium trioxide-water system, J. Inorg. Nucl. Chem., **35** (1973) 761-779.

[73HUL/DES] Hultgren, R., Desai, P.D., Hawkins, D.T., Gleiser, M., Kelley, K.K., Wagman, D.D., Selected values of the thermodynamic properties of the elements, American Society for Metals, Metals Park, Ohio, USA, 1973, 636p.

[73INA/NAI] Inaba, H., Naito, K., Specific heat anomaly of non-stoichiometric U_4O_{9-y}, J. Nucl. Mater., **49** (1973) 181-188.

[73KOZ/BLO] Kozlov, Yu.A., Blokhin, V.V., Shurukhin, V.V., Mironov, V.E., Thermodynamics of hydrogen fluoride in aqueous salt solutions, Russ. J. Phys. Chem., **47** (1973) 1343-1344.

[73KRU/ROD] Krunchak, E.G., Rodichev, A.G., Khvorostin, Ya.S., Krumgal'z, B.S., Krunchak, V.G., Yusova, Yu.I., Determination of the dissociation constants of hydrogen sulphite ions at various temperatures, Zh. Neorg. Khim., **18(10)** (1973) 2859-2860, in Russian; Engl. transl.: Russ. J. Inorg. Chem., **18** (1973) 1519-1520.

[73MAJ] Majchrzak, K., Equilibria in solutions and anion exchange of uranium(VI) complexes: IV. Stability constants and sorption of uranyl sulphate complexes, Nucleonika, **8(3)** (1973) 105-119, in Polish.

[73MAV] Mavrodin-Tărăbic, M., L'étude des espèces ioniques formées par l'hydrolyse de l'ion d'uranium hexavalent, Rev. Roum. Chim., **18** (1973) 73-88, in French.

[73MAV2] Mavrodin-Tărăbic, M., Contribution à l'étude de l'hydrolyse de l'ion UO_2^{2+} en milieu perchlorique, Rev. Roum. Chim., **18** (1973) 609-621, in French.

[73MOS] Moskvin, A.I., Some thermodynamic characteristics of the processes of formation of actinide compounds in a solid form: I. Energy and entropy of the crystal lattice, heats of formation and heats of solution, Sov. Radiochem., **15** (1973) 356-363.

[73MOS2] Moskvin, A.I., Some thermodynamic characteristics of the processes of formation of actinide compounds in a solid form: II. Heat capacity and linear and bulk thermal-expansion coefficients, Sov. Radiochem., **15** (1973) 364-367.

[73OHA/HOE2] O'Hare, P.A.G., Hoekstra, H.R., Thermochemistry of uranium compounds: II. Standard enthalpy of formation of α-sodium uranate (α-Na_2UO_4), J. Chem. Thermodyn., **5** (1973) 769-775.

[73PIT/MAY] Pitzer, K.S., Mayorga, G., Thermodynamics of electrolytes. II. Activity and osmotic coefficients for strong electrolytes with one or both ions univalent, J. Phys. Chem., **77** (1973) 2300-2308.

[73PIT] Pitzer, K.S., Thermodynamics of electrolytes: I. Theoretical basis and general equations, J. Phys. Chem., **77** (1973) 268-277.

[73PRI] Prins, G., Investigation on uranyl chloride, its hydrates, and basic salts, Ph.D. thesis in natural sciences, University of Amsterdam, The Netherlands, 1973, 119p.

[73STA/CEL] Stankov, V., Čeleda, J., Jedináková, V., Contributions to the chemistry of highly concentrated aqueous electrolyte solutions: 32. Spectrophotometric study of nitrito- and chloro complexes of UO_2^{2+} ion, Scientific papers of the institute of chemical technology, Prague, Sborník VŠCHT Praha, **B17** (1973) 109-143, in Czech.

[73TSV/SEL] Tsvetkov, A.A., Seleznev, V.P., Sudarikov, B.N., Gromov, B.V., Uranyl hydroxide fluorides, Zh. Neorg. Khim., **18(1)** (1973) 12-15, in Russian; Engl. transl.: Russ. J. Inorg. Chem., **18(1)** (1973) 5-7.

[73VDO/SUG] Vdovenko, V.M., Suglobova, I.G., Chirkst, D.E., Complex formation in the system uranium halide-alkali metal halide: Enthalpy of formation of Rb_2UBr_6 and Cs_2UBr_6, Sov. Radiochem., **15** (1973) 56-58.

[73VER/KHA] Verma, V.P., Khandelwal, B.L., Electrometric studies on the interaction between selenous acid and uranyl ions, Indian J. Chem., **11** (1973) 602-603.

[74BIN/SCH] Binnewies, M., Schäfer, H., Gasförmige Halogenidkomplexe und ihre Stabilität, Z. Anorg. Allg. Chem., **407** (1974) 327-344, in German.

[74BUN] Bunus, F.T., An ion exchange study of the uranium (U^{4+} and UO_2^{2+}) complex species with Cl^- as ligand, J. Inorg. Nucl. Chem., **36** (1974) 917-920.

[74BUN2] Bunus, F.T., Determinarea speciilor complexe ale U(IV) în prezenţa ionului Cl^- ca ligand şi a funcţiilor termodinamice care intervin, prin metoda schimbului ionic, Rev. Chim. (Bucharest), **25(5)** (1974) 367-370, in Rumanian.

[74COR/PRI] Cordfunke, E.H.P., Prins, G., Equilibria involving volatile UO_2Cl_2, J. Inorg. Nucl. Chem., **36** (1974) 1291-1293.

[74GIG] Giggenbach, W.F., Equilibria involving polysulfide ions in aqueous sulfide solutions up to 240°C, Inorg. Chem., **13** (1974) 1724-1730.

[74JED] Jedináková, V., Contributions to the chemistry of highly concentrated aqueous electrolyte solutions: 37. Densimetric study of complex formation of the UO_2^{2+} ion in isomolar series perchlorate-halide, Scientific Papers of the Institute of Chemical Technology, Prague, Sborník VŠCHT Praha, **B18** (1974) 113-125, in Czech.

[74KAK/ISH] Kakihana, H., Ishiguro, S., Potentiometric and spectrophotometric studies of fluoride complexes of uranium(IV), Bull. Chem. Soc. Japan, **47** (1974) 1665-1668.

[74KAN/ZAR] Kanevskii, E.A., Zarubin, A.T., Rengevich, V.B., Pavlovskaya, G.R., Heat of solution of anhydrous uranyl nitrate in water and certain organic solvents, Sov. Radiochem., **16** (1974) 407-408.

[74KAN] Kanno, M., Thermodynamics of uranium intermetallic compounds by solid electrolyte EMF method, J. Nucl. Mater., **51** (1974) 24-29.

[74MAV] Mavrodin-Tărăbic, M., Contribution à l'étude de l'hydrolyse de l'uranium(VI) en milieu azotique, Rev. Roum. Chim., **19** (1974) 1461-1470, in French.

[74MES/BAE] Mesmer, R.E., Baes, C.F., Jr., Phosphoric acid dissociation equilibria in aqueous solutions to 300°C, J. Solution Chem., **3** (1974) 307-322.

[74MIL] Mills, K.C., Thermodynamic data for inorganic sulphides, selenides and tellurides, London: Butterworths, 1974, 845p.

[74MUK/SEL] Mukhametshina, Z.B., Selenznev, V.P., Suponitskii, Yu.L., Bodrov, V.G., Karapet'yants, M.K., Sudarikov, B.N., Standard enthalpies of formation of potassium, rubidium and caesium nonafluorotetroxodiuranates, Russ. J. Phys. Chem., 48(2) (1974) 293.

[74NIK] Nikolaeva, N.M., Oxidation-reduction potential of the U(VI)/U(IV) system at the elevated temperature, Izv. Sib. Otd. Akad. Nauk SSSR Ser Khim. Nauk, **1** (1974) 14-20, in Russian.

[74OHA/HOE] O'Hare, P.A.G., Hoekstra, H.R., Thermochemistry of molybdates: II. Standard enthalpies of formation of rubidium molybdate and the aqueous molybdate ion, J. Chem. Thermodyn., **6** (1974) 117-122.

[74OHA/HOE2] O'Hare, P.A.G., Hoekstra, H.R., Thermochemistry of uranium compounds: III. Standard enthalpy of formation of cesium uranate (Cs_2UO_4), J. Chem. Thermodyn., **6** (1974) 251-258.

[74OHA/HOE3] O'Hare, P.A.G., Hoekstra, H.R., Thermochemistry of uranium compounds: IV. Standard enthalpy of formation of sodium uranium(V) trioxide ($NaUO_3$), J. Chem. Thermodyn., **6** (1974) 965-972.

[74OHA/HOE4] O'Hare, P.A.G., Hoekstra, H.R., Thermochemistry of uranium compounds: V. Standard enthalpies of formation of the monouranates of lithium, potassium and rubidium, J. Chem. Thermodyn., **6** (1974) 1161-1169.

[74OSB/FLO] Osborne, D.W., Flotow, H.E., Hoekstra, H.R., Cesium molybdate, Cs_2MoO_4: Heat capacity and thermodynamic properties form 5 to 350 K, J. Chem. Thermodyn., **6** (1974) 179-183.

[74PIT/KIM] Pitzer, K.S., Kim, J.J., Thermodynamics of electrolytes. IV. Activity and osmotic coefficients for mixed electrolytes, J. Am. Chem. Soc., **96** (1974) 5701-5707.

[74PIT/MAY] Pitzer, K.S., Mayorga, G., Thermodynamics of electrolytes. III. Activity and osmotic coefficients for 2-2 electrolytes, J. Solution Chem., **3** (1974) 539-546.

[74SAM/KOS] Samorukov, O.P., Kostryukov, V.N., Kostylev, F.A., Tumbakov, V.A., Low-temperature specific heat and thermodynamic functions of uranium beryllide, Sov. J. At. Energy, **37** (1974) 705.

[74SUP/SEL] Suponitskii, Yu.L., Seleznev, V.P., Makhametshina, Z.B., Bodrov, V.G., Karapet'yants, M.K., Sudarikov, B.N. Standard enthalpies of formation of alkali metal and ammonium pentafluorodiuranylates, Sov. Radiochem., **16** (1974) 84-87.

[74TAG] Tagawa, H., Phase relations and thermodynamic properties of the uranium-nitrogen system, J. Nucl. Mater., **51** (1974) 78-89.

[74VDO/VOL] Vdovenko, V.M., Volkov, V.A., Suglobova, I.G., Complex formation in the systems uranium halide-alkali metal halide: Enthalpy of formation of alkali metal chlorouranates(IV), Sov. Radiochem., **16** (1974) 364-368.

[75ALY/ABD] Aly, H.F., Abdel-Rassoul, A.A., Zakareia, N., Use of zirconium phosphate for stability constant determination of uranium and antimony chlorocomplexes, Z. Phys. Chem. (Frankfurt/Main), **94** (1975) 11-18.

[75BAR/SEN] Bartušek, M., Senkyr, J., Study of thiocyanate complexes with a silver thiocyanate membrane electrode, Scr. Fac. Sci. Natur. Univ. Purkynianae Brun., **5(5)** (1975) 61-69, in German, Chem. Abstr. 85:84716d.

[75BRO] Brookins, D.G., Coffinite-uraninite stability relations in Grant Mineral Belt, New Mexico, Abs. Bull. Am. Assoc. Geologist, **59(5-8)** (1975) 905.

[75CHE] Chen, C.H., A method of estimation of standard free energies of formation of silicate minerals at 298.15 K, Am. J. Sci., **275** (1975) 801-817.

[75CIN/SCA] Cinnéide, S.Ó., Scanlan, J.P., Hynes, M.J., Equilibria in uranyl carbonate systems: I. The overall stability constant of $UO_2(CO_3)_3^{4-}$, J. Inorg. Nucl. Chem., **37** (1975) 1013-1018.

[75COR/OUW] Cordfunke, E.H.P., Ouweltjes, W., Prins, G., Standard enthalpies of formation of uranium compounds: I. β-UO_3 and γ-UO_3, J. Chem. Thermodyn., **7** (1975) 1137-1142.

[75COR] Cordfunke, E.H.P., Personal communication, Reactor Centrum Nederland, 1975, cited by Fuger, J., Oetting, F.L., The chemical thermodynamics of the actinide elements and compounds: Part 2, Vienna: International Atomic Energy Agency, 1976, 65p.

[75COR2] Cordfunke, E.H.P., Personal communication, Reactor Centrum Nederland, 1975, cited by Cordfunke, E.H.P., O'Hare, P.A.G., The chemical thermodynamics of the actinide elements and compounds: Part 3. Miscellaneous actinide compounds, Vienna: International Atomic Energy Agency, 1978, 83p.

[75DAV/EFR] Davydov, Yu.P., Efremenkov, V.M., Investigation of the hydrolytic properties of tetravalent uranium: II. Conditions of formation of mononuclear and polynuclear hydroxo complexes of U(IV), Sov. Radiochem., **17** (1975) 160-164.

[75FLO/OSB] Flotow, H.E., Osborne, D.W., Fried, S.M., Malm, J.G., Experimental heat capacities of ^{242}-PuF_3 and ^{242}-PuF_4 from 10 to 350 K and of ^{244}-PuO_2 from 4 to 25 K, in: Thermodynamics of nuclear materials 1974, Proc. Symp. held 21-25 October, 1974, in Vienna, Vienna: International Atomic Energy Agency, 1975, *pp.*477-488.

[75HIL/CUB] Hildenbrand, D.L., Cubicciotti, D.D., Mass spectrometric study of the molecular species formed during vaporization of UCl_4 in Cl_2 and $AlCl_3$, Lawrence Livermore National Laboratory, Livermore, California, USA, 1975, 22p.

[75HUT] Hutin, M.F., Etudes des complexes uranylperiodates en solution aqueuse, Bull. Soc. Chim. Fr., (1975) 463-466, in French.

[75JOH] Johnson, I., Report ANL-RDP-36, Argonne National Laboratory, Argonne, Illinois, USA, 1975.

[75LAG/SUS] Lagnier, R., Suski, W., Wojakowski, A., Heat capacity of the uranium sesquiselenide, Phys. Status Solidi, **A29** (1975) K51-K54.

[75NUG] Nugent, L.J., Standard electrode potentials and enthalpies of formation of some lanthanide and actinide aquo-ions, J. Inorg. Nucl. Chem., **37** (1975) 1767-1770.

[75OHA/ADE] O'Hare, P.A.G., Ader, M., Hubbard, W.N., Johnson, G.K., Settle, J.L., Calorimetric studies on actinide compounds and materials of interest in reactor technology, in: Thermodynamics of nuclear materials 1974, Proc.

Symp. held 21-25 October, 1974, in Vienna, Vienna: International Atomic Energy Agency, 1975, *pp*.439-453.

[75OHA/HOE2] O'Hare, P.A.G., Hoekstra, H.R., Thermochemistry of uranium compounds: VI. Standard enthalpy of formation of $Cs_2U_2O_7$, thermodynamics of formation of cesium and rubidium uranates at elevated temperatures, J. Chem. Thermodyn., **7** (1975) 831-838.

[75PAL/CIR] Palenzona, A., Cirafici, S., Thermodynamic properties of Y, Th and U, MX_3 compounds with IIIA and IVA elements, Thermochim. Acta, **13** (1975) 357-360.

[75PIT] Pitzer, K.S., Thermodynamics of electrolytes. V. Effects of higher-order electrostatic terms, J. Solution Chem., **4** (1975) 249-265.

[75SAW/SIE] Sawlewicz, K., Siekierski, S., Determination of the vapour pressure of uranium tetrachloride by the gas saturation method: Association of uranium tetrachloride in the gas phase, Radiochem. Radioanal. Letters, **23** (1975) 71.

[75STE] Steinhaus, D.W., Personal communication, Los Alamos Scientific Laboratory, Los Alamos, N.M., 1975, cited by Fuger, J., Oetting, F.L., The chemical thermodynamics of actinide elements and compounds: Part 2. The actinides aqueous ions, Vienna: International Atomic Energy Agency, 1976, 65p.

[75WES/MAG] Westrum, E.F., Jr., Magadanz, H., (University of Michigan), private communication, 1975, cited by Cordfunke, E.H.P., O'Hare, P.A.G., The chemical thermodynamics of actinide elements and compounds: Part 3. Miscellaneous actinide compounds, Vienna: International Atomic Energy Agency, 1978, 83p.

[75WES/SOM] Westrum, E.F., Jr., Sommers, J.A., Downie, D.B., Grønvold, F., Cryogenic heat capacity and Neel transitions of two uranium pnictides: UAs_2 and USb_2, in: Thermodynamic of nuclear materials 1974, Proc. Symp. held 21-25 October, 1974, in Vienna, Vienna: International Atomic Energy Agency, 1975, *pp*.409-418.

[75YOK/TAK] Yokokawa, H., Takahashi, Y., Mukaibo, T., The heat capacity of uranium monophosphide from 80 to 1080 K and the electronic contribution, in: Thermodynamcis of nuclear materials 1974, Proc. Symp. held 21-25 October, 1974, in Vienna, Vienna: International Atomic Energy Agency, 1975, *pp*.419-430.

[76BAE/MES] Baes, C.F., Jr., Mesmer, R.E., The hydrolysis of cations, New York: Wiley & Sons, 1976, 489p.

[76BOE/OHA] Boerio, J., O'Hare, P.A.G., Standard enthalpies of solution and formation at 298.15 K of an- hydrous magnesium and calcium nitrates, J. Chem. Thermodyn., 8 (1976) 725-729.

[76BOU/BON] Bousquet, J., Bonnetot, B., Claudy, P., Mathurin, D., Turck, G., Enthalpies de formation standard des formiates d'uranyle anhydre et hydrate par calorimétrie de réaction, Thermochim. Acta, **14** (1976) 357-367, in French.

[76CHO/UNR] Choppin, G.R., Unrein, P.J., Thermodynamic study of actinide fluoride complexation, in: Transplutonium 1975, Proc. Symp. held 13-17 September, 1975, in Baden-Baden, Amsterdam: North-Holland, 1976, *pp.*97-107.

[76COR/OUW] Cordfunke, E.H.P., Ouweltjes, W., Prins, G., Standard enthalpies of formation of uranium compounds: II. UO_2Cl_2 and UCl_4, J. Chem. Thermodyn., **8** (1976) 241-250.

[76DEW/PER] Dewally, D., Perrot, P., Contribution à l'étude thermodynamique des oxydes mixtes d'uranium et de nickel, Rev. Chim. Minérale, **13** (1976) 605-613, in French.

[76FRE/OHA] Fredrickson, D.R., O'Hare, P.A.G., Enthalpy increments for α- and β-Na_2UO_4 and Cs_2UO_4 by drop calorimetry; The enthalpy of the α to β transition in Na_2UO_4, J. Chem. Thermodyn., **8** (1976) 353-360.

[76FRO/ITO] Frondel, C., Ito, J., Honea, R.M., Weeks, A.M., Mineralogy of the zippeite group, Canadian Mineral., **14** (1976) 429-436.

[76FUG/OET] Fuger, J., Oetting, F.L., The chemical thermodynamics of actinide elements and compounds: Part **2**. The actinide aqueous ions, Vienna: International Atomic Energy Agency, 1976, 65p.

[76GHO/MUK] Ghosh, N.N., Mukhopadhyay, S.K., Studies on some *n*-acyl *n*-phenylhydroxylamines as metal complexing ligands: Part IX. Formation constants of UO_2^{2+} complexes of some *n*-hydroxysuccinamic acids, J. Indian Chem. Soc. **53** (1976) 233-237.

[76IKE/TAM] Ikeda, Y., Tamaki, M., Matsumoto, G., Vaporization of uranium-carbon-nitrogen system, J. Nucl. Mater., **59** (1976) 103-111.

[76MAS/DES] Masson, J.P., Desmoulin, J.P., Charpin, P., Bougon, R., Synthesis and characterization of a new uranium(V) compound: $H_3OUF_6^-$, Inorg. Chem., **15** (1976) 2529-2531.

[76MUE] Münstermann, E., Massenspektrometrische Untersuchungen zur thermischen Dissoziation von Uranylfluorid, Uranylchlorid und Thoriumtetrajodid, Ph.D. thesis, Technische Hochschule Aachen, Aachen, Federal Republic of Germany, 1976, in German.

[76NIK] Nikolaeva, N.M., Investigation of hydrolysis and complexing at elevated temperatures, Proc. Int. Corr. Conf. Ser. 1973, NACE-4 (1976) 146-152.

[76NIK2] Nikolaeva, N.M., Solubility product UO_2CO_3 at elevated temperature, Izv. Sib. Otd. Akad. Nauk SSSR, **6** (1976) 30-31, in Russian.

[76OET/RAN] Oetting, F.L., Rand, M.H., Ackermann, R.J., The chemical thermodynamics of actinide elements and compounds: Part 1. The actinide elements, Vienna: International Atomic Energy Agency, 1976, 111p.

[76OHA/BOE] O'Hare, P.A.G., Boerio, J., Hoekstra, H.R., Thermochemistry of uranium compounds: VIII. Standard enthalpies of formation at 298.15 K of the uranates of calcium ($CaUO_4$) and barium ($BaUO_4$). Thermodynamics of the behavior of barium in nuclear fuels, J. Chem. Thermodyn., **8** (1976) 845-855.

[76OHA/HOE] O'Hare, P.A.G., Hoekstra, H.R., Fredrickson, D.R., Thermochemistry of uranium compounds: VII. Solution calorimetry of α- and β-Na_2UO_4, standard enthalpy of formation of β-Na_2UO_4, and the enthalpy of the α to β transition at 298.15 K, J. Chem. Thermodyn., **8** (1976) 255-258.

[76OSB/BRL] Osborne, D.W., Brletic, P.A., Hoekstra, H.R., Flotow, H.E., Cesium uranate, Cs_2UO_4: Heat capacity and entropy from 5 to 350 K and standard Gibbs energy of formation at 298.15 K, J. Chem. Thermodyn., **8** (1976) 361-365.

[76PAT/RAM] Patil, S.K., Ramakrishna, V.V., Sulphate and fluoride complexing of U(VI), Np(VI) and Pu(VI), J. Inorg. Nucl. Chem., **38** (1976) 1075-1078.

[76PIR/NIK] Pirozhkov, A.V., Nikolaeva, N.M., Determination the stability constants UO_2CO_3 at the temperature from 25 to 150°C, Izv. Sib. Otd. Akad. Nauk SSSR, **5** (1976) 55-59, in Russian.

[76PIT/SIL] Pitzer, K.S., Silvester, L.F., Thermodynamics of electrolytes. VI. Weak electrolytes including H_3PO_4, J. Solution Chem., **5(4)** (1976) 269-278.

[76SCA] Scatchard, G., Equilibrium in solution: Surface and colloid chemistry, Cambridge, Massachusetts: Harvard University Press, 1976, 306p.

[76SMI/MAR] Smith, R.M., Martell, A.E., Critical stability constants: Volume 4. Inorganic complexes, New York: Plenum Press, 1976, 257p.

[76SOU/SHA] Souka, N., Shabana, R., Farah, K., Adsorption behaviour of some actinides on zirconium phosphate stability constant determinations, J. Radioanal. Chem., **33** (1976) 215-222.

[76WEI/HOF] Weigel, F., Hoffmann, G., The phosphates and arsenates of hexavalent actinides: Part I. Uranium, J. Less-Common Met., **44** (1976) 99-123.

[77ALL/FAL] Alles, A., Falk, B.G., Westrum, E.F., Jr., Grønvold, F., Zaki, M.R., Actinoid pnictides: I. Heat capacities from 5 to 950 K and magnetic transitions of U_3As_4 and U_3Sb_4: Ferromagnetic transitions, J. Inorg. Nucl. Chem., **39** (1977) 1993-2000.

[77BAR/KNA] Barin, I., Knacke, O., Kubaschewski, O., Thermochemical properties of inorganic substances (supplement), Berlin: Springer-Verlag, 1977, 861p.

[77BUS/MES] Busey, R.H., Mesmer, R.E., Ionization equilibria of silicic acid and polysilicate formation in aqueous sodium chloride solutions to 300°C, Inorg. Chem., **16** (1977) 2444-2450.

[77COR/OUW] Cordfunke, E.H.P., Ouweltjes, W., Standard enthalpies of formation of uranium compounds: III. UO_2F_2, J. Chem. Thermodyn., **9** (1977) 71-74.

[77COR/OUW2] Cordfunke, E.H.P., Ouweltjes, W., Standard enthalpies of formation of uranium compounds: IV. α-UO_2SO_4, β-UO_2SeO_4, and UO_2SeO_3, J. Chem. Thermodyn., **9** (1977) 1057-62.

[77COR/PRI] Cordfunke, E.H.P., Prins, G., van Vlaanderen, P., $(UO_2)_2Cl_3$: A new oxide chloride of uranium, J. Inorg. Nucl. Chem., **39** (1977) 2189-2190.

[77FER] Fernelius, W.C. (chairman), How to name an inorganic substance, IUPAC Commission on Nomenclature of Inorganic Chemistry, Oxford: Pergamon Press, 1977, 36p.

[77FLO/OSB] Flotow, H.E., Osborne, D.W., Lyon, W.G., Grandjean, F., Fredrickson, D.R., Hastings, I.J., Heat capacity and thermodynamic properties of triuranium silicide from 1 to 1203 K, J. Chem. Thermodyn., **9** (1977) 473-481.

[77GAB] Gabelica, Z., Sur l'existence des thiosulfates d'uranyle, de zirconyle et de terres rares, J. Less-Common Met., **51** (1977) 343-351, in French.

[77HIL] Hildenbrand, D.L., Thermochemistry of gaseous UF_5 and UF_4, J. Chem. Phys., **66** (1977) 4788-4794.

[77INA/SHI] Inaba, H., Shimizu, H., Naito, K., λ-type heat capacity anomalies in U_3O_8, J. Nucl. Mater., **64** (1977) 66-70.

[77ISH/KAO] Ishiguro, S., Kao, C.F., Kakihana, H., Formation constants of HF_m^{1-m} and $UO_2F_n^{2-n}$ complexes in 1 mol·dm^{-3} (NaCl) medium, Denki Kagaku Oyobi Kogyo Butsuri Kagaku, **45(10)** (1977) 651-653.

[77LYO/OSB] Lyon, W.G., Osborne, D.W., Flotow, H.E., Hoekstra, H.R., Sodium uranium(V) trioxide, $NaUO_3$; Heat capacity and thermodynamic properties from 5 to 350 K, J. Chem. Thermodyn., **9** (1977) 201-210.

[77NIK/TSV] Nikolaeva, N.M., Tsvelodub, L.D., Complexing of uranium(IV) with sulfate(-2) ions in aqueous solutions at elevated temperatures, Izv. Sib. Otd. Akad. Nauk SSSR Ser. Khim. Nauk, **5** (1977) 38-43, in Russian, Chem. Abstr. 88:28492c.

[77NIK] Nikolaeva, N.M., Study of uranium(IV) complexing with chlorides at elevated temperatures, Izv. Sib. Otd. Akad. Nauk SSSR Ser. Khim. Nauk, **3** (1977) 114-120, in Russian.

[77NIK2] Nikolaeva, N.M., Complexing in uranyl chloride solutions at elevated temperatures, Izv. Sib. Otd. Akad. Nauk SSSR Ser. Khim. Nauk, **1** (1977) 56-59, in Russian.

[77OHA/BOE] O'Hare, P.A.G., Boerio, J., Fredrickson, D.R., Hoekstra, H.R., Thermochemistry of uranium compounds: IX. Standard enthalpy of formation and high-temperature thermodynamic functions of magnesium uranate ($MgUO_4$). A comment on the non-existence of beryllium uranate, J. Chem. Thermodyn., **9** (1977) 963-972.

[77OHA] O'Hare, P.A.G., Thermochemistry of uranium compounds: X. Standard enthalpies of formatiom of uranyl oxalate, uranyl acetate, and their hydrates. Thermodynamics of the $UO_2C_2O_4+H_2O$ and $UO_2(CH_3COO)_2+H_2O$ systems, J. Chem. Thermodyn., **9** (1977) 1077-1086.

[77PIT/ROY] Pitzer, K.S., Roy, R.N., Silvester, L.F., Thermodynamics of electrolytes: 7. Sulfuric acid, J. Am. Chem. Soc., **99** (1977) 4930-4936.

[77ROS/RYA] Rosenzweig, A., Ryan, R.R., Kasolite, $PbUO_2SiO_4 \cdot H_2O$, Crystal Structure Communications, **6** (1977) 617-621.

[77SCA] Scanlan, J.P., Equilibria in uranyl carbonate systems: II. The overall stability constant of $UO_2(CO_3)_2^{2-}$ and the third formation constant of $UO_2(CO_3)_3^{4-}$, J. Inorg. Nucl. Chem., **39** (1977) 635-639.

[77VOL/BEL] Volk, V.I., Belikov, A.D., Investigation of the hydrolysis of the uranyl ion in solutions of uranyl pertechnetate and uranyl perchlorate by the method of two phase potentiometric titration, Radiokhimiya, **19** (1977) 811-816; Engl. transl: Sov. Radiochem., **19** (1977) 676-681.

[78ALL/BEA] Allard, B., Beall, G.W., Predictions of actinide species in the groundwater, in: Workshop on the environmental chemistry and research of the actinides elements, Proc. Workshop held 8-12 October, 1978, in Warrenton, Virginia, USA, 1978.

[78BED/FID] Bednarczyk, L., Fidelis, I., Determination of stability constants of U(VI), Np(VI) and Pu(VI) with chloride ions by extraction chromatography, J. Radioanal. Chem., **45** (1978) 325-330.

[78BER/HEI] Berg, L., Heibel, B., Hinz, I., Karl, W., Keller-Rudek, H., Leonard, A., Ruprecht, S., Steiss, P., Uran und Sauerstoff, in: Gmelin Handbuch der Anorganischen Chemie, U-Ergänzungsband C2, Berlin: Springer-Verlag, 1978, 312p., in German.

[78BLA/FOU] Blaise, A., Fournier, J.-M., Lagnier, R., Mortimer, M.J., Schenkel, R., Henkie, Z., Wojakowski, A., Physical properties of uranium dipnictides, in: Rare Earths and Actinides 1977, Conf. Ser. Inst. Phys., **37** (1978) 184-189, Chem. Abstr. **89** (1978) 83868a.

[78BRA] Brandenburg, N.P., Thesis, University of Amsterdam, 1978, 141p., cited by Fuger, J., Chemical thermodynamic properties—selected values, in: Gmelin Handbook of inorganic chemistry, 8*th* ed., Uranium, Supp. Vol. **A6**, Berlin: Springer-Verlag, 1983, *pp.*165-192.

[78COD] CODATA recommended Key Values for thermodynamics 1977, Committee on Data for Science and Technology (CODATA), Bulletin Nr. 28, Paris: International Council of Scientific Unions, 1978, 6p.

[78COR/OHA] Cordfunke, E.H.P., O'Hare, P.A.G., The chemical thermodynamics of actinide elements and compounds: Part 3. Miscellaneous actinide compounds, Vienna: International Atomic Energy Agency, 1978, 83p.

[78GLU/GUR] Glushko, V.P., Gurvich, L.V., Bergman, G.A., Veits, I.V., Medvedev, V.A., Khachkuruzov, G.A., Yungman, V.S., Thermodynamic properties of individual substances: Elements O, H(D, T), F, Cl, Br, I, He, Ne, Ar, Kr, Xe, Rn, S, N, P, and their compounds, Vol **1**, Moscow, 1978, in Russian.

[78GRO/ZAK] Grønvold, F., Zaki, M.R., Westrum, E.F., Jr., Sommers, J.A., Downie, D.B., Actinoid pnictides: II. Heat capacities of UAs_2 and USb_2 from 5 to 750 K and antiferromagnetic transitions, J. Inorg. Nucl. Chem., **40** (1978) 635-645.

[78JOH/OHA] Johnson, G.K., O'Hare, P.A.G., Thermochemistry of uranium compounds: XI. Standard enthalpy of formation of γ-UO_3, J. Chem. Thermodyn., **10** (1978) 577-580.

[78KOB/KOL] Kobets, L.V., Kolevich, T.A., Umreiko, D.S., Crystalline hydrated forms of triuranyl diorthophosphate, Russ. J. Inorg. Chem., **23** (1978) 501-505.

[78KUD/SUG] Kudryashov, V.L., Suglobova, I.G., Christ, D.E., Enthalpies of formation of hexafluorouranates(V) of the alkali metals, Sov. Radiochem., **20** (1978) 320-324.

[78LAN] Langmuir, D., Uranium solution - mineral equilibria at low temperatures with applications to sedimentary ore deposits, Geochim. Cosmochim. Acta, **42** (1978) 547-569.

[78MEI/HEI] Meixner, D., Heintz, A., Lichtenthaler, R.N., Dampfdrucke und Phasenumwandlungsenthalpien der Fluoride IF_5, IF_7, MoF_6, WF_6 und UF_6: Experimentelle und flüssigkeitstheoretische Untersuchungen, Ber. Bunsen-Ges. Phys. Chem., **82** (1978) 220-225, in German.

[78MON/AMI] Monk, C.B., Amira, M.F., Electromotive force studies of electrolyte dissociation: Part 12. Dissociation constants of some strongly ionising acids at zero ionic strength and 25°C, J. Chem. Soc. Dalton Trans., **74** (1978) 1170-1178.

[78NIK/PIR] Nikolaeva, N.M., Pirozhkov, A.V., The solubility product of U(IV) hydroxide at the elevated temperatures, Izv. Sib. Otd. Akad. Nauk SSSR Ser. Khim. Nauk, **5** (1978) 82-88, in Russian.

[78NIK] Nikolaeva, N.M., The hydrolysis of U^{4+} ions at elevated temperatures, Izv. Sib. Otd. Akad. Nauk SSSR, Ser. Khim. Nauk, **9(4)** (1978) 91-96, in Russian; Engl. transl.: Report ORNL/TR-88/2, Oak Ridge National Laboratory, Oak Ridge, Tennessee, USA, 1988, 14p.

[78PIT/PET] Pitzer, K.S., Peterson, J.R., Silvester, L.F., Thermodynamics of electrolytes: IX. Rare earth chlorides, nitrates, and perchlorates, J. Solution Chem., **7** (1978) 45-56.

[78PRI/COR] Prins, G., Cordfunke, E.H.P., Ouweltjes, W., Standard enthalpies of formation of uranium compounds: V. UO_2Br_2, $UO_2Br_2{\cdot}H_2O$, $UO_2Br_2{\cdot}3H_2O$, and $UO_2(OH)Br{\cdot}2H_2O$, a comment on the stability of uranyl iodide and hydrates, J. Chem. Thermodyn., **10** (1978) 1003-1010.

[78ROB/HEM2] Robie, R.A., Hemingway, B.S., Fisher, J.R., Thermodynamic properties of minerals and related substances at 298.15 K and 1 bar pressure and at higher temperatures, US Geological Survey Bulletin, **1452** (1978) 456p.

[78SCH/SUL] Schmidt, K.H., Sullivan, J.C., Gordon, S., Thompson, R.C., Determination of hydrolysis constants of metal cations by a transient conductivity method, Inorg. Nucl. Chem. Letters, **14** (1978) 429-434.

[78SIN/PRA] Singh, Z., Prasad, R., Venugopal, V., Sood, D.D., The vaporization thermodynamics of uranium tetrachloride, J. Chem. Thermodyn., **10** (1978) 129-134.

[78SUG/CHI] Suglobova, I.G., Chirkst, D.E., Enthalpy of formation of uranium tetrachloride, Sov. Radiochem., **20** (1978) 175-177.

[78SUG/CHI2] Suglobova, I.G., Chirkst, D.E., Structural-energetic interpretation of the competition of forms of complexes, with the systems UBr_3-MBr as an example, Radiokhimiya, **20(3)** (1978) 352-360, in Russian; Engl. transl.: Sov. Radiochem., **20** (1978) 302-309.

[78THA/AHM] Thakur, L., Ahmad, F., Prasad, R., Heats of solution of uranium metal and uranium tetrachloride and of neptunium iodide in aqueous hydrochloric acid, Indian J. Chem., **16A** (1978) 661-664.

[78WAL] Walenta, K., Uranospathite and arsenuranospathite, Mineral. Mag., **42** (1978) 117-128, Abstract in Am. Mineral., **64** (1979) 465.

[79BAU/ARM] Bauman, J.E., de Arman, S.K., Thermodynamics of uranium(VI) aquo ions, Abstracts od the 34th Annual Calorimetry Conference held 24-28 July, 1979, Kent, Ohio, USA, cited by Ullman, W.J., Schreiner, F., Calorimetric determination of the enthalpies of the carbonate complexes of U(VI), Np(VI), and Pu(VI) in aqueous solution at 25°C, Radiochim. Acta, **43** (1988) 37-44.

[79BLA/LAG] Blaise, A., Lagnier, R., Mulak, J., Zolnierek, Z., Magnetic susceptibility and heat capacity anomalies of $U(OH)_2SO_4$ at 21 K, J. Phys. Colloq. C4 (Supp. 4), **40** (1979) 176-178.

[79BLA] Blaise, A., The specific heat of actinide compounds: Are those measurements useful? J. Phys. Colloq. C4 (Supp. 4), **40** (1979) 49-61.

[79CHI/MAS] Chiotti, P., Mason, J.T., Lee, T.S., Thermodynamic properties of the mercury-uranium system, J. Less-Common Met., **66** (1979) 41-50.

[79CIA/FER] Ciavatta, L., Ferri, D., Grimaldi, M., Palombari, R., Salvatore, F., Dioxouranium(VI) carbonate complexes in acid solution, J. Inorg. Nucl. Chem., **41** (1979) 1175-1182.

[79COR/MUI] Cordfunke, E.H.P., Muis, R.P., Prins, G., High-temperature enthalpy increments of uranyl dihalides UO_2X_2 (X = F, Cl, Br), and thermodynamic functions, J. Chem. Thermodyn., **11** (1979) 819-823.

[79COR/OUW] Cordfunke, E.H.P., Ouweltjes, W., Investigations on uranium carbonitrides: II. Phase relationships, J. Nucl. Mater., **79** (1979) 271-276.

[79DEL/PIR] Deliens, M., Piret, P., Uranyl aluminum phosphates from Kobokobo: II. New minerals phuralumite and upalite, Bull. Minéral. (Soc. Fr. Minéral. Cristallogr.), **102** (1979) 333-337, in French, Abstract in Am. Mineral., **65** (1980) 208.

[79FRO/RYK] Frolov, A.A., Rykov, A.G., Interaction of ions of pentavalent actinides with multiply charged cations, Sov. Radiochem., **21** (1979) 281-292.

[79FUG] Fuger, J., Alkali metal actinide complex halides: Thermochemical and structural considerations, J. Physique Colloq. C4 (Supp. 4), **40** (1979) 207-213.

[79GOL] Goldberg, R.N., Evaluated activity coefficients and osmotic coefficients for aqueous solutions: Bi-univalent compounds of lead, copper, manganese and uranium, J. Phys. Chem. Ref. Data, **8** (1979) 1005-1050.

[79GUP/GIN] Gupta, S.K., Gingerich, K.A., Observation and atomization energies of the gaseous uranium carbides UC, UC_2, UC_3, UC_4, UC_5 and UC_6 by high temperature mass spectrometry, J. Chem. Phys., **71** (1979) 3072-3080.

[79HAA/WIL] Haacke, D.F., Williams, P.A., The aqueous chemistry of uranium minerals: Part I. Divalent cation zippeite, Mineral. Mag., **43** (1979) 539-541.

[79JOH/PYT] Johnson, K.S., Pytkowicz, R.M., Ion association and activity coefficients in multicomponent solutions, in: Activity coefficients in electrolyte solutions (Pytkowicz, R.M., *ed.*), Vol. **II**, Boca Raton, Florida: CRC Press, 1979, *pp.*1-62.

[79JOH] Johnson, G.K., The enthalpy of formation of uranium hexafluoride, J. Chem. Thermodyn., **11** (1979) 483-490.

[79KLE/HIL] Kleinschmidt, P.D., Hildenbrand, D.L., Thermodynamics of the dimerization of gaseous UF_5, J. Chem. Phys., **71** (1979) 196-201.

[79KUB/ALC] Kubaschewski, O., Alcock, C.B., Metallurgical thermochemistry, 5*th* ed., Oxford: Pergamon Press, 1979, 449p.

[79LAJ/PAR] Lajunen, L.H.J., Parhi, S., A note on the hydrolysis of uranyl(VI) ions in 0.5 M $NaClO_4$ medium, Finn. Chem. Letters, (1979) 143-144.

[79MAD/GUI] Madic, C., Guillaume, B., Morisseau, J.C., Moulin, J.P., "Cation-cation" complexes of pentavalent actinides: I. Spectrophotometric study of complexes between neptunium(V) and UO_2^{2+} and NpO_2^{2+} ions in aqueous perchloric and nitric acid solution, J. Inorg. Nucl. Chem., **41** (1979) 1027-1031.

[79MIL/ELK] Milic, N.B., El Kass, G., Hydrolysis of the uranyl ion in a potassium chloride medium, Bull. Soc. Chim. Beograd, **44(4)** (1979) 275-279.

[79MIL] Millero, F.J., Effects of pressure and temperature on activity coefficients, in: Activity coefficients in electrolyte solutions (Pytkowicz, R.M., *ed.*), Vol. **II**, Boca Raton, Florida: CRC Press, 1979, *pp.*63-151.

[79PIT] Pitzer, K.S., Theory: Ion interaction approach, in: Activity coefficients in electrolyte solutions (Pytkowicz, R.M., *ed.*), Vol. **I**, Boca Raton: CRC Press, 1979, *pp.*157-208.

[79PRA/NAG] Prasad, R., Nagarajan, K., Singh, Z., Bhupathy, M., Venugopal, V., Sood, D.D., Thermodynamics of the vaporization of thorium and uranium halides, Proc. International Symp. on thermodynamics of nuclear materials, Vol. **1**, Jülich, Federal Republic of Germany, 1979, *p.*45.

[79PYT] Pytkowicz, R.M., Activity coefficients, ionic media, and equilibria in solutions, in: Activity coefficients in electrolyte solutions (Pytkowicz, R.M., *ed.*), Vol. **II**, Boca Raton, Florida: CRC Press, 1979, *pp.*301-305.

[79SHI/CRI] Shin, C., Criss, C.M., Standard enthalpies of formation of anhydrous and aqueous magnesium chloride at 298.15 K, J. Chem. Thermodyn., **11** (1979) 663-666.

[79SPI/ARN] Spiess, B., Arnaud-Neu, F., Schwing-Weill, M.J., Behaviour of uranium(VI) with some cryptands in aqueous solution, Inorg. Nucl. Chem. Letters, **15** (1979) 13-16.

[79STO] Stokes, R.H., Thermodynamics of solutions, in: Activity coefficients in electrolyte solutions (Pytkowicz, R.M., *ed.*), Vol. I, Boca Raton: Florida, CRC Press, 1979, *pp.*1-28.

[79SYL/DAV] Sylva, R.N., Davidson, M.R., The hydrolysis of metal ions: Part 1. Copper(II), J. Chem. Soc. Dalton Trans., (1979) 232-235.

[79SYL/DAV2] Sylva, R.N., Davidson, M.R., The hydrolysis of metal ions: Part 2. Dioxouranium(VI), J. Chem. Soc. Dalton Trans., (1979) 465-471.

[79TAG/FUJ] Tagawa, H., Fujino, T., Yamashita, T., Formation and some chemical properties of alkaline-earth metal monouranates, J. Inorg. Nucl. Chem., **41** (1979) 1729-1735.

[79WHI] Whitfield, M., Activity coefficients in natural waters, in: Activity coefficients in electrolyte solutions (Pytkowicz, R.M., *ed.*), Vol. **II**, Boca Raton, Florida: CRC Press, 1979, pp.153-299.

[79WHI2] Whiffen, D.H. (former chairman), Manual of symbols and terminology for physicochemical quantities and units, IUPAC Commission on physicochemical symbols, terminology and units, Oxford: Pergamon Press, 1979, 41p.

[79WOL] Wolery, T.J., Calculation of chemical equilibrium between aqueous solution and minerals: The EQ3/6 software package, Report UCRL-52658, Lawrence Livermore National Laboratory, Livermore, California, USA, 1979.

[80ALL/KIP] Allard, B., Kipatsi, H., Liljenzin, J.O., Expected species of uranium, neptunium and plutonium in neutral aqueous solutions, J. Inorg. Nucl. **42** (1980) 1015-1027.

[80ALW/WIL] Alwan, A.A., Williams, P.A., The aqueous chemistry of uranium minerals: Part II. Minerals of the liebigite group, Mineral. Mag., **43** (1980) 665-667.

[80BAR/COR2] Barten, H., Cordfunke, E.H.P., The formation and stability of hydrated and anhydrous uranyl phosphates, Thermochim. Acta, **40** (1980) 357-365.

[80BEN/TEA] Benson, L.V., Teague, L.S., A tabulation of thermodynamic data for chemical reactions involving 58 elements common to radioactive waste package systems, Report LBL-11448, Lawrence Berkeley Laboratory, Berkeley, USA, 1980, 97p.

[80BLA/LAG] Blaise, A., Lagnier, R., Troć, R., Henkie, Z., Markowski, P.J., The heat capacities of U_3As_4 and Th_3As_4, J. Low Temp. Phys., **39** (1980) 315-328.

[80BLA/LAG2] Blaise, A., Lagnier, R., Wojakowski, A., Zygmunt, A., Mortimer, M.J., Low-temperature specific heats of some uranium ternary compounds UAsY (Y = S, Se, Te), J. Low Temp. Phys., **41** (1980) 61-72.

[80BLA/TRO] Blaise, A., Troć, R., Lagnier, R., Mortimer, M.J., The heat capacity of uranium monoarsenide, J. Low Temp. Phys., **38** (1980) 79-92.

[80BOE/BOO] De Boer, F.R., Boom, R., Miedema, A.R., Enthalpies of formation of liquid and solid binary alloys based on 3d metals: I. Alloys of scandium, titanium and vanadium, Physica, **101B** (1980) 294-319.

[80BON/HEF] Bond, A.M., Hefter, G.T., Critical survey of stability constants and related thermodynamic data of fluoride complexes in aqueous solution, IUPAC Chemical Data Series **27**, Oxford: Pergamon Press, 1980, 67p.

[80BRO/HUA] Brooker, M.H., Huang, C.H., Sylwestrowicz, J., Raman spectroscopic studies of aqueous uranyl nitrate and perchlorate systems, J. Inorg. Nucl. Chem., **42** (1980) 1431-1440.

[80CIA] Ciavatta, L., The specific interaction theory in evaluating ionic equilibria, Ann. Chim. (Rome), (1980) 551-567.

[80COD] CODATA tentative set of key values for thermodynamics Part VIII, Committee on Data for Science and Technology (CODATA) Special Report 8, Paris: International Council of Scientific Unions, 1980, 48p.

[80COR/WES] Cordfunke, E.H.P., Westrum, E.F., Jr., Investigations on caesium uranates: VII. Thermochemical properties of $Cs_2U_4O_{12}$, in: Thermodynamics of nuclear materials 1979, Proc. Symp. held 29 January to 2 February, 1979, in Jülich, Federal Republic of Germany, Vienna: International Atomic Energy Agency, 1980, *pp.*125-141.

[80DON/LAN] Dongarra, G., Langmuir, D., The stability of UO_2OH^+ and $UO_2[HPO_4]_2^{2-}$ complexes at 25°C, Geochim. Cosmochim. Acta, **44** (1980) 1747-1751.

[80DOU/PON] Double, G., Pongi, N.K., Hurwic, J., Complexes carbonates de U(VI) et carbonates doubles d'uranium et de sodium, Bull. Soc. Chim. Fr., (1980) 354-359, in French.

[80EDW/STA] Edwards, J.G., Starzynski, J.S., Peterson, D.E., Sublimation thermodynamics of the URu_3 compound, J. Chem. Phys., **73** (1980) 908-912.

[80GAL/KAI] Galloway, W.E., Kaiser, W.R., Catahoula formation of the Texas coastal plain: Origin, geochemical evolution, and characteristic of uranium deposits, Report of investigation No. 100, Bureau of Econ. Geol., University of Texas, Austin, Texas, USA, 1980, 81p.

[80GHO/GHO] Ghosh, A.K., Ghosh, J.C., Prasad, B., Third dissociation constant of phosphoric acid from 283.15 K to 323.15 K, J. Indian Chem. Soc., **57** (1980) 1194-1199.

[80HU/NI] Hu, R., Ni, X., Bai, M., Liu, K., Hu, C., Sui, D., Zhang, M., Huang, W., Determination of the standard enthalpy of formation of uranium tetrafluoride by solution calorimetry, Sci. Sin., **23** (1980) 1386-1395.

[80JOH/PAP] Johnson, G.K., Papatheodorou, G.N., Johnson, C.E., The enthalpies of formation and high-temperature thermodynamic functions of As_4S_4 and As_2S_3, J. Chem. Thermodyn., **12** (1980) 545-557.

[80KHR/LYU] Khromov, Yu.F., Lyutikov, R.A., Certain thermodynamic properties of uranium nitride UN_y, Sov. Atom. Energy, **49** (1980) 448-452.

[80KOB/KOL] Kobets, L.V., Kolevich, T.A., Ksenofontova, N.M., Spectroscopic investigation of uranyl bis(dihydrogen orthophosphate), Russ. J. Inorg. Chem., **25** (1980) 732-735.

[80LAN/CHA] Langmuir, D., Chatham, J.R., Groundwater prospecting for sandstone-type uranium deposits: A preliminary comparison of the merits of mineral-solution equilibria, and single element tracer methods, J. Geochem. Explor., **13** (1980) 201-219.

[80LEM/TRE] Lemire, R.J., Tremaine, P.R., Uranium and plutonium equilibria in aqueous solutions to 200°C, J. Chem. Eng. Data, **25** (1980) 361-370.

[80MIE/CHA] Miedema, A.R., de Châtel, P.F., de Boer, F.R., Cohesion in alloys-Fundamentals of a semi-empirical model, Physica, **100B** (1980) 1-28.

[80NAG/BHU] Nagarajan, K., Bhupathy, M., Prasad, R., Singh, Z., Venugopal, V., Sood, D.D., Vaporization behaviour of uranium tetrafluoride, J. Chem. Thermodyn., **12** (1980) 329-333.

[80OHA/FLO] O'Hare, P.A.G., Flotow, H.E., Hoekstra, H.R., Heat capacity (5 to 350 K) and recommended thermodynamic properties of barium uranate ($BaUO_4$) to 1100 K, J. Chem. Thermodyn., **12** (1980) 1003-1008.

[80PAR] Parker, V.B., The thermochemical properties of the uranium halogen containing compounds, Report NBSIR80-2029, US National Bureau of Standards, Washington, D.C., USA, 1980, 160p.

[80PON/DOU] Pongi, N.K., Double, G., Hurwic, J., Hydrolyse de U(VI) et uranates de sodium, Bull. Soc. Chim. Fr., **9-10** (1980) 347-353, in French.

[80PRI/COR] Prins, G., Cordfunke, E.H.P., Depaus, R., Investigations of uranium carbonitrides III. Nitrogen vapour pressures and thermodynamic properties, J. Nucl. Mater., **89** (1980) 221-228.

[80SUG/CHI] Suglobova, I.G., Chirkst, D.E., On the formation enthalpy of uranium tetrachloride, Radiokhimia, **22** (1980) 779-781, in Russian.

[80VOL/SUG] Volkov, V.A., Suglobova, D.E., Chirkst, D.E., Koord. Khim., **6** (1980) 417-422, cited by Fuger, J., Parker, V.B., Hubbard, W.N., Oetting, F.L., The chemical thermodynamics of the actinide elements: Part 8. The actinide halides, Vienna: International Atomic Energy Agency, 1983, 267p.

[80WES/ZAI] Westrum, E.F., Jr., Zainel, H.A., Jakes, D., Alkaline earth uranates: An exploration of their thermophysical stabilities, in: Thermodynamics nuclear materials 1979, Proc. Symp. held 29 January to 2 February, 1979, in Jülich, Federal Republic of Germany, Vienna: International Atomic Energy Agency, 1980, *pp.*143-154.

[81AWA/SUN] Awasthi, S.P., Sundaresan, M., Spectrophotometric and calorimetric study of uranyl cation/chloride anion system in aqueous solution, Indian J. Chem., **20A** (1981) 378-381.

[81BAE/MES] Baes, C.F., Jr., Mesmer, R.E., The thermodynamics of cation hydrolysis, Am. J. Sci., **281** (1981) 935-962.

[81CHI/AKH] Chiotti, P., Akhachinskiy, V.V., Ansara, I., Rand, M.H., The chemical thermodynamics of actinide elements and compounds: Part 5. The actinide binary alloys, Vienna: International Atomic Energy Agency, 1981, 275p.

[81CIA/FER2] Ciavatta, L., Ferri, D., Grenthe, I., Salvatore, F., The first acidification step of the tris (carbonato) dioxouranate(VI) ion, $UO_2(CO_3)_3^{4-}$, Inorg. Chem., **20** (1981) 463-467.

[81COR/OUW] Cordfunke, E.H.P., Ouweltjes, W., Standard enthalpies of formation of uranium compounds: VI. MUO_3 (M = Li, Na, K, and Rb), J. Chem. Thermodyn., **13** (1981) 187-192.

[81COR/OUW2] Cordfunke, E.H.P., Ouweltjes, W., Standard enthalpies of formation of uranium compounds: VII. UF_3 and UF_4 (by solution calorimetry), J. Chem. Thermodyn., **13** (1981) 193-197.

[81FER/GRE] Ferri, D., Grenthe, I., Salvatore, F., Dioxouranium(VI) carbonate complexes in neutral and alkaline solutions, Acta Chem. Scand., **A35** (1981) 165-168.

[81FUJ/TAG] Fujino, T., Tagawa, H., Adachi, T., On some factors affecting the non-stoichiometry in U_3O_8, J. Nucl. Mater., **97** (1981) 93-103.

[81GOL/TRE] Gol'tsev, V.P., Tretyakov, A.A., Kerko, P.F., Barinov, V.I., Thermodynamic characteristics of calcium uranates, Vest. Akad. Navuk Belaruskai SSR Ser. Fiz. Ener., (1981) 24-29, in Russian.

[81GUI/HOB] Guillaume, B., Hobart, D.E., Bourges, J.Y., "Cation-cation" complexes of pentavalent actinides: II. Spectrophotometric study of complexes of Am(V) with UO_2^{2+} and NpO_2^{2+} in aqueous perchlorate solution, J. Inorg. Nucl. Chem., **43** (1981) 3295-3299.

[81HEL/KIR] Helgeson, H.C., Kirkham, D.H., Flowers, G.C., Theoretical prediction of the thermodynamic behavior of aqueous electrolytes at high pressures and temperatures: IV. Calculation of activity coefficients, osmotic coefficients, and apparent molal and standard and relative partial molal properties to 600°C and 5 kb, Am. J. Sci., **281** (1981) 1249-1516.

[81JOH/COR] Johnson, G.K., Cordfunke, E.H.P., The enthalpies of formation of uranium mononitride and alpha and beta uranium sesquinitride by fluorine bomb calorimetry, J. Chem. Thermodyn., **13** (1981) 273-282.

[81JOH/STE3] Johnson, G.K., Steele, W.V., The standard enthalpy of formation of uranium dioxide by fluorine bomb calorimetry, J. Chem. Thermodyn., **13** (1981) 717-723.

[81LIN/BES] Lindemer, T.B., Besmann, T.M., Johnson, C.E., Thermodynamic review and calculations: Alkali-metal oxide systems with nuclear fuels, fission products, and structural materials, J. Nucl. Mater., **100** (1981) 178-226.

[81MAY/BEG] Maya, L., Begun, G.M., A Raman spectroscopy study of hydroxo and carbonato species of the uranyl(VI) ion, J. Inorg. Nucl. Chem., **43** (1981) 2827-2832.

[81MOR/FAH] Morss, L.R., Fahey, J.A., Gens, R., Fuger, J., Jenkins, H.D.B., Stabilities of ternary alkaline earth-actinide(VI) oxides M_3AnO_6, in: Actinides-1981 (Abstracts), held 10-15 September, 1981, in Pacific Grove, California, USA, *p.*264.

[81NIE/BOE] Niessen, A.K., de Boer, F.R., The enthalpy of formation of solid borides, carbides, nitrides, silicides and phosphides of transition and noble metals, J. Less-Common Met., **82** (1981) 75-80.

[81OBR/WIL] O'Brien, T.J., Williams, P.A., The aqueous chemistry of uranium minerals: Part 3. Monovalent cation zippeites, Inorg. Nucl. Chem. Letters, **17** (1981) 105-107.

[81OHA/FLO] O'Hare, P.A.G., Flotow, H.E., Hoekstra, H.R., Cesium diuranate ($Cs_2U_2O_7$): Heat capacity (5 to 350 K) and thermodynamic functions to 350 K; A re-evaluation of the standard enthalpy of formation and the thermodynamics of (cesium + uranium + oxygen), J. Chem. Thermodyn., **13** (1981) 1075-1080.

[81SAD/LIN] Sadiq, M., Lindsay, W.L., Arsenic supplement to Technical Bulletin 134: Selection of standard free energies of formation for use in soil chemistry, Colorado State University Experiment Station, Fort Collins, Colorado, USA, 1981, 39p.

[81SJO/NOR] Sjöberg, S., Nordin, A., Ingri, N., Equilibrium and structural studies of silicon(IV) and aluminium(III) in aqueous solution: II. Formation constants for the monosilicate ions $SiO(OH)_3^-$ and $SiO_2(OH)_2^{2-}$. A precision study at 25°C in a simplified seawater medium, Mar. Chem., **10** (1981) 521-532.

[81STO/SMI] Stohl, F.V., Smith, D.K., The crystal chemistry of the uranyl silicates minerals, Am. Mineral., **66** (1981) 610-625.

[81STU/MOR] Stumm, W., Morgan, J.J., Aquatic Chemistry, New York: John Wiley and Sons, 1981, 780p.

[81SUG/CHI] Suglobova, I.G., Chirkst, D.E., Enthalpy of formation of alkali metal chlorouranates(III), Sov. Radiochem., **23** (1981) 316-320.

[81TRE/CHE] Tremaine, P.R., Chen, J.D., Wallace, G.J., Boivin, W.A., Solubility of uranium (IV) oxide in alkaline aqueous solutions to 300°C, J. Solution Chem., **10** (1981) 221-230.

[81TUR/WHI] Turner, D.R., Whitfield, M., Dickson, A.G., The equilibrium speciation of dissolved components in freshwater and seawater at 25°C and 1 atm pressure, Geochim. Cosmochim. Acta, **45** (1981) 855-881.

[81VAI/MAK] Vainiotalo, A., Makitie, O., Hydrolysis of dioxouranium(VI) ions in sodium perchlorate media, Finn. Chem. Letters, (1981) 102-105.

[81WIJ] Wijbenga, G., Thermochemical investigations on intermetallic UMe_3 compounds (Me = Ru, Rh, Pd), Ph.D. thesis, Report ECN-102, Netherlands Research Foundation, Petten, The Netherlands, 1981.

[82BOE/BOO] De Boer, F.R., Boom, R., Miedema, A.R., Enthalpies of formation of liquid and solid binary alloys based on 3d metals: II. Alloys of chromium and manganese, Physica, **113B** (1982) 18-41.

[82COR/MUI] Cordfunke, E.H.P., Muis, R.P., Ouweltjes, W., Flotow, H.E., O'Hare, P.A.G., The thermodynamic properties of Na_2UO_4, $Na_2U_2O_7$, and $NaUO_3$, J. Chem. Thermodyn., **14** (1982) 313-322.

[82COR/OUW] Cordfunke, E.H.P., Ouweltjes, W., Prins, G., Standard enthalpies of formation of uranium compounds: VIII. UCl_3, UCl_5 and UCl_6, J. Chem. Thermodyn., **14** (1982) 495-502.

[82COR/OUW2] Cordfunke, E.H.P., Ouweltjes, W., Barten, H., Standard enthalpies of formation of uranium compounds: IX. Anhydrous uranyl arsenates, J. Chem. Thermodyn., **14** (1982) 883-886.

[82DEV/YEF] Devina, O.A., Yefimov, M.Ye., Medvedev, V.A., Khodakovsky, I.L., Thermodynamic properties of the uranyl ion in aqueous solution at elevated temperatures, Geochem. Int., **19(5)** (1982) 161-172.

[82FEN/ZHA] Feng, S., Zhang, Z., Zhang, S., The vacuum adiabatic calorimetry of anhydrous uranyl sulphate and uranyl nitrate trihydrate, Chem. J. Chinese Univ., **3** (1982) 374-380, in Chinese.

[82FUG] Fuger, J., Thermodynamic properties of the actinides: Current perspectives, in: Actinides in perspective, Oxford: Pergamon Press, 1982, 409-431.

[82GLU/GUR] Glushko, V.P., Gurvich, L.V., Bergman, G.A., Veits, I.V., Medvedev, V.A., Khachkuruzov, G.A., Yungman, V.S., Thermodynamic properties of individual substances: Vol. **4**, Moscow: Nauka, 1982, in Russian.

[82GRE/HAS] Greis, O., Haschke, J.M., Rare earth fluorides, in: Handbook on the physics and chemistry of rare earths, Vol. **5** (Gschneidner, K.A., Jr., Eyring, L., *eds.*), Amsterdam: Elsevier Science Publishers BV, 1982, *pp.*387-460.

[82GUI/BEG] Guillaume, B., Begun, G.M., Hahn, R.L., Raman spectrometric studies of "cation-cation" complexes of pentavalent actinides in aqueous perchlorate solutions, Inorg. Chem., **21** (1982) 1159-1166.

[82HEM] Hemingway, B.S., Thermodynamic properties of selected uranium compounds and aqueous species at 298.15 K and 1 bar and at higher temperatures. Preliminary models for the origin of coffinite deposits, Open File Report 82-619, US Geological Survey, 1982, 89p.

[82JEN] Jensen, B.S., Migration phenomena of radionuclides into the geosphere, CEC Radioactive Waste Management Series No. 5, Chur, Switzerland: Harwood Academic Publishers, 1982, 197p.

[82LAF] Laffitte, M. (chairman), A report of IUPAC commission I.2 on thermodynamics: Notation for states and processes, significance of the word "standard" in chemical thermodynamics, and remarks on commonly tabulated forms of thermodynamic functions, J. Chem. Thermodyn., **14** (1982) 805-815.

[82LAU/HIL] Lau, K.H., Hildenbrand, D.L., Thermochemical properties of the gaseous lower valent fluorides of uranium, J. Chem. Phys., **76** (1982) 2646-2652.

[82MAR/SMI] Martell, A.E., Smith, R.M., Critical stability constants, Vol. 5, First supplement, New York: Plenum Press, 1982.

[82MAY] Maya, L., Hydrolysis and carbonate complexation of dioxouranium(VI) in the neutral-pH range at 25°C, Inorg. Chem., **21** (1982) 2895-2898.

[82MAY2] Maya, L., Detection of hydroxo and carbonato species of dioxouranium(VI) in aqueous media by differential pulse polarography, Inorg. Chim. Acta, **65** (1982) L13-L16.

[82MIL/SUR] Milic, N.B., Suranji, T.M., Hydrolysis of the uranyl ion in sodium nitrate medium, Z. Anorg. Allg. Chem., **489** (1982) 197-203.

[82MOR/FUG] Morss, L.R., Fuger, J., Jenkins, H.D.B., Thermodynamics of actinide perovskite-type oxides: I. Enthalpy of formation of Ba_2MgUO_6 and Ba_2MgNpO_6, J. Chem. Thermodyn., **14** (1982) 377-384.

[82MOR] Morss, L.R., Complex oxide systems of the actinides, in: Actinides in perspective, Oxford: Pergamon Press, 1982, *pp.*381-407.

[82NAI/INA] Naito, K., Inaba, H., Takahashi, S., Phase transitions in U_3O_{8-z}: I. Heat capacity measurements, J. Nucl. Mater., **110** (1982) 317-323.

[82OHA/MAL] O'Hare, P.A.G., Malm, J.G., Eller, P.G., Thermochemistry of uranium compounds: XII. Standard enthalpies of formation of the α and β modifications of UF_5. Enthalpy of the β-to-α transition at 298.15 K, J. Chem. Thermodyn., **14** (1982) 323-330.

[82OHA/MAL2] O'Hare, P.A.G., Malm, J.G., Thermochemistry of uranium compounds: XIII. Standard enthalpies of formation of α-UOF_4 and UO_2F_2, thermodynamics of (U+O+F), J. Chem. Thermodyn., **14** (1982) 331-336.

[82OVE/LUN] Overvoll, P.A., Lund, W., Complex formation of uranyl ions with polyaminopolycarboxylic acids, Anal. Chim. Acta, **143** (1982) 153-161.

[82PAN] Pankratz, L.B., Thermodynamic properties of elements and oxides, US Bureau of Mines Bulletin, **672** (1982) 509p.

[82PER] Perrin, D.D., Ionisation constants of inorganic acids and bases in aqueous solution, 2*nd* ed., IUPAC Chemical Data Series **29**, Oxford: Pergamon Press, 1982, 180p.

[82ROY/PRA] Roy, K.N., Prasad, R., Venugopal, V., Singh, Z., Sood, D.D., Studies on ($2\ UF_4 + H_2 = 2\ UF_3 + 2\ HF$) and vapour pressure of UF_3, J. Chem. Thermodyn., **14** (1982) 389.

[82SET/MAT] Seta, K., Matsui, T., Inaba, H., Naito, K., Heat capacity and electrical conductivity of non-stoichiometric $U4O_{9-y}$ from 300 to 1200 K, J. Nucl. Mater., **110** (1982) 47-54.

[82TAY] Taylor, J.R., An introduction to error analysis: The study of uncertainties in physical measurements, Mill Valley, California: University Science Books, 1982, 270p.

[82WAG/EVA] Wagman, D.D., Evans, W.H., Parker, V.B., Schumm, R.H., Halow, I., Bailey, S.M., Churney, K.L., Nuttall, R.L., The NBS tables of chemical thermodynamic properties: Selected values for inorganic and C_1 and C_2 organic substances in SI units, J. Phys. Chem. Ref. Data, **11** suppl. No. **2** (1982) 1-392.

[82WIJ/COR] Wijbenga, G., Cordfunke, E.H.P., Determination of standard Gibbs energies of formation of URh_3 and URu_3 by solid state e.m.f. measurements, J. Chem. Thermodyn., **14** (1982) 409-417.

[82WIJ] Wijbenga, G., The enthalpy of formation of UPd_3 by fluorine bomb calorimetry, J. Chem. Thermodyn., **14** (1982) 483-493.

[83ABE/FER] Aberg, M., Ferri, D., Glaser, J., Grenthe, I., Studies of metal carbonate equilibria. VIII. Structure of the hexakis(carbonato)tri[dioxouranate(VI)] ion in aqueous solution. An X-ray diffraction and ^{13}C NMR study, Inorg. Chem., **22** (1983) 3981-3985.

[83ALL] Allard, B., Actinide solution equilibria and solubilities in geologic systems, Report TR-83-35, SKBF/KBS, Stockholm, Sweden, 1983, 48p.

[83AUR/CHI] Aurov, N.A., Chirkst, D.E., Enthalpies of formation of hexahalide complexes of uranium(III), Sov. Radiochem., **25** (1983) 445-449.

[83BAG/JAC] Bagnall, K.W., Jacob, E., Potter, P.E., Uranium, carbonates, cyanides, alkoxides, carboxylates compounds with silicon in: Gmelin Handbook of Inorganic Chemistry, Supp. C13, 1983.

[83BRO/ELL] Brown, P.L., Ellis, A.J., Sylva, R.N., The hydrolysis of metal ions: Part 5. Thorium(IV), J. Chem. Soc. Dalton Trans., (1983) 31-34.

[83CAC/CHO2] Caceci, M.S., Choppin, G.R., The first hydrolysis constant of uranium(VI), Radiochim. Acta, **33** (1983) 207-212.

[83CHI/NEV] Chierice, G.O., Neves, E.A., The stepwise formation of complexes in the uranyl/azide system in aqueous medium, Polyhedron, **2** (1983) 31-35.

[83CHI] Chirkst, D.E., Energetics of complex uranium halides in the light of D.I. Mendeleev's periodic system, Sov. Radiochem., **24** (1983) 620-645.

[83CIA/FER] Ciavatta, L., Ferri, D., Grenthe, I., Salvatore, F., Spahiu, K., Studies on metal carbonate equilibria: 4. Reduction of the tris(carbonato)dioxouranate(VI) ion, $UO_2(CO_3)_3^{4-}$, in hydrogen carbonate solutions, Inorg. Chem., **22** (1983) 2088-2092.

[83COR/OUW] Cordfunke, E.H.P., Ouweltjes, W., Prins, G., van Vlaanderen, P., Standard enthalpies of formation of UCl_5, UCl_6, and UOCl: Correction of errors in previous papers, J. Chem. Thermodyn., **15** (1983) 1103-1104.

[83COR/OUW2] Cordfunke, E.H.P., Ouweltjes, W., van Vlaanderen, P., Standard enthalpies of formation of uranium compounds: X. Uranium oxide chlorides, J. Chem. Thermodyn., **15** (1983) 237-243.

[83DAV/EFR] Davydov, Yu.P., Efremenkov, V.M., The hydrolysis of uranium(VI) in solution, Russ. J. Inorg. Chem., **28** (1983) 1313-1316.

[83DAV/YEF] Davina, O.A., Yefimov, M.Ye., Medvedev, V.A., Khodakovsky, I.L., Thermochemical determination of the stability constant of $UO_2(CO_3)_3^{4-}$(sol) at 25-200°C, Geochem. Int., **20(3)** (1983) 10-18.

[83DEV/YEF] Devina, O.A., Yefimov, M.Ye., Medvedev, V.A., Khodakovsky, I.L., Thermochemical determination of the stability constant of $UO_2(CO_3)_3^{4-}$ at 25/200°C, Geokhimiya, **5** (1983) 677-684, in Russian; Engl. transl.: Geochem. Int., **20(3)** (1983) 10-18.

[83FER/GRE] Ferri, D., Grenthe, I., Salvatore, F., Studies on metal carbonate equilibria. 7. Reduction of the tris(carbonato)dioxouranate(VI) ion, $UO_2(CO_3)_3^{4-}$, in carbonate solutions, Inorg. Chem., **22** (1983) 3162-3165.

[83FLE] Fleischer, M., Glossary of mineral species 1983, Tucson, Arizona: Mineralogical Record, 1983, 202p.

[83FUG/PAR] Fuger, J., Parker, V.B., Hubbard, W.N., Oetting, F.L., The chemical thermodynamics of actinide elements and compounds: Part 8. The actinide halides, Vienna: International Atomic Energy Agency, 1983, 267p.

[83FUG] Fuger, J., Chemical thermodynamic properties—selected values, in: Gmelin Handbook of inorganic chemistry, 8*th* ed., Uranium, Supp. Vol. **A6**, Berlin: Springer-Verlag, 1983, *pp.*165-192.

[83GIR/PET] Girichev, G.V., Petrov, V.M., Giricheva, N.I., Zasorin, E.Z., Krasnov, K.S., Kiselev, Yu.M., An electron diffraction study of the structure of the uranium tetrafluoride molecule, Zh. Strukt. Khim., **24** (1983) 70-74; Engl. transl.: J. Struct. Chem., **24** (1983) 61-64.

[83KAG/KYS] Kaganyuk, D.S., Kyskin, V.I., Kazin, I.V., Calculation of enthalpies of formation for radioelement compounds, Sov. Radiochem., **25** (1983) 65-69.

[83KHA] Khattak, G.D., Specific heat of uranium dioxide (UO_2) between 0.3 and 50 K, Phys. Status Solidi, **A75** (1983) 317-321.

[83KOH] Kohli, R., Heat capacity and thermodynamic properties of alkali metal compounds: II. Estimation of the thermodynamic properties of cesium and rubidium zirconates, Thermochim. Acta, **65** (1983) 285-293.

[83KOV/CHI] Kovba, V.M., Chikh, I.V., IR spectra of vapors over UF_4, UCl_4, UBr_4, and in the UCl_4-Cl_2 system, Zh. Strukt. Khim., **24** (1983) 172-173; Engl. transl.: J. Struct. Chem., **24** (1983) 326-327.

[83LEI] Leitnaker, J.M., Thermodynamic data for uranium fluorides, K/PS-352, Union Carbide Corporation, Oak Ridge Diffusion Plant, Oak Ridge, Tennessee, USA, 1983, 43p.

[83MAR/PAV] Marković, M., Pavković, N., Solubility and equilibrium constants of uranyl(2+) in phosphate solutions, Inorg. Chem., **22** (1983) 978-982.

[83MEY/WAR] Meyer, B., Ward, K., Koshlap, K., Peter, L., Second dissociation constant of hydrogen sulfide, Inorg. Chem., **22** (1983) 2345-2346.

[83MOR/WIL2] Morss, L.R., Williams, C.W., Choi, I.K., Gens, R., Fuger, J., Thermodynamics of actinide perkovskite-type oxides: II. Enthalpy of formation of Ca_3UO_6, Sr_3UO_6, Ba_3UO_6, Sr_3NpO_6, and Ba_3NpO_6, J. Chem. Thermodyn., **15** (1983) 1093-1102.

[83NAI/TSU] Naito, K., Tsuji, T., Ohya, F., Phase transitions in U_3O_{8-z}: II. Electrical conductivity measurement, J. Nucl. Mater., **114** (1983) 136-142.

[83NAS/CLE2] Nash, K.L., Cleveland, J.M., Free energy, enthalpy, and entropy of plutonium(IV)-sulfate complexes, Radiochim. Acta, **33** (1983) 105-111.

[83OBR/WIL] O'Brien, T.J., Williams, P.A., The aqueous chemistry of uranium minerals: 4. Schröckingerite, grimselite, and related alkali uranyl carbonates, Mineral. Mag., **47** (1983) 69-73.

[83ROB] Robel, X., Chemical thermodynamics, in: Gmelin handbook of inorganic chemistry, 8*th* ed., Uranium, Supp. Vol. **A6**, Berlin: Springer-Verlag, 1983.

[83RYA/RAI] Ryan, J.L., Rai, D., The solubility of uranium(IV) hydrous oxide in sodium hydroxide solutions under reducing conditions, Polyhedron, **2** (1983) 947-952.

[83SJO/HAE] Sjöberg, S., Hägglund, Y., Nordin, A., Ingri, N., Equilibrium and structural studies of silicon(IV) and aluminium(III) in aqueous solution: V. Acidity constants of silicic acid and the ionic product of water in the medium range 0.05-2.0 M Na(Cl) at 25°C, Mar. Chem., **13** (1983) 35-44.

[83SPA] Spahiu, K., Carbonate complex formation in lanthanoid and actinoid systems, Ph.D. thesis, The Royal Institute of Technology, Stockholm, Sweden, 1983.

[83WOL] Wolery, T.J., EQ3NR: A computer program for geochemical aqueous speciation-solubility calculations: User's guide and documentation, UCRL-53414, Lawrence Livermore National Laboratory, Livermore California, 1983, 191p.

[84BUS/PLU] Busenberg, E., Plummer, L.N., Parker, V.B., The solubility of strontianite ($SrCO_3$) in CO_2-H_2O solutions between 2 and 91°C, the association constants of $SrHCO_3^+$ and $SrCO_3$(aq) between 5 and 80°C, and an evaluation of the thermodynamic properties of Sr^{2+} and $SrCO_3$(cr) at 25°C and 1 atm total pressure, Geochim. Cosmochim. Acta, **48** (1984) 2021-2055.

[84CHO/RAO] Choppin, G.R., Rao, L.F., Complexation of pentavalent and hexavalent actinides by fluoride, Radiochim. Acta, **37** (1984) 143-146.

[84COR/KUB] Cordfunke, E.H.P., Kubaschewski, O., The thermochemical properties of the system uranium-oxygen-chlorine, Thermochim. Acta, **74** (1984) 235-245.

[84FLO/HAS] Flotow, H.E., Haschke, J.M., Yamauchi, S., The chemical thermodynamics of actinide elements and compounds: Part 9. The actinide hydrides, Vienna: International Atomic Energy Agency, 1984, 115p.

[84FRE] Freeman, R.D., Conversion of standard (1 atm) thermodynamic data to the new standard-state pressure, 1 bar (10^5 Pa), J. Chem. Eng. Data, **29** (1984) 105-111.

[84FUL/SMY] Fuller, E.L., Jr., Smyrl, N.R., Condon, J.B., Eager, M.H., Uranium oxidation: Characterization of oxides formed by reaction with water by infrared and sorption analyses, J. Nucl. Mater., **120** (1984) 174-194.

[84GEN/WEI] Van Genderen, A.C.G., van der Weijden, C.H., Prediction of Gibbs energies of formation and stability constants of some secondary uranium minerals containing the uranyl group, Uranium, **1** (1984) 249-256.

[84GOR/SMI] Gorokhov, L.N., Smirnov, V.K., Khodeev, Yu.S., Thermochemical characterization of uranium UF_n molecules, Russ. J. Phys. Chem., **58** (1984) 980.

[84GRE/FER] Grenthe, I., Ferri, D., Salvatore, F., Riccio, G., Studies on metal carbonate equilibria. Part 10. A solubility study of the complex formation in the uranium(VI) water-carbonate dioxide system at 25°C, J. Chem. Soc. Dalton Trans., (1984) 2439-2443.

[84GRE/SPA] Grenthe, I., Spahiu, K., Olofsson, G., Studies on metal carbonate equilibria: 9. Calorimetric determination of the enthalpy and entropy changes for the formation of uranium(IV) and (V) carbonate complexes at 25°C in a 3 M (Na,H)ClO_4, Inorg. Chim. Acta, **95** (1984) 79-84.

[84GRO/DRO] Grønvold, F., Drowart, J., Westrum, E.F., Jr., The chemical thermodynamics of actinide elements and compounds: Part 4. The actinide chalcogenides (excluding oxides), Vienna: International Atomic Energy Agency, 1984, 265p.

[84GUR/DOR] Gurvich, L.V., Dorofeeva, O.V., Personal communication, Inst. for High Temperatures, Moscow, 1984, cited by Hildenbrand, D.L., Gurvich, L.V., Yungman, L.V., The chemical thermodynamics of actinide elements and compounds, Part 13. The gaseous actinide ions, Vienna: International Atomic Energy Agency, 1985, 187p.

[84HOL/RAN] Holley, C.E., Jr., Rand, M.H., Storms, E.K., The chemical thermodynamics of actinide elements and compounds: Part 6. The actinide carbides, Vienna: International Atomic Energy Agency, 1984, 101p.

[84KOT/EVS] Kotvanova, M.K., Evseev, A.M., Borisova, A.P., Torchenkova, E.A., Zakharov, S.V., Study of uranyl ion hydrolysis reaction equilibria in potassium nitrate solutions, Vest. Mosk. Univ. Khim., **39(6)** (1984) 551-554, in Russian; Engl. transl.: Moscow Univ. Chem. Bull., **39(6)** (1984) 37-40.

[84LAU/HIL] Lau, K.H., Hildenbrand, D.L., Thermochemical studies of the gaseous uranium chlorides, J. Chem. Phys., **80** (1984) 1312-1317.

[84MUL/DUD] Muller, A.B., Duda, L.E., The U-H_2O system: Behavior of dominant aqueous and solid components, Report SAND83-0105, Sandia National Laboratories, Albuquerque, New Mexico, USA, 1984, 167p.

[84NRI] Nriagu, J.O., Phosphate minerals: Their properties and general modes of occurrence, in: Phosphate minerals (Nriagu, J.O., Moore, P.B., *eds.*), Berlin: Springer-Verlag, 1984, *pp.*1-136.

[84NRI2] Nriagu, J.O., Formtion and stability of base metal phosphates in soils and sediments (Nriagu, J.O., Moore, P.B., *eds.*), Berlin: Springer-Verlag, 1984, *pp.*318-329.

[84OHA/MAL] O'Hare, P.A.G., Malm, J.G., Thermochemistry of uranium compounds: XIV. Standard molar enthalpy of formation at 298.15 K of $AgUF_6$ and its relation to the thermodynamics of AgF and AgF_2, J. Chem. Thermodyn., **16** (1984) 753-759.

[84SET/OHA] Settle, J.L., O'Hare, P.A.G., Thermochemistry of inorganic sulfur compounds: III. The standard molar enthalpy of formation at 298.15 K of $US_{1.992}$ (β-uranium disulfide) by fluorine bomb calorimetry, J. Chem. Thermodyn., **16** (1984) 1175-1180.

[84SID/PYA] Sidorova, I.V., Pyatenko, A.T., Gorokhov, L.N., Smirnov, V.K., Determination of the enthalpy of formation of ions UOF_5^- and $UO_2F_3^-$ by the method of ionic-molecular equilibrium, High Temp., **22** (1984) 857-861.

[84SMI/GOR] Smirnov, V.K., Gorokhov, L.N., Thermodynamics of the sublimation and decomposition of uranyl fluoride, Russ. J. Phys. Chem., **58** (1984) 346-348.

[84TOT/FRI] Toth, L.M., Friedman, H.A., Begun, G.M., Dorris, S.E., Raman study of uranyl ion attachment to thorium(IV) hydrous polymer, J. Phys. Chem., **88** (1984) 5574-5577.

[84TRI] Tripathi, V.S., Uranium(VI) transport modeling: Geochemical data and submodels, Ph.D. thesis, Stanford University, Palo Alto, California, 1984.

[84VIE/TAR] Vieillard, P., Tardy, Y., Thermochemical properties of phosphates, in: Phosphate minerals (Nriagu, J.O., Moore, P.B., *eds.*), Berlin: Springer-Verlag, 1984, *pp.*171-198.

[84VOC/GOE] Vochten, R., Goeminne, A., Synthesis, crystallographic data, solubility and electrokinetic properties of meta-zeunerite, meta-kirchheimerite and nickel-uranylarsenate, Phys. Chem. Minerals, **11** (1984) 95-100.

[84WAN] Wanner, H., Bildung und Hydrolyse von Palladium(II)-Komplexen mit Pyridin, 2,2′-Bipyridyl und 1,10-Phenanthrolin, Ph.D. thesis in technical sciences, Swiss Federal Institute of Technology, Zürich, Switzerland, 1984, 183p., in German.

[84WIL/MOR] Williams, C.W., Morss, L.R., Choi, I.K., Stability of tetravalent actinides in perovskites, in: Geochemical behavior of disposed radioactive waste, ACS Symp. Ser., **216** (1984) 323-334.

[85ALD/BRO] Aldridge, J.P., Brock, E.G., Filip, H., Flicker, H., Fox, K., Galbraith, H.W., Holland, R.F., Kim, K.C., Krohn, B.J., Magnuson, D.W., Maier II, W.B., McDowell, R.S., Patterson, C.W., Person, W.B., Smith, D.F., Werner, G.K., Measurement and analysis of the infrared-active stretching fundamental (ν_3) of UF_6, J. Chem. Phys., **83** (1985) 34-48.

[85BAR/COR] Barten, H., Cordfunke, E.H.P., Thermochemical properties of uranyl arsenates: I. Arsenic oxide dissociation pressures, Thermochim. Acta, **90** (1985) 177-187.

[85BRE/REI] Breitung, W., Reil, K.O., Report KfK 3939, Kernforschungszentrum Karlsruhe, Karlsruhe, Federal Republic of Germany, 1985.

[85COR/OUW] Cordfunke, E.H.P., Ouweltjes, W., Standard enthalpies of formation of uranium compounds: XII. Anhydrous phosphates, J. Chem. Thermodyn., **17** (1985) 465-471.

[85COR/OUW2] Cordfunke, E.H.P., Ouweltjes, W., Prins, G., Standard enthalpies of formation of uranium compounds: XI. Lithium uranates(VI), J. Chem. Thermodyn., **17** (1985) 19-22.

[85COR/VLA] Cordfunke, E.H.P., van Vlaanderen, P., Goubitz, K., Loopstra, B.O., Pentauranium(V) chloride dodecaoxide $U_5O_{12}Cl$, J. Solid State Chem., **56** (1985) 166-170.

[85FER/GRE] Ferri, D., Grenthe, I., Hietanen, S., Néker-Neumann, E., Salvatore, F., Studies on metal carbonate equilibria: 12. Zinc(II) carbonate complexes in acid solution, Acta Chem. Scand., **A39** (1985) 347-353.

[85FUG] Fuger, J., Thermochemistry of the alkali metal and alkaline earth-actinide complex oxides, J. Nucl. Mater., **130** (1985) 253-265.

[85GEN/FUG] Gens, R., Fuger, J., Morss, L.R., Williams, C.W., Thermodynamics of actinide perovskite-type oxides: III. Molar enthalpies of formation of Ba_2MAnO_6 (M = Mg, Ca, or Sr; An = U, Np, or Pu) and of M_3PuO_6 (M = Ba or Sr), J. Chem. Thermodyn., **17** (1985) 561-573.

[85GUO/LIU] Guorong, M., Liufang, Z., Chengfa, Z., Investigation of the redox potential UO_2^{2+}/U^{4+} an the complex formation between U^{4+} and NO_3^- in nitric acid, Radiochim. Acta, **38** (1985) 145-147.

[85HIL/GUR] Hildenbrand, D.L., Gurvich, L.V., Yungman, V.S., The chemical thermodynamics of actinide elements and compounds: Part **13**. The gaseous actinide ions, Vienna: International Atomic Energy Agency, 1985, 187p.

[85JOH] Johnson, G.K., The enthalpy of formation of UF_4 by fluorine bomb calorimetry, J. Nucl. Mater., **130** (1985) 102-108.

[85LAU/BRI] Lau, K.H., Brittain, R.D., Hildenbrand, D.L., Complex sublimation/decomposition of uranyl fluoride: Thermodynamics of gaseous UO_2F_2 and UOF_4, J. Phys. Chem., **89** (1985) 4369-4373.

[85LAU/HIL] Lau, K.H., Hildenbrand, D.L., Unpublished data, cited by Hildenbrand, D.L., Gurvich, L.V., Yungman, L.V., The chemical thermodynamics of the actinide elements: Part 13. The gaseous actinide ions, Vienna: International Atomic Energy Agency, 1985, 187p.

[85MAR/FUG] Martinot, L., Fuger, J., The actinides, in: Standard potentials in aqueous solution (Bard, A.J., Parsons, R., Jordan, J., *eds.*), New York: Marcel Dekker, 1985, *pp.*631-674.

[85MOR/SON] Morss, L.R., Sonnenberger, D.C., Enthalpy of formation of americium sesquioxide; systematics of actinide sesquioxide thermochemistry, J. Nucl. Mater., **130** (1985) 266-272.

[85MOR2] Morss, L.R., Thermodynamic systematics of oxides of americium, curium, and neighboring elements, in: Americium and curium chemistry and technology (Edelstein, N.M., Navratil, J.D., Schulz, W.W., *eds.*), Dardrecht, The Netherlands: D. Reidel Publishing Company, 1985, *pp.*147-158.

[85MUL4] Muller, A.B., International chemical thermodynamic data base for nuclear applications, Radioact. Waste Manage. Nucl. Fuel Cycle, **6(2)** (1985) 131-141.

[85OHA2] O'Hare, P.A.G., Thermochemistry of uranium compounds: XV. Calorimetric measurements on UCl_4, UO_2Cl_2, and UO_2F_2, and the standard molar enthalpy of formation at 298.15 K of UCl_4, J. Chem. Thermodyn., **17** (1985) 611-622.

[85OHS/BAB] Ohse, R.W., Babelot, J.F., Cercignani, C., Hiernaut, J.P., Hoch, M., Hyland, G.J., Magill, J., Equation of state of uranium oxide, J. Nucl. Mat., **130** (1985) 165.

[85PAR/POH] Parks, G.A., Pohl, D.C., Hydrothermal solubility of uraninite, Report DOE/ER 12016-1, Stanford University, Stanford, California, USA, 1985, 43p.

[85PAS] Past, V., Antimony, in: Standard potentials in aqueous solution (A.J. Bard, R. Parsons and J. Jordan, eds.), New York: Marcel Dekker, 1985, 172-176.

[85PHI/PHI] Phillips, S.L., Phillips, C.A., Skeen, J., Hydrolysis, formation and ionization constants at 25°C, and at high temperature-high ionic strength, Report LBL-14996, Lawrence Berkeley Laboratory, Berkeley, California, USA, 1985.

[85RAR] Rard, J.A., Chemistry and thermodynamics of ruthenium and some of its inorganic compounds and aqueous species, Chem. Rev., **85(1)** (1985) 1-39.

[85RUD/OTT] Rudigier, H., Ott, H.R., Vogt, O., Low-temperature specific heat of uranium monopnictides and monochalcogenides, Phys. Rev. B, **32** (1985) 4584-4591.

[85RUD/OTT2] Rudigier, H., Ott, H.R., Vogt, O., Low-temperature specific heat of uranium monopnictides and monochalcogenides, Physica, **130B** (1985) 538-540.

[85SAW/CHA] Sawant, R.M., Chaudhuri, N.K., Rizvi, G.H., Patil, S.K., Studies on fluoride complexing of hexavalent actinides using a fluoride ion selective electrode, J. Radioanal. Nucl. Chem., **91** (1985) 41-58.

[85SAW/RIZ] Sawant, R.M., Rizvi, G.H., Chaudhuri, N.K., Patil, S.K., Determination of the stability constant of Np(V) fluoride complex using a fluoride ion selective electrode, J. Radioanal. Nucl. Chem., Articles, **89(2)** (1985) 373-378.

[85SCH/FRI] Schreiner, F., Friedman, A.M., Richards, R.R., Sullivan, J.C., Microcalorimetric measurement or reaction enthalpies in solutions of uranium and neptunium compounds, J. Nucl. Mater., **130** (1985) 227-233.

[85SJO/OEH] Sjöberg, S., Öhman, L.O., Ingri, N., Equilibrium and structural studies of silicon(IV) and aluminium(III) in aqueous solution: 11. Polysilicate formation in alkaline aqueous solution. A combined potentiometric and ^{29}Si NMR study, Acta Chem. Scand., **A39** (1985) 93-107.

[85TAC] Tachikawa, H., Lithium, sodium, potassium, rubidium, cesium and francium, in: Standard potentials in aqueous solution (A.J. Bard, R. Parsons and J. Jordan, eds.), New York: Marcel Dekker, 1985, 727-762.

[85TSO/BRO] Tso, T.C., Brown, D., Judge, A.I., Halloway, J.H., Fuger, J., Thermodynamics of the actinoid elements: Part 6. The preparation and heats of formation of some sodium uranates(VI), J. Chem. Soc. Dalton Trans., (1985) 1853-1858.

[85WEI] Weigel, F., The carbonates, phosphates and arsenates of the hexavalent and pentavalent actinides, in: Handbook on the Physics and Chemistry of the Actinides, Vol. **3** (Freeman, A.J., Keller, C., *eds.*), Amsterdam: Elsevier Science Publishers 1985, *pp.*243-288.

[86BAR] Barten, H., Thermochemical investigations on uranyl phosphates and arsenates, Ph.D. thesis, ECN-188, Netherlands Energy Research Foundation, Petten, The Netherlands, 1986, 145p.

[86BRA/LAG] Bratsch, S.G., Lagowski, J.J., Actinide thermodynamic predictions: 3. Thermodynamics of compounds and aquo ions of the 2+, 3+ and 4+ oxidation states and standard electrode potentials at 298.15 K, J. Phys. Chem., **90** (1986) 307-312.

[86BRU/FER] Bruno, J., Ferri, D., Grenthe, I., Salvatore, F., Studies on metal carbonate equilibria: 13. On the solubility of uranium(IV) dioxide, $UO_2(s)$, Acta Chem. Scand., **40** (1986) 428-434.

[86BRU/GRE] Bruno, J., Grenthe, I., Muñoz, M., Studies on the radionuclide coprecipitation-solid solution formation: The $UO_2(s)$-$La(OH)_3(s)$ coprecipitation as an analog for the $UO_2(s)$-$Pu(OH)_3(s)$ system, Sci. Basis Nucl. Waste Management IX, held 9-11 September, 1985, in Stockholm, Mat. Res. Soc. Symp. Proc., **50** (1986) 717-728.

[86BRU/GRE2] Bruno, J., Grenthe, I., Lagerman, B., Redox processes and $UO_2(s)$ solubility: The determination of the UO_2^{2+}/U^{4+} redox potential at 25°C in $HClO_4$ media of different strength, Sci. Basis Nucl. Waste Management IX, held 9-11 September, 1985, in Stockholm, Mat. Res. Soc. Symp. Proc., **50** (1986) 299-308.

[86BRU] Bruno, J., Stoichiometric and structural studies on the Be^{2+}-H_2O-$CO_2(g)$ system, Ph.D. thesis, The Royal Institute of Technology, Stockholm, Sweden, 1986.

[86BUS/PLU] Busenberg, E., Plummer, L.N., The solubility of $BaCO_3(cr)$ (witherite) in CO_2-H_2O solutions between 0 and 90°C, evaluation of the association constants of $BaHCO_3^+(aq)$ and $BaCO_3(aq)$ between 5 and 80°C, and a preliminary evaluation of the thermodynamic properties of $Ba^{2+}(aq)$, Geochim. Cosmochim. Acta, **50** (1986) 2225-2233.

[86COD] CODATA recommended values for the fundamental constants 1986, Committee on Data for Science and Technology (CODATA), Newsletter Nr. 38, Paris: International Council of Scientific Unions, 1986, 12p.

[86COR/OUW] Cordfunke, E.H.P., Ouweltjes, W., Prins, G., Standard enthalpies of formation of uranium compounds: XIII. Cs_2UO_4, J. Chem. Thermodyn., **18** (1986) 503-509.

[86COR/PRI] Cordfunke, E.H.P., Prins, G., The standard enthalpy of formation of UP_2O_7: A correction, J. Chem. Thermodyn., **18** (1986) 501.

[86DES/KRA] Deschenes, L.L., Kramer, G.H., Monserrat, K.J., Robinson, P.A., The potentiometric and laser raman study of the hydrolysis of uranyl chloride under physiological conditions and the effect of systematic and random errors on the hydrolysis constants, Report AECL-9266, Atomic Energy of Canada Ltd., Chalk River, Ontario, Canada, 1986, 23p.

[86GRE/RIG] Grenthe, I., Riglet, C., Vitorge, P., Studies of metal-carbonate complexes: 14. Composition and equilibria of trinuclear neptunium(VI)- and plutonium(VI)-carbonates complexes, Inorg. Chem., **25** (1986) 1679-1684.

[86HIS/BEN] Hisham, M.W.M., Benson, S.W., Thermochemistry of inorganic solids: 3. Enthalpies of formation of solid metal oxyhalide compounds, J. Phys. Chem., **90** (1986) 885-888.

[86LIC/HOD] Licht, S., Hodes, G., Manassen, J., Numerical analysis of aqueous polysulfide solutions and its application to cadmiun chalcogen/polysulfide photoelectrochemical solar cells, Inorg. Chem., **25** (1986) 2486-2489.

[86MOR] Morss, L.R., Thermodynamic properties, in: The chemistry of the actinide elements, 2*nd* ed. (Katz, J.J., Seaborg, G.T., Morss, L.R., *eds.*), Vol. **2**, London: Chapman and Hall, 1986, *pp.*1278-1360.

[86MYE] Myers, R.J., The new low value for the second dissociation constant for H_2S, J. Chem. Educ., **63** (1986) 687-690.

[86POW] Powder Diffraction File, Alphabetic index: Inorganic phases, International Centre for Diffraction Data, Swarthmore, Pennsylvania, USA, 1986.

[86SHE/TOT] Sherrow, S.A., Toth, L.M., Begun, G.M., Raman spectroscopic studies of UO_2^{2+} association with hydrolytic species of Group IV (4) metals, Inorg. Chem., **25** (1986) 1992-1996.

[86SUG/CHI] Suglobova, I.G., Chirkst, D.E., Enthalpy of formation of binary uranium bromides, Radiokhimiya, **28(5)** (1986) 569-572, in Russian; Engl. transl.: Sov. Radiochem., **28(5)** (1986) 513-516.

[86ULL/SCH] Ullman, W.J., Schreiner, F., Calorimetric determination of the stability of U(VI)-, Np(VI)-, and Pu(VI)-SO_4^{2-} complexes in aqueous solution at 25°C, Radiochim. Acta, **40** (1986) 179-183.

[86VOC/GRA] Vochten, R., de Grave, E., Pelsmaekers, J., Synthesis, crystallographic and spectroscopic data, solubility and electrokinetic properties of metakahlerite and its Mn analogue, Am. Mineral., **71** (1986) 1037-1044.

[86WAN] Wanner, H., Modelling interaction of deep groundwaters with bentonite and radionuclide speciation, EIR-Bericht Nr. 589 and Nagra NTB 86-21, National Cooperative for the Storage of Radioactive Waste (Nagra), Baden, Switzerland, 1986, 103p.

[86WEI] Weigel, F., Uranium, in: The chemistry of the actinide elements, 2*nd* ed. (Katz, J.J., Seaborg, G.T., Morss, L.R., *eds.*), Vol. **1**, London: Chapman and Hall, 1986, *pp.*169-442.

[87BAR] Barten, H., Chemical stability of arsenates of uranium with a valency lower than six, with emphasis on uranium(V) arsenic oxide $UAsO_5$, Thermochim. Acta, **118** (1987) 107-110.

[87BOB/SIN] Bober, M., Singer, J., Vapor pressure determination of liquid UO_2 using a boiling point technique, Nucl. Sci. Eng., **97** (1987) 344.

[87BON/KOR] Bondarenko, A.A., Korobov, M.V., Sidorov, L.N., Karasev, N.M., Enthalpy of formation of gaseous uranium pentafluoride, Russ. J. Phys. Chem., **61** (1987) 1367-1370.

[87BRO/WAN] Brown, P.L., Wanner, H., Predicted formation constants using the unified theory of metal ion complexation, Paris: OECD Nuclear Energy Agency, 1987, 102p.

[87BRU/CAS] Bruno, J., Casas, I., Lagerman, B., Muñoz, M., The determination of the solubility of amorphous $UO_2(s)$ and the mononuclear hydrolysis constants of uranium(IV) at 25°C, Sci. Basis Nucl. Waste Management X, held 1-4 December, 1986, in Boston, Massachusetts, Mat. Rec. Soc. Symp. Proc., **84** (1987) 153-160.

[87BRU/CAS2] Bruno, J., Casas, I., Grenthe, I., Lagerman, B., Studies on metal carbonate equilibria: 19. Complex formation in the Th(IV)-H_2O-$CO_2(g)$ system, Inorg. Chim. Acta, **140** (1987) 299-301.

[87BRU/GRE] Bruno, J., Grenthe, I., Sandström, M., Studies of metal-carbonate equilibria: Part 15. The Be^{2+}-H_2O-$CO_2(g)$ system in acidic 3.0 mol/dm^3 media, J. Chem. Soc. Dalton Trans., (1987) 2439-2444.

[87BRU] Bruno, J., Beryllium(II) hydrolysis in 3.0 mol·dm^{-3} perchlorate, J. Chem. Soc. Dalton Trans., (1987) 2431-2437.

[87CIA/IUL] Ciavatta, L., Iuliano, M., Porto, R., The hydrolysis of the La(III) ion in aqueous perchlorate solution at 60°C, Polyhedron, **6** (1987) 1283-1290.

[87DUB/RAM] Dubessy, J., Ramboz, C., Nguyen-Trung, C., Cathelineau, M., Charoy, B., Cuney, M., Leroy, J., Poty, B., Weisbrod, A., Physical and chemical controls (f_{O_2}, T, pH) of the opposite behaviour of U and Sn-W as exemplified by hydrothermal deposits in France and Great-Britain, and solubility data, Bull. Mineral., **110** (1987) 261-281.

[87FIS] Fischer, E.A., Evaluation of the urania equation of state based on recent vapour pressure measurements, Report KfK 4087, Kernforschungszentrum Karlsruhe, Karlsruhe, Federal Republic of Germany, 1987.

[87GAR/PAR] Garvin, D., Parker, V.B., White, H.J., Jr., CODATA thermodynamic tables: Selections for some compounds of calcium and related mixtures: A prototype set of tables, Washington, D.C.: Hemisphere Publishing Corp., 1987, 356p.

[87GUR/SER] Gurevich, V.M., Sergeyeva, E.I., Gavrichev, K.S., Gorbunov, V.E., Khodakovsky, I.L., Low-temperature specific heat of UO_2CO_3, Russ. J. Phys. Chem., **61(6)** (1987) 856-857.

[87HIS/BEN] Hisham, M.W.M., Benson, S.W., Thermochemistry of inorganic solids: 7. Empirical relations among enthalpies of formation of halides, J. Phys. Chem., **91** (1987) 3631-3637.

[87LAU/HIL] Lau, K.H., Hildenbrand, D.L., Thermochemistry of the gaseous uranium bromides UBr through UBr_5, J. Phys. Chem., **86** (1987) 2949-2954.

[87LIC/MAN] Licht, S., Manassen, J., The second dissociation constant of H_2S, J. Electrochem. Soc., **134** (1987) 918-921.

[87MAT/OHS] Matsui, T., Ohse, R.W., Thermodynamic properties of uranium nitride, plutonium nitride and uranium-plutonium mixed nitride, High Temp. High Pressures, **19** (1987) 1-17.

[87RIG/VIT] Riglet, C., Vitorge, P., Grenthe, I., Standard potentials of the (MO_2^{2+}/ MO_2^+) systems for uranium and other actinides, Inorg. Chim. Acta, **133** (1987) 323-329.

[87SUG/CHI] Suglobova, I.G., Chirkst, D.E., Enthalpies of formation for binary uranium chlorides, Sov. Radiochem., **29** (1987) 125-132.

[87THE] Thermodata, Evaluated data, Domaine Universitaire de Grenoble, St. Martin d'Hères, France, 1987.

[87VIL] Viljoen, C.L., Hydroxo-species of uranium(VI) in aqueous media, M.Sc. dissertation, University of Port Elizabeth, South Africa, 1987, 104p.

[87WIL] Willis, B.T.M., Crystallographic studies of anion-excess uranium oxides, J. Chem. Soc. Faraday Trans. 2, **83** (1987) 1073-1081.

[88ALI/MAL] Alikhanyan, A.S., Malkerova, I.P., Sevast'yanov, V.G., Yuldashev, F., Gorgoraki, V.I., Thermodynamic properties of lower bromides and chlorobromides of uranium UBr_3, UBr_4, UCl_3Br, UCl_2Br_2, $UClBr_3$, High Temp., trans. from Vys. Vesh., **1** (1988) 85-90.

[88BAR/BUR] Barnes, D.S., Burkinshaw, P.M., Mortimer, C.T., Enthalpy of combustion of triphenylarsine oxide, Thermochim. Acta, **131** (1988) 107-113.

[88BAR] Barten, H., Comparison of the thermochemistry of uranyl/uranium phosphates and arsenates, Thermochim. Acta, 124 (1988) 339-344.

[88BAR2] Barten, H., Thermodynamical properties of uranyl arsenates: II. Derivations of standard entropies, Thermochim. Acta, **126** (1988) 385-391.

[88BAR3] Barten, H., The thermochemistry of uranyl phosphates, Thermochim. Acta, **126** (1988) 375-383.

[88BOR/BOL] Borshchevskii, A.Ya., Boltalina, O.V., Sorokin, I.D., Sidorov, L.N., Thermochemical quantities for gas-phase iron, uranium, and molybdenum fluorides, and their negative ions, J. Chem. Thermodyn., **20** (1988) 523-537.

[88BRU] Bruno, J., Private communication, The Royal Institute of Technology, Stockholm, Sweden, 1988.

[88BUR/TO] Burriel, R., To, M.W.K., Zainel, H.A., Westrum, E.F., Jr., Thermodynamics of uranium intermetallic compounds: II. Heat capacity of UPd_3 from 8 to 850 K, J. Chem. Thermodyn., **20** (1988) 815-823.

[88CIA] Ciavatta, L., Private communication, Università de Napoli, Naples, Italy, June 1988.

[88COL/BES] Colinet, C., Bessoud, A., Pasturel, A., Enthalpies of formation of rare earth and uranium tin compounds, J. Less-Common Met., **143** (1988) 265-278.

[88COR/IJD] Cordfunke, E.H.P., IJdo, D.J.W., $Ba_2U_2O_7$: crystal structure and phase relationships, J. Phys. Chem. Solids, **49** (1988) 551-554.

[88COR/OUW] Cordfunke, E.H.P., Ouweltjes, W., Standard enthalpies of formation of uranium compounds: XIV. BaU_2O_7 and $Ba_2U_2O_7$, J. Chem. Thermodyn., **20** (1988) 235-238.

[88COR/WES] Cordfunke, E.H.P., Westrum, E.F., Jr., The thermodynamic properties of β-UO_3 and γ-UO_3, Thermochim. Acta, **124** (1988) 285-296.

[88COW] Cowan, C.E., Review of selenium thermodynamic data, Report EPRI EA-5655, Batelle, Pacific Northwest Laboratories, Richland, Washington, USA, 1988.

[88CRO/EWA] Cross, J.E., Ewart, F.T., Harwell/Nirex Thermodynamic Database for Chemical Equilibrium Studies, a compiled data base using the Ashton-Tate DBASE(III)plus software on a PC-DOS/MS-DOS computer, Harwell Laboratory, Didcot, United Kingdom, 1988.

[88DUB/LEL] Dubois, P., Lelieur, J.P., Lepoutre, G., Identification and characterization of ammonium polysulfides in solution in liquid ammonia, Inorg. Chem., **27** (1988) 1883-1890.

[88EZH/KOM] Ezhov, Yu.S., Komarov, S.A., Mikulinskaya, N.M., Electron diffraction by uranium tetrachloride, Zh. Strukt. Khim., **29** (1988) 42-45; Engl. transl.: J. Struct. Chem., **29** (1989) 692-694.

[88FER/GLA] Ferri, D., Glaser, J., Grenthe, I., Confirmation of the structure of $(UO_2)_3(CO_3)_6^{6-}$ by ^{17}O-NMR, Inorg. Chim. Acta, **148** (1988) 133-134.

[88HED] Hedlund, T., Studies of complexation and precipitation equilibria in some aqueous aluminium(III) systems, Ph.D. thesis, University of Umeå, Umeå, Sweden, 1988.

[88JAC/WOL] Jackson, K.J., Wolery, T.J., Bourcier, W.L., Delany, J.M., Moore, R.M., Clinnick, M.L., Lundeen, S.R., MCRT user's guide and documentation, UCID-21406 Rev. 1, Lawrence Livermore National Laboratory, Livermore, California, USA, 1988, 91p.

[88KHA/CHO] Khalili, F.I., Choppin, G.R., Rizkalla, E.N., The nature of U(VI) complexation by halates and chloroacetates, Inorg. Chim. Acta, **143** (1988) 131-135.

[88LEM] Lemire, R.J., Effects of high ionic strength groundwaters on calculated equilibrium concentrations in the uranium-water system, Report AECL-9549, Atomic Energy of Canada Ltd., Pinawa, Manitoba, Canada, 1988, 40p.

[88OHA/LEW] O'Hare, P.A.G., Lewis, B.M., Nguyen, S.N., Thermochmemistry of uranium compounds: XVII. Standard molar enthalpy of formation at 298.15 K of dehydrated schoepite $UO_3{\cdot}0.9H_2O$. Thermodynamics of (schoepite + dehydrated schoepite + water), J. Chem. Thermodyn., **20** (1988) 1287-1296.

[88PAR/POH] Parks, G.A., Pohl, D.C., Hydrothermal solubility of uraninite, Geochim. Cosmochim. Acta, **52** (1988) 863-875.

[88PHI/HAL] Phillips, S.L., Hale, F.V., Silvester, L.F., Siegel, M.D., Thermodynamic tables for nuclear waste isolation: Aqueous solution database, Vol. **1**, Report NUREG/CR-4864, US Nuclear Regulatory Commission, Washington, D.C., Report LBL-22860, Lawrence Berkeley Laboratory, Berkeley, California, USA, SAND87-0323, Sandia National Laboratories, Albuquerque, New Mexico, USA, 1988, 181p.

[88PHI/HAL2] Phillips, S.L., Hale, F.V., Silvester, L.F., Siegel, M.D., Thermodynamic tables for nuclear waste isolation, an aqueous solutions database, retrieval using on-line services, Lawrence Berkeley Laboratory, Berkeley, California, USA,1988.

[88SCH/BAR] Schoonen, M.A.A., Barnes, H.L., An approximation of the second dissociation constant for H_2S, Geochim. Cosmochim. Acta, **52** (1988) 649-654.

[88SHO/HEL] Shock, E.L., Helgeson, H.C., Calculation of the thermodynamic and transport properties of aqueous species at high pressures and temperatures: Correlation algorithms for ionic species and equation of state predictions to 5 kb and 1000°C, Geochim. Cosmochim. Acta, **52** (1988) 2009-2036.

[88TAN/HEL] Tanger IV, J.C., Helgeson, H.C., Calculation of the thermodynamic and transport properties of aqueous species at high pressures and temperatures: Revised equations of state for the standard partial molal properties of ions and electrolytes, Am. J. Sci., **288** (1988) 19-98.

[88TAS/OHA] Tasker, I.R., O'Hare, P.A.G., Lewis, B.M., Johnson, G.K., Cordfunke, E.H.P., Thermochemistry of uranium compounds: XVI. Calorimetric determination of the standard molar enthalpy of formation at 298.15 K, low-temperature heat capacity, and high-temperature enthalpy increments of $UO_2(OH)_2 \cdot H_2O$ (schoepite), Can. J. Chem., **66** (1988) 620-625.

[88ULL/SCH] Ullman, W.J., Schreiner, F., Calorimetric determination of the enthalpies of the carbonate complexes of U(VI), Np(VI), and Pu(VI) in aqueous solution at 25°C, Radiochim. Acta, **43** (1988) 37-44.

[88WAN] Wanner, H., The NEA Thermochemical Data Base Project, Radiochim. Acta, **44/45** (1988) 325-329.

[89BAZ/KOM] Bazhanov, V.I., Komarov, S.A., Ezhov, Yu.S., Structure of the uranium tetrachloride molecule, Zh. Fiz. Khim., **64** (1989) 2247-2249; Engl. transl.: Russ. J. Phys. Chem., **63** (1989) 1234-1235.

[89BRU/GRE] Bruno, J., Grenthe, I., Robouch, P., Studies of metal carbonate equilibria: 20. Formation of tetra(carbonato)uranium(IV) ion, $U(CO_3)_4^{4-}$, in hydrogen carbonate solutions, Inorg. Chim. Acta, **158** (1989) 221-226.

[89BRU/PUI] Bruno, J., Puigdomenech, I., Validation of the SKBU1 uranium thermodynamic data base for its use in geochemical calculations with EQ3/6,

Sci. Basis Nucl. Waste Management XII, held 10-13 October, 1988, in Berlin (West), Mat. Res. Soc. Symp. Proc., **127** (1989) 887-896.

[89BRU/SAN] Bruno, J., Sandino, A., The solubility of amorphous and crystalline schoepite in neutral to alkaline solutions, Sci. Basis Nucl. Waste Management XII, held 10-13 October, 1988, in Berlin (West), Mat. Res. Soc. Symp. Proc., **127** (1989) 871-878.

[89COR/KON] Cordfunke, E.H.P., Konings, R.J.M., Westrum, E.F., Jr., Recent thermochemical research on reactor materials and fission products, J. Nucl. Mat., **167** (1989) 205.

[89COX/WAG] Cox, J.D., Wagman, D.D., Medvedev, V.A., CODATA Key Values for Thermodynamics, New York: Hemisphere Publishing Corp., 1989, 271p.

[89EZH/BAZ] Ezhov, Yu.S., Bazhanov, V.I., Komarov, S.A., Popik, M.S., Sevast'yanov, V.G., Yuldachev, F., Molecular structure of uranium tetrabromide, Zh. Fiz. Khim., **63(11)** (1989) 3094-3097, in Russian.

[89GRE/BID] Grenthe, I., Bidoglio, G., Omenetto, N., Use of thermal lensing spectrophotometry (TLS) for the study of mononuclear hydrolysis of uranium(IV), Inorg. Chem., **28** (1989) 71-74.

[89GRE/WAN] Grenthe, I., Wanner, H., Guidelines for the extrapolation to zero ionic strength, TDB-2.1 (Revision 1), OECD Nuclear Energy Agency, Data Bank, Gif-sur-Yvette, France, September 1989, 23p.

[89HOV/NGU] Hovey, J.K., Nguyen-Trung, C., Tremaine, P.R., Thermodynamics of aqueous uranyl ion: Apparent and partial molar heat capacities and columes of aqueous uranyl perchlorate from 10 to 55°C, Geochim. Cosmochim. Acta, **53** (1989) 1503-1509.

[89RED/SAV] Red'kin, A.F., Savelyeva, N.I., Sergeyeva, E.I., Omelyanenko, B.I., Ivanov, I.P., Khodakovsky, I.L., Investigation of uraninite (UO_2(c)) solubility under hydrothermal conditions, Sci. Geol. Bull., **42(4)** (1989) 329-334.

[89RIG/ROB] Riglet, C., Robouch, P., Vitorge, P., Standard potentials of the (MO_2^{2+}/ MO_2^{+}) and (M^{4+}/M^{3+}) redox systems for neptunium and plutonium, Radiochim. Acta, **46** (1989) 85-94.

[89SCH/GEN] Schoebrechts, J.P., Gens, R., Fuger, J., Morss, L.R., Thermochemical and structural studies of complex actinide chlorides $Cs_2NaAnCl_6$ (An = U-Cf); ionization potentials and hydration enthalpies of An^{3+} ions, Thermochim. Acta, **139** (1989) 49-66.

[89SHO/HEL] Shock, E.L., Helgeson, H.C., Corrections to [88SHO/HEL], Geochim. Cosmochim. Acta, **53** (1989) 215.

[89SHO/HEL2] Shock, E.L., Helgeson, H.C., Sverjensky, D.A., Calculation of the thermodynamic and transport properties of aqueous species at high pressures and temperatures: Standard partial molal properties of inorganic neutral species, Geochim. Cosmochim. Acta, **53** (1989) 2157-2183.

[89WAN2] Wanner, H., Guidelines for the review procedure and data selection, TDB-1, OECD Nuclear Energy Agency, Data Bank, Gif-sur-Yvette, France, March 1989, 5p.

[89WAN3] Wanner, H., Guidelines for the assignment of uncertainties, TDB-3, OECD Nuclear Energy Agency, Data Bank, Gif-sur-Yvette, France, March 1989, 14p.

[89WAN4] Wanner, H., Standards and conventions for TDB publications, TDB-5.1 (Revision 1), OECD Nuclear Energy Agency, Data Bank, Gif-sur-Yvette, France, September 1989, 24p.

[89YAN/PIT] Yang, J., Pitzer, K.S., Thermodynamics of aqueous uranyl sulfate to 559 K, J. Solution Chem., **18** (1989) 189-199.

[90AHR/HEF] Ahrland, S., Hefter, G.T., Norén, B., A calorimetric study of the mononuclear fluoride complexes of zirkonium(IV), hafnium(IV), thorium(IV) and uranium(IV), Acta Chem. Scand., **44** (1990) 1-7.

[90BRU/GRE] Bruno, J., Grenthe, I., Lagerman, B., The redox potential of U(VI)/U(IV), Acta Chem. Scand., **44** (1990) 896-901.

[90BRU/SAN] Bruno, J., Sandino, A., Fredlund, F., The solubility of $(UO_2)_3(PO_4)_2 \cdot 4H_2O(s)$ in neutral and alkaline media, Radiochim. Acta, (1990) in press.

[90CAP/VIT] Capdevila, H., Vitorge, P., Temperature and ionic strength influence on U(VI/V) and U(IV/III) redox potentials in aqueous acidic and carbonate solutions, J. Radioanal. Nucl. Chem., Articles, **143** (1990) 403-414.

[90CIA] Ciavatta, L., The specific interaction theory in equilibrium analysis: Some empirical rules for estimating interaction coefficients of metal ion complexes, Ann. Chim. Rome, **80** (1990) 255-263.

[90COR/KON] Cordfunke, E.H.P., Konings, R.J.M., Potter, P.E., Prins, G., Rand, M.H., Thermochemical data for reactor materials and fission products (Cordfunke, E.H.P., Konings, R.J.M., *eds.*), Amsterdam: North-Holland, 1990, 695p.

[90COR/KON2] Cordfunke, E.H.P., Konings, R.J.M., Ouweltjes, W., The standard enthalpies of formation of MO(s), $MCl_2(s)$, and $M^{2+}(aq, \infty)$, (M = Ba, Sr), J. Chem. Thermodyn., **22** (1990) 991-996.

[90DIC/WES] Dickson, A.G., Wesolowski, D.J., Palmer, D.A., Mesmer, R.E., Dissociation constant of bisulfate ion in aqueous sodium chloride solutions to 250°C, J. Phys. Chem., **94** (1990) 7978-7985.

[90EIK] Eikenberg, J., On the problem of silica solubility at high pH, PSI-Bericht Nr. 74, Paul Scherrer Institut, Würenlingen, Switzerland, 1990, 54p.

[90ELY/BRI] Elyahyaoui, A., Brillard, L., Boulhassa, S., Hussonnois, M., Guillaumont, R., Complexes of thorium with phosphoric acid, Radiochim. Acta, **49** (1990) 39-44.

[90GUR] Gurvich, L.V., Reference books and data banks on the thermodynamic properties of inorganic substances, High Temp. Sci., **26** (1990) 197-206.

[90HAY/THO] Hayes, S.L., Thomas, J.K., Peddicord, K.L., Material property correlations for uranium mononitride IV. Thermodynamic properties, J. Nucl. Mater., **171** (1990) 300-318.

[90LYC/MIK] Lychev, A.A., Mikhalev, V.A., Suglobov, D.N., Phase equilibria in the system uranyl fluoride-water-acetone at 20°C, Radiokhimiya, **32(3)** (1990) 25-43, in Russian; Engl. transl.: Sov. Radiochem., **32** (1990) 158-164.

[90NGU/HOV] Nguyen-Trung, C., Hovey, J.K., Thermodynamics of complexed aqueous uranyl species. 1. Volume and heat capacity changes associated with the formation of uranyl sulfate from 10 to 55°C and calculation of the ion-pair equilibrium constant to 175°C, J. Phys. Chem., **94** (1990) 7852-7865.

[90RAI/FEL] Rai, D., Felmy, A.R., Ryan, J.L., Uranium(IV) hydrolysis constants and solubility product of $UO_2 \cdot xH_2O(am)$, Inorg. Chem., **29** (1990) 260-264.

[90RIG] Riglet, C., Chimie du neptunium et autres actinides en milieu carbonate, Report CEA-R-5535, Commissariat à l'Energie Atomique, Gif-sur-Yvette, France, 1990, 267p., in French.

[90WAN] Wanner, H., The NEA Thermochemical Data Base project, TDB-0.1 (Revision 1), OECD Nuclear Energy Agency, Data Bank, Gif-sur-Yvette, France, January 1990, 13p.

[90WAN2] Wanner, H., Guidelines for the independent peer review of TDB reports, TDB-6, OECD Nuclear Energy Agency, Data Bank, Gif-sur-Yvette, France, January 1990, 6p.

[90ZAR/NEM] Zarubin, D.P., Nemkina, N.V., The solubility of amorphous silica in alkaline aqueous solutions at constant ionic strength, Zh. Neorg. Khim., **35** (1990) 31-38.

[91BID/CAV] Bidoglio, G., Cavalli, P., Grenthe, I., Omenetto, N., Qi, P., Tanet, G., Studies on metal carbonate equilibria, Part 20: Study of the U(VI)-H_2O-CO_2(g) system by thermal lensing spectrophotometry, Talanta, **38** (1991) 433-437.

[91CHO/MAT] Choppin, G.R., Mathur, J.N., Hydrolysis of actinyl(VI) cations, Radiochim. Acta, **53/54** (1991) 25-28.

[91COR/KON] Cordfunke, E.H.P., Konings, R.J.M., The vapour pressure of UCl_4, J. Chem. Thermodyn., **23** (1991) 1121-1124.

[91GRE/LAG] Grenthe, I., Lagerman, B., Studies on the metal carbonate equilibria: 22. A coulometric study of the uranium(VI)-carbonate system, the composition of the mixed hydroxide carbonate species, Acta Chem. Scand., **45** (1991) 122-128.

[91GRE/LAG2] Grenthe, I., Lagerman, B., Studies on metal carbonate equilibria: 23. Complex formation in the Th(IV)-H_2O-CO_2(g) system, Acta Chem. Scand., **45** (1991) 231-238.

[91HIL/LAU] Hildenbrand, D.L., Lau, K.H., Brittain, R.D., The entropies and probable symmetries of the gaseous thorium and uranium tetrahalides, J. Chem. Phys., (1991), in press.

[91HIL/LAU2] Hildenbrand, D.L., Lau, K.H., Redermination of the thermochemistry of gaseous UF_5, UF_2 and UF, J. Chem. Phys., **94** (1991) 1420-1425.

[91MAT] Mathur, J.N., Complexation and thermodynamics of the uranyl ion with phosphate, Polyhedron, **10** (1991) 47-53.

[91PUI/RAR] Puigdomenech, I., Rard, J.A., Wanner, H., Guidelines for temperature corrections, TDB-4, OECD Nuclear Energy Agency, Data Bank, Gif-sur-Yvette, France, Draft of June 1991, 39p.

[91SAN] Sandino, A., Processes affecting the mobility of uranium in natural waters, Ph.D. thesis in inorganic chemistry, The Royal Institute of Technology, Stockholm, Sweden, 1991.

Chapter VIII

Authors list

This chapter contains an alphabetical list of the authors of the references cited in this book, *cf.* Chapter VII. The reference codes given with each name corresponds to the publications of which the person is the author or a co-author. Note that inconsistencies may occur due to different transliterations. For example, Efimov, M.E., and Yefimov, M.Y., refer to the same person. Most of these inconsistencies are not corrected in this review.

Author	References
Abbrent, M.	[65VES/PEK]
Abdel-Rassoul, A.A.	[75ALY/ABD]
Aberg, M.	[70ABE], [71ABE], [83ABE/FER]
Abraham, B.M.	[55ABR/FLO], [59FLO/LOH], [60ABR/OSB]
Accary, A.	[68DOG/VAL]
Achachinsky, V.V.	[63ACH/ALC]
Ackermann, R.J.	[76OET/RAN]
Adachi, T.	[81FUJ/TAG]
Ader, M.	[75OHA/ADE]
Agafonova, A.L.	[50PAM/AGA]
Agron, P.A.	[58AGR]
Ahmad, F.	[78THA/AHM]
Ahrland, S.	[49AHR], [49AHR2], [51AHR], [51AHR2], [54AHR/HIE], [54AHR/LAR], [54AHR/LAR2], [56AHR/LAR], [67AHR], [68AHR], [71AHR/KUL], [71AHR/KUL2], [71AHR/KUL3], [90AHR/HEF]
Akhachinskiy, V.V.	[81CHI/AKH]
Akishin, P.A.	[61AKI/KHO], [69EZH/AKI]
Alcock, C.B.	[61ALC/GRI], [62ALC/GRI], [63ACH/ALC], [79KUB/ALC]
Alcock, K.	[56DAW/WAI]
Aldridge, J.P.	[85ALD/BRO]
Alexander, C.A.	[69ALE/OGD], [71ALE/OGD]
Alikhanyan, A.S.	[88ALI/MAL]
Aling, P.	[63COR/ALI], [65COR/ALI]
Allard, B.	[78ALL/BEA], [80ALL/KIP], [83ALL]
Allen, K.A.	[58ALL]
Alles, A.	[77ALL/FAL]
Almagro, V.	[73ALM/GAR]
Aloy, J.	[1896ALO]
Alter, H.W.	[51ZEB/ALT]
Altman, D.	[43ALT], [43ALT/LIP], [43ALT2], [44ALT]
Alwan, A.A.	[80ALW/WIL]
Aly, H.F.	[75ALY/ABD]
Alyamovskaya, K.V.	[61CHU/ALY]
Amaya, T.	[72MAE/AMA]
Amira, M.F.	[78MON/AMI]
Amis, E.S.	[60HEF/AMI], [63MEL/AMI]
Amphlett, C.B.	[41AMP/THO], [48AMP/MUL]
Anderson, O.E.	[42JEN/AND]
Anderson, P.D.	[63HUL/ORR]
Andon, R.J.L.	[64AND/COU]
Ansara, I.	[81CHI/AKH]
Antipina, V.A.	[68NIK/ANT]
Appleman, D.E.	[57COL/APP]
Appleman, E.H.	[70JOH/SMI]
Argue, G.R.	[61ARG/MER]
Arnaud-Neu, F.	[79SPI/ARN]

Author	References
Arnek, R.	[68ARN/SCH], [72ARN/BAR]
Asmussen, R.W.	[51MUU/REF]
Aurov, N.A.	[83AUR/CHI]
Awad, A.M.	[61SHE/AWA], [62SHE/AWA], [62SHE/AWA2]
Awasthi, S.P.	[81AWA/SUN]
Babelot, J.F.	[85OHS/BAB]
Babko, A.K.	[60BAB/KOD]
Baes, C.F., Jr.	[53BAE/SCH2], [54SCH/BAE], [55SCH/BAE], [56BAE], [56BAE2], [62BAE/MEY], [65BAE/MEY], [74MES/BAE], [76BAE/MES], [81BAE/MES]
Bagnall, K.W.	[83BAG/JAC]
Bai, M.	[80HU/NI]
Bailey, A.R.	[71BAI/LAR]
Bailey, S.M.	[68WAG/EVA], [82WAG/EVA]
Baker, F.B.	[65NEW/BAK]
Bale, W.D.	[57BAL/DAV]
Bandyopadhyay, N.	[72CHA/BAN]
Banerjea, D.	[61BAN/TRI]
Baran, V.	[70BAR]
Baranova, L.I.	[52KAP/BAR]
Barber, C.M.	[66WES/BAR]
Barcza, L.	[71BAR/SIL], [72ARN/BAR]
Barin, I.	[73BAR/KNA], [77BAR/KNA]
Barinov, V.I.	[81GOL/TRE]
Barkelew, C.	[45BAR]
Barnes, D.S.	[88BAR/BUR]
Barnes, H.L.	[88SCH/BAR]
Barnes, R.D.	[70FRE/BAR]
Barnett, M.K.	[56ORB/BAR]
Barten, H.	[80BAR/COR2], [82COR/OUW2], [85BAR/COR], [86BAR], [87BAR], [88BAR], [88BAR2], [88BAR3]
Bartušek, M.	[63BAR/SOM], [64BAR/SOM], [67BAR], [71POL/BAR], [75BAR/SEN]
Baskin, Y.	[70BAS/SMI]
Battles, J.E.	[72BAT/SHI]
Bauer, A.A.	[64BUG/BAU]
Bauman, J.E.	[79BAU/ARM]
Bazhanov, V.I.	[89BAZ/KOM], [89EZH/BAZ]
Beall, G.W.	[78ALL/BEA]
Beck, G.	[28BEC]
Bednarczyk, L.	[78BED/FID]
Beezer, A.E.	[65BEE/MOR]
Begun, G.M.	[81MAY/BEG], [82GUI/BEG], [84TOT/FRI], [86SHE/TOT]
Behl, W.K.	[66BEH/EGA]
Belikov, A.D.	[77VOL/BEL]
Belova, L.N.	[58BEL], [73BEL]
Benson, L.V.	[80BEN/TEA]

Author	References
Benson, S.W.	[86HIS/BEN], [87HIS/BEN]
Bereznikova, I.A.	[60LEO/REZ]
Berg, L.	[78BER/HEI]
Bergman, G.A.	[78GLU/GUR], [82GLU/GUR]
Besmann, T.M.	[81LIN/BES]
Bessoud, A.	[88COL/BES]
Betts, R.H.	[49BET/MIC], [50BET/LEI], [55BET]
Bevington, P.R.	[69BEV]
Bhupathy, M.	[79PRA/NAG], [80NAG/BHU]
Bianco, P.	[68HAL/SAB]
Bidoglio, G.	[89GRE/BID], [91BID/CAV]
Biegański, Z.	[63STA/NIE], [66STA/BIE], [67STA/BIE]
Biltz, W.	[28BIL/FEN]
Binnewies, M.	[74BIN/SCH]
Birky, M.M.	[63WU/BIR]
Bjerrum, N.	[29BJE/UNM]
Blackburn, P.E.	[72BAT/SHI]
Blair, A.	[73BLA/IHL]
Blaise, A.	[78BLA/FOU], [79BLA], [79BLA/LAG], [80BLA/LAG], [80BLA/LAG2], [80BLA/TRO]
Blake, C.A.	[56BLA/COL]
Blankenship, F.F.	[60LAN/BLA]
Block, F.E.	[63WIC/BLO]
Blokhin, V.V.	[73KOZ/BLO]
Blumenthal, B.	[60BLU]
Bober, M.	[87BOB/SIN]
Bodrov, V.G.	[74MUK/SEL], [74SUP/SEL]
Boerio, J.	[76BOE/OHA], [76OHA/BOE], [77OHA/BOE]
Boivin, W.A.	[81TRE/CHE]
Boltalina, O.V.	[88BOR/BOL]
Bond, A.M.	[80BON/HEF]
Bondarenko, A.A.	[87BON/KOR]
Bones, R.L.	[67MAR/BON]
Bonham, R.W.	[63CAV/BON]
Bonnetot, B.	[76BOU/BON]
Boom, R.	[80BOE/BOO], [82BOE/BOO]
Borisova, A.P.	[84KOT/EVS]
Borshchevskii, A.Ya.	[88BOR/BOL]
Bougon, R.	[76MAS/DES]
Boulhassa, S.	[90ELY/BRI]
Bourcier, W.L.	[88JAC/WOL]
Bourges, J.Y.	[81GUI/HOB]
Bousquet, J.	[76BOU/BON]
Boyle, J.S.	[56ORB/BAR]
Brand, J.R.	[70BRA/COB]
Brandenburg, N.P.	[78BRA]
Bratsch, S.G.	[86BRA/LAG]

Author	References
Breitung, W.	[85BRE/REI]
Brewer, L.	[45BRE/BRO], [58BRE/BRO], [61PIT/BRE]
Brickwedde, F.G.	[48BRI/HOD], [51BRI/HOD]
Briggs, G.G.	[58BRI]
Brillard, L.	[90ELY/BRI]
Brittain, R.D.	[85LAU/BRI], [91HIL/LAU]
Brletic, P.A.	[76OSB/BRL]
Brock, E.G.	[85ALD/BRO]
Broene, H.H.	[47BRO/VRI]
Bromley, L.A.	[45BRE/BRO], [58BRE/BRO], [63WIL/BRO]
Brønsted, J.N.	[22BRO], [22BRO2]
Brooker, M.H.	[80BRO/HUA]
Brookins, D.G.	[75BRO]
Brown, D.	[68BRO], [71FUG/BRO], [73FUG/BRO], [85TSO/BRO]
Brown, K.B.	[56BLA/COL]
Brown, P.L.	[83BRO/ELL], [87BRO/WAN]
Brown, R.D.	[54BRO/BUN]
Browne, J.C.	[60SAV/BRO]
Bruno, J.	[86BRU], [86BRU/FER], [86BRU/GRE], [86BRU/GRE2], [87BRU], [87BRU/CAS], [87BRU/CAS2], [87BRU/GRE], [88BRU], [89BRU/GRE], [89BRU/PUI], [89BRU/SAN], [90BRU/GRE], [90BRU/SAN]
Brusilovskii, S.A.	[58BRU]
Buchacek, R.	[72BUC/GOR]
Bugl, J.	[64BUG/BAU]
Bukhtiyarova, T.N.	[67MOS/ESS]
Bullwinkel, E.P.	[56MCC/BUL]
Bunger, W.B.	[54BRO/BUN]
Bunus, F.T.	[74BUN], [74BUN2]
Burkinshaw, P.M.	[88BAR/BUR]
Burns, J.H.	[60BUR/OSB]
Burriel, R.	[88BUR/TO]
Busenberg, E.	[84BUS/PLU], [86BUS/PLU]
Busey, R.H.	[77BUS/MES]
Butler, T.	[58JOH/BUT]
Byakhova, N.I.	[65VID/BYA]
Bynyaeva, M.K.	[63OSI/BYN]
Caceci, M.S.	[83CAC/CHO2]
Caillat, R.	[53CAI/COR]
Caja, J.	[68CUK/CAJ], [69CAJ/PRA]
Campanella, D.A.	[50WAZ/CAM]
Campi, E.	[68OST/CAM]
Capdevila, H.	[90CAP/VIT]
Caramazza, R.	[57CAR]
Casas, I.	[87BRU/CAS], [87BRU/CAS2]
Cathelineau, M.	[87DUB/RAM]
Cavalli, P.	[91BID/CAV]

Author	References
Cavett, A.D.	[63CAV/BON]
Cercignani, C.	[85OHS/BAB]
Chakravorti, M.C.	[72CHA/BAN]
Chandrasekharaiah, M.S.	[68TVE/CHA]
Charoy, B.	[87DUB/RAM]
Charpin, P.	[76MAS/DES]
Chasanov, M.G.	[70FRE/BAR], [72FRE/CHA]
Chatham, J.R.	[80LAN/CHA]
Chaudhuri, N.K.	[85SAW/CHA], [85SAW/RIZ]
Chen, C.H.	[75CHE]
Chen, J.D.	[81TRE/CHE]
Cheney, R.K.	[47WAC/CHE]
Chengfa, Z.	[85GUO/LIU]
Chernyayev, I.I.	[66CHE]
Chierice, G.O.	[83CHI/NEV]
Chikh, I.V.	[83KOV/CHI]
Chilton, D.R.	[56DAW/WAI]
Chiotti, P.	[69CHI/KAT], [79CHI/MAS], [81CHI/AKH]
Chirkst, D.E.	[73VDO/SUG], [78SUG/CHI], [78SUG/CHI2], [80SUG/CHI], [80VOL/SUG], [81SUG/CHI], [83AUR/CHI], [83CHI], [86SUG/CHI], [87SUG/CHI]
Choi, I.K.	[83MOR/WIL2], [84WIL/MOR]
Choporov, D.Ya.	[68CHO/CHU], [70CHU/CHO]
Choppin, G.R.	[76CHO/UNR], [83CAC/CHO2], [84CHO/RAO], [88KHA/CHO], [91CHO/MAT]
Chottard, G.	[70CHO]
Chrétien, A.	[37CHR/KRA], [37CHR/KRA2]
Christ, D.E.	[78KUD/SUG]
Chudinov, É.G.	[70CHU/CHO]
Chudinov, É.T.	[68CHO/CHU]
Chukhlantsev, V.G.	[56CHU/STE], [61CHU/ALY]
Churney, K.L.	[82WAG/EVA]
Ciavatta, L.	[79CIA/FER], [80CIA], [81CIA/FER2], [83CIA/FER], [87CIA/IUL], [88CIA], [90CIA]
Cinnéide, S.Ó.	[75CIN/SCA]
Ciopec, M.	[70MAR/CIO], [71MAR/CIO]
Cirafici, S.	[75PAL/CIR]
Claudy, P.	[76BOU/BON]
Clayton, H.	[62GRO/HAY]
Cleveland, J.M.	[83NAS/CLE2]
Clifton, J.R.	[69CLI/GRU]
Clinnick, M.L.	[88JAC/WOL]
Cobble, J.W.	[61ARG/MER], [64CRI/COB], [64CRI/COB2], [70BRA/COB]
Cole, D.L.	[67COL/EYR]
Coleman, C.F.	[56BLA/COL]
Coleman, R.G.	[57COL/APP]
Colinet, C.	[88COL/BES]

Author	References
Condon, J.B.	[84FUL/SMY]
Connick, R.E.	[58CON/WAH], [61CON/PAU]
Cooper, G.R.	[41VOS/COO]
Cordfunke, E.H.P.	[63COR/ALI], [64COR], [65COR/ALI], [66COR], [66LOO/COR], [67COR/LOO], [68COR/PRI], [69COR], [71COR/LOO], [72COR], [74COR/PRI], [75COR], [75COR/OUW], [75COR2], [76COR/OUW], [77COR/OUW], [77COR/OUW2], [77COR/PRI], [78COR/OHA], [78PRI/COR], [79COR/MUI], [79COR/OUW], [80BAR/COR2], [80COR/WES], [80PRI/COR], [81COR/OUW], [81COR/OUW2], [81JOH/COR], [82COR/MUI], [82COR/OUW], [82COR/OUW2], [82WIJ/COR], [83COR/OUW], [83COR/OUW2], [84COR/KUB], [85BAR/COR], [85COR/OUW], [85COR/OUW2], [85COR/VLA], [86COR/OUW], [86COR/PRI], [88COR/IJD], [88COR/OUW], [88COR/WES], [88TAS/OHA], [89COR/KON], [90COR/KON], [90COR/KON2], [91COR/KON]
Coriou, H.	[53CAI/COR]
Corruccini, R.J.	[47GIN/COR]
Coughlin, J.P.	[54COU]
Coulter, L.V.	[40COU/PIT]
Counsell, J.F.	[64AND/COU], [66COU/DEL], [67COU/DEL], [68COU/MAR]
Covington, A.K.	[65COV/DOB]
Cowan, C.E.	[88COW]
Cox, J.D.	[89COX/WAG]
Crego, A.	[62FAU/CRE]
Criss, C.M.	[64CRI/COB], [64CRI/COB2], [79SHI/CRI]
Crist, R.H.	[48WEI/CRI]
Cromer, D.T.	[56STA/CRO]
Cross, J.E.	[88CRO/EWA]
Cubicciotti, D.D.	[75HIL/CUB]
Cukman, D.	[68CUK/CAJ]
Cuney, M.	[87DUB/RAM]
Datyuk, Yu.V.	[63VDO/SUG]
Davidson, M.R.	[79SYL/DAV], [79SYL/DAV2]
Davidson, P.	[45DAV/MAC]
Davidson, P.H.	[45DAV], [45DAV/STR]
Davies, C.W.	[62DAV]
Davies, E.W.	[57BAL/DAV], [57DAV/MON]
Davina, O.A.	[83DAV/YEF]
Davydov, Yu.P.	[73DAV/EFR], [75DAV/EFR], [83DAV/EFR]
Dawson, J.K.	[56DAW/WAI]
Day, R.A., Jr.	[54DAY/POW], [55DAY/WIL]
De Arman, S.K.	[79BAU/ARM]
De Boer, F.R.	[80BOE/BOO], [80MIE/CHA], [81NIE/BOE], [82BOE/BOO]
De Bruyn, P.L.	[60LIE/BRU]
De Châtel, P.F.	[80MIE/CHA]
De Conink, O.	[01CON]
De Crescente, M.A.	[65VOZ/CRE]

Author	References
De Forcrand, R.	[15FOR]
De Grave, E.	[86VOC/GRA]
De Vries, T.	[47BRO/VRI]
Deichman, E.N.	[61TAN/DEI]
Delany, J.M.	[88JAC/WOL]
Deliens, M.	[79DEL/PIR]
Dell, R.M.	[66COU/DEL], [67COU/DEL], [68COU/MAR]
Denotkina, R.G.	[60DEN/MOS]
Depaus, R.	[80PRI/COR]
Desai, P.D.	[73HUL/DES]
Deschenes, L.L.	[86DES/KRA]
Desmoulin, J.P.	[76MAS/DES]
Devina, O.A.	[82DEV/YEF], [83DEV/YEF]
Dewally, D.	[76DEW/PER]
Dickson, A.G.	[81TUR/WHI], [90DIC/WES]
Dobson, J.V.	[65COV/DOB]
Dogu, A.	[68DOG/VAL]
Dongarra, G.	[80DON/LAN]
Doody, E.	[58LI/DOO]
Dorofeeva, O.V.	[84GUR/DOR]
Dorris, S.E.	[84TOT/FRI]
Double, G.	[80DOU/PON], [80PON/DOU]
Downer, B.	[72TAY/KEL]
Downie, D.B.	[75WES/SOM], [78GRO/ZAK]
Drobnic, M.	[66DRO/KOL]
Drowart, J.	[84GRO/DRO]
Drozdova, V.M.	[58SHC/VAS], [59SHC/VAS]
Drăgulescu, C.	[65DRA/JUL], [70DRA/VIL], [73DRA/VIL]
Dubessy, J.	[87DUB/RAM]
Dubois, P.	[88DUB/LEL]
Duda, L.E.	[84MUL/DUD]
Duflo-Plissonnier, M.	[72DUF/SAM]
Dunaeva, K.M.	[71SAN/VID], [72SAN/VID]
Dunkl, F.	[37SCH/MAR]
Dunn, H.W.	[56DUN]
Dunsmore, H.S.	[63DUN/HIE], [63DUN/SIL], [64DUN/NAN]
Eager, M.H.	[84FUL/SMY]
Edwards, J.G.	[80EDW/STA]
Efimov, A.I.	[56SHC/VAS]
Efremenkov, V.M.	[73DAV/EFR], [75DAV/EFR], [83DAV/EFR]
Egan, J.J.	[66BEH/EGA], [66HEU/EGA]
Eikenberg, J.	[90EIK]
El Kass, G.	[79MIL/ELK]
Eldridge, E.	[68FAR/ELD]
Eller, P.G.	[82OHA/MAL]
Ellert, G.V.	[71SEV/SLO]
Ellinger, F.H.	[54MUL/ELL]

Author	References
Ellis, A.J.	[63ELL], [71ELL/GIG], [83BRO/ELL]
Elyahyaoui, A.	[90ELY/BRI]
Engel, T.K.	[69ENG]
Erdös, E.	[62ERD], [62ERD2]
Ermolaev, N.P.	[62ERM/KRO]
Essen, L.N.	[67MOS/ESS]
Evans, H.T., Jr.	[63EVA]
Evans, W.H.	[68WAG/EVA], [71PAR/WAG], [82WAG/EVA]
Evseev, A.M.	[84KOT/EVS]
Ewart, F.T.	[88CRO/EWA]
Eyring, E.M.	[67COL/EYR]
Ezhov, Yu.S.	[69EZH/AKI], [88EZH/KOM], [89BAZ/KOM], [89EZH/BAZ]
Fahey, J.A.	[81MOR/FAH]
Falk, B.G.	[77ALL/FAL]
Farah, K.	[76SOU/SHA]
Farkas, M.	[68FAR/ELD]
Faucherre, J.	[48FAU], [54FAU], [62FAU/CRE]
Faustova, D.G.	[61FAU/IPP], [61IPP/FAU]
Favre, P.A.	[1853FAV/SIL]
Feder, H.M.	[63SET/FED], [68OHA/SET]
Felmy, A.R.	[90RAI/FEL]
Fendius, C.	[28BIL/FEN]
Feng, S.	[82FEN/ZHA]
Feng, S.Y.	[65FEN/LI]
Ferguson, W.J.	[44FER/PRA], [44FER/RAN]
Fernelius, W.C.	[77FER]
Ferraro, J.R.	[52KAT/SIM]
Ferri, D.	[79CIA/FER], [81CIA/FER2], [81FER/GRE], [83ABE/FER], [83CIA/FER], [83FER/GRE], [84GRE/FER], [85FER/GRE], [86BRU/FER], [88FER/GLA]
Fidelis, I.	[78BED/FID]
Filip, H.	[85ALD/BRO]
Fischer, E.A.	[87FIS]
Fisher, J.R.	[78ROB/HEM2]
Fitzgibbon, G.C.	[67FIT/PAV], [71FIT/PAV]
Fleischer, M.	[83FLE]
Flicker, H.	[85ALD/BRO]
Flis, I.E.	[58FLI], [63OSI/BYN], [64FLI/MIS]
Flotow, H.E.	[55ABR/FLO], [59FLO/LOH], [60ABR/OSB], [60FLO/LOH], [66FLO/OSB], [67FLO/OSB], [68FLO/OSB], [69FLO/OSB], [72OSB/FLO], [74OSB/FLO], [75FLO/OSB], [76OSB/BRL], [77FLO/OSB], [77LYO/OSB], [80OHA/FLO], [81OHA/FLO], [82COR/MUI], [84FLO/HAS]
Flowers, G.C.	[81HEL/KIR]
Fontana, B.J.	[47FON], [58FON]
Fournier, J.-M.	[78BLA/FOU]
Fournier, R.O.	[62MOR/FOU]

Author	References
Fox, K.	[85ALD/BRO]
Fredlund, F.	[90BRU/SAN]
Fredrickson, D.R.	[70FRE/BAR], [72FRE/CHA], [76FRE/OHA], [76OHA/HOE], [77FLO/OSB], [77OHA/BOE]
Freeman, R.D.	[84FRE]
Frei, V.	[70FRE/WEN]
Frey, P.R.	[29GER/FRE]
Fried, S.M.	[75FLO/OSB]
Friedman, A.M.	[85SCH/FRI]
Friedman, H.A.	[84TOT/FRI]
Frolov, A.A.	[79FRO/RYK]
Fronæus, S.	[51FRO]
Frondel, C.	[58FRO], [76FRO/ITO]
Frydman, M.	[58FRY/NIL], [68SCH/FRY]
Fuchs, L.H.	[59FUC/HOE]
Fuger, J.	[71FUG/BRO], [72FUG], [73FUG/BRO], [76FUG/OET], [79FUG], [81MOR/FAH], [82FUG], [82MOR/FUG], [83FUG], [83FUG/PAR], [83MOR/WIL2], [85FUG], [85GEN/FUG], [85MAR/FUG], [85TSO/BRO], [89SCH/GEN]
Fujino, T.	[79TAG/FUJ], [81FUJ/TAG]
Fuller, E.L., Jr.	[84FUL/SMY]
Fuoss, R.M.	[33FUO/KRA]
Gabelica, Z.	[77GAB]
Gagarinskii, Yu.V.	[50POP/GAG], [59GAG/MAS], [60MAL/GAG], [67GAG/KHA]
Galbraith, H.W.	[85ALD/BRO]
Gal'chenko, G.L.	[58POP/GAL], [59POP/GAL]
Galkin, N.P.	[60GAL/STE], [60STE/GAL], [62STE/GAL], [71GAL]
Gallagher, F.X.	[70HOE/SIE]
Galloway, W.E.	[80GAL/KAI]
Garcia, F.S.	[73ALM/GAR]
Garrels, R.M.	[62HOS/GAR]
Garretson, H.H.	[41TAR/GAR]
Garvin, D.	[87GAR/PAR]
Gavrichev, K.S.	[87GUR/SER]
Gayer, K.H.	[55GAY/LEI], [57GAY/LEI]
Gennaro, M.C.	[68OST/CAM]
Gens, R.	[81MOR/FAH], [83MOR/WIL2], [85GEN/FUG], [89SCH/GEN]
Germann, F.E.E.	[29GER/FRE]
Ghosh, A.K.	[80GHO/GHO]
Ghosh, J.C.	[80GHO/GHO]
Ghosh, N.N.	[76GHO/MUK]
Ghosh, R.	[70GHO/NAI]
Giggenbach, W.F.	[71ELL/GIG], [71GIG], [74GIG]
Gilles, P.W.	[45BRE/BRO], [58BRE/BRO]
Gillikland, A.A.	[71MIN/GIL]
Gingerich, K.A.	[69GIN], [70GIN], [79GUP/GIN]
Ginnings, D.C.	[47GIN/COR]

Author	References
Girdhar, H.L.	[68GIR/WES]
Girichev, G.V.	[83GIR/PET]
Giricheva, N.I.	[83GIR/PET]
Gladtchikova, J.N.	[38SHI/GLA]
Glaser, J.	[83ABE/FER], [88FER/GLA]
Gleiser, M.	[73HUL/DES]
Glevob, V.A.	[72GLE/KLY]
Gliolitti, F.	[06GLI/LIB]
Glukova, L.P.	[68NEM/PAL]
Glushko, V.P.	[78GLU/GUR], [82GLU/GUR]
Goeminne, A.	[84VOC/GOE]
Goglev, V.S.	[67RYZ/NAU]
Goldberg, R.N.	[79GOL]
Gol'tsev, V.P.	[81GOL/TRE]
Goodman, R.M.	[66GOO/WES]
Gorbunov, V.E.	[87GUR/SER]
Gordon, G.	[64GOR/KER], [72BUC/GOR]
Gordon, J.	[52JON/GOR]
Gordon, S.	[78SCH/SUL]
Gorgoraki, V.I.	[88ALI/MAL]
Gorokhov, L.N.	[84GOR/SMI], [84SID/PYA], [84SMI/GOR]
Gotoo, K.	[65GOT/NAI]
Goubitz, K.	[85COR/VLA]
Grady, H.F.	[58YOU/GRA]
Grandjean, F.	[77FLO/OSB]
Greenberg, E.	[56GRE/WES], [56GRE/WES3]
Gregory, N.W.	[45GRE], [46GRE], [46GRE2], [46GRE3]
Greis, O.	[82GRE/HAS]
Grenthe, I.	[69GRE/VAR], [81CIA/FER2], [81FER/GRE], [83ABE/FER], [83CIA/FER], [83FER/GRE], [84GRE/FER], [84GRE/SPA], [85FER/GRE], [86BRU/FER], [86BRU/GRE], [86BRU/GRE2], [86GRE/RIG], [87BRU/CAS2], [87BRU/GRE], [87RIG/VIT], [88FER/GLA], [89BRU/GRE], [89GRE/BID], [89GRE/WAN], [90BRU/GRE], [91BID/CAV], [91GRE/LAG], [91GRE/LAG2]
Grieveson, P.	[61ALC/GRI], [62ALC/GRI]
Grimaldi, M.	[79CIA/FER]
Grisard, J.W.	[53OLI/MIL]
Gromov, B.V.	[71SUP/TSV], [71TSV/SEL], [72TSV/SEL], [73TSV/SEL]
Grønvold, F.	[59WES/GRO], [62WES/GRO], [62WES/GRO2], [65WES/TAK], [70GRO/KVE], [70WES/GRO], [75WES/SOM], [77ALL/FAL], [78GRO/ZAK], [84GRO/DRO]
Gross, P.	[62GRO/HAY], [71GRO/HAY]
Gruen, D.M.	[69CLI/GRU], [69GRU/MCB]
Gryzin, Yu.I.	[67GRY/KOR]
Guggenheim, E.A.	[35GUG], [66GUG]
Guillaume, B.	[79MAD/GUI], [81GUI/HOB], [82GUI/BEG]
Guillaumont, R.	[90ELY/BRI]
Guillemin, C.	[59GUI/PRO]

Author	References
Guiter, H.	[47GUI]
Gundersen, G.E.	[67REI/GUN]
Guorong, M.	[85GUO/LIU]
Gupta, S.K.	[79GUP/GIN]
Gurevich, A.M.	[43KHL/GUR], [57GUR]
Gurevich, V.M.	[87GUR/SER]
Gurvich, L.V.	[78GLU/GUR], [82GLU/GUR], [84GUR/DOR], [85HIL/GUR], [90GUR]
Gustafson, R.L.	[60GUS/RIC]
Haacke, D.F.	[79HAA/WIL]
Haebler, H.	[32NEU/KRO]
Hägglund, Y.	[83SJO/HAE]
Hagisawa, H.	[40HAG]
Hahn, R.L.	[82GUI/BEG]
Haladjan, J.	[68HAL/SAB]
Hale, F.V.	[88PHI/HAL], [88PHI/HAL2]
Halloway, J.H.	[85TSO/BRO]
Halow, I.	[68WAG/EVA], [82WAG/EVA]
Hamilton, F.D.	[55DAY/WIL]
Harris, W.E.	[47HAR/KOL]
Hartl, K.	[57WEI/HAR]
Haschke, J.M.	[82GRE/HAS], [84FLO/HAS]
Hastings, I.J.	[77FLO/OSB]
Hawkins, D.T.	[73HUL/DES]
Hayes, S.L.	[90HAY/THO]
Hayman, C.	[62GRO/HAY], [71GRO/HAY]
Hayman, H.J.G.	[67HAY]
Hearne, J.A.	[57HEA/WHI]
Hedger, H.J.	[64AND/COU]
Hedlund, T.	[88HED]
Hefley, J.D.	[60HEF/AMI]
Hefter, G.T.	[80BON/HEF], [90AHR/HEF]
Heibel, B.	[78BER/HEI]
Heiks, J.R.	[56ORB/BAR]
Heintz, A.	[78MEI/HEI]
Heinzelmann, A.	[11RUF/HEI]
Helgeson, H.C.	[69HEL], [81HEL/KIR], [88SHO/HEL], [88TAN/HEL], [89SHO/HEL], [89SHO/HEL2]
Hemingway, B.S.	[78ROB/HEM2], [82HEM]
Henkie, Z.	[78BLA/FOU], [80BLA/LAG]
Hepler, L.G.	[63WU/BIR]
Heumann, F.K.	[51ZEB/ALT]
Heus, R.J.	[66HEU/EGA]
Hiernaut, J.P.	[85OHS/BAB]
Hietanen, S.	[54AHR/HIE], [56HIE], [56HIE2], [59HIE/SIL], [63DUN/HIE], [63HIE/ROW], [85FER/GRE]

Author	References
Hildenbrand, D.L.	[75HIL/CUB], [77HIL], [79KLE/HIL], [82LAU/HIL], [84LAU/HIL], [85HIL/GUR], [85LAU/BRI], [85LAU/HIL], [87LAU/HIL], [91HIL/LAU], [91HIL/LAU2]
Hill, D.G.	[56BLA/COL]
Hindman, J.C.	[49KRI/HIN], [52SUL/HIN], [54HIN], [54SUL/HIN], [59SUL/HIN], [61SUL/HIN]
Hinz, I.	[78BER/HEI]
Hirono, S.	[65MUT/HIR]
Hisham, M.W.M.	[86HIS/BEN], [87HIS/BEN]
Hobart, D.E.	[81GUI/HOB]
Hoch, M.	[85OHS/BAB]
Hodes, G.	[86LIC/HOD]
Hodge, H.J.	[48BRI/HOD], [51BRI/HOD]
Hoekstra, H.R.	[54HOE/KAT], [59FUC/HOE], [61HOE/SIE], [70HOE/SIE], [73HOE/SIE], [73OHA/HOE2], [74OHA/HOE], [74OHA/HOE2], [74OHA/HOE3], [74OHA/HOE4], [74OSB/FLO], [75OHA/HOE2], [76OHA/BOE], [76OHA/HOE], [76OSB/BRL], [77LYO/OSB], [77OHA/BOE], [80OHA/FLO], [81OHA/FLO]
Hoenig, C.L.	[71HOE]
Hoffmann, G.	[72HOF], [76WEI/HOF]
Hofmann, U.	[57WEI/HAR]
Holden, R.B.	[64LAP/HOL]
Holland, R.F.	[85ALD/BRO]
Holley, C.E., Jr.	[52HUB/HOL], [62HUB/HOL], [63ACH/ALC], [67FIT/PAV], [68HOL/STO], [69HUB/HOL], [69HUB/HOL3], [71FIT/PAV], [84HOL/RAN]
Honea, R.M.	[76FRO/ITO]
Hostetler, P.B.	[62HOS/GAR]
Hovey, J.K.	[89HOV/NGU], [90NGU/HOV]
Hu, C.	[80HU/NI]
Hu, R.	[80HU/NI]
Huang, C.H.	[80BRO/HUA]
Huang, W.	[80HU/NI]
Hubbard, W.N.	[62HUB], [63SET/FED], [68OHA/SET], [69FLO/OSB], [70JOH/SMI], [75OHA/ADE], [83FUG/PAR]
Huber, E.J., Jr.	[52HUB/HOL], [62HUB/HOL], [67STO/HUB], [69HUB/HOL], [69HUB/HOL3]
Huggins, J.C.	[56MCC/BUL]
Hultgren, R.	[63HUL/ORR], [73HUL/DES]
Hunt, P.D.	[71TET/HUN]
Huntzicker, J.J.	[71HUN/WES]
Hurwic, J.	[80DOU/PON], [80PON/DOU]
Hussonnois, M.	[90ELY/BRI]
Hutin, M.F.	[71HUT], [75HUT]
Hyland, G.J.	[85OHS/BAB]
Hynes, M.J.	[75CIN/SCA]
Ihle, H.	[73BLA/IHL]

Author	References
IJdo, D.J.W.	[88COR/IJD]
Ikeda, Y.	[76IKE/TAM]
Inaba, H.	[73INA/NAI], [77INA/SHI], [82NAI/INA], [82SET/MAT]
Ingri, N.	[67ING/KAK], [81SJO/NOR], [83SJO/HAE], [85SJO/OEH]
Inouye, H.	[68INO/LEI]
Ippolitova, E.A.	[61IPP/FAU], [65VID/BYA], [71SAN/VID], [71VID/IPP], [72SAN/VID]
Ippolitova, V.I.	[61FAU/IPP]
Ishiguro, S.	[74KAK/ISH], [77ISH/KAO]
Israeli, Y.J.	[65ISR]
Ito, J.	[76FRO/ITO]
Iuliano, M.	[87CIA/IUL]
Ivanov, I.P.	[89RED/SAV]
Ivanov, M.I.	[57POP/IVA], [58IVA/TUM], [59IVA/TUM]
Ivantsov, A.I.	[67SOL/TSV]
Jacksic, M.M.	[69STO/JAC]
Jackson, K.J.	[88JAC/WOL]
Jacob, E.	[83BAG/JAC]
Jain, D.V.S.	[66MIT/MAL]
Jakes, D.	[67JAK/SCH], [80WES/ZAI]
Jedináková, V.	[73STA/CEL], [74JED]
Jenkins, F.A.	[42JEN/AND]
Jenkins, H.D.B.	[81MOR/FAH], [82MOR/FUG]
Jensen, B.S.	[82JEN]
Jensen, K.A.	[71JEN]
Jensen, R.P.	[67COL/EYR]
Job, P.	[28JOB]
Johnson, C.E.	[80JOH/PAP], [81LIN/BES]
Johnson, G.K.	[70JOH/SMI], [75OHA/ADE], [78JOH/OHA], [79JOH], [80JOH/PAP], [81JOH/COR], [81JOH/STE3], [85JOH], [88TAS/OHA]
Johnson, G.L.	[49KRA/NEL]
Johnson, I.	[72JOH/STE], [75JOH]
Johnson, J.S.	[62RUS/JOH], [63RUS/JOH]
Johnson, K.S.	[79JOH/PYT]
Johnson, O.	[58JOH/BUT]
Johnsson, K.O.	[47JOH]
Johnstone, H.F.	[34JOH/LEP]
Jones, E.V.	[66MAR/JON]
Jones, L.V.	[56ORB/BAR]
Jones, W.M.	[52JON/GOR]
Judge, A.I.	[85TSO/BRO]
Julean, I.	[65DRA/JUL], [70DRA/VIL]
Junkison, A.R.	[67COU/DEL], [68COU/MAR]
Kaganyuk, D.S.	[83KAG/KYS]
Kaiser, W.R.	[80GAL/KAI]
Kakihana, H.	[70MAE/KAK], [72MAE/AMA], [74KAK/ISH], [77ISH/KAO]

Author	References
Kakolowicz, W.	[67ING/KAK]
Kandykin, Yu.M.	[60RAF/KAN]
Kanevskii, E.A.	[74KAN/ZAR]
Kangro, W.	[63KAN]
Kanno, M.	[74KAN]
Kao, C.F.	[77ISH/KAO]
Kapitonov, V.I.	[68LEB/NIC2]
Kapustinsky, A.F.	[52KAP/BAR]
Karapet'yants, M.K.	[68KAR/KAR], [71SUP/TSV], [74MUK/SEL], [74SUP/SEL]
Karapet'yants, M.L.	[68KAR/KAR]
Karasev, N.M.	[87BON/KOR]
Karl, W.	[78BER/HEI]
Karpov, V.I.	[61KAR]
Karpova, T.F.	[57POP/KOS]
Karyakin, A.V.	[68NEM/PAL]
Kateley, J.A.	[69CHI/KAT]
Katz, J.J.	[51KAT/RAB], [54HOE/KAT], [57KAT/SEA]
Katzin, L.I.	[52KAT/SIM]
Kazin, I.V.	[83KAG/KYS]
Keder, W.E.	[61WIL/KED]
Keller-Rudek, H.	[78BER/HEI]
Kelley, C.M.	[56KEL/TAR]
Kelley, K.K.	[32MAI/KEL], [40KEL], [47MOO/KEL], [60KEL], [61KEL/KIN], [63HUL/ORR], [73HUL/DES]
Kelly, J.W.	[72TAY/KEL]
Kelmy, M.	[32ROS/KEL]
Kerko, P.F.	[81GOL/TRE]
Kern, D.M.H.	[49KER/ORL], [64GOR/KER]
Khachkuruzov, G.A.	[78GLU/GUR], [82GLU/GUR]
Khalili, F.I.	[88KHA/CHO]
Khanaev, E.I.	[67GAG/KHA], [68KHA], [70KHA/KRI]
Khandelwal, B.L.	[73VER/KHA]
Khattak, G.D.	[83KHA]
Khitarov, N.I.	[68RYZ/KHI]
Khlopin, V.G.	[43KHL/GUR]
Khodadad, P.	[57KHO]
Khodakovsky, I.L.	[71NAU/RYZ], [72NIK/SER], [72SER/NIK], [82DEV/YEF], [83DAV/YEF], [83DEV/YEF], [87GUR/SER], [89RED/SAV]
Khodeev, Yu.S.	[61AKI/KHO], [84GOR/SMI]
Khomakov, K.G.	[61KHO/SPI]
Khromov, Yu.F.	[80KHR/LYU]
Khvorostin, Ya.S.	[73KRU/ROD]
Kielland, J.	[37KIE]
Kigoshi, K.	[50KIG]
Kim, J.J.	[74PIT/KIM]
Kim, K.C.	[85ALD/BRO]
King, A.J.	[57KIN/PFE]

Author	References
King, E.G.	[61KEL/KIN], [71KIN/SIE]
Kipatsi, H.	[80ALL/KIP]
Kirkham, D.H.	[81HEL/KIR]
Kirslis, S.S.	[54WEN/KIR]
Kiselev, Yu.M.	[83GIR/PET]
Kleinschmidt, P.D.	[79KLE/HIL]
Klygin, A.E.	[59KLY/KOL], [59KLY/SMI], [59KLY/SMI2], [60KLY/KOL], [70KLY/KOL], [72GLE/KLY]
Knacke, O.	[69KNA/LOS], [69KNA/LOS2], [72KNA/MUE], [73BAR/KNA], [77BAR/KNA]
Knoch, W.	[70LAH/KNO]
Kobets, L.V.	[78KOB/KOL], [80KOB/KOL]
Koch, G.	[66KOC/SCH]
Kodenskaya, V.S.	[60BAB/KOD]
Kohli, R.	[83KOH]
Kolar, D.	[66DRO/KOL]
Kolevich, T.A.	[78KOB/KOL], [80KOB/KOL]
Kolthoff, I.M.	[47HAR/KOL]
Kolyada, N.S.	[59KLY/KOL], [60KLY/KOL], [70KLY/KOL], [72GLE/KLY]
Kolychev, V.B.	[62NIK/PAR]
Komar, N.P.	[55KOM/TRE]
Komarov, E.V.	[68KRY/KOM], [68KRY/KOM3]
Komarov, S.A.	[88EZH/KOM], [89BAZ/KOM], [89EZH/BAZ]
Konings, R.J.M.	[89COR/KON], [90COR/KON], [90COR/KON2], [91COR/KON]
Korobov, M.V.	[87BON/KOR]
Koryttsev, K.Z.	[67GRY/KOR]
Koshlap, K.	[83MEY/WAR]
Kostryukov, V.N.	[74SAM/KOS]
Kostylev, F.A.	[57POP/KOS], [59POP/KOS], [74SAM/KOS]
Kothari, P.S.	[71KOT]
Kotvanova, M.K.	[84KOT/EVS]
Kovaleva, T.V.	[57VDO/KOV]
Koval'skaya, M.P.	[57VDO/KOV]
Kovba, V.M.	[83KOV/CHI]
Kozlov, Yu.A.	[73KOZ/BLO]
Kraft, J.	[37CHR/KRA], [37CHR/KRA2]
Kramer, G.H.	[86DES/KRA]
Krasnov, K.S.	[83GIR/PET]
Kraus, C.A.	[22KRA/PAR], [33FUO/KRA]
Kraus, K.A.	[49KRA/NEL], [49KRA/NEL2], [50KRA/NEL], [50MOO/KRA], [51NEL/KRA], [55KRA/NEL], [56KRA/MOO], [62RUS/JOH]
Krestov, G.A.	[72KRE]
Kripin, L.A.	[70KHA/KRI]
Kritchevsky, E.S.	[49KRI/HIN]
Kröger, C.	[32NEU/KRO]
Krohn, B.J.	[85ALD/BRO]
Krot, N.N.	[62ERM/KRO]

Author	References
Kruger, O.L.	[67MOS/KRU]
Krumgal'z, B.S.	[73KRU/ROD]
Krunchak, E.G.	[73KRU/ROD]
Krunchak, V.G.	[72PAL/LVO], [73KRU/ROD]
Krylov, V.N.	[68KRY/KOM], [68KRY/KOM3]
Ksenofontova, N.M.	[80KOB/KOL]
Kubaschewski, O.	[63ACH/ALC], [63RAN/KUB], [77BAR/KNA], [79KUB/ALC], [84COR/KUB]
Kudryashov, V.L.	[78KUD/SUG]
Kudryashova, Z.P.	[68MAR/KUD]
Kullberg, L.	[71AHR/KUL], [71AHR/KUL2], [71AHR/KUL3]
Kumok, V.N.	[67MER/SKO]
Kurata, H.	[65MUT/HIR]
Kuteinikov, A.F.	[61KUT]
Kveseth, N.J.	[70GRO/KVE]
Kyskin, V.I.	[83KAG/KYS]
Laffitte, M.	[82LAF]
Lagerman, B.	[86BRU/GRE2], [87BRU/CAS], [87BRU/CAS2], [90BRU/GRE], [91GRE/LAG], [91GRE/LAG2]
Lagerström, G.	[59LAG]
Lagnier, R.	[75LAG/SUS], [78BLA/FOU], [79BLA/LAG], [80BLA/LAG], [80BLA/LAG2], [80BLA/TRO]
Lagowski, J.J.	[86BRA/LAG]
Lahr, H.	[70LAH/KNO]
Lajunen, L.H.J.	[79LAJ/PAR]
Langer, S.	[60LAN/BLA]
Langmuir, D.	[78LAN], [80DON/LAN], [80LAN/CHA]
Lapat, P.E.	[64LAP/HOL]
Larson, J.W.	[71BAI/LAR]
Larsson, R.	[54AHR/LAR], [54AHR/LAR2], [56AHR/LAR]
Latimer, W.M.	[38LAT/PIT], [40COU/PIT], [52LAT]
Lau, K.H.	[82LAU/HIL], [84LAU/HIL], [85LAU/BRI], [85LAU/HIL], [87LAU/HIL], [91HIL/LAU], [91HIL/LAU2]
Le Doux, R.A.	[67OTE/DOU]
Lebedev, V.A.	[68LEB/NIC2]
Lee, T.S.	[79CHI/MAS]
Leider, H.	[55GAY/LEI], [57GAY/LEI]
Leigh, R.M.	[50BET/LEI]
Leist, M.	[55LEI]
Leitnaker, J.M.	[68INO/LEI], [72OET/LEI], [83LEI]
Lelieur, J.P.	[88DUB/LEL]
Lemire, R.J.	[80LEM/TRE], [88LEM]
Leonard, A.	[78BER/HEI]
Leonidov, V.Ya.	[60LEO/REZ]
Lepoutre, G.	[88DUB/LEL]
Leppla, P.W.	[34JOH/LEP]
Leroy, J.	[87DUB/RAM]

Author	References
Lesser, J.M.	[53BAE/SCH2]
Lewis, B.M.	[88OHA/LEW], [88TAS/OHA]
Li, H.Y.	[65FEN/LI]
Li, N.C.	[41LI/LO], [58LI/DOO]
Liberi, G.	[06GLI/LIB]
Licht, S.	[86LIC/HOD], [87LIC/MAN]
Lichtenthaler, R.N.	[78MEI/HEI]
Liebhafsky, H.A.	[34LIE]
Lietzke, M.H.	[60LIE/STO]
Liljenzin, J.O.	[80ALL/KIP]
Lindemer, T.B.	[81LIN/BES]
Lindsay, W.L.	[81SAD/LIN]
Linke, W.F.	[58LIN]
Lipilina, I.I.	[54LIP/SAM]
Lipkin, D.	[43ALT/LIP]
Liu, K.	[80HU/NI]
Liufang, Z.	[85GUO/LIU]
Livey, D.T.	[63ACH/ALC]
Llewellyn, D.R.	[53LLE]
Lo, Y.T.	[41LI/LO]
Lofgren, N.L.	[45BRE/BRO], [58BRE/BRO]
Lohr, H.R.	[55OSB/WES], [57OSB/WES], [59FLO/LOH], [60FLO/LOH]
Long, E.A.	[52JON/GOR]
Longsworth, L.G.	[47MAC/LON]
Lonsdale, H.K.	[65WES/SUI]
Loopstra, B.O.	[66LOO/COR], [67COR/LOO], [71COR/LOO], [85COR/VLA]
Lossmann, G.	[69KNA/LOS], [69KNA/LOS2]
Lowrie, R.S.	[56BLA/COL]
Luk'yanychev, Yu.A.	[61NIK/LUK], [62LUK/NIK], [63LUK/NIK], [63LUK/NIK2]
Lumme, P.	[68TUM/LUM]
Lund, W.	[82OVE/LUN]
Lundeen, S.R.	[88JAC/WOL]
Lundgren, G.	[53LUN]
Luther, R.	[08LUT/MIC]
L'vova, T.I.	[72PAL/LVO]
Lychev, A.A.	[90LYC/MIK]
Lynch,.	[41WIL/LYN]
Lyon, W.G.	[77FLO/OSB], [77LYO/OSB]
Lyutikov, R.A.	[80KHR/LYU]
MacInnes, D.A.	[47MAC/LON]
MacLeod, A.C.	[69MAC]
MacWood, G.E.	[45DAV/MAC], [58MAC]
Madic, C.	[79MAD/GUI]
Maeda, M.	[70MAE/KAK], [72MAE/AMA]
Magadanz, H.	[75WES/MAG]
Magill, J.	[85OHS/BAB]
Magnuson, D.W.	[85ALD/BRO]

Author	References
Maier II, W.B.	[85ALD/BRO]
Maier, C.G.	[32MAI/KEL]
Majchrzak, K.	[73MAJ]
Makhametshina, Z.B.	[74SUP/SEL]
Makitie, O.	[81VAI/MAK]
Malhotra, H.C.	[66MIT/MAL]
Malinin, G.V.	[70VDO/ROM]
Malkerova, I.P.	[88ALI/MAL]
Mallett, M.W.	[55MAL/SHE], [55MAL/SHE2]
Malm, J.G.	[75FLO/OSB], [82OHA/MAL], [82OHA/MAL2], [84OHA/MAL]
Mal'tsev, V.A.	[60MAL/GAG]
Mal'tsev, Y.G.	[58SHC/VAS2]
Mal'tseva, L.A.	[58VDO/MAL]
Manassen, J.	[86LIC/HOD], [87LIC/MAN]
Marchgraber, R.	[37SCH/MAR]
Marchidan, D.I.	[70MAR/CIO], [71MAR/CIO]
Marcus, R.B.	[60ABR/OSB]
Marcus, Y.	[58MAR], [65SHI/MAR]
Markétos, M.	[12MAR]
Markin, T.L.	[67MAR/BON]
Markov, V.K.	[64MAR/VER]
Marković, M.	[83MAR/PAV]
Markowski, P.J.	[80BLA/LAG]
Marshall, W.L.	[54BRO/BUN], [66MAR/JON]
Martell, A.E.	[60GUS/RIC], [64SIL/MAR], [71SIL/MAR], [76SMI/MAR], [82MAR/SMI]
Martin, A.E.	[72OHA/SHI]
Martin, J.F.	[64AND/COU], [66COU/DEL], [67COU/DEL], [68COU/MAR]
Martinot, L.	[85MAR/FUG]
Martynova, N.S.	[58SHC/VAS2], [59SHC/VAS], [68MAR/KUD]
Mashirev, V.P.	[59GAG/MAS]
Masi, J.F.	[49MAS]
Maslov, P.G.	[64MAS]
Mason, J.T.	[79CHI/MAS]
Masson, J.P.	[76MAS/DES]
Mathur, J.N.	[91CHO/MAT], [91MAT]
Mathurin, D.	[76BOU/BON]
Matsui, T.	[82SET/MAT], [87MAT/OHS]
Matsumoto, G.	[76IKE/TAM]
Matsuo, S.	[60MAT]
Mavrodin-Tărăbic, M.	[73MAV], [73MAV2], [74MAV]
Maya, L.	[81MAY/BEG], [82MAY], [82MAY2]
Mayer, S.W.	[64OWE/MAY]
Mayorga, G.	[73PIT/MAY], [74PIT/MAY]
McBeth, R.L.	[69GRU/MCB]
McClaine, L.A.	[56MCC/BUL]
McDowell, R.S.	[85ALD/BRO]

Author	References
McKay, H.A.C.	[64MCK/WOO]
Medvedev, V.A.	[78GLU/GUR], [82DEV/YEF], [82GLU/GUR], [83DAV/YEF], [83DEV/YEF], [89COX/WAG]
Meierkord, E.H.	[52HUB/HOL]
Meixner, D.	[78MEI/HEI]
Melton, S.L.	[63MEL/AMI]
Mercer, E.E.	[61ARG/MER]
Merkusheva, S.A.	[67MER/SKO]
Mesmer, R.E.	[74MES/BAE], [76BAE/MES], [77BUS/MES], [81BAE/MES], [90DIC/WES]
Meyer, B.	[83MEY/WAR]
Meyer, N.J.	[62BAE/MEY], [65BAE/MEY]
Michels, R.K.	[49BET/MIC]
Michie, A.C.	[08LUT/MIC]
Miedema, A.R.	[80BOE/BOO], [80MIE/CHA], [82BOE/BOO]
Mikhalev, V.A.	[90LYC/MIK]
Mikulinskaya, N.M.	[88EZH/KOM]
Milic, N.B.	[79MIL/ELK], [82MIL/SUR]
Millero, F.J.	[79MIL]
Mills, K.C.	[74MIL]
Milton, H.T.	[53OLI/MIL]
Minc, S.	[61SOB/MIN]
Minor, J.I.	[71MIN/GIL]
Mironov, V.E.	[73KOZ/BLO]
Mishchenko, K.P.	[63OSI/BYN], [64FLI/MIS]
Mitra, R.R.	[66MIT/MAL]
Mixter, W.G.	[12MIX]
Monk, C.B.	[57BAL/DAV], [57DAV/MON], [78MON/AMI]
Monserrat, K.J.	[86DES/KRA]
Montignie, E.	[37MON]
Moore, G.E.	[47MOO/KEL], [50MOO/KRA], [56KRA/MOO]
Moore, R.M.	[88JAC/WOL]
Morey, G.W.	[62MOR/FOU]
Morgan, J.J.	[81STU/MOR]
Morgans, D.B.	[57BAL/DAV]
Morisseau, J.C.	[79MAD/GUI]
Morozumi, T.	[67OHA/MOR], [67OHA/MOR2]
Morris, J.C.	[66MOR]
Morss, L.R.	[81MOR/FAH], [82MOR], [82MOR/FUG], [83MOR/WIL2], [84WIL/MOR], [85GEN/FUG], [85MOR/SON], [85MOR2], [86MOR], [89SCH/GEN]
Mortimer, C.T.	[65BEE/MOR], [88BAR/BUR]
Mortimer, M.J.	[78BLA/FOU], [80BLA/LAG2], [80BLA/TRO]
Moser, J.B.	[67MOS/KRU]
Moskvin, A.I.	[60DEN/MOS], [67MOS/ESS], [67MOS/SHE], [69MOS], [73MOS], [73MOS2]
Moulin, J.P.	[79MAD/GUI]

Author	References
Mrazek, F.C.	[69FLO/OSB], [72OHA/SHI]
Muis, R.P.	[79COR/MUI], [82COR/MUI]
Mukaibo, T.	[62MUK/NAI], [62MUK/NAI2], [63ACH/ALC], [75YOK/TAK]
Mukhametshina, Z.B.	[74MUK/SEL]
Mukhina, P.S.	[64VAS/MUK]
Mukhopadhyay, S.K.	[76GHO/MUK]
Mulak, J.	[79BLA/LAG]
Mulford, R.N.R.	[54MUL/ELL], [63OLS/MUL]
Muller, A.B.	[84MUL/DUD], [85MUL4]
Müller, F.	[69KNA/LOS], [69KNA/LOS2], [72KNA/MUE]
Müller, M.E.	[48MUE]
Mullinger, L.W.	[48AMP/MUL]
Muñoz, M.	[86BRU/GRE], [87BRU/CAS]
Münstermann, E.	[76MUE]
Muradova, G.A.	[68NEM/PAL]
Musikas, C.	[72MUS]
Muto, T.	[65MUT], [65MUT/HIR]
Muus, J.	[51MUU/REF]
Myers, R.J.	[86MYE]
Nagarajan, K.	[79PRA/NAG], [80NAG/BHU]
Naidich, S.	[39NAI/RIC]
Nair, B.K.S.	[61NAI/PRA]
Nair, V.S.K.	[58NAI/NAN], [64NAI], [70GHO/NAI]
Naito, K.	[62MUK/NAI], [62MUK/NAI2], [65GOT/NAI], [73INA/NAI], [77INA/SHI], [82NAI/INA], [82SET/MAT], [83NAI/TSU]
Nancollas, G.H.	[58NAI/NAN], [64DUN/NAN]
Nash, K.L.	[83NAS/CLE2]
Naumov, G.B.	[67RYZ/NAU], [71NAU/RYZ], [72NIK/SER], [72SER/NIK]
Navratil, J.D.	[72OET/NAV]
Nazarov, V.K.	[66SHI/NAZ]
Neckel, A.	[63ACH/ALC]
Neckrasova, Z.A.	[58NEC]
Néker-Neumann, E.	[85FER/GRE]
Nelson, F.	[49KRA/NEL], [49KRA/NEL2], [50KRA/NEL], [51NEL/KRA], [55KRA/NEL], [56KRA/MOO]
Nemkina, N.V.	[90ZAR/NEM]
Nemodruk, A.A.	[68NEM/PAL]
Neumann, B.	[32NEU/KRO]
Neves, E.A.	[83CHI/NEV]
Newton, A.S.	[58JOH/BUT]
Newton, T.W.	[65NEW/BAK]
Nguyen, S.N.	[88OHA/LEW]
Nguyen-Trung, C.	[87DUB/RAM], [89HOV/NGU], [90NGU/HOV]
Ni, X.	[80HU/NI]
Nichkov, I.F.	[68LEB/NIC2]
Niemic, J.	[63STA/NIE]
Niessen, A.K.	[81NIE/BOE]

Author	References
Nikitin, A.A.	[72NIK/SER], [72SER/NIK]
Nikolaev, N.S.	[61NIK/LUK], [62LUK/NIK], [63LUK/NIK], [63LUK/NIK2]
Nikolaeva, N.M.	[62NIK/PAR], [62PAR/NIK], [62PAR/NIK2], [68NIK/ANT], [70NIK], [71NIK], [74NIK], [76NIK], [76NIK2], [76PIR/NIK], [77NIK], [77NIK/TSV], [77NIK2], [78NIK], [78NIK/PIR]
Nikolic, B.Z.	[69STO/JAC]
Nikol'skaya, N.A.	[59KLY/SMI2]
Nikol'skii, B.P.	[62PAR/NIK2]
Nilsson, G.	[58FRY/NIL], [58NIL/REN]
Nordin, A.	[81SJO/NOR], [83SJO/HAE]
Norén, B.	[69NOR], [90AHR/HEF]
Norman, J.H.	[64NOR/WIN]
Nottorf, R.	[44NOT/POW]
Nowotny, H.	[63ACH/ALC]
Nriagu, J.O.	[84NRI], [84NRI2]
Nugent, L.J.	[75NUG]
Nuttall, R.L.	[82WAG/EVA]
O'Brien, T.J.	[81OBR/WIL], [83OBR/WIL]
Oetting, F.L.	[72OET/LEI], [72OET/NAV], [76FUG/OET], [76OET/RAN], [83FUG/PAR]
Ogden, J.S.	[69ALE/OGD], [71ALE/OGD]
O'Hare, P.A.G.	[68OHA/SET], [69FLO/OSB], [72OHA/SHI], [73OHA/HOE2], [74OHA/HOE], [74OHA/HOE2], [74OHA/HOE3], [74OHA/HOE4], [75OHA/ADE], [75OHA/HOE2], [76BOE/OHA], [76FRE/OHA], [76OHA/BOE], [76OHA/HOE], [77OHA], [77OHA/BOE], [78COR/OHA], [78JOH/OHA], [80OHA/FLO], [81OHA/FLO], [82COR/MUI], [82OHA/MAL], [82OHA/MAL2], [84OHA/MAL], [84SET/OHA], [85OHA2], [88OHA/LEW], [88TAS/OHA]
Ohashi, H.	[67OHA/MOR], [67OHA/MOR2]
Öhman, L.O.	[85SJO/OEH]
Ohse, R.W.	[85OHS/BAB], [87MAT/OHS]
Ohtaki, H.	[72MAE/AMA]
Ohya, F.	[83NAI/TSU]
Oliver, G.D.	[53OLI/MIL]
Olofsson, G.	[84GRE/SPA]
Olson, W.M.	[63OLS/MUL]
Omelyanenko, B.I.	[89RED/SAV]
Omenetto, N.	[89GRE/BID], [91BID/CAV]
Orban, E.	[56ORB/BAR]
Orlemann, E.F.	[49KER/ORL]
Orlova, M.M.	[67ZAK/ORL]
Orr, R.L.	[63HUL/ORR]
Osborne, D.W.	[55OSB/WES], [57OSB/WES], [59FLO/LOH], [60ABR/OSB], [60BUR/OSB], [66FLO/OSB], [67FLO/OSB], [68FLO/OSB], [69FLO/OSB], [72OSB/FLO], [74OSB/FLO], [75FLO/OSB], [76OSB/BRL], [77FLO/OSB], [77LYO/OSB]

Author	References
Osinska-Tanevska, S.M.	[63OSI/BYN]
Ostacoli, G.	[68OST/CAM], [69VAN/OST]
Otey, M.G.	[67OTE/DOU]
Ott, H.R.	[85RUD/OTT], [85RUD/OTT2]
Ouweltjes, W.	[75COR/OUW], [76COR/OUW], [77COR/OUW], [77COR/OUW2], [78PRI/COR], [79COR/OUW], [81COR/OUW], [81COR/OUW2], [82COR/MUI], [82COR/OUW], [82COR/OUW2], [83COR/OUW], [83COR/OUW2], [85COR/OUW], [85COR/OUW2], [86COR/OUW], [88COR/OUW], [90COR/KON2]
Overbeek, J.T.G.	[60LIE/BRU]
Overvoll, P.A.	[82OVE/LUN]
Owens, B.B.	[64OWE/MAY]
Pai, S.A.	[69RAO/PAI]
Pal'chevskii, V.V.	[72PAL/LVO]
Palei, P.N.	[68NEM/PAL]
Palenzona, A.	[75PAL/CIR]
Palmer, D.A.	[90DIC/WES]
Palombari, R.	[79CIA/FER]
Pamfilov, A.V.	[50PAM/AGA]
Pankratz, L.B.	[82PAN]
Papatheodorou, G.N.	[80JOH/PAP]
Paramonova, V.I.	[62NIK/PAR], [62PAR/NIK], [62PAR/NIK2]
Pardue, W.M.	[69ALE/OGD], [71ALE/OGD]
Parhi, S.	[79LAJ/PAR]
Parker, H.C.	[22KRA/PAR]
Parker, V.B.	[65PAR], [68WAG/EVA], [71PAR/WAG], [80PAR], [82WAG/EVA], [83FUG/PAR], [84BUS/PLU], [87GAR/PAR]
Parks, G.A.	[85PAR/POH], [88PAR/POH]
Pascal, P.	[61PAS], [67PAS]
Past, V.	[85PAS]
Pastukova, E.D.	[68NIK/ANT]
Pasturel, A.	[88COL/BES]
Patil, S.K.	[76PAT/RAM], [85SAW/CHA], [85SAW/RIZ]
Patterson, C.W.	[85ALD/BRO]
Paul, A.D.	[61CON/PAU]
Pavković, N.	[83MAR/PAV]
Pavlovskaya, G.R.	[74KAN/ZAR]
Pavone, D.	[67FIT/PAV], [71FIT/PAV]
Peddicord, K.L.	[90HAY/THO]
Pekárek, V.	[65PEK/VES], [65VES/PEK]
Peligot, E.	[1848PEL]
Pelsmaekers, J.	[86VOC/GRA]
Perio, P.	[53CAI/COR]
Perminov, P.S.	[67MOS/SHE]
Perrin, D.D.	[82PER]
Perrot, P.	[76DEW/PER]

Author	References
Person, W.B.	[85ALD/BRO]
Peter, L.	[83MEY/WAR]
Peterson, A.	[61PET]
Peterson, D.E.	[80EDW/STA]
Peterson, J.R.	[78PIT/PET]
Pethybridge, A.D.	[67PET/PRU]
Petrov, V.M.	[83GIR/PET]
Petrova, A.E.	[63POZ/STE]
Pfeiffer, R.	[57KIN/PFE]
Phillips, C.A.	[85PHI/PHI]
Phillips, L.R.	[56SCH/PHI]
Phillips, S.L.	[85PHI/PHI], [88PHI/HAL], [88PHI/HAL2]
Piret, P.	[79DEL/PIR]
Pirozhkov, A.V.	[76PIR/NIK], [78NIK/PIR]
Pissarjewsky, L.	[00PIS]
Pitzer, K.S.	[38LAT/PIT], [40COU/PIT], [61PIT/BRE], [73PIT], [73PIT/MAY], [74PIT/KIM], [74PIT/MAY], [75PIT], [76PIT/SIL], [77PIT/ROY], [78PIT/PET], [79PIT], [89YAN/PIT]
Plummer, L.N.	[84BUS/PLU], [86BUS/PLU]
Podolshaia, N.S.	[58IVA/TUM]
Pohl, D.C.	[85PAR/POH], [88PAR/POH]
Polasek, M.	[71POL/BAR]
Pongi, N.K.	[80DOU/PON], [80PON/DOU]
Popik, M.S.	[89EZH/BAZ]
Popov, M.M.	[50POP/GAG], [57POP/IVA], [57POP/KOS], [58POP/GAL], [59POP/GAL], [59POP/KOS], [60MAL/GAG]
Porter, R.A.	[71POR/WEB]
Porto, R.	[87CIA/IUL]
Posey, J.C.	[65WOL/POS]
Potter, P.E.	[83BAG/JAC], [90COR/KON]
Poty, B.	[87DUB/RAM]
Powell, J.	[44NOT/POW]
Powers, R.M.	[54DAY/POW]
Pozharskii, B.G.	[63POZ/STE]
Prabhu, L.H.	[61NAI/PRA]
Prasad, B.	[80GHO/GHO]
Prasad, R.	[78SIN/PRA], [78THA/AHM], [79PRA/NAG], [80NAG/BHU], [82ROY/PRA]
Prather, J.	[44FER/PRA]
Pravdic, V.	[68CUK/CAJ], [69CAJ/PRA]
Prins, G.	[68COR/PRI], [73PRI], [74COR/PRI], [75COR/OUW], [76COR/OUW], [77COR/PRI], [78PRI/COR], [79COR/MUI], [80PRI/COR], [82COR/OUW], [83COR/OUW], [85COR/OUW2], [86COR/OUW], [86COR/PRI], [90COR/KON]
Protas, J.	[59GUI/PRO]
Prue, J.E.	[67PET/PRU]

Author	References
Puigdomenech, I.	[89BRU/PUI], [91PUI/RAR]
Pusenok, G.I.	[64FLI/MIS]
Pushlenkov, M.F.	[68KRY/KOM], [68KRY/KOM3]
Pyatenko, A.T.	[84SID/PYA]
Pytkowicz, R.M.	[79JOH/PYT], [79PYT]
Qi, P.	[91BID/CAV]
Quist, A.S.	[61VAN/QUI]
Rabinowitch, E.	[51KAT/RAB]
Rafalskii, R.P.	[60RAF/KAN]
Rai, D.	[83RYA/RAI], [90RAI/FEL]
Ramakrishna, V.V.	[76PAT/RAM]
Rambidi, N.G.	[69EZH/AKI]
Ramboz, C.	[87DUB/RAM]
Rammelsberg, C.F.	[1872RAM]
Rammelsberg, S.	[1842RAM]
Rampton, D.T.	[67COL/EYR]
Rand, M.H.	[63ACH/ALC], [63ACH/ALC], [63RAN/KUB], [76OET/RAN], [81CHI/AKH], [84HOL/RAN], [90COR/KON]
Rand, R.D.	[44FER/RAN]
Rao, C.L.	[69RAO/PAI]
Rao, L.F.	[84CHO/RAO]
Rapp, K.E.	[65WOL/POS]
Rard, J.A.	[85RAR], [91PUI/RAR]
Raspopin, S.P.	[68LEB/NIC2]
Red'kin, A.F.	[89RED/SAV]
Refn, I.	[51MUU/REF]
Reil, K.O.	[85BRE/REI]
Reishus, J.W.	[67REI/GUN]
Rengemo, T.	[58FRY/NIL], [58NIL/REN]
Rengevich, V.B.	[74KAN/ZAR]
Reschetnikov, F.	[63ACH/ALC]
Rezukhina, T.N.	[60LEO/REZ]
Ricci, J.E.	[39NAI/RIC]
Riccio, G.	[84GRE/FER]
Richard, C.	[60GUS/RIC]
Richards, R.R.	[85SCH/FRI]
Richardson, D.W.	[71KIN/SIE]
Riglet, C.	[86GRE/RIG], [87RIG/VIT], [89RIG/ROB], [90RIG]
Rizkalla, E.N.	[88KHA/CHO]
Rizvi, G.H.	[85SAW/CHA], [85SAW/RIZ]
Robel, X.	[83ROB]
Roberts, C.E.	[65BAE/MEY]
Robie, R.A.	[68ROB/WAL], [78ROB/HEM2]
Robins, R.G.	[66ROB]
Robinson, P.A.	[86DES/KRA]
Robinson, R.A.	[59ROB/STO]
Robouch, P.	[89BRU/GRE], [89RIG/ROB]

Author	References
Rodichev, A.G.	[73KRU/ROD]
Roletto, E.	[69VAN/OST]
Romanov, G.A.	[63VDO/ROM], [63VDO/ROM2], [63VDO/ROM3], [66VDO/ROM], [67VDO/ROM], [69VDO/ROM], [70VDO/ROM], [70VDO/ROM2]
Ron, A.	[69CLI/GRU]
Rosengren, K.	[56AHR/LAR]
Rosenheim, A.	[22ROS/TRE], [32ROS/KEL]
Rosenzweig, A.	[77ROS/RYA]
Ross, V.	[55ROS]
Rossotti, F.J.C.	[61ROS/ROS]
Rossotti, H.	[61ROS/ROS]
Row, B.R.L.	[63HIE/ROW]
Rowe, J.J.	[62MOR/FOU]
Roy, K.N.	[82ROY/PRA]
Roy, R.N.	[77PIT/ROY]
Rudigier, H.	[85RUD/OTT], [85RUD/OTT2]
Ruff, O.	[11RUF/HEI]
Ruprecht, S.	[78BER/HEI]
Rush, R.M.	[62RUS/JOH], [63RUS/JOH]
Ryan, J.L.	[60RYA], [83RYA/RAI], [90RAI/FEL]
Ryan, R.R.	[77ROS/RYA]
Rydberg, J.	[55RYD]
Rykov, A.G.	[79FRO/RYK]
Ryzhenko, B.N.	[67RYZ/NAU], [68RYZ/KHI], [71NAU/RYZ]
Sabbah, R.	[68HAL/SAB]
Sadiq, M.	[81SAD/LIN]
Salvatore, F.	[79CIA/FER], [81CIA/FER2], [81FER/GRE], [83CIA/FER], [83FER/GRE], [84GRE/FER], [85FER/GRE], [86BRU/FER]
Samhoun, K.	[72DUF/SAM]
Samoilov, O.Ya.	[54LIP/SAM]
Samorukov, O.P.	[74SAM/KOS]
Sancho, J.	[73ALM/GAR]
Sandino, A.	[89BRU/SAN], [90BRU/SAN], [91SAN]
Sandström, M.	[87BRU/GRE]
Santalova, N.A.	[71SAN/VID], [72SAN/VID]
Sato, K.	[62MUK/NAI], [62MUK/NAI2]
Savage, A.W., Jr.	[60SAV/BRO]
Savelyeva, N.I.	[89RED/SAV]
Sawant, R.M.	[85SAW/CHA], [85SAW/RIZ]
Sawlewicz, K.	[75SAW/SIE]
Scanlan, J.P.	[75CIN/SCA], [77SCA]
Scatchard, G.	[36SCA], [76SCA]
Schäfer, H.	[74BIN/SCH]
Schauer, V.	[67JAK/SCH]
Schedin, U.	[68SCH/FRY]
Schelberg, A.	[42SCH/THO], [42THO/SCH], [48THO/SCH]
Schenkel, R.	[78BLA/FOU]
Schlyter, K.	[68ARN/SCH]

Author	References
Schmid, H.	[37SCH/MAR]
Schmidt, K.H.	[78SCH/SUL]
Schmitt, J.M.	[56BLA/COL]
Schoebrechts, J.P.	[89SCH/GEN]
Schoonen, M.A.A.	[88SCH/BAR]
Schott, H.F.	[28SCH/SWI]
Schreiner, F.	[85SCH/FRI], [86ULL/SCH], [88ULL/SCH]
Schreyer, J.M.	[53BAE/SCH2], [54SCH/BAE], [55SCH], [55SCH/BAE], [56SCH/PHI]
Schumm, R.H.	[68WAG/EVA], [82WAG/EVA]
Schwind, E.	[66KOC/SCH]
Schwing-Weill, M.J.	[79SPI/ARN]
Scott, R.B.	[48BRI/HOD], [51BRI/HOD]
Seaborg, G.T.	[57KAT/SEA]
Secoy, C.H.	[54BRO/BUN]
Selenznev, V.P.	[74MUK/SEL]
Seleznev, V.P.	[71SUP/TSV], [71TSV/SEL], [72TSV/SEL], [73TSV/SEL], [74SUP/SEL]
Senin, M.D.	[58POP/GAL], [59POP/GAL]
Senkyr, J.	[75BAR/SEN]
Serebrennikov, V.V.	[67MER/SKO]
Sergeyeva, E.I.	[72NIK/SER], [72SER/NIK], [87GUR/SER], [89RED/SAV]
Seta, K.	[82SET/MAT]
Settle, J.L.	[63SET/FED], [68OHA/SET], [69FLO/OSB], [75OHA/ADE], [84SET/OHA]
Sevast'yanov, V.G.	[71SEV/SLO], [88ALI/MAL], [89EZH/BAZ]
Shabana, R.	[76SOU/SHA]
Shcherbakov, V.A.	[63VDO/ROM], [63VDO/ROM2], [63VDO/ROM3], [66VDO/ROM]
Shchukarev, S.A.	[56SHC/VAS], [58SHC/VAS], [58SHC/VAS2], [59SHC/VAS]
Sheipline, V.M.	[55MAL/SHE], [55MAL/SHE2]
Shelyakina, A.M.	[67MOS/SHE]
Sherif, F.G.	[61SHE/AWA], [62SHE/AWA], [62SHE/AWA2]
Sherrow, S.A.	[86SHE/TOT]
Shevchenko, V.B.	[60DEN/MOS]
Shilin, I.V.	[66SHI/NAZ]
Shiloh, M.	[65SHI/MAR]
Shilov, E.A.	[34SHI], [38SHI/GLA]
Shimizu, H.	[77INA/SHI]
Shin, C.	[79SHI/CRI]
Shinn, W.A.	[72BAT/SHI], [72OHA/SHI]
Shock, E.L.	[88SHO/HEL], [89SHO/HEL], [89SHO/HEL2]
Shurukhin, V.V.	[73KOZ/BLO]
Sidorov, L.N.	[87BON/KOR], [88BOR/BOL]
Sidorova, I.V.	[84SID/PYA]
Siegel, M.D.	[88PHI/HAL], [88PHI/HAL2]
Siegel, S.	[61HOE/SIE], [70HOE/SIE], [73HOE/SIE]
Siekierski, S.	[75SAW/SIE]

Author	References
Siemans, R.E.	[71KIN/SIE]
Silbermann, J.T.	[1853FAV/SIL]
Sillén, L.G.	[54AHR/HIE], [54SIL], [54SIL2], [58FRY/NIL], [58NIL/REN], [59HIE/SIL], [63DUN/HIE], [63DUN/SIL], [63HIE/ROW], [64SIL/MAR], [67ING/KAK], [69SIL/WAR], [71BAR/SIL], [71SIL/MAR]
Silvester, L.F.	[76PIT/SIL], [77PIT/ROY], [78PIT/PET], [88PHI/HAL], [88PHI/HAL2]
Silzars, A.	[67COL/EYR]
Simon, D.M.	[52KAT/SIM]
Singer, J.	[87BOB/SIN]
Singh, Z.	[78SIN/PRA], [79PRA/NAG], [80NAG/BHU], [82ROY/PRA]
Sjöberg, S.	[81SJO/NOR], [83SJO/HAE], [85SJO/OEH]
Skeen, J.	[85PHI/PHI]
Skorik, N.A.	[67MER/SKO]
Slovyanskikh, V.K.	[71SEV/SLO]
Smirnov, V.K.	[84GOR/SMI], [84SID/PYA], [84SMI/GOR]
Smirnova, I.D.	[59KLY/SMI], [59KLY/SMI2], [70KLY/KOL], [72GLE/KLY]
Smith, B.C.	[69SMI/THA]
Smith, D.F.	[85ALD/BRO]
Smith, D.K.	[81STO/SMI]
Smith, E.R.	[44TAY/SMI]
Smith, P.N.	[70JOH/SMI]
Smith, R.M.	[76SMI/MAR], [82MAR/SMI]
Smith, S.D.	[70BAS/SMI]
Smith, W.H.	[68SMI]
Smith, W.V.	[38LAT/PIT]
Smyrl, N.R.	[84FUL/SMY]
Sobkowski, J.	[61SOB], [61SOB/MIN]
Sokolov, A.P.	[59VDO/SOK]
Solntseva, L.V.	[69VDO/ROM], [70VDO/ROM], [70VDO/ROM2]
Solovkin, A.S.	[67SOL/TSV]
Sommer, L.	[63BAR/SOM], [64BAR/SOM]
Sommers, J.A.	[75WES/SOM], [78GRO/ZAK]
Sonnenberger, D.C.	[85MOR/SON]
Sood, D.D.	[78SIN/PRA], [79PRA/NAG], [80NAG/BHU], [82ROY/PRA]
Sorokin, I.D.	[88BOR/BOL]
Souka, N.	[76SOU/SHA]
Spahiu, K.	[83CIA/FER], [83SPA], [84GRE/SPA]
Spiess, B.	[79SPI/ARN]
Spitsyn, V.I.	[61FAU/IPP], [61IPP/FAU], [61KHO/SPI]
Stabrovskii, A.I.	[60STA2], [70STA]
Staliński, B.	[63STA/NIE], [66STA/BIE], [67STA/BIE]
Stankov, V.	[73STA/CEL]
Staritzky, E.	[56STA/CRO]
Starý, J.	[60STA]
Starzynski, J.S.	[80EDW/STA]

Author	References
Stebunov, O.B.	[69VDO/STE], [69VDO/STE2]
Steele, W.V.	[81JOH/STE3]
Steidl, D.V.	[72JOH/STE]
Steinhaus, D.W.	[75STE]
Steiss, P.	[78BER/HEI]
Stepanov, M.A.	[60GAL/STE], [60STE/GAL], [62STE/GAL]
Stepanov, S.I.	[56CHU/STE]
Sterlingova, T.N.	[63POZ/STE]
Stohl, F.V.	[81STO/SMI]
Stojkovic, D.	[69STO/JAC]
Stokes, R.H.	[59ROB/STO], [79STO]
Storms, E.K.	[62STO], [67STO], [67STO/HUB], [68HOL/STO], [84HOL/RAN]
Stoughton, R.W.	[60LIE/STO]
Streeter, I.	[45DAV/MAC], [45DAV/STR]
Stumm, W.	[81STU/MOR]
Sudarikov, B.N.	[71SUP/TSV], [71TSV/SEL], [72TSV/SEL], [73TSV/SEL], [74MUK/SEL], [74SUP/SEL]
Suglobov, D.N.	[63VDO/SUG], [90LYC/MIK]
Suglobova, D.E.	[80VOL/SUG]
Suglobova, I.G.	[63VDO/SUG], [73VDO/SUG], [74VDO/VOL], [78KUD/SUG], [78SUG/CHI], [78SUG/CHI2], [80SUG/CHI], [81SUG/CHI], [86SUG/CHI], [87SUG/CHI]
Sui, D.	[80HU/NI]
Suits, E.	[65WES/SUI]
Sullivan, J.C.	[52SUL/HIN], [54SUL/HIN], [59SUL/HIN], [61SUL/HIN], [78SCH/SUL], [85SCH/FRI]
Sundaresan, M.	[81AWA/SUN]
Suponitskii, Yu.L.	[71SUP/TSV], [74MUK/SEL], [74SUP/SEL]
Suranji, T.M.	[82MIL/SUR]
Suski, W.	[75LAG/SUS]
Sutton, J.	[47SUT], [49SUT]
Sveen, A.	[70GRO/KVE]
Sverjensky, D.A.	[89SHO/HEL2]
Swift, E.H.	[28SCH/SWI]
Sylva, R.N.	[79SYL/DAV], [79SYL/DAV2], [83BRO/ELL]
Sylwestrowicz, J.	[80BRO/HUA]
Tachikawa, H.	[85TAC]
Tagawa, H.	[74TAG], [79TAG/FUJ], [81FUJ/TAG]
Takahashi, S.	[82NAI/INA]
Takahashi, Y.	[65TAK/WES], [65WES/TAK], [75YOK/TAK]
Tamaki, M.	[76IKE/TAM]
Tanayev, I.V.	[61TAN/DEI]
Tanet, G.	[91BID/CAV]
Tanger IV, J.C.	[88TAN/HEL]
Tardy, Y.	[84VIE/TAR]
Tartar, H.V.	[41TAR/GAR], [56KEL/TAR]

Author	References
Tasker, I.R.	[88TAS/OHA]
Taylor, J.C.	[72TAY/KEL]
Taylor, J.K.	[44TAY/SMI]
Taylor, J.R.	[82TAY]
Teague, L.S.	[80BEN/TEA]
Tetenbaum, M.	[71TET/HUN]
Thakur, L.	[69SMI/THA], [78THA/AHM]
Thamer, B.J.	[57THA]
Thomas, J.K.	[90HAY/THO]
Thomas, L.F.	[41AMP/THO], [48AMP/MUL]
Thompson, R.C.	[78SCH/SUL]
Thompson, R.W.	[42SCH/THO], [42THO/SCH], [48THO/SCH]
Thorn, R.J.	[63ACH/ALC]
Tichy, J.	[70GRO/KVE]
Titlestad, N.	[10TIT]
To, M.W.K.	[88BUR/TO]
Torchenkova, E.A.	[84KOT/EVS]
Toth, L.M.	[84TOT/FRI], [86SHE/TOT]
Tremaine, P.R.	[80LEM/TRE], [81TRE/CHE], [89HOV/NGU]
Tretyak, Z.A.	[55KOM/TRE]
Tretyakov, A.A.	[81GOL/TRE]
Trewendt, G.	[22ROS/TRE]
Tripathi, K.K.	[61BAN/TRI]
Tripathi, V.S.	[84TRI]
Troć, R.	[66STA/BIE], [67STA/BIE], [80BLA/LAG], [80BLA/TRO]
Tso, T.C.	[85TSO/BRO]
Tsuji, T.	[83NAI/TSU]
Tsvelodub, L.D.	[77NIK/TSV]
Tsvetkov, A.A.	[71SUP/TSV], [71TSV/SEL], [72TSV/SEL], [73TSV/SEL]
Tsvetkova, Z.N.	[67SOL/TSV]
Tsymbal, C.	[69TSY]
Tumbakov, V.A.	[58IVA/TUM], [59IVA/TUM], [74SAM/KOS]
Tummavuori, J.	[68TUM/LUM]
Turck, G.	[76BOU/BON]
Turner, D.R.	[81TUR/WHI]
Tveekrem, J.O.	[68TVE/CHA]
Tyler, E.G.	[65BEE/MOR]
Uchijima, T.	[62MUK/NAI], [62MUK/NAI2]
Ullman, W.J.	[86ULL/SCH], [88ULL/SCH]
Umreiko, D.S.	[78KOB/KOL]
Unmack, A.	[29BJE/UNM]
Unrein, P.J.	[76CHO/UNR]
Vainiotalo, A.	[81VAI/MAK]
Val, C.	[68DOG/VAL]
Van der Weijden, C.H.	[84GEN/WEI]
Van Genderen, A.C.G.	[84GEN/WEI]
Van Lier, J.A.	[60LIE/BRU]

Author	References
Van Rensen, E.	[72KNA/MUE]
Van Vlaanderen, P.	[68COR/PRI], [77COR/PRI], [83COR/OUW], [83COR/OUW2], [85COR/VLA]
Van Wazer, J.R.	[50WAZ/CAM]
Vanderzee, C.E.	[61VAN/QUI]
Vanni, A.	[69VAN/OST]
Varfeldt, J.	[69GRE/VAR]
Vartak, D.G.	[61NAI/PRA]
Vasil'ev, V.P.	[62VAS], [64VAS/MUK]
Vasil'kova, I.V.	[56SHC/VAS], [58SHC/VAS], [58SHC/VAS2], [59SHC/VAS]
Vasilkova, I.V.	[68MAR/KUD]
Čeleda, J.	[73STA/CEL]
Vdovenko, V.M.	[57VDO/KOV], [59VDO/SOK], [63VDO/ROM], [63VDO/ROM2], [63VDO/ROM3], [63VDO/SUG], [66VDO/ROM], [67VDO/ROM], [69VDO/ROM], [69VDO/STE], [69VDO/STE2], [70VDO/ROM], [70VDO/ROM2], [73VDO/SUG], [74VDO/VOL]
Veits, I.V.	[78GLU/GUR], [82GLU/GUR]
Venugopal, V.	[78SIN/PRA], [79PRA/NAG], [80NAG/BHU], [82ROY/PRA]
Verma, V.P.	[73VER/KHA]
Vernyi, E.A.	[64MAR/VER]
Veselý, V.	[65PEK/VES], [65VES/PEK]
Vidavskii, L.M.	[65VID/BYA], [71SAN/VID], [71VID/IPP], [72SAN/VID]
Vieillard, P.	[84VIE/TAR]
Vîlceanu, N.	[65DRA/JUL], [70DRA/VIL], [73DRA/VIL]
Viljoen, C.L.	[87VIL]
Vinogradov, A.V.	[64MAR/VER]
Vitorge, P.	[86GRE/RIG], [87RIG/VIT], [89RIG/ROB], [90CAP/VIT]
Vochten, R.	[84VOC/GOE], [86VOC/GRA]
Vodvenko, V.M.	[58VDO/MAL]
Vogt, O.	[85RUD/OTT], [85RUD/OTT2]
Volk, V.I.	[77VOL/BEL]
Volkov, V.A.	[74VDO/VOL], [80VOL/SUG]
Vosburgh, W.C.	[41VOS/COO]
Vozzella, P.A.	[65VOZ/CRE]
Wacker, P.F.	[47WAC/CHE]
Wagman, D.D.	[68WAG/EVA], [71MIN/GIL], [71PAR/WAG], [73HUL/DES], [82WAG/EVA], [89COX/WAG]
Wahl, A.C.	[58CON/WAH]
Wait, E.	[56DAW/WAI]
Waldbaum, D.R.	[68ROB/WAL]
Walenta, K.	[78WAL]
Walker, J.B.	[70WAL]
Wallace, G.J.	[81TRE/CHE]
Wallace, R.M.	[67WAL]
Wanner, H.	[84WAN], [86WAN], [87BRO/WAN], [88WAN], [89GRE/WAN], [89WAN2], [89WAN3], [89WAN4], [90WAN], [90WAN2], [91PUI/RAR]

Author	References
Ward, K.	[83MEY/WAR]
Warnquist, B.	[67ING/KAK], [69SIL/WAR]
Wassef, M.A.	[69SMI/THA]
Weber, Jr.	[71POR/WEB]
Webster, R.A.	[43WEB]
Weeks, A.M.	[76FRO/ITO]
Weigel, F.	[76WEI/HOF], [85WEI], [86WEI]
Weinstock, B.	[48WEI/CRI]
Weisbrod, A.	[87DUB/RAM]
Weiss, A.	[57WEI/HAR]
Weissman, S.	[43ALT/LIP]
Wendolkowski, W.S.	[54WEN/KIR]
Wendt, H.	[70FRE/WEN]
Werner, G.K.	[85ALD/BRO]
Wesolowski, D.J.	[90DIC/WES]
Westrum, E.F., Jr.	[55OSB/WES], [56GRE/WES], [56GRE/WES3], [57OSB/WES], [59WES/GRO], [60BUR/OSB], [62WES/GRO], [62WES/GRO2], [63ACH/ALC], [65TAK/WES], [65WES/SUI], [65WES/TAK], [66GOO/WES], [66WES], [66WES/BAR], [68FLO/OSB], [68GIR/WES], [70WES/GRO], [71HUN/WES], [75WES/MAG], [75WES/SOM], [77ALL/FAL], [78GRO/ZAK], [80COR/WES], [80WES/ZAI], [84GRO/DRO], [88BUR/TO], [88COR/WES], [89COR/KON]
Wheeler, V.J.	[67MAR/BON]
Whiffen, D.H.	[79WHI2]
White, A.G.	[57HEA/WHI]
White, H.J., Jr.	[87GAR/PAR]
White, J.M.	[58LI/DOO]
Whitfield, M.	[79WHI], [81TUR/WHI]
Wicks, C.E.	[63WIC/BLO]
Wijbenga, G.	[81WIJ], [82WIJ], [82WIJ/COR]
Wilcox, D.E.	[63WIL/BRO]
Wilhite, R.N.	[55DAY/WIL]
Williams, C.W.	[83MOR/WIL2], [84WIL/MOR], [85GEN/FUG]
Williams, P.A.	[79HAA/WIL], [80ALW/WIL], [81OBR/WIL], [83OBR/WIL]
Williamson, A.T.	[41WIL/LYN]
Willis, B.T.M.	[87WIL]
Wilson, A.S.	[61WIL/KED]
Wilson, G.L.	[71GRO/HAY]
Winchell, P.	[64NOR/WIN]
Witteman, W.G.	[69HUB/HOL3]
Wojakowski, A.	[75LAG/SUS], [78BLA/FOU], [80BLA/LAG2]
Wolery, T.J.	[79WOL], [83WOL], [88JAC/WOL]
Wolf, A.S.	[65WOL/POS]
Woodhead, J.L.	[64MCK/WOO]
Wooster, C.B.	[38WOO]
Wu, C.H.	[63WU/BIR]

Author	References
Yamashita, T.	[79TAG/FUJ]
Yamauchi, S.	[84FLO/HAS]
Yang, J.	[89YAN/PIT]
Yefimov, M.Ye.	[82DEV/YEF], [83DAV/YEF], [83DEV/YEF]
Yokokawa, H.	[75YOK/TAK]
Yost, D.M.	[28SCH/SWI]
Young, H.S.	[58YOU/GRA]
Yuldachev, F.	[89EZH/BAZ]
Yuldashev, F.	[88ALI/MAL]
Yungman, V.S.	[78GLU/GUR], [82GLU/GUR], [85HIL/GUR]
Yusova, Yu.I.	[73KRU/ROD]
Zachariasen, W.H.	[54MUL/ELL]
Zainel, H.A.	[80WES/ZAI], [88BUR/TO]
Zakareia, N.	[75ALY/ABD]
Zakharov, S.V.	[84KOT/EVS]
Zakharova, F.A.	[67ZAK/ORL]
Zaki, M.R.	[77ALL/FAL], [78GRO/ZAK]
Zarubin, A.T.	[74KAN/ZAR]
Zarubin, D.P.	[90ZAR/NEM]
Zasorin, E.Z.	[83GIR/PET]
Zebroski, E.L.	[51ZEB/ALT]
Zeek, W.	[57KIN/PFE]
Zhang, M.	[80HU/NI]
Zhang, S.	[82FEN/ZHA]
Zhang, Z.	[82FEN/ZHA]
Zhvanko, S.A.	[61KHO/SPI]
Zielen, A.J.	[59ZIE], [61SUL/HIN]
Zmbov, K.F.	[69ZMB]
Zolnierek, Z.	[79BLA/LAG]
Zubova, N.V.	[59POP/KOS]
Zygmunt, A.	[80BLA/LAG2]

Chapter IX

Formula list

This chapter presents, in standard order of arrangement (*cf.* Section II.1.7) a list of formulae of uranium containing species. The meaning of the phase designators, (aq), (cr), *etc.*, is explained in Section II.1.4.

The formulae for which selected data are presented in Chapter III, as well as those which are discussed but for which no data are recommended by this review, are marked correspondingly. The formulae that are not discussed in the present review are provided with information on references that may contain thermodynamic data on these compounds or complexes (for complete citations see Chapter VII).

The inclusion of these formulae in this chapter is to be understood as information on the existence of published material on these formulae. It in no way implies that the present review gives any credit to either the thermodynamic data or the chemical composition or existence of these species.

Formula	References
β-U	[47MOO/KEL], [87THE]
U(cr)	data selected in the present review
U(g)	data selected in the present review
γ-U	[47MOO/KEL], [87THE]
U(l)	[87THE]
U^{2+}	discussed in the present review
U^{3+}	data selected in the present review
U^{4+}	data selected in the present review
U^{5+}	[82FUG]
U^{6+}	[82FUG]
UD_3(cr)	[60ABR/OSB]
UO(cr)	[51KAT/RAB], [54HOE/KAT], [58BRE/BRO], [62WES/GRO], [69HUB/HOL3], [72KRE], [86BRA/LAG], [86MOR]
UO(g)	data selected in the present review
UO^{2+}	[60STE/GAL]
$UO_{1.5}$(cr)	[86BRA/LAG]
UO_2(am)	[78LAN], [80GAL/KAI], [84MUL/DUD], [88PHI/HAL2], [89BRU/PUI]
UO_2(cr)	data selected in the present review
UO_2(g)	data selected in the present review
γ-UO_2	[67FIT/PAV], [71NAU/RYZ]
UO_2(l)	[87THE]
UO_2(s)	[87THE], [89BRU/PUI]
UO_2^+	data selected in the present review
UO_2^{2+}	data selected in the present review
$UO_{2.017}$(s)	[70GRO/KVE]
$UO_{2.023}$(s)	[81JOH/STE3]
$UO_{2.125}$(cr)	[70MAR/CIO]
$UO_{2.20}$(s)	[82SET/MAT]
$UO_{2.228}$(s)	[73INA/NAI]
$UO_{2.235}$(s)	[82SET/MAT]
$UO_{2.240}$(s)	[73INA/NAI]
β-$UO_{2.25}$	data selected in the present review
$UO_{2.25}$(cr)	data selected in the present review
$UO_{2.25}$(s)	[65GOT/NAI], [71NAU/RYZ]
$UO_{2.254}$(s)	[70GRO/KVE]
α-$UO_{2.3333}$	data selected in the present review
β-$UO_{2.3333}$	data selected in the present review
$UO_{2.62}$(s)	[58BRE/BRO]
$UO_{2.63}$(cr)	[70MAR/CIO]
$UO_{2.640}$(s)	[82NAI/INA]
$UO_{2.656}$(s)	[82NAI/INA]
$UO_{2.663}$(s)	[82NAI/INA]
$UO_{2.6667}$(cr)	data selected in the present review
$UO_{2.86}\cdot 0.5H_2O$(cr)	data selected in the present review
$UO_{2.86}\cdot 1.5H_2O$(cr)	data selected in the present review
$UO_{2.9}$(s)	[82HEM]
$UO_{2.92}$(s)	[64COR], [72FUG]

Formula	References
α-UO_3	data selected in the present review
UO_3(am)	data selected in the present review
β-UO_3	data selected in the present review
UO_3(cr)	[12MIX]
δ-UO_3	data selected in the present review
ε-UO_3	data selected in the present review
UO_3(g)	data selected in the present review
γ-UO_3	data selected in the present review
UO_3(hexa)	[54COU], [57KAT/SEA]
UO_4^{2-}	[55GAY/LEI]
U_2O_3(cr)	[63WIL/BRO], [72KRE], [85MOR/SON], [85MOR2], [86MOR]
U_2O_5(cr)	[71MAR/CIO], [72KRE], [86MOR]
$U_2O_5^{2+}$	[62NIK/PAR]
UO_2UO_4(cr)	[83KAG/KYS]
U_2O_9(cr)	[69KNA/LOS2]
α-U_3O_7	[62WES/GRO2], [62WES/GRO], [71NAU/RYZ], [82HEM], [82WAG/EVA], [83FUG], [86MOR]
β-U_3O_7	[62WES/GRO2], [62WES/GRO], [71NAU/RYZ], [72FUG], [82FUG], [82HEM], [82MOR], [82WAG/EVA], [83FUG], [86MOR], [88LEM]
U_3O_7(cr)	[62MUK/NAI], [89BRU/PUI]
α-U_3O_8	[61KHO/SPI], [88PHI/HAL2]
U_3O_8(cr)	[12MIX], [79TAG/FUJ], [89BRU/PUI], [89COX/WAG]
U_4O_8(cr)	[72KRE]
U_4O_9(cr)	[65MUT/HIR], [65MUT], [66WES], [80LEM/TRE], [89BRU/PUI]
U_4O_{17}(cr)	[69KNA/LOS2]
β-UH_3	data selected in the present review
UH_3(cr)	[52LAT], [54HOE/KAT], [55ABR/FLO], [55MAL/SHE2], [57KAT/SEA], [58MAC], [59FLO/LOH], [61KEL/KIN], [63RAN/KUB], [63SET/FED], [67FLO/OSB], [72KRE], [73BAR/KNA], [79KUB/ALC], [82WAG/EVA], [83FUG], [84FLO/HAS], [86MOR]
UOH^{2+}	[78ALL/BEA], [80ALL/KIP]
UOH^{3+}	data selected in the present review
$U(OH)_2^+$	[78ALL/BEA], [80ALL/KIP]
$U(OH)_2^{2+}$	[78LAN], [80BEN/TEA], [80GAL/KAI], [86WAN], [89BRU/PUI]
$UO_3\cdot0.393H_2O$(cr)	data selected in the present review
UO_2OH(aq)	[78ALL/BEA], [80ALL/KIP], [83ALL], [86WAN], [87BRO/WAN]
UO_2OH(cr)	[78ALL/BEA], [80ALL/KIP], [83ALL], [86WAN], [87BRO/WAN]
UO_2OH^+	data selected in the present review
$UO_3\cdot0.5H_2O$(cr)	[72SAN/VID]
$UO_3\cdot0.648H_2O$(cr)	data selected in the present review
$UO_3\cdot0.64H_2O$(cr)	[88OHA/LEW]
α-$UO_3\cdot0.85H_2O$	data selected in the present review
$UO_3\cdot0.85H_2O$(cr)	[88OHA/LEW]
$UO_3\cdot0.8H_2O$(cr)	[66ROB]
α-$UO_3\cdot0.9H_2O$	data selected in the present review
$UO_3\cdot0.9H_2O$(cr)	[88OHA/LEW]

Formula	References
$UO(OH)_2(cr)$	[60STE/GAL]
$U(OH)_3(aq)$	[78ALL/BEA], [80ALL/KIP]
$U(OH)_3(cr)$	[52LAT], [63WIL/BRO], [72KRE], [73MOS], [80ALL/KIP], [86MOR]
$U(OH)_3^+$	[78LAN], [80BEN/TEA], [86WAN], [89BRU/PUI]
HUO_4^-	[55GAY/LEI], [62HOS/GAR], [65MUT/HIR], [65MUT]
α-$UO_2(OH)_2$	[71NAU/RYZ]
$UO_2(OH)_2(aq)$	data selected in the present review
β-$UO_2(OH)_2$	data selected in the present review
$UO_2(OH)_2(cr)$	[82JEN]
γ-$UO_2(OH)_2$	data selected in the present review
$UO_2(OH)_2^-$	[78ALL/BEA], [80ALL/KIP], [87BRO/WAN]
α-$UO_3{\cdot}H_2O$	[82HEM], [82WAG/EVA], [83FUG]
β-$UO_3{\cdot}H_2O$	[64COR], [72FUG], [78BER/HEI], [78COR/OHA], [82HEM], [82WAG/EVA], [83FUG], [86MOR]
$UO_3{\cdot}H_2O(cr)$	[88OHA/LEW]
ε-$UO_3{\cdot}H_2O$	[64COR], [72FUG], [78COR/OHA], [82WAG/EVA], [86MOR]
$H_3UO_4^-$	[57GAY/LEI]
$U(OH)_4(aq)$	data selected in the present review
$U(OH)_4(cr)$	[52LAT], [60STE/GAL], [63WIL/BRO], [72KRE], [73MOS2], [73MOS], [78ALL/BEA], [80ALL/KIP], [82JEN], [83ALL], [86WAN]
$U(OH)_4^-$	[78ALL/BEA], [80ALL/KIP]
$UO_2{\cdot}2H_2O(am)$	[82HEM]
$UO_2{\cdot}2H_2O(cr)$	[82HEM], [83FLE]
$UO_2(OH)_3^-$	data selected in the present review
$UO_2(OH)_3^{2-}$	[87BRO/WAN]
$UO_2(OH)_2{\cdot}H_2O(cr)$	[89BRU/PUI]
α-$UO_3{\cdot}2H_2O$	[82HEM]
$UO_3{\cdot}2H_2O(am)$	[89BRU/SAN]
$UO_3{\cdot}2H_2O(cr)$	data selected in the present review
ε-$UO_3{\cdot}2H_2O$	[82HEM]
$U(OH)_5^-$	data selected in the present review
$UO_2(OH)_4^{2-}$	data selected in the present review
$UO_2(OH)_4^{3-}$	[87BRO/WAN]
$UO_4{\cdot}2H_2O(cr)$	data selected in the present review
$U(OH)_6^{2-}$	[87BRO/WAN]
$UO_2(OH)_5^{3-}$	[87BRO/WAN]
$UO_2(OH)_5^{4-}$	[87BRO/WAN]
$UO_4{\cdot}4H_2O(cr)$	data selected in the present review
$UO_2{\cdot}833.2H_2O(cr)$	discussed in the present review
$U_2(OH)_2^{4+}$	[78ALL/BEA], [80ALL/KIP]
$U_2(OH)_2^{6+}$	[78ALL/BEA], [80ALL/KIP], [83ALL]
$U_2(OH)_3^{5+}$	[80ALL/KIP]
$U_2(OH)_4^{4+}$	[80ALL/KIP]
$(UO_2)_2OH^{3+}$	data selected in the present review
$U_2(OH)_5^{3+}$	[80ALL/KIP]

Formula	References
$(UO_2)_2(OH)_2^{2+}$	data selected in the present review
$U_3O_8OH^+$	[54HIN], [62NIK/PAR]
$(UO_2)_3(OH)_4^{2+}$	data selected in the present review
$(UO_2)_3(OH)_5^+$	data selected in the present review
$(UO_2)_3(OH)_6(aq)$	discussed in the present review
$(UO_2)_3(OH)_7^-$	data selected in the present review
$(UO_2)_3(OH)_8^{2-}$	discussed in the present review
$(UO_2)_4(OH)_2^{6+}$	[56ORB/BAR]
$U_4(OH)_{12}^{4+}$	[87BRO/WAN]
$(UO_2)_4(OH)_6^{2+}$	discussed in the present review
$(UO_2)_4(OH)_7^+$	data selected in the present review
$U_6(OH)_{15}^{9+}$	data selected in the present review
$U_6O_{17}\cdot 10H_2O(cr)$	see $UO_2(UO_3)_5\cdot 10H_2O(cr)$
$(UO_2)_9(OH)_{19}^-$	discussed in the present review
$UF(g)$	data selected in the present review
UF^{3+}	data selected in the present review
$UF_2(cr)$	[86BRA/LAG]
$UF_2(g)$	data selected in the present review
UF_2^{2+}	data selected in the present review
$UF_3(cr)$	data selected in the present review
$UF_3(g)$	data selected in the present review
$UF_3(s)$	[87THE]
UF_3^+	data selected in the present review
$UF_4(aq)$	data selected in the present review
$UF_4(cr)$	data selected in the present review
$UF_4(g)$	data selected in the present review
$UF_{4.25}(cr)$	see $U_4F_{17}(cr)$
$UF_{4.5}(cr)$	see $U_2F_9(cr)$
α-UF_5	data selected in the present review
β-UF_5	data selected in the present review
$UF_5(cr)$	[45BRE/BRO], [52LAT], [54HOE/KAT], [58BRE/BRO], [63RAN/KUB], [63WIC/BLO], [68BRO], [69KNA/LOS2], [72FUG], [72KRE], [73BAR/KNA], [73MOS], [79KLE/HIL], [79KUB/ALC], [87BON/KOR], [87THE]
$UF_5(g)$	data selected in the present review
UF_5^-	data selected in the present review
$UF_6(cr)$	data selected in the present review
$UF_6(g)$	data selected in the present review
$UF_6(l)$	[79KUB/ALC]
UF_6^{2-}	data selected in the present review
$U_2F_9(cr)$	data selected in the present review
$(UF_5)_2(g)$	see $U_2F_{10}(g)$
$U_2F_{10}(g)$	data selected in the present review
$U_4F_{17}(cr)$	data selected in the present review
$UOF(g)$	[85HIL/GUR]
$UOF_2(cr)$	data selected in the present review
$UOF_2(g)$	[85HIL/GUR]

Formula	References
$UOF_2(s)$	[86HIS/BEN]
$UOF_3(cr)$	discussed in the present review
$UOF_3(g)$	[85HIL/GUR]
$UOF_4(cr)$	data selected in the present review
$UOF_4(g)$	data selected in the present review
UOF_5^-	[84SID/PYA]
$UO_2F(aq)$	[78ALL/BEA], [87BRO/WAN]
$UO_2F(g)$	[85HIL/GUR]
UO_2F^+	data selected in the present review
$UO_2F_2(aq)$	data selected in the present review
$UO_2F_2(cr)$	data selected in the present review
$UO_2F_2(g)$	data selected in the present review
$UO_2F_2^-$	[78ALL/BEA], [87BRO/WAN]
$UO_2F_3^-$	data selected in the present review
$UO_2F_3^{2-}$	[78ALL/BEA], [87BRO/WAN]
$UO_2F_4^{2-}$	data selected in the present review
$UO_2F_4^{3-}$	[78ALL/BEA], [87BRO/WAN]
$UO_2F_5^{3-}$	[87BRO/WAN]
$UO_2F_5^{4-}$	[87BRO/WAN]
$U_2O_3F_6(cr)$	data selected in the present review
$U_3O_5F_8(cr)$	data selected in the present review
$UF_4{\cdot}1.333H_2O(cr)$	[67GAG/KHA]
$UF_4{\cdot}H_2O(cr)$	[51KAT/RAB]
$UF_4{\cdot}1.5H_2O(cr)$	[67GAG/KHA]
$H_3OUF_6(cr)$	data selected in the present review
$UOFOH(cr)$	data selected in the present review
$UOFOH{\cdot}0.5H_2O(cr)$	data selected in the present review
$UOF_2{\cdot}H_2O(cr)$	data selected in the present review
$UF_4{\cdot}2.5H_2O(cr)$	data selected in the present review
$UO_2FOH(cr)$	[73TSV/SEL], [75FLO/OSB]
$UO_2F_2{\cdot}H_2O(cr)$	[72TSV/SEL], [73TSV/SEL], [80PAR]
$UO_2F_2{\cdot}1.6H_2O(cr)$	[71SUP/TSV], [71TSV/SEL]
$UO_2FOH{\cdot}H_2O(cr)$	data selected in the present review
$UO_2F_2{\cdot}2H_2O(cr)$	[71SUP/TSV], [71TSV/SEL], [72TSV/SEL], [73TSV/SEL], [83KAG/KYS]
$UO_2FOH{\cdot}2H_2O(cr)$	data selected in the present review
$UO_2F_2{\cdot}3H_2O(cr)$	data selected in the present review
$UO_2F_2{\cdot}4H_2O(cr)$	[71SUP/TSV], [71TSV/SEL], [72TSV/SEL], [73TSV/SEL], [83KAG/KYS]
$UO_2(HF_2)_2{\cdot}4H_2O(cr)$	discussed in the present review
$UCl(g)$	data selected in the present review
UCl^{2+}	discussed in the present review
UCl^{3+}	data selected in the present review
$UCl_2(cr)$	[72KRE], [86BRA/LAG]
$UCl_2(g)$	data selected in the present review
UCl_2^{2+}	discussed in the present review
$UCl_3(cr)$	data selected in the present review
$UCl_3(g)$	data selected in the present review

Formula	References
UCl_3^+	[78ALL/BEA], [87BRO/WAN]
$UCl_4(aq)$	[80PAR], [82WAG/EVA], [87BRO/WAN]
$UCl_4(cr)$	data selected in the present review
$UCl_4(g)$	data selected in the present review
$UCl_4(l)$	[59POP/GAL]
$UCl_5(cr)$	data selected in the present review
$UCl_5(g)$	data selected in the present review
UCl_5^-	[87BRO/WAN]
$UCl_6(cr)$	data selected in the present review
$UCl_6(g)$	data selected in the present review
UCl_6^{2-}	[79FUG]
$U_2Cl_8(g)$	data selected in the present review
$(UCl_5)_2(g)$	see $U_2Cl_{10}(g)$
$U_2Cl_{10}(g)$	data selected in the present review
$UOCl(cr)$	data selected in the present review
$UOCl_2(cr)$	data selected in the present review
$UOCl_2(s)$	[86HIS/BEN]
$UOCl_3(cr)$	data selected in the present review
$UO_2Cl(aq)$	[87BRO/WAN]
$UO_2Cl(cr)$	data selected in the present review
UO_2Cl^+	data selected in the present review
$UO_2Cl_2(aq)$	data selected in the present review
$UO_2Cl_2(cr)$	data selected in the present review
$UO_2Cl_2(g)$	data selected in the present review
$UO_2Cl_2^-$	[87BRO/WAN]
$UO_2Cl_3^-$	[87BRO/WAN]
$UO_2Cl_3^{2-}$	[87BRO/WAN]
$UO_2Cl_4^{2-}$	[87BRO/WAN]
$UO_2Cl_4^{3-}$	[87BRO/WAN]
$UO_2Cl_5^{3-}$	[87BRO/WAN]
$UO_2Cl_5^{4-}$	[87BRO/WAN]
$UClO_4^{3+}$	discussed in the present review
$UO_2ClO_2^+$	discussed in the present review
$UO_2ClO_3^+$	data selected in the present review
$U_2O_2Cl_5(cr)$	data selected in the present review
$(UO_2)_2Cl_3(cr)$	data selected in the present review
$U_5O_{12}Cl(cr)$	data selected in the present review
$UOClOH^{4+}$	discussed in the present review
$UO_2Cl_2 \cdot H_2O(cr)$	data selected in the present review
$UO_2ClOH \cdot 2H_2O(cr)$	data selected in the present review
$UO_2Cl_2 \cdot 3H_2O(cr)$	data selected in the present review
$UCl_3F(cr)$	data selected in the present review
$UCl_2F_2(cr)$	data selected in the present review
$UClF_3(cr)$	data selected in the present review
$UBr(g)$	data selected in the present review
UBr^{2+}	discussed in the present review
UBr^{3+}	data selected in the present review

Formula	References
$UBr_2(cr)$	[72KRE], [86BRA/LAG]
$UBr_2(g)$	data selected in the present review
UBr_2^{2+}	[87BRO/WAN]
$UBr_3(cr)$	data selected in the present review
$UBr_3(g)$	data selected in the present review
UBr_3^+	[87BRO/WAN]
$UBr_4(aq)$	[87BRO/WAN]
$UBr_4(cr)$	data selected in the present review
$UBr_4(g)$	data selected in the present review
$UBr_5(cr)$	data selected in the present review
$UBr_5(g)$	data selected in the present review
UBr_5^-	[87BRO/WAN]
$UBr_6(cr)$	[72KRE]
UBr_6^{2-}	[79FUG]
$UOBr(cr)$	discussed in the present review
$UOBr_2(cr)$	data selected in the present review
$UOBr_2(s)$	[86HIS/BEN]
$UOBr_3(cr)$	data selected in the present review
$UO_2Br(aq)$	[87BRO/WAN]
UO_2Br^+	data selected in the present review
$UO_2Br_2(aq)$	[52LAT], [87BRO/WAN]
$UO_2Br_2(cr)$	data selected in the present review
$UO_2Br_2^-$	[87BRO/WAN]
$UO_2Br_3^-$	[87BRO/WAN]
$UO_2Br_3^{2-}$	[87BRO/WAN]
$UO_2Br_4^{2-}$	[87BRO/WAN]
$UO_2Br_4^{3-}$	[87BRO/WAN]
$UO_2Br_5^{3-}$	[87BRO/WAN]
$UO_2Br_5^{4-}$	[87BRO/WAN]
$UO_2BrO_3^+$	data selected in the present review
$UO_2Br_2\cdot H_2O(cr)$	data selected in the present review
$UO_2Br_2\cdot 2H_2O(cr)$	[82HEM]
$UO_2BrOH\cdot 2H_2O(cr)$	data selected in the present review
$UO_2Br_2\cdot 3H_2O(cr)$	data selected in the present review
$UBr_3\cdot 6H_2O(cr)$	[73MOS2], [73MOS]
$UBr_2Cl(cr)$	data selected in the present review
$UBr_3Cl(cr)$	data selected in the present review
$UBr_3Cl(g)$	discussed in the present review
$UClBr_3(g)$	[88ALI/MAL]
$UBrCl_2(cr)$	data selected in the present review
$UBr_2Cl_2(cr)$	data selected in the present review
$UBr_2Cl_2(g)$	discussed in the present review
$UCl_2Br_2(g)$	[88ALI/MAL]
$UBrCl_3(cr)$	data selected in the present review
$UBrCl_3(g)$	discussed in the present review
$UCl_3Br(g)$	[88ALI/MAL]
$UI(g)$	data selected in the present review

Formula	References
UI^{3+}	data selected in the present review
$UI_2(cr)$	[86BRA/LAG]
$UI_2(g)$	data selected in the present review
UI_2^{2+}	[87BRO/WAN]
$UI_3(cr)$	data selected in the present review
$UI_3(g)$	data selected in the present review
UI_3^+	[87BRO/WAN]
$UI_4(aq)$	[87BRO/WAN]
$UI_4(cr)$	data selected in the present review
$UI_4(g)$	data selected in the present review
$UI_4(l)$	[59POP/GAL]
$UI_5(cr)$	[72KRE], [87HIS/BEN]
UI_5^-	[87BRO/WAN]
$UI_6(cr)$	[72KRE]
UI_6^{2-}	[87BRO/WAN]
$UOI_3(cr)$	[72KRE]
$UO_2I(aq)$	[87BRO/WAN]
UO_2I^+	discussed in the present review
$UO_2I_2(aq)$	[87BRO/WAN]
$UO_2I_2(cr)$	[78PRI/COR], [82HEM], [86MOR]
$UO_2I_2^-$	[87BRO/WAN]
$UO_2I_3^-$	[87BRO/WAN]
$UO_2I_3^{2-}$	[87BRO/WAN]
$UO_2I_4^{2-}$	[87BRO/WAN]
$UO_2I_4^{3-}$	[87BRO/WAN]
$UO_2I_5^{3-}$	[87BRO/WAN]
$UO_2I_5^{4-}$	[87BRO/WAN]
UIO_3^{3+}	[87BRO/WAN]
$UO_2IO_3(aq)$	[87BRO/WAN]
$UO_2IO_3^+$	data selected in the present review
$U(IO_3)_2^{2+}$	[87BRO/WAN]
$UO_2(IO_3)_2(aq)$	data selected in the present review
$UO_2(IO_3)_2(cr)$	data selected in the present review
$UO_2(IO_3)_2^-$	[87BRO/WAN]
$U(IO_3)_3^+$	[87BRO/WAN]
$UO_2(IO_3)_3^-$	[87BRO/WAN]
$UO_2(IO_3)_3^{2-}$	[87BRO/WAN]
$U(IO_3)_4(aq)$	[87BRO/WAN]
$UO_2(IO_3)_4^{2-}$	[87BRO/WAN]
$UO_2(IO_3)_4^{3-}$	[87BRO/WAN]
$U(IO_3)_5^-$	[87BRO/WAN]
$UO_2(IO_3)_5^{3-}$	[87BRO/WAN]
$UO_2(IO_3)_5^{4-}$	[87BRO/WAN]
$U(IO_3)_6^{2-}$	[87BRO/WAN]
$H(UO_2)_2IO_6(cr)$	discussed in the present review
$UClI_3(cr)$	data selected in the present review
$UCl_2I_2(cr)$	data selected in the present review

Formula	References
$UCl_3I(cr)$	data selected in the present review
$UBrI_3(cr)$	data selected in the present review
$UBr_2I_2(cr)$	data selected in the present review
$UBr_3I(cr)$	data selected in the present review
$US(cr)$	data selected in the present review
$US(g)$	[74MIL], [83FUG], [84GRO/DRO], [86MOR], [87THE]
$US_{1.011}(s)$	[68OHA/SET], [72FUG]
$US_{1.5}(cr)$	see $U_2S_3(cr)$
$US_{1.67}(s)$	[83FUG]
$US_{1.90}(cr)$	data selected in the present review
$US_{1.992}(cr)$	[75OHA/ADE], [84SET/OHA]
$US_2(cr)$	data selected in the present review
$US_2(g)$	[74MIL], [84GRO/DRO]
$US_3(cr)$	data selected in the present review
$(US)_2(g)$	[83FUG]
$U_2S_2(g)$	[84GRO/DRO]
$U_2S_3(cr)$	data selected in the present review
$U_2S_5(cr)$	data selected in the present review
$U_3S_5(cr)$	data selected in the present review
USO_4^{2+}	data selected in the present review
$UO_2SO_3(aq)$	data selected in the present review
$UO_2SO_3(cr)$	data selected in the present review
$UO_2S_2O_3(aq)$	data selected in the present review
$UO_2SO_4(aq)$	data selected in the present review
$UO_2SO_4(cr)$	data selected in the present review
$UO_2SO_4^-$	[78ALL/BEA], [87BRO/WAN]
$U(SO_3)_2(cr)$	data selected in the present review
$U(SO_4)_2(aq)$	data selected in the present review
$U(SO_4)_2(cr)$	data selected in the present review
$UO_2(SO_3)_2^{2-}$	discussed in the present review
$UO_2(SO_4)_2^{2-}$	data selected in the present review
$UO_2(SO_4)_2^{3-}$	[87BRO/WAN]
$UO_2(SO_3)_3^{4-}$	discussed in the present review
$U(SO_4)_3^{2-}$	[78ALL/BEA], [87BRO/WAN]
$UO_2(SO_4)_3^{4-}$	[65MUT/HIR], [78ALL/BEA], [81TUR/WHI], [82JEN], [86WAN], [87BRO/WAN]
$UO_2(SO_4)_3^{5-}$	[87BRO/WAN]
$U(SO_4)_4^{4-}$	[87BRO/WAN]
$UO_2(SO_4)_4^{6-}$	[87BRO/WAN]
$UO_2(SO_4)_4^{7-}$	[87BRO/WAN]
$U(SO_4)_5^{6-}$	[87BRO/WAN]
$UO_2(SO_4)_5^{8-}$	[87BRO/WAN]
$U(SO_4)_6^{8-}$	[87BRO/WAN]
$U_2(SO_4)_3(cr)$	discussed in the present review
$U(OH)_2SO_4(cr)$	data selected in the present review
$UO_2S_2O_3 \cdot H_2O(cr)$	discussed in the present review
$UO_2SO_4 \cdot H_2O(cr)$	[63RAN/KUB], [64OWE/MAY], [82WAG/EVA]

Formula	References
$UO_2SO_4{\cdot}2.5H_2O(cr)$	data selected in the present review
$UO_2SO_4{\cdot}3H_2O(cr)$	data selected in the present review
$UO_2SO_4{\cdot}3.5H_2O(cr)$	data selected in the present review
$UO_2SO_3{\cdot}4.5H_2O(cr)$	discussed in the present review
$U(SO_4)_2{\cdot}4H_2O(cr)$	data selected in the present review
$U(SO_4)_2{\cdot}8H_2O(cr)$	data selected in the present review
$(UO_2)_3(OH)_2(SO_4)_2{\cdot}8H_2O(cr)$	[82HEM]
$(UO_2)_6SO_4(OH)_{10}{\cdot}5H_2O(cr)$	[83FLE]
$(UO_2)_6SO_4(OH)_{10}{\cdot}12H_2O(cr)$	[83FLE]
USe(cr)	data selected in the present review
USe(g)	[74MIL], [83FUG], [84GRO/DRO]
$USe_{1.33}(cr)$	see $U_3Se_4(cr)$
$USe_{1.5}(cr)$	see $U_2Se_3(cr)$
$USe_{1.667}(cr)$	see $U_3Se_5(cr)$
$USe_{1.90}(s)$	[62WES/GRO], [83FUG]
α-USe_2	data selected in the present review
β-USe_2	data selected in the present review
$USe_2(cr)$	[66WES], [70BAS/SMI], [70WES/GRO], [72FUG], [74MIL], [82HEM]
$USe_2(g)$	[84GRO/DRO]
$USe_3(cr)$	data selected in the present review
$U_2Se_3(cr)$	data selected in the present review
$U_3Se_4(cr)$	data selected in the present review
$U_3Se_5(cr)$	data selected in the present review
$UO_2SeO_3(aq)$	discussed in the present review
$UO_2SeO_3(cr)$	data selected in the present review
$UO_2SeO_4(cr)$	data selected in the present review
$(UO_2)_2(OH)_2SeO_3(aq)$	discussed in the present review
$Te_{1.33}U(cr)$	see $U_3Te_4(cr)$
UTe(cr)	[62WES/GRO], [70BAS/SMI], [72KRE], [74MIL], [82HEM], [82WAG/EVA], [83FUG], [84GRO/DRO], [86MOR]
UTe(g)	[74MIL], [83FUG], [84GRO/DRO]
$UTe_2(cr)$	[62WES/GRO], [70BAS/SMI], [72KRE], [74MIL], [82HEM], [83FUG], [84GRO/DRO]
$UTe_2(g)$	[84GRO/DRO]
$UTe_3(cr)$	[62WES/GRO], [72KRE], [74MIL], [79BLA], [82WAG/EVA], [83FUG], [84GRO/DRO], [86MOR]
$U_2Te_3(cr)$	[62WES/GRO], [72KRE], [74MIL], [83FUG], [84GRO/DRO]
$U_2Te_5(cr)$	[62WES/GRO], [72KRE], [83FUG], [84GRO/DRO]
$Te_4U_3(cr)$	see $U_3Te_4(cr)$
$U_3Te_4(cr)$	[62WES/GRO], [70BAS/SMI], [74MIL], [82WAG/EVA], [83FUG], [84GRO/DRO], [86MOR]
$U_3Te_5(cr)$	[74MIL], [83FUG]
$U_3Te_7(cr)$	[74MIL]
UOTe(cr)	data selected in the present review
$UOTe_3(cr)$	data selected in the present review
$UO_2TeO_3(cr)$	data selected in the present review

Formula	References
$UN_{0.965}(s)$	[68OHA/SET], [72FUG]
$UN_{0.997}(s)$	[81JOH/COR], [82WAG/EVA], [83FUG]
$UN(cr)$	data selected in the present review
$UN(s)$	[68INO/LEI]
β-$UN_{1.466}$	data selected in the present review
$UN_{1.466}(s)$	[81JOH/COR], [83FUG]
$UN_{1.5}(cr)$	see $U_2N_3(cr)$
$UN_{1.51}(s)$	[68OHA/SET], [72FUG]
$UN_{1.54}(cr)$	[86MOR]
α-$UN_{1.59}$	data selected in the present review
$UN_{1.59}(s)$	[72FUG], [83FUG]
α-$UN_{1.606}$	data selected in the present review
$UN_{1.606}(s)$	[81JOH/COR], [83FUG]
$UN_{1.65}(cr)$	[45BRE/BRO]
α-$UN_{1.674}$	data selected in the present review
$UN_{1.674}(s)$	[81JOH/COR]
$UN_{1.69}(s)$	[68OHA/SET], [72FUG]
α-$UN_{1.73}$	data selected in the present review
$UN_{1.73}(s)$	[72FUG], [83FUG]
$UN_2(cr)$	[72FUG], [81NIE/BOE]
$U_2N(cr)$	[81NIE/BOE]
$U_2N_3(cr)$	[45BRE/BRO], [52LAT], [54HOE/KAT], [57KAT/SEA], [58BRE/BRO], [63RAN/KUB], [63WIC/BLO], [72KRE], [79COR/OUW], [79KUB/ALC], [82WAG/EVA], [83FUG], [86MOR]
$UO_2N_3^+$	data selected in the present review
$UO_2(N_3)_2(aq)$	data selected in the present review
$UO_2(N_3)_3^-$	data selected in the present review
$UO_2(N_3)_4^{2-}$	data selected in the present review
UNO_3^{3+}	data selected in the present review
$UO_2NO_3(aq)$	[78ALL/BEA], [87BRO/WAN]
$UO_2NO_3^+$	data selected in the present review
$U(NO_3)_2^{2+}$	data selected in the present review
$UO_2(NO_3)_2(aq)$	discussed in the present review
$UO_2(NO_3)_2(cr)$	data selected in the present review
$UO_2(NO_3)_2^-$	[87BRO/WAN]
$U(NO_3)_3^+$	discussed in the present review
$UO_2(NO_3)_3^-$	discussed in the present review
$UO_2(NO_3)_3^{2-}$	[87BRO/WAN]
$U(NO_3)_4(aq)$	discussed in the present review
$UO_2(NO_3)_4^{2-}$	[87BRO/WAN]
$UO_2(NO_3)_4^{3-}$	[87BRO/WAN]
$U(NO_3)_5^-$	[87BRO/WAN]
$UO_2(NO_3)_5^{3-}$	[87BRO/WAN]
$UO_2(NO_3)_5^{4-}$	[87BRO/WAN]
$U(NO_3)_6^{2-}$	[87BRO/WAN]
UNH_3^{4+}	[87BRO/WAN]
$U(NH_3)_2^{4+}$	[87BRO/WAN]

Formula	References
$U(NH_3)_3^{4+}$	[87BRO/WAN]
$U(NH_3)_4^{4+}$	[87BRO/WAN]
$U(NH_3)_5^{4+}$	[87BRO/WAN]
$U(NH_3)_6^{4+}$	[87BRO/WAN]
$UO_2NH_3^{+}$	[87BRO/WAN]
$UO_2NH_3^{2+}$	[87BRO/WAN]
$UO_2(NH_3)_2^{+}$	[87BRO/WAN]
$UO_2(NH_3)_2^{2+}$	[87BRO/WAN]
$UO_2(NH_3)_3^{+}$	[87BRO/WAN]
$UO_2(NH_3)_3^{2+}$	[87BRO/WAN]
$UO_2(NH_3)_4^{+}$	[87BRO/WAN]
$UO_2(NH_3)_4^{2+}$	[87BRO/WAN]
$UO_2(NH_3)_5^{+}$	[87BRO/WAN]
$UO_2(NH_3)_5^{2+}$	[87BRO/WAN]
$UO_3NH_3{\cdot}3H_2O(cr)$	[83FUG]
$UO_2(NO_3)_2{\cdot}H_2O(cr)$	data selected in the present review
$UO_2(NO_3)_2{\cdot}2H_2O(cr)$	data selected in the present review
$UO_2(NO_3)_2{\cdot}3H_2O(cr)$	data selected in the present review
$UO_2(NO_3)_2{\cdot}6H_2O(cr)$	data selected in the present review
$(UO_3)_2NH_3{\cdot}3H_2O(cr)$	[64COR], [78COR/OHA], [82WAG/EVA], [86MOR]
$(NH_4)_2U_2O_7(cr)$	[64COR]
$(UO_3)_3NH_3{\cdot}5H_2O(cr)$	[64COR], [78COR/OHA], [82WAG/EVA], [83FUG], [86MOR]
$(UO_3)_3(NH_3)_2{\cdot}4H_2O(cr)$	[64COR], [78COR/OHA], [82WAG/EVA], [83FUG], [86MOR]
$NOUF_6(cr)$	[76MAS/DES], [86MOR]
$(NH_4)_3UO_2F_5(cr)$	[80PAR], [82WAG/EVA], [83FUG/PAR], [83FUG], [86MOR]
$NH_4(UO_2)_2F_5(cr)$	[74SUP/SEL], [80PAR], [82WAG/EVA], [83FUG/PAR], [83FUG], [86MOR]
$NH_4(UO_2)_2F_5{\cdot}3H_2O(cr)$	[71TSV/SEL], [80PAR], [82WAG/EVA], [83FUG/PAR], [83FUG], [86MOR]
$NH_4(UO_2)_2F_5{\cdot}4H_2O(cr)$	[71TSV/SEL], [80PAR], [82WAG/EVA], [83FUG/PAR], [83FUG], [86MOR]
$(NH_4)_4(UO_2)_6(SO_4)_3(OH)_{10}{\cdot}4H_2O(cr)$	[81OBR/WIL]
$UP_{0.992}(s)$	[68OHA/SET], [72FUG]
$UP(cr)$	data selected in the present review
$UP_{1.007}(s)$	[72FUG]
$UP_{1.33}(cr)$	see $U_3P_4(cr)$
$UP_{1.35}(s)$	[72FUG]
$UP_2(cr)$	data selected in the present review
$U_2P(cr)$	discussed in the present review
$U_3P_4(cr)$	data selected in the present review
UPO_4^{+}	[87BRO/WAN]
$UPO_5(cr)$	data selected in the present review
$UO_2PO_4^{-}$	data selected in the present review
$UO_2PO_4^{2-}$	[87BRO/WAN]

Formula	References
$UP_2O_7(aq)$	discussed in the present review
$UP_2O_7(cr)$	data selected in the present review
$U(PO_4)_2^{2-}$	[87BRO/WAN]
$UO_2(PO_3)_2(cr)$	[85COR/OUW], [88BAR3], [88BAR]
$UO_2(PO_4)_2^{4-}$	[87BRO/WAN]
$UO_2(PO_4)_2^{5-}$	[87BRO/WAN]
$U(PO_4)_3^{5-}$	[87BRO/WAN]
$UO_2(PO_4)_3^{7-}$	[87BRO/WAN]
$UO_2(PO_4)_3^{8-}$	[87BRO/WAN]
$U(PO_4)_4^{8-}$	[87BRO/WAN]
$UO_2(PO_4)_4^{10-}$	[87BRO/WAN]
$UO_2(PO_4)_4^{11-}$	[87BRO/WAN]
$U(PO_4)_5^{11-}$	[87BRO/WAN]
$UO_2(PO_4)_5^{13-}$	[87BRO/WAN]
$UO_2(PO_4)_5^{14-}$	[87BRO/WAN]
$U(PO_4)_6^{14-}$	[87BRO/WAN]
$(UO_2)_2P_2O_7(cr)$	data selected in the present review
$(UO_2)_3(PO_4)_2(cr)$	data selected in the present review
$U_3(PO_4)_4(cr)$	discussed in the present review
$UHPO_4^{2+}$	[78LAN], [80BEN/TEA], [86WAN]
$UH_2PO_4^{2+}$	[69MOS]
$UH_2PO_4^{3+}$	[78ALL/BEA], [87BRO/WAN]
$UO_2HPO_3(cr)$	discussed in the present review
$UOHPO_4(cr)$	discussed in the present review
$UO_2HPO_4(aq)$	data selected in the present review
$UO_2HPO_4(cr)$	[78ALL/BEA], [78LAN], [80BEN/TEA], [83MAR/PAV], [86WAN]
$UO_2HPO_4^-$	[87BRO/WAN]
$UO_2H_2PO_4(aq)$	[87BRO/WAN]
$UO_2H_2PO_4^+$	data selected in the present review
$UO_2H_3PO_4^{2+}$	data selected in the present review
$UO_2(H_2PO_2)_2(cr)$	discussed in the present review
$U(HPO_4)_2(aq)$	[78LAN], [80BEN/TEA], [86WAN]
$U(HPO_4)_2(cr)$	discussed in the present review
$U(H_2PO_4)_2^+$	[69MOS]
$U(H_2PO_4)_2^{2+}$	[87BRO/WAN]
$UO_2HPO_3{\cdot}3H_2O(cr)$	discussed in the present review
$U(H_2PO_2)_4(s)$	discussed in the present review
$UO_2H_2P_2O_7(cr)$	discussed in the present review
$UO_2(HPO_4)_2^{2-}$	[67MOS/SHE]
$UO_2(HPO_4)_2^{3-}$	[87BRO/WAN]
$UO_2(H_2PO_4)_2(aq)$	data selected in the present review
$UO_2(H_2PO_4)_2^-$	[87BRO/WAN]
$UO_2(H_2PO_4)(H_3PO_4)^+$	data selected in the present review
$UO_2H_5(PO_4)_2^+$	[84TRI]
$U(HPO_4)_2{\cdot}2H_2O(cr)$	discussed in the present review
$UO_2HPO_4{\cdot}4H_2O(cr)$	data selected in the present review

Formula	References
$U(HPO_3)_2 \cdot 4H_2O(cr)$	discussed in the present review
$U(H_2PO_2)_4 . H_3PO_2(s)$	discussed in the present review
$UO_2(H_2PO_4)_2 \cdot H_2O(cr)$	discussed in the present review
$UO_2HPO_4 \cdot 5H_2O(cr)$	[84VIE/TAR]
$U(HPO_4)_3^{2-}$	[78LAN], [80BEN/TEA], [86WAN]
$U(H_2PO_4)_3(aq)$	[69MOS]
$U(H_2PO_4)_3^{+}$	[87BRO/WAN]
$U(HPO_4)_2 \cdot 4H_2O(cr)$	data selected in the present review
$U(HPO_4)_2H_3PO_4 \cdot H_2O(cr)$	discussed in the present review
$UO_2(H_2PO_4)_2 \cdot 3H_2O(cr)$	discussed in the present review
$UP_2O_7 \cdot 6H_2O(cr)$	discussed in the present review
$UO_2(HPO_4)_3^{4-}$	[87BRO/WAN]
$UO_2(HPO_4)_3^{5-}$	[87BRO/WAN]
$UO_2(H_2PO_4)_3^{-}$	[86WAN], [87BRO/WAN]
$UO_2(H_2PO_4)_3^{2-}$	[87BRO/WAN], [88CRO/EWA]
$UO_2H_3PO_4(H_2PO_4)_2(aq)$	[83MAR/PAV]
$U(HPO_4)_2 \cdot 6H_2O(cr)$	discussed in the present review
$U(H_2PO_3)_2HPO_3 \cdot 5H_2O(cr)$	discussed in the present review
$U(HPO_4)_4^{4-}$	[78LAN], [80BEN/TEA]
$U(H_2PO_4)_4(aq)$	[87BRO/WAN]
$UO_2(HPO_4)_4^{6-}$	[87BRO/WAN]
$UO_2(HPO_4)_4^{7-}$	[87BRO/WAN]
$UO_2(H_2PO_4)_4^{2-}$	[87BRO/WAN]
$U(HPO_4)_5^{6-}$	[87BRO/WAN]
$U(H_2PO_4)_5^{-}$	[87BRO/WAN]
$UO_2(HPO_4)_5^{8-}$	[87BRO/WAN]
$UO_2(HPO_4)_5^{9-}$	[87BRO/WAN]
$U(HPO_4)_6^{8-}$	[87BRO/WAN]
$U(H_2PO_4)_6^{2-}$	[87BRO/WAN]
$UP_2O_7 \cdot 20H_2O(cr)$	discussed in the present review
$(UO_2)_2(HPO_4)_2(cr)$	[80LEM/TRE], [88PHI/HAL2]
$H_2(UO_2)_2(PO_4)_2(cr)$	discussed in the present review
$H_2(UO_2)_2(PO_4)_2 \cdot 2H_2O(cr)$	discussed in the present review
$(UO_2)_2P_2O_7 \cdot 4H_2O(cr)$	discussed in the present review
$H_2(UO_2)_2(PO_4)_2 \cdot 4H_2O(cr)$	discussed in the present review
$H_2(UO_2)_2(PO_4)_2 \cdot 8H_2O(cr)$	discussed in the present review
$H_2(UO_2)_2(PO_4)_2 \cdot 10H_2O(cr)$	discussed in the present review
$H_{11}(UO_2)_2(PO_4)_5(cr)$	discussed in the present review
$U_3(H_2PO_4)_2^{10+}$	[78ALL/BEA]
$(UO_2)_3(PO_4)_2 \cdot 4H_2O(cr)$	data selected in the present review
$(UO_2)_3(PO_4)_2 \cdot 6H_2O(cr)$	data selected in the present review
$(UO_2)_3(PO_4)_2 \cdot 8H_2O(cr)$	discussed in the present review
$H_2(UO_2)_3(HPO_3)_4 \cdot 12H_2O(cr)$	discussed in the present review
$U(UO_2)_3(PO_4)_2(OH)_6 \cdot 2H_2O(cr)$	[83FLE]
$U(UO_2)_3(PO_4)_2(OH)_6 \cdot 4H_2O(cr)$	[83FLE]
$U_2SP(cr)$	[68COU/MAR]
$U_4SP_3(cr)$	[68COU/MAR]

Formula	References
$U_4S_3P(cr)$	[68COU/MAR]
$NH_4UO_2PO_4(cr)$	discussed in the present review
$NH_4UO_2PO_4(s)$	[56CHU/STE]
$NH_4UO_2PO_4 \cdot 3H_2O(cr)$	[71NAU/RYZ], [83FLE], [84VIE/TAR], [88PHI/HAL2]
$(NH_4)_2(UO_2)_2(PO_4)_2(cr)$	[88PHI/HAL2]
UAs(cr)	data selected in the present review
$UAs_{1.33}(cr)$	see $U_3As_4(cr)$
$UAs_2(cr)$	data selected in the present review
$U_3As_4(cr)$	data selected in the present review
$UAsO_5(cr)$	data selected in the present review
$UO_2(AsO_3)_2(cr)$	data selected in the present review
$(UO_2)_2As_2O_7(cr)$	data selected in the present review
$(UO_2)_3(AsO_4)_2(cr)$	data selected in the present review
$UO_2HAsO_4(cr)$	[84GEN/WEI]
$(UO_2)_3(AsO_4)_2 \cdot 12H_2O(cr)$	[83FLE]
UAsS(cr)	data selected in the present review
UAsSe(cr)	data selected in the present review
UAsTe(cr)	data selected in the present review
USb(cr)	data selected in the present review
$USb_2(cr)$	data selected in the present review
$U_3Sb_4(cr)$	data selected in the present review
$U_4Sb_3(cr)$	data selected in the present review
BiU(cr)	[63RAN/KUB], [68LEB/NIC2], [70BAS/SMI], [77BAR/KNA], [79KUB/ALC], [82WAG/EVA], [83FUG], [86MOR]
$Bi_2U(cr)$	[63RAN/KUB], [68LEB/NIC2], [75WES/SOM], [78GRO/ZAK], [83FUG], [86MOR]
$Bi_4U_3(cr)$	[63RAN/KUB], [68LEB/NIC2], [70BAS/SMI], [77BAR/KNA], [79KUB/ALC], [82WAG/EVA], [83FUG], [86MOR]
$Bi_2U_2O_9 \cdot 3H_2O(cr)$	[83FLE]
$Bi_4UO_2(AsO_4)_2O_4 \cdot 2H_2O(cr)$	[83FLE]
$UC_{0.96}(s)$	[67STO], [68HOL/STO], [72FUG]
$UC_{0.996}(s)$	[67STO], [68HOL/STO], [72FUG]
β-UC	[84HOL/RAN]
UC(cr)	data selected in the present review
UC(g)	[79GUP/GIN], [83FUG], [86MOR]
$UC_{1.032}(s)$	[67STO], [72FUG]
$UC_{1.039}(s)$	[68HOL/STO]
$UC_{1.1}(s)$	[72FUG]
$UC_{1.86}(s)$	[62HUB/HOL]
$UC_{1.9}(cr)$	[65WES/SUI], [87THE]
$UC_{1.9}(s)$	[63RAN/KUB], [64AND/COU], [67STO/HUB], [67STO], [68HOL/STO], [69MAC], [72FUG], [73BAR/KNA], [79KUB/ALC], [86MOR]
α-$UC_{1.94}$	data selected in the present review

Formula	References
α-UC_2	[45BRE/BRO], [52LAT], [54HOE/KAT], [55MAL/SHE], [57KAT/SEA], [58BRE/BRO], [62ALC/GRI], [62MUK/NAI2], [62STO], [63ACH/ALC], [63WIC/BLO], [66BEH/EGA], [67STO], [72KRE], [72OET/NAV], [79COR/OUW], [81NIE/BOE], [86MOR]
$UC_2(g)$	[62STO], [64NOR/WIN], [79GUP/GIN], [83FUG], [86MOR]
$UC_3(g)$	[79GUP/GIN], [83FUG], [86MOR]
$UC_4(g)$	[79GUP/GIN], [83FUG], [86MOR]
$UC_5(g)$	[79GUP/GIN], [83FUG], [86MOR]
$UC_6(cr)$	[79GUP/GIN]
$UC_6(g)$	[83FUG], [86MOR]
$U_2C(cr)$	[81NIE/BOE]
$U_2C_3(cr)$	data selected in the present review
UCO_3^{2+}	[87BRO/WAN]
$UO_2CO_3(aq)$	data selected in the present review
$UO_2CO_3(cr)$	data selected in the present review
$UO_2CO_3^-$	[82HEM], [83ALL], [87BRO/WAN]
$U(CO_3)_2(aq)$	[87BRO/WAN]
$U(CO_3)_2(cr)$	[87BRO/WAN]
$UO_2C_2O_4(cr)$	[77OHA], [82WAG/EVA], [83FUG], [86MOR]
$UO_2(CO_3)_2^{2-}$	data selected in the present review
$UO_2(CO_3)_2^{3-}$	[83ALL], [87BRO/WAN]
$U(C_2O_4)_2(cr)$	[78COR/OHA], [83FUG], [85MAR/FUG], [86MOR]
$U(CO_3)_3^{2-}$	[87BRO/WAN]
$UO_2(CO_3)_3^{4-}$	data selected in the present review
$UO_2(CO_3)_3^{5-}$	data selected in the present review
$U(CO_3)_4^{4-}$	data selected in the present review
$UO_2(CO_3)_4^{6-}$	[87BRO/WAN]
$UO_2(CO_3)_4^{7-}$	[87BRO/WAN]
$U(CO_3)_5^{6-}$	data selected in the present review
$UO_2(CO_3)_5^{8-}$	[87BRO/WAN]
$UO_2(CO_3)_5^{9-}$	[87BRO/WAN]
$U(CO_3)_6^{8-}$	[87BRO/WAN]
$(UO_2)_3(CO_3)_6^{6-}$	data selected in the present review
$HCUO_2^-$	[52LAT]
$UHCO_3^{3+}$	[87BRO/WAN]
$UO_2HCO_3(aq)$	[87BRO/WAN]
$UO_2HCO_3^+$	[87BRO/WAN]
$UCO_3(OH)_2(aq)$	discussed in the present review
$UO_2CO_3OH^-$	discussed in the present review
$UO_2(OH)_2CO_2(aq)$	see $UO_2CO_3(aq)$
$UO_2CO_3{\cdot}H_2O(cr)$	discussed in the present review
$U(HCO_3)_2^{2+}$	[87BRO/WAN]
$UO_2(HCO_2)_2(cr)$	[76BOU/BON], [78COR/OHA], [82WAG/EVA], [83FUG], [86MOR]
$U(OH)_3CO_3^-$	[83ALL], [86WAN], [87BRO/WAN]
$UO_2(C_2H_3O_2)_2(aq)$	see $UO_2(CH_3CO_2)_2(aq)$

Formula	References
$UO_2(CH_3CO_2)_2(aq)$	[52LAT], [54HOE/KAT]
$UO_2(CH_3CO_2)_2(cr)$	[73MOS], [77OHA], [78COR/OHA], [82WAG/EVA], [83FUG], [86MOR]
$UO_2C_2H_3O_2{\cdot}2H_2O(cr)$	[52LAT]
$UO_2(OH)_2CO_3^{2-}$	[78ALL/BEA], [87BRO/WAN]
$UO_2C_2O_4{\cdot}H_2O(cr)$	[77OHA], [82WAG/EVA], [83FUG], [86MOR]
$UO_2CO_3{\cdot}2.2H_2O(cr)$	discussed in the present review
$UO_2CO_3{\cdot}2H_2O(cr)$	[83FLE]
$UO_2(HCO_2)_2{\cdot}H_2O(cr)$	[76BOU/BON], [78COR/OHA], [82WAG/EVA], [83FUG], [86MOR]
$UO_2CO_3{\cdot}2.5H_2O(cr)$	discussed in the present review
$U(CO_3)_2(OH)_2^{2-}$	discussed in the present review
$UO_2(HCO_3)_2(aq)$	[87BRO/WAN]
$UO_2(HCO_3)_2^-$	[87BRO/WAN]
$UO_2(CH_3CO_2)_3^-$	[58CON/WAH]
$UO_2(CH_3CO_2)_2{\cdot}2H_2O(cr)$	[54HOE/KAT], [77OHA], [78COR/OHA], [82WAG/EVA], [83FUG], [86MOR]
$U(CH_3CO_2)_4(cr)$	[73MOS], [86MOR]
$U(HCO_3)_3^+$	[87BRO/WAN]
$UO_2C_2O_4{\cdot}3H_2O(cr)$	[73MOS], [77OHA], [82WAG/EVA], [83FUG], [86MOR]
$UO_2(CO_3)_2(H_2O)_2^{2-}$	[56MCC/BUL]
$U(CO_3)_3(OH)_2^{4-}$	discussed in the present review
$UO_2(HCO_3)_3^-$	[87BRO/WAN]
$UO_2(HCO_3)_3^{2-}$	[87BRO/WAN]
$U(HCO_3)_4(aq)$	[87BRO/WAN]
$UO_2(CH_3CO_2)_2{\cdot}6H_2O(cr)$	[52LAT]
$U(CO_3)_4(OH)_2^{6-}$	discussed in the present review
$UO_2(HCO_3)_4^{2-}$	[87BRO/WAN]
$UO_2(HCO_3)_4^{3-}$	[87BRO/WAN]
$U(C_2O_4)_2{\cdot}6H_2O(cr)$	[73MOS]
$U(HCO_3)_5^-$	[87BRO/WAN]
$UO_2(HCO_3)_5^{3-}$	[87BRO/WAN]
$U(HCO_3)_6^{2-}$	[87BRO/WAN]
$(UO_2)_2CO_3(OH)_3^-$	data selected in the present review
$(UO_2)_3CO_3(OH)_3^+$	[88LEM]
$(UO_2)_3O(OH)_2(HCO_3)^+$	data selected in the present review
$(UO_2)_3(OH)_5CO_2^+$	see $(UO_2)_3O(OH)_2(HCO_3)^+$
$(UO_2)_{11}(CO_3)_6(OH)_{12}^{2-}$	data selected in the present review
$(UO_2)_{11}(OH)_{25}(CO_2)_6^{3-}$	see $(UO_2)_{11}(OH)_{13}(CO_3)_6^{3-}$
$(UO_2)_{13}(OH)_{30}(CO_2)_7^{4-}$	see $(UO_2)_{13}(OH)_{16}(CO_3)_7^{4-}$
$NH_4UO_2(CO_3)OH{\cdot}3H_2O(cr)$	discussed in the present review
$NH_4UO_2(CH_3CO_2)_3{\cdot}6H_2O(cr)$	[83FUG]
$UO_2NH_4(CH_3CO_2)_3{\cdot}6H_2O(cr)$	[78COR/OHA], [86MOR]
$NH_4(UO_2)_2(CO_3)_3OH{\cdot}5H_2O(cr)$	discussed in the present review
$(NH_4)_2UF_4(HCO_3)_2(cr)$	discussed in the present review
$USCN^{3+}$	data selected in the present review
$U(SCN)_2^{2+}$	data selected in the present review

Formula	References
$U(SCN)_3^+$	discussed in the present review
$U(SCN)_4(aq)$	[87BRO/WAN]
$U(SCN)_5^-$	[87BRO/WAN]
$U(SCN)_6^{2-}$	[87BRO/WAN]
$UO_2SCN(aq)$	[87BRO/WAN]
UO_2SCN^+	data selected in the present review
$UO_2(SCN)_2(aq)$	data selected in the present review
$UO_2(SCN)_2^-$	[87BRO/WAN]
$UO_2(SCN)_3^-$	data selected in the present review
$UO_2(SCN)_3^{2-}$	[87BRO/WAN]
$UO_2(SCN)_4^{2-}$	[87BRO/WAN]
$UO_2(SCN)_4^{3-}$	[87BRO/WAN]
$UO_2(SCN)_5^{3-}$	[87BRO/WAN]
$UO_2(SCN)_5^{4-}$	[87BRO/WAN]
USi(cr)	[62GRO/HAY], [63RAN/KUB], [69HUB/HOL3], [77BAR/KNA], [82WAG/EVA], [83FUG]
$USi_{1.88}(s)$	[83FUG]
$USi_2(cr)$	[62GRO/HAY], [63RAN/KUB], [73BAR/KNA], [77BAR/KNA], [82WAG/EVA]
$USi_3(cr)$	[62GRO/HAY], [63RAN/KUB], [73BAR/KNA], [75PAL/CIR], [77BAR/KNA], [82WAG/EVA], [83FUG]
$U_2Si(cr)$	[72KRE], [81NIE/BOE]
$U_3Si(cr)$	[63RAN/KUB], [63SET/FED], [73BAR/KNA], [75OHA/ADE], [77BAR/KNA], [77FLO/OSB], [79KUB/ALC], [82WAG/EVA], [83FUG], [86MOR]
$U_3Si_2(cr)$	[61ALC/GRI], [62GRO/HAY], [63RAN/KUB], [72KRE], [77BAR/KNA], [79KUB/ALC], [82WAG/EVA], [83FUG], [86MOR]
$U_3Si_5(cr)$	[61ALC/GRI], [63RAN/KUB], [72KRE], [77BAR/KNA], [79KUB/ALC], [83FUG], [86MOR]
$USiO_4(am)$	discussed in the present review
$USiO_4(cr)$	data selected in the present review
$UO_2SiO(OH)_3^+$	discussed in the present review
$(UO_2)_2SiO_4 \cdot 2H_2O(cr)$	discussed in the present review
$UH_6(UO_2)_6(SiO_4)_6 \cdot 30H_2O(cr)$	[83FLE]
GeU(cr)	[61ALC/GRI], [63RAN/KUB], [77BAR/KNA], [79KUB/ALC], [83FUG], [86MOR]
$Ge_2U(cr)$	[61ALC/GRI], [63RAN/KUB], [77BAR/KNA], [79KUB/ALC], [83FUG], [86MOR]
$Ge_3U(cr)$	[61ALC/GRI], [63RAN/KUB], [75PAL/CIR], [77BAR/KNA], [79KUB/ALC], [83FUG], [86MOR]
$Ge_5U_3(cr)$	[61ALC/GRI], [63RAN/KUB], [77BAR/KNA], [79KUB/ALC], [83FUG], [86MOR]
$Ge_3U_5(cr)$	[61ALC/GRI], [63RAN/KUB], [77BAR/KNA], [79KUB/ALC], [83FUG], [86MOR]
$Sn_3U(cr)$	[61ALC/GRI], [63RAN/KUB], [82WAG/EVA], [83FUG], [86MOR], [88COL/BES]
$Sn_2U_3(cr)$	[61ALC/GRI], [63RAN/KUB], [83FUG], [86MOR]

Formula	References
$Sn_5U_3(cr)$	[61ALC/GRI], [63RAN/KUB], [83FUG], [86MOR]
$PbU(cr)$	[61ALC/GRI], [63RAN/KUB], [83FUG], [86MOR]
$Pb_3U(cr)$	[61ALC/GRI], [63RAN/KUB], [75PAL/CIR], [82WAG/EVA], [83FUG], [86MOR]
$PbU_4O_{13}\cdot 4H_2O(cr)$	[83FLE]
$Pb_2U_5O_{17}\cdot 4H_2O(cr)$	[83FLE]
$PbU_7O_{22}\cdot 12H_2O(cr)$	[83FLE]
$PbUO_2(TeO_3)_2(cr)$	[83FLE]
$Pb_2UO_2(PO_4)_2(cr)$	[84GEN/WEI]
$Pb(UO_2)_2(PO_4)_2(cr)$	[78LAN], [85PHI/PHI], [88PHI/HAL2]
$Pb_2UO_2(PO_4)_2\cdot 2H_2O(cr)$	[83FLE], [84NRI2]
$Pb(UO_2)_2(PO_4)_2\cdot 3H_2O(cr)$	[83FLE]
$Pb(UO_2)_2(PO_4)_2\cdot 4H_2O(cr)$	[83FLE], [84NRI2]
$Pb(UO_2)_2(PO_4)_2\cdot 10H_2O(cr)$	[65MUT/HIR]
$Pb(UO_2)_3(PO_4)_2(OH)_2(cr)$	[84GEN/WEI]
$Pb_2(UO_2)_3(PO_4)_2(OH)_4\cdot 3H_2O(cr)$	[83FLE], [84NRI2]
$Pb(UO_2)_4(PO_4)_2(OH)_4\cdot 7H_2O(cr)$	[84NRI2], [84NRI]
$Pb(UO_2)_4(PO_4)_2(OH)_4\cdot 8H_2O(cr)$	[84NRI2]
$Pb_2UO_2(AsO_4)_2(cr)$	[83FLE], [84GEN/WEI]
$Pb_2UO_2(CO_3)_3(cr)$	[83FLE]
$PbUO_2SiO_4\cdot H_2O(cr)$	discussed in the present review
$UB(cr)$	[81NIE/BOE]
$UB(g)$	[70GIN], [83FUG]
$UB_{1.98}(s)$	[82WAG/EVA], [83FUG]
$UB_2(cr)$	[62ALC/GRI], [63RAN/KUB], [69FLO/OSB], [70FRE/BAR], [77BAR/KNA], [79KUB/ALC], [81CHI/AKH], [81NIE/BOE], [82WAG/EVA], [83FUG], [86MOR]
$UB_2(g)$	[70GIN], [83FUG]
$UB_4(cr)$	[62ALC/GRI], [63RAN/KUB], [77BAR/KNA], [79KUB/ALC], [82WAG/EVA], [83FUG], [86MOR]
$UB_{12}(cr)$	[62ALC/GRI], [63RAN/KUB], [77BAR/KNA], [79KUB/ALC], [82WAG/EVA], [83FUG], [86MOR]
$UB_{13}(cr)$	[68FAR/ELD]
$U_2B(cr)$	[81NIE/BOE]
$Al_2U(cr)$	[58IVA/TUM], [63RAN/KUB], [69CHI/KAT], [77BAR/KNA], [79KUB/ALC], [82WAG/EVA], [83FUG], [86MOR]
$Al_3U(cr)$	[58IVA/TUM], [63RAN/KUB], [69CHI/KAT], [75PAL/CIR], [77BAR/KNA], [79KUB/ALC], [82WAG/EVA], [83FUG], [86MOR]
$Al_4U(cr)$	[58IVA/TUM], [63RAN/KUB], [69CHI/KAT], [77BAR/KNA], [79KUB/ALC], [82WAG/EVA], [83FUG], [86MOR]
$HAlUO_2PO_4(OH)_3\cdot 4H_2O(cr)$	[83FLE]
$Al_2UO_2(PO_4)_2(OH)_2\cdot 8H_2O(cr)$	[83FLE]
$Al(UO_2)_2(PO_4)_2OH\cdot 8H_2O(cr)$	[83FLE]
$Al(UO_2)_3(PO_4)_2(OH)_3(cr)$	[79DEL/PIR]

Formula	References
$Al(UO_2)_3(PO_4)_2(OH)_3 \cdot 5.5H_2O(cr)$	[83FLE]
$Al_2(UO_2)_3(PO_4)_2(OH)_6 \cdot 10H_2O(cr)$	[79DEL/PIR], [83FLE]
$HAl(UO_2)_4(PO_4)_4(cr)$	[84GEN/WEI]
$HAl(UO_2)_4(PO_4)_4 \cdot 16H_2O(cr)$	[83FLE]
$HAl(UO_2)_4(PO_4)_4 \cdot 40H_2O(cr)$	[83FLE]
$HAl(UO_2)_4(AsO_4)_4 \cdot 40H_2O(cr)$	[78WAL]
$GaU(cr)$	[82WAG/EVA], [86MOR]
$Ga_2U(cr)$	[82WAG/EVA], [83FUG], [86MOR]
$Ga_3U(cr)$	[63RAN/KUB], [75PAL/CIR], [82WAG/EVA], [83FUG], [86MOR]
$Ga_3U_2(cr)$	[83FUG]
$In_3U(cr)$	[63RAN/KUB], [75PAL/CIR], [82WAG/EVA], [83FUG], [86MOR]
$Tl_3U(cr)$	[63RAN/KUB], [75PAL/CIR], [82WAG/EVA], [83FUG], [86MOR]
$TlUO_2(CO_3)OH \cdot 3H_2O(cr)$	discussed in the present review
$Tl(UO_2)_2(CO_3)_3OH \cdot 5H_2O(cr)$	discussed in the present review
$U_2Zn_{17}(cr)$	[63RAN/KUB], [83FUG], [86MOR]
$Zn_2(UO_2)_6(SO_4)_3(OH)_{10} \cdot 8H_2O(cr)$	discussed in the present review
$Zn(UO_2)_2(AsO_4)_2(cr)$	[84GEN/WEI]
$Zn(UO_2)_2(AsO_4)_2 \cdot 10H_2O(cr)$	[83FLE]
$Cd_{11}U(cr)$	[63RAN/KUB], [73BAR/KNA], [79KUB/ALC], [81CHI/AKH], [83FUG], [86MOR]
$HgU(cr)$	[79CHI/MAS], [83FUG]
$Hg_2U(cr)$	[63RAN/KUB], [79CHI/MAS], [83FUG], [86MOR]
$Hg_3U(cr)$	[63RAN/KUB], [79CHI/MAS], [83FUG], [86MOR]
$Hg_4U(cr)$	[63RAN/KUB], [79CHI/MAS], [83FUG], [86MOR]
$Cu_5U(cr)$	[74KAN]
$CuUO_2(OH)_4(cr)$	[83FLE]
$Cu_2(UO_2)_3(OH)_{10} \cdot 5H_2O(cr)$	[83FLE]
$CuUCl_5(g)$	[74BIN/SCH]
$Cu_2UCl_6(g)$	[74BIN/SCH]
$Cu(UO_2)_2(SO_4)_2(OH)_2 \cdot 8H_2O(cr)$	[83FLE]
$Cu_4UO_2(SeO_3)_2(OH)_6 \cdot H_2O(cr)$	[83FLE]
$Cu(UO_2)_3(SeO_3)_3(OH)_2 \cdot 7H_2O(cr)$	[83FLE]
$Cu(UO_2)_2(PO_4)_2(cr)$	[78LAN], [85PHI/PHI], [88PHI/HAL2]
$Cu(UO_2)_2(PO_4)_2 \cdot 8H_2O(cr)$	[83FLE], [84NRI2], [84VIE/TAR]
$Cu(UO_2)_2(PO_4)_2 \cdot 10H_2O(cr)$	[65MUT/HIR]
$Cu(UO_2)_2(AsO_4)_2(cr)$	[84GEN/WEI]
$Cu(UO_2)_2(AsO_4)_2 \cdot 8H_2O(cr)$	[83FLE], [84VOC/GOE]
$Cu(UO_2)_2Si_2O_6(OH)_2 \cdot 6H_2O(cr)$	[82HEM], [83FLE]
$AgUF_6(cr)$	[84OHA/MAL], [86MOR]
$Ag_2UF_6(cr)$	[84OHA/MAL]
$AgUO_2(CO_3)OH \cdot 3H_2O(cr)$	discussed in the present review
$Ag(UO_2)_2(CO_3)_3OH \cdot 5H_2O(cr)$	discussed in the present review
$Ni_2U(cr)$	[83FUG]
$Ni_5U(cr)$	[83FUG]

Formula	References
$NiUO_4(cr)$	[83FUG]
$NiU_2O_6(cr)$	[76DEW/PER], [83FUG], [86MOR]
$NiU_3O_8(cr)$	[82MOR], [85PHI/PHI]
$NiU_3O_{10}(cr)$	[76DEW/PER], [83FUG], [86MOR], [88PHI/HAL2]
$Ni_2(UO_2)_6(SO_4)_3(OH)_{10}\cdot 8H_2O(cr)$	discussed in the present review
$Ni(UO_2)_2(PO_4)_2\cdot 7H_2O(cr)$	[84VIE/TAR]
$Co_{11}U_2(cr)$	[74KAN]
$Co_2(UO_2)_6(SO_4)_3(OH)_{10}\cdot 8H_2O(cr)$	discussed in the present review
$Co(UO_2)_2(PO_4)_2\cdot 7H_2O(cr)$	[84VIE/TAR]
$Co(UO_2)_2(AsO_4)_2(cr)$	[84GEN/WEI]
$Co(UO_2)_2(AsO_4)_2\cdot 7H_2O(cr)$	[84VOC/GOE]
$Co(UO_2)_2(AsO_4)_2\cdot 8H_2O(cr)$	[83FLE]
$Fe_2U(cr)$	[74KAN]
$FeU_6(cr)$	[63RAN/KUB], [83FUG]
$Fe(UO_2)_2(PO_4)_2(cr)$	[78LAN], [85PHI/PHI], [86WAN], [88PHI/HAL2]
$Fe(UO_2)_2(PO_4)_2\cdot 8H_2O(cr)$	[83FLE]
$Fe(UO_2)_2(PO_4)_2\cdot 10H_2O(cr)$	[65MUT/HIR]
$Fe(UO_2)_2(AsO_4)_2(cr)$	[84GEN/WEI]
$Fe(UO_2)_2(AsO_4)_2\cdot 8H_2O(cr)$	[83FLE], [86VOC/GRA]
$Pd_3U(cr)$	[82WIJ], [88BUR/TO]
$Rh_3U(cr)$	[73BAR/KNA], [79KUB/ALC], [82WIJ/COR]
$Ru_3U(cr)$	[77BAR/KNA], [79KUB/ALC], [82WIJ/COR], [83FUG], [85RAR]
$URu_3(cr)$	[80EDW/STA]
$Ru_3UC_{0.7}(s)$	[85RAR]
$MnU(cr)$	[82BOE/BOO]
$Mn_2U(cr)$	[82BOE/BOO]
$Mn_5U(cr)$	[82BOE/BOO]
$MnU_2(cr)$	[82BOE/BOO]
$MnU_5(cr)$	[82BOE/BOO]
$MnUO_4(cr)$	[78COR/OHA], [83FUG]
$Mn(UO_2)_2(AsO_4)_2\cdot 8H_2O(cr)$	[86VOC/GRA]
$CrU(cr)$	[82BOE/BOO]
$Cr_2U(cr)$	[82BOE/BOO]
$Cr_5U(cr)$	[82BOE/BOO]
$CrU_2(cr)$	[82BOE/BOO]
$CrU_5(cr)$	[82BOE/BOO]
$UO_2CrO_4(aq)$	[54HOE/KAT]
$UO_2CrO_4(cr)$	[78COR/OHA], [83FUG]
$UO_2CrO_4\cdot 5.5H_2O(cr)$	[54HOE/KAT], [78COR/OHA], [82WAG/EVA], [83FUG]
$U(MoO_4)_2(cr)$	[83FLE]
$UO_2MoO_4\cdot 4H_2O(cr)$	[57COL/APP]
$UO_2Mo_2O_7\cdot 3H_2O(cr)$	[83FLE]
$UMo_5O_{12}(OH)_{10}(cr)$	[83FLE]
$H_4U(UO_2)_3(MoO_4)_7\cdot 18H_2O(am)$	[83FLE]
$UV(cr)$	[80BOE/BOO]

Formula	References
$UV_2(cr)$	[80BOE/BOO]
$UV_5(cr)$	[80BOE/BOO]
$U_2V(cr)$	[80BOE/BOO]
$U_5V(cr)$	[80BOE/BOO]
$Pb(UO_2)_2(VO_4)_2(cr)$	[84GEN/WEI]
$Pb(UO_2)_2(VO_4)_2{\cdot}5H_2O(cr)$	[83FLE]
$Al(UO_2)_2(VO_4)_2OH(cr)$	[84GEN/WEI]
$Al(UO_2)_2(VO_4)_2OH{\cdot}8H_2O(cr)$	[83FLE]
$Al(UO_2)_2(VO_4)_2OH{\cdot}11H_2O(cr)$	[83FLE]
$CuUO_2VO_4(cr)$	[84GEN/WEI]
$Cu_2(UO_2)_2(VO_4)_2(OH)_2{\cdot}6H_2O(cr)$	[83FLE]
$Mn(UO_2)_2(VO_4)_2{\cdot}10H_2O(cr)$	[83FLE]
$TiU(cr)$	[80BOE/BOO]
$Ti_2U(cr)$	[80BOE/BOO]
$Ti_5U(cr)$	[80BOE/BOO]
$TiU_2(cr)$	[80BOE/BOO]
$TiU_5(cr)$	[80BOE/BOO]
$U_2Ti_4O_{12}(OH)_2(cr)$	[83FLE]
$ScU(cr)$	[80BOE/BOO]
$Sc_2U(cr)$	[80BOE/BOO]
$Sc_5U(cr)$	[80BOE/BOO]
$ScU_2(cr)$	[80BOE/BOO]
$ScU_5(cr)$	[80BOE/BOO]
$U_4PuN_5(s)$	[71ALE/OGD]
$(UO_2)_2(PuO_2)(CO_3)_6^{6-}$	data selected in the present review
$UNpO_4^{3+}$	[61SUL/HIN]
$(UO_2)_2(NpO_2)(CO_3)_6^{6-}$	data selected in the present review
$Be_{13}U(cr)$	data selected in the present review
$BeUO_4(cr)$	[77OHA/BOE]
$MgUO_4(cr)$	data selected in the present review
$MgU_2O_6(cr)$	discussed in the present review
$MgU_2O_7(cr)$	discussed in the present review
$MgU_3O_{10}(cr)$	data selected in the present review
$Mg_3U_3O_{10}(cr)$	discussed in the present review
$Mg_2(UO_2)_6(SO_4)_3(OH)_{10}{\cdot}8H_2O(cr)$	discussed in the present review
$Mg(UO_2)_2(PO_4)_2(cr)$	[78LAN], [85PHI/PHI], [86WAN], [88PHI/HAL2]
$Mg(UO_2)_2(PO_4)_2{\cdot}10H_2O(cr)$	[65MUT/HIR]
$Mg(UO_2)_2(AsO_4)_2(cr)$	[84GEN/WEI]
$Mg(UO_2)_2(AsO_4)_2{\cdot}12H_2O(cr)$	[83FLE]
$Mg_2UO_2(CO_3)_3{\cdot}18H_2O(cr)$	discussed in the present review
$Mg(UO_2)_2Si_2O_6(OH)_2{\cdot}5H_2O(cr)$	[82HEM], [83FLE]
$MgU_2Mo_2O_{13}{\cdot}6H_2O(cr)$	[83FLE]
α-$CaUO_4$	[71NAU/RYZ]
β-$CaUO_4$	data selected in the present review
$CaUO_4(cr)$	data selected in the present review
$Ca_3UO_6(cr)$	data selected in the present review
$CaU_2O_6(cr)$	discussed in the present review

Formula	References
$CaU_2O_7(cr)$	discussed in the present review
$CaU_6O_{19}\cdot 11H_2O(cr)$	[83FLE]
$CaU(PO_4)_2(cr)$	[86WAN]
$Ca(UO_2)_2(PO_4)_2(cr)$	[78LAN], [85PHI/PHI], [86WAN], [88PHI/HAL2]
$CaU(PO_4)_2\cdot 2H_2O(cr)$	[65MUT/HIR], [65MUT], [78LAN], [84VIE/TAR], [88PHI/HAL2]
$Ca(UO_2)_2(PO_4)_2\cdot 10H_2O(cr)$	[65MUT/HIR]
$H_4Ca(UO_2)_2(PO_4)_4\cdot 9H_2O(cr)$	[84NRI]
$Ca_2(UO_2)_3(PO_4)_2(OH)_4\cdot 4H_2O(cr)$	[83FLE]
$Ca(UO_2)_3(PO_4)_2(OH)_2\cdot 6H_2O(cr)$	[83FLE]
$Ca_2(UO_2)_3(PO_4)_2(OH)_4\cdot 5.5H_2O(cr)$	[84NRI]
$Ca(UO_2)_4(PO_4)_2(OH)_4(cr)$	[84GEN/WEI]
$Ca(UO_2)_2(AsO_4)_2(cr)$	[84GEN/WEI]
$Ca(UO_2)_2(AsO_4)_2\cdot 8H_2O(cr)$	[83FLE]
$Ca(UO_2)_2(AsO_4)_2\cdot 10H_2O(cr)$	[83FLE]
$Ca(UO_2)_4(AsO_4)_2(OH)_4(cr)$	[84GEN/WEI]
$Ca(UO_2)_4(AsO_4)_2(OH)_4\cdot 6H_2O(cr)$	[58BEL]
$CaUO_2(CO_3)_2\cdot 3H_2O(cr)$	discussed in the present review
$CaUO_2(CO_3)_2\cdot 5H_2O(cr)$	discussed in the present review
$CaUO_2(CO_3)_2\cdot 10H_2O(cr)$	discussed in the present review
$Ca_2UO_2(CO_3)_3\cdot 10H_2O(cr)$	discussed in the present review
$Ca(UO_2)_2(SiO_3OH)_2(cr)$	[78LAN], [80BEN/TEA], [88LEM], [88PHI/HAL2]
$Ca(UO_2)_2(SiO_3OH)_2\cdot 5H_2O(cr)$	[82HEM]
$Ca(UO_2)_2Si_6O_{15}\cdot 5H_2O(cr)$	[82HEM], [83FLE]
$Ca_2CuUO_2(CO_3)_4\cdot 6H_2O(cr)$	[82HEM], [83FLE]
$Ca(UO_2)_3(MoO_4)_3(OH)_2\cdot 11H_2O(cr)$	[83FLE]
$Ca(UO_2)_2(VO_4)_2(cr)$	[78LAN], [80BEN/TEA], [85PHI/PHI], [88PHI/HAL2]
$Ca(UO_2)_2V_{10}O_{28}\cdot 16H_2O(cr)$	[83FLE]
$CaMgUO_2(CO_3)_3\cdot 12H_2O(cr)$	discussed in the present review
$Ca_3Mg_3(UO_2)_2(CO_3)_6(OH)_4\cdot 18H_2O(cr)$	discussed in the present review
α-$SrUO_4$	data selected in the present review
β-$SrUO_4$	data selected in the present review
$SrUO_4(cr)$	[83KOH]
$Sr_2UO_5(cr)$	data selected in the present review
$Sr_3UO_6(cr)$	data selected in the present review
$Sr_2U_3O_{11}(cr)$	data selected in the present review
$SrU_4O_{13}(cr)$	data selected in the present review
$Sr(UO_2)_2(PO_4)_2(cr)$	[78LAN], [85PHI/PHI], [86WAN], [88PHI/HAL2]
$Sr(UO_2)_2(PO_4)_2\cdot 10H_2O(cr)$	[65MUT/HIR]
$BaUO_3(cr)$	data selected in the present review
$BaUO_4(cr)$	data selected in the present review
$Ba_3UO_6(cr)$	data selected in the present review
$BaU_2O_7(cr)$	data selected in the present review
$Ba_2U_2O_7(cr)$	data selected in the present review
$BaU_2O_7\cdot 4H_2O(cr)$	[83FLE]
$BaU_6O_{19}\cdot 11H_2O(cr)$	[83FLE]

Formula	References
$Ba(UO_2)_3(SeO_3)_2(OH)_4\cdot 3H_2O(cr)$	[83FLE]
$Ba(UO_2)_2(PO_4)_2(cr)$	[78LAN], [84GEN/WEI], [85PHI/PHI], [88PHI/HAL2]
$Ba(UO_2)_2(PO_4)_2\cdot 8H_2O(cr)$	[83FLE]
$Ba(UO_2)_2(PO_4)_2\cdot 10H_2O(cr)$	[65MUT/HIR]
$Ba(UO_2)_2(PO_4)_2\cdot 12H_2O(cr)$	[83FLE]
$Ba_2(UO_2)_3(PO_4)_2(OH)_4\cdot 5.5H_2O(cr)$	[84NRI]
$Ba(UO_2)_2(AsO_4)_2(cr)$	[84GEN/WEI]
$Ba(UO_2)_2(AsO_4)_2\cdot 8H_2O(cr)$	[83FLE]
$Ba_{0.5}UO_2(CO_3)OH\cdot 3H_2O(cr)$	discussed in the present review
$BaUO_2(CO_3)_2\cdot 2H_2O(cr)$	discussed in the present review
$Ba_{0.5}(UO_2)_2(CO_3)_3OH\cdot 5H_2O(cr)$	discussed in the present review
$BaUO_2(CO_3)F_2\cdot 2H_2O(cr)$	discussed in the present review
$Ba(UO_2)_2(VO_4)_2(cr)$	[84GEN/WEI]
$Ba_6Dy_2(UO_6)_3(cr)$	[82MOR]
$Ba_2MgUO_6(cr)$	data selected in the present review
$Ba_2CaUO_6(cr)$	data selected in the present review
$BaCa(UO_2)_3(PO_4)_2(OH)_4\cdot 5.5H_2O(cr)$	[84NRI]
$Ba_2SrUO_6(cr)$	data selected in the present review
$LiUO_3(cr)$	data selected in the present review
$Li_2UO_4(cr)$	data selected in the present review
$Li_3UO_4(cr)$	[81LIN/BES]
$Li_4UO_5(cr)$	data selected in the present review
$Li_6UO_6(cr)$	[81LIN/BES]
$Li_7UO_6(cr)$	[81LIN/BES]
$Li_2U_2O_7(cr)$	data selected in the present review
$Li_2U_3O_{10}(cr)$	data selected in the present review
$Li_2U_4O_{11}(cr)$	[81LIN/BES]
$Li_2U_4O_{13}(cr)$	[81LIN/BES]
$Li_6U_5O_{18}(cr)$	[81LIN/BES]
$Li_2U_6O_{19}(cr)$	[81LIN/BES]
$LiUF_6(cr)$	[83KAG/KYS]
$Li_2UF_6(cr)$	[83KAG/KYS]
$Li_2UCl_6(cr)$	[74VDO/VOL], [79FUG], [80PAR], [82WAG/EVA], [83CHI], [83FUG/PAR], [83FUG], [83KAG/KYS], [86MOR]
$Li_2UBr_6(cr)$	[83KAG/KYS]
$Li_2UI_6(cr)$	[83KAG/KYS]
$LiUO_2AsO_4(cr)$	[82WAG/EVA]
$NaUO_3(cr)$	data selected in the present review
α-Na_2UO_4	data selected in the present review
β-Na_2UO_4	data selected in the present review
$Na_2UO_4(cr)$	[52LAT], [57KAT/SEA], [61FAU/IPP], [61IPP/FAU], [63RAN/KUB], [72BAT/SHI], [72KRE], [73OHA/HOE2], [76FRE/OHA], [82HEM], [83DAV/YEF], [83KAG/KYS], [83KOH], [85TAC], [86WAN], [88LEM], [89BRU/PUI]

Formula	References
$Na_3UO_4(cr)$	data selected in the present review
$Na_4UO_5(cr)$	data selected in the present review [82COR/MUI], [82MOR], [83DAV/YEF], [83FUG], [85PHI/PHI]
$Na_2U_2O_7(cr)$	data selected in the present review
$Na_6U_7O_{21}(cr)$	[78COR/OHA]
$Na_6U_7O_{24}(cr)$	data selected in the present review
$Na_4O_4UO_4{\cdot}9H_2O(cr)$	[82WAG/EVA]
$Na_2U_2O_7{\cdot}1.5H_2O(cr)$	discussed in the present review
α-$NaUF_6$	[78KUD/SUG], [83FUG/PAR], [83FUG], [86MOR]
β-$NaUF_6$	[78KUD/SUG], [80PAR], [83CHI], [83FUG/PAR], [83FUG], [86MOR]
$NaUF_6(cr)$	[82WAG/EVA], [83KAG/KYS], [84OHA/MAL]
$Na_2UF_6(cr)$	[83CHI], [83KAG/KYS]
$NaUF_7(cr)$	[80PAR], [83FUG/PAR], [83FUG], [86MOR]
$Na_3UF_7(cr)$	[83CHI], [83FUG/PAR], [83FUG]
$Na_2UF_8(cr)$	[80PAR], [83FUG/PAR], [83FUG], [86MOR]
$Na_3UF_8(cr)$	[83CHI], [86MOR]
$Na_3UO_2F_5(cr)$	[80PAR], [82WAG/EVA], [83FUG/PAR], [83FUG], [86MOR]
$Na(UO_2)_2F_5(cr)$	[74SUP/SEL], [80PAR], [82WAG/EVA], [83FUG/PAR], [83FUG], [86MOR]
α-$NaUCl_6$	[80PAR], [82WAG/EVA], [83CHI], [83FUG/PAR], [83FUG], [86MOR]
β-$NaUCl_6$	[80PAR], [82WAG/EVA], [83CHI], [83FUG/PAR], [83FUG], [86MOR]
$Na_2UCl_6(cr)$	[68MAR/KUD], [72FUG], [74VDO/VOL], [79FUG], [80PAR], [82WAG/EVA], [83CHI], [83FUG/PAR], [83FUG], [83KAG/KYS], [86MOR]
$NaUBr_6(cr)$	[83CHI]
$Na_2UBr_6(cr)$	[79FUG], [83CHI], [83FUG/PAR], [83FUG], [83KAG/KYS], [86MOR]
$Na_2UI_6(cr)$	[83KAG/KYS]
$Na_4(UO_2)_6(SO_4)_3(OH)_{10}{\cdot}4H_2O(cr)$	[81OBR/WIL], [82HEM]
$NaUO_2PO_4(cr)$	discussed in the present review
$Na_2(UO_2)_2(PO_4)_2(cr)$	[88PHI/HAL2]
$NaUO_2PO_4{\cdot}4H_2O(cr)$	[83FLE]
$Na_2(UO_2)_2(PO_4)_2{\cdot}10H_2O(cr)$	[65MUT/HIR]
$NaUO_2AsO_4{\cdot}4H_2O(cr)$	[71NAU/RYZ]
$Na_4UO_2(CO_3)_3(cr)$	data selected in the present review
$Na_2UF_4(HCO_3)_2(cr)$	discussed in the present review
$NaUO_2VO_4(cr)$	[84GEN/WEI]
$NaUO_2VO_4{\cdot}3H_2O(cr)$	[83FLE]
$CaNa_2UO_2(CO_3)_3{\cdot}6H_2O(cr)$	discussed in the present review
$Ca_3NaUO_2(CO_3)_3F(SO_4){\cdot}4H_2O(cr)$	discussed in the present review
$Ca_3NaUO_2(CO_3)_3F(SO_4){\cdot}10H_2O(cr)$	discussed in the present review
$KUO_3(cr)$	data selected in the present review
$K_2UO_4(cr)$	data selected in the present review

Formula	References
$K_4UO_5(cr)$	[81LIN/BES]
$K_2U_2O_7(cr)$	data selected in the present review
$K_2U_4O_{13}(cr)$	[81LIN/BES]
$K_2U_7O_{22}(cr)$	[81LIN/BES]
$K_2U_6O_{19}\cdot 11H_2O(cr)$	[83FLE]
$KUF_6(cr)$	[78KUD/SUG], [80PAR], [82WAG/EVA], [83CHI], [83FUG/PAR], [83FUG], [83KAG/KYS], [84OHA/MAL], [86MOR]
$K_2UF_6(cr)$	[83CHI], [83KAG/KYS]
$K_2UF_7(cr)$	[83CHI], [86MOR]
$K_3UF_7(cr)$	[83CHI]
$K_3UF_8(cr)$	[83CHI], [86MOR]
$K_4UF_9(cr)$	[83CHI], [86MOR]
$K_3UO_2F_5(cr)$	[80PAR], [82WAG/EVA], [83FUG/PAR], [83FUG], [86MOR]
$K(UO_2)_2F_5(cr)$	[74SUP/SEL], [80PAR], [82WAG/EVA], [83FUG/PAR], [83FUG], [86MOR]
$K_5(UO_2)_2F_9(cr)$	[80PAR], [83FUG/PAR], [83FUG], [86MOR]
$K_5(UO_2)_3F_9(cr)$	[74MUK/SEL]
$KUCl_5(cr)$	[68MAR/KUD], [72FUG], [74VDO/VOL], [79FUG], [80PAR], [82WAG/EVA], [83CHI], [83FUG/PAR], [83FUG], [86MOR]
$K_2UCl_5(cr)$	[81SUG/CHI], [83CHI]
$KUCl_6(cr)$	[80PAR], [82WAG/EVA], [83CHI], [83FUG/PAR], [83FUG], [86MOR]
$K_2UCl_6(cr)$	[68MAR/KUD], [72FUG], [74VDO/VOL], [79FUG], [80PAR], [82WAG/EVA], [83CHI], [83FUG/PAR], [83FUG], [83KAG/KYS], [86MOR]
$K_3UCl_8(cr)$	[83CHI]
$K_2UBr_5(cr)$	[78SUG/CHI2], [83CHI], [83FUG/PAR], [83FUG], [86MOR]
$KUBr_6(cr)$	[83CHI]
$K_2UBr_6(cr)$	[79FUG], [83CHI], [83FUG/PAR], [83FUG], [83KAG/KYS], [86MOR]
$K_2UI_6(cr)$	[83KAG/KYS]
$K_4(UO_2)_6(SO_4)_3(OH)_{10}\cdot 4H_2O(cr)$	[81OBR/WIL], [82HEM]
$KUO_2PO_4(cr)$	discussed in the present review
$KUO_2PO_4(s)$	[56CHU/STE]
$K_2(UO_2)_2(PO_4)_2(cr)$	[88PHI/HAL2]
$KUO_2PO_4\cdot 3H_2O(cr)$	[71NAU/RYZ], [84VIE/TAR], [85PHI/PHI], [88PHI/HAL2]
$K_2(UO_2)_2(PO_4)_2\cdot 6H_2O(cr)$	[83FLE]
$K_2(UO_2)_2(PO_4)_2\cdot 10H_2O(cr)$	[65MUT/HIR]
$KUO_2AsO_4(cr)$	[71NAU/RYZ], [82WAG/EVA]
$KUO_2AsO_4\cdot 4H_2O(cr)$	[83FLE]
$K_3UO_2(CO_3)F_3(cr)$	discussed in the present review
$H_3OKUO_2SiO_4(cr)$	[82HEM], [83FLE]

Formula	References
$K_2(UO_2)_2Si_6O_{15}\cdot 4H_2O(cr)$	[82HEM], [83FLE]
$K_2(UO_2)_2(VO_4)_2(cr)$	[78LAN], [80BEN/TEA], [85PHI/PHI], [88PHI/HAL2]
$K_2(UO_2)_2(VO_4)_2\cdot 3H_2O(cr)$	[62HOS/GAR], [83FLE]
$K_2CaU_6O_{20}\cdot 9H_2O(cr)$	[83FLE]
$KNaUCl_6(cr)$	[68MAR/KUD], [72FUG], [82WAG/EVA], [83FUG]
$NaKUCl_6(cr)$	[79FUG], [80PAR], [83FUG/PAR], [86MOR]
$K_2NaUCl_6(cr)$	[83AUR/CHI], [83CHI], [86MOR]
$K_2NaUBr_6(cr)$	[83AUR/CHI], [83CHI], [86MOR]
$K_3NaUO_2(CO_3)_3\cdot H_2O(cr)$	[83FLE], [83OBR/WIL]
$Na_{0.7}\cdot K_{0.3}H_3OUO_2SiO_4\cdot H_2O(cr)$	[82HEM]
$RbUO_3(cr)$	data selected in the present review
$Rb_2UO_3(cr)$	[75OHA/ADE]
$Rb_2UO_4(cr)$	data selected in the present review
$Rb_4UO_5(cr)$	[81LIN/BES]
$Rb_2U_2O_7(cr)$	data selected in the present review
$Rb_2U_4O_{13}(cr)$	[81LIN/BES]
$Rb_2U_7O_{22}(cr)$	[81LIN/BES]
$RbUF_6(cr)$	[78KUD/SUG], [80PAR], [82WAG/EVA], [83CHI], [83FUG/PAR], [83FUG], [83KAG/KYS], [84OHA/MAL], [86MOR]
$Rb_2UF_6(cr)$	[83CHI], [83KAG/KYS]
$Rb_2UF_7(cr)$	[83CHI], [86MOR]
$Rb_3UF_7(cr)$	[83CHI], [83FUG/PAR], [83FUG], [86MOR]
$Rb_3UF_8(cr)$	[83CHI], [86MOR]
$Rb_4UF_9(cr)$	[83CHI], [86MOR]
$Rb_3UO_2F_5(cr)$	[80PAR], [82WAG/EVA], [83FUG/PAR], [83FUG], [86MOR]
$Rb(UO_2)_2F_5(cr)$	[74SUP/SEL], [80PAR], [82WAG/EVA], [83FUG/PAR], [83FUG], [86MOR]
$Rb_5(UO_2)_2F_9(cr)$	[74MUK/SEL], [80PAR], [82WAG/EVA], [83FUG/PAR], [83FUG], [86MOR]
$Rb_{0.5}UCl_{3.5}(cr)$	[81SUG/CHI], [83CHI]
$RbUCl_5(cr)$	[74VDO/VOL], [79FUG], [80PAR], [82WAG/EVA], [83CHI], [83FUG/PAR], [83FUG], [86MOR]
$Rb_2UCl_5(cr)$	[81SUG/CHI], [83CHI]
$RbUCl_6(cr)$	[80PAR], [82WAG/EVA], [83CHI], [83FUG/PAR], [83FUG], [86MOR]
$Rb_2UCl_6(cr)$	[74VDO/VOL], [79FUG], [80PAR], [82WAG/EVA], [83CHI], [83FUG/PAR], [83FUG], [83KAG/KYS], [86MOR]
$Rb_4UCl_8(cr)$	[74VDO/VOL], [79FUG], [80PAR], [82WAG/EVA], [83FUG/PAR], [83FUG], [86MOR]
$Rb_{0.43}UBr_{3.43}(cr)$	[83CHI]
$Rb_2UBr_5(cr)$	[78SUG/CHI2], [83CHI], [83FUG/PAR], [83FUG], [86MOR]
$RbUBr_6(cr)$	[83CHI]

Formula	References
Rb_2UBr_6(cr)	[73VDO/SUG], [79FUG], [80PAR], [82WAG/EVA], [83CHI], [83FUG/PAR], [83FUG], [83KAG/KYS], [86MOR]
Rb_2UI_6(cr)	[83KAG/KYS]
$RbUO_2PO_4$(cr)	discussed in the present review
Rb_2NaUCl_6(cr)	[83AUR/CHI], [83CHI], [86MOR]
$CsUO_3$(cr)	[83KAG/KYS]
$Cs_2UO_{3.56}$(s)	[81LIN/BES]
Cs_2UO_4(cr)	data selected in the present review
$Cs_2U_2O_7$(cr)	data selected in the present review
$Cs_2U_4O_{12}$(cr)	data selected in the present review
$Cs_2U_4O_{13}$(cr)	discussed in the present review
$Cs_2U_5O_{16}$(cr)	discussed in the present review
$Cs_4U_5O_{17}$(cr)	discussed in the present review
$Cs_2U_6O_{18}$(cr)	[81LIN/BES]
$Cs_2U_7O_{22}$(cr)	discussed in the present review
$Cs_2U_9O_{27}$(cr)	[81LIN/BES]
$Cs_2U_{15}O_{46}$(cr)	[81LIN/BES]
$Cs_2U_{16}O_{49}$(cr)	discussed in the present review
$CsUF_6$(cr)	[78KUD/SUG], [80PAR], [82WAG/EVA], [83CHI], [83FUG/PAR], [83FUG], [83KAG/KYS], [84OHA/MAL], [86MOR]
Cs_2UF_6(cr)	[83CHI], [83FUG], [83KAG/KYS], [86MOR]
$CsUF_7$(cr)	[83CHI]
α-Cs_2UF_7	[86MOR]
β-Cs_2UF_7	[86MOR]
Cs_2UF_7(cr)	[83CHI]
Cs_3UF_7(cr)	[83CHI]
Cs_3UF_8(cr)	[83CHI], [86MOR]
$Cs_3UO_2F_5$(cr)	[80PAR], [82WAG/EVA], [83FUG/PAR], [83FUG], [86MOR]
$Cs(UO_2)_2F_5$(cr)	[74SUP/SEL], [80PAR], [82WAG/EVA], [83FUG/PAR], [83FUG], [86MOR]
$Cs_5(UO_2)_2F_9$(cr)	[74MUK/SEL], [80PAR], [82WAG/EVA], [83FUG/PAR], [83FUG], [86MOR]
$CsUCl_4$(cr)	[81SUG/CHI], [83CHI]
$CsUCl_5$(cr)	[74VDO/VOL], [79FUG], [80PAR], [82WAG/EVA], [83CHI], [83FUG/PAR], [83FUG], [86MOR]
Cs_2UCl_5(cr)	[81SUG/CHI], [83CHI]
$CsUCl_6$(cr)	[80PAR], [82WAG/EVA], [83CHI], [83FUG/PAR], [83FUG], [86MOR]
Cs_2UCl_6(cr)	[71FUG/BRO], [72FUG], [74VDO/VOL], [79FUG], [80PAR], [82WAG/EVA], [83CHI], [83FUG/PAR], [83FUG], [83KAG/KYS], [86MOR]
Cs_2UCl_7(cr)	[83CHI]
Cs_3UCl_8(cr)	[83CHI]
CsU_2Cl_9(cr)	[74VDO/VOL], [79FUG], [80PAR], [82WAG/EVA], [83FUG/PAR], [83FUG], [86MOR]

Formula	References
$Cs_2UO_2Cl_4(cr)$	[79FUG], [83FUG/PAR], [83FUG], [86MOR]
$Cs_{0.59}UBr_{3.59}(cr)$	[83CHI]
$Cs_2UBr_5(cr)$	[78SUG/CHI2], [83CHI], [83FUG/PAR], [83FUG], [86MOR]
$CsUBr_6(cr)$	[83CHI]
$Cs_2UBr_6(cr)$	[73VDO/SUG], [79FUG], [80PAR], [82WAG/EVA], [83CHI], [83FUG/PAR], [83FUG], [83KAG/KYS], [86MOR]
$Cs_3UBr_6(cr)$	[78SUG/CHI2], [83AUR/CHI], [83CHI], [86MOR]
$Cs_2UO_2Br_4(cr)$	[79FUG], [83FUG/PAR], [83FUG], [86MOR]
$Cs_2UI_6(cr)$	[83KAG/KYS]
$CsUO_2PO_4(cr)$	discussed in the present review
$CsUO_2(CO_3)F{\cdot}H_2O(cr)$	discussed in the present review
$Cs_2NaUCl_6(cr)$	[89SCH/GEN]
$Cs_2NaUBr_6(cr)$	[83AUR/CHI], [83CHI], [86MOR]
$FrUO_3(cr)$	[83KAG/KYS]
$Fr_2UO_4(cr)$	[83KAG/KYS]
$FrUF_6(cr)$	[83KAG/KYS]
$Fr_2UF_6(cr)$	[83KAG/KYS]
$Fr_2UCl_6(cr)$	[83KAG/KYS]
$Fr_2UBr_6(cr)$	[83KAG/KYS]
$Fr_2UI_6(cr)$	[83KAG/KYS]

Chapter X

Alphabetical name list

This chapter presents a list of names of uranium containing mineral compounds with the corresponding chemical formulae. For the minerals which are discussed or mentioned in the present review, the corresponding section number of Chapter V is indicated. If data are selected, the relevant section number is marked in **bold** typeface. More information on published material on minerals that are not discussed in the present review, may be found under the corresponding formula in Chapter IX.

Very few precise thermodynamic data exist on uranium containing minerals, and the information in the literature is to a large extent based on predictions whose quality cannot be estimated reliably. Some data evaluators were aware of this fact, *e.g.*, the data base provided with the code EQ3/6 [83WOL, 88JAC/WOL] contains the comment "quality unspecified" for the data on most of these minerals. Modellers who need to include data on uranium containing minerals in their calculations are advised to use the compilation by Langmuir [78LAN] as a guideline.

Name	Formula	Section
abernathyite	$KUO_2AsO_4 \cdot 4H_2O(cr)$	
andersonite	$CaNa_2UO_2(CO_3)_3 \cdot 6H_2O(cr)$	V.7.1.2.2.b, V.10.2.4.1
andersonite, anhydrous	$CaNa_2UO_2(CO_3)_3(cr)$	
arsenuranospathite	$HAl(UO_2)_4(AsO_4)_4 \cdot 40H_2O(cr)$	
arsenuranylite	$Ca(UO_2)_4(AsO_4)_2(OH)_4 \cdot 6H_2O(cr)$	
arsenuranylite, anhydrous	$Ca(UO_2)_4(AsO_4)_2(OH)_4(cr)$	
autunite	$Ca(UO_2)_2(PO_4)_2 \cdot 10H_2O(cr)$	V.6.2.1.1.b, V.6.2.2.10.c
autunite, Cs, anhydrous	$CsUO_2PO_4(cr)$	V.6.2.1.1.b
autunite, H	$UO_2HPO_4 \cdot 4H_2O(cr)$	V.6.2.1.1.a, **V.6.2.1.1.b**, **V.6.2.2.10.c**
autunite, H, anhydrous	$H_2(UO_2)_2(PO_4)_2(cr)$	V.6.2.2.10.c
autunite, H, anhydrous	$UO_2HPO_4(cr)$	
autunite, H, octahydrated	$H_2(UO_2)_2(PO_4)_2 \cdot 8H_2O(cr)$	V.6.2.2.10.c
autunite, K	$K_2(UO_2)_2(PO_4)_2(cr)$	
autunite, K, anhydrous	$KUO_2PO_4(cr)$	V.6.2.1.1.b
autunite, NH_4, anhydrous	$NH_4UO_2PO_4(cr)$	V.6.2.1.1.b
autunite, Na	$NaUO_2PO_4 \cdot 4H_2O(cr)$	
autunite, Na	$Na_2(UO_2)_2(PO_4)_2(cr)$	
autunite, Na, anhydrous	$NaUO_2PO_4(cr)$	V.6.2.1.1.b
autunite, Rb, anhydrous	$RbUO_2PO_4(cr)$	V.6.2.1.1.b
autunite, Sr, anhydrous	$Sr(UO_2)_2(PO_4)_2(cr)$	
autunite, anhydrous	$Ca(UO_2)_2(PO_4)_2(cr)$	
bassetite	$Fe(UO_2)_2(PO_4)_2 \cdot 8H_2O(cr)$	
bassetite, anhydrous	$Fe(UO_2)_2(PO_4)_2(cr)$	
bauranoite	$BaU_2O_7 \cdot 4H_2O(cr)$	
bayleyite	$Mg_2UO_2(CO_3)_3 \cdot 18H_2O(cr)$	V.7.1.2.2.b, V.9.2.3
bayleyite, anhydrous	$Mg_2UO_2(CO_3)_3(cr)$	
becquerelite	$CaU_6O_{19} \cdot 11H_2O(cr)$	
bergenite, Ba	$Ba_2(UO_2)_3(PO_4)_2(OH)_4 \cdot 5.5H_2O(cr)$	
bergenite, Ca	$Ca_2(UO_2)_3(PO_4)_2(OH)_4 \cdot 5.5H_2O(cr)$	
bergenite, Ca-Ba	$BaCa(UO_2)_3(PO_4)_2(OH)_4 \cdot 5.5H_2O(cr)$	
billietite	$BaU_6O_{19} \cdot 11H_2O(cr)$	
boltwoodite	$H_3OKUO_2SiO_4(cr)$	
calcurmolite	$Ca(UO_2)_3(MoO_4)_3(OH)_2 \cdot 11H_2O(cr)$	
carnotite	$K_2(UO_2)_2(VO_4)_2 \cdot 3H_2O(cr)$	
carnotite, anhydrous	$K_2(UO_2)_2(VO_4)_2(cr)$	
coffinite	$USiO_4(cr)$	**V.7.2.2.2**, V.7.2.2.3
compreignacite	$K_2U_6O_{19} \cdot 11H_2O(cr)$	
cousinite	$MgU_2Mo_2O_{13} \cdot 6H_2O(cr)$	
cuprosklodowskite	$Cu(UO_2)_2Si_2O_6(OH)_2 \cdot 6H_2O(cr)$	
curienite	$Pb(UO_2)_2(VO_4)_2 \cdot 5H_2O(cr)$	
curienite, anhydrous	$Pb(UO_2)_2(VO_4)_2(cr)$	
curite	$Pb_2U_5O_{17} \cdot 4H_2O(cr)$	
derriksite	$Cu_4UO_2(SeO_3)_2(OH)_6 \cdot H_2O(cr)$	
dewindtite	$Pb(UO_2)_2(PO_4)_2 \cdot 3H_2O(cr)$	

Name	Formula	Section
dumontite	$Pb_2(UO_2)_3(PO_4)_2(OH)_4\cdot3H_2O(cr)$	
fourmarierite	$PbU_4O_{13}\cdot4H_2O(cr)$	
francevillite	$Ba(UO_2)_2(VO_4)_2\cdot5H_2O(cr)$	
francevillite, anhydrous	$Ba(UO_2)_2(VO_4)_2(cr)$	
fritzcheite	$Mn(UO_2)_2(VO_4)_2\cdot10H_2O(cr)$	
furongite	$Al_2UO_2(PO_4)_2(OH)_2\cdot8H_2O(cr)$	
grimselite	$K_3NaUO_2(CO_3)_3\cdot H_2O(cr)$	
grimselite, anhydrous	$K_3NaUO_2(CO_3)_3(cr)$	
guilleminite	$Ba(UO_2)_3(SeO_3)_2(OH)_4\cdot3H_2O(cr)$	
gummite	$UO_3(am)$	**V.3.3.1.3**, V.3.3.1.5
haiweeite	$Ca(UO_2)_2Si_6O_{15}\cdot5H_2O(cr)$	
hallimondite	$Pb_2UO_2(AsO_4)_2(cr)$	
hydrogen spinite	$UO_2HAsO_4\cdot4H_2O(cr)$	
ianthinite	$UO_{2.833}\cdot2H_2O(cr)$	V.3.3.3.5
iriginite	$UO_2Mo_2O_7\cdot3H_2O(cr)$	
johannite	$Cu(UO_2)_2(SO_4)_2(OH)_2\cdot8H_2O(cr)$	
joliotite	$UO_2CO_3\cdot2H_2O(cr)$	V.7.1.2.2.a
kahlerite, anhydrous	$Fe(UO_2)_2(AsO_4)_2(cr)$	
kasolite	$PbUO_2SiO_4\cdot H_2O(cr)$	V.7.3
liebigite	$Ca_2UO_2(CO_3)_3\cdot10H_2O(cr)$	V.7.1.2.2.b, V.9.3.2.2
liebigite, anhydrous	$Ca_2UO_2(CO_3)_3(cr)$	
marthozite	$Cu(UO_2)_3(SeO_3)_3(OH)_2\cdot7H_2O(cr)$	
masuyite	$UO_2\cdot2H_2O(cr)$	
meta-ankoleite	$K_2(UO_2)_2(PO_4)_2\cdot6H_2O(cr)$	
meta-uranocircite	$Ba(UO_2)_2(PO_4)_2\cdot8H_2O(cr)$	
meta-uranopilite	$(UO_2)_6SO_4(OH)_{10}\cdot5H_2O(cr)$	
meta-uranospinite	$Ca(UO_2)_2(AsO_4)_2\cdot8H_2O(cr)$	
metaheinrichite	$Ba(UO_2)_2(AsO_4)_2\cdot8H_2O(cr)$	
metaheinrichite, anhydrous	$Ba(UO_2)_2(AsO_4)_2(cr)$	
metakahlerite	$Fe(UO_2)_2(AsO_4)_2\cdot8H_2O(cr)$	
metakirchheimerite	$Co(UO_2)_2(AsO_4)_2\cdot8H_2O(cr)$	
metakirchheimerite, anhydrous	$Co(UO_2)_2(AsO_4)_2(cr)$	
metakirchheimerite, heptahydrate	$Co(UO_2)_2(AsO_4)_2\cdot7H_2O(cr)$	
metalodevite	$Zn(UO_2)_2(AsO_4)_2\cdot10H_2O(cr)$	
metalodevite, anhydrous	$Zn(UO_2)_2(AsO_4)_2(cr)$	
metatorbernite	$Cu(UO_2)_2(PO_4)_2\cdot8H_2O(cr)$	
metavanmeersscheite	$U(UO_2)_3(PO_4)_2(OH)_6\cdot2H_2O(cr)$	
metavanuralite	$Al(UO_2)_2(VO_4)_2OH\cdot8H_2O(cr)$	
metazellerite	$CaUO_2(CO_3)_2\cdot3H_2O(cr)$	V.7.1.2.2.b
metazeunerite	$Cu(UO_2)_2(AsO_4)_2\cdot8H_2O(cr)$	
metazeunerite, anhydrous	$Cu(UO_2)_2(AsO_4)_2(cr)$	
moctezumite	$PbUO_2(TeO_3)_2(cr)$	
moluranite	$H_4U(UO_2)_3(MoO_4)_7\cdot18H_2O(am)$	
mourite	$UMo_5O_{12}(OH)_{10}(cr)$	
mundite	$Al(UO_2)_3(PO_4)_2(OH)_3\cdot5.5H_2O(cr)$	

Name	Formula	Section
ningyoite	$CaU(PO_4)_2 \cdot 2H_2O(cr)$	
novacekite	$Mg(UO_2)_2(AsO_4)_2 \cdot 12H_2O(cr)$	
novacekite, anhydrous	$Mg(UO_2)_2(AsO_4)_2(cr)$	
orthobrannerite	$U_2Ti_4O_{12}(OH)_2(cr)$	
parsonite	$Pb_2UO_2(PO_4)_2 \cdot 2H_2O(cr)$	
parsonite, anhydrous	$Pb_2UO_2(PO_4)_2(cr)$	
phosphuranylite	$Ca(UO_2)_3(PO_4)_2(OH)_2 \cdot 6H_2O(cr)$	
phuralumite	$Al_2(UO_2)_3(PO_4)_2(OH)_6 \cdot 10H_2O(cr)$	
phurcalite	$Ca_2(UO_2)_3(PO_4)_2(OH)_4 \cdot 4H_2O(cr)$	
przhevalskite	$Pb(UO_2)_2(PO_4)_2 \cdot 4H_2O(cr)$	
przhevalskite, anhydrous	$Pb(UO_2)_2(PO_4)_2(cr)$	
pseudoautunite	$H_4Ca(UO_2)_2(PO_4)_4 \cdot 9H_2O(cr)$	
rabbittite	$Ca_3Mg_3(UO_2)_2(CO_3)_6(OH)_4 \cdot 18H_2O(cr)$	V.9.3.3.2
rameauite	$K_2CaU_6O_{20} \cdot 9H_2O(cr)$	
ranunculite	$HAlUO_2PO_4(OH)_3 \cdot 4H_2O(cr)$	
rauvite	$Ca(UO_2)_2V_{10}O_{28} \cdot 16H_2O(cr)$	
renardite	$Pb(UO_2)_4(PO_4)_2(OH)_4 \cdot 7H_2O(cr)$	
roubaultite	$Cu_2(UO_2)_3(OH)_{10} \cdot 5H_2O(cr)$	
rutherfordine	$UO_2CO_3(cr)$	V.7.1.2.2
sabugalite	$HAl(UO_2)_4(PO_4)_4 \cdot 16H_2O(cr)$	
sabugalite, anhydrous	$HAl(UO_2)_4(PO_4)_4(cr)$	
saleeite	$Mg(UO_2)_2(PO_4)_2 \cdot 10H_2O(cr)$	
saleeite, anhydrous	$Mg(UO_2)_2(PO_4)_2(cr)$	
schmitterite	$UO_2TeO_3(cr)$	**V.5.3.2.2**
schoepite	$UO_3 \cdot 2H_2O(cr)$	V.3.2.1.3, V.3.2.1.4, **V.3.3.1.5**
schroeckingerite	$Ca_3NaUO_2(CO_3)_3F(SO_4) \cdot 10H_2O(cr)$	V.7.1.2.2.b
sedovite	$U(MoO_4)_2(cr)$	
sengierite	$Cu_2(UO_2)_2(VO_4)_2(OH)_2 \cdot 6H_2O(cr)$	
sharpite	$UO_2CO_3 \cdot H_2O(cr)$	V.7.1.2.2.a
sklodowskite	$Mg(UO_2)_2Si_2O_6(OH)_2 \cdot 5H_2O(cr)$	
soddyite	$(UO_2)_2SiO_4 \cdot 2H_2O(cr)$	V.7.2.2.1
strelkinite	$NaUO_2VO_4 \cdot 3H_2O(cr)$	
strelkinite, anhydrous	$NaUO_2VO_4(cr)$	
studtite	$UO_4 \cdot 4H_2O(cr)$	**V.3.3.1.4**
swamboite	$UH_6(UO_2)_6(SiO_4)_6 \cdot 30H_2O(cr)$	
swartzite	$CaMgUO_2(CO_3)_3 \cdot 12H_2O(cr)$	V.7.1.2.2.b, V.9.3.3.1
swartzite, anhydrous	$CaMgUO_2(CO_3)_3(cr)$	
threadgoldite	$Al(UO_2)_2(PO_4)_2OH \cdot 8H_2O(cr)$	
torbernite	$Cu(UO_2)_2(PO_4)_2 \cdot 10H_2O(cr)$	
torbernite, anhydrous	$Cu(UO_2)_2(PO_4)_2(cr)$	
troegerite	$(UO_2)_3(AsO_4)_2 \cdot 12H_2O(cr)$	
tyuyamunite	$Ca(UO_2)_2(VO_4)_2 \cdot 8H_2O(cr)$	
tyuyamunite, anhydrous	$Ca(UO_2)_2(VO_4)_2(cr)$	
umohoite	$UO_2MoO_4 \cdot 4H_2O(cr)$	

Name	Formula	Section
upalite	$Al(UO_2)_3(PO_4)_2(OH)_3(cr)$	
uramphite	$NH_4UO_2PO_4\cdot 3H_2O(cr)$	
uraninite	$UO_2(cr)$	V.3.2.3.2, V.3.2.3.3, **V.3.3.2.1**, V.3.3.2.2, V.4.1.3.4.a, V.4.2.1.3.e, V.4.3.1.3.e, V.5.1.2.2.c, V.7.2.2.2, V.7.2.2.3
uranocircite I	$Ba(UO_2)_2(PO_4)_2\cdot 12H_2O(cr)$	
uranocircite II	$Ba(UO_2)_2(PO_4)_2\cdot 10H_2O(cr)$	
uranocircite, anhydrous	$Ba(UO_2)_2(PO_4)_2(cr)$	
uranophane	$Ca(UO_2)_2(SiO_3OH)_2\cdot 5H_2O(cr)$	
uranophane, anhydrous	$Ca(UO_2)_2(SiO_3OH)_2(cr)$	
uranopilite	$(UO_2)_6SO_4(OH)_{10}\cdot 12H_2O(cr)$	
uranospathite	$HAl(UO_2)_4(PO_4)_4\cdot 40H_2O(cr)$	
uranosphaerite	$Bi_2U_2O_9\cdot 3H_2O(cr)$	
uranospinite	$Ca(UO_2)_2(AsO_4)_2\cdot 10H_2O(cr)$	
uranospinite, anhydrous	$Ca(UO_2)_2(AsO_4)_2(cr)$	
vandenbrandeite	$CuUO_2(OH)_4(cr)$	
vandendriesscheite	$PbU_7O_{22}\cdot 12H_2O(cr)$	
vanmeersscheite	$U(UO_2)_3(PO_4)_2(OH)_6\cdot 4H_2O(cr)$	
vanuralite	$Al(UO_2)_2(VO_4)_2OH\cdot 11H_2O(cr)$	
vanuralite, anhydrous	$Al(UO_2)_2(VO_4)_2OH(cr)$	
voglite	$Ca_2CuUO_2(CO_3)_4\cdot 6H_2O(cr)$	
voglite, anhydrous	$Ca_2CuUO_2(CO_3)_4(cr)$	
walpurgite	$Bi_4UO_2(AsO_4)_2O_4\cdot 2H_2O(cr)$	
weeksite	$K_2(UO_2)_2Si_6O_{15}\cdot 4H_2O(cr)$	
widenmannite	$Pb_2UO_2(CO_3)_3(cr)$	
zellerite	$CaUO_2(CO_3)_2\cdot 5H_2O(cr)$	V.7.1.2.2.b, V.9.3.2.1
zellerite, anhydrous	$CaUO_2(CO_3)_2(cr)$	
zippeite, Co	$Co_2(UO_2)_6(SO_4)_3(OH)_{10}\cdot 8H_2O(cr)$	V.5.1.3.2.c
zippeite, K	$K_4(UO_2)_6(SO_4)_3(OH)_{10}\cdot 4H_2O(cr)$	
zippeite, Mg	$Mg_2(UO_2)_6(SO_4)_3(OH)_{10}\cdot 8H_2O(cr)$	V.5.1.3.2.c, V.9.2.2
zippeite, Na	$Na_4(UO_2)_6(SO_4)_3(OH)_{10}\cdot 4H_2O(cr)$	
zippeite, Ni	$Ni_2(UO_2)_6(SO_4)_3(OH)_{10}\cdot 8H_2O(cr)$	V.5.1.3.2.c
zippeite, Zn	$Zn_2(UO_2)_6(SO_4)_3(OH)_{10}\cdot 8H_2O(cr)$	V.5.1.3.2.c

Appendix A

Discussion of publications

This appendix is comprised of discussions relating to a number of publications which contain experimental information cited in this review. While these discussions are fundamental in explaining the accuracy of the data concerned and the interpretation of the experiments, they are either too lengthy or are related to too many different sections to be included in the main text. The notation used in this appendix is consistent with that used in the other parts of this book, and not necessarily with that used in the publication under discussion.

[15FOR]

De Forcrand, R., Etude des hydrates du nitrate d'uranyle et de l'anhydride uranique, Ann. de Chim. (Paris), 9(3) (1915) 5-48, in French.

De Forcrand [15FOR] reported a value for the enthalpies of solution of hydrated and anhydrous dioxouranium(VI) nitrates in water at 12°C. Analysis of the data for the higher hydrates is described in Section V.6.1.3.2. For the monohydrate, the reported enthalpy of solution is $-49.7\,\mathrm{kJ\cdot mol^{-1}}$. From Kopp's Law, $C^{\circ}_{p,\mathrm{m}}(\mathrm{UO_2(NO_3)_2{\cdot}H_2O}$, cr, 298.15 K) is estimated to be approximately $36\,\mathrm{J\cdot K^{-1}\cdot mol^{-1}}$ less than the value for the solid dihydrate (*i.e.*, $242\,\mathrm{J\cdot K^{-1}\cdot mol^{-1}}$). $C^{\circ}_{p,\mathrm{m}}(\mathrm{UO_2(NO_3)_2}$, cr, 298.15 K) = $206\,\mathrm{J\cdot K^{-1}\cdot mol^{-1}}$ is estimated in a similar way. From these, $C^{\circ}_{p,\mathrm{m}}(\mathrm{UO_2(NO_3)_2{\cdot}6H_2O}$, cr, 298.15 K), $C^{\circ}_{p,\mathrm{m}}(\mathrm{H_2O}$, l, 298.15 K) and $\Delta_{\mathrm{sol}}H^{\circ}_{\mathrm{m}}(\mathrm{UO_2(NO_3)_2{\cdot}6H_2O}$, cr, 298.15 K), the enthalpy differences at 25°C are calculated:

$$\begin{aligned}
\Delta_{\mathrm{sol}}H^{\circ}_{\mathrm{m}}(\mathrm{UO_2(NO_3)_2{\cdot}6H_2O}) - \Delta_{\mathrm{sol}}H^{\circ}_{\mathrm{m}}(\mathrm{UO_2(NO_3)_2{\cdot}H_2O}) &= 74.5\,\mathrm{kJ\cdot mol^{-1}}\\
\Delta_{\mathrm{sol}}H^{\circ}_{\mathrm{m}}(\mathrm{UO_2(NO_3)_2{\cdot}6H_2O}) - \Delta_{\mathrm{sol}}H^{\circ}_{\mathrm{m}}(\mathrm{UO_2(NO_3)_2}) &= 105.2\,\mathrm{kJ\cdot mol^{-1}}
\end{aligned}$$

Calculations using the selected value of $\Delta_{\mathrm{f}}H^{\circ}_{\mathrm{m}}(\mathrm{UO_2(NO_3)_2{\cdot}6H_2O}$, cr, 298.15 K) and the CODATA value for $\Delta_{\mathrm{f}}H^{\circ}_{\mathrm{m}}(\mathrm{H_2O}$, l, 298.15 K) result in the values $\Delta_{\mathrm{f}}H^{\circ}_{\mathrm{m}}(\mathrm{UO_2(NO_3)_2{\cdot}H_2O}$, cr, 298.15 K) = $-1663.8\,\mathrm{kJ\cdot mol^{-1}}$ and $\Delta_{\mathrm{f}}H^{\circ}_{\mathrm{m}}(\mathrm{UO_2(NO_3)_2}$, cr, 298.15 K) = $-1347.3\,\mathrm{kJ\cdot mol^{-1}}$.

[29GER/FRE]

Germann, F.E.E., Frey, P.R., The hydrates of uranyl nitrate, their respective vapor pressures, and the relative stability of the lower hydrate, J. Colorado-Wyoming Acad. Sci., **34** (1929) 54-55.

This is an abstract, not a full paper. The vapour pressures for transformation of the dioxouranium(VI) nitrate hydrates were obtained by an isopiestic method at an unspecified temperature. Wendelkowski and Kirslis [54WEN/KIR] deduced that the measurements were done at 23.4°C, and this was presumably based on the reported vapour pressures and corresponding concentrations of aqueous sulphuric acid. No reference was given as to the source for the water vapour pressures over sulphuric acid. Therefore, the data from this paper are not used in the recalculation in this review of the temperature dependence of the transformation water vapour pressures for the hydrates.

[40COU/PIT]

Coulter, L.V., Pitzer, K.S., Latimer, W.M., Entropies of large ions: The heat capacity, entropy and heat of solution of K_2PtCl_6, $(CH_3)_4NI$ and $UO_2(NO_3)_2{\cdot}6H_2O$, J. Am. Chem. Soc., **62** (1940) 2845-2851.

The authors reported experimental heat capacities for $UO_2(NO_3)_2 \cdot 6H_2O(cr)$ from 13.67 to 298.68 K, and "smoothed values" at selected temperatures from 15 to 300 K, but not specifically at 298.15 K. For calculation of the entropy at 298.1 K, contributions for 0 to 14 K (Debye and Einstein extrapolation) and for 14 to 298 K (from graphical integration) were summed. The compilation of Kelley and King [61KEL/KIN] used exactly the same numerical values for these contributions, but took the sum as the value for 298.15 K (but quotes a slightly reduced uncertainty). However, the heat capacity values reported by Kelley and King [61KEL/KIN] are somewhat lower than those reported in the original paper. Their result for $C^\circ_{p,m}(UO_2(NO_3)_2 \cdot 6H_2O$, cr, 298.15) is $(466.9 \pm 2.1)\ J \cdot K^{-1} \cdot mol^{-1}$, although the table of smoothed values in Coulter *et al.* [40COU/PIT] would suggest a value slightly greater than $468\ J \cdot K^{-1} \cdot mol^{-1}$.

In this review the raw data are replotted for temperatures greater than 260 K. From this, and considering the authors' comments concerning the relative weightings of the data points, the value selected is $C^\circ_{p,m}(UO_2(NO_3)_2{\cdot}6H_2O$, cr, 298.15 K) = $(468.0 \pm 2.5)\ J \cdot K^{-1} \cdot mol^{-1}$. It should also be mentioned that attempts to fit the heat capacity data for the temperature range 260 to 300 K to the Maier-Kelley equation [32MAI/KEL] by the method of least squares lead to systematic differences between the calculated and (sparse) experimental values.

[47FON]

Fontana, B.J., The heat of solution of uranium tetrachloride in aqueous perchloric acid solutions, MDDC-1452, US Atomic Energy Commission, Oak Ridge, Tennessee, USA, 1947, 2p.

As noted by Betts [55BET] the primary difficulty with this work is the assumption that UOH^{3+} is the sole hydrolysis product of U^{4+} when UCl_4 is added to water. As discussed in Chapter V this assumption is questionable for a pH near 2.2. Also, as noted by Fontana, the measured heat may have included contributions from precipitation, as the final solution was supersaturated with latex antUO_2(am).

[47GUI]

Guiter, H., Hydrolyse du nitrate d'uranyle, Bull. Soc. Chim. France, (1947) 64-67, in French.

This paper includes hydroxy-nitrate complexes of dioxouranium(VI) in the hydrolysis model. Such a model might, of course, be used. However, the ion interaction coefficient model used in this review is preferred to the calculation of formation constants between weakly complexing anionic ligands and the cationic hydrolysis species. Because of the apparent lack of temperature control in the experiments and the imprecise colorimetric method used to measure the pH values no attempt is made to reanalyse the data, and the reported constants are not credited in this review.

[47HAR/KOL]

Harris, W.E., Kolthoff, I.M., The polarography of uranium: III. Polarography in very weakly acid, neutral or basic solution, J. Am. Chem. Soc., **69** (1947) 446-451.

Possible polymer formation was ignored. A rough reanalysis of the data gives $\log_{10}{}^*\beta_{2,2} = -5.2$, in poor agreement with the results of other authors. This value is not used in the evaluation of ${}^*\beta^\circ_{2,2}$ in this review.

[47MAC/LON]

MacInnes, D.A., Longsworth, L.G., The measurement and interpretation of pH and conductance values of aqueous solutions of uranyl salts, MDDC-9911, US Atomic Energy Commission, Oak Ridge, Tennessee, USA, 1947, 10p.

This is a nice early piece of work, but the lack of control of temperature and ionic strength precludes its use in this review.

[48FAU]

Faucherre, J., Sur la condensation des ions basiques dans l'hydrolyse des nitrates d'aluminium et d'uranyle, C. R. Hébd. Séances Acad. Sci., **227** (1948) 1367-1369, in French.

This is an early study. The values reported for ${}^*\beta_{2,2}$ at $I = 0.6$ M and $I = 0.06$ M ($Ba(NO_3)_2$ supporting electrolyte) are in good agreement with later values. The values are assigned a weight of zero in this review because the temperature of the measurements was not reported.

[49BET/MIC]

Betts, R.H., Michels, R.K., Ionic association in aqueous solutions of uranyl sulphate and uranyl nitrate, J. Chem. Soc., **S58** (1949) 286-294.

Betts and Michels [49BET/MIC] did a spectrophotometric study of the complex formation of UO_2^{2+} with sulphate and nitrate ions at 25°C. The ionic medium was 2.00 M $HClO_4$. The ionic strengths varied from 5.38 M to 7.05 M ($NaNO_3$ + U(VI)) in the nitrate studies. In the sulphate studies the ionic strength was 2.56 M.

The authors studied the reactions

$$UO_2^{2+} + HSO_4^- \rightleftharpoons UO_2SO_4(aq) + H^+ \quad (A.1)$$
$$UO_2^{2+} + NO_3^- \rightleftharpoons UO_2NO_3^+ \quad (A.2)$$

and reported $\log_{10}{}^*\beta_1(A.2) = (0.70 \pm 0.06)$ and $\log_{10}\beta_1(A.2) = -(0.68 \pm 0.04)$, respectively. This review multiplies the reported uncertainties by two.

There was no evidence for the formation of complexes with HSO_4^-. The HSO_4^- concentration used to calculate the equilibrium constant of the first reaction was so low (< 0.08 M) that the formation of $UO_2(SO_4)_2^{2+}$ was disregarded. This was not the case in the Job-plot, *cf.* [49BET/MIC, *p.*289], nevertheless, there is no doubt that the stoichiometry of the complex is UO_2SO_4. Correction of the value of $\log_{10} K_1$ from $I = 2.65$ M (*i.e.*, $I_m = 3.05$ m) to $I = 0$ with $\Delta\varepsilon = -(0.31 \pm 0.04)$, *cf.* Appendix B, gives $\log_{10}{}^*\beta_1^\circ(A.2) = (0.74 \pm 0.13)$. Combination of this value with the equilibrium constant $\log_{10}\beta_1^\circ(HSO_4^-) = (1.98 \pm 0.05)$ for the reaction $H^+ + SO_4^{2-} \rightleftharpoons HSO_4^-$ gives $\log_{10}\beta_1^\circ(UO_2SO_4(aq)) = (2.72 \pm 0.14)$.

The nitrate system was studied at a much higher ionic strength than the sulphate system. Hence the correction to $I = 0$ is more uncertain. This review therefore does not include this value in the evaluation of the dioxouranium(VI) nitrate system.

[49KRI/HIN]

Kritchevsky, E.S., Hindman, J.C., The potentials of uranium three-four and five-six couples in perchloric and hydrochloric acids, J. Am. Chem. Soc., **71** (1949) 2096-2106.

This is a careful polarographic study. However, the authors used an SCE reference electrode connected to the test solution via a NH_4Cl (or KCl) agar bridge. It is unlikely that this arrangement eliminates all diffusion potentials, *cf.* [49KER/ORL, *p.*2104]. This review does not consider these data to be as precise as those of Kraus, Nelson and Johnson [49KRA/NEL] or Kern and Orleman [49KER/ORL]. From the reported formal potentials, the standard potential is calculated by using the specific ion interaction theory with $\varepsilon_{(UO_2^+,ClO_4^-)} = (0.28 \pm 0.05)$. Using this value and $E^\circ = 0.2415$ V for the saturated calomel reference electrode, this review obtains the values listed in Table V.2 of Section V.2.2 for Ref. [49KRI/HIN]. These values are in reasonably good agreement with those of Refs. [49KRA/NEL] and [49KER/ORL].

[49SUT]

Sutton, J., The hydrolysis of the uranyl ion: Part I, J. Chem. Soc., **S58** (1949) 275-286.

This is a report of a careful study. However, the ionic strength was not properly controlled. It also was not shown that the minor species assumed were the correct minor species. The raw data are not available for reanalysis.

[50BET/LEI]

Betts, R.H. Leigh, R.M., Ionic species of tetravalent uranium in perchloric and sulphuric acids, Can. J. Res., **28** (1950) 514-525.

Betts and Leigh studied the formation of uranium(IV) sulphate complexes by using an extraction method. The measurements were made at 25°C in 2 M $HClO_4$. The experimental work is satisfactory but not the computation procedure. The authors reported $^*K_1 = 338$ and $^*K_2 = 250$ for the reactions $U^{4+} + nHSO_4^- \rightleftharpoons U(SO_4)_n^{4-2n}$ where $n = 1$ and 2. The data re-evaluated later by Sullivan and Hindman [52SUL/HIN] are accepted in this review.

[50KRA/NEL]

Kraus, K.A., Nelson, F., Hydrolytic behavior of metal ions: I. The acid constants of uranium (IV) and plutonium(IV), J. Am. Chem. Soc., 72 (1950) 3901-3906.

Kraus and Nelson measured the hydrolysis of U^{4+} in perchlorate and chloride solutions by a spectrophotometric method in combination with pH measurements. The authors estimated the experimental errors to be about 10% of the measured values. In this review the uncertainties are estimated as 20% of the reported values and double this for cases in which the uranium salts contributed more than 10% to the total ionic strength. Analysis of the results for the chloride solutions is complicated by the need to use the same data to calculate the hydrolysis constant, $^*\beta_1^\circ$, the equilibrium constant for the formation of UCl^{3+}, K_1°, and the extinction coefficient for UCl^{3+}. The calculation of K_1° by assuming the same ionic strength dependence of the hydrolysis in perchlorate and chloride solutions is equivalent to assuming the same ion interaction coefficients for U^{4+} in the different media. The calculation by Kraus and Nelson also appears to use an extinction coefficient equal to that for U^{4+} in solutions containing only small amounts of chloride, but equal to the much smaller extinction coefficient for UOH^{3+} in the more concentrated chloride solutions. Reanalysis of the data is not attempted because only the extrema of the acid concentrations were provided for each ionic strength. All of the hydrolysis results in chloride media are assigned a weight of zero in the present evaluation.

Data on the formation of chloride complexes were also obtained. The authors estimated $\beta_1^\circ = (7 \pm 2)$ for the reaction $U^{4+} + Cl^- \rightleftharpoons UCl^{3+}$. From the concentration quotients, Kraus and Nelson calculated the activity quotients, *i.e.*, the equilibrium constants at $I = 0$, by using an extended Debye-Hückel method. The Debye-Hückel term differs somewhat from the one used in the specific ion interaction theory. The numerator there is $(1 + 2.5\sqrt{I})$ rather than $(1 + 1.5\sqrt{I})$. This value also fits the experimental data at rather low ionic strengths, where the difference between the Debye-Hückel method and the specific ion interaction theory is small. The spectral changes between a perchlorate solution and a 2 M Cl^- solution where 42% of the uranium is present as UCl^{3+} (according to the stability constant) are very small, leading to a large uncertainty in the equilibrium constant. The experiments performed by Kraus and Nelson were not optimized for the determination of the formation constants of the chloride complexes. Furthermore, at very low acidities hydrolysis species other than UOH^{3+} may also be significant. Thus, it is not surprising that the value for the equilibrium constant, $\log_{10}\beta_1^\circ$, obtained by Kraus and Nelson for the formation of UO_2Cl^+, is different from the value selected in this review.

[50MOO/KRA]
Moore, G.E., Kraus, K.A., Anionic exchange studies of uranium(VI) in HCl-HF mixtures: Separation of Pa(VI) and U(VI), ORNL-795, Oak Ridge National Laboratory, Oak Ridge, Tennessee, USA, 1950, *pp.*16-17.

This is an anion exchange study made at 25°C, but at varying ionic strength. It is difficult to evaluate the accuracy of the reported equilibrium constant, $\log_{10}\beta_1 = 4.32$ for the reaction $UO_2^{2+} + F^- \rightleftharpoons UO_2F^+$, and this review therefore gives zero weight to these data.

[50WAZ/CAM]
Van Wazer, J.R., Campanella, D.A., Structure and properties of the condensed phosphates: IV. Complex ion formation in polyphosphate solutions, J. Am. Chem. Soc., 72 (1950) 655-663.

This is a qualitative study of the complex formation between a polyphosphate with an average chain length of five phosphate groups and a number of metal ions including uranium. The experimental study indicated clearly that strong complexes were formed, but the analysis of the data does not permit conclusions to be drawn about the composition of the complexes or their stability constants. The equilibrium constants proposed by the authors are not considered reliable by this review.

[51AHR]
Ahrland, S., On the complex chemistry of the uranyl ion: VI. The complexity of uranyl chloride, bromide and nitrate, Acta Chem. Scand., 5 (1951) 1271-1282.

Ahrland measured the equilibrium constants for the formation of U(VI) chloride, bromide and nitrate complexes in 1 M $NaClO_4$ at 20°C. He used ligand displacement with acetate as the experimental method. The chloride system was also studied spectrophotometrically. The experimental procedures used are satisfactory. All complexes formed are very weak, and the author reported $\beta_1(UO_2Cl^+) = (0.8 \pm 0.2)$ (emf) and (0.5 ± 0.3) (sp), $\beta_1(UO_2Br^+) = (0.5 \pm 0.2)$ and $\beta_1(UO_2NO_3^+) = (0.5 \pm 0.2)$. The constants for the chloride system are corrected to 25°C using the enthalpy value selected in this review. The selected value of $\beta_1^\circ(UO_2Cl^+)$ is obtained from a linear extrapolation to $I = 0$ of values from Refs. [51AHR, 57DAV/MON, 78BED/FID, 81AWA/SUN] by using the specific ion interaction theory, *cf.* Section V.4.2.1.2.a). The value obtained from the emf determination by Ahrland seems to be an outlier and is therefore not included in the linear regression. From the $\Delta\varepsilon$ value obtained from this extrapolation this review calculates $\varepsilon_{(UO_2Cl^+,ClO_4^-)} = (0.33 \pm 0.04)$. The constants $\beta_1(UO_2Br^+)$ and $\beta_1(UO_2NO_3^+)$ are corrected to $I = 0$ assuming that $\varepsilon_{(UO_2Br^+,ClO_4^-)} \approx \varepsilon_{(UO_2NO_3^+,ClO_4^-)} \approx \varepsilon_{(UO_2Cl^+,ClO_4^-)} = (0.33 \pm 0.04)$.

[51AHR2]
Ahrland, S., On the complex chemistry of the uranyl ion: V. The complexity of uranyl sulfate, Acta Chem. Scand., 5 (1951) 1151-1167.

Ahrland used a spectrophotometric and a potentiometric method to study equilibria in the dioxouranium(VI) sulphate system at 20°C and $I = 1$ M $NaClO_4$ according to Eq. (A.3).

$$UO_2^{2+} + qSO_4^{2-} \rightleftharpoons UO_2(SO_4)_q^{2-2q} \quad (A.3)$$

The potentiometric method (using pH measurements and acetate as a competing ligand) is less precise than the spectrophotometric study (especially for the formation constants of the complexes $UO_2(SO_4)_2^{2-}$ and $UO_2(SO_4)_3^{4-}$). In fact, the evidence for the presence of a third complex $UO_2(SO_4)_3^{4-}$ is not conclusive. This complex is not included in the recommended data base. This choice is supported by a later calorimetric study by Ahrland and Kullberg [71AHR/KUL3] where no evidence for $UO_2(SO_4)_3^{4-}$ was found. The experimental data at 20°C are corrected to 25°C by using the enthalpy of reaction selected in this review, $\Delta_r H_m^\circ(A.3, q = 1) = (19.5 \pm 1.6)\,kJ \cdot mol^{-1}$ and $\Delta_r H_m^\circ(A.3, q = 2) = (35.1 \pm 1.0)\,kJ \cdot mol^{-1}$, leading to a correction of +0.06 in $\log_{10}\beta_1(A.3, q = 1)$ and +0.10 in $\log_{10}\beta_1(A.3, q = 2)$.

[51NEL/KRA]

Nelson, F., Kraus, K.A., Chemistry of aqueous uranium(V) solutions: III. The uranium(IV)-(V)-(VI) equilibrium in perchlorate and chloride solutions, J. Am. Chem. Soc., **73** (1951) 2157-2161.

Nelson and Kraus studied the disproportionation equilibrium $2UO_2^+ + 4H^+ \rightleftharpoons UO_2^{2+} + U^{4+} + 2H_2O(l)$ in perchlorate and chloride solutions. The medium dependence of the equilibrium constant was interpreted in terms of complex formation between chloride and U^{4+} and UO_2^{2+}, respectively. The authors made the reasonable assumption that only the monochloride complexes were formed. The concentration equilibrium constants were recalculated to activity constants, *i.e.*, they referred to the reference state of $I = 0$ by using a Debye-Hückel expression with the denominator $(1 + 2.46\sqrt{I})$ for U(IV) and $(1 + 1.97\sqrt{I})$ for U(VI). The authors used the value of $\log_{10}\beta_1^\circ$ determined by Nelson and Kraus [51NEL/KRA] for the formation of UCl^{3+} to determine $\beta_1^\circ(UO_2Cl^+)$.

The numerical value of $\log_{10}\beta_1^\circ(UO_2Cl^+)$ is directly dependent on the corresponding value for UCl^{3+}, which is rather uncertain. Hence, the authors considered their proposed value $\log_{10}\beta_1^\circ(UO_2Cl^+) = 0.38$ as a rough estimate only. In view of this, zero weight is given to this value.

[52KAP/BAR]

Kapustinsky, A.F., Baranova, L.I., Free energy and heat of dehydration of crystal-hydrates of uranyl nitrate and chloride, Izv. Akad. Nauk SSSR, Engl. Transl., (1952) 981-983.

Water vapour pressures were measured for the equilibrium $UO_2(NO_3)_2 \cdot 6H_2O \rightleftharpoons UO_2(NO_3)_2 \cdot 3H_2O + 3H_2O(g)$. The vapour pressures are systematically lower than those found by Cordfunke [66COR] or Vdovenko and Sokolov [59VDO/SOK]. The uncertainties in the measured vapour pressure values are estimated in this review as 100 Pa for measurements at all temperatures. The measurements for the trihydrate/dihydrate equilibrium were not considered reproducible by the authors, and these measurements are not used in calculations done in this review.

[52SUL/HIN]
Sullivan, J.C., Hindman, J.C., An analysis of the general mathematical formulations for the calculation of association constants on complex ion systems, J. Am. Chem. Soc., **74** (1952) 6091-6096.

Sullivan and Hindman re-evaluated the experimental data of Betts and Leigh [50BET/LEI]. For the reaction $U^{4+} + q HSO_4^- \rightleftharpoons U(SO_4)_q^{4-2q}$ they found, as the average value of several evaluations with different methods, $\log_{10}{}^*\beta_1 = (2.41 \pm 0.05)$ and $\log_{10}{}^*\beta_2 = (3.72 \pm 0.25)$, where the uncertainties are estimated by this review. These values agree very well with those of Day, Wilhite and Hamilton [55DAY/WIL]. Correction to $I = 0$ is done using the following estimates for the ion interaction coefficients: $\varepsilon_{(U^{4+},ClO_4^-)} = (0.76 \pm 0.06)$, $\varepsilon_{(USO_4^{2+},ClO_4^-)} = (0.3 \pm 0.1)$, and $\varepsilon_{(HSO_4^-,H^+)} = -(0.01 \pm 0.02)$. The resulting values are $\log_{10}{}^*\beta_1^\circ = (4.45 \pm 0.31)$ and $\log_{10}{}^*\beta_2^\circ = (6.37 \pm 0.40)$, and substituting for the protonation of sulphate (see Chapter VI), $\log_{10}\beta_1^\circ = (6.43 \pm 0.32)$ and $\log_{10}\beta_2^\circ = (10.33 \pm 0.41)$.

[53BAE/SCH2]
Baes, C.F., Jr., Schreyer, J.M., Lesser, J.M., The chemistry of uranium(VI) orthophosphate solutions: Part I. A spectrophotometric investigation of uranyl phosphate complex formation in perchloric acid solutions, AECD-3596, US Atomic Energy Commission, Oak Ridge, Tennessee, USA, 1953.

See comments under Baes [56BAE].

[54AHR/LAR]
Ahrland, S., Larsson, R., The complexity of uranyl fluoride, Acta Chem. Scand., **8** (1954) 354-366.

See comments under Ahrland, Larsson, and Rosengren [56AHR/LAR].

[54AHR/LAR2]
Ahrland, S., Larsson, R., The complexity of uranium(IV) chloride, bromide and thiocyanate, Acta Chem. Scand., **8** (1954) 137-150.

Ahrland and Larsson studied the formation of Cl^-, Br^- and SCN^- complexes of U(IV) by using an emf-method (the determination of the redox potential of U(VI)/U(IV) as a function of the ligand concentration). The ionic medium was 0.6 M $HClO_4$ and 0.4 M $NaClO_4$ and the temperature 20°C. The experimental work is satisfactory even though there were fairly large changes in the ionic medium throughout the experimental situations. Corrections to $I = 0$ are made by using the specific ion interaction theory as described in Chapter V. Corrections to 25°C are made for the chloride system by using the enthalpy data selected in this review.

[54BRO/BUN]
Brown, R.D., Bunger, W.B., Marshall, W.L., Secoy, C.H., The electrical conductivity of uranyl sulfate in aqueous solutions, J. Am. Chem. Soc., **76** (1954) 1532-1535.

Brown *et al.* studied the conductivity of dioxouranium(VI) sulphate in aqueous solutions at 0, 25, 50, 90, 125, and 200°C. The concentration range studied was

5×10^{-5} to 3.64 M. The experimental data were corrected for hydrolysis of U(VI) and protonation of SO_4^{2-} (the authors used $K = 100$ for the latter reaction). The ionic conductivities of UO_2^{2+} and $U_2O_5^{2+}$ ($\triangleq (UO_2)_2(OH)_2^{2+}$) were estimated from the conductance data together with the dissociation constant of $UO_2SO_4(aq)$ at zero ionic strength. This was done by using an extended Debye-Hückel method with a distance of closest approach in the ion pair of 7×10^{-10} m. This corresponds to $Ba_j = 2.3$ compared to 1.5 used by this review in the specific ion interaction method. The dissociation constant of $UO_2SO_4(aq)$ has a value of $(5.9 \pm 0.1) \times 10^{4-}$ at $I = 0$. In view of the many corrections necessary to obtain this value, this review increases this uncertainty to ± 1.0. However, this value is more an estimate of the accuracy than the precision of this well-performed study. The complex formation constant is $\log_{10} \beta_1^\circ = (3.23 \pm 0.07)$, which is in good agreement with other values at $I = 0$, *cf.* Table V.28.

The high-temperature data cannot be used to estimate equilibrium constants because of lack of information on the ionic conductivities of H^+, HSO_4^- and $U_2O_5^{2+}$ at these temperatures.

[54DAY/POW]

Day, R.A., Jr., Powers, R.M., Extraction of uranyl ion from some aqueous salt solutions with 2-thenoyltrifluoroacetone, J. Am. Chem. Soc., **76** (1954) 3895-3897.

Day and Powers measured the equilibrium constants for the reactions

$$UO_2^{2+} + HF(aq) \rightleftharpoons UO_2F^+ + H^+ \quad (A.4)$$
$$UO_2^{2+} + Cl^- \rightleftharpoons UO_2Cl^+ \quad (A.5)$$
$$UO_2^{2+} + NO_3^- \rightleftharpoons UO_2NO_3^+ \quad (A.6)$$
$$UO_2^{2+} + HSO_4^- \rightleftharpoons UO_2SO_4(aq) + H^+ \quad (A.7)$$
$$UO_2^{2+} + 2HSO_4^- \rightleftharpoons UO_2(SO_4)_2^{2-} + 2H^+ \quad (A.8)$$

at 10, 25 and 40°C in 2.0 M $NaClO_4$. The fluoride system was also studied at $I = 0.05$, 0.25, 0.50 and 1.00 M. The experimental method was based on an extraction method with 2-thenoyltrifluoroacetone as the extractant. The authors estimated the errors to be at least $\pm 10\%$. From the experimental data published $\pm 20\%$ seems more reasonable, and this estimate is used here.

The enthalpy values calculated from the temperature variation of the equilibrium constants differ substantially from the corresponding data of Wallace [67WAL], Ahrland and Kullberg [71AHR/KUL2, 71AHR/KUL3], and Bailey and Larson [71BAI/LAR]. For this reason, and because of the lack of proper error estimates, this review gives these data zero weight. However, the equilibrium constant data at 10 and 40°C are used to recalculate the values at 25°C by using the enthalpy data selected in this review.

Day and Powers [54DAY/POW] used $\log_{10} K = 0.084$ for the reaction $HSO_4^- \rightleftharpoons H^+ + SO_4^{2-}$ at $I = 2.0$ M, a value that agrees fairly well with $\log_{10} K = 0.10$ obtained by recalculating the $I = 0$ value to $I = 2.0$ M. The equilibrium constants for the sulphate complexes at 10 and 40°C are recalculated to 25°C using the enthalpy values selected in this review. For the chloride, nitrate, and sulphate systems, only the 25°C data are used.

[54FAU]
Faucherre, J., Sur la constitution des ions basiques métalliques, Bull. Soc. Chim. France, Ser. 5, **21(2)** (1954) 253-267, in French.

See comments under Faucherre [48FAU].

[54SCH/BAE]
Schreyer, J.M., Baes, C.F., Jr., The solubility of uranium(VI) orthophosphates in phosphoric acid solutions, J. Am. Chem. Soc., **76** (1954) 354-357.

The solubilities of the solids $(UO_2)_3(PO_4)_2{\cdot}4H_2O$(cr), $UO_2HPO_4{\cdot}4H_2O$(cr) and $UO_2(H_2PO_4)_2{\cdot}3H_2O$(cr) were determined at varying concentrations of phosphoric acid. The data were obtained at 25°C in solutions of varying ionic strength. All solid phases were well characterized. The solubilities of the two first phases *increased* with increasing concentration of phosphoric acid, indicating the formation of soluble complexes with at least two phosphate groups per dioxouranium(VI). The solubility of the third phase *decreased* with increasing concentration of phosphoric acid, indicating that the concentration of soluble phosphate complexes with more than two phosphate groups per dioxouranium(VI) was small. The invariant points where two phases are in equilibrium were located at a total phosphate concentration of 0.014 M for the two phases $(UO_2)_3(PO_4)_2 \cdot 4H_2O$(cr) and $UO_2HPO_4 \cdot 4H_2O$(cr), and at a total phosphate concentration of ~6 M for the two phases $UO_2HPO_4{\cdot}4H_2O$(cr) and $UO_2(H_2PO_4)_2{\cdot}3H_2O$(cr). The location of the invariance points gives a relationship between the solubility products of the two phases involved.

[54WEN/KIR]
Wendolkowski, W.S, Kirslis, S.S., Thermal decomposition of uranyl nitrate hexahydrate, K-1086, K-25 Plant, Carbide and Carbon Chemicals Company, Union Carbide and Carbon Corporation, Oak Ridge, Tennessee, USA, 1954, 23p.

The results for 41.8°C from this report are used (together with values obtained by others) to calculate the variation of the water vapour pressure over hydrated dioxouranium(VI) nitrates as a function temperature. The authors noted slight decomposition even at this low temperature, and results for higher temperatures are not used. The uncertainty in the reported water vapour pressures is estimated in this review as 100 Pa.

[55BET]
Betts, R.H., Heat of hydrolysis of uranium(IV) in perchloric acid solutions, Can. J. Chem., **33** (1955) 1775-1779.

Some problems were encountered in establishing the values of the extinction coefficients in this spectrophotometric study. In this review the values accepted for $\log_{10}{}^*\beta_1$ from this reference are those reported with the extinction coefficients for U^{4+}, denoted as E_4 by the authors, considered to be temperature dependent. This is in agreement with the observations of Kraus and Nelson [55KRA/NEL]. The uranium(IV) perchlorate contributes a substantial fraction to the total ionic strength (0.19 M), and therefore, rather large uncertainties in the values of $\log_{10}{}^*\beta_1$ are estimated (± 0.20).

The method of least-squares is used to refit the reported values for $\log_{10}{}^*\beta_1$ from 15.2 to 24.7°C as a function of temperature. Values of $\Delta_r H_m$ ($44.3\,\text{kJ}\cdot\text{mol}^{-1}$) and $\Delta_r S_m$ ($126\,\text{J}\cdot\text{K}^{-1}\cdot\text{mol}^{-1}$) are calculated ($I = 0.19$ M (ClO_4^-), $\Delta_r C_{p,m}$ assumed equal to 0). The enthalpy of reaction differs only slightly from the reported $44.4\,\text{J}\cdot\text{K}^{-1}\cdot\text{mol}^{-1}$. The fitted value at 298.15 K is $\log_{10}{}^*\beta_1 = -(1.161 \pm 0.19)$. Because the reaction entropy and enthalpy are calculated from measurements done over a very narrow temperature range, uncertainties in $\Delta_r H_m$ and $\Delta_r S_m$ are expected to be large. However, these uncertainties are unlikely to be as large as those calculated if the uncertainties in the values of $\log_{10}{}^*\beta_1$ (± 0.20) are estimated as described above. Errors in the values for the three temperatures are likely to have occurred in a systematic fashion. The uncertainty in $\Delta_r H_m$ is estimated as $\pm 6\,\text{kJ}\cdot\text{mol}^{-1}$ and the uncertainty in $\Delta_r S_m$ as $\pm 20\,\text{J}\cdot\text{K}^{-1}\cdot\text{mol}^{-1}$.

[55DAY/WIL]

Day, R.A., Jr., Wilhite, R.N., Hamilton, F.D., Stability of complexes of uranium(IV) with chloride, sulfate and thiocyanate, J. Am. Chem. Soc., 77 (1955) 3180-3182.

Day, Wilhite and Hamilton studied the formation of complexes between U^{4+} and Cl^-, SO_4^{2-} and SCN^-. They also estimated the equilibrium constants for the reactions $U^{4+} + qHF(aq) \rightleftharpoons UF_q^{4-q} + H^+$, where $q = 1$ and 2. The experiments were performed at $I = 2$ M and at 10, 25 and 40°C by using a liquid-liquid extraction method with 2-thenoyltrifluoroacetone as the extracting ligand. The test solutions had the composition 1 M H^+, $(2 - c_A)$ M ClO_4^-, where c_A is the total concentration of the ligand. The ionic medium was changed from 0 to 2 M Cl^-, from 0 to 0.3 M SCN^-, and from 0 to 0.1 M SO_4^{2-}. There was no experimental information given on the fluoride system, and these data are therefore given zero weight.

The authors reported stability constants for the formation of UCl^{3+} ($\beta_1 = 1.21$) and UCl_2^{2+} ($\beta_2 = 1.14$). No uncertainty estimates were given. The formation of UCl_2^{2+} might well be an artifact brought about by the large change in the composition of the ionic medium. This review does not consider the existence of this complex to be proven by these data. The equilibrium constants calculated by omitting the formation of UCl_2^{2+} are $\beta_1 = 1.5$, 1.8 and 3.3 at 40, 25 and 10°C, respectively. From the temperature variation of these constants $\Delta_r H_m = -(19 \pm 9)\,\text{kJ}\cdot\text{mol}^{-1}$ is obtained. The uncertainties in the $\log_{10}\beta_1$ values are estimated to be ± 0.15.

The equilibrium constants in the sulphate system referred to the reactions

$$U^{4+} + qHSO_4^- \rightleftharpoons U(SO_4)_q^{4-2q} + qH^+ \tag{A.9}$$

were reported as $\log_{10}{}^*\beta_1(\text{A.9}, q = 1) = 2.52$ and $\log_{10}{}^*\beta_2(\text{A.9}, q = 1) = 3.87$. This review estimates the uncertainties in $\log_{10}{}^*\beta_1$ and $\log_{10}{}^*\beta_2$ to be ± 0.05 and ± 0.15, respectively. The constants are corrected to $I = 0$ by using the ion interaction coefficients $\varepsilon_{(U^{4+},ClO_4^-)} = (0.76 \pm 0.06)$, $\varepsilon_{(USO_4^{2+},ClO_4^-)} = (0.3 \pm 0.1)$ and $\varepsilon_{(HSO_4^-,H^+)} = -(0.01 \pm 0.02)$.

This review uses the constants ${}^*\beta_1$ and ${}^*\beta_2$ reported at 10, 25 and 40°C to obtain the enthalpy changes of Reaction (A.9), giving $\Delta_r H_m(\text{A.9}, q = 1, I = 2) = -(14.4 \pm$

2.5) $kJ \cdot mol^{-1}$ and $\Delta_r H_m(A.9, q = 2, I = 2) = -(12.1 \pm 1.8)\, kJ \cdot mol^{-1}$, respectively. The uncertainties are estimated by this review. Assuming the ionic strength effect can be disregarded (which seems reasonable for the enthalpy of protonation of sulphate, *cf.* Chapter VI), these two values are combined with the enthalpy of protonation of SO_4^{2-} selected in Chapter VI to obtain the enthalpy values for the formation reactions

$$U^{4+} + qSO_4^{2-} \rightleftharpoons U(SO_4)_q^{4-2q}, \qquad (A.10)$$

assumed to be valid at $I = 0$: $\Delta_r H_m^\circ(A.10, q = 1) = (8.0 \pm 2.7)\, kJ \cdot mol^{-1}$ and $\Delta_r H_m^\circ(A.10, q = 2) = (32.7 \pm 2.8)\, kJ \cdot mol^{-1}$. These values differ considerably from those given by the authors, $\Delta_r H_m(A.9, q = 1, I = 2) = -13.4\, kJ \cdot mol^{-1}$ and $\Delta_r H_m(A.9, q = 2, I = 2) = -9.6\, kJ \cdot mol^{-1}$, obviously because the authors did not take into account the enthalpy change in the protonation reaction of sulphate, and their values thus refer to Reaction (A.9), rather than to Reaction (A.10).

[55GAY/LEI]
Gayer, K.H., Leider, H., The solubility of uranium trioxide, $UO_3 \cdot H_2O$, in solutions of sodium hydroxide and perchloric acid at 25°C, J. Am. Chem. Soc., **77** (1955) 1448-1550.

Baes and Mesmer [76BAE/MES] reanalysed the data from this study and calculated $\log_{10} {}^*K_{s,0}^\circ = (5.6 \pm 0.15)$. Naumov and coworkers [71NAU/RYZ, 72NIK/SER] noted that the solid phase in Gayer and Leider's solubility experiments in acidic solutions was probably $UO_3 \cdot 2H_2O$, not $UO_3 \cdot H_2O$, as the dihydrate is the stable form at 25°C [66ROB]. Gayer and Leider [55GAY/LEI] noted that their hydrated oxide was dried over drierite before it was analysed, and Naumov's conclusion is probably correct. Thus, Baes and Mesmer's analysis [76BAE/MES] is justified, but applies to $UO_3 \cdot 2H_2O$, not $UO_3 \cdot H_2O$. This leads to a value of $\Delta_f G_m^\circ(UO_2 \cdot 2H_2O, cr, 298.15\,K)$ more than $4\, kJ \cdot mol^{-1}$ more negative than the value determined by Tasker *et al.* [88TAS/OHA]. The data are too poor to use to calculate hydrolysis constants. The solid in basic solutions was probably a "uranate", not the oxide. Polymeric species were mentioned but not incorporated in the data analysis. The shape of the solubility curve in acidic media is not compatible with currently accepted hydrolysis schemes.

[55KRA/NEL]
Kraus, K.A., Nelson, F., Hydrolytic behavior of metal ions: IV. The acid constant of uranium (IV) as a function of temperature, J. Am. Chem. Soc., **77** (1955) 3721-3722.

This study reports spectrophotometrically determined values of the first hydrolysis constant of U^{4+} at 10 and 43°C, $I = 0.5$ M (ClO_4^-). The method of least-squares is used to refit values reported for $\log_{10} {}^*\beta_1$ as a function of temperature. Values of $\Delta_r H_m$ ($46.9\, kJ \cdot mol^{-1}$) and $\Delta_r S_m$ ($129\, J \cdot K^{-1} \cdot mol^{-1}$) are calculated ($\Delta_r C_{p,m}$ assumed equal to 0). The enthalpy of reaction is essentially the same as the reported $46.7\, J \cdot K^{-1} \cdot mol^{-1}$. The fitted value of $\log_{10} {}^*\beta_1$ at 298.15 K is $-(1.46_7 \pm 0.05)$. Because the reaction entropy and enthalpy are calculated from measurements done over a fairly narrow temperature range, uncertainties in $\Delta_r H_m$ and $\Delta_r S_m$ are expected to be large. However, these uncertainties are unlikely to be as large as those calculated if the uncertainties in the values of $\log_{10}{}^*\beta_1$ are estimated as described above (see discussion of Ref. [50KRA/NEL]). Errors in the values of $^*\beta_1$ for the three temperatures are likely to have occurred in a systematic fashion. The uncertainty in $\Delta_r H_m$ is estimated as $\pm 3\,kJ \cdot mol^{-1}$ and the uncertainty in $\Delta_r S_m$ as $\pm 10\,J \cdot K^{-1} \cdot mol^{-1}$. The value used by Kraus and Nelson for $^*\beta_1$ at 25°C (0.0342) was derived from values in Ref. [50KRA/NEL], but was not simply a weighted average of the values reported in perchlorate media with $I \approx 0.5$ M (0.0316). Since 25°C is near the middle of the temperature range used, the difference in the value at 25°C makes essentially no difference to the calculated values of $\Delta_r H_m$ and $\Delta_r S_m$. The authors attempted to correct the entropy and enthalpy of reaction to $I = 0$ by applying an extended Debye-Hückel correction to the equilibrium constants. The values of $\Delta_r H_m^\circ$ ($49.4\,kJ \cdot mol^{-1}$) and $\Delta_r S_m^\circ$ ($153\,J \cdot K^{-1} \cdot mol^{-1}$) at $I = 0$ calculated by this method are larger than the values at $I = 0.5$ M. Use of the specific ion interaction theory (see Appendix B) leads to a similar conclusion.

[55SCH]

Schreyer, J.M., The solubility of uranium(IV) orthophosphates in phosphoric acid solutions, J. Am. Chem. Soc., 77 (1955) 2972-2974.

The solubility of $U(HPO_4)_2{\cdot}6H_2O$ was studied at 25°C in solutions of phosphoric acid with concentrations from 1.5 to 15.3 M. The hexahydrate was the stable phase at phosphoric acid concentrations less than 9.8 M, at higher concentrations $U(HPO_4)_2{\cdot}H_3PO_4{\cdot}H_2O(s)$ was formed. The solubility of $U(HPO_4)_2{\cdot}6H_2O$ increased with increasing phosphate concentration indicating that negatively charged complexes containing more than two phosphate groups per U(IV) are formed. The hydrogen ion concentration and the phosphate concentration are not independent under the experimental conditions used. Hence, the data cannot be used to deduce the compositions of the complexes formed.

[56AHR/LAR]

Ahrland, S., Larsson, R., Rosengren, K., On the complex chemistry of the uranyl ion: VIII. The complexity of uranyl fluoride, Acta Chem. Scand., 10 (1956) 705-718.

Ahrland and Larsson [54AHR/LAR], and Ahrland, Larsson and Rosengren [56AHR/LAR] determined the equilibrium constants for the reactions

$$UO_2^{2+} + qF^- \rightleftharpoons UO_2F_q^{2-q}, \qquad (A.11)$$

where $q = 1$ to 4, at 20°C in a 1 M $NaClO_4$ medium. Ref. [54AHR/LAR] is a preliminary communication of the investigations discussed more extensively in Ref. [56AHR/LAR]. The measurements were made by using a potentiometric method, *viz.*, measurement of $[H^+]$ by using a quinhydrone electrode. The equilibrium constants are recalculated to 25°C by using the enthalpy data of Ahrland and Kullberg [71AHR/KUL2]. The correction term is 0.006 logarithmic units at most.

Ahrland and Kullberg [71AHR/KUL] studied the dioxouranium(VI) fluoride system by potentiometric measurements of $[H^+]$ and $[F^-]$ using a quinhydrone and a fluoride membrane electrode, respectively. The temperature was 25°C and the ionic medium 1.00 M $NaClO_4$. This review assigns an uncertainty of ± 0.05 to $\log_{10}\beta_q$ in both Ref. [56AHR/LAR] and Ref. [71AHR/KUL].

The enthalpies of reaction for the complex formation were determined calorimetrically by Ahrland and Kullberg [71AHR/KUL2]. These are found to be the best data available. It is assumed here that the enthalpies of reaction do not change significantly with the ionic strength and that the $\Delta_r H_m$(A.11) values of Ahrland and Kullberg [71AHR/KUL2] are valid also at $I = 0$. The stepwise values given in Ref. [71AHR/KUL2] are recalculated to correspond to the overall complexation reactions, $UO_2^{2+} + qF^- \rightleftharpoons UO_2F_q^{2-q}$. This results in $\Delta_r H_m^\circ$ values of (1.70 ± 0.08), (2.10 ± 0.19), (2.35 ± 0.31), and $(0.29 \pm 0.47)\,kJ \cdot mol^{-1}$ for $q = 1$, 2, 3, and 4, respectively. The uncertainties, representing "three standard deviations" as pointed out in the paper, are taken as the selected uncertainties in this review.

[56BAE]

Baes, C.F., Jr., A spectrophotometric investigation of uranyl phosphate complex formation in perchloric acid solution, J. Phys. Chem., **60** (1958) 878-883.

This study and the paper of Baes, Schreyer and Lesser [53BAE/SCH2] describe the complex formation between dioxouranium(VI) and phosphoric acid in solutions of constant acidity, 0.1 M and 1 M $HClO_4$, respectively, and a temperature of (25 ± 2)°C. The equilibria (A.12) and (A.13) were studied.

$$UO_2^{2+} + H_3PO_4(aq) \rightleftharpoons UO_2H_rPO_4^{r-1} + (3-r)H^+ \qquad (A.12)$$
$$UO_2^{2+} + 2H_3PO_4(aq) \rightleftharpoons UO_2H_y(PO_4)_2^{y-4} + (6-y)H^+ \qquad (A.13)$$

$UO_2H_rPO_4^{r-1}$ denotes the sum of all the protonated mononuclear phosphate complexes with the ratio U:P = 1:1. $UO_2H_y(PO_4)_2^{y-4}$ is the corresponding sum for species with the ratio U:P = 1:2.

The experimental studies are precise as is the data analysis. The authors reported, for the 1 M $HClO_4$ medium, K_1(A.12) = (11 ± 2) and K_2(A.13) = (24 ± 5), and for the medium (0.1 M $HClO_4$/1 M $NaClO_4$) they reported K_1(A.12) = (58 ± 4) and K_2(A.13) = (470 ± 20). These are conditional constants that can be expressed as

$$K_1(A.12) = {}^*\beta_1(A.14)/[H^+] + \beta_1(A.15)$$
$$K_2(A.13) = {}^*\beta_2(A.16)/[H^+]^2 + {}^*\beta(A.17)/[H^+].$$

The various β values refer to the reactions

$$UO_2^{2+} + H_3PO_4(aq) \rightleftharpoons UO_2H_2PO_4^+ + H^+ \qquad (A.14)$$
$$UO_2^{2+} + H_3PO_4(aq) \rightleftharpoons UO_2H_3PO_4^{2+} \qquad (A.15)$$
$$UO_2^{2+} + 2H_3PO_4(aq) \rightleftharpoons UO_2(H_2PO_4)_2(aq) + 2H^+ \qquad (A.16)$$
$$UO_2^{2+} + 2H_3PO_4(aq) \rightleftharpoons UO_2(H_2PO_4)(H_3PO_4)^+ + H^+ \qquad (A.17)$$

From the values given for K_1(A.12) and K_2(A.13) this review obtains

$$^*\beta_1(A.14) = 5.2 \pm 0.5$$
$$\beta_1(A.15) = 5.8 \pm 2.0$$
$$^*\beta_2(A.16) = 2.56 \pm 0.59$$
$$^*\beta(A.17) = 21.4 \pm 5.0,$$

where the uncertainties are derived from those given by the authors. Baes used an experimental value of $\log_{10} K = -1.68$ for the first dissociation constant of phosphoric acid. This value differs somewhat from $\log_{10} K = -(1.77 \pm 0.05)$, the selected value after correction to $I = 1$ M by using the ion interaction coefficients in Appendix B. However, the difference is small, and this review accepts Baes' value without recalculation.

It should be mentioned that, although this is a precise study, the identification of the stoichiometries of the complexes is not unambiguous. Thamer [57THA] made a similar study and found excellent agreement with Baes in 1 M $HClO_4$. His data at 0.5 M $HClO_4$ do not agree with Baes' model, while the study of Marcus [58MAR] does. This review considers the evidence in favour of complex formation with both $H_2PO_4^-$ and H_3PO_4 sufficient to adopt the model proposed by Baes [56BAE].

The study of Baes, Schreyer and Lesser [53BAE/SCH2] contains additional spectrophotometric data and also quantitative information on the equilibrium constant for the formation of $UO_2H_rPO_4^{r-1}$ in 1 M $HClO_4$. The calculations are less precise than those presented by Baes [56BAE] because the formation of complexes of the composition $UO_2H_y(PO_4)_2^{y-4}$ was neglected. The reported equilibrium constant is therefore given zero weight.

[56BAE2]
Baes, C.F., Jr., The reduction of uranium(VI) by iron(II) in phosphoric acid solution, J. Phys. Chem., 60 (1958) 805-806.

The author studied the redox properties of the Fe(III)/Fe(II) and U(VI)/U(IV) couples in acidic sulphate solution containing phosphoric acid of various concentrations, and a temperature of 25°C. Equilibrium constants for the reaction $2Fe(II) + U(VI) \rightleftharpoons 2Fe(III) + U(IV)$ were reported, where the chemical symbols denote the sums of the species found in the given oxidation states. The data indicate clearly that Fe(III) and U(IV) species are stabilized at higher phosphate concentrations (1 M, or higher) and that U(VI) then is reduced by Fe(II). It is not possible to use these data to obtain quantitative information on equilibrium constants and/or the chemical composition of the species formed. However, the graph of E' *vs.* $\log_{10}[H_3PO_4]$, where E' is the formal U(IV)/U(VI) potential [56BAE2, Figure 1] shows that U(IV) contains more phosphate ligands than U(VI) in the entire phosphate concentration range investigated (1.5 to 6 M).

[56BLA/COL]
Blake, C.A., Coleman, C.F., Brown, K.B., Hill, D.G., Lowrie, R.S., Schmitt, J.M., Studies in the carbonate-uranium system, J. Am. Chem. Soc., 78 (1956) 5978-5983.

Solubility and spectrophotometric data were used in this study to deduce the composition of two dioxouranium(VI) species, $UO_2(CO_3)_2^{2-}$ and $UO_2(CO_3)_3^{4-}$. The solubility measurements were made in a dioxouranium(VI) concentration range where essentially only the trimer $(UO_2)_3(CO_3)_6^{6-}$ should exist. This review reinterprets the solubility data by using the EQ3/6 code [79WOL]. The solid line shown in Figure A.1 is calculated by using the values of $\log_{10} K_{s,0} = -14.20$, $\log_{10} \beta_{6,3} = 51$, $\log_{10} \beta_2 = 16.7$ and $\log_{10} \beta_3 = 20.0$, the dashed line is calculated for the same model

Figure A.1: Experimental measurements of the solubility of UO_2CO_3 in solutions of different Na_2CO_3 concentrations [56BLA/COL] (o), compared with two different models: The solid line is calculated assuming the formation of a trinuclear complex, $(UO_2)_3(CO_3)_6^{6-}$, the dashed line is calculated by assuming the formation of mononuclear complexes only.

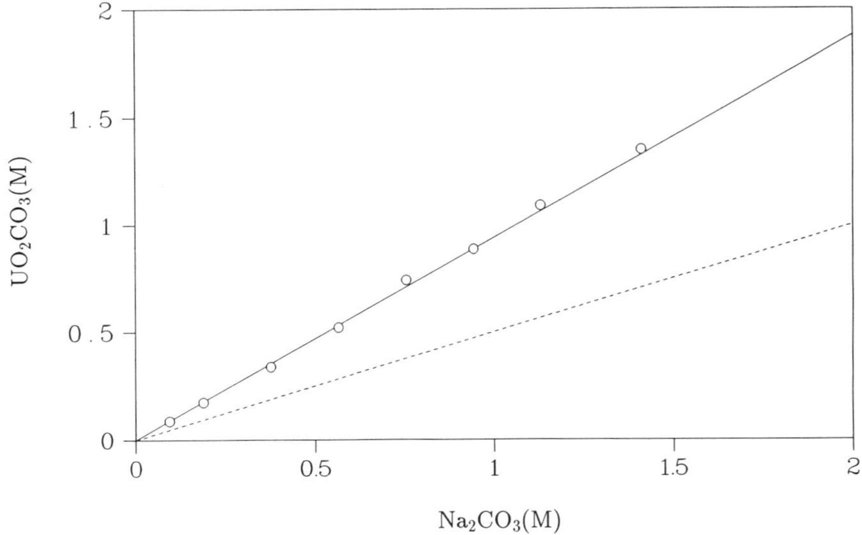

excluding the trinuclear complex. The activity factors are estimated using the specific ion interaction theory. It is clear that experimental data cannot be described by a set of mononuclear complexes. The data are in good agreement with the trinuclear model. Because of the very high and variable ionic strengths of the solutions, the numerical values of the equilibrium constants are not judged precise enough to be used for the calculation of the selected constants. There is no doubt that the data of Blake *et al.* [56BLA/COL] give additional experimental support to the formation of $(UO_2)_3(CO_3)_6^{6-}$.

The spectrophotometric measurements were also performed at such a high dioxouranium(VI) concentration (0.035 M) that the triuranyl complexes should be predominant. The authors used the method of continuous variation to provide information about the chemical composition of the species formed. This method cannot be applied in systems with several equilibria, *cf.* Rossotti and Rossotti [61ROS/ROS], hence, the proposal of the formation of a mixed hydroxide/carbonate complex, which the authors tentatively formulated as $U_2O_5OHCO_3^-$, cannot be considered proven. However, this complex, which is identical with $(UO_2)_2CO_3(OH)_3^-$, was later proposed by Maya [82MAY] and discussed by Grenthe and Lagerman [91GRE/LAG].

[56CHU/STE]

Chukhlantsev V.G., Stepanov S.P., Solubility of uranyl and thorium phosphates, J. Inorg. Chem. USSR, 1 (1956) 135-141.

Chukhlantsev and Stepanov reported solubility data, at 19 to 20°C, for three solid phases considered as $UO_2NH_4PO_4$, UO_2KPO_4 and UO_2HPO_4 in H_2SO_4 and HNO_3 solutions. Apparently, these three compounds were not characterized before and after their equilibration with aqueous acid solutions. Also it is not clear from Ref. [56CHU/STE] that these three compounds were anhydrous.

The solubility of the three compounds in H_2SO_4 seems to be consistent with but slightly higher than that in HNO_3 probably because of the formation of uranium sulphate complexes.

In the calculation of the solubility products, only UO_2^{2+} was taken into account by the authors. In fact, complexes such as $UO_2H_2PO_4^+$, $UO_2(H_2PO_4)_2(aq)$, $UO_2SO_4(aq)$ and $UO_2(SO_4)_2^{2-}$ could be present in the pH range 1.8 to 2.3.

The reported values of solubility products,

$$\begin{aligned} K_{s,0} &= [UO_2^{2+}][NH_4^+][PO_4^{3-}] &= 4.36\times10^{-27} \\ K_{s,0} &= [UO_2^{2+}][K^+][PO_4^{3-}] &= 7.76\times10^{-24} \\ K_{s,0} &= [UO_2^{2+}][HPO_4^{2-}] &= 2.14\times10^{-11}, \end{aligned}$$

increase significantly at lower pH. This could be interpreted, in some of the experiments, by the fact that the equilibrium was not attained.

It is well known that the compound $UO_2HPO_4(cr)$ is stable in the pH range 0 to 2. Therefore, in such a low pH range 1.8 to 2.3, it is very probable that the two solid phases $UO_2NH_4PO_4(cr)$ and $UO_2KPO_4(cr)$ would be partially tansformed to $UO_2HPO_4(cr)$.

For these reasons, this review considers that the solubility products of $UO_2NH_4PO_4(cr)$, $UO_2KPO_4(cr)$ and $UO_2HPO_4(cr)$ reported by Chukhlantsev and Stepanov [56CHU/STE] are not reliable enough to be accepted.

[56HIE]

Hietanen, S., Studies on the hydrolysis of metal ions: 17. The hydrolysis of the uranium(IV) ion, U^{4+}, Acta Chem. Scand., 10 (1956) 1531-1546.

This is a careful potentiometric study, but the data analysis is inadequate. Baes and Mesmer [76BAE/MES] recalculated the results of this study and obtained $\log_{10}{}^*\beta_1 = -(2.1 \pm 0.15)$ and $\log_{10}{}^*\beta_{15,6} = -(16.9 \pm 0.3)$. In this review the uncertainties in the logarithms of the formation constants are estimated as twice those selected in Ref. [76BAE/MES]. Baes and Mesmer [76BAE/MES] reported their data analysis indicated the formation of at least one additional species, and comparison with the thorium system [83BRO/ELL, 65BAE/MEY] suggests that further investigation of this system would be useful.

[56HIE2]

Hietanen, S., The mechanism of thorium(IV) and uranium(IV) hydrolysis, J. R. Neth. Chem. Soc., **75** (1956) 711-715.

This is a less complete treatment of the data from Hietanen [56HIE].

[56KRA/MOO]

Kraus, K.A., Moore, G.E., Nelson, F., Anion-exchange studies: XXI. Th(IV) and U(IV) in hydrochloric acid, separation of thorium, protactinium and uranium, J. Am. Chem. Soc., **78** (1956) 2692-2694.

This is a qualitative anion exchange study which shows that anionic U(IV) chloride complexes are formed only at very high (7.55 M) chloride concentrations.

[56MCC/BUL]

McClaine, L.A., Bullwinkel, E.B., Huggins, J.C., The carbonate chemistry of uranium: Theory and applications, in: Proc. International Conf. on the Peaceful Uses of Atomic Energy, held 8-20 August, 1955, in Geneva, Switzerland, Vol. **VIII**, New York: United Nations, 1956, *pp.*26-37.

Two different experimental methods (*viz.* spectrophotometry and solubility technique) were used in the study. The authors reported the equilibrium constants $K(\mathrm{A.18}) = 65$ and $K(\mathrm{A.19}) = 26$ for the reactions

$$UO_2(CO_3)_2^{2-} + 2HCO_3^- \rightleftharpoons UO_2(CO_3)_3^{4-} + H_2O(l) + CO_2(g) \qquad (A.18)$$

$$UO_2CO_3(cr) + 2HCO_3^- \rightleftharpoons UO_2(CO_3)_2^{2-} + H_2O(l) + CO_2(g). \qquad (A.19)$$

These equilibrium constants were obtained from the spectrophotometric concentration determination of the uranium complexes and the known total concentrations of carbonate and did not involve any determination of the hydrogen ion concentration. The concentrations were recalculated to activities by using Kieland's [37KIE] activity data. Hence, the constants reported refer to $I = 0$.

McClaine, Bullwinkel and Huggins [56MCC/BUL] do not give any information about the composition of the solutions used to determine the equilibrium constants. However, it seems as if the total concentrations of uranium used is around 10^{-2} M. Hence, the concentration of $(UO_2)_3(CO_3)_6^{6-}$ should predominate over that of $UO_2(CO_3)_2^{2-}$. The presence of the trinuclear complex is also indicated by the published absorption spectrum for "$UO_2(CO_3)_2(H_2O)_2^{2-}$", *cf.* discussion of Ref. [77SCA].

From the known protolysis constants of Reactions (A.20) and (A.21), as well as from the equilibrium constants of Reactions (A.18) and (A.19),

$$CO_2(g) + H_2O(l) \rightleftharpoons HCO_3^- + H^+ \qquad (A.20)$$

$$CO_3^{2-} + H^+ \rightleftharpoons HCO_3^- \qquad (A.21)$$

at $I = 0$, $\log_{10} K^\circ(\mathrm{A.20}) = -7.83$ and $\log_{10} K^\circ(\mathrm{A.21}) = 10.32$, one can obtain the stepwise formation constants $\log_{10} K_3^\circ(\mathrm{A.22}) = 4.3$ and $\log_{10} K_2^\circ(\mathrm{A.23}) = 3.9$.

$$UO_2(CO_3)_2^{2-} + CO_3^{2-} \rightleftharpoons UO_2(CO_3)_3^{4-} \qquad (A.22)$$

$$UO_2CO_3(cr) + CO_3^{2-} \rightleftharpoons UO_2(CO_3)_2^{2-} \qquad (A.23)$$

It is likely that not negligible amounts of $(UO_2)_3(CO_3)_6^{6-}$ are present in the test solutions studied in Ref. [56MCC/BUL]. Hence, $UO_2(CO_3)_2^{2-}$ in Eqs. (A.18) to (A.23) refers to a mixture of $UO_2(CO_3)_2^{2-}$ and the trimer. By combining Eqs. (A.22) and (A.23) the constant $\log_{10} K^\circ_{s,3} = 8.2$ is obtained for the reaction $UO_2CO_3(cr) + 2CO_3^{2-} \rightleftharpoons UO_2(CO_3)_3^{4-}$. This value is not in agreement with the value of 7.13 calculated from the values selected in this review.

The equilibrium constants for the reactions

$$UO_2^{2+} + 2CO_3^{2-} \rightleftharpoons UO_2(CO_3)_2^{2-} \quad (A.24)$$

$$UO_2^{2+} + 3CO_3^{2-} \rightleftharpoons UO_2(CO_3)_3^{4-} \quad (A.25)$$

proposed by the authors, $\log_{10} K^\circ(A.24) = 14.6$ and $\log_{10} K^\circ(A.25) = 18.3$, are based on standard free energies of formation which are slightly in error, *cf.* [56MCC/BUL, Table III].

[57BAL/DAV]

Bale, W.D., Davies, E.W., Morgans, D.B., Monk, C.B., The study of some ion-pairs by spectrophotometry, Discuss. Faraday Soc., 24 (1957) 94-102.

Bale *et al.* studied the formation of UO_2SCN^+ in 20% methanol/water from 15 to 45°C and the formation of UO_2Cl^+ in water. Spectrophotometry was used as the experimental method. Some previous spectrophotometric results for the dioxouranium(VI) chloride system in different ionic media were reviewed and the authors pointed out that the measured difference in molar absorbtivity was dependent on the ionic medium, especially at high ionic strengths. These observations might be due to changes in the outer-sphere UO_2^{2+}-Cl complexes with the very large changes in water activity that take place at high ionic strengths. Nevertheless, for moderate ionic strengths the effect is not serious. The authors reported a dissociation constant, which this review recalculates to $\beta_1^\circ(UO_2Cl^+) = (1.67 \pm 0.03)$ at 25°C. The authors used a Debye-Hückel expression which differs from the one used in this review to calculate activity factors. Hence, this review estimates a somewhat larger uncertainty of ± 0.10 in β_1°. The value of $\log_{10} \beta_1^\circ$ is in good agreement with the results of other studies.

The dioxouranium(VI) thiocyanate system was studied at six temperatures between 15 and 45°C in a 20% methanol/water solvent. The authors reported $\log_{10} \beta_1(UO_2SCN^+) = (1.00 \pm 0.01)$ and $\Delta_r H_m = 4.8\,kJ \cdot mol^{-1}$. It is impossible to estimate the activity corrections to the pure water reference state from these data. They are therefore given zero weight.

[57DAV/MON]

Davies, E.W., Monk, C.B., Spectrophotometric studies of electrolytic dissociation: Part 4. Some uranyl salts in water, Trans. Faraday Soc., 53 (1957) 442-449.

Davies and Monk made a spectrophotometric study of the complex formation between UO_2^{2+} and Cl^-, Br^-, SCN^- and SO_4^{2-}. The ionic strength is known and the temperature is 25°C. The experimental method used is satisfactory. However, the authors calculated the activity coefficients of the ionic species by using a simplified

Debye-Hückel expression (Davies equation). This method is doubtful, particularly in media of high ionic strength.

In this review the preliminary data in Ref. [57DAV/MON] are used to calculate concentration equilibrium constants. The data for the chloride system are used, together with others, for extrapolation to $I = 0$. The value of $\Delta\varepsilon = -(0.25 \pm 0.02)$ results from the linear regression. This value is also used as an estimate for the corresponding Br^- and SCN^- complexes. There is a printing error in Table 2 of Ref. [57DAV/MON]: the chloride concentrations range from 0.8 to 0.2 M, not from 0.08 to 0.02 M.

[57HEA/WHI]

Hearne, J.A., White, A.G., Hydrolysis of the uranyl ion, J. Chem. Soc., (1957) 2168-2174.

This is reasonably careful work, but no polymers beyond the dimer were considered in the data analysis. The "constants" may be mixed constants incorporating both concentrations and activities. In view of these problems the reported values for the formation constant of $(UO_2)_2(OH)_2^{2+}$ are not credited in this review. The uranium concentrations ranged to considerably lower values than used in many studies of uranium hydrolysis. Therefore, in this review the reported values of $^*\beta_1$ are used (with estimated uncertainties of ± 0.8 in $\log_{10}{^*\beta_1}$) in the evaluation of $^*\beta_1^\circ$ (see Chapter V).

[57KIN/PFE]

King, A.J., Pfeiffer, R., Zeek, W., Thermal stability of the hydrates of uranyl nitrate, Final Report NYO-6313, US Atomic Energy Commission, 1957.

The results obtained in this work do not show good internal agreement nor is the reported vapour pressure for 50.1°C consistent with data from several other groups. Data from King *et al.* [57KIN/PFE] are given zero weight in the recalculation of the temperature dependence of the hexahydrate/trihydrate equilibrium vapour pressures done in this review.

[57THA]

Thamer, B.J., Spectrophotometric and solvent-extration studies of uranyl phosphate complexes, J. Am. Chem. Soc., **79** (1957) 4298-4305.

The experimental study was carried out at approximately 25°C in a perchlorate medium with $I = 1.06$ M. Two different constant acidities, $[H^+] = 0.51$ and 1.004 M, were used. This is a precise study with a careful analysis of experimental and model uncertainties. The author used an experimental value of the first dissociation constant of phosphoric acid, $\log_{10} K_1 = -(1.72 \pm 0.05)$. This is in excellent agreement with the $\log_{10} K_1^\circ$ value selected in this review, corrected to $I = 1.06$ M using the corresponding ion interaction coefficients: $\log_{10} K_1 = -(1.77 \pm 0.05)$.

The author reported $^*\beta_1(\text{A.26}, q = 1) = (15.5 \pm 1.6)$ for the reaction

$$UO_2^{2+} + qH_3PO_4(aq) \rightleftharpoons UO_2(H_2PO_4)_q^{2-q} + qH^+, \qquad (\text{A.26})$$

using spectrophotometry. Thamer found no evidence for the formation of a mixture of $UO_2H_2PO_4^+$ and $UO_2H_3PO_4^{2+}$ as found previously by Baes [56BAE]. The value of

K(A.12) = (11 ± 2) found by Baes [56BAE] agrees fairly well with the one of Thamer [57THA].

By using solvent extraction, Thamer found K(A.13) = (21.8 ± 2.0) which is in good agreement with the value of (24 ± 5) in 1 M $HClO_4$ [56BAE]. At high phosphoric acid concentrations ($[H_3PO_4] > 0.5$ M) Thamer found evidence for the formation of $UO_2(H_2PO_4)_2(H_3PO_4)(aq)$ with K(A.27) = (10.2 ± 0.5) for the reaction

$$UO_2^{2+} + 3H_3PO_4(aq) \rightleftharpoons UO_2(H_2PO_4)_2(H_3PO_4)(aq) + 2H^+. \qquad (A.27)$$

It is not likely that the activity of H_3PO_4 is constant at these high concentrations, and this review finds there is not sufficient evidence for the formation of $UO_2(H_2PO_4)_2(H_3PO_4)(aq)$ to be included in the selected data table. Furthermore, complexes containing more than two phosphate groups per U(VI) are not compatible with the solubility data of Schreyer and Baes [54SCH/BAE], see comments under Ref. [54SCH/BAE].

[58ALL]
Allen, K.A., The uranyl sulfate complexes form tri-n-octylamine extraction equilibria, J. Am. Chem. Soc., **80** (1958) 4133-4137.

Allen [58ALL] studied the extraction of U(VI) from aqueous solutions of H_2SO_4 and Na_2SO_4 into an organic phase of benzene and tri-n-octylamine. The author estimated the activity coefficients by using a Debye-Hückel expression. The method used seems to be satisfactory. The presence of two aqueous complexes $UO_2SO_4(aq)$ and $UO_2(SO_4)_2^{2-}$ was established, but the overall formation constants for these species cannot be determined with precision, due to the low free concentration of UO_2^{2+}. The author made an estimate of β_1 and β_2 at $I = 1$. He found $\beta_1 = 34$, (17, 135) and $\beta_2 = 200$ (100, 260) where the numbers within parentheses indicate the lowest and highest estimate of the corresponding constant. This review gives these constants zero weight. However, the equilibrium constant $K_2 = (6 \pm 1)$ for the stepwise reaction $UO_2SO_4(aq) + SO_4^{2-} \rightleftharpoons UO_2(SO_4)_2^{2-}$ is well determined, and in fair agreement with the value $K_2 = (10.0 \pm 0.8)$ selected in this review.

[58BRU]
Brusilovskii, S.A., Investigation of the precipitation of hexavalent uranium hydroxide Proc. Acad. Sci. USSR, **120** (1958) 343-347.

Polymer formation was neglected. Tripathi [84TRI] calculated $\log_{10}{}^*\beta_1 = -4.06$ from these data. Neither the reported constant nor the recalculated value are credited in this review.

[58LI/DOO]
Li, N.C., Doody, E., White, J.M., Copper(II), nickel and uranyl complexes of some amino acids, Am. J. Sci., **80** (1958) 5901-5903.

UO_2OH^+ was not considered for the extraction solutions in which it might be expected to form, but this should not affect the calculated hydrolysis constants that were determined from potentiometric titrations at higher uranium concentrations.

The data were analysed considering a "core-and-links" scheme. Only a few of the raw data were published, therefore the hydrolysis constants cannot be recalculated on the basis of another model. Consideration of the reanalysis of the data of Ahrland [49AHR] by Rush *et al.* [62RUS/JOH] indicates that the value of $^*\beta_{2,2}$ should be reasonably model independent, and this review uses $\log_{10}(kk_\circ) = \log_{10}{^*\beta_{2,2}} = -6.05$ with an estimated uncertainty of ±0.4 in its evaluation of $\log_{10}{^*\beta^\circ_{2,2}}$.

[58MAR]
Marcus, Y., Anion exchange of metal complexes: The uranyl phosphate system, in: Proc. 2*nd* United Nations International Conf. on the Peaceful Uses of Atomic Energy, held 1-13 September, 1958, in Geneva, Switzerland, Vol. **3**, Geneva: United Nations, 1958, *pp*.465-471.

Marcus used an anion exchange method to determine the complex formation between dioxouranium(VI) and H_3PO_4. The method indicated the formation of a number of mononuclear complexes and their equilibrium constants. The method used required an estimate of the activity factors of the reactants and products both in the aqueous and the resin phase. The equilibrium constants are "mixed" constants obtained by using the activity of the ligand and the concentrations of the uranium bearing species. The author considered the constants as approximate, and this review therefore gives them zero weight. Marcus proposed complexes which contain more than two phosphate groups per uranium at high phosphoric acid concentrations. These stoichiometries do not agree with the solubility data of Schreyer and Baes [54SCH/BAE], *cf.* discussion under Ref. [54SCH/BAE].

[59HIE/SIL]
Hietanen, S., Sillén, L.G., Studies on the hydrolysis of metal ions: 24. Hydrolysis of the uranyl ion, UO_2^{2+}, in perchlorate self-medium, Acta Chem. Scand., **13** (1959) 1828-1838.

High uranium concentrations and high ionic strengths used in this work do not permit it to be used in the present analysis.

[59KLY/KOL]
Klygin, A.E., Kolyada, N.S., A study of the UO_2SO_3-$(NH_4)_2SO_3$-H_2O system by the solubility method, Russ. J. Inorg. Chem., **4** (1959) 101-103.

The only quantitative equilibrium studies available on the aqueous U(VI) sulphite system are those of Klygin and Kolyada [59KLY/KOL] and of Zakharova and Orlova [67ZAK/ORL].

Klygin and Kolyada did their solubility study in solutions of varying ionic strength. Their experiment was not well designed for the determination of equilibrium constants. The first ten experimental points, *cf.* [59KLY/KOL, Table 1], had very similar values of the free sulphite concentration and consequently represent approximately the same solubility. This review uses the points of the highest hydrogen ion concentrations (where the hydrolysis is negligible) to estimate $K_{s,0}$ and β_1. The results obtained are $\log_{10} K_{s,0} = -(8.6 \pm 0.2)$ and $\log_{10}\beta_1 = (6.0 \pm 0.2)$. These values refer to an "average" ionic strength of $I \approx 0.02$ M. Recalculation to $I = 0$ by neglecting the specific ion interaction term due to the low ionic strength gives $\log_{10} K^{\circ}_{s,0} = -(9.1 \pm 0.3)$ and $\log_{10} \beta^{\circ}_1 = (6.5 \pm 0.3)$. The uncertainty is increased in order to account for the varying ionic strength.

The value of $\log_{10} \beta_2$ reported in Ref. [59KLY/KOL] is erroneous as discussed in Ref. [67ZAK/ORL]. This review evaluates $\log_{10}(\beta_2 K_{s,0})$ from the seven last experimental points of Table 1 in Ref. [59KLY/KOL] and finds $\log_{10}(\beta_2 K_{s,0}) = -(1.5 \pm 0.1)$. These data are assumed to be valid for the average ionic strength $I = (0.15 \pm 0.04)$ M. These data fall on a smooth curve of slope +1, as expected, but the numerical value differs very much from the value given in Ref. [67ZAK/ORL]. This review prefers the solubility data of Ref. [59KLY/KOL] over those of Ref. [67ZAK/ORL], but neither of the experimental sets is really satisfactory.

$\log_{10}(\beta_2 K_{s,0})$ should be very near independent of the ionic strength, because the contribution of the Debye-Hückel term for the reaction $UO_2SO_3 \cdot 4.5H_2O(cr) + SO_3^{2-} \rightleftharpoons UO_2(SO_3)_2^{2-} + 4.5H_2O(l)$ is zero and $\Delta\varepsilon$ of this reaction is close to zero. Hence, $\log_{10}(\beta^{\circ}_2 K^{\circ}_{s,0}) = -(1.5 \pm 0.1)$.

Zakharova and Orlova [67ZAK/ORL] studied the formation of U(VI) sulphite complexes at (23 ± 1)°C by using three different experimental methods. In the potentiomtric study the authors used a standard titration technique (measurements of $[H^+]$). They did not keep the ionic strength constant, but the variations were rather small from 0.02 to 0.04 M. The titration procedure seems more or less satisfactory as judged by the determination of the first protonation constant of sulphite. Zakharova and Orlova [67ZAK/ORL] reported $\log_{10} K_1(HSO_3^-) = 6.55$ compared to $\log_{10} K^{\circ}_1(HSO_3^-) = (6.94 \pm 0.08)$ recalculated to $I = 0.03$ M from the value selected in this review (*cf.* Chapter VI) using the specific ion interaction theory but neglecting the $\Delta\varepsilon\, m$ term. Zakharova and Orlova [67ZAK/ORL] reported $\log_{10} \beta_1 = (5.85 \pm 0.03)$ for the reaction $UO_2^{2+} + SO_3^{2-} \rightleftharpoons UO_2SO_3(aq)$. A better estimate from their data in Table 1 of Ref. [67ZAK/ORL] would be $\log_{10} \beta_1 = (5.8 \pm 0.3)$ which, converted to a molality constant and corrected from $I = 0.03$ M to $I = 0$, corresponds to $\log_{10} \beta^{\circ}_1 = (6.4 \pm 0.3)$ (only the Debye-Hückel term needs to be taken into account).

The spectrophotometric study in Ref. [67ZAK/ORL] was done at $I = 0.1$ M (NH_4ClO_4). When interpreting their data Zakharova and Orlova [67ZAK/ORL] assumed that "half the uranyl ions present are combined in the complex". This is an approximation. Based on the reported pH values this review reinterprets the spectrophotometric data and finds that approximately 60% of U(VI) is present as $UO_2SO_3(aq)$ and 40% as UO_2^{2+}. This results in $\log_{10} \beta_1 = (6.2 \pm 0.2)$, which is slightly different from the value proposed in Ref. [67ZAK/ORL], $\log_{10} \beta_1 = (6.01 \pm 0.01)$. Conversion to molality units and recalculation of $\log_{10} \beta_1 = (6.2 \pm 0.2)$ from $I = 0.1$ M to $I = 0$ (neglecting the Debye-Hückel term) results in $\log_{10} \beta^{\circ}_1 = (7.0 \pm 0.2)$, which is not consistent with the potentiometric value.

The solubility data of $UO_2SO_3 \cdot 4.5H_2O(cr)$ in Ref. [67ZAK/ORL] were obtained in a 1 M NaCl ionic medium. These data are rather unprecise as can be seen when $\log_{10} S$ (S is the measured solubility) is plotted *vs.* $\log_{10}[SO_3^{2-}]$. There is a significant scatter of the experimental points (points 10, 11 and 12 in Table 3 of Ref. [67ZAK/ORL] must be affected by some experimental error). The uncharged complex $UO_2SO_3(aq)$ has a broad range of existence as indicated by the rather shallow solubility minimum. Hence, it is reasonable to assume that the minimum solubility is equal to $[UO_2SO_3(aq)]$ as proposed by Zakharova and Orlova [67ZAK/ORL]. The equilibrium constant for the reaction

$$UO_2SO_3 \cdot 4.5H_2O(cr) \rightleftharpoons UO_2SO_3(aq) + 4.5H_2O(l) \qquad (A.28)$$

is then $\log_{10} K(A.28) = \log_{10}(\beta_1 K_{s,0}) = -(2.49 \pm 0.04)$. This equilibrium constant is independent of the ionic strength. Hence, $\log_{10}(\beta_1^\circ K_{s,0}^\circ) = -(2.49 \pm 0.04)$.

The latter part of the solubility data of Ref. [67ZAK/ORL] could in principle give information on the reaction $UO_2SO_3 \cdot 4.5H_2O(l) + SO_3^{2-} \rightleftharpoons UO_2(SO_3)_2^{2-} + 4.5H_2O(l)$. However, there is a significant scatter in the data, and they are therefore discarded in favour of those of Ref. [59KLY/KOL] which are more precise.

[59KLY/SMI]
Klygin, A.E., Smirnova, I.D., The dissociation constant of the $UO_2(CO_3)_3^{4-}$ ion, Russ. J. Inorg. Chem., 4 (1959) 16-18.

The solubility of $UO_2R_2{\cdot}HR(s)$, where HR is hydroxyquinoline, was measured as a function of the carbonate concentration. The measurements were made in a concentration range where $UO_2(CO_3)_3^{4-}$ is the predominant complex ($[CO_3^{2-}] = 33.5 \times 10^{-5}$ M to 1.5×10^{-2} M). The stoichiometry of the complex was confirmed in the measurements. The carbonate concentration was calculated from experimental measurements of the free hydrogen ion concentration by using a glass-electrode. The electrode was calibrated by using buffers of known pH, but it is not clear whether the system is calibrated in terms of activities or concentrations. The constants used for the protonation of CO_3^{2-} ($CO_3^{2-} + 2H^+ \rightleftharpoons CO_2(aq) + H_2O(l)$, $\log_{10} K = 16.67$) seems to refer to $I = 0$, rather than the 1 M NH_4Cl medium used.

The equilibrium constants reported by the authors [59KLY/SMI, Table 1] show a systematic increase with increasing concentration of $(NH_4)_2CO_3$. This increase is large, a factor of approximately five, and the model proposed does not properly explain the experimental observations. The equilibrium constant obtained is of the right order of magnitude, but for the reasons stated above, the measurements are not reliable enough to warrant their use in the set of selected equilibrium constants.

[59KLY/SMI2]
Klygin, A.E., Smirnova, I.D., Nikol'skaya, N.A., Equilibria in the the $(UO_2(IO_3)_2$-KIO_3-H_2O system, Russ. J. Inorg. Chem., 4 (1959) 754-756.

The solubility of $UO_2(IO_3)_2(cr)$ was studied at 25°C and 60°C in a 0.2 M NH_4Cl ionic medium. The procedures used by the authors are satisfactory, and hydrolysis did not seriously influence the experimental results. The authors reported for the solubility product $\log_{10} K_{s,0} = \log_{10}([UO_2^{2+}][IO_3^-]^2) = -(7.0 \pm 0.1)$ at 25°C and $\log_{10} K_{s,0} = -(6.65 \pm 0.10)$ at 60°C. The uncertainties are estimated by this review. The other equilibria reported are

$$UO_2^{2+} + 2IO_3^- \rightleftharpoons UO_2(IO_3)_2(aq) \qquad (A.29)$$
$$UO_2^{2+} + 3IO_3^- \rightleftharpoons UO_2(IO_3)_3^- \qquad (A.30)$$

with the constants reported for 25°C, $\log_{10}\beta_2 = 2.73$ and $\log_{10}\beta_3 = 3.67$, and for 60°C, $\log_{10}\beta_2 = 2.74$ and $\log_{10}\beta_3 = 3.44$, Klygin, Smirnova and Nikol'skaya did not take the formation of $UO_2IO_3^+$ into account. This review reinterprets the experimental data at (25°C) by adding the complex $UO_2IO_3^+$ found by Khalili, Choppin and Rizkalla [88KHA/CHO]. The solubility data in Ref. [59KLY/SMI2] do not allow an independent determination of the equilibrium constants. This review uses an alternative graphical evaluation of the 25°C data in Ref. [59KLY/SMI2]. From the first two experimental points in each series, $\log_{10} K_{s,0} = -(7.10 \pm 0.10)$ and $\log_{10}\beta_1 = (1.48 \pm 0.14)$ are obtained, the latter value being in good agreement with the value reported in Ref. [88KHA/CHO]. Continuing the graphical fitting yields $\log_{10}\beta_2 = (2.80 \pm 0.15)$. Klygin, Smirnova and Nikol'skaya proposed the formation of $UO_2(IO_3)_3^-$. However, the observed increase in the solubility at higher iodate concentration does not fit this chemical model very well. In this region the solubility may be affected by the large changes in the ionic medium. Up to $[IO_3^-] \approx 0.1$ M the solubility data are well described by the following constants: $\log_{10} K_{s,0} = -(7.10 \pm 0.10)$, $\log_{10}\beta_1 = (1.48 \pm 0.14)$, and $\log_{10}\beta_2 = (2.80 \pm 0.15)$. This review does not consider the formation of $UO_2(IO_3)_3^-$ as proven.

All the equilibrium constants given by the authors—and those re-evaluated by this review—are conditional constants. This review corrects the re-evaluated constants for the formation of UO_2Cl^+ by using $\log_{10}\beta_1(UO_2Cl^+) = -(0.33 \pm 0.02)$ calculated for $I = 0.2$ M NH_4Cl from the selected value of $\log_{10}\beta_1^\circ(UO_2Cl^+)$ and the $\Delta\varepsilon$ value obtained for this reaction by extrapolation to $I = 0$. The corrected values are $\log_{10} K_{s,0} = -(7.14 \pm 0.10)$, $\log_{10}\beta_1 = (1.52 \pm 0.14)$, and $\log_{10}\beta_2 = (2.84 \pm 0.15)$. Correction of the equilibrium data to $I = 0$ is made by using the ion interaction coefficients given in Appendix B, with $\varepsilon_{(IO_3^-,NH_4^+)} \approx \varepsilon_{(BrO_3^-,Na^+)} = -(0.06 \pm 0.02)$, $\varepsilon_{(UO_2Cl^+,Cl^-)} \approx \varepsilon_{(UO_2Cl^+,ClO_4^-)} = (0.46 \pm 0.03)$ and $\varepsilon_{(UO_2IO_3^+,Cl^-)} \approx \varepsilon_{(UO_2Cl^+,ClO_4^-)} = (0.33 \pm 0.04)$. The equilibrium constants obtained in the standard state are $\log_{10} K^\circ_{s,0} = -(7.88 \pm 0.10)$, $\log_{10}\beta_1^\circ = (2.05 \pm 0.14)$ and $\log_{10}\beta_2^\circ = (3.59 \pm 0.15)$. The data reported in Ref. [59KLY/SMI2] at 25°C and 60°C are not sufficiently precise to allow an estimate of $\Delta_r H_m$.

[59SUL/HIN]

Sullivan, J.C., Hindman, J.C., The hydrolysis of neptunium(IV), J. Phys. Chem., 63 (1959) 1332- 1333.

This is a spectrophotometric study. The uncertainty estimated in this review for $^*\beta_1$ of the U^{4+} hydrolysis is $\pm 0.1 \times 10^{-2}$, twice the uncertainty selected by the authors.

[59VDO/SOK]

Vdovenko, V.M., Sokolov, A.P., Crystallisation of uranyl nitrate, Radiokhimiya, 1 (1959) 117-120, in Russian.

Water vapour pressures were measured for the two following equilibria:

$$UO_2(NO_3)_2 \cdot 6H_2O(cr) \rightleftharpoons UO_2(NO_3)_2 \cdot 3H_2O(cr) + 3H_2O(g) \quad (A.31)$$

$$UO_2(NO_3)_2 \cdot 3H_2O(cr) \rightleftharpoons UO_2(NO_3)_2 \cdot 2H_2O(cr) + H_2O(g). \quad (A.32)$$

It is apparent from the scatter in the data that the uncertainties in the measured values rose with increasing temperature, at least in part because of increasing content of nitrogen oxides in the gas phase. In this review the uncertainties in the vapour pressures are estimated as 40 Pa for measurements at temperatures less than or equal to 30°C, 60 Pa for measurements between 30 and 50°C and 100 Pa for measurements at 50°C or above.

[59WES/GRO]
Westrum, E.F., Jr., Grønvold, F., Low temperature heat capacity and thermodynamic functions of triuranium octoxide, J. Am. Chem. Soc., 81 (1959) 1777-1780.

The heat capacity of U_3O_8 was measured by low temperature adiabatic calorimetry from 4.8-346.9 K. The authors estimated the errors in the experimental heat capacities for temperatures above 25 K to be 0.1%. In this review, for the purposes of calculating a function describing the temperature dependence of the heat capacity, the uncertainty (95% confidence level) of each measurement (232.9 to 346.9 K) is estimated as 0.2% of the reported value, and the data are weighted accordingly.

[60BAB/KOD]
Babko, A.K., Kodenskaya, V.S., Equilibria in solutions of uranyl carbonate complexes, Russ. J. Inorg. Chem., 5 (1960) 1241-1244.

This investigation was discussed in detail by Sergeyeva *et al.* [72SER/NIK]. The solubility of $UO_2(OH)_2(cr)$ was measured as a function of the carbonate concentration in the range $7.25 < pH < 9.40$, but the solid in equilibrium with the carbonate solutions was not characterized. By using the solubility product of $UO_2(OH)_2(cr)$ given by the authors and the solubility product of $UO_2CO_3(cr)$ selected in this review, one can calculate the equilibrium constant, $\log_{10}{}^*K_p(A.33) \approx 2.5$.

$$UO_2(OH)_2(cr) + CO_2(g) \rightleftharpoons UO_2CO_3(cr) + H_2O(l) \qquad (A.33)$$

Hence, at a partial pressure of $p_{CO_2} > 10^{-2.5}$ atm, $UO_2(OH)_2(cr)$ should be transformed into $UO_2CO_3(cr)$. This is in good agreement with the direct measurement of Sergeyeva *et al.* [72SER/NIK, Table 4], $\log_{10}{}^*K_p(A.33) = 2.22$. Such a transformation is also thermodynamically possible in the range where the complex $UO_2(CO_3)_2^{2-}$ is formed [60BAB/KOD, Table 2]. The claim of the authors that "uranyl carbonate can exist as a solid phase only under high pressures of CO_2" is not correct.

The authors gave no information about the method used to calibrate the glass-electrode, and it is unclear whether or not they were measuring H^+ concentrations. The value of $\log_{10}\beta_2 = 18.16$ used for the equilibrium $CO_3^{2-} + 2H^+ \rightleftharpoons CO_2(g) + H_2O(l)$ seems to be the constant referring to $I = 0$, and not to the ionic medium used. These uncertainties in the pH measurement procedure result in an uncertainty in the absolute values of the equilibrium constants but not in the proposed composition of the complexes.

In view of these facts, this review assigns zero weight to the equilibrium constants reported by these authors.

[60GUS/RIC]

Gustafson, R.L., Richard, C., Martell, A.E., Polymerization of uranyl-tiron chelates, J. Am. Chem. Soc., **82** (1960) 1526-1534.

The experiments and data analysis were quite well done. Baes and Mesmer [76BAE/MES] noted the lack of detail on calibration of the pH meter. The constants obtained were certainly sufficient for the authors' purposes, but obtaining these constants was not their primary aim. Calculation of a formation constant for $(UO_2)_4(OH)_6^{2+}$ rather than the more likely $(UO_2)_3(OH)_5^+$ probably had a minimal effect on the values calculated for $^*\beta_1$ and $^*\beta_{2,2}$. This review accepts the values for these two constants, but increases the estimated uncertainty in $\log_{10}{}^*\beta_{2,2}$ to ±0.30 and estimates the uncertainty in $\log_{10}{}^*\beta_1$ as ±0.6.

[60HEF/AMI]

Hefley, J.D., Amis, E.S., A spectrophotometric study of the complexes formed between uranyl and chloride ions in water and water-ethanol solvents, J. Phys. Chem., **64** (1960) 870-872.

Hefley and Amis studied the formation of dioxouranium(VI) chloride complexes in aqueous solutions at 25°C by using spectrophotometry. The data in dilute solution 0.02 to 0.18 M could be satisfactorily explained by assuming the formation of UO_2Cl^+ only. The Job method [28JOB, 61ROS/ROS] used to evaluate the composition of the complex is then satisfactory. When the same measurements were made on more concentrated solutions (0.237 to 1.58 M) the authors reported four maxima and one minimum in Job plots at various λ. The authors assigned these to the complexes $UO_2Cl^+, UO_2Cl_2(aq)$ and $UO_2Cl_3^-$. This interpretation is erroneous. It is well known that the Job method cannot be used to determine the composition of complexes in solutions where more than two absorbing species are present [61ROS/ROS]. The numerical data from these solutions are not reliable. The value reported for the formation constants of UO_2Cl^+ at $I = 1.238$ M is equal to $\log_{10}\beta_1 = 1.64$. This value is much larger than those reported by other authors. This discrepancy might well be due to the erroneous procedure used by Hefley and Amis [60HEF/AMI] to interpret their data. The data from this study are given zero weight.

[60LIE/STO]

Lietzke, M.H., Stoughton, R.W., The solubility of silver sulfate in electrolyte solutions: Part 7. Solubility in uranyl sulfate solutions, J. Phys. Chem., **64** (1960) 816-820.

Formation constants for $UO_2SO_4(aq)$, $UO_2(SO_4)_2^{2-}$ and $(UO_2)_2(OH)_2^{2+}$ were calculated from the measured solubility of $AgSO_4$ in aqueous solution as a function of the concentration of dioxouranium(VI) sulphate from 25 to 200°C. The total concentrations of uranium were high (0.1 to 1.35 m) and the authors assumed the only hydrolysis product formed is $(UO_2)_2(OH)_2^{2+}$. The authors claimed only "that the hydrolysis quotient is not larger than that calculated by about an order of magnitude". The value of $\log_{10}{}^*\beta_{2,2} = -5.50$ determined by Lietzke and Stoughton at 25°C differs by 0.12 from the value selected in this review, $-(5.62\pm0.04)$, and is based on a value of -6.00 (1 M $NaClO_4$, 20°C) supposedly from Ref. [54AHR/HIE] wherein a value of -6.05 was given. The reported equilibrium constants [60LIE/STO, Table 1 and

Eq. (9)] are smoothed values based on the 25, 100 and 150°C data (extrapolated to $I = 0$). In this review the uncertainties in the values of $\log_{10}{}^*\beta_{2,2}$ are estimated as ±1.4 for 100 and 150°C. The smoothed values for other temperatures are assigned a weight of zero.

The data cannot be used for an independent determination of the formation constants of $UO_2SO_4(aq)$ and $UO_2(SO_4)_2^{2-}$. The authors used the 25°C data of Kraus and Nelson [51NEL/KRA] to estimate the equilibrium constants and the enthalpy changes at higher temperatures. The data must be corrected for hydrolysis and protonation of SO_4^{2-} and this cannot be done with any precision at temperatures higher than 25°C. The enthalpy values calculated by the authors seem very doubtful and the numerical values depend strongly on the expression used to fit the $\log_{10} K(T)$ data. These thermodynamic quantities are given zero weight.

[60MAT]

Matsuo, S., Sulphate complexes of uranium(VI), J. Chem. Soc. Japan, **81** (1960) 833-836, in Japanese.

Matsuo made a spectrophotometric study of the complex formation between UO_2^{2+} and SO_4^{2-} at 25°C in 1 M $NaClO_4$. The experimental method and the calculation procedure are satisfactory. The author started his experiments from a solution containing UO_2SO_4 and $HClO_4$, then added Na_2SO_4. In order to calculate the sulphate concentration, the protonation constant of SO_4^{2-} is needed. The author used the value 50 M^{-1} rather than 15 M^{-1} which would be a more correct choice. This introduces an error of at most around 10% in the sulphate concentration. This error makes the equilibrium constant $\beta_1 = 65$ reported by Matsuo [60MAT] at most 10% too low and the equilibrium constant $\beta_2 = 195$ at most around 30% too low. β_1 and β_2 refer to the formation of $UO_2SO_4(aq)$ and $UO_2(SO_4)_2^{2-}$, respectively. The uncertainties reported by the author are too small. This review estimates the uncertainties in β_1 and β_2 to $\pm10\%$, and $\pm30\%$, respectively. After multiplication of β_1 and β_2 by 1.1 and 1.3 to give $\log_{10}\beta_1 = (1.85 \pm 0.10)$ and $\log_{10}\beta_2 = (2.40 \pm 0.20)$, respectively, the constants agree fairly well with other determinations.

[60SAV/BRO]

Savage, A.W., Jr., Browne, J.C., The solubility of uranium(IV) fluoride in aqueous fluoride solutions, J. Am. Chem. Soc., **82** (1960) 4817-4821.

Savage and Browne determined the solubility of $UF_5\cdot2.5H_2O(cr)$ as a function of [HF] in solutions of $HClO_4$. The temperature in the experiment was 25°C and $[HClO_4]$ varied between 0.1212 M and 0.0646 M. The concentration equilibrium constant for $H^+ + F^- \rightleftharpoons HF(aq)$ used was $\log_{10} K = 2.98$, which is in good agreement with the value of 2.95 in 0.1 m $HClO_4$, obtained from the specific ion interaction theory. The solid is well characterized and the authors reported its properties and chemical analysis. This seems to be a precise study. The authors reported "lowest" and "highest" values for the equilibrium constants of the reactions below. This review averages these values and takes the spans between the lowest and highest values as the uncertainty intervals.

$$\begin{array}{lcll}
UF_4{\cdot}2.5H_2O(cr) & \rightleftharpoons & UF_2^{2+} + 2F^- & \log_{10} K_{s,2} = -12.45 \pm 0.04 \\
UF_4{\cdot}2.5H_2O(cr) & \rightleftharpoons & UF_3^+ + F^- & \log_{10} K_{s,3} = -8.24 \pm 0.09 \\
UF_4{\cdot}2.5H_2O(cr) & \rightleftharpoons & UF_4(aq) & \log_{10} K_{s,4} = -3.96 \pm 0.03 \\
UF_4{\cdot}2.5H_2O(cr) + F^- & \rightleftharpoons & UF_5^- & \log_{10} K_{s,5} = -2.39 \pm 0.25 \\
UF_4{\cdot}2.5H_2O(cr) + 2\,F^- & \rightleftharpoons & UF_6^{2-} & \log_{10} K_{s,6} = -0.08 \pm 0.08
\end{array}$$

These constants are converted to molality units and corrected from $I = 0.12$ m to $I = 0$ by using the specific ion interaction theory but neglecting the $\Delta\varepsilon\, m_{HClO_4}$ terms because of the low ionic strength. This review obtains

$$\begin{array}{lcl}
\log_{10} K^\circ_{s,2} & = & -13.15 \pm 0.04 \\
\log_{10} K^\circ_{s,3} & = & -8.47 \pm 0.09 \\
\log_{10} K^\circ_{s,4} & = & -3.96 \pm 0.03 \\
\log_{10} K^\circ_{s,5} & = & -2.39 \pm 0.25 \\
\log_{10} K^\circ_{s,6} & = & -0.32 \pm 0.08.
\end{array}$$

The uncharged complex is dominant over the broad range of HF(aq) concentration, $10^{-4} < [HF(aq)] < 10^{-2}$, *cf.* the behaviour of the corresponding hydroxide system. Other authors who studied uranium fluoride complexes were not able to reach the anionic region due to the use of a much higher acidity in their test solutions.

The uncertainties quoted above are maximum uncertainties given by the authors. This review uses the values reported here for $\log_{10} K^\circ_{s,q}$, $q = 2$ to 4, and the selected values of $\log_{10} \beta^\circ_q$, $q = 2$ to 4, to calculate the solubility product $\log_{10} K^\circ_{s,0} = \log_{10}(a_{U^{4+}} \times a^4_{OH^-})$. The values obtained are $\log_{10} K^\circ_{s,0} = -(29.38 \pm 0.16)$ $(q = 2)$, $-(30.07 \pm 1.00)$ $(q = 3)$, and $-(29.56 \pm 1.00)$ $(q = 4)$. The weighted average value is $\log_{10} K^\circ_{s,0} = -(29.40 \pm 0.18)$. This value is used to calculate the selected overall formation constants $\log_{10} \beta^\circ_5 = (27.01 \pm 0.30)$ and $\log_{10} \beta^\circ_6 = (29.08 \pm 0.18)$.

It should be mentioned that the solubility product calculated from the standard Gibbs energies of formation of $UF_4{\cdot}2.5H_2O(cr)$ selected in this review, *cf.* Section V.4.1.3.1.b), is $\log_{10} K^\circ_{s,0} = -(33.5 \pm 1.2)$. This value differs significantly from the direct experimental determination of Savage and Browne. There may be several reasons for this discrepancy:

- The solid phase used in Ref. [60SAV/BRO] contained lower hydrates and thus gave a larger solubility.
- The particle size of the solid used in Ref. [60SAV/BRO] was so small that the solubility was increased through the influence of surface forces.
- The estimation of the entropy of $UF_4{\cdot}2.5H_2O(cr)$ made by Fuger *et al.* [83FUG/PAR] is in error.
- The standard Gibbs energy of formation of U^{4+}(aq) selected in this review is in error.

The first two error sources are judged less likely due to the experimental procedures used in Ref. [60SAV/BRO]. An error in the estimated entropy of either $UF_4{\cdot}2.5H_2O(cr)$ or $\Delta_f G^\circ_m(U^{4+}$, aq, 298.15 K) can hardly explain the difference in the standard Gibbs energy of formation of $UF_4{\cdot}2.5H_2O(cr)$. It seems most likely that the discrepancy is due to a combination of errors. On the basis of available experimental information it is not possible to make a selection between the two sets of values.

[60STA]

Starý, J., Untersuchungen über die Extraktion des U(VI)-Komplexes mit Benzoylaceton, Coll. Czechoslov. Chem. Commun., 25 (1960) 890-896, in German.

The authors made the assumption that at uranium concentrations below 2×10^{-4} M only monomeric hydrolysis species are formed. This is probably not true, and in particular $(UO_2)_3(OH)_5^+$ should be significant near pH 4. The extraction results indicate the formation of cationic, neutral and anionic species, but it is not clear if the species are monomeric or polymeric. It is also not clear that the data analysis unambiguously separated the constants for simple hydrolysis from those for formation of benzoylacetone (1-phenyl-1,3-butanedione) complexes.

[61BAN/TRI]

Banerjea, D., Tripathi, K.K., Association of uranium(VI) with anions in aqueous perchloric acid medium, J. Inorg. Nucl. Chem., 18 (1961) 199-206.

Banerjea and Tripathi did a cation exchange study of the complex formation of UO_2^{2+} with Cl^-, NO_3^-, SCN^- and SO_4^{2-} in 1 M $NaClO_4$ and at pH = 3.0. The temperature was 32°C. The authors assumed that only UO_2^{2+} was sorbed on the exchanger. This approximation is doubtful in any system where UO_2X^+ is formed ($X^- = Cl^-, NO_3^-, ...$). The neglection of the UO_2X^+ sorption may lead to systematic errors in the equilibrium constants of the higher complexes. These are determined in ligand concentration ranges where $[UO_2^{2+}] \ll [UO_2X^+]$.

The equilibrium constant for the formation of UO_2Cl^+ is in fair agreement with other data. The data for NO_3^- and SCN^- seem very doubtful, and it is certainly not possible to obtain equilibrium constants for complexes higher than the first. This review does not credit the study of the nitrate complexes.

The sulphate data are not influenced by systematic errors due to the absorption of $UO_2SO_4(aq)$, hence these data should be more reliable than those for the other ligands. The formation constant obtained, $\beta_1(A.34) = 43$,

$$UO_2^{2+} + SO_4^{2-} \rightleftharpoons UO_2SO_4(aq) \qquad (A.34)$$

is in fair agreement with other data, while the value $\beta_2(A.35) = 6\times10^3$ seems too high.

$$UO_2^{2+} + 2SO_4^{2-} \rightleftharpoons UO_2(SO_4)_2^{2-} \qquad (A.35)$$

The linearity of $1/\lambda$ *vs.* $[SO_4^{2-}]$ in Ref. [61BAN/TRI, Figure 1] indicates that no appreciable amounts of $UO_2(SO_4)_2^{2-}$ are formed, and this review gives therefore zero weight to the value found in this paper for the formation of $\beta_2(UO_2(SO_4)_2^{2-})$.

The authors did not make any error estimate, but $\log_{10}\beta_1(A.35) = (1.63 \pm 0.20)$ seems to be reasonable. This value is corrected to 25°C by using $\Delta_rH_m^\circ(A.34) =$

$(19.5 \pm 1.6)\,\mathrm{kJ \cdot mol^{-1}}$ selected by this review, leading to a correction of -0.08.

[61CHU/ALY]
Chukhlantsev, V.G., Alyamovskaya, K.V., Solubility product of uranyl, beryllium and cerium phosphates, Izv. Vuz. Khim. Tekhnol., **4** (1961) 359-363, in Russian.

This is a solubility study made at 19 to 20°C using $(UO_2)_3(PO_4)_2(s)$ as the solid phase. The study was carried out in the range $1.6 < pH < 2.4$, and the authors reported $\log_{10} K_{s,0}(A.36) = -(49.09 \pm 0.29)$ for the reaction

$$(UO_2)_3(PO_4)_2(s) \rightleftharpoons 3UO_2^{2+} + 2PO_4^{3-}. \qquad (A.36)$$

Dilute solutions of $I \approx 0.01\,M$ were used. No corrections for the complex formation between $H_rPO_4^{r-3}$ and UO_2^{2+} were made, and the experimental information is scarce. Recalculation of $\log_{10} K_{s,0}$ to $I = 0$ using the specific ion interaction method results in $\log_{10} K^\circ_{s,0}(A.36) = -(50.41 \pm 0.29)$, which is not in agreement with the value selected in this review from experimental solubility studies of the tetrahydrate, $\log_{10} K^\circ_{s,0} = -(49.36 \pm 0.30)$, *cf.* Section V.6.2.1.1.b). In view of the limited experimental information this review gives zero weight to this study.

The authors also investigated the solubility product of $Ce_3(PO_4)_4(s)$ and reported $\log_{10} K_{s,0}(A.37) = -(90.14 \pm 0.32)$ for the reaction

$$Ce_3(PO_4)_4(s) \rightleftharpoons 3Ce^{4+} + 4PO_4^{3-}. \qquad (A.37)$$

No corrections were made for the hydrolysis of Ce(IV) or complex formation with $H_rPO_4^{r-3}$. Hence, the solubility product is too large and thus certainly in error. This solid phase seems to have been used as a model for $U_3(PO_4)_4(s)$ by Allard and Beall [78ALL/BEA]. There are large chemical differences between Ce(IV) and U(IV), and such an analogy is therefore not very reliable. This fact, together with the experimental shortcomings in this solubility study of $Ce_3(PO_4)_4(s)$, results in a very uncertain estimate of the solubility of $U_3(PO_4)_4(s)$, *cf.* Section V.6.2.2.5.b).

[61CON/PAU]
Connick, R.E., Paul, A.D., The fluoride complexes of silver and stannous ions in aqueous solution, J. Phys. Chem., **65** (1961) 1216-1220.

Apart from the fluoride complexation of silver(I) and tin(II), Connick and Paul determined the equilibrium constant for the reaction $UO_2^{2+} + HF(aq) \rightleftharpoons UO_2F^+ + H^+$ by using a spectrophotometric method. This involves a simultaneous determination of the molar absorption of the complex. No error estimate was given, but the error is probably large, due to the covariation of K_1 and the molar absorption of UO_2F^+. The authors reported $^*\beta_1 = 15$, from which this review calculates $\log_{10} \beta_1^\circ = 4.77$ for the reaction $UO_2^{2+} + F^- \rightleftharpoons UO_2F^+$. In view of the limited information available on the calculation procedure used by the authors and the lack of an error estimate, this value of the equilibrium constants is given zero weight.

[61KAR]

Karpov, V.I., The solubility of triuranyl phosphate, Russ. J. Inorg. Chem., 6 (1961) 271-272.

Karpov determined the solubilty of $(UO_2)_3(PO_4)_2 \cdot 6H_2O(cr)$ in nitric acid solutions. The solid phase was properly characterized, but no corrections of the measured solubilities for complex formation were made. A reinterpretation of the experimental data of [61KAR, Table 3] (using the dioxouranium(VI) phosphate equilibrium constants and the dissociation constants of phosphoric acid selected in this review and corrected to the ionic strength used by Karpov) yields $\log_{10}{}^*K_s(A.38) = -5.24, -5.16$ and -5.42, respectively, for the reaction

$$(UO_2)_3(PO_4)_2 \cdot 6H_2O(cr) + 6H^+ \rightleftharpoons 3UO_2^{2+} + 2H_3PO_4(aq) + 6H_2O(l). \quad (A.38)$$

The corresponding values corrected to $I = 0$ are $\log_{10}{}^*K_s^\circ(A.38) = -5.52, -5.34$ and -5.40, respectively. The uncertainties in these constants are estimated by this review to ± 0.3. The average value is $\log_{10}{}^*K_s^\circ(A.38) = -(5.42 \pm 0.30)$, which is in fair agreement with the value found by Veselý *et al.* [65VES/PEK], $\log_{10}{}^*K_s^\circ(A.50) = -(5.96 \pm 0.30)$, for the tetrahydrate. However, this review does not find sufficient evidence that the two phases (tetra- and hexahydrate) are different.

[61KUT]

Kuteinikov, A.F., Spectroscopic investigation of the stability of the complex compound of uranium(VI) fluoride, Radiokhimiya, 3 (1961) 706-711, in Russian.

No translation of this paper is available for this review. The author used a spectrophotometric method, competition between arsenazo reagent and F^-, to determine the equilibrium constant for the reaction $UO_2^{2+} + F^- \rightleftharpoons UO_2F^+$. The measurements appear to have been performed at room temperature and at low ionic strength. This review is not able to extract sufficient experimental details from the abstract to use these data. However, the equilibrium constant, $\log_{10}\beta_1^\circ = 4.77$, is close to the selected value.

[61NAI/PRA]

Nair, B.K.S., Prabhu, L.H., Vartak, D.G., Spectrophotometric investigation of uranyl azide complex in aqueous solution, J. Sci. Indust. Res., Sect. B, 20 (1961) 489-492.

Nair, Prabhu and Vartak [61NAI/PRA] determined the complexes formed in the dioxouranium(VI) azide system spectrophotometrically in solutions of $I = 0.3$ M $(NaClO_4)$. It is assumed that the authors worked at "room temperature" (~25°C). Both the experimental method and the data treatment are satisfactory and the authors reported $\log_{10}\beta_1 = (3.49 \pm 0.01)$. When calculating $\log_{10}\beta_1$, the authors used $\log_{10}K = -4.62$ for the dissociation constant $HN_3(aq)$, which is in good agreement with the value selected in Chapter VI.

Extrapolation to $I = 0$ gives $\log_{10}\beta_1^\circ = (4.05 \pm 0.04)$. This value is much higher than the value obtained by other authors. This might be due to decomposition of $HN_3(aq)$ which is present in not negligible amounts in the test solutions. For this reason $\log_{10}\beta_1^\circ$ in [61NAI/PRA] is given zero weight.

[61NIK/LUK]
Nikolaev, N.S., Luk'yanychev, Yu.A., The hydrolysis of uranium tetrafluoride, Sov. Atomic Energy, 11 (1961) 704-706.

Solutions saturated with $UF_4 \cdot 2.5H_2O(cr)$ were diluted with water and the pH measured. From the uranium concentrations and pH, values for the hydrolysis constants $^*\beta_n$ ($n = 1, 4$) were proposed. In the more concentrated (low pH) solutions it is probable that fluoride complexation of uranium(IV) interfered with the measurements. There are insufficient data to identify (or justify the equilibrium constants for) the more highly hydrolysed species. It is not clear if any UO_2 was precipitated on dilution of the solutions.

[61PET]
Peterson, A., Studies on the hydrolysis of metal ions: 32. The uranyl ion, UO_2^{2+}, in Na_2SO_4 medium, Acta Chem. Scand., 15 (1961) 101-120.

See comments under Dunsmore and Sillén [63DUN/SIL] and the footnote in Section V.5.1.3.1.a.

[61SHE/AWA]
Sherif, F.G., Awad, A.M., The structure of uranyl azide: Its instability constant in aqueous solutions, J. Inorg. Nucl. Chem., 19 (1961) 94-100.

Sherif and Awad studied the dioxouranium(VI) azide system by using spectrophotometry. The only quantitative equilibrium information refers to the reaction $UO_2^{2+} + N_3^- \rightleftharpoons UO_2N_3^+$ studied at 25°C and at $I = 0.096$ M ($NaClO_4$). The experimental method and the calculaton procedure are satisfactory. The authors reported $\log_{10}\beta_1 = (2.31 \pm 0.04)$. This review recalculates this value to $I = 0$ using the estimated ion interaction coefficients $\varepsilon_{(UO_2N_3^+,ClO_4^-)} = (0.3 \pm 0.1)$ and $\varepsilon_{(N_3^-,Na^+)} = (0.0 \pm 0.1)$, *i.e.*, $\Delta\varepsilon = -(0.16 \pm 0.14)$. This leads to $\log_{10}\beta_1^\circ = (2.58 \pm 0.04)$. The uncertainty of $\log_{10}\beta_1^\circ$ is estimated by this review.

[61SOB]
Sobkowski, J., The oxidation-reduction potential of UO_2^{2+}-U^{4+} system: II. The influence of $HClO_4$, HCl, H_2SO_4 and of temperature on the oxidation potential of UO_2^{2+}-U^{4+}, J. Inorg. Nucl. Chem., 23 (1961) 81-90.

Sobkowski studied the variation of the U(VI)/U(IV) redox potential in perchloric, hydrochloric, and sulphuric acid solutions of varying ionic strength. A hydrogen electrode was used as the reference electrode of the galvanic cell. The temperature dependence of the emf was studied in the range of 20 to 50°C. The experimental data are accurate, but there are some shortcomings: *i)* the U(VI)/U(IV) ratio was not varied and the Nernst behaviour not tested, *ii)* the total concentrations of U(VI) and U(IV) were high and made a significant contribution to the ionic strength of the solutions and to the diffusion potential. No corrections for the latter were made by the author.

The interpretation of the experimental results given by the author is not satisfactory for two reasons: *i)* the hydrogen ion activity was estimated from the value in the pure acid, and *ii)* the extrapolation procedure to $I = 0$ is erroneous because the simple Debye-Hückel law of the type $\log_{10}\gamma_i = -z_i^2 A\sqrt{I}$ cannot be used at these high ionic strengths. Bruno, Grenthe and Lagerman [90BRU/GRE] recently reinterpreted the data of Sobkowski, *cf.* Section V.2.3. They found that the value of $\log_{10} K^\circ$(A.39)

$$UO_2^{2+} + 2H^+ + H_2(g) \rightleftharpoons U^{4+} + 2H_2O(l) \qquad (A.39)$$

is $\log_{10} K^\circ(A.39) = (9.16 \pm 0.14)$ after correction for chloride complex formation, *i.e.*, $E^\circ = (0.271 \pm 0.004)$ V, which is in good agreement with the corresponding value in $HClO_4$ media $\log_{10} K^\circ(A.39) = (9.04 \pm 0.05)$, $E^\circ = (0.2690 \pm 0.0015)$ V. Bruno, Grenthe and Lagerman [90BRU/GRE] used the selected values for the uranium chloride complexes and the corresponding ion interaction coefficients from this review. This results confirm the choices made. This review accepts the reasoning in Ref. [90BRU/GRE].

The sulphate experiments were not well designed, and this review considers it impossible to extract accurate equilibrium constants from them. For Reaction (A.40)

$$U^{4+} + SO_4^{2-} \rightleftharpoons USO_4^{2+} \qquad (A.40)$$

Sobkowski reported $K(A.40) = 413$ at $I = 1$ M. This constant probably refers to Reaction (A.41)

$$U^{4+} + HSO_4^- \rightleftharpoons USO_4^{2+} + H^+ \qquad (A.41)$$

which is the actual reaction studied. The equilibrium constant, $K = 413$, is of the correct order of magnitude for Reaction (A.41), but in view of the remarks made above, this review gives zero weight to this quantity.

Sobkowski also determined the entropy, $S_m(U^{4+}$, 1 M $HClO_4$, 298.15 K$) = -416\ J\cdot K^{-1}\cdot mol^{-1}$, from the temperature coefficient $(\partial E/\partial T)_p$ which varied from $-1.2\ mV\cdot K^{-1}$ to $-0.8\ mV\cdot K^{-1}$ between $I = 2.34$ and 0.234 M. Hence, it is difficult to estimate the entropy at zero ionic strength.

[61TAN/DEI]
Tanayev, I.V., Deichman, E.N., The properties of uranium hydroxyfluoride in solutions, Radiokhimiya, **3** (1961) 712-718, in Russian.

The authors investigated the solubility of PbClF(cr) and CaF_2(cr) in dioxouranium(VI) nitrate solutions and the reaction between UO_2^{2+} and HF(aq) using pH measurements. The measurements seem to refer to "room temperature". No constant ionic medium was used.

From the solubility of PbClF(cr) the authors reported $\log_{10}\beta_1 = (4.0 \pm 0.1)$ and $\log_{10}\beta_2 = (8.7 \pm 0.2)$. These constants have the correct magnitude, however, in view of the varying ionic medium used, this review gives these data zero weight.

The solubility of CaF_2(cr) in dioxouranium(VI) nitrate solutions gave a value of $\log_{10}\beta_1$ which varied from 4.1 to 4.7. The data at the lowest total concentrations of UO_2^{2+} agree reasonably well with the PbClF(cr) data. These data are given zero weight.

The potentiometric studies of the reaction $UO_2^{2+} + qHF(aq) \rightleftharpoons UO_2F_q^{2-q} + qH^+$ were not interpreted quantitatively. The results only indicate that UO_2F^+ and $UO_2F_2(aq)$ are formed.

[61WIL/KED]
Wilson, A.S., Keder, W.E., The extraction of uranium(IV) from nitric acid by tri-*n*-octylamine, J. Inorg. Nucl. Chem., 18 (1961) 259-262.

The extraction of U(IV) was measured at 25°C from HNO_3 solutions of concentrations varying from 1 to 12 M. The extracted complex is the diaminsalt of $U(NO_3)_6^{2-}$ and the extraction curve has a maximum around 6 M HNO_3 indicating that the average charge of the complexes is zero, *e.g.*, by the predominance of $U(NO_3)_4(aq)$. This paper gave only qualitative information about the composition and equilibrium constants of the uranium(IV) nitrate complexes.

[62BAE/MEY]
Baes, C.F., Jr., Meyer, N.J., Acidity measurements at elevated temperatures: I. Uranium(VI) hydrolysis at 25 and 94°C, Inorg. Chem., 1 (1962) 780-789.

This is a good, careful study. Perhaps the only problem is the lack of statistical support for not including hydrolysis schemes incorporating minor species such as $(UO_2)_4(OH)_7^+$. The reported uncertainties in the hydrolysis constants were estimated from the "probable" uncertainties in the concentrations and potentials, and are not simply the statistical uncertainties in the curve fitting parameters. In this review the uncertainties (95% confidence level) for $\log_{10}{}^*\beta_1$ and $\log_{10}{}^*\beta_{2,2}$ are estimated as twice the uncertainties estimated in the original paper. For $\log_{10}{}^*\beta_{5,3}$ the possible effects of model error increase the uncertainties to ±0.3 at 25°C and ±0.6 at 94.4°C.

[62ERM/KRO]
Ermolaev, N.P., Krot, N.N., Some data on the behavior of uranium(IV) in nitric acid solutions, Sov. Radiochem., 4 (1962) 600-606.

This is a carefully performed spectrophotometric study of the nitrate system,

$$U^{4+} + qNO_3^- \rightleftharpoons U(NO_3)_q^{4-q}, \qquad (A.42)$$

using spectrophotometry and cation exchange. The experiments were done at four different ionic strengths (I = 2.0, 2.5, 3.0 and 3.5 M) in $H(ClO_4,NO_3)$ media. The spectrophotometric study was discussed by McKay and Woodhead [64MCK/WOO]. This discussion indicates that the interpretation of the spectrophotometric data in Ref. [62ERM/KRO] at higher nitrate concentrations might be erroneous. The procedure used by Ermolaev and Krot should not influence the equilibrium constant for the formation of UNO_3^{3+} and $U(NO_3)_2^{2+}$. This is verified by the good agreement between their spectrophotometric and cation exchange results. No uncertainties have been given by the authors, and this review has therefore made an uncertainty estimate for the values at $I \leq 3.5$ m used in the linear extrapolation to $I = 0$, based on experience from the type of experimental techniques used.

The equilibrium constants reported for the reactions (A.42, $q > 2$) are not reliable (*cf.* Ref. [64MCK/WOO]) and were therefore given zero weight. The cation exchange values are also very uncertain because of the very large covarition between κ_i and K_i (*cf.* [62ERM/KRO, *p.*604]). The calculations procedure used by the authors does not seem to have taken this into account. The uncertainty in κ_i should have a fairly small influence on the value of $\log_{10}\beta_2$.

[62FAU/CRE]

Faucherre, J., Crego, A., Détermination cryoscopique de constantes de dissociation de complexes peu stable, Bull. Soc. Chim. France, (1962) 1820-1824, in French.

Faucherre and Crego used a cryoscopic method to determine the stability constant of UO_2Cl^+. The experiments were made in saturated KNO_3 at approximately $-3°C$. The authors reported $\beta_1(UO_2Cl^+) = (1.8 \pm 0.4)$. Saturated KNO_3 is approximately 2 M at that temperature. The data are recalculated to $I = 0$ and 25°C by assuming the $\Delta\varepsilon$ for the complex formation reaction is the same in KNO_3 as in $NaClO_4$. By using $\Delta_r H_m = (8 \pm 2)\,kJ \cdot mol^{-1}$ as the enthalpy change for the reaction $UO_2^{2+} + Cl^- \rightleftharpoons UO_2Cl^+$ at -3 to $+25°C$, the value $\log_{10}\beta_1^\circ = 0.7$ can be obtained. This value seems high in comparison with the results of other investigations. No correction for the formation of $UO_2NO_3^+$ is made, and hence the recalculation is very uncertain. This equilibrium constant is therefore given zero weight.

[62LUK/NIK]

Luk'yanychev, Yu.A., Nikolaev, N.S., Solubility of tetravalent uranium hydroxide in hydrofluoric acid solutions, Sov. At. Energy, **13** (1962) 779-781.

The authors determined the solubility of $U(OH)_4(s)$ in hydrofluoric acid solutions at $(25 \pm 1)°C$. No inert ionic medium was used, and the composition of the solid phase was not checked. The experiments seem to have been performed in the pH range 5.7 to 7, but no information about the pH is given. The authors proposed that the dominant complexation reaction is $U^{4+} + HF(aq) \rightleftharpoons UF^{3+} + H^+$. From the known hydrolysis constant of U(IV) it is quite clear that this assumption cannot be correct. The value of $^*\beta_1$ is in error by at least three orders of magnitude. This study is given zero weight.

[62NIK/PAR]

Nikolaeva, N.M., Paramonova, V.I., Kolychev, V.B., Hydrolysis of uranyl in nitrate solutions, Izv. Sib. Otd. Akad. Nauk. SSSR, **3** (1962), translated from the Russian: ORNL-TR-417, Oak Ridge National Laboratory, Tennessee, 1965, 15p.

Apparent lack of temperature control precludes use of the values in this paper.

[62PAR/NIK2]

Paramonova, V.I., Nikol'skii, B.P., Nikolaeva, N.M., Interaction of uranyl nitrate solutions and alkali-metal carbonates, Russ. J. Inorg. Chem., **7** (1962) 528-532.

The authors made ion-exchange investigations of the dioxouranium(VI) carbonate system at a uranium concentration of about 10^{-3} M, where there should be significant amounts of the trimer $(UO_2)_3(CO_3)_6^{6-}$ formed. Only the anion exchange data are used to derive equilibrium constant information. In their calculations, the authors assumed that only the limiting complex $UO_2(CO_3)_3^{4-}$ is significantly sorbed by the ion exchanger, while $UO_2(CO_3)_2^{2-}$ is not sorbed at all because of its lower charge. In precise work, one can determine the individual sorption coefficients of all charged species but this has not been attempted here. Hence, the procedure used is not state of the art.

There is no information about the calibration procedure used to convert emf readings to H^+ concentrations. A reference is given to the values of the protonation constants of CO_3^{2-} used, but it is not clear whether these figures refer to $I = 0$ or to 0.5 M $NaNO_3$. These data were given zero weight.

[62RUS/JOH]

Rush, R.M., Johnson, J.S., Kraus, K.A., Hydrolysis of uranium(VI): Ultracentrifugation and acidity measurements in chloride solutions, Inorg. Chem., **1** (1962) 378-386.

The authors claim that addition of the species UO_2OH^+ to their hydrolysis scheme did not noticeably improve the fit to the data. UO_2OH^+ would not be expected to be significant at the high uranium concentrations used (0.001 to 0.1 M), and the curve-fit value for $^*\beta_1$ was not considered significant by the authors nor has it been credited in this review. However, it should be noted that the value is consistent with values reported from other potentiometric studies. Uncertainties in the equilibrium constants were not reported. In this review the uncertainties in values of the equilibrium constants from this study are estimated as follows: ± 0.2 for $\log_{10}{}^*\beta_{2,2}$, and ± 0.4 for both $\log_{10}{}^*\beta_{4,3}$ and $\log_{10}{}^*\beta_{5,3}$. These have been selected primarily to reflect the possibility that minor species omitted in the hydrolysis model may influence the calculated values of the constants.

[62SHE/AWA]

Sherif, F.G., Awad, A.M., Spectrophotometric determination of uranium(VI) by the azide reaction, Anal. Chim. Acta, **26** (1962) 235-241.

This study confirms the stoichiometry $UO_2N_3^+$ found by the authors in a previous study [61SHE/AWA]. There are indications of the formation of higher complexes when the ratio $[N_3^-]$:$[UO_2^{2+}]$ exceeds 2. The authors reported a value of $\log_{10}\beta_1 = (2.64 \pm 0.07)$ at $I = 0.04$ M. This seems to be based on one experimental point only, *cf.* Ref. [62SHE/AWA, Figure 4], hence this value was given zero weight.

[62SHE/AWA2]

Sherif, F.G., Awad, A.M., Re-evaluation of the continuous variation method as applied to the uranyl azide complexes, J. Inorg. Nucl. Chem., **24** (1962) 179-182.

This is a discussion of the method of continuous variation which gives very little new information. The authors suggested that three different dioxouranium(VI) azide complexes can be formed. No quantitative information on the equilibria was given.

[62STE/GAL]

Stepanov, M.A., Galkin, N.P., Solubility product of basic uranium(IV) sulphate, Russ. J. Inorg. Chem., **7** (1962) 506-508.

The authors used a $U(OH)_2SO_4(cr)$ crystalline phase to determine the solubility product and reported $\log_{10} K_{s,0}(A.43) = -(31.17\pm0.03)$ for the hypothetical reaction

$$U(OH)_2SO_4(cr) \rightleftharpoons U^{4+} + 2OH^- + SO_4^{2-}. \qquad (A.43)$$

The authors corrected for the formation of sulphate and hydroxide complexes of U(IV). They also reported to have corrected for a deviation from ideality by using "the Debye-Hückel theory". Whatever form was chosen, the differences may not be large due to the low ionic strength of 0.0188 M. The numerical values of the equilibrium constants used for the formation of the sulphate and hydroxide complexes are in fair agreement with the ones selected in this review. The solubility product was derived by solving a set of mass balance equations. This method is not very precise in the present case, and the uncertainty in the solubility product proposed by the authors is underestimated according to this review. There is no information about the analysis or other methods used to characterize the solid phase used in Ref. [62STE/GAL]. However, the authors claimed that the slope of the solubility curve is consistent with a solid of the composition "$U(OH)_2SO_4$". This review considers the reported solubility product as tentative and assigns, somewhat arbitrarily, an uncertainty of ±0.5 in $\log_{10} K_{s,0}(A.43)$.

[63BAR/SOM]

Bartusek, M., Sommer, L., Ph.D. Thesis, Institut für Analytische Chemie der J.E. Purkyne Universität, Brno, Czechoslovakia, 1963, cited by Baes, C.F., Jr., Mesmer, R.F., The hydrolysis of cations, New York: Wiley & Sons, 1976, 177p.

This paper was unavailable to the reviewers. See comments under Bartusek and Sommer [64BAR/SOM].

[63DUN/HIE]

Dunsmore, H.S., Hietanen, S., Sillén, L.G., Studies on the hydrolysis of metal ions: 46. Uranyl ion, UO_2^{2+}, in chloride, perchlorate, nitrate and sulfate media, Acta Chem. Scand., **17** (1963) 2644-2656.

See comments under Dunsmore and Sillén [63DUN/SIL].

[63DUN/SIL]

Dunsmore, H.S., Sillén, L.G., Studies on the hydrolysis of metal ions: 47. Uranyl ion in 3 M (Na)Cl medium, Acta Chem. Scand., **17** (1963) 2657-2663.

This discussion also concerns Refs. [61PET, 63DUN/HIE, 63HIE/ROW]. The experiments were done with extreme care. The work in $Mg(ClO_4)_2$, $Ca(ClO_4)_2$ and Na_2SO_4 media has not been used in this review in the calculation of interaction parameters because of the high ionic strengths used in these studies. Even in the rest of these studies the high ionic strengths may have given junction potential problems at high dioxouranium(VI) concentrations. Uncertainties in the equilibrium constants are not reported. Instead values of "3σ" are provided for the parameters (equilibrium constants) obtained by curve-fitting of the data to models. These do not take into account uncertainties resulting from model error and from inaccuracies in solution compositions or potentiometer calibration, factors that will be more important for calculation of the formation constants of the minor species. In this review the uncertainties in values of the equilibrium constants from these studies are estimated as follows: ±0.10 for $\log_{10}{}^*\beta_{2,2}$, ±0.3 for $\log_{10}{}^*\beta_{5,3}$ (perchlorate and nitrate media), ±1.0 for $\log_{10}{}^*\beta_{4,3}$ (perchlorate and nitrate media), ±0.4 for both $\log_{10}{}^*\beta_{4,3}$ (3 M NaCl) and $\log_{10}{}^*\beta_{5,3}$ (3 M NaCl). Uncertainties in the constants for other species are discussed in Chapter V.

[63HIE/ROW]

Hietanen, S., Row, B.R.L., Sillén, L.G., Studies on the hydrolysis of metal ions: 48. The uranyl ion in sodium, magnesium and calcium perchlorate medium, Acta Chem. Scand., 17 (1963) 2735-2749.

See comments under Dunsmore and Sillén [63DUN/SIL].

[63LUK/NIK]

Luk'yanychev, Yu.A., Nikolaev, N.S., The solubility of uranium tetrafluoride in aqueous solutions of acids, Sov. At. Energy, 15 (1963) 1184-1187.

The authors measured the solubility of $UF_4 \cdot 2.5H_2O(cr)$ in solutions of H_2SO_4, HCl, and $HClO_4$ at 25°C. The ionic strength was not kept constant. The data were used to estimate the solubility product of the solid, according to the reaction

$$UF_4 \cdot 2.5H_2O(cr) \rightleftharpoons U^{4+} + 4F^- + 2.5H_2O(l) \qquad (A.44)$$

The authors used their own value for the formation constant of UF^{3+} which is the predominant complex under the experimental conditions used. As this value is seriously in error, *cf.* comments to Ref. [62LUK/NIK], the calculated solubility product, $K_{s,0}(A.44) = 6 \times 10^{-22}$, is also in error. This review has recalculated the solubility product using the experimental data in Ref. [63LUK/NIK] and $\log_{10}\beta_1(UF^{3+}) \approx 9$. In this way $\log_{10} K^\circ_{s,0}(A.44) = -(25.7 \pm 0.3)$ is obtained, which is still in poor agreement with the results of Savage and Browne [60SAV/BRO]. There is no characterization made of the solid used in Ref. [63LUK/NIK], and this review also considers the experimental procedure used as less precise than that of Savage and Browne [60SAV/BRO]. For this reason the recalculated value of $\log_{10} K^\circ_{s,0}(A.44)$ is given zero weight.

[63LUK/NIK2]

Luk'yanychev, Yu.A., Nikolaev, N.S., Solubility product of uranium tetrafluoride, Russ. J. Inorg. Chem., 8 (1963) 927-928.

The authors studied the solubility of $UF_4 \cdot 2.5H_2O(cr)$ in the presence of Al^{3+}. They proposed that the observed increase in the solubility of the solid is due to the reaction $UF_4(s) + Al^{3+} \rightleftharpoons U^{4+} + AlF^{2+}$. This assumption seems unjustified as the equilibrium constant for the reaction $U^{4+} + F^- \rightleftharpoons UF^{3+}$ is larger than those for the reactions $Al^{3+} + F^- \rightleftharpoons AlF^{2+}$ and $AlF^{2+} + F^- \rightleftharpoons AlF_2^+$. This study was also discussed by Vdovenko and Romanov [67VDO/ROM] and is given zero weight in this review.

[63MEL/AMI]

Melton, S.L., Amis, E.S., A spectrophotometric study of the complexes formed between uranyl and thiosulfate ions in water and water-ethanol solvents, Anal. Chem., **25** (1963) 1626-1630.

This is a spectrophotometric study of the complexes formed in the dioxouranium(VI) thiosulphate system at 15, 20 and 25°C. The equilibria were studied in water and in 30, 60 and 90% ethanol-water mixtures. Two different constant ionic strengths were used: 0.0638 and 0.133 M. The data from the ethanol-water mixtures will not be discussed here. The data at low ionic strength have been interpreted by assuming the reaction

$$UO_2^{2+} + S_2O_3^{2-} \rightleftharpoons UO_2S_2O_3(aq). \qquad (A.45)$$

The 1:1 stoichiometry of the dioxouranium(VI) thiosulphate complex was confirmed by using Job's method of continuous variation [28JOB, 41VOS/COO]. This is not normally a satisfactory method to study stepwise complex formation, but the results are justified in this case. The authors reported $\log_{10}\beta_1(15°C) = (2.47 \pm 0.01)$, $\log_{10}\beta_1(20°C) = (2.29 \pm 0.03)$ and $\log_{10}\beta_1(25°C) = (2.04 \pm 0.07)$. They did not correct the constant to $I = 0$. Using the Debye-Hückel term ($8D = 0.29$ at $I = 0.0638$ M) for correction to $I = 0$, this review obtains $\log_{10}\beta_1^\circ(25°C) = (2.8 \pm 0.3)$, where the uncertainty is an estimate. From the temperature variation of $\log_{10}\beta_1$ the authors deduce $\Delta_r H_m(A.45) = -77.4\,kJ\cdot mol^{-1}$. This value is not credited here because of the small temperature range involved (15 to 25°C). The authors proposed the formation of a triuranyl complex $(UO_2)_3(S_2O_3)_2^{2+}$ at higher ionic strength. This composition is also deduced from a Job plot. This method is only satisfactory when one complex is formed. There is no reason why $UO_2S_2O_3(aq)$ should not be present at high ionic strength. The formation of $(UO_2)_3(S_2O_3)_2^{2+}$ has not been proven.

[63POZ/STE]

Pozharskii, B.G., Sterlingova, T.N., Petrova, A.E., Hydrolysis and complex formation of uranyl in mineral acid solutions, Russ. J. Inorg. Chem., 8 (1963) 831-839.

Pozharskii, Sterlingova, and Petrova studied the formation of dioxouranium(VI) sulphate complexes by using pH-measurements at 24.7°C. The ionic strength varies between 10^{-3} M and 3×10^{-2} M. Only four experiments were made in the sulphate system. There is no information available about the calibration procedure used by the authors and it is not clear whether the electrode was calibrated in activity or concentration units. The experimental data were interpreted in terms of the formation of UO_2OH^+, $UO_2OHSO_4^-$. The equilibrium constant for the hydrolysis reaction was determined in a separate experiment. Both the experimental data and the mathematical analysis are unsatisfactory. Other hydrolysis complexes should be formed, and there is no proof for the proposed $UO_2OHSO_4^-$ complex. The numerical value reported for the equilibrium constant for the reaction $UO_2^{2+} + SO_4^{2-} \rightleftharpoons UO_2SO_4(aq)$

is $\log_{10}\beta_1 = 3.8$ or $\log_{10}\beta_1^\circ \approx 4.4$. This is much higher than the values reported by other investigations. In view of the shortcomings of this study, the reported equilibrium constants were given zero weight.

[63RUS/JOH]
Rush, R.M., Johnson, J.S., Hydrolysis of uranium(VI): Absorption spectra of chloride and perchlorate solutions, J. Phys. Chem., **67** (1963) 821-825.

Only eight experimental potentiometric measurements were reported and, thus, only a low weight can be assigned to this work. In this review the uncertainties for $\log_{10}{}^*\beta_{2,2}$ and $\log_{10}{}^*\beta_{5,3}$ are estimated as ± 0.3 and ± 0.6, respectively. The results from the spectrophotometric study are considered to be less accurate and are not used in this review in the evaluation of the hydrolysis constants and interaction parameters.

[63VDO/ROM]
Vdovenko, V.M., Romanov, G.A., Shcherbakov, V.A., Hydrolysis of the ion U(IV), Sov. Radiochem, **5** (1963) 119-120.

The results (obtained using the spin-lattice relaxation time method) indicate the formation of an initial hydrolysis species followed by the formation of polymeric species at higher pH. No values are reported for the formation constants of the polymeric species. The data were reanalysed in Ref. [69VDO/STE].

[63VDO/ROM2]
Vdovenko, V.M., Romanov, G.A., Shcherbakov, V.A., Study of the complex-formation of uranium(IV) with the fluoride ion by the proton resonance method, Sov. Radiochem., **5** (1963) 538-541.

The authors measured the proton relaxation time, $1/T_1$, as a function of the total concentration of added fluoride in solutions of uranium(IV). The measurements were done at (20 ± 1)°C in a 2 M $HClO_4$ medium. The analysis of the experimental data is analogous to the conventional spectrophotometric methods and requires the determination of both the equilibrium constants and a set of relaxation parameters, one for each complex. The authors have not described the calculation procedure used in sufficient detail to make an estimate of the accuracy of their results. This review is not confident that the solution of the equation systems, as proposed by the authors, is a good method when strong complexes are formed. However, it is interesting to see that the authors found equilibrium constants for the reactions $U^{4+} + qF^- \rightleftharpoons UF_q^{4-q}$, $q = 1$ to 3, that are in fair agreement with other determinations for $q = 1$ and 2. β_1 is probably the most precise constant in the set. In view of the lack of computational details, this review has given zero weight to this study.

[63VDO/ROM3]
Vdovenko, V.M., Romanov, G.A., Shcherbakov, V.A., Proton resonance study of complex formation of U(IV) with halide, sulphate and perchlorate anions, Sov. Radiochem., **5** (1963) 624-627.

Vdovenko, Romanov and Shcherbakov studied the complex formation of U(IV) with Cl^-, Br^-, I^-, SO_4^{2-} and ClO_4^- by using a NMR-method (measurements of the proton relaxation time T_1 in U^{4+} as a function of the ligand concentration). The measurements were made at 20°C and in the presence of 2.00 M $HClO_4$ as a background electrolyte. However, the ionic strength varied considerably throughout the experiments, from 2.2 to 2.7 M for Cl^-, Br^- and I^-, and from 2.38 to 3.38 M for SO_4^{2-}. The authors reported the following stability constants:

$$\begin{array}{lcllcl} U^{4+} + Cl^- & \rightleftharpoons & UCl^{3+} & \beta_1 & = & 6 \pm 1.5 \\ U^{4+} + Br^- & \rightleftharpoons & UBr^{3+} & \beta_1 & = & 2 \pm 0.3 \\ U^{4+} + I^- & \rightleftharpoons & UI^{3+} & \beta_1 & = & 1.5 \pm 0.5 \\ U^{4+} + ClO_4^- & \rightleftharpoons & UClO_4^{3+} & \beta_1 & = & 0.14 \pm 0.02 \\ U^{4+} + SO_4^{2-} & \rightleftharpoons & USO_4^{2+} & \beta_1 & = & 50 \end{array}$$

The sulphate constant is in error because the authors did not take the protonation of SO_4^{2-} into account. However, even after correcting for the protonation equilibrium, the constant is considerably lower than other values reported in the literature, and the authors did not report any error estimate. This review considers the sulphate data unreliable. The other equilibrium constants seem more precise. The uncertainties given by the authors are probably underestimated due to the variations in the ionic strength. This review assumes that the constants refer to I = 2.5 M and doubles the above uncertainties given by the authors.

[64BAR/SOM]
Bartusek, M., Sommer, L., Über die Hydrolyse von UO_2^{2+} in verdünnten Lösungen, Z. Phys. Chem. Leipzig, 226 (1964) 309-332, in German.

The omission of $(UO_2)_3(OH)_5^+$ in the hydrolysis scheme may have affected the constants reported for $(UO_2)_2OH^{3+}$ and $(UO_2)_2(OH)_2^{2+}$. The values used in Baes and Mesmer's review [76BAE/MES] are based on a thesis [63BAR/SOM] (unavailable to the reviewers) and apparently the data given in the thesis are more extensive than in the paper [64BAR/SOM]. Nevertheless, in neither reference [64BAR/SOM], nor a later publication [67BAR] by Bartusek, is there reference to data obtained by Bartusek and Sommer in a nitrate medium. In this review only $\log_{10}{}^*\beta_{2,2} = 6.09$ from the potentiometric study (0.1 M $NaClO_4$) is accepted, and an uncertainty of ±0.4 in $\log_{10}{}^*\beta_{2,2}$ is assigned.

[64COR]
Cordfunke, E.H.P., Heats of formation of some hexavalent uranium compounds, J. Phys. Chem., 68 (1964) 3353-3356.

The author indicated that the uncertainties in his heat of solution measurements are 0.21 kJ·mol^{-1} (0.05 kcal·mol^{-1}). Several heats were reported to the nearest hundredth of a kilocalorie. For these, an uncertainty (95% confidence level) of $\pm 0.4\,kJ \cdot mol^{-1}$ appears to be reasonable. Other heats are only reported to the nearest tenth of a kilocalorie. In these cases, the uncertainties are estimated in this review to be $\pm 1.0\,kJ \cdot mol^{-1}$. As noted by Loopstra and Cordfunke [66LOO/COR], α-UO_3 has a disordered structure and samples may not have the exact stoichiometry $UO_{3.00}$. In view of the possible effects of this problem, the uncertainty in the enthalpy of formation of α-UO_3 is estimated in the present review as $\pm 3\,kJ \cdot mol^{-1}$.

The enthalpy of formation of δ-UO_3 is based on the heat of solution of a mixture of δ-UO_3 and α-UO_3, and the uncertainty in $\Delta_f H_m^\circ(\delta\text{-}UO_3)$ is estimated in this review as $\pm 5\,kJ\cdot mol^{-1}$. The reaction describing the oxidation of $UO_{2.92}$ in the aqueous 6 M HNO_3 was apparently not investigated, and the enthalpy of solution given for this oxide cannot be used to determine its enthalpy of formation.

[64GOR/KER]
Gordon, G., Kern, D.M.H., Observations on the complex between uranyl and chlorite ions, Inorg. Chem., **3** (1964) 1055-1056.

This is a spectrophotometric study made in a (Na,H)ClO_4 ionic medium with I = 1.0 M at 25°C. The authors give indications of the formation of a complex $UO_2ClO_2^+$, but it is not possible to make an estimate of the equilibrium constant from the measured absorptivity data. The authors claimed that the equilibrium constant for the reaction

$$UO_2^{2+} + ClO_2^- \rightleftharpoons UO_2ClO_2^+ \qquad (A.46)$$

is $\beta_1 > 0.02$ which is not very informative. They also claimed that at most 5% of the chlorite ions present are bonded to U(VI). This would mean $\beta_1 \approx 0.3$ or less, *i.e.*, the complexes are very weak.

[64MCK/WOO]
McKay, H.A.C., Woodhead, J.L., A spectrophotometric study of the nitrate complexes of uranium(IV), J. Chem. Soc., (1964) 717-723.

McKay and Woodhead determined the equilibrium constants in the uranium(IV) nitrate system at 20°C by using a spectrophotometric method. The experiments were performed at constant acidity (1 M $HClO_4$) and in different ionic media (1 M $HClO_4$ + 1 M $LiClO_4$, 1 M $HClO_4$ + 2 M $LiClO_4$ and 1 M $HClO_4$ + 3 M $LiClO_4$). Formation of the two nitrate complexes UNO_3^{3+} and $U(NO_3)_2^{2+}$ were well established. The evidence for the formation of higher complexes $U(NO_3)_3^+$ and $U(NO_3)_4(aq)$ is not conclusive. The authors thoroughly discussed the earlier study of Ermolaev and Krot [62ERM/KRO]. At very high acidities (6-14 M HNO_3) there is evidence for the formation of higher uranium(IV) nitrate complexes, possibly $U(NO_3)_6^{2-}$ (*cf.* Ryan [60RYA]). The study of McKay and Woodhead is precise, and the authors have used a proper method for data evaluation. The data at $I \leq 3.5$ m were used by this review along with those of Ermolaev and Krot [62ERM/KRO] to determine $\log_{10}\beta_1^\circ$, $\Delta\varepsilon_1$, $\log_{10}\beta_2^\circ$, and $\Delta\varepsilon_2$ for the formation of UNO_3^{3+} and $U(NO_3)_2^{2+}$, *cf.* Figs. V.13 and V.14.

The value of the hydrolysis constant $^*\beta_1$ is, within the experimental uncertainties, equal to the value obtained at 25°C by Kraus and Nelson in a similar medium. Rather than attempt to correct this single value from 20 to 25°C, the value from McKay and Woodhead [64MCK/WOO] is not used in this review in the evaluation of $^*\beta_1^\circ$ at 25°C.

[64VAS/MUK]

Vasil'ev, V.P., Mukhina, P.S., Equilibria of uranyl complexes in aqueous thiocyanate solutions, Izv. Vyssh. Ucheb. Zaved. Khim., 7 (1964) 711-714, in Russian, Chem. Abstr. 62 (1965) 7172c.

The original publication was not available to the reviewers, only the abstract. The authors used a spectrophotometric method with a blue filter to define the wavelength region. The concentration of KSCN varied from 0.1 to 3.8 M, which results in large changes in the ionic medium. The authors did not report the formation of $UO_2(SCN)_3^-$, which is a well-established species from the investigations of Ahrland [49AHR2] and Ahrland and Kullberg [71AHR/KUL3]. This might result in a small error in β_2. The equilibrium data of Vasil'ev and Mukhina [64VAS/MUK] agree well with the results of other investigations. The values reported for $I = 2.5$ M ($NaNO_3$) were used, after correction for the formation of $UO_2NO_3^+$ using the formation constant selected in this review, and to $I = 0$ using the specific ion interaction theory, to evalulate $\log_{10} \beta_1^\circ$ and $\log_{10} \beta_2^\circ$.

[65DRA/JUL]

Drăgulescu, C., Julean, I., Vîlceanu, N., Sur les pyrophosphates complexes d'uranyle, Rev. Roum. Chim., 10 (1965) 809-819, in French.

The authors reviewed some of the older literature and performed potentiometric, conductometric, spectrophotometric, radiometric and radioelectrophporesis measurements [65DRA/JUL, 70DRA/VIL]. Neither the experimental methods nor the interpretation of the measurements are satisfactory. No inert ionic medium was used, there is no information on electrode calibration, and the stoichiometry of the species are based on discontinuities in the various titration curves.

In the first study [65DRA/JUL] the authors proposed the formation of the species $(UO_2)_5P_2O_7^{6+}$, $(UO_2)_4P_2O_7^{4+}$, $(UO_2)_3P_2O_7^{2+}$, $UO_2P_2O_7^{2-}$, and $UO_2(P_2O_7)_2^{6-}$, but no attempt was made to obtain a quantitavive model. This review considers it highly unlikely that species of the type $(UO_2)_5P_2O_7^{6+}$ and $(UO_2)_4P_2O_7^{4+}$ are formed as they are not compatible with the known coordination geometry of U(VI) and $P_2O_7^{4-}$.

In the second paper [70DRA/VIL] the authors proposed the formation of protonated pyrophosphate complexes, but there is no support from the experiments.

There is no doubt that U(VI) pyrophosphate complexes are formed that probably also contain protons. However, the stoichiometries proposed in the two studies [65DRA/JUL, 70DRA/VIL] are not credited in this review.

[65ISR]

Israeli, Y.J., Le mécanisme d'hydrolyse de l'ion uranyle, Bull. Soc. Chim. France, (1965) 193-196, in French.

The values for the constants were based on a single titration at an unspecified temperature, and alternate hydrolysis schemes were not discussed. These constants were assigned a weight of zero in this review.

[65MUT]

Muto, T., Thermochemical stability of ningyoite, Mineral. J., 4 (1965) 245-274.

The author measured the solubility of four synthetic ningyoite samples. Their chemical analysis indicated a highly valiable composition with a Ca/U ratio between 1.17 and 1.60.

The solubility experiments were made at 25°C and 100°C in solutions of pH values in the range 0.92 to 1.50. The author made no correction for U(IV) hydrolysis and phosphate complexation when he calculated the solubility product. In view of these and the variable composition of the solid samples, the solubility product $\log_{10}{}^*K_{s,0}(298.15\text{ K}) = -17.1$, corresponding to the reaction

$$CaU(PO_4)_2 \cdot 2H_2O(cr) + 6H^+ \rightleftharpoons Ca^{2+} + U^{4+} + 2H_3PO_4(aq) + 2H_2O(l),$$

is not credited in this review.

Muto also reported solubility measurements at 25°C on natural autunite, $Ca(UO_2)_2(PO_4)_2{\cdot}10H_2O(cr)$, as well as on a synthetic hydrogen autunite, $H_2(UO_2)_2(PO_4)_2{\cdot}10H_2O(cr)$, in hydrochloric acid solutions of a pH between 1.77 and 4.78. These data are not considered reliable by this review for the following reasons: The composition of the solid phases was not well established; the presence of $10H_2O$ in the formula is based on an assumption without analytical support; the composition of the hydrogen autunite sample changed during the equilibration with the aqueous phase. Furthermore, no correction of the measured solubilities for the formation of U(VI) phosphate complexes was made. Hence the reported solubility products, $\log_{10}{}^*K_{s,0}(298.15\text{ K}) = -7.4$ for the reaction

$$Ca(UO_2)_2(PO_4)_2{\cdot}10H_2O(cr) + 6H^+ \rightleftharpoons Ca^{2+} + 2UO_2^{2+} + 2H_3PO_4(aq) + 10H_2O(l),$$

and $\log_{10}{}^*K_{s,0}(298.15\text{ K}) = -10.3$ for the reaction

$$H_2(UO_2)_2(PO_4)_2 \cdot 10H_2O(cr) + 4H^+ \rightleftharpoons 2UO_2^{2+} + 2H_3PO_4(aq) + 10H_2O(l),$$

are not credited by this review, nor are the Gibbs energies of formation of the compounds $Ca(UO_2)_2(PO_4)_2 \cdot 10H_2O(cr)$ and $H_2(UO_2)_2(PO_4)_2 \cdot 10H_2O(cr)$ derived from these constants.

[65SHI/MAR]

Shiloh, M., Marcus, Y., A spectrophotometric study of trivalent actinide complexes in solution: I. Uranium, Israel J. Chem., 3 (1965) 123-131.

Shiloh and Marcus studied the spectra of uranium(III) in solutions of sulphuric acid and lithium chloride or lithium bromide. There is evidence for the formation of complexes, and the authors reported $\beta_1 = (1.3 \pm 0.3) \times 10^{-3}\text{M}^{-1}$ and $\beta_2 = (1.1 \pm 0.2) \times 10^{-4}\text{M}^{-1}$ for UCl^{2+} and UBr^{2+}, respectively. The complexes are thus very weak, and noticeable changes in the spectra occur only at ligand concentrations larger than 6 M. The constants reported by the authors refer to 25°C and solutions of varying and very high ionic strength. Hence, it is not possible to estimate equilibrium constants at $I = 0$. However, the data clearly indicate that complex formation between U^{3+} and Cl^- or Br^- is negligible except in brine systems. It should also be mentioned that no quantitative estimate of the possible formation of U(III) sulphate complexes was made.

[65VES/PEK]
Veselý, V., Pekárek, V., Abbrent, M., A study on uranyl phosphates: III. Solubility products of uranyl hydrogen phosphate, uranyl orthophosphate and some alkali uranyl phosphates, J. Inorg. Nucl. Chem., 27 (1965) 1159-1166.

This is a careful solubility study performed with properly characterized solid phases. The measurements have been carried out at (20 ± 1)°C in acid $NaNO_3$ media. The main solubility reaction of uranyl hydrogen phosphate is

$$UO_2HPO_4 \cdot 4H_2O(cr) + 2H^+ \rightleftharpoons UO_2^{2+} + H_3PO_4(aq) + 4H_2O(l). \quad (A.47)$$

However, a part of the dissolved uranium is present as phosphate complexes, while a part of the dissolved phosphate is present as $H_2PO_4^-$ and $H_3PO_4(aq)$. For the calculation of $\log_{10}{}^*K_s(A.47)$, the authors corrected their experimental total concentrations of U(VI) and phosphate for the presence of these species, using the data of Baes [56BAE] that refer to $I = 1$ M and $t = 25$°C, rather than to the experimental conditions used by Veselý, Pekárek and Abrent. A recalculation of these data using the equilibrium constants selected in this review for the formation of $UO_2H_2PO_4^+$ and $UO_2H_3PO_4^{2+}$ and the dissociation constants of phosphoric acid, corrected to the proper ionic strength ($I = 0.32$ M), yields $\log_{10}{}^*K_s(A.47) = -(2.14 \pm 0.05)$. Correction to $I = 0$ using the ion interaction coefficients listed in the tables of Appendix B, results in $\log_{10}{}^*K_s^\circ(A.47) = -(2.42 \pm 0.05)$. This corresponds to $\log_{10} K_s(A.48) = -(11.77 \pm 0.06)$ and $\log_{10} K_s(A.49) = -(24.12 \pm 0.07)$.

$$UO_2HPO_4 \cdot 4H_2O(cr) \rightleftharpoons UO_2^{2+} + HPO_4^{2-} + 4H_2O(l) \quad (A.48)$$

$$UO_2HPO_4 \cdot 4H_2O(cr) \rightleftharpoons UO_2^{2+} + H^+ + PO_4^{3-} + 4H_2O(l) \quad (A.49)$$

The value of $\log_{10}{}^*K_s(A.47)$ differs somewhat from $\log_{10}{}^*K_s(A.47) = -2.59$, recalculated here from the data of Moskvin *et al.* [67MOS/SHE]. The value of the derived constant, $\log_{10} K_s(A.49)$, differs significantly from $\log_{10} K_s(A.49) = -25.0$ proposed by Tripathi [84TRI], who has used equilibrium data referring to a different ionic strength.

Veselý, Pekárek and Abrent [65VES/PEK] have also determined the solubility product of $(UO_2)_3(PO_4)_2 \cdot 4H_2O(cr)$, according to the reaction

$$(UO_2)_3(PO_4)_2 \cdot 4H_2O(cr) + 6H^+ \rightleftharpoons 3UO_2^{2+} + 2H_3PO_4(aq) + 4H_2O(l). \quad (A.50)$$

Their data were recalculated by this review using the equilibrium constants selected by this review (as above). The equilibrium constant calculated from the second experimental point [65VES/PEK, Table 2] was more than a factor of ten larger than the others and was therefore omitted. The average constant calculated from the remaining four points is $\log_{10}{}^*K_s(A.50) = -(5.13 \pm 0.30)$. Correction to $I = 0$

using the specific ion interaction theory results in $\log_{10}{}^{*}K^{\circ}_{s}(\text{A.50}) = -(5.96 \pm 0.30)$. $\log_{10} K^{\circ}_{s}(\text{A.51}) = -(49.36 \pm 0.30)$ is the derived constant for the reaction

$$(UO_2)_3(PO_4)_2 \cdot 4H_2O(cr) \rightleftharpoons 3UO_2^{2+} + 2PO_4^{3-} + 4H_2O(l) \qquad (A.51)$$

and differs somewhat from the one proposed by Veselý, Pekárek and Abrent [65VES/PEK] and by Tripathi [84TRI].

The solubility constants for the reactions

$$MUO_2PO_4 \cdot xH_2O(cr) + 3H^+ \rightleftharpoons M^+ + UO_2^{2+} + H_3PO_4(aq) + xH_2O(l) \quad (A.52)$$

were recalculated for $M^+ = Na^+$, K^+, Rb^+, Cs^+ and NH_4^+, using equilibrium constants corrected to $I = 0.2$ M. This review obtains $\log_{10}{}^{*}K_s(\text{A.52}, M = Na) = -(1.66 \pm 0.09)$, $\log_{10}{}^{*}K_s(\text{A.52}, M = K) = -(3.00 \pm 0.15)$, $\log_{10}{}^{*}K_s(\text{A.52}, M = Rb) = -(3.23 \pm 0.20)$, $\log_{10}{}^{*}K_s(\text{A.52}, M = Cs) = -(2.90 \pm 0.20)$, and $\log_{10}{}^{*}K_s(\text{A.52}, M = NH_4) = -(3.82 \pm 0.10)$. There is a systematic drift in some of the solubility constants which might indicate lack of equilibrium or some other systematic error. The corresponding constants at $I = 0$ are $\log_{10}{}^{*}K^{\circ}_{s}(\text{A.52}, M = Na) = -(1.66 \pm 0.09)$, $\log_{10}{}^{*}K^{\circ}_{s}(\text{A.52}, M = K) = -(3.00 \pm 0.15)$, $\log_{10}{}^{*}K^{\circ}_{s}(\text{A.52}, M = Rb) = -(3.23 \pm 0.20)$, $\log_{10}{}^{*}K^{\circ}_{s}(\text{A.52}, M = Cs) = -(2.90 \pm 0.20)$, and $\log_{10}{}^{*}K^{\circ}_{s}(\text{A.52}, M = NH_4) = -(3.82 \pm 0.10)$.

[65VID/BYA]

Vidavskii, L.M., Byakhova, N.I., Ippolitova, E.A., Enthalpy of the reaction of γ-UO_3 with hydrofluoric acid and enthalpy of formation of γ-UO_3, Russ. J. Inorg. Chem., **10** (1965) 953-954.

See comments under Santalova *et al.* [71SAN/VID].

[66COR]

Cordfunke, E.H.P., Thermodynamic properties of hexavalent uranium compounds, in: Thermodynamics, Vol. **2**, Proc. Symp. held 22-27 July, 1965, in Vienna, Austria, Vienna: International Atomic Energy Agency, 1966, *pp.*483-495.

Water vapour pressures were measured for the two following equilibria:

$$UO_2(NO_3)_2 \cdot 6H_2O(cr) \rightleftharpoons UO_2(NO_3)_2 \cdot 3H_2O(cr) + 3H_2O(g) \qquad (A.53)$$
$$UO_2(NO_3)_2 \cdot 3H_2O(cr) \rightleftharpoons UO_2(NO_3)_2 \cdot 2H_2O(cr) + H_2O(g). \qquad (A.54)$$

It is apparent from the scatter in the data that the uncertainties in the measured values rose with increasing temperature, at least in part because of increasing content of nitrogen oxides in the gas phase. In this review the uncertainties in the vapour pressures are estimated as 40 Pa for measurements at temperatures less than or equal to 30°C, 60 Pa for measurements between 30 and 50°C and 100 Pa for measurements at 50°C or above. Values for the enthalpy and entropy of reaction, as calculated from the variation of p_{H_2O} with temperature, are strongly correlated:

$$\begin{aligned} p_{H_2O} = \; & \exp\{-\tfrac{1}{3RT}(\Delta_r H^{\circ}_m(298.15\,K) - T\Delta_r S^{\circ}_m(298.15\,K) \\ & +(T - 298.15)\Delta_r C_{p,m} - T\Delta_r C_{p,m}\ln(T/298.15))\}. \end{aligned} \qquad (A.55)$$

A similar correlation is also found in values calculated from the water vapour pressures for dehydration of $UO_2Cl_2 \cdot 3H_2O(cr)$ to the monohydrate (see Section V.4.2.1.3.h), and for that reason the vapour pressure measurements are not used in this review in the calculation of the enthalpy of formation of either of the chloride hydrates.

[66DRO/KOL]
Drobnic, M., Kolar, D., Calorimetric determination of enthalpy of hydration of UO_3, J. Inorg. Nucl. Chem., 28 (1966) 2833-2835.

The enthalpy of hydration of amorphous UO_3 is reported as a function of the temperature at which the sample was treated. The authors selected the maximum heat per mole of UO_3 as the true enthalpy of hydration. Results for samples prepared at lower temperatures were discounted by assuming the samples were not completely dehydrated. Results above 350°C were discounted because of possible decomposition of the amorphous solid. In this review, the uncertainty (95% confidence level) in the enthalpy of hydration from reference [66DRO/KOL] is estimated as twice the difference in the enthalpies of hydration found for samples treated at 300°C and at 350°C. This is markedly larger than the authors' estimate of 2% precision for their measurements.

[66KOC/SCH]
Koch, G., Schwind, E., Extraktion von Uran(IV)-Nitrat durch Tricaprylmethylammoniumnitrat, J. Inorg. Nucl. Chem., 28 (1966) 571-575, in German.

This is an extraction study of U(IV) from HNO_3 solutions of varying concentrations. The distribution curve of U(IV) has a maximum around 3 M HNO_3 indicating a zero average change of the complexes in the aqueous phase. The organic phase contains the diamine salt of $U(NO_3)_6^{2-}$ throughout the investigated concentration range. This study gives a qualitative confirmation of Ref. [61WIL/KED].

[66SHI/NAZ]
Shilin, I.V., Nazarov, V.K., Complex formation of neptunium(IV) with nitrate and chloride ions, Sov. Radiochem., 8 (1966) 474-478.

This paper contains an experimental study of the complex formation between Np(IV) and Cl^- and NO_3^-. The authors also described the extrapolation procedure (Vasil'ev [62VAS]) used to obtain the equilibrium constant at $I = 0$. They made these extrapolations also for Pu(IV) and U(IV). Vasil'ev's equation is identical with the specific ion interaction theory approach except that the factor $(1 + 1.5\sqrt{I})$ in the denominator of the Debye-Hückel term of the specific ion interaction theory is replaced by $(1 + 1.6\sqrt{I})$ in Vasil'ev's equation. The authors reported $\log_{10} \beta_1^\circ = 1.55$ for the formation of UNO_3^{3+}. The value of b was the same for Th(IV), U(IV) and Pu(IV) and equal to 0.19. This value differs from the value obtained by using the specific ion interaction theory, because the specific ion interaction theory uses molal concentration units while Vasil'ev's equation uses molar units. The UNO_3^{3+} data are based on the experiments of Ermolaev and Krot [62ERM/KRO]. This review prefers to use the original data reported by Ermolaev and Krot in the specific ion interaction extrapolation to $I = 0$, and the value of $\log_{10} \beta_1^\circ = 1.55$ reported by Shilin and

Nazarov [66SHI/NAZ] is therefore not considered, even though it is consistent with the resulting value of $\log_{10}\beta_1^\circ = (1.47 \pm 0.13)$selected in this review.

[66VDO/ROM]

Vdovenko, V.M., Romanov, G.A., Shcherbakov, V.A., A study of fluoro-complexes of uranium(IV) in aluminium salt solutions, Russ. J. Inorg. Chem., 11 (1966) 139-141.

The equilibrium constant of the reaction $UF^{3+} + Al^{3+} \rightleftharpoons U^{4+} + AlF^{2+}$ was studied at 20°C by using spectrophotometry. The ionic medium varied, but an estimate of the ratio $\beta_1(UF^{3+})/\beta_1(AlF^{2+}) = 440$ can be made. From this ratio the authors obtain $\log_{10}\beta_1(UF^{3+}) = 8.8$, valid at $I \approx 0.5$ M. This value corresponds to $\log_{10}\beta_1^\circ(UF^{3+}) = 10.2$, which is significantly higher than the value selected in this review, $\log_{10}\beta_1^\circ = (9.25 \pm 0.11)$. The equilibrium constant proposed in Ref. [66VDO/ROM] is less precise than the values published in other papers (see Section V.4.1.2.3), mainly because of the varying ionic strength. Therefore the value of $\log_{10}\beta_1$ in Ref. [66VDO/ROM] is given zero weight.

[67COL/EYR]

Cole, D.L. Eyring, E.M., Rampton, D.T., Silzars, A., Jensen, R.P., Rapid reaction rates in a uranyl ion hydrolysis equilibrium, J. Phys. Chem., 71 (1967) 2771-2775.

Baes and Mesmer [76BAE/MES] suggest the data are consistent with $\log_{10}{}^*\beta_1 \leq -5.3$. The data are quite scattered and could probably be considered consistent with values in the range $-5.0 > \log_{10}{}^*\beta_1 > -6.0$.

[67GRY/KOR]

Gryzin, Yu.I., Koryttsev, K.Z., A study of the behaviour of UO_3 and its hydrates in solutions with an electrode of the third kind, Russ. J. Inorg. Chem., 12 (1967) 50-53.

As pointed out by O'Hare *et al.* [88OHA/LEW], the fact that the temperature dependence of the potentials infers a negative entropy for the solid uranium trioxide monohydrate suggests a serious problem in the results of this paper.

[67MER/SKO]

Merkusheva, S.A., Skorik, N.A., Kumok, V.N., Serebrennikov, V.V., Thorium and uranium(IV) pyrophosphates, Sov. Radiochem., 9 (1967) 683-685.

Merkusheva *et al.* measured the solubility of $UP_2O_7 \cdot 20H_2O(cr)$ (and also those of $ThP_2O_7 \cdot 4H_2O(cr)$ and $CeP_2O_7 \cdot 9H_2O(cr)$) in $HClO_4/NaClO_4$ solutions of $I =$ 0.1 M ionic strength and at (25.00 ± 0.05)°C. The uranium solid phase used for the measurements was prepared by the authors by precipitation from an oversaturated solution of $U(SO_4)_2$ and the calculated amount of a 2.5% aqueous solution of $K_4P_2O_7$. The precipitate was then dried in air. The analysis yielded a proportion of 20 H_2O per uranium(IV).

The authors reported the solubility of uranium(IV) pyrophosphate at ten different pH values, from 1.00 to 2.68. They verified the absence of dioxouranium(VI) in solution. In order to avoid oxidation of uranium(IV) to dioxouranium(VI) by atmoshperic oxygen, they carried out the experiment in a nitrogen atmosphere. The solubility measurements for uranium(IV) pyrophosphate were interpreted as the sum of two species, U^{4+} and $UP_2O_7(aq)$, according to Eqs. (A.56) and (A.57).

$$UP_2O_7(cr) \rightleftharpoons U^{4+} + P_2O_7^{4-} \quad (A.56)$$

$$U^{4+} + P_2O_7^{4-} \rightleftharpoons UP_2O_7(aq) \quad (A.57)$$

Using the solubility measurements above and the protonation constants of pyrophosphate determined at the same ionic strength [66MIT/MAL], the authors calculated $K_{s,0}(A.56) = (1.35 \pm 0.32) \times 10^{-24}$, and $\beta_1(A.57) = (1.17 \pm 0.34) \times 10^{19}$. These constants are not considered reliable by this review for three reasons:

- The authors did not provide any X-ray data to prove that the solid phase was effectively $UP_2O_7 \cdot 20H_2O(cr)$ before and after the experiments.

- The authors suggested the predominance of the neutral species $UP_2O_7(aq)$, probably because the solubility of uranium(IV) pyrophosphate is apparently independent of pH in the pH range from 1.00 to 2.68. This argument is logical but incomplete. Although there is no reliable analytical study (potentiometry, spectroscopy, *etc.*) on aqueous uranium(IV) pyrophosphate, it is probable, by similarity with phosphate, that pyrophosphate ligands form charged and uncharged uranium(IV) complexes other than the $UP_2O_7(aq)$.

[67MOS/ESS]
Moskvin, A.I., Essen, L.N., Bukhtiyarova, T.N., The formation of thorium(IV) and uranium(IV) complexes in phosphate solutions, Russ. J. Inorg. Chem., **12** (1967) 1794-1795.

The authors reported experimental solubility data of $Th(HPO_4)_2(s)$ in 0.35 M $HClO_4$ solutions with varying concentrations of phosphoric acid (10^{-4} to 1 M H_3PO_4). The solubilities were small, hence the hydrogen ion concentration remained essentially constant. The authors used earlier experimental data of Moiseev [64MAR/VER] (which are not available to this review) to calculate equilibrium constants for the corresponding U(IV) phosphate complexes. It is assumed that the conditions were the same as for the Th(IV) system. The Th(IV) data indicate clearly that a limiting complex with four phosphate groups per Th(IV) is formed. As the experiments were performed at constant hydrogen ion concentration, it is not possible to determine the composition of the ligand. It could be H_3PO_4, $H_2PO_4^-$, HPO_4^{2-}, or a mixture of them. However, Zebroski *et al.* [51ZEB/ALT] showed that complexes with both H_3PO_4 and $H_2PO_4^-$ are formed for Th(IV) at $[H^+] = 0.35$ M. Additional data at different hydrogen ion concentrations are necessary. The data of Moskvin *et al.* [67MOS/ESS] clearly showed that complexes of the form $UH_r(PO_4)_q^{4-3q+r}$, $q = 1$ to 4, were formed. As the value of q of these complexes is unknown, no stability constants can be credited. The constants given in Ref. [67MOS/ESS] are thus conditional equilibrium constants valid only in 0.35 M $HClO_4$.

These constants can be used to correct the measured solubilities for complex formation, which in turn makes it possible to determine a correct value of the solubility product of $U(HPO_4)_2(s)$. This review recalculates the solubility product given in Ref. [67MOS/ESS] to refer to the reaction

$$U(HPO_4)_2(s) + 4H^+ \rightleftharpoons U^{4+} + 2H_3PO_4(aq), \quad (A.58)$$

yielding $\log_{10}{}^*K_s(A.58) = -(9.96 \pm 0.15)$. This value is somewhat affected by hydrolysis of U(IV). Hence, this review assigns a fairly large uncertainty. Correction of $\log_{10}{}^*K_s(A.58)$ to $I = 0$ results in $\log_{10}{}^*K_s^\circ(A.58) = -(11.79 \pm 0.15)$.

By using the dissociation constants of phosphoric acid selected by this review, $\log_{10} K_s^\circ(A.59) = -(30.49 \pm 0.16)$ and $\log_{10}{}^*K_s^\circ(A.60) = -(55.19 \pm 0.17)$ are obtained.

$$U(HPO_4)_2(s) \rightleftharpoons U^{4+} + 2HPO_4^{2-} \quad (A.59)$$

$$U(HPO_4)_2(s) \rightleftharpoons U^{4+} + 2PO_4^{3-} + 2H^+ \quad (A.60)$$

$$(A.61)$$

[67MOS/SHE]

Moskvin, A.I., Shelyakina, A.M., Perminov, P.S., Solubility product of uranyl phosphate and the composition and dissociation constants of uranyl phosphato-complexes, Russ. J. Inorg. Chem., 12 (1967) 1756-1760.

The solubility product of $UO_2HPO_4 \cdot 4H_2O(cr)$ was determined in solutions of different ionic strength and at different temperatures. The solid phase is reasonably well characterized.

Only the solubility data in HNO_3 [67MOS/SHE, Table 2] are considered reliable by this review. However, they are reanalyzed here because the authors have not taken the complex formation between UO_2^{2+} and $H_2PO_4^-/H_3PO_4$ into account, nor have they made corrections for the medium dependence of the dissociation constants of phosphoric acid. The following procedure is used by this review:

- An approximate ionic strength was calculated by assuming that the predominant uranium species are UO_2^{2+} and $UO_2H_3PO_4^{2+}$. The free hydrogen ion concentration was calculated from the mass balance.

- The concentration constants for the equilibria $H_3PO_4(aq) \rightleftharpoons H^+ + H_2PO_4^-$, $UO_2^{2+} + H_3PO_4(aq) \rightleftharpoons UO_2H_3PO_4^{2+}$ and $UO_2^{2+} + H_3PO_4(aq) \rightleftharpoons UO_2H_2PO_4^+ + H^+$ were calculated from the values selected in this review, using the specific ion interaction theory. The ion interaction coefficients used were for the nitrate medium, hence no corrections were made for the formation of nitrate complexes of UO_2^{2+}.

- The concentrations of UO_2^{2+}, $UO_2H_3PO_4^{2+}$, $UO_2H_2PO_4^+$, $H_3PO_4(aq)$ and H^+ were calculated from the known equilibrium conditions and the mass balance equations. Then the concentration constants for the reaction

  $$UO_2HPO_4 \cdot 4H_2O(cr) + 2H^+ \rightleftharpoons UO_2^{2+} + H_3PO_4(aq) + 4H_2O(l) \quad (A.62)$$

  was calculated for the points 1 to 7 [67MOS/SHE, Table 2]: $\log_{10}{}^*K_s(A.62) = -2.37, -2.29, -2.30, -2.36, -2.39, -2.43$, and -2.54, respectively.

- The extrapolation to $I = 0$ using the specific ion interaction theory yields $\log_{10}{}^*K_s^\circ(A.62) = -(2.59 \pm 0.05)$ and $\Delta\varepsilon(A.62) = (0.13 \pm 0.03)$. The value of $\Delta\varepsilon$ agrees well with the value calculated from the values of Appendix B, (0.10 ± 0.04).

From the value of $\log_{10}{}^*K_s^\circ(\text{A.62}) = -(2.59 \pm 0.05)$ and the selected dissociation constants of phosphoric acid this review obtains $\log_{10} K_s^\circ(\text{A.63}) = -(11.94 \pm 0.06)$ and $\log_{10} K_s^\circ(\text{A.64}) = -(24.29 \pm 0.07)$ for the reactions

$$UO_2HPO_4 \cdot 4H_2O(cr) \rightleftharpoons UO_2^{2+} + HPO_4^{2-} + 4H_2O(l) \quad (A.63)$$
$$UO_2HPO_4 \cdot 4H_2O(cr) \rightleftharpoons UO_2^{2+} + H^+ + PO_4^{3-} + 4H_2O(l). \quad (A.64)$$

These values differ somewhat from those selected by Tripathi [84TRI].

Moskvin *et al.* have studied the temperature variation of the solubility product. However, these data are not sufficiently precise for the determination of enthalpy and entropy changes.

The numerical value of the solubility product is not strongly dependent on the model used for the soluble dioxouranium(VI) phosphate complexes. By assuming that only $UO_2H_2PO_4^+$ is formed with $\log_{10}{}^*K_1^\circ = 1.45$, values of $\log_{10}{}^*K_s^\circ(\text{A.62}) = -(2.63 \pm 0.05)$ and $\Delta\varepsilon(\text{A.62}) = (0.17 \pm 0.03)$ are obtained.

The composition and stability constants of the U(VI) phosphate complexes given by Moskvin *et al.* are erroneous, as pointed out, *e.g.*, by Tripathi [84TRI]. The experimental data given in [67MOS/SHE, Table 5] were reinterpreted by this review. They indicate that the solubility reaction is

$$UO_2HPO_4 \cdot 4H_2O(cr) + H_3PO_4(aq) \rightleftharpoons UO_2(H_rPO_4)_2(aq). \quad (A.65)$$

The solubility data were measured at constant acidity (0.5 M HNO_3), hence, it is not possible to determine the proton content in the complex. The conditional equilibrium constant calculated by using the experimental points 2 to 5 was $\log_{10}{}^*K_s(\text{A.65}) = -(0.75 \pm 0.07)$. By combination with Eq. (A.62), a value of $\log_{10}{}^*\beta_2(\text{A.66}) = (1.54 \pm 0.09)$ at $I = 0.5$ M is obtained for the equilibrium

$$UO_2^{2+} + 2H_3PO_4(aq) \rightleftharpoons UO_2(H_rPO_4)_2(aq). \quad (A.66)$$

This value is in fair agreement with the conditional constants found by Thamer [57THA] and Baes [56BAE], $\log_{10}{}^*\beta_2(\text{A.66}) = (1.36 \pm 0.09)$ and $\log_{10}{}^*\beta_2(\text{A.66}) = 1.57$, respectively.

The data in Ref. [67MOS/SHE] confirm the formation of $UO_2(H_2PO_4)_2(aq)$ and $UO_2(H_2PO_4)(H_3PO_4)^+$ rather than $UO_2HPO_4(aq)$ and $UO_2(HPO_4)_2^{2-}$.

[67OHA/MOR]
Ohashi, H., Morozumi, T., Electrometric determination of stability constants of uranyl chloride and uranyl nitrate complexes with pCl-Stat, J. At. Energy Soc. Japan, **9(2)** (1967) 65-71, in Japanese, Chem. Abstr. **67** (1967) 111876.

Ohashi and Morozumi [67OHA/MOR] determined the equilibrium constants of UO_2Cl^+ and $UO_2NO_3^+$ in aqueous solutions to KNO_3 and $NaNO_3/NaClO_4$ containing minor amounts of Cl^-, at 25°C. The stability constant $\beta_1(UO_2Cl^+)$ at various ionic strengths was estimated from the amount of UO_2Cl^+ ion obtained by integration of the current-time curve of the pCl, after the addition of a given amount of dioxouranium(VI) salt. The authors reported $\beta_1 = 2.42$, 1.40 and 0.89 at $I = 0.54$, 0.82 and 1.06 M, respectively.

The stability constant of $UO_2NO_3^+$ was determined by measuring the variation of $\beta_1(UO_2Cl^+)$ in mixed solutions of $NaNO_3$ and $NaClO_4$. The authors reported $\beta_1(UO_2NO_3^+) = 0.37$, 0.20 and 0.19 at $I = 0.54$, 0.82 and 1.06 M, respectively.

It is remarkable that the authors used a free chloride concentration of only 1.1×10^{-3} M. Very little UO_2Cl^+ is present in such solutions. The experimental procedure seems questionable and the chloride data are therefore given zero weight. The chloride data are also significantly different from the results found in other investigations.

It must be very difficult to determine the equilibrium constants of the much weaker nitrate complexes using this method (essentially ligand completion). The nitrate data are also given zero weight.

[67OHA/MOR2]
Ohashi, H., Morozumi, T., Temperature dependence of the stability constants of uranyl-chloro complexes, J. At. Energy Soc. Japan, 9(4) (1967) 200-201, in Japanese.

The authors used the same experimental method as used by Ohashi and Morozumi [67OHA/MOR]. They reported equilibrium constants for the reaction $UO_2^{2+} + Cl^- \rightleftharpoons UO_2Cl^+$ determined at a constant free ligand concentration of 10^{-3} M. The ionic strength varied from 0.05 to 0.5 M (KNO_3 as ionic medium). Measurements were made at 15 and 35°C. By extrapolation to $I = 0$ the authors found $\beta_1^\circ(UO_2Cl^+) =$ 19.6, 15.4, and 13.0 M^{-1} at 15, 25, and 35°C, respectively. This value is much higher than those given by other authors. This review is not confident that the experimental approach used by the authors is satisfactory. Only one free ligand concentration was investigated, and the concentration was very low (10^{-3} M). This means that only about 1% of U(VI) was present as UO_2Cl^+. The equilibrium data and the derived enthalpy data are given zero weight.

[67RYZ/NAU]
Ryzhenko, B.N., Naumov, G.B., Goglev, V.S., Hydrolysis of uranyl ions at elevated temperatures, Geochem. Internat., 4 (1967) 363-367.

The results of this conductance study provide values for the formation constant of $(UO_2)_2(OH)_2^{2+}$ at 100, 150 and 200°C. Except at 100°C the results are in good agreement with the results of Lietzke and Stoughton. The value of $\log_{10}{}^*\beta_{2,2}$ at 100°C is more negative by 0.6 than the value obtained by Lietzke and Stoughton, and by 0.4 than the value found by Baes and Meyer [62BAE/MEY]. The primary difficulty is the use of $Ba(NO_3)_2$ as a model salt for both fully dissociated $UO_2(NO_3)_2$ and $(UO_2)_2(OH)_2)(NO_3)_2$. Ryzhenko *et al.* [67RYZ/NAU] felt their results confirm those of Lietzke and Stoughton [324], but preferred to use the fitted values of the latter study. In this review an uncertainty of ± 2.0 in $\log_{10}{}^*\beta_{2,2}$ is estimated for the values from the conductance study.

[67SOL/TSV]

Solovkin, A.S., Tsvetkova, Z.N., Ivantsov, A.I., Thermodynamic constants for complex formation by U^{4+} with OH^- ions and by Zr^{4+} with OH^- and NO_3^- ions, Russ. J. Inorg. Chem., **12** (1967) 326-330.

This is a reanalysis of the data in Ref. [50KRA/NEL] but no new data were provided.

[67VDO/ROM]

Vdovenko, V.M., Romanov, G.A., Solubility of uranium tetrafluoride in solutions of aluminium salts, Sov. Radiochem., **9** (1967) 375-377.

This is a spectrophotometric study in which the authors point out that UF_2^{2+} is formed when Al^{3+} is added to a suspension of $UF_4 \cdot 2.5H_2O(cr)$ in water. This proves that the assumption made by Luk'yanychev and Nikolaev [63LUK/NIK2] when determining the solubility product of $UF_4 \cdot 2.5H_2O(cr)$ was in error.

[67WAL]

Wallace, R.M., Determination of stability constants by Donnan membrane equilibrium: The uranyl sulphate complexes, J. Phys. Chem., **71** (1967) 1271-1276.

Wallace studied the formation of dioxouranium(VI) sulphate complexes according to the reactions

$$UO_2^{2+} + SO_4^{2-} \rightleftharpoons UO_2SO_4(aq) \quad (A.67)$$
$$UO_2^{2+} + 2SO_4^{2-} \rightleftharpoons UO_2(SO_4)_2^{2-}. \quad (A.68)$$

He measured the distribution of UO_2^{2+} between complexing and non-complexing solutions separated by a permselective membrane. Measurements were made at 10, 25, and 40°C in solutions where the ionic strength varied from 0.010 to 0.150 M using NH_4ClO_4. The author reports concentration constants and also recalculated these to $I = 0$ by using a Debye-Hückel equation for the estimation of the activity coefficients. This is a very reliable study. However, this review has preferred to use the specific ion interaction theory with $\Delta\varepsilon(A.67) = -(0.34 \pm 0.07)$ for the correction to $I = 0$. The uncertainties of the constant $\log_{10}\beta_1$ at $I \neq 0$ were increased to ± 0.05 logarithmic units. The difference between Wallace's value and the one selected in this review is very small (< 0.05 logarithmic units). Wallace's value for the formation constant of $UO_2(SO_4)_2^{2-}$ was assigned an uncertainty of ± 0.10 logarithmic units and corrected to $I = 0$ by estimating $\varepsilon_{(UO_2(SO_4)_2^{2-},NH_4^+)} \approx \varepsilon_{(SO_4^{2-},NH_4^+)} \approx \varepsilon_{(SO_4^{2-},Na^+)} = -(0.12 \pm 0.06)$.

The enthalpies of reaction are $\Delta_r H_m(A.67) = (21 \pm 2)\,kJ \cdot mol^{-1}$ and $\Delta_r H_m(A.68) = (29 \pm 4)\,kJ \cdot mol^{-1}$, respectively, as evaluated from the temperature dependence of the corresponding equilibrium constants. These data are in good agreement with the calorimetric data of Ahrland and Kullberg [71AHR/KUL3] and Bailey and Larson [71BAI/LAR].

[67ZAK/ORL]
Zakharova, F.A., Orlova, M.M., The formation of uranyl sulphito-complexes, Russ. J. Inorg. Chem., **12** (1967) 1596-1599.

See comments under Klygin and Kolyada 73sta [59KLY/KOL].

[68ARN/SCH]
Arnek, R., Schlyter, K., Thermochemical studies of hydrolytic reactions: 8. A recalculation of calorimetric data on uranyl hydrolysis, Acta Chem. Scand., **22** (1968) 1331-1333.

Formation constants have been reported only for $(UO_2)_2(OH)_2^{2+}$ and $(UO_2)_3(OH)_5^+$. The data are insufficient (in quantity and accuracy) to evaluate hydrolysis constants for the minor species. The reported standard deviations only reflect the uncertainties in the curve-fitting procedure and this review estimates $\log_{10}{}^*\beta_{2,2} = 0.15$ and $\log_{10}{}^*\beta_{5,3} = 0.3$. This is an important study as it is one of the few calorimetric studies on dioxouranium(VI) complexes. The uncertainties in the reaction enthalpies have been accepted as reported. The uncertainties in the reaction entropies also reflect the estimated uncertainties in the equilibrium constants.

[68COR/PRI]
Cordfunke, E.H.P., Prins, G., van Vlaanderen, P., Preparation and properties of the violet "U_3O_8 hydrate", J. Inorg. Nucl. Chem., **30** (1968) 1745-1750.

This paper describes the synthesis of two mixed oxidation state oxide hydrates, $UO_{2.86} \cdot 1.5H_2O(cr)$ and $UO_{2.86} \cdot 0.5H_2O(cr)$, and reports the enthalpies of formation of these solids. As discussed by Cordfunke and O'Hare [78COR/OHA], the paper suffers from errors in the calculations. Recalculation using the reported heat of solution of $UO_{2.86} \cdot 1.5H_2O(cr)$ in aqueous 0.057 M $Ce(SO_4)_2$/1.5 M H_2SO_4, the CODATA Key Values for $\Delta_f H_m^\circ(U_3O_8$, cr, 298.15 K) and $\Delta_f H_m^\circ(H_2O$, l, 298.15 K), and a molar enthalpy of solution for a $UO_{2.028}/UO_{2.993}$ mixture with an O:U ratio of 2.86 from Fitzgibbon *et al.* [67FIT/PAV] results in $\Delta_f H_m^\circ(UO_{2.86}\cdot 1.5H_2O$, cr, 298.15 K) $= -1666\,kJ\cdot mol^{-1}$ [78COR/OHA]. The value for $\Delta_{trs}H_m[H_2O, H_2O(l) \rightarrow Ce(SO_4)_2/1.5\ M\ H_2SO_4]$ is assumed to be zero within the experimental error.

The heat of solution of U_3O_8(cr) reported in Ref. [68COR/PRI] differs by less than $1\,kJ\cdot mol^{-1}$ from the value obtained for solution into the same medium by Fitzgibbon *et al.* [67FIT/PAV]. However, the heat of solution of $UO_{2.24}$(cr) differs by more than $20\,kJ\cdot mol^{-1}$ from the value for dissolution of $UO_{2.25}$ reported by Fitzgibbon *et al.* [67FIT/PAV], and would lead to the unlikely conclusion that U_4O_9(cr) can disproportionate to UO_2 and U_3O_7 (or U_3O_8). The calculated value of $\Delta_f H_m(UO_3, \alpha, 298.15\,K)$ based on the heat of solution of α-UO_3 in the cerium sulphate/sulphuric acid solution is about $8\,kJ\cdot mol^{-1}$ more positive than the value selected in Section V.3.3.1.2 of this review. An uncertainty of $\pm 10\,kJ\cdot mol^{-1}$ in $\Delta_f H_m^\circ(UO_{2.86}\cdot 1.5H_2O, cr, 298.15\,K)$ is estimated.

The enthalpy of dehydration of $UO_{2.86}\cdot 1.5H_2O$(cr) to $UO_{2.86}\cdot 0.5H_2O$(cr) ($\Delta_{dehyd}H^\circ = (57.53 \pm 0.84)\,kJ\cdot mol^{-1}$) as recalculated by Cordfunke and O'Hare [78COR/OHA] is accepted in this review. The uncertainty in the value $\Delta_f H_m^\circ(UO_{2.86}\cdot 0.5H_2O$, cr, 298.15 K) $= -(1367 \pm 10)\,kJ\cdot mol^{-1}$ selected in this review is based on the estimated uncertainty in $\Delta_f H_m^\circ(UO_{2.86}\cdot 1.5H_2O$, cr, 298.15 K), as discussed above.

[68GIR/WES]
Girdhar, H.L., Westrum, E.F., Jr., λ-type thermal anomaly in triuranium octaoxide at 482.7 K, J. Chem. Eng. Data, 13 (1968) 531-533.

The heat capacity of U_3O_8 was measured by adiabatic calorimetry from 300 to 500 K. The authors estimated the errors in the experimental heat capacities to be 0.1%. In this review, for the purposes of calculating a function describing the temperature dependence of the heat capacity, the uncertainty of each measurement (95% confidence level) is estimated as 0.2% of the reported value, and the data are weighted accordingly.

[68KRY/KOM3]
Krylov, V.N., Komarov, E.V., Pushlenkov, M.F., Complex formation of U(VI) with the fluoride ion in solutions of $HClO_4$, Sov. Radiochem., 10(6) (1968) 708-710.

Krylov, Komarov and Pushlenkov measured the equilibrium constants for

$$UO_2^{2+} + HF(aq) \rightleftharpoons UO_2F^+ + H^+ \qquad (A.69)$$

in $HClO_4$ solutions at concentrations of 2.10, 1.04, 0.51 and 0.20 M, using the cation exchange method of Fronæus [51FRO]. There is no information given about the temperature of the experiment and it has been assumed that the data refer to 25°C. The experimental method is satisfactory. The authors have not made an error estimate. This review has therefore estimated the error to be $\pm 15\%$ in the value of $^*K_1$ for Reaction (A.69). This estimate is based on the published $\pm 10\%$ error in ϕ, *cf.* Ref. [68KRY/KOM3, Table 1], and an additional 5% due to the omission of l_1 [68KRY/KOM, Eq. (1)].

The data in 0.20 M $HClO_4$ have been used to calculate the equilibrium constants for the formation of $UO_2F_2(aq)$.

$$UO_2^{2+} + 2HF(aq) \rightleftharpoons UO_2F_2(aq) + 2H^+ \qquad (A.70)$$

The numerical value of this constant is strongly dependent on the precision in the distribution coefficients and the error in $(\beta_{1H} - l_1)$ and l_1 [68KRY/KOM, Eq. (1)]. This review has re-evaluated the published data and finds for the reaction (A.70) $\log_{10}{}^*\beta_2(A.70) = (1.8 \pm 0.2)$ if we assume $l_1 = (5 \pm 5)$ and $(\beta_1 - l_1) = (180 \pm 10)$.

The data cannot be used to obtain a reliable value of the equilibrium constant for the formation of $UO_2F_3^-$.

[68NEM/PAL]
Nemodruk, A.A., Palei, P.N., Glukova, L.P., Karyakin, A.V., Muradova, G.A., The decomposition of the uranyl group in strong acid solutions, Dokl. Chem. (Engl. transl.), 180 (1968) 510-513.

Nemodruk *et al.* reported that the UV-visible spectrum for acidic media containing dioxouranium(VI) varied as a function of the total acid concentration at constant chloride concentration for chloride concentrations from 1 to 11 M. Infrared absorption spectra were similarly found to change. The authors interpeted these observations in terms of formation of UO^{4+} according to Reaction (A.71)

$$UO_2^{2+} + 2H^+ \rightleftharpoons UO^{4+} + H_2O(l) \quad (A.71)$$

and give an equilibrium constant of 5×10^{-5} to 6×10^{-5}. The speciesUO^{4+} is not credited in this review as it has not been otherwise confirmed. It is probable the observed spectral changes are related to changes in structure or composition of chloro complexes of dioxouranium(VI) [49SUT], and activity effects in the high ionic strength media.

[68NIK/ANT]
Nikolaeva, N.M., Antipina, V.A., Pastukova, E.D., Investigation of hydrolysis of uranyl nitrate at elevated temperatures, Deposited document No. 395, VINITI, Moscow, USSR, 1968, in Russian.

This paper was not available to the reviewers. See comments under Nikolaeva [71NIK].

[68OST/CAM]
Ostacoli, G., Campi, E., Gennaro, M.C., Complessi di alcuni acidi diidrossibenzoici con lo ione uranile in soluzione acquosa, Gazzetta Chim. Ital., **98** (1968) 301-315, in Italian.

The calibration procedures are not clear. The discussion of the data analysis gives no information on whether minor species were considered. In this review the uncertainties in the hydrolysis constants are estimated as ±0.3 in $\log_{10}{}^*\beta_{2,2}$ and ±0.6 in $\log_{10}{}^*\beta_{5,3}$.

[68SCH/FRY]
Schedin, U., Frydman, M., Studies on the hydrolysis of metal ions: 59. The uranyl ion in magnesium nitrate medium, Acta Chem. Scand., **22** (1968) 115-127.

Values in this paper were obtained in solutions of very high ionic strength. These data, therefore, cannot be used to obtain simple interaction coefficients.

[69GRE/VAR]
Grenthe, I., Varfeldt, J., A potentiometric study of fluoride complexing of uranium(IV) and uranium(VI) using the U(VI)/U(IV) redox couple, Acta Chem. Scand., **23** (1969) 988-998.

These measurements were made at 25°C in a mixed $(Na,H)ClO_4$ medium at $I =$ 4 M. The equilibrium constants were calculated from the measured variation of the U(VI)/U(IV) redox potential with [HF]. Norén [69NOR] compared this method with the fluoride membrane electrode method and found good agreement up to $\overline{q} \approx 1.4$. The highest value of $\overline{q}$ measured by Grenthe and Varfeldt is 1.5. Hence, only a small amount of UF_3^+ can be present in solution. For this reason, and because of a systematic error in the data at values of $\overline{q} > 1.4$, this review gave zero weight to $\log_{10}\beta_3$. The error estimated by Grenthe and Varfeldt seems too small. This review has selected error limits which are the same as those given by Norén [69NOR].

Grenthe and Varfeldt also determined the equilibrium constant for the reaction $UO_2^{2+} + HF(aq) \rightleftharpoons UO_2F^+ + H^+$ by measuring the redox potential U(VI)/U(IV) as a function of the total concentrations of HF, U(VI) and U(IV), giving $^*\beta_1(UO_2F^+) =$ (30 ± 6). This corresponds to $\log_{10}{}^*\beta_1 = (1.48 \pm 0.09)$. The uncertainty given by the authors seems reliable. However, the specific ion interaction theory used in this review for extrapolation to zero ionic strength is at the limit of its validity for such a high ionic strength. This value has therefore not been used in the linear extrapolation to $I = 0$.

[69NOR]
Norén, B., A solvent extraction and potentiometric study of fluoride complexes of thorium(IV) and uranium(IV), Acta Chem. Scand., **23** (1969) 931-942.

The equilibrium data of Norén for the reactions

$$U^{4+} + HF(aq) \rightleftharpoons UF^{3+} + H^+ \qquad (A.72)$$
$$U^{4+} + 2HF(aq) \rightleftharpoons UF_2^{2+} + 2H^+ \qquad (A.73)$$
$$U^{4+} + 3HF(aq) \rightleftharpoons UF_3^{+} + 3H^+ \qquad (A.74)$$

were obtained in 4 M $HClO_4$ at 20°C by using a fluoride membrane electrode which has an excellent precision. Norén also checked the data of Grenthe and Varfeldt [69GRE/VAR] and found good agreement between the two different methods, at least up to a value of $\overline{q} \approx 1.4$, where $\overline{q}$ is the average number of bonded F^- per U(IV). The uncertainties of the constants as listed in Table V.17 of Section V.4.1.2.3 are estimated based on the error in the average complexation degree, $\overline{q}$, given by the author. For the complexes UF_q^{4-q}, this results in $\log_{10}{}^*\beta_1(\text{A.72, 293.15 K}) =$ (5.54 ± 0.02), $\log_{10}{}^*\beta_2(\text{A.73, 293.15 K}) = (8.72 \pm 0.03)$, and $\log_{10}{}^*\beta_3(\text{A.74, 293.15 K})$ $= (10.68 \pm 0.05)$. The corrections to 298.15 K were done using the $\Delta_rH_m^\circ$ values selected in this review, assuming they are independent of the ionic strength. The resulting values at 25°C are $\log_{10}{}^*\beta_1(\text{A.72, 298.15 K}) = (5.49 \pm 0.02)$, $\log_{10}{}^*\beta_2(\text{A.73,}$ $\text{298.15 K}) = (8.64 \pm 0.03)$, and $\log_{10}{}^*\beta_3(\text{A.74, 293.15 K}) = (10.57 \pm 0.05)$.

[69RAO/PAI]
Rao, C.L., Pai, S.A., A study of nitrate and sulphate complexes of uranium(IV), Radiochim. Acta, **12** (1969) 135-140.

The authors used TTA as the auxiliary ligand. The ionic medium was 3.8 M (H, Na)ClO_4 with $[H^+] = 1.0$ M. There is no information given about the temperature, which was assumed to be around 25°C. The authors find that there is some systematic error in the TBP-data for the nitrate system and these are therefore disregarded. Several different computational methods were tested and these gave concordant results. The highest nitrate concentration was 1.2 M, and the amounts of $U(NO_3)_2^{2+}$ formed in this concentration range were negligible. This finding is not in agreement with the results in Refs. [70LAH/KNO], [65VID/BYA] and [62ERM/KRO]. There seems to be some experimental ambiguity in this study, and the ionic strength is a bit too high for a reliable extrapolation to $I = 0$ using the specific ion interaction method. Hence the data from this study are not used in the final selection, even though they are in fair agreement with the results of other studies.

The sulphate equilibria refer to the reactions $U^{4+} + qHSO_4^- \rightleftharpoons U(SO_4)_q^{4-2q} + qH^+$, where $q = 1$ and 2. The highest HSO_4^- concentration investigated is 0.066 M. This is sufficiently high to ensure the formation of $U(SO_4)_2(aq)$. The authors reported $\log_{10} K_1 = (2.32 \pm 0.05)$ and $\log_{10} K_2 = (3.81 \pm 0.08)$. The uncertainties are underestimated. This review has made a graphical re-evaluation of the data in Ref. [69RAO/PAI] and finds the conditional equilibrium constants $^*\beta_1 = (103 \pm 20)$ and $^*\beta_2 = (2440 \pm 500)$ for the reactions $U^{4+} + HSO_4^- \rightleftharpoons USO_4^{2+} + H^+$ and $U^{4+} + 2HSO_4^- \rightleftharpoons U(SO_4)_2(aq) + 2H^+$, respectively. Correction to $I = 0$ is uncertain due to the mixed inert salt medium used in this study. This review has therefore given zero weight to this study.

[69TSY]
Tsymbal, C., Contribution à la chimie de l'uranium(VI) en solution, Ph.D. thesis, CEA-R-3479, Université de Grenoble, Grenoble, France, 1969, 97p, in French.

Based on the experimental procedure, this looks like it should be a reliable study. However, the results for $\log_{10}{^*\beta_1}$ and $\log_{10}{^*\beta_{1,2}}$ are inconsistent with many other reported values. The value of $\log_{10}{^*\beta_{7,3}} = -24.03$ does not appear to be correct. Tripathi [84TRI] reported that reanalysis of Tsymbal's data leads to $\log_{10}{^*\beta_{7,3}} \approx -28$, but even this value is too large as discussed in Section V.3.2.1.3. In view of the apparently large systematic errors, this review assigns a weight of zero to all the hydrolysis constants reported in this thesis.

The hydroxide complexes are minor species in the presence of carbonate, and this review therefore tries to reinterpret the carbonate data of Tsymbal. Some details about the calculations done by this review are given in Table A.1.

The potentiometric measurements seem carefully performed and precautions were taken to standardize the glass electrode and to control liquid junction potentials. The experiments were titrations in which acid was added to solutions of known total concentrations of dioxouranium(VI) (concentration range 0.5 to 5 mM) and carbonate (concentration range 3 to 30 mM). The total concentration of carbonate was kept constant by working in a closed cell in which there was a continuous increase of p_{CO_2} as CO_3^{2-}/HCO_3^- were transformed to $CO_2(g)$. There is a risk of precipitation of $UO_2CO_3(cr)$ around pH = 5, but no mention of the formation of such a precipitate is given by the author. The constants given by Tsymbal [69TSY] have been criticized, for example, by Tripathi [84TRI]. Tripathi [84TRI] has also done some re-interpretations of Tsymbal's data assuming "that titrations were carried out rapidly enough to prevent equilibration with atmospheric CO_2". This is not the procedure used by Tsymbal, so a different re-interpretation was found necessary by this review.

The system is considered as a two-phase system, one solution and one gas-phase, with the possibility of transfer of $CO_2(g)$ between the two phases. A special subroutine of the LETAGROP procedure [69SIL/WAR] for distribution measurements (LETAGROP DISTR) was used in the least-squares analysis of various equilibrium models. The total concentration of protons was chosen as the error-carrying variable. Note that the total concentration of protons in a c M CO_3^{2-} solution is equal to $-2c$ M. The hydrolysis constants of Tsymbal were used as fixed parameters in

Table A.1: Least-squares analysis of equilibrium models for data from Tsymbal [69TSY] for the formation of complexes $(UO_2)_m(CO_3)_q(OH)_n$ according to the reaction $mUO_2^{2+} + qCO_3^{2-} + nH_2O(l) \rightleftharpoons (UO_2)_m(CO_3)_q(OH)_n + nH^+$.

$q:n:m$	$\log_{10} K$			
	Ref. [69TSY]	Model 1[(a)]	Model 2[(a)]	Model 3[(a)]
0:2:2	−6.09	−6.05	−5.70	−5.70
0:5:3	−15.64	−16.24	−15.60	−15.60
1:0:1		9.02±0.18	9.3±0.2	9.5±0.2
2:0:1	16.18±0.06	16.4±0.1	16.1±0.2	16.0±0.2
3:0:1	21.57±0.02	21.8±0.2	21.7±0.1	21.3±0.3
6:0:3		54.9±0.1	55.0±0.2	54.4±0.1
1:1:1	4.10±0.03			
"6:24:11" [(c)]			34.7 ± 0.2[(d)]	32.9 ± 0.5[(d)]
$\sum_i s_i^2 \times 10^3$		86.4	82.3	40.3
$s(H) \times 10^3$[(b)]		0.0202	0.0195	0.0137

(a) Tsymbal denoted the least-square data in the original publication [69TSY]. Models 1 and 2 are the least-squares calculations of this review using Tsymbal's data. Model 3 denotes the least-squares calculations of this review using Tsymbal's data corrected for a systematic error in $[H]_T$ of the U(VI) stock solutions.

(b) The standard deviation of the error-carrying variable H (the total concentration of dissociable protons), in $mol \cdot dm^{-3}$, is reasonable in view of the concentration range investigated, $[H]_T = 6$ to $60 \times 10^{-3}\, mol \cdot dm^{-3}$.

(c) Refers to the complex $(UO_2)_{11}(OH)_{24}(CO_2)_6^{2-}$ for the reaction $11UO_2^{2+} + 24H_2O(l) + 6CO_2(g) \rightleftharpoons (UO_2)_{11}(OH)_{24}(CO_2)_6^{2-} + 24H^+$.

(d) The value from Ciavatta *et al.* [79CIA/FER] is $\log_{10} K = (33.6 \pm 0.1)$.

these refinements. The value of $\log_{10}{}^*\beta_{7,3}$ selected in this review was used in these calculations. It was difficult to obtain convergence in the least-squares procedure when using Tsymbal's published data. There is also a strong correlation between the equilibrium constants for $UO_2CO_3(aq)$ and $(UO_2)_{11}(OH)_{24}CO_2^{2-}$. The data indicated the presence of a systematic error in the total acidity of the dioxouranium(VI) stock solution used by the author. He assumed the total acidity to be zero but a ratio H/U = 0.20 is found to give a much better fit of all the experimental data and the error square sum decreased significantly. The indicated error is surprisingly large and one cannot be sure if there is not some other error in the experiments.

The results of the least squares refinements show the following:

- The equilibrium constants of the carbonate complexes are not strongly dependent on the equilibrium constants of the hydroxide species because these are present in fairly low concentrations in most of the pH-ranges investigated.

- There is a significant amount of $(UO_2)_3(CO_3)_6^{6-}$ formed. Tsymbal's data thus confirm the studies in Refs. [81CIA/FER2, 81FER/GRE, 84GRE/FER]. The equilibrium constant of $\log_{10}{}^*\beta_{6,3} = (54.4 \pm 0.1)$ for the reaction $3UO_2^{2+} + 6CO_3^{2-} \rightleftharpoons (UO_2)_3(CO_3)_6^{6-}$ is in fair agreement with the value of Grenthe *et al.* [84GRE/FER], $\log_{10}{}^*\beta_{6,3} = 54.2$, recalculated to 0.1 M $NaClO_4$.

- The agreement between observed and calculated data is improved if the chemical model includes the large polynuclear species $(UO_2)_{11}(OH)_{24}(CO_2)_6^{2-}$ proposed by Ciavatta *et al.* [79CIA/FER]. Tsymbal has few experimental data in the pH-range where this complex predominates. Hence, the refinement is less sensitive for this species than was the case with the data of Ciavatta *et al.* [79CIA/FER], and Grenthe and Lagerman [91GRE/LAG].

- By using Tsymbal's data it is not possible to distinguish between the various mixed hydroxide/carbonate complexes proposed. The equilibrium constant for the formation of both $(UO_2)_2CO_3(OH)_3^-$ and $(UO_2)_3(OH)_5CO_2^+$ was set equal to zero when the least-squares refinements was made. This is understandable because only 1% of the total uranium concentration is present as $(UO_2)_3(OH)_5CO_2^+$ (by using the data of Ciavatta *et al.* [79CIA/FER]). However, by using the data of Maya [82MAY] this review found that about 20% of uranium should be present as $(UO_2)_2CO_3(OH)_3^-$. This model obviously does not fit Tsymbal's data.

However, the study confirms the main features of the U(VI)-CO_2-H_2O system observed by other investigations. This review therefore uses the equilibrium constants of the major species but assigned fairly large uncertainties.

[69VAN/OST]
Vanni, A., Ostacoli, G., Roletto, E., Complessi formati dallo ione uranile in soluzione acquosa con acidi bicarbossilici della seri satura, Ann. Chim. Rome, 59 (1969) 847-859, in Italian.

This is a study done at 25°C under the same conditions used by Baes and Meyer [62BAE/MEY] and yielded essentially the same results. According to the authors evidence was found for UO_2OH^+. However, considering all solutions had relatively high total uranium concentrations (2×10^{-3} to 6×10^{-3} M), $^*\beta_1$ from this study is not credited in this review. The uncertainties in the other hydrolysis constants are estimated in this review to be the same as those in the Baes and Meyer study, ± 0.2 in $\log_{10}{}^*\beta_{2,2}$ and ± 0.3 in $\log_{10}{}^*\beta_{5,3}$.

[69VDO/STE]
Vdovenko, V.M., Stebunov, O.B., Relaxation processes during complex formation: IV. Determination of the stability constants of the complexes from the data of the proton relaxation method, Sov. Radiochem., 11 (1969) 625-629.

This is a reanalysis of the data in Ref. [63VDO/ROM]. Lack of detail on solution preparation precludes use of these data. The temperature is not indicated.

[69VDO/STE2]
Vdovenko, V.M., Stebunov, O.B., Relaxation processes during complex formation: V. Determination of the stability constants of strong complexes by the NMR method in the case of unknown equilibrium concentrations of the ligand, Sov. Radiochem., 11 (1969) 630-632.

Vdovenko and Stebunov studied the formation of UO_2F^+ at $I = 0.5$ M ($NaClO_4$?) and $t = 20$°C, by measuring proton relaxation times. A competitive method was used where first the proton relaxation time in the presence of paramagnetic Fe^{3+} was measured at different fluoride concentrations. These data were used to determine the stability constants for the Fe^{3+}-F^- complexes. The uncertainties vary from $\pm 10\%$ for $\beta_1(FeF^{2+})$ to $\pm 50\%$ for FeF_2^+ and $FeF_3(aq)$. The constants are in good agreement with the literature values as given, *e.g.*, by Sillén and Martell [64SIL/MAR]. From the measurements of the proton relaxation time in the presence of UO_2^{2+} and Fe^{3+}, and from the known equilibrium constants of the Fe^{3+}-F^- system, the equilibrium constant for the UO_2^{2+}-F^- system was calculated using the method of corresponding solutions, *cf.* Ref. [61ROS/ROS]. The value obtained for the formation constant of UO_2F^+ was $\beta_1 = (4.5 \pm 0.5) \times 10^4$. This uncertainty seems small in view of the uncertainties in the Fe^{3+}-F^- constants. The predominant Fe(III) fluoride complex in the UO_2^{2+}-F^- study is FeF^{2+}, hence this review proposes $\beta_1 = (4.5 \pm 1.0) \times 10^4$, *i.e.*, $\log_{10}\beta_1 = (4.66 \pm 0.10)$, as a more reasonable uncertainty estimate. The constant has been corrected to $t = 25$°C with the data given by Ahrland and Kullberg [71AHR/KUL2]. Conversion to molality units and correction to $I = 0$ with the ion interaction coefficients given in Appendix B, this review obtains $\log_{10}\beta_1^\circ = (5.25 \pm 0.10)$.

[70BAS/SMI]
Baskin, Y., Smith, S.D., Enthalpy-of-formation data on compounds of uranium with groups VA and VIA elements, J. Nucl. Mater., **37** (1970) 209-222.

This paper provides enthalpy of formation data for a number of binary uranium compounds using a direct reaction method. The enthalpy of formation value for U_3As_4(cr) depends only slightly on an estimated value for $\Delta_f H_m^\circ(UAs_2$, cr, 298.15 K) which is only 3 mole-% of the final product. The estimate for $\Delta_f H_m^\circ(UAs_2$, cr, 298.15 K) appears to be reasonable within the uncertainty noted by the authors, and the value is accepted in this review.

[70DRA/VIL]
Drăgulescu, C., Vîlceanu, N., Julean, I., Komplexe Uranylpyrophosphate: II. Untersuchung der Bildung anionischer Komplexe, Z. Anorg. Allg. Chem., **376** (1970) 196-204, in German.

See comments under Drăgulescu, Julean and Vîlceanu [65DRA/JUL].

[70FRE/WEN]
Frei, V., Wendt, H., Fast ionic reactions in solution: VIII. The fission and formation kinetics of the dimeric isopolybase $(UO_2OH)_2^{2+}$, Ber. Bunsen-Ges., **74** (1970) 593-599.

This is a useful kinetic study corroborating the existence of $(UO_2)_2OH^{3+}$. The value determined for $^*\beta_{2,2}$ has a very large uncertainty, and it is assigned a weight of zero in this review.

[70KLY/KOL]
Klygin, A.E., Kolyada, N.S. Smirnova, I.D., Spectrophotometric investigation of complex formation in the uranyl nitrate-nitric acid-water system, Russ. J. Inorg. Chem., **15** (1970) 1719-1721.

This is a spectrophotometric study at 20°C of U(VI) nitrate complex formation in nitric acid solutions with concentrations varying from 0.59 M to 11.1 M.

The measurements indicate clearly the formation of several different nitrate complexes, but it is difficult to estimate the accuracy of the equilibrium constants proposed. The authors report $\log_{10}\beta_2 = -(1.66\pm0.04)$ for the reaction $UO_2^{2+} + 2NO_3^- \rightleftharpoons UO_2(NO_3)_2(aq)$, and $\log_{10}{}^*K = (1.66 \pm 0.16)$ for the reaction $UO_2^{2+} + 3NO_3^- + H^+ \rightleftharpoons UO_2(NO_3)_2(HNO_3)(aq)$.

No evidence for the formation of $UO_2NO_3^+$ was found. One observes fairly large errors in the least squares calculated quantities given by Klygin, Kolyada and Smirnova [70KLY/KOL, *p*.1720]. Correlations betwen molar absorptivities and equilibrium constants often have these effects. This review does not consider the quantitative information given by Klygin, Kolyada and Smirnova as sufficiently precise to be included in this database.

[70LAH/KNO]

Lahr, H., Knoch, W., Bestimmung von Stabilitätskonstanten einiger Aktinidenkomplexe: II. Nitrat- und Chloridkomplexe von Uran, Neptunium, Plutonium und Americium, Radiochim. Acta, **13** (1970) 1-5, in German.

Lahr and Knoch studied the formation of nitrate and chloride complexes of U(IV), U(VI) and other actinides in 8 M H^+ at 20°C. They used an extraction method based on tri-*n*-octylamin dissolved in xylene. For the nitrate complexes they find for the reaction $U^{4+} + qNO_3^- \rightleftharpoons U(NO_3)_q^{4-q}$: $\beta_1 = 0.84$, $\beta_2 = 3.76$, $\beta_3 = 0.37$, and $\beta_4 = 0.50$, and for the reaction $UO_2^{2+} + qNO_3^- \rightleftharpoons UO_2(NO_3)_q^{2-q}$: $\beta_1 = 2.94$ and $\beta_2 = 0.03$. The nitrate concentration varies from 0 to 8 M. This should result in large changes in the activity factors of the various species. The "equilibrium constants" given by the authors are better regarded as empirical parameters than as true stability constants. It is not possible to extrapolate data from this high ionic strength to $I = 0$. Hence, the nitrate data were given zero weight.

For the chloride system the authors reported for the reaction $UO_2^{2+} + qCl^- \rightleftharpoons UO_2Cl_q^{2-q}$: $\beta_1 = 0.82$ and $\beta_2 = 0.05$. These constants cannot be recalculated to $I = 0$ because of lack of experimental details. They were therefore given zero weight.

The publication contains a survey of literature data and a discussion of systematics of chloride and nitrate complexes among the actinides.

[70MAE/KAK]

Maeda, M., Kakihana, H., The hydrolysis of the uranyl ion in heavy water, Bull. Chem. Soc. Japan, **43** (1970) 1097-1100.

See comments under Maeda *et al.* [72MAE/AMA].

[71AHR/KUL]

Ahrland, S., Kullberg, L., Thermodynamics of metal complex formation in aqueous solution: I. A potentiometric study of fluoride complexes of hydrogen, uranium(VI) and vanadium(IV), Acta Chem. Scand., **25** (1971) 3457-3470.

See comments under Ahrland, Larsson, and Rosengren [56AHR/LAR].

[71AHR/KUL2]

Ahrland, S., Kullberg, L., Thermodynamics of metal complex formation in aqueous solution: II. A calorimetric study of fluoride complexes of hydrogen, uranium(VI) and vanadium(IV), Acta Chem. Scand., **25** (1971) 3471-3483.

See comments under Ahrland, Larsson, and Rosengren [56AHR/LAR].

[71AHR/KUL3]

Ahrland, S., Kullberg, L., Thermodynamics of metal complex formation in aqueous solution: III. A calorimetric study of hydrogen sulphate and uranium(VI) sulphate, acetate and thiocyanate complexes, Acta Chem. Scand., **25** (1971) 3677-3691.

This is a precise calorimetric study of the enthalpy changes for a number of equilibria, including

$$UO_2^{2+} + qSO_4^{2-} \rightleftharpoons UO_2(SO_4)_q^{2-2q} \quad (A.75)$$

$$UO_2^{2+} + qSCN^- \rightleftharpoons UO_2(SCN)_q^{2-q} \quad (A.76)$$

at 25.0°C and in 1 M $NaClO_4$. The authors reported $\Delta_r H_m(A.75, q = 1) = (18.23 \pm 0.17)$ kJ · mol^{-1}, $\Delta_r H_m(A.75, q = 2) = (35.11 \pm 0.40)$ kJ · mol^{-1}, $\Delta_r H_m(A.76, q = 1) = (3.22 \pm 0.06)$ kJ · mol^{-1}, $\Delta_r H_m(A.76, q = 2) = (8.9 \pm 0.6)$ kJ · mol^{-1}, and $\Delta_r H_m(A.76, q = 3) = (6.0 \pm 1.2)$ kJ · mol^{-1}. The authors could not confirm the formation of $UO_2(SO_4)_3^{4-}$ [51AHR2].

[71BAI/LAR]

Bailey, A.R., and Larson, J.W., Heats of dilution and the thermodynamics of dissociation of uranyl and vanadyl sulfates, J. Phys. Chem., 75 (1971) 2368-2372.

This is a careful calorimetric study of the thermodynamics of Reaction A.77.

$$UO_2^{2+} + SO_4^{2-} \rightleftharpoons UO_2SO_4(aq) \quad (A.77)$$

Two sets of data were used:

1. Heats of dilution of $UO_2SO_4(aq)$ which gave $\Delta_r H_m^\circ(A.78) = -(20.84 \pm 0.42)$ kJ · mol^{-1}.

   $$UO_2SO_4(aq) \rightleftharpoons UO_2^{2+} + SO_4^{2-} \quad (A.78)$$

   This value was obtained by using data from ionic strengths ranging from 0.0652 to 0.240 M and a value of $\log_{10} \beta_1^\circ(A.77) = 2.72$, which is lower than the constant proposed in this review. The equilibrium concentrations at the various ionic strengths were obtained by estimating the activity factors of the ions, by using an extended Debye-Hückel expression. This procedure seems satisfactory as judged by the near constancy of the calculated value of $\Delta_r H_m^\circ$ (*cf.* Ref. [71BAI/LAR, Table I].

2. Heats of Reaction (A.77). The authors reported $\Delta_r H_m^\circ(A.77) = (20.79 \pm 0.29)$ kJ · mol^{-1} at $I = 0$.

There is excellent agreement between scheme 1 and scheme 2. This value is in fair agreement with the data of Ahrland and Kullberg [71AHR/KUL3], indicating that the ionic strength dependence of $\Delta_r H_m$ is small (about 10% difference over the ionic strength range investigated). This review accepts $\Delta_r H_m^\circ(A.77) = (19.5 \pm 1.6)$ kJ · mol^{-1}, *cf.* Section V.5.1.3.1.a.

[71COR/LOO]

Cordfunke, E.H.P., Loopstra, B.O., Sodium uranates: Preparation and thermochemical properties, J. Inorg. Nucl. Chem., 33 (1971) 2427-2436.

Cordfunke and Loopstra determined the enthalpies of dissolution of α-Na_2UO_4, β-Na_2UO_4, Na_4UO_5, $Na_2U_2O_7$ and $Na_6U_7O_{24}$ in 6 M HNO_3 to be $-(192.8 \pm 1.4)$, $-(199.0 \pm 1.0)$, $-(418.9 \pm 2.1)$, $-(184.35 \pm 1.13)$, and $-(85.1 \pm 0.3)$ kJ · mol^{-1}, respectively.

$$Na_xU_yO_{0.5x+3y} + (x + 2y)HNO_3(sln) \rightleftharpoons yUO_2(NO_3)_2(sln) + xNaNO_3(sln) + (0.5x + y)H_2O(sln) \quad (A.79)$$

These authors reported the following values:

$$\begin{aligned}
\Delta_f H_m^\circ(Na_2UO_4, \alpha, 298.15\,K) &= -1874\,kJ\cdot mol^{-1}\\
\Delta_f H_m^\circ(Na_2UO_4, \beta, 298.15\,K) &= -1866\,kJ\cdot mol^{-1}\\
\Delta_f H_m^\circ(Na_4UO_5, cr, 298.15\,K) &= -2414\,kJ\cdot mol^{-1}\\
\Delta_f H_m^\circ(Na_2U_2O_7, cr, 298.15\,K) &= -3180\,kJ\cdot mol^{-1}\\
\Delta_f H_m^\circ(Na_6U_7O_{24}, cr, 298.15\,K) &= -10799\,kJ\cdot mol^{-1}.
\end{aligned}$$

Recalculations by Cordfunke and O'Hare [78COR/OHA] using the new value of the enthalpy of solution of $NaNO_3$, $(15.44 \pm 0.21)\,kJ\cdot mol^{-1}$ in 6 M HNO_3, yielded

$$\begin{aligned}
\Delta_f H_m^\circ(Na_2UO_4, \alpha, 298.15\,K) &= -(1889.9 \pm 2.1)\,kJ\cdot mol^{-1}\\
\Delta_f H_m^\circ(Na_2UO_4, \beta, 298.15\,K) &= -(1883.6 \pm 2.1)\,kJ\cdot mol^{-1}\\
\Delta_f H_m^\circ(Na_4UO_5, cr, 298.15\,K) &= -(2450.6 \pm 2.1)\,kJ\cdot mol^{-1}\\
\Delta_f H_m^\circ(Na_2U_2O_7, cr, 298.15\,K) &= -(3194.5 \pm 2.1)\,kJ\cdot mol^{-1}\\
\Delta_f H_m^\circ(Na_6U_7O_{24}, cr, 298.15\,K) &= -(11355 \pm 15)\,kJ\cdot mol^{-1}.
\end{aligned}$$

In this review, the values are again recalculated using the following auxiliary data: $\Delta_f H_m^\circ(NaNO_3$, cr, 298.15 K$) = -(467.58 \pm 0.41)\,kJ\cdot mol^{-1}$ based on the CODATA values for NO_3^- and Na^+ [89COX/WAG] and the weighted average of the values for the enthalpy of solution of $NaNO_3$ from Table II-1 of the same reference; the enthalpy of solution of $NaNO_3$(cr) in 6 M HNO_3 as cited above; $\Delta_f H_m^\circ(HNO_3$, 6 M HNO_3, 298.15 K$) = -(200.34 \pm 0.41)\,kJ\cdot mol^{-1}$ based on the CODATA [89COX/WAG] value for the enthalpy of formation of NO_3^- and enthalpy of dilution data from Ref. [65PAR]; $\Delta_f H_m^\circ(H_2O$, 6 M $HNO_3) = -(286.37 \pm 0.08)\,kJ\cdot mol^{-1}$ [65PAR, 89COX/WAG]; $\Delta_f H_m^\circ(UO_2(NO_3)_2$, 6 M HNO_3, 298.15 K$) = -(1409.5 \pm 1.5)\,kJ\cdot mol^{-1}$ based on the CODATA value for $\Delta_f H_m^\circ UO_2(NO_3)_2{\cdot}6H_2O(cr) = -(3167.5 \pm 1.5)\,kJ\cdot mol^{-1}$ [89COX/WAG, footnote 119 to Table 5], $\Delta_f H_m^\circ(H_2O$, 6 M $HNO_3)$ (as above), and the enthalpy of solution of $UO_2(NO_3)_2{\cdot}6H_2O$ in 6 M HNO_3, $39.8\,kJ\cdot mol^{-1}$. The uncertainties are estimated considering the strong correlation of the uncertainties in the nitrate species.

$$\begin{aligned}
\Delta_f H_m^\circ(Na_2UO_4, \alpha, 298.15\,K) &= -(1892.4 \pm 2.3)\,kJ\cdot mol^{-1}\\
\Delta_f H_m^\circ(Na_2UO_4, \beta, 298.15\,K) &= -(1886.1 \pm 2.3)\,kJ\cdot mol^{-1}\\
\Delta_f H_m^\circ(Na_4UO_5, cr, 298.15\,K) &= -(2456.2 \pm 3.0)\,kJ\cdot mol^{-1}\\
\Delta_f H_m^\circ(Na_2U_2O_7, cr, 298.15\,K) &= -(3196.1 \pm 3.9)\,kJ\cdot mol^{-1}\\
\Delta_f H_m^\circ(Na_6U_7O_{24}, cr, 298.15\,K) &= -(11351 \pm 14)\,kJ\cdot mol^{-1}
\end{aligned}$$

There is an indication in Ref. [71COR/LOO] that the sample of β-Na_2UO_4 in this work was contaminated with an unspecified amount of $Na_2U_2O_7$. For this reason, and because the difference in the enthalpies of solution of the α- and β-forms is only $6.3\,kJ\cdot mol^{-1}$ as opposed to the $12.6\,kJ\cdot mol^{-1}$ [85TSO/BRO] and $13.9\,kJ\cdot mol^{-1}$ [76OHA/HOE] found in other studies, the value for β-Na_2UO_4 from this study ([71COR/LOO]) is not considered reliable. This conclusion was reached despite the fact the enthalpy of formation value is in reasonably good agreement with the values from the other studies, and the value for the α-form is not.

The reported values for the enthalpies of solution of $NaUO_3$(cr), β-UO_3 and U_3O_8(cr) in 1.7 M H_2SO_4/0.057 M $Ce(SO_4)_2$ were also reported. The value of

$\Delta_f H_m^\circ(NaUO_3, cr, 298.15 K)$ was recalculated (including correction of a major error in one of the auxiliary data) by O'Hare and Hoekstra [74OHA/HOE3], who obtained $\Delta_f H_m^\circ(NaUO_3, cr, 298.15 K) = -(1492.4 \pm 7.9)\,kJ \cdot mol^{-1}$. Cordfunke and O'Hare [78COR/OHA] did a further recalculation based on more recent auxiliary data, and the results of their recalculation are accepted in this review with the exception of a minor change to ensure consistency with the CODATA [89COX/WAG] value for $\Delta_f H_m^\circ(SO_4^{2-}, aq, 298.15 K)$, and use of the value $\Delta_f H_m^\circ(Na_2SO_4, cr, 298.15 K) = -(1387.61 \pm 0.84)\,kJ \cdot mol^{-1}$. These changes lead to a value of $\Delta_f H_m^\circ(NaUO_3, cr, 298.15 K) = -(1497.6 \pm 4.9)\,kJ \cdot mol^{-1}$.

[71HUT]
Hutin, M.F., Chimie analytique: Sur l'existence des complexes uranyl-periodiques. Application analytique, C. R. Hébd. Séances Acad. Sci., Sér. C, **273** (1971) 739-740, in French.

See comments under Hutin [75HUT].

[71NIK]
Nikolaeva, N.M., The study of hydrolysis and complexing of uranyl ions in sulphate solutions at elevated temperatures, Izv. Sib. Otd. Akad. Nauk SSSR, Ser. Khim. Nauk, 7(3) (1971) 61-67, translated from the Russian: ORNL/TR-88/1, Oak Ridge National Laboratory, Oak Ridge, Tennessee, USA.

Nikolaeva performed two studies of the dissociation constant of $UO_2SO_4(aq)$, the first one [70NIK] was done by using conductivity measurements and the second [71NIK] by using potentiometric data H^+-determinations. The first study was made at 25, 50, 70 and 90°C at low ionic strength < 0.02 M. Corrections for hydrolysis and protonation of SO_4^{2-} have to be made. Interpretation of the conductivity data also requires λ_o values for all ionic species. The value reported in Ref. [70NIK] for the reaction $UO_2^{2+} + SO_4^{2-} \rightleftharpoons UO_2SO_4(aq)$, $\log_{10} \beta_1^\circ = 3.35$, is in fair agreement with the value at $I = 0$ selected here. In view of the various corrections that have to be made, this review has not taken this constant into account for the evaluation of the selected value.

The potentiometric data [71NIK] were obtained at 50, 70, 90, 100, 125 and 150°C in solutions where the total concentrations of U(VI) varied from 7×10^{-5} M to 1.2×10^{-3} M, (*i.e.*, the ionic strength is low). The experimental data have to be corrected for hydrolysis (studied separately) and for the protonation of SO_4^{2-}. The authors do not report equilibrium constants at the various temperatures, only a temperature function for the dissociation of $UO_2SO_4(aq)$. The value calculated from this function does not agree very well with the direct determinations in [70NIK]. The calculated concentration ratios $[UO_2SO_4(aq)]:([UO_2^{2+}][SO_4^{2-}])$ based on the data given in Ref. [71NIK, Table X.2] are the following:

$T(°C)$	$\frac{[UO_2SO_4]}{[UO_2^{2+}][SO_4^{2-}]}(\times 10^3)$	$T(°C)$	$\frac{[UO_2SO_4]}{[UO_2^{2+}][SO_4^{2-}]}(\times 10^3)$	$T(°C)$	$\frac{[UO_2SO_4]}{[UO_2^{2+}][SO_4^{2-}]}(\times 10^3)$
50	1.54	90	4.13	125	99.1
	1.55		5.56		149
	5.3		13.6	150	294
70	1.96	100	21.8		596
	3.39		16.0		253
	8.64		21.6		
			25.6		
			62.9		
			59.3		

It is obvious that the ratio corresponding to the formation constant of $UO_2SO_4(aq)$ is not a constant at a given temperature. These data are not considered reliable and are therefore given zero weight. The derived enthalpy information is even more uncertain and is also given zero weight. The data in Ref. [71NIK] indicate that composition of the solutions at high temperture is uncertain. The equilibrium data at 50, 70 and 90°C in Ref. [70NIK] are therefore also given zero weight.

The only experimental hydrolysis data reported in this paper [71NIK] are for 100, 125 and 150°C. The formation constants for $UO_2(OH)_2(aq)$ were calculated by assuming values for the formation constants for UO_2OH^+ and $(UO_2)_2(OH)_2^{2+}$ that appear to have been calculated from the temperature dependent functions described as Eqs. (4) and (5) in Ref. [71NIK]. These constants were apparently based on data in Ref. [68NIK/ANT]. Even if $UO_2(OH)_2(aq)$ is significant above 100°C (and the evidence for this is not conclusive) the species $(UO_2)_3(OH)_5^+$ would be expected to be significant for $T \leq 100°C$ in the more concentrated uranium solutions. However, no hydrolysis scheme including this species appears to have been used to try to explain the data. The values from these references for $\log_{10}{}^*\beta_{2,2}$, $\log_{10}{}^*\beta_2$ and $\log_{10}{}^*\beta_1$ extrapolated to 25°C are assigned a weight of zero in this review. Nikitin *et al.* [72NIK/SER] noted that activity effects had not been properly taken into account in Ref. [68NIK/ANT], and provided temperature dependent equations for $\log_{10}{}^*\beta_1$ and $\log_{10}{}^*\beta_{2,2}$ based on the data of Ref. [68NIK/ANT].

[71POL/BAR]

Polasek, M., Bartusek, M., Thiocyanate complexes of bivalent metals in dilute aqueous solutions, Scr. Fac. Sci. Natur. Univ. Purkynianae Brun., 1(9) (1971) 109-118.

The original publication was not available to the reviewers, only an abstract (CA **77** (1972) 131395v). Polasek and Barbusek have used an emf-method (Ag, AgSCN(s) and Cd-Hg electrodes) to study the complex formation between UO_2^{2+} and SCN^-. Such emf-methods are usually precise. The value of $\log_{10}\beta_1$ is also in good agreement with other data if corrections are made for the composition of the ionic medium (0.1M KNO_3). The value of $\log_{10}\beta_2$ reported here differs considerably from those found in other studies. This might be due to changes in the ionic medium when a large part of NO_3^- is replaced by SCN^-. The $\log_{10}\beta_2$ value was given zero weight. Corrections for the formation of nitrate complexes of UO_2^{2+} are negligible.

[71SAN/VID]

Santalova, N.A., Vidavskii, L.M., Dunaeva, K.M., Ippolitova, E.A., Enthalpy of formation of uranium trioxide dihydrate, Sov. Radiochem., 13 (1971) 608-612.

The enthalpies of solution of $UO_3 \cdot 2H_2O(cr)$ in dilute aqueous hydrofluoric acid as determined by Santalova *et al.* [71SAN/VID], $-72.7\,kJ \cdot mol^{-1}$, and by Tasker *et al.* [88TAS/OHA], $-76.25\,kJ \cdot mol^{-1}$, are markedly different. The same is true for the reported enthalpies of dissolution of γ-UO_3 [65VID/BYA, 88TAS/OHA]. However, the difference $[\Delta_r H_m^\circ(UO_3 \cdot 2H_2O) - \Delta_r H_m^\circ(\gamma\text{-}UO_3)]$ is essentially the same for both groups. In this review the assumption has been made that each group accurately measured the differences in the enthalpies of solution, but that a some systematic error exists in one of the sets of results, most probably in the results of the Moscow group [71SAN/VID].

[71VID/IPP]

Vidavskii, L.M., Ippolitova, E.A., Standard enthalpy of formation of the amorphous modification of uranium trioxide, Sov. Radiochem., 13 (1971) 306-308.

Vidavskii and Ippolitova reported enthalpies of solution of γ-UO_3 and of a mixture of UO_3(am) and α-UO_3. They corrected their result for the content of α-UO_3 using the difference between the enthalpies of solution of α-UO_3 and UO_3(am) determined by Hoekstra and Siegel [61HOE/SIE]. In this review a somewhat different procedure has been followed in that the difference in the selected values for the enthalpies of formation of α-UO_3 and γ-UO_3 ($\Delta = (6.26 \pm 0.53)\,kJ \cdot mol^{-1}$) has been used to allow for the α-UO_3 content. After correction for the small amount of hydrated solid, the difference $[\Delta_f H_m^\circ(UO_3, am, 298.15\,K) - \Delta_f H_m^\circ(UO_3, \gamma, 298.15\,K)] = -(13.92 \pm 0.81)\,kJ \cdot mol^{-1}$ is obtained. Although not significantly different from values in other studies, the derived value was assigned a weight of zero in this review because of the large α-UO_3 content of the UO_3(am) sample.

[72BUC/GOR]

Buchacek, R., Gordon, G., The kinetics and mechanism of the oxidation-reduction reaction between uranium(IV) and chlorine(III) in the presence of phenol in aqueous solution, Inorg. Chem., 11 (1972) 2154-2160.

The kinetics of oxidation of U(IV) by chlorite ions was studied at 25°C in a 2.0 M (Na,H)ClO_4 ionic medium. From the observed rate law (an inverse hydrogen ion dependency and the overall inhibition of the reaction by $HClO_2$) the authors suggest the following mechanism:

$$U^{4+} + HClO_2(aq) \overset{K}{\rightleftharpoons} UOClOH^{4+}$$
$$H^+ + ClO_2^- \overset{K_3}{\rightleftharpoons} HClO_2(aq)$$
$$U^{4+} + ClO_2^- \overset{k_1}{\rightleftharpoons} \text{products}$$

From the experimental rate data the authors obtained $K = (4.13 \pm 0.25)\,M^{-1}$. This seems to be a careful study, and the constants would probably not vary strongly with the ionic strength because the reactants and products have the same charge. The value of $\log_{10} K^\circ = (0.62 \pm 0.10)$ would thus be a reasonable guess, where the uncertainty has been re-estimated by this review. However, due to the lability of the complex $UOClOH^{4+}$ in systems in which it is formed, this review has not taken this equilibrium constant into account.

[72CHA/BAN]

Chakravorti, M.C., Bandyopadhyay, N., Fluoro complexes of hexavalent uranium-IV: Complexes of the series $UO_2F_3^-$ and $(UO_2)_2F_5^-$, J. Inorg. Nucl. Chem., **34** (1972) 2867-2874.

Chakravorti and Bandyopadhyay reported the protonation constant for the formation of $HUO_2F_3(aq)$. No evidence is presented for the existence of such a complex in solution. If the solid $HUO_2F_3 \cdot H_2O(cr)$ is dissolved in an aqueous solvent it will form a whole set of $UO_2F_q^{2-q}$, $q = 1$ to 4 depending on $[F^-]$. This was not taken into account by the authors. The experimental method is also unsatisfactory, as the use of KCl in the salt bridge in contact with $NaClO_4$ will result in precipitation of $KClO_4$ and ill-defined liquid junction potentials. The electrode calibration should have been made in concentration units, rather than activities.

[72GLE/KLY]

Glevob, V.A., Klygin, A.E., Smirnova, I.D., Kolyada, N.S., Hydrolysis of quadrivalent uranium, Russ. J. Inorg. Chem., **17** (1972) 1740-1742.

A value of 0.052 is reported for ${}^*\beta_1$ from a spectrophotometric study at an unspecified temperature. The ionic strength was allowed to vary with the pH of the solutions, but the reported constant is an average value. The results from this study are not used in this review.

[72MAE/AMA]

Maeda, M., Amaya, T., Ohtaki, H., Kakihana, H., Mixed solvent deuterium isotope effects on the hydrolysis of UO_2^{2+} and Be^{2+} ions, Bull. Chem. Soc. Japan, **45** (1972) 2464-2468.

It would have been useful if more details on the selection of the hydrolysis scheme had been given. The experimental work looks reasonable. Uncertainties in the equilibrium constants are not reported. Instead values of "3σ" are provided for the parameters (equilibrium constants) obtained by curve-fitting of the data to a model. These do not take into account uncertainties resulting from model error and from inaccuracies in solution compositions or potentiometer calibration. In this review the uncertainties in values of the equilibrium constants from this study are estimated as follows: ± 0.2 in $\log_{10}{}^*\beta_{2,2}$, ± 1.0 in $\log_{10}{}^*\beta_{4,3}$, and ± 0.4 in $\log_{10}{}^*\beta_{5,3}$. These estimates are rather large, but have been chosen to reflect the omission of minor species from the model. The differences between the values reported for the solvolysis in H_2O and D_2O are probably more accurate than the actual values of the solvolysis constants. In this review the solvolysis constants for D_2O were estimated by using

$$\begin{aligned}\log_{10}{}^*\beta^\circ_{n,m}(D_2O) &= \log_{10}{}^*\beta^\circ_{n,m}(H_2O, \text{ this review}) \\ &\quad - \log_{10}{}^*\beta_{n,m}(3\,M\ NaClO_4, H_2O, [72MAE/AMA]) \\ &\quad + \log_{10}{}^*\beta_{n,m}(3\,M\ NaClO_4, D_2O, [72MAE/AMA])\end{aligned}$$

and the uncertainties in the $\log_{10}{}^*\beta^\circ_{n,m}$ (D_2O) solvolysis constants were estimated to be the same as estimated above for $\log_{10}{}^*\beta_{n,m}$ (3 M $NaClO_4$, H_2O).

[72NIK/SER]

Nikitin, A.A., Sergeyeva, E.I., Khodakovsky, I.L., Naumov, G.B., Hydrolysis of uranyl in the hydrothermal region, Geokhimiya, 3 (1972) 297-307; translated from the Russian: Atomic Energy of Canada Ltd., Pinawa, Manitoba, Canada, 21p.

The values for the formation constants of UO_2OH^+ and $(UO_2)_2(OH)_2^{2+}$ at 25°C are (a) based on data reported by Nikolaeva but "corrected" to $I = 0$ and (b) smoothed values to conform with data obtained for higher temperatures. The species $(UO_2)_3(OH)_5^+$ was omitted in the analysis, although, based on the work of Baes and Meyer [62BAE/MEY], it would be expected that this species would be significant in the solutions studied—at least for temperatures $\leq$ 100°C. The constants at 25°C are not credited in this review. Nikitin *et al.* also report solubility measurements for $UO_3 \cdot 2H_2O(cr)$ at 25°C (pH 6.2 to 8.2). The experimental solubilities at 25°C are lower than those calculated using the solubility product for $UO_3 \cdot 2H_2O(cr)$ provided in the same reference if $(UO_2)_3(OH)_5^+$ is included in the database. Conversely, unless a fairly high stability is assigned to $UO_2(OH)_2(aq)$ (or a polymer thereof), the experimental solubilities are higher than those calculated if the recently reported value of $\Delta_f G^\circ_m(UO_3{\cdot}2H_2O$, cr, 298.15 K)[88TAS/OHA] is used (also see Section V.3.3.1.5).

Nikitin *et al.* also use their solubility data (25 to 200°C) and their reanalysis of the data from Ref. [68NIK/ANT] to estimate values for the second monomeric hydrolysis constant $^*K_2$. It is difficult to assign an uncertainty to these values, and it is not even certain $UO_2(OH)_2(aq)$ forms in these solutions. Other hydrolysis schemes have not been considered, yet the total dioxouranium(VI) concentrations in the solutions are quite low and monomeric species may well predominate, especially at the higher temperatures. The fact that the stable solid changes within the range of the measurements was apparently not addressed in the calculation of the thermodynamic parameters for Reaction (A.80) (see Ref. [72NIK/SER, Eq. (2)]).

$$UO_2(OH)_2(cr) \rightarrow UO_2(OH)_2(aq) \qquad (A.80)$$

The stable solid is probably the dihydrate even at 100°C [88TAS/OHA]. Also, pH values for the high temperature equilibrium solutions were not reported. Thus the justification for the extrapolated constant for Reaction (A.80) at 25°C ($\log_{10} K = -6.15$) is weak, and for the justification for the reported enthalpy of reaction is even weaker. The 25°C solubility data suggest that $\log_{10} K(A.81) \leq -5.5$ probably represents an upper limit on the stablity of $UO_2(OH)_2(aq)$.

$$UO_3 \cdot 2H_2O(cr) \rightarrow UO_2(OH)_2(aq) + H_2O(l) \qquad (A.81)$$

[72SER/NIK]

Sergeyeva, E.I., Nikitin, A.A., Khodakovsky, I.L., Naumov, G.B., Experimental investigation of equilibria in the system UO_3-CO_2-H_2O in 25-200°C temperature interval, Geochim. Int., 11 (1972) 900-910.

Sergeyeva *et al.* made the largest number of measurements in the pH range where the uncharged complex $UO_2CO_3(aq)$ is predominant. The measurements at 25°C were made at only one value of $p_{CO_2} \approx 1$ atm. An independent assessment of the chemical composition of the system is thus not possible. There are only two experimental points in the range where UO_2^{2+} is predominant, *i.e.*, where the solubility product is more accurately determined, and three experimental points where the two species $UO_2(CO_3)_2^{2-}$ and $UO_2(CO_3)_3^{4-}$ are formed. The authors made corrections for the hydrolysis of U(VI). The number of experimental data are too few to allow an independent selection of a chemical model. However, in combination with the other evidence presented in this survey, these data allow a good estimate of equilibrium constants at $I = 0$. The experimental study was made at such low ionic strength that the activity coefficient calculations made by the authors should be quite precise. It is difficult to make a proper estimation of uncertainties with the limited experimental data available, hence, the uncertainties quoted by the authors have been increased to ± 0.3 logarithmic units by this review.

The solubility data were discussed by Langmuir [78LAN] and Tripathi [84TRI]. These authors propose equilibrium constants that are somewhat different from those of Sergeyeva *et al.* [72SER/NIK].

[73ALM/GAR]

Almagro, V., Garcia, F.S., Sancho, J., Estudio polarografico del ion UO_2^{2+} en medio carbonato sodico, An. Quim., 69 (1973) 709-716, in Spanish.

This is a polarographic study made in the range $9 < pH < 12$ and with the ratio carbonate/uranium $10 < (C:U) < 180$. The equilibrium constants reported for the formation of $UO_2CO_3(aq)$, $UO_2(CO_3)_2^{2-}$ and $UO_2(CO_3)_3^{4-}$ are 6×10^2, 1.08×10^4 and 5.0×10^7, respectively. These values are obviously in error. The authors seem to have misinterpreted their data. The process studied, at least at $pH > 10$, is the reduction of U(VI) to U(V) (*cf.* Harris and Kolthoff [47HAR/KOL]). The complex formation of dioxouranium(V) must thus be taken into account.

[73DAV/EFR]

Davydov, Yu.P., Efremenkov, V.M., Investigation of the hydrolysis properties of hexavalent uranium, Vest. Akad. Nauk Belaruskai SSR Ser. Fiz. Energ., (4) (1973) 21-25, in Russian.

See comments under Davydov and Efremenkov [83DAV/EFR].

[73DRA/VIL]
Drăgulescu, C., Vîlceanu, N., Uranylpyrophosphate complexes: III. The monopyrophosphate-pentauranyl cation complex, Rev. Roum. Chim., **18** (1973) 49-52.

This is a spectrophotometric study, where the authors reported an equilibrium constant of $K = 0.0344$ for the reaction

$$5UO_2^{2+} + H_2PO_7^{2-} \rightleftharpoons (UO_2)_5H_2PO_7^{8+}. \qquad (A.82)$$

The present review accepts neither the stoichiometry proposed, nor the value of the equilibrium constant for the following reasons:

- The experimental absorption spectra reported in Figures 1 and 2 of Ref. [73DRA/VIL] clearly show the formation of complexes. However, the authors' assumption that only the complex contributes to the measured absorbance is not correct.
- There is no proof of the authors' assumption that only one equilibium reaction takes place. Their proposal of a discontinuity in the absorbance *vs.* the total concentration of $Na_2H_2P_2O_7$ is not justified by their own data in Figure 2 of [73DRA/VIL].
- If the assumption that only one coloured complex contributed to the measured absorbance had been correct, then the limiting absorbance should be constant and not increasing as in [73DRA/VIL, Figure 2].
- The coordination of five dioxouranium(VI) ions to a ligand containing only four unprotonated oxygen donors (and one bridging) is not likely.

[73MAJ]
Majchrzak, K., Equilibria in solutions and anion exchange of uranium(VI) complexes: IV. Stability constants and sorption of uranyl sulphate complexes, Nucleonika, **8(3)** (1973) 105-119, in Polish.

Majchrzak made an anion exchange study of the complex formation between UO_2^{2+} and SO_2^{2-}. The temperature is not given and the ionic strength varies with the concentration of Na_2SO_4. For the reactions $UO_2^{2+} + qSO_4^{2-} \rightleftharpoons UO_2(SO_4)_q^{2-2q}$, $q = 1$ to 3, the author reports the following equilibrium constant: $\beta_1 = 8500$ $\beta_2 = 25000$ and $\beta_3 = 2500$. These constant are based on the assumption that the ratio between the activity factors in the aqueous and anion exchange phases are constant, *i.e.*,

$$\frac{\overline{\gamma}_{(UO_2(SO_4)_q^{2-2q})}\gamma_{(SO_4^{2-})}}{\gamma_{(UO_2(SO_4)_q^{2-2q})}\overline{\gamma}_{(SO_4^{2-})}} = \text{const.}$$

This assumption is not verified, and it is likely that the ratio varies with the invasion of sulphate into the resin phase. It is quite clear from this study that the formation of anionic sulphate complexes were proven but the numerical values of the equilibrium constants are not reliable. Hence, they were given zero weight.

[73MAV]
Mavrodin-Tărăbic, M.. L'étude des espèces ioniques formées par l'hydrolyse de l'ion d'uranium hexavalent, Rev. Roum. Chim., **18** (1973) 73-88, in French.
See comments under Mavrodin-Tărăbic [74MAV].

[73MAV2]
Mavrodin-Tărăbic, M., Contribution à l'étude de l'hydration de l'ion UO_2^{2+} en milieu perchlorique, Rev. Roum. Chim., **18** (1973) 609-621, in French.
See comments under Mavrodin-Tărăbic [74MAV].

[73OHA/HOE2]
O'Hare, P.A.G., Hoekstra, H.R., Thermochemistry of uranium compounds: II. Standard enthalpy of formation of α-sodium uranate (α-Na_2UO_4), J. Chem. Thermodyn., **5** (1973) 769-775.

In this work the enthalpy of solution of α-Na_2UO_4 in dilute hydrochloric acid (0.1 M) was measured and reported as $-(175.5 \pm 0.7)\,\mathrm{kJ \cdot mol^{-1}}$. The main problem with this work has been the shifting values of the auxiliary data required to calculate the enthalpy of formation of α-Na_2UO_4. Thus, the original estimate of $-1864.2\,\mathrm{kJ{\cdot}mol^{-1}}$ was revised to $-1889.0\,\mathrm{kJ{\cdot}mol^{-1}}$ based on a revised $\Delta_f H_m^\circ(UO_2Cl_2$, cr), then to $-1895.9\,\mathrm{kJ \cdot mol^{-1}}$ based on 1) a further revision of that quantity, 2) a new determination of the enthalpy of solution of UO_2Cl_2 in 0.1 M HCl(aq) and 3) revised values for the enthalpies of formation of Cl^-(aq) and NaCl(cr). The auxiliary values used in Ref. [82COR/MUI] are reasonably close to being consistent with the CODATA compilation and the auxiliary data selected in this review. A recalculation of the enthalpy of formation of HCl, in 0.1 M HCl(aq), from the elements, a slightly revised value for the enthalpy of formation of Cl^-(aq), and a minor change in $\Delta_f H_m^\circ(UO_2Cl_2$, cr) (to $-1243.6\,\mathrm{kJ \cdot mol^{-1}}$) lead to the value $\Delta_f H_m^\circ(Na_2UO_4, \alpha, 298.15\,\mathrm{K}) = -(1896.8 \pm 2.3)\,\mathrm{kJ \cdot mol^{-1}}$, recalculated in this review.

[73STA/CEL]
Stankov, V., Čeleda, J., Jedináková, V., Contributions to the chemistry of highly concentrated aqueous electrolyte solutions: 32. Spectrophotometric study of nitrito- and chloro complexes of UO_2^{2+} ion, Scientific papers of the Institute of Chemical Technology, Prague, Sborník VSCHT Praha, **B17** (1973) 109-143, in Czech.

Stankov, Čeleda and Jedináková present a qualitative study of the formation of nitrate and chloride complexes of U(VI) using spectrophotometry, electrophoresis and apparent molar volumes. Only the abstract has been translated into English. The qualitative results are in agreement with the known behaviour of these systems, indicating that UO_2Cl^+ is predominant at chloride concentrations less than 5 M.

[74BUN]
Bunus, F.T., An ion exchange study of the uranium (U^{4+} and UO_2^{2+}) complex species with Cl^- as ligand, J. Inorg. Nucl. Chem., **36** (1974) 917-920.

This is a cation exchange study of the chloride complexes of UO_2^{2+} and U^{4+} over a range of temperatures and ionic strengths. The U(VI) data are satisfactory. The U(IV) data are surprising in two ways: 1) The complex UCl^{3+} turned out weaker than UO_2Cl^+ (for "electrostatic" reasons one would expect the inverse); 2) The enthalpy change for the reaction $U^{4+} + Cl^- \rightleftharpoons UCl^{3+}$ is strongly negative ($\Delta_r H_m = -30\,kJ \cdot mol^{-1}$), and the entropy change is negative as well ($\Delta_r S_m = -100\,J \cdot K^{-1} \cdot mol^{-1}$). Interactions between hard donors and acceptors always lead to positive $\Delta_r S_m$ values, *cf.* Ahrland [67AHR, 68AHR]. These chemical "anomalies" may result from experimental shortcomings. Even a very small oxidation of U(IV) will lead to large errors in the measured distribution coefficients, ϕ, for the high ϕ values encountered in the U(IV)-Cl^- system. This review has given zero weight to this study.

[74BUN2]
Bunus, F.T., Determinarea speciilor complexe ale U(IV) în prezența ionului Cl^- ca ligand și a funcțiilor termodinamice care intervin, prin metoda schimbului ionic, Rev. Chim. (Bucharest), **25(5)** (1974) 367-370, in Rumanian.

This seems to be the same study as Ref. [74BUN].

[74JED]
Jedináková, V., Contributions to the chemistry of highly concentrated aqueous electrolyte solutions: 37. Densimetric study of complex formation of the UO_2^{2+} ion in isomolar series perchlorate-halide, Scientific papers of the Institute of Chemical Technology, Prague, Sborník VSCHT Praha, **B18** (1974) 113-125, in Czech.

Jedináková studied the formation of UO_2^{2+} complexes with Cl^-, Br^- and I^- using the determination of apparent molar volumes from experimental densities. The experiments were done at several different ionic strengths and presumably at room temperature. The details in the paper could not be followed by this review because of language problems. However, the orders of magnitude of the equilibrium constants are in good agreement with other studies. Since no uncertainty estimate was given by the author, and because of the insufficient insight into the experimental details, this study is given zero weight.

[74KAK/ISH]
Kakihana, H., Ishiguro, S., Potentiometric and spectrophotometric studies of fluoride complexes of uranium(IV), Bull. Chem. Soc. Japan, **47** (1974) 1665-1668.

Kakihana and Ishiguro studied the U^{4+}-HF(aq) system by using a fluoride membrane electrode. The measurements were made at 25°C in a 1 M mixed $(Na^+, H^+)Cl^-$ medium. They also made some spectrophotometric measurements which were used to determine the molar absorbtivities of the various complexes. The complex formation was studied in a fluoride concentration range where $\overline{q}$, the average number of bonded F^- per U(IV), varied between 1.3 and 2.6. Hence, the concentration of free U^{4+}

should be very low, even at the lowest values of $[F^-]$. This indicates that the value of β_1 is rather uncertain. The stepwise stability constants $UF_q^{4-q} + F^- \rightleftharpoons UF_{q+1}^{3-q}$ should be more accurate. The uncertainties quoted by Kakihana and Ishiguro for $\log_{10}\beta_2$, $\log_{10}\beta_3$ and $\log_{10}\beta_4$ probably refer to the precision in the stepwise constants. The uncertainties in the overall $\log_{10}\beta_q$ values cannot be smaller than the one in $\log_{10}\beta_1$ with the method used. Hence, this review assigns new uncertainties for $\log_{10}\beta_q$, $q > 1$, as shown below. The authors gave little information about $[H^+]$, only that it was larger than 0.1 M. However, hydrolysis is not expected to influence the complex formation curve in the concentration range studied. It may, however, influence the value of β_1. The experimental data must be corrected for the formation of U(IV) chloride complexes. This is done in this review by assuming that only UCl^{3+} is formed, with $\log_{10}\beta_1(UCl^{3+}) \approx (0.38 \pm 0.15)$ valid in this solution. The protonation constant of F^- used, $\log_{10} K = 2.84$, agrees well with the value of 2.86 calculated at $I = 1$ M from the value 3.18 at $I = 0$ selected in this review (*cf.* Chapter VI). The authors reported equilibrium constants for both reactions of the type (A.83) and (A.84).

$$U^{4+} + qF^- \rightleftharpoons UF_q^{4-q} \quad \text{(A.83)}$$

$$U^{4+} + qHF(aq) \rightleftharpoons UF_q^{4-q} + qH^+ \quad \text{(A.84)}$$

This review uses the latter set, corrected for chloride complexation and corrected to $I = 0$. For $q = 1$ it is assumed that $\Delta\varepsilon$ is the same as for the reaction $U^{4+} + H_2O(l) \rightleftharpoons UOH^{3+} + H^+$ in perchlorate medium (see Section V.4.1.2.3), $\Delta\varepsilon(A.84, q = 1) = -(0.14 \pm 0.10)$. The uncertainty is increased more than for the evaluations of Refs. [69GRE/VAR, 69NOR, 76CHO/UNR] because a chloride medium was used here rather than a perchlorate medium. Similarly, for $q \geq 2$ it is assumed that the ionic interactions involving chloride are similar to those involving perchlorate. Hence, the following analogies are made: $\varepsilon_{(U^{4+},Cl^-)} \approx \varepsilon_{(U^{4+},ClO_4^-)} = (0.76 \pm 0.06)$, $\varepsilon_{(UF_2^{2+},Cl^-)} \approx \varepsilon_{(M^{2+},ClO_4^-)} = (0.3 \pm 0.1)$ and $\varepsilon_{(UF_3^+,Cl^-)} \approx \varepsilon_{(M^+,ClO_4^-)} = (0.1 \pm 0.1)$, and an estimated uncertainty of ± 0.10 in each value of $\Delta\varepsilon$. Using $\varepsilon_{(H^+,ClO_4^-)}$ rather than $\varepsilon_{(H^+,Cl^-)}$ for consistency reasons, the following values are obtained: $\Delta\varepsilon(A.84, q = 2) = -(0.18 \pm 0.10)$, $\Delta\varepsilon(A.84, q = 3) = -(0.24 \pm 0.10)$, and $\Delta\varepsilon(A.84, q = 4) = -(0.24 \pm 0.10)$, *cf.* [77ISH/KAO]. The results are shown in Table A.2. The values are in fairly good agreement with those reported by other authors and are used for the calculation of the weighted average values selected in this review for $q = 1$, 2 and 3. For $q = 4$, this paper is the only direct determination, and $\log_{10}{}^*\beta_4^\circ$ listed above is accepted by this review.

[74MAV]

Mavrodin-Tărăbic, M., Contribution à l'étude de l'hydrolyse de l'uranium(VI) en milieu azotique, Rev. Roum. Chim., 19 (1974) 1461-1470, in French.

The apparent lack of temperature control in Refs. [73MAV, 73MAV2, 74MAV] makes it pointless to try to calculate proper hydrolysis constants from these data.

Table A.2: Re-evaluation of the uranium(IV) fluoride complexation data of Ref. [74KAK/ISH], according to the reactions $U^{4+} + q\mathrm{HF(aq)} \rightleftharpoons \mathrm{UF}_q^{4-m} + q\mathrm{H}^+$.

	$\log_{10}{}^*\beta_1$	$\log_{10}{}^*\beta_2$	$\log_{10}{}^*\beta_3$	$\log_{10}{}^*\beta_4$
Reported [74KAK/ISH]	4.49 ± 0.04	7.45 ± 0.02	8.95 ± 0.05	10.11 ± 0.17
Corrected for Cl^- complex formation	5.02 ± 0.11	7.98 ± 0.11	9.48 ± 0.12	10.64 ± 0.20
Corrected to $I = 0$	6.10 ± 0.15	9.80 ± 0.15	11.69 ± 0.16	12.89 ± 0.22

[74OHA/HOE2]
O'Hare, P.A.G., Hoekstra, H.R., Thermochemistry of uranium compounds: III. Standard enthalpy of formation of cesium uranate (Cs_2UO_4), J. Chem. Thermodyn., 6 (1974) 251-258.

The value from this paper for $\Delta_f H^\circ_m(\mathrm{Cs_2UO_4}$, cr, 298.15 K) is recalculated using the values $\Delta_f H^\circ_m(\mathrm{UO_2Cl_2}$, cr, 298.15 K) $= -(1243.6\pm1.3)\,\mathrm{kJ\cdot mol^{-1}}$ and $\Delta_{sol} H_m(\mathrm{UO_2Cl_2}$, 1 M HCl(aq)) $= -(101.7\pm1.7)\,\mathrm{kJ\cdot mol^{-1}}$ selected in the present review, the CODATA [89COX/WAG] values for $\Delta_f H^\circ_m(\mathrm{Cs^+}$, aq, 298.15 K) and $\Delta_f H^\circ_m(\mathrm{Cl^-}$, aq, 298.15 K), and the heat of solution, $\Delta_{sol} H^\circ_m$, of CsCl(cr) in H_2O(l) based on those reported in Annex II of the CODATA review [89COX/WAG]. Hence, $\Delta_f H^\circ_m(\mathrm{Cs_2UO_4}$, cr, 298.15 K) $= -(1929.4 \pm 2.5)\,\mathrm{kJ \cdot mol^{-1}}$ is calculated.

[74OHA/HOE3]
O'Hare, P.A.G., Hoekstra, H.R., Thermochemistry of uranium compounds: IV. Standard enthalpy of formation of sodium uranium(V) trioxide ($NaUO_3$), J. Chem. Thermodyn., 6 (1974) 965-972.

This appears to be a careful study of the enthalpy of formation of $NaUO_3$(cr). The procedure used involved oxidation of the compound with XeO_3 in aqueous (approximately 1.5 M) HCl solution. The enthalpy of formation as recalculated by Cordfunke and O'Hare [78COR/OHA], and with CODATA compatible data in this review ($\Delta_f H^\circ_m(\mathrm{NaUO_3, cr, 298.15\,K}) = -1520.5\,\mathrm{kJ \cdot mol^{-1}}$), is 20 to 25 $\mathrm{kJ \cdot mol^{-1}}$ more negative than the values from two other careful studies [71COR/LOO, 81COR/OUW]. There is no apparent reason for the large discrepancy. The recalculations were done using the value of the enthalpy of solution of UO_2Cl_2(cr) in the XeO_3/HCl(aq) mixture as reported in the original publication. Based on more recent information on the enthalpy of solution of UO_2Cl_2(cr) [83FUG/PAR], there may be reason to suspect better auxiliary data would give an even more negative value. The value of $\Delta_f H^\circ_m(\mathrm{NaUO_3})$ from the work of O'Hare and Hoekstra [74OHA/HOE3] is rejected in this review.

[74OHA/HOE4]
O'Hare, P.A.G., Hoekstra, H.R., Thermochemistry of uranium compounds: V. Standard enthalpies of formation of the monouranates of lithium, potassium and rubidium, J. Chem. Thermodyn., **6** (1974) 1161-1169.

The values from this paper for $\Delta_f H_m^\circ(M_2UO_4$, cr, M = Li, K, Rb) are recalculated in a manner similar to that described in this Appendix for the recalculation of the results for Cs_2UO_4(cr) from Ref. [74OHA/HOE2]. The values obtained are:

$$\begin{aligned}\Delta_f H_m^\circ(Li_2UO_4, \text{cr}, 298.15\text{ K}) &= -(1971.2 \pm 3.0)\,\text{kJ}\cdot\text{mol}^{-1}\\ \Delta_f H_m^\circ(K_2UO_4, \text{cr}, 298.15\text{ K}) &= -(1920.7 \pm 2.2)\,\text{kJ}\cdot\text{mol}^{-1}\\ \Delta_f H_m^\circ(Rb_2UO_4, \text{cr}, 298.15\text{ K}) &= -(1922.7 \pm 2.2)\,\text{kJ}\cdot\text{mol}^{-1}.\end{aligned}$$

[75ALY/ABD]
Aly, H.F., Abdel-Rassoul, A.A., Zakareia, N., Use of zirconium phosphate for stability constant determination of uranium and antimony chlorocomplexes, Z. Phys. Chem. (Frankfurt/Main), **94** (1975) 11-18.

Aly, Abdel-Rassoul and Zakareia studied the formation of dioxouranium(VI) chloride complexes by using a zirconium phosphate ion-exchanger. The measurements were made in HCl solutions, the concentrations of which varied from 10^{-2} to 4.0 M. The technique is the same as used by Souka, Shabana and Farah [76SOU/SHA]. No account was taken of varying activity coefficients of reactants and products. This seriously impairs the usefulness of the results obtained. It is quite clear that the value of $\beta_1 = (39 \pm 4)$ for the formation of UO_2Cl^+ given by the authors is much too large. It means that UO_2Cl^+ would be the dominating complex in 30 mM Cl^-, a result that contradicts a number of other experimental observations. This value is given zero weight.

[75BAR/SEN]
Bartusek, M., Senkyr, J., Study of thiocyanate complexes with a silver thiocyanate membrane electrode, Scr. Fac. Sci. Natur. Univ. Purkyniznae Brun., **5(5)** (1975) 61-69, in German, Chem. Abstr. 85:84716d.

Only an abstract is available to the reviewers. Barbusek and Senkyr [75BAR/SEN] used an emf-method for determining the equilibrium constants of UO_2SCN^+. The authors used a procedure that, in general, gives precise data. The value of $\log_{10} K_1$ is not available to this review.

[75CIN/SCA]
Cinnéide, S.O., Scanlan, J.P., Hynes, M.J., Equilibria in uranyl carbonate systems: I. The overall stability constant of $UO_2(CO_3)_3^{4-}$, J. Inorg. Nucl. Chem., **37** (1975) 1013-1018.

This study was made in a carbonate concentration range where the concentration of all uranium species except $UO_2(CO_3)_3^{4-}$ are negligible.

The experimental measurements was made by using a two-phase distribution method. It is not clear from the paper whether the glass-electrode is calibrated in terms of concentrations or activities. The same protonation constants as in [77SCA] were used to calculate CO_3^{2-} from the experimental data. However, in this study the free carbonate concentration is so high (2 to 10 mM) that an error in the protolysis constants does not greatly influence the equilibrium constants.

The data of this paper are corrected from 20°C to 25°C using the enthalpy value selected in this review. In view of the uncertainties in the procedures this review doubles the uncertainties given by the authors and assigns an uncertainty of ± 0.06 to the logarithmic equilibrium constants.

[75DAV/EFR]
Davydov, Yu.P., Efremenkov, V.M., Investigation of the hydrolytic properties of tetravalent uranium: II. Conditions of formation of mononuclear and polynuclear hydroxo complexes of U(IV), Sov. Radiochem., **17** (1975) 160-164.

This paper reports on one of the few attempts to obtain hydrolysis data for U^{4+} at concentrations as low as 10^{-5} M. Values for β_1 and β_2 were reported for perchlorate and chloride media (I = 0.5 M). The effect of complexation with chloride was not discussed. Neither the temperature nor the value used for the ion product of water are specified, but calculations suggest the value of $^*\beta_1$ found in perchlorate medium is similar to those reported by Kraus and Nelson [50KRA/NEL]. There is no evidence provided that the second stage of hydrolysis actually involves the formation of $U(OH)_2^{2+}$. If the dihydroxo species is formed, the paper suggests the formation constant is about an order of magnitude smaller than the formation constant for UOH^{3+}. The reported formation constants for UOH^{3+} are given a weight of zero in this review.

[75HUT]
Hutin, M.F., Etudes des complexes uranylperiodates en solution aqueuse, Bull. Soc. Chim. France, (1975) 463-466, in French.

Hutin [71HUT, 75HUT] made a calorimetric study of complexes between U(VI) and $H_3IO_6^{2-}$ at pH > 13 and concluded that a complex containing four anions per U(VI) be formed. The evaluation was done using the method of continuous variation (the Job method) which is useful only when one complex is formed. The author did not address the question of other complexes being formed, nor did he discuss the extensive hydrolysis that is expected at these high pH values. This review finds no reason to support the claims that $UO_2(H_3IO_6)_4^{6-}$ should be formed.

Hutin also did a potentiometric study of complex formation between U(VI) and periodates at low pH (1.10 < pH < 1.65) where hydrolysis is negligible. The author proposed that the following reactions take place:

$$IO_4^- + 2H_2O + 2UO_2^{2+} \rightleftharpoons (UO_2)_2HIO_6(aq) + 3H^+$$
$$H^+ + (UO_2)_2HIO_6(aq) + IO_4^- + 2H_2O \rightleftharpoons 2H_3UO_2IO_6(aq).$$

The author seems to have based the idea of the stoichiometry of the complexes on an analysis of solids that are formed in the same pH region, *e.g.*, $H(UO_2)_2IO_6 \cdot 8H_2O(cr)$ and $NaH_2UO_2IO_6 \cdot 7H_2O(cr)$. The stoichiometric composition of the solids does not inherently give any information about the composition of the complexes present in solution, and the potentiometric study only gives a qualitative indication that dioxouranium(VI) forms complexes with periodate. No information about the ionic medium, temperature or calibration procedures is given in Refs. [71HUT, 75HUT].

[75JOH]
Johnson, I., Report ANL-RDP-36, Argonne National Laboratory, Argonne, Illinois, USA, 1975.

This author plotted the Gibbs energies of formation of the caesium uranates from their oxides per gram-mole of oxide *vs.* the mole fraction of $Cs_2O(cr)$, x_{Cs_2O}, in the compound. The Gibbs energies of formation of $Cs_2UO_4(cr)$ ($-1775.34\,kJ\cdot mol^{-1}$, $x_{Cs_2O}=\frac{1}{2}$) and $Cs_2U_2O_7(cr)$ ($-2960.76\,kJ\cdot mol^{-1}$, $x_{Cs_2O}=\frac{1}{3}$) were used as reference data. The value of $\Delta_f G^\circ_m$ of $Cs_4U_5O_{17}(cr)$ ($x_{Cs_2O}=\frac{2}{7}$) was estimated to be about $0.8\,kJ\cdot mol^{-1}$ less negative than that obtained by a linear extrapolation using the values for $Cs_2UO_4(cr)$ and $Cs_2U_2O_7(cr)$. This estimation makes the compound $Cs_2U_2O_7(cr)$ stable with respect to decomposition into $Cs_2UO_4(cr)$ and $Cs_4U_5O_{17}(cr)$. The Gibbs energy of formation of five other compounds, $Cs_2U_4O_{13}(cr)$, $Cs_2U_5O_{16}(cr)$, $Cs_2U_7O_{22}(cr)$, $Cs_2U_{16}O_{49}(cr)$, and $Cs_2U_4O_{12}(cr)$, were estimated in a similar way.

Johnson [75JOH] used a similar procedure to estimate the enthalpies of formation of the six compounds. The enthalpies of formation of $Cs_2UO_4(cr)$ ($-1897.28\,kJ\cdot mol^{-1}$) and $Cs_2U_2O_7(cr)$ ($-3155.99\,kJ\cdot mol^{-1}$) were used as reference data. From the estimated values of $\Delta_f G^\circ_m$ and $\Delta_f H^\circ_m$, the entropy of formation and the absolute entropy of each compounds were computed. Results obtained by Johnson [75JOH] are presented in Table A.3.

Table A.3: Thermodynamic functions of caesium polyuranates estimated by Johnson [75JOH] at 298.15 K and 0.1 MPa. These data are not selected in this review.

Compound	$\Delta_f G^\circ_m$ ($kJ\cdot mol^{-1}$)	$\Delta_f H^\circ_m$ ($kJ\cdot mol^{-1}$)	$\Delta_f S^\circ_m$ ($J\cdot K^{-1}\cdot mol^{-1}$)	S°_m ($J\cdot K^{-1}\cdot mol^{-1}$)
$Cs_4U_5O_{17}(cr)$	−7099.2[(a)]	−7562.6	−1554.4	770.0
$Cs_2U_4O_{13}(cr)$	−5314.4[(a)]	−5659.7	−1158.1	543.5
$Cs_2U_5O_{16}(cr)$	−6480.8[(a)]	−6899.8	−1405.4	654.8
$Cs_2U_7O_{22}(cr)$	−8797.9[(a)]	−9372.6	−1927.6	848.5
$Cs_2U_{16}O_{49}(cr)$	−19218.5[(a)]	−20479.8	−4230.4	1767.3
$Cs_2U_4O_{12}(cr)$	−5196.2[(a)]	−5517.4	−1077.4	522.6

(a) calculated in this review.

For this graphical method, Johnson [75JOH] used unreliable thermodynamic data of $Cs_2UO_4(cr)$ and $Cs_2U_2O_7(cr)$ as reference data to estimate thermodynamic functions of six caesium polyuranate compounds. Furthermore, the choice of the value of $\Delta_f G^\circ_m$ of each compound to be about $0.8\,kJ\cdot mol^{-1}$ less negative than that obtained by a linear extrapolation of the two values of $Cs_2UO_4(cr)$ and $Cs_2U_2O_7(cr)$, was purely arbitrary. For these two reasons, this review does not select the thermodynamic values estimated by Johnson [75JOH].

[75OHA/HOE2]
O'Hare, P.A.G., Hoekstra, H.R., Thermochemistry of uranium compounds: VI. Standard enthalpies of $Cs_2U_2O_7$, thermodynamics of formation of cesium and rubidium uranates at elevated temperatures, J. Chem. Thermodyn., 7 (1975) 831-838.

The value from this paper for $\Delta_f H^\circ_m(Cs_2U_2O_7$, cr, 298.15 K) is recalculated using the values $\Delta_f H^\circ_m(UO_2Cl_2$, cr, 298.15 K) = $-(1243.6 \pm 1.3)$ kJ · mol^{-1} and $\Delta_{sol}H(UO_2Cl_2$, 1 M HCl(aq)) = $-(101.7 \pm 1.7)$ kJ · mol^{-1} selected in the present review, the CODATA [89COX/WAG] values for $\Delta_f H^\circ_m(Cs^+$, aq, 298.15 K) and $\Delta_f H^\circ_m(Cl^-$, aq, 298.15 K), and the heat of solution, $\Delta_{sol}H^\circ_m$, of CsCl(cr) in H_2O(l) based on those reported in Annex II of the CODATA review [89COX/WAG]. Hence, $\Delta_f H^\circ_m(Cs_2U_2O_7$, cr, 298.15 K) = $-(3220.1 \pm 4.5)$ kJ · mol^{-1} is calculated.

A slightly less negative value would result if $\Delta_{sol}H(UO_2Cl_2$, 1 M HCl(aq)) = $-(100.2 \pm 1.7)$ kJ · mol^{-1} were used, as is done elsewhere by Cordfunke *et al.* [90COR/KON]. This heat of solution value would be more consistent with experimental work done by Cordfunke and coworkers [82COR/OUW, 88COR/OUW], but slightly inconsistent with the value used elsewhere in this review, and by CODATA as part of their analysis for $\Delta_f H^\circ_m(UO_2^{2+}$, aq, 298.15 K). The values agree within the stated uncertainties, and the selection of the heat of dilution from Fuger *et al.* [83FUG/PAR] is done for no reason other than consistency (also see the Appendix A discussion for Ref. [88COR/OUW]).

[75WES/SOM]
Westrum, E.F., Jr., Sommers, J.A., Downie, D.B., Grønvold, F., Cryogenic heat capacities and Neel temperatures of two uranium pnictides, UAs_2 and USb_2, Thermodynamics of Nuclear Materials 1974, Proc. Symp. held 1974 in Vienna, Austria, Vol. II, Vienna: International Atomic Energy Agency, 1975, *pp.*409-418.

See comments under Grønvold *et al.* [78GRO/ZAK].

[76BAE/MES]
Baes, C.F., Jr., Mesmer, R.F., The hydrolysis of cations, New York: Wiley & Sons, 1976, 489p.

See comments under Hietanen [56HIE].

[76CHO/UNR]
Choppin, G.R., Unrein, P.J., Thermodynamic study of actinide fluoride complexation, in: Transplutonium 1975, Proc. Symp. held 13-17 September, 1975, in Baden-Baden, Federal Republic of Germany, Amsterdam: North-Holland, 1976, *pp.*97-107.

Choppin and Unrein studied the thermodynamic parameters of formation of the mono-fluoride complexes of Am(III), Cm(III), Bk(III), Cf(III), Th(IV), U(IV) and Np(IV) at 25°C and in 1 M perchlorate medium. The U(IV) studies were performed in 1 M $HClO_4$ by using a fluoride membrane electrode. The equilibrium constant for the reaction

$$U^{4+} + HF(aq) \rightleftharpoons UF^{3+} + H^+ \qquad (A.85)$$

was determined at 3, 25 and 47°C. The experimental method is satisfactory, as can be seen from the good agreement between the potentiometric and extraction data for Th(IV). The uncertainties reported by the authors represent "one standard deviation", *i.e.*, the 68% confidence level (*cf.* Ref. [82TAY, Table A]). As mentioned in Appendix C, this review assigns uncertainties that represent the 95% confidence level, and therefore doubles the uncertainties reported in this paper.

From the temperature dependence of this reaction and the enthalpy of protonation of F^- the authors reported $\Delta_r H_m(A.86) = (0.75 \pm 4.00)$.

$$U^{4+} + F^- \rightleftharpoons UF^{3+} \qquad (A.86)$$

The authors used the value of Walker [70WAL], $\Delta_r H_m(A.87) = (13.56 \pm 0.42)\,kJ \cdot mol^{-1}$ to correct for the protonation reaction.

$$F^- + H^+ \rightleftharpoons HF(aq) \qquad (A.87)$$

This review prefers the more precise value of Ahrland and Kullberg [71AHR/KUL2] of $\Delta_r H_m(A.87) = (12.17 \pm 0.04)\,kJ \cdot mol^{-1}$. A direct error estimate of $\Delta_r H_m(A.87)$ from the authors published data shows that $\Delta_r H_m(A.85) = -(13 \pm 3)\,kJ \cdot mol^{-1}$. This results in $\Delta_r H_m(A.86) = (0.8 \pm 4.0)\,kJ \cdot mol^{-1}$.

[76NIK]

Nikolaeva, N.M., Investigation of hydrolysis and complexing at elevated temperatures, Proc. Int. Corr. Conf. Ser. 1973, NACE-4 (1976) 146-152.

Values given in Table 4 of this paper appear to have been calculated from the equations reported in Ref. [71NIK] (although more data points are plotted in Ref. [76NIK, Figure 5] than in Ref. [71NIK]). This is contrary to the statement in the footnote of Table 5 [76NIK] that the pK values are those "experimentally determined" except at 25 and 200°C. Also, contrary to the discussion in the text, the hydrolysis scheme is not the same as that used in Ref. [62NIK/PAR].

[76NIK2]

Nikolaeva, N.M., Solubility product of UO_2CO_3 at elevated temperature, Izv. Sib. Otd. Akad. Nauk SSSR, 6 (1976) 30-31, in Russian.

Both the studies by Nikolaeva [76NIK2] and by Piroshkov and Nikolaeva [76PIR/NIK] concern solubility investigations of $UO_2CO_3(cr)$. The experiments were carried out in the pH-range 3.67 to 4.18 at partial pressures of $CO_2(g)$ from 0.1 to 2 atm. The investigations cover the temperature range 25° to 150°C. The investigations were made in a pH-range where UO_2^{2+} and $UO_2CO_3(aq)$ will be the predominant species. However, the authors made corrections for the hydrolysis (which was in fact fairly small). The experimental technique and the analytical procedures are satisfactory and the reported error estimates are fairly small. The uncertainty in pH, p_{CO_2}, total uranium and total carbonate concentrations are ±0.02, ±0.02 atm, $\pm3\%$ and $\pm0.1\%$, respectively. The values reported for the solubility product and the formation constant of $UO_2CO_3(aq)$ are the averages of three to four experimental determinations. There are no details of the calibration of the pH-electrode and the protonation constants of carbonate seem to refer to $I = 0$. Therefore, this review doubles the uncertainties reported by Nikolaeva.

As no details are given about the experimental procedures, it is difficult to evaluate the high-temperature data, and this review therefore gives them zero weight. However, from the solubility products at 50, 75 and 100°C an enthalpy of $\Delta_r H_m(A.88) = -4.8\,kJ \cdot mol^{-1}$ can be derived for the reaction

$$UO_2CO_3(cr) \rightleftharpoons UO_2^{2+} + CO_3^{2-}, \qquad (A.88)$$

which is close to the value calculated at 25°C using the enthalpies of formation selected in this review, $\Delta_r H_m^\circ(A.88) = -(4.6\pm2.4)\,kJ\cdot mol^{-1}$. This agreement may be a coincidence as the values at 25 and 150°C deviate considerably from the experimental $\log_{10} K(1/T)$ function.

This review estimates that $I \approx 10^{-2}$ M and corrects the constants reported to $I = 0$ using the specific ion interaction theory. The uncertainty in this correction is estimated to be ±0.12 in $\log_{10} K_s$.

[76OHA/BOE]
O'Hare, P.A.G., Boerio, J., Hoekstra, H.R., Thermochemistry of uranium compounds: VIII. Standard enthalpies of formation at 298.15 K of the uranates of calcium ($CaUO_4$) and barium ($BaUO_4$). Thermodynamics of the behavior of barium in nuclear fuels, J. Chem. Thermodyn., 8 (1976) 845-855.

The enthalpy of reaction of $CaUO_4(cr)$ with 3 M $HNO_3(aq)$ from this paper iss used to recalculate $\Delta_f H_m^\circ(CaUO_4, cr, 298.15\ K) = -(2002.3 \pm 2.3)\,kJ \cdot mol^{-1}$ using auxiliary data from the CODATA compilations [87GAR/PAR, 89COX/WAG].

The enthalpy of reaction of $BaUO_4(cr)$ with 1 M HCl(aq) from this paper is used to recalculate $\Delta_f H_m^\circ(BaUO_4, cr, 298.15\ K) = -(1993.8 \pm 3.3)\,kJ \cdot mol^{-1}$ using $\Delta_f H_m^\circ(BaCl_2, s, 298.15\ K) = -(855.15 \pm 2.5)\,kJ \cdot mol^{-1}$ (see Section VI.5.2), the selected value $\Delta_f H_m^\circ(UO_2Cl_2, cr, 298.15\ K) = -(1243.6 \pm 1.3)\,kJ \cdot mol^{-1}$ (Section V.4.2.1.3.h), and $-(101.7 \pm 1.7)\,kJ \cdot mol^{-1}$ as the enthalpy of solution of $UO_2Cl_2(cr)$ in 1 M HCl(aq).

[76OHA/HOE]
O'Hare, P.A.G., Hoekstra, H.R., Fredrickson, D.R., Thermochemistry of uranium compounds: VII. Solution calorimetry of α- and β-Na_2UO_4, standard enthalpy of formation of β-Na_2UO_4, and the enthalpy of the α to β transition at 298.15 K, J. Chem. Thermodyn., 8 (1976) 255-258.

See discussion under Tso *et al.* [85TSO/BRO].

[76PAT/RAM]
Patil, S.K., Ramakrishna, V.V., Sulphate and fluoride complexing of U(VI), Np(VI) and Pu(VI), J. Inorg. Nucl. Chem., 38 (1976) 1075-1078.

Patil and Ramakrishna studied the complex formation of UO_2^{2+} with SO_4^{2-} and F^- at 25°C in a 2.00 M $HClO_4$ solution using an extraction method with dinonyl naphtalene sulphonic acid as the extractant. The maximum $[HSO_4^-]$ was sufficiently high, 0.864 M, to ensure the formation of $UO_2(SO_4)_2^{2-}$. This resulted in large changes in the composition of the ionic medium and the numerical value of the equilibrium constant for the formation of $UO_2(SO_4)_2^{2-}$ is much more uncertain than that for the formation of $UO_2SO_4(aq)$.

The authors reported $^*\beta_1 = 7.6$ and $^*\beta_2 = 17$ for the following reactions:

$$UO_2^{2+} + HSO_4^- \rightleftharpoons UO_2SO_4(aq) + H^+ \quad (A.89)$$
$$UO_2^{2+} + 2HSO_4^- \rightleftharpoons UO_2(SO_4)_2^{2-} + 2H^+. \quad (A.90)$$

No uncertainty estimates are given. This review estimates the uncertainties from the published experimental data to ±1.0 and ±5 to $^*\beta_1$ and $^*\beta_2$, respectively.

Correction to $I = 0$ is performed using the ion interaction coefficients $\varepsilon_{(UO_2^{2+},ClO_4^-)} = (0.46\pm0.03)$, $\varepsilon_{(HSO_4^-,H^+)} = -(0.01\pm0.02)$, $\varepsilon_{(H^+,ClO_4^-)} = (0.14\pm0.02)$, and an estimated $\varepsilon_{(UO_2(SO_4)_2^{2-},Na^+)} \approx \varepsilon_{(SO_4^{2-},Na^+)} = -(0.12 \pm 0.06)$, leading to $\Delta\varepsilon_1 = -(0.31 \pm 0.04)$ and $\Delta\varepsilon_2 = -(0.28 \pm 0.09)$. The corrected constants are $\log_{10}{}^*K_1^\circ = (1.13 \pm 0.11)$ and $\log_{10}{}^*K_2^\circ = (0.61 \pm 0.24)$. By combining these data with the protonation constant of SO_4^{2-}, $\log_{10} K_1^\circ = (1.98 \pm 0.05)$ (see Chapter VI, the values obtained are $\log_{10} \beta_1^\circ = (3.11 \pm 0.12)$ and $\log_{10} \beta_2^\circ = (4.57 \pm 0.26)$.

The data for the fluoride system refer to the reaction $UO_2^{2+} + HF(aq) \rightleftharpoons UO_2F^+ + H^+$. These data were obtained in a HF(aq) concentration range where the amount of $UO_2F_2(aq)$ is negligible.

[76SOU/SHA]
Souka, N., Shabana, R., Farah, K., Adsorption behaviour of some actinides on zirconium phosphate stability constant determinations, J. Radioanal. Chem., 33 (1976) 215-222.

Souka, Shabana and Farah used cation exchange on zirconium phosphate to determine equilibrium constants for the formation of dioxouranium(VI) chloride and nitrate complexes. The measurements were made at (25 ± 3)°C. The authors did not use a medium of constant ionic strength, instead they varied the concentration of HNO_3 or NH_4NO_3 and of HCl. By assuming that only UO_2^{2+} was sorbed by the ion-exchanger the equilibrium constants for the formation of UO_2Cl^+ and $UO_2NO_3^+$ were determined. The values obtained were $\beta_1(UO_2Cl^+) = 2.2$, $\beta_1(UO_2NO_3^+) = 3.3$. It is not clear from the publication how the authors handled the large variation in activity coefficients. They seem to have used the mean activity of HNO_3 and HCl as a measure of the activities of NO_3^- and Cl^-, respectively. Nothing is mentioned about the activities of UO_2^{2+}, UO_2Cl^+, and $UO_2NO_3^+$. The data treatment used by the authors is considered unsatisfactory, and the results are given zero weight.

[77ALL/FAL]
Alles, A., Falk, B.G., Westrum, E.F., Jr., Grønvold, F., Zaki, M.R., Actinoid pnictides: I. Heat capacities from 5 to 950 K and magnetic transitions of U_3As_4 and U_3Sb_4: Ferromagnetic transitions, J. Inorg. Nucl. Chem., **39** (1977) 1993-2000.

This paper provides good quality heat capacity data for U_3As_4(cr) and U_3Sb_4(cr). The value for $S^\circ_m(U_3As_4, cr, 298.15\,K)$ in Table 6 of this paper was incorrectly copied from Table 4.

[77COR/OUW2]
Cordfunke, E.H.P., Ouweltjes, W., Standard enthalpies of formation of uranium compounds: IV. α-UO_2SO_4, β-UO_2SeO_4, and UO_2SeO_3, J. Chem. Thermodyn., **9** (1977) 1057-1062.

The phase designations of α-UO_2SO_4 and β-UO_2SeO_4 are transposed in the title of this work.

[77INA/SHI]
Inaba, H., Shimizu, H., Naito, K., λ-type heat capacity anomalies in U_3O_8, J. Nucl. Mat., **64** (1977) 66-70.

The heat capacity of U_3O_8 was measured by scanning adiabatic calorimetry from 310-970 K, and reported for intervals of 5 K. The authors estimated the precision in the experimental heat capacities for temperatures below 700 K to be 1%. In this review, for the purposes of calculating a function describing the temperature dependence of the heat capacity, the uncertainty (95% confidence level) of each measurement (310 to 600 K) is estimated as 2% of the reported value, and the data are weighted accordingly. It should be noted that the multiplicative factors given in Ref. [77INA/SHI, Eq. (1)] are incorrect. The equation should read:

$$C^\circ_{p,m}(J{\cdot}K^{-1}{\cdot}mol^{-1}) \;=\; 88.08 + 1.5462 \times 10^{-2}\,T - 1.0434 \times 10^{6}\,T^{-2}.$$

[77ISH/KAO]
Ishiguro, S., Kao, C.F., Kakihana, H., Formation constants of hydrogen fluoride ion (HF_m^{1-m}) and uranyl fluoride ion ($UO_2F_n^{2-n}$) complexes in 1 mol·dm^{-3} (sodium chloride) medium, Denki Kagaku Oyobi Kogyo Butsuri Kagaku, **45(10)** (1977) 651-653.

The authors studied the formation of $UO_2F_q^{2-q}$, $q = 1$ to 4, at 25°C in a 1 M NaCl medium using a fluoride electrode. This is a precise method, but the equilibrium constants reported must be corrected for complexation with Cl^-. At $[Cl^-] = 1.0$ M only UO_2Cl^+ is present whose formation constant is $\log_{10}\beta_1(1.0\text{ MCl}^-) = -(0.47 \pm 0.06)$. This value is calculated from the selected data of this review assuming the same $\Delta\varepsilon$ in chloride media as in perchlorate media. Hence, the corrected fluoride equilibrium constants are $\log_{10}\beta_1 = (4.52 \pm 0.06)$, $\log_{10}\beta_2 = (7.90 \pm 0.06)$, $\log_{10}\beta_3 = (9.96 \pm 0.07)$, and $\log_{10}\beta_4 = (10.94 \pm 0.12)$. For the extrapolation of these data to $I = 0$, the values $\varepsilon_{(F^-,Na^+)} = (0.02 \pm 0.02)$, $\varepsilon_{(UO_2F^+,Cl^-)} \approx \varepsilon_{(UO_2F^+,ClO_4^-)} = (0.29 \pm 0.05)$ and $\varepsilon_{(UO_2^{2+},Cl^-)} \approx \varepsilon_{(UO_2^{2+},ClO_4^-)} = (0.46 \pm 0.03)$ (see Appendix B), as well as the estimates $\varepsilon_{(UO_2F_3^-,Na^+)} = (0.00 \pm 0.05)$ and $\varepsilon_{(UO_2F_4^{2-},Na^+)} = -(0.08 \pm 0.06)$ are used. These approximations thus lead to the same $\Delta\varepsilon$ values in both sodium chloride and sodium perchlorate media. (The use of the coefficient $\varepsilon_{(UO_2^{2+},Cl^-)} = (0.21 \pm 0.02)$, as proposed by Ciavatta [80CIA], would lead to inconsistent results because it was obtained by interpreting experimental results exclusively in terms of ionic interaction between UO_2^{2+} and Cl^-, whereas this review in addition considers ion pair formation between UO_2^{2+} and Cl^-.)

With these assumptions and after conversion of the constants to molality units, this review obtains $\log_{10}\beta_1^\circ = (5.13 \pm 0.09)$, $\log_{10}\beta_2^\circ = (8.60 \pm 0.08)$, $\log_{10}\beta_3^\circ = (10.64 \pm 0.11)$ and $\log_{10}\beta_4^\circ = (11.11 \pm 0.15)$. It should be mentioned that the accuracy of $\log_{10}\beta_4$ as determined by the authors cannot be very high because only three points of one titration curve out of four had an average complexation degree $\overline{q}$ markedly over 3, and the maximum value is only $\overline{q} = 3.45$ (see Figure 2 of Ref. [77ISH/KAO]). It is possible that this is part of the reason for the considerable discrepancy between $\log_{10}\beta_4^\circ$ of Ref. [77ISH/KAO] compared with those from Refs. [56AHR/LAR] and [71AHR/KUL], where higher average complexation degrees were achieved. Nevertheless, this argument is not sufficient to discard $\log_{10}\beta_4^\circ$ from Ref. [77ISH/KAO] in the determination of the selected value.

[77NIK]
Nikolaeva, N.M., Study of uranium(IV) complexing with chlorides at elevated temperatures, Izv. Sib. Otd. Akad. Nauk SSSR, Ser. Khim. Nauk, **3** (1977) 114-120, in Russian.

Only an abstract of this paper is available to the reviewers. Nikolaeva reported a potentiometric study (presumably using a chloride selective electrode) of the formation of UCl^{3+} in the temperature range 25 to 150°C. Several different ionic strengths in the range 0.5 to 4.0 M were investigated. The author reported $\log_{10} K^\circ = 21.61 - 11302.3/T + 1680720/T^2$, where K° referred to the reaction $U^{4+} + Cl^- \rightleftharpoons UCl^{3+}$ at $I = 0$. No information was given in the abstract about the extrapolation procedure or the composition of the ionic medium. This review assumes that the ionic medium is $HClO_4$ which is used to reduce the hydrolysis. From the $\log_{10} K^\circ$ function this review calculates $\log_{10} K^\circ(25°C) = 2.6$. This value differs considerably from other determinations and is therefore given zero weight.

[77NIK2]
Nikolaeva, N.M., Complexing in uranyl chloride solutions at elevated temperatures, Izv. Sib. Otd. Akad. Nauk SSSR, Ser. Kim. Nauk, **1** (1977) 56-59, in Russian.

Only an abstract of this publication is available to the reviewers. The author reported equilibrium constants for the reaction $UO_2^{2+} + Cl^- \rightleftharpoons UO_2Cl^+$ in the temperature range 50 to 150°C. The ionic strength was 0.03, and no information was given about the ionic medium. The experimental method was potentiometry, presumably using a chloride electrode. The author reported

$$\log_{10} K = 5.775 - 0.032473\,T + 6.259 \times 10^{-5}\,T^2 \qquad \text{(A.91)}$$

From this function, the values $\log_{10} K(50°C) = 1.8$ and $\log_{10} K(25°C) = 1.7$ were obtained. The extrapolated value seems high. There is no information available about the pH and hydrolysis might at these elevated temperatures affect the data. In view of the limited experimental information available to us we have given these data zero weight. Also, the derived enthalpy data were given zero weight. It is interesting to compare the enthalpy data of Ohashi and Morozumi [67OHA/MOR2] and Nikolaeva [77NIK]. The former work gives $\Delta_r H_m^\circ = -15.4\,\mathrm{kJ\cdot mol^{-1}}$, while Nikolaeva's data indicate a positive $\Delta_r H_m^\circ = 15.9\,\mathrm{kJ\cdot mol^{-1}}$ at 50°C. The two sets of data obviously do not agree very well with one another. The enthalpy value of Nikolaeva [77NIK] is in better agreement with the value of Awasthi and Sundaresan [81AWA/SUN], $\Delta_r H_m^\circ = (8 \pm 2)\,\mathrm{kJ\cdot mol^{-1}}$, selected by this review.

[77NIK/TSV]
Nikolaeva, N.M., Tsvelodub, L.D., Complexing of uranium(IV) with sulfate(2–) ions in aqueous solutions at elevated temperatures, Izv. Sib. Otd. Akad. Nauk SSSR, Ser. Khim. Nauk, 5 (1977) 38-43, in Russian, Chem. Abstr. 88:28492c.

An abstract in English is available to the reviewers. The authors made a potentiometric study of the formation of uranium(IV) sulphate complexes at different temperatures in the range 25 to 150°C and at ionic strengths of 0.4, 1.1, 1.4, 1.6, 1.8, and 2.0 M. The ionic medium was $NaClO_4$. The authors measured the redox potential U(VI)/U(IV) as a function of the concentration of HSO_4^- and made corrections for the formation of HSO_4^- and uranium(IV) sulphate complexes. The hydrolysis of U(IV) was not taken into account. This assumption is doubtful at the higher temperatures. The authors reported values of the equilibrium constants for Reactions (A.92) and (A.93).

$$U^{4+} + SO_4^{2-} \rightleftharpoons USO_4^{2+} \qquad (A.92)$$
$$U^{4+} + 2SO_4^{2-} \rightleftharpoons U(SO_4)_2(aq) \qquad (A.93)$$

They found for $I = 2$ M:

$$\log_{10}\beta_1(A.92) = -2.23 + 1.558\times10^{-2}\,T + 1.552\times10^{-5}\,T^2$$
$$\log_{10}\beta_2(A.93) = -5.92 + 3.819\times10^{-2}\,T - 52765.7\,T^{-2}$$

for $I = 1.1$ M:

$$\log_{10}\beta_1(A.92) = -3.58 + 2.424\times10^{-2}\,T + 9.413\times10^{-6}\,T^2$$
$$\log_{10}\beta_2(A.93) = -0.14 + 2.991\times10^{-2}\,T - 272547.9\,T^{-2}$$

and corrected to $I = 0$:

$$\log_{10}\beta_1(A.92) = 3.098 - 2.208\times10^{-3}\,T + 6.272\times10^{-5}\,T^2$$
$$\log_{10}\beta_2(A.93) = 4.06 + 3.618\times10^{-2}\,T - 351300.6\,T^{-2}$$

This review corrects the values at 25°C of $\log_{10}\beta_1$ and $\log_{10}\beta_2$ to $I = 0$ by using the specific ion interaction theory with $\Delta\varepsilon(A.92) = -(0.38 \pm 0.15)$ and $\Delta\varepsilon(A.93) = -(0.56 \pm 0.16)$, after conversion of the constants to molality units. For $\log_{10}\beta_1^\circ$ the deviations from the values obtained from Refs. [52SUL/HIN] and [55DAY/WIL] are larger for the values measured at the lower ionic strengths; for $\log_{10}\beta_2^\circ$ these deviations are larger at higher ionic strengths. It is difficult to judge the quality of this study. The authors had to make a large number of corrections and also used auxiliary data that are not known to the reviewers. Moreover, the Debye-Hückel term is very large for reactions involving multicharged ions. The numerical values at $I = 0$ are thus critically dependent on the accuracy of the extrapolation procedure. This review does not find it justified to give this study any weight in the data selection procedure.

[77OHA/BOE]
O'Hare, P.A.G., Boerio, J., Fredrickson, D.R., Hoekstra, H.R., Thermochemistry of uranium compounds: IX. Standard enthalpy of formation and high-temperature thermodynamic functions of magnesium uranate ($MgUO_4$). A comment on the non-existence of beryllium uranate, J. Chem. Thermodyn., 9 (1977) 963-972.

The experimental data from this paper are recalculated using CODATA auxiliary data [89COX/WAG] and $\Delta_{sol}H_m^\circ(Mg(NO_3)_2, cr, 298.15\ K) = -(90.1 \pm 0.3)\ kJ \cdot mol^{-1}$ as determined by Boerio and O'Hare [76BOE/OHA]. This results in $\Delta_f H_m^\circ(MgUO_4, cr, 298.15\ K) = -(1859.4 \pm 2.4)\ kJ \cdot mol^{-1}$ from the heat of solution of $MgUO_4$(cr) in 12 M HNO_3, and $\Delta_f H_m^\circ(MgUO_4, cr, 298.15\ K) = -(1856.1 \pm 1.9)\ kJ \cdot mol^{-1}$ from the heat of solution in an HF/HNO_3 mixture. The weighted average is $\Delta_f H_m^\circ(MgUO_4, cr, 298.15\ K) = -(1857.3 \pm 1.5)\ kJ \cdot mol^{-1}$.

[77SCA]
Scanlan, J.P., Equilibria in uranyl carbonate systems: II. The overall stability constant of $UO_2(CO_3)_2^{2-}$ and the third formation constant of $UO_2(CO_3)_3^{4-}$, J. Inorg. Nucl. Chem., 39 (1977) 635-639.

The author determined the overall equilibrium constants for the formation of $UO_2(CO_3)_2^{2-}$ and $UO_2(CO_3)_3^{4-}$ by using a solvent extraction method (for $UO_2(CO_3)_2^{2-}$) and a spectrophotometric method (for the determination of the equilibrium constant for the reaction $UO_2(CO_3)_2^{2-} + CO_3^{2-} \rightleftharpoons UO_2(CO_3)_3^{4-}$). The measurements were made in the presence of CO_2(g) of atmospheric pressure and at such low total concentrations of uranium (1 to 2×10^{-4} M) that no significant amounts of $(UO_2)_3(CO_3)_6^{6-}$ should be present. This is supported by the fact that the authors reported a different absorption spectrum for this species than McClaine *et al.* [56MCC/BUL] and Ferri *et al.* [81FER/GRE]. The test solutions used had a concentration of $NaHCO_3$ in the range 10 to 80 mM. The author worked at a constant ionic strength of 0.1 M ($NaNO_3$) but used an equilibrium constant for $CO_3^{2-} + 2H^+ \rightleftharpoons CO_2(g) + H_2O(l)$ which refers to $I = 0$ ($\log_{10} K = 18.18$).

There are no details presented about the calibration procedure used in the emf measurements, hence, it is not known whether the electrode was calibrated in concentration or activity units. This is a precise study, but the numerical value of the equilibrium constant might be subject to a slight systematic error because of the use of slightly erroneous values of the protonation constants of CO_3^{2-}. Nevertheless, these data are used in this review in the evaluation of the selected value for $\log_{10}\beta_2^\circ$ and $\log_{10}\beta_3^\circ$, but with an increased uncertainty of ± 0.3 in the $\log_{10}\beta$ values reported.

[78BED/FID]
Bednarczyk, L., Fidelis, I., Determination of stability constants of U(VI), Np(VI) and Pu(VI) with chloride ions by extraction chromatography, J. Radioanal. Chem., **45** (1978) 325-330.

Bednarczyk and Fidelis used extraction chromatography with HDEHP as the stationary phase for the determination of equilibrium constants for the formation of $MO_2Cl_q^{2-q}$ complexes, where M is U, Np and Pu. The measurements were made in 2 M $H^+(ClO_4^-, Cl^-)$ at (25 ± 1)°C. The experimental method is satisfactory and the authors report $\beta_1(UO_2Cl^+) = (0.75 \pm 0.04)$ and $\beta_2(UO_2Cl_2(aq)) = (0.23 \pm 0.03)$. The uncertainty in $\log_{10} \beta_1$ is estimated to ± 0.1 by this review. The value of $\log_{10} \beta_1 = -(0.12 \pm 0.10)$ is used, after correction from molarity to molality units, in the extrapolation to $I = 0$. The value of β_2 is more uncertain because of the large changes in the ionic medium necessary in situations where this complex is formed. The uncertainty in this quantity is thus much larger than claimed by the authors. This review assigns an uncertainty of ± 0.30 to $\log_{10} \beta_2$ and uses it, together with data from Ref. [81AWA/SUN], in the extrapolation procedure to $I = 0$.

[78BLA/FOU]
Blaise, A.., Fournier, J.M., Lagnier, R., Mortimer, M.J., Schenkel, R., Henkie, Z., Wojakowski, A., Physical properties of uranium dipnictides, Conf. Ser. Inst. Phys. (Rare Earths and Actinides 1977), **37** (1978), 184-189.

The heat capacity results for UAs_2(cr) in this paper differ by several percent from those reported by Grønvold *et al.* [78GRO/ZAK] for reasons that are not clear to the reviewer. The raw data are not given, and the data appear to be of lower precision than the data in Refs. [75WES/SOM, 78GRO/ZAK].

[78GRO/ZAK]
Grønvold, F., Zaki, M.R., Westrum, E.F., Jr., Sommers, J.A., Downie, D.B., Actinoid pnictides: II. Heat capacities of UAs_2 and USb_2 from 5 to 750 K and antiferromagnetic transitions, J. Inorg. Nucl. Chem., **40** (1978) 635-645.

The data for UAs_2(cr) are the same as in Ref. [75WES/SOM]. However, Grønvold *et al.* [78GRO/ZAK] gave complete details of measurements outlined in the earlier conference proceedings [75WES/SOM]. The agreement in values obtained using two different techniques near 300 K is good.

[78KOB/KOL]
Kobets, L.V., Kolevich, T.A., Umreiko, D.S., Crystalline hydrated forms of triuranyl diorthophosphate, Russ. J. Inorg. Chem., **23** (1978) 501-505.

The authors gave evidence that the tetra- and hexa-hydrates are stable in air at room temperature. This indicates that the equilibrium constant for the transition reaction $(UO_2)_3(PO_4)_2 \cdot 6H_2O(cr) \rightleftharpoons (UO_2)_3(PO_4)_2 \cdot 4H_2O(cr) + 2H_2O(g)$ can be estimated to $\log_{10} K = \log_{10} p^2_{H_2O} \approx \log_{10}(0.017)^2 = -3.54$, assuming a "room temperature" of 20°C and a relative humidity of 70%. No enthalpy data are available for this transition reaction.

[78NIK]

Nikolaeva, N.M., The hydrolysis of U^{4+} ions at elevated temperatures, Izv. Sib. Otd. Akad. Nauk SSSR, Ser. Khim. Nauk, (4) (1978) 91-95, translated from the Russian: ORNL/TR-88/2, Oak Ridge National Laboratory, Oak Ridge, Tennessee, USA, 1988, 14p.

Values are reported for the first hydrolysis constant of U^{4+}, for the temperature range 25 to 150°C, in 0.25 to 1.68 m aqueous $HClO_4$. This is the only major study of the hydrolysis of U^{4+} in acidic media over a wide temperature range. For 25°C the equilibrium constants for the reaction

$$U^{4+} + H_2O(l) \rightleftharpoons UOH^{3+} + H^+ \qquad (A.94)$$

from this spectrophotometric study are slightly smaller than those reported by Kraus and Nelson [55KRA/NEL].

Nikolaeva reported values for the equilibrium constants corrected to zero ionic strength. However, the procedure used was not consistent with the specific ion interaction method used in this review. If the method described in Appendix B is used for the higher temperature data by simply changing the Debye-Hückel A factor appropriately with temperature [59ROB/STO], the extrapolated equilibrium constants are slightly more positive than those reported by Nikolaeva. The resulting values of $\Delta\varepsilon$ are badly scattered [(0.4 ± 0.6), (0.6 ± 0.1), (0.4 ± 0.2), (0.0 ± 0.2), (0.1 ± 0.1), (0.2 ± 0.2) at 25, 50, 75, 100, 125, and 150°C, respectively], but all are substantially more positive than the value for 25°C selected in this review, -0.14. This value is based on all the available experimental data, not just Nikolaeva's. (The value of $\Delta\varepsilon$ might be expected to increase slightly with increasing temperature only because the Debye-Hückel B term becomes larger with increasing temperature.)

If the value for $\Delta\varepsilon$ is fixed at the value found for 25°C, the equilibrium constants extrapolated to zero ionic strength are smaller than Nikolaeva's values by factors ranging from 0.3 to 0.7, and the fit to the experimental data is substantially poorer. This suggests that Nikolaeva's data has rather large uncertainties, but also may reflect failings in the specific ion interaction treatment. In this review the uncertainties in the reported experimental values of $\log_{10} K(A.94)$ from Nikolaeva's paper are estimated as ± 0.2, based primarily on the internal consistency of the data.

If the equilibrium constants for uranium(IV) in 0.5 m aqueous $HClO_4$ are considered, the value $\Delta_r H_m(A.94) = (42 \pm 7)\,kJ \cdot mol^{-1}$ is calculated for the temperature range 25 to 100°C. A similar enthalpy, $(45 \pm 7)\,kJ \cdot mol^{-1}$, is calculated using the data from the 0.99 m $HClO_4$ solutions (50 to 125°C). The extrapolated ($I = 0$) equilibrium constants (specific ion interaction method) can be used to calculate average values (25 to 150°C) of $\Delta_r H_m(A.94) = (46 \pm 4)\,kJ \cdot mol^{-1}$ and $\Delta_r S_m(A.94) = (143 \pm 14)\,J \cdot K^{-1} \cdot mol^{-1}$ if $\log_{10} {}^*\beta_1(A.94)$ for 25°C is fixed at the value selected in this review and $\Delta_r C^\circ_{p,m}$ is assigned a value of $0.0\,J \cdot K^{-1} \cdot mol^{-1}$. The calculated uncertainty does not properly reflect the model error. For example, if the ($I = 0$) equilibrium constants calculated with the $\Delta\varepsilon$ value fixed (at -0.14) are used, a value of $\Delta_r H_m(A.94) = 38\,kJ \cdot mol^{-1}$ is obtained.

[78NIK/PIR]
Nikolaeva, N.M., Pirozhkov, A.V., The solubility product of U(IV) hydroxide at the elevated temperatures, Izv. Sib. Otd. Akad. Nauk SSSR, Ser. Khim. Nauk, **5** (1978) 82-88, in Russian.

The method used for these measurements of the solubility product of "U(IV) hydroxide" is essentially the same as was later used by Bruno *et al.* [86BRU/FER]. Neglect of the probable formation of hydrolysis species, particularly at the higher temperatures, introduces uncertainty in the calculated solubility products [86BRU/FER, 88PAR/POH]. The solubility product at 25°C is substantially smaller than that obtained by Bruno *et al.* [86BRU/FER], but it is still greater than the "thermodynamic" value. The extent of the crystallinity of the UO_2 samples was not reported, although the presence of two different solids was noted for temperatures $\geq 100°C$.

[78SCH/SUL]
Schmidt, K.H., Sullivan, J.C., Gordon, S., Thompson, R.C., Determination of hydrolysis constants of metal cations by a transient conductivity method, Inorg. Nucl. Chem. Letters, **14** (1978) 429-434.

No mention is made of whether the temperature of the experimental solutions was controlled and, if it was, at what temperature. Other work discussed in this review indicates neither UO_2^{2+} nor UO_2OH^+ are expected to be the predominant dioxouranium(VI) species in solutions with pH > 4.4. However, reactions other than

$$UO_2OH^+ + H_3O^+ \rightleftharpoons UO_2^{2+} + 2H_2O(l) \qquad (A.95)$$

were apparently not considered in the calculations.

[79CIA/FER]
Ciavatta, L., Ferri, D., Grimaldi, M., Palombari, R., Salvatore, F., Dioxouranium(VI) carbonate complexes in acid solution, J. Inorg. Nucl. Chem., **41** (1979) 1175-1182.

This is a precise study using emf technique (measurements of H^+ by using a glass electrode). All necessary experimental precautions were taken and the electrodes were calibrated in concentration units. The concentration and p_{CO_2} ranges investigated were $3.2 < -\log_{10} H^+ < 4.8$, 2.5×10^{-4} M < U(VI) < 10^{-2} M and 0.195 atm < p_{CO_2} < 0.97 atm. The medium was 3 M $NaClO_4$.

The authors proposed an equilibrium model involving the species $(UO_2)_2(OH)_2^{2+}$, $(UO_2)_3(OH)_5^+$, $UO_2(OH)_2CO_2(aq) \triangleq UO_2CO_3(aq)$,
$(UO_2)_3(OH)_5CO_2^+$ and a large polynuclear complex of a composition equal, or close to $(UO_2)_{11}(OH)_{24}(CO_2)_6^{2-}$.

The equilibrium constants for the hydroxide complexes are in good agreement with other determinations (in Table V.5, Section V.3.2.1) indicating that the experimental procedure is satisfactory.

A calculation of the relative amounts of the various complexes reveals that only very small amounts of $(UO_2)_3(OH)_5CO_2^+$ are present, at most, 5% of the total concentration of uranium. This review performs some tests of a model without this complex. The mixed complexes proposed by Maya [82MAY] and Tsymbal [69TSY], *viz.* $(UO_2)_2CO_3(OH)_5^-$ and $UO_2CO_3(OH)_3^-$ are also tested.

The results are shown in Table A.4. The error-carrying variable is the emf, E. The equilibrium constants refer to reaction written as

$$m\text{UO}_2^{2+} + n\text{H}_2\text{O(l)} + q\text{CO}_2\text{(g)} \rightleftharpoons (\text{UO}_2)_m(\text{OH})_n(\text{CO}_2)_q + n\text{H}^+$$

The least squares refinements are made by using the subroutine E-TITER of the LETAGROP procedure [69SIL/WAR].

The results indicate that the complexes (0,2,2), (0,5,3), (1,2,1) and (6,24,11) are well determined. The complex (1,2,1) is present in fairly small amounts (~10%) in the solutions with the highest metal ion concentrations, but reaches about 50% at the lowest total concentrations of uranium. The large polynuclear complex is predominant in the region immediately before precipitation. It is not possible from these data to get unequivocal evidence for the composition of the mixed hydroxide/carbonate of lower nuclearity, because, at most, 5% of the total uranium can be present in the form of this complex. The mixed complexes proposed by Maya [82MAY] and Tsymbal [69TSY] are not present in noticeable amounts in the concentration range studied by Ciavatta *et al.* [79CIA/FER].

A mixed hydroxide/carbonate complex appears to be formed, so the composition $(\text{UO}_2)_3(\text{OH})_5\text{CO}_2^+$ is preferred on chemical grounds. This stoichiometry is equivalent to $(\text{UO}_2)_3\text{O(OH)}_2\text{HCO}_3^+$ which may be formed by addition of CO_2 to one of the bound hydroxide groups in $(\text{UO}_2)_3\text{O(OH)}_3^+ \triangleq (\text{UO}_2)_3(\text{OH})_5^+$, *cf.* [70ABE].

The equilibrium constant for the reaction

$$3\text{UO}_2^{2+} + \text{CO}_2\text{(g)} + 5\text{H}_2\text{O(l)} \rightleftharpoons (\text{UO}_2)_3(\text{OH})_5\text{CO}_2^+ + 5\text{H}^+ \qquad \text{(A.96)}$$

at $I = 0$ was calculated from the specific ion interaction theory by assuming that the ion interaction coefficient for the large cation $(\text{UO}_2)_3(\text{OH})_5\text{CO}_2^+$ is equal to $\varepsilon_{((\text{UO}_2)_3(\text{OH})_5\text{CO}_2^+,\text{ClO}_4^-)} = (0.0 \pm 0.1)$. This results in $\log_{10} K^\circ(\text{A.96}) = -(17.3 \pm 0.5)$.

Reanalysis of the data including the final species proposed by the authors, except the large polymer but including $\text{UO}_2(\text{CO}_3)_2^{2-}$, leads to $\log_{10}{}^*\beta_{2,2} = -6.00$ and $\log_{10}{}^*\beta_{5,3} = -16.7$. Inclusion of $(\text{UO}_2)_{11}(\text{CO}_3)_6(\text{OH})_{12}^{2-}$ significantly improves the fit to the data, and this species is included in the reanalysis. The value of ${}^*\beta_{5,3}$ is particularly sensitive to the choice of minor species, especially $(\text{UO}_2)_4(\text{OH})_7^+$. UO_2OH^+ and $(\text{UO}_2)_3(\text{OH})_4^{2+}$ do not appear to be important in the solutions studied. However, for certain solutions, 3% of the uranium would be $\text{UO}_2(\text{CO}_3)_2^{2-}$ even if $\log_{10} \beta_{1,2}$ is as small as 16.1. The fitted value of $\log_{10} \beta_{1,2}$ is 16.6. The value of $\log_{10}{}^*\beta_{2,2}$ in this study is relatively insensitive to the model selected and is assigned an uncertainty of ±0.10 in this review. The value for $\log_{10}{}^*\beta_{5,3}$ is very dependent on the model and is assigned an uncertainty of ±0.6.

[79HAA/WIL]

Haacke, D.F., Williams, P.A., The aqueous chemistry of uranium minerals: Part I. Divalent cation zippeite, Mineral. Mag., **43** (1979) 539-541.

The authors reported the Gibbs energies of formation of divalent metal ion zippeites, $\text{M}_2(\text{UO}_2)_6(\text{SO}_4)_3(\text{OH})_{10}\cdot 8\text{H}_2\text{O}$, M = Mg, Co, Ni and Zn from solubility measurements of the corresponding compounds. The solubilities were corrected for complex

Table A.4: Four aqueous models used by this review to model the experimental data presented by Ciavatta *et al.* [79CIA/FER] on the carbonate-hydroxide complexes of dioxouranium(VI).

Model	Complexes[(a)] (q, n, m)	$\log_{10}{}^*\beta_{q,n,m}$	Standard deviation in the error-carrying variable E (mV)
[79CIA/FER]	(0,2,2)	–6.01 ± 0.01	0.497
	(0,5,3)	–16.63 ± 0.07	
	(1,2,1)	–9.00 ± 0.02	
	(6,24,11)	–71.9 ± 0.1	
	(1,5,3)	–16.8[(b)]	
[79CIA/FER] model without (1,5,3)	(0,2,2)	–6.01 ± 0.01	0.505
	(0,5,3)	–16.56 ± 0.06	
	(1,2,1)	–9.01 ± 0.01	
	(6,24,11)	–71.9 ± 0.1	
[79CIA/FER] model but with [82MAY] mixed complex	(0,2,2)	–5.99 ± 0.01	0.477
	(0,5,3)	–16.68 ± 0.07	
	(1,2,1)	–9.01 ± 0.01	
	(6,24,11)	–72.0 ± 0.1	
	(1,5,2)	–20.1 ± 0.1	
[79CIA/FER] but with [69TSY] mixed complex	(0,2,2)	–6.00 ± 0.01	0.492
	(0,5,3)	–16.68 ± 0.07	
	(1,2,1)	–9.02 ± 0.01	
	(6,24,11)	–71.9 ± 0.1	
	(1,3,1)	–14.7 ± 0.2	

(a) (q, n, m) designates complexes of the form $(UO_2)_m(OH)_n(CO_2)_q^{2m-n}$.
(b) Maximum value –16.5.

formation using published stability constants. The only complexes that influence the solubility in the investigated pH range 4.95 to 5.15 are the hydroxide complexes of UO_2^{2+}. The authors made corrections for activity factors and presumably also used the proper hydrolysis constants from Baes and Mesmer [76BAE/MES]. The experimental information is scarce. No variation of pH or other parameters was made which would allow a confirmation of the results. The authors reported $\Delta_f G_m^\circ$ values for the four zippeites equal to $-13506\,kJ \cdot mol^{-1}$ (M = Mg), $-12695\,kJ \cdot mol^{-1}$ (M = Co), $-12683\,kJ \cdot mol^{-1}$ (M = Ni), and $-12870\,kJ \cdot mol^{-1}$ (M = Zn). These values were obtained from the dissolution reactions

$$M_2(UO_2)_6(SO_4)_3(OH)_{10} \cdot 8H_2O(cr) \rightleftharpoons 2M^{2+} + 6UO_2^{2+} + 3SO_4^{2-} + 10OH^- + 8H_2O(l), \quad (A.97)$$

for which the authors reported solubility products $\log_{10} K_{s,0}^\circ$ equal to -146.1, -145.9, -145.6 and -153.0 for M^{2+} equal to Mg, Co, Ni and Zn, respectively. From these solubility products, this review derives the $\Delta_f G_m^\circ$ values of $-13161\,kJ \cdot mol^{-1}$ (M = Mg), $-12356\,kJ \cdot mol^{-1}$ (M = Co), $-12337\,kJ \cdot mol^{-1}$ (M = Ni), and $-12584\,kJ \cdot mol^{-1}$ (M = Zn), which are very different from those reported by the authors. The reason for this is unclear, since the discrepancy is too large to be attributed to differences in the auxiliary data used. Further experimental studies are needed to obtain conclusive evidence on the behaviour of zippeites in aqueous solutions.

[79LAJ/PAR]

Lajunen, L.H.J., Parhi, S., A note on the hydrolysis of uranyl(VI) ions in 0.5 M $NaClO_4$ medium, Finn. Chem. Letters, (1979) 143-144.

This looks like good work although not enough detail was given for a thorough evaluation. Uncertainties in the equilibrium constants were not reported. Instead, values of "standard deviation" were provided for the parameters (equilibrium constants) obtained by curve-fitting of the data to a model. These did not take into account uncertainties resulting from model error and from inaccuracies in solution compositions or potentiometer calibration. In this review the uncertainties in values of the equilibrium constants from this study are estimated as follows: ± 0.1 in $\log_{10}{}^*\beta_{2,2}$, ± 0.5 in $\log_{10}{}^*\beta_{4,3}$, and ± 0.3 in $\log_{10}{}^*\beta_{5,3}$. Uncertainties in the constants for other minor species are discussed in Chapter V, but it is unfortunate that no information is provided about the other models tested.

[79MIL/ELK]

Milic, N.B., El Kass, G., Hydrolysis of the uranyl ion in a potassium chloride medium, Bull. Soc. Chim. Beograd, 44(4) (1979) 275-279.

There is no explanation given as to why this unusual set of hydrolysis species was chosen. From the diagram, $(UO_2)_2(OH)_3^+$ looks like a species incorporated to correct some type of systematic error. Also, Viljoen [87VIL] noted that equilibrium constants for $(UO_2)_2(OH)_3^+$ and $(UO_2)_3(OH)_5^+$ are not mathematically distinguishable when calculated exclusively by fitting data for low values of the ratio $[OH^-]:[U]_T$. The data of Milic and El Kass were obtained only from solutions with ratios ≤ 0.4.

The raw data were not supplied in the paper, thus reanalysis is not possible. The hydrolysis constants from the study are not credited in this review.

[79SPI/ARN]
Spiess, B., Arnaud-Neu, F., Schwing-Weill, M.J., Behaviour of uranium(VI) with some cryptands in aqueous solution, Inorg. Nucl. Chem. Letters, **15** (1979) 13-16.

The experiments and data analysis used in this work were not designed to obtain precise hydrolysis constants. A limited number of measurements was done and minor species have not been considered in the data analysis. Uncertainties in the equilibrium constants were not reported. Instead values of "σ" were provided for the parameters (equilibrium constants) obtained by curve-fitting of the data to a model. These did not take into account uncertainties resulting from model error and from inaccuracies in solution compositions or potentiometer calibration. In this review the uncertainties in values of the equilibrium constants from this study are estimated as follows: ±0.3 in $\log_{10}{}^*\beta_{2,2}$ and ±0.6 in $\log_{10}{}^*\beta_{5,3}$.

[79SYL/DAV2]
Sylva, R.N., Davidson, M.R., The hydrolysis of metal ions: Part 2. Dioxouranium(VI), J. Chem. Soc. Dalton Trans., (1979) 465-471.

This is one of the most thorough studies of dioxouranium(VI) hydrolysis to date with a good discussion of the problems encountered and the hydrolysis schemes considered. For the most concentrated uranium solutions the ionic strength may have changed by $> 4\%$ during titrations. This would mainly affect the values for the formation constants of the minor species $(UO_2)_3(OH)_4^{2+}$, UO_2OH^+ and, therefore, the species with the highest ratio $n : m$, $(UO_2)_4(OH)_7^+$ (and/or $(UO_2)_3(OH)_7^-$). Uncertainties in the equilibrium constants were not reported. Instead, values of the estimated standard deviations were provided for the parameters (equilibrium constants) obtained by curve-fitting of the data to models. These did not take into account uncertainties resulting from model error and from inaccuracies in solution compositions or potentiometer calibration, factors that will be more important for calculation of the formation constants of the minor species. In this review the uncertainties in values of the equilibrium constants from this study are estimated as follows: ±0.06 in $\log_{10}{}^*\beta^\circ_{2,2}$, ±0.2 in $\log_{10}{}^*\beta^\circ_{5,3}$, and ±0.4 in both $\log_{10}{}^*\beta^\circ_{4,3}$ and $\log_{10}{}^*\beta^\circ_1$. Uncertainties in the constants for other species are discussed in Chapter V.

[80BLA/LAG]
Blaise, A., Lagnier, R., Troć, R., Henkie, Z., Markowski, P.J., Mortimer, M.J., The heat capacities of U_3As_4 and Th_3As_4, J. Low Temp. Phys., **39** (1980) 315-328.

This heat capacity study covered the temperature range 5 to 300 K. The authors indicated that the data of Alles *et al.* [77ALL/FAL] are to be preferred above 200 K. The data in Ref. [77ALL/FAL] also show less scatter than those of Ref. [80BLA/LAG].

[80BLA/LAG2]

Blaise, A., Lagnier, R., Wojakowski, A., Zygmunt, A., Mortimer, M.J., Low-temperature specific heats of some uranium ternary compounds UAsY (Y = S, Se, Te), J. Low Temp. Phys., **41** (1980) 61-72.

The authors estimated the "errors" (uncertainties) in the heat capacities as "about 3%" above 15 K and as ±3% for the measurements on the samples in the low-temperature region (1 to 15 K, done using a different calorimeter).

[80BLA/TRO]

Blaise, A., Troć, R., Lagnier, R., Mortimer, M.J., The heat capacity of uranium monoarsenide, J. Low Temp. Phys., **38** (1980) 79-92.

This heat capacity study covered the temperature range 5 to 300 K. The authors estimated errors in the thermal capacity values as 5% at 6 K, and less than 1% above 15 K. As the compounds only comprised half the mass on which the measurements were made, the uncertainties in the heat capacities at higher temperatures were estimated to be 2%. These rather large uncertainties seem reasonable considering values for other uranium arsenides as determined by this group [80BLA/LAG] and by Alles *et al.* [77ALL/FAL].

[80BRO/HUA]

Brooker, M.H., Huang, C.H., Sylwestrowicz, J., Raman spectroscopy studies of aqueous uranyl nitrate and perchlorate systems, J. Inorg. Nucl. Chem., **42** (1980) 1431-1440.

This study of equilibria in the dioxouranium(VI) nitrate system was made by using Raman spectroscopy. Some studies were made in melts of $UO_2(NO_3)_2{\cdot}6H_2O$(cr). Others were made in aqueous solution. Only the latter will be discussed here. The aqueous studies were made at 25°C in solutions of high ionic strength (0.75 M U(VI) and $NaClO_4$+ $NaNO_3$ = 4.00 M). The Raman data clearly indicate that only one nitrate complex is formed and that this has a concentration equilibrium constant equal to (0.15 ± 0.04). Because of the high ionic strength it is impossible to extrapolate this value to $I = 0$. This value is therefore given zero weight. This study is valuable because it gives direct spectroscopic evidence that only one NO_3^- is coordinated to UO_2^{2+} in aqueous solution. Higher complexes seem only to be formed in very concentrated electrolyte solutions where the solute/solvent ratio is high.

[80COR/WES]

Cordfunke, E.H.P., Westrum, E.F., Jr., Investigations on caesium uranates: VII. Thermochemical properties of $Cs_2U_4O_{12}$, in: Thermochemistry of Nuclear Materials 1979, Proc. Symp. held 1979 in Jülich, Federal Republic of Germany, Vienna: International Atomic Energy Agency, 1980, *pp.*125-141.

In this conference paper, values were reported for the enthalpy of reaction of α-$Cs_2U_4O_{12}$(cr) with an aqueous solution of (1.505 M H_2SO_4 + 0.0350 M $Ce(SO_4)_2$). The weighted average of the heats of reaction is $-(479.94 \pm 1.42)\,kJ \cdot mol^{-1}$. These values were used in a cycle involving dissolution of UO_2 and γ-UO_3 in the same solvent to obtain a value for $\Delta_f H_m^\circ(Cs_2U_4O_{12}, \alpha, 298.15\text{ K})$. However, Cordfunke and Ouweltjes [81COR/OUW] measured the heat of solution of U_3O_8(cr) in the same mixture, and a cycle involving $U_3O_8(cr)$ rather than $UO_2(cr)$ is less likely to be subject to errors rooted in experimental difficulties.

Therefore, the cycle in Ref. [80COR/WES] is modified by replacing Reaction (2) with the reaction

$$U_3O_8(cr) + [2H_2SO_4 + Ce(SO_4)_2](sln) \rightarrow [3UO_2SO_4 + 3Ce_2(SO_4)_3 + 2H_2O](sln)$$

and Reaction (5) with the reaction

$$3U(s) + 4O_2(g) \rightarrow U_3O_8(cr).$$

$\Delta_r H_m(2) = -(351.25 \pm 1.34)\,kJ \cdot mol^{-1}$ is from [81COR/OUW], and $\Delta_r H_m(5)$ is a CODATA value [89COX/WAG]. From CODATA [89COX/WAG] values for $\Delta_f H_m^\circ(Cs^+$, aq, 298.15 K) and $\Delta_f H_m^\circ(SO_4^{2-}$, aq, 298.15 K), and the weighted average of the heats of solution, $\Delta_{sol} H_m^\circ$, of $Cs_2SO_4(cr)$ in $H_2O(l)$ reported in Annex II of Ref. [89COX/WAG], $(17.04 \pm 0.15)\,kJ \cdot mol^{-1}$, a value of $\Delta_f H_m^\circ(Cs_2SO_4$, cr, 298.15 K) $= -(1442.4 \pm 1.1)\,kJ \cdot mol^{-1}$ is calculated. Using these values and other auxiliary data from the original paper, $\Delta_f H_m^\circ(Cs_2U_4O_{12}$, cr, 298.15 K) $= -(5571.8 \pm 3.6)\,kJ \cdot mol^{-1}$ is calculated.

[80DON/LAN]

Dongarra, G., Langmuir, D., The stability of UO_2OH^+ and $UO_2[HPO_4]_2^{2-}$ complexes at 25°C, Geochim. Cosmochim. Acta, 44 (1980) 1747-1751.

Dongarra and Langmuir performed a potentiometric study of the U(VI) phosphate system in the range $4 < -\log_{10}[H^+] < 5$ and reported equilibrium constants for the formation of $UO_2HPO_4(aq)$ and $UO_2(HPO_4)_2^{2-}$ and hydroxide complexes. This study was discussed and criticized by Tripathi [84TRI]. The present review independently examines the data, involving least-squares refinement of different chemical models, based on the published primary experimental data. The present review finds:

- that, for the solutions containing phosphate, it is not possible to refine either the hydrolysis constants above, or the hydrolysis and hydrogen phosphate constants together.
- that $(UO_2)_2(OH)_2^{2+}$ is the predominant hydrolysis complex under the experimental conditions used, not UO_2OH^+ as the authors claim.
- it is not possible to refine the equilibrium constants for the formation of $UO_2(HPO_4)_2^{2-}$, even when the hydroxide constants are not varied in the least-squares refinement. The $UO_2(HPO_4)_2^{2-}$ constant converges to zero, hence, this complex cannot be predominant as was suggested by Dongarra and Langmuir [80DON/LAN].
- the best fit to the experimental data is obtained with a model that contains $UO_2HPO_4(aq)$ and $UO_2PO_4^-$, but even this model is too poor to be accepted by this review. The published data are either too few and/or of insufficient quality to be interpreted by a unique chemical model. Hence, this review agrees with Tripathi [84TRI] and gives this study zero weight.

Regardless of the hydrolysis scheme selected, the data in this paper lead to $\log_{10}{}^*\beta_1^\circ = -(5.1 \pm 0.1)$. Reanalysis with ${}^*\beta_{2,2}$ and ${}^*\beta_{5,3}$ as variables (rather than fixed to the Baes and Mesmer [76BAE/MES] values), gives essentially no changes in the values of ${}^*\beta_1$, ${}^*\beta_{2,2}$ and ${}^*\beta_{5,3}$. The method of extrapolation to $I = 0$ also has little effect. However, uncertainties of a few hundredths of a pH unit in either the measurements or the electrode calibrations would introduce substantial uncertainties in ${}^*\beta_1$. There are only seven solutions for which the pH has been measured in the absence of phosphate. (The total uranium concentration in each initial solution in the absence of phosphate can be calculated using the data for the solutions containing phosphate.) In this review the uncertainty is estimated as ± 0.4 in $\log_{10}{}^*\beta_1^\circ$.

[80PON/DOU]
Pongi, N.K., Double, G., Hurwic, J., Hydrolyse de U(VI) et uranates de sodium, Bull. Soc. Chim. France, **9-10** (1980) 347-353, in French.

The authors indicated that their work supports the existence of $(UO_2)_3(OH)_4^+$, $(UO_2)_3(OH)_6(aq)$ and $(UO_2)_3(OH)_7^+$, in addition to UO_2OH^+, $(UO_2)_2(OH)_2^{2+}$ and $(UO_2)_3(OH)_5^+$, for which they calculated hydrolysis constants. Tripathi [84TRI] digitized the titration curve in one of the figures in this paper, and obtained a reasonable value for $\log_{10}{}^*\beta_{7,3}$. The value for ${}^*\beta_{5,3}$ is much smaller than values reported by other workers. This may be the result of a poor choice of model for the hydrolysis sequence.

[81AWA/SUN]
Awasthi, S.P., Sundaresan, M., Spectrophotometric and calorimetric study of uranyl cation/chloride anion system in aqueous solution, Indian J. Chem., **20A** (1981) 378-381.

Awasthi and Sundaresan determined the equilibrium constants and the enthalpy changes for the reactions

$$UO_2^{2+} + q\mathrm{Cl}^- \rightleftharpoons UO_2Cl_q^{2-q}, \qquad (A.98)$$

where $q = 1$ to 3. The equilibrium measurements were done at 25, 35 and 45°C. Several different ionic strengths up to $I = 4.9$ M were investigated. The ionic medium was $HClO_4$. A spectrophotometric method was used and both this and the calculation procedure are satisfactory. The authors reported their results as empirical functions of $\log_{10}\beta_q(I)$:

$$\log_{10}\beta_1(25°\mathrm{C}) = (0.23 \pm 0.02) - \frac{2.04\sqrt{I}}{1 + 1.6\sqrt{I}} + (0.25 \pm 0.02)I \qquad (A.99)$$

$$\log_{10}\beta_2(25°\mathrm{C}) = -(1.2 \pm 0.03) - \frac{3.06\sqrt{I}}{1 + 1.6\sqrt{I}} + (0.62 \pm 0.02)I \qquad (A.100)$$

At the highest ionic strength (4.9 M) there was evidence for the formation of $UO_2Cl_3^-$.

Eqs. (A.99) and (A.100) have the same form as expected from the specific ion interaction theory (apart from the factor 1.6 instead of 1.5, in the numerator of the Debye-Hückel term). The ionic strength is given in molarity instead of molality, which may affect the $\Delta\varepsilon$ slightly, but not the equilibrium constant at $I = 0$. This review uses Eq. (A.99) to calculate $\log_{10}\beta_1$ at $I = 0.1, 0.2, 0.5, 1, 1.5, 2, 2.5$ and 3 M. These data are then used in the extrapolation procedure to $I = 0$, after conversion to molality units. Data from other studies are also included. The uncertainty given in Eqs. (A.99) seems too optimistic. This review assigns an uncertainty of ± 0.1 in $\log_{10}\beta_1$.

For $\log_{10}\beta_2$ this review calculates values at $I = 0.1, 0.5, 1, 2$ and 3 M and assigns an uncertainty of ± 0.5 in $\log_{10}\beta_2$. After conversion to molality units, these data for $\log_{10}\beta_2$ are used, together with the value of Bednarczyk and Fidelis [78BED/FID], for an extrapolation to $I = 0$. The resulting $\Delta\varepsilon$ value (see Section V.4.2.1.2.a) of $-(0.62 \pm 0.17)$ is compatible with the value calculated for a $HClO_4$ medium using the ion interaction coefficients listed in Appendix B, $-(0.70 \pm 0.04)$. This indicates that the presence of large proportions of chloride ion does not significantly affect the extrapolation procedure.

Awashi and Sundaresan reported $\log_{10}\beta_3 = -1.5$ and -0.1 at $I = 4.0$ and 4.9 M, respectively. It is difficult to interpret these results quantitatively because of the large variations in the ionic medium and the activity of water. The evidence for the formation of a third complex is not found conclusive by this review. Zero weight is therefore assigned to the proposed equilibrium constant.

The calorimetric procedure used is satisfactory but not up to the state of the art. It is assumed here that "subtraction of the blank" [81AWA/SUN, *p*.379] refers to the heat of dilution of the 0.9 M $UO_2(ClO_4)_2$ solution in 4.9 M perchloric acid. The authors report enthalpies of reaction which vary with the ionic strength. This review does not consider it possible to extrapolate these data to $I = 0$ and has therefore somewhat arbitrarily selected $\Delta_r H_m^\circ(\text{A.98}, q = 1) = (8 \pm 2)\,\text{kJ}\cdot\text{mol}^{-1}$ and $\Delta_r H_m^\circ(\text{A.98}, q = 2) = (15 \pm 6)\,\text{kJ}\cdot\text{mol}^{-1}$ as tentative values for the enthalpy changes of the overall formation reactions of UO_2Cl^+ and $UO_2Cl_2(aq)$. These values are in agreement with the temperature dependence of the equilibrium constants at $I = 0$. The proposed value for $\Delta_r H_m(\text{A.98}, q = 3) = 16\,\text{kJ}\cdot\text{mol}^{-1}$ is given zero weight. Confirmation of the formation of anionic chloride complexes of UO_2^{2+} is given in the anion exchange study of Kraus, Moore and Nelson [56KRA/MOO]. These authors observe a maximum in the U(VI) distribution curve between anion and exchanger at a HCl concentration around 8 to 9 M. At this maximum the uncharged complex predominates. These data indicate that $UO_2Cl_3^-$ is a very weak complex.

[81CIA/FER2]

Ciavatta, L., Ferri, D., Grenthe, I., Salvatore, F., The first acidification step of the tris(carbonato)dioxouranate(VI) ion, $UO_2(CO_3)_3^{4-}$, Inorg. Chem., **20** (1981) 463-467.

See comments under Ferri *et al.* [81FER/GRE].

[81COR/OUW]

Cordfunke, E.H.P., Ouweltjes, W., Standard enthalpies of formation of uranium compounds: VI. MUO_3 (M = Li, Na, K, and Rb), J. Chem. Thermodyn., **13** (1981) 187-192.

In this review, using slightly different values for the enthalpies of formation of the sulphate species (in an attempt to maintain consistency with the CODATA assessment [89COX/WAG]—see the comments for Ref. [82COR/MUI] in this Appendix), the value $\Delta_f H_m^\circ(NaUO_3, cr, 298.15\,K) = -(1494.5 \pm 1.6)\,kJ \cdot mol^{-1}$ is calculated from the data in this reference. The value of $\Delta_f H_m^\circ$(NaUO$_3$, cr, 298.15 K) from this paper is in good agreement with an earlier value based on the work of Cordfunke and Loopstra [71COR/LOO]. The values of $\Delta_f H_m^\circ$ for $LiUO_3$(cr), KUO_3(cr) and $RbUO_3$(cr) are similarly recalculated.

It appears from later papers from this group (*e.g.*, [85COR/OUW2]) that the value reported for the heat of Reaction 10 in the reaction cycle is really the sum of the heats for Reactions 10 and 11. The enthalpies of formation of the alkali metal sulphates are calculated from the CODATA values of $\Delta_f H_m^\circ$(M^+, aq, 298.15 K) and $\Delta_f H_m^\circ$(SO_4^{2-}, aq, 298.15 K) and the heats of solution for each alkali metal sulphate in water [89COX/WAG, Annex II].

$$\begin{aligned}\Delta_f H_m^\circ(LiUO_3, cr, 298.15\,K) &= -(1522.3 \pm 1.8)\,kJ \cdot mol^{-1}\\ \Delta_f H_m^\circ(KUO_3, cr, 298.15\,K) &= -(1522.9 \pm 1.7)\,kJ \cdot mol^{-1}\\ \Delta_f H_m^\circ(RbUO_3, cr, 298.15\,K) &= -(1520.9 \pm 1.8)\,kJ \cdot mol^{-1}\end{aligned}$$

[81FER/GRE]

Ferri, D., Grenthe, I., Salvatore, F., Dioxouranium(VI) carbonate complexes in neutral and alkaline solutions, Acta Chem. Scand., A35 (1981) 165-168.

In the studies [81CIA/FER2] and [81FER/GRE] the authors investigated the equilibrium between the species $(UO_2)_3(CO_3)_6^{6-}$ and $UO_2(CO_3)_3^{4-}$ in the presence of CO_2(g), according to the reaction

$$3UO_2(CO_3)_3^{4-} + 3CO_2(g) \rightleftharpoons (UO_2)_3(CO_3)_6^{6-} + 6HCO_3^-.$$

The concentration ranges investigated were $p_{CO_2} = 0.1$ to 0.97 atm, $[U^{VI}]_T = 1$ to 30 mM and $-\log_{10}[H^+] = 5.3$ to 11.8. The emf study is the more precise of the two investigations as indicated by the estimated errors. The glass-electrodes used have been calibrated in solutions of known hydrogen ion concentrations. There are some small deviations between the chemical model suggested by the authors and the experimental data at the lowest total concentration of dioxouranium(VI) used (1.00×10^{-3} M). These deviations occur in the range where the average number of carbonate ions per UO_2^{2+} is close to two and are probably due to the presence of small amounts of $UO_2(CO_3)_2^{2-}$. The inclusion of this complex in the chemical model does not change the equilibrium constant suggested by the authors.

The equilibrium constant obtained in these studies can also be calculated from the solubility investigation reported by Grenthe *et al.* [84GRE/FER]. However, the precision here is much lower than in the emf and the spectrophotometric studies. The data in Refs. [81CIA/FER2] and [81FER/GRE] were used to verify the chemical model, *i.e.*, the formation of $(UO_2)_3(CO_3)_6^{6-}$ and to interpret the solubility data, *cf.* [84GRE/FER, *p.*2441].

The errors in the overall constants are determined by the errors in the solubility study, while the error in the stepwise reaction above is very much lower.

[81MAY/BEG]
Maya, L., Begun, G.M., A Raman spectroscopic study of hydroxo and carbonato species of the uranyl(VI) ion, J. Inorg. Nucl. Chem., 43 (1981) 2827-2832.

The authors studied the frequency of the UO_2-ν_1 symmetric stretching vibration in solutions of varying composition and used this as additional evidence for the speciation in the U(VI)-CO_3^{2-}-OH^- system. The sensitivity of the Raman method is rather low, hence the authors had to use total concentrations of uranium around 0.01 M. Under the experimental conditions used the concentration of $UO_2(CO_3)_2^{2-}$ is negligible in comparison with $(UO_2)_3(CO_3)_6^{6-}$, hence the chemical model used by the authors is not correct. The evidence for the formation of $(UO_2)_2CO_3(OH)_3^-$ presented in this study (the stoichiometry of a solid formed in the same pH/p_{CO_2} range as the complex, and the sharpening of a vibration band) are not convincing to this review. However, the emf study by Maya [82MAY] showed that this complex indeed is formed. The solutions used in the Raman study contain several complexes and not one predominant. A much more systematic study of the variations of band positions and intensities is necessary in order to make any statements about the stoichiometry, *cf.* Rossotti and Rossotti [61ROS/ROS]. The ^{13}C-NMR data indicate the presence of two different carbonate sites, which do not seem to exchange at room temperature. There is no information on integrals and the composition of the solutions recorded. Hence no stoichiometric information is obtained from these data either.

[81VAI/MAK]
Vainiotalo, A., Makitie, O., Hydrolysis of dioxouranium(VI) ions in sodium perchlorate media, Finn. Chem. Letters, (1981) 102-105.

The total uranium concentration was varied only slightly in these experiments. Thus, it is not surprising that the data analysis only provides evidence for the complexes $(UO_2)_2(OH)_2^{2+}$ and $(UO_2)_3(OH)_5^+$. The neglect of the minor species probably slightly distorts the values of the formation constants of these major species, particularly $^*\beta_{5,3}$. Values of three times the standard deviation are provided for the parameters (equilibrium constants) obtained by curve-fitting of the data to the model, but these deviations do not adequately reflect the uncertainties in the constants. In this review the uncertainties in the constants ($I \geq 0.1$ M) are estimated as ± 0.3 in $\log_{10}{}^*\beta_{2,2}$ and ± 0.8 in $\log_{10}{}^*\beta_{5,3}$. For $I = 0.024$ M the estimated uncertainties are larger (± 0.4 in $\log_{10}{}^*\beta_{2,2}$ and ± 0.8 in $\log_{10}{}^*\beta_{5,3}$) because the ionic strength changes during the titration were relatively large.

[82COR/MUI]
Cordfunke, E.H.P., Muis, R.P., Ouweltjes, W., Flotow, H.E., O'Hare, P.A.G., The thermodynamic properties of Na_2UO_4, $Na_2U_2O_7$, and $NaUO_3$, J. Chem. Thermodyn., 14 (1982) 313-322.

The data in this paper are recalculated using $\Delta_f H_m^\circ(UO_2SO_4) = -(1845.14 \pm 0.84)\,kJ\cdot mol^{-1}$, the value selected in this review, and, based on values in the CODATA assessment (including [89COX/WAG, Table II-1]), $\Delta_f H_m^\circ(\beta\text{-}Na_2SO_4) = -(1387.61 \pm 0.84)\,kJ\cdot mol^{-1}$ and $\Delta_f H_m^\circ(H_2SO_4, 1.505H_2SO_4) = -(884.98 \pm 0.04)\,kJ\cdot mol^{-1}$. The recalculated values are $\Delta_f H_m^\circ(Na_2UO_4, \alpha, 298.15\,K) = -(1897.4 \pm 1.1)\,kJ\cdot mol^{-1}$ and

$\Delta_f H_m^\circ(Na_2U_2O_7, \alpha, 298.15\,K) = -(3194.5 \pm 1.8)\,kJ \cdot mol^{-1}$.

[82COR/OUW2]
Cordfunke, E.H.P., Ouweltjes, W., Barten, H., Standard enthalpies of formation of uranium compounds: IX. Anhydrous uranyl arsenates, J. Chem. Thermodyn., **14** (1982) 883-886.

In this paper values were reported for the enthalpies of solution of $(UO_2)_3(AsO_4)_2(cr)$, $(UO_2)_2As_2O_7(cr)$ and $UO_2(AsO_3)_2(cr)$ in 1.505 M aqueous H_2SO_4. These values together with the heat of solution of γ-UO_3 in the same medium [77COR/OUW2], the heats of solution of $(As_2O_5)_3 \cdot 5H_2O(cr)$ in water and in aqueous sulphuric acid, and $\Delta_f H_m^\circ$("$H_3AsO_4(aq)$") from the US NBS compilation [82WAG/EVA] are used to calculate the enthalpies of formation of the dioxouranium(VI) arsenates.

The use of the value for "$H_3AsO_4(aq)$" is ambiguous in that it is dependent on the final total concentration of arsenic in solution. Cordfunke *et al.* [82COR/OUW2] appear to have used the US NBS value, but a proper weighted average based on the values $\Delta_f H_m^\circ$("$H_3AsO_4(ao)$") and $\Delta_f H_m^\circ(H_2AsO_4^-$, aq, 298.15 K) is only $0.1\,kJ \cdot mol^{-1}$ more negative. The experimental heat of solution of $(As_2O_5)_3 \cdot 5H_2O(cr)$ into water, when analysed in terms of $\Delta_f H_m^\circ$("$H_3AsO_4(ao)$") and $\Delta_f H_m^\circ(H_2AsO_4^-, aq, 298.15\,K)$, is not in agreement with the US NBS value for $\Delta_f H_m^\circ((As_2O_5)_3 \cdot 5H_2O, cr, 298.15\,K)$. The heat of solution of $(As_2O_5)_3 \cdot 5H_2O(cr)$ in the aqueous sulphuric acid suggests an unusually large heat, $6\,kJ \cdot mol^{-1}$, for the transfer of "$H_3AsO_4(ao)$" from water to 1.505 M aqueous H_2SO_4.

A revised set of calculations was presented in Barten's thesis [86BAR], but the corrected values, $20.7\,kJ \cdot mol^{-1}$ more negative for each of the anhydrous dioxouranium(VI) arsenates, appear themselves to be in error (see comments under Ref. [86BAR]).

In this review the enthalpies of formation for $(UO_2)_3(AsO_4)_2(cr)$, $(UO_2)_2As_2O_7(cr)$ and $UO_2(AsO_3)_2(cr)$ from Ref. [82COR/OUW2] are accepted, but the uncertainties are estimated as $\pm 8\,kJ \cdot mol^{-1}$ based on uncertainties in the arsenic database, and especially uncertainties in $\Delta_f H_m^\circ((As_2O_5)_3{\cdot}5H_2O$, cr, 298.15 K) and $\Delta_f H_m^\circ(As_2O_5$, cr, 298.15 K).

[82MAY]
Maya, L., Hydrolysis and carbonate complexation of dioxouranium(VI) in the neutral-pH range at 25°C, Inorg. Chem., **21** (1982) 2895-2898.

The experimental emf measurements were made at constant, but very low partial pressures of $CO_2(g)$ (3×10^{-4} to 10^{-2} atm). The pH and dioxouranium(VI) concentration ranges were $5 < pH < 8.5$ and $10^{-3}\,M < U(VI) < 2 \times 10^{-3}\,M$, respectively. The measurements were made with necessary precautions but nevertheless do not seem to be of the highest precision. This judgement is based on the drift reported by the author ($< 1\,mV/2h$) and the standard deviation quoted (2.4 mV or 0.04 pH units). The protolysis constants of carbonate were determined experimentally, and the value $\log_{10}\beta_{p,1,2} = (17.67 \pm 0.03)$ at $I = 0.1$ M ($NaClO_4$) for the reaction $CO_3^{2-} + 2H^+ \rightleftharpoons CO_2(g) + H_2O(l)$ is in fair agreement with the value

$\log_{10}\beta_{p,1,2}$ = (17.52 ± 0.05) calculated by using the specific ion interaction theory and the selected value of $\log_{10}\beta_{p,1,2}$ = (18.15 ± 0.04) (see Chapter VI). A test of Maya's published experimental data against three different chemical models is shown in Table A.5.

Table A.5: Test of the experimental data (I = 0.1 M) published by Maya [82MAY] using three different chemical models. The values not given in the columns of "Model 1", "Model 2" and "Model 3" are the selected values used as fixed parameters, while the values indicated are used as variables in a least squares refinement using the LETAGROP [69SIL/WAR] procedure.

$(UO_2)_m(OH)_n(CO_2)_q$			$\log_{10}{}^*\beta_{q,n,m}$ (a)			
q	n	m	Values selected in the present review (b)	Model 1	Model 2	Model 3
0	2	2	−5.89			
0	5	3	−16.19			
1	2	1	−8.65			
2	4	1	−18.88			
6	12	3	−51.17			
3	6	1	−30.91			−31.19 ± 0.19
0	7	3	−31.81			
1	5	2	−18.63 (c)	−18.83 ± 0.10	−18.87 ± 0.11	−18.90 ± 0.10
1	1	0	−7.616		−7.85 ± 0.09	−7.78 ± 0.06
2	1	0	−17.52			
Error square sum				1.44 × 10³	0.41 × 10³	0.31 × 10³
Standard deviation in measured emf (mV)				6.3	3.4	2.95

(a) Constants according to the reactions $mUO_2^{2+} + nH_2O(l) + qCO_2(g) \rightleftharpoons (UO_2)_m(OH)_n(CO_2)_q^{2m-n} + nH^+$.
(b) The values listed on the selected values recalculated to $I = 0.1$ M.
(c) This is the value that was reported by Maya [82MAY].

All models give the same numerical value of $\log_{10}{}^*\beta_{1,5,2}$. However, the values are somewhat smaller than the one reported by Maya [82MAY]. This may be due to the inclusion of more (unpublished) experimental data in Maya's refinement. There is a small, but significant difference between the formation constant of $UO_2CO_3(aq)$ selected by this review and the one calculated from Maya's experimental data, indicating a small systematic error in the experiments. However, this error does not influence the chemical model or the numerical value of $\log_{10}{}^*\beta_{1,5,2}$.

Grenthe and Lagerman [91GRE/LAG] confirmed the formation of the complex $(UO_2)_2CO_3(OH)_3^-$ ($\hat{=}$ $(UO_2)_2(OH)_5CO_2^-$) but obtained an equilibrium constant that is smaller than the one reported by Maya [82MAY]. However, this constant is in better agreement with the value obtained by this review when reinterpreting Maya's data ($\log_{10}{}^*\beta_{1,5,2} = -(19.21 \pm 0.10)$ [91GRE/LAG] *vs.* $\log_{10}{}^*\beta_{1,5,2} = -(18.90 \pm 0.10)$ in this review). The precision is higher in the study of Grenthe and Lagerman than in the one by Maya (the standard deviations in the measured emf are ± 0.60 mV [91GRE/LAG] and ± 2.40 mV [82MAY], respectively). The equilibrium constant $\log_{10}{}^*\beta_{3,6,1} = -(31.19 \pm 0.19)$ is in fair agreement with the value calculated from the data selected in this review, $\log_{10}{}^*\beta_{3,6,1} = -(30.91 \pm 0.10)$, indicating that the accuracy of the constant is satisfactory.

The structure of the complex $(UO_2)_2CO_3(OH)_3^-$ is not trivial, and it would be of interest to establish the number of magnetically distinct OH^- groups by ^{17}O-NMR. Such a study should be feasible because of the rather broad range of predominance of this complex.

The uncertainty in the chemical model used to describe the carbonate species also introduces uncertainty in the value for the hydrolysis constant ${}^*\beta_{0,5,3}$. The extremely sparse data at low pH lead to low precision in the calculated values of both ${}^*\beta_{0,2,2}$ and ${}^*\beta_{0,5,3}$ (reflected in the large σ values reported for the constants in Ref. [82MAY]). The reported hydrolysis constants are not used in the evaluation in this review.

[82MIL/SUR]

Milic, N.B., Suranji, T.M., Hydrolysis of the uranyl ion in sodium nitrate medium, Z. Anorg. Allg. Chem., **489** (1982) 197-203.

The constants may be mixed constants as the analysis procedure is not clear. In the experimental range considered (using the authors' constants and extrapolated constants from their figure), the species $(UO_2)_3(OH)_4^{2+}$ and $(UO_2)_3(OH)_5^+$ are roughly of equal importance at all selected $NaNO_3$ concentrations between pH 2 and pH 3.8. Both species should have been included in the analysis for all solutions. Part of the difficulty in the data analysis probably results from using titrations with a single total uranium concentration (0.01 M) for most of the media. In this review only the constants based on all the titrations in 3.0 M $NaNO_3$ are credited. The uncertainties are estimated as ± 0.3 in $\log_{10}{}^*\beta_{2,2}$ and ± 0.6 in $\log_{10}{}^*\beta_{5,3}$.

[82MOR/FUG]

Morss, L.R., Fuger, J., Jenkins, H.D.B., Thermodynamics of actinide perovskite-type oxides: I. Enthalpy of formation of Ba_2MgUO_6 and Ba_2MgNpO_6, J. Chem. Thermodyn., **14** (1982) 377-384.

The enthalpy of formation of $MgCl_2$ in 1 M HCl(aq) is calculated from the CODATA [89COX/WAG] values for Mg^{2+}(aq) and Cl^-(aq) and the difference ($5.86\,kJ \cdot mol^{-1}$) in the enthalpies of solution of $MgCl_2$(cr) in 1 M HCl(aq) and in water reported by Shin and Criss [79SHI/CRI]. The reported enthalpy of reaction of Ba_2MgUO_6 with 1 M HCl(aq) is used with this and other auxiliary data selected in the present review (see Chapter VI and the entry for [83MOR/WIL2] in Appendix A) to obtain

$\Delta_f H_m^\circ(Ba_2MgUO_6, cr, 298.15\ K) = -(3245.9 \pm 6.5)\ kJ \cdot mol^{-1}$.

[82OVE/LUN]
Overvoll, P.A., Lund, W., Complex formation of uranyl ions with polyaminopolycarboxylic acids, Anal. Chim. Acta, 143 (1982) 153-161.

This seems to be reasonably careful work. The evidence for $(UO_2)_2(OH)_3^+$ in these solutions is tenuous. However, it is perhaps surprising that no evidence of UO_2OH^+ could be found in titrations containing 5×10^{-4} M uranium. The reported uncertainties in the formation constants appear to have been deduced with considerable thought. In this review the 95% confidence level uncertainties have been estimated as twice the uncertainties selected by the authors except for $\log_{10}{}^*\beta_{5,3}$ for which a larger value of ±0.30 has been estimated to reflect possible model error (omission of other species with a high hydroxide to metal ratio).

[83CAC/CHO2]
Caceci, M.S., Choppin, G.R., The first hydrolysis constant of uranium(VI), Radiochim. Acta, 33 (1983) 207-212.

This is a careful piece of work, but the data analysis suffers from the same flaw as many extraction studies, *i.e.*, there is a lack of consideration of alternate hydrolysis schemes. The triuranyl species $(UO_2)_3(OH)_5^+$ and, perhaps, $(UO_2)_4(OH)_7^+$, or even $(UO_2)_3(OH)_7^-$, should be among the major dioxouranium(VI) species in the pH and concentration range proposed by the authors—even if their own values for ${}^*\beta_1$ are used. Using the values for ${}^*\beta_{5,3}$ and $\varepsilon_{((UO_2)_3(OH)_5^+,ClO_4^-)}$ calculated in this review, more than 20% of the ($\sim 2\times10^{-7}$ M) uranium in the authors' 1 M NaClO4 solutions would be in the form of $(UO_2)_2(OH)_2^{2+}$ or $(UO_2)_3(OH)_5^+$ at pH 6. The polymeric species are even more stable at lower ionic strengths (perchlorate media) and acidities. The reason the values obtained for ${}^*\beta_1$ are smaller than those from potentiometric studies is not clear. The reported hydrolysis constants are assigned a weight of zero in this review.

[83CHI/NEV]
Chierice, G.O., Neves, E.A., The stepwise formation of complexes in the uranyl/azide system in aqueous medium, Polyhedron, 2 (1983) 31-35.

Chierice and Neves [83CHI/NEV] used a potentiometric method to determine the equilibria in the dioxouranium(VI) azide system. The measurements were performed at 25°C in a 2 M $NaClO_4$ medium. The glass electrode was properly standardized in concentration units. The authors report equilibrium constants for the reactions $UO_2^{2+} + qN_3^- \rightleftharpoons UO_2(N_3)_q^{2-q}$, where $q = 1$ to 6. The data at higher ligand concentrations are of lower accuracy. There are two reasons for this:

- The difference between the total ligand concentration and the total concentration of uncomplexed ligands, *i.e.*, $[HN_3(aq)] + [N_3^-]$, is small. This results in large errors in $\bar{q}$ (the average number of bonded ligands). This is a general problem with this experimental method when applied to systems where weak complexes are formed.

- At high ligand concentrations one necessarily obtains large changes in the medium. Hence, the assumption of constant activity factors may not be correct.

A reinterpretation of the experimental data of Chierice and Neves results in somewhat different values for the equilibrium constant. This review does not consider the existence of the higher complexes, $UO_2(N_3)_5^{3-}$ and $UO_2(N_3)_6^{4-}$, as adequately proven.

[83CIA/FER]
Ciavatta, L., Ferri, D., Grenthe, I., Salvatore, F., Spahiu, K., Studies on metal carbonate equilibria: 4. Reduction of the tris(carbonato)dioxouranate(VI) ion, $UO_2(CO_3)_3^{4-}$, in hydrogen carbonate solutions, Inorg. Chem., 22 (1983) 2088-2092.

The normal potential and the chemical composition of the chemical species participating in the reaction

$$UO_2(CO_3)_3^{4-} + 2e^- + 2CO_2(g) \rightleftharpoons U(CO_3)_5^{6-} \qquad (A.101)$$

were determined by a combination of redox, H^+ and spectrophotometric measurements. The complexes are the limiting species of dioxouranium(VI) and uranium(IV), respectively. This is due to the fairly high concentrations of HCO_3^- which had to be used in the study (100 to 300 m M), in order to avoid precipitation of $UO_2(cr)$. These concentrations are much higher than those found in most ground water systems, and caution should be used when modelling such systems at lower carbonate concentrations, where carbonate species of quite different stoichiometries and equilibrium constants may be present.

The authors determined the normal potential for Reaction (A.101) against NHE and reported a value of $-(0.279 \pm 0.001)$ V, *i.e.*, $\log_{10} K = -(9.44 \pm 0.03)$. By combination of Reaction (A.101) with the equilibria

$$UO_2^{2+} + 2e^- + 4H^+ \rightleftharpoons U^{4+} + 2H_2O(l) \qquad (A.102)$$

$$UO_2^{2+} + 3CO_3^{2-} \rightleftharpoons UO_2(CO_3)_3^{4-} \qquad (A.103)$$

$$CO_2(g) + H_2O(l) \rightleftharpoons CO_3^{2-} + 2H^+ \qquad (A.104)$$

whose selected constants are corrected to $I = 3$ M ($NaClO_4$), $\log_{10} K(A.102) = (11.54 \pm 0.40)$, $\log_{10} K(A.103) = (22.40 \pm 0.11)$ and $\log_{10} K(A.104) = -(17.72 \pm 0.18)$, this review obtains for $\log_{10} K(3\text{ M NaClO}_4) = (36.86 \pm 0.55)$ for the reaction

$$U^{4+} + 5CO_3^{2-} \rightleftharpoons U(CO_3)_5^{6-}. \qquad (A.105)$$

The authors used the value $\log_{10} K = 24$ for the equilibrium (A.103), a value that was later revised [79CIA/FER]. The value for Reaction (A.102) used by the authors ($\log_{10} K = 9.2$) refers to $I = 0$. In this review the selected values of $\log_{10} K(A.102)$ and $\log_{10} K(A.103)$, recalculated to 3 M, are preferred.

No information is available on the ionic strength dependence of $\log_{10} K(A.105)$. However, the Debye-Hückel term cancels out and the ionic strength dependence is determined by $\Delta\varepsilon$ only. This review estimates $\varepsilon_{(U(CO_3)_5^{6-},Na^+)} = -(0.27 \pm 0.15)$ and then obtains $\Delta\varepsilon = -(0.82 \pm 0.19)$ and $\log_{10} \beta_5^\circ = (34.0 \pm 0.9)$.

[83DAV/EFR]
Davydov, Yu.P., Efremenkov, V.M., The hydrolysis of uranium(VI) in solution, Russ. J. Inorg. Chem., **28** (1983) 1313-1316.

This paper is based on the same experimental data as Ref. [73DAV/EFR]. If the results of these papers were correct, $(UO_2)_2(OH)_2^{2+}$ (or other polymeric species) would be a minor species except for total uranium concentrations much greater than 0.1 M, a result in disagreement with more than thirty other studies done in the last forty years.

[83FER/GRE]
Ferri, D., Grenthe, I., Salvatore, F., Studies on metal carbonate equilibria: 7. Reduction of the tris(carbonato)dioxouranate(VI) ion, $UO_2(CO_3)_3^{4-}$, in carbonate solution, Inorg. Chem., **22** (1983) 3162-3165.

The authors measured the normal potential for Reaction (A.106) and the chemical composition of the limiting complex of U(V).

$$UO_2(CO_3)_3^{4-} + e^- \rightleftharpoons UO_2(CO_3)_3^{5-} \qquad (A.106)$$

and obtained $E^\circ = -(0.5236 \pm 0.0003)$ V, *i.e.*, $\log_{10} K = -(8.851 \pm 0.005)$. It is difficult to extrapolate this value to $I = 0$ because of the strong ion pairing between $UO_2(CO_3)_3^{5-}$ and Na^+, resulting in a deviation from the linear relationship in the specific ion interaction extrapolation at $I > 2$ M, *cf.* Riglet [90RIG, *p.*105]. By using $\varepsilon_{(UO_2(CO_3)_3^{5-},Na^+)} = -(0.69 \pm 0.14)$ derived from $\Delta\varepsilon$(A.106) [90RIG] and $\varepsilon_{(UO_2(CO_3)_3^{4-},Na^+)}$ evaluated in this review (*cf.* Appendix B), this review obtains $\log_{10} K^\circ(A.106) = -(13.60 \pm 0.49)$. By combining this value with the selected values of $\log_{10} K^\circ$ of Eqs. (A.107) and (A.108),

$$UO_2^{2+} + 3CO_3^{2-} \rightleftharpoons UO_2(CO_3)_3^{4-} \qquad (A.107)$$

$$UO_2^{2+} + e^- \rightleftharpoons UO_2^+ \qquad (A.108)$$

a value of $\log_{10} K(A.109) = (6.54 \pm 0.49)$ can be calculated for the reaction

$$UO_2^+ + 3CO_3^{2-} \rightleftharpoons UO_2(CO_3)_3^{5-}. \qquad (A.109)$$

This value is in fair agreement with $\log_{10} K^\circ(A.109) = (7.1 \pm 0.2)$ proposed by Riglet [90RIG]. This review accepts the uncertainty reported by Riglet [90RIG]. The value of $\log_{10} K(A.109)$ reported by Ferri *et al.* [83FER/GRE] refers to 3.0 M $NaClO_4$ medium. The value of $\log_{10} K(A.107)$ was suggested by the authors, based on the value of $\log_{10} K(A.108)$ calculated from the experimental data of Kern and Orleman [49KER/ORL].

The experimental study was made in solutions of high carbonate concentration $(0.51 < -\log_{10}[CO_3^{2-}] < 1.08)$. Hence, different dioxouranium(V) complexes might form at lower carbonate concentrations.This fact should be kept in mind when modelling systems under these conditions.

[83MAR/PAV]
Marković, M., Pavković, N., Solubility and equilibrium constants of uranyl(2+) in phosphate solutions, Inorg. Chem., 22 (1983) 978-982.

This solubility study was made at 25°C and an ionic strength in the range $3 \cdot 10^{-3}\,\mathrm{M} < I < 39 \cdot 10^{-3}\,\mathrm{M}$ using $UO_2HPO_4 \cdot 4H_2O(cr)$ and $(UO_2)_3(PO_4)_2 \cdot 8H_2O(cr)$ as solid phases. By measuring the solubility of $UO_2HPO_4 \cdot 4H_2O(cr)$ in solutions with $1.41 < \mathrm{pH} < 2.49$ and varying concentrations of phosphate, the authors suggested a chemical model for the soluble species which are formed.

The experimental technique is satisfactory and the solid phases were adequately characterized. The authors' analytical technique and the pH measurements are also state of the art. It is unfortunate that the glass electrode is calibrated in activity units and not in concentrations. However, by using the information about the pH in 0.0100 M HCl + 0.090 M KCl, this review recalculates the pH values [83MAR/PAV, Table I] to the corresponding $-\log_{10}[H^+]$ quantities and uses these data in a reinterpretation of the experimental data.

Marković and Pavković reported equilibrium data for the complexes $UO_2H_2PO_4^+$, $UO_2H_3PO_4^{2+}$, $UO_2(H_2PO_4)_2(aq)$ and $UO_2(H_2PO_4)_2(H_3PO_4)(aq)$. The chemical model is accepted by this review, with the exception of $UO_2(H_2PO_4)_2(H_3PO_4)(aq)$. However, the equilibrium constants differ considerably from the ones selected in this review. One possible reason for this discrepancy might be the iterative method used in Ref. [83MAR/PAV] to determine the equilibrium constants. The authors stated in their caption for Figure 1 [83MAR/PAV] that "the experimental pH values in the samples deviate from the theoretical iso-pH curve" and these deviations are between 0.1 and 0.2 pH units, indicating that the chemical model proposed might not be the "best" one. The data in Ref. [83MAR/PAV] are used to test the model selected in this review. The model includes the species $UO_2H_3PO_4^{2+}$, $UO_2H_2PO_4^+$, $UO_2(H_2PO_4)_2(aq)$ and $UO_2(H_2PO_4)(H_3PO_4)^+$ with the equilibrium constants selected by this review and recalculated to an average ionic strength of 0.02 M. From the reported total concentrations in [83MAR/PAV, Table I] and the recalculated values of $-\log_{10}[H^+]$, this review calculates the concentrations of all species using the program HALTAFALL [67ING/KAK]. From the concentrations of free H^+, UO_2^{2+} and $H_3PO_4(aq)$ thus obtained, the solubility produced for the reaction

$$UO_2HPO_4 \cdot 4H_2O(cr) + 2H^+ \rightleftharpoons UO_2^{2+} + H_3PO_4(aq) + 4H_2O(l) \quad (A.110)$$

is calculated to be $\log_{10}{}^*K_s = -(2.41 \pm 0.13)$, where the uncertainty is chosen so that all the 13 experimental values of $\log_{10}{}^*K_s$ are included. This value refers to an average ionic strength of 0.02 and the corresponding value at $I = 0$ is $\log_{10}{}^*K_s = -(2.53 \pm 0.13)$. This value is virtually identical with the one selected in this review, $\log_{10}{}^*K_s = -(2.50 \pm 0.09)$. Hence, the data of Marković and Pavković are in excellent agreement with the chemical model proposed by this review.

The solubility product of $(UO_2)_3(PO_4)_2 \cdot 8H_2O(cr)$ is in good agreement with the value proposed in this review.

[83MOR/WIL2]
Morss, L.R., Williams, C.W., Choi, I.K., Gens, R., Fuger, J., Thermodynamics of actinide perovskite-type oxides: II. Enthalpy of formation of Ca_3UO_6, Sr_3UO_6, Ba_3UO_6, Sr_3NpO_6, and Ba_3NpO_6, J. Chem. Thermodyn., 15 (1983) 1093-1102.

The enthalpy of reaction of Ca_3UO_6(cr) with 1 M HCl(aq) and the reaction cycles from this paper are used to recalculate $\Delta_f H^\circ_m(Ca_3UO_6$, cr, 298.15 K$) = -(3305.4 \pm 4.1)\,kJ\cdot mol^{-1}$ using auxiliary data from the CODATA compilations [87GAR/PAR, 89COX/WAG]. Using the enthalpy of reaction of Ca_3UO_6(cr) with 2.73 M HNO_3(aq) and CODATA auxiliary data [87GAR/PAR], $\Delta_f H^\circ_m(Ca_3UO_6$, cr, 298.15 K$) = -(3300.2 \pm 4.1)\,kJ\cdot mol^{-1}$ is calculated. The two results agree within the overlap of the stated uncertainties, and the average, $\Delta_f H^\circ_m(Ca_3UO_6$, cr, 298.15 K$) = -(3302.8 \pm 2.9)\,kJ\cdot mol^{-1}$, is accepted in the present review.

The enthalpy of formation of Ba_3UO_6(cr) is recalculated to be $-(3210.4 \pm 8.0)\,kJ\cdot mol^{-1}$ using $\Delta_f H^\circ_m(BaCl_2$, s, 298.15 K$) = -(855.15\pm2.5)\,kJ\cdot mol^{-1}$ (see Section VI.5.2) and the enthalpy of solution of $BaCl_2$(cr) in 1.0 M aqueous HCl as determined (1.0109 M) by Cordfunke, Konings and Ouweltjes [90COR/KON2].

[84GRE/FER]
Grenthe, I., Ferri, D., Salvatore, F., Riccio, G., Studies on metal carbonate equilibria: Part 10. A solubility study of the complex formation in the uranium(VI) water-carbon dioxide system at 25°C, J. Chem. Soc. Dalton Trans., (1984) 2439-2443.

These measurements were carried out at different partial pressures of CO_2(g) (0.051 atm $< p_{CO_2} <$ 0.97 atm). The solubility $\log_{10} S$ (equal to $\log_{10} p_{CO_2} - 2\log_{10}[H^+]$) is independent of p_{CO_2}, hence, no ternary complexes containing OH^- or HCO_3^- are formed in the pH region investigated. The formation of such ternary complexes was reported in Refs. [79CIA/FER] and [82MAY]. However, there is no contradiction between these results as the two latter investigations were made at quite different values of the total concentration of U(VI), pH and p_{CO_2}. The solubility data of Grenthe *et al.* [84GRE/FER] are corrected for hydrolysis using the selected hydrolysis constants for U(VI). The data at two different ionic strengths are used to verify and determine the ion interaction coefficients $\varepsilon_{(UO_2^{2+},ClO_4^-)}$, $\varepsilon_{(UO_2(CO_3)_2^{2-},Na^+)}$ and $\varepsilon_{(UO_2(CO_3)_3^{4-},Na^+)}$ as described in the original publication. The calculations refer to reactions of the type

$$m UO_2CO_3(cr) + q CO_2(g) + q H_2O(l) \rightleftharpoons (UO_2)_m(CO_3)_{m+q}^{-2q} + 2q H^+ \quad (A.111)$$

where $p = 1$ or 3 and $q = -1$, 0, 1, 2, 3. The equilibrium constants according to Reaction (A.111) given in Ref. [84GRE/FER, Table 2] are used, after recalculation using $\log_{10} K^\circ(A.112) = (18.15 \pm 0.04)$, for the reaction

$$CO_3^{2-} + 2H^+ \rightleftharpoons CO_2(g) + H_2O(l) \quad (A.112)$$

as selected in Section VI.4.1.1, and corrected to the appropriate ionic strength, to calculate the formation constants of the various complexes that are used for the evaluation in Section V.7.1.2.

The measured equilibrium data [84GRE/FER, Table 2] are satisfactory, some of the derived constants given in [84GRE/FER, Table 3] are not. For the reaction $CO_2(g)+H_2O(l) \rightleftharpoons CO_3^{2-}+2H^+$ this review prefers using the value $\log_{10} K(I = 0.5\,M) = -(17.25 \pm 0.04)$ calculated from the CODATA Key Values selected in this review (*cf.* Section VI.4.1.1) at $I = 0$ and the specific ion interaction theory, rather than the value $\log_{10} K^\circ = -(17.15 \pm 0.03)$ given by Grenthe *et al.* [84GRE/FER]. The values thus obtained are listed in Table A.6. The various ε values are also re-evaluated by this review as described in Section V.7.1.2.1.a).

The uncertainty estimates given by Grenthe *et al.* [84GRE/FER] are accepted in the present review.

Table A.6: Equilibrium constants reported for the dioxouranium(VI)-carbonate system by Grenthe *et al.* [84GRE/FER].

Reaction			$\log_{10} K(I = 0.5\,M)$	$\log_{10} K(I = 3.0\,M)$
$UO_2CO_3(cr)$	$\rightleftharpoons$	$UO_2^{2+} + CO_3^{2-}$	-13.31 ± 0.03	-13.94 ± 0.03
$UO_2^{2+} + CO_3^{2-}$	$\rightleftharpoons$	$UO_2CO_3(aq)$	8.39 ± 0.10	8.89 ± 0.10
$UO_2^{2+} + 2CO_3^{2-}$	$\rightleftharpoons$	$UO_2(CO_3)_2^{2-}$	15.56 ± 0.15	16.20 ± 0.15
$UO_2^{2+} + 3CO_3^{2-}$	$\rightleftharpoons$	$UO_2(CO_3)_3^{4-}$	21.76 ± 0.15	22.61 ± 0.15
$3UO_2^{2+} + 6CO_3^{2-}$	$\rightleftharpoons$	$(UO_2)_3(CO_3)_6^{6-}$	54.33 ± 0.30	56.23 ± 0.30

[84KOT/EVS]
Kotvanova, M.K., Evseev, A.M., Borisova, A.P., Torchenkova, E.A., Zakharov, S.V., Study of uranyl ion hydrolysis reaction equilibria in potassium nitrate solutions, Moscow Univ. Chem. Bull., **39(6)** (1984) 37-40.

The potentiometric titration(s?) was apparently of a solution 0.001 M in dioxouranium(VI) nitrate ($I = 0.1$ M (KNO_3)). There is no indication what method was used to calibrate the electrodes or how the uranium stock solution was checked for acid content. The data analysis was based on first assuming two hydrolysis species, $(UO_2)_2(OH)_2^{2+}$ and $(UO_2)_4(OH)_7^+$ and fitting the data to hydrolysis constants for these species. Then, with these constants fixed, the authors attempted to add other species to the hydrolysis model that would improve agreement between the experimental and calculated pH values. None of the parameters (equilibrium constants) calculated by such a procedure is likely to be correct unless further refined, and they are not credited in this review.

[85COR/OUW2]

Cordfunke, E.H.P., Ouweltjes, W., Prins, G., Standard enthalpies of formation of uranium compounds: XI. Lithium uranates, J. Chem. Thermodyn., 17 (1985) 19-22.

Cordfunke, Ouweltjes and Prins used a value for $\Delta_f H_m^\circ(SO_4^{2-}$, aq, 298.15 K) from the 1975 tentative CODATA set. In the present recalculation of these results, CODATA [89COX/WAG] values for $\Delta_f H_m^\circ(Li^+$, aq, 298.15 K) and $\Delta_f H_m^\circ(SO_4^{2-}$, aq, 298.15 K) are used. Also, using the weighted average of the heats of solution, $\Delta_{sol} H_m^\circ$, of Li_2SO_4(cr) in H_2O(l) reported in Ref. [89COX/WAG, Annex II], $-(29.86\pm0.15)$ kJ · mol^{-1}, a value of $\Delta_f H_m^\circ(Li_2SO_4$, cr, 298.15 K) $= -(1436.4 \pm 0.5)$ kJ · mol^{-1} is calculated. Using these minor corrections, the recalculated enthalpies of formation of the uranates are:

$$\begin{aligned}\Delta_f H_m^\circ(Li_2UO_4, \text{cr}, 298.15\text{ K}) &= -(1967.4 \pm 1.5)\text{ kJ}\cdot\text{mol}^{-1}\\ \Delta_f H_m^\circ(Li_4UO_5, \text{cr}, 298.15\text{ K}) &= -(2639.4 \pm 1.7)\text{ kJ}\cdot\text{mol}^{-1}\\ \Delta_f H_m^\circ(Li_2U_3O_{10}, \text{cr}, 298.15\text{ K}) &= -(4437.4 \pm 4.1)\text{ kJ}\cdot\text{mol}^{-1}\end{aligned}$$

[85FUG]

Fuger, J., Thermochemistry of the alkali metal and alkaline earth-actinide complex oxides, J. Nucl. Mater., 130 (1985) 253-265.

The enthalpies of reaction of the diuranates with 1 M HCl(aq) from this paper are used with auxiliary data consistent with the present review (see Appendix A comments on Refs. [74OHA/HOE2] and [74OHA/HOE4], including heats of solution of the alkali metal chlorides in 1 M HCl(aq) from these references) to calculate the enthalpies of formation:

$$\begin{aligned}\Delta_f H_m^\circ(Li_2U_2O_7, \text{cr}, 298.15\text{ K}) &= -(3213.6 \pm 5.3)\text{ kJ}\cdot\text{mol}^{-1}\\ \Delta_f H_m^\circ(K_2U_2O_7, \text{cr}, 298.15\text{ K}) &= -(3250.5 \pm 4.5)\text{ kJ}\cdot\text{mol}^{-1}\\ \Delta_f H_m^\circ(Rb_2U_2O_7, \text{cr}, 298.15\text{ K}) &= -(3232.0 \pm 4.3)\text{ kJ}\cdot\text{mol}^{-1}\\ \Delta_f H_m^\circ(Cs_2U_2O_7, \text{cr}, 298.15\text{ K}) &= -(3225.5 \pm 4.5)\text{ kJ}\cdot\text{mol}^{-1}\end{aligned}$$

[85GEN/FUG]

Gens, R., Fuger, J., Morss, L.R., Williams, C.W., Thermodynamics of actinide perovskite-type oxides: III. Molar enthalpies of formation of Ba_2MAnO_6 (M = Mg, Ca or Sr); An = U, Np or Pu) and of M_3PuO_6 (M = Ba or Sr), J. Chem. Thermodyn., 17 (1985) 561-573.

The enthalpies of reaction are used with auxiliary data selected in the present review (see Chapter VI and the entry for [83MOR/WIL2] in this Appendix) to obtain $\Delta_f H_m^\circ(Ba_2CaUO_6$, cr, 298.15 K) $= -(3295.8 \pm 5.9)$ kJ · mol^{-1} and $\Delta_f H_m^\circ(Ba_2SrUO_6$, cr, 298.15 K) $= -(3257.3 \pm 5.7)$ kJ · mol^{-1}.

[85GUO/LIU]

Guorong, M., Liufang, Z., Chengfa, Z., Investigation of the redox potential UO_2^{2+}/U^{4+} and the complex formation between U^{4+} and NO_3^- in nitric acid, Radiochim. Acta, 38 (1985) 145-147.

Guorung, Liufang and Chengfa studied the complex formation between U^{4+} and NO_3^- in $H(ClO_4,NO_3)$ medium with constant ionic strength and constant acidity, both equal to 3.0 M. The temperature was (30 ± 1)°C. The authors measured the redox potential of the UO_2^{2+}/U^{4+} complex. The experimental procedure is not satisfactory as the investigators used a Ag, AgCl|K^+,Cl^- 1.00 M reference electrode. This will result in a precipitation of $KClO_4$ in the fritted glass disk used to connect the reference and the test solutions, as well as in a large unreproducible diffusion potential due to the large difference in acidity and ionic strength between test and reference solutions.

This review has redone the graphical extrapolation used in Ref. [85GUO/LIU]. The plot of ψ_1 [85GUO/LIU, *p*.146] is satisfctory, but ψ_2 and ψ_3 reveal that the experimental data are not satisfactory at nitrate concentrations higher than 0.6 M. The existence of UNO_3^{3+} and $U(NO_3)_2^{2+}$ certified by this study and the equilibrium constants quoted seem, of the correct magnitude. This review has estimated the uncertainties in β_1 and β_2 to ±0.15 and ±0.25 , respectively. The data do not have the quality necessary for the determination of higher complexes. In view of the experimental shortcomings given above, this review has given zero weight to this study, even though the values for β_1 and β_2 agree well with other experimental data.

[85PAR/POH]
Parks, G.A., Pohl, D.C., Hydrothermal solubility of uraninite, DOE/ER/12016-1, Stanford University, Standford, California, USA, Dec. 1985, 43p.

See comments under Parks and Pohl [88PAR/POH]. There are a number of errors in the data analysis in this report that were corrected in the subsequent paper [88PAR/POH]. Markedly different experimental results are reported for samples 11-12 and 9-5 in Ref. [88PAR/POH].

[85RUD/OTT]
Rudigier, H., Ott, H.R. and Vogt, O., Low-temperature specific heat of uranium monopnictides and monochalcogenides, Phys. Rev. B, **32** (1985) 4584-4591.

Ref. [85RUD/OTT2] appears to be a preliminary presentation of the experiments described in detail in this paper [85RUD/OTT]. Data for US(cr) is given only in Ref. [85RUD/OTT].

[85RUD/OTT2]
Rudigier, H., Ott, H.R. and Vogt, O., Specific heat of uranium monopnictides and monochalcogenides, Physica, **130B** (1985) 538-540.

See comments under Rudigier, Ott and Vogt [85RUD/OTT].

[85SAW/CHA]
Sawant, R.M., Chaudhuri, N.K., Rizvi, G.H., Patil, S.K., Studies on fluoride complexing of hexavalent actinides using a fluoride ion selective electrode, J. Radioanal. Nucl. Chem., **91** (1985) 41-58.

Sawant *et al.* studied the complex-formation reactions of the type $UO_2^{2+} + qF^- \rightleftharpoons UO_2F_q^{2-q}$ in 1 M $NaClO_4$ at 21°C. A fluoride selective electrode was used, and all necessary experimental precautions were taken to control liquid-junction potentials and to calibrate the electrodes. This review recalculates the equilibrium constants to 25°C by using the enthalpy data of Ahrland and Kullberg [71AHR/KUL2]. The corrections terms are not larger than 0.004 logarithmic units and consequently do not affect the magnitudes of the constants: $\log_{10}\beta_1 = (4.56 \pm 0.05)$, $\log_{10}\beta_2 = (7.99 \pm 0.05)$, $\log_{10}\beta_3 = (10.34 \pm 0.05)$. Since the authors did not make an uncertainty estimate, this review estimates their uncertainties to be similar to those of Ahrland and Kullberg [71AHR/KUL] who used a similar technique. The constants are converted to molality constants and corrected to $I = 0$ by using the ion interactions coefficients as listed in Appendix B. This results in $\log_{10}\beta_1^\circ = (5.17 \pm 0.08)$, $\log_{10}\beta_2^\circ = (8.66 \pm 0.07)$ and $\log_{10}\beta_3^\circ = (10.96 \pm 0.10)$.

[85SCH/FRI]
Schreiner, F., Friedman, A.M., Richards, R.R., Sullivan, J.C., Microcalorimetric measurement of reaction enthalpies in solutions of uranium and neptunium compounds, J. Nucl. Mat., 130 (1985) 227-233.

This is a calorimetric study of the enthalpy changes in the dioxouranium(VI) carbonate system in the presence of Na_2SO_4. The authors reported enthalpy changes for Reactions (A.113) and (A.114),

$$UO_2^{2+} + 2CO_3^{2-} \rightleftharpoons UO_2(CO_3)_2^{2-} \quad (A.113)$$

$$UO_2^{2+} + 3CO_3^{2-} \rightleftharpoons UO_2(CO_3)_3^{4-}, \quad (A.114)$$

$\Delta_r H_m^\circ(A.113) = -(39.6 \pm 1)\,kJ\cdot mol^{-1}$ and $\Delta_r H_m^\circ(A.114) = -(57.5 \pm 1.5)\,kJ\cdot mol^{-1}$. The technical quality of the enthalpy data seems to be satisfactory. However, the reactions studied are not the ones proposed by the authors. The dioxouranium(VI) which is not complexed to carbonate is present as sulphate complexes and not as UO_2^{2+}. The two main reactions taking place under the conditions used by Schreiner *et al.* are

$$UO_2SO_4(aq) + 2CO_3^{2-} \rightarrow \tfrac{1}{3}(UO_2)_3(CO_3)_6^{6-} + SO_4^{2-} \quad (A.115)$$

$$UO_2SO_4(aq) + 3CO_3^{2-} \rightarrow UO_2(CO_3)_3^{4-} + SO_4^{2-} \quad (A.116)$$

At the rather high uranium concentrations used ($\approx$ 30 mM) no $UO_2(CO_3)_2^{2-}$ is formed, but rather $(UO_2)_3(CO_3)_6^{6-}$. By combining Eq. (A.116) with the selected enthalpy of reaction for

$$UO_2^{2+} + SO_4^{2-} \rightleftharpoons UO_2SO_4(aq), \quad (A.117)$$

$\Delta_r H_m^\circ(A.117) = (19.5 \pm 1.6)\,kJ\cdot mol^{-1}$, this review calculates, for the reaction

$$UO_2^{2+} + 3CO_3^{2-} \rightleftharpoons UO_2(CO_3)_3^{4-}, \quad (A.118)$$

$\Delta_r H_m^\circ(A.118) = -(38.0 \pm 2.2)\,kJ\cdot mol^{-1}$, which is used for the evaluation of the value selected in this review.

In both $UO_2(CO_3)_3^{4-}$ and $(UO_2)_3(CO_3)_6^{6-}$, the UO_2^{2+} groups are entirely surrounded by coordinated CO_3^{2-}, hence, the formation of mixed CO_3^{2-}/SO_4^{2-} is unlikely. By taking the difference between the enthalpy values reported, this review obtains $\Delta_r H_m^\circ(A.119) = -57.5 + 39.6 = -(17.9 \pm 1.8)\,kJ\cdot mol^{-1}$.

$$\tfrac{1}{3}(UO_2)_3(CO_3)_6^{6-} + CO_3^{2-} \rightleftharpoons UO_2(CO_3)_3^{4-} \quad (A.119)$$

This value is in good agreement with $\Delta_r H_m^\circ(A.119) = -(15.4 \pm 0.9)\,kJ \cdot mol^{-1}$ reported by Grenthe *et al.* [84GRE/SPA] in 3 M $NaClO_4$. The large difference between the enthalpy value of Reaction (A.114) above and the results of other investigators is probably due to the formation of sulphate complexes.

[85TSO/BRO]

Tso, T.C., Brown, D., Judge, A.I., Halloway, J.H., Fuger, J., Thermodynamics of the actinoid elements: Part 6. The preparation and heats of formation of some sodium uranates(VI), J. Chem. Soc. Dalton Trans., (1985) 1853-1858.

The auxiliary data used in this paper appear to be consistent with the CODATA compilation [89COX/WAG] and with values used elsewhere in this review. The auxiliary data were used directly with the enthalpies of solution for α-Na_2UO_4 and β-Na_2UO_4 from [76OHA/HOE] to calculate $\Delta_f H_m^\circ(Na_2UO_4, \alpha, 298.15\,K) = -(1900.5 \pm 2.3)\,kJ \cdot mol^{-1}$ and $\Delta_f H_m^\circ(Na_2UO_4, \beta, 298.15\,K) = -(1886.7 \pm 2.3)\,kJ \cdot mol^{-1}$.

[86BAR]

Barten, H., Thermochemical investigations on uranyl phosphates and arsenates, Ph.D. thesis, ECN-188, The Netherlands Energy Research Foundation, Petten, The Netherlands, 1986, 145p.

Much of the content of this thesis was published in papers which are discussed elsewhere in Appendix A of this review, *cf.* [82COR/OUW2, 87BAR, 88BAR2].

The values $\Delta_f H_m^\circ((UO_2)_3(AsO_4)_2, cr, 298.15\,K) = -(4689.4 \pm 2.8)\,kJ \cdot mol^{-1}$, $\Delta_f H_m^\circ((UO_2)_2As_2O_7, cr, 298.15\,K) = -(3426.0 \pm 2.1)\,kJ \cdot mol^{-1}$ and $\Delta_f H_m^\circ(UO_2(AsO_3)_2, cr, 298.15\,K) = -(2156.6 \pm 1.3)\,kJ \cdot mol^{-1}$ [82COR/OUW2], based on the value of $\Delta_f H_m^\circ$("H_3AsO_4(aq)") tabulated by the US NBS [68WAG/EVA], were decreased by $20.7\,kJ \cdot mol^{-1}$ in this dissertation [86BAR]. It appears that the "correction" is itself in error, or that a different thermodynamic cycle was used to obtain the new result. The details of the calculation of the "correction" are not provided.

Calculations of $\Delta_f H_m^\circ(As_2O_5, cr, 298.15\,K)$ based on recent publications by Beezer *et al.* [65BEE/MOR, 88BAR/BUR] and Minor, *et al.* [71MIN/GIL] lead to the conclusion that the US NBS value, if in error, is probably too negative. Thus, the values of $\Delta_f H_m^\circ$("H_3AsO_4(aq)") and $\Delta_f H_m^\circ((As_2O_5)_3 \cdot 5H_2O, cr, 298.15\,K)$ are also probably too negative, even though Barten's "correction" would indicate they are too positive.

Equations are provided for the enthalpy content functions for $(UO_2)_3(AsO_4)_2$, $(UO_2)_2As_2O_7$, $UO_2(AsO_3)_2$ and $UAsO_5$ (from drop calorimetry results, constrained to give $[H_m^\circ(T) - H_m^\circ(298.15\,K)] = 0$ exactly at 298.15 K), and heat capacity functions obtained by differentiation of the enthalpy content functions. The heat capacity functions are accepted in this review. The calculated heat capacities at any particular temperature are probably accurate within a few percent, however, considering the lack of experimental detail, the heat capacities at 298.15 K are estimated in this review to have an uncertainty of $30\,J \cdot K^{-1} \cdot mol^{-1}$.

[86BRU/FER]

Bruno, J., Ferri, D., Grenthe, I., Salvatore, F., Studies on metal carbonate equilibria: 13. On the solubility of uranium(IV) dioxide, $UO_2(s)$, Acta Chem. Scand., **40** (1986) 428-434.

These were obviously careful measurements, although the nature of the oxide surfaces was not completely established. The "amorphous" material apparently showed some crystallinity, but could be reproducibly synthesized. As suggested by the authors, the high value determined for the solubility product of the ceramic UO_2 probably results from control of the solubility by a small quantity of fine-grained material.

[86COR/OUW]

Cordfunke, E.H.P., Ouweltjes, W., Prins, G., Standard enthalpies of formation of uranium compounds: XIII. Cs_2UO_4, J. Chem. Thermodyn., **18** (1986) 503-509.

$\Delta_f H_m^\circ(Cs_2UO_4$, cr, 298.15 K) $= -(1928.5 \pm 2.4)$ kJ · mol^{-1} is calculated for the HF cycle using the value for $\Delta_f H_m^\circ(UO_3$, γ, 298.15 K) selected in this review, and the CODATA values [89COX/WAG] for $\Delta_f H_m^\circ(Cs^+$, aq, 298.15 K) and $\Delta_f H_m^\circ(F^-$, aq, 298.15 K).

For the sulphate cycle, CODATA [89COX/WAG] values are used for $\Delta_f H_m^\circ(Cs^+$, aq, 298.15 K) and $\Delta_f H_m^\circ(SO_4^{2-}$, aq, 298.15 K) as was the weighted average of the heats of solution, of Cs_2SO_4(cr) in H_2O(l), $\Delta_{sol} H_m^\circ = (17.04 \pm 0.15)$ kJ·mol^{-1} [89COX/WAG, Annex II]. This value is markedly less positive than the experimental value, (17.68 ± 0.17) kJ · mol^{-1}, reported by Cordfunke, Ouweltjes and Prins [86COR/OUW], but the latter enthalpy was not corrected to $I = 0$. $\Delta_f H_m^\circ(Cs_2UO_4$, cr, 298.15 K) $= -(1927.1 \pm 1.7)$ kJ · mol^{-1} is obtained.

[86DES/KRA]

Deschenes, L.L., Kramer, G.H., Monserrat, K.J., Robinson, P.A., The potentiometric and laser raman study of the hydrolysis of uranyl chloride under physiological conditions and the effect of systematic and random errors on the hydrolysis constants, AECL-9266, Atomic Energy of Canada Ltd., Chalk River, Ontario, Canada, 1986, 23p.

This is a reasonable study, but it is difficult to make a direct comparison of the results with those of other authors because no other values have been reported for this temperature. It is unfortunate that neither the raw data nor the initial concentrations of the reagents in the titrations were supplied. The information provided concerning the models used for the data analysis is excellent.

[86ULL/SCH]

Ullman, W.J., Schreiner, F., Calorimetric determination of the stability of U(VI)-, Np(VI)-, and Pu(VI)-SO_4^{2-} complexes in aqueous solution at 25°C, Radiochim. Acta, **40** (1986) 179-183.

This is a precise microcalorimetric study of the enthalpies of reaction for M = U, Np and Pu, according to the reactions

$$MO_2^{2+} + SO_4^{2-} \rightleftharpoons MO_2SO_4(aq) \qquad (A.120)$$

$$MO_2^{2+} + 2SO_4^{2-} \rightleftharpoons MO_2(SO_4)_2^{2-}. \qquad (A.121)$$

Only the uranium data are discussed here.

The experiments were carried out at 25°C in solutions where the ionic strength varied from approximately 0.005 to 0.015 M. In order to calculate the composition of the test solutions the authors used equilibrium constants derived from the literature, $\log_{10}\beta_1^\circ(A.120) = (3.22\pm0.04)$ and $\log_{10}\beta_2^\circ(A.121) = (4.4\pm0.1)$. These values are in fair agreement with the ones selected in this review, (3.15 ± 0.02) and (4.14 ± 0.07), respectively. The authors reported a value of $\Delta_r H_m(A.120) = (19.6\pm0.7)\,kJ\cdot mol^{-1}$, which is in good agreement with the values reported in Refs. [71BAI/LAR] and [71AHR/KUL3]. The uncertainty is based on an estimated uncertainty of 10% in β_1, which is accepted by this review. The value of $\Delta_r H_m(A.121) = (37.8\pm2.0)\,kJ\cdot mol^{-1}$ is uncertain, and this review gives this value zero weight. However, it is close to the more accurate value reported by Ahrland and Kullberg [71AHR/KUL3].

[87BAR]

Barten, H., Chemical stability of arsenates of uranium with a valency lower than six, with emphasis on uranium(V) arsenic oxide $UAsO_5$, Thermochim. Acta, 118 (1987) 107-110.

In this paper experimental values for $\Delta_f H_m^\circ(298.15\text{ K})$ and $[H_m^\circ(T) - H_m^\circ(298.15\text{ K})]$ for $UAsO_5$(cr) are reported (note: in line 1 of page 110 [87BAR] the value of the parameter "C" should be 10.542×10^5 (not 10.542×10^{-5})). The values tabulated at higher temperatures may be suspect because the transition temperatures involving As(cr), As_4(g) and As_2(g) in the paper are not consistent with those in the thesis, *e.g.*, [86BAR, Tables V.7 and B6-9]. Furthermore, the value of $\Delta_f H_m^\circ(UAsO_5, cr, 298.15\text{ K})$ probably is subject to the same difficulty with auxiliary data described for the anhydrous dioxouranium(VI) arsenates in Section V.6.3.2.1. However, for $UAsO_5$ no proper cycle is presented to describe how the enthalpy of formation was calculated. Thus, the value of $\Delta_f H_m^\circ(UAsO_5, cr, 298.15\text{ K})$ cannot be corrected to be consistent with the arsenic database selected in this review. The entropy is an estimate. In view of these problems, the results of Barten for $UAsO_5$, except for the the corrected function for $C_{p,m}^\circ$ [86BAR, Appendix B], are not accepted in this review.

[87BRU/CAS]

Bruno, J., Casas, I., Lagerman, B., Muñoz, M., The determination of the solubility of amorphous UO_2(s) and the mononuclear hydrolysis constants of uranium(IV) at 25°C, Sci. Basis Nucl. Waste Management X, held Dec. 1986 in Boston, Massachusetts, USA, Mat. Rec. Soc. Symp. Proc., 84 (1987) 153-160.

Bruno *et al.* determined the solubility of "amorphous" UO_2 at 25°C in 0.5 M $NaClO_4$ solution. The measured solubility was $10^{-(4.4\pm0.4)}$ (where the uncertainty is estimated in this review as twice the uncertainty reported by the authors), independent of pH between pH values of 5.5 and 10. The pH independence over such a large pH range is a strong indication that material dissolving from the surface of the solid is uranium(IV), and that there is little oxidation of the solution species to dioxouranium(VI) species during the duration of the experiment [88BRU].

The solubility of UO_2("am") ($I = 0.5$ M, $NaClO_4$) increased slowly with decreasing pH between pH values of 5 and 2. This increase was interpreted as indicating that for Reaction (A.122)

$$UO_2(am) + H^+ + H_2O(l) \rightleftharpoons U(OH)_3^+ \qquad (A.122)$$

the equilibrium constant is $\log_{10} K(A.122) = -(0.5 \pm 0.1)$ (although it was not possible to exclude the possibility that polymeric species were present). For the reaction as written, the only charged species involved are one singly charged cationic reactant and one singly charged cationic product. Thus the equilibrium constant should vary only slightly with ionic strength. Bruno *et al.* [87BRU/CAS] calculated $\log_{10}{}^*\beta_3 = -(1.1 \pm 0.1)$ in 0.5 M $NaClO_4$ using the solubility product of the "amorphous" solid [86BRU/FER] corrected to this medium. Similarly, a value of $\log_{10}{}^*\beta_3^\circ = -(0.6 \pm 0.7)$ can be estimated using $\log_{10}{}^*K_{s,0}^\circ = 10^{(0.1\pm0.7)}$ for the solubility product of UO_2("am") (see Section V.3.3.2.2) and using the same correction for the solubility product that was used for "crystalline UO_2" in Ref. [86BRU/FER].

For a medium of 0.5 M $NaClO_4$, the value accepted in this review (based primarily on the results of Kraus and Nelson [50KRA/NEL]) is $\log_{10}{}^*\beta_1 = -(1.5 \pm 0.2)$. If both this and the calculated value for ${}^*\beta_3$ in the medium were correct, the concentration of $U(OH)_3^+$ would be greater than that of UOH^{3+} for acid concentrations < 1.5 M, *i.e.*, for all solutions in which UOH^{3+} was considered to be the sole hydrolysis product of U^{4+} in the experimental hydrolysis studies [50KRA/NEL]. If $U(OH)_3^+$ is not predominant for acid concentrations > 0.01 M (a much more reasonable assumption considering also Refs. [55BET] and [89GRE/BID]) the value of $\log_{10}{}^*\beta_3$ would be < -5.5. The hydrolysis of uranium(IV) below pH 5 is undoubtedly more complicated than assumed by Bruno *et al.* [87BRU/CAS], and their value for ${}^*\beta_3$ (and the derived value for ${}^*\beta_3^\circ$) is not accepted in this review.

[87VIL]
Viljoen, C.L., Hydroxo-species of uranium(VI) in aqueous media, M.Sc. dissertation, University of Port Elizabeth, South Africa, 1987, 104p.

This thesis raises some interesting questions concerning the species for uranium hydrolysis that have been selected in a large number of potentiometric studies of the hydrolysis of dioxouranium(VI). In particular, it is proposed that $(UO_2)_2(OH)_3^+$, not $(UO_2)_3(OH)_5^+$, and $(UO_2)_9(OH)_{19}^-$ rather than $(UO_2)_3(OH)_7^-$ are the important species in hydrolysis of 10^{-4}-10^{-3} M solutions of UO_2Cl_2.

The main result of Viljoen's work is the establishment of distinct breaks in the titration curves (of aqueous uranyl chloride with sodium hydroxide solution) at $[OH^-]/[U]_T$ ratios of 1.50 and 2.11. These values are in marked disagreement with those found by Sutton [49SUT] (1.65 and 2.34, respectively). Tsymbal [69TSY] also reported breaks in the titration curves at $[OH^-]/[U]_T$ ratios between 1.6 and 1.8 but, as discussed in Appendix A of this review, the ratios reported by Tsymbal are suspect.

The study by Viljoen was not designed to provide good numerical values of the hydrolysis constants, and there is some doubt as to the excess acid content of the stock uranium solutions. The values obtained are not considered precise by the author and have accordingly not been used in this review.

It would be useful if other groups were able to confirm or refute the data in this thesis. If the conclusions of Viljoen are correct, extensive re-evaluation of earlier work would be required to obtain proper hydrolysis constants.

The plots shown in a paper from Sillen's group [63DUN/HIE] suggest the 3:2 species is not significant. However, reanalysis of their data [63DUN/SIL, 63HIE/ROW] clearly does not, on a statistical basis, rule out the newly proposed hydrolysis scheme, particularly if other species such as $(UO_2)_4(OH)_7^+$ and $(UO_2)_3(OH)_4^{2+}$ are also included. Similarly, reanalysis of the data of Ciavatta *et al.* [79CIA/FER] does not lead to unambiguous identification of the hydrolysis species.

However, even if Viljoen's model were correct, most calculated total uranium concentrations at 298.15 K would not be significantly affected for pH values less than 5 to 6. At higher pH values the uncertainties are very large both in Viljoen's constants and in the values presented in this review.

[88BAR2]

Barten, H., Thermodynamical properties of uranyl arsenates: II. Derivations of standard entropies, Thermochim. Acta, 126 (1988) 385-391.

In this paper experimental values of the enthalpy content functions $[H_m^\circ(T) - H_m^\circ(298.15\,K)]$ for $(UO_2)_3(AsO_4)_2(cr)$, $(UO_2)_2As_2O_7(cr)$ and $UO_2(AsO_3)_2(cr)$ were reported. These were used with the corresponding enthalpies of formation [82COR/OUW2, 86BAR], the measured vapour pressure of $As_4O_{10}(g)$ over the arsenates (801 to 1273 K) [85BAR/COR], and the value of $\Delta_f H_m^\circ(As_4O_{10}, g, 298.15\,K)$ derived from the experimentally determined vapour pressures over $As_2O_5(cr)$ at high temperatures (865 to 1009 K), to calculate entropies for the dioxouranium(VI) arsenates.

The entropy values for the uranium arsenates are very dependent on the thermodynamic parameters for other arsenic compounds. The value of $\Delta_f H_m^\circ(As_4O_{10}, g, 298.15\,K)$ was used directly in the calculations. $\Delta_f H_m^\circ(As_4O_{10}, g, 298.15\,K) = -(1514 \pm 12)\,kJ \cdot mol^{-1}$ was reported in the paper, but in the thesis [86BAR] the value $-(1565 \pm 16)\,kJ \cdot mol^{-1}$ was preferred. Tables 5, 6 and 7 of Ref. [88BAR2] are identical to Tables B6, B7 and B8 of the thesis [86BAR], and it appears $\Delta_f H_m^\circ(As_4O_{10}, g, 298.15\,K) = -(1565 \pm 16)\,kJ \cdot mol^{-1}$ was also used in the paper. If the US NBS [82WAG/EVA] auxiliary data are used, the value of $\Delta_f H_m^\circ(As_4O_{10}, g, 298.15\,K)$ (based on the estimate of $-1565\,kJ \cdot mol^{-1}$) is decreased by $13\,kJ \cdot mol^{-1}$ to $-1578\,kJ \cdot mol^{-1}$. The enthalpies of formation for the uranium arsenates were taken from the thesis, not Ref. [82COR/OUW2], and thus differ by $20.7\,kJ \cdot mol^{-1}$ from the values selected in this review (see comments under Refs. [82COR/OUW2] and [86BAR]). Finally, it should be mentioned that use of the calculated entropies for $P_4O_6(g)$ and $P_4O_{10}(g)$ from Glushko *et al.* [78GLU/GUR] would lead to a slightly lower estimate for $S_m^\circ(As_4O_{10}, g, 298.15\,K)$.

If the auxiliary data from this review are accepted, together with the enthalpies of formation from Ref. [82COR/OUW2], and the heat capacity functions [86BAR] are assumed to be satisfactory over the entire range of temperature of the vapour pressure measurements, entropy values are obtained that are about $20\,J \cdot K^{-1} \cdot mol^{-1}$ greater than those reported in Ref. [88BAR2]. In this review the values $S_m^\circ((UO_2)_3(AsO_4)_2,$

cr, 298.15 K) = (387 ± 30) J · K^{-1} · mol^{-1}, S°_{m}((UO_2)$_2$$As_2O_7$, cr, 298.15 K) = (307 ± 30) J · K^{-1} · mol^{-1}, and S°_{m}(UO_2(AsO_3)$_2$, cr, 298.15 K) = (231 ± 30) J · K^{-1} · mol^{-1} are selected. The large uncertainties are chosen to reflect the uncertainties in the values for $\Delta_f H^{\circ}_{m}$(As_4O_{10}, g, 298.15 K), $\Delta_f H^{\circ}_{m}$(As_2O_5, cr, 298.15 K), $\Delta_f H^{\circ}_{m}$((UO_2)$_3$(AsO_4)$_2$, cr, 298.15 K), $\Delta_f H^{\circ}_{m}$((UO_2)$_3$$As_2O_7$, cr, 298.15 K), $\Delta_f H^{\circ}_{m}$(UO_2(AsO_3)$_2$, cr, 298.15 K), and S°_{m}(As_4O_{10}, g, 298.15 K).

[88COR/OUW]
Cordfunke, E.H.P., Ouweltjes, W., Standard enthalpies of formation of uranium compounds: XIV. BaU_2O_7 and $Ba_2U_2O_7$, J. Chem. Thermodyn., **20** (1988) 235-238.

The enthalpies of formation of $Ba_2U_2O_7$ and BaU_2O_7 are recalculated using $\Delta_f H^{\circ}_{m}$($BaCl_2$, s, 298.15 K) = $-(855.15 \pm 2.5)$ kJ · mol^{-1} (see Section VI.5.2) and the selected value $\Delta_f H^{\circ}_{m}$(UO_2Cl_2, s, 298.15 K) = $-(1243.6 \pm 1.3)$ kJ · mol^{-1} (Section V.4.2.1.3.h).

Cordfunke, Ouweltjes and Prins [82COR/OUW] reported $-(102.68 \pm 0.64)$ kJ · mol^{-1} for the enthalpy of solution of UO_2Cl_2(s) in the same medium used in Ref. [88COR/OUW]. Later, a correction factor was reported [83COR/OUW] for selected experimental values in Ref. [82COR/OUW], but specifically not for the enthalpy of solution of UO_2Cl_2(s) in aqueous HCl. Nevertheless, in a subsequent paper [88COR/OUW] an identical correction factor seems to have been applied to obtain -102.18 kJ · mol^{-1}. If estimated according to the method of Fuger *et al.* [83FUG/PAR], the value of the enthalpy of solution would be $-(103.3 \pm 1.7)$ kJ · mol^{-1}. Elsewhere in the present review, the enthalpy of solution values from Fuger *et al.* [83FUG/PAR] are used consistently. It would be preferable to use an experimental value. However, in view of the ambiguities in the literature, the value $-(103.3 \pm 1.7)$ kJ · mol^{-1} is used in the present review to recalculate the enthalpies of formation of $Ba_2U_2O_7$(cr) and BaU_2O_7(cr) to be $-(3740.0 \pm 6.3)$ kJ · mol^{-1} and $-(3237.2 \pm 5.0)$ kJ · mol^{-1}, respectively. The uncertainty assigned to the enthalpy of solution encompasses either value reported to have been based on the experiments described by Cordfunke, Ouweltjes and Prins [82COR/OUW].

[88KHA/CHO]
Khalili, F.I., Choppin, G.R., Rizkalla, E.N., The nature of U(VI) complexation by halates and chloroacetates, Inorg. Chim. Acta, **143** (1988) 131-135.

The formation of $UO_2ClO_3^+$, $UO_2BrO_3^+$ and $UO_2IO_3^+$ was studied at 25°C in a 0.1 M $NaClO_4$ medium by extraction with HDEHP. The enthalpies of reaction for the three systems were measured calorimetrically. The experimental procedures are satisfactory. For Reaction (A.123)

$$UO_2^{2+} + XO_3^- \rightleftharpoons UO_2XO_3^+ \qquad X = Cl, Br, I \qquad (A.123)$$

the authors reported the following values: β_1(A.123, X = Cl) = (1.2 ± 0.2), β_1(A.123, X = Br) = (1.6 ± 0.3), β_1(A.123, X = I) = (37.6 ± 1.0), $\Delta_r H_m$(A.123, X = Cl) = $-(3.9 \pm 0.1)$ kJ·mol^{-1}, $\Delta_r H_m$(A.123, X = Br) = (0.1 ± 0.1) kJ·mol^{-1}, and $\Delta_r H_m$(A.123, X = I) = (9.8 ± 0.3) kJ · mol^{-1}. The equilibrium constants were converted to molality units and corrected to $I = 0$ by this review, assuming that $\varepsilon_{(UO_2XO_3^+, ClO_4^-)} \approx$

$\varepsilon_{(UO_2Cl^+,ClO_4^-)} = (0.33 \pm 0.04)$ and $\varepsilon_{(IO_3^-,Na^+)} \approx \varepsilon_{(BrO_3^-,Na^+)} = -(0.06 \pm 0.02)$, resulting in $\log_{10} \beta_1^\circ(X = Cl) = (0.50 \pm 0.07)$, $\log_{10} \beta_1^\circ(X = Br) = (0.63 \pm 0.08)$, and $\log_{10} \beta_1^\circ(X = I) = (2.00 \pm 0.01)$. This review assumes that the uncertainties reported by the authors represent a measure for the precision of the experiment. Since the changes in the ionic medium are not negligible during the experiment, this review has somewhat arbitrarily doubled the uncertainties reported by the authors and considers them as better estimates for the overall accuracy of the resulting values.

The enthalpies of reaction were assumed to have the same value at $I = 0$ as in 0.1 M $NaClO_4$ solution, but the uncertainties of the selected $\Delta_r H_m^\circ$ values were increased from ± 0.1 to $\pm 0.3\,kJ \cdot mol^{-1}$ for X = Cl, Br and from ± 0.3 to $\pm 0.9\,kJ \cdot mol^{-1}$ for X = I for this reason. The equilibrium constant for the iodate complex differs significantly from the results reported by Klygin, Smirnova and Nikol'skaya [59KLY/SMI2]. This review re-interprets the data of Klygin, Smirnova and Nikol'skaya [59KLY/SMI2] and finds better agreement with the observations made by Khalili, Choppin and Rizkalla, who worked at such low concentrations of iodate that only $UO_2IO_3^+$ is formed in significant amounts.

[88PAR/POH]
Parks, G.A., Pohl, D.C., Hydrothermal solubility of uraninite, Geochim. Cosmochim. Acta, **52** (1988) 863-875.

These appear to be careful measurements. The experimental solubilities were found to be essentially pH and temperature independent for pH > 4. The authors concluded the only important solution species for these conditions is $U(OH)_4(aq)$. Their temperature independent value for the equilibrium constant of

$$UO_2(cr) + 2H_2O(l) \;=\; U(OH)_4(aq), \qquad (A.124)$$

$\log_{10} K(A.124) = -(9.47 \pm 0.56)$, is consistent with a value of $\Delta_r S_m^\circ$(A.124, 298.15 K) of approximately $-181\,J \cdot K^{-1} \cdot mol^{-1}$ and $\Delta_r C_{p,m}^\circ$ defined as $0\,J \cdot K^{-1} \cdot mol^{-1}$. This would lead to $S_m^\circ(U(OH)_4$, aq, 298.15 K$) = 36\,J \cdot K^{-1} \cdot mol^{-1}$ and $C_{p,m}^\circ(U(OH)_4$, aq, 298.15 K$) \approx 220\,J \cdot K^{-1} \cdot mol^{-1}$. Such values are not unreasonable for a neutral aqueous species, although the entropy is a bit small. A larger entropy value for this species would lead to a calculated solubility for UO_2 at 298.15 K that is lower than those found experimentally at the higher temperatures.

If the values of $\log_{10}[U]_T$ are fitted to a function of temperature,

$$\log_{10}[U]_T = \left(\frac{298.15}{T}\right) \log_{10}[U]_{298.15} + \Delta_r S_m^\circ \left(\frac{T - 298.15}{2.303\,RT}\right)$$

for solutions with pH > 4.0 (allowing $\Delta_r S_m^\circ$ to vary, but again setting $\Delta_r C_{p,m}^\circ$ to zero) an extrapolated value of $10^{-(9.8\pm1.3)}$ M is obtained for the solubility at 298.15 K, and $S_m^\circ(U(OH)_4, aq, 298.15\,K) = (49 \pm 44)\,J \cdot K^{-1} \cdot mol^{-1}$ is calculated. Justification for a second fitting parameter is marginal.

There was considerable variation in the ionic strength of the acidic solutions, which adds uncertainty to the interpretation of the solubilities in terms of the highly charged aqueous species, UOH^{3+} and $U(OH)_2^{2+}$. Furthermore, it is possible that uranium(IV) fluoro complexes were also formed in the experimental solutions [88PAR/POH]. In view of these difficulties no attempt has been made in this review to calculate Gibbs energies of formation (at 25°C) from these high-temperature data for any species less hydrolysed than $U(OH)_4(aq)$.

[88TAS/OHA]
Tasker, I.R., O'Hare, P.A.G., Lewis, B.M., Johnson, G.K., Cordfunke, E.H.P., Thermochemistry of uranium compounds: XVI. Calorimetric determination of the standard molar enthalpy of formation at 298.15 K, low-temperature heat capacity, and high-temperature enthalpy increments of $UO_2(OH)_2{\cdot}H_2O$ (schoepite), Can. J. Chem., **66** (1988) 620-625.

The values reported for $\Delta_f G^\circ_m(UO_3 \cdot 2H_2O, cr, 298.15\,K)$ appear to have been calculated using the 1 bar standard state entropies for $O_2(g)$ and $H_2(g)$ [82WAG/EVA] rather than (as stated in the paper) the CODATA 1 atm standard state entropies [78COD].

[88ULL/SCH]
Ullman, W.J., Schreiner, F., Calorimetric determination of the enthalpies of the carbonate complexes of U(VI), Np(VI) and Pu(VI) in aqueous solution at 25°C, Radiochim. Acta, **43** (1988) 37-44.

This is essentially the same type of experimental study as the one described by Schreiner *et al.* [85SCH/FRI]. The technique and method of data interpretation is described in Ref. [86ULL/SCH]. The experiments were made at 25°C in a Na_2SO_4 medium with varying ionic strength (from 0.27 to 1.08 M). The methods are satisfactory, but it is unfortunate that the activities for the MO_2^{2+} ions were calculated with the Davies equation, which is not accurate at the high ionic strengths used. However, the errors thus introduced will hardly affect the precision in the measured enthalpy values.

The authors reported $\Delta_r H_m(A.125) = -(42.1 \pm 1.3)\,kJ \cdot mol^{-1}$ and $\Delta_r H_m(A.126) = -(23 \pm 2)\,kJ \cdot mol^{-1}$ for the reactions

$$UO_2^{2+} + 3CO_3^{2-} \rightleftharpoons UO_2(CO_3)_3^{4-} \qquad (A.125)$$
$$UO_2^{2+} + 2CO_3^{2-} \rightleftharpoons UO_2(CO_3)_2^{2-}. \qquad (A.126)$$

The first value is in fair agreement with the one reported by Schreiner *et al.* [85SCH/FRI] after correction for U(VI) sulphate complex formation ($\Delta_r H_m(A.125) = -57.5 + 10.5 = -38.0\,kJ \cdot mol^{-1}$).

The second reaction studied is not the one proposed by Ullman and Schreiner, because at the total concentration of U(VI) used the predominant complex will be $(UO_2)_3(CO_3)_6^{6-}$, rather than $UO_2(CO_3)_2^{2-}$. The value of $\Delta_r H_m(A.126)$ thus refers to the reaction

$$UO_2^{2+} + 2CO_3^{2-} \rightleftharpoons \tfrac{1}{3}(UO_2)_3(CO_3)_6^{6-}, \qquad (A.127)$$

and the experimental value agrees well with the one reported by Grenthe, Spahiu and Olofsson [84GRE/SPA], $\Delta_r H_m(A.127) = -(20.5 \pm 0.9)\,kJ \cdot mol^{-1}$. Ullman and Schreiner [88ULL/SCH] "know of no evidence that polymeric species are formed at ionic strengths and ligand concentrations similar to the ones used in our experiments". Such evidence is presented, *e.g.*, in Ref. [84GRE/FER].

[89BRU/GRE]
Bruno, J., Grenthe, I., Robouch, P., Studies of metal carbonate equilibria: 20. Formation of tetra(carbonato)uranium(IV) ion, $U(CO_3)_4^{4-}$, in hydrogen carbonate solutions, Inorg. Chim. Acta, **158** (1989) 221-226.

This is a spectrophotometric study of the dissociation reaction $U(CO_3)_5^{6-} \rightleftharpoons U(CO_3)_4^{4-} + CO_3^{2-}$ performed at 25°C in solutions of varying concentration of $NaClO_4$ (0.5 to 3 M). The concentrations of U(IV) are around 4×10^{-4} M, and the pH range investigated is $6.86 \leq \log_{10}[H^+] \leq 9.31$. The data indicated both the formation of $U(CO_3)_4^{4-}$ and a strong ionic pairing between Na^+ and $U(CO_3)_4^{4-}$. The equilibrium constant for the addition of CO_3^{2-} to $U(CO_3)_4^{4-}$ is rather small, indicating that K_1 to K_4 for the reaction $U(CO_3)_p^{4-2p} + CO_3^{2-} \rightleftharpoons U(CO_3)_{p+1}^{2-2p}$, $p = 1$ to 3, must be large.

[89BRU/SAN]
Bruno, J., Sandino, A., The solubility of amorphous and crystalline schoepite in neutral to alkaline solutions, Sci. Basis Nucl. Waste Management **XII**, held Oct. 1988 in Berlin, Federal Republic of Germany, Mat. Res. Soc. Symp. Proc., **127** (1989) 871-878.

The paper reports preliminary values for the solubility of amorphous and microcrystalline forms of schoepite for pH values between 7 and 9. The solubilities, particularly of the amorphous form, provide confirmation of the amphoteric behaviour of dioxouranium(VI) in aqueous solutions. The reported values for ${}^*\beta_{7,3}$ and ${}^*\beta_3$ are consistent with the values selected in this review, but are inconsistent with each other if the reported solubility products for the two solids are also considered. Indeed, calculations show that if $\log_{10}{}^*\beta_{7,3} = -31.9$ and $\log_{10}{}^*\beta_3 = -19.69$ in 0.5 M $NaClO_4$, the (7,3) species will be predominant over both solids at saturation for pH values above 8, and the calculated solubilities of the microcrystalline material are much higher than the experimental values. For this reason, and because calculations including other possible species such as $UO_2(OH)_2(aq)$ and $(UO_2)_4(OH)_7^+$ were not reported, the values of the equilibrium constants from this paper have not been accepted directly in this review.

[89GRE/BID]
Grenthe, I., Bidoglio, G., Omenetto, N., Use of thermal lensing spectrophotometry (TLS) for the study of mononuclear hydrolysis of uranium(IV), Inorg. Chem., **28** (1989) 71-74.

Thermal lensing spectrophotometry (TLS) was used so that hydrolysis could be studied in solutions with low (1 to 2×10^{-5} M) total uranium(IV) concentrations. The hydrolysis species formed in 3 M $NaClO_4$ were found to be monomeric. As noted by the authors, the interpretation of the data is complicated by correlation between the calculated values of the molar TLS signal for the hydrolysed species and the values for the association constants. The limiting value for $\log_{10}{}^*\beta_2$ is based on selection of reasonable values for the molar TLS signal for both UOH^{3+} and $U(OH)_2^{2+}$, and the maximum value of the ratio ${}^*\beta_2/{}^*\beta_1^2$ that still gives reasonable agreement with the experimental data. However, the existence of $U(OH)_2^{2+}$, as opposed to more highly hydrolysed species, was not proven. In this review the uncertainty in the value of $\log_{10}{}^*\beta_1$ has been increased to ± 0.10 to reflect the difficulties in selecting a model.

[89YAN/PIT]
Yang, J., Pitzer, K.A., Thermodynamics of aqueous uranyl sulfate to 559 K, J. Sol. Chem., **18** (1989) 189-199.

The authors used earlier isopiestic data for the dioxouranium(VI) sulphate system and Pitzer's ion interaction equations to perform a consistent modelling of the osmotic coefficients of the solutions over a large concentration and temperature range. The authors concluded that the accuracy of the equations at low molality is subject to appreciable uncertainty, which thus will affect the activity coefficients. The result is that calculated activity coefficients within the investigated molality range at a given temperature will be subject to correction by a constant factor. The treatment is incomplete as no complex formation between dioxouranium(VI) and sulphate was taken into account.

[90AHR/HEF]
Ahrland, S., Hefter, G., Norén, B., A calorimetric study of the mononuclear fluoride complexes of zirconium(IV), hafnium(IV), thorium(IV) and uranium(IV), Acta. Chem. Scand., **44** (1990) 1-7.

This is a precise calorimetric study of the enthalpies of reaction of the listed M(IV) ions with HF(aq) in a 4 M $HClO_4$ ionic medium, and at a temperature of 25°C. This review accepts the enthalpies of reaction proposed by the authors, but doubles their uncertainties, which are given as one standard deviation of the least squares refined experimental data in Ref. [90AHR/HEF]. The enthalpies of reaction vary between the different M(IV) ions, while the corresponding entropies of reaction are very nearly the same. The enthalpies of reaction between M(IV) and the fluoride ion was calculated from the experimental data and the experimental enthalpy of reaction for the protonation of fluoride. The latter value was not determined by the authors, but taken from Kozlov *et al.* [73KOZ/BLO]. The value $-11.9\,\mathrm{kJ\cdot mol^{-1}}$ is in good agreement with values at other ionic strengths listed in Chapter VI of this review. This review estimates the uncertainty in the enthalpy of protonation of fluoride in 4 M $HClO_4$ to be $0.5\,\mathrm{kJ\cdot mol^{-1}}$, making the uncertainties in the derived enthalpies of the reaction

$$M^{4+} + qF^- \rightarrow MF_q^{4-q} \qquad \text{(A.128)}$$

considerably larger than for the corresponding reactions with HF(aq).

The enthalpies and entropies of the stepwise complexation reactions, obtained by using the uncertainties estimated in this review and error propagation analysis, are summarized in Table A.7. The enthalpy of reaction for the formation of UF^{3+} is much more precise than the value proposed by Choppin and Unrein [76CHO/UNR], and therefore preferred in this review.

Table A.7: Equilibrium thermodynamic data for the U(IV) fluoride system as reported in Ref. [90AHR/HEF]. The data refer to a solution of 4 M $HClO_4$ at 298.15 K.

Equilibrium			$\Delta_r H_m$ ($kJ \cdot mol^{-1}$)	$\Delta_r S_m$ ($J \cdot K^{-1} \cdot mol^{-1}$)
$U^{4+} + HF(aq)$	$\rightleftharpoons$	$UF^{3+} + H^+$	-17.52 ± 0.12	46.2 ± 0.4
$UF^{3+} + HF(aq)$	$\rightleftharpoons$	$UF_2^{2+} + H^+$	-9.8 ± 0.4	27.5 ± 1.2
$UF_2^{2+} + HF(aq)$	$\rightleftharpoons$	$UF_3^+ + H^+$	-8 ± 4	11 ± 10
$U^{4+} + F^-$	$\rightleftharpoons$	UF^{3+}	-5.6 ± 0.5	154 ± 2
$UF^{3+} + F^-$	$\rightleftharpoons$	UF_2^{2+}	2.1 ± 0.6	136 ± 2
$UF_2^{2+} + F^-$	$\rightleftharpoons$	UF_3^+	4 ± 4	119 ± 13

[90CAP/VIT]
Capdevila, H., Vitorge, P., Temperature and ionic strength influence on U(VI/V) and U(IV/III) redox potentials in aqueous acidic and carbonate solutions, J. Radioanal. Nucl. Chem., Articles, **143** (1990) 403-414.

This is an electrochemical study done by using cyclic voltammetry, *cf.* Refs. [89RIG/ROB, 90RIG]. The experiments in acid solutions were carried out at 1.00 M $HClO_4$ over a temperature range from 5 to 55°C. The redox potential obtained for the UO_2^{2+}/UO_2^+ couple, E(1 M $HClO_4$, 298.15 K) = (0.060 ± 0.004) V, is in very good agreement with the data reported in other studies, *cf.* Table V.2. The reported temperature coefficients are not incompatible with the entropy values selected in the present review, but the uncertainties reported appear too low by a factor of two.

The redox potential reported for the U^{4+}/U^{3+} couple, $E = -0.632$ V at 1 M $HClO_4$ and -0.553 V at $I = 0$, differs slightly from the value given by Riglet, Robouch and Vitorge [89RIG/ROB]. As for the UO_2^{2+}/UO_2^+ couple, the reported temperature coefficients are not incompatible with the entropy values selected in the present review, but the uncertainties reported appear too low by a factor of two.

The data for the UO_2^{2+}/UO_2^+ couple in carbonate media were obtained at four different ionic strengths between 5 and 55°C. The method is the same as the one reported by Riglet [90RIG], but the evaluation was done using $\log_{10} \beta_3^\circ(UO_2(CO_3)_3^{4-}$, 298.15 K) = 21.3 rather than 21.6 as selected in the present review. The reported equilibrium constant, $\log_{10} \beta_3^\circ(UO_2(CO_3)_3^{5-}, 298.15\,K) = (6.6 \pm 0.3)$ (6.9 when using the value selected in the present review), differs from that reported by Riglet [90RIG], $\log_{10} \beta_3^\circ(UO_2(CO_3)_3^{5-}, 298.15\,K) = 7.1$ (7.4 when using the value selected in the present review). The reason for this difference is not clear from Ref. [90CAP/VIT]. This review prefers the value of Riglet [90RIG] which is better documented. It should be mentioned that the ion interaction coefficient obtained by Capdevila and Vitorge, $\varepsilon_{(UO_2(CO_3)_3^{5-},Na^+)} = -0.83$, differs from the one derived by Riglet [90RIG] and adopted in this review, $\varepsilon_{(UO_2(CO_3)_3^{5-},Na^+)} = -0.66$.

[90COR/KON]
Cordfunke, E.H.P., Konings, R.J.M., Potter, P.E., Prins, G., Rand, M.H., Thermochemical data for reactor materials and fission products (Cordfunke, E.H.P., Konings, R.J.M., *eds.*), Amsterdam: North-Holland, 1990, 695p.

These authors reported $\Delta_f H_m^\circ(Cs_2U_2O_7, cr, 298.15\ K) = -(3215.3 \pm 1.9)\ kJ \cdot mol^{-1}$ based on a sulphate cycle. The sulphate cycle is less prone to difficulties with the values for auxiliary data than is the cycle using HCl (*e.g.*, [75OHA/HOE2]). The use of auxiliary data consistent with the present review (E.H.P. Cordfunke, private communication, 1991) leads to $\Delta_f H_m^\circ(Cs_2U_2O_7, cr, 298.15\ K) = -(3214.3 \pm 2.1)\ kJ \cdot mol^{-1}$. The main source of the difference is the different value used for the enthalpy of formation of Cs_2SO_4(cr): $-(1442.4 \pm 1.1)\ kJ \cdot mol^{-1}$ instead of $-(1443.4 \pm 0.5)\ kJ \cdot mol^{-1}$.

[90RAI/FEL]
Rai, D., Felmy, A.R., Ryan, J.L., Uranium(IV) hydrolysis constants and solubility product of $UO_2 \cdot xH_2O$(am), Inorg. Chem., **29** (1990) 260-264.

This study of the solubility of "$UO_2 \cdot xH_2O$(am)" appears to have been carefully designed and executed. The exact slope of the change in solubility with increasing pH is claimed to be three orders of magnitude per pH unit. Although the scatter in the data for pH values ≤ 4 results in a large uncertainty in the slope, it is clear that species such as $U(OH)_2^{2+}$ and $U(OH)_3^+$ can have little range of predominance, if any. The pH independence of the solubility for pH values greater than 4 is in good qualitative agreement with the work of Parks and Pohl [88PAR/POH], who reported the solubility of a crystalline form of UO_2, uraninite, at somewhat higher temperatures. However, the estimated values for the values of $\Delta_f G_m^\circ(U(OH)_4, aq, 298.15\ K)$ that can be roughly derived from the two papers differ by approximately $40\ kJ \cdot mol^{-1}$.

[91GRE/LAG]
Grenthe, I., Lagerman, B., Studies on metal carbonate equilibria: 22. A coulometric study of the uranium(VI)-carbonate system, the composition of the mixed hydroxide carbonate species, Acta Chem. Scand., **45** (1991) 122-128.

This is an experimental study designed to confirm or reject the possible formation of $(UO_2)_2CO_3(OH)_3^-$ and $(UO_2)_{11}(OH)_{24}(CO_2)_6^{2-} \triangleq (UO_2)_{11}(CO_3)_6(OH)_{12}^{2-}$. The experiments were made coulometrically at 25°C, $I = 0.500$ m ($NaClO_4$), and covered the following ranges: $3.5 < -\log_{10}[H^+] < 5.2$; 0.2 mM $< [U^{VI}]_T <$ 1.0 mM; and 0.30 atm $< p_{CO_2} <$ 0.97 atm. An example of a species distribution diagram is given in Figure 1 of Ref. [91GRE/LAG]. The authors confirmed the formation of the two species mentioned above and proposed a chemical model with the following eight complexes and formation constants, $\log_{10}{}^*\beta(0.5\ m\ NaClO_4)$:

$$\begin{array}{llll}
UO_2^{2+} + 2H_2O(l) + CO_2(g) & \rightleftharpoons & UO_2CO_3(aq) + 2H^+ & -8.71 \pm 0.05 \\
2UO_2^{2+} + 2H_2O(l) & \rightleftharpoons & (UO_2)_2(OH)_2^{2+} + 2H^+ & -6.07 \pm 0.15 \\
3UO_2^{2+} + 5H_2O(l) & \rightleftharpoons & (UO_2)_3(OH)_5^+ + 5H^+ & -16.40 \pm 0.22 \\
3UO_2^{2+} + 6H_2O(l) + 6CO_2(g) & \rightleftharpoons & (UO_2)_3(CO_3)_6^{6-} + 12H^+ & -49.68 \pm 0.17
\end{array}$$

$11UO_2^{2+} + 18H_2O(l) + 6CO_2(g)$	$\rightleftharpoons$	$(UO_2)_{11}(CO_3)_6(OH)_{12}^{2-} + 24H^+$	-72.48 ± 0.31
$2UO_2^{2+} + 4H_2O(l) + CO_2(g)$	$\rightleftharpoons$	$(UO_2)_2CO_3(OH)_3^- + 5H^+$	-19.40 ± 0.11
$UO_2^{2+} + 3H_2O(l) + 3CO_2(g)$	$\rightleftharpoons$	$UO_2(CO_3)_3^{4-} + 6H^+$	-29.29 ± 0.09
$UO_2^{2+} + 2H_2O(l) + 2CO_2(g)$	$\rightleftharpoons$	$UO_2(CO_3)_2^{2-} + 4H^+$	-19.57 ± 0.30

The equilibrium constants of the major species are in excellent agreement with the ones selected by this review and recalculated to 0.5 m $NaClO_4$. The value for the complex $(UO_2)_3(CO_3)_6^{6-}$ is also in good agreement with the value selected in this review. The complexes $UO_2(CO_3)_2^{2-}$ and $UO_2(CO_3)_3^{4-}$ occur only in very small amounts in the experiments carried out by Grenthe and Lagerman [91GRE/LAG]. Their constants can therefore not be very precise, and they are not used in the evaluation of the selected formation constants in the present review. Equally, the values of $^*\beta_{2,2}$ and $^*\beta_{5,3}$ from this study are not used in the overall evaluation of $^*\beta^\circ_{2,2}$ and $^*\beta^\circ_{5,3}$, as these complexes are present in fairly small amounts in the test solutions. The present review doubles the uncertainties given by Grenthe and Lagerman to evaluate the recommended constants. The complexes $(UO_2)_2CO_3(OH)_3^-$ and $(UO_2)_{11}(CO_3)_6(OH)_{12}^{2-}$ were formed in substantial amounts. Hence, the study of Grenthe and Lagerman gives additional support to the earlier studies of Ciavatta *et al.* [79CIA/FER] and Maya [82MAY]. Grenthe and Lagerman also discussed previous studies and proposed two structural models in support of the proposed mixed complex $(UO_2)_{11}(OH)_{24}(CO_2)_6^{2-}$, *cf.* Figure A.2.

Figure A.2: Structure models for the complex $(UO_2)_{11}(OH)_{24}(CO_2)_6^{2-}$ proposed by Grenthe and Lagerman [91GRE/LAG]: The model on the left-hand side is based on the rutherfordine ($UO_2CO_3(cr)$) structure. The complex is planar with the UO_2^{2+} groups perpendicular to the plane formed by the ligands: CO_3^{2-}, OH^- (striped) and H_2O (shaded). The two inner UO_2^{2+} ions are hexacoordinated in the plane, while the external ones may be either five or six coordinate. The model on the right-hand side is based on the known structure for $(UO_2)_2CO_3(OH)_3^-$ [83ABE/FER]. The symbols have the same significance as in both models.

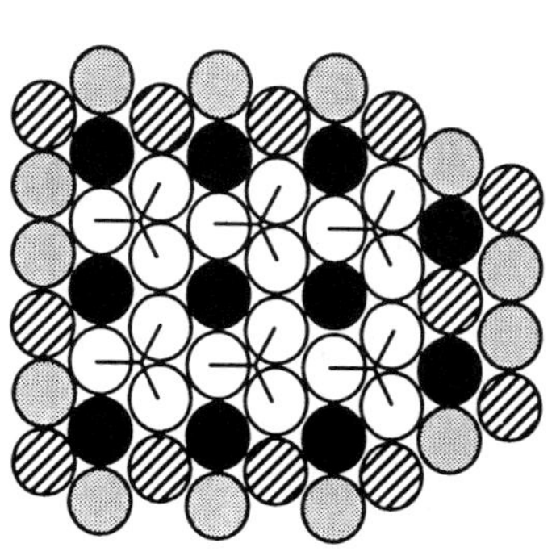

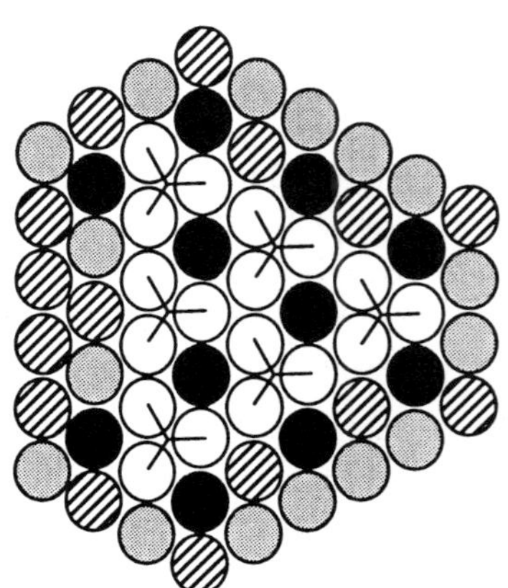

[91MAT]
Mathur, J.N., Complexation and thermodynamics of the uranyl ion with phosphate, Polyhedron, 10 (1991) 47-53.

The equilibrium constants for the reactions

$$UO_2^{2+} + qH_2PO_4^- \rightleftharpoons UO_2(H_2PO_4)_q^{2-q}, \qquad (A.129)$$

were determined at five different ionic strengths varying from 0.100 to 1.00 M ($NaClO_4$). The equilibrium constants were determined at $I = 1.00$ M and at four different temperatures ranging from 10°C to 50°.

The equilibrium constants measured at different ionic strengths were extrapolated to $I = 0$ using an extended Debye-Hückel expression. The reported values are $\log_{10}\beta_1^\circ$(A.129, $q = 1$, 298.15 K) = (3.47 ± 0.02) and $\log_{10}\beta_2^\circ$(A.129, $q = 2$, 298.15 K) = (5.77 ± 0.02). The second value differs significantly from previous determinations. The values selected in Section V.6.2.1.1.a, recalculated to refer to the reactions (A.129) above, are $\log_{10}\beta_1^\circ$(A.129, $q = 1$, 298.15 K) = (3.26 ± 0.07) and $\log_{10}\beta_2^\circ$(A.129, $q = 2$, 298.15 K) = (4.92 ± 0.13). The author indicated that these differences are "due to uncertainty about the quality of experimental conditions such as ionic strength, high concentrations of H^+ ion, very high concentrations of phosphoric acid, uncertainty in pK_{a1} values or the mixed nature of equilibrium constants". In contrast, the present review concludes that Mathur used an erroneous chemical model to interpret the experimental findings.

The experimental technique used by Mathur is satisfactory. The fact that the calibration of the pH electrode was done in activity instead of concentration units, has no significant influence on the concentration of $H_2PO_4^-$. The reason for part of the differences between the equilibrium constants from Mathur and those from previous studies is an incomplete chemical model, *i.e.*, Mathur did not take the complexes $UO_2H_3PO_4^{2+}$, $UO_2HPO_4(aq)$ and $UO_2(H_rPO_4)_2^{2r-4}$ into account. By working at a constant pH it is not possible to obtain information on the degree of protonation of the bonded phosphate, and the equilibrium constants reported are therefore conditional constants.

The results of this study are therefore not necessarily contradictory with the values selected in the present review.

[91SAN]
Sandino, A., Processes affecting the mobility of uranium in natural waters, Ph.D. thesis in inorganic chemistry, The Royal Institute of Technology, Stockholm, Sweden, 1991.

This thesis contains a discussion of both hydroxide and phosphate complexation of dioxouranium(VI). The experimental data were obtained at 25°C in 0.5 M $NaClO_4$ by measuring the solubility of amorphous and crystalline schoepite, as well as $(UO_2)_3(PO_4)_2 \cdot 4H_2O$. The solid phases used were well characterized. The phosphate study was carried out at $[PO_4]_T = 0.01$ M by varying the hydrogen ion concentration in the range $3 < -\log_{10}[H^+] < 9$. The low total phosphate concentration was used to avoid the phase transformations observed by Schreyer and Baes [54SCH/BAE]; it would have been desirable to use different total concentrations in order to confirm the chemical model proposed. Even at the highest H^+ concentration used, the concentration of uncomplexed UO_2^{2+} is very small. For the evaluation of the solubility product of $(UO_2)_3(PO_4)_2 \cdot 4H_2O$, the author used the stability constants selected in the present review for the complexes formed in the most acid range. The solubility product thus obtained, $\log_{10} K_s(0.5 \text{ M NaClO}_4) = -(48.48 \pm 0.12)$, was recalculated to $I = 0$, resulting in $\log_{10} K_s^\circ = -(53.28 \pm 0.13)$. This value is considerably lower than the one selected in the present review, indicating that Sandino used a more crystalline phase than the previous investigators.

The experimental solubility data in the range $5.5 < -\log_{10}[H^+] < 6.5$ indicate the formation of predominantly uncharged phosphate complexes, most probably $UO_2HPO_4(aq)$. In the range $6.5 < -\log_{10}[H^+] < 9$ there is an indication that negatively charged complexes are formed, and that at least one of them contains phosphate. The present review accepts the arguments in favour of $UO_2PO_4^-$ and $UO_2(OH)_3^-$. For the equilibria

$$UO_2^{2+} + HPO_4^{2-} \rightleftharpoons UO_2HPO_4(aq) \quad (A.130)$$

$$UO_2^{2+} + PO_4^{3-} \rightleftharpoons UO_2PO_4^- \quad (A.131)$$

$$UO_2^{2+} + 3H_2O(l) \rightleftharpoons UO_2(OH)_3^- + 3H^+ \quad (A.132)$$

the author reported equilibrium constants which they recalculated to $I = 0$ by using the ion interaction coefficients $\varepsilon_{(UO_2PO_4^-,Na^+)} = \varepsilon_{(UO_2(OH)_3^-,Na^+)} = -(0.09 \pm 0.05)$. This resulted in $\log_{10} \beta_1^\circ(\text{A.130, 298.15 K}) = (7.24 \pm 0.17)$, $\log_{10} \beta_1^\circ(\text{A.131, 298.15 K}) = (13.23 \pm 0.10)$ and $\log_{10} {}^*\beta_3^\circ(\text{A.132, 298.15 K}) = -(19.18 \pm 0.29)$. The uncertainties reported by Sandino represent two times the standard deviations obtained from a least-squares refinement. This review accepts these values but prefers to increase the uncertainties to be equal to three times the standard deviation.

Appendix B

Ionic strength corrections

Thermodynamic data always refer to a selected standard state. The definition given by IUPAC [82LAF] is adopted in this review as outlined in Section II.3.1. According to this definition, the standard state for a solute B in a solution is a hypothetical solution, at the standard state pressure, in which $m_B = m^\circ = 1\,\text{mol}\cdot\text{kg}^{-1}$, and in which the activity coefficient γ_B is unity. However, for many reactions, measurements cannot be made accurately (or at all) in dilute solutions from which the necessary extrapolation to the standard state would be simple. This is invariably the case for reactions involving ions of high charge. Precise thermodynamic information for these systems can only be obtained in the presence of an inert electrolyte of sufficiently high concentration, ensuring that activity factors are reasonably constant throughout the measurements. This appendix describes and illustrates the method used in this review for the extrapolation of experimental equilibrium data to zero ionic strength.

The activity factors of all the species participating in reactions in high ionic strength media must be estimated in order to reduce the thermodynamic data obtained from the experiment to the standard state ($I = 0$). Two alternative methods can be used to describe the ionic medium dependence of equilibrium constants:

- One method takes into account the individual characteristics of the ionic media by using a medium dependent expression for the activity coefficients of the species involved in the equilibrium reactions. The medium dependence is described by virial or ion interaction coefficients as used in the Pitzer equations and in the specific ion interaction theory.

- The other method uses an extended Debye-Hückel expression in which the activity coefficients of reactants and products depend only on the ionic charge and the ionic strength, but it accounts for the medium specific properties by introducing ionic pairing between the medium ions and the species involved in the equilibrium reactions. Earlier, this approach has been used extensively in marine chemistry, *cf.* Refs. [79JOH/PYT, 79MIL, 79PYT, 79WHI].

The activity factor estimates are thus based on the use of Debye-Hückel type equations. The "extended" Debye-Hückel equations are either in the form of specific ion interaction methods or the Davies equation [62DAV]. However, the Davies equation should in general not be used at ionic strengths larger than 0.1 mol·kg^{-1}. The method preferred in the NEA Thermochemical Data Base review is a medium-dependent expression for the activity coefficients, which is the specific ion interaction theory in the form of the Brønsted-Guggenheim-Scatchard approach. Other forms of specific ion interaction methods (the Pitzer and Brewer "B-method" and the Pitzer virial coefficient method) are described in the NEA Guidelines for the extrapolation to zero ionic strength [89GRE/WAN].

The specific ion interaction methods are reliable for intercomparison of experimental data in a given concentration range. In many cases this includes data at rather low ionic strengths, $I = 0.01$ to 0.1 M, *cf.* Figure B.1, while in other cases, notably for cations of high charge ($\geq +4$ and ≤ -4), the lowest ionic strength may be 0.2 M or higher, *cf.* Figures V.12 and V.13. It is reasonable to assume that the extrapolated equilibrium constants at $I = 0$ are more precise in the former than in the latter cases. The extrapolation error is composed of two parts, one due to experimental errors, the other due to model errors. The model errors seem to be rather small for many systems, less than 0.1 units in $\log_{10} K^\circ$. For reactions involving ions of high charge, which are extensively hydrolyzed, one cannot perform experiments at low ionic strengths. Hence, it is impossible to estimate the extrapolation error. This is true for all methods used to estimate activity corrections. Systematic model errors of this type are not included in the uncertainties assigned to the selected data in this review.

It should be emphasized that the specific ion interaction model is *approximate.* Modifying it as recently suggested by Ciavatta [90CIA] obtained have slightly different ion interaction coefficients and equilibrium constants, *cf.* Footnotes 10-14 in Sections V.6.1.3.1, V.7.1.2.1.a and V.7.1.4.1.a and V.7.1.4.1.c, respectively. Both methods provide an internally consistent set of values. However, their absolute values may differ somewhat. This review *estimates* that these differences in general are less than 0.2 units in $\log_{10} K^\circ$, *i.e.*, approximately 1 kJ·mol^{-1} in derived $\Delta_f G^\circ_m$ values.

B.1. The specific ion interaction theory

B.1.1. Background

The Debye-Hückel term, which is the dominant term in the expression for the activity coefficients in dilute solution, accounts for electrostatic, non-specific long-range interactions. At higher concentrations short range, non-electrostatic interactions have to be taken into account. This is usually done by adding ionic strength dependent terms to the Debye-Hückel expression. This method was first outlined by Brønsted [22BRO, 22BRO2], and elaborated by Scatchard [36SCA] and Guggenheim [66GUG]. The two basic assumptions in the specific ion interaction theory are described below.

Assumption 1: The activity coefficient γ_j of an ion j of charge z_j in the solution of ionic strenght I_m may be discribed by Eq. (B.1).

$$\log_{10} \gamma_j = -z_j^2 D + \sum_k \varepsilon_{(j,k,I_m)} m_k \qquad \text{(B.1)}$$

D is the Debye-Hückel term:

$$D = \frac{A\sqrt{I_m}}{1 + Ba_j\sqrt{I_m}} \tag{B.2}$$

A and B are constants which are temperature dependent, and a_j is the effective diameter of the hydrated ion j. The values of A and B as a function of temperature are listed in Table B.1.

Table B.1: Debye-Hückel constants, adapted from Helgeson, Kirkham and Flowers [81HEL/KIR].

t(°C)	A	$B(\times 10^{-8})$
0	0.4913	0.3247
5	0.4943	0.3254
10	0.4976	0.3261
15	0.5012	0.3268
20	0.5050	0.3276
25	0.5091	0.3283
30	0.5135	0.3291
35	0.5182	0.3299
40	0.5231	0.3307
45	0.5282	0.3316
50	0.5336	0.3325
55	0.5392	0.3334
60	0.5450	0.3343
65	0.5511	0.3352
70	0.5573	0.3362
75	0.5639	0.3371

The term Ba_j in the denominator of the Debye-Hückel term has been assigned a value of $Ba_j = 1.5$, as proposed by Scatchard [76SCA] and accepted by Ciavatta [80CIA]. This value has been found to minimize, for several species, the ionic strength dependence of $\varepsilon_{(j,k,I_m)}$ between $I_m = 0.5$ m and $I_m = 3.5$ m. It should be mentioned that some authors have proposed different values for Ba_j, ranging from $Ba_j = 1.0$ [35GUG] to $Ba_j = 1.6$ [62VAS]. However, the parameter Ba_j is empirical and as such correlated to the value of $\varepsilon_{(j,k,I_m)}$. Hence, this variety of values for Ba_j does not represent an uncertainty range, but rather indicates that several different sets of Ba_j and $\varepsilon_{(j,k,I_m)}$ may describe equally well the experimental mean activity coefficients of a given electrolyte. The ion interaction coefficients listed in Tables B.3 through B.5 have thus to be used with $Ba_j = 1.5$.

The summation in Eq. (B.1) extends over all ions k present in solution. Their molality is denoted m_k. The concentrations of the ions of the ionic medium is often very much larger than those of the reacting species. Hence, the ionic medium ions will make the main contribution to the value of $\log_{10} \gamma_j$ for the reacting ions. This fact often makes it possible to simplify the summation $\sum_k \varepsilon_{(j,k,I_m)} m_k$ so that only ion interaction coefficients between the participating ionic species and the ionic medium ions are included, as shown in Eqs. (B.5) to (B.9).

Assumption 2: The ion interaction coefficients $\varepsilon_{(j,k,I_m)}$ are zero for ions of the same charge sign and for uncharged species. The rationale behind this is that ε, which describes specific short-range interactions, must be small for ions of the same charge since they are usually far from one another due to electrostatic repulsion. This holds to a lesser extent also for uncharged species.

Eq. (B.1) will allow fairly accurate estimates of the activity coefficents in mixtures of electrolytes if the ion interaction coefficients are known. Ion interaction coefficients for simple ions can be obtained from tabulated data of mean activity coefficients of strong electrolytes or from the corresponding osmotic coefficients. Ion interaction coefficients for complexes can either be estimated from the charge and size of the ion or determined experimentally from the variation of the equilibrium constant with the ionic strength. Ion interaction coefficients are not strictly constant but may vary slightly with the ionic strength. The extent of this variation depends on the charge type and is small for 1:1, 1:2 and 2:1 electrolytes for molalities less than 3.5 m. The concentration dependence of the ion interaction coefficients can thus often be neglected. This point was emphasized by Guggenheim [66GUG], who has presented a considerable amount of experimental material supporting this approach. The concentration dependence is larger for electrolytes of higher charge. In order to accurately reproduce their activity coefficient data, concentration dependent ion interaction coefficients have to be used, *cf.* Pitzer and Brewer [61PIT/BRE], Baes and Mesmer [76BAE/MES], or Ciavatta [80CIA]. By using a more elaborate virial expansion, Pitzer and co-workers [73PIT, 73PIT/MAY, 74PIT/KIM, 74PIT/MAY, 75PIT, 76PIT/SIL, 78PIT/PET, 79PIT] have managed to describe measured activity coefficients of a large number of electrolytes with high precision over a large concentration range. Pitzer's model generally contains three parameters as compared to one in the specific ion interaction theory. The use of the theory requires the knowledge of all these parameters. The derivation of Pitzer coefficients for many complexes such as those of the actinides would require a very large amount of additional experimental work, since no data of this type are currently available.

The way in which the activity coefficient corrections are performed in this review according to the specific ion interaction theory is illustrated below for a general case of a complex formation reaction. Charges are omitted for brevity.

$$m\mathrm{M} + q\mathrm{L} + n\mathrm{H_2O(l)} \rightleftharpoons \mathrm{M}_m\mathrm{L}_q(\mathrm{OH})_n + n\mathrm{H}^+ \quad \text{(B.3)}$$

The formation constant of $\mathrm{M}_m\mathrm{L}_q(\mathrm{OH})_n$, ${}^*\beta_{q,n,m}$, determined in an ionic medium (1:1 salt NX) of the ionic strenght I_m, is related to the corresponding value at zero ionic strenght, ${}^*\beta^\circ_{q,n,m}$, by Eq. (B.4).

$$\begin{aligned}\log_{10}{}^*\beta_{q,n,m} &= \log_{10}{}^*\beta^\circ_{q,n,m} + m\log_{10}\gamma_{\mathrm{M}} + q\log_{10}\gamma_{\mathrm{L}} + n\log_{10}a_{\mathrm{H_2O}} \\ &\quad - \log_{10}\gamma_{q,n,m} - n\log_{10}\gamma_{\mathrm{H^+}}\end{aligned} \tag{B.4}$$

The subscript (q,n,m) denotes the complex ion $\mathrm{M}_m\mathrm{L}_q(\mathrm{OH})_n$. If the concentrations of N and X are much greater than the concentrations of M, L, $\mathrm{M}_m\mathrm{L}_q(\mathrm{OH})_n$ and H^+, only the molalities m_{N} and m_{X} have to be taken into account for the calculation of the term $\sum_k \varepsilon_{(j,k,I_m)} m_k$ in Eq. (B.1). For example, for the activity coefficient of the metal cation M, γ_{M}, Eq. (B.5) is obtained.

$$\log_{10}\gamma_{\mathrm{M}} = \frac{-z^2_{\mathrm{M}}0.5091\sqrt{I_m}}{1+1.5\sqrt{I_m}} + \varepsilon_{(\mathrm{M,X},I_m)}m_{\mathrm{X}} \tag{B.5}$$

Under these conditions, $I_m \approx m_{\mathrm{X}} = m_{\mathrm{N}}$. Substituting the $\log_{10}\gamma_j$ values in Eq. (B.4) with the corresponding forms of Eq. (B.5) and rearranging leads to

$$\log_{10}{}^*\beta_{q,n,m} - \Delta z^2 D - n\log_{10}a_{\mathrm{H_2O}} = \log_{10}{}^*\beta^\circ_{q,n,m} - \Delta\varepsilon I_m \tag{B.6}$$

where

$$\Delta z^2 = (mz_{\mathrm{M}} - qz_{\mathrm{L}} - n)^2 + n - mz^2_{\mathrm{M}} - qz^2_{\mathrm{L}} \tag{B.7}$$

$$D = \frac{0.5091\sqrt{I_m}}{1+1.5\sqrt{I_m}} \tag{B.8}$$

$$\Delta\varepsilon = \varepsilon_{(q,n,m,\mathrm{N\ or\ X})} + n\varepsilon_{(\mathrm{H,X})} - q\varepsilon_{(\mathrm{N,L})} - m\varepsilon_{(\mathrm{M,X})} \tag{B.9}$$

Here $(mz_{\mathrm{M}} - qz_{\mathrm{L}} - n)$, z_{M} and z_{L} are the charges of the complex $\mathrm{M}_m\mathrm{L}_q(\mathrm{OH})_n$, the metal ion M and the ligand L, respectively. Equilibria involving $\mathrm{H_2O(l)}$ as a reactant or product require a correction for the activity of water, $a_{\mathrm{H_2O}}$. The activity of water in an electrolyte mixture can be calculated as

$$\log_{10}a_{\mathrm{H_2O}} = \frac{-\Phi\sum_k m_k}{\ln(10)\times 55.51} \tag{B.10}$$

where Φ is the osmotic coefficient of the mixture and the summation extends over all ions k with molality m_k present in the solution. In the presence of an ionic medium NX in dominant concentration, Eq. (B.10) can be simplified by neglecting the contributions of all minor species, *i.e.*, the reacting ions. Hence, for a 1:1 electrolyte of ionic strength $I_m \approx m_{\mathrm{NX}}$, Eq. (B.10) becomes

$$\log_{10}a_{\mathrm{H_2O}} = \frac{-2m_{\mathrm{NX}}\Phi}{\ln(10)\times 55.51}. \tag{B.11}$$

Values of osmotic coefficients for single electrolytes have been compiled by various authors, *e.g.*, Robinson and Stokes [59ROB/STO]. The activity of water can also be calculated from the known activity coefficients of the dissolved species.

In the presence of an ionic medium NX of a concentration much larger than those of the reacting ions, the osmotic coefficient can be calculated according to Eq. (B.12) [79STO].

$$1 - \Phi = \frac{A\ln(10)|z_+ z_-|}{m_{NX} \times (1.5)^3}\left[1 + 1.5\sqrt{m_{NX}} - 2\log_{10}(1 + 1.5\sqrt{m_{NX}}) - \frac{1}{1 + 1.5\sqrt{m_{NX}}}\right] + \frac{1}{\ln(10)}\varepsilon_{(N,X)} m_{NX} \quad (B.12)$$

The activity of water is obtained by inserting Eq. (B.12) into Eq. (B.11). It should be mentioned that in mixed electrolytes with several components at high concentrations, it is necessary to use Pitzer's equation to calculate the activity of water. On the other hand, a_{H_2O} is near constant (and equal to 1) in most experimental studies of equilibria in dilute aqueous solutions, where an ionic medium is used in large excess with respect to the reactants. The medium electrolyte thus determines the osmotic coefficient of the solvent.

In natural waters the situation is similar; the ionic strength of most surface waters is so low that the activity of $H_2O(l)$ can be set equal to unity. A correction may be necessary in the case of seawater, where a sufficiently good approximation for the osmotic coefficient may be obtained by considering NaCl as the dominant electrolyte.

In more complex solutions of high ionic strengths with more than one electrolyte at significant concentrations, *e.g.*, (Na^+, Mg^{2+}, Ca^{2+})(Cl^-, SO_4^{2-}), Pitzer's equation (*cf.* [89GRE/WAN]) may be used to estimate the osmotic coefficient; the necessary interaction coefficients are known for most systems of geochemical interest.

Note that in all ion interaction approaches, the equation for mean activity coefficients can be split up to give equations for conventional single ion activity coefficients in mixtures, *e.g.*, Eq. (B.1). The latter are strictly valid only when used in combinations which yield electroneutrality. Thus, while estimating medium effects on standard potentials, a combination of redox equilibria with $H^+ + e^- \rightleftharpoons \frac{1}{2}H_2(g)$ is necessary (*cf.* Example B.3).

B.1.2. *Estimation of ion interaction coefficients*

B.1.2.1. *Estimation from mean activity coefficient data*

Example B.1:

The ion interaction coefficient $\varepsilon_{(H^+,Cl^-)}$ can be obtained from published values of $\gamma_{\pm,HCl}$ *vs.* m_{HCl}.

$$\begin{aligned} 2\log_{10}\gamma_{\pm,HCl} &= \log_{10}\gamma_{+,H^+} + \log_{10}\gamma_{-,Cl^-} \\ &= -D + \varepsilon_{(H^+,Cl^-)} m_{Cl^-} - D + \varepsilon_{(Cl^-,H^+)} m_{H^+} \\ \log_{10}\gamma_{\pm,HCl} &= -D + \varepsilon_{(H^+,Cl^-)} m_{HCl} \end{aligned}$$

By plotting $\log_{10}\gamma_{\pm,HCl} + D$ *vs.* m_{HCl} a straight line with the slope $\varepsilon_{(H^+,Cl^-)}$ is obtained. The degree of linearity should in itself indicate the range of validity of the specific ion interaction approach. Osmotic coefficient data can be treated in an analogous way.

B.1.2.2. Estimations based on experimental values of equilibrium constants at different ionic strengths

Example B.2:

Equilibrium constants are given in Table B.2 for the reaction

$$UO_2^{2+} + Cl^- \rightleftharpoons UO_2Cl^+. \tag{B.13}$$

Table B.2: The preparation of the experimental equilibrium constants for the extrapolation to $I = 0$ with the specific ion interaction method, according to Reaction (B.13). The linear regression of this set of data is shown in Figure B.1.

I_m	$\log_{10}\beta_1(\exp)^{(a)}$	$\log_{10}\beta_{1,m}^{(b)}$	$\log_{10}\beta_{1,m} + 4D$
0.1	-0.17 ± 0.10	-0.174	0.264 ± 0.100
0.2	-0.25 ± 0.10	-0.254	0.292 ± 0.100
0.26	-0.35 ± 0.04	-0.357	0.230 ± 0.040
0.31	-0.39 ± 0.04	-0.397	0.220 ± 0.040
0.41	-0.41 ± 0.04	-0.420	0.246 ± 0.040
0.51	-0.32 ± 0.10	-0.331	0.371 ± 0.100
0.57	-0.42 ± 0.04	-0.432	0.288 ± 0.040
0.67	-0.34 ± 0.04	-0.354	0.395 ± 0.040
0.89	-0.42 ± 0.04	-0.438	0.357 ± 0.400
1.05	-0.31 ± 0.10	-0.331	0.491 ± 0.100
1.05	-0.277 ± 0.260	-0.298	0.525 ± 0.260
1.61	-0.24 ± 0.10	-0.272	0.618 ± 0.100
2.21	-0.15 ± 0.10	-0.193	0.744 ± 0.100
2.21	-0.12 ± 0.10	-0.163	0.774 ± 0.100
2.82	-0.06 ± 0.10	-0.021	0.860 ± 0.100
3.5	0.04 ± 0.10	-0.021	0.974 ± 0.100

(a) Equilibrium constants for Reaction (B.13) with assigned uncertainties, corrected to 25°C where necessary.
(b) Equilibrium constants corrected from molarity to molarity units, as described in Chapter II.2.

The following formula is deduced from Eq. (B.6) for the extrapolation to $I = 0$:

$$\log_{10}\beta_1 + 4D = \log_{10}\beta_1^{\circ} - \Delta\varepsilon I_m \tag{B.14}$$

The linear regression is done as described in Appendix C. The following results are obtained:

Figure B.1: Plot of $\log_{10}\beta_1 + 4D$ *vs.* I_m for Reaction (B.13), $UO_2^{2+} + Cl^- \rightleftharpoons UO_2Cl^+$. The straight line shows the result of the weighted linear regression, and the shaded area represents the uncertainty range obtained by propagating the resulting uncertainties at $I = 0$ back to $I = 4$ m.

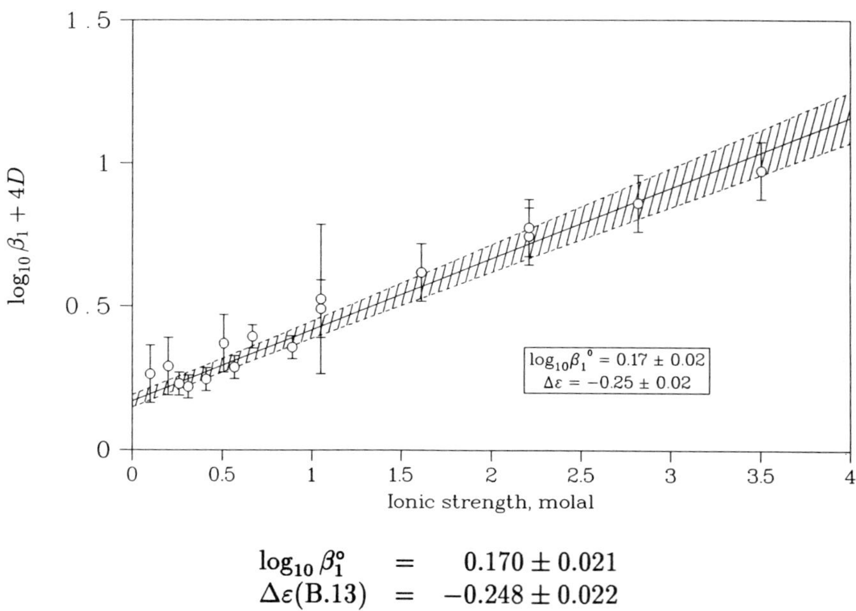

$$\begin{aligned} \log_{10}\beta_1^\circ &= 0.170 \pm 0.021 \\ \Delta\varepsilon(\text{B.13}) &= -0.248 \pm 0.022 \end{aligned}$$

The experimental data are depicted in Figure B.1, where the dashed area represents the uncertainty range that is obtained by using the results in $\log_{10}\beta_1^\circ$ and $\Delta\varepsilon$ and correcting back to $I \neq 0$.

Example B.3:

When using the specific ion interaction theory, the relationship between the redox potential of the couple UO_2^{2+}/U^{4+} in a medium of ionic strength I_m and the corresponding quantity at $I = 0$ should be calculated in the following way. The reaction in the galvanic cell

$$\text{Pt}, \text{H}_2 \mid \text{H}^+ \parallel \text{UO}_2^{2+}, \text{U}^{4+} \mid \text{Pt} \tag{B.15}$$

is

$$\text{UO}_2^{2+} + \text{H}_2(\text{g}) + 2\text{H}^+ \rightleftharpoons \text{U}^{4+} + 2\text{H}_2\text{O}(\text{l}). \tag{B.16}$$

For this reaction

$$\log_{10} K^\circ = \log_{10}\left(\frac{a_{\text{U}^{4+}} \times a_{\text{H}_2\text{O}}^2}{a_{\text{UO}_2^{2+}} \times a_{\text{H}^+}^2 \times f_{\text{H}_2}}\right), \tag{B.17}$$

$$\begin{aligned} \log_{10} K^\circ = {} & \log_{10} K + \log_{10}\gamma_{\text{U}^{4+}} - \log_{10}\gamma_{\text{UO}_2^{2+}} - 2\log_{10}\gamma_{\text{H}^+} - \log_{10}\gamma_{f,\text{H}_2} \\ & +2\log_{10} a_{\text{H}_2\text{O}}, \end{aligned} \tag{B.18}$$

$f_{H_2} \approx p_{H_2}$ at reasonably low partial pressure of $H_2(g)$, $a_{H_2O} \approx 1$, and

$$\log_{10} \gamma_{U^{4+}} = -16D + \varepsilon_{(U^{4+},ClO_4^-)} m_{ClO_4^-} \quad (B.19)$$

$$\log_{10} \gamma_{UO_2^{2+}} = -4D + \varepsilon_{(UO_2^{2+},ClO_4^-)} m_{ClO_4^-} \quad (B.20)$$

$$\log_{10} \gamma_{H^+} = -D + \varepsilon_{(H^+,ClO_4^-)} m_{ClO_4^-} \,. \quad (B.21)$$

Hence,

$$\log_{10} K^\circ = \log_{10} K - 10D + \left(\varepsilon_{(U^{4+},ClO_4^-)} - \varepsilon_{(UO_2^{2+},ClO_4^-)} - 2\varepsilon_{(H^+,ClO_4^-)}\right) m_{ClO_4^-} \,. \quad (B.22)$$

The relationship between the equilibrium constant and the redox potential is

$$\ln K = \frac{nF}{RT} E \quad (B.23)$$

$$\ln K^\circ = \frac{nF}{RT} E^\circ. \quad (B.24)$$

E is the redox potential in a medium of ionic strength I, E° is the corresponding standard potential at $I = 0$, and n is the number of transferred electrons in the reaction considered. Combining Eqs. (B.22), (B.23) and (B.24) and rearranging them leads to Eq. (B.25).

$$E - 10D\left(\frac{RT\ln(10)}{nF}\right) = E^\circ - \Delta\varepsilon m_{ClO_4^-}\left(\frac{RT\ln(10)}{nF}\right) \quad (B.25)$$

For $n = 2$ in the present example and $T = 298.15\,K$, Eq. (B.25) becomes

$$E[mV] - 295.8D = E^\circ[mV] - 29.58\Delta\varepsilon m_{ClO_4^-} \quad (B.26)$$

where

$$\Delta\varepsilon = \left(\varepsilon_{(U^{4+},ClO_4^-)} - \varepsilon_{(UO_2^{2+},ClO_4^-)} - 2\varepsilon_{(H^+,ClO_4^-)}\right). \quad (B.27)$$

B.1.3. *On the magnitude of ion interaction coefficients*

Ciavatta [80CIA] made a compilation of ion interaction coefficients for a large number of electrolytes. Similar data for complexations of various kinds were reported by Spahiu [83SPA] and Ferri, Grenthe and Salvatore [83FER/GRE]. These and some other data have been collected and are listed in Tables B.3 to B.5. It is obvious from the data in these tables that the charge of an ion is of great importance for the magnitude of the ion interaction coefficient. Ions of the same charge type have similar ion interaction coefficients with a given counter-ion. Based on the tabulated data, this review judges it possible to estimate, with an error of at most ± 0.1 in ε, ion interaction coefficients for cases where there are insufficient experimental data for an extrapolation to $I = 0$. The error that is made by this approximation is estimated to ± 0.1 in $\Delta\varepsilon$ in most cases, based on comparison with $\Delta\varepsilon$ values of various reactions of the same charge type.

B.2. Ion interaction coefficients and equilibrium constants for ion pairs

It can be shown that the virial type of activity coefficient equations and the ionic pairing model are equivalent provided that the ionic pairing is weak. In these cases it is in general difficult to distinguish between complex formation and activity coefficient variations unless independent experimental evidence for complex formation is available, *e.g.*, from spectroscopic data, as is the case for the weak uranium(VI) chloride complexes. It should be noted that the ion interaction coefficients evaluated and tabulated by Ciavatta [80CIA] were obtained from experimental mean activity coefficient data without taking into account complex formation. However, it is known that many of the metal ions listed by Ciavatta form weak complexes with chloride and nitrate ion. This fact is reflected by ion interaction coefficients that are smaller than those for the non-complexing perchlorate ion, *cf.* Table B.3. This review takes chloride and nitrate complex formation into account when these ions are part of the ionic medium and uses the value of the ion interaction coefficient $\varepsilon_{(M^{n+},ClO_4^-)}$ for $\varepsilon_{(M^{n+},Cl^-)}$ and $\varepsilon_{(M^{n+},NO_3^-)}$. In this way, the medium dependence of the activity coefficients is described with a combination of a specific ion interaction model and an ion pairing model. It is evident that the use of NEA recommended data with ionic strength correction models that differ from those used in the evaluation procedure can lead to inconsistencies in the results of the speciation calculations.

It should be mentioned that complex formation may also occur between negatively charged complexes and the cation of the ionic medium. An example is the stabilization of the complex ion $UO_2(CO_3)_3^{5-}$ at high ionic strength, *cf.* Section V.7.1.2.1.d.

B.3. Tables of ion interaction coefficients

Tables B.3 through B.5 contain the selected specific ion interaction coefficients used in this review, according to the specific ion interaction theory described. Table B.3 contains cation interaction coefficients with Cl^-, ClO_4^- and NO_3^-, Table B.4 anion interaction coefficients with Li^+, with Na^+ or NH_4^+ and with K^+. The coefficients have the units of $kg \cdot mol^{-1}$ and are valid for 298.15 K. The species are ordered by charge and appear, within each charge class, in standard order of arrangement, *cf.* Ref. [82WAG/EVA].

In some cases, the ionic interaction can be better described by assuming ion interaction coefficients as functions of the ionic strength rather than as constants. Ciavatta [80CIA] proposed the use of Eq. (B.28) for cases where the uncertainties in Tables B.3 and B.4 are ± 0.03 or greater.

$$\varepsilon = \varepsilon_1 + \varepsilon_2 \log_{10} I_m \qquad \text{(B.28)}$$

For these cases, and when the uncertainty can be improved with respect to the use of a constant value of ε, the values ε_1 and ε_2 given in Table B.5 should be used.

It should be noted that ion interaction coefficients tabulated in Tables B.3 through B.5 may also involve ion pairing effects, as described in Section B.2. In direct comparisons of ion interaction coefficients, or when estimates are made by analogy, this aspect must be taken into account.

Table B.3: Ion interaction coefficients $\varepsilon_{j,k}$ for cations j with $k = Cl^-$, ClO_4^- and NO_3^-, taken from Ciavatta [80CIA, 88CIA] unless indicated otherwise. The uncertainties represent the 95% confidence level. The ion interaction coefficients marked with † can be described more accurately with an ionic strength dependent function, listed in Table B.5. The coefficients $\varepsilon_{(M^{n+},Cl^-)}$ and $\varepsilon_{(M^{n+},NO_3^-)}$ reported by Ciavatta [80CIA] were evaluated without taking chloride and nitrate complexation into account, as discussed in Section B.2.

j $k \rightarrow$ ↓	Cl^-	ClO_4^-	NO_3^-
H^+	0.12 ± 0.01	0.14 ± 0.02	0.07 ± 0.01
NH_4^+	-0.01 ± 0.01	-0.08 ± 0.04†	-0.06 ± 0.03†
H_2gly^+	-0.06 ± 0.02		
Tl^+		-0.21 ± 0.06†	
$ZnHCO_3^+$	0.2(h)		
$CdCl^+$		0.25 ± 0.02	
CdI^+		0.27 ± 0.02	
$CdSCN^+$		0.31 ± 0.02	
$HgCl^+$		0.19 ± 0.02	
Cu^+		0.11 ± 0.01	
Ag^+		0.00 ± 0.01	-0.12 ± 0.05†
YCO_3^+		0.17 ± 0.04(d)	
UO_2^+		0.26 ± 0.03(c)	
UO_2OH^+		-0.06 ± 3.7(c)	0.51 ± 1.4(c)
$(UO_2)_3(OH)_5^+$	0.81 ± 0.17(c)	0.45 ± 0.15(c)	0.41 ± 0.22(c)
UF_3^+	0.1 ± 0.1(f)	0.1 ± 0.1(f)	
UO_2F^+	0.04 ± 0.07(b)	0.29 ± 0.05(c)	
UO_2Cl^+		0.33 ± 0.04(c)	
$UO_2ClO_3^+$		0.33 ± 0.04(f)	
UO_2Br^+		0.24 ± 0.04(f)	
$UO_2BrO_3^+$		0.33 ± 0.04(f)	
$UO_2IO_3^+$		0.33 ± 0.04(f)	
$UO_2N_3^+$		0.3 ± 0.1(f)	
$UO_2NO_3^+$		0.33 ± 0.04(f)	
UO_2SCN^+		0.22 ± 0.04(f)	
NpO_2^+		0.25 ± 0.05(b)	
PuO_2^+		0.17 ± 0.05(b)	
$AlOH^{2+}$	0.09(a)	0.31(a)	

Table B.3 (continued)

j $k \rightarrow$ ↓	Cl^-	ClO_4^-	NO_3^-
$Al_2CO_3(OH)_2^{2+}$	0.26[(a)]		
Pb^{2+}		0.15 ± 0.02	$-0.20 \pm 0.12^\dagger$
Zn^{2+}		0.33 ± 0.03	0.16 ± 0.02
$ZnCO_3^{2+}$	0.35 ± 0.05[(h)]		
Cd^{2+}			0.09 ± 0.02
Hg^{2+}		0.34 ± 0.03	$-0.1 \pm 0.1^\dagger$
Hg_2^{2+}		0.09 ± 0.02	$-0.2 \pm 0.1^\dagger$
Cu^{2+}	0.08 ± 0.01	0.32 ± 0.02	0.11 ± 0.01
Ni^{2+}	0.17 ± 0.02		
Co^{2+}	0.16 ± 0.02	0.34 ± 0.03	0.14 ± 0.01
$FeOH^{2+}$		0.38[(d)]	
$FeSCN^{2+}$		0.45[(d)]	
Mn^{2+}	0.13 ± 0.01		
$YHCO_3^{2+}$		0.39 ± 0.04[(d)]	
UO_2^{2+}	0.21 ± 0.02[(i)]	0.46 ± 0.03	0.24 ± 0.03[(i)]
$(UO_2)_2(OH)_2^{2+}$	0.69 ± 0.07[(c)]	0.57 ± 0.07[(c)]	0.49 ± 0.09[(c)]
$(UO_2)_3(OH)_4^{2+}$	0.50 ± 0.18[(c)]	0.89 ± 0.23[(c)]	0.72 ± 1.0[(c)]
UF_2^{2+}		0.3 ± 0.1[(f)]	
USO_4^{2+}		0.3 ± 0.1[(f)]	
$U(NO_3)_2^{2+}$		0.49 ± 0.14[(g)]	
Mg^{2+}	0.19 ± 0.02	0.33 ± 0.03	0.17 ± 0.01
Ca^{2+}	0.14 ± 0.01	0.27 ± 0.03	0.02 ± 0.01
Ba^{2+}	0.07 ± 0.01	0.15 ± 0.02	-0.28 ± 0.03
Al^{3+}	0.33 ± 0.02		
Fe^{3+}		0.56 ± 0.03	0.42 ± 0.08
Cr^{3+}	0.30 ± 0.03		0.27 ± 0.02
La^{3+}	0.22 ± 0.02	0.47 ± 0.03	
$La^{3+} \rightarrow Lu^{3+}$		$0.47 \rightarrow 0.52$[(d)]	
UOH^{3+}		0.48 ± 0.08[(g)]	
UF^{3+}		0.48 ± 0.08[(f)]	
UCl^{3+}		0.59 ± 0.10[(g)]	
UBr^{3+}		0.52 ± 0.10[(f)]	
UI^{3+}		0.55 ± 0.10[(f)]	
UNO_3^{3+}		0.62 ± 0.08[(g)]	
Be_2OH^{3+}		0.50 ± 0.05[(e)]	

Table B.3 (continued)

j ↓ $k \rightarrow$	Cl^-	ClO_4^-	NO_3^-
$Be_3(OH)_3^{3+}$	0.30 ± 0.05[e]	0.51 ± 0.05[e]	0.29 ± 0.05[e]
$Al_3CO_3(OH)_4^{4+}$	0.41[a]		
$Fe_2(OH)_2^{4+}$		0.82[d]	
$Y_2CO_3^{4+}$		0.80 ± 0.04[d]	
Pu^{4+}		1.03 ± 0.05[b]	
Np^{4+}		0.82 ± 0.05[b]	
U^{4+}		0.76 ± 0.06[f]	
Th^{4+}	0.25 ± 0.03		0.11 ± 0.02
$Al_3(OH)_4^{5+}$	0.66[a]	1.30[a]	

(a) Taken from Hedlund [88HED].
(b) Taken from Riglet, Robouch and Vitorge [89RIG/ROB], where the following assumptions were made : $\varepsilon_{(Np^{3+},ClO_4^-)} \approx \varepsilon_{(Pu^{3+},ClO_4^-)} = 0.49$ as for other (M^{3+}, ClO_4^-) interactions, and $\varepsilon_{(NpO_2^{2+},ClO_4^-)} \approx \varepsilon_{(PuO_2^{2+},ClO_4^-)} \approx \varepsilon_{(UO_2^{2+},ClO_4^-)} = 0.46$.
(c) Evaluated in this present review.
(d) Taken from Spahiu [83SPA].
(e) Taken from Bruno [86BRU], where the following assumptions were made: $\varepsilon_{(Be^{2+},ClO_4^-)} = 0.30$ as for other $\varepsilon_{(M^{2+},ClO_4^-)}$, $\varepsilon_{(Be^{2+},Cl^-)} = 0.17$ as for other $\varepsilon_{(M^{2+},Cl^-)}$, and $\varepsilon_{(Be^{2+},NO_3^-)} = 0.17$ as for other $\varepsilon_{(M^{2+},NO_3^-)}$.
(f) Estimated in this review.
(g) Evaluated in this present review using $\varepsilon_{(U^{4+},ClO_4^-)} = (0.76 \pm 0.06)$.
(h) Taken from Ferri *et al.* [85FER/GRE].
(i) It is recalled that these coefficients are not used in the present review because they were evaluated by Ciavatta [80CIA] without taking chloride and nitrate complexation into account.

Table B.4: Ion interaction coefficients $\varepsilon_{j,k}$ for anions j with $k = Li^+, Na^+$ and K^+, taken from Ciavatta [80CIA, 88CIA] unless indicated otherwise. The uncertainties represent the 95% confidence level. The ion interaction coefficients marked with † can be described more accurately with an ionic strength dependent function, listed in Table B.5.

j ↓ k →	Li^+	Na^+	K^+
OH^-	$-0.02 \pm 0.03^\dagger$	0.04 ± 0.01	0.09 ± 0.01
F^-		0.02 ± 0.02 (a)	0.03 ± 0.02
HF_2^-		-0.11 ± 0.06 (a)	
Cl^-	0.10 ± 0.01	0.03 ± 0.01	0.00 ± 0.01
ClO_3^-		-0.01 ± 0.02	
ClO_4^-	0.15 ± 0.01	0.01 ± 0.01	
Br^-	0.13 ± 0.02	0.05 ± 0.01	0.01 ± 0.02
BrO_3^-		-0.06 ± 0.02	
I^-	0.16 ± 0.01	0.08 ± 0.02	0.02 ± 0.01
IO_3^-		-0.06 ± 0.02 (b)	
HSO_4^-		-0.01 ± 0.02	
N_3^-		0.0 ± 0.1 (b)	
NO_2^-	$0.06 \pm 0.04^\dagger$	0.00 ± 0.02	-0.04 ± 0.02
NO_3^-	0.08 ± 0.01	$-0.04 \pm 0.03^\dagger$	$-0.11 \pm 0.04^\dagger$
$H_2PO_4^-$		$-0.08 \pm 0.04^\dagger$	$-0.14 \pm 0.04^\dagger$
HCO_3^-		-0.03 ± 0.02	
SCN^-		0.05 ± 0.01	-0.01 ± 0.01
$HCOO^-$		0.03 ± 0.01	
CH_3COO^-	0.05 ± 0.01	0.08 ± 0.01	0.09 ± 0.01
$SiO(OH)_3^-$		-0.08 ± 0.03 (a)	
$Si_2O_2(OH)_5^-$		-0.08 ± 0.04 (b)	
$B(OH)_4^-$		$-0.07 \pm 0.05^\dagger$	
$UO_2(OH)_3^-$		-0.09 ± 0.05 (b)	
$UO_2F_3^-$		0.00 ± 0.05 (b)	
$UO_2(N_3)_3^-$		0.0 ± 0.1 (b)	
$(UO_2)_2CO_3(OH)_3^-$		0.00 ± 0.05 (b)	
SO_3^{2-}		$-0.08 \pm 0.05^\dagger$	
SO_4^{2-}	$-0.03 \pm 0.04^\dagger$	$-0.12 \pm 0.06^\dagger$	-0.06 ± 0.02
$S_2O_3^{2-}$		$-0.08 \pm 0.05^\dagger$	

Table B.4 (continued)

j $k\rightarrow$ ↓	Li^+	Na^+	K^+
HPO_4^{2-}		$-0.15 \pm 0.06^{\dagger}$	$-0.10 \pm 0.06^{\dagger}$
CO_3^{2-}		-0.05 ± 0.03	0.02 ± 0.01
$SiO_2(OH)_2^{2-}$		-0.10 ± 0.07 [(a)]	
$Si_2O_3(OH)_4^{2-}$		-0.15 ± 0.06 [(b)]	
CrO_4^{2-}		$-0.06 \pm 0.04^{\dagger}$	$-0.08 \pm 0.04^{\dagger}$
$UO_2F_4^{2-}$		-0.08 ± 0.06 [(b)]	
$UO_2(SO_4)_2^{2-}$		-0.12 ± 0.06 [(b)]	
$UO_2(N_3)_4^{2-}$		-0.1 ± 0.1 [(b)]	
$UO_2(CO_3)_2^{2-}$		0.04 ± 0.09 [(a)]	
PO_4^{3-}		$-0.25 \pm 0.03^{\dagger}$	-0.09 ± 0.02
$Si_3O_6(OH)_3^{3-}$		-0.25 ± 0.03 [(b)]	
$Si_3O_5(OH)_5^{3-}$		-0.25 ± 0.03 [(b)]	
$Si_4O_7(OH)_5^{3-}$		-0.25 ± 0.03 [(b)]	
$P_2O_7^{4-}$		-0.26 ± 0.05	-0.15 ± 0.05
$Fe(CN)_6^{4-}$			-0.17 ± 0.03
$U(CO_3)_4^{4-}$		0.09 ± 0.10 [(b)]	
$UO_2(CO_3)_3^{4-}$		0.08 ± 0.11 [(a)]	
$UO_2(CO_3)_3^{5-}$		-0.66 ± 0.14 [(a)]	
$U(CO_3)_5^{6-}$		-0.27 ± 0.15 [(a)]	
$(UO_2)_3(CO_3)_6^{6-}$		0.55 ± 0.11 [(a)]	

(a) Evaluated in this review.
(b) Estimated in this review.

Table B.5: Ion interaction coefficients $\varepsilon_{(1,j,k)}$ and $\varepsilon_{(2,j,k)}$ for cations j with k = Cl^-, ClO_4^- and NO_3^- (first part), and for anions j with k = Li^+, Na^+ and K^+ (second part), according to the relationship $\varepsilon = \varepsilon_1 + \varepsilon_2 \log_{10} I_m$. The data are taken from Ciavatta [80CIA, 88CIA]. The uncertainties represent the 95% confidence level.

j ↓ k →	Cl^-		ClO_4^-		NO_3^-	
	ε_1	ε_2	ε_1	ε_2	ε_1	ε_2
NH_4^+			-0.088 ± 0.002	0.095 ± 0.012	-0.075 ± 0.001	0.057 ± 0.004
Ag^+					-0.1432 ± 0.0002	0.0971 ± 0.0009
Tl^+			-0.18 ± 0.02	0.09 ± 0.02		
Hg_2^{2+}					-0.2300 ± 0.0004	0.194 ± 0.002
Hg^{2+}					-0.145 ± 0.001	0.194 ± 0.002
Pb^{2+}					-0.329 ± 0.007	0.288 ± 0.018

j ↓ k →	Li^+		Na^+		K^+	
	ε_1	ε_2	ε_1	ε_2	ε_1	ε_2
OH^-	-0.039 ± 0.002	0.072 ± 0.006				
NO_2^-	0.02 ± 0.01	0.11 ± 0.01				
NO_3^-			-0.049 ± 0.001	0.044 ± 0.002	-0.131 ± 0.002	0.082 ± 0.006
$B(OH)_4^-$			-0.092 ± 0.002	0.103 ± 0.005		
$H_2PO_4^-$			-0.109 ± 0.001	0.095 ± 0.003	-0.1473 ± 0.0008	0.121 ± 0.004
SO_3^{2-}			-0.125 ± 0.008	0.106 ± 0.009		
SO_4^{2-}	-0.068 ± 0.003	0.093 ± 0.007	-0.184 ± 0.002	0.139 ± 0.006		
$S_2O_3^{2-}$			-0.125 ± 0.008	0.106 ± 0.009		
HPO_4^{2-}			-0.19 ± 0.01	0.11 ± 0.03	-0.152 ± 0.007	0.123 ± 0.016
CrO_4^{2-}			-0.090 ± 0.005	0.07 ± 0.01	-0.123 ± 0.003	0.106 ± 0.007
PO_4^{3-}			-0.29 ± 0.02	0.10 ± 0.01		

Appendix C

Assigned uncertainties

One of the objectives of the NEA Thermochemical Data Base (TDB) project is to provide an idea of the uncertainties associated with the data selected in this review. As a rule, the uncertainties define the range within which the corresponding data can be reproduced with a probability of 95% at any place and by any appropriate method. In many cases, statistical treatment is limited or impossible due to the availability of only one or few data points. A particular problem has to be solved when significant discrepancies occur between different source data. This appendix outlines the statistical procedures which were used for fundamentally different problems and explains the philosophy used in this review when statistics were inapplicable. These rules are followed consistently throughout the series of reviews within the TDB Project. Four fundamentally different cases are considered:

1. One source datum available
2. Two or more independent source data available
3. Several data available at different ionic strengths
4. Data at non-standard conditions: Procedures for data correction and recalculation.

C.1. One source datum

The assignment of an uncertainty to a selected value that is based on only one experimental source is a highly subjective procedure. In some cases, the number of data points the selected value is based on allows the use of the "root mean square" [82TAY] deviation of the data points X_i to describe the standard deviation s_X associated with the average $\overline{X}$:

$$s_X = \sqrt{\frac{1}{N-1}\sum_{i=1}^{N}(X_i - \overline{X})^2}\,. \qquad (C.1)$$

The standard deviation s_X is thus calculated from the dispersion of the equally weighted data points X_i around the average $\overline{X}$, and the probability is 95% that an

X_i is within $\overline{X} \pm 1.96\, s_X$), see Taylor [82TAY, *pp.*244-245]. The standard deviation s_X is a measure of the precision of the experiment and does not include any systematic errors.

Many authors report standard deviations s_X calculated with Eq. (C.1) (but often not multiplied by 1.96), but these do not represent the quality of the reported values in absolute terms. It is thus important not to confuse the standard deviation s with the uncertainty σ. The latter reflects the reliability and reproducibility of an experimental value and also includes all kinds of systematic errors s_j that may be involved. The uncertainty σ can be calculated with Eq. (C.2), assuming that the systematic errors are independent.

$$\sigma_{\overline{X}} = \sqrt{s_X^2 + \sum_j (s_j^2)} \qquad (C.2)$$

The estimation of the systematic errors s_j (which, of course, have to relate to $\overline{X}$ and be expressed in the same unit) can only be made by a person who is familiar with the experimental method. The uncertainty σ corresponds to the 95% confidence level preferred in this review. It should be noted that for all the corrections and recalculations that were made (*e.g.*, temperature or ionic strength corrections) the rules of the propogation of errors have been followed, as outlined in Section C.4.2.

More often, the determination of s_X is not possible because either only one or two data points are available, or the authors did not report the individual values. The uncertainty σ in the resulting value can still be estimated using Eq. (C.2) assuming that s_X^2 is much smaller than $\sum_j (s_j^2)$, which is usually the case anyway.

C.2. Two or more independent source data

Frequently, two or more experimental data sources are available, reporting experimental determinations of the desired thermodynamic data. In general, the quality of these determinations varies widely, and the data have to be weighted accordingly for the calculation of the mean. Instead of assigning weight factors, the individual source data X_i are provided with an uncertainty σ_i that also includes all systematic errors and represents the 95% confidence level, as described in Section C.1. The weighted mean $\overline{X}$ and its uncertainty $\sigma_{\overline{X}}$ are then calculated according to Eqs. (C.3) and (C.4).

$$\overline{X} \equiv \frac{\sum_{i=1}^{N} \left(\frac{X_i}{\sigma_i^2}\right)}{\sum_{i=1}^{N} \left(\frac{1}{\sigma_i^2}\right)} \qquad (C.3)$$

$$\sigma_{\overline{X}} = \sqrt{\frac{1}{\sum_{i=1}^{N} \left(\frac{1}{\sigma_i^2}\right)}} \qquad (C.4)$$

Eqs. (C.3) and (C.4) may only be used if all the X_i belong to the same parent distribution. If there are serious discrepancies among the X_i, one proceeds as described below under "Discrepancies". It can be seen from Eq. (C.4) that $\sigma_{\overline{X}}$ is directly dependent on the absolute magnitude of the σ_i values, and not on the dispersion of the data points around the mean. This is reasonable because there are no discrepancies among the X_i, and because the σ_i values already represent the 95% confidence level. The selected uncertainty $\sigma_{\overline{X}}$ will therefore also represent the 95% confidence level.

In cases where all the uncertainties are equal $\sigma_i = \sigma$, Eqs. (C.3) and (C.4) reduce to Eqs. (C.5) and (C.6).

$$\overline{X} = \frac{1}{N}\sum_{i=1}^{N} X_i \tag{C.5}$$

$$\sigma_{\overline{X}} = \frac{\sigma}{\sqrt{N}} \tag{C.6}$$

Example C.1:

Five data sources report values for the thermodynamic quantity X. The reviewer has assigned uncertainties that represent the 95% confidence level as described in Section C.1.

i	X_i	σ_i
1	25.3	0.5
2	26.1	0.4
3	26.0	0.5
4	24.85	0.25
5	25.0	0.6

According to Eqs. (C.3) and (C.4), the following result is obtained:

$$\overline{X} = 25.3 \pm 0.2.$$

The calculated uncertainty $\sigma_{\overline{X}} = 0.2$ appears relatively small but is statistically correct, for the values are assumed to follow a Gaussian distribution. As a consequence of Eq. (C.4), $\sigma_{\overline{X}}$ will always come out smaller than the smallest σ_i. Assuming $\sigma_4 = 0.10$ instead of 0.25 would yield $\overline{X} = (25.0 \pm 0.1)$, and $\sigma_4 = 0.60$ would result in $\overline{X} = (25.6 \pm 0.2)$. In fact, the values $(X_i \pm \sigma_i)$ in this example are at the limit of consistency, that is, the range $(X_4 \pm \sigma_4)$ does not overlap with the ranges $(X_2 \pm \sigma_2)$ and $(X_3 \pm \sigma_3)$. There might be a better way to solve this problem. Three possible alternatives seem more reasonable:

i. The uncertainties σ_i are reassigned because they appear too optimistic after further consideration. Some assessments may have to be reconsidered and the uncertainties reassigned. For example, multiplying all the σ_i by 2 would yield $\overline{X} = (25.3 \pm 0.3)$.

ii. If reconsideration of the previous assessments gives no evidence for reassigning the X_i and σ_i (95% confidence level) values listed above, the statistical conclusion will be that all the X_i do not belong to the same parent distribution and cannot therefore be treated in the same group (*cf.* item iii below for a non-statistical explanation). The values for i = 1, 4 and 5 might be considered as belonging to Group A and the values for i = 2 and 3 to Group B. The weighted average of the values in Group A is $X_A(i = 1,4,5) = (24.95 \pm 0.21)$ and of those in Group B $X_B(i = 2,3) = (26.06 \pm 0.31)$, the second digit after the decimal point being carried over to avoid loss of information. The selected value is now determined as described below under "Discrepancies" (Case I). X_A and X_B are averaged (straight average, there is no reason for giving X_A a larger weight than X_B), and $\sigma_{\overline{X}}$ is chosen in such a way that it covers the complete ranges of expectancy of X_A and X_B. The selected value is then $\overline{X} = (25.5 \pm 0.9)$.

iii Another explanation could be that unidentified systematic errors are associated with some values. If this seems likely to be the case, there is no reason for splitting the values up into two groups. The correct way of proceeding would be to calculate the unweighted average of all the five points and assign an uncertainty that covers the whole range of expectancy of the five values. The resulting value is then $\overline{X} = (25.45 \pm 1.05)$, which is rounded according to the rules in Section C.4.3 to $\overline{X} = (25.4 \pm 1.1)$.

Discrepancies:

Two data are called discrepant if they differ significantly, *i.e.*, their uncertainty ranges do not overlap. In this context, two cases of discrepancies are considered. Case I: Two significantly different source data are available. Case II: Several, mostly consistent source data are available, one of them being significantly different, *i.e.*, an "outlier".

Case I: This is a particularly difficult case because the number of data points is obviously insufficient to allow the preference of one of the two values. If there is absolutely no way of discarding one of the two values and selecting the other, the only solution is to average the two source data in order to obtain the selected value, because the underlying reason for the discrepancy must be unrecognized systematic errors. There is no point in calculating a weighted average, even if the two source data have been given different uncertainties, because there is obviously too little information to give even only limited preference to one of the values. The uncertainty $\sigma_{\overline{X}}$ assigned to the selected mean $\overline{X}$ has to cover the range of expectation of both source data X_1, X_2, as shown in Eq. (C.7),

$$\sigma_{\overline{X}} = \left|X_i - \overline{X}\right| + \sigma_{\max}, \tag{C.7}$$

where $i = 1,2$, and $\sigma_{\max}$ is the larger of the two uncertainties σ_i, see Example C.1.ii and Example C.2.

Example C.2:

The following credible source data are given:

$$X_1 = 4.5 \pm 0.3$$
$$X_2 = 5.9 \pm 0.5.$$

The uncertainties have been assigned by the reviewer. Both experimental methods are satisfactory, and there is no justification to discard one of the data. The selected value is then:

$$\overline{X} = 5.2 \pm 1.2.$$

Illustration for Example C.2:

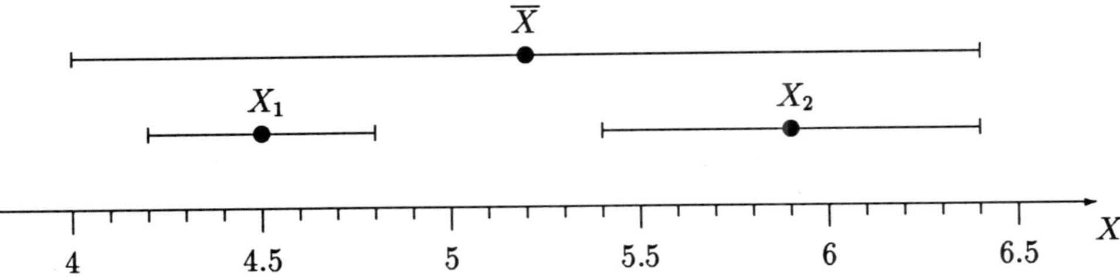

Case II: This problem can often be solved by either discarding the outlying data point, or by providing it with a large uncertainty to lower its weight. If, however, the outlying value is considered to be of high quality and there is no reason to discard all the other data, this case is treated in a way similar to Case I. Example C.3 illustrates the procedure.

Example C.3:

The following data points are available. The reviewer has assigned the uncertainties and sees no justification for any change.

i	X_i	σ_i
1	4.45	0.35
2	5.9	0.5
3	5.7	0.4
4	6.0	0.6
5	5.2	0.4

There are two sets of data that, statistically, belong to different parent distributions A and B. According to Eqs. (C.3) and (C.4), the following average values are found for the two groups: $X_A(i = 1) = (4.45 \pm 0.35)$ and $X_B(i = 2, 3, 4, 5) = (5.62 \pm 0.23)$. The selected value will be the straight average of X_A and X_B, analogous to Example C.1:

$$\overline{X} = 5.0 \pm 0.9.$$

C.3. Several data at different ionic strengths

The extrapolation procedure used in this review is the specific ion interaction theory outlined in Appendix B. The objective of this review is to provide selected data sets at standard conditions, *i.e.*, among others, at infinite dilution for aqueous species. Equilibrium constants determined at different ionic strengths can, according to the specific ion interaction theory, be extrapolated to $I = 0$ with a linear regression model, yielding as the intercept the desired equilibrium constant at $I = 0$, and as the slope the stoichiometric sum of the ion interaction coefficients, $\Delta\varepsilon$. The ion interaction coefficient of the target species can usually be extracted from $\Delta\varepsilon$ and is listed in the corresponding table of Appendix B.

The available source data may sometimes be sparse or may not cover a sufficient range of ionic strengths to allow a proper linear regression. In this case, the correction to $I = 0$ should be carried out according to the procedure described in Section C.4.1.

If sufficient data are available at different ionic strengths and in the same inert salt medium, a weighted linear regression will be the appropriate way to obtain both the constant at $I = 0$, $\overline{X}^{\circ}$, and $\Delta\varepsilon$. The first step is the conversion of the ionic strength from the frequently used molar ($\text{mol} \cdot \text{dm}^{-3}$, M) to the molal ($\text{mol} \cdot \text{kg}^{-1}$, m) scale, as described in Section II.2. The second step is the assignment of an uncertainty σ_i, representing the 95% confidence level, to each data point X_i at the molality $m_{k,i}$, according to the rules described in Section C.1. A large number of commercial and public domain computer programs and routines exist for weighted linear regressions. The subroutine published by Bevington [69BEV, *pp.*104-105] has been used for the calculations in the examples of this appendix. Eqs. (C.8) through (C.12) present the equations that are used for the calculation of the intercept $\overline{X}^{\circ}$ and the slope $\Delta\varepsilon$:

$$\overline{X}^{\circ} = \frac{1}{\Delta}\left(\sum_{i=1}^{N}\frac{m_{k,i}^2}{\sigma_i^2}\sum_{i=1}^{N}\frac{X_i}{\sigma_i^2} - \sum_{i=1}^{N}\frac{m_{k,i}}{\sigma_i^2}\sum_{i=1}^{N}\frac{m_{k,i}X_i}{\sigma_i^2}\right) \quad (C.8)$$

$$\Delta\varepsilon = \frac{1}{\Delta}\left(\sum_{i=1}^{N}\frac{1}{\sigma_i^2}\sum_{i=1}^{N}\frac{m_{k,i}X_i}{\sigma_i^2} - \sum_{i=1}^{N}\frac{m_{k,i}}{\sigma_i^2}\sum_{i=1}^{N}\frac{X_i}{\sigma_i^2}\right) \quad (C.9)$$

$$\sigma_{\overline{X}^{\circ}} = \sqrt{\frac{1}{\Delta}\sum_{i=1}^{N}\frac{m_{k,i}^2}{\sigma_i^2}} \quad (C.10)$$

$$\sigma_{\Delta\varepsilon} = \sqrt{\frac{1}{\Delta}\sum_{i=1}^{N}\frac{1}{\sigma_i^2}}, \quad (C.11)$$

where

$$\Delta = \sum_{i=1}^{N}\frac{1}{\sigma_i^2}\sum_{i=1}^{N}\frac{m_{k,i}^2}{\sigma_i^2} - \left(\sum_{i=1}^{N}\frac{m_{k,i}}{\sigma_i^2}\right)^2. \quad (C.12)$$

In this way, the uncertainties σ_i are not only used for the weighting of the data in Eqs. (C.8) and (C.9), but also for the calculation of the uncertainties $\sigma_{\overline{X}^{\circ}}$ and $\sigma_{\Delta\varepsilon}$ in Eqs. (C.10) and (C.11). If the σ_i represent the 95% confidence level, $\sigma_{\overline{X}^{\circ}}$ and $\sigma_{\Delta\varepsilon}$ will also do so. In other words, the uncertainties of the intercept and the slope do not depend on the dispersion of the data points around the straight line but rather directly on their absolute uncertainties σ_i.

Example C.4:

Ten independent determinations of the equilibrium constant $\log_{10}{}^*\beta$ for the reaction

$$UO_2^{2+} + HF(aq) \rightleftharpoons UO_2F^+ + H^+ \qquad (C.13)$$

are available in $HClO_4/NaClO_4$ media at different ionic strengths. Uncertainties that represent the 95% confidence level have been assigned by the reviewer. A weighted linear regression, ($\log_{10}{}^*\beta + 2D$) *vs.* m_k, according to the formula $\log_{10}{}^*\beta(C.13) + 2D = \log_{10}{}^*\beta^\circ(C.13) - \Delta\varepsilon\, m_k$, will yield the correct values for the intercept $\log_{10}{}^*\beta^\circ(C.13)$ and the slope $\Delta\varepsilon$. In this case, m_k corresponds to the molality of ClO_4^-. D is the Debye-Hückel term, *cf.* Appendix B.

i	$m_{ClO_4^-,i}$	$\log_{10}{}^*\beta_i(C.13) + 2D$	σ_i
1	0.05	1.88	0.10
2	0.25	1.86	0.10
3	0.51	1.73	0.10
4	1.05	1.84	0.10
5	2.21	1.88	0.10
6	0.52	1.89	0.11
7	1.09	1.93	0.11
8	2.32	1.78	0.11
9	2.21	2.03	0.10
10	4.95	2.00	0.32

The results of the linear regression are:

$$\begin{aligned} \text{intercept} &= 1.837 \pm 0.054 = \log_{10}{}^*\beta^\circ(C.13) \\ \text{slope} &= 0.029 \pm 0.036 = -\Delta\varepsilon. \end{aligned}$$

Calculation of the ion interaction coefficient $\varepsilon_{(UO_2F^+,ClO_4^-)} = \Delta\varepsilon + \varepsilon_{(UO_2^{2+},ClO_4^-)} - \varepsilon_{(H^+,ClO_4^-)}$: From $\varepsilon_{(UO_2^{2+},ClO_4^-)} = (0.46 \pm 0.04)$, $\varepsilon_{(H^+,ClO_4^-)} = (0.14 \pm 0.01)$ (see Appendix B) and the slope of the linear regression, $\Delta\varepsilon = (-0.03 \pm 0.04)$, it follows that $\varepsilon_{(UO_2F^+,ClO_4^-)} = (0.29 \pm 0.07)$. Note that the uncertainty (± 0.07) is obtained based on the rules of error propagation as described in Section C.4.2:

$$\sigma = \sqrt{(0.04)^2 + (0.04)^2 + (0.01)^2}\,.$$

The resulting selected values are thus

$$\begin{aligned} \log_{10}{}^*\beta^\circ(C.13) &= 1.84 \pm 0.05 \\ \varepsilon_{(UO_2F^+,ClO_4^-)} &= 0.29 \pm 0.07. \end{aligned}$$

Discrepancies or insufficient number of data points:

Discrepancies are principally treated as described in Section C.2. Again, two cases can be defined. Case I: Only two data are available. Case II: An "outlier" cannot be discarded. If only one data point is available, the procedure for correction to zero ionic strength outlined in Section C.4 should be followed.

Case I: If only two source data are available, there will be no straightforward way to decide whether or not these two data points belong to the same parent distribution unless either the slope of the straight line is known or the two data refer to the same ionic strength. Drawing a straight line right through the two data points is an inappropriate procedure because all the errors associated with the two source data would accumulate and may lead to highly erroneous values of $\log_{10} K^\circ$ and $\Delta\varepsilon$. In this case, an ion interaction coefficient for the key species in the reaction in question may be selected by anology (charge is the most important parameter), and a straight line with the slope $\Delta\varepsilon$ as calculated may then be drawn through each data point. If there is no reason to discard one of the two data points based on the quality of the underlying experiment, the selected value will be the unweighted average of the two standard state data obtained by this procedure, and its uncertainty must cover the entire range of expectancy of the two values, analogous to Case I in Section C.2. It should be mentioned that the ranges of expectancy of the corrected values at $I = 0$ are given by their uncertainties which are based on the uncertainties of the source data at $I \neq 0$ and the uncertainty in the slope of the straight line. The latter uncertainty is not an estimate but is calculated from the uncertainties in the ion interaction coefficients involved, according to the rules of error propagation outlined in Section C.4.2. The ion interaction coefficients estimated by analogy are listed in the table of selected ion interaction coefficients (Appendix B), but they are flagged as estimates.

Example C.5:

Three determinations of the equilibrium constant $\log_{10}{}^*\beta$ of the reaction

$$\mathrm{Pu^{4+} + H_2O(l) \rightleftharpoons PuOH^{3+} + H^+} \qquad \text{(C.14)}$$

are available in $NaClO_4$ solutions at different ionic strengths:

$$\begin{array}{ll} I_c = 0.5\ \mathrm{M\ (NaClO_4)}: & \log_{10}{}^*\beta(\mathrm{C.14}) = -1.60 \pm 0.18 \\ I_c = 1.05\ \mathrm{M\ (NaClO_4)}: & \log_{10}{}^*\beta(\mathrm{C.14}) = -1.48 \pm 0.10 \\ I_c = 2.0\ \mathrm{M\ (NaClO_4)}: & \log_{10}{}^*\beta(\mathrm{C.14}) = -1.72 \pm 0.12. \end{array}$$

The uncertainties have been assigned by the reviewer and represent the 95% confidence level. The data do not appear too inconsistent at first sight, but a linear regression would be an inappropriate procedure with only three data points. The ion interaction coefficients for Pu^{4+} and H^+, and an estimate for $PuOH^{3+}$ based on the values of other species with charge +3, are taken from the tables in Appendix B:

$$\begin{aligned}\varepsilon_{(Pu^{4+},ClO_4^-)} &= 1.03 \pm 0.05\\ \varepsilon_{(H^+,ClO_4^-)} &= 0.14 \pm 0.01\\ \varepsilon_{(PuOH^{3+},ClO_4^-)} &= 0.50 \pm 0.05\end{aligned}$$

This leads to $\Delta\varepsilon = -(0.39 \pm 0.07)$. The specific ion interaction theory requires the use of molal (mol · kg^{-1}, m) units and leads to a linear equation, Eq. (C.15), where D is the Debye-Hückel term.

$$\log_{10}{}^*\beta(\text{C.14}) + 6D \;=\; \log_{10}{}^*\beta^\circ(\text{C.14}) - \Delta\varepsilon\, m_{ClO_4^-} \tag{C.15}$$

The results for $\log_{10}{}^*\beta^\circ$(C.14) are shown below. The ionic strength values I_i have been converted from $I_{i,\text{M}}$ (molar units) to $I_{i,\text{m}}$ (molal units) [76BAE/MES, *p*.439], assuming that $I_{i,m} = m_{ClO_4^-}$. The uncertainties have been obtained based on the rules of error propagation, as described in Section C.4.2.

i	$I_{i,m}$	$\log_{10}{}^*\beta_i(\text{C.14}) + 6D_i$	$\log_{10}{}^*\beta_i^\circ(\text{C.14})$
1	0.51	$-0.54_7 \pm 0.18$	-0.75 ± 0.18
2	1.05	$-0.24_6 \pm 0.10$	-0.66 ± 0.12
3	2.21	$-0.32_4 \pm 0.12$	-1.19 ± 0.20

Obviously, the corrected values $\log_{10}{}^*\beta_i^\circ$(C.14) are inconsistent and have therefore to be treated as described in Case I of Section C.2. That is, the selected value will be the unweighted average of $\log_{10}{}^*\beta_i^\circ$(C.14), and its uncertainty will cover the entire range of expectancy of the three values. A weighted average would only be justified if the three data $\log_{10}{}^*\beta_i^\circ$(C.14) turned out consistent. The result is $\log_{10}{}^*\beta^\circ(\text{C.14}) = -(0.87 \pm 0.52)$ and may be rounded to one significant digit (see Section C.4), yielding

$$\log_{10}{}^*\beta^\circ(\text{C.14}) \;=\; -0.9 \pm 0.5.$$

It should be noted that a linear regression according to Eq. (C.15), which is not the appropriate way but theoretically feasible, would yield $\log_{10}{}^*\beta^\circ(\text{C.14}) = -(0.38 \pm 0.17)$ and $\Delta\varepsilon = (0.04 \pm 0.11)$, giving $\varepsilon_{(PuOH^{3+},ClO_4^-)} = (0.93 \pm 0.12)$, a value which is considerably larger than the typical range for species with a charge +3.

Case II: This case includes situations where it is difficult to decide whether or not a large number of points belong to the same parent distribution. There is no general rule on how to solve this problem, and decisions are left to the judgement of the reviewer. For example, if eight data points follow a straight line reasonably well and two lie way out, it may be justified to discard the "outliers". If, however, the eight points are scattered considerably and two points are just a bit further out, one can probably not consider them as "outliers". It depends on the particular case and on the judgement of the reviewer whether it is reasonable to increase the uncertainties of the data to reach consistency, or whether the slope $\Delta\varepsilon$ of the straight line should be estimated by analogy.

C.4. Procedures for data handling

C.4.1. Correction to zero ionic strength

The correction of experimental data to zero ionic strength is necessary in all cases where a linear regression is impossible or appears inappropriate. The method used throughout the review is the specific ion interaction theory. This theory is described in detail in Appendix B. Two variables are needed for this correction, and both have to be provided with an uncertainty at the 95% confidence level: the experimental source value, $\log_{10} K$ or $\log_{10} \beta$, and the stoichiometric sum of the ion interaction coefficients, $\Delta\varepsilon$. The ion interaction coefficients (see Tables B.3, B.4 and B.5 of Appendix B) required to calculate $\Delta\varepsilon$ may not all be known. Missing values therefore need to be estimated by analogy, and it is recalled that the electric charge of an ion has the most significant influence on the magnitude of the ion interaction coefficients. The uncertainty of the corrected value at $I = 0$ is calculated by taking into account the propagation of errors, as described below. It is recalled that the ionic strength is frequently given in moles per dm^3 of solution (molar, M) and has to be converted to moles per kg H_2O (molal, m), as the theory requires. A table of conversion factors for the most common inert salts given by Baes and Mesmer [76BAE/MES, *p.*439] is represented in Table II.5.

Example C.6:

For the equilibrium constant of the reaction

$$Fe^{3+} + 2H_2O(l) \rightleftharpoons Fe(OH)_2^+ + 2H^+ \qquad (C.16)$$

only one credible determination in 3 M $NaClO_4$ solution is known, $\log_{10}{}^*\beta(C.16) = -6.31$, to which an uncertainty of ± 0.12 has been assigned. The ion interaction coefficients are as follows:

$$\begin{aligned} \varepsilon_{(Fe^{3+},ClO_4^-)} &= 0.56 \pm 0.07 \\ \varepsilon_{(Fe(OH)_2^+,ClO_4^-)} &= 0.26 \pm 0.11 \\ \varepsilon_{(H^+,ClO_4^-)} &= 0.14 \pm 0.01. \end{aligned}$$

The values of $\Delta\varepsilon$ and $\sigma_{\Delta\varepsilon}$ can be obtained readily:

$$\begin{aligned} \Delta\varepsilon &= \varepsilon_{(Fe(OH)_2^+,ClO_4^-)} + 2\varepsilon_{(H^+,ClO_4^-)} - \varepsilon_{(Fe^{3+},ClO_4^-)} = -0.02 \\ \sigma_{\Delta\varepsilon} &= \sqrt{(0.11)^2+(2\times 0.01)^2+(0.07)^2} = 0.13. \end{aligned}$$

The two variables are thus:

$$\begin{aligned} \log_{10}{}^*\beta(C.16) &= -6.31 \pm 0.12 \\ \sigma_{\Delta\varepsilon} &= -0.02 \pm 0.13 \end{aligned}$$

According to the specific ion interaction theory the following equation is used to correct for ionic strength for the reaction considered here:

$$\log_{10}{}^*\beta(\text{C.16}) + 6D \quad = \quad \log_{10}{}^*\beta^\circ(\text{C.16}) - \Delta\varepsilon\, m_{\text{ClO}_4^-}$$

D is the Debye-Hückel term: $D = 0.5091\sqrt{I_m}/(1+1.5\sqrt{I_m})$. The ionic strength I_m and the molality $m_{\text{ClO}_4^-}$ ($I_m \approx m_{\text{ClO}_4^-}$) have to be expressed in molal units, 3 M $NaClO_4$ corresponding to 3.5 m $NaClO_4$ (see Section II.2), giving $D = 0.25$. This results in

$$\log_{10}{}^*\beta^\circ(\text{C.16}) \quad = \quad -4.88.$$

The uncertainty in $\log_{10}{}^*\beta^\circ$ is calculated from the uncertainties in $\log_{10}{}^*\beta$ and $\Delta\varepsilon$:

$$\sigma_{\log_{10}{}^*\beta^\circ} = \sqrt{\sigma_{\log_{10}{}^*\beta} + \left(m_{\text{ClO}_4^-}\,\sigma^2_{\Delta\varepsilon}\right)^2} = \sqrt{(0.12)^2 + (3.5\times 0.12)^2} = 0.44.$$

The selected, rounded value is

$$\log_{10}{}^*\beta^\circ(\text{C.16}) \quad = \quad -4.9 \pm 0.4.$$

C.4.2. Propagation of errors

Whenever data are converted or recalculated, or other algebraic manipulations are performed that involve uncertainties, the propagation of these uncertainties has to be taken into account in a correct way. A clear outline of the propagation of errors is given by Bevington [69BEV]. A simplified form of the general formula for error propagation is given by Eq. (C.17), supposing that X is a function of $Y_1, Y_2, ..., Y_N$.

$$\sigma_X^2 \quad = \quad \sum_{i=1}^{N}\left(\frac{\partial X}{\partial Y_i}\sigma_{Y_i}\right)^2 \qquad \text{(C.17)}$$

Eq. (C.17) can be used only if the variables Y_1, Y_2, ..., Y_N are independent or if their uncertainties are small, that is the covariances can be disregarded. One of these two assumptions can almost always be made in chemical thermodynamics, and Eq. (C.17) can thus almost universally be used in this review. Eqs. (C.18) through (C.22) present explicit formulas for a number of frequently encountered algebraic expressions, where c, c_1, c_2 are constants.

$$X = c_1Y_1 \pm c_2Y_2: \qquad \sigma_X^2 = (c_1\sigma_{Y_1})^2 + (c_2\sigma_{Y_2})^2 \qquad \text{(C.18)}$$

$$X = \pm cY_1Y_2 \text{ and } X = \pm\frac{cY_1}{Y_2}: \qquad \left(\frac{\sigma_X}{X}\right)^2 = \left(\frac{\sigma_{Y_1}}{Y_1}\right)^2 + \left(\frac{\sigma_{Y_2}}{Y_2}\right)^2 \qquad \text{(C.19)}$$

$$X = c_1Y^{\pm c_2}: \qquad \frac{\sigma_X}{X} = c_2\frac{\sigma_Y}{Y} \qquad \text{(C.20)}$$

$$X = c_1\, e^{\pm c_2 Y}: \qquad \frac{\sigma_X}{X} = c_2\sigma_Y \qquad \text{(C.21)}$$

$$X = c_1\ln(\pm c_2 Y): \qquad \sigma_X = c_1\frac{\sigma_Y}{Y} \qquad \text{(C.22)}$$

Example C.7:

A few simple calculations illustrate how these formulas are used. The values have not been rounded.

$$\text{Eq. (C.18)}: \quad \Delta_r G_m = 2[-(277.4 \pm 4.9)]\,\text{kJ}\cdot\text{mol}^{-1} - [-(467.3 \pm 6.2)]\,\text{kJ}\cdot\text{mol}^{-1}$$

$$\Delta_r G_m = -(87.5 \pm 11.6)\,\text{kJ}\cdot\text{mol}^{-1}$$

$$\text{Eq. (C.19)}: \quad K = \frac{(0.038 \pm 0.002)}{(0.0047 \pm 0.0005)} = (8.09 \pm 0.92)$$

$$\text{Eq. (C.20)}: \quad K = 4(3.75 \pm 0.12)^3 = (210.9 \pm 20.3)$$

$$\text{Eq. (C.21)}: \quad K^\circ = e^{\frac{-\Delta_r G^\circ_m}{RT}}; \qquad \begin{aligned} \Delta_r G^\circ_m &= -(2.7 \pm 0.3)\,\text{kJ}\cdot\text{mol}^{-1} \\ R &= 8.3145\,\text{J}\cdot\text{K}^{-1}\cdot\text{mol}^{-1} \\ T &= 298.15\,\text{K} \end{aligned}$$

$$K^\circ = 2.97 \pm 0.36$$

Note that powers of 10 have to be reduced to powers of e, *i.e.*, the variable has to be multiplied by $\ln(10)$, *e.g.*,

$$\log_{10} K = (2.45 \pm 0.10); \quad K = 10^{\log_{10} K} = e^{(\ln(10)\,\log_{10} K)} = (282 \pm 65).$$

$$\text{Eq. (C.22)}: \quad \Delta_r G^\circ_m = -RT \ln K^\circ; \qquad \begin{aligned} K^\circ &= (8.2 \pm 1.2) \times 10^6 \\ R &= 8.3145\,\text{J}\cdot\text{K}^{-1}\cdot\text{mol}^{-1} \\ T &= 298.15\,\text{K} \end{aligned}$$

$$\begin{aligned} \Delta_r G^\circ_m &= -(39.46 \pm 0.36)\,\text{kJ}\cdot\text{mol}^{-1} \\ \ln K^\circ &= 15.92 \pm 0.15 \\ \log_{10} K^\circ &= \ln K^\circ / \ln(10) = 6.91 \pm 0.06 \end{aligned}$$

Again, it can be seen that the uncertainty in $\log_{10} K^\circ$ cannot be the same as in $\ln K^\circ$. The constant conversion factor of $\ln(10) = 2.303$ is also to be applied to the uncertainty.

C.4.3. Rounding

The standard rules to be used for rounding are:

1. When the digit following the last digit to be retained is less than 5, the last digit retained is kept unchanged.

2. When the digit following the last digit to be retained is greater than 5, the last digit retained is increased by 1.

3. When the digit following the last digit to be retained is 5 and

a) there are no digits (or only zeroes) beyond the 5, an odd digit in the last place to be retained is increased by 1 while an even digit is kept unchanged.

b) other non-zero digits follow, the last digit to be retained is increased by 1, whether odd or even.

This procedure avoids introducing a systematic error from always dropping or not dropping a 5 after the last digit retained.

When adding or subtracting, the result is rounded to the number of decimal places (not significant digits) in the term with the least number of places. In multiplication and division, the results are rounded to the number of significant digits in the term with the least number of significant digits.

In general, all operations are carried out in full, and only the final results are rounded, in order to avoid the loss of information from repeated rounding. For this reason, several additional digits are carried in all calculations until the final selected set of data is developed, and only then are data rounded.

C.4.4. Significant digits

The uncertainty of a value basically defines the number of significant digits a value should be given.

Examples: 3.478 ± 0.008
3.48 ± 0.01
2.8 ± 0.4

In the case of auxiliary data or values that are used for later calculations, it is often not convenient to round to the last significant digit. In the value (4.85 ± 0.26), for example, the "5" is close to being significant and should be carried along a recalculation path in order to avoid loss of information. In particular cases, where the rounding to significant digits could lead to slight internal inconsistencies, digits with no significant meaning in absolute terms are nevertheless retained. The uncertainty of a selected value always contains the same number of digits after the decimal point as the value itself.

Appendix D

The estimation of entropies

Methods of entropy estimation for solids have been determined by Latimer [52LAT]. Based on later experimental data, Naumov *et al.* [71NAU/RYZ] prepared a revised table of parameters to be used with Latimer's method.

Langmuir [78LAN] described an improved method of estimating entropy values for solid compounds containing the UO_2^{2+} moiety. In his procedure, the contributions of "UO_2^{2+}" to the entropy of γ-UO_3, β-$UO_2(OH)_2$ and schoepite were used to estimate the contribution of "UO_2^{2+}" to the entropies of other uranium compounds. Although Latimer [52LAT] and Naumov *et al.* [71NAU/RYZ] suggested a value of $39.3\ J \cdot K^{-1} \cdot mol^{-1}$ for each water of hydration in a crystalline solid hydrate, Langmuir [78LAN] considered the value of $44.7\ J \cdot K^{-1} \cdot mol^{-1}$, the value for the entropy of ice-I from the compilation of Robie and Waldbaum [68ROB/WAL], to be more appropriate.

Thus, the entropies in the literature, based on Latimer's method, may vary considerably, depending on the exact set of parameters used in the estimation. Also, for some compounds, especially binary solids, better entropy values may be estimated by comparison with values for closely related solids, than by Latimer's method. In the present review an attempt has been made to ensure that entropies estimated for solids have been calculated using consistent procedures.

Three methods are used. For the most part, entropies of solids for which experimental data are unavailable, or unreliable, are calculated using Latimer's method with the parameters from Naumov *et al.* [71NAU/RYZ]. The entropy contribution per water of hydration is the Robie and Waldbaum value for ice-I [68ROB/WAL]. Entropy values at 298.15 K, based on low-temperature heat capacity data have been selected in this review for the following solid compounds of uranium(VI): UO_2Cl_2(cr), UO_2CO_3(cr), α-UO_3, β-UO_3, γ-UO_3, $UO_3 \cdot 2H_2O$ (*i.e.*, $UO_2(OH)_2 \cdot H_2O$), $UO_2(NO_3)_2 \cdot 6H_2O$. These were used in the manner described by Langmuir to calculate a contribution of $94.7\ J \cdot K^{-1} \cdot mol^{-1}$ for UO_2^{2+} in uranium(VI) solids—as opposed to the value $93.3\ J \cdot K^{-1} \cdot mol^{-1}$ proposed by Langmuir [78LAN].

Other parameters required, but not available from Ref. [71NAU/RYZ] have been estimated. All values used are provided in Table D.1. Uncertainties are estimated as $\pm 4\ J \cdot K^{-1} \cdot mol^{-1}$ per contributing group other than water of hydration. For each water of hydration the uncertainty is estimated as $\pm 6\ J \cdot K^{-1} \cdot mol^{-1}$. This increased uncertainty reflects the fact that the parameters from Naumov *et al.* [71NAU/RYZ] were derived by considering entropies from many compounds, some of which could have been hydrates. Thus the potential exists for an internal inconsistency.

In estimating entropies for solids containing U(III), U(IV) and U(V) no allowance has been made for contributions from the electronic degeneracy of these uranium ions. This is obviously incorrect, however, the sources (and weighting) of the data used to derive the Latimer parameters [71NAU/RYZ] are not clear. Therefore, it is not certain if corrections should actually be applied, or if they are already included in suggested parameters (or trends in parameters). It would appear that in any future systematic attempt to improve the Latimer parameters only data for compounds containing ions without such electronic contributions should be used.

Second, for some solids the entropy is known for one or more hydrates, but is not known for another. In such cases the unknown entropies are estimated using the known values and the recommended contribution of $44.7\,J \cdot K^{-1} \cdot mol^{-1}$ per water of hydration.

Finally, for some binary metal systems for which the entropy values are known for two (or more) compounds, but are not well established for others, the entropies are estimated by assuming the same entropy contribution for each of the two types of metal atoms in the set of binary compounds.

Table D.1: Contributions to entropies of solids ($J \cdot K^{-1} \cdot mol^{-1}$). These are mainly from Ref. [71NAU/RYZ] as discussed in the text. The values in parentheses were not directly based on experimental values, but were estimated by Naumov, Ryzhenko and Khodakovsky [71NAU/RYZ].

Anion	Average cation charge			
	+1	+2	+3	+4
OH^-	30.5	19.2	17.5	(19.2)
O^{2-}	8.4	2.5	2.1	1.3
F^-	23.0	17.6	16.1	20.1
Cl^-	43.9	32.6	29.3	33.1
Br^-	56.1	45.2	41.8	49.8
I^-	63.2	54.4	55.2	51.0
IO_3^-	104.6	(92)		
CO_3^{2-}	64.9	49.4		
NO_3^-	90.8	73.2		
SO_3^{2-}	83.3	62.3	50.2	(46)[(a)]
SO_4^{2-}	92.9	67.8	57.3	(50)
PO_4^{3-}	79.5	62.8	57.3	(50)[(a)]
PO_3^-	66.9	54.0	(50.0)	(48)[(a)]
HPO_4^-	87.9	72.4	66.9	(63)
H_2O	44.7			
UO_2^{2+}[(b)]	94.7			
U[(c)]	66.9			

(a) Estimated in the present review.
(b) Treated as a dipositive ion for the purpose of selecting anion contributions.
(c) For uranium compounds not containing $U^{VI}O_2$.

ORGANISATION FOR ECONOMIC CO-OPERATION AND DEVELOPMENT

Pursuant to Article 1 of the Convention signed in Paris on 14th December 1960, and which came into force on 30th September 1961, the Organisation for Economic Co-operation and Development (OECD) shall promote policies designed:

— to achieve the highest sustainable economic growth and employment and a rising standard of living in Member countries, while maintaining financial stability, and thus to contribute to the development of the world economy;
— to contribute to sound economic expansion in Member as well as non-member countries in the process of economic development; and
— to contribute to the expansion of world trade on a multilateral, non-discriminatory basis in accordance with international obligations.

The original Member countries of the OECD are Austria, Belgium,Canada, Denmark, France, Germany, Greece, Iceland, Ireland, Italy, Luxembourg, the Netherlands, Norway, Portugal, Spain, Sweden, Switzerland, Turkey, the United Kingdom and the United States. The following countries became Members subsequently through accession at the dates indicated hereafter: Japan (28th April 1964), Finland (28th January 1969), Australia (7th June 1971) and New Zealand (29th May 1973). The Commission of the European Communities takes part in the work of the OECD (Article 13 of the OECD Convention). Yugoslavia has a special status at OECD (agreement of 28th October 1961).

The ECD Nuclear Energy Agency (NEA) was established on 1st February 1958 under the name of the OEEC European Nuclear Energy Agency. It received its present designation on 20th April 1972, when Japan became its first non-European full Member. NEA membership today consists of all European Member countries of OECD as well as Australia, Canada, Japan and the United States. The Commission of the European Communities takes part in the work of the Agency.

The primary objective of NEA is to promote co-operation among the governments of its participating countries in furthering the development of nuclear power as a safe, environmentally acceptable and economic energy source.

This is achieved by:

— *encouraging harmonization of national regulatory policies and practices, with particular reference to the safety of nuclear installations, protection of man against ionising radiation and preservation of the environment, radioactive waste management, and nuclear third party liability and insurance;*
— *assessing the contribution of nuclear power to the overall energy supply by keeping under review the technical and economic aspects of nuclear power growth and forecasting demand and supply for the different phases of the nuclear fuel cycle;*
— *developing exchanges of scientific and technical information particularly through participation in common services;*
— *setting up international research and development programmes and joint undertakings.*

In these and related tasks, NEA works in close collaboration with the International Atomic Energy Agency in Vienna, with which it has concluded a Co-operation Agreement, as well as with other international organisations in the nuclear field.